A Practical Guide to Ore Microscopy—Volume 1

Ricardo Castroviejo

A Practical Guide to Ore Microscopy—Volume 1

Mineral Identification

Springer

Ricardo Castroviejo
E.T.S. de Ingenieros de Minas y Energía
Universidad Politécnica de Madrid
Madrid, Spain

ISBN 978-3-031-12656-7 ISBN 978-3-031-12654-3 (eBook)
https://doi.org/10.1007/978-3-031-12654-3

This Springer imprint is published by the registered company Springer Nature Switzerland AG
The registered company address is: Gewerbestrasse 11, 6330 Cham, Switzerland

Dedico este libro, fruto de un esfuerzo compartido, a Antonia, mi amiga, esposa y apoyo abnegado en momentos difíciles.

(This book, the fruit of a shared effort, is dedicated to Antonia, my friend, wife and selfless support in difficult times.)

Foreword

Congratulations on acquiring the most useful book on ore microscopy! You'll find the book packed with information relevant to geologists, metallurgists, and others interested in not only ore minerals but all common minerals that are opaque in transmitted light.

Following the adage "when all else fails, look at the rock", this book is a first-rate text on how to maximize the use of a polarizing reflected-light microscope (covered in Part I), identify more than 150 ore minerals (Part II) and more than 50 gangue minerals with which they are commonly associated (Part III), interpret intergrowths of minerals for geological interpretations of ore-forming processes and for applications in geometallurgy and environmental considerations (Part IV) and use the light microscope for image-analysis automation in a much less expensive but complementary approach to scanning electron microscope (SEM)-based techniques (Part V).

The photographs of ore and gangue minerals in all chapters are as accurate in colour as possible and provide excellent examples culled from worldwide collections. Each ore mineral in Part II is photographed under the four types of illumination commonly used in reflected-light petrography (plane polarized light in air then with oil plus crossed polarized light in both air and oil). The photographs in Part IV are illustrative of the intergrowths that are essential to accurate characterization of mineral resources and reserves, including potential difficulties in ore processing and waste disposal.

The book comes in two volumes, with Parts I through III in Volume 1 and Parts IV and V plus references and annexes (appendices) in Volume 2. The annexes are exceptionally useful. They address common associations of ore and gangue minerals (#2), techniques for the preparation of polished sections (#3), tables of mineral properties (#4), and well-proven determinative tables (#5).

Professor Castroviejo has spent his career using ore microscopy in research and applications to industry. He demonstrates that reflected-light microscopy remains the most cost-effective approach to characterize and interpret ores in terms of genesis, mineral processing, and environmental concerns during and after production. Although SEM, electron microprobe, and other chemical analytical techniques are essential, the ore microscope remains the first and best approach to look at the rocks.

Jonathan G. Price Ph.D.
(Emeritus) State Geologist
Nevada Bureau of Mines and Geology
University of Nevada
Reno, USA

Acknowledgements

This work has benefited from the generous support of numerous individuals and institutions. Being impossible to name them all, I would like to express my gratitude to all of them. The description of minerals, the main body of the work, is based on direct observations, mainly on the author's ore collections and on those of the School of Mines. The former, gathered throughout the author's professional life, have been supplemented with replicas from the *Rehwald-Ramdohr Collection* in Heidelberg, kindly loaned at the time by Prof. G. C. Amstutz. The latter comprise both those of the Department of Geological and Mining Engineering (collected successively by Profs. I. Roso de Luna, T. Febrel, F. Vázquez and J. Samper) and those loaned for study by the Museum of Mineralogy (Dir. Prof. B. Calvo) at the Escuela Técnica Superior de Ingenieros de Minas y Energía, **ETSIME**, of the Universidad Politécnica de Madrid, **UPM**.

I would like to acknowledge the generous contribution of Dr. Heinz-Jürgen Bernhardt, *Institut für Geologie, Mineralogie und Geophysik der Ruhr-Universität Bochum*, **RUB**. He has carried out the EMP analyses, shared his own ore collections and, above all, has continuously enriched the author's approach with his advice and discussions. If there is any merit in this work, it is shared with him. Dr. Hans-Peter Schertl, *curator of* the Museum of Mineralogy of the RUB, has provided additional samples. Ms. Sabine Weisel, Ms. Ellen Kessler and Ms. Tanja Westphal, specialists of the *Sample Preparation* Lab, RUB, have produced polished sections of excellent quality.

Dr. Jonathan G. Price, *Nevada State Geologist* and Director (*Emeritus*) of the **NBMG**, *Nevada Bureau of Mines and Geology* and *McKay School of Mines* (Reno, NV, USA), generously helped me to complete the collection of studied ores by accessing the *Stanford Collection-Mckay School of Mines* and several active gold and silver mines in Nevada. To Prof. Hartmut Beurlen, *Universidade Federal de Pernambuco*, **UFPE**, Recife (Brazil), I owe the Nb, Ta and REE samples, documented with his precise *EMPA* analyses; to Profs. Eric Pirard (*Université de Liège*, Belgium) and J. M. Charlet (*University of Mons*, Belgium), some CopperBelt ores (DR Congo); and to Dr. María José Correa, *Universidad Nacional de La Plata* (Argentina), the romanechite samples.

The text layout and execution profited from the selfless support of Dr. J. C. Catalina (*author of the* Chap. 2, vol. 2, on automated microscopy) and Ing. U. Grunwald-Romera (*author of Annex 3* on polished sections preparation). Besides, they have contributed, together with Ings. P. Romero and D. Alarcón, to data compilation for DB and to digital design. Dr. César Cánepa, *Universidad Nacional Mayor de San Marcos*, **UNMSM**, Lima (Peru), a master for many generations of mineralogists, has given generous and constant help throughout the process. Professors Soledad Fernández-Santín and Alfredo Hernández-Pacheco, Universidad Complutense de Madrid, **UCM**, have enriched the work with their samples, discussions and critical reviews of the manuscript. Encouraging support has been provided by the opinions of many colleagues consulted, who have taken the trouble to read and critically comment on early drafts: Profs. Emil Mackowycki (University of Copenhagen), José Antonio Espí (UPM), Colombo G Tassinari (University of São Paulo, **USP**), Eulogio Pardo and Juan Locutura (IGME/UPM), Joaquín Proenza (University of Barcelona) and Fernando Gervilla (University of Granada), Drs. Rolando Lastra (CanMet, Ottawa), Laura Pérez-Barnuevo (*COREM*, Quebec) and Thomas Aiglsperger (Luleå), Ing. Jorge Acosta, M.Sc. (*INGEMMET*, Lima), among others.

The inspiration of Prof. Ana García Moreno, UCM, creator and Director of the digital magazine **REDUCA**—*Recursos Educativos* (http://www.revistareduca.es), and the early guidance of Dr. Joan Roca, mathematician and photography researcher, have been very helpful for the digital image processing (to *sRGB colour space*, *IEC standard*) and for designing the virtual text.

Dr. Alexis Vizcaíno, *Springer*'s editor, has provided encouraging and patient support in the laborious editing process. The *Hudson Institute of Mineralogy* is acknowledged for the open access mineral database Mindat.org, source of credited data referred to for some ores. **EIT** (*European Institute for Innovation and Technology*) funded, through *EIT-RawMaterials* Project no. **15039** (2016–2018), the innovative **AMCO** System (*Automated Microscopic Characterisation of Ores*). The support of the project's colleagues (*University of Liège*, Belgium; *ThinSectionLab*, France; *KGHM*, Poland; *Cobre Las Cruces*, Spain, **UPM**-*Consortium leader*) is gratefully acknowledged.

Last but not least, I wish to thank Prof. T. Febrel (†) for his motivating teaching of ore microscopy, at the ETSI Minas-Madrid, as well as the *Deutscher Akademischer Austausch-dienst* (**DAAD**, German Academic Exchange Service) for the fellowship that made possible my early stay at the *Mineralogisch-Petrographisches Institut der Universität Heidelberg* (Germany), my research training with professors like P. Ramdohr (†) and G. C. Amstutz (†) and the enriching contact with postgraduate colleagues from all over the world. All this is at the root of this work. I am also grateful to UPM, RUB, USP and NBMG-McKay *School of Mines* for their support of my *sabbatical stay* (2010/2011), which was decisive for this project.

Contents

Part I

Introduction and Methodology

1 Introduction to Ore Microscopy

1.1 State of the Art and Purpose of the Text

Optical microscopy of metallic ores, based on the polarising reflected light microscope and initially known as *mineragraphy*, developed rapidly in the second half of the last century. It has become an indispensable tool for the study and beneficiation of mineral deposits and, in many cases, the technique par excellence for metallogenic research, to such an extent that the optical reflection microscope came to be known as the *metallogenic microscope*.

The literature on the subject is abundant but, from the point of view of methodology and mineralogical description and interpretation, some milestones[1] such as Schneiderhöhn and Ramdohr (1931), Schneiderhöhn (1952), Ramdohr (1955, 1960, 1980), Picot and Johan (1977, 1982), as well as the determinative tables of Schouten (1962) and Uytenbogaardt and Burke (1971) should be highlighted. These are fundamental contributions in a rich and varied set, which should not be ignored and which the interested reader can document in the excellent synthesis by Criddle (1998), e.g., Spry and Gedlinske (1987), Ineson (1989), Tarkian (1974), Tarkian and Liessmann (1991), and more recently, Cabri (2002), Marshall, Anglin and Mumin (2004), Pracejus (2008), Paar et al. (2016) and Neumann (2019). Virtual texts can also be consulted, such as Ixer and Duller (1998), in addition to the mineralogical information contained in websites such as *mindat* (cf. References/*WEBSITES*), *IMA/COM* (Commission on Ore Mineralogy of the International Mineralogical Association: cf. References/*WEBSITES*) or RRUFF (cf. References/*WEBSITES*). In Spanish, the pioneering work of Febrel (1970) stands out; the determinative tables of Novitzky (1957) should also be mentioned, as well as the texts of Hagel (1979) and Fenoll Hach-Alí and Gervilla (2005). From the point of view of mineral optics, the work of Peckett (1992) is noteworthy, while Sanderson (2019) provides a comprehensive and well-documented treatment of optical microscopy.

There are also interesting collective contributions of synthesis and updating of the technique and its industrial application to processing mineralogy (*ore processing* and concentration), e.g., Amstutz (1962), Freund (1966), Jambor and Vaughan (1990), Cabri and Vaughan (1998) and, in the present century, Cabri (2002), Becker, Wightman and Evans (2016). If we compare the evolution over time of the techniques applied according to these contributions, we can see how the initial prominence of optical microscopy has been diluted by the spread of microanalytical techniques (**EMP** or electron microprobe, **SEM** or electron microscopy), which provide complementary information. The trend is accentuated when these are combined with image analysis to achieve automation, with the spread in the present century of commercial systems offering so-called *automated mineralogy* (e.g., QemScan, MLA, Tescan Tima, Zeiss Mineralogic…).

As a consequence of these developments, SEM techniques have become increasingly popular on the market since the last quarter of the twentieth century, partly taking over tasks that had previously been the exclusive domain of optical microscopy. This is a trend driven by the recent forced reduction of content in university earth science curricula, which makes learning difficult. However, the interest in (optical) microscopy, both from the point of view of science (ore genesis) and of industrial applications, including mining exploration, process control, and environmental impact, is not only alive but has been enhanced by recent technological advances (Pirard 2016). It retains some important advantages over electron microscopy, such as its simplicity, flexibility, lower requirements for infrastructure and specialised personnel, and its considerably lower price. Moreover, it can be automated, as demonstrated by the

[1] From a historical point of view, pioneering contributions date back more than two centuries with Berzelius and continue with, among others, Sorby, Baumhauer and, above all, Campbell from 1905 onwards, who is considered by Orcel, in his documented Historical Note (1972), as the founder of ore microscopy. The reader interested in the historical aspects can find a valuable source of information in this short summary.

R. Castroviejo, *A Practical Guide to Ore Microscopy—Volume 1*,
https://doi.org/10.1007/978-3-031-12654-3_1

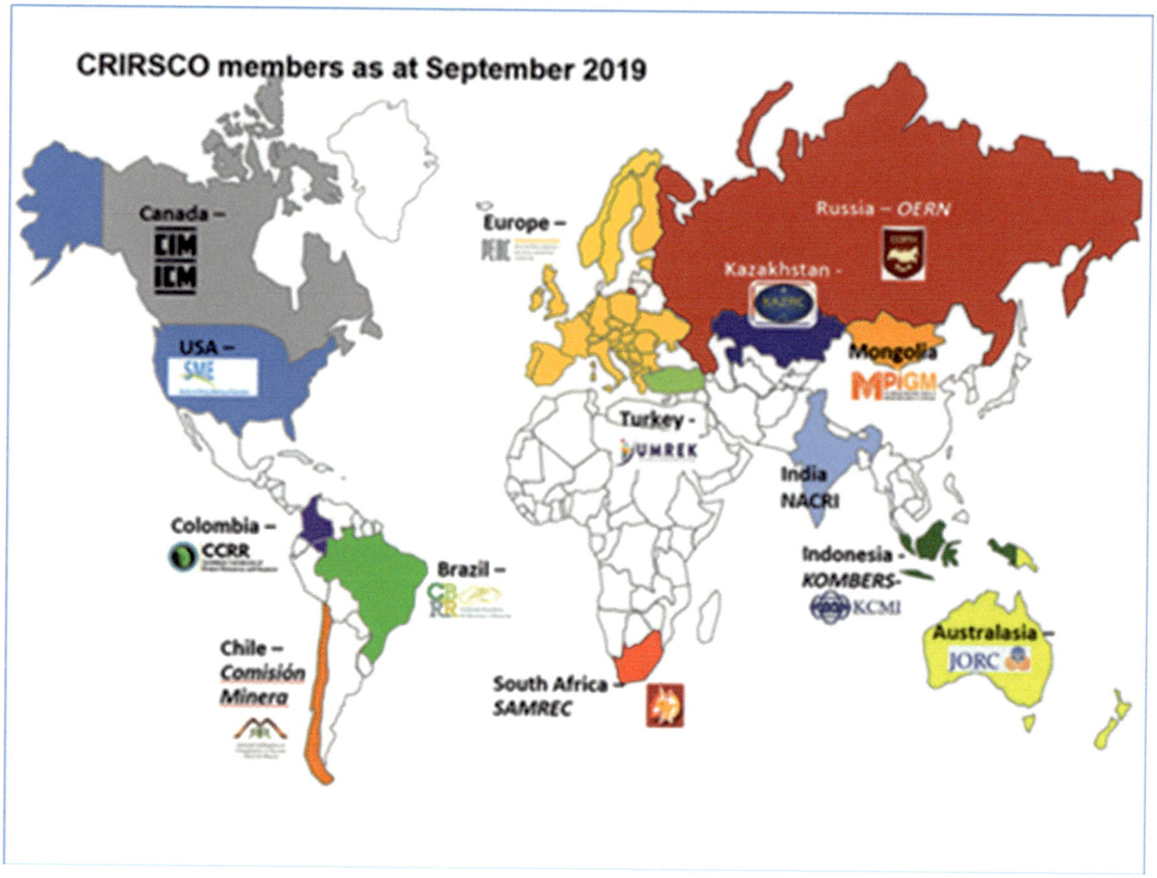

Fig. 1.1 Countries present in CRIRSCO

AMCO System (*Automated Microscopic Characterisation of Ores*: cf description in ***Part II***, vol. 2), developed at the *Applied Microscopy Laboratory* (**LMA**, ETSI Minas y Energía, UPM). AMCO offers an alternative to the aforementioned commercial *automated (electron) microscopy* systems at a fraction of the price. This makes it affordable for small-scale mining, thus putting the application of mineralogical characterisation of ores within reach for exploration and, above all, for the design and optimisation of the treatment process, as an indispensable component in the modern integrated concept of *Geometallurgy*.

In this respect, it is interesting to analyse the role attributed to microscopy (petrography and mineragraphy) in the industrial qualification of mineral resources and reserves, according to current quality *standards* for project certification. Certification is essential for investors, after the Busang scandal.[2] Today, it is unthinkable to invest in a mining project without adhering to a *quality standard* that qualifies the resources and reserves obtained in the exploration and research phase for mining development. The most important *standards* come from the Australian JORC (*Joint Ore Reserves Committee*), which is developed under different acronyms and variants, often national (Chilean Code, South African SAMREC, Canadian National Instrument **NI 43** 101, IMEB in Spain and Portugal, among others) or even continental (European PERC), as shown in the accompanying map[3] (Fig. 1.1). All of them are under the control of CRIRSCO (*Committee for Mineral Reserves International Reporting Standards*).

The purpose of these requirements is to ensure that the project is carried out according to a code of good practice, which makes it credible to the interested investor. Petrographic and mineragraphic studies are necessary to ensure quality in the qualification of resources and reserves. In fact, mineralogical-petrographic characterisation studies for geometallurgical development and their conclusions are

[2] This multimillion-dollar scam was uncovered in 1997, when analytical results certifying high grades and huge reserves at the alleged Busang epithermal gold deposit (Kalimantan, Indonesia) were found to be fraudulent. This led to the ruin of many investors who, relying on *Bre X Minerals'* reports and the involvement of major international corporations, had been putting money into the project.

[3] J.A. Espí, Universidad Politécnica de Madrid, *personal communication*.

found in Chap. 13 (*Mineral Processing and Metallurgical Testing*) of all Canadian NI 43, for example.

In view of the above, it is of the utmost importance to facilitate learning and familiarisation of professionals and *junior* engineers with the working methodology of optical ore microscopy (always a great source of information affordable for modest investments), as well as with its automation, combining machine vision and image analysis.

The present text aims to respond to these urgent needs in two ways. The main focus is on observation and recognition of ores with the "classic" reflection microscope, an indispensable tool for the investigation and processing of mineral resources. This is the subject of ***Part II*** (Chaps. 2–132), which describes the ores considered to be of greatest interest ($\approx$ 150 cf. Sect. 1.2), following the methodology described (Sects. 1.3–1.6) and illustrating their properties with the photomicrographs necessary for learning. Gangue minerals ($\approx$ 50) also deserve attention, ***Part III*** (Chaps. 133–138). Textural aspects and their significance, which are decisive for research (e.g., genetic interpretation) as well as for utilisation (e.g., optimisation of in-plant processing) of mineral resources are addressed in ***Part I***, vol. 2.

The needs of the industry, with its implicit demand for automation, should not be neglected. Indeed, automated analysis has become indispensable to effectively respond to the massive demand for mineralogical information in today's concentration plants. This demand far exceeds, in terms of volume and complexity, the capacity of a human operator, whose eyes could not possibly provide the millions of measurements required. Several automated systems based on electron microscopy (SEM) are commercially available and operational but, for optical microscopy, there has been no comparable development,[4] although similar performance to SEM systems can be achieved at a fraction of the cost. ***Part II***, vol. 2 describes this alternative, as a first in optical microscopy of ores, including conception, fundamentals, working methodology and applications (with *case studies*, in supplement) of the AMCO System, whose recently developed prototype is in use at the Applied Microscopy Laboratory (LMA/UPM).

It is thus expected to contribute to the lifelong learning, LLL, objectives that have become indispensable in today's society for students, professionals and researchers in earth, mining and materials sciences, in line with the strategy and *LLE, Lifelong Education,* priorities of the EU, *EIT-RawMaterials*.

[4] The collective synthesis *Process Mineralogy* (Becker et al. 2016) provides an update on application possibilities and developments in the field of optical microscopy (Pirard 2016). It includes a comparative table (Appendix 2) of the various analytical techniques available today, with their limitations and prospects, which shows the potential for automation of optical microscopy.

1.2 Mineral Selection Criteria

1.2.1 Metallic Ores

Among the more than 5500 mineral species recognised by the **IMA**, *International Mineralogical Association*, about one fifth could be considered as metallic ores (absorbing or opaque minerals, typically including those of economic interest because of their metallic content). Although Criddle (1998) puts the number of minerals that could be considered truly "opaque" at $\approx$660, their number could exceed a thousand if minerals that are not so absorbent but can be characterised by reflected light[5] are taken into account. However, most of them are scarce and, from an industrial point of view, irrelevant in terms of their economic interest. If abundance and industrial importance are taken into account, this number is considerably reduced, for example, 37 *common* minerals for Bowie and Simpson (1980), around 50 for other texts, $\approx$70 for CAMEVA and for the most elementary of the AMCO databases (VNIR Multispectral Reflectances, Castroviejo et al. 2010–2014).

The purpose of this text is practical—to facilitate learning regarding typically encountered ore species. Practitioners (student or professional) will acquire knowledge of the descriptive methodology that can be applied to the recognition of any other species, even those that are unknown, facilitating identification through appropriate consultation of specific sources.

The selected species are, therefore, those that can be considered *common* in the practical sense of the term, those that the mineral resources professional or researcher should recognise, either because of their *abundance* or because of their *economic or industrial importance*. For example, precious metals such as Au or Pt are not abundant but are undoubtedly important for industry. Pyrite is abundant in many deposits and is not desired from an economic point of view, but it is essential to recognise it and often to quantify it in order to control the environmental impact of its oxidation (acid mine drainage, acid rain, pollution by minor elements associated with sulphide and released when it oxidises, etc.) or to remedy it, in the case of uncontrolled discharge. Arsenopyrite is not usually abundant, but it is frequent even as an accessory. Although it is sometimes important as a gold carrier, its *industrial interest is generally negative*. Like

[5] According to the inventory compiled by Tarkian and Liessmann (1991), **1337 ores** are recorded, which is a considerable extension of the figures previously reported. However, most of them are classified as *rare* or *very rare* and, in some cases, the ore designation seems debatable (e.g., *johannsenite* pyroxene). Among them, he has selected 262 for optical determination and characterisation on R-VHN diagrams, which is also an extension of his previous selection of ***100 common ores*** also characterised by their R (monochromatic)-VHN values (Tarkian 1974).

in the case of pyrite, it must be detected in order to control As pollution of the ore or plant effluents.

Practicality, understood as a suitable approach for industrial reality, also imposes a criterion which may be *heterodox* from the strict point of view of scientific **IMA/COM** (*IMA-Commission on Ore Mineralogy*) definitions but which is inevitable if the mineralogical reports are to be useful for industry. There are terms such as limonite, psilomelane, melnicovite, bravoite, electrum, and amalgam, among others, which are rejected as mineral species by IMA/COM but which remain undoubtedly valid for the miners, because they correspond to the materials they actually encounter and must treat, concentrate or separate. It is of little use to the miner or plant engineer to be told that limonite has no mineralogical entity because it is a mixture that includes finely intergrown goethite, lepidocrocite, hematite, jarosite, silica, clay, etc. if they cannot separate and treat these species individually, because what is found in the mine is precisely the inextricable mixture, i.e., limonite. This is particularly important for automated microscopy applied as geometallurgical support, which must document and quantify the actually processed product. Therefore, not only *s.str.* mineral species but also other materials (or *mineraloids*) and varieties are described when their industrial or practical significance justifies it.

In conclusion, the ores selected are *common* ores, understood as:

- the most abundant (pyrite, hematite, magnetite), or
- the usual companions of these (either in primary or in secondary paragenesis, both associated in alteration assemblages) even if they are not abundant, or
- the most frequent, although present in minor amounts (they are usually only accessory: arsenopyrite, pyrrhotite, lepidocrocite, etc.), or
- those of major economic importance, even if scarce (Au, Pt, other precious metal species, REE ‹Rare-earth element› minerals such as monazite, etc.) or
- varieties or mineraloids of industrial interest (limonite, psilomelane, amalgam), as well as industrial minerals such as graphite.

In total, out of some 300 options (cf. ***Annex 1.1***), ≈200 minerals (*s.l*) have been selected for this text, of which ≈150 ores and ≈50 gangue components are described (cf. ***Annexes 1.2*** and ***1.3***). It is clear in this relatively large assemblage that not all minerals are of equal importance or occur with equal frequency. For this reason, specific categories have been established according to their properties (***Annexes 2*** and ***4***), and priority groups of ores have been defined for their orderly recognition. These groups have been marked in ***Annex 1.2*** (*right column*) as **Priority1 (indispensable)** or *Priority2* (*important*), while those that are not marked in this column but are nevertheless described as ores are considered as lower priority, i.e., minerals of limited interest and not frequently encountered. For descriptive purposes and to lighten the text, especially in the explanations of the photomicrographs, the photomicrographs and the corresponding ores are identified by **abbreviations**, as indicated in **Annex 1**.

The learning strategy (cf. Sect. 1.6) aims to motivate observation skills and to facilitate the progress of the beginners, including the *determinative tables* (***Annexes 4*** and ***5***), to guide the identification of an unknown ore from their own observations. If, in addition to facilitating learning, this text contributes to fostering the reader's curiosity and awakening his or her joyful capacity for wonder at the fascinating spectacle some ores offer under the microscope—an excellent motivation for this sometimes arduous and always rigorous work of observation—the author would feel amply rewarded.

1.2.2 Gangue Minerals

It must never be forgotten that an ore is a rock and, as such, all the minerals that compose it require attention, including the *gangue* components. These are almost always transparent minerals (quartz, carbonates, phyllosilicates, barite, anhydrite, epidote, etc.), and they must therefore be studied with transmitted light (petrographic microscope). However, since the technique commonly used in deposit research and geometallurgical applications is reflected light microscopy, it is of interest to characterise by this technique, as far as possible, all the minerals present, including gangue minerals. In general, the most obvious property that distinguishes the two classes is transparency of the gangue. This translates, when the sample is observed with reflected light, into a *low brightness*, and this provides the first criterion to differentiate them. The gangue appears dark, in contrast to the remarkable lustre of most ores. As brightness is a subjective judgement, an alternative that can be rigorously measured and quantified is needed: *specular reflectance*.

Reflectance is the proportion of light reflected by the surface of a mineral compared to the total light incident on that surface. This proportion, referring to light rays orthogonal to the surface of the mineral, is called **specular reflectance, R**, and it is mathematically expressed[6] as:

$$R = \frac{intensity\ of\ reflected\ light}{intensity\ of\ incident\ light}$$

[6] Since this ratio is always ≤ 1, the value of R is usually expressed in %, i.e., R (%) = 100× reflected light intensity/incident light intensity.

In the most general case (transparent minerals), the reflectance **R** can be determined experimentally, or it can be calculated as a function of the refringence (or refractive indices: **n** of the mineral; **N** of the medium) by simplifying the Fresnel equation:

$$R = \frac{(n-N)^2 + k^2}{(n+N)^2 + k^2}$$

for non-absorbent minerals (**k** ≈ **0**):

$$R = \frac{(n-N)^2}{(n+N)^2}$$

and, if the medium is air (*dry* objective), N≈1 and

$$R = \frac{(n-1)^2}{(n+1)^2}$$

This equation has been plotted in Fig. 1.2, which shows the reflectance values R of transparent minerals as a function of their refractive index n for observation in air (N≈1) and in oil immersion (N≈1.5).

Consultation of the graph (or an easy calculation, entering in each case the value of n in the formula) shows that the reflectance in air of the most common gangue minerals (n ≈ 1.5–1.6: quartz, feldspars, phyllosilicates), as well as that of the inlay resin of the polished section, is around R ≈ 5%. Other minerals differ slightly depending on the value of n, resulting, for example, in being somewhat lower for fluorite (n ≈ 1.43, R ≈ 3%) and somewhat higher for barite, inosilicates, epidote, etc. (n ≈ 1.6–1.8, R ≈ 6–8%). There is a noticeable change with orientation for minerals with high birefringence, such as jarosite and, above all, carbonates (*vbgr.* calcite: n ≈ 1.5^- to 1.7^-, cerussite: n ≈ 1.8^+ to 2.1^-), whose R values range from ≈ 4% to 12%.

The reflectances actually measured in gangue minerals usually slightly exceed these values due to the effect of internal reflections, which add to the mirror reflectance, thus increasing the amount of light reflected. However, even these values remain clearly below those typical for metallic ores (R > 12%, with very few exceptions; often >50%), which explains the characteristic grey colour of gangue and its dark appearance compared to ores.

The contrast between ore and gangue is extreme with *oil immersion objectives*. Substitute in the Fresnel equation N for its value (oil: commonly N = 1.518, for λ = 546.1 nm), very close to the n≈1.5 of the gangue, and see that

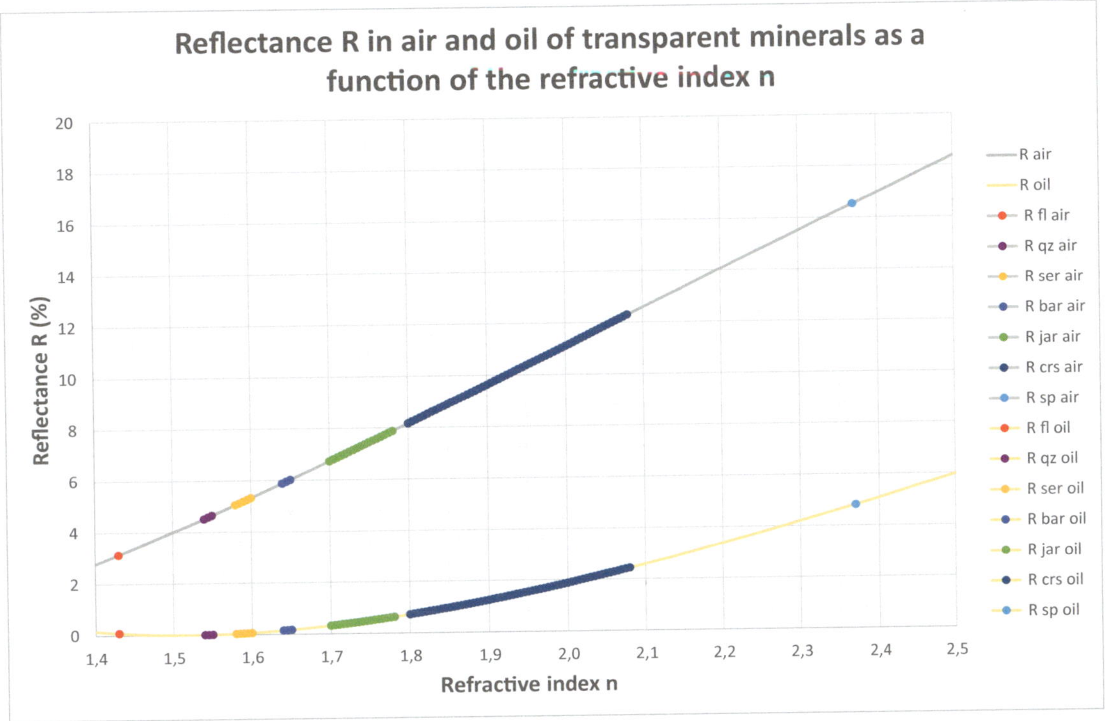

Fig. 1.2 Specular reflectance, in air and in oil, of non-absorbing (transparent) minerals, calculated as a function of their refractive index n, according to the simplified Fresnel equation (for values: k = 0, N = 1 air, N = 1.5 oil). *Minerals*: *fl-fluorite*; *qz-quartz*; *ser-sericite* (muscovite); *bar-barite*; *jar-jarosite*; *crs-cerussite*; *sp-sphalerite*. In air, the R-values of typical gangue minerals are below 12%, while sphalerite, with R ≈ 16.5–17% (depending on composition) is already projected into the range of metallic ores. In oil, R quartz falls to 0 (not visible or black) and sphalerite to R ≈ *5%*

$$R = \frac{(n-N)^2}{(n+N)^2} \approx \frac{(1.5-1.5)^2}{(1.5+1.5)^2} = 0$$

This means that the most typical silicate gangue, such as quartz, disappears in immersion (it appears black or looks like a hole). However, carbonates retain the contrast between their different orientations due to their high birefringence, so calcite disappears if aligned with R_1, but it is still visible, although very faintly, in the R_2 position. In addition, in coloured minerals such as malachite, the flashes of their internal reflections intensify the typical colours, reinforcing the light reflected on the surface. The ores, including translucent ores such as cassiterite or sphalerite, do not disappear, even if they become darker.

These data show that, with reflected light, it is possible to make observations that provide interesting information on the nature of the gangue. It is therefore considered reasonable to include among the minerals described some characteristic of the gangue, such as quartz, feldspars, sericite/muscovite, calcite, malachite, cerussite, etc.

1.3 Systematic Description Plan

The optical study of metallic ores, most of which are characterised as absorbing or opaque media, usually requires the use of reflected light (**RL**) microscopy. For this purpose, polished sections of hand samples or milling concentrates of the ore must be prepared, either in polished specimens or blocks, or in polished thin sections (for preparation techniques, see *Annex 3*).

Reflected light microscopy is based on theoretical approaches that do not coincide exactly with those of mineral optics applied to transmitted light (**TL**) microscopy for the study of common rocks in thin sections, widely used in the world of geology. Even if the comparison of the two techniques (Gribble and Hall 1992) makes their distinction apparent, it should not be forgotten that both refer to materials that are, in principle, anisotropic (minerals) and, therefore, share the need to work with polarisation. This translates into two essential features common to both types of microscopy:

- use of polarised light, involving two components—the polariser (always inserted in the light beam and in a fixed position) and the analyser (mobile), and
- rotating stage, to check and change the orientation of the minerals.

Both designs clearly differentiate them from non-polarising microscopes (*vbgr.* biological microscopes (TL) or metallographic microscopes (RL), used respectively in medicine, histology, biology or in metallography—material sciences, all of them with no polarisers) and therefore require the application of specific theoretical principles, based on polarisation optics.

These principles, as far as RL on absorbing media is concerned, have been developed theoretically in important research works, which show the complexity of the phenomena involved. The reader interested in their theoretical foundation (mineral optics) can find a rigorous but accessible synthesis in Galopin and Henry (1972) and in more succinct or informative presentations, but sufficient for basic work in ore microscopy, in the aforementioned texts by Criddle (1998) and Gribble and Hall (1992), as well as in Craig and Vaughan (1994) and, in Spanish, Febrel (1970). More detailed presentations can be found in Peckettt (1992) and Sanderson (2019).

In what follows, these principles, generally taught in the university teachings of the subject, are assumed to be known, and readers are referred to the recommended bibliography, should they need to complete their knowledge. However, the procedures applied to achieve a reliable identification, which depends first and foremost on rigorous observation and systematic and objective description of the relevant properties, will be briefly discussed.

In order to facilitate systematic work, these properties have been arranged according to a descriptive model, with an identical card format for all minerals. Each description is enriched with original photomicrographs, in which special care has been taken to ensure fidelity to the observed colours. This follows the path taken by Picot and Johan (1982), who showed the diagnostic value of polarisation colours, illustrating them for the first time with high quality, reliable and very useful colour photomicrographs.

Now, taking advantage of the possibilities of *Internet* presentation, the graphic information is extended by numerous digital photomicrographs, whose printing on paper requires special care to ensure reproducibility in print of the true colours. Even if consulted or downloaded as virtual images, their reliability requires some basic computer requirements as well (cf. Sect. 1.5.3). Not only the polarisation colours (polariser plus analyser) are documented graphically, but also the natural colour, brightness and pleochroism (observation with polariser only) and, where appropriate, peculiarities such as light etching, tarnish (*patina*) or alterations, micro-structures, etc.

The systematic approach involves orderly observation of the visible properties:

- with plane (or linear) polarised light: using *one polariser only*, abbreviated **sp** (**single polariser**, a setting formerly known as parallel *nicols* or N//, designation now obsolete, and equivalent to *Plane Polarised Light* PPL)

- in air (dry objective, *dry*, abbrev. **d**) and
- in oil (oil immersion objective; *oil*, abbrev. **o**)

• with double polarisation, i.e., with polariser and analyser crossed at 90° (abbrev: **+p**; formerly *crossed nicols* or NX; in English, *crossed polars, observation between two polars*: XPL or +nic)
  - in air (dry objective, *dry*, abbrev. **d**) and
  - in oil (oil immersion objective; *oil*, abbrev. **o**).

This results in the four basic combinations that will be used in all descriptions. They are indicated by the abbreviations **dsp**, **osp**, **d+p**, **o+p**. Occasionally, observations are made with a slightly rotated analyser (e.g., crossed at 87° of the polariser instead of the usual 90°), which makes it easier to characterise some polarisation effects (e.g., in weakly anisotropic minerals) and is indicated as **xp** (**dxp**, **oxp**). When interference contrast is required, the *Wollaston Prism* is inserted at the +p position, and this setting is denoted as **+p.w** (resulting in: **d+p.w** or **o+p.w**).

The properties thus observed are described following a constant model (card), grouping them under the two headings: **sp** (appearance, morphology, colour, reflectance, polishing, exfoliation, hardness) and **+p** (extinction, anisotropism, internal reflections, twinning). Many authors follow a different order of presentation, for example, distinguishing physical and optical properties (Criddle 1998) or describing them independently (Febrel 1970) or grouped in different categories (e.g., *polishing properties, reflection behaviour, etching, physico-chemistry, fabric;* Ramdohr 1980; *Colour, Reflectance, Anisotropism, Texture, Associated Minerals*; Picot and Johan 1982). The choice of this text is to follow the natural order of observation according to the setting of the microscope. So the observer always knows how to look for a property and, above all, is facilitated in working with the microscope. The aim is to pay attention to each and every one of the properties listed. These are to be described systematically, in order to achieve identification. For the sake of clarity, a brief practical guide summarising and graphically showing the main properties to be observed is presented (Sect. 1.4).

The photomicrographs illustrating any of the properties are always *identified*, both in the image (label in the *lower righthand corner*) and in the caption or explanatory legend, with a code containing the symbol or abbreviation of the mineral (cf. explanation and list in Annex 1). The microscope setting is also reported with its code (*dsp , osp , d+p , o+p*). In addition, the *graphic scale* is integrated in each image by means of a horizontal bar (100 μm, *upper lefthand corner*), and another label (*lower lefthand corner*) refers to the sample (polished section or polished thin section).

Finally, the characterisation is updated with the spectral information available for each ore: specular reflectance (**R**) determined with the AMCO System (cf. Chap. 140). This information, obtained from multispectral images composed of 20 reflectance bands in the VNIR range (from 370 to 1000 nm), is applied here for the first time in a systematic way and provides an objective and precise complement to the qualitative observations described. It is presented in the form of R–λ (Reflectance–Wavelength) tables and graphs and is also compared with the available published information, in the visible range (400–700 nm), usually by Criddle and Stanley (1993) in QDF3 and also by Picot and Johan (1982).

In contrast to the classical texts, the techniques of chemical attack and microchemical identification (Short 1940), which (with some exceptions) are rarely used today, have been omitted here. More precise compositional information is obtained with currently available microanalytical techniques such as the electron microprobe or electron microscope, whose complementary use to optical microscopy is documented by Tarkian and Liessmann (1991). However, although the high price of the microhardness tester has greatly restricted its use and it is no longer found in most mineralogical laboratories, *Vickers microhardness* (*VHN*) information is retained as relevant. Its combination with reflectance data (R %—VHN tables, McLeod and Chamberlain 1968; Tarkian 1974; Tarkian and Liessmann 1991) can effectively aid identification, based on quantitative determinations, free of subjective components.

The methodology used and reflected in the descriptive sheets—Boxes 1.1 and 1.2—corresponds to the most widespread identification strategy (e.g. short courses endorsed by the IMA Commission on Ore Mineralogy, Jambor and Vaughan 1990; Cabri and Vaughan 1998), but it is not the only possible methodology.

Cameron (1961) described phenomena linked to conoscopic properties and crystalline symmetry, which could give rise to an alternative strategy. This has had few followers, perhaps because it requires a deeper understanding of optical theory, so its applications are not widespread. The interested reader can find more up-to-date information in Pecket (1992). Cervelle and Moëlo (1990) offer a synthesis, rigorous but accessible to the non-specialist, of the relationships between optical properties, crystalline structure and type of polarisation (plane or linear, circular, elliptical). This is a field in which the observer can undoubtedly go even deeper and take advantage of the observation of angles and properties of extinction, rotation of reflection, polarisation colours at 45°, etc. in relation to the structure and orientation of the mineral. It would be highly recommended to do so for a more complete characterisation. However, this has not been the choice of the present text, nor of the most widespread systematic works (e.g., Ramdohr 1980; Picot and Johan 1982; Uytenbogaardt and Burke 1971; Schouten 1962). Priority has been given to the simplicity and agility of the method, to avoid overloading the descriptions. Consequently, the identification strategy chosen focuses on observation of the properties recorded in the description sheet (explained in Sect. 1.5, Box 1.1 and Box 1.2).

For some of these properties, although of a qualitative nature (polishing hardness, anisotropism), their intensity has been estimated by means of a numerical scale, in view of the potential of computer resources based on the digitisation of optical parameters for automated identification (*vbgr.* Bernhardt 1979, 1987; Atkin and Harvey 1979; Kühnel et al. 1980: *Delft System*, reworked for MS-DOS by Glass and Voncken, at TU Delft, The Netherlands; Gerlitz et al. 1989; Shoji and Kaneda 1994, among others; *cf* discussion: Bernhardt 1990). However, the automation used in the present text is based only on multispectral specular reflectance measurements (AMCO System 2018, Chap. 2, vol. 2).

1.4 Practical Guide for Microscopic Characterisation

A prerequisite for work with the RL microscope is the careful preparation of representative polished sections of the ore to be studied. These can be polished blocks, polished transparent slides or even double-polished slides (those used for studies of fluid inclusions); the slides must never have coverslips. The ore may come from hand sample (embedded in a block or specimen with *epoxy* resin) or from concentrate (e.g., from pan or minerallurgical plant). Good *polishing quality* is essential, as poor polishing may not only remove some species from the ore, but even those that are visible may be unrecognisable. Excessive relief, due to differences in hardness, must be avoided, especially when polishing by hand with soft cloths. This requires continuous control of the preparation process (cf. *Annex 3*).

The study should start with a low *magnification* objective (2.5× or 4×, even a *stereomicroscope*) and then progressively move to objectives of increasing magnification, as required by the *grain size* of the ore[7] and the properties to be observed. A complete study usually requires at least 20× magnification objective (both in air and in immersion), although the 10× objective is perhaps the most suitable for most properties. The 5× objective is also very useful and the 50× objective, preferably in immersion, may be necessary for very fine particle sizes (e.g., precious metals, often ≤ 10 μm) but for polarisation it is unreliable. Wide field lenses (as the 2.5× magnification objective), which are optimal for an overview, to appreciate textures, interpretation of paragenesis etc., also tend to introduce chromatic aberrations with polarised light and are, therefore, unreliable for defining anisotropism.

[7] *Table 1.1* in chap. 1, vol. 2, defines the grain size categories used.

Observation in *immersion* (*oil* = **o**) complements that carried out with a *dry* objective (air, **d** = *dry*), with important advantages: by generally decreasing the brightness (more precisely, the absolute value of the reflectance, **R**, according to Fresnel's formula), the relative differences between low R minerals (such as sphalerite and magnetite) increase, making it easier to distinguish them. Typical gangue minerals such as quartz, whose reflectance goes from ≈5% in air to 0% in oil, “disappear”. For the same reason, the contrast between the different R values in anisotropic minerals (R_1–R_2; R_a–R_b–R_c) increases. This facilitates the definition of bireflectance and pleochroism. Moreover, it improves the sharpness of the image and increases the intensity or saturation of the colour. This translates into a higher value of the **Pe** parameter, or *excitation purity*, as can be seen in the descriptions (cf. § *Box 1.1*, *Summary-Table*, comprising descriptive data and QDF3 parameters, after Criddle and Stanley 1993) and allows a better appreciation of anisotropism and internal reflections. Finally, the mere observation in oil can be diagnostic (cf. description and photomicrographs **F cv**, covellite).

For all these reasons, it is useful to compare both types of observations (*d* and *o*). Therefore, a first set of four images (or *Basic Photos*, d and o photomicrographs, with and without analyser) enlighten each ore description. While recognising the inconvenience of changing lenses for immersion and the need to carefully clean the oil afterwards, users will be able to see by themselves the usefulness of immersion, e.g., to define a difficult paragenesis or to study (and not overlook!) fine-grained precious metals; they will also soon discover that it is highly recommended for common ores.

The description must be based on an orderly and systematic determination of the properties (cf*:* § *Box 1.1* and § *Box 1.2*), including observations:

- with **polariser only** (or single polariser, **sp**, formerly *N//*, or *parallel nicols*, refering to *Nichol prisms*, which are no longer in use)
- and with **polariser and analyser crossed at 90°** (**+p**, formerly *NX*; sometimes it is useful to uncross them slightly, up to ≈**87°**: **xp**);
- in some cases, *interference contrast*, ***i.c.*** , is used in order to better observe properties such as zoning, the presence of inclusions or the quality of the polishing (e.g., Fig. 1.3, Fig. 63.4, etc.). The surface quality can also be affected by tarnishing but, as shown by Figs. 1.3a to 1.3c, this can be corrected by slightly re-polishing, and *i.c.* can be used to assess the result, as well as other properties, like relief, scratches, etc. For this setting, which is indicated as **+p. w.**, the +p setting is modified by intercalating the *Wollaston Prism* and rotating the analyser until the optimum contrast is obtained.

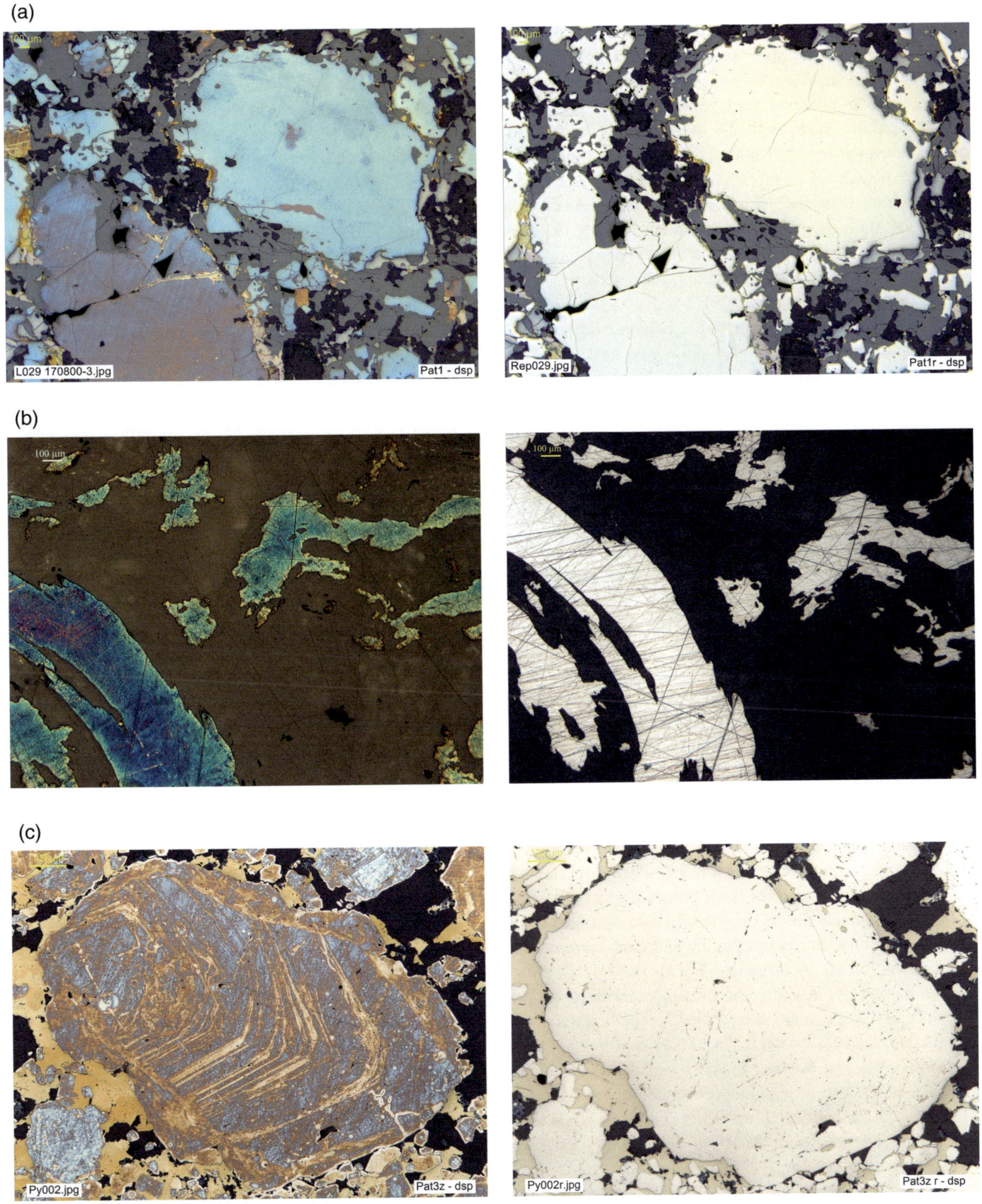

Fig. 1.3 Series of photomicrographs (1.3a to 1.3c from top to bottom) to compare tarnished sections (lefthand column) and same sections clean (righthand column) after repolishing (0.25 µm diamond paste). Figure 1.3a (obj 2.5x, *dsp*). Pyrite (yellowish white), chalcopyrite (yellow), sphalerite (grey), pyrrhotite (pinkish brown). Figure 1.3b (obj 5x, *dsp*). Native silver. Figure 1.3c (obj 10x, *dsp*). Pyrite and chalcopyrite: zoning in pyrite microclast evidenced by tarnishing patina, which disappears on repolishing.

Finally, it should be remembered that the study can be completed with the observation by transmission, **T**, by means of the **petrographic or transmitted light microscope**. This can be done on the same section studied by reflection, when polished thin sections (*LTP*) or double polished sections (*LDP*, typically used to study fluid inclusions) are available. Transmitted light microscopy is not among the objectives of this text but, in some cases where the mineralogy (especially for low reflectance minerals) or the type of study (e.g., the definition of textures or alterations) makes it advisable, T characterisation has been added to the usual RL description, indicating the settings:

- **TSP**: transmitted light, with single polariser (**sp**),
- **T+P:** transmitted light, with **polariser and analyser crossed at 90°** (**+p**).

1.4.1 Properties Observed with Polariser Only (Single Polariser, sp)

These properties (*sp*) include appearance or visual perception, determined by the response to polishing (or surface quality, relief, pits or pull-outs, scratches, etc.), colour, gloss or brightness, reflectance (which quantifies gloss), bireflectance and pleochroism:

- The perception and description of colour is inevitably subjective and depends not only on the visual qualities of the observer but also on the medium (d, o) and the accompanying minerals.
- The colour can change, sometimes very rapidly (e.g., in Ag, As, bornite), due to surface alteration or spontaneous oxidation of the polished surface (*patina* formation or *tarnishing*) after polishing and it is, therefore, advisable to carry out the study with the fresh polish in order to not be obliged to repolish the section; in any case, this alteration can provide an identifying criterion and must therefore be recognised (cf. series of images Fig. 1.3).
- *Light etching*, typical of some minerals and in particular of several of the main silver ores, is both a help—it facilitates diagnosis—and a hindrance, especially if it is instantaneous and intense. It alters and decomposes the surface of the ore, as can be seen in the images accompanying the description of acanthite (Fig. 3.8), stephanite (Fig. 117.3), polybasite-pearceite (Figs. 95.3–95.5) or myargirite (Fig. 83.7).
- The proliferation of colour names and the subjectivity of their definition has been a source of confusion in the literature. To avoid this, *12* well-known *colours* -or 13 if black is called a colour- have been selected for descriptions: white, grey, black, blue, green, yellow, orange, red, purple, violet, pink, olive, brown. Only these and their combinations are used in the text for all purposes. They are listed, with their abbreviations, at the end of the descriptive scheme (cf. § *Box 1.1*).
- To achieve the required objectivity, constant illumination must be maintained (in this text, the reference is CIE *illuminant C*). This is approximated if the source is, as the one used in this work, a halogen lamp at its nominal power, with a blue or daylight filter interspersed. LED sources, which are becoming more and more widespread, must be adapted in each case. The intensity must be kept constant and, if it is necessary to reduce it to avoid being dazzled by very bright ores, neutral (grey) filters can be used.
- As far as the spectral response is concerned, the optimum sensitivity of the human eye for brightness comparison, i.e., for estimating the reflectance value, corresponds to $\lambda \approx 546$ nm. This is achieved by inserting a green filter usually offered by the manufacturers. In this way, R values corresponding to *white light*, such as those offered by Picot and Johan (1982), can be compared.
- The multispectral measurement of *reflectance*, **R**, requires specific equipment, but even if this is not available, R information is always of interest and can be of diagnostic value. Multispectral R values are available in the visible spectrum, 400–700 nm (Picot and Johan 1982; *QDF3*: Criddle and Stanley 1993). In the VNIR spectrum, 350–1000 nm (visible and near infra-red), they have been measured with the CAMEVA (Castroviejo et al. 2014) and AMCO (Chap. 2, vol. 2) Systems and compiled in databases. These allow comparative multispectral plots (displayed for each of the ores at the end of the description), which are instrumental for ore characterisation and provide, as well, a powerful tool for automated identification, supported by computer vision.
- R remains constant in optically isotropic minerals, but it varies with orientation in anisotropic ores. The latter are said to be *bireflective* if what changes is only the magnitude of R (the same grain gains or loses luster with no colour change when the stage is rotated, Fig. 1.7-top, left, rutile) and *pleochroic*, if the colour also changes (Fig. 1.7-bottom, left, berthierite). This occurs because the slopes of the multispectral curves, R_1 and R_2 or R_a, R_b and R_c, etc., are parallel in the first case and changing—even intersecting sometimes—in the second.

Other properties observed with single polariser (*sp*) to be noted are:

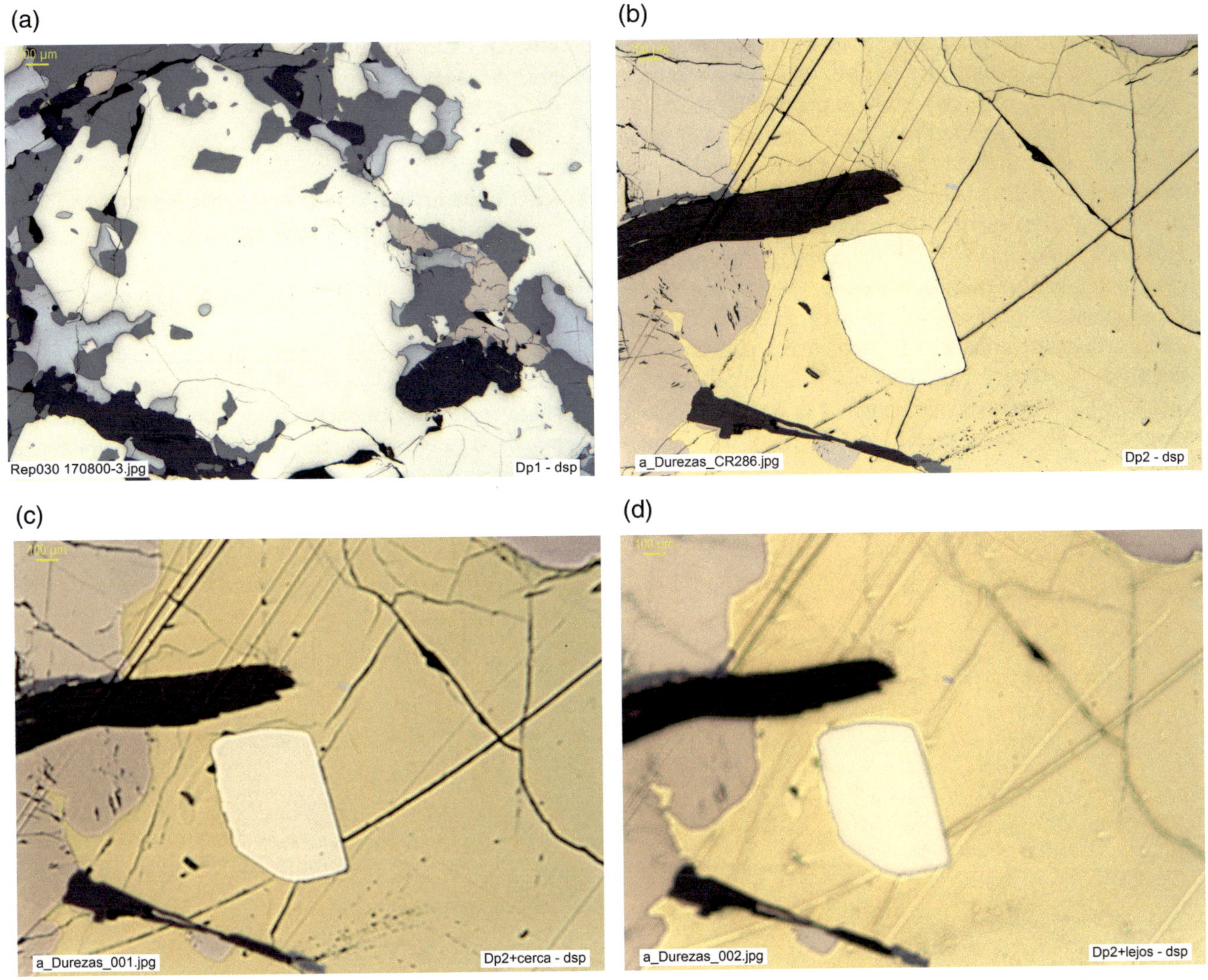

Fig. 1.4 (obj. 5×, dsp) Polishing hardnesses, PolH, of pyrite (yellowish white), chalcopyrite (yellow), sphalerite (grey), galena (light grey), pyrrhotite (pinkish brown) and mica (black), compared. By the relief of pyrite (hard) on galena (soft), evident in the contacts Fig. 1.4-Dp1. By the displacement of the white line when changing the focus (Fig. 1.4-Dp2; zooming in; Fig. 1.4+cerca=bottom left, zooming out; Fig. 1.4+lejos=bottom right) with respect to the focused position (Fig. 1.4-Dp2), resulting in PolH pyrite > pyrrhotite > chalcopyrite > mica. The polishing scratches (well visible in Fig. 1.4, increasingly thinner in the order mica-chalcopyrite-pyrrhotite until their final disappearance in pyrite, indicate a scratch hardness coincident with the PolH data)

- *Polishing hardness*, ***PolH*** (or **Dp**[8]): if there is much difference between the PolH of adjacent minerals, those that are harder stand out (*relief*, Fig. 1.4). Excessive relief is an impediment to study, but moderate relief is an aid, because it allows comparing the PolH (Dp) of two adjacent minerals, observing the luminous contour (white line or *Kalb's line*) that forms at the contact edge when the contrast is increased (closing the diaphragm slightly, obj 10×) and the field is gently defocused. When zoomed out, the white line moves towards the softer mineral and when zoomed in, the opposite happens (cf. **F** series Fig. 1.4). Careful observation of the polishing scratches also provides a useful criterion: the scratches are thicker on soft minerals and become thinner or disappear on harder minerals (cf. Fig. 1.4);
- *Cleavage*: the cleavage (or *exfoliation*) observed in hand samples is not always visible under the microscope. It depends to a large extent on the polishing and is, for some species, a relevant feature (*vbgr.* prismatic 100, cubic, octahedral, rhombohedral, etc.). Sometimes it is enhanced by characteristic intersections (*vbgr. triangular pits* of galena: Fig. 1.6).

[8] The photomicrographs Fig. 1.4 showing the Kalb's Line estimation of Polishing Hardness (abridged *PolH*) are labelled **Dp** (from the original Spanish equivalent term for PolH, *Dureza de Pulido*).

- The *morphology* of the grains is also of interest. Some minerals, such as cassiterite, have a marked tendency towards idiomorphic development, while with others, such as galena, the opposite occurs. Still others are colloform, cryptocrystalline, etc. The morphological characterisation includes several aspects: occurrence in masses or aggregates—and, in this case, crystallinity—or in isolated crystals, growth tendency (idiomorphic or euhedral, allotriomorphic or anhedral, pseudomorphic, etc.), habit, zoning, among others.
- Zoning usually responds to slight compositional differences which are enhanced by chemical attack to make them visible. Occasionally they can be observed without attack, e.g., when there are contrasts in hardness, colour (**Fig. 1.3a, vol. 2**) or alteration (Fig. 1.5a), weathering, tarnish (or patina, Fig. 1.3), and also when there are differences in the effects of anisotropism (visible in +p position, Fig. 1.5b–c).

1.4.2 Properties Observed with Polariser and Analyser Crossed at 90° (+p)

By polariser and analyser crossed at 90° (**+p** position),

- the optically *isotropic* (cubic minerals and amorphous materials) or *anisotropic* (all crystalline materials except cubic materials) character is determined first and foremost, as evidenced respectively by the permanent

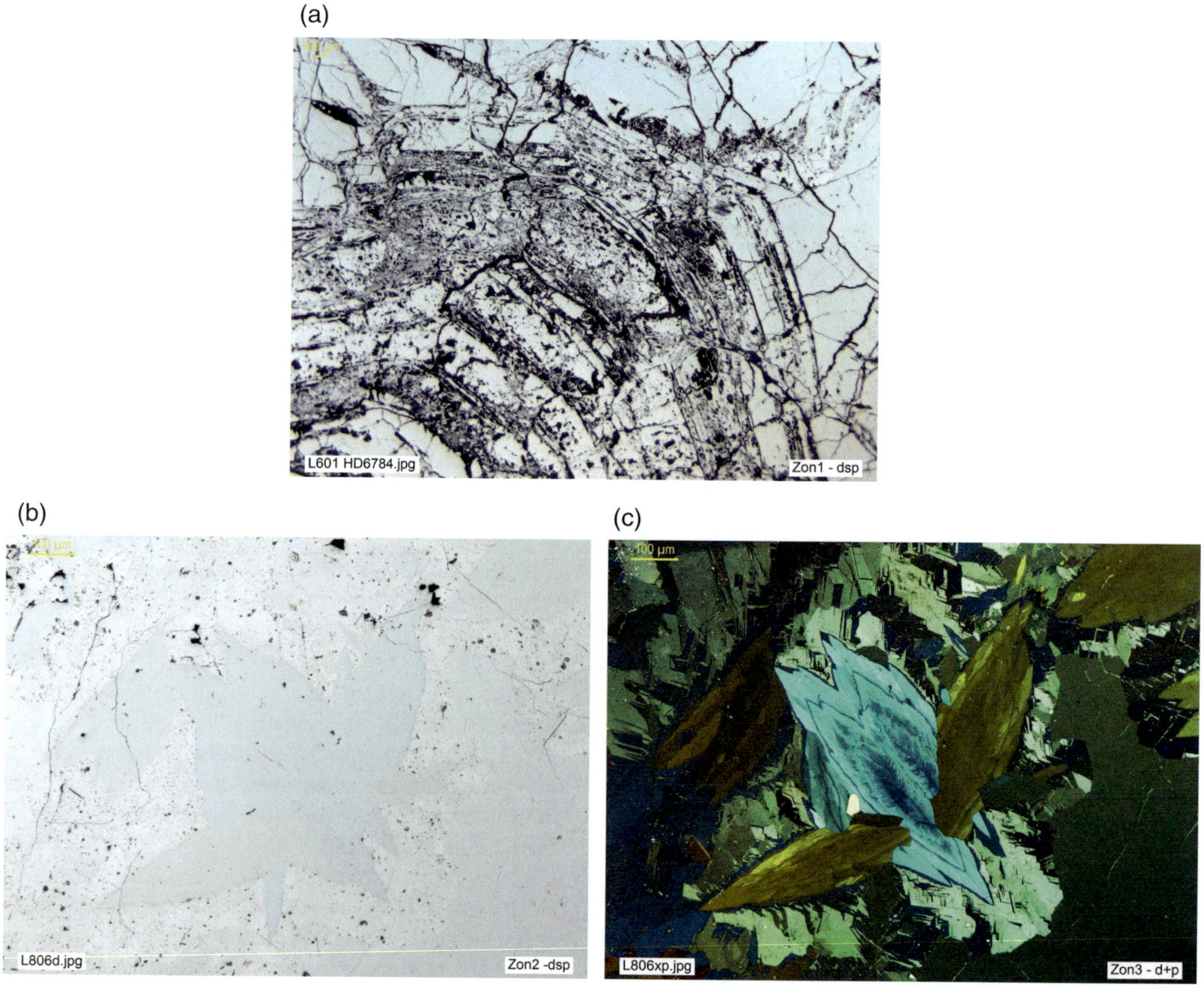

Fig. 1.5 **a** Zoned skutterudite, partially altered and replaced by rammelsbergite, both white (obj 2.5×, dsp). Figure 1.5 **b–c** Zoning in löllingite, visible only with *+p* setting (Fig. 1.5c). Löllingite is intergrown with rammelsbergite (polysynthetic twins, greenish anisotropism), which comprises the groundmass (obj 10×, **dsp** - **d+p**, resp.)

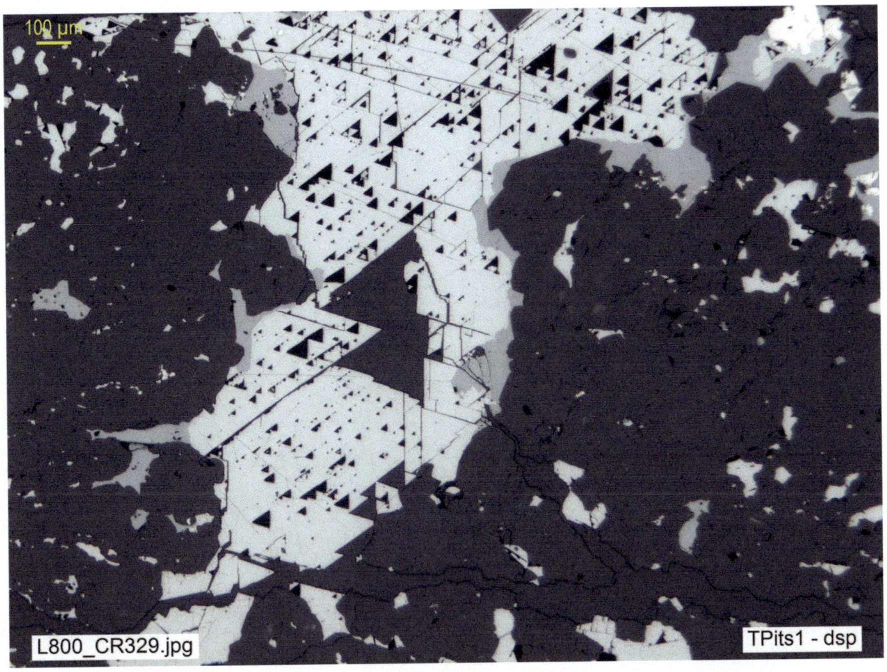

Fig. 1.6 Cubic exfoliation (100) and triangular *pits* in galena (obj. 5×, dsp)

extinction when rotating the stage in the first case and by the alternation of extinction positions (repeated every 90°) and illumination (maximum at 45° of the former), in the second case;

- other data such as the *type of extinction*, the *extinction angle* or the intensity and colours of anisotropism or polarisation tints must be specified. There are supposedly isotropic ores, such as native metals, whose extinction is *incomplete*, an effect accentuated by polishing scratches. Others, such as pyrite or magnetite, may occasionally show *anomalous anisotropism*;
- there are anisotropic minerals whose anisotropism effects are so weak that they could easily be confused with cubic ores. That is why pinpointing the *intensity of anisotropism* is important, as well as the *polarisation tints or anisotropism colours*, which provide a very important identification criterion, as Picot and Johan (1982) have shown. Sometimes, it is useful to slightly uncross (by rotating ≈ 2°–3°) the analyser (***xp*** setting) to observe weak anisotropism or to obtain additional information on polarisation colours; in special cases ***+p.w***.
- *Twins*[9] are regular intergrowths of two or more individuals of the same species. As these differ in orientation, even though they belong to the same grain, they are evident in anisotropic minerals by their extinction in different positions (Fig. 1.7). In isotropic minerals, as they are all in permanent extinction, it is not easy to distinguish them, unless they can be identified by morphology, grain contours or other features. Some minerals have such a strong bireflectance that it is possible to see the twins even without an analyser (Fig. 1.7-bottom left). The twins can be characterised by different criteria: contact, penetration; lamellar, sectored, irregular; single, multiple, polysynthetic. By their origin, *growth twins* are formed in equilibrium as the grain grows, producing usually simple forms; *inversion twins* are due to transformation, e.g., as the primary phase cools and becomes unstable (potentially useful as a geothermometer) and *deformation or pressure* twins are induced by post-crystalline stress.
- The more weakly absorbing ores (e.g. sphalerite or cassiterite), as well as all transparent minerals, do not reflect all the incident light but a part penetrates into the crystal and is reflected from within the crystal in microcracks or reticular planes. It is perceived by the eye as an additional luminosity, showing the mineral's own colour, which interferes with the polarisation colours and can even mask them (Fig. 1.7-top right). This *diffuse luminosity* is also referred to as *internal reflections*, **IR**, and can vary with the composition of the mineral (e.g., from colourless or yellowish and abundant in pure sphalerite to red or brownish and scarce in the iron-rich variety, marmatite, or green in gallium-bearing varieties; from scarce in ferberite or Fe-wolframite to massive and luminous in the Mn term, hübnerite). The abundance, colour and intensity of the RI are important data.

[9] Also called Macles, a common term in various languages, as abridged in the photomicrographs: Fig. 1.7.

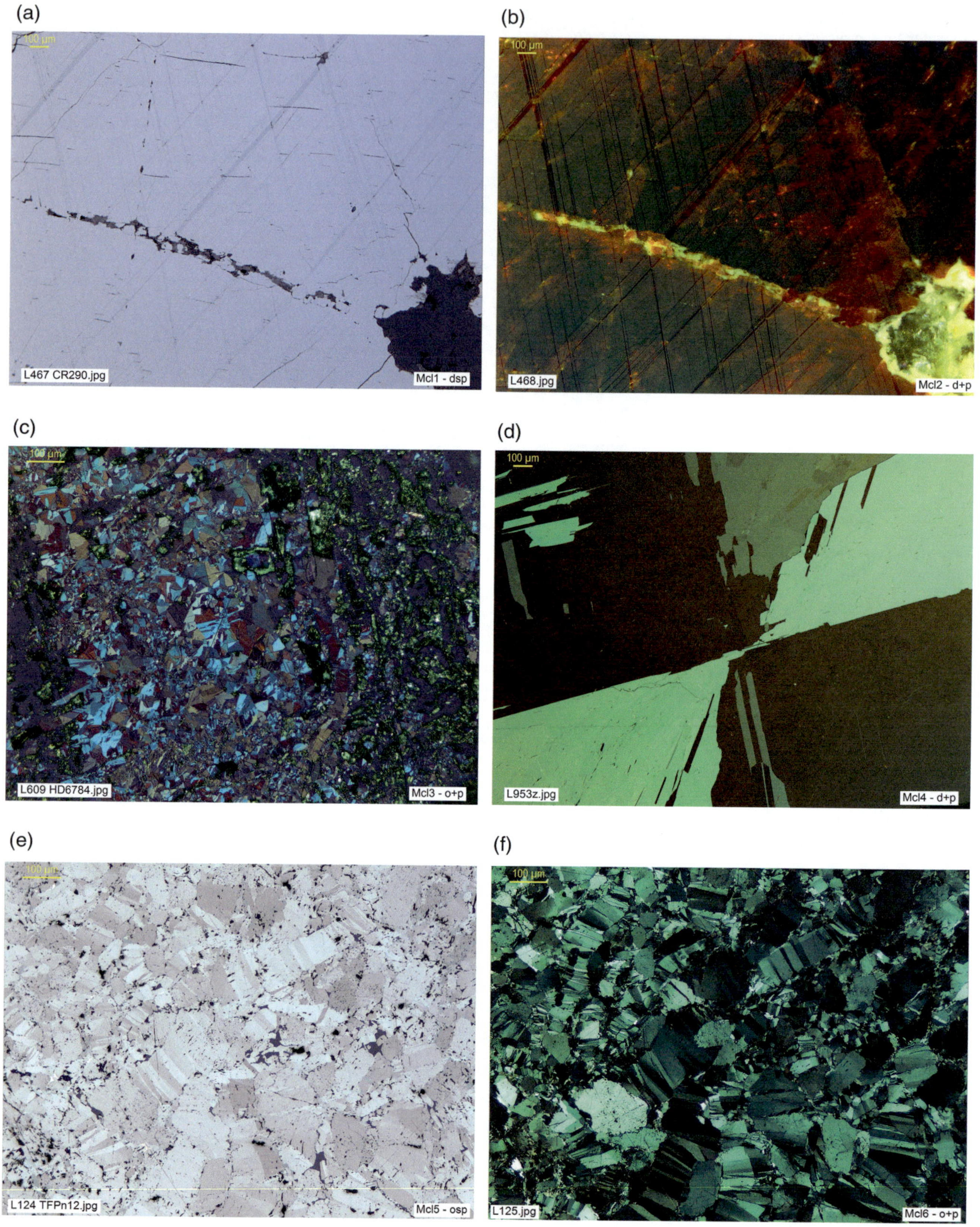

Fig. 1.7 Twinning. Rutile twins (Fig. 1.7-top, left/right=dsp/d+p, laminar, visible even with no analyser or *dsp setting*, due to its high bireflectance), rammelsbergite (Fig. 1.7-center, left=o+p, polysynthetic), marcasite (Fig. 1.7-center, right=d+p in sectors) and berthierite (Fig. 1.7-bottom, left/right=osp/o+p deformation, visible already *osp* due to strong pleochroism)

Box 1.1 General Outline Followed for Description of ORES

(Summary Table of Properties, Descriptive Text and Definitions)

Summary Table with Basic Information

MINERAL *(name, abbreviation/synonyms)*	**Formula**	**Symmetry**

Mineral		R_{546} (%)	Y (%)	λ_d (nm)	P_e (%)	Colour sensation
Air	R_1					
	R_2					
Oil	R_1					
	R_2					
Bir: Bireflectance (**1–5**)				*Pleoc*: Pleochroism (1–5)		*Tarn*: Tarnish (=*Pat*: Pátina)
Anis + P: Anisotropism + P (**1–5**)				*IR*: Internal Reflections		*Twg*: Twinning (=Mcl: Maclas)
Mor: Morphology				*Clv*: Cleavage (=*Exf*: Exfoliation)		*PolQ*: Polishing quality
Hardness	*PolH*: Polishing Hardness (**1–5**)			*Mohs*:	*Talm*:	*VHN* $_{(100\ g)}$: Vickers Microhardness$_{(weight)}$
Obs **:** Specific remarks/Spectral information and VHN sources/*Abbreviations*						

Parameters, Symbols and Concepts Used

R = Specular reflectance (%), defined in Sect. 1.2.2, refers to the symmetry of the mineral:

- In cubic minerals, **R** = single value
- For anisotropic minerals, $\mathbf{R_1}$ = minimum, $\mathbf{R_2}$ = maximum (or $\mathbf{R_o}$, $\mathbf{R_e}$, **R'**, $\mathbf{R_a}$, $\mathbf{R_b}$, $\mathbf{R_c}$, etc. if applicable)
- In possibly anisotropic minerals, whose R measurement did not allow safe separation of the two curves R_1 and R_2 because these are very close or uncertain, e.g. due to scarcity of the ores, **R′** = value provisionally accepted as representative

Air/oil = observation with dry objective, **d** (air)/in oil immersion (**o** oil), *standard ISO 8036-1*.

$\mathbf{R_{546}}$ (%) = reflectance measured for wavelength 546 nm.

Y (%) = luminance or luminous reflectance, equals the *integrated* brightness or reflectance in the vision of a *standard* or statistically representative observer, with CIE *illuminant C*.

$\boldsymbol{\lambda_d}$ (nm) = dominant wavelength, suggests the subjective perception of colour (*hue*) by a *standard* observer, with CIE *illuminant C*.

$\mathbf{P_e}$ (%) = excitation purity, corresponds to the saturation or intensity of the colour compared to white light (CIE *illuminant* C).

Colour or chromatic sensation = description of the subjective perception of the colour and brilliance of the mineral as seen by an observer, as far as possible statistically representative, with polarised light (polariser only) and CIE *illuminant C*. The *colour names* accepted are exclusively those selected from the **list** (and abbreviations) below.

Bir = characterisation of bireflectance (observed with polariser alone, without analyser, *sp* setting) by updating with numerical code (**1 to 5**) the categories of Galopin and Henry (1972):

- *strong* (**S** = **5**): striking and always visible because of the strong contrast between R_1 and R_2 (e.g., covellite, graphite, molybdenite)
- *medium* (**M** = **4**): visible even in isolated grains although the contrast is moderate (e.g., bismuthite, hematite, pyrrhotite, ilmenite)
- *weak* (**W** = **3**): visible by comparison with adjacent grains but with difficulty in isolated grains (e.g., enargite, stannite, stephanite)
- *faint to absent* (**F–A** = **2**): not perceptible to the eye, even when comparing neighbouring grains of different orientation but demonstrable by inserting the analyser (+p setting), revealing the anisotropic character of the mineral (e.g., acanthite, arsenopyrite, chalcopyrite)
- *null* (**N** = **1**): non-existent due to cubic symmetry of the mineral (e.g., chromite, pyrite)

The same numerical categories apply to the characterisation of **pleochroism**, which can be considered as a particular case of bireflectance with colour change.

Hardness (H) = qualitative H (**PolH** = Polishing H, **Talm** = Talmage H) or quantitative H (**Mohs** = Mohs H; **VHN** = Vickers Microhardness Number). The polishing hardness, **PolH**, is estimated according to the categories (coded from **1** to **5**, mod. Schneiderhöhn 1952):

- *very low* (**1** = very soft minerals, PolH < antimonite),
- *low* (**2** = soft minerals, PolH < galena),
- *medium* (**3** = *PolH* between galena and pyrrhotite),
- *high* (**4** = hard minerals, PolH > pyrrhotite)
- *very high* (**5** = very hard *minerals*, PolH > pyrite).

It is defined by direct comparison, using Kalb's line, with adjacent minerals, e.g., for alabandite, medium PolH, it might be constrained as $PolH_{alb}$ >> $PolH_{ac}$ (acanthite); $PolH_{alb}$ > $PolH_{ccp}$ (chalcopyrite); $PolH_{alb}$ < $PolH_{sp}$ (sphalerite); $PolH_{alb}$ << $PolH_{py}$ (pyrite). Ores are sorted by PolH in *Annex 4.*

IR = internal reflections, also described by some authors as diffuse reflectance or scattered light (reflected from lattice planes, microcracks or defects from inside the crystal).

Anis +P = description of the polarisation colours or *anisotropic rotation tints* (Criddle 1998), observed with polariser and analyser crossed at 90° (*+p* setting), when rotating the stage from extinction to maximum illumination positions, according to the five categories of Galopin and Henry (1972), identified with a numerical code (from 1 to 5)*:*

- **1** = *isotropic minerals with good extinction* (typically low to medium R, e.g., sphalerite, chromite, alabandite)
- **2** = *isotropic* minerals *with imperfect extinction* (typically medium–high R, e.g., pyrite, skutterudite, native metals)
- **3** = *weakly anisotropic* minerals (bireflectance not visible with *sp setting*, e.g. chalcocite, acanthite, cobaltite)
- **4** = *clearly anisotropic* minerals (clearly defined extinction positions with +p; with sp, visible bireflectance; e.g. pyrrhotite, bismuthinite, arsenopyrite)
- **5** = *strongly anisotropic* minerals (very clearly defined extinction positions and the colours in between are very bright, sometimes very vivid and varied, e.g. covellite, marcasite, nickelite, graphite)

In order to avoid misunderstandings and ambiguities, the *colours shall be* described using only the 13 terms listed in the following list of abbreviations and their possible combinations.

Colour abbreviations

- **Colour names used**: *WHITE–GREY–BLACK–BLUE–GREEN–YELLOW–ORANGE–RED–PURPLE–VIOLET–PINK–OLIVE–BROWN*
- **Colour abbreviations** (in the same order as the names): *Wh , Gy, Bk, Blu , Gn , Yl, Or, Rd, Pu, Vi, Pi, Ol, Br*
- **Combinations**: *BluWh , YlGn , GyBr ,* etc.—*bluish white, yellowish green, greyish brown,* etc.

Mineral abbreviations *: see* **Annex 1**

1.5 Data Presentation and Identification Methodology

According to the properties described above, characterisation of the different mineral species is based on their systematic observation. This involves four possible basic microscope settings: **dsp , d+p , osp , o+p** (***d***, *dry*: observation in air, i.e. with dry objective; ***o***, *oil*: observation in oil immersion; ***sp***: single polariser; ***+p***: polariser and analyser crossed at 90°; ***xp***: *íd.* ≈ 87°; ***+p.w***: *íd.* with Wollaston Prism insertion, for *interference contrast*, *i.c.*).

These properties are summarised for each species in a record which will be the basic tool for recognition and whose format contains, successively, the following types of information:

1.5.1 Table-Summary of Properties

Basic data on the mineral, such as name, composition and crystalline system, followed by:

a. quantitative information (optical parameters and VHN, based on IMA/COM data taken from QDF3, Criddle and Stanley 1993; alternatively, from Picot and Johan 1982, or Uytenbogaardt and Burke 1971).
b. the optical properties linked to anisotropy and polishing hardness, also estimated as approximate numerical categories, which can be digitally processed (for automated identification).
c. *remarks* or *observations* in the last line of the table, containing any additional information and, in particular, an explanation of the ***mineral abbreviations*** used in the description of this ore or the corresponding

photomicrographs. (For a complete list of all mineral abbreviations or symbols in the text, *see* **Annex 1**).

This table (§ Box 1.1) is followed by a brief explanatory text (*parameters, symbols and concepts used*), which is important to ensure precision and avoid misunderstandings. For example, it specifies the 13 **accepted colours** (including black, although in reality it is an absence of colour: *white, grey, black, blue, green, yellow, orange, red, purple, violet, pink, olive, brown*) and their **abbreviations** (*Wh , Gy, Bk, Blu , Gn , Yl, Or, Rd, Pu, Vi, Pi, Ol, Br*) and combinations. These or their combinations will be the only colours used.

Box 1.2
Format Used for Microscopic Description

General
Chemical composition, varieties, family, abundance, deposit, etc.

Microscopic Description
SP Observation *(properties observed with polariser only—appearance, shape, colour, polishing properties).*

- *Optical impression—R (brightness), colour, bireflectance/pleochroism*
- *Morphology (eu-/sub-/anhedral tendency, habit, aggregates), zoning, cleavage (exfoliation), polishing properties (quality, scratches, pits, hardness)*
- *Other—tarnishing (patina), light etching, very significant alterations* (e.g., *native arsenic, manganite, argentite, pyrrhotite)*

Observation + P *(properties observed with polariser + analyser or "crossed nicols" at* 90°. When observing by slightly uncrossing the analyser, usually to an angle of ≈87°, ***xp*** is indicated. If the Wollaston Prism is inserted, for interference contrast, ***+p.w is*** indicated).

- *Anisotropism (intensity/polarisation colours), extinction*
- *Internal reflections*
- *Twinning (visible twins)*

Paragenesis
Accompanying minerals, natural association, deposits, substitutions, alterations.
As a complementary support, *Annex* 2 shows in several tables common ores and parageneses or, more generally, their associations, sorted by deposit types.

Identification
Diagnostic properties and recognition criteria, possible confusions, distinction from similar minerals.

Photomicrographs (*see below—scheme, description, discussion*)

Spectral Information (*VNIR and visible values in air, sketch below*)
....

Photomicrographs
The graphic documentation strategy uses two groups of photographs. First, in order to provide homogeneous, objectively comparable information for the different mineral species, a set of four photomicrographs (identified as **Basic Photomicrographs**) is offered for each of them, usually taken with a 10 × objective and always with the same combination of microscope settings, successively and in the same order, namely:

- **dsp** (*air, d = dry; sp = polariser only*)
- ***osp*** (*oil, o = oil; polariser only*)
- ***d + p*** (*air, d = dry; +p = polariser + analyser at* 90°. In some cases, the polarisation should also be observed by slightly reducing the angle between polariser and analyser, e.g., by 1° to 5°. In this case **xp** is indicated; use of the Wollaston Prism for interference contrast, *i.c., is* indicated **+p.w**)
- ***o + p*** (*oil, o = oil; +p = polariser + analyser* 90°)

However, in many cases it is necessary to additionally document other properties or peculiarities of each species (varieties, microstructures, different paragenesis or associations, alterations, etc.). For this purpose, a second group of images appropriate to each case, identified as **Supplementary Photomicrographs**, is provided if necessary. As explained (Sect. 1.4), ***Tsp*** or ***T + p*** is indicated in the case of *sp* or + *p* observations with **transmitted light** (*see* Fig. 1.53, vol. 2).

All photomicrographs have a **graphic scale** in the **upper left-hand corner** (bar with length indicated, always 100 μm, except in the case of very high magnification images, e.g., microprobe or electron microscope images, in which case this is indicated as appropriate). All of them also have a visible **label**, with the identification code used in the text when referring to each ore, in the **lower right-hand corner**, which should avoid any confusion in following the descriptions.

Leica DMLP and LeicaDM2500P microscopes were used, with the magnification and setting indicated in each case. The photographs were taken with a *Leica DFC500* digital camera, LAS software, *Leica Application System v3.8*, for image acquisition, graphic scale insertion, white balance control, etc.

In order to optimise colour fidelity when downloading or browsing the *Internet*, the selected images have had the **sRGB IEC61966-2.1 colour space** information added to them using *Adobe Photoshop software*, which is the IEC standard for use on monitors, printers and the Internet. If the equipment and the display of the reader are correctly calibrated, the reproducibility of the colours on the screen should be assured.

Basic Photomicrographs F m1-4.
(m = mineral abbreviation. Same scene, observed with: dsp, osp / d+p, o+p).

dsp (*air: dry; polariser only*)	**osp** (*oil: oil; polariser only*)
d+p (*air: dry; polariser + analyser at* 90°)	**o+p** (*oil: oil; polariser + analyser* 90°)

F m1—*Description.* Sample: y, Mine z (**dsp**). **F m2**—Same scene as Fm1 (osp…), etc

Supplementary Photomicrographs: F m5, 6…

F m5—*Description*…Sample…, Mine… (**dsp**)/(**osp**)/(**d+p**)/(**o+p**), etc

....

Spectral Information

(*Presentation scheme, applied as an example to the case of löllingite, compared with possible accompanying minerals as measured in air in the AMCO System and compared with QDF3 spectra, after* Criddle and Stanley 1993; Picot and Johan 1982).

Explanation:

- *λ = wavelength (nm), spanning VNIR and near-UV ranges in AMCO*
- *μ = mean specular reflectance value, R, as computed from the AMCO D.B*
- P_5 *to* P_{95}*—R-values corresponding to the 5th, 20th, 80th and 95th percentiles*

Löllingite: Multi-spectral Reflectance R (in air, without polariser) (%)

λ (nm)	370	400	425	450	475	500	525	550	575	600	625	650	675	700	750	800	850	900	950	1000
μ	51.8	52.6	53.1	53.1	53.6	53.4	53.6	53.3	53.3	53.0	52.7	52.4	51.9	51.7	51.0	50.6	50.3	50.3	50.3	50.1
P_5	50.3	51.0	51.8	51.8	52.2	52.2	52.7	52.3	52.3	52.1	51.8	51.6	51.2	51.0	50.3	49.9	49.5	49.6	49.5	49.4
P_{20}	50.10	51.8	52.5	52.4	53.2	52.8	53.2	52.8	52.8	52.6	52.3	52.1	51.6	51.3	50.6	50.3	49.9	49.9	49.8	49.6
P_{80}	52.7	53.6	53.9	54.1	54.3	54.3	54.3	54.0	53.9	53.7	53.3	53.0	52.4	52.1	51.4	51.2	50.9	50.9	50.8	50.7
P_{95}	53.8	54.5	54.7	54.7	54.9	54.7	54.6	54.3	54.3	54.1	53.8	53.5	52.9	52.6	51.9	51.5	51.4	51.3	51.4	51.1

Comparison of Specular Reflectance values (VNIR, AMCO data) from löllingite, native bismuth, rammelsbergite and magnetite

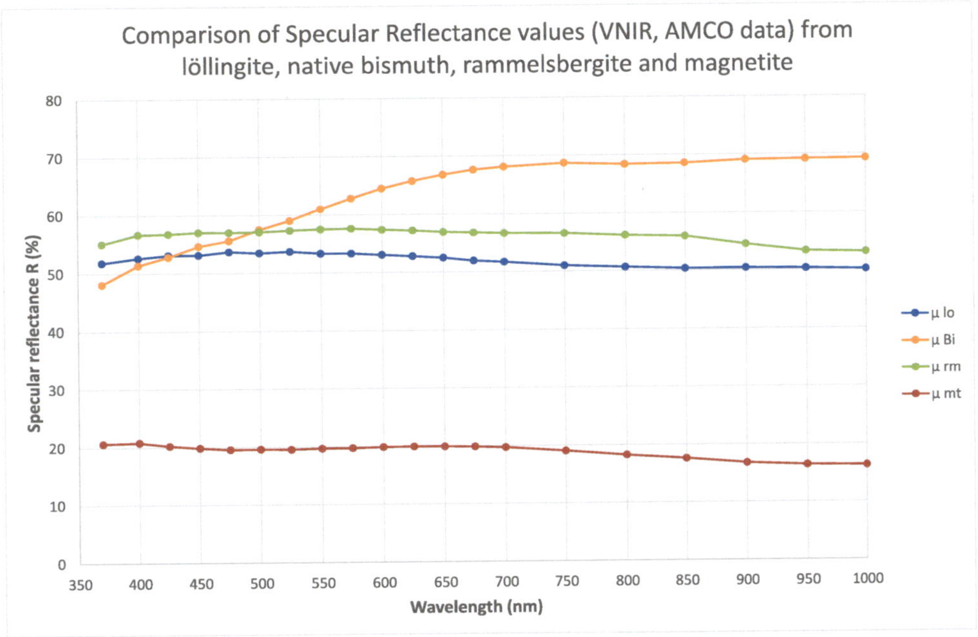

Comparison of Specular Reflectance VNIR values from löllingite (AMCO), löllingite R_1 & löllingite R_2 (QDF3), and löllingite R_{max} and löllingite R_{min} (Picot and Johan 1982)**.**

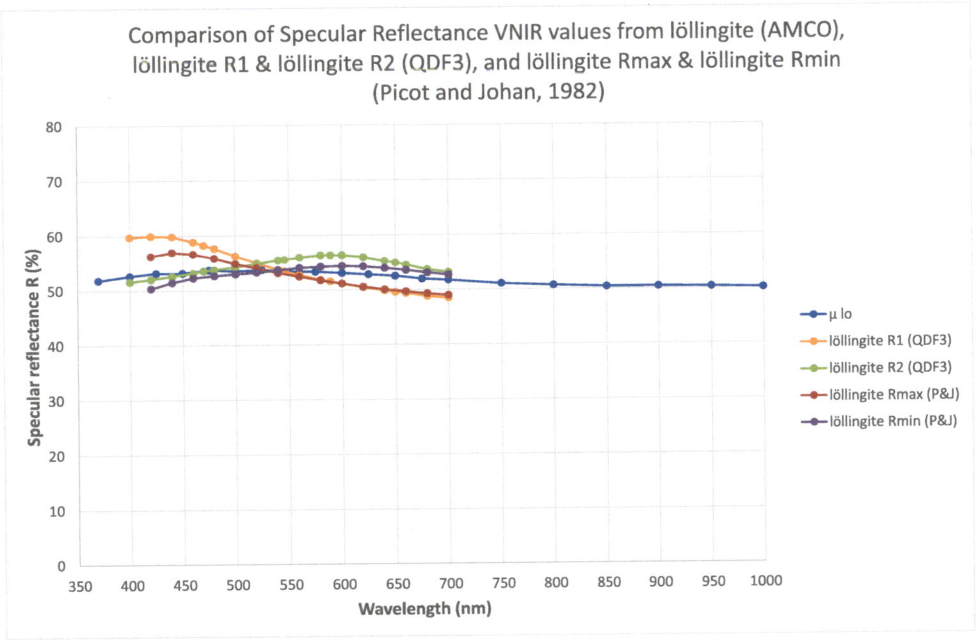

1.5.2 Descriptive Text and Diagnostic Properties

A systematic description, in fixed format (§ Box 1.2) is provided for each ore. After some general considerations, the different properties are successively described, referring to the required microscope setting (sp, +p), and their diagnostic value for recognition (ore identification) is discussed.

1.5.3 Photomicrographs

The photomicrographs (F, cf. § Box 1.2) illustrate the Appearance and Properties of the Mineral (*xx*):

a. There is always a **basic group** (*F xx1 to F xx4*) of four photos with fixed format, to illustrate the observation in air (***d***) and in immersion oil (***o***), both with polariser alone (***sp***) and with polariser and analyser crossed at 90° (**+ *p***). Sometimes a slightly uncrossed analiser is required, departing a little from perpendicularity, e.g., 87° or ca 90°, whichever is indicated, ***xp***.
b. In most cases, there is also a **complementary group** (*Fxx5 ff*) to complete the graphic documentation, offering images of the described ore in different associations and illustrating the change in the subjective perception of colour and lustre depending on the accompanying mineral.

All photomicrographs have a **graphic scale** in the **upper left-hand corner** (bar with length indicated, always 100 µm, except in the case of high magnification images, e.g., microprobe or electron microscope images, in which case the appropriate is indicated). All of them also have a visible **label**, with the identification code used in the text when referring to each ore, in the **lower right-hand corner**. This should avoid any confusion in following the descriptions.

The reliable reproduction of the colours obtained with a digital camera is difficult due to the usual methods of printing on paper. However, with rigorous acquisition like the one carried out (CIE iluminant C, with a stable halogen source of constant intensity, high quality camera, white balance of each image, etc.) and with conversion into a sRGB colour space profile of the images thus acquired, a realistic reproduction is possible in the digital version offered, if the appropriate equipment and a calibrated monitor are available (cf. Explanation of *PHOTOMICROGRAPHS* in ***Box*** 1.2, for more details).

1.5.4 Spectral Information

This Comprises:

a. a table with the multispectral R-values, specular reflectance, measured in air with the AMCO System (Chap. 2, vol. 2) in the range 370–1000 nm (visible and near-infrared, VNIR, and near-UV) of the considered ore (as an example, löllingite has been represented). These values include mean and percentiles 5, 20, 80, 95;
b. a graph plotting the R-values (mean; percentiles are usually omitted from the graphs for clarity) as a function of wavelength, λ, and also, for reference, in comparison with other selected species (VNIR range, AMCO) and with QDF3 values (only available in the visible range) and/or, sometimes, Picot and Johan (1982), also in the visible range.

This information is structured schematically, according to the model summarised in ***Boxes*** 1.1 ***and*** 1.2 (*General Outline* and *Model or Format of Record* for the systematic description).

1.6 Learning Strategy

Learning ore microscopy involves many hours of observation. The guidance of an expert is also an invaluable help. Discussion—or at least documentation—is another essential support. Indeed, this is a task in which experience is very important. This introduction therefore concludes with some *recommendations to the beginner*, to benefit from the experience accumulated in more than a century of efforts by different researchers.

The starting point is not the same for the expert and the beginner. Experts usually get a general idea of the ore and its components right away by a quick examination of the sample. Even if the actual study then requires much more time and attention for precise determinations and correction of possible errors of a hasty assessment, their experience has given them a habit of observation and a spontaneous memorisation of data that leads to diagnosis, just as a known coordinate system allows them to mark the location without hesitation. Beginners, on the other hand, first must internalise this system of coordinates, to develop a capacity of observation allowing them a reliable orientation in the space of diagnostic properties, so as to work confidently without confusing the fundamental and the accessory. To achieve this, they need not only a basic theoretical knowledge of mineral optics, but also to become familiar with the handling and the different settings of the microscope, enabling them to move freely from one type of vision to another (*d/o*, *sp/+p*, *xp*, etc.). These are basic requirements to understand the optical phenomena observed in each case and to define the corresponding properties without bias or error.

Therefore, learning must firstly insist on practice, on the training necessary to develop this ability. The identification of each mineral species by name comes next, as a second step that builds on the first. The two steps are, in fact, inseparable, as the practice is done with real ores, but it is only from precise observation and description of the properties that the identity can be determined. In the first instance, it is more important to describe correctly than to name the mineral. Recognition is achieved once the necessary observation skills have been acquired.

Galopin and Henry (1972) were no doubt well aware of this when they proposed a didactic progression from qualitative observation to quantitative determination. In the first mode, they set four exercises to practice with progressively more complex and varied ores and observations, in which beginners tests their observation skills. In the second, they apply the measurement of quantitative properties (in their case Vickers microhardness, *VHN*, and reflectance in white light, *R%*, 546 nm) to confirm qualitative observations or, if not, to resolve or narrow down their indeterminacy. This was largely achieved by means of **R%-VHN diagrams**, which represented a breakthrough at the time (McLeod and Chamberlain 1968; Tarkian 1974; Tarkian and Liessmann 1991).

The method retains all its value, although the use of quantitative properties has evolved significantly:

- The VHN number is still valuable, but almost none of today's mineralogical laboratories have the expensive microhardness tester which was used to measure it, so this application to ores has declined sharply;
- The determination of reflectances, on the other hand, has been greatly enriched and diversified from monochromatic or white light measurements to systematic multispectral measurements in the visible (Criddle and Stanley 1993; Picot and Johan 1982) or visible and near-infrared (VNIR, this text) ranges.

The disuse of the microhardness meter has made R%-VHN diagrams less popular, but reflectance measurements have increased their usefulness and have even become an essential tool for automated recognition. This is important and will be dealt with in due course (cf. Sect. 1.5) but not before due attention has been given to basic recognition by direct observation under the microscope. Both forms of recognition are important, not exclusive but complementary.

This introduction concludes with recommendations for optimising the learning strategy of ore microscopy with reflected light. Although the book is primarily intended as a reference text or practical guide (*LLE* support) for *senior* professionals in the sector (geological surveys, industrial or university laboratories, consultants, a.o.), it is also intended to be accessible to beginners, and it is hoped that it can be useful to students and *junior* graduates in mineral resources. It is precisely for them that these *recommendations for the beginner* are addressed, with the aim of facilitating learning through progressive steps, from the simple to the complex. It is advisable to start writing down the own observations orderly, on a *card* for each mineral. This card is a basic personal document, which will allow practitioners to retrieve, follow up, correct and, finally, memorise the descriptions, as well as to check and assess their own progress.

It is therefore recommended that, starting from the collections available in each case, ore samples are selected and studied according to the following sequence:

1. Start with easy samples, containing one or, at best, several priority minerals (listed as *Learning Priority 1* in the ore compilations of the *Mineral Indices*, as well as in *Annex 1.2*, *e.g.*, pyrite, magnetite, sphalerite, galena, native silver or gold), and observe them with dry objective (in air, no analyser, ***dsp***), to compare and order by brightness and colour the species found, estimating relative values of their reflectance (for a start, gangue can be used for reference as R ≈ 5% and pyrite, R ≈ 55%):
   a. also in a sample containing magnetite (R ≈ 20%);
   b. in a sample containing galena (R ≈ 43%) and/or sphalerite (R ≈ 17%);
   c. in a sample containing either native silver (reference as R ≈ 90%) or native gold (reference as R ≈ 70%) or both;
   d. observe also, on all of them, their polishing hardnesses (*PolH*); compare, in particular, PolH of pyrite, magnetite, sphalerite and galena; describe cleavage where visible; identify *triangular pits* of galena.
2. Observe the same samples with the analyser crossed at 90° (***d* + *p***), and distinguish isotropic from anisotropic species (if any). Compare and describe the internal reflections, **IR**, of sphalerite and gangue, and describe the effects of anisotropism, if anisotropic species are present.
3. Observe the same samples in immersion (**osp**) and note changes in appearance (brightness, colour). Compare the effect of immersion on gangue, magnetite and sphalerite.
4. *Id.* with 90° crossed analyser (***o* + *p***), compare the effect of immersion on IR of gangue and sphalerite.
5. Compare observations with the descriptions of each mineral given in the text. Resolve discrepancies with new observations or by consulting an expert.
6. Move now—still *Priority* 1—to samples with anisotropic minerals (such as hematite, covellite, chalcopyrite, arsenopyrite, marcasite, nickelite, pyrrhotite, molybdenite, graphite) and observe them, again following steps 1–5 for all of them (never forgetting any of the four basic settings: ***dsp***, ***d* + *p***, **osp**, ***o* + *p***; and using the most appropriate objective in each case). Fulfil the observations until all the properties outlined are defined (*Boxes* 1.1 *and* 1.2), for example, describing and comparing the

bireflectance of hematite and molybdenite, the pleochroism of covellite, pyrrhotite and graphite, the twinning of chalcopyrite, and the anisotropism and polarisation tints of pyrrhotite, marcasite, molybdenite or graphite.
7. Continue with the same observation methodology until all *Priority 1* and then *Priority 2* ores are familiarised.

This continued work of observation until the process is complete should provide the necessary fluency to tackle the microscopic study of new ores (including the non-priority ores described in the text), with the help of the summary tables (*Annex 4*) and all the documentation published over decades by specialists. Now that they have acquired the required observation skills, beginners will be able to resort, when necessary, to the determinative tables (*Annex 5*). These are designed as a guide or orientation for a gradual advance towards definition of the unknown materials. Progress in identification is achieved through observations and comparisons that focus the diagnosis by eliminating opposites. Data supported by observation are highlighted to exclude inconsistent alternatives. It should be noted, once again, that these tables presuppose that one's own observations are reliable; if this is not proven in practice (as indicated, cf. *Annex 5.2*), the results will always be questionable.

References

Amstutz GC (1962) How microscopy can increase recovery in your milling circuit. Mining World 24:19–23

Atkin BP, Harvey PK (1979) Nottingham interactive system for opaque mineral identification: NISOMI. Trans Inst Mining Metall 88:1324–1327

Becker M, Wightman EM, Evans CL (eds) (2016) Processs mineralogy. SMI-JKMRC Monograft Series in Mining and Mineral Processing No. 6. University of Queensland, Australia, 470p

Bernhardt H-J (1979) Computer-gestuetzte Erzmineral-diagnose mittels reflexionsspektren in sichtbarem Licht. N Jb Miner Mh H.9:403–407 (Stuttgart)

Bernhardt H-J (1987) A simple, fully-automated system for ore mineral identification. Mineral Petrol 36:241–245

Bernhardt H-J (1990) Microscopic identification, and identification schemes, of ore minerals.In: Jambor JL, Vaughan DJ (eds) Advanced microscopic studies of ore minerals, pp 189–212. Short course handbook 17, Mineralogical Association of Canada, Ottawa, 426p

Bowie SHU, Simpson PR (1978, 1980) The Bowie-Simpson system for the microscopic determination of ore minerals. First students issue. Applied Mineralogy Group of the Mineralogical Society, McCrone Research Associates, London, 10p

Cabri LJ (2002) The platinum-group minerals. In: Cabri LJ (ed) CIM Special. Canad Inst Mining, Metall and Petroleum, vol 54, pp 13–129. Calgary, Alberta, Canada, 852p

Cabri LJ, Vaughan DJ (eds) (1998) Modern approaches to ore and environmental mineralogy. COM/IMA Short Course Series, vol 27, Ottawa, Ontario, 421p

Cameron EN (1961) Ore microscopy. Wiley, New York, 293p

Castroviejo R, Catalina JC, Bernhardt HJ, Pirard E, Segundo F, Brea C, Pérez Barnuevo L (2010) A fully automated system for multispectral ore microscopy. IMA2010 (20th General Meet Internat Mineralogical Association). Acta Mineral Petrogr Abstr Ser (HU ISSN 0324-6523/HU ISSN 1589-4835), Szeged, Budapest, p 281. (Proc CD_ROM, Session EG53_Geometallurgy and Process mineralogy, Code 4NF15)

Castroviejo R, Catalina JC, Bernhardt HJ, Pirard E, Brea C, Perez-Barnuevo L, Segundo F (2014) Multispectral (visible and near infrared, 400–1000 nm range) reflectance data file from common ore minerals. International Mineralogical Association/Commission on Ore Mineralogy. Disponible en Internet en: http://projects.gtk.fi/com/results/reflectance_data.html

Cervelle B, Moello Y (1990) Reflected-light optics. In: Jambor JL, Vaughan DJ (eds) Advanced microscopic studies of ore minerals, pp 87–108. Short Course Handbook 17, Mineralogical Association of Canada, Ottawa, 426p

Craig J, Vaughan D (1994) Ore microscopy and ore petrography, 2nd edn. Wiley, New York, p 434

Criddle AJ (1998) Ore microscopy and photometry (1890–1998). In: Cabri LJ, Vaughan DJ (eds) Modern approaches to ore and environmental mineralogy. COM/IMA Short Course Series, vol 27, Ottawa, Ontario, 421p

Criddle AJ, Stanley CJ (eds) (1993) Quantitative data file for ore minerals, 3rd edn. Chapman & Hall, London, p 635

Febrel T (1970) Microscopía de Minerales Opacos. Publs. ETSI Minas y Energía, Madrid, p 102

Fenoll Hach-Alí P, Gervilla F (2005) Identificación práctica de minerales opacos mediante microscopía de luz reflejada. Dpto. Mineralogía y Petrología. Universidad de Granada, 121p

Freund H (ed) (1966) Applied ore microscopy. In: Theory and technique. Collier-Macmillan, New York, 607pp

Galopin R, Henry NFM (1972) Microscopic study of opaque minerals. Heffer, Cambridge, p 322

Gerlitz CN, Leonard BF, Criddle AJ (1989) QDF database system, version 1.0: reflectance of ore minerals—a search-and-match identification system for IBM and compatible microcomputers using the IMA/COM quantitative data file for ore minerals, second issue. US Geol Surv Open-File Report 89-0306A-E

Gribble CD, Hall AJ (1992) Optical mineralogy: principles & practice. Routledge, London, p 303

Hagel E (1979) Microscopía de Menas. Inst° Geología Económica Aplicada, Universidad de Chile, Santiago, 205p

Ineson PR (1989) Introduction to practical ore microscopy. Longman, 197p

Ixer RA, Duller PR (1998) Virtual atlas of opaque and ore minerals in their associations. http://www.atlas-of-ore-minerals.com (visit. 20201111)

Jambor JL, Vaughan DJ (eds) (1990) Advanced microscopic studies of ore minerals. MAC Short Course, Ottawa, Canada, 426p

Kühnel RA, Prins JJ, Roorda HJ (1980) The 'Delft' system for mineral identification. Delft University Press, Delft, The Netherlands, 204p

Marshall D, Anglin D, Mumin H (2004) Ore mineral atlas. Geol Assoc Canada Mineral Deposits Division, 112p

McLeod CR, Chamberlain JA (1968) Reflectivity and Vickers microhardness. Paper 68-64, Geol Surv Can

Neumann U (2019) Guide for the microscopical identification of ore and gangue minerals: mineral profiles with photomicrographs. Tübingen University Press, 319p

Novitzky A (1957) Tablas para la determinación microscópica y Rayos X de minerales opacos, 2ª. Univ. Técnica de Oruro, Bolivia

Orcel J (1972) Historical note. App VII. In: Galopin R, Henry NFM (eds) Microscopic study of opaque minerals, pp 301–305. Heffer, Cambridge, 322p

Paar WH, Brodtkorb MK, Putz H, Martin RF (2016) Atlas of ore minerals: focus on epithermal deposits of Argentina. Canadian Mineralogist, Spec Pub 11, Quebec, 402p
Peckett A (1992) The colours of opaque minerals. Wiley, New York, p 471
Picot P, Johan Z (1977) Atlas des Mineraux Metalliques. Mem BRGM 90. Paris, 406p
Picot P, Johan Z (1982) Atlas of ore minerals. BRGM/Elsevier
Pirard E (2016) Optical microscopy. Chap. 5. In: Becker M, Wightman EM, Evans CL (eds) Process mineralogy, pp 51–66. JKMRC Monograph Series in Mining & Mineral Processing: No. 6. The University of Queensland, Australia, 470p
Pracejus B (2008) The ore minerals under the microscope. In: Atlases in geoscience, 3rd edn. Elsevier, 875p
Ramdohr P (1955) Die Erzmineralien und ihre Verwachsungen. Akademie Verlag, Berlin, p 875
Ramdohr P (1960) The ore minerals and their intergrowths. Akademie Verlag, Berlin, p 875
Ramdohr P (1980) The ore minerals and their intergrowths, 2nd edn. Pergamon Press, Oxford, 2 volumes, 1205p
Sanderson J (2019) Understanding light microscopy. Royal Microsc Soc (Wiley, London). ISBN 978-0-470-97375-2
Schneiderhöhn H (1952) Erzmikroskopisches Praktikum. E. Schweizerbart'sche Verlagsbuchhandlung, Stuttgart, p 274
Schneiderhöhn H, Ramdohr P (1931) Lehrbuch der Erzmikroskopie, I & II Band, mit Anhang: Erzmikroskopische Bestimmungstafeln. Gebr. Borntraeger, Berlin, 714, 313 & 47p
Schouten C (1962) Determination tables for ore microscopy. Elsevier Publishing Company, Amsterdam, p 242
Shoji T, Kaneda H (1994) An interactive system to assist mineral identification in ore microscopy. Math Geol 26(8):961–972
Short MN (1940) Microscopic determination of the ore minerals, 2nd edn. US Geol Survey Bull 914, 314p
Spry PG, Gedlinske BL (1987) Tables for the determination of common opaque minerlas. Economic Geology Publ Co Littleton, Colo, USA, 52p
Tarkian M (1974) A key-diagram for the optical determination of common ore minerals. Minerals Sci Engng 6(2):101–105
Tarkian M, Liessmann W (1991) A guide for optical and analytical identification of ore minerals. Clausthaler Geol Abh 47, 78p (Verlag Sven von Loga, Köln)
Uytenbogaardt W, Burke EAJ (1971) Tables for microscopic identification of ore minerals, 2nd edn. Elsevier, Amsterdam

Websites

Commission on Ore Mineralogy of the International Mineralogical Association (COM/IMA). https://www.ima-mineralogy.org/COM_Strategy.htm (visit. 20220221)
Mindat. https://www.mindat.org/ (visit. 20220221)
RRUFF: https://rruff.info/ (visit. 20220221)

Part II

Description of Ores

2 Introduction to Ore Description

2.1 Mineral Selection

The minerals selected are metallic ores of general interest either because of their abundance or frequent occurrence in nature or because of their economic or scientific interest for mining, exploration or mineral resource research. This heterogeneous group of minerals will be referred to by the conventional term ***common ores*** (or *frequent ores*, even if they are sometimes rare, as in the platinum group minerals). The criteria applied have been outlined in Chap. 1, resulting in the selection of ore and gangue minerals set out in the lists in Annex 1 (cf § *Annex* 1.2 and 1.3). In order to facilitate the work (presentation of data in tables and graphs, description of photomicrographs, etc. in a reduced space, in addition to the rapid annotation of data during microscopic observation), the selected species or mineral groups are identified by means of abbreviations or symbols (cf § *Annex* 1).

Metallic ores generally have in common the fact that they are absorbent to light radiation, i.e., they have little or no transparency. For this reason, they are frequently designated as *opaque minerals* and are studied by reflection microscopy with polarised light, usually known as *ore microscopy* (also *opaque microscopy*, *mineragraphy*, *chalcography* or *ore petrography*). However, what defines an ore is its economic value, not only its colour or opacity. For this reason, the student of metallic ores will also come across some that are transparent or semi-transparent, such as cassiterite or sphalerite, which can be studied perfectly well with a classic petrographic (or transmitted light) microscope.

Ores are associated with common, non-economic minerals (*gangue*), which are usually transparent and therefore studied by transmitted light (**TL**). However, their characterization by reflected light (**RL**) will be addressed as far as possible, describing and illustrating with photomicrographs the properties that can be determined with this technique. While being aware that the appropriate technique in these cases is TL (usual setting for petrography), the attentive observer can obtain useful information and, not infrequently, an accurate or approximate recognition of the gangue through the RL study. And, in any case, he should make the most of the observation possibilities that the technique offers.

The study of ores is dealt with below (Part II, Chaps. 3–132). It will be complemented in Part III by RL characterisation/discussion of gangue minerals.

2.2 Photomicrographs and Colours

The RL microscopic description and discussion of each ore species follows in alphabetical order, comprising Chaps. 3–132. The optical properties and the application of diagnostic recognition criteria are illustrated by means of appropriate photomicrographs (*F*). In order to avoid confusion due to their large number, they are always identified in the same way: a **label**, numbered and *visible in the lower right corner* of the image, with the abbreviation of the described ore as the first piece of information (e.g., Figs. 99.1a, b and 28.1 = py1, py2, and ccp1, respectively, referring to pyrite and chalcopyrite)). If the scope of the image is of a general character and not the illustration of a particular ore, the mineral abbreviation is replaced by a number referring to the related topic, or chapter.

As explained in Chap. 1, colours and brightness are very important data for RL characterisation. All photomicrographs, taken with a DFC500—LAS digital camera on Leica DMLP or DM2500 microscopes, have been processed and defined in *sRGB colour space*, in order to guarantee unalterable and high quality images. When using or viewing these virtual images online, the user must ensure that the **colour calibration of the monitor** is reproducible. Otherwise, if using an uncontrolled monitor or printer, the colours may be substantially altered.

2.3 Description of Selected Ores

The individual descriptions of the ores are presented in alphabetical order, followed by their *abbreviations*, e.g., pyrite (py)". The members of isomorphic series are usually

R. Castroviejo, *A Practical Guide to Ore Microscopy—Volume 1*,
https://doi.org/10.1007/978-3-031-12654-3_2

described with the related main ore mineral and are therefore not listed in the chapter headings of the Table of Contents. This applies also to some ore species with particular relationships, as well as to some common mineral varieties. However, their descriptions are discussed with those of the corresponding main ores. They can be retrieved by their own names from the **Analytical Index of Minerals**, where a reference to the main ore is reported (stating the ore's symbol or abbreviation[1]). The information should be easily found (e.g., for *melnicovite*, cf *py* = pyrite; for briartite and germanite, cf. *ren* = renierite; for each of the many terms comprising the linnaeite or the *fahlore* families, cf. *ln* = linnaeite or cf *td* = tetrahedrite, *resp*).

Other abbreviations used in the descriptions, particularly those of the 13 **colours** selected to define optical properties, are explained in Chap. 1 (§ *Box* 1.1).

[1] Compiled in *Annex 1*, see *Annex 1.2.a* or *1.1.a* (ores or minerals listed by their symbols).

3 Acanthite (*ac*)

Ag_2S: Monoclinic (pseudocubic)

Acanthite[a]		R'_{546} (%)	Y (%)	λ_d (nm)	P_e (%)	Colour sensation
Air		31.0	30.6	483	5.1	Grey. Often polishing scratches and dark tarnishing
Oil		15.7	15.5	482	8.7	
Bir: Very weak to absent (2)				*Pleoc*: Very weak (2)		*Tarn*: Darkening. Light etching
Anis + P: Weak, BluGy-GnGy (3)				*IR*: No		*Twg*: Coarse lamellar twins
Mor: Aggregates, microcracks				*Clv*: No		*PolQ*: Difficult (scratches)
Hardness	*PolH*: very low, < gn, td (1)			*Mohs*: 2	*Talm*: A	*VHN* $_{(50\ g)}$: 21–25

[a]Often confused in the past with its cubic polymorph argentite, which is only stable above 177 °C and always transforms into acanthite at lower temperatures. However, in this case the inherited pseudocubic symmetry of acanthite is revealed by the closeness of its R1 and R2 reflectance values (Picot and Johan 1977), reduced to the single R′ value in QDF3

Obs: Common Ag ore; often paramorphic, inheriting cubic forms from argentite; very soft (unavoidable polishing scratches); sensitive to oxidation (dark surface) and light etching

Spectral information: cf. Table and R plots, multispectral data (VNIR, AMCO) and visible range (QDF3)

Mineral abbreviations: ac—acanthite; asp—arsenopyrite; gg—gangue; mc—marcasite; sp—sphalerite; stp—stephanite

Colour abbrev.: BluGy—Bluish Grey; GnGy—Greenish Grey

3.1 General

Acanthite, *ac*, is the only Ag_2S species directly observable in nature, as it is the only silver sulphide stable under surface conditions. Its paramorph argentite is only stable at high temperature and, on cooling, its cubic structure becomes monoclinic (acanthite).

This transition is unavoidable (*not quenchable*) and occurs at 177 °C, a temperature which also marks the upper stability limit of acanthite, whose structure reverses to cubic above 177 °C (for more details, cf. comments on *Argentite*). The transition provides a *geological thermometer* when it can be detected, e.g., by the observation of pseudomorphs of cubic forms or by the presence of polysynthetic twins. These, according to Picot (1982), would result from inversion of the cubic lattice by cooling, indicating an original precipitation above 179 °C. Conversely, evidence of primary acanthite indicates deposition below that temperature.

It is not an abundant mineral, but it is common, although minor, in silver deposits, where it is usually accompanied by native silver, silver sulphosalts, galena, etc.

3.2 Microscopic Description

3.2.1 SP Observation (*Single Polariser*)

The optical impression (Figs. 3.1a, b and 3.2) is primarily conditioned by the difficult polishing, with abundant scratches resulting in a dark surface appearance. Grey colour (with a slight greenish or bluish–green tinge, noticeable especially in oil, in direct comparison with silver or galena). Moderate reflectance (≈31%, white light), on the order of tetrahedrite or freibergite in freshly polished samples but, with time, darkens markedly by spontaneous oxidation of the polished surface (also, very rapid darkening under the microscope: *light etching*). Pleochroism is imperceptible in air and weak in oil (Fig. 3.1a, b).

No cleavage, nor zoning commonly. However, clearly visible zoning enhanced by light etching has been cited (Petrulian in Ramdohr 1980, p. 472). Very soft, softer than its usual companions (galena, sulphosalts, native silver), except jalpaite, which shows a similar polishing hardness. Difficult polishing, unavoidable scratches (Fig. 3.2). It

R. Castroviejo, *A Practical Guide to Ore Microscopy—Volume 1*,
https://doi.org/10.1007/978-3-031-12654-3_3

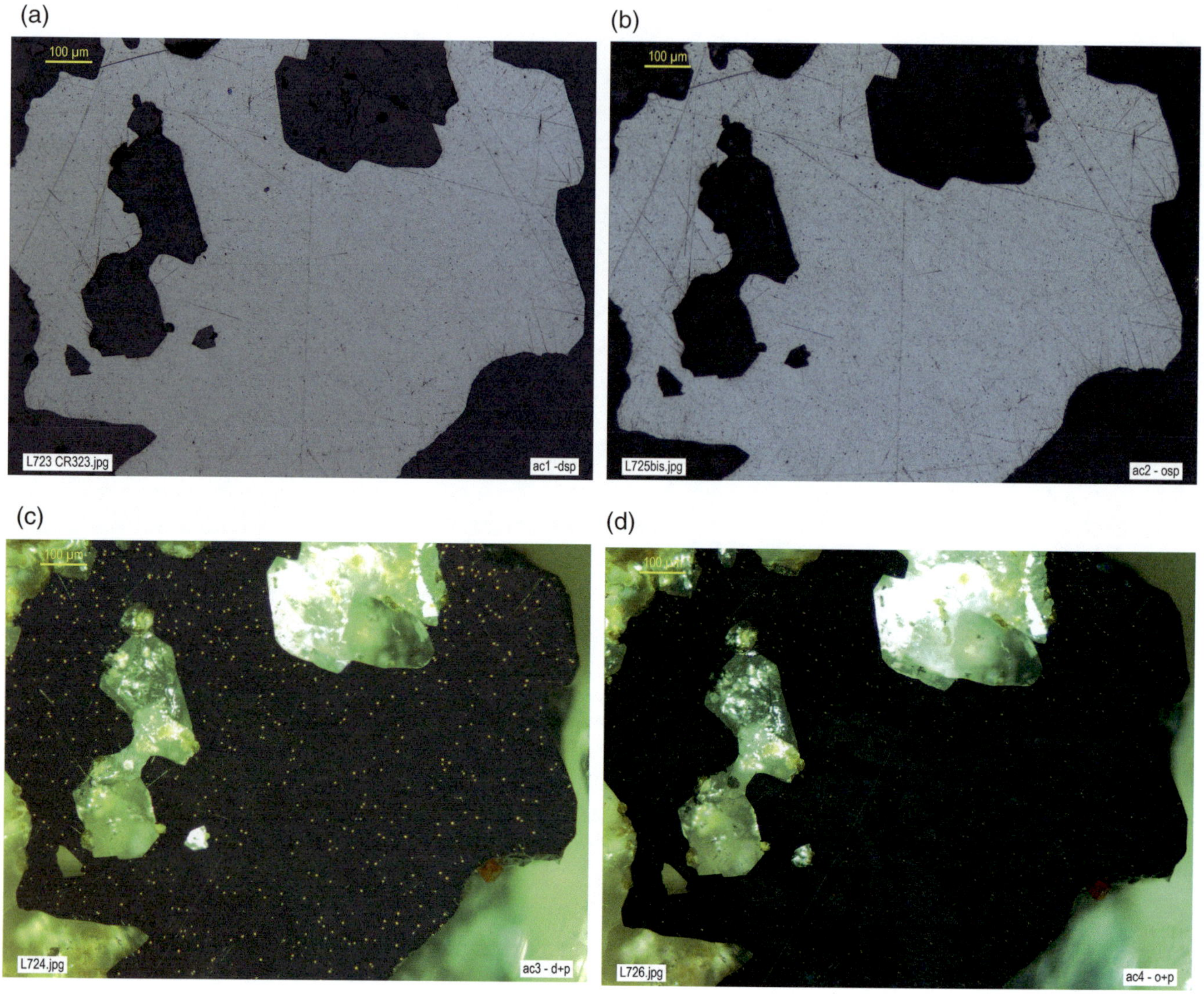

Fig. 3.1 **a** Acanthite (light grey) and silicate gangue (dark grey). Sample CR323, Mina Baliolu, Falset, Tarragona (**dsp**), **b** Same scene as ac1 (**osp**). Note the higher contrast between ac and gg, **c** Same scene as ac1 (**d+p**). Weak anisotropism, but sufficient to recognise polysynthetic twinning. Incipient light etching (unavoidable when removing filters for **+p** observation), evidenced by the numerous light spots marking the surface of ac, **d** Same scene as ac1 (**o+p**). Anisotropism colours more vivid than in air, with greyish blue-green shades

tarnishes and darkens due to spontaneous oxidation of the polished surface over time. Fast but irregular *light etching* under the RL microscope (Figs. 3.1c, 3.5 and 3.6). Very rapid under electron beam (EPMA or SEM).

It occurs disseminated, in anhedral inclusions or in microcracks and occasionally in crystalline aggregates with a subhedral tendency. It forms usually as paramorph from argentite.

3.2.2 Observation +P (*Polariser + Analyser 90°*)

Weak anisotropism but usually perceptible (except on heavily scratched surfaces), with grey and blue–green tones, sharper in oil (Fig. 3.1c, d). No internal reflections.

Frequent thick lamellar twins (Figs. 3.1c and d), interpreted by most authors as inversion twins of the high-temperature polymorph (argentite) and, therefore, evidence of a deposition of the primary ore above the inversion temperature (≈177 °C, cf. argentite), considered as a geological thermometer. According to Ramdohr (1980), the absence of such twins indicates a primary formation of acanthite, below the inversion temperature (≈177 °C).

3.2.3 Paragenesis

Typical ore from argentiferous deposits, both hypogene—usually in late phases—and supergene, upgraded in the

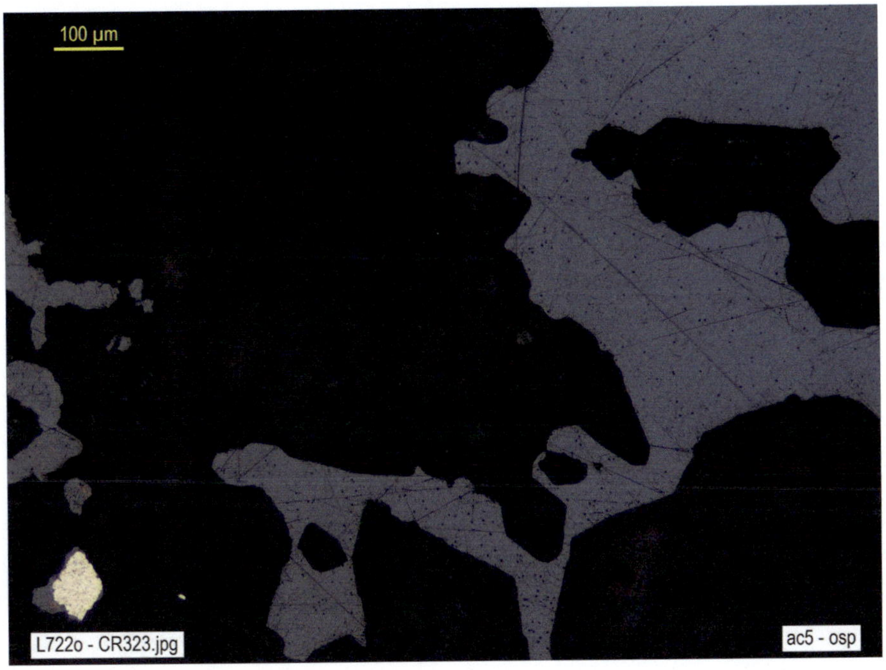

Fig. 3.2 Acanthite (light grey) and native Ag (yellowish white), in silicate gangue (black). Immersion in oil accentuates the contrast of reflectances between Ag (Y% 90.9), ac (Y% 15.5) and gg (practically 0). The black dots on ac are the effect of light etching, after 15 min of exposure to the microscope lamp under normal observation conditions. Sample CR323, Mina Baliolu, Falset, Tarragona (**osp**)

sulphide blanket or supergene sulphide enrichment zone (more rarely, in oxidized ores).

Its formation as a late mineral in hypogene deposits (as impregnation, replacement, pore filling, cementing microcracks) implies that the accompanying minerals are not always paragenetic with acanthite and can be very varied (pyrite, skutterudite, maucherite, etc. in Ni–Co–Bi–Ag–U or *5-element type* deposits; chalcopyrite, sphalerite, pyrite, galena, etc. in epithermal formations). As assemblage companions, acanthite is generally associated with native silver (Fig. 3.2), silver sulphosalts (polybasite, pearceite, stephanite, pyrargyrite, jalpaite, etc.), tetrahedrite/freibergite, galena (Figs. 3.3 and 3.4).

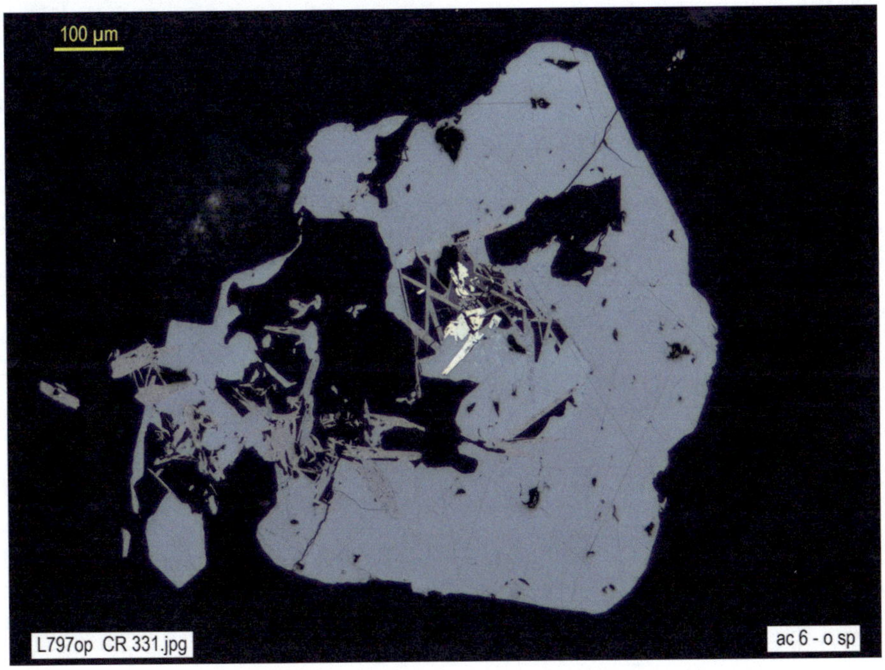

Fig. 3.3 Crystal aggregate of ac (bluish–grey) with intergrown laths of stephanite, stp (grey with brownish–violet tones); stp forms scattered tabular or acicular crystals, marginal or associated with pores, accompanied by asp (white), mc (faint pinkish white shade) and sp (dark grey, sparse); gg black. CR331: Fresnillo, Zacatecas, Mexico (**osp**)

Fig. 3.4 (Same scene as Fig. 3.3; **oxp**). In oil immersion and with partially crossed polars, the ores are clearly distinguished by their polarization tints: dull ac, in violet–grey to bluish–green tones; bright stp, in yellowish–brown to greenish tones; bright asp and mc, while sp remains dark (isotropic), with barely perceptible internal reflections

Fig. 3.5 *Light etching* on ac, diffuse and spontaneous (due to microscope illumination, as in Fig. 3.2), as black dots scattered over the whole mass of ac (visible mainly on the righthand side); concentrated (enhanced by closing diaphragm for 45 min) on the field appearing as a darker grey spot, in the upper central part of the image (**dsp**)

3.2.4 Identification

Common Ag ore, often paramorphic, inheriting cubic forms from argentite, and occurring in argentiferous paragenesis. It is distinguished from polybasite by its lack of internal reflections and its polarisation tints (anisotropism). Additional distinctive features for acanthite are its:

- grey colour, inconspicuous due to its moderate reflectance and with a greenish hue on well-polished surfaces,
- difficult polishing (abundant scratches), extremely low polishing hardness,
- gradual darkening by tarnishing or rapid darkening under the microscope (light etching), weak anisotropism with bluish-green tints,
- lamellar twinning.

Fig. 3.6 (Same scene as Fig. 3.5: **d+p**). With crossed polars, luminous effects are observed on some etched areas, due to the decomposition of the mineral, with the formation of unpolished silver corpuscles

3.3 Spectral Information

3.3.1 Acanthite: Multispectral Reflectance R (in Air, Without Polariser) (%)

λ (nm)	370	400	425	450	475	500	525	550	575	600	625	650	675	700	750	800	850	900	950	1000
μ	28.9	29.3	29.6	30.1	30.5	30.1	29.4	28.3	27.7	27.2	26.9	26.8	26.7	26.6	26.5	26.3	26.1	25.5	25.3	25.0
P_5	24.5	25.4	25.5	26.2	26.5	26.4	25.9	25.2	24.8	24.5	24.3	24.3	24.3	24.2	24.3	24.2	24.1	23.6	23.5	23.2
P_{20}	27.3	27.9	28.1	28.7	29.1	28.8	28.2	27.2	26.7	26.3	26.0	25.9	25.8	25.7	25.7	25.5	25.4	24.7	24.5	24.2
P_{80}	30.6	30.9	31.2	31.7	32.0	31.4	30.7	29.5	28.8	28.3	28.0	27.8	27.7	27.5	27.3	27.1	26.9	26.3	26.1	25.8
P_{95}	31.4	31.7	31.9	32.4	32.7	32.1	31.4	30.1	29.4	28.8	28.5	28.3	28.2	28.0	27.8	27.6	27.4	27.0	26.9	26.8

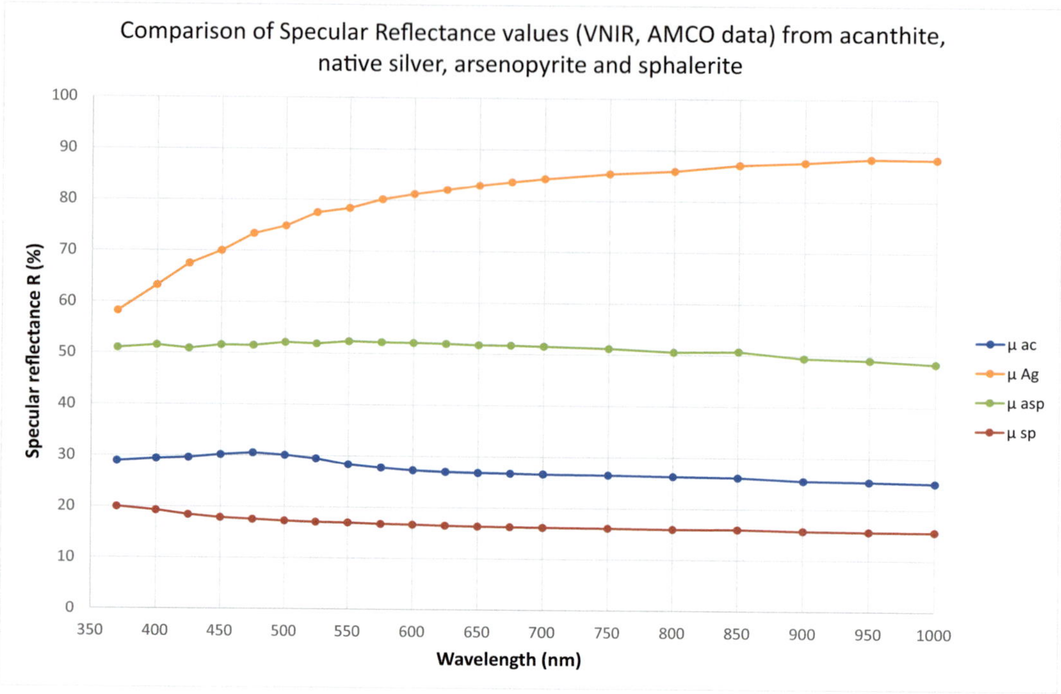
Comparison of Specular Reflectance values (VNIR, AMCO data) from acanthite, native silver, arsenopyrite and sphalerite
Specular reflectance R (%)
Wavelength (nm)
μ ac
μ Ag
μ asp
μ sp

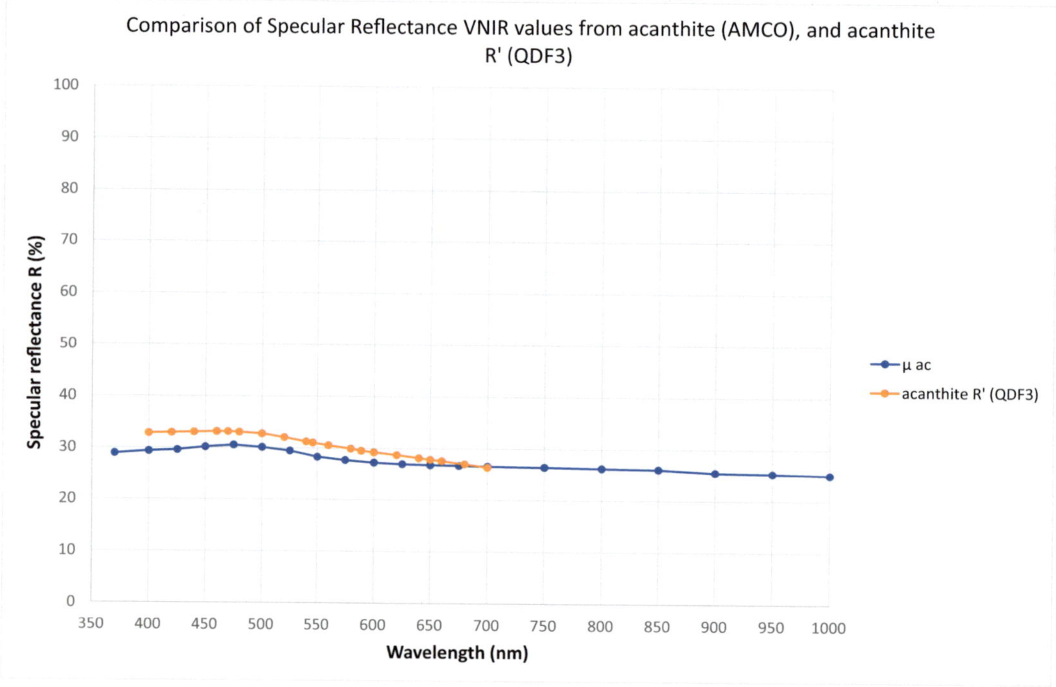
Comparison of Specular Reflectance VNIR values from acanthite (AMCO), and acanthite R' (QDF3)
Specular reflectance R (%)
Wavelength (nm)
μ ac
acanthite R' (QDF3)

References

Picot P, Johan Z (1977) Atlas des Mineraux Metalliques. Mem B R G M 90, 406p (Paris)

Ramdohr P (1980) The ore minerals and their intergrowths, 2nd edn. Pergamon Press, Oxford. 2 volumes, 1205p

Aguilarite (*agl*)

4

Ag_4SeS: Monoclinic

Aguilarite		R_{546} (%)	Y (%)	λ_d (nm)	P_e (%)	Colour sensation
Air	R_1	33.9	33.4	487	3.2	Grey
	R_2	35.4	34.8	492	1.9	
Oil	R_1	18.6	18.1	486	5.9	
	R_2	19.8	19.3	490	3.5	
Bir: Very weak to absent (2)				*Pleoc*: Very weak (2)		*Tarn*: nd
Anis + P: Weak (3)				*IR*: No		*Twg*: nd
Mor: Disseminated, massive, skeletal crystals				*Clv*: No		*PolQ*: Delicate (soft, scratches)
Hardness	*PolH*: Very low (1)			*Mohs*: 2½	*Talm*: nd	*VHN* $_{(25\ g)}$: 20.5–23.8

Obs: cf. Table and R plots, multispectral data (VNIR, AMCO) and visible range (QDF3, Picot and Johan 1982)
Mineral abbreviations: ac—acanthite; Ag—native silver; agl, (agu)—aguilarite; ccp—chalcopyrite; el—electrum; gg—gangue; gn—galena; my—miargyrite; nau—naumannite; py—pyrite; sp—sphalerite; td—tetrahedrite

4.1 General

Aguilarite, *agl*, is isostructural with acanthite, *ac*, but it differs from ac by incorporating Se. Composition: Ag_4SeS, with possible Cu impurities. Initially it was supposed to be orthorhombic. Apparently, in the Ag–Se–S system, aguilarite is part of two solid solution series—a monoclinic series with acanthite (Ag_2S) and an orthorhombic series with naumannite (Ag_2Se).

As in the case of acanthite and argentite, the inversion of a high-temperature cubic form of Ag_4SeS to aguilarite at 133 °C is detected by the appearance of polysynthetic twins (Picot and Johan 1982), suggesting its possible application as a geological thermometer.

4.2 Microscopic Description

4.2.1 SP Observation (*Single Polariser*)

Grey *color*, very similar to acanthite. Like acanthite, it can be polished with difficulty because it is very soft, but with aguilarite a better *polishing* quality is achieved.

Moderate *reflectance*, R agl ≈ my; slightly > ac, td and slightly < nau; < ccpy, gn, py; << Ag, el (Figs. 4.1a, b and 4.2). R quantitative data differ slightly according to sources. Lacking a better explanation, this might be due to compositional variations (*cf* comparative graphs).

Bireflectance and pleochroism are imperceptible.

It occurs in small disseminated anhedral aggregates or as an impregnation associated with acanthite, silver, electrum, naumannite, pyrite, chalcopyrite, etc. (Figs. 4.1a, b and 4.2). Occasionally found in skeletal crystals.

4.2.2 Observation +P (*Polariser + Analyser at 90°*)

Weakly *anisotropic*, like acanthite (Fig. 4.1c, d). Occasional lamellar *twinning*, interesting for indicating inversion of cubic form (133 °C) (Fig. 4.2).

4.2.3 Paragenesis

It is a rare mineral, typical of low temperature hydrothermal or *epithermal argentiferous* paragenesis, of low sulfidation but rich in Se.

R. Castroviejo, *A Practical Guide to Ore Microscopy—Volume 1*,
https://doi.org/10.1007/978-3-031-12654-3_4

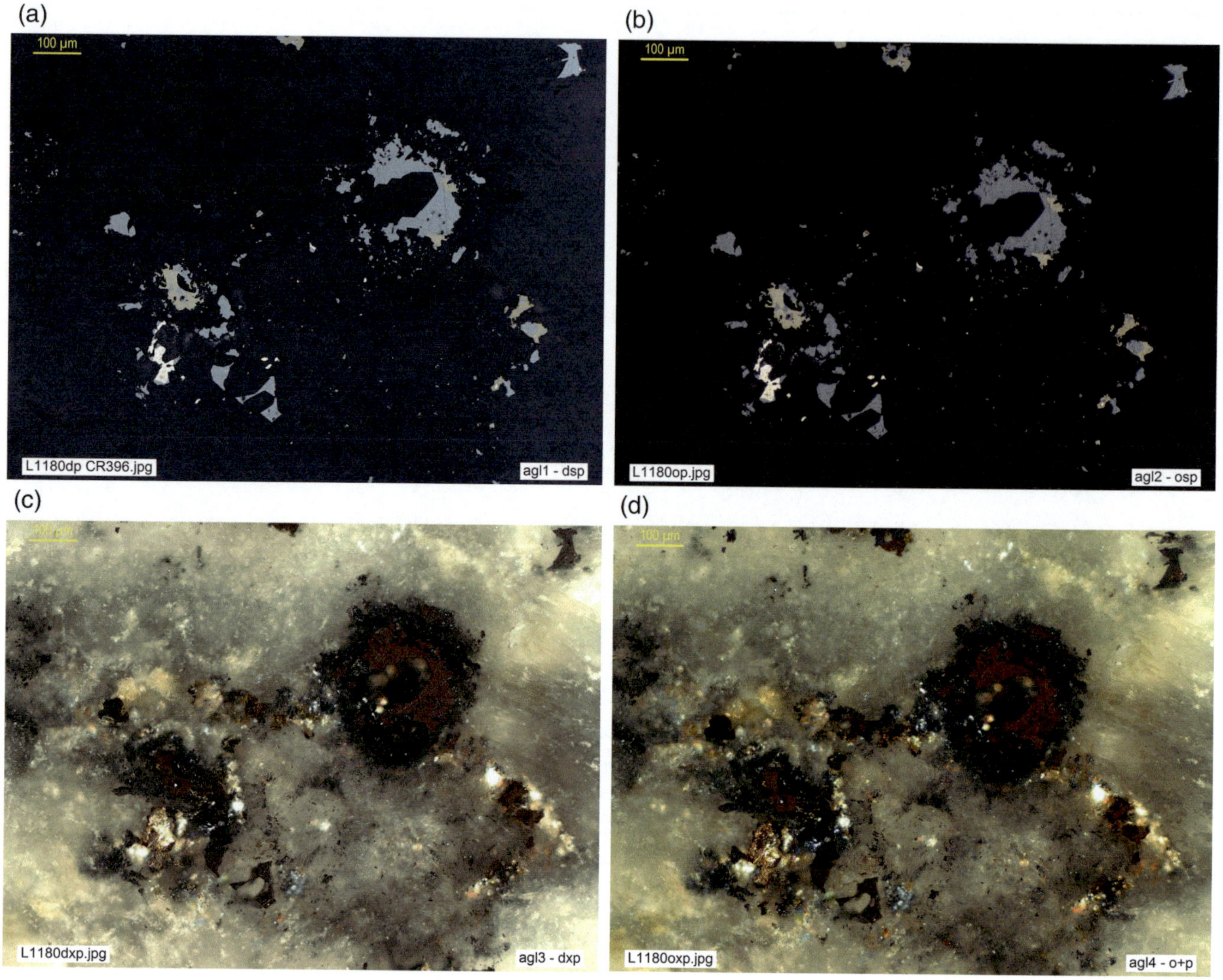

Fig. 4.1 **a**, **b** (**dsp** and **osp**, respectively). Quartz vein (dark background) with interstitial impregnation of agl (gray), el (yellowish white, very bright) and ccp (yellow, with a brownish tint suggesting the presence of Se in the crystal lattice of ccp). CR396, Midas Mine, NV, USA. **c**, **d** Same scene as Fig. 4.1a (**d+p** and **oxp**, respectively), weak anisotropism of agl and ccp

Accompanying ores: acanthite, silver, electrum, naumannite, pyrite, chalcopyrite, galena, sphalerite and other sulfides or sulfosalts of Ag–Se.

4.2.4 Identification

The paragenesis is typical, yet aguilarite could be confused with acanthite at first. The absence of light etching in aguilarite, an attack to which acanthite is very sensitive, is a decisive difference, to which is added the better polishing of aguilarite and its somewhat more coloured anisotropism.

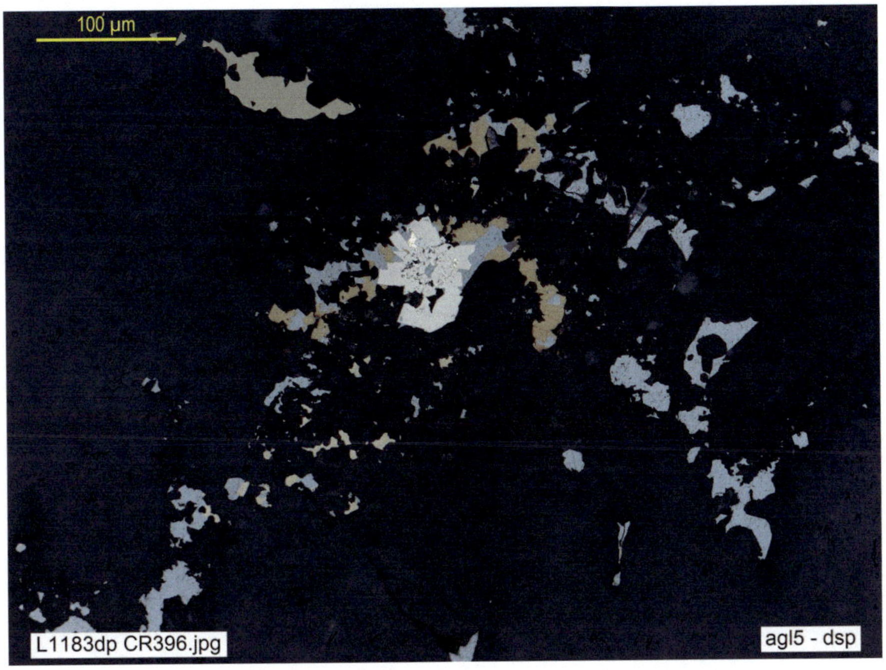

Fig. 4.2 Impregnation of agl (light grey), ccp, py, el and sp (dark grey) in the same vein as Fig. 4.1a. When contiguous with agl, ccp shows a brownish hue, indicating its Se contamination, which is not the case with the isolated ccp (surrounded only by q, upper left), whose color is closer to its normal yellow. CR396, Midas Mine, NV, USA (**dsp**)

4.3 Spectral Information

4.3.1 Aguilarite: Multispectral Reflectance R (in Air, Without Polariser) (%)

λ (nm)	370	400	425	450	475	500	525	550	575	600	625	650	675	700	750	800	850	900	950	1000
μ	34.6	34.8	34.9	35.0	34.8	34.6	34.0	33.3	32.8	32.3	31.7	31.2	30.7	30.2	29.4	28.7	27.9	27.5	27.3	27.5
P_5	34.4	34.4	34.4	34.4	34.0	33.6	33.0	32.4	32.0	31.6	31.0	30.4	30.0	29.6	28.8	28.0	27.2	26.8	26.8	26.8
P_{20}	34.2	34.4	34.6	34.8	34.4	34.0	33.4	32.8	32.4	32.0	31.4	30.8	30.4	30.0	29.2	28.4	27.6	27.2	27.2	27.2
P_{80}	35.2	35.2	35.2	35.2	35.2	35.2	34.4	33.6	33.0	32.4	32.0	31.6	31.0	30.4	29.6	29.2	28.4	27.6	27.6	27.6
P_{95}	35.0	35.2	35.4	35.6	35.6	35.6	34.8	34.0	33.4	32.8	32.2	31.6	31.2	30.8	30.0	29.2	28.4	28.0	27.6	28.0

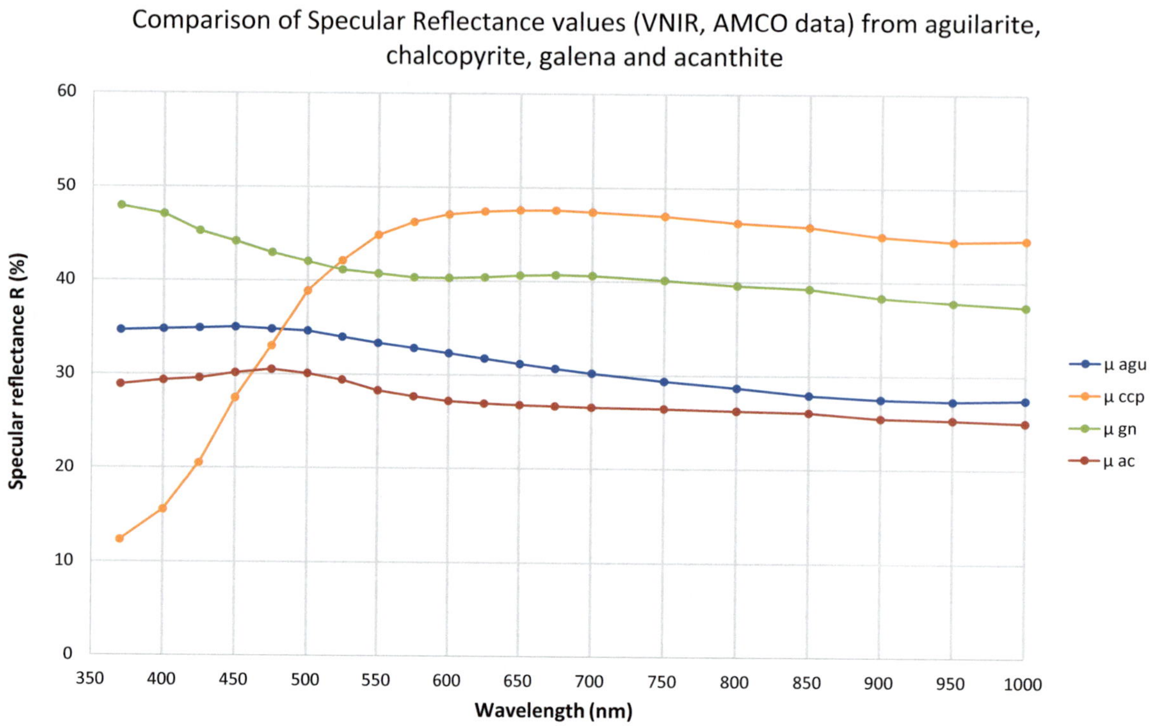
Comparison of Specular Reflectance values (VNIR, AMCO data) from aguilarite, chalcopyrite, galena and acanthite
Specular reflectance R (%)
Wavelength (nm)
μ agu
μ ccp
μ gn
μ ac

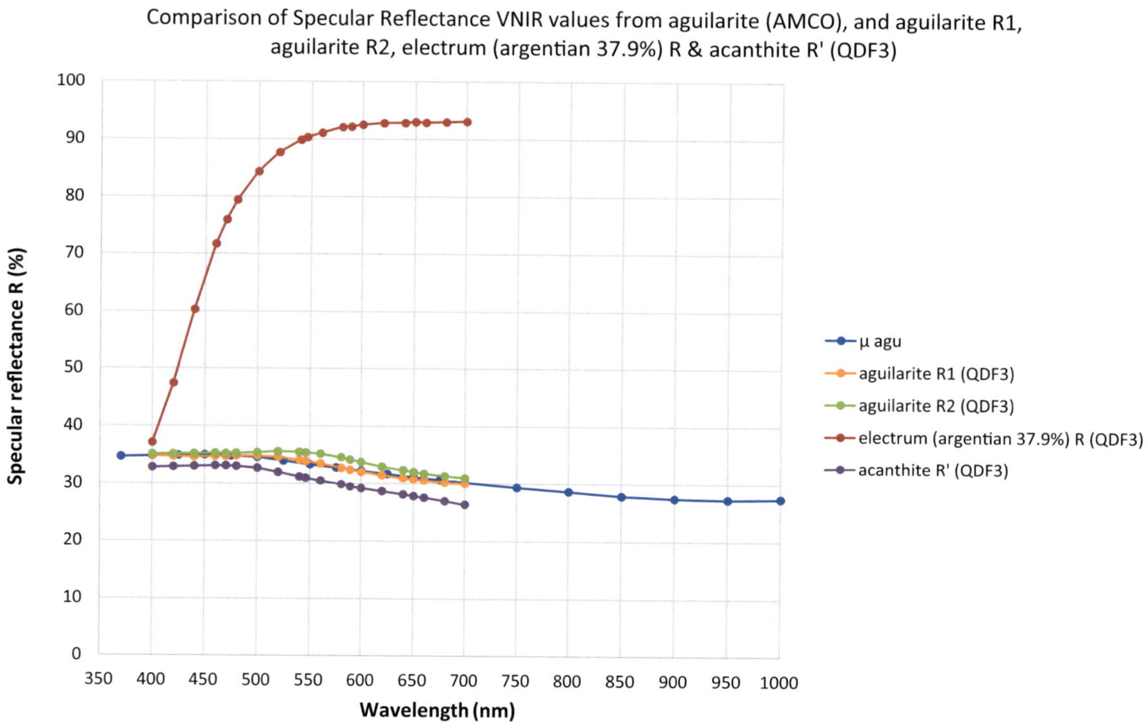
Comparison of Specular Reflectance VNIR values from aguilarite (AMCO), and aguilarite R1, aguilarite R2, electrum (argentian 37.9%) R & acanthite R' (QDF3)
Specular reflectance R (%)
Wavelength (nm)
μ agu
aguilarite R1 (QDF3)
aguilarite R2 (QDF3)
electrum (argentian 37.9%) R (QDF3)
acanthite R' (QDF3)

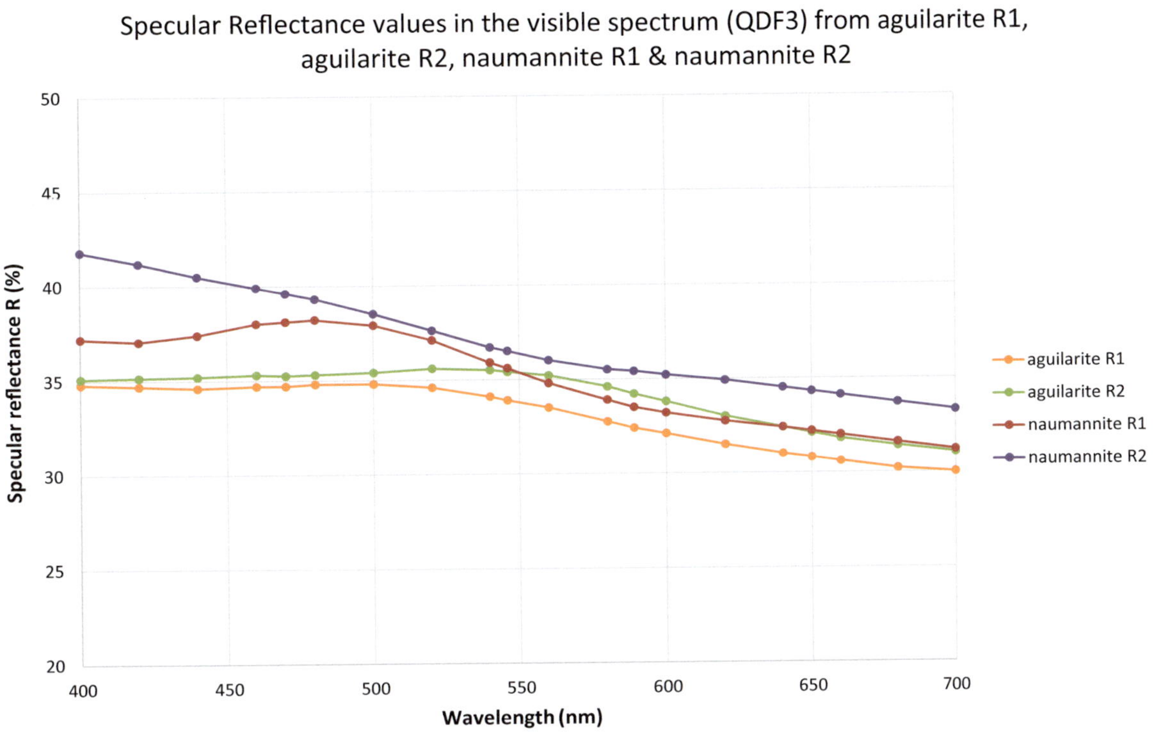
Specular Reflectance values in the visible spectrum (QDF3) from aguilarite R1, aguilarite R2, naumannite R1 & naumannite R2
Specular reflectance R (%)
Wavelength (nm)
aguilarite R1
aguilarite R2
naumannite R1
naumannite R2
50
45
40
35
30
25
20
400
450
500
550
600
650
700

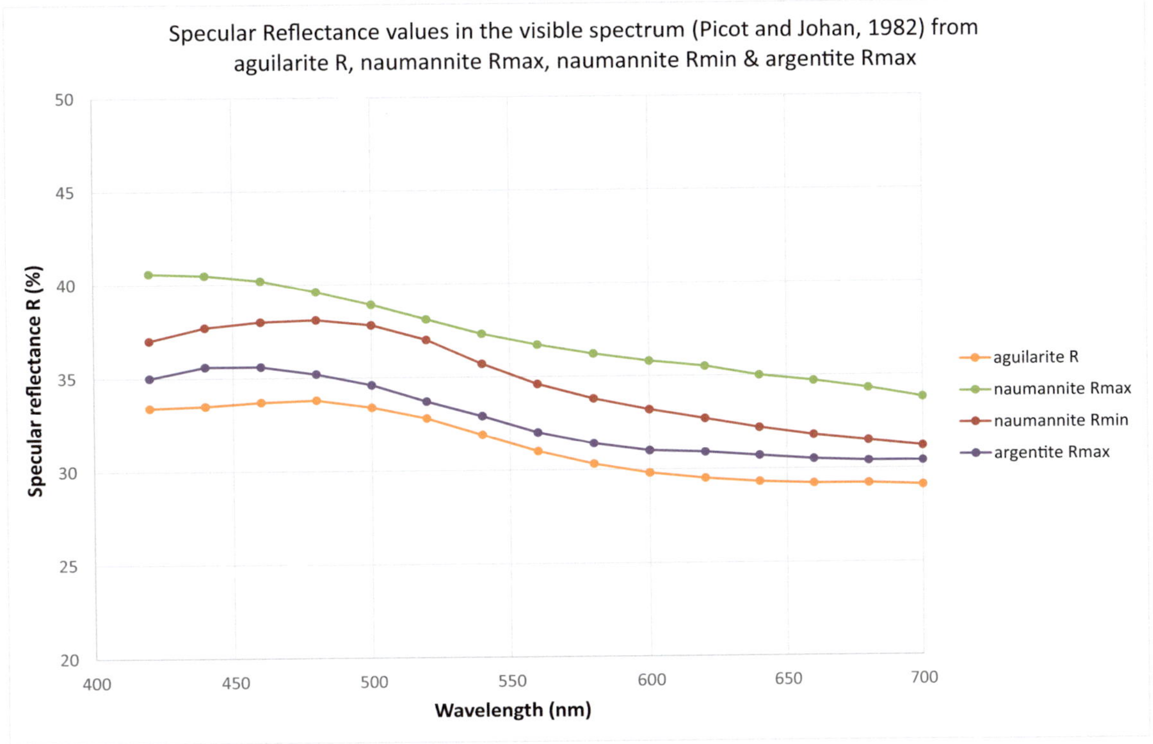
Specular Reflectance values in the visible spectrum (Picot and Johan, 1982) from aguilarite R, naumannite Rmax, naumannite Rmin & argentite Rmax
Specular reflectance R (%)
Wavelength (nm)
aguilarite R
naumannite Rmax
naumannite Rmin
argentite Rmax
50
45
40
35
30
25
20
400
450
500
550
600
650
700

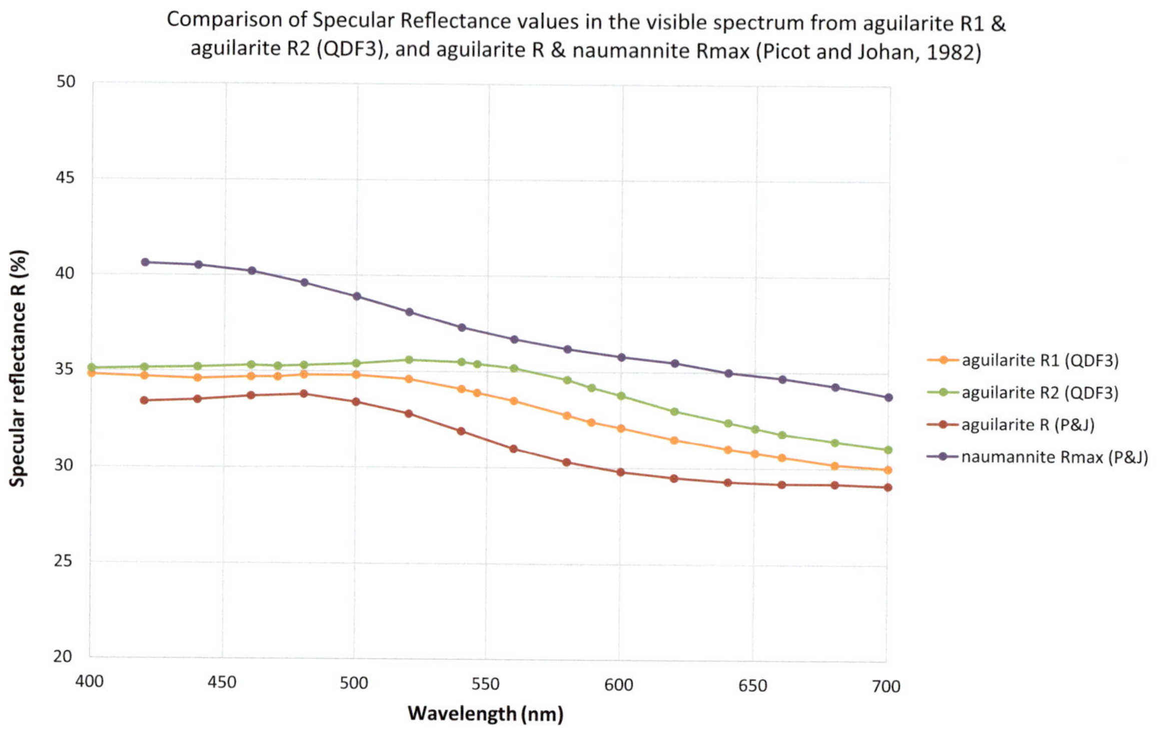

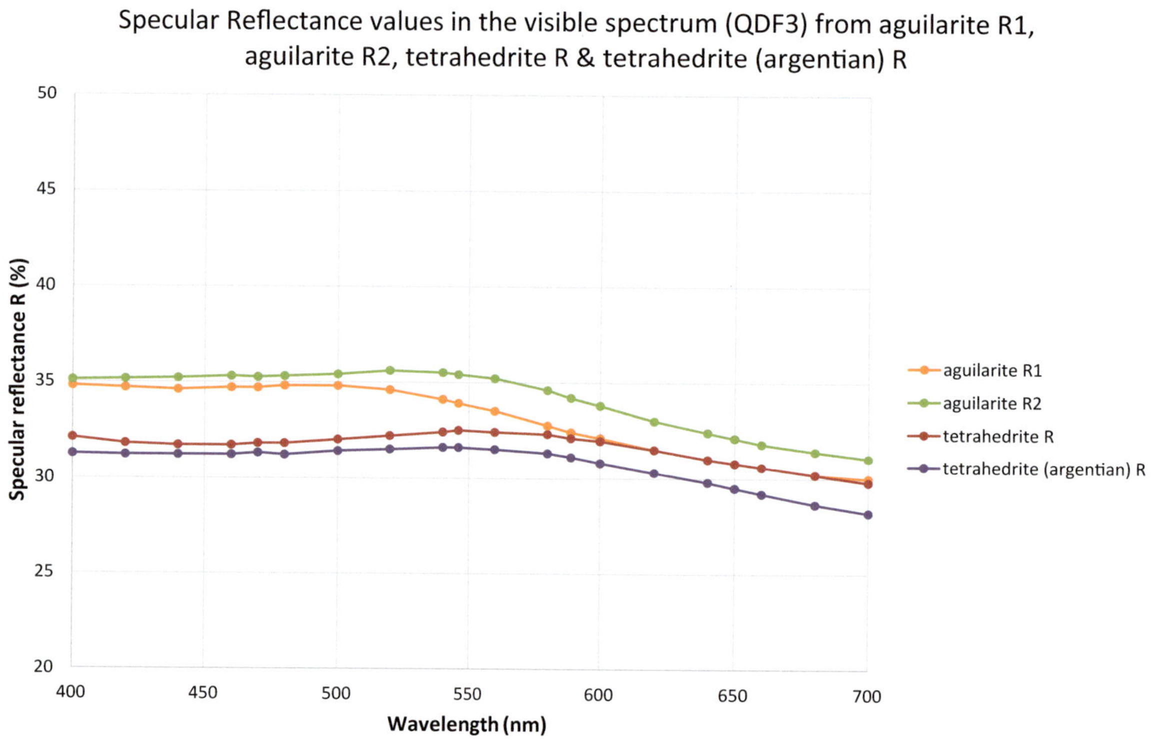

Reference

Picot P, Johan Z (1982) Atlas of ore minerals. B.R.G.M./Elsevier

5 Alabandite (*alb*)

MnS: Cubic

Alabandite	R_{546} (%)	Y (%)	λ_d (nm)	P_e (%)	Colour sensation
Air	22.8	22.8	475	4.8	Grey, lighter than sp
Oil	9.0	9.0	474	6.2	Grey
Bir: No (1)			*Pleoc*: No (1)		*Tarn*: Frequent
Anis +P: Isotropic, total extinction +P (1)			*IR:* Green or brown		*Twg*: Thin lamellar twins (111)
Mor: Varied, crystalline aggregates			*Clv*: // (100) occasional		*PolQ*: Good quality polishing
Hardness	*PolH*: >ccp, <sp (3)		*Mohs*: 3.5+	*Talm*: C	*VHN* $_{(100\ g)}$: 240–251

Obs: Rare mineral. Epithermal paragenesis with Ag/Au ores
Spectral information: cf. Table and R plots, multispectral data (VNIR, AMCO) and visible range (QDF3)
Mineral abbreviations: alb—alabandite; gg—gangue; gn—galena; ln—linneite; mc—marcasite; mt—magnetite; po—pyrrhotite; py—pyrite; sp—sphalerite

5.1 General

Composition: (Mn,Fe)S. Fe contents variable from ≈ 0% (pure MnS) to values close to 50% (MnS:FeS ≈ 1:1, *Eisenalabandin*, Ramdohr 1957). The Fe content seems to depend on the formation temperature (only for FeS contents > 40 mol %) but is not reliable as a geological thermometer (for more information and other data on the effects of Fe content on the properties of alabandite, cf. Törnroos 1982).

5.2 Microscopic Description

5.2.1 SP Observation (*Single Polariser*)

Grey *color*, similar to sphalerite but lighter (Figs. 5.1a, b, 5.2, 5.3 and 5.4). It is also lighter than magnetite which, in direct comparison, shows a brownish grey tone; darker than freibergite (Y fbg > alb > mt > sp). The contrast with these minerals increases in immersion (oil).

Good *polish*, but rapid tarnishing. In a short time, a superficial alteration *patina* usually appears. *Polishing hardness* between that of sphalerite and chalcopyrite ($PolH_{alb}$ < sp; $PolH_{alb}$ > ccp). Vickers Microhardness, VHN, increases with Fe content from ≈170 to ≈250, for % mol FeS from 0 to 40, respectively (Törnroos 1982). Mohs hardness: 3.5–4, Talmage: C.

Low *reflectance*, lower than freibergite or tetrahedrite, acanthite, hematite but higher than sphalerite, magnetite or manganite (cf. *Table and spectral graphs R*). R increases with Fe content (λ = 546 nm, from R = 23.2% for 1 mol % FeS, to 26.0% for 10.6 mol % FeS). Isotropic.

Occurs in crystalline aggregates and encrustations, sometimes pseudomorphic. Occasional zoning. Cleavage // (100) visible in sections with defective polish.

5.2.2 Observation +P (*Polariser + Analyser at 90°*)

Isotropic, always in extinction (+P).

Sporadic *internal reflections*, more visible in oil (Figs. 5.1c, d, 5.3, 5.4 and 5.5). Typical IR are usually described as dark green but more often show brown to reddish-brown shades. This can be explained as a compositional effect. The green colour is restricted to pure terms

R. Castroviejo, *A Practical Guide to Ore Microscopy—Volume 1*,
https://doi.org/10.1007/978-3-031-12654-3_5

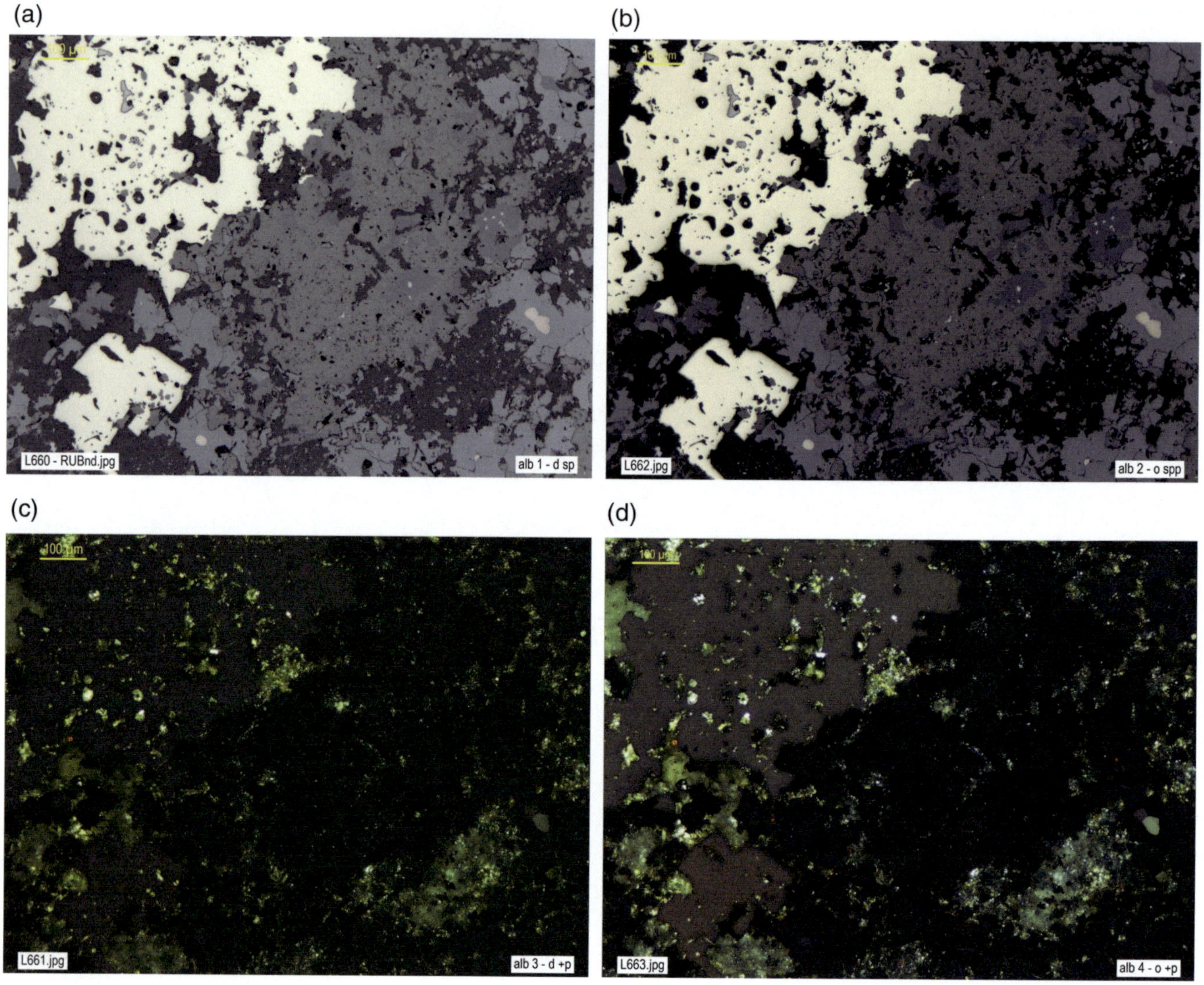

Fig. 5.1 **a** Aggregate of py (upper left zone, with gn inclusions), mt (center) and alb (right); mt with sp inclusions. Compare the three grey minerals, alb, mt, sp, in this order, from lightest to darkest (R alb > mt > sp). Mt shows a brownish hue that is lacking in alb and sp, and both alb and sp exhibit IR that mt lacks. Lower right border shows inclusion of po (light pinkish brown) in alb. R py > po > alb (**dsp**, sample RUB, n/d), **b** Same scene as Fig. 5.1a (**osp**). Increased R contrast and colour saturation make it easier to distinguish sulfides from each other and from mt, **c** Same scene as Fig. 5.1a (**d+p**). Total extinction of alb, mt and sp (isotropic), as well as py and gn, in contrast to po (coloured, greenish and brownish, anisotropism), **d** Same scene as Fig. 5.1a (**o+p**). Immersion allows more clearly distinguishing the sporadic IR of alb (in this section, the IR are not very visible in sp because of their small size, being included in mt)

(MnS), becoming brownish or reddish in the abundant ferriferous terms (the colour change is apparently dominant from 4 to 5% Fe).

Very fine laminar *twinning* (111?) similar to that of sphalerite, occasionally visible and enhanced by polishing relief.

5.2.3 Paragenesis

Epithermal paragenesis with native Ag and/or Au, electrum, sphalerite, marcasite, galena, chalcopyrite, pyrite, pyrrhotite, magnetite, stannite, Au tellurides, Ag sulfosalts, etc., as well as rhodochrosite and/or rhodonite gangue.

It is more rarely found in *basalts*. It has also been seen in *meteorites*. It appears, as a product of *metallurgical processes*, in many *mattes*.

5.2.4 Identification

Typical for alabandite are its grey color, good polish, low reflectance (between sphalerite or magnetite and freibergite),

Fig. 5.2 Hydrothermal infill of *alb* (light grey), with *sp* (darker grey) and fine aggregates of *mc* (greenish white), *ln* (pinkish white) and *py* (yellowish white) scattered in the quartz (almost black) and carbonate gangue (different shades of grey, due to the strong birefringence but always darker than sp). CR 334, Inakuraischi Mine. Hokkaido, Japan (**dsp**)

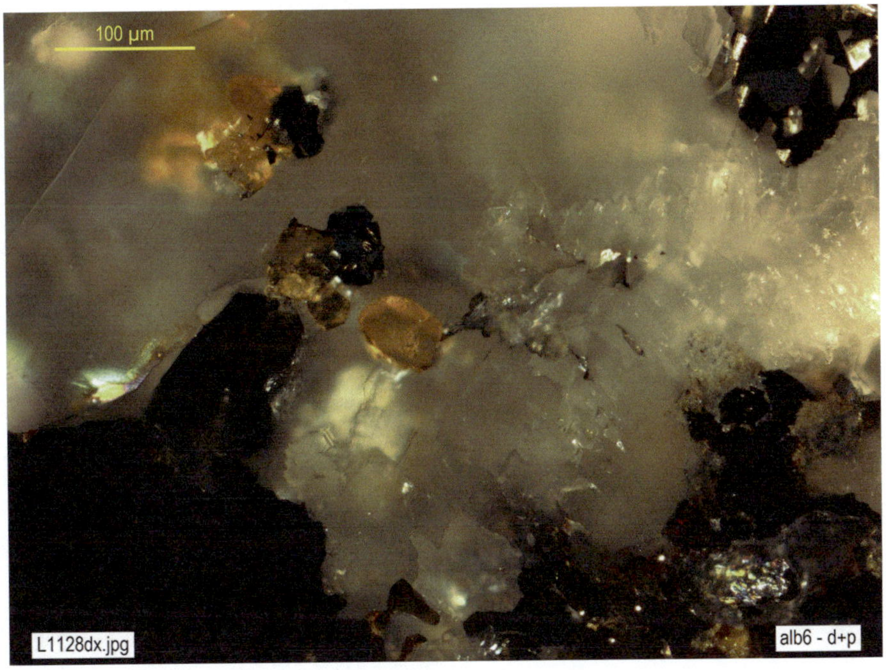

Fig. 5.3 Same scene as Fig. 5.2 (**d+p**), widespread IR in quartz-carbonate *gg* and *sp*, much scarcer in *alb*

isotropic character, occasional IR and association with manganiferous gangue (rhodochrosite).

It can be confused with sphalerite (sp), from which alb can be distinguished by its higher reflectance and lower polishing hardness. Sphalerite shows much more frequent and luminous IR. Compared to magnetite, mt is darker and lacks IR, shows higher polishing hardness and a brownish tone that is not found in alabandite. Against freibergite and minerals of the tetrahedrite-tennantite series, it appears distinctly darker.

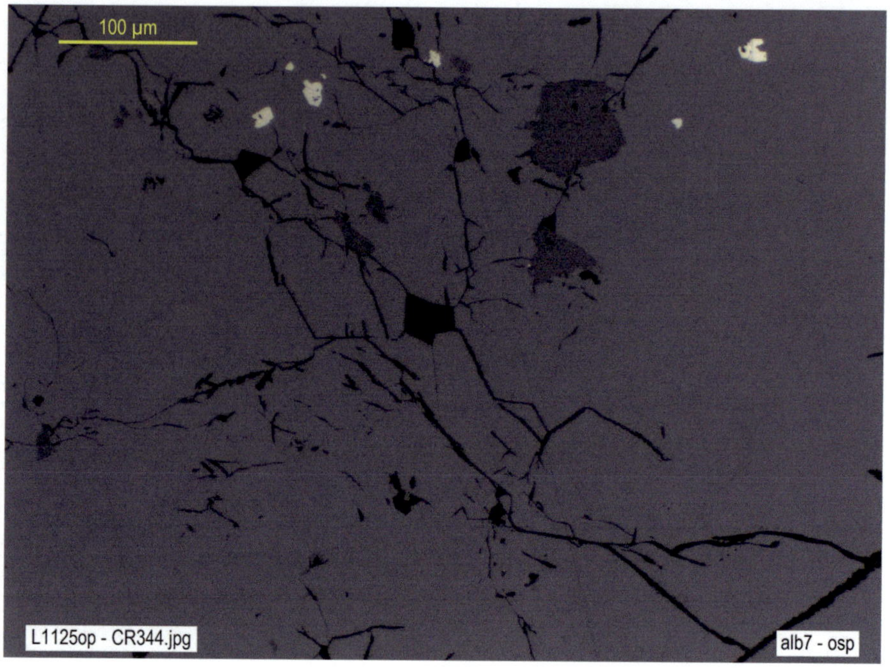

Fig. 5.4 Crystal aggregate of alb (light grey), with inclusions of sp (darker grey) and py+mc (white/cream); quartz (black in immersion: R ≈ 0%) in intergranular films and microcracks, or interstitial in alb. CR 334, Inakuraischi Mine. Hokkaido, Japan (**osp**)

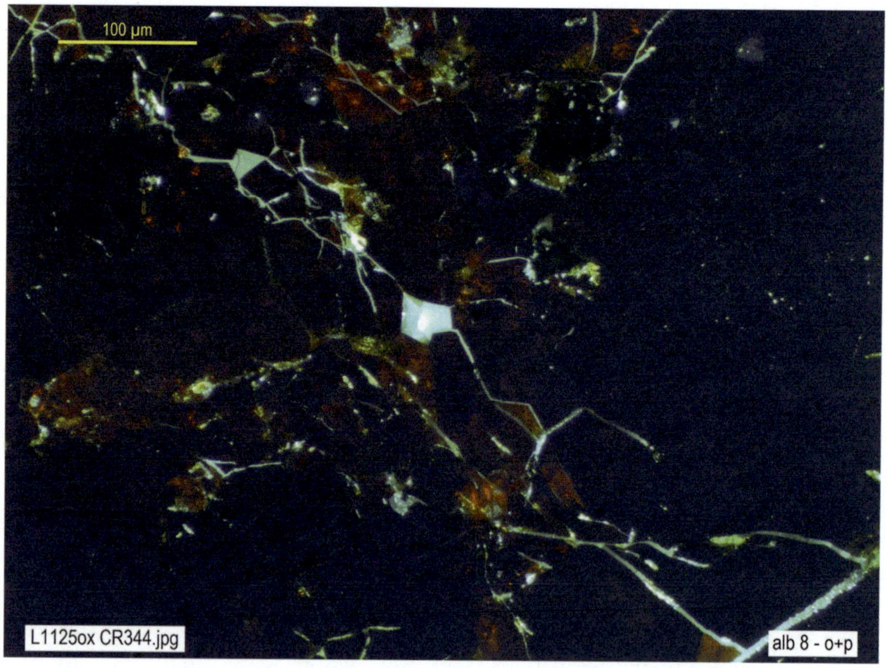

Fig. 5.5 Same scene as Fig. 5.4, +P, greenish and reddish-brownish IR of alb, clearly visible in immersion. Widespread, colourless and much more luminous IR of interstitial quartz (**o+p**)

5.3 Spectral Information

5.3.1 Alabandite: Multispectral Reflectance R (in Air, Without Polariser) (%)

λ (nm)	370	400	425	450	475	500	525	550	575	600	625	650	675	700	750	800	850	900	950	1000
μ	28.6	27.6	26.6	25.7	25.0	24.2	23.7	23.2	22.8	22.5	22.3	22.1	22.0	21.8	21.6	21.4	21.3	21.1	20.9	20.7
P_5	28.1	27.2	26.1	25.2	24.6	23.8	23.3	22.8	22.4	22.1	21.9	21.8	21.6	21.5	21.3	20.9	20.8	20.5	20.3	20.1
P_{20}	28.3	27.5	26.4	25.5	24.8	24.0	23.5	23.0	22.6	22.3	22.1	21.9	21.8	21.7	21.4	21.1	21.0	20.7	20.5	20.3
P_{80}	28.8	27.8	26.8	25.9	25.2	24.5	23.9	23.4	23.0	22.7	22.5	22.3	22.1	22.0	21.8	21.6	21.5	21.4	21.2	21.0
P_{95}	29.1	28.0	27.0	26.1	25.5	24.7	24.2	23.6	23.2	22.9	22.7	22.5	22.3	22.2	22.1	22.0	22.0	22.0	21.8	21.6

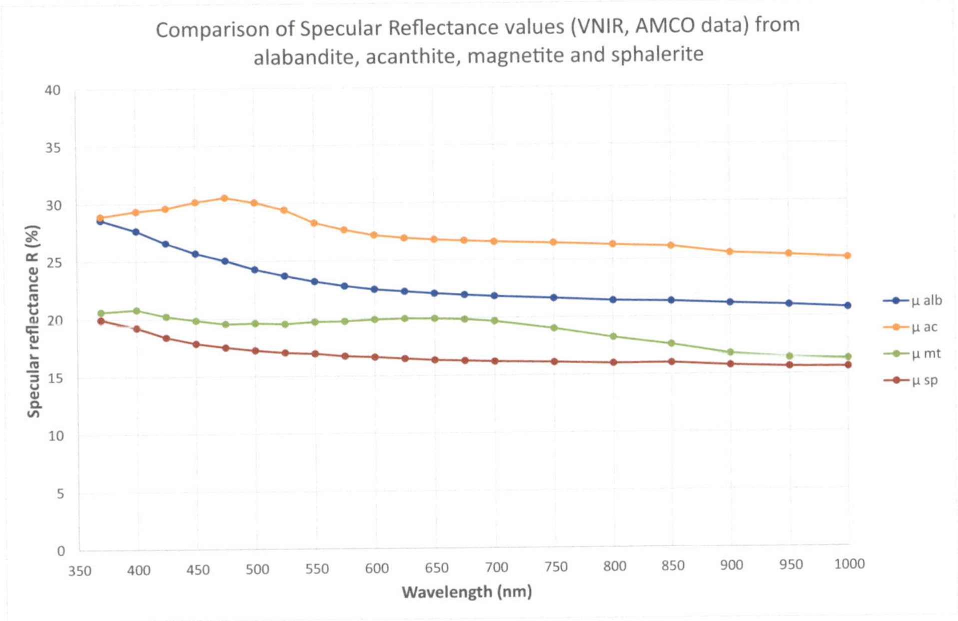

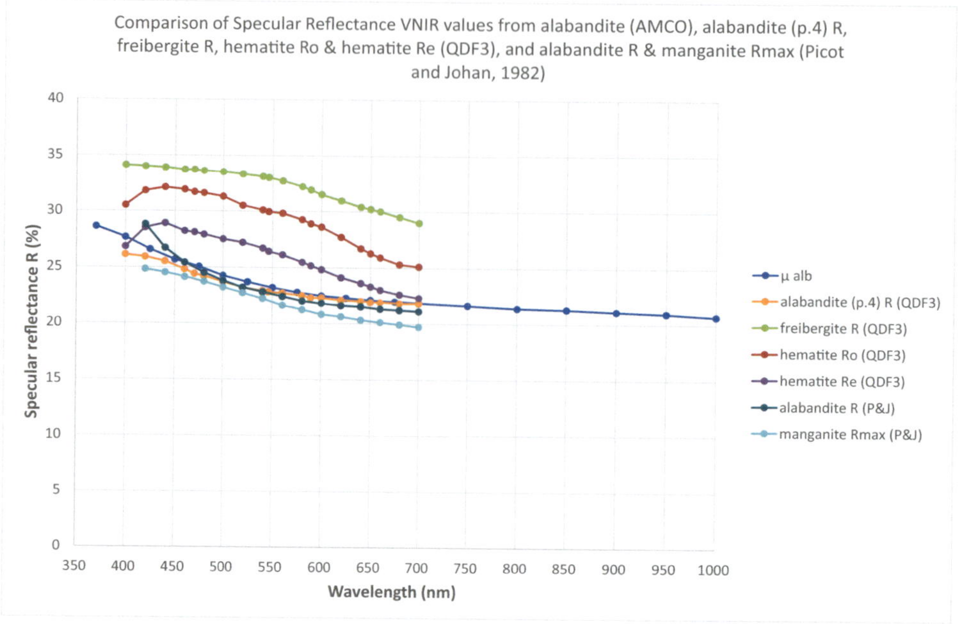

References

Ramdohr P (1957) Eisenalabandin, ein merkwürdiger natürlicher Hochtemperatur-Mischkristall. Neues Jhb Mineral Abh 91:89–93

Törnroos R (1982) Properties of alabandite; alabandite from Finland. N Jb Minere Abh 144(1):107–123

Allargentum (*aag*)

6

$Ag_{1-x}Sb_x$ (x ≈ 0.09–0.16): Hexagonal

Allargentum		R_{546} (%)	Y (%)	λ_d (nm)	P_e (%)	Colour sensation
Air	R_1	69.5	nd	nd	nd	Bright white, grey versus Ag
	R_2	70.9	nd	nd	nd	
Oil	R_1	nd	nd	nd	nd	
	R_2	nd	nd	nd	nd	
Bir: Imperceptible (2)				*Pleoc*: Imperceptible (2)		*Tarn*: No
Anis +P: Weak (3)				*IR*: No		*Twg*: Lamellar/Polysynthetic
Mor: Finely intergrown with Ag–Sb				*Clv*: No		*PolQ*: Good, but inevitably scratched
Hardness	*PolH*: Medium (3)			*Mohs*: 3.5	*Talm*: B	*VHN*$_{(100\ g)}$: 172–203

Obs: complex textures, due to phase relations in the Ag–Sb binary system; cf. Table and R-plots, multispectral data (VNIR, AMCO) and visible range (Picot and Johan 1982)
Mineral abbreviations: aag—allargentum; dy—discrasite; eug—eugenite; gg—gangue; gn—galena; lo—löllingite; nc—nickelite; prg—pyrgyrite; rm—rammelsbergite; sk—skutterudite

6.1 General

Mineral first described by Ramdohr and redefined by IMA as proposed by Petruk et al. (1969, 1970). The currently accepted composition, $Ag_{1-x}Sb_x$ (x ≈ 0.09–0.16), implies a remarkable range of variation from the original, Ag_6Sb. This would correspond, in the new formula, to a value x ≈ 0.14 (with 84.4% Ag and 15.6% Sb), i.e., it lies towards the more Sb-rich end of the range now admitted for allargentum (between ≈ 82.3% Ag-17.7% Sb for x = 0.16 and 90% Ag-10% Sb for x = 0.09). It is abundant, associated with *antimonial silver* (Ag, with Sb ≤ 5 wt%) in deposits at Cobalt, Ontario, where it had initially been mistaken for dyscrasite (cf. discussion *dy*). Its intergrowths are evidence of the complexity of phase relationships in the ternary S–Ag–Sb system and in the binary Ag–Sb system (Keighin and Honea 1969). For a more detailed discussion of the allargentum problem, cf. Uytenbogaardt and Burke (1971, p. 83). Frost and Raynor (1950) have shown experimentally that Ag is insoluble in Sb, which explains why the term Sb, native antimony, does not contain Ag. Allargentum may contain traces of Hg, Ni, As.

This mineral does not appear in QDF3, so numerical information on most of the optical parameters, marked (**nd**) in the summary table (above), is lacking. However, the reflectance data published by Picot and Johan (1977) are presented.

6.2 Microscopic Description

6.2.1 SP Observation (*Single Polariser*)

White *colour*, which looks grey in contact with silver, with a slight violet-brown hue in immersion. Good *polish*, although scratches are unavoidable. Unlike silver, it does not tarnish (Figs. 6.1a, b, 6.2 and 6.3).

High *reflectance*, although distinctly lower than silver and antimonial silver (cf. R plots); higher than dyscrasite and much higher than skutterudite, löllingite, etc.; very close (although slightly lower) to native antimony (Figs. 6.1a, b; comp. Fig. 6.8). Although slightly bireflective, bireflectance and pleochroism are usually not perceived. No cleavage is observed. Occurs in fine anhedral or lanceolate aggregates, intergrown with other argentiferous phases, particularly

R. Castroviejo, *A Practical Guide to Ore Microscopy—Volume 1*,
https://doi.org/10.1007/978-3-031-12654-3_6

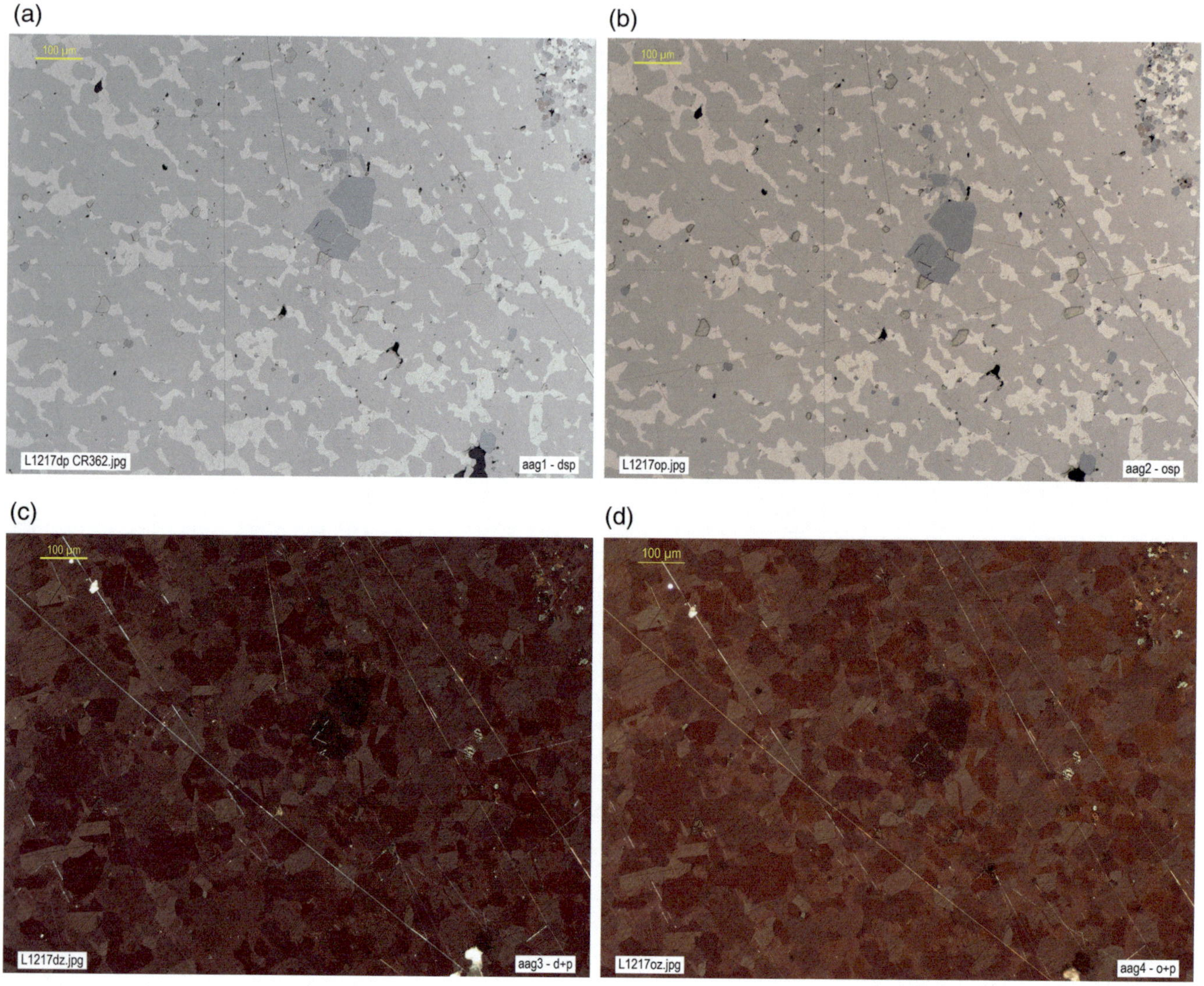

Fig. 6.1 **a** Symplectitic intergrowth of aag (grey) and antimonial silver (white), with microclasts of sk (dark grey), fine dissemination of Sb (tiny grey corpuscles, somewhat lighter than aag) and residual aggregate of nc (pink) + sk in up right corner. CR362: Nipissing Mine, Cobalt, Ontario, Canada (**dsp**). **b** Same scene as aag1 (**osp**): note the higher contrast, which makes it easier to distinguish the tiny Sb inclusions. **c** Same scene as aag1 (**d+p**): aag shows occasional lamellar twins and distinct but weak anisotropism, which contrasts with the strong anisotropism of Sb. Sb shows luminous tones and typical, very thin, polysynthetic twins. The aag twins are best seen in coarser grain sizes (comp. Figs. 7.1 and 7.2: cf. amalgam description). **d** Same scene as aag1 (**o+p**): aag and Sb anisotropism reinforced in immersion, particularly in Sb, with more visible twins, sharp, in spite of the small size

silver, antimonial silver, dyscrasite, *amalgam* (Ag–Hg, *eugenite*, $Ag_{11}Hg_2$, Figs. 6.4, 6.5, 6.6 and 6.7), occasionally with arsenides (sk, rm, nc, lo) Figs. 6.1a, b and 6.6 (Fig. 6.9).

Medium *polishing hardness*, similar to but somewhat lower than silver; superior to antimony, dyscrasite and galena.

6.2.2 Observation +P (*Polariser + Analyser at 90°*)

Weakly (but clearly) *anisotropic*, with polarization tints in grey and dull reddish-brown tones (Figs. 6.1c, d). No *internal reflections* (Fig. 6.9). Occasionally, in some of the

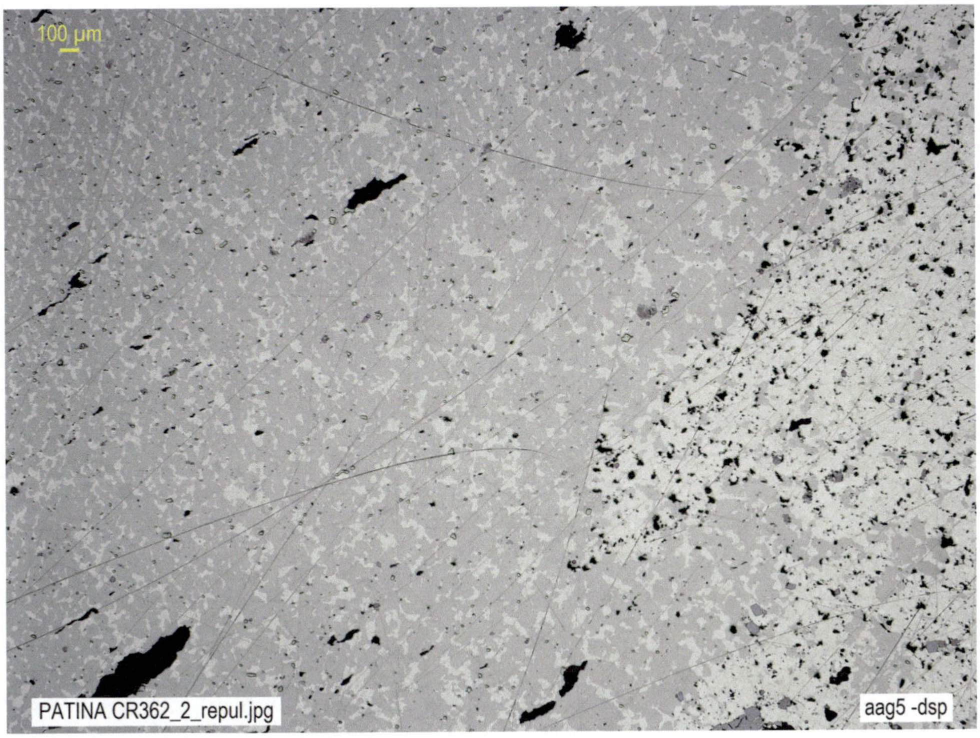

Fig. 6.2 Fine aag-antimonial Ag intergrowth, similar to Fig. 6.1, with Ag band (white) on the right. Despite the similarity between the two Ag varieties, Ag and antimonial Ag, note the slightly lower reflectance of the latter, as well as its greater resistance to patina formation, as Fig. 6.3 illustrates. CR362: Nipissing Mine, Cobalt, Ont.; **dsp**

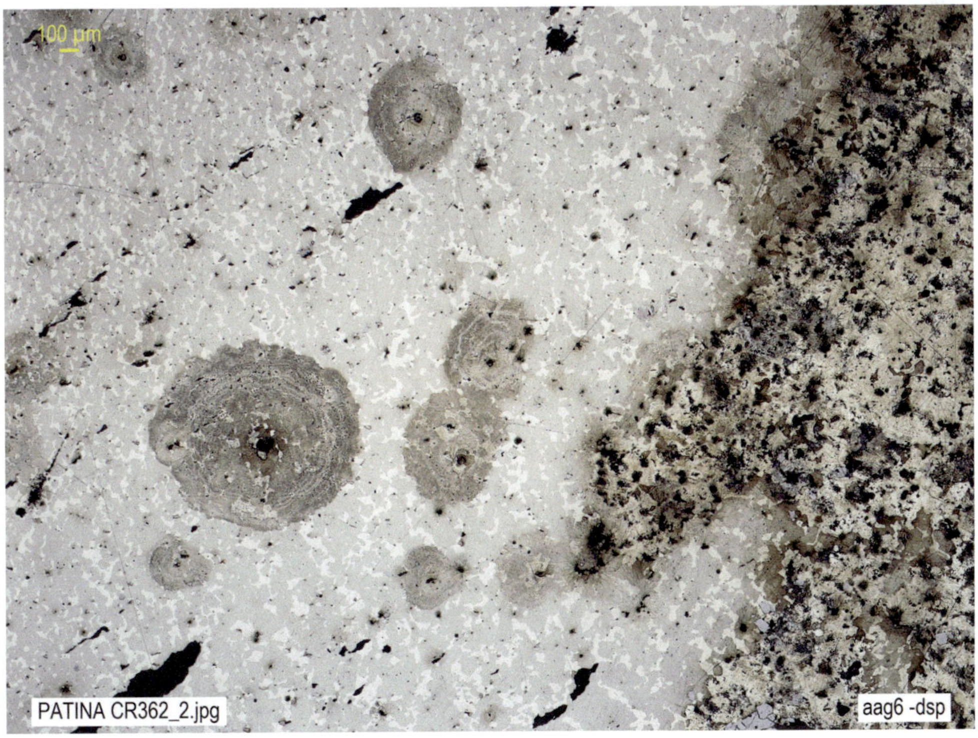

Fig. 6.3 Same scene as Fig. 6.2, several weeks after polishing. Note the darkening of Ag by tarnishing, which does not affect aag nor antimonial Ag (the globular stains on these phases are due to exsudation of the polishing lubricant retained in pores—black dots in the centre of the stain; **dsp**)

grains of microgranular aggregates, fine isolated lamellar *twins* are observed, well visible in spite of the polishing scratches (Figs. 6.1c, d). They should not be confused with the laminar or lanceolate exsolutions that can occur when included or in exsolution in other argentiferous phases (Figs. 6.6 and 6.7).

6.2.3 Paragenesis

Paragenesis characteristic of argentiferous hydrothermal deposits, frequent in the Cobalt, Ontario type, associated with silver, antimonial silver, dyscrasite, *amalgam-Ag* (vbgr. eugenite), breithauptite, skutterudite, etc.

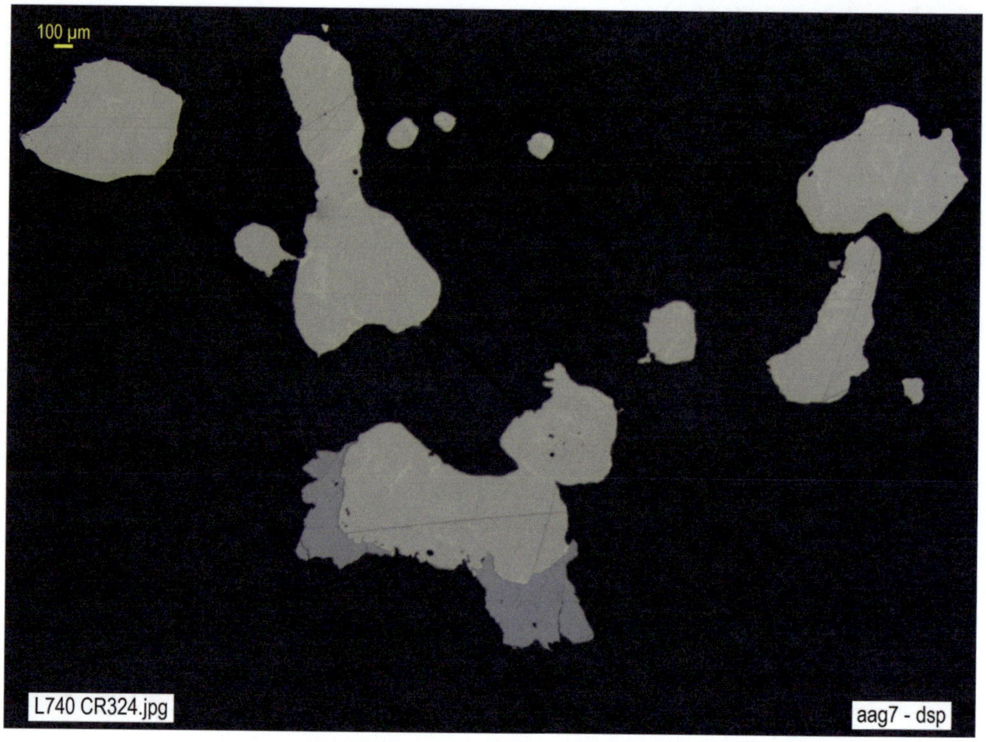

Fig. 6.4 Impregnation of aag (brownish grey) intergrown with *amalgam* (*eugenite*, white), with a border of lo (grey), in carbonate gangue (dark). CR324: Bouismas Mine, Bou Azzer, Morocco (**dsp**)

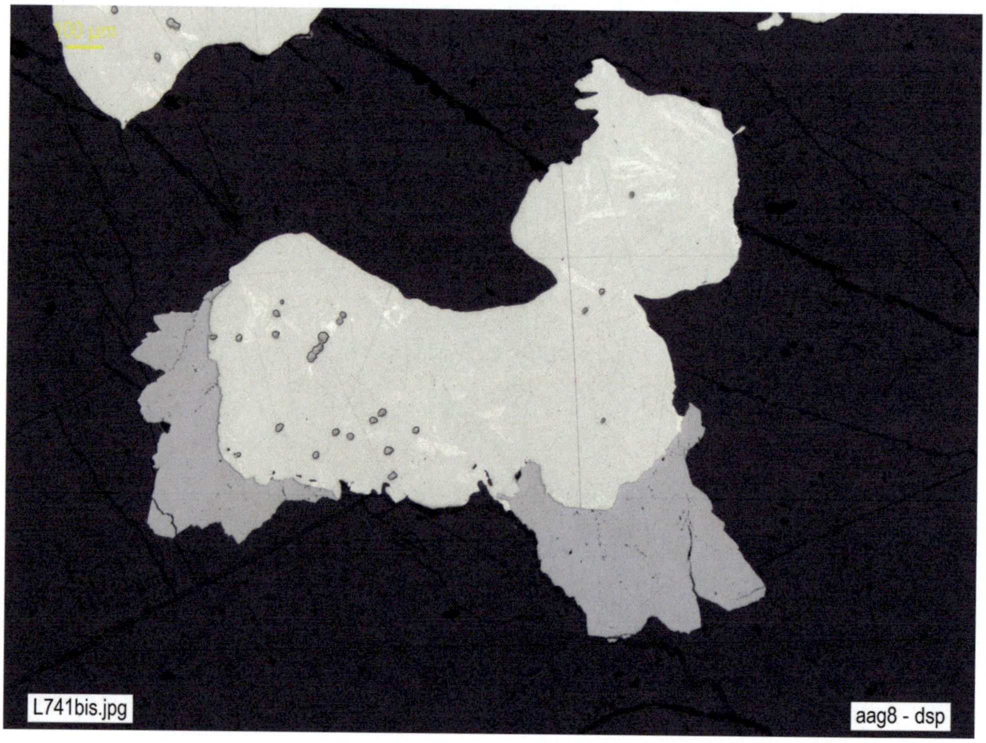

Fig. 6.5 Detail of Fig. 6.4, after EMP (electron microprobe) Analysis of the Ag–Hg amalgam. The points hit by the electron beam are marked by decomposition to eugenite ($Ag_{11}Hg_2$). The identities of lo and aag have also been confirmed by EMPA, showing that the subtle variations in colour hue and reflectance of aag are explained by slight changes in the Ag:Sb ratio (**dsp**)

6.2.4 Identification

Because of its properties and paragenesis, aag can be confused with few minerals, being dyscrasite the most similar. Its colour, high reflectance and anisotropism allow us to distinguish it from silver (much more reflective and isotropic), from antimony (slightly more reflective and with a much stronger and luminous anisotropism), and from dyscrasite (clearly less reflective, slightly more anisotropic, never laminar or lanceolate). Additionally, it is distinguished from all of them by the absence of tarnishing in aag.

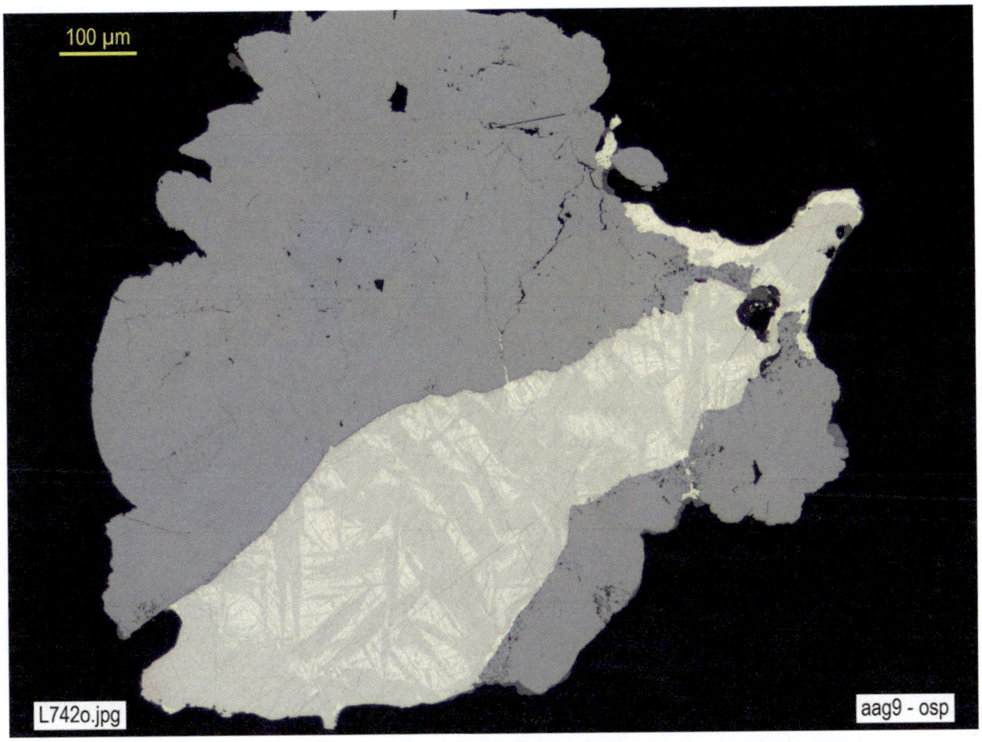

Fig. 6.6 Lamellar and lanceolate intergrowths (probable exsolutions) of aag (purplish grey) and eug (white), filling microcracks in lo aggregate (darker grey, slightly pleochroic in different shades). CR324: Bouismas Mine, Bou Azzer, Morocco (**osp**)

Fig. 6.7 Same scene as Fig. 6.6 (**o+p**). The strong anisotropism of lo, with bright colours, contrasts with the weak anisotropism of aag and the imperfect extinction of eug

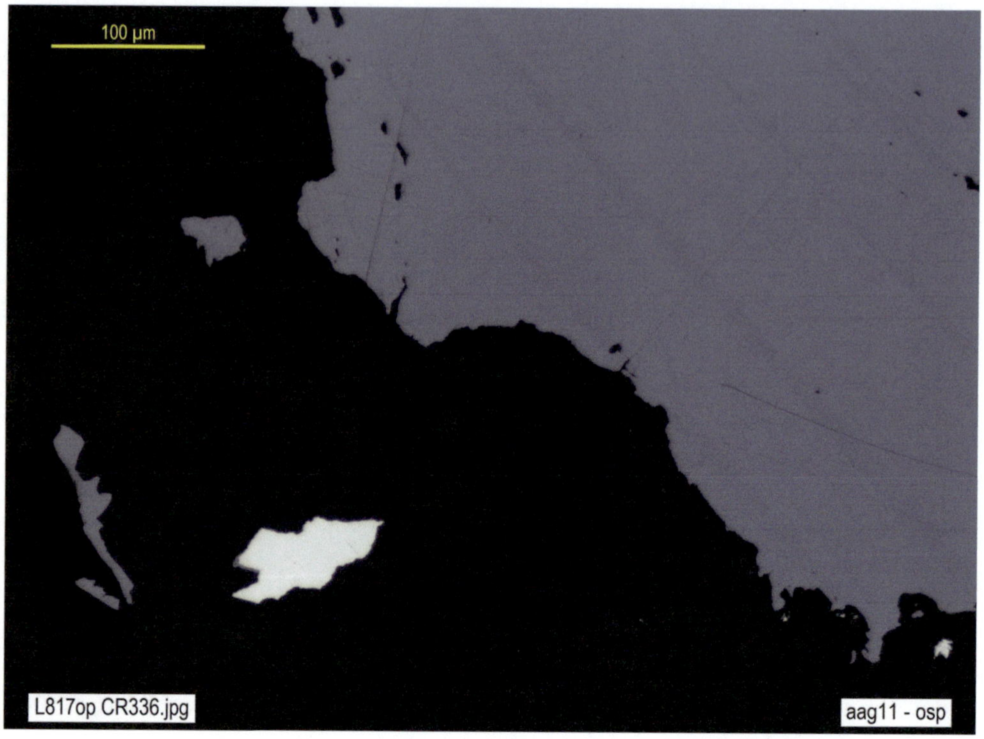

Fig. 6.8 Particle of aag (white, bright) in gangue, with pyrargyrite (bluish grey). Unlike the previous images, aag appears here only in the company of non-colored minerals of low-very low reflectance, which reinforces the subjective sensation of high brightness and white color of aag, resulting in a visual perception very similar to that of Ag. CR336: Freiberg, Germany; **osp**

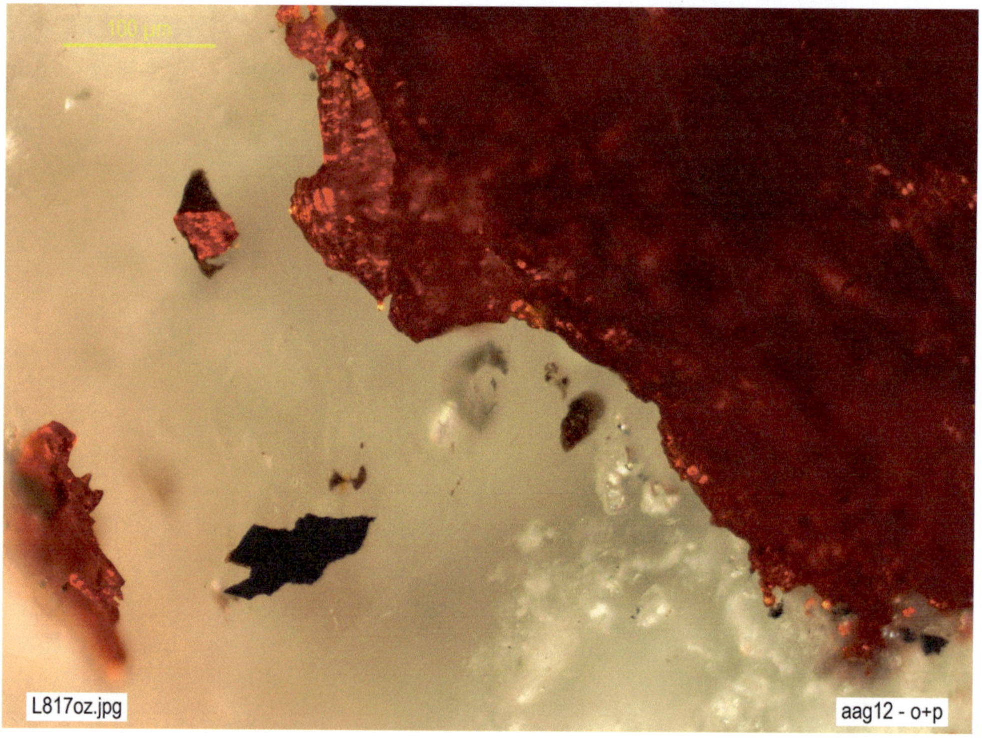

Fig. 6.9 Same scene as Fig. 6.8 (**o+p**). The weak anisotropism of aag is practically imperceptible, masked by the abundant and luminous internal reflections of gangue and prg (the latter, twinned)

6.2.5 Allargentum: Multispectral Reflectance R (in Air, Without Polariser) (%)

λ (nm)	370	400	425	450	475	500	525	550	575	600	625	650	675	700	750	800	850	900	950	1000
μ	56.0	59.1	61.8	63.0	65.7	66.2	68.4	68.7	70.1	70.8	71.4	72.1	72.5	73.0	73.8	74.2	75.2	75.9	76.6	76.9
P_5	53.2	56.3	58.8	59.9	62.4	63.0	65.2	65.6	67.1	67.9	68.6	69.3	69.9	70.5	71.4	72.0	73.1	74.0	74.8	75.1
P_{20}	54.9	58.1	60.7	61.9	64.6	65.1	67.3	67.6	69.1	69.8	70.5	71.1	71.6	72.2	72.9	73.4	74.5	75.3	76.0	76.3
P_{80}	57.2	60.3	63.1	64.3	67.1	67.5	69.7	69.9	71.3	72.0	72.5	73.2	73.6	74.1	74.7	75.2	76.1	76.8	77.5	77.6
P_{95}	58.0	61.0	63.9	65.2	67.9	68.4	70.5	70.7	72.0	72.7	73.3	73.9	74.3	74.8	75.4	75.8	76.7	77.3	78.0	78.1

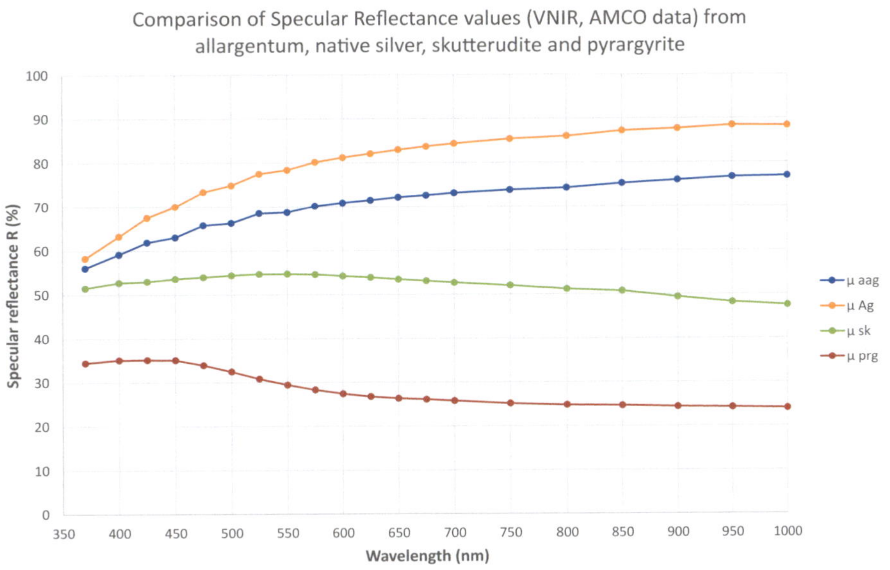

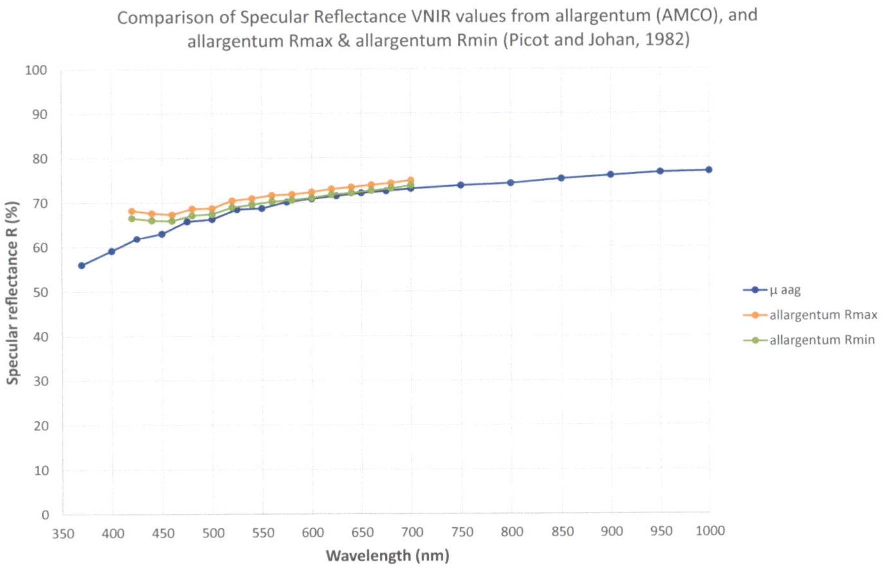

References

Frost BRT, Raynor GV (1950) The system silver-magnesium-antimony, with reference to the theory of alloy formation. Proc Roy Soc (Lond) A-203:134–135

Keighin CW, Honea RM (1969) The Ag-Sb-S system from 600 °C to 200 °C. Mineral Deposita 4:153–171

Petruk W, Harris DC, Cabri LJ, Stewart JM (1969) Native silver and silver-antimony minerals in the Cobalt-Gowganda ores. Can Mineral 10:144ff

Petruk W, Cabri LJ, Harris DC, Stewart JM (1970) Allargentum redefined. Can Mineral 10:163–172

Picot P, Johan Z (1977) Atlas des Mineraux Metalliques. Mem. B.R.G. M., 90, Paris, 406 p

Picot P, Johan Z (1982) Atlas of ore minerals. B.R.G.M., Elsevier

Uytenbogaardt W, Burke EAJ (1971) Tables for microscopic identification of ore minerals, 2nd edn. Elsevier, Amsterdam

7 Amalgam (*am, mercurial silver*)

(**Ag**, **Hg**), also (**Au**, Ag)Hg and (Ag, Hg, **Cu**)

7.1 General

Common amalgam or mercurial silver is a compound of variable composition (Ag+Hg, with Hg < 55%), defined by Dana as the α-phase of the Ag–Hg system, but it is discredited today as a mineral species by IMA-COM, unlike the γ-phase of the same system, which is accepted as Moschellandsbergite, Ag_2Hg_3. Nevertheless, because it is still referenced in use (e.g., it appears in the authoritative compilation of Hey 1975) and especially because of its mineral significance, it deserves at least a brief comment.

It should be noted that Hey (1975) presents the group of amalgams, also known as *native amalgam, natural pellet, Colombianite,* etc., essentially as natural **Ag–Hg** compounds, but in practice it is customary to also include gold-bearing compounds in this group. From this point of view, the group comprises as varieties the Cu and Au compounds, *cuproarcherite*, and *gold amalgam* (or *mercurial gold*), *respectively*. However, **gold amalgam** has recently acquired greater relevance as an industrial product, generated in gold concentration processes by amalgamation.

7.2 Description

According to Hey (1975, Chemical Index, File 1.10), there are the following varieties of amalgam:

- Kongsbergite, Hg ≈ 5% (File 1.11)
- Archerite, Hg ≈ 13% (File 1.12)
- Bordosite, Hg ≈ 30% (File 1.13)
- Cuprokerite, (Ag, Hg, Cu), Hg ≈ 13% (File 1.15).
- *Goldamalgam,* (Au,Ag)Hg, Hg ≈ 60% (File 1.16)

However, since this nomenclature no longer corresponds to accredited mineral species, it is preferable when adequate chemical information is available (usually precise compositions at mineral grain scale by electron microprobe, EMPA) to try and identify recognized species such as **Eugenite** ($Ag_{11}Hg_2$, cubic), **Moschellandsbergite** (Ag_2Hg_3, cubic) or **Schachnerite** ($Ag_{1.1}Hg_{0.9}$, hexagonal), among others.

Although these minerals are very rare and beyond the scope of this book, some images of the first one, eugenite (described by Kucha 1986), the only one of these three minerals that is not mentioned by Ramdohr (1980), are given as illustration. These images (Figs. 7.1 and 7.2) are complemented by those illustrating the electron microprobe analyses and the explanation of the associated *allargentum* (Figs. 6.4, 6.5, 6.6 and 6.7, cf. description aag).

Further information on these and related species can be found in the IMA/COM mineral species databases, http://rruff.info/ima.

Unlike the purely argentiferous, white varieties, the **gold amalgam** is more yellowish, as corresponds to its gold content (Figs. 7.3a, b, c, d.

R. Castroviejo, *A Practical Guide to Ore Microscopy—Volume 1*,
https://doi.org/10.1007/978-3-031-12654-3_7

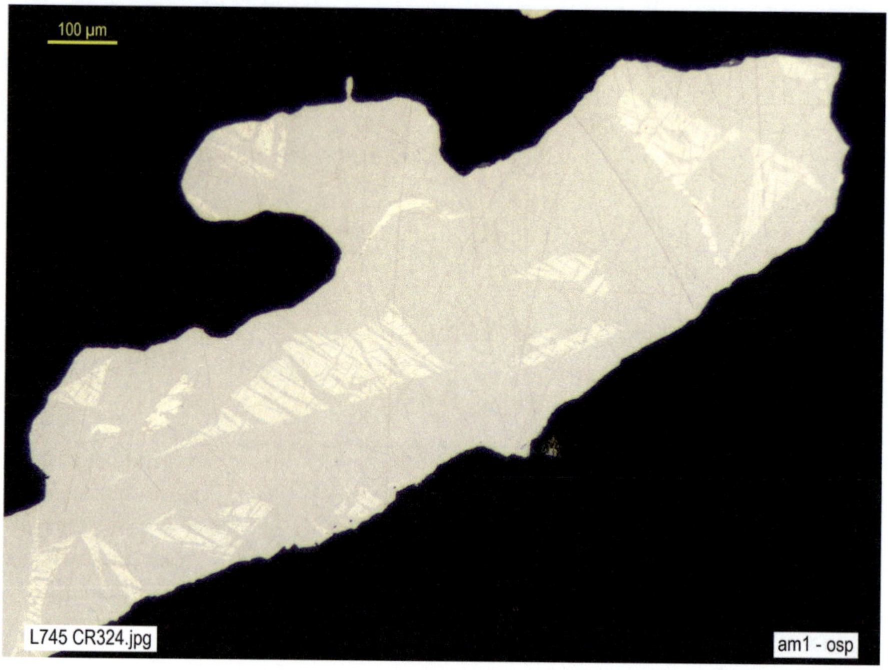

Fig. 7.1 Aggregate of allargentum (brownish grey) intergrown with *amalgam* (*eugenite*, white, very reflective), in carbonate gangue (CR324: Bouismas Mine, Bou Azzer, Morocco; **osp**)

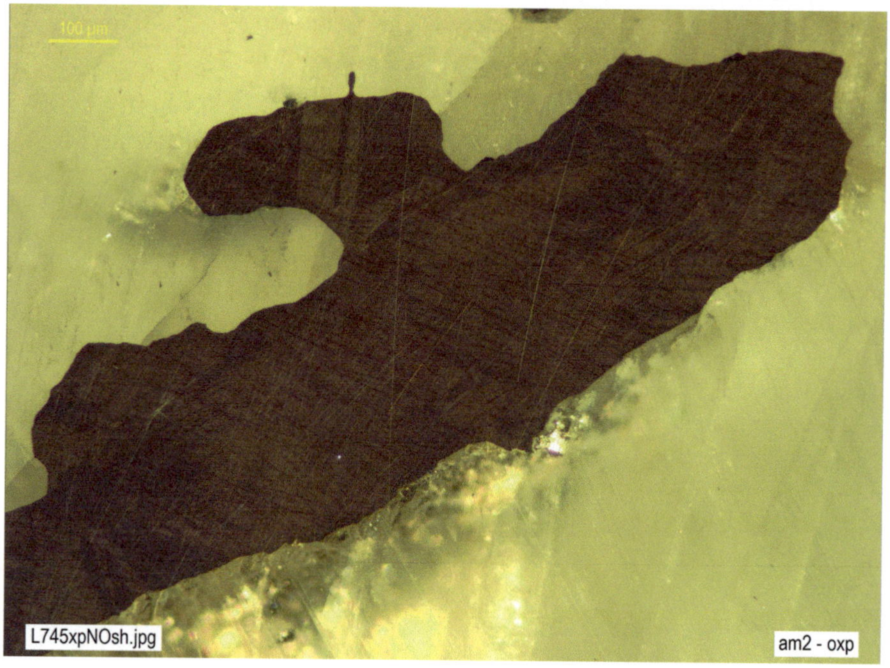

Fig. 7.2 Same scene as Fig. 7.1 (**o+p**): lamellar or polysynthetic twins of aag (allargentum), clearly visible in spite of their weak anisotropism, while eug (eugenite), although cubic, shows the incomplete extinction of native metals

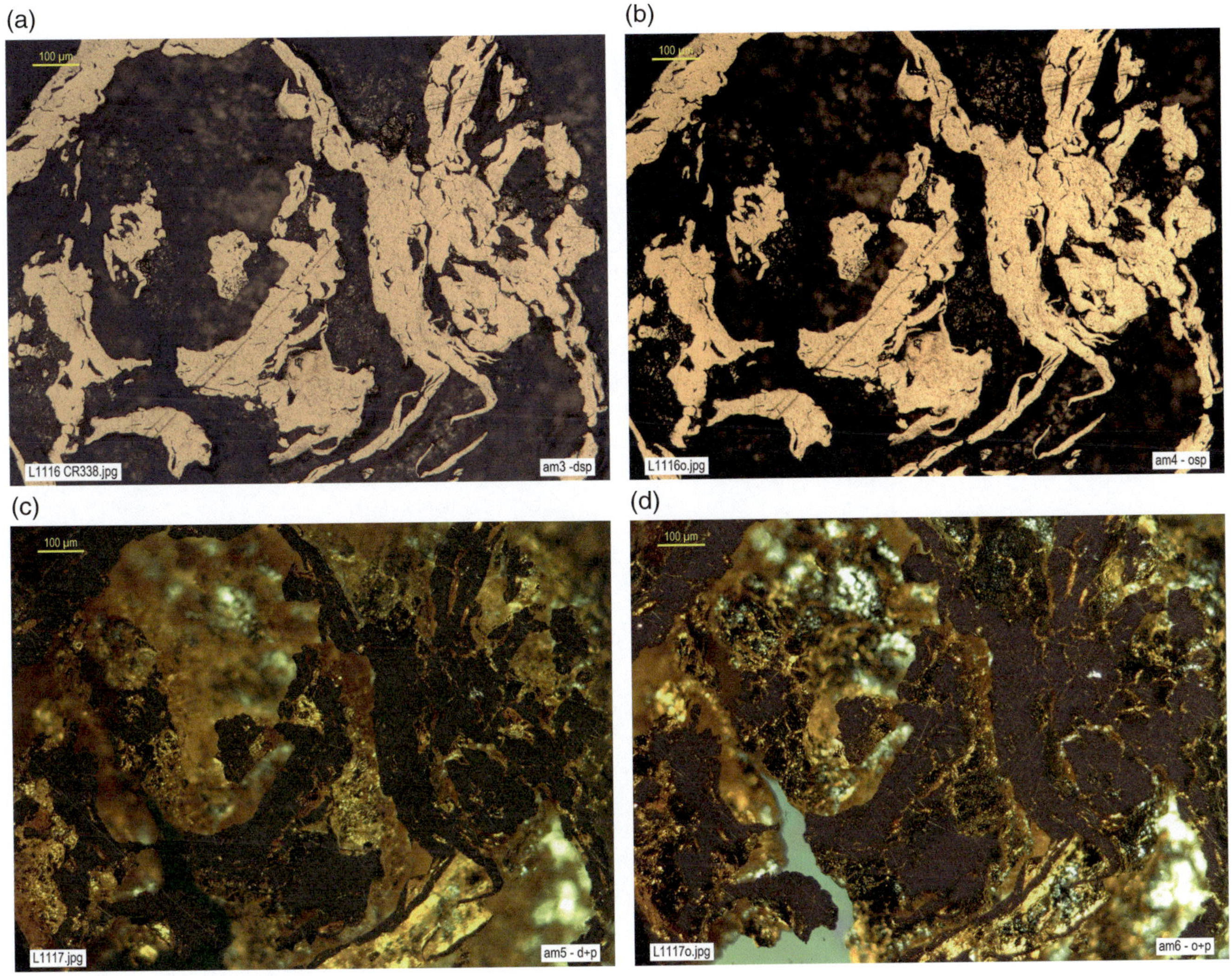

Fig. 7.3 **a** Gold amalgam chips (**dsp**). Sample CR338: Au bullion obtained by amalgamation from ores from Colombia, at ETSIME Madrid, UPM (Ph.D. F. Pantoja, 1999). **b** Same scene as Fig. 7.3 (**osp**). **c** Same scene as Fig. 7.3 (**d+p**). **d** Same scene as Fig. 7.3 (**o+p**)

References

Hey MH (1975) An index of mineral species & varieties arranged chemically (Chemical index of minerals), 3 vols. British Museum (Nat. History), London

Kucha H (1986) Eugenite, $Ag_{11}Hg_2$—a new mineral from Zechstein copper deposits in Poland. Mineral Pol 17:3–10

8 Anatase (*ana/octaedrite*)

TiO_2: Tetragonal

Anatase		R_{546} (%)	Y (%)	λ_d (nm)	P_e (%)	Colour sensation
Air	R_1	19.5	19.5	476	4.9	Grey
	R_2	19.2	19.1	476	5.6	
Oil	R_1	6.8	6.8	475	8.5	
	R_2	6.6	6.6	475	9.3	
Bir: Very weak/Absent (2)				*Pleoc*: Very weak/A. (2)		*Tarn*: No
Anis +P: Imperceptible by RI (3)				*IR*: Very abundant		*Twg*: No
Mor: Euhedral, pseudom, critpox				*Clv*: {001} and {011}		*PolQ*: Good
Hardness	*PolH*: High (4)			*Mohs*: 6−	*Talm*: nd	*VHN* $_{(100\ g)}$: 616–698

Obs: cf. R plots (visible range, QDF3)
Mineral abbreviations: ana—anatase; brk—brookite; ccp—chalcopyrite; gg—gangue; ht—hematites; lcx—leucoxene; mt—magnetite; py—pyrite; rt—rutile; sp—sphalerite; tit—titanite

8.1 General

Composition: TiO_2, with frequent impurities of Al and Fe^{3+}, occasionally traces of V, Sn, Ca, Nb. Unlike rutile, it rarely has important contents of Nb or Ta. A transparent mineral, it can also be studied with transmitted light. Polymorph of *rutile* and *brk, brookite*. Elongated bipyramidal habit, to which it owes its second name, *octaedrite*. Usually secondary to other titaniferous species. In very fine or cryptocrystalline aggregates, sometimes associated with *titanite,* in the leucoxene variety, also known as anatase gel. Ti ore.

8.2 Microscopic Description

8.2.1 SP Observation (*Single Polariser*)

Grey *colour*, slightly more reflective than sphalerite, which it resembles.

Moderate *reflectance* (cf. multispectral R chart and Figs. 8.1a, b and 8.2) but relatively high for a transparent mineral, due to its high refringence (refractive indices, $n\omega = 2.56$, $n\varepsilon = 2.49$). In air $R_{ana} \approx$ mt, or slightly lower than mt (but in immersion distinctly $R_{ana} <$ mt Fig. 8.1b); always $R_{ana} <$ rt, ht; and R_{ana} distinctly > gg. Practically not bireflective or pleochroic (Fig. 8.1c, d).

Polishes well. High *polishing hardness* (relief). Exfoliation {001} and {011} occasionally visible.

It occurs well crystallized, *vbgr*. elongated bipyramidal forms in alpine veins or hydrothermal filling,but also, as a secondary product, pseudomorphous, often in cryptocrystalline or colomorph aggregates (*leucoxene*, *anatase gel*).

8.2.2 Observation +P (*Polariser + Analyser at 90°*)

Anisotropic, but anisotropism effects are hardly noticeable due to the massive and luminous *internal reflections*, ***IR***, colourless, brownish–yellow, bluish–green; very abundant and lighter than those of rutile (Figs. 8.1c, d and 8.3). Its pseudo-isotropy is very noticeable in leucoxene.

8.2.3 Paragenesis

Low temperature polymorph of TiO_2, it can be found as an accessory component of igneous and metamorphic rocks, pegmatites, carbonatites, among others and also in detrital

R. Castroviejo, *A Practical Guide to Ore Microscopy—Volume 1*,
https://doi.org/10.1007/978-3-031-12654-3_8

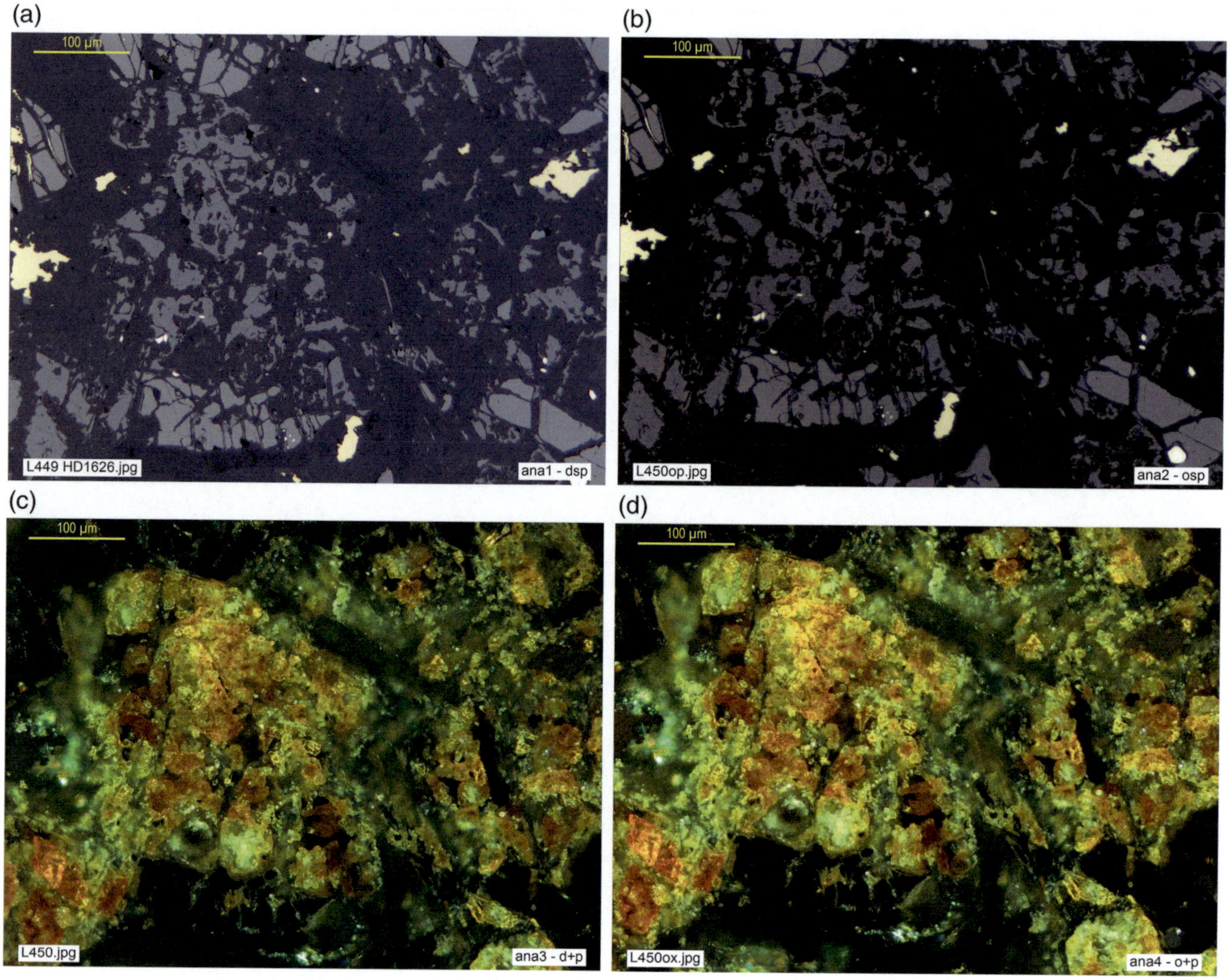

Fig. 8.1 **a**, **b** Disseminated ore of ccp (yellow) and mt (brownish grey), with ana (grey, slightly bluish against mt and a little darker than mt, especially in oil, but distinctly lighter than gg) secondary to mt and intergrown with gg. HD 1626, Bayernland Mine, Germany (**dsp**) and (**osp**), respectively. **c**, **d** Same scene as ana1, (**d+p**) and (**o+p**), respectively. The abundant RI of ana contrast with the total extinction of mt

concentrations. As a secondary mineral, it frequently appears as a pseudomorph of titaniferous phases, including Ti-mt, from which it can assimilate Fe^{3+}. The iron impurities may cause IR to become coloured in brownish and yellowish tones, otherwise they would be typically colourless. Frequent component of leucoxene, associated with titanite. Companions are titanite, magnetite, ilmenite, rutile.

8.2.4 Identification

It resembles rutile, but anatase is darker, pseudo-isotropic, shows lighter IR and lacks twins.

It is distinguished from sphalerite because sp is slightly darker, is associated with sulphides and lacks the bluish or colourless IR of anatase.

Because of its inconspicuous appearance, anatase often goes unnoticed in many studies, but it is very common, particularly as an accessory resulting from hydrothermal alteration bound to the genesis of many metallic deposits. A clue to its recognition can be given by the presence of scattered grey particles or aggregates, distinctly lighter than gangue, apparently isotropic and with massive colourless IR. Its higher reflectance distinguishes it from titanite, with which it can be associated in leucoxene.

Fig. 8.2 Same sample as Fig. 8.1a, comprising a wider field (Fig. 8.1a is central zone of Fig. 8.2 in detail). Ore of mt, ccp and scarce py (white) and sp (grey, darker than mt, associated with ccp, upper right corner), with alteration of mt to ana (intergrown with gg), partly pseudomorphic after mt (**dsp**)

Fig. 8.3 Same scene as Fig. 8.2 (**d+p**). The alteration from mt to ana is perfectly delimited by the high luminosity of ana's IR (colourless, yellowish and brownish–yellow to blue–green); sp is distinguished from ana by its more coloured (reddish) internal reflections, as well as by its association with ccp, instead of the mt with which ana is associated

8.3 Spectral Information

8.3.1 Anatase: Multispectral Reflectance R (in Air, Visible Range, QDF3) (%)

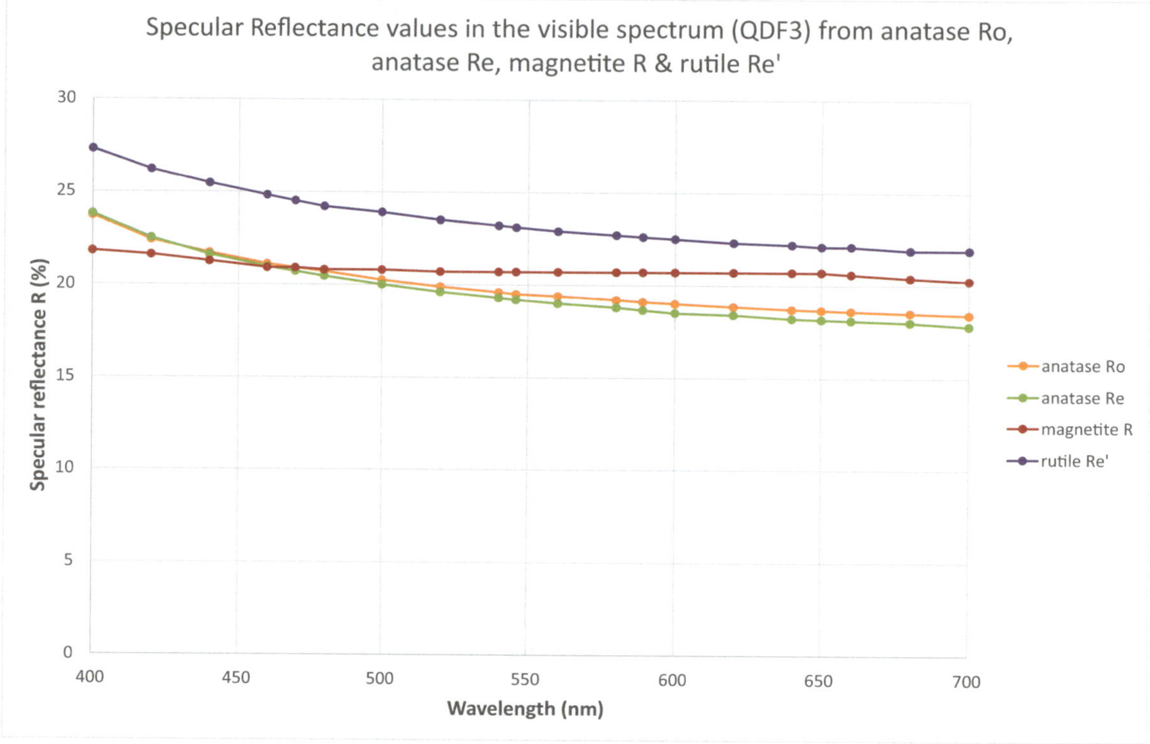

Antimony (*Sb/Native Antimony*)

9

Sb: Trigonal

Mineral		R_{546} (%)	Y (%)	λ_d (nm)	P_e (%)	Colour sensation
Air	*R_1*	74.4	73.6	500	0.8	Bright white, highly reflective
	R_2	77.9	77.3	535	0.6	
Oil	*R_1*	64.7	63.6	493	1.8	
	R_2	69.7	69.0	505	0.6	
Bir: Very weak (*immersion*) (2)				*Pleoc*: Weak (*imm.*) (3)		*Tarn*: No
Anis +P: Distinct (orange-brown) (4)				*IR*: No		*Twg*: Common, lam-polysynthetic
Mor: Xenomorphic aggreg., botryoidal				*Clv*: Occasional		*PolQ*: Polishes well
Hardness	*PolH*: Low (<dy, gn; >cv, Bi) (2)			*Mohs*: 3^+	*Talm*: –	*VHN* $_{(100\ g)}$: 50–69

Obs: It is not common but occurs occasionally with precious metals (Ag, Au)
Spectral information: cf. Table and R plots, multispectral data (VNIR, AMCO) and visible range (QDF3)
Mineral abbreviations: Ag—native silver; As—native arsenic; Au—native gold; Bi—native bismuth; cin—cinnabar; cv—covellite; dy—dyscrasite; gn—galena; prg—pyrargyrite; Sb—native antimony; sba—stibarsen; sbt—stibnite (antimonite)

9.1 General

Composition: Sb, with occasional As impurities. Combines with As to form the mineral stibarsen (SbAs, formerly *allemontite*).

9.2 Microscopic Description

9.2.1 SP Observation (*Single Polariser*)

Brilliant white *colour* (with a very faint yellowish tone). Very good *polish*, although frequent scratches may appear due to its low hardness (Figs. 9.1a, b, and 9.4). Unlike native arsenic, it does not tarnish.

Very high *reflectance*, only surpassed by native silver and moschellandsbergite (Ag_2Hg_3). Although it is anisotropic, its bireflectance is practically negligible and its pleochroism is very weak (perceptible in immersion on adjacent grains).

Usually occurs in fine-grained xenomorphic aggregates, also colomorphic or botryoidal. Cleavage occasionally observable.

Low *polishing hardness*: $PolH_{Sb}$ < sba, As, dy, cin, gn; however, $PolH_{Sb}$ > cv, Bi.

Vickers microhardness, 50–69 (load 100 g). *Mohs hardness* 3–3.5.

9.2.2 Observation +P (*Polariser + Analyser at 90°*)

Anisotropic, with vivid polarization colors, showing brownish and orange tints, especially in oil (Fig. 9.1c and d). It shows no *internal reflections*, but frequent polysynthetic or lamellar *twins* in several directions.

9.2.3 Paragenesis

Hydrothermal, particularly argentiferous epithermal *paragenesis*, with native arsenic, stibnite, stibarsen, dyscrasite, Co–Ni minerals, etc., as well as in stibnite-Au seams. In the oxidation zone, it is altered to antimony ochres such as *valentinite* and *senarmontite* (Sb_2O_3, Figs. 9.1a, b and 9.4).

R. Castroviejo, *A Practical Guide to Ore Microscopy—Volume 1*,
https://doi.org/10.1007/978-3-031-12654-3_9

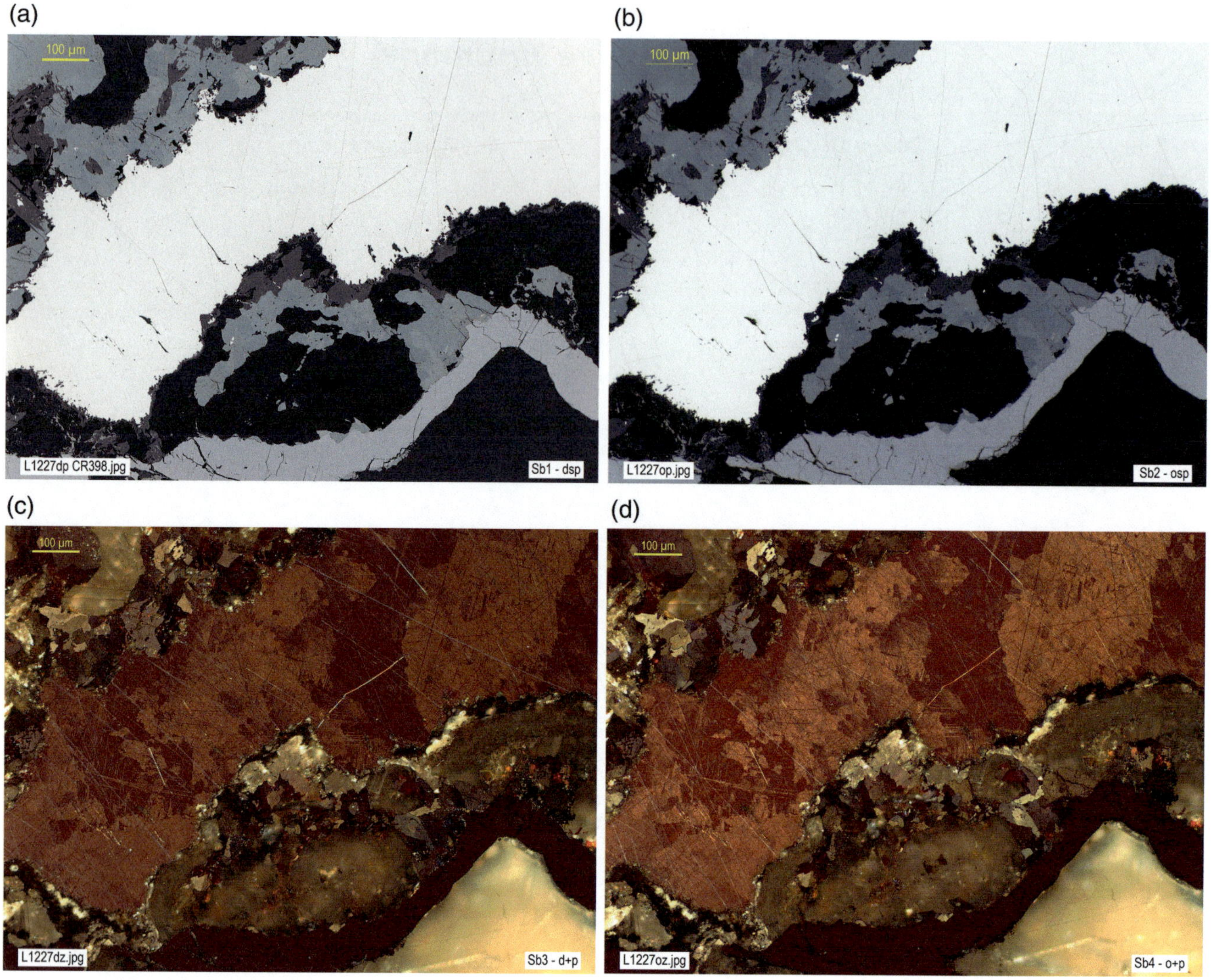

Fig. 9.1 **a** Native Sb (bright white, diagonal band NE-SW), surrounded by pyrargyrite (bluish gray, pleochroic) intergrown with Sb ochre (dark gray, but lighter than quartz) and gangue (quartz), with an outermost rim of galena (light gray, lower band). CR-398, Hiendelaencina, Guadalajara (**dsp**). **b** Same scene as Sb1, with **osp** setting, greater contrast, quartz disappears, while Sb ochre becomes much darker. **c** Same scene as Fig. 9.1 (**d+p**), anisotropism of Sb (brownish-orange polarization tints) and of prg, polysynthetic twins (Sb) and internal reflections, IR, of prg (red) and of Sb ochre and quartz (colorless and more abundant). **d** Same scene as Fig. 9.1 (**o+p**), enhanced IR (prg, Sb-ochre and quartz) and effects of anisotropism (Sb) and twinning (Sb)

9.2.4 Identification

Native antimony is easily characterized by its extreme reflectance, white color, good polish, distinct anisotropism in orange-brown shades, thin polysynthetic twinning and paragenesis.

It is less reflective than silver, but more so than arsenic and bismuth. It is also distinguished from silver and arsenic by the absence of tarnish. In addition, silver is isotropic, whereas arsenic and bismuth are more strongly anisotropic than antimony and show different polarisation tints.

However, when it occurs isolated in gangue, it is at first easily confused with silver because of its color and brightness, but the anisotropism of Sb clearly differentiates them (Figs. 9.2 and 9.3).

Against stibarsen, antimony is more reflective.

Against dyscrasite, antimony is more reflective and shows stronger anisotropism and thinner, polysynthetic twins but no tarnish.

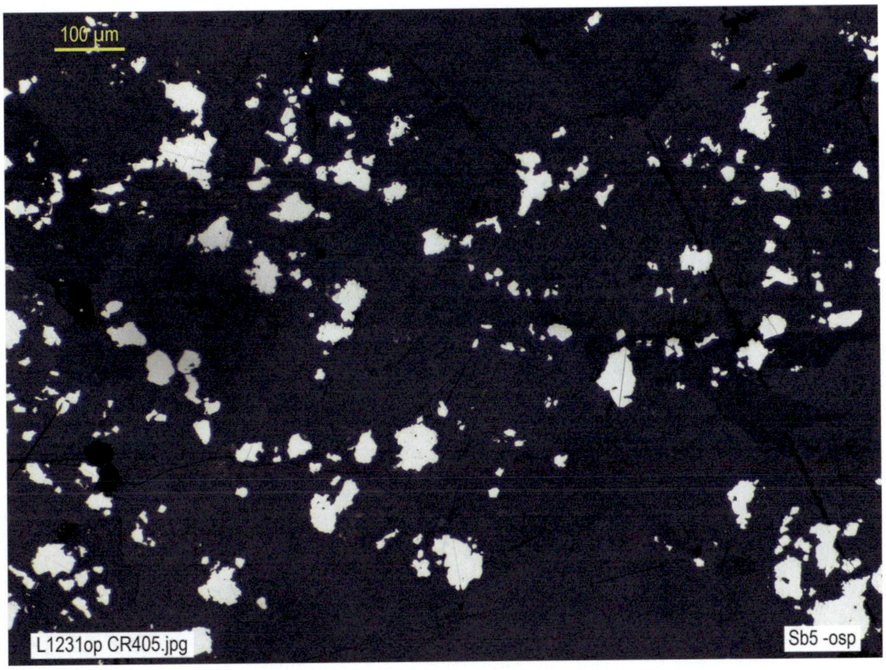

Fig. 9.2 Sb disseminated in silicate gangue (quartz). CR 405, West Gore, Hants, Nova Scotia, Canada (**osp**)

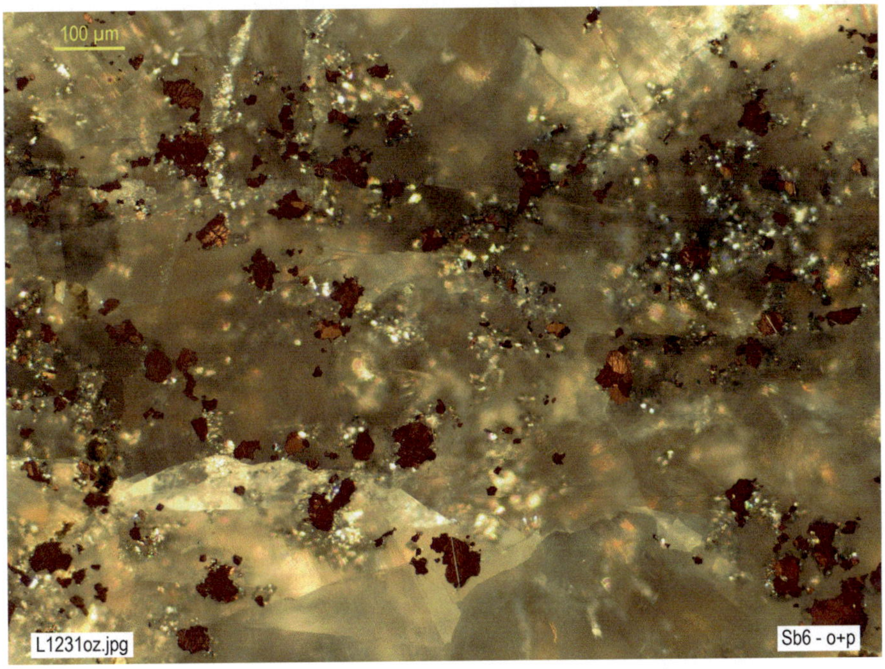

Fig. 9.3 Same scene as Fig. 9.2 (**o+p**), note anisotropism colors and Sb twinning

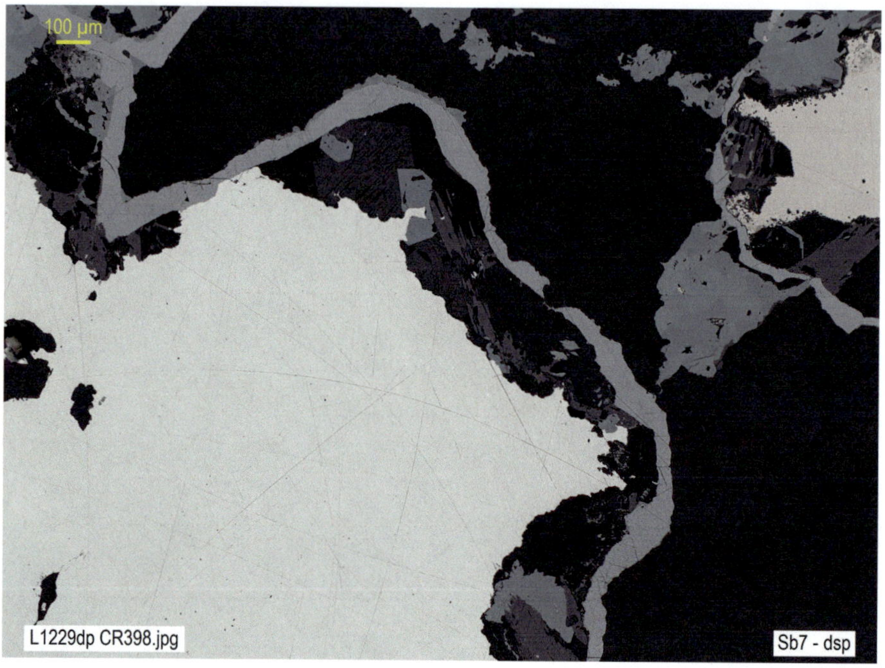

Fig. 9.4 Native antimony (massive allotriomorphic, white, polishing scratches) with a crown of Sb ochres (dark grey) intergrown with prg (bluish grey, slightly pleochroic) and, finally, gn (pure grey, lighter than prg), all sealed by a final episode of quartz (very dark grey) and second generation of prg, which close the hydrothermal infill. The *antimony ochres* (*Sb_2O_3*, partly secondary to native Sb as their corroded contacts suggest, are distinguished by their dark grey colour (but lighter than quartz, whose reflectance is clearly lower) and by their morphology. They are *valentinite* (fibrous habit) and its dimorph *senarmontite* (cubic, idiomorphic, darker than valentinite, and very scratched due to its scarce hardness). CR 398, Hiendelaencina, Guadalajara (**dsp**)

9.3 Spectral Information

9.3.1 Antimony: Multispectral Reflectance R (in Air, Without Polariser) (%)

λ (nm)	370	400	425	450	475	500	525	550	575	600	625	650	675	700	750	800	850	900	950	1000
μ	63.2	65.0	66.7	67.7	68.4	68.8	69.0	68.9	68.6	68.4	68.2	67.8	67.6	67.5	67.4	67.2	67.5	67.4	67.5	67.3
P_5	60.7	62.5	64.2	65.4	66.2	66.9	67.2	67.3	67.2	67.0	66.9	66.5	66.5	66.4	66.4	66.2	66.5	66.4	66.6	66.3
P_{20}	61.9	63.8	65.5	66.6	67.3	67.9	68.1	68.1	67.9	67.7	67.5	67.2	67.0	67.0	66.9	66.6	66.9	66.8	67.0	66.7
P_{80}	64.4	66.3	67.9	68.9	69.5	69.8	69.9	69.7	69.4	69.1	68.8	68.4	68.3	68.1	68.0	67.8	68.1	68.0	68.1	67.9
P_{95}	65.3	67.4	69.2	70.1	70.6	70.9	70.9	70.6	70.2	69.8	69.6	69.1	68.9	68.8	68.6	68.3	68.5	68.5	68.6	68.3

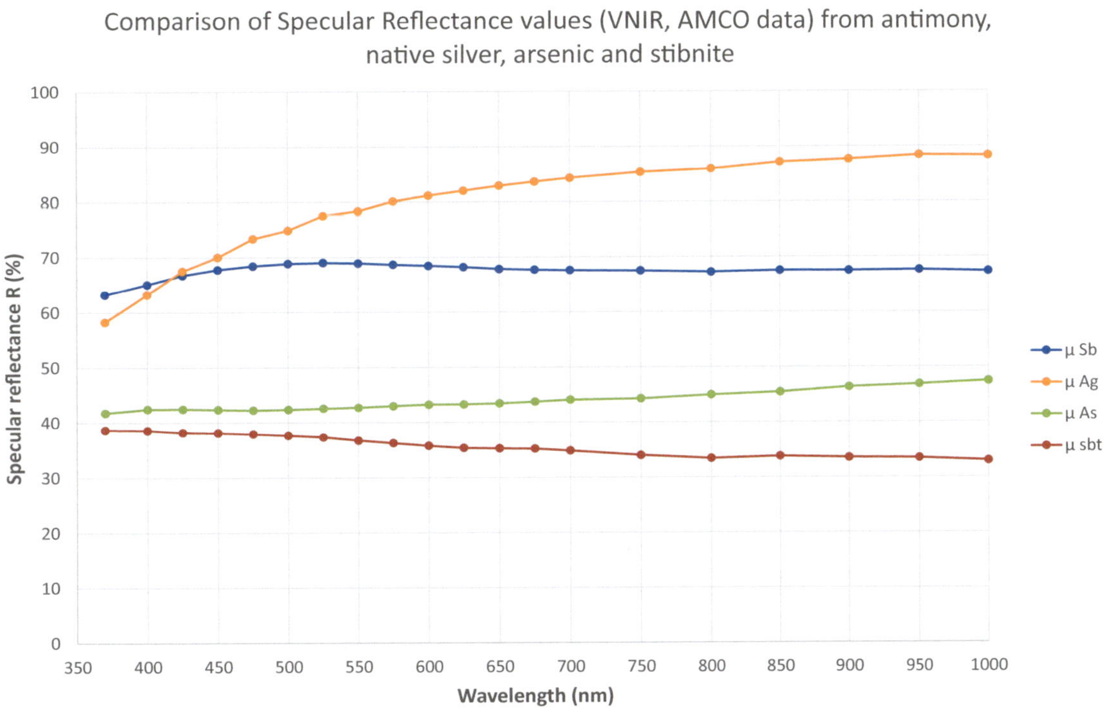
Comparison of Specular Reflectance values (VNIR, AMCO data) from antimony, native silver, arsenic and stibnite
Specular reflectance R (%)
Wavelength (nm)
µ Sb
µ Ag
µ As
µ sbt

Comparison of Specular Reflectance VNIR values from antimony (AMCO), antimony R1, antimony R2, stibnite Rb & stibnite Rc (QDF3), and antimony Rmax & antimony Rmin (Picot and Johan, 1982)
Specular reflectance R (%)
Wavelength (nm)
µ Sb
antimony R1 (QDF3)
antimony R2 (QDF3)
stibnite Rb (QDF3)
stibnite Rc (QDF3)
antimony Rmax (P&J)
antimony Rmin (P&J)

Argentite (*arg*)

Ag_2S: Cubic

10

10.1 An Enigmatic Sulphide

Argentite, ***arg***, was traditionally presented as a typical silver mineral in the sulfide class, cubic in symmetry and Ag_2S in composition. However, over time it has been proven that the mineral found in nature is always monoclinic. In reality, it often shows a *pseudocubic habit*, which had led to its original characterization as cubic.

It is now understood that, in these cases, what has happened is a paramorphic transformation of the Ag_2S sulphide on cooling, from the high-temperature form (argentite proper, cubic) to the only form stable at low temperature (monoclinic, which has been given the name *acanthite*, **ac**).

10.2 Experimental Data

Modern experimental data explain this apparent contradiction, as do Craig and Scott (1976), by the existence of three polymorphs of $Ag_{2\pm x}S$, whose composition progressively departs from stoichiometry at increasing temperatures, according to experimental results (Kracek 1946; Craig and Scott 1976, CS-102). These polymorphs, in pure Ag–S systems, cannot be *quenched*.

Therefore, only the low-temperature form (monoclinic) can survive at room temperature. This is the only species found in nature (acanthite, ac), although most often ac is not the primary phase, since in many cases it formed by modification of the higher temperature polymorphs, especially of argentite (cubic).

10.3 Discussion and Consequences

The transition temperature between acanthite and argentite depends on the composition and varies between 177.8 °C in the S-saturated limit and 176.3 °C in the Ag-saturated limit. This transition can be used as a geothermometer, if some evidence of it is found in the acanthite, such as pseudocubic morphology[1] or polysynthetic twins (cf. Acanthite, ac, description).

The microscopic observations and descriptions of argentite so far must, therefore, be referred to acanthite, taking into account the experimental data cited. For this reason, it has been decided not to describe argentite here, since it is not observable under normal conditions of temperature and pressure. The reader is referred to what is said for acanthite.

10.4 Conclusion—A Ghost Mineral?

Taking into account these data and applying the official criteria of IMA-COM regarding nomenclature, all the samples identified so far as argentite, including those existing in collections and museums, as well as all those that can be studied under the ore microscope, would not be argentite but, in fact, acanthite.

However, not all mineralogists agree with the application of criteria which, in this case, make a clean sweep of the classical nomenclature. It is worth mentioning, for example, the authorised opinion of Paul Ramdohr, who comments in this respect:

«*Argentite is today often used as a name for the high temperature form of* Ag_2S. *I think that this is unfortunate because all other experimental high forms would consequently merit an own name or* vice versa *in the observable mineral world real argentite would actually not be present at all*». (Ramdohr 1980, p. 471).

It can be argued, but it is hard to deny, that this is a reasonable objection, even if it has not been widely heard. As things are now, argentite will never be seen by men's eyes on earth, apparently it became just a *ghost*!

[1] A case of *paramorphism*, i.e., pseudomorphism between polymorphs.

R. Castroviejo, *A Practical Guide to Ore Microscopy—Volume 1*,
https://doi.org/10.1007/978-3-031-12654-3_10

References

Craig JR, Scott SD (1976) Sulphide Phase Equilibria. In: Ribbe PH (ed) Sulfide Mineralogy. Mineralogical Soc. Am Rev Mineralogy 1 (5):CS1–110

Kracek (1946) Craig and Scott 1974, CS-102

Ramdohr P (1980) The ore minerals and their intergrowths, vol 2, 2nd edn. Pergamon Press, Oxford, 1205p

Argentopyrite (*agp/argentopyrites with stb*) 11

$AgFe_2S_3$: Monoclinic

Argentopyrite		R_{546} (%)	Y (%)	λ_d (nm)	P_e (%)	Colour sensation
Air	R_1	26.4	26.8	579	8.6	Pinkish-brown, bireflective
	R_2	34.3	34.8	580	6.7	
Oil	R_1	13.7	14.0	580	10.8	
	R_2	20.8	21.1	582	7.7	
Bir: Medium (4)				Pleoc: Apparent (4)		Tanr: -
Anis + P: Strong and coloured (5)				IR: No		Twg: Complex, pseudo-hexagonal
Mor: Tabular, prismatic				Clv: (001) visible		PolQ: Good
Hardness	PolH: Very low (1)			Mohs: 3½-4	Talm:	VHN $_{(100g)}$: 225–242

Spectral information: cf. R plots, visible range (QDF3, and Picot and Johan 1982)
Mineral abbreviations: aag—allargentum, ac—acanthite, Ag—native silver, agp—argentopyrite, bth—breithauptite, cc—calcocite, ccp—chalcopyrite, fbg—freibergite, gg—gangue, gn—galena, lo—löllingite, mc—marcasite, nc—nickelite, my—miargyrite, pc—pearceite, plb—polybasite, po—pyrrhotite, pru—proustite, prg—pyrargyrite, py—pyrite, q—quartz, sbt—stibnite (antimonite), sf—safflorite, sk—skutterudite, sp—sphalerite, stb—sternbergite, stp—stephanite

11.1 General

Argentopyrite is a mineral with a complicated history. Known since ancient times as *Silberkies* in Germany, its definition was very imprecise and sometimes contradictory because of frequent intergrowths and alterations (pseudomorphized by pyrite, marcasite, iron oxides). It is now a defined species and admitted by IMA, composition $AgFe_2S_3$, sometimes with traces of Cu, As. Following Picot and Johan (1982), we should speak rather of *Argentopyrites* (in plural) including, in addition to the species argentopyrite, its orthorhombic dimorph *sternbergite*, **stb**, with very similar optical properties.

11.2 Microscopic Description

11.2.1 SP Observation (*Single Polariser*)

Pinkish-brown *colour* (Figs. 11.1, 11.2 and 11.3a), like pyrrhotite, but somewhat darker and more *bireflective*. In appearance, clearly pleochroic, although the dispersion of its R1-R2 reflectance curves is similar and, strictly speaking, it should be called bireflectance (*cf.* R-plot: agp). However, it is usually described as a pleochroic ore, since the difference in brightness on rotation is perceived as a change in colour from intense dark brown to light yellowish-brown.

Moderate *reflectance*, slightly lower than pyrrhotite (*cf* comparative R-plots). According to Picot and Johan (1982), in white light R agp (35%) ≈ R stb (34.5%) and, compared to other ores:

- R agp and stb > ac (32%), cc (31.5%), plb (31%), prg (31%), pc (30.5%);
- R agp and stb < my (36%) < po (39%), << nc (50.5%), py (51%), sk (54%); <<< aag (68%) and Ag (85%).

These quantitative data should be kept in mind because the subjective visual perception can be very different depending on the accompanying minerals, which can make the same ore appear pinkish brown or brownish grey to dark brown, as can be seen by comparing Figs. 11.1, 11.2 and 11.3a, in which argentopyrite appears very dark accompanied by silver, and Fig. 83.4a, b (*cf* description *my*), in which

R. Castroviejo, *A Practical Guide to Ore Microscopy—Volume 1*,
https://doi.org/10.1007/978-3-031-12654-3_11

Fig. 11.1 Argentiferous ore: Ag (bright white), allargentum (light grey next to Ag, upper right corner of the image), prg (dark grey) and agp (dark brown, included in prg), with sk (grey, idiomorph), nc (deep pink), bth (very fine, automorphic, pale pinkish-grey crystals or aggregates, included in nc and/or sk) and black gg. CR 362, Nipissing Mine, Cobalt, Ont, Canada (**osp**)

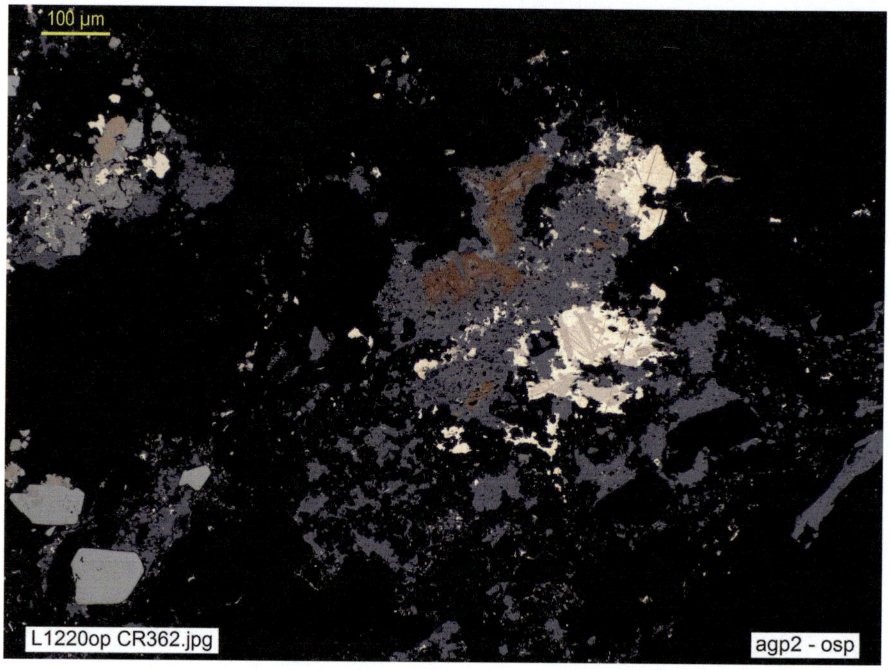

Fig. 11.2 Vein fill with silver ores composed of prg (grey, porous), agp (brownish, included in prg), Ag (bright white) and aag (included in Ag, distinguished by its brownish grey tone, less bright than Ag), permeating and cementing the carbonate gg (black). For reference, compare with aggregates (upper and lower left corners) of sk (light grey, idiomorphic), nc (light pink) and associated crystallites of bth (grey with dull pink hue). CR 362, Nipissing Mine, Cobalt, Ont, Canada (**osp**)

sternbergite appears much lighter, accompanied by pyrargyrite and miargyrite only.

Good *polish*. Very low *polishing hardness* (≈ prg). Distinct cleavage (001).

It occurs in tabular, prismatic or anhedral crystals, generally forming small inclusions in argentiferous ores (with proustite, pyrargyrite, polybasite, among others) and sometimes forming oriented intergrowths with sternbergite.

11.2.2 Observation +P (*Polariser + Analyser at 90°*)

Clearly anisotropic, with effects varying from sternbergite (multicoloured tones) to argentopyrite (duller, greenish-grey to brownish, Fig. 11.3b). No *internal reflections*.

Complex *twinning*, sometimes with pseudo-hexagonal forms like aragonite.

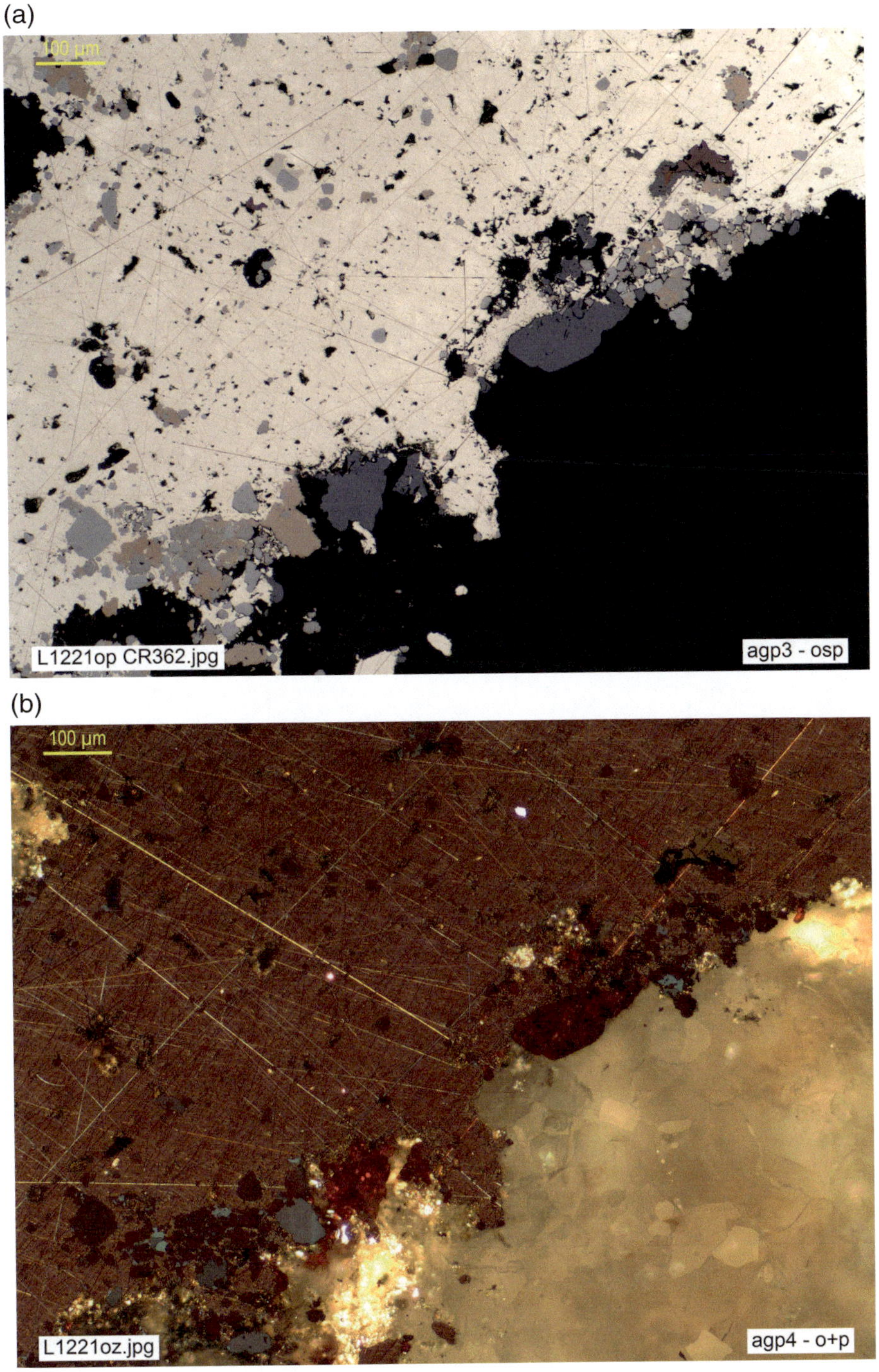

Fig. 11.3 **a**, **b** (Same scene, **osp** and **o + p**, *resp*ectively), *Ag* veinlets (white, slightly yellowish, very bright, isotropic, polishing scratches), cementing carbonate *gg* (black, osp; massive colourless **IR**, internal reflections, *o + p*). Minor ores, associated with Ag: *sk* (grey, isotropic), *nc* (intense pink, vivid anisotropism) and traces of *bth* (pinkish grey), disseminated; prg: two dark grey crystals (in contact between Ag and Carb, well characterized by their abundant red, IR; *o + p F agp4*), and some aggregates disseminated in the Ag mass, one of which (upper right corner) comprises *agp* (dark brown, anisotropic but without the lively colours of *nc*). CR 362, Nipissing Mine, Cobalt, Ont, Canada

11.2.3 Paragenesis

Late hypogene mineral in *five element*-type hydrothermal *paragenesis* (Ni–Co–Bi–Ag–U), accompanied by Ag and Ag sulfosalts, Ni–Co arsenides, pitchblende, pyrite, marcasite, chalcopyrite, galena, sphalerite. Occasionally also in epithermal paragenesis.

11.2.4 Identification

Although rare, the identification of argentopyrite and sternbergite should not be problematic. Its resemblance to pyrrhotite is typical, but in argentiferous paragenesis, it is also distinguished from pyrrhotite by its lower reflectance and by its more marked bireflectance and anisotropism. In the face of Ni-Co minerals and, in particular, the silver sulfosalts that usually accompany them, the distinction is very easy by colour and reflectance (*cf.* comparative graphs R).

Argentopyrite resembles its dimorph sternbergite, which is distinguished by its more vivid anisotropism, more perfect cleavage and worse polish.

11.3 Spectral Information

11.3.1 Argentopyrite Reflectance R (in Air, Visible Spectrum) (%)

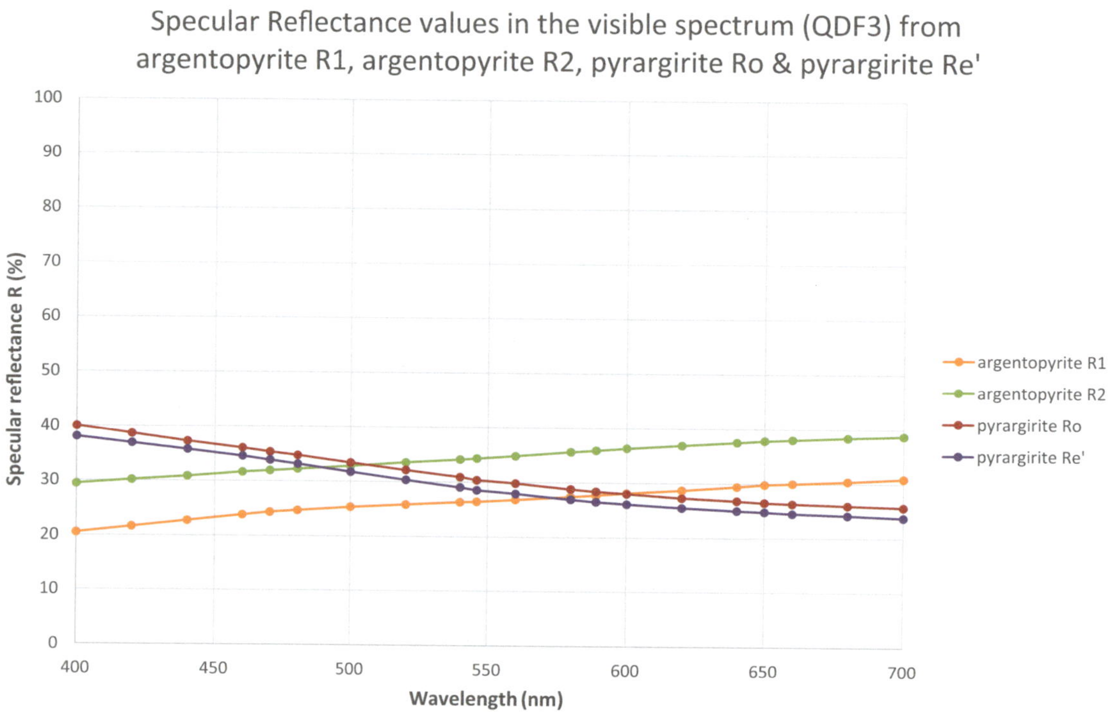

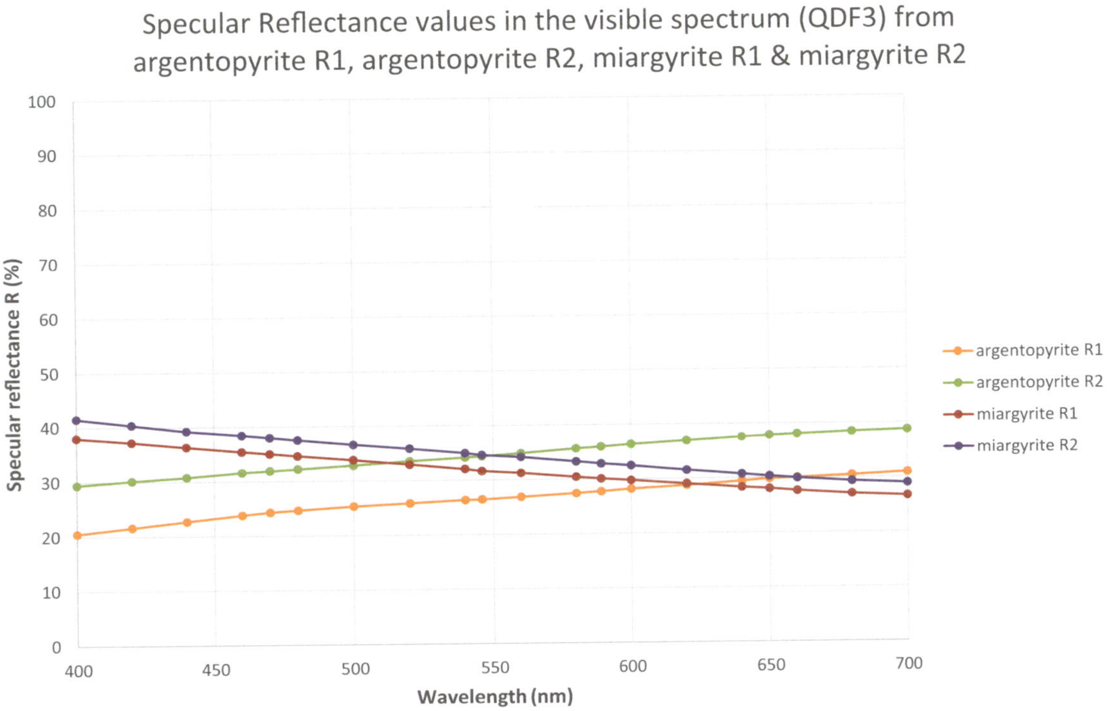

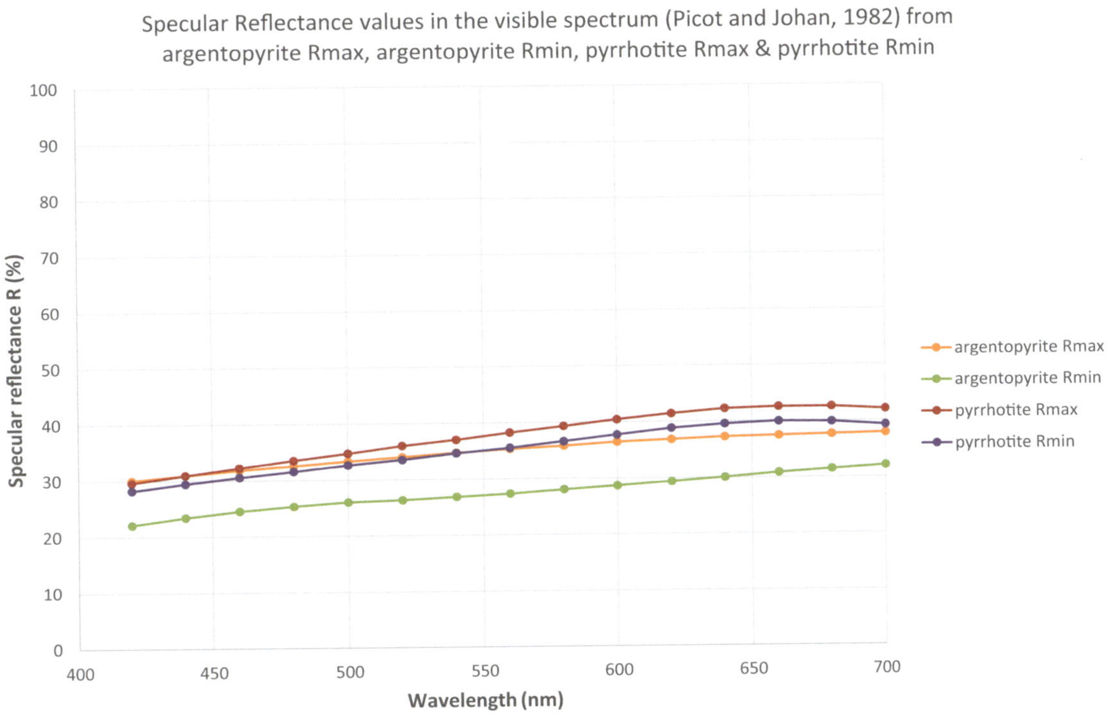

Reference

Picot P, Johan Z (1982) Atlas of ore minerals. B.R.G.M./Elsevier

12 Arsenic (*As, Native Arsenic*)

As: Trigonal

Arsenic		R_{546} (%)	Y (%)	λ_d (nm)	P_e (%)	Colour sensation
Air	R_1	51.7	51.6	476	2.1	Bright white
	R_2	55.7	55.5	483	0.6	
Oil	R_1	37.1	37.0	475	3.3	Bright white (slight yellowish hue)
	R_2	42.1	41.9	480	1.0	
Bir: Weak (3)				*Pleoc*: Weak (3)		*Tarn*: Darkens quickly
Anis + P: Distinct (4)				*IR.*: No		*Twg*: Frequent and varied
Mor: Fine-sized aggr, collom, botryoidal				*Clv*: // (0001)		*PolQ*: Polishes well, but soon tarnishes
Hardness	*PolH*: Low (>Sb; < gn) (2)			*Mohs*: 3–4	*Talm*: C	*VHN* $_{(100\ g)}$: 72–173

Obs: Uncommon; waning hydrothermal stages of Ni–Co–Bi–Ag–U and shallow epithermal parageneses
Spectral information: cf. Table and R plots, multispectral data (VNIR, AMCO) and visible range (QDF3; Picot and Johan 1982)
Mineral abbreviations: Ag—native silver, As—native arsenic, asp—arsenopyrite, Au—native gold, Bi—native bismuth, bm—bismuthinite, cc—chalcocite, cin—cinnabar, cv—covellite, dg—digenite, gn—galena, lo—löllingite, nc—nickelite, Sb—native antimony, sba—stibarsen, sbt—stibnite (antimonite), td—tetrahedrite

12.1 General

Composition: As, commonly containing some Sb, with which it is frequently associated, forming peculiar unmixing structures from Sb–As gels (cf. *Sb* and stibarsen, *sba*).

12.2 Microscopic Description

12.2.1 SP Observation (*Single Polariser*)

Bright white *colour*, with a slight yellowish tinge, more noticeable in immersion (compare Figs. 12.1a, b, 12.2, and 12.4). High *reflectance*, $R_{As} > R_{galena}$ but noticeably $< R_{Ag,\ Sb,\ Au}$; of the order of löllingite or pyrite, although rapid tarnishing may give As a darker appearance. Barely perceptible *bireflectance* or *pleochroism*.

As shows an excellent *polish* but then *tarnishes* rapidly, darkens noticeably after a few hours and becomes completely black in a few months (Figs. 12.6, 12.7 and 12.8). Low *polishing hardness*, $PolH_{As}$ < gn, cc, bm, dg or cin, but > Sb, Bi, sbt or cv.

Cleavage // (0001) visible in crystals of sufficient size, which may also show occasional zoning.

Commonly occurs in microcrystalline aggregates with rhythmic or banded colloform textures (Figs. 12.1a–d, 12.2, 12.3, 12.4 and 12.5). Often intergrown with *stibarsen*, in symplectitic or unmixing aggregates (cf. **F sba**), traditionally known as *allemontite*. In these, native arsenic is easily distinguished by its lower reflectance, higher polishing hardness and tarnishing.

12.2.2 Observation +P (*Polariser + Analyser at 90°*)

Clearly *anisotropic*, but with faintly coloured polarisation tints, more lively in immersion (Figs. 12.1c, d, 12.3, and 12.5), greyish-brown to yellowish, sometimes with a slight greenish tinge (occasionally somewhat orange, though not as much as native antimony).

No internal reflections. Frequent polysynthetic twins $(01\bar{1}2)$, also deformation and saw-toothed twinning.

R. Castroviejo, *A Practical Guide to Ore Microscopy—Volume 1*,
https://doi.org/10.1007/978-3-031-12654-3_12

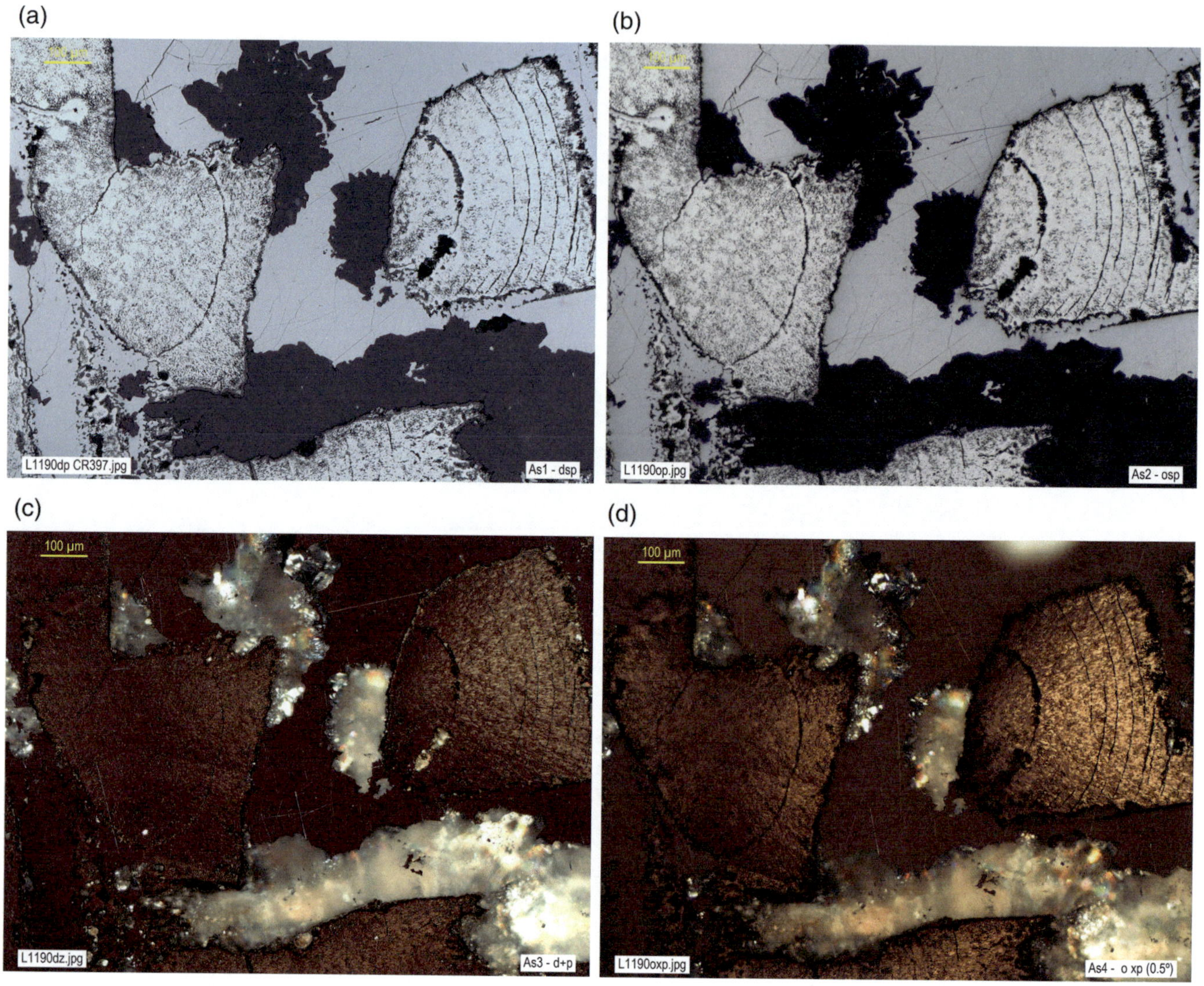

Fig. 12.1 Concentric, collomorphic mass of As (white) brecciated and cemented by galena (grey) and quartz. CR 397: Hiendelaencina, Guadalajara (**dsp**). **b** Same scene as As1 (**osp**), the slightly yellowish tone of As and the greater contrast with gn is more clearly perceived. **c** Same scene as Fig. 12.1 (**d+p**). **d** Same scene as Fig. 12.1 (**o+p**)

12.2.3 Paragenesis

Rare mineral, typical of late episodes or waning stages of hydrothermal activity in Ni–Co–Bi–Ag–U deposits and in precious metal, especially epithermal deposits.

It may be accompanied by arsenopyrite, löllingite, nickeline, Ni and Co sulphides and arsenides, galena, stibarsen, Ag sulphosalts, etc.

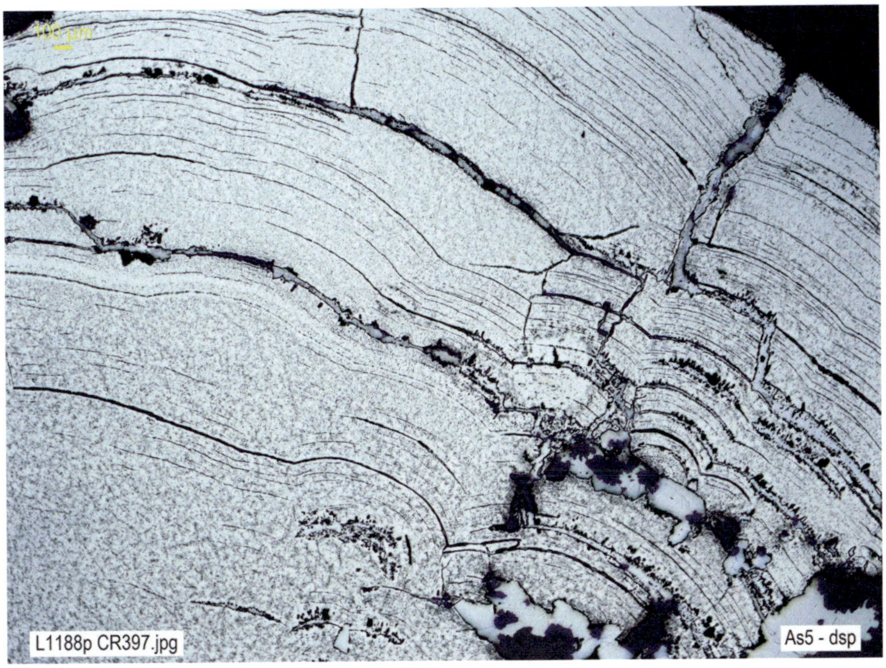

Fig. 12.2 Banded colloform aggregate of As (white), with microcracks cemented by gn (light grey), td (darker grey) and quartz (very dark grey). CR 397: Hiendelaencina, Guadalajara (**dsp**)

Fig. 12.3 Same scene as Fig. 12.2 (**d+p**)

12.2.4 Identification

Rapid darkening by alteration (tarnishing), colloform, concentric and banded textures, relatively low reflectance for a native element, and paragenesis characterize native arsenic. Native antimony is more reflective and softer, shows weaker anisotropism (but more coloured with orange tones) and no tarnishing. Native bismuth is much brighter and softer.

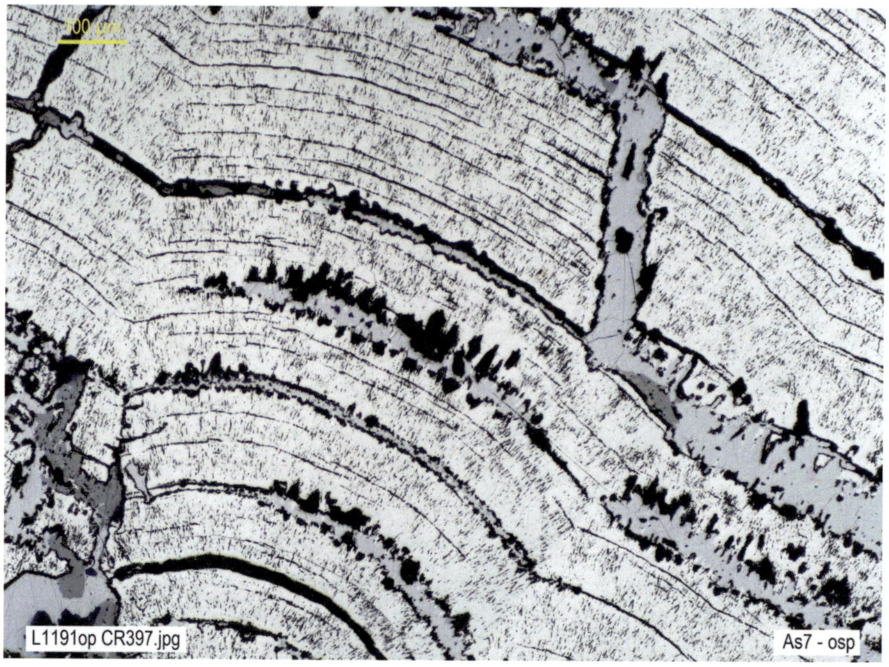

Fig. 12.4 Banded colloform aggregate of As (yellowish white), brecciated and partially replaced by gn (light grey), td (dark grey) and quartz (black). CR 397: Hiendelaencina, Guadalajara (**osp**)

Fig. 12.5 Same scene as As7 (**o +p**)

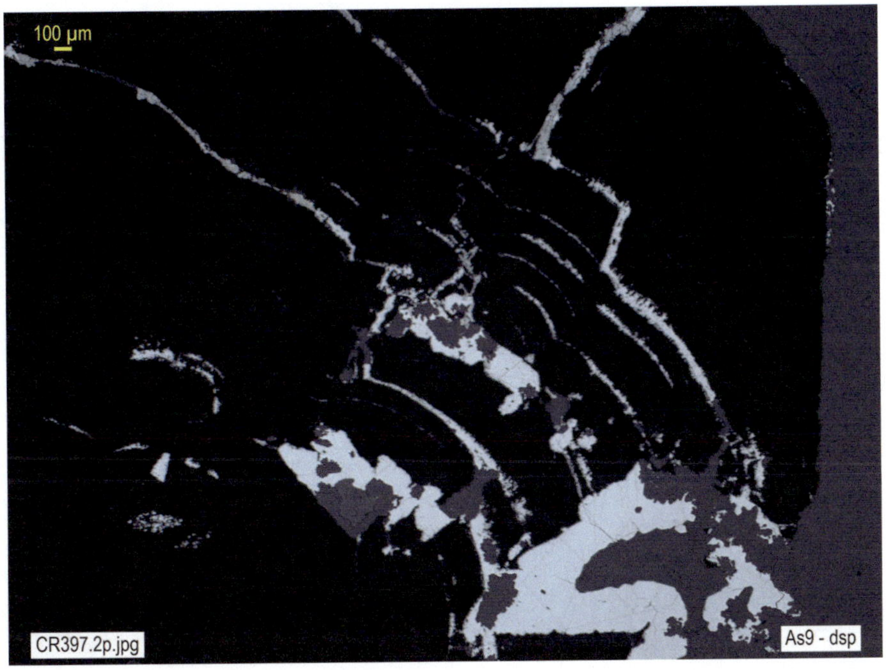

Fig. 12.6 Massive As, darkened by tarnishing: As became totally black, in contrast with gn (almost white) and quartz (grey), not affected by surface alteration

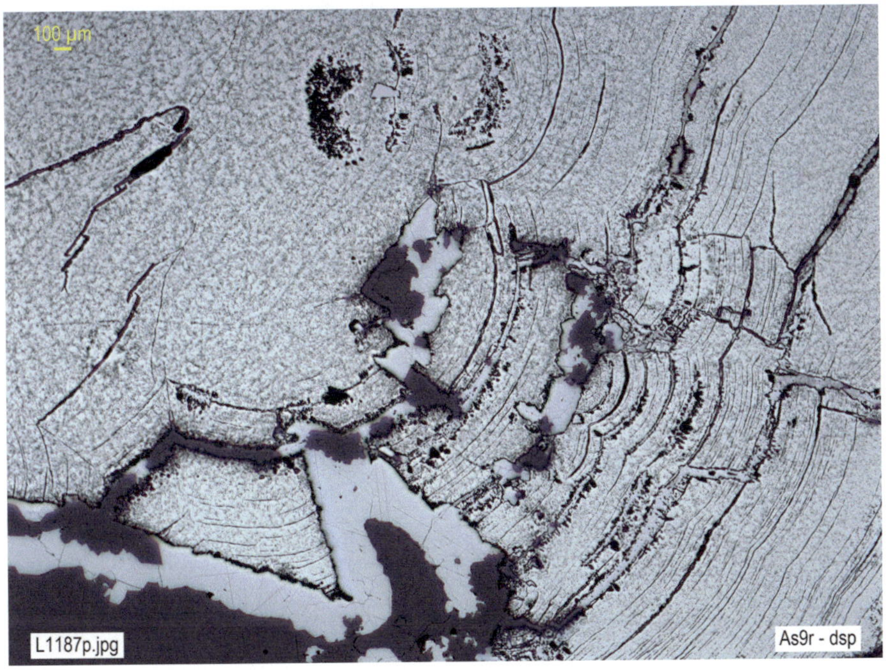

Fig. 12.7 Same scene as Fig. 12.6, after cleaning tarnish by repolishing with 0.25 µm diamond powder. The colour, reflectance and colloform texture of As were totally hidden by tarnishing (CR 397: Hiendelaencina, Guadalajara; both photos **dsp**)

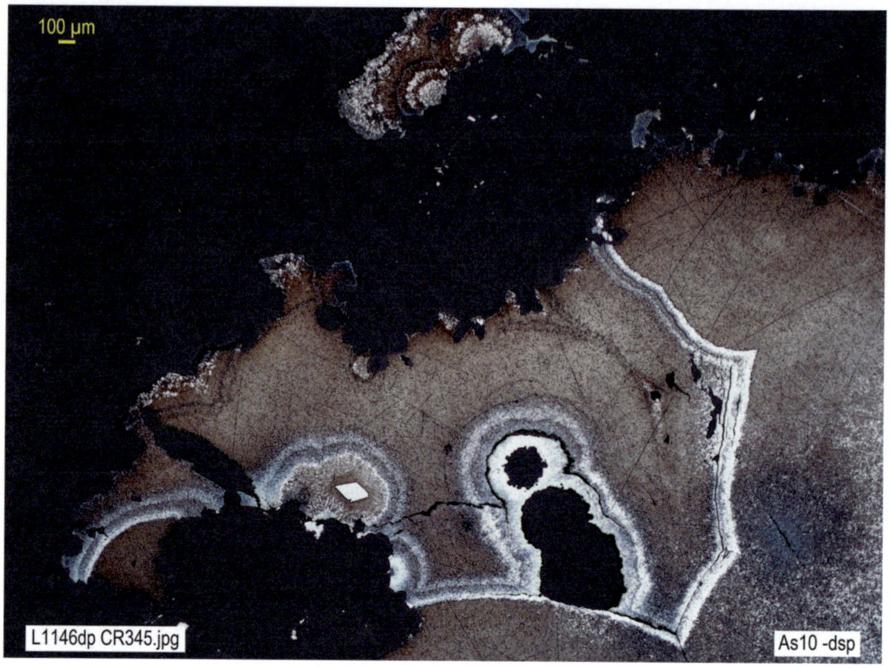

Fig. 12.8 Quartz vein (q very dark, massive, showing hexagonal and prismatic crystals) with colloform, banded intergrowths of As (brownish gray mass) and stibarsen (white bands), and scattered sparse inclusions of lo (white, idiomorphic). Reflectance contrast between sba and As enhanced by tarnishing (*patina* developed, *only on As*, in one month). CR 345: Engineer Mine, Atlin District, BC, Canada (**dsp**)

12.3 Spectral Information

12.3.1 Arsenic: Multispectral Reflectance R (in Air, Without Polariser) (%)

λ (nm)	370	400	425	450	475	500	525	550	575	600	625	650	675	700	750	800	850	900	950	1000
μ	41.8	42.4	42.5	42.3	42.3	42.3	42.5	42.7	43.0	43.2	43.3	43.4	43.7	44.1	44.3	44.9	45.4	46.3	46.8	47.4
P_5	35.1	36.1	36.0	36.3	36.1	36.7	36.9	37.6	38.0	38.5	38.7	39.0	39.6	40.2	40.7	41.6	42.3	43.5	44.2	45.0
P_{20}	37.8	38.7	38.9	39.0	38.9	39.3	39.5	39.9	40.3	40.7	40.8	41.0	41.5	42.0	42.3	43.1	43.8	44.9	45.5	46.2
P_{80}	45.1	45.7	45.6	45.4	45.2	45.1	45.3	45.2	45.3	45.5	45.5	45.6	45.8	46.0	46.1	46.6	47.0	47.7	48.1	48.6
P_{95}	48.6	48.6	48.5	48.0	47.9	47.6	47.6	47.4	47.6	47.6	47.4	47.5	47.4	47.7	47.5	48.0	48.2	48.8	49.1	49.4

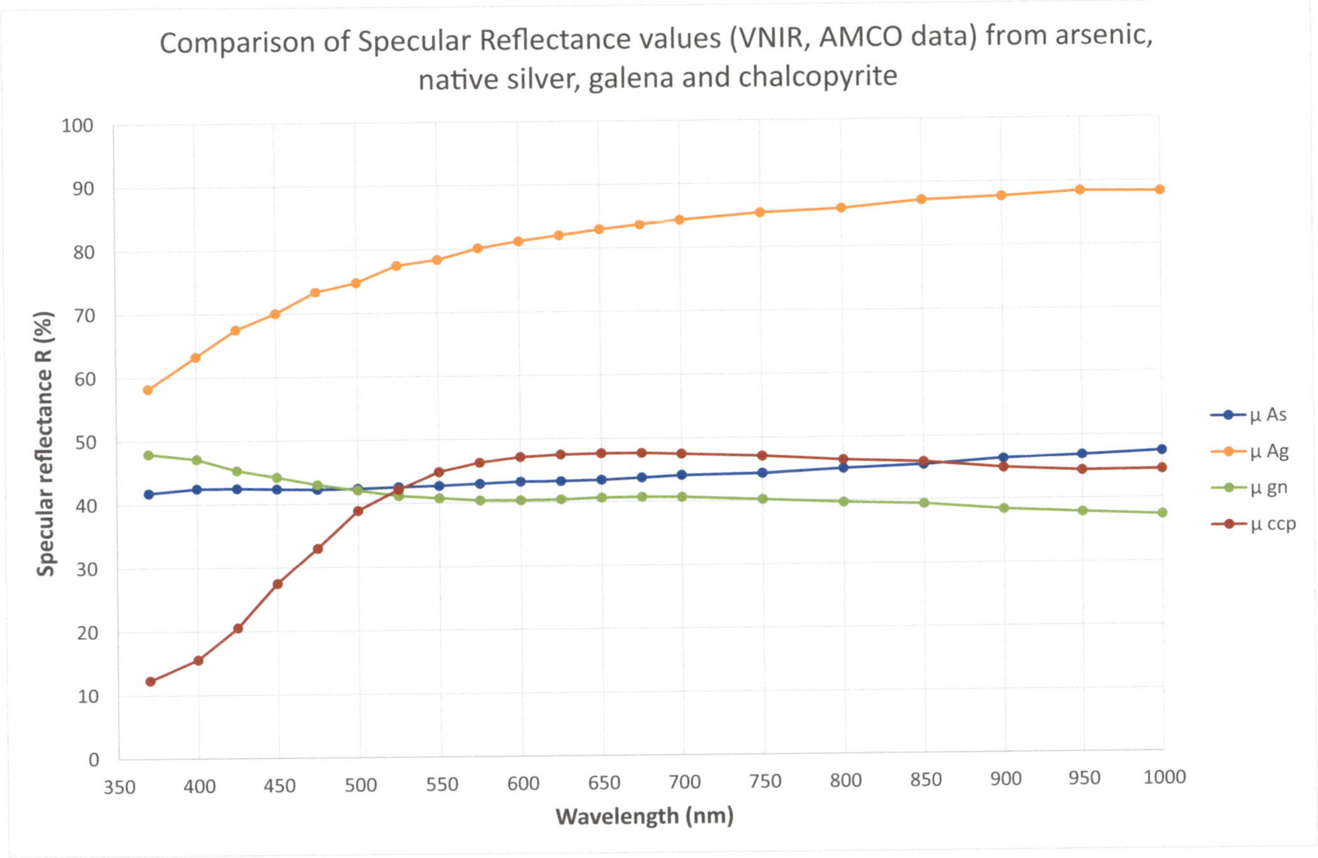
Comparison of Specular Reflectance values (VNIR, AMCO data) from arsenic, native silver, galena and chalcopyrite
Specular reflectance R (%)
Wavelength (nm)
µ As
µ Ag
µ gn
µ ccp

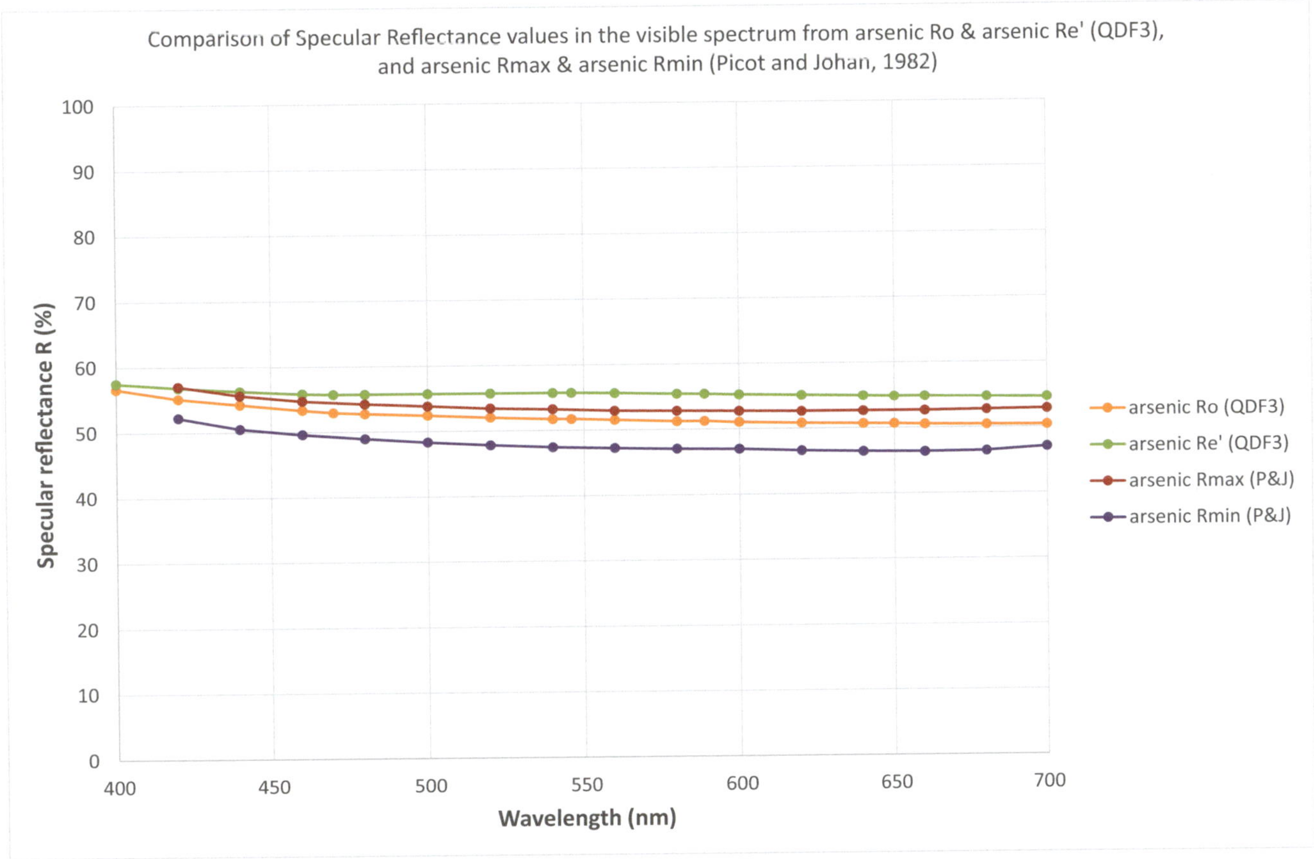
Comparison of Specular Reflectance values in the visible spectrum from arsenic Ro & arsenic Re' (QDF3), and arsenic Rmax & arsenic Rmin (Picot and Johan, 1982)
Specular reflectance R (%)
Wavelength (nm)
arsenic Ro (QDF3)
arsenic Re' (QDF3)
arsenic Rmax (P&J)
arsenic Rmin (P&J)

Reference

Picot P, Johan Z (1982) Atlas of Ore Minerals. B.R.G.M./Elsevier

13 Arsenopyrite (*asp/mispickel*)

FeAsS: Monoclinic (Pseudorhombic)

Arsenopyrite		R_{546} (%)	Y (%)	λ_d (nm)	P_e (%)	Colour sensation
Air	R_a	52.3	52.5	579	2.4	White to YlWh, glossy
	R_b	51.8	51.8	576	4.5	
	R_b	51.9	51.8	498	0.2	
Oil	R_a	37.6	37.6	577	5.5	Bright white, slightly pleochroic
	R_b	37.3	37.3	576	7.0	
	R_c	37.2	37.1	562	0.4	
Bir: Imperceptible (2)				*Pleoc*: Weak (3)		*Tarn*: Occasional (brownish tones)
Anis +P: Distinct and coloured (4)				*IR*: –		*Twg*: Frequent, irregular shapes
Mor: Idiom (typical lozenges), massive				*Clv*: Occasional //(101)		*PolQ*: Good, but slow (hard min.)
Hardness	*PolH*: High, but < py (4)			*Mohs*: 6–	*Talm*: F	*VHN* $_{(100\ g)}$: 1081

Obs: Mineral frequent but minor. When it is abundant, its presence in the ore processed in the plant or in the tailings is problematic due to the environmental impact of the As it contains

Spectral information: cf. Table and R plots, multispectral data (VNIR, AMCO) and visible range (QDF3)

Mineral abbreviations: ac—acanthite, asp—arsenopyrite, bn—bornite, ccp—chalcopyrite, cob—cobaltite, cst—cassiterite, cv—covellite, ln—linneitite, lo—löllingite, mc—marcasite, my—miargyrite, po—pyrrhotite, py—pyrite, rm—rammelsbergite, sch—scheelite, sf—safflorite, sk—skutterudite, sp—sphalerite, wf—wolframite

Colour abbreviations: YlWh—yellowish white

13.1 General

Composition: FeAsS, usually with Co replacing Fe in variable proportions up to Fe:Co = 2:1 (*danaite* variety, ≈ 12% Co), reaching Fe:Co = 1:6 in *glaucodot*.

13.2 Microscopic Description

13.2.1 SP Observation (*Single Polariser*)

White *colour*, with a very faint yellowish tinge (arsenopyrite looks white in contact with pyrite but yellowish against löllingite). Weak pleochroism in bluish, yellowish and reddish tones, more perceptible in immersion. It does not show the greenish and pinkish tones of marcasite (Fig. 13.1a, b). The hardness of the mineral requires patient *polishing*, but careful work provides an excellent polish. It can darken, sometimes quickly, by *tarnishing* or surface alteration, in irregular brownish spots. With time, alteration becomes invasive in humid atmospheres and can totally destroy the sample, oxidizing it to *scorodite* dust, even in collections kept in closed rooms.

High *reflectance*, although against native Ag and Au it appears relatively dark (Figs. 13.4, 56.1 and 56.2), of the order of löllingite or cobaltite, slightly lower than pyrite but R_{asp} > ccp and gn; $R_{asp} \gg$ po, my, sp; $R_{asp} \ggg$ cst (Figs. 13.1a, b, 13.2 and 13.6; cf. multispectral comparison plots).

High *polishing hardness*, between po and py: $PolH_{asp} > PolH_{po}$; $PolH_{asp}$ < PolH cob, py or cst.

It frequently shows idiomorphic development when it is disseminated (crystals typically show rhombic and prismatic shapes, Figs. 13.4 and 13.5), but it also occurs in anhedral masses and aggregates (Fig. 13.1a, b).

R. Castroviejo, *A Practical Guide to Ore Microscopy—Volume 1*,
https://doi.org/10.1007/978-3-031-12654-3_13

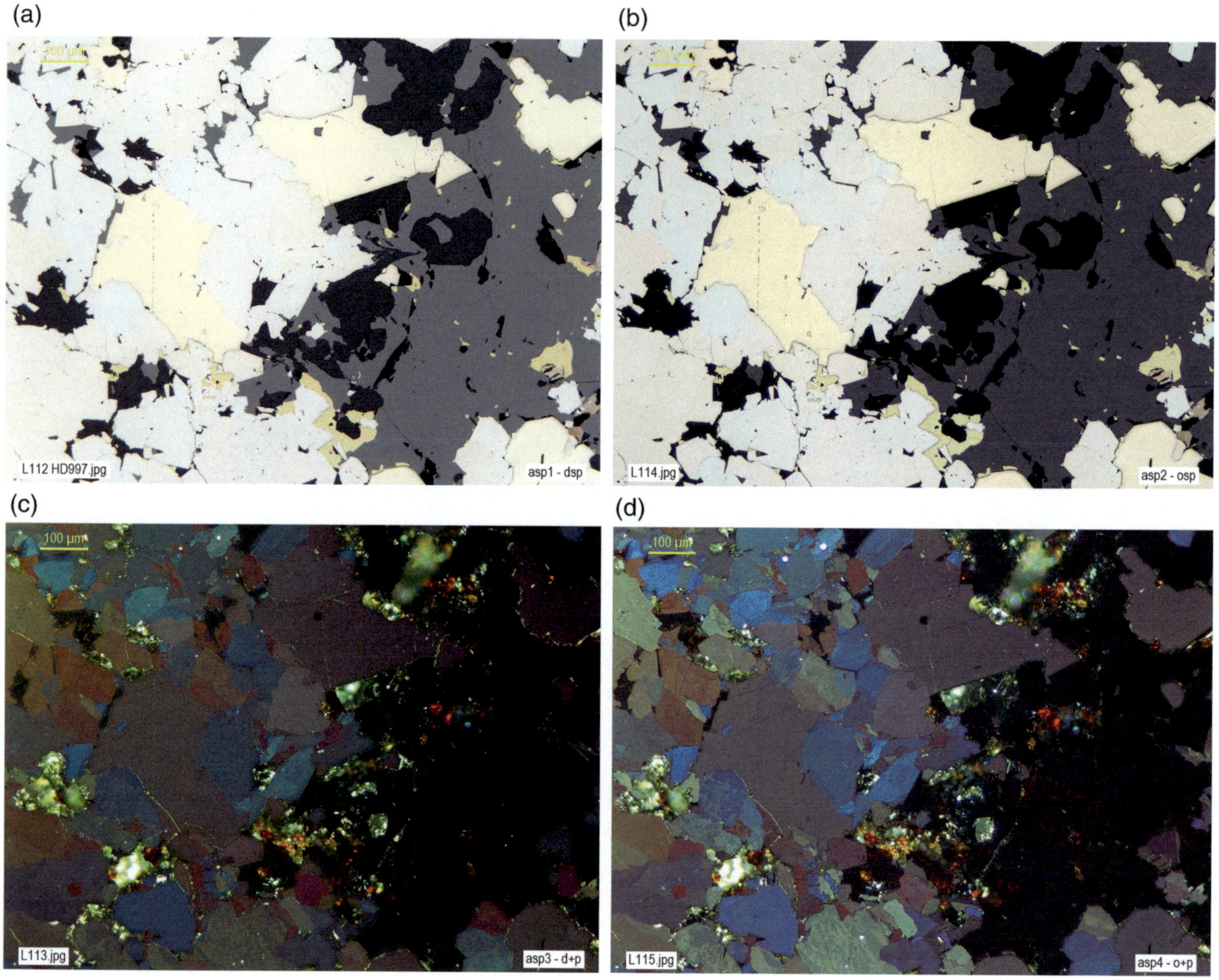

Fig. 13.1 **a** Massive asp (white, with faint bluish pleochroism) intergrown with py (yellow), sp (grey) and gg (dark grey), with small inclusions of ccp (intense yellow) and po (pinkish brown, lower right corner). HD 997, Boliden, Sweden (**dsp**). **b** Same scene as (**a**) (**osp**). Note the enhanced pleochroism of asp and the stronger contrast of reflectances with sp due to the immersion effect. **c** Same scene as Fig. 13.1a (**d+p**). **d** Same scene as Fig. 13.1a (**o+p**). Note the typical polarization colours of asp, brighter and more striking in immersion

13.2.2 Observation +P (*Polariser + Analyser at 90°*)

Distinctly *anisotropic*, with varied colours (bluish, greenish, brownish and violet tones), which may be striking, especially in immersion (Figs. 13.1c, d and 13.3). The anisotropism can make visible occasional *zoning*.

No *internal reflections*. Frequently *twinned*: varied twins, laminar, hourglass or with characteristic irregular shapes, clearly visible in immersion (Figs. 13.1c and 13.3).

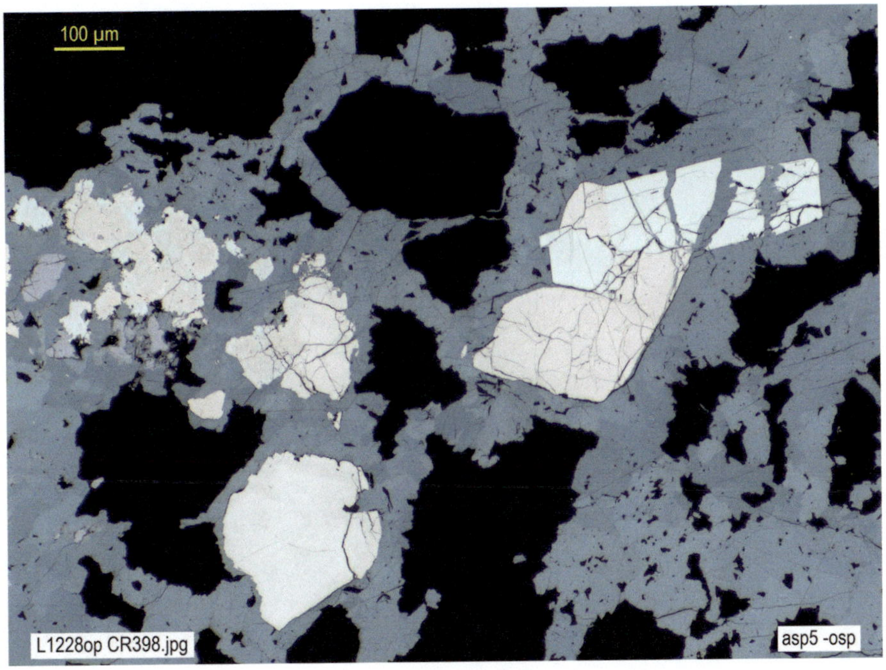

Fig. 13.2 Aggregates of asp (white) and carbonate gg (dolomite, black), fragmented, corroded and cemented by my (bluish grey, pleochroic). Sample CR398, Hiendelaencina Mine, Guadalajara (**osp**)

Fig. 13.3 Same scene as Fig. 13.2 (**o+p**). Note the irregular asp twins, internal reflections of my and dolomite, and lamellar twins of dolomite

13.2.3 Paragenesis

Almost ubiquitous as a trace mineral, but abundant only in some types of deposits, such as pegmatitic-pneumatolitic Sn-W formations and "five-element" (Ni–Co–Bi–Ag–U) deposits. Also sometimes found in volcano-sedimentary massive sulphides, in skarn formations and in orogenic Au deposits. As typical companions, apart from the common sulphides, can be found according to the corresponding *paragenesis*: cassiterite, wolframite, scheelite, Ag and Bi

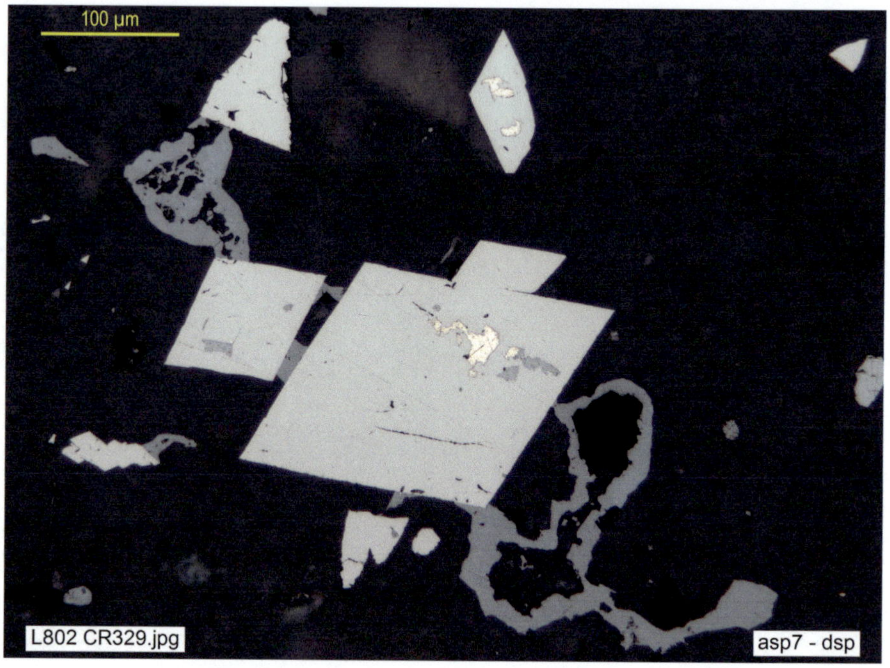

Fig. 13.4 Inclusions of native Ag (bright white) in idiomorphic asp (rhombic, white but greyish compared to Ag), with posthumous ac (grey) lining microgeodes. CR329, Imiter (Morocco) (**dsp**)

Fig. 13.5 Hydrothermal infill of asp (white) and sp (grey, lower right) and dissemination of idiomorphic asp in schistose wallrock with disseminated ru (grey acicular crystals). PET2_In409, Rio Gibranzos, Cáceres (**dsp**)

sulphosalts, native gold, *white arsenides* (löllingite, safflorite, rammelsbergite), etc.

By weathering, especially in humid environments, it is easily altered to the oxidized phase ***scorodite*** ($Fe^{3+}AsO_4 \cdot 2H_2O$), which can be recognized by its low reflectance ($R \approx 9\%$) and replacement textures (Figs. 13.6 and 13.7), although its proper identification must be done with transmitted light, which allows checking its high relief and birefringence. It can appear impregnated with limonite, in which case it acquires a reddish tone.

Fig. 13.6 Holocrystalline aggregate of cst (grey), with microgeode filled with asp and py; asp partially altered to scorodite (grey, darker than cst, corrosion rims in contact with asp); sparse cv and bn (included in asp and py, respectively). HJB_RUB607, Geevor Mine, Cornwall, UK (**dsp**)

Fig. 13.7 Same scene as Fig. 13.6 (**d+p**). Anisotropism of cst, cv and asp, as well as internal reflections of cst and scorodite, characteristic of these mineral phases

13.2.4 Identification

Its bright white colour, idiomorphic tendency, good polish, high hardness, twinning and coloured anisotropism allow identification. However, it can be easily confused with some of the so-called *white arsenides*:

- It can be distinguished from *löllingite*, *safflorite* and *rammelsbergite* because these show a purer white colour, lower hardness and more vivid anisotropism;
- *Glaucodot* shows a weaker anisotropism and is usually accompanied by cobaltiferous minerals (cobaltite, linneite, safflorite, skutterudite);

- *Gudmundite*, whose appearance is very similar to arsenopyrite, is softer, more reflective, more pleochroic and shows more vivid polarization tints, with orange tones different from those of arsenopyrite.

13.3 Spectral Information

13.3.1 Arsenopyrite Multispectral Reflectance R (in Air, Without Polariser) (%)

λ (nm)	370	400	425	450	475	500	525	550	575	600	625	650	675	700	750	800	850	900	950	1000
μ	51.0	51.5	50.9	51.5	51.5	52.1	51.9	52.3	52.2	52.1	52.0	51.7	51.7	51.5	51.2	50.6	50.7	49.4	49.1	48.4
P_5	46.7	47.7	47.8	48.5	49.2	49.7	50.3	50.7	50.9	50.9	50.6	50.3	50.2	50.0	49.6	49.1	49.3	48.0	47.7	47.0
P_{20}	49.3	50.1	49.6	50.3	50.4	51.3	51.1	51.6	51.5	51.4	51.3	50.9	50.8	50.6	50.2	49.7	49.8	48.6	48.2	47.6
P_{80}	52.6	53.0	51.9	52.8	52.5	52.9	52.6	52.9	52.7	52.6	52.4	52.3	52.3	52.1	51.8	51.3	51.1	50.0	49.4	48.8
P_{95}	54.6	54.9	53.4	53.9	53.4	53.6	53.9	54.1	54.0	54.1	54.1	54.1	54.3	54.2	53.9	52.8	52.9	51.4	51.4	50.4

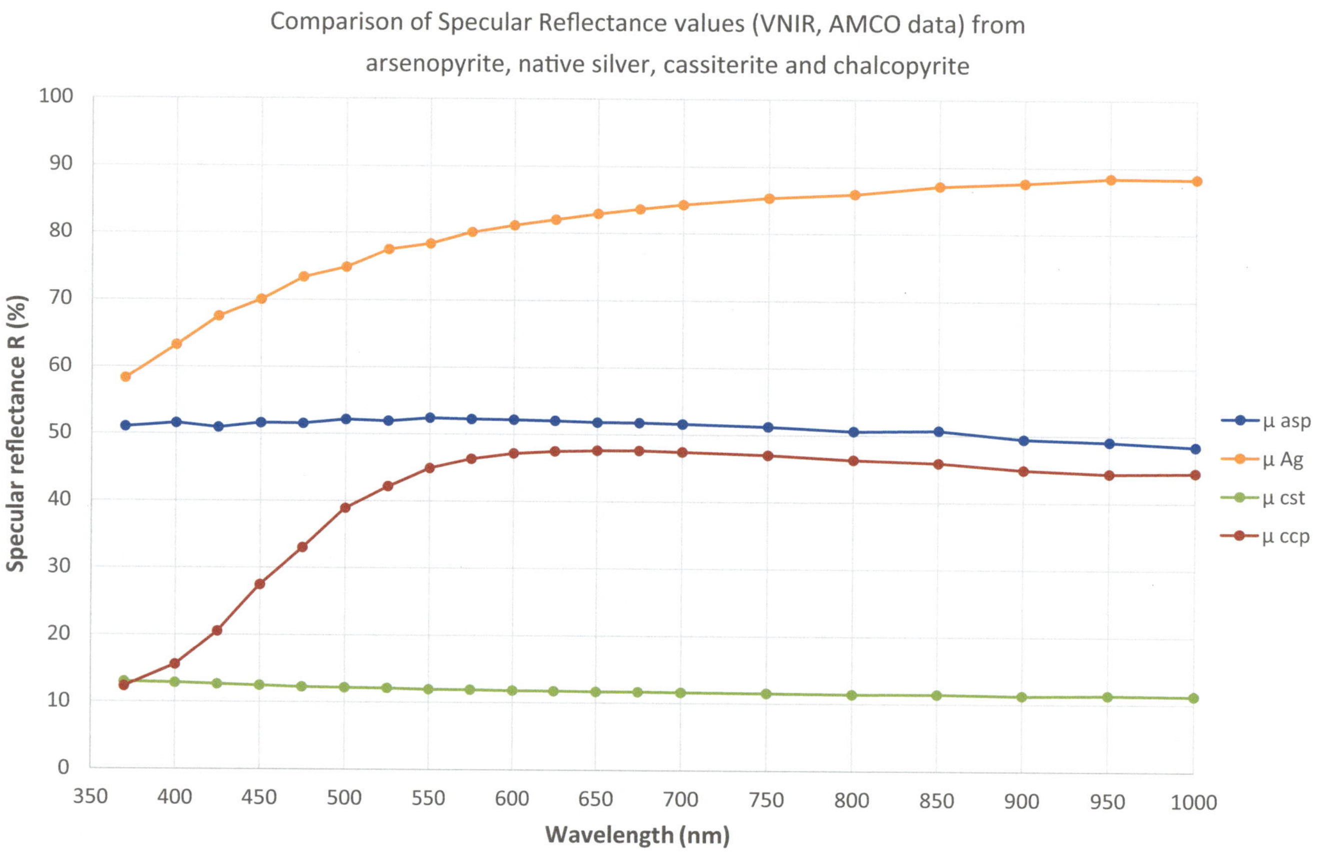

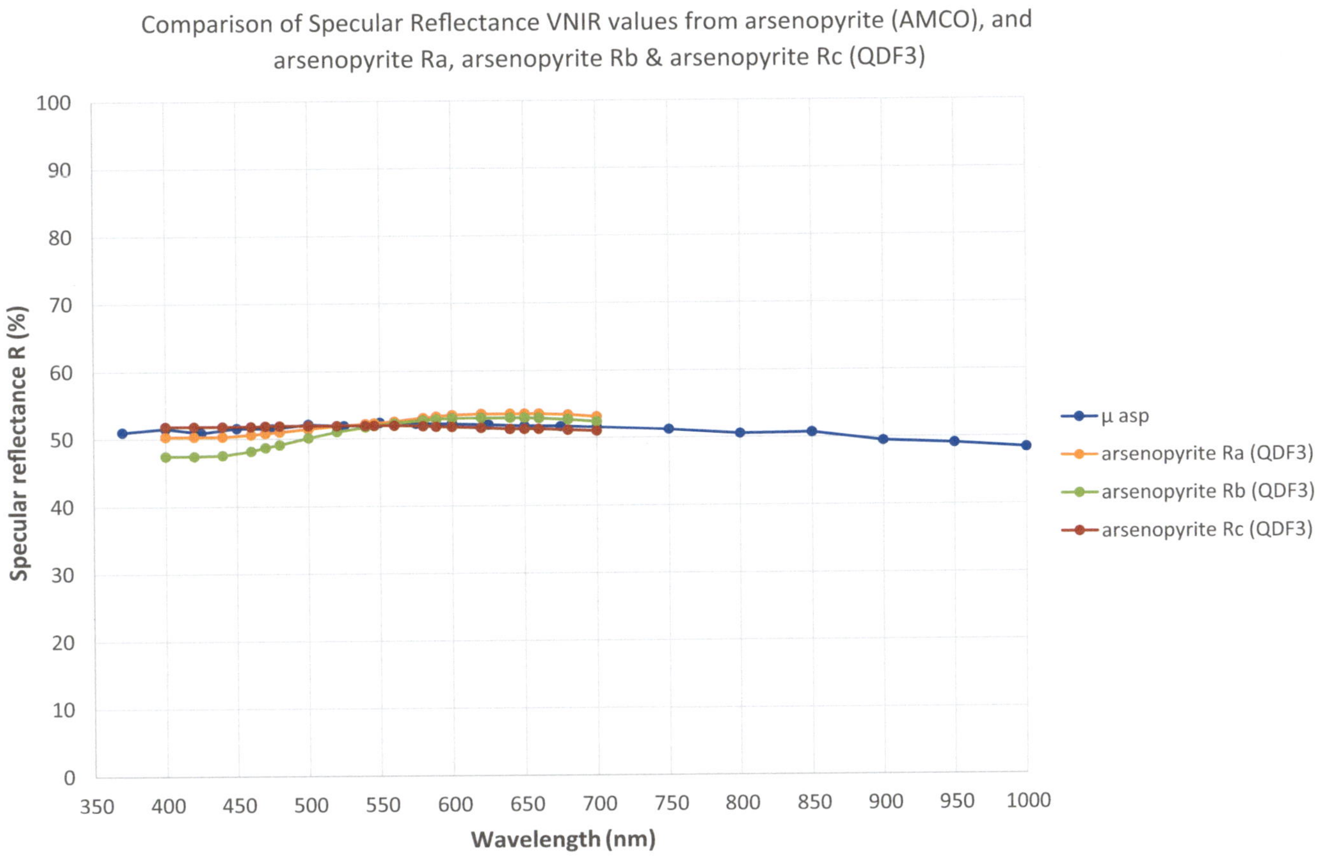
Comparison of Specular Reflectance VNIR values from arsenopyrite (AMCO), and arsenopyrite Ra, arsenopyrite Rb & arsenopyrite Rc (QDF3)
Specular reflectance R (%)
100
90
80
70
60
50
40
30
20
10
0
350
400
450
500
550
600
650
700
750
800
850
900
950
1000
Wavelength (nm)
μ asp
arsenopyrite Ra (QDF3)
arsenopyrite Rb (QDF3)
arsenopyrite Rc (QDF3)

14 Azurite (*az*)

$Cu_3(CO_3)_2(OH)_2$: Monoclinic

Azurite		R_{546} (%)	Y (%)	λ_d (nm)	P_e (%)	Colour sensation
Air	R_1	8.8	8.8	479	6.2	Grey
	R_2	9.2	9.3	475	5.7	Light grey
Oil	R_1	nd	nd	nd	nd	Dark grey
	R_2	nd	nd	nd	nd	Grey
Bir: Notable (4)				*Pleoc*: Weak/very weak (2)		*Tarn*: No
Anis +P: Strong but hidden by IR (4)				*IR*: Intense blue, massive		*Twg*: Rare
Mor: Various crystalline aggregates				*Clv*: No		*PolQ*: Good
Hardness	*PolH*: Mean, ≈mch, <td (3)			*Mohs*: 3.5+	*Talm*: –	*VHN* $_{(100\ g)}$: nd

Obs: The blue colour, so typical in hand samples, is hardly perceptible under the microscope (without analyzer); however, it is detected by the reflectance spectrum, R, with higher values towards blue (*cf. multispectral information graph VNIR*). The chromatic sensation (grey to bluish grey) changes dramatically with the analyzer crossed at 90° (+p). The colour of the mineral is then perceived as massive and intense blue internal reflections

Spectral information: cf. Table and R plots, multispectral data (VNIR, AMCO) and visible range (QDF3)

Mineral abbreviations: Ag—native silver, Au—native gold, az—azurite, cc—calcocite, ccp—chalcopyrite, Cu—native copper, cv—covellite, dg—digenite, en—enargite, fm—famatinite, gn—galena, lz—luzonite, mch—malachite, sp—sphalerite, td—tetrahedrite

14.1 General

$Cu_3(CO_3)_2(OH)_2$, a very frequent mineral, as a typical secondary product, in the oxidation zone of copper deposits.

14.2 Microscopic Description

14.2.1 SP Observation (*Single Polariser*)

Grey *colour*, relatively dark (Figs. 14.1a, b). The blue colour of the mineral—intense and striking in hand specimen and evident with transmitted light because it is a transparent mineral—is not clearly perceived with reflected light; only sometimes, due to the effect of internal reflections, it can also be perceived without the analyzer, especially in immersion, resulting in a bluish grey tone (Fig. 14.2). Clearly bireflective, in polycrystalline aggregates it shows a variable level of grey depending on the orientation of the grains but, due to the absence of colour, the pleochroism is hardly or not at all visible.

Polishes well. Medium *polishing hardness*: $PolH_{az} \approx$ mch; $PolH_{az} >$ Au, Ag, calcite, ccp, Cu, gn; and $PolH_{az} <$ dolomite, td, sp, siderite.

Low *reflectance compared* to sulphide ores, but clearly higher than common gangue minerals.

Strong *bireflectance* (especially in immersion Fig. 14.2), as befits a carbonate.

However, *pleochroism* is usually hardly noticeable. In general, only a change in light/dark shades of grey is observed.

Azurite occurs in crystalline aggregates, with euhedral tendency and varied forms, frequently tabular or prismatic. As a supergene mineral, it often contains sulphide relics (cc, cv, etc. Figs. 14.1a, b). It is sometimes massive, also botryoidal or intergrown with malachite, which can pseudomorphize it.

R. Castroviejo, *A Practical Guide to Ore Microscopy—Volume 1*,
https://doi.org/10.1007/978-3-031-12654-3_14

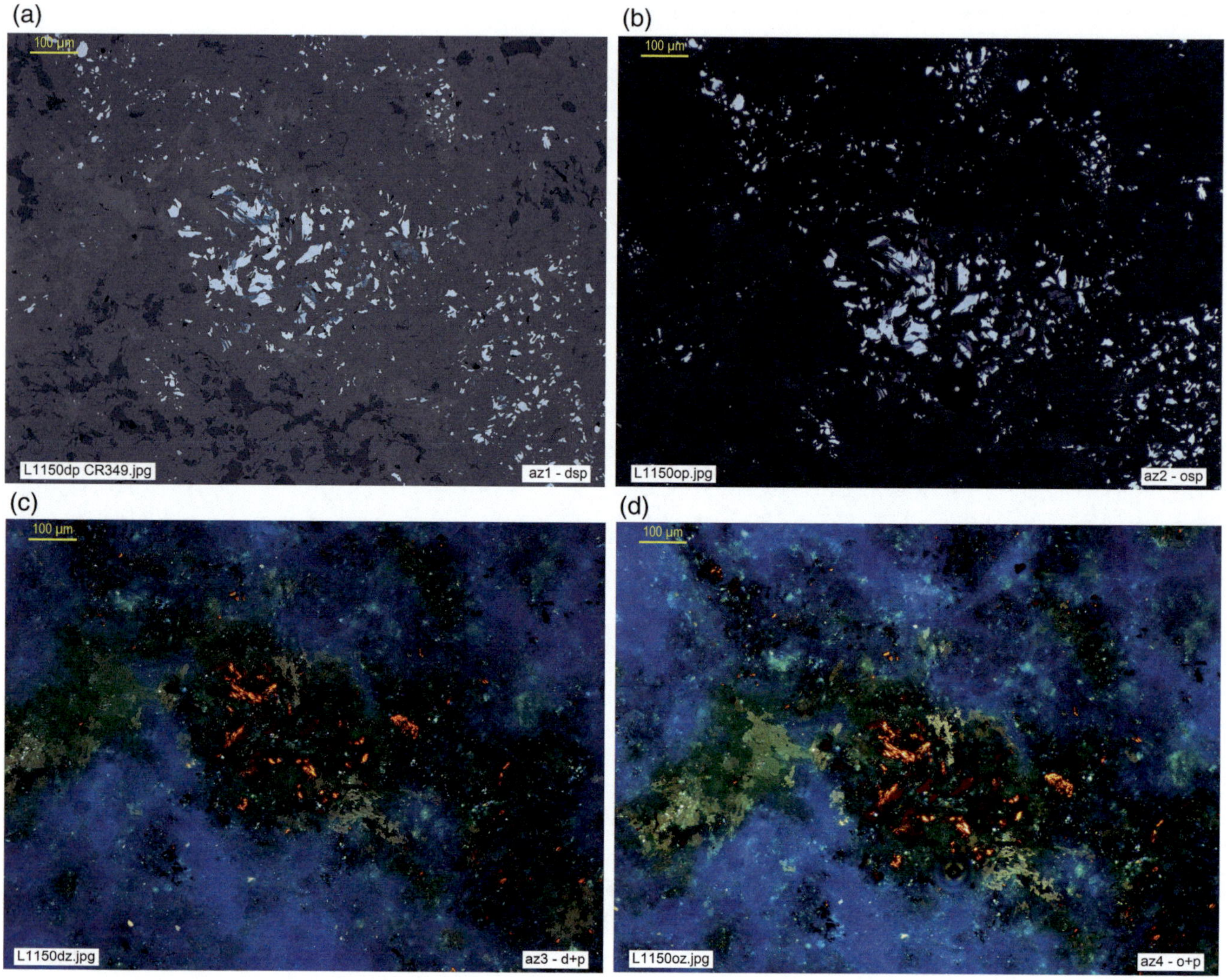

Fig. 14.1 **a** Aggregate of az and mch (grey, bireflective), with scattered relicts of cc (bluish-white) and cv (deep blue). CR349, Grand Gulch, Coconino County, Ariz. USA (**dsp**). **b** Same scene as Fig. 14.1 (**osp**). Note the sharp decrease in reflectance (darkening) of az and mch and the increased contrast with the relict sulfides, which now appear brighter. **c**, **d** Same scene as Fig. 14.1 (**d+p**) and (**o+p**), respectively. The anisotropism of the carbonates, mch and az, is now masked by the massive internal reflections, green and blue respectively, while among the sulphides only cv can be clearly seen, due to its strong and bright orange anisotropism

14.2.2 Observation +P (*Polariser + Analyser at 90°*)

Distinctly *anisotropic*, but the anisotropism is masked by the intense blue colour of its massive and luminous *internal reflections* (Figs, 14.1c, d, and 14.3).

Rarely *twinned*.

14.2.3 Paragenesis

It appears as a secondary product of supergene alteration, typically in the oxidation zone, in practically all copper deposits with the presence of carbonates.

Its presence on the surface can serve as an exploration guide for Cu, although it is less frequent than malachite and usually appears *only associated with the alteration of ores containing enargite, luzonite, famatinite or tetrahedrite as primary minerals*.

14.2.4 Identification

Its optical properties, particularly its bireflectance, its crystallinity and its blue internal reflections are so typical that they make its identification easy, taking into account its association with cupriferous deposits. However, in case of doubt, it can be confirmed by an attack with dilute acid: az releases CO_2, with effervescence.

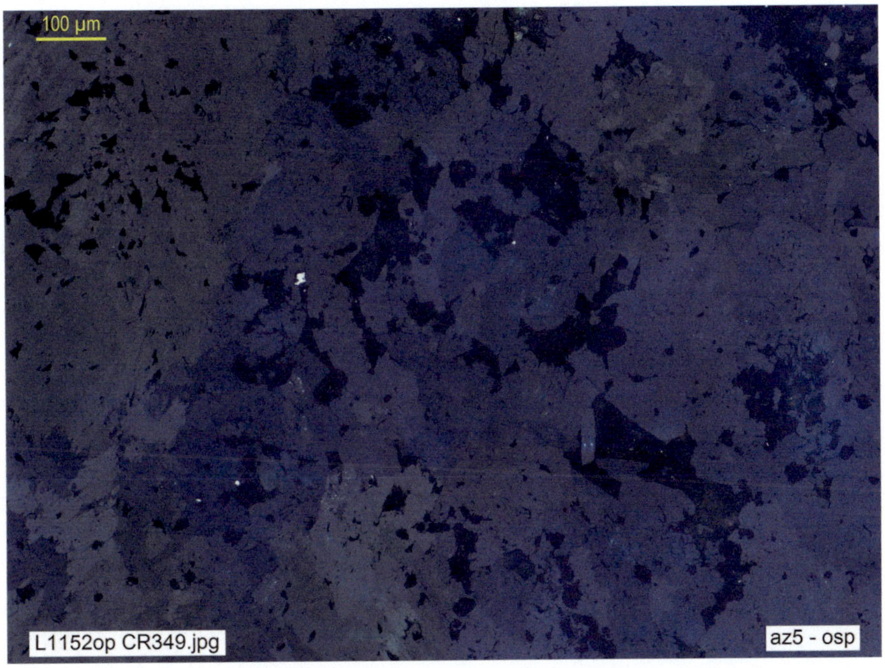

Fig. 14.2 Aggregate of az and mch (grey, bireflective), with sparse relics of cc (white). Strong bireflectance of carbonates, az and mch. Euhedral, prismatic and tabular forms of az. The colour of az is perceived here by reflection as bluish-grey because of internal reflections. CR349, Grand Gulch, Coconino County, Ariz. USA (**osp**)

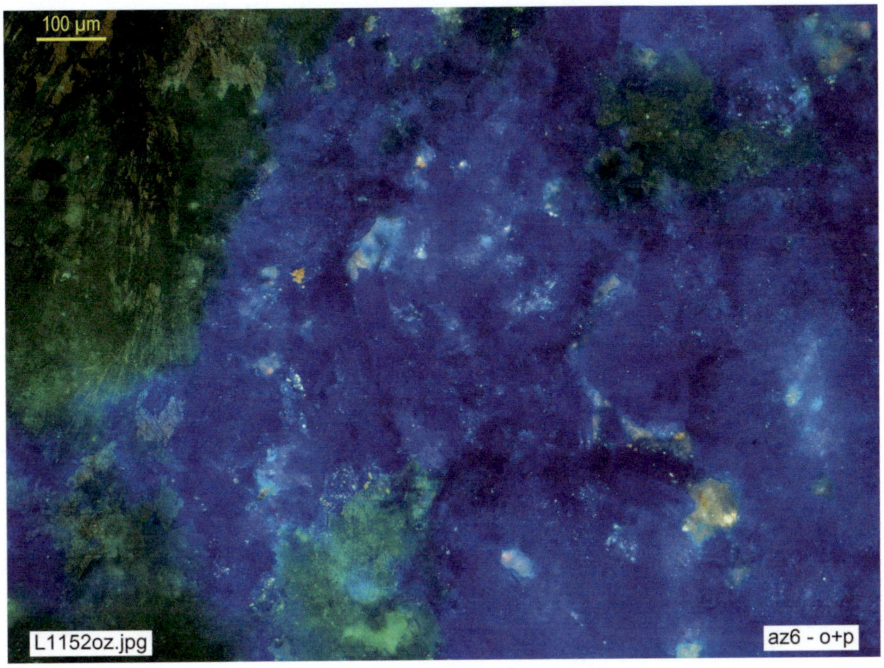

Fig. 14.3 Same scene as Fig. 14.2 (**o+p**). Massive internal reflections of az and mch, blue and green respectively, which partly hide their anisotropism

14.3 Spectral Information

14.3.1 Azurite Multispectral Reflectance R (in Air, Without Polariser) (%)

λ (nm)	370	400	425	450	475	500	525	550	575	600	625	650	675	700	750	800	850	900	950	1000
μ	9.9	9.9	10.4	10.4	9.9	8.9	8.3	8.0	7.8	7.7	7.7	7.6	7.6	7.6	7.6	7.6	7.6	7.5	7.6	7.9
P_5	9.4	9.3	9.3	9.1	8.9	8.4	7.9	7.6	7.4	7.3	7.3	7.2	7.2	7.2	7.2	7.2	7.2	7.1	7.3	7.5
P_{20}	9.7	9.6	9.7	9.5	9.3	8.6	8.1	7.7	7.6	7.5	7.4	7.4	7.4	7.4	7.4	7.4	7.4	7.3	7.4	7.7
P_{80}	10.1	10.2	11.0	11.1	10.4	9.1	8.4	8.1	8.0	7.9	7.9	7.8	7.8	7.8	7.8	7.8	7.8	7.7	7.8	8.0
P_{95}	10.3	10.3	11.2	11.3	10.6	9.5	9.2	9.0	8.8	8.7	8.6	8.6	8.6	8.5	8.5	8.5	8.5	8.5	8.5	8.5

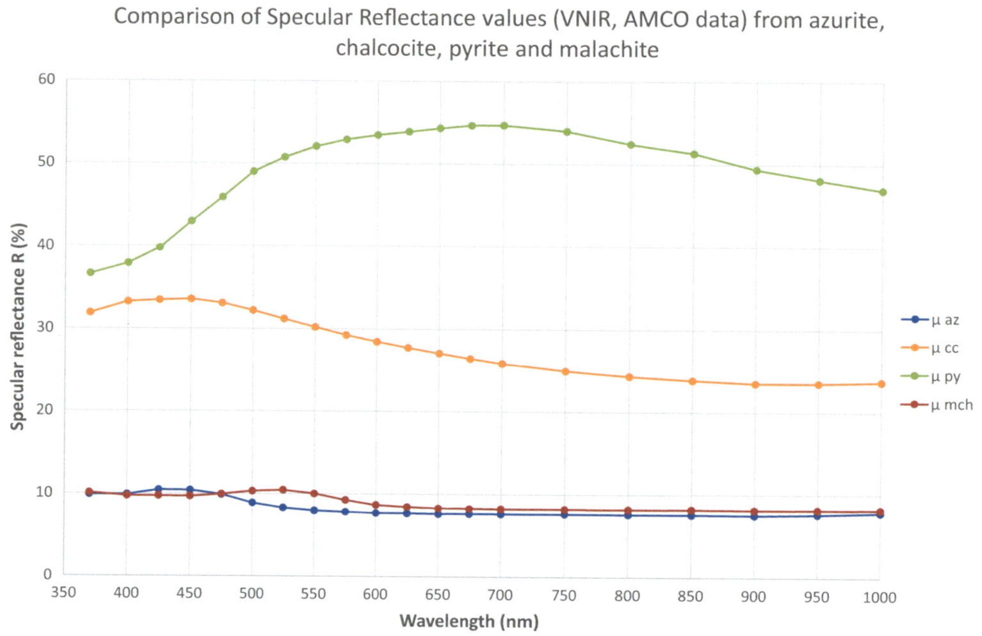

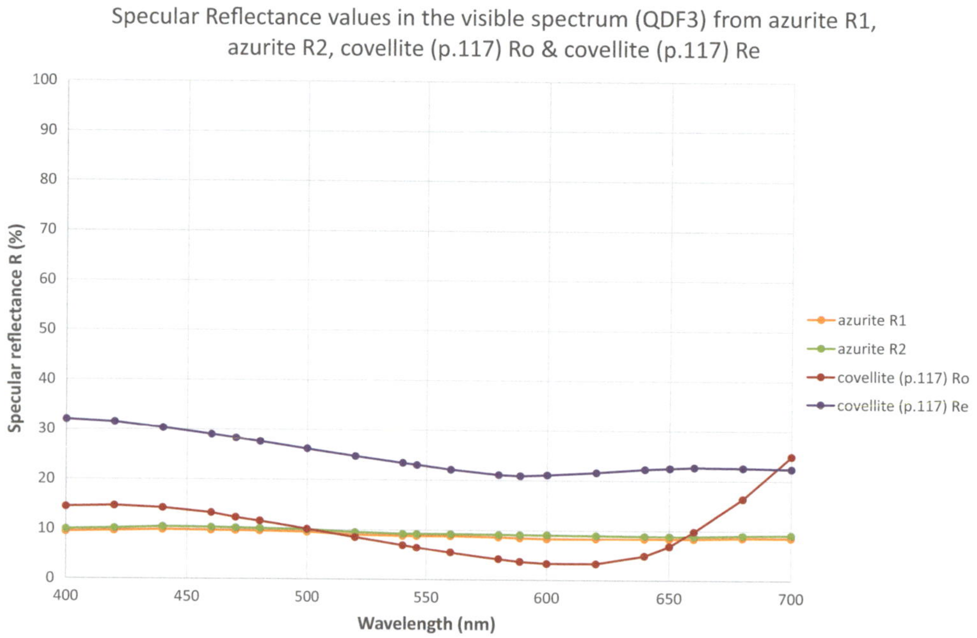

15 Bastnäsite (*bst*)

$Ce(CO_3)F$: Hexagonal

Bastnäsite		R_{546} (%)	Y (%)	λ_d (nm)	P_e (%)	Colour sensation
Air	R_1	≈7.2	nd	nd	nd	Grey
	R_2	≈8.5	nd	nd	nd	
Oil	R_1	nd	nd	nd	nd	Dark grey
	R_2	nd	nd	nd	nd	
Bir: Weak (3)				*Pleoc*: Very weak (2)		*Tarn*: No
Anis +P: Partially veiled by IR (3)				*IR*: Frequent, clear		*Twg*: nd
Mor: Tabular, massive				*Clv*: Faintly visible {1010}		*PolQ*: Variable
Hardness	*PolH*: Medium (3)			*Mohs*: 4–4½	*Talm*: nd	*VHN* $_{(100\ g)}$: nd

Obs: No published data on reflectance or description of properties with reflected light have been found
Mineral abbreviations: bar—barite, bst—bastnäsite, cst—cassiterite, gg—gangue, gn—galena, py—pyrite, q—quartz, tit—titanite

15.1 General

Composition: $Ce(CO_3)F$ or Bastnäsite-(Ce). In reality, the composition can be varied since it can incorporate Y, Nd or La in substitution of Ce, or hydroxyl (OH) instead of F. Currently, IMA has defined the different compositions that make up the *Bastnäsite Group* as independent species. Among them, the most frequent is the one described here, Bastnäsite-(Ce), which forms a series with Hydroxyl-bastnäsite-(Ce) or $Ce(CO_3)(OH)$.

The other species are named in the same way, indicating the corresponding metal: Bastnäsite-(Y), Bastnäsite-(La) or Bastnäsite-(Nd). However, the exact determination of the composition requires additional information (e.g., chemical analysis, electron microprobe). The microscopist who does not have this information can only characterize **bst**, bastnäsite, in a broad sense (sensu lato, implying any composition). This is what optical microscopy allows us to determine. Since it is a transparent mineral, it is also useful to characterize it by **TL** (transmitted light or petrographic microscope).

15.2 Microscopic Description

15.2.1 SP Observation (*Single Polariser*)

Grey *colour* (Figs. 15.1a, b and 15.2a), resulting in a very light grey tone against quartz and not so light against barite; in any case, much darker than usual ores such as pyrite or other sulphides. Lacking specular reflectance measurements and given that it is a transparent mineral (Fig. 15.3a), the *Reflectance R* has been estimated by the Fresnel equation, taking into account the minimum and maximum refractive indices (n_ω = 1.717–1.722, n_ε = 1.818–1.823), as explained (cf. Chap. 1, §1.2.2 and Fig. 1.2). The resulting values for R ($R_\omega \approx 7.2$ and $R_\varepsilon \approx 8.5$) are, as expected, distinctly higher than those for barite (R ≈ 6%) and quartz (R ≈ 4.5%). In the sample described, the bireflectance is not as clearly discernible as one would expect in a carbonate, possibly because of alteration and the many inclusions.

It occurs in tabular crystals, idiomorphic but sometimes fragmented (Fig. 15.2a), and in massive aggregates.

R. Castroviejo, *A Practical Guide to Ore Microscopy—Volume 1*,
https://doi.org/10.1007/978-3-031-12654-3_15

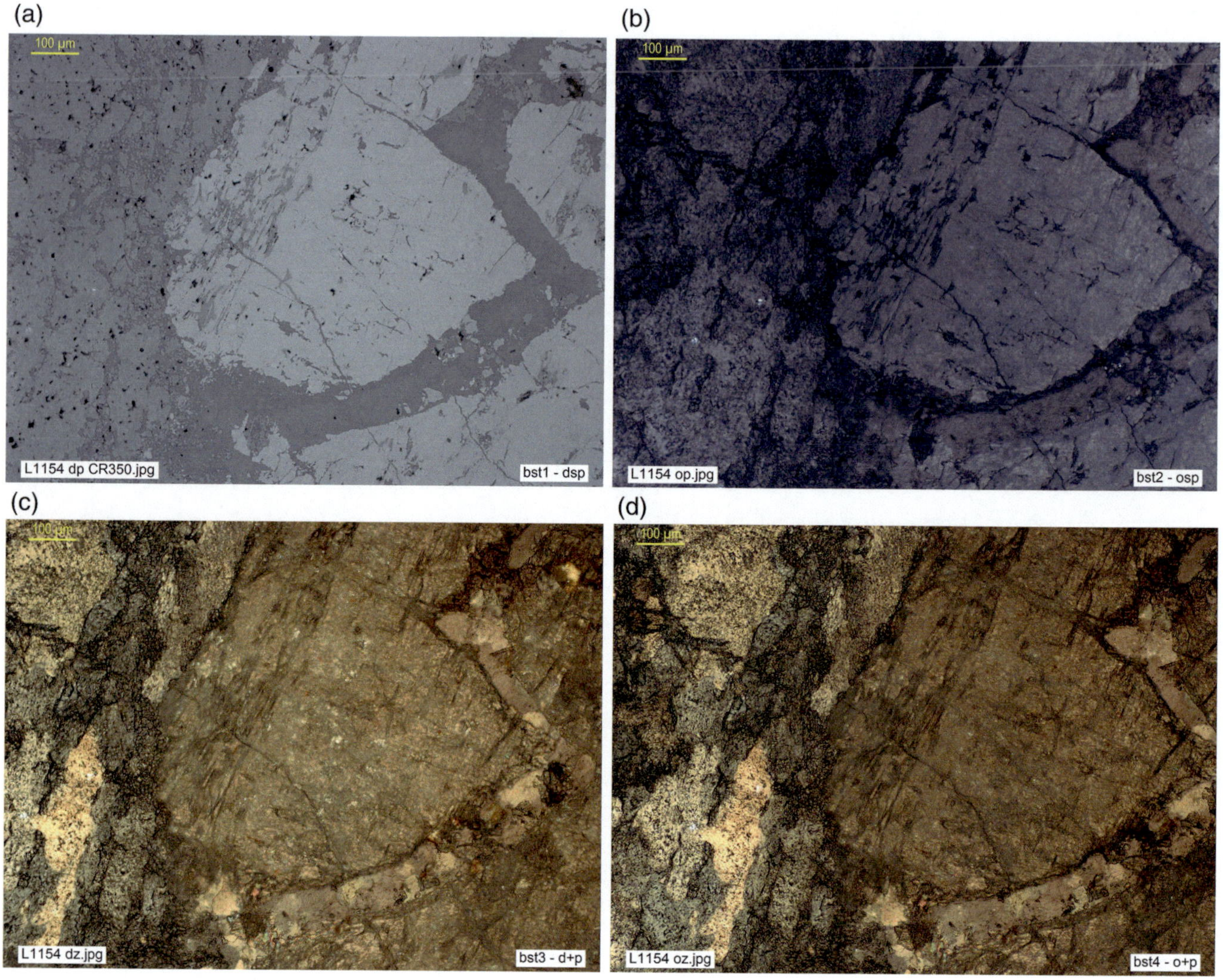

Fig. 15.1 **a**, **b** (Same scene as **dsp** and **osp**, *respectively*). Ore of bst (light grey) and bar (medium grey) brecciated and cemented by q (darker grey). The range of grey levels, from lightest to darkest, corresponds to the sequence of reflectances: R bst > R bar > R q. Idiomorphic development of bst, which forms large fragmented prismatic crystals, as seen in wider-field images of the same sample (Figs. 15.2a, b and 15.3a, b), and shows a poorly visible cleavage, as sparse aligned or parallel microcracks. CR350, Mountain Pass, Calif., USA. **c**, **d** Same scene as Fig. 15.1a (**d+p** and **o+p**, respectively). The anisotropism of bst, which should be more visible than that of bar or q, is masked by the colourless or yellowish IR and could go unnoticed

15.2.2 Observation +P (*Polariser + Analyser at 90°*)

Anisotropic, but with inconspicuous effects. Its anisotropism is largely obliterated by *internal reflections*, **IR**, which are colourless (Fig. 15.1c, d) but may appear coloured by transparency or transmission of the high-order TL polarization colour of the mineral (Figs. 15.2b and 15.3b, with low magnification objective).

15.2.3 Observation with TL (*Transmitted Light*)

Observation with **transmitted light**, *TL*, can help with recognition, so the main properties are summarized briefly. As in the case described here, to observe them is enough to make a polished thin section, with which the information (RL and TL) can be duplicated. In transmitted light (**Tsp**, no analyser) bastnäsite is colourless or yellowish, weakly pleochroic, shows high relief (refractive indices ≈ 1.72 to

Fig. 15.2 **a**, **b** (Same scene, **dsp** and **d+p**, *respectively*; both are wide-field images of Fig. 15.1a, 2.5 × objective, **RL**). Vein with brecciated holocrystalline hydrothermal filling of bst, whose prismatic shape can be recognized despite fragmentation, and bar, with subordinate q. In Fig. 15.2b (**d+p**), false impression of coloured anisotropism of bst. What is seen in colour are actually internal reflections, producing interference colours resulting from their transmission through the transparent mineral (comp. **TL +p**, Fig. 15.3b). CR350, Mountain Pass, Calif., USA.

1.82), which highlights it on barite and quartz, and a poorly visible {1010} cleavage Fig. 15.3a. With analyser (**T+p**), it shows straight extinction and positive uniaxial character, strong birefringence (up to 0.101) and very high polarization colours (third order and higher), although the remarkable absorption ($\varepsilon > \omega$) detracts from its transparency and, therefore, luminosity Fig. 15.3b.

15.2.4 Paragenesis

Bastnäsite is an important RE (Rare Earth) ore. It is the most frequent ore among RE minerals, occuring in several types of deposits: carbonatites, alkaline rocks, granites and syenites, pegmatites, peralkaline gneisses, skarn and contact metamorphic formations. It may be seen in detrital

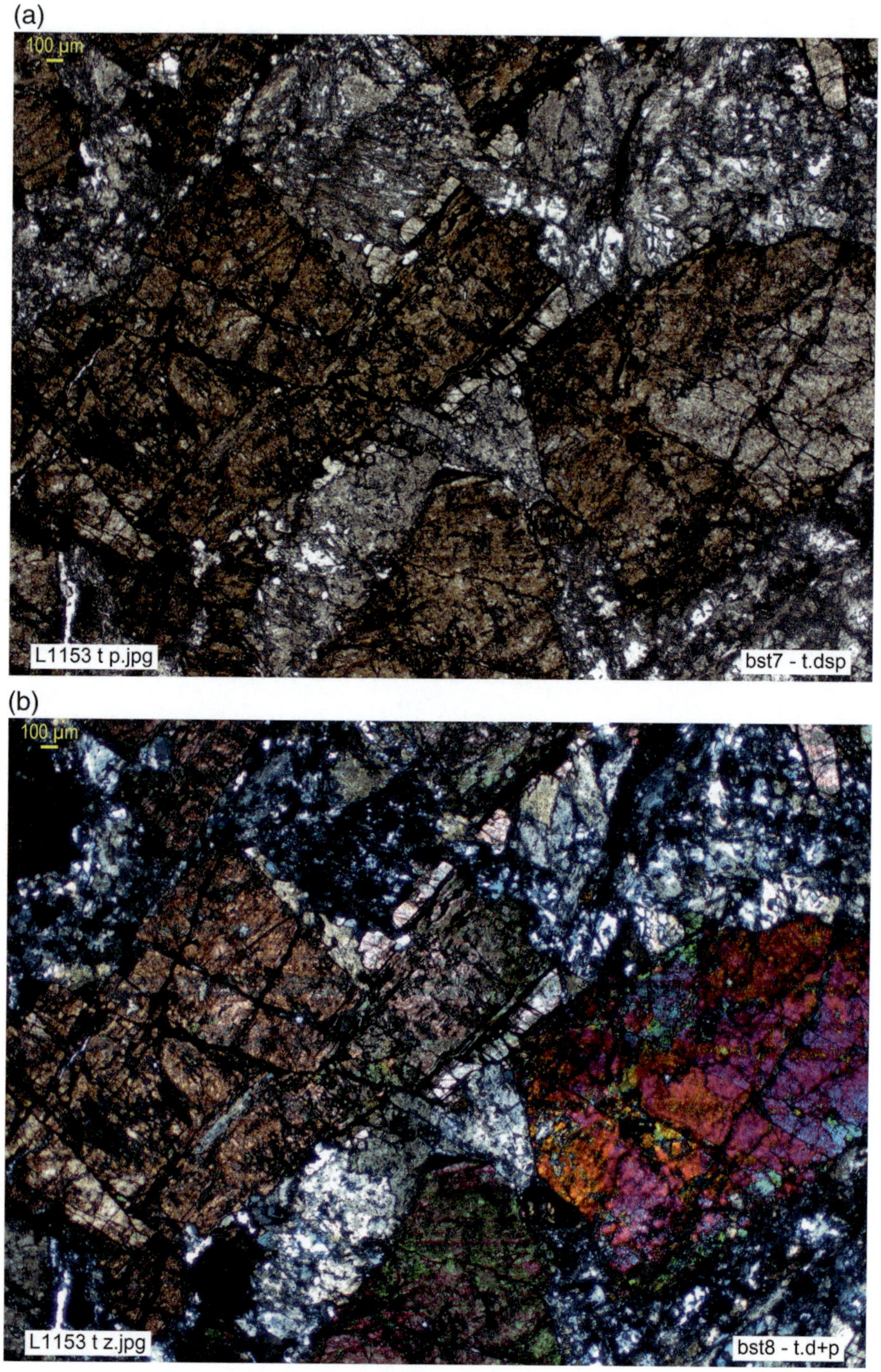

Fig. 15.3 **a**, **b** Same scene as Fig. 15.2a, seen with transmitted light, TL (**Tsp** and **T+p**, *respectively*). Strong relief and birefringence of bst, reminiscent of ores like tit or cst

concentrations as well, but these have little significance as mineral resource, due to the low hardness and to the solubility of the mineral in surface environments, as is typical for carbonates.

As companions, depending on each paragenesis, other carbonates (calcite, dolomite), fluorite, barite, allanite, egirine, quartz, microcline, pyrite, etc. are found.

15.2.5 Identification

The paragenesis of REE is very important data. These are typical bastnäsite features:

- grey colour, with reflectance much lower than that of common metallic ores (pyrite, chalcopyrite, galena, etc.)

but notably higher than that of the usual gangue, such as quartz, feldspars, etc.

- transparency, which gives rise to ubiquitous, generally colourless IRs (unless they transmit polarization colours by transparency Fig. 15.2b); IRs largely mask anisotropism
- the properties observed with **transmitted light** (cf. **Observations TL**)

Berthierite (*brt*)

16

$FeSb_2S_4$: Orthorhombic

Berthierite		R_{546} (%)	Y (%)	λ_d (nm)	P_e (%)	Colour sensation
Air	R_1	30.3	30.6	c548	1.7	Grey-pink-brown
	R_2	42.3	42.1	480	4.1	
Oil	R_1	15.5	15.7	c544	2.6	
	R_2	26.4	26.2	479	6.9	
Bir: Medium (4)				*Pleoc*: Moderate (4)		*Tarn*: Frequent, dark
Anis +P: Net (4)				*IR*: No		*Twg*: Occasional
Mor: Prismatic, microgranular				*Clv*: Weak		*PolQ*: Good
Hardness	*PolH*: Low, $\geq$ sbt (2)			*Mohs*: 3	*Talm*: –	$VHN_{(100\ g)}$: 168–228

Spectral information: cf. Reflectance plots, visible range (Picot and Johan 1982)
Mineral abbreviations: asp—arsenopyrite, Au—native gold, bm—bismuthinite, brt—berthierite, ccp—chalcopyrite, cin—cinnabar, gd—gudmundite, gn—galena, mc—marcasite, po—pyrrhotite, prg—pyrargyrite, pru—proustite, py—pyrite, q—quartz, sbt—stibnite (antimonite), sp—sphalerite

16.1 General

Composition: $FeSb_2S_4$, with possible incorporation of cations (Mn, Ag, Cu), facilitated by its unique structure, with channels oriented //c.

It decomposes easily into stibnite + pyrite (melnicovite).

16.2 Microscopic Description

16.2.1 SP Observation (*Single Polariser*)

Greyish-white to brownish-grey/pinkish-grey, *pleochroic* (Fig. 16.1a, b). Good *polish*, but its usual darkening by tarnishing may soon require cleaning.

Moderate *reflectance*, R slightly < stibnite (between Rmax and Rmin of stibnite, cf comparative graphs and Fig. 16.2a).

Low *polishing hardness*, but somewhat higher than that of stibnite, on which it appears in relief: PolH between that of stibnite and that of galena or bismuthinite, clearly < sphalerite. More resistant than stibnite to deformation, showing a more brittle behaviour, which contrasts with the ductility of stibnite (Fig. 16.2a, b).

It occurs in granular aggregates and in crystals with a prismatic to fibrous habit and cleavage usually imperceptible. It tends to decompose into stibnite + melnicovitic pyrite.

16.2.2 Observation +P (*Polariser + Analyser at 90°*)

Strongly anisotropic, similar to stibnite but with light greenish tones and without the brownish-pink tones of stibnite (Figs. 16.1c, d and 16.2b).

R. Castroviejo, *A Practical Guide to Ore Microscopy—Volume 1*,
https://doi.org/10.1007/978-3-031-12654-3_16

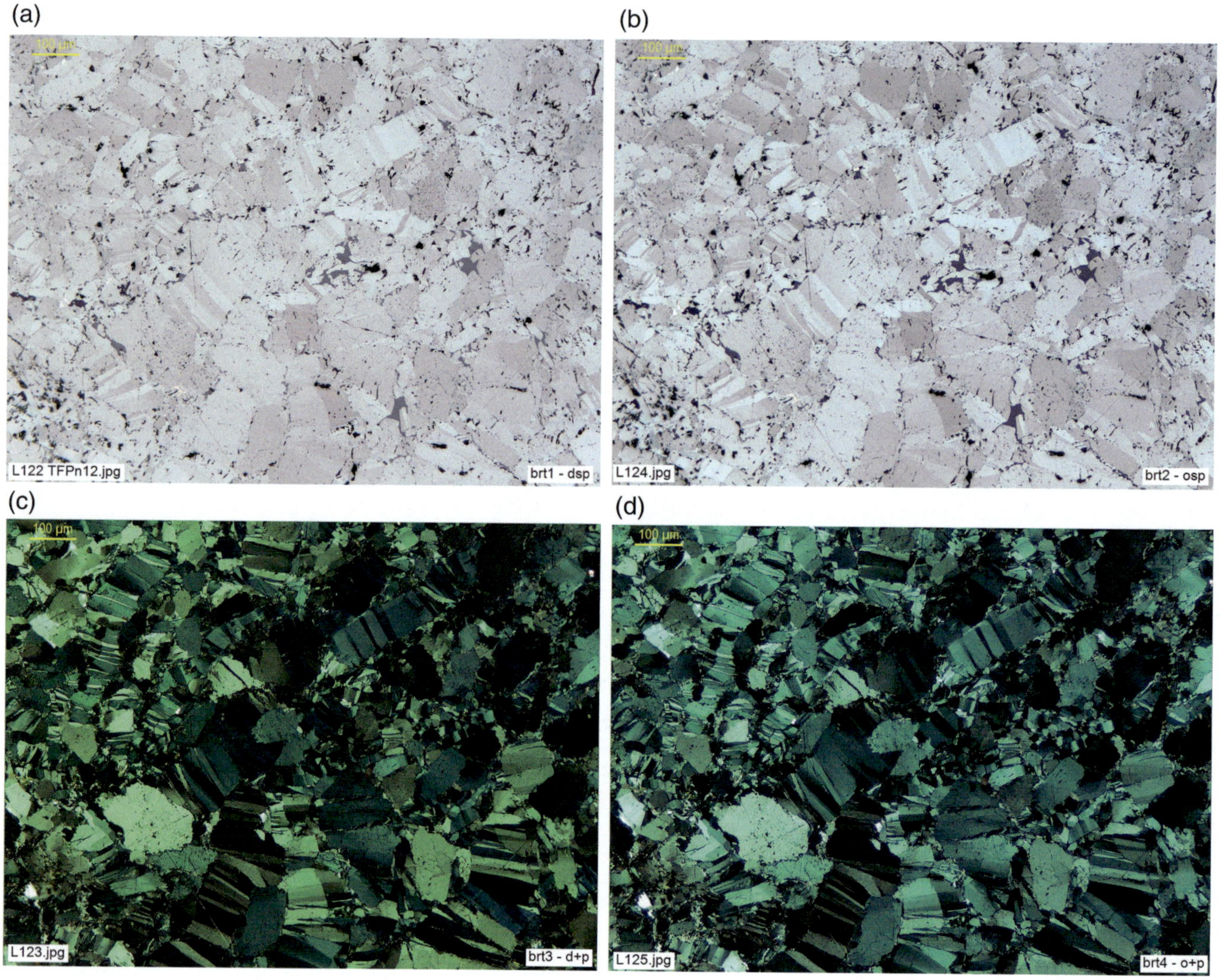

Fig. 16.1 **a**, **b** Same scene, **dsp** and **osp**, *respectively*. Microgranular aggregate of twinned brt, showing its remarkable pleochroism (enough to make the twins visible without analyser); interstitial sphalerite (grey), sbt and quartz (dark grey). TF Pn12, Almuradiel, Ciudad Real. **c**, **d** Same scene as Fig. 16.1 (**d+p** and **o+p**, respectively). Twinning and anisotropism (in greenish to bluish-green tones) of brt, in deformed ore

Occasional *twinning*, less frequent than in stibnite. It does not usually show the pressure twins typical of stibnite, which deforms more easily (Figs. 16.1b–d and 16.2b).

16.2.3 Paragenesis

Typically, hydrothermal *parageneses*; brt is generally related to waning, low-temperature stages of vein filling, like stibnite.

Accompanying minerals: stibnite, pyrite, marcasite, sphalerite, chalcopyrite, cinnabar, native gold, proustite-pyrgyrite and other silver sulphosalts, arsenopyrite, gudmundite, etc.

16.2.4 Identification

Its reflectance, pleochroism, anisotropism and paragenesis are characteristic.

It can be confused with stibnite, which it resembles.

- However, the following properties of berthierite differ from sbt: its pinkish-brown colour (reminiscent of pyrrhotite) in one of the positions (//axis a, while for b and c the colours are very similar), its greater hardness and more brittle behaviour, its anisotropism with greenish tones and without the pinkish tones of stibnite.
- On the other hand, stibnite is whiter, somewhat lighter, more fibrous and more easily deformed, with frequent pressure twins that are not so common in berthierite.

(a)

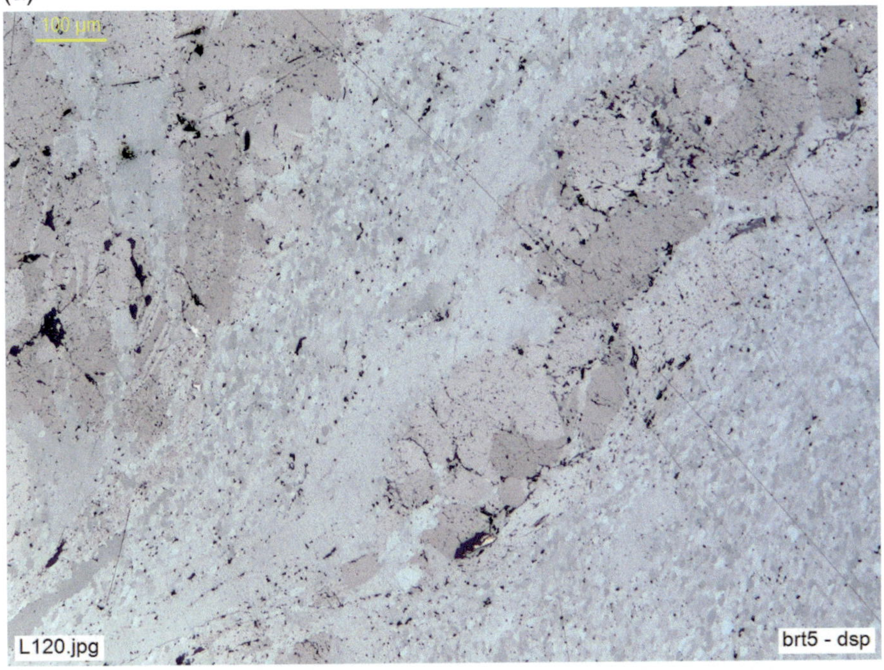

(b)

Fig. 16.2 **a**, **b** (**dsp** and **d+p**, resp) Ore of brt and sbt (with traces of interstitial q, sp and mc) deformed in differentiated bands, of more ductile behaviour (sbt, white–grey, shear microbands, very fine-grained, with incipient recrystallization) or more brittle (brt, white–grey–brown, microclast fragments or aggregates). TF Pn12, Almuradiel, Ciudad Real

16.3 Spectral Information

16.3.1 Berthierite Reflectance in the Visible Range (in Air) (%)

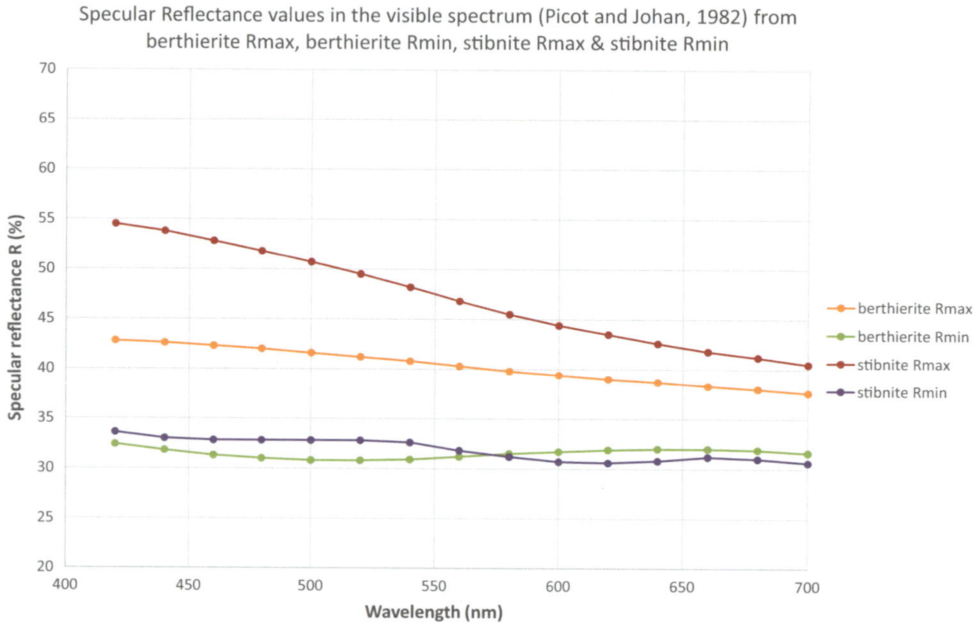

Reference

Picot P, Johan Z (1982) Atlas of ore minerals. B.R.G.M./Elsevier

17 Bismuth (*Bi*)

Bi: Trigonal

Bismuth		R_{546} (%)	Y (%)	λ_d (nm)	P_e (%)	Colour sensation
Air	R_1	59.8	59.9	577	8.2	Bright white or yellowish-W to light pinkish-grey
	R_2	67.2	67.4	578	6.6	
Oil	R_1	47.8	48.1	578	10.1	
	R_2	55.7	56.2	582	6.9	
Bir: Weak, barely perceptible (3)				*Pleoc*: Weak (oil) (3)		*Tarn*: Yellowish to reddish
Anis +P: Distinct, green tones (4)				*IR*: No		*Twg*: Frequent. Typical polysynthetic
Mor: Fine intergrowths				*Clv*: //(0001) occasional		*PolQ*: Easy, but often scratched
Hardness	*PolH*: Low: >sbt, <bm (2)			*Mohs*: 2	*Talm*: A+	*VHN* $_{(100\ g)}$: 16–18

Obs: Avoid heating when preparing the sample because of possible transformation at 75 °C
Spectral information: cf. Table and R plots, multispectral data (VNIR, AMCO) and visible range (QDF3)
Mineral abbreviations: Bi-native bismuth, bm-bismuthinite, dy-dyscrasite, lo-löllingite, py-pyrite, rm-rammelsbergite, sbt-stibnite (antimonite), sf-safflorite, sk-skutterudite

17.1 General

Composition: Bi practically pure, with occasional traces of As or Te.Ú

17.2 Microscopic Description

17.2.1 SP Observation (*Single Polariser*)

Very bright white *colour*, *in* spite of the almost inevitable polishing scratches and pits. After polishing, it soon *tarnishes* and gradually acquires yellowish to reddish tones.

High *reflectance* (>As, but <Sb), particularly noticeable in immersion. Weak *bireflectance* and *pleochroism* hardly visible in air, but noticeable in immersion (white or yellowish-white, greyish and pinkish tones, Fig. 17.3).

Low *polishing hardness* (it is the softest of all the minerals that usually accompany it, including bismuthinite). It polishes reasonably well, but due to low hardness it scratches a lot, especially if it is accompanied by hard minerals (Fig. 17.1a, b).

It occurs intergrown with different minerals, such as sulfides, sulfosalts and particularly Ni-Co-Bi-Ag minerals of the so-called *five element formation*, sometimes developing typical skeletal, dendritic and myrmektitic textures, occasionally rimmed by safflorite or löllingite crusts (Figs. 17.2 and 17.3). Also as a late mineral, disseminated, interstitial or in microcracks.

17.2.2 Observation +P (*Polariser + Analyser at 90°*)

Distinctly *anisotropic*, though in dull tones, with characteristic green colours (Figs. 17.1c, d, 17.4 and 17.5). Incomplete extinction, as usually happens with native metals, accentuated by the numerous polishing scratches.

Almost constant *twinning*, dominated by polysynthetic or lamellar lanceolate twins in parquetry-patterns or in various directions. Also penetration, hourglass (Fig. 17.5), sectored, and deformation twins.

R. Castroviejo, *A Practical Guide to Ore Microscopy—Volume 1*,
https://doi.org/10.1007/978-3-031-12654-3_17

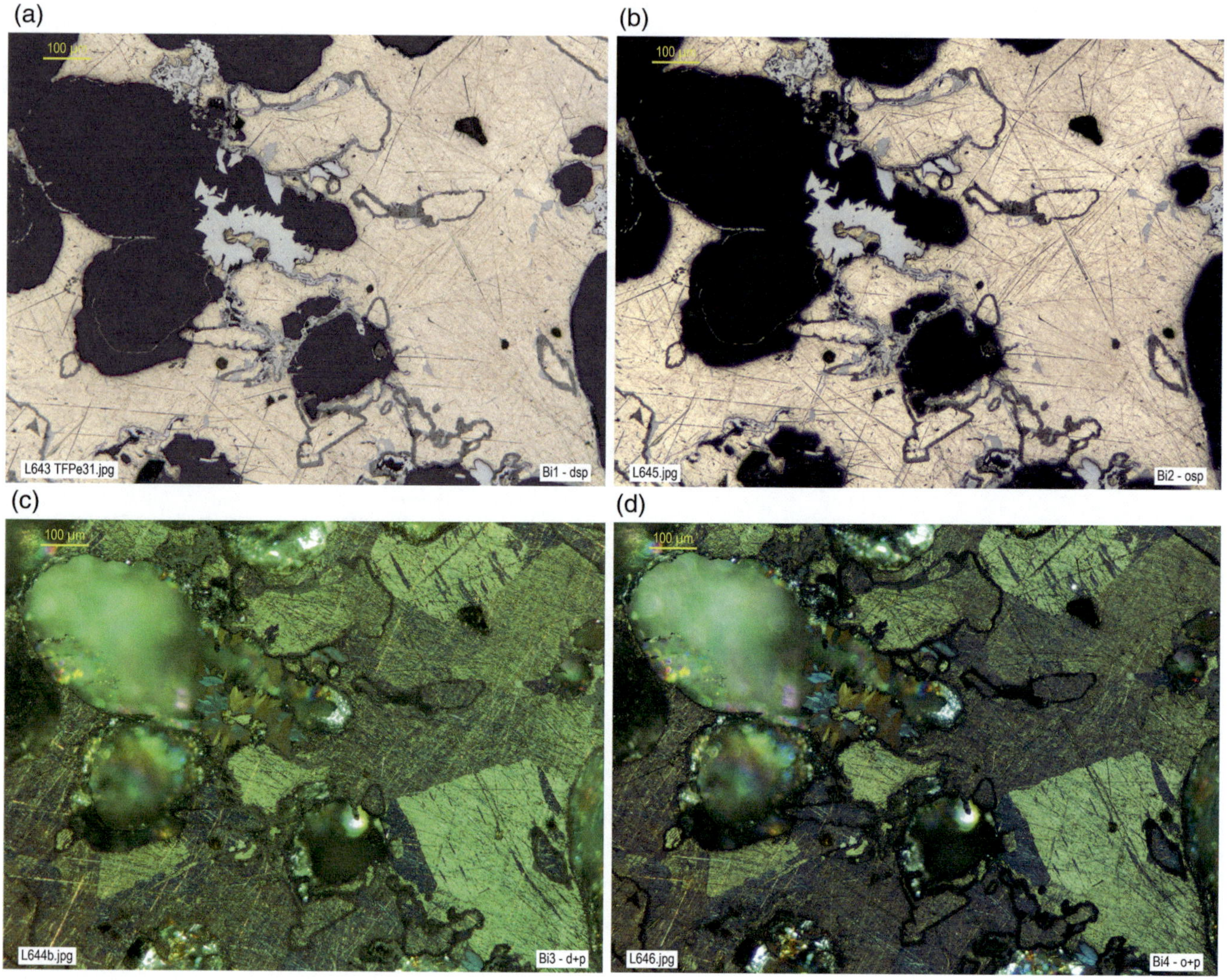

Fig. 17.1 **a**, **b** (Same scene: **dsp** and **osp** , *respectively*). Aggregate of native Bi (yellowish white, scratched and with black pits due to polishing), occasionally zoned and with thin rims of bm (grey), gangue (dark grey) and sf (light grey); *corona* or *cockade* on Bi grain (towards the center) of euhedral tabular sf crystals in radial disposition around Bi core. TF Pe31: Schneeberg, Saxen, Germany. **c**, **d** (Same scene as Fig. 17.1a: **d+p** and **o+p** , *respectively*). Note the dull anisotropism but also the brighter green tones of Bi, in contrast with the lively bluish and orange colours of sf

17.2.3 Paragenesis

Paragenesis relatively varied. It appears in pegmatitic-pneumatolytic and hydrothermal deposits, either of magmatic or metamorphogenic derivation but not in those of the sedimentary cycle. Its presence is typical, particularly in hydrothermal deposits of the Ni–Co–Bi–Ag–U type (*five-element formation*).

Despite its scarcity, it is the most ubiquitous of the native metals. It can appear accompanied by very diverse minerals, such as Co–Ni arsenides (safflorite, skutterudite, etc.), löllingite, bismuthinite, wolframite, scheelite, molybdenite, galena, pyrite.

17.2.4 Identification

Bi can look at first like Ag, Sb or dyscrasite, but it is easily identified by its:

- colour and luster (high reflectance but <Sb and Ag), rapid tarnishing
- low hardness (PolH lower than these ores and lower than all its likely paragenetic ores)
- abundant scratches, including sometimes polishing pits (Fig. 17.1a, b)
- distinct anisotropism with characteristic green tints; frequent and typical twinning.

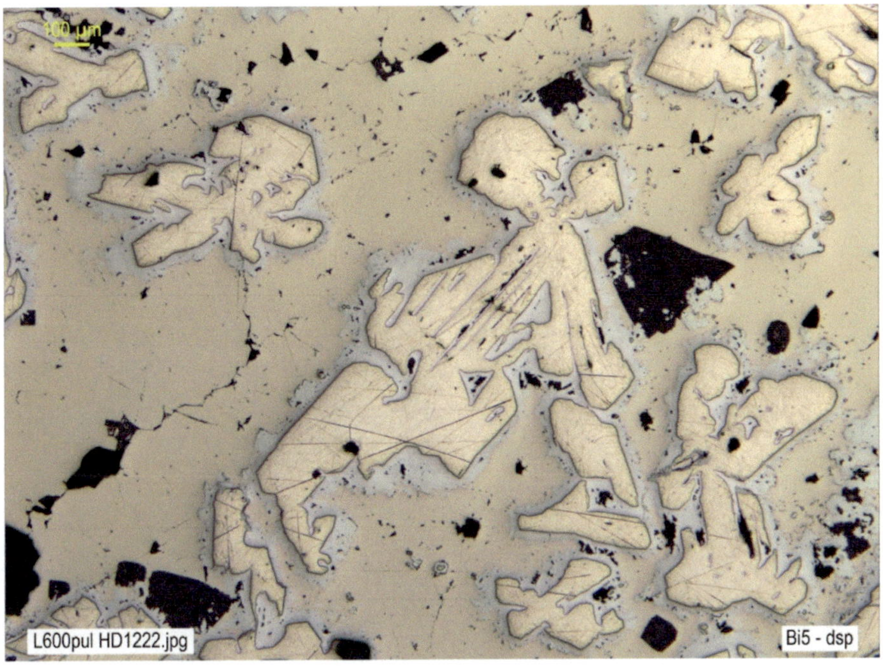

Fig. 17.2 *Skeletal* aggregates of Bi (yellowish white, scratched) with *corona* of rm, sf, lo, sk (finely intergrown, as thin grey crusts rimming Bi), all included in py mass. The perception of the colours is conditioned by the high reflectance of the adjoining Bi, against which the other species appear dull or darker; however, the slightly brownish tone of py may be influenced by its As content (1%, EMP analysis). HD 1222: Eisleben, Germany (**dsp**)

Fig. 17.3 Prismatic-radial aggregate of lo (light grey, slightly pleochroic as shown by the variation in its shades of grey, well perceptible in immersion) with relic skeletal Bi inclusions and scarce minute aggregates of bm (dark grey). The Bi inclusions (bright white, faintly bireflectant and slightly pinkish by tarnishing) show irregular, corroded borders. Rub1070: Hüttenberger Erzrevier, Lölling, Schacht Hauptlager, Kärnten, Austria (**osp**)

Fig. 17.4 Same scene as Fig. 17.3 (**o+p**). The polarization tints of Bi seem to be dulled by the strong anisotropism of lo, whose radial/palmated texture is clearly evidenced

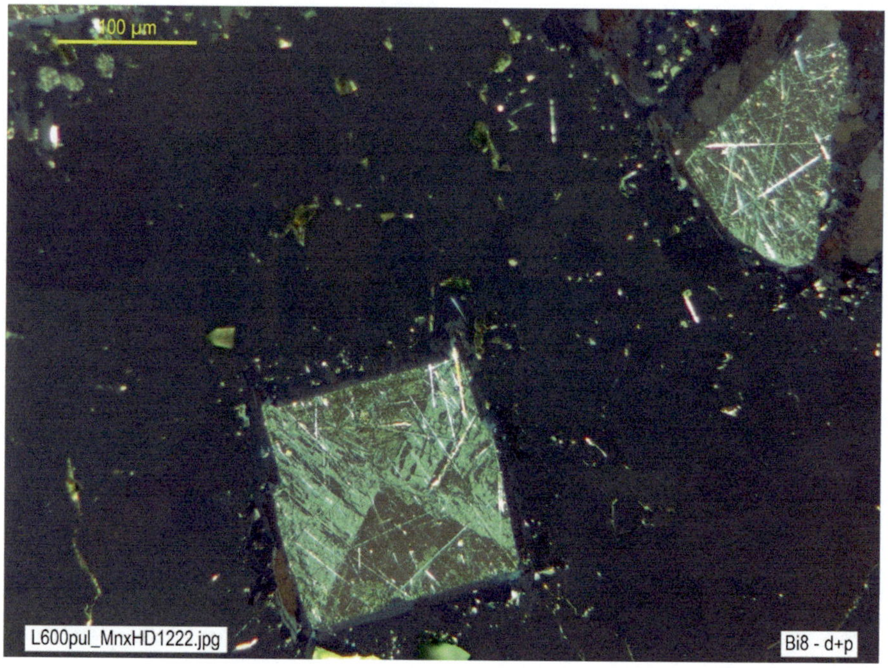

Fig. 17.5 Twins and greenish anisotropism colours of native Bi, disseminated in py (dark background). Hourglass twin and tiny subordinate lamellar twins on euhedral Bi grain of square section, lower center. HD 1222: Eisleben, Germany Mine (**d+p**)

17.3 Spectral Information

17.3.1 Bismuth: Multi-spectral Reflectance R (in Air, Without Polariser) (%)

λ (nm)	370	400	425	450	475	500	525	550	575	600	625	650	675	700	750	800	850	900	950	1000
μ	48.1	51.4	52.8	54.6	55.5	57.4	59.0	61.0	62.8	64.5	65.7	66.8	67.6	68.1	68.7	68.4	68.5	69.0	69.1	69.2
P_5	42.8	46.2	47.8	49.5	50.7	52.8	54.6	56.7	58.7	60.5	61.8	63.0	63.8	64.4	64.9	64.6	64.8	64.8	65.0	64.5
P_{20}	45.6	49.0	50.5	52.3	53.3	55.3	56.9	59.0	60.9	62.7	64.0	65.2	66.0	66.5	67.0	66.8	67.1	67.3	67.6	67.3
P_{80}	50.8	54.0	55.3	57.2	58.0	59.9	61.4	63.4	65.1	66.8	68.0	69.0	69.8	70.1	70.7	70.3	70.3	71.1	71.0	71.4
P_{95}	53.5	56.5	57.5	59.5	60.0	61.9	63.2	65.2	66.8	68.3	69.5	70.5	71.3	71.5	72.0	71.5	71.5	72.3	72.2	72.5

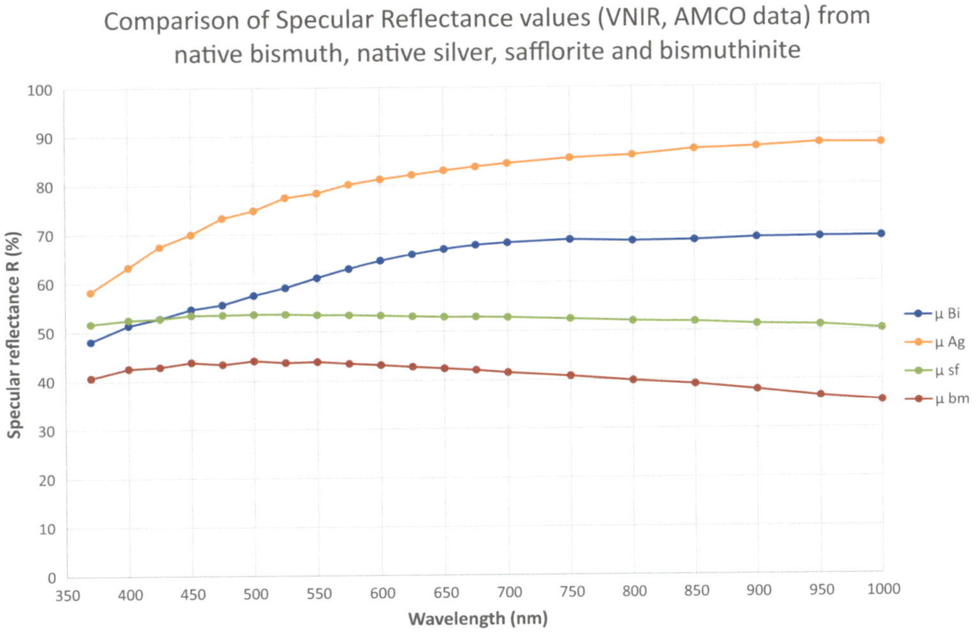

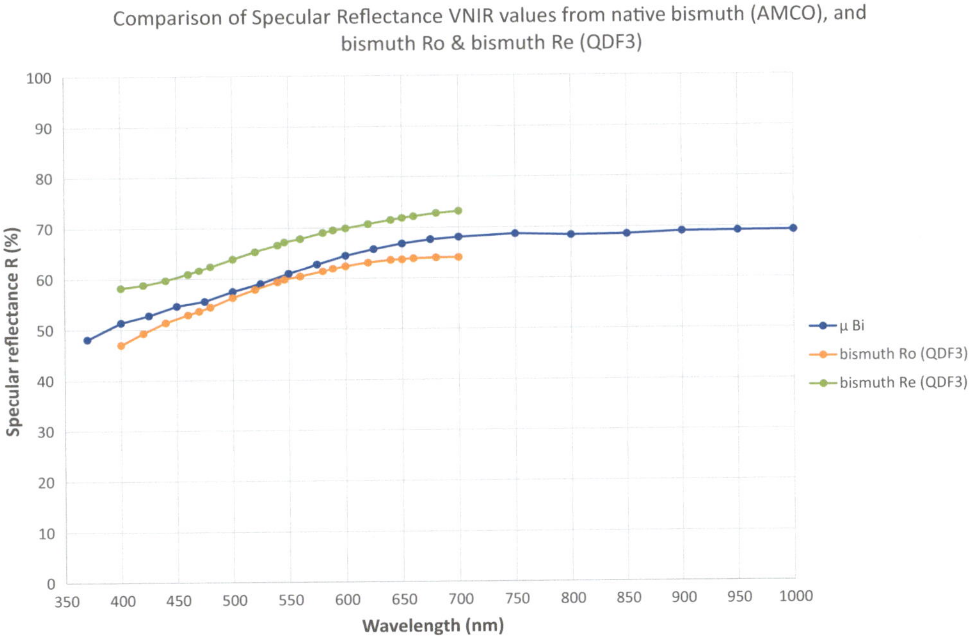

Bismuthinite (*bm*)

18

Bi_2S_3: Orthorhombic

Bismuthinite			R546 (%)	Y (%)	λ_d (nm)	P_e (%)	Colour sensation
Air	R_a		43.8	43.7	482	1.0	White-YlWh to light grey
	R_b		37.1	37.0	488	0.9	
	R_b		49.0	48.6	507	0.6	
Oil	R_a		29.2	29.0	481	2.6	
	R_b		22.2	22.1	483	2.8	
	R_c		33.6	33.2	510	0.9	
Bir: Medium (4)					*Pleoc*: Moderate (3)		*Tarn*: No
Anis +P: Distinct, greenish tones (4)					*IR.*: No		*Twg*: Occasional
Mor: Prismatic, acicular					*Clv*: (010)		*PolQ*: Good
Hardness		*PolH*: Low, ≈ gn, cc, dg (2)			*Mohs*: 2	*Talm*: B+	*VHN* $_{(50\ g)}$: 68–190

Obs: cf. Table and R charts, multispectral data (VNIR, AMCO) and visible range (QDF3); VHN: Uytenbogaardt and Burke (1971)
Mineral abbreviations: asp—arsenopyrite, Au—native gold, Bi—native bismuth, bm—bismuthinite, bn—bornite, cc—chalcocite, ccp—chalcopyrite, cin—cinnabar, dg—digenite, gg—gangue, gn—galena, mol—molybdenite, mt—magnetite, po—pyrrhotite, py—pyrite, q—quartz, Sb—native antimony, sbt—stibnite (antimonite), td—tetrahedrite

18.1 General

Composition: Bi_2S_3. Possible Sb contents, *bismuthinite-stibnite series* (Bi_2S_3–Sb_2S_3) and *horobetsuite* variety (intermediate compositions between Sb-rich bismuthinite and Bi-rich stibnite). There are also compositional variations in the *aikinite-bismuthinite series* ($CuPbBiS_3$- Bi_2S_3). Other impurities such as Fe, As, Te, exceptionally with notable S-Se substitution.

18.2 Microscopic Description

18.2.1 SP Observation (*Single Polariser*)

White *colour*, with variable shades depending on the environment, slightly yellowish against galena, greyish white or very faintly bluish grey against chalcopyrite or native bismuth (Fig. 18.1a, b).

Relatively high *reflectance*, close to that of galena (cf comparative multispectral plots) and higher than tetrahedrite or stibnite but distinctly lower than native bismuth. *Bireflectance* and *pleochroism* (yellowish white to greyish white), moderate but noticeable.

Polishes well. Occasionally, in fibrous ores, visible (010) cleavage. Low *polishing hardness* (as galena, chalcocite and digenite, in the low-medium limit) > cin, Sb, Bi, sbt; < bn, mol.

As an early mineral it shows an idiomorphic tendency, with prismatic or acicular habit (Fig. 18.1a, b). It also occurs, as a secondary mineral, in allotriomorphic aggregates.

18.2.2 Observation +P (*Polariser + Analyser at 90°*)

Clearly anisotropic (Fig. 18.1c, d), not as anisotropic as stibnite but with characteristic and striking bright green shades (brownish in weakly anisotropic varieties, greyish in

R. Castroviejo, *A Practical Guide to Ore Microscopy—Volume 1*,
https://doi.org/10.1007/978-3-031-12654-3_18

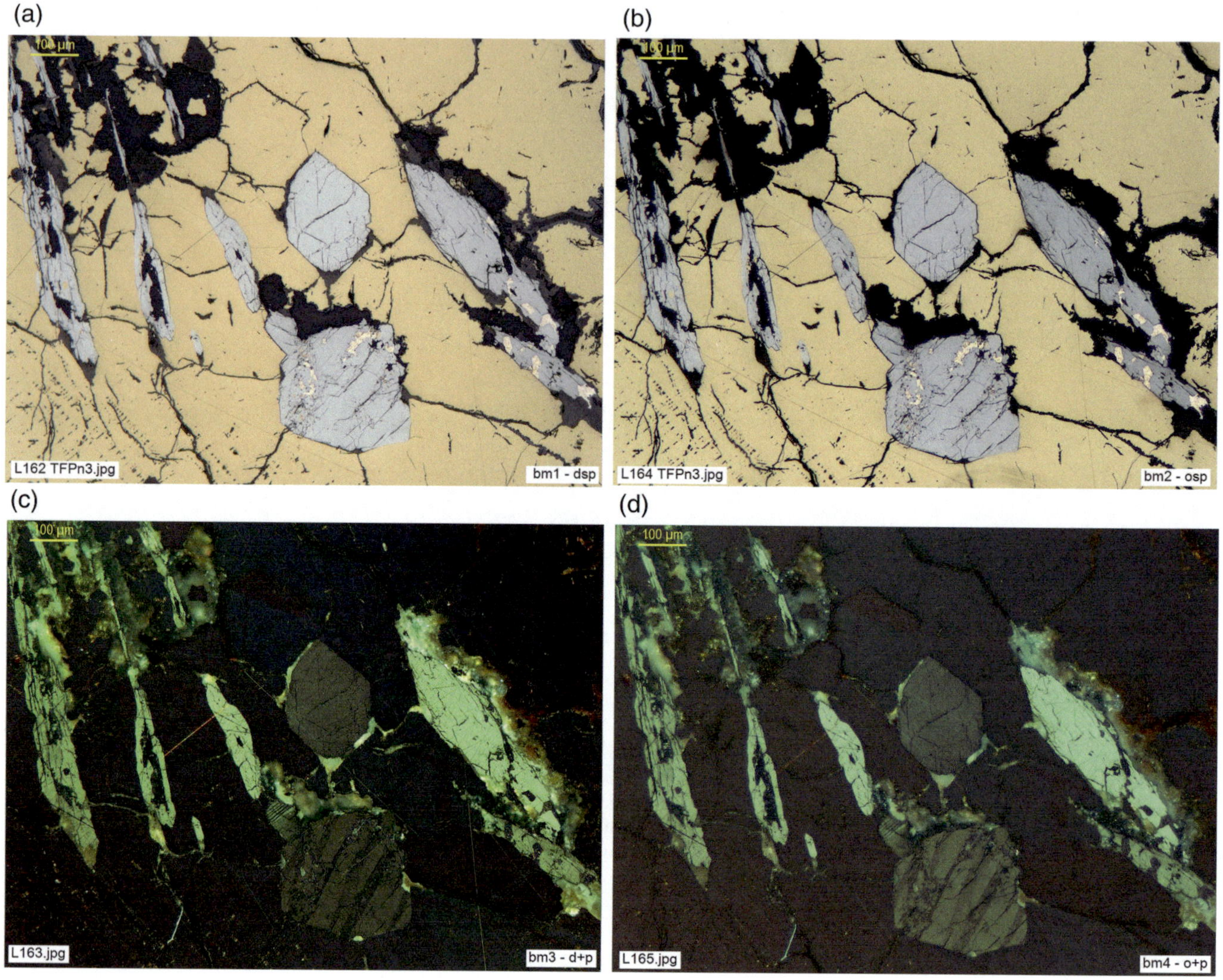

Fig. 18.1 **a**, **b** (Same scene: **dsp** and **osp**, *respectively*) Massive ccp ore (yellow) with idiomorphic inclusions of bm (light grey) which, in turn, include Bi (white, bright) and q gg. TF-Pn3, Lozoyuela, Madrid. **c**, **d** Same scene as Fig. 18.1a (**d+p** and **o+p**, *respectively*), anisotropism and twinning of bm, ccp in extinction position

diagonal position). Extinction straight. Occasional polysynthetic deformation *twinning*, as stibnite but much less frequent in bm.

18.2.3 Paragenesis

Varied hydrothermal *parageneses*. Typical in hydrothermal formations of the *five-element* class (Ni–Co–Bi–Ag–U). In volcanogenic formations, particularly in the roots (*feeder* or stockwork) of VMS deposits, etc. In high temperature concentrations (Sn, W, Au ores), bm may occur mostly as a late mineral, undesirable for the miner because it interferes with the metallurgical process.

Accompanying ores: native bismuth and Bi sulphosalts, cassiterite, wolframite, native gold, Co–Ni ores, arsenopyrite, pyrrhotite, magnetite.

18.2.4 Identification

It is recognized by its colour and pleochroism, its relatively high reflectance, its striking anisotropism, with generally luminous green tones, as well as its paragenesis and the company of native bismuth or other bismuth minerals. It is more reflective, less pleochroic and less anisotropic than stibnite.

18.3 Spectral Information

18.3.1 Bismutinite Multispectral Reflectance R (in Air, Without Polariser) (%)

λ(nm)	370	400	425	450	475	500	525	550	575	600	625	650	675	700	750	800	850	900	950	1000
μ	40.7	42.6	42.9	43.8	43.4	44.1	43.7	43.9	43.5	43.1	42.8	42.4	42.1	41.5	40.8	39.8	39.1	37.8	36.5	35.6
P_5	36.0	38.3	39.4	40.4	40.5	40.9	41.1	41.2	40.9	40.7	40.3	40.0	39.7	39.1	38.3	37.3	36.5	35.3	34.0	33.3
P_{20}	37.2	39.5	40.6	41.5	41.6	41.9	42.0	42.1	41.8	41.5	41.1	40.7	40.4	39.8	39.0	37.9	37.0	35.9	34.6	33.8
P_{80}	43.5	45.2	45.0	46.0	45.2	46.2	45.4	45.6	45.1	44.7	44.3	43.9	43.6	43.0	42.3	41.4	40.7	39.5	38.0	37.0
P_{95}	44.0	45.6	45.6	46.4	45.7	46.6	45.9	46.1	45.5	45.1	44.7	44.3	43.9	43.4	42.6	41.6	40.9	39.7	38.3	37.2

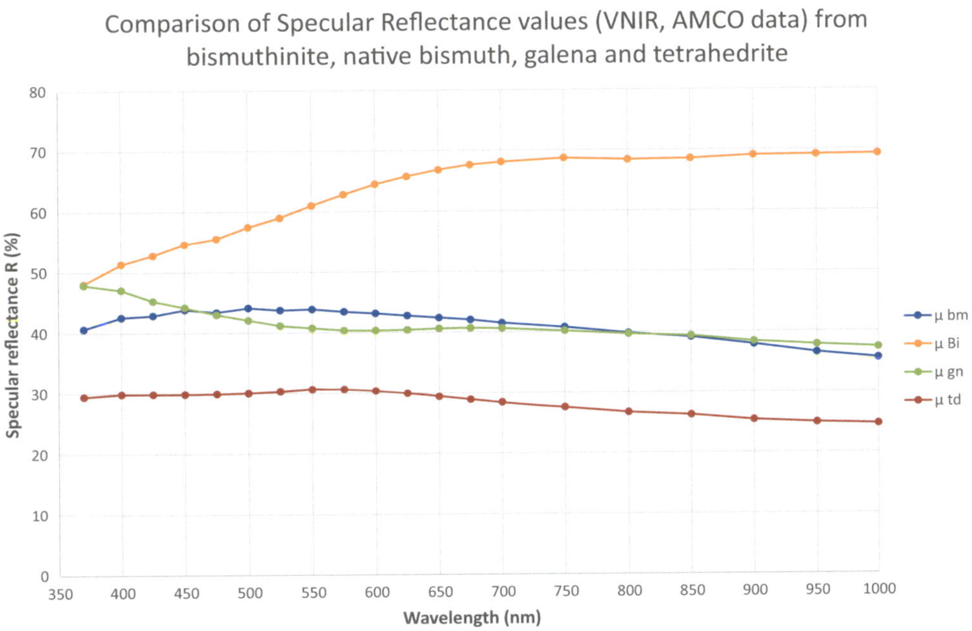

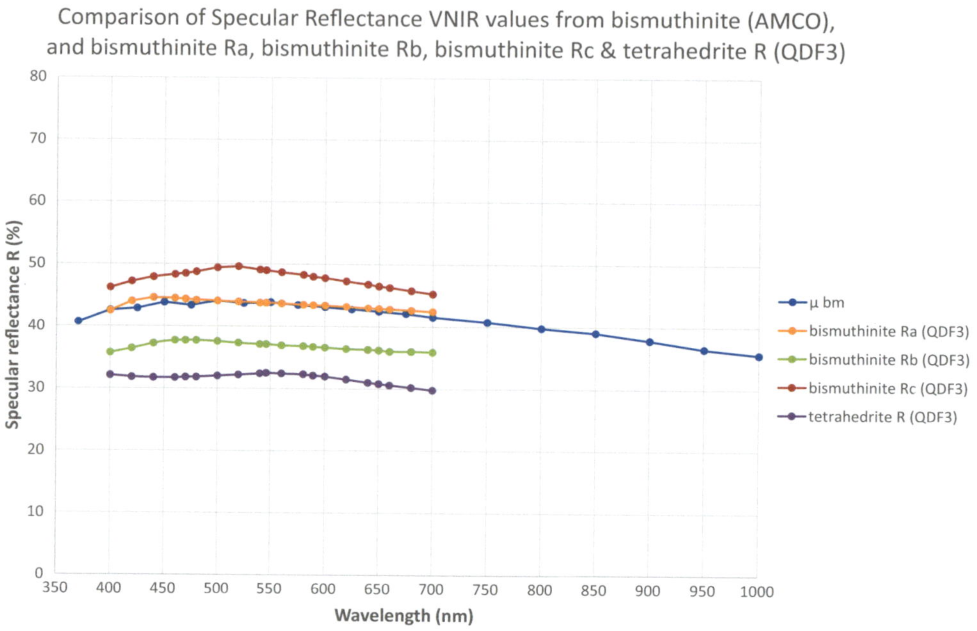

Reference

Uytenbogaardt W, Burke EAJ (1971) Tables for microscopic identification of ore minerals, 2nd edn. Elsevier, Amsterdam

Bornite (*bn*) 19

Cu_5FeS_4: Orthorhombic (*Pseudotetragonal/Pseudocubic*)

Bornite		R′$_{546}$ (%)	Y (%)	λ_d (nm)	P_e (%)	Colour sensation
Air	*R*	21.7	22.8	586	15.9	Brownish-pink color
Oil	*R*	10.4	11.3	584	22.6	Intense brownish-pink colour
Bir: Very weak (2)				*Pleoc*: Very weak (2)		*Tarn*: Fast, practically unavoidable
Anis + P: Weak, sometimes net (3)				*IR*: No		*Twg*: Lamellar, lanceolate, etc
Mor: Anhedral aggregates				*Clv*: Occasional (111)		*PolQ*: Easily good
Hardness	*PolH*: Med: > gn, < ccp (3)			*Mohs*: 3	*Talm*: D$^+$	*VHN* $_{(100\ g)}$: Vickers Microhardness$_{(weight)}$

Spectral information: cf. Table and R plots, multispectral data (VNIR, AMCO) and visible range (QDF3)
Mineral abbreviations: Ag—native silver, Au—native gold, ccp—chalcopyrite, cc—calcocite, cv—covellite, gn—galena, po—pyrrhotite, py—pyrite, sp—sphalerite, stn—stannite, td—tetrahedrite

19.1 General

Although the formula Cu_5FeS_4 is accepted, in nature there are remarkable compositional variations, some of which characterize defined varieties such as the *argentiferous bornite*, $(Cu,Ag)_5FeS_4$, an important source of Ag, or the Fe-deficient varieties described in the Polish Kupferschiefer deposits (Kucha et al. 1981), *half-bornite* (50% Fe deficit) and *quatr-bornite* (75% Fe deficit). Possible contents of Pb, Bi, Ge, In have also been mentioned.

The structure of bornite is temperature dependent, becoming cubic above 190 °C (Frueh 1950, in: Ramdohr 1980). During cooling, modifications such as transformation twins and exsolutions (chalcopyrite, chalcocite) may occur. Substitutions are also frequent (chalcopyrite, chalcocite, digenite, pyrite).

19.2 Microscopic Description

19.2.1 SP Observation (*Single Polariser*)

Light pinkish-brown *color*, similar to pyrrhotite (Figs. 19.1a, b and 19.2), but it darkens very quickly by surface alteration, presenting a dark *patina* of varied colors (reddish-brown to violet or blue-violet: Fig. 19.3), tarnishing sometimes in less than an hour. It should be noted that color and reflectance, as always, depend not only on the minerals in the immediate environment but can also vary considerably with composition, and, of course, with the time elapsed after polishing (tarnishing), as shown by the reflectance data (graph 1: bn-T).

Moderate *reflectance* in fresh sections changes with time due to tarnishing, as evidenced by the multispectral comparative graphs (at the end of the description), when comparing the curve of bornite (fresh) with bornite_T (T = *Tarnished*).

Weakly anisotropic (*pseudocubic*) but *pleochroism* may occasionally be seen.

Cleavage occasionally visible, with two possible orientations, // (100) and (111).

Medium–low *polishing hardness*, $PolH_{bn}$ > gn, but < ccp, Ag, Au, stn, td and << py.

It usually occurs in anhedral aggregates with rounded grains or soft contacts. However, reaction edges or indented contacts with other sulfides are also frequent, due to disequilibrium and replacement (Figs. 19.4, 19.5 and 19.7).

R. Castroviejo, *A Practical Guide to Ore Microscopy—Volume 1*,
https://doi.org/10.1007/978-3-031-12654-3_19

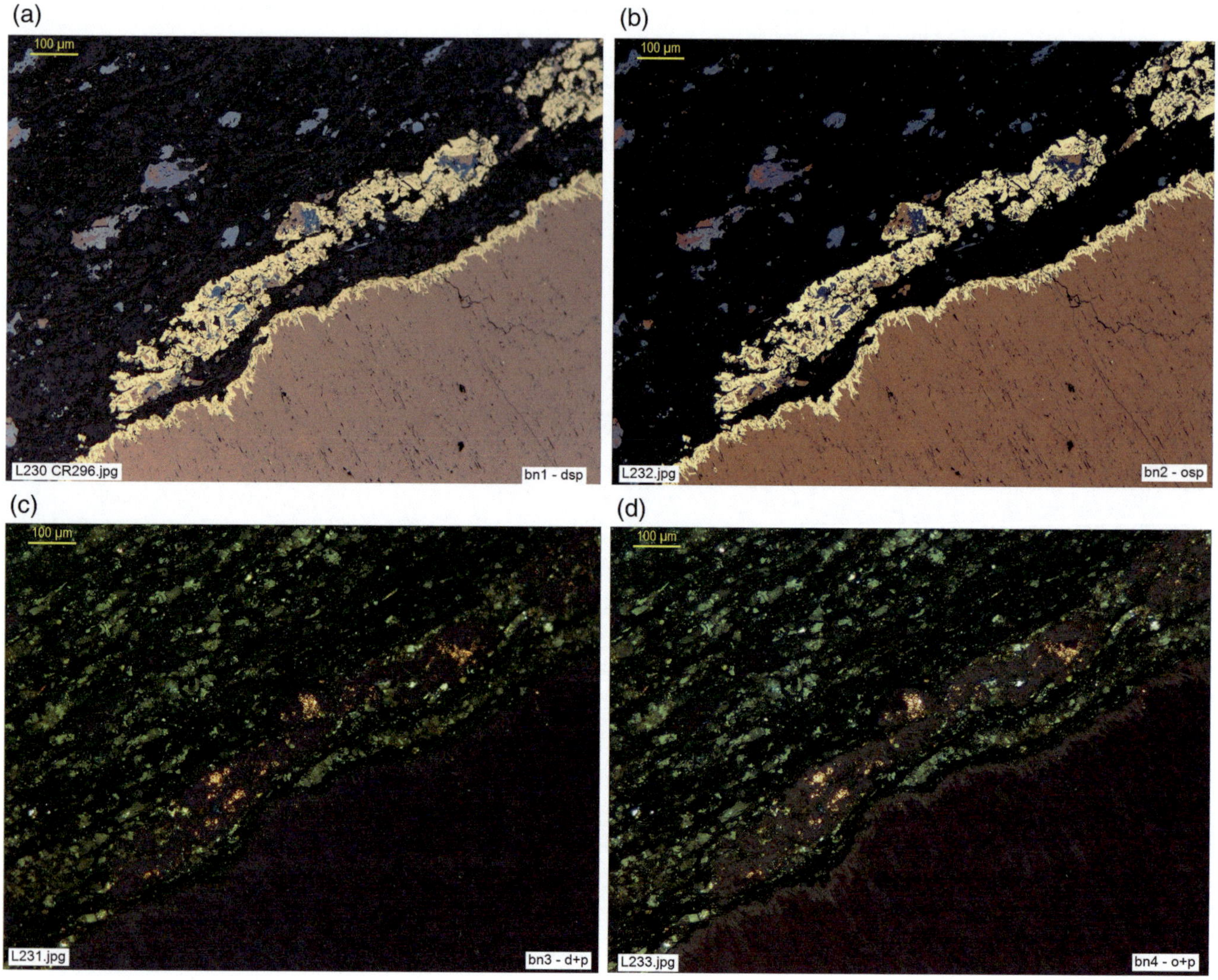

Fig. 19.1 **a** Banded ore of bn (brownish-pink), with marginal ccp (bright yellow). Intergrown, microlenticular bn, ccp, dg (bluish grey) and scanty cc (white) and cv (deep blue), scattered in the schist. CR296, Kupferschiefer, Poland (**dsp**). **b** Same scene as bn1 (**osp**). **c**, **d** Same scene as bn1 (**d+p**) and (**o+p**), respectively. Very weak anisotropism of bn, perceptible in immersion, and very strong anisotropism of cv, with very bright yellow tones

19.2.2 Observation +P (*Polariser + Analyser at 90°*)

Variable *anisotropism*. Normally weak to very weak (Fig. 19.1c, d), visible, although not very coloured when the analyser is partially uncrossed (Fig. 19.6), showing the grain contours. However, it is sometimes relatively strong and coloured (greyish-brown to violet-brown tones), net even with perfectly crossed analyser (90° Fig. 19.8).

Frequent lamellar *twinning*, with polysynthetic and lanceolate twins of various laws, well visible in the more anisotropic sections (Fig. 19.8).

Fig. 19.2 Banded texture of bn, ccp, dg (cc, cv) ore, *Kupferschiefer* type (same sample as F bn1, wide field objective, **dsp**)

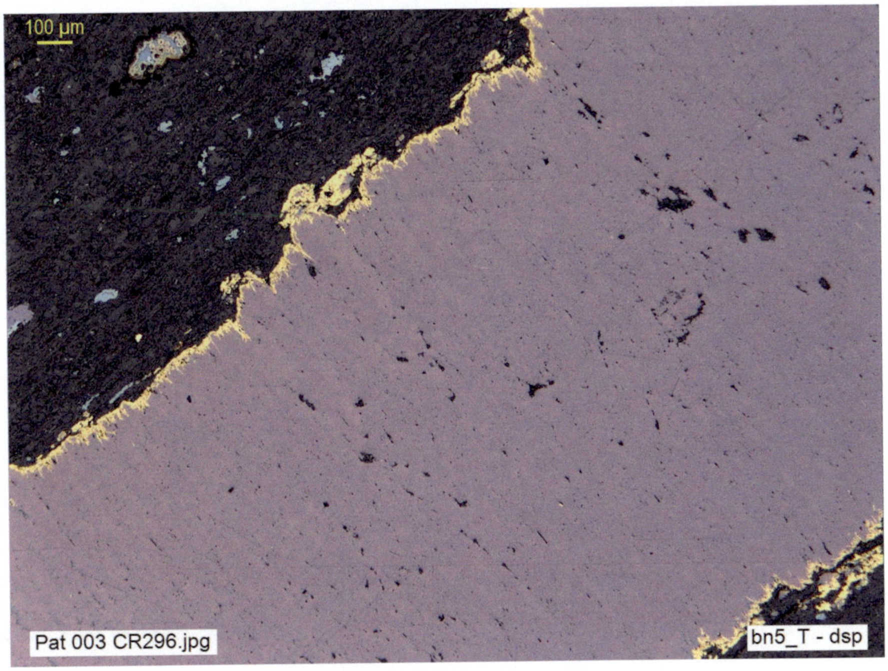

Fig. 19.3 Detail of Fig. 19.2, showing color and R change of bn by tarnishing, two months after polishing (**dsp**). Compare R spectra (bn and bn_T) in graph of multispectral measurements

19.2.3 Paragenesis

Very varied *paragenesis*, hypogene and supergene. Deposits of the magmatic series, e. g., pegmatitic, pneumatolitic and hydrothermal formations of different types, copper porphyries, basic rocks. Metamorphic concentrations. Stratabound concentrations, Kupferschiefer type, and supergene, descending. Associated with chalcopyrite, digenite, covellite, chalcocite, tetrahedrite (*s.l.*), sphalerite, pyrite, among others.

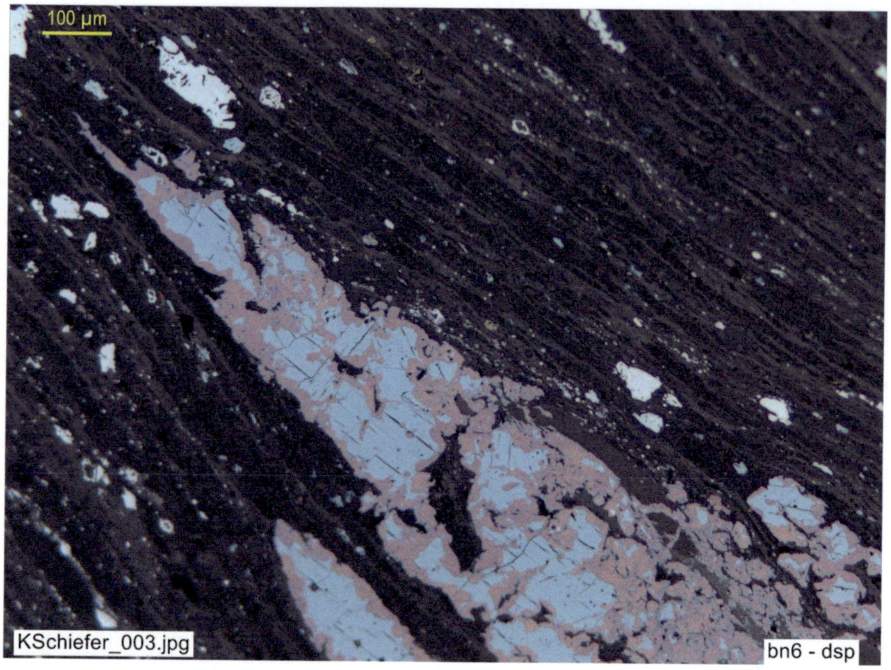

Fig. 19.4 Microlenticular intergrowth of bn and dg and dissemination of cc, dg, bn; dg replaces cc; and bn, dg. *Kupferschiefer-003*, Rudna-Lubin, Poland (**dsp**)

Fig. 19.5 Replacement of bn by cc, with relics (*phantoms*) of zoned cubic py. Sample 26hjb, Tsumeb, Namibia (**dsp**)

19.2.4 Identification

Typical is its pink color, which darkens quickly, showing a purplish patina. Its twinning and weak anisotropism, as well as its low hardness, distinguish it from pyrrhotite. Luzonite is strongly anisotropic and less frequent (paragenesis with As–Sb). Germanite is isotropic but more reflective and much rarer than bornite (cf. renierite).

Fig. 19.6 Same scene as bn7 (**dxp**: *Air, polarizer + analyzer at ≈87°)*. By slightly uncrossing the polarizers, the anisotropism of cc and bn and the lamellar or lanceolate twins of bn are clearly seen

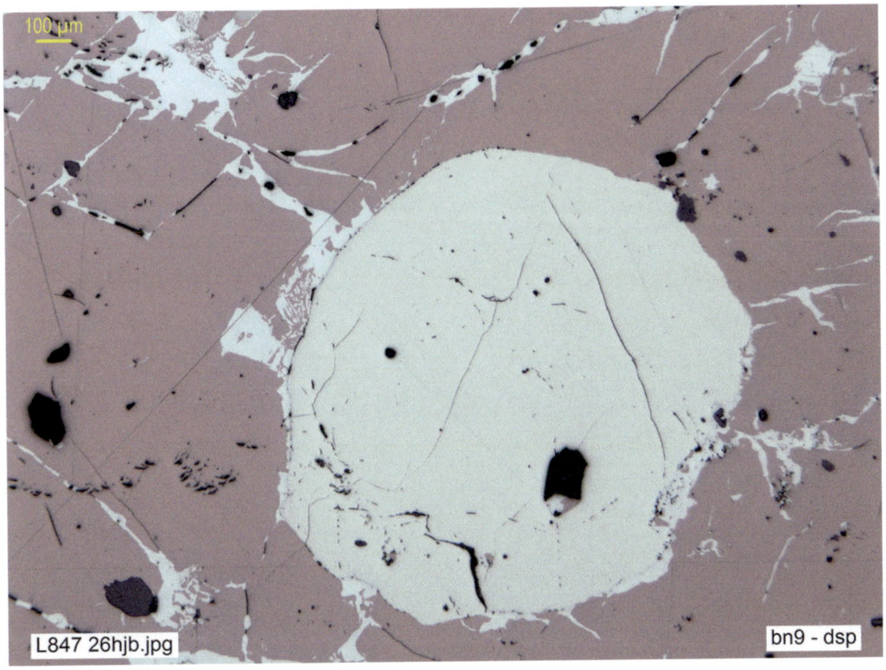

Fig. 19.7 Massive bn ore (partially replaced by cc, enhancing microcracks and lattice planes), with early td inclusion (large rounded grey crystal in the centre), slightly corroded by bn and cc. Sample 26hjb, Tsumeb, Namibia (**dsp**)

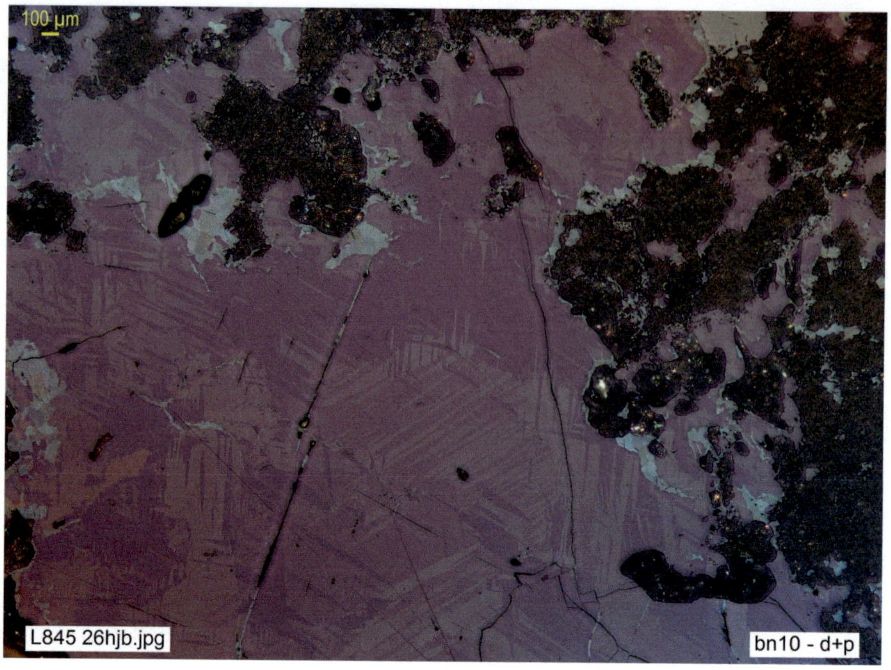

Fig. 19.8 Clearly anisotropic bn aggregate, with polysynthetic and orthogonal twins, sparse secondary cc. Sample 26hjb, Tsumeb, Namibia (**d+p**)

19.3 Spectral Information

19.3.1 Bornite Multi-spectral Reflectance R (in Air, Without Polarizer) (%)

λ (nm)	370	400	425	450	475	500	525	550	575	600	625	650	675	700	750	800	850	900	950	1000
μ	22.2	21.2	19.4	17.8	17.3	17.1	17.7	18.7	20.0	21.6	23.0	24.6	26.1	27.6	30.2	32.3	33.9	35.0	35.5	36.1
P_5	19.7	19.2	17.7	16.3	15.6	14.8	14.8	15.4	16.4	17.7	18.9	20.5	22.0	23.5	26.2	28.3	30.2	31.9	32.9	33.8
P_{20}	20.4	19.7	18.2	17.0	16.4	16.0	16.2	16.9	18.0	19.5	20.8	22.5	24.0	25.6	28.4	30.5	32.2	33.6	34.5	35.4
P_{80}	23.6	22.3	20.3	18.5	18.1	18.2	19.2	20.5	22.0	23.7	25.1	26.7	28.2	29.7	32.2	34.2	35.6	36.5	36.5	36.8
P_{95}	24.3	22.9	20.8	18.9	18.4	18.6	19.8	21.1	22.6	24.3	25.7	27.4	28.9	30.3	32.7	34.6	36.0	36.8	36.8	37.1

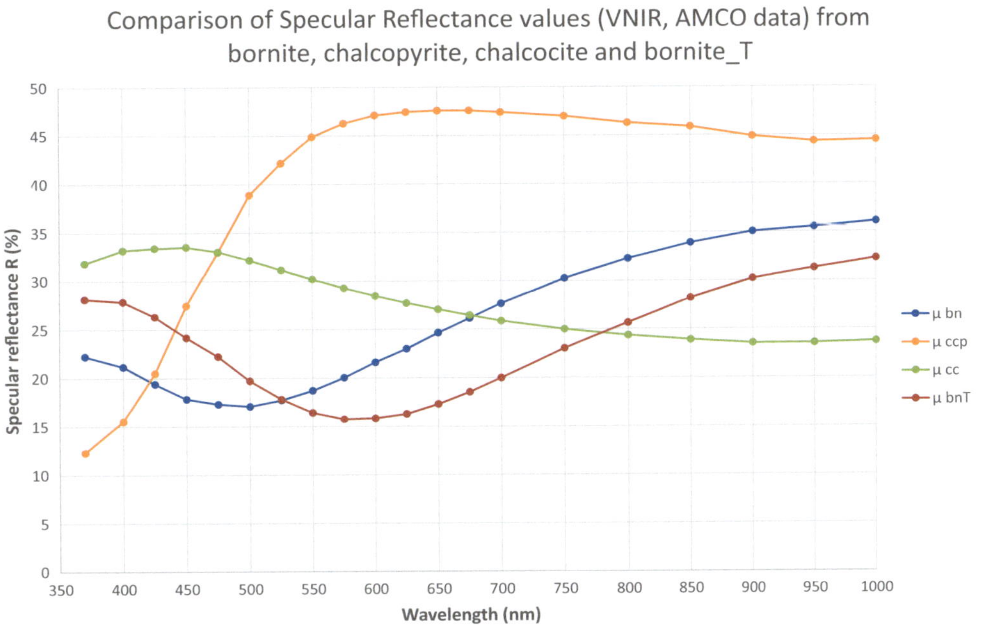

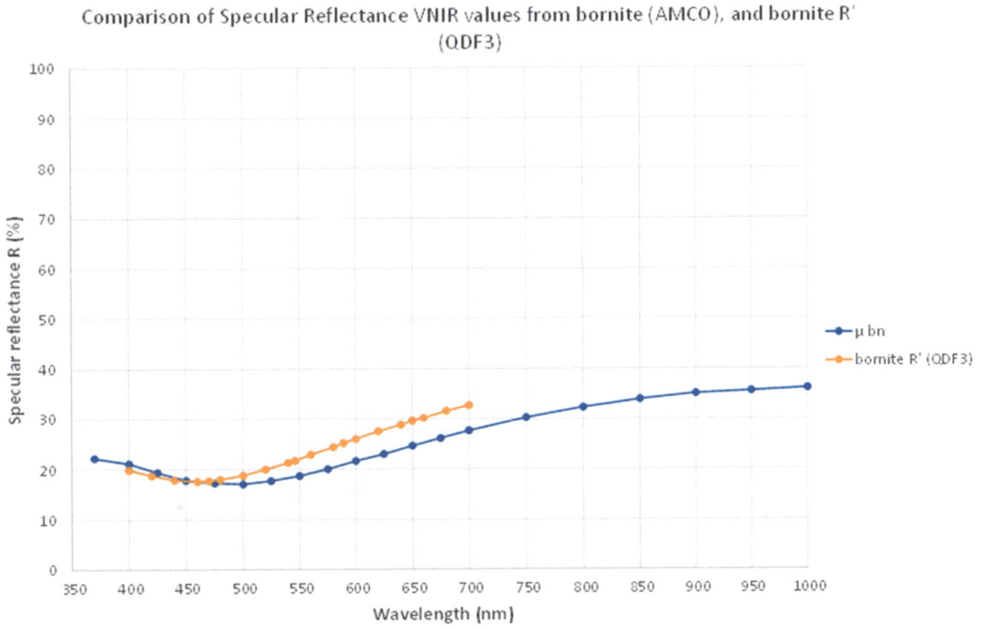

References

Kucha H, Mayer W, Piestrzyński A, Wieczorek A (1981) The replacement of rutile by chalcocite in the Zechstein copper ores of the Fore-Sudetic Monocline. Mineral Pol 12(1):69–76

Ramdohr P (1980) The ore minerals and their intergrowths, 2nd edn. Pergamon Press, Oxford. 2 volumes, 1205p

20 Boulangerite (*bl*)

$Pb_5Sb_4S_{11}$: Monoclinic

Boulangerite		R_{546} (%)	Y (%)	λ_d (nm)	P_e (%)	Colour sensation
Air	R_1	37.4	37.2	481	3.3	Greyish-white, slightly pleochroic
	R_2	41.8	41.5	484	3.1	
Oil	R_1	23.0	22.9	480	6.4	
	R_2	27.6	27.2	482	6.4	
Bir: Medium (4)				*Pleoc*: Perceptible (4)		*Tarn*: No
Anis + P: Distinct, greenish tints (4)				*IR*: Scarce, red		*Twg*: No
Mor: Acicular, fibrous				*Clv*: Visible (// 100)		*PolQ*: Good
Hardness	*PolH*: Low (2)			*Mohs*: 3	*Talm*: B +	*VHN* $_{(50\ g)}$: 116–182

Obs: cf. Table and R charts, multispectral data (VNIR, AMCO) and visible range (QDF3). VHN (Uytenbogaardt and Burke 1971)
Mineral abbreviations: bl—boulangerite, bnn—bournonite, gg—gangue, gn—galena, flk—falkmanite, jm—jamesonite, po—pyrrhotite, py—pyrite, q—quartz, sbt—stibnite (antimonite), sp—sphalerite, td—tetrahedrite

20.1 General

Composition: $Pb_5Sb_4S_{11}$, with possible traces of Fe, Zn, Sn, Cu.

20.2 Microscopic Description

20.2.1 SP Observation (*Single Polariser*)

White to greyish-white *colour*, with a slightly greenish or bluish tone (comp. Figs. 20.1a, b and 20.2a), *pleochroic*, which distinguishes it from galena.

Good *polish. Polishing hardness* low, slightly < gn and slightly > cin.

Relatively high *reflectance*, slightly < gn and slightly > bnn (between galena and tetrahedrite, between Rmax and Rmin of stibnite, cf multispectral comparison plots).

It typically occurs in crystals with acicular or fibrous habit and cleavage parallel to elongation; also with laminar habit and in aggregates.

20.2.2 Observation +P (*Polariser + Analyser at 90°*)

Clearly *anisotropic* but poorly coloured, in grey tones, although greenish shades are often discernible (Figs. 20.1c, d, 20.2b). Extinction //.

Reddish *internal reflections*, scarce, visible in oil. Not twinned or zoned.

20.2.3 Paragenesis

Typical of hydrothermal *paragenesis* in varied environments, although generally as a late mineral, accompanied by galena, other Sb–Pb sulphosalts, sphalerite, pyrite, pyrrhotite, etc.

20.2.4 Identification

Boulangerite, **bl**, is a relatively frequent mineral, although it is not very abundant. Possibly it has not been cited in many cases because of inadvertence (it resembles galena, which is

R. Castroviejo, *A Practical Guide to Ore Microscopy—Volume 1*,
https://doi.org/10.1007/978-3-031-12654-3_20

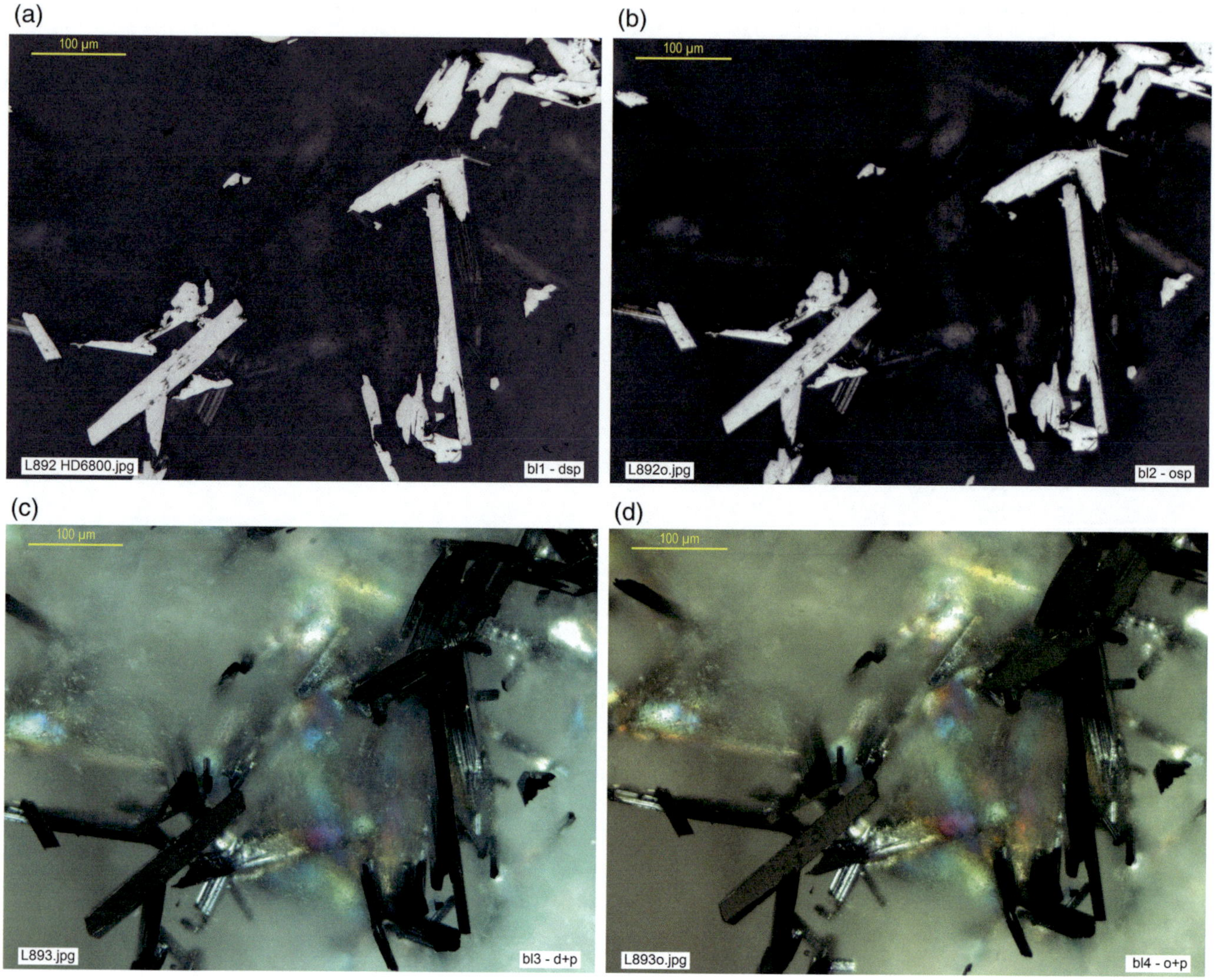

Fig. 20.1 **a**, **b** (Same scene: **dsp** and **osp**, *respectively*). Dissemination of acicular bl in q. Cleavage parallel to the prism axis, visible as striations on faces not affected by polishing, perceived through the q gangue, especially in Fig. 20.1c, d. HD 6800, Ramsbeck, Westphalia, Germany. **c**, **d** Same scene as Fig. 20.1 (**d+p** and **o+p**, *respectively*). The typical anisotropism (moderate, with greenish tones of bl) is, in this case, somewhat obscured by the effect of the ubiquitous IR of q but is well perceived in Fig. 20.2a

much more common) or because of confusion with other sulfosalts such as bournonite, jamesonite or falkmanite.

It is easy to avoid confusion with *galena*, recalling the pleochroism, the anisotropism, and the IR of bl, as well as its lower R, compared to gn, which is a cubic, isotropic mineral.

Although its appearance is similar to that of *stibnite*, sbt has higher reflectance and bireflectance and stronger anisotropism than bl. The absence of twins in boulangerite helps to distinguish it from *jamesonite*, which is very similar, and from *bournonite* (less reflective and not as strongly anisotropic as boulangerite).

(a)

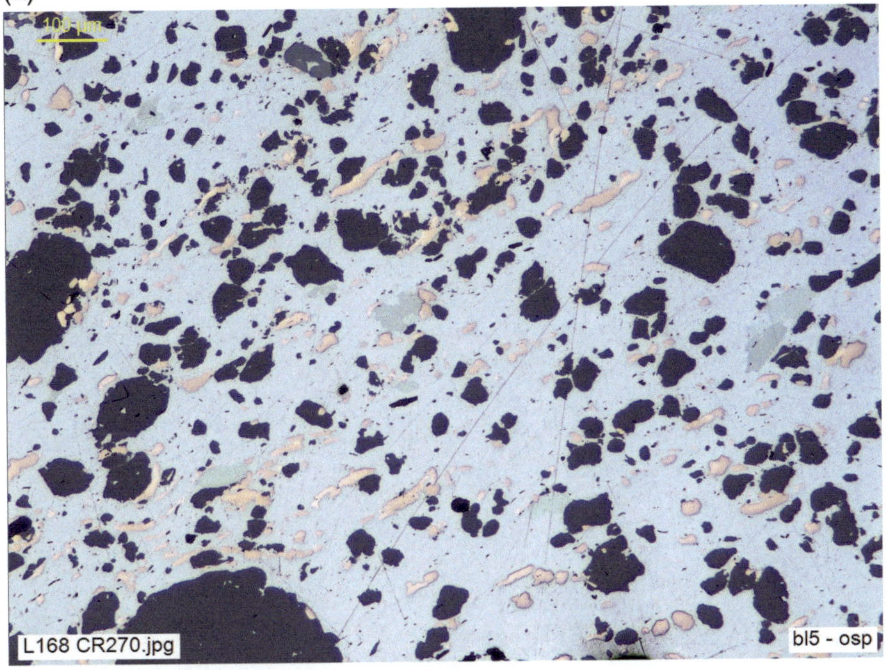

(b)

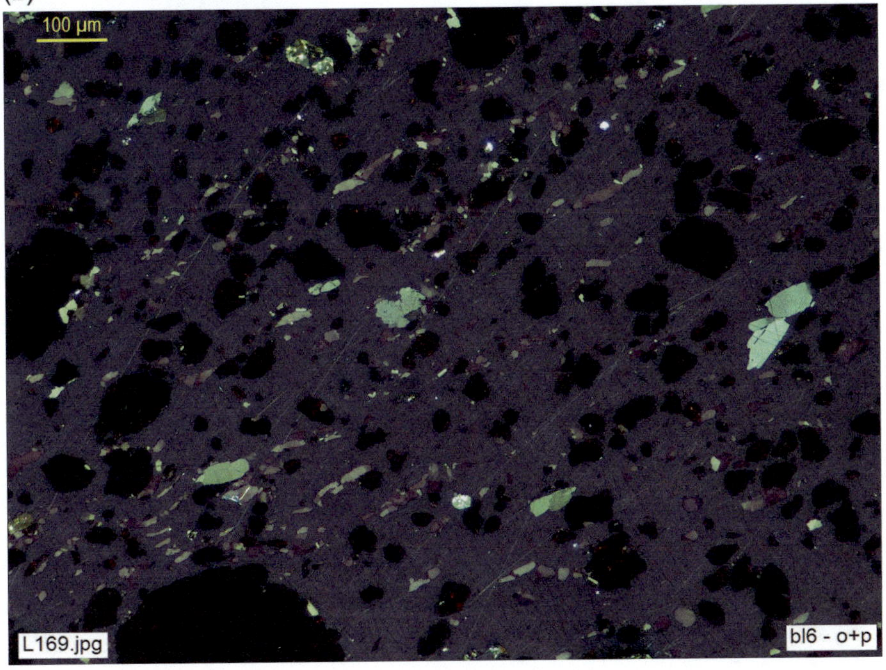

Fig. 20.2 **a**, **b** (Same scene: **dsp** and **osp**, *respectively*). Massive ore of gn (bluish-white), with inclusions of sp (very dark grey), po (pinkish-brown) and bl (similar to gn, but somewhat less reflective and with light grey to greenish-white pleochroism; bl does not show its typical acicular habit here, but its identity has been confirmed with EMPA). With (+p, Fig. 20.2b), gn and sp (cubic) are in extinction. In sp some IR are seen, despite the small size of the inclusions; po and bl are anisotropic but bl's anisotropism is more lively and greenish. CR 270, Sullivan Mine, BC, Canada

Falkmanite, ***flk***, a disputed mineral, $Pb_3Sb_2S_6$, or a variety of bl according to Picot and Johan (1982), provisionally enjoys the status of *grandfathered* species, credited as such by IMA until an accepted solution to its crystal structure is found to take a decision.[1] Under the RL microscope flk occurs in thicker crystals, not showing the fibrous habit of boulangerite and is less vividly anisotropic than bl.

[1] Further information: https://www.mindat.org/min-1446.html, *consult. 2021/11/16.*

20.3 Spectral Information

20.3.1 Boulangerite Multispectral Reflectance R (in Air, Without Polariser) (%)

λ (nm)	370	400	425	450	475	500	525	550	575	600	625	650	675	700	750	800	850	900	950	1000
μ	41.3	41.6	41.1	40.7	40.4	40.1	39.7	39.5	38.9	38.3	37.6	36.8	36.1	35.4	34.3	33.1	32.7	32.7	32.8	32.5
P_5	39.3	39.7	39.3	38.8	38.6	38.2	37.9	37.7	37.2	36.6	36.0	35.3	34.7	34.1	33.0	31.9	31.5	30.8	30.2	29.8
P_{20}	40.0	40.4	40.0	39.5	39.3	39.0	38.6	38.4	37.9	37.3	36.6	35.8	35.2	34.6	33.5	32.4	32.0	32.2	31.6	31.2
P_{80}	43.1	43.0	42.6	41.8	41.8	41.1	41.0	40.6	40.0	39.3	38.5	37.7	37.0	36.2	34.9	33.9	33.4	33.7	34.3	34.0
P_{95}	43.7	43.5	43.1	42.4	42.3	41.7	41.5	41.1	40.5	39.7	38.9	38.0	37.3	36.5	35.3	34.1	33.6	34.3	35.3	35.1

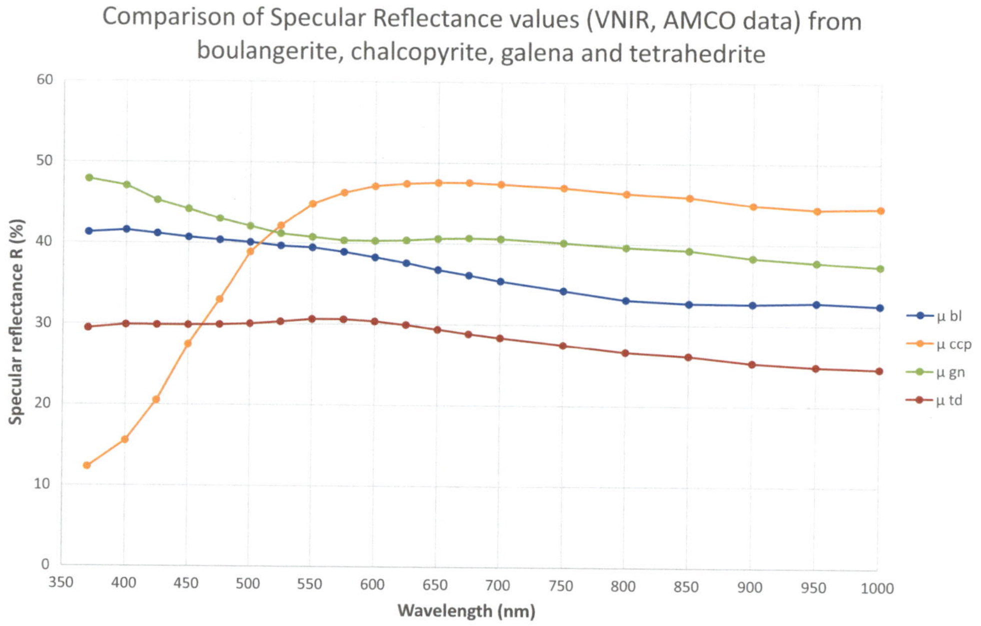

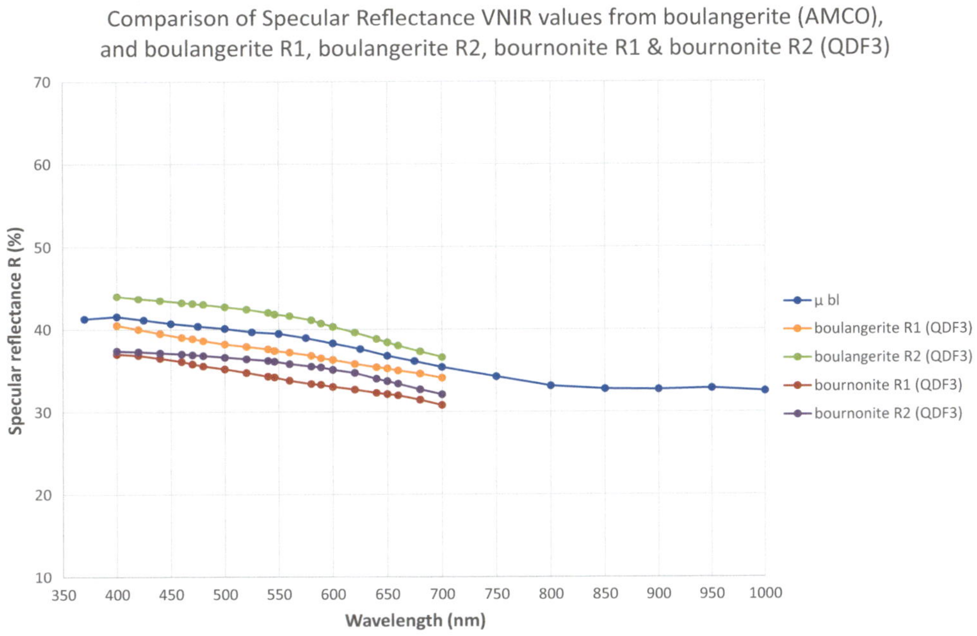

References

Picot P, Johan Z (1982) Atlas of Ore Minerals. B.R.G.M./Elsevier.

Uytenbogaardt W, Burke EAJ (1971) Tables for microscopic identification of ore minerals, 2nd edn. Elsevier, Amsterdam

21 Bournonite (*bnn*)

$PbCuSbS_3$: Orthorhombic

Bournonite		R_{546} (%)	Y (%)	λ_d (nm)	P_e(%)	Colour sensation
Air	R_1	34.2	34.0	480	3.8	Light grey, slightly greenish
	R_2	36.1	35.8	484	2.3	
Oil	R_1	18.9	18.7	479	6.6	
	R_2	20.5	20.3	483	4.2	
Bir: Very weak (2)				*Pleoc*: Very weak (2)		*Tarn*: No
Anis +P: Weak (3)				*IR*: Only exceptional		*Twg*: Polysynthetic *parquetry* typical
Mor: Prismatic. Inclusions in gn alotriom				*Clv*: Not visible		*PolQ*: Very good, little scratched
Hardness	*PolH*: Medium (3)			*Mohs*: 3	*Talm*: C	*VHN* $_{(100\ g)}$: 163–194

Obs: cf. Table and R charts, multispectral data (VNIR, AMCO) and visible range (QDF3)
Mineral abbreviations: bl–boulangerite, bn—bornite, bnn—bournonite, ccp—chalcopyrite, flb—freieslebenite, gg—gangue, gn—galena, py—pyrite, sbt—stibnite (antimonite), sp—sphalerite, td—tetrahedrite

21.1 General

Composition: $PbCuSbS_3$, occasionally with variable contents of As replacing Sb, forms a series with seligmannite ($PbCuAsS_3$). Other possible impurities: Fe, Ag, Zn, Mn, Ni.

21.2 Microscopic Description

21.2.1 SP Observation (*Single Polariser*)

Colour white to light grey, with greenish to olive tones, very weak *pleochroism*. Medium *reflectance* (Fig. 21.1a, b), R slightly > td but clearly < gn (cf. multispectral comparison plots).

Very good *polish*, less scratched than galena. Medium *polishing hardness*, between that of galena and chalcopyrite (slightly > gn; < bn, ccp).

It forms well-developed crystals, idiomorphic, with prismatic or tabular habit, if it grows in cavities or fractures (open spaces that facilitate crystallization). Otherwise, it occurs intergrown in aggregates or in small allotriomorphic inclusions.

21.2.2 Observation +P (*Polariser + Analyser at 90°*)

Weakly *anisotropic*, with dull greenish to bluish or purplish colours.

Only exceptionally (twice) have *internal reflections been* observed (Ramdohr 1980).

Characteristic is the *polysynthetic twinning in parquetry* (two series of practically orthogonal polysynthetic twins Fig. 21.1c, d). Although they are generally growth twins, they sometimes appear deformed and can also be generated as pressure twins, as has been experimentally demonstrated.

21.2.3 Paragenesis

It is found as an accessory mineral, typically in intermediate temperature *hydrothermal paragenesis*, especially in Pb–Zn deposits. Frequently intergrown with galena or tetrahedrite.

R. Castroviejo, *A Practical Guide to Ore Microscopy—Volume 1*,
https://doi.org/10.1007/978-3-031-12654-3_21

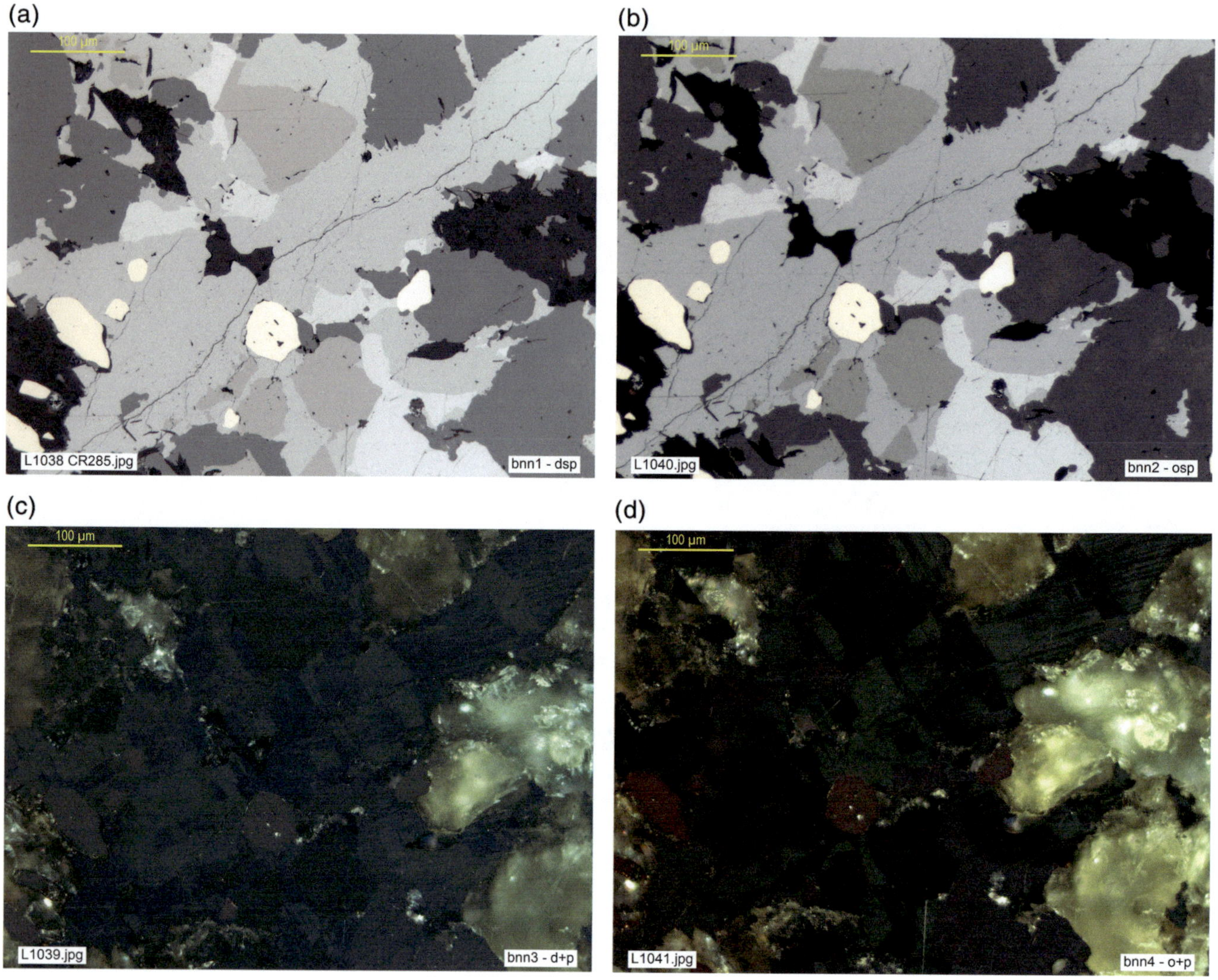

Fig. 21.1 **a**, **b** (Same scene, **dsp** and **osp**, respectively) Vein filling of Pb–Zn ore: bnn, sp, gn, td, py, gg. Sequence of increasing reflectances, gg (dark grey, black in oil immersion Fig. 21.1b), sp (grey, dark grey in oil immersion Fig. 21.1b), td, bnn, gn, py. CR285, Linares (Jaén). **c**, **d** Same scene as bnn1 (**d+p** and **o+p**, *respectively*), bnn shows weak anisotropism and polysynthetic twins in two orthogonal orientations. The other sulfides are in extinction (cubic), but sp is illuminated by its numerous colourless IR (as befits its composition, pure ZnS, confirmed EPMA

21.2.4 Identification

Because of its unremarkable appearance and properties, it is easy to overlook bnn in a quick survey. However, its lamellar twins *in parquet* provide a very useful criterion. In addition:

- it is not difficult to avoid confusion with their frequent associates, *galena* and *tetrahedrite* (both isotropic and not twinned). Bournonite is distinguished from galena by being harder and less reflective, and from tetrahedrite by being softer and more reflective;
- it is distinguished from *freieslebenite* by the more vivid and coloured anisotropism of the latter, which can also show polysynthetic twins but not in parquetry;
- confusion with *boulangerite* is avoided, considering that boulangerite is not twinned and also shows an acicular habit not found in bournonite;
- there are also properties that clearly differentiate it from *stibnite* (*antimonite*). Stibnite is much softer and has much stronger bireflectance, pleochroism and anisotropism than bournonite.

21.3 Spectral Information

21.3.1 Bournonite Multispectral Reflectance R (in Air, Without Polariser) (%)

λ (nm)	370	400	425	450	475	500	525	550	575	600	625	650	675	700	750	800	850	900	950	1000
μ	37.2	37.0	36.8	36.6	36.3	36.0	35.6	35.2	34.8	34.3	33.7	33.0	32.4	31.8	30.7	29.8	28.9	28.6	28.3	28.4
P_5	36.0	36.0	36.0	36.0	35.6	35.2	34.8	34.4	34.0	33.6	33.0	32.4	31.8	31.2	30.0	29.2	28.0	28.0	27.6	27.6
P_{20}	36.4	36.4	36.4	36.4	36.0	35.6	35.2	34.8	34.4	34.0	33.4	32.8	32.0	31.2	30.4	29.6	28.4	28.0	28.0	28.0
P_{80}	37.4	37.2	37.0	36.8	36.6	36.4	36.0	35.6	35.2	34.8	34.2	33.6	32.8	32.0	31.2	30.0	29.2	28.8	28.8	28.8
P_{95}	38.2	38.0	37.8	37.6	37.2	36.8	36.4	36.0	35.6	35.2	34.4	33.6	33.0	32.4	31.6	30.4	29.6	29.2	29.2	29.2

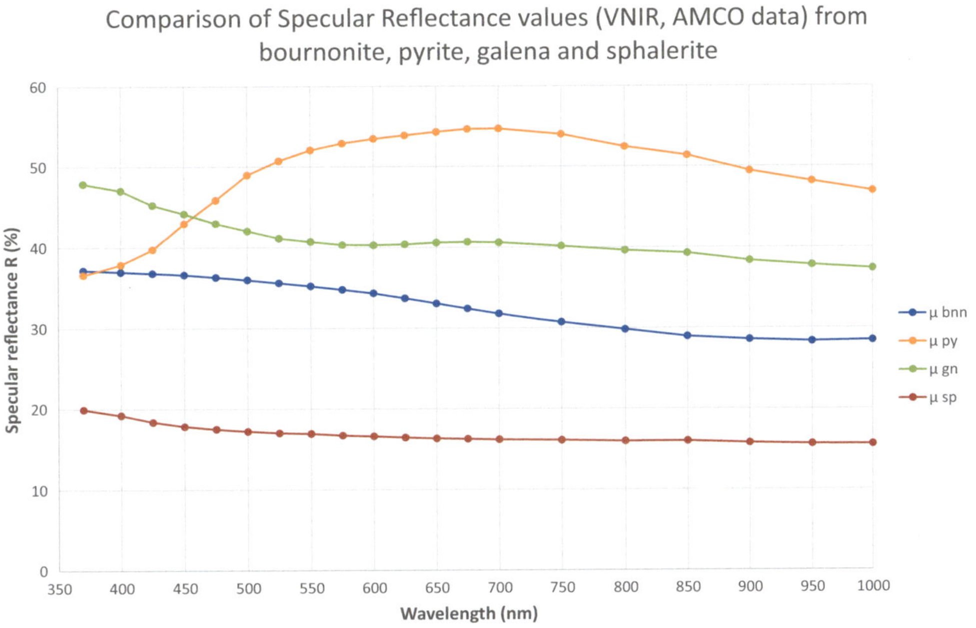

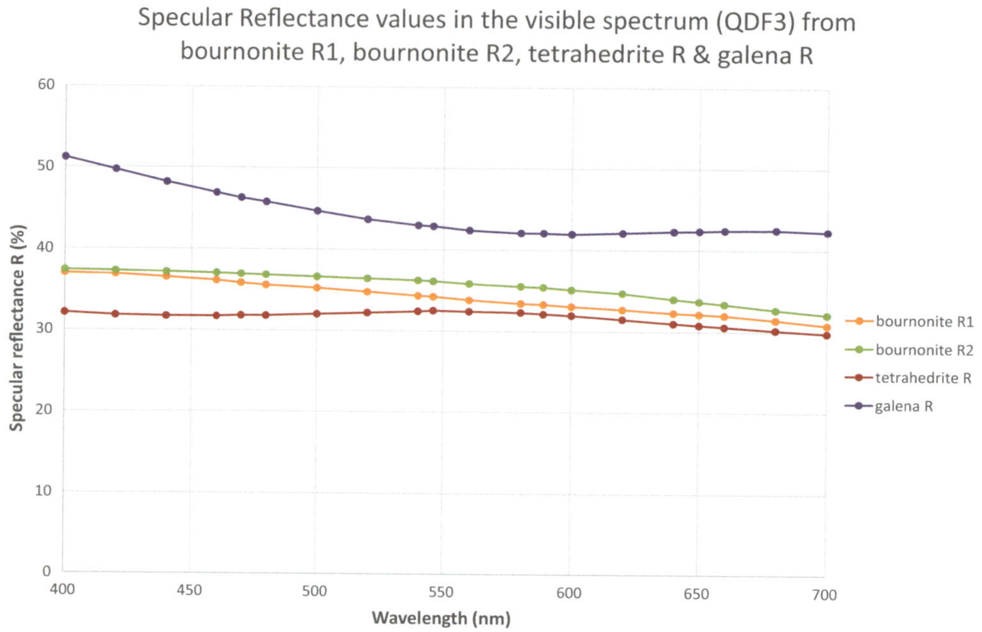

Reference

Ramdohr P (1980) The ore minerals and their intergrowths, 2nd edn, 2 vols. Pergamon Press, Oxford, 1205 p

22 Braunite (*bra*)

$Mn^{2+}Mn^{3+}{}_6SiO_{12}$: Tetragonal

Braunite		R_{546} (%)	Y (%)	λ_d (nm)	P_e (%)	Colour sensation
Air	R_o	18.9	18.8	477	4.7	Grey
	R_e	19.9	19.8	478	4.9	
Oil	R_o	6.4	6.3	476	8.7	
	R_e	7.0	7.0	477	8.9	
Bir: Very weak, imperceptible (2)				*Pleoc*: Imperceptible (2)		*Tarn*: No
Anis +P: Weak (3)				*IR*: Very rare		*Twg*: Exceptional
Mor: Hypidiomorphic aggregates				*Clv*: Not visible under the microscope		*PolQ*: Good
Hardness	*PolH*: High, > ht, mt; < py (4)			*Mohs*: 6–6½	*Talm*: E+	*VHN* $_{(100\ g)}$: 920-1097pinacoid, 1025-1196basal

Obs: cf. Table and R charts, visible range (QDF3 and Picot and Johan 1982)
Mineral abbreviations: bra—braunite, gth—goethite, ht—hematites, fkl—franklinite, gg—gangue, jac—jacobsite, mng—manganite, mt—magnetite, pie—piemontite, prl—pyrolusite, psi—psilomelane, py—pyrite, rom—romanechite, sp—sphalerite

22.1 General

Composition: IMA formula $Mn^{2+}Mn^{3+}{}_6(SiO_4)O_8$, although it may contain Fe (in the lattice as Si, after Ramdohr 1982) and/or minor contents of other metals such as Al, Mg, Ca, Ti, Ba.

High *polish hardness*: $PolH_{bra} \approx$ hausmannite. $PolH_{bra}$ slightly > ht. $PolH_{bra}$ > mt, but $PolH_{bra}$ < py and distinctly < prl.

It usually occurs crystallized, in dispersed euhedral (pseudo-octahedral) grains, in hypidiomorphic aggregates or in microcrystalline masses of fine- to medium-grain size. Substitution relationships with other Mn oxides are not uncommon.

22.2 Microscopic Description

22.2.1 SP Observation (*Single Polariser*)

Grey *colour*, good *polish*.

Low-moderate *reflectance*: $R_{bra} \approx$ jac, fkl, mt; R_{bra} < psi, rom, prl; R_{bra} > sp, gth (cf. multispectral comparative charts and Figs. 22.1a, b, 22.2, 22.3 and 22.4a). Bireflectance or pleochroism imperceptible.

22.2.2 Observation +P (*Polariser + Analyser at 90°*)

Weakly anisotropic, with dull and poorly coloured shades (grey to slightly bluish-greenish-grey) but well visible, especially in immersion (Figs. 22.1c, d and 22.4b). Exceptionally, (112) *twins* can be observed under the microscope.

R. Castroviejo, *A Practical Guide to Ore Microscopy—Volume 1*,
https://doi.org/10.1007/978-3-031-12654-3_22

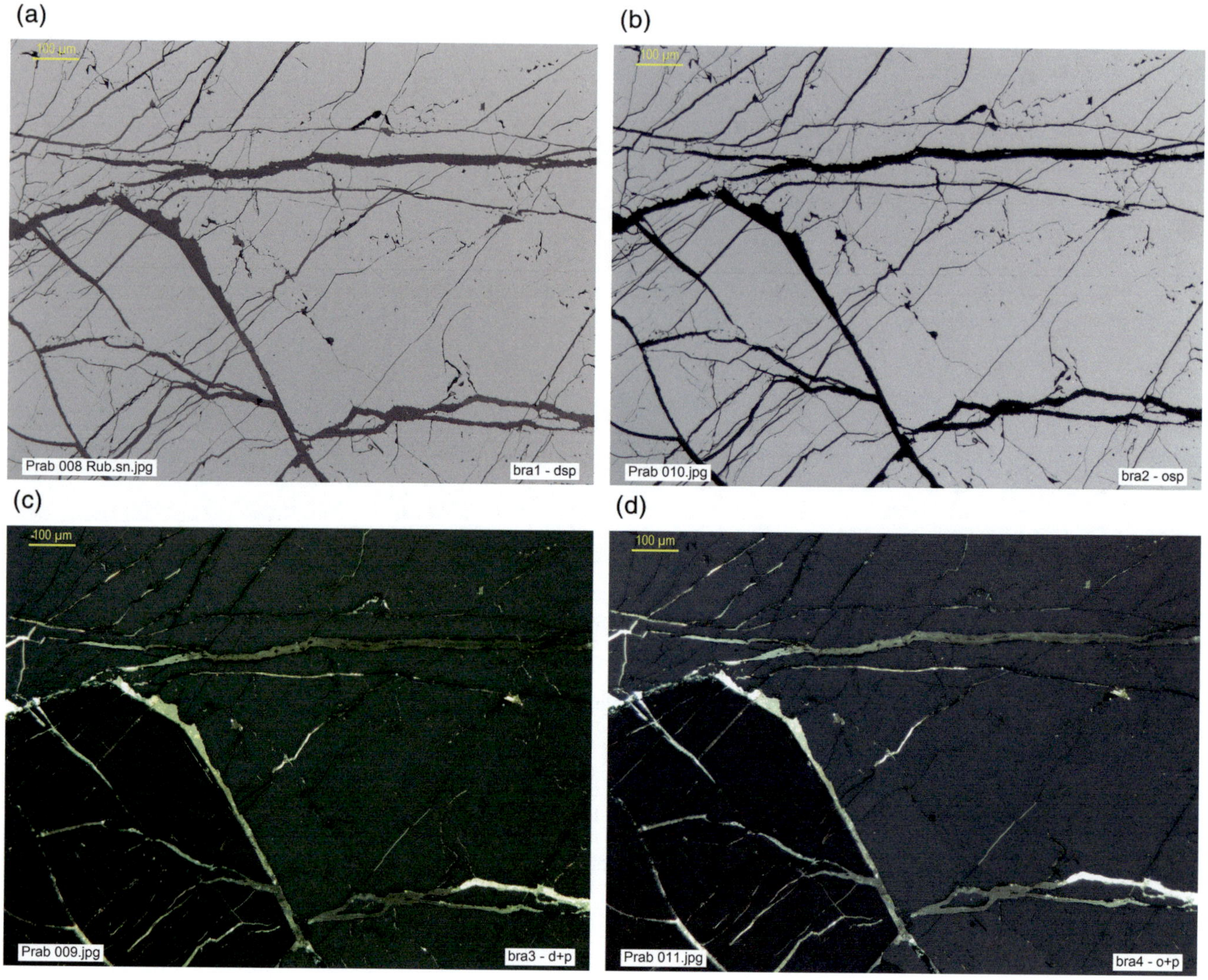

Fig. 22.1 **a**, **b** (Same scene: **dsp** and **osp**, *respectively*) Holocrystalline aggregate of bra (grey) and, separated by thin intergranular film of q, jac (also grey, cubic automorph, bottom left corner). Microcracks cemented by q (detail of Fig. 22.2). The colour and luster of bra and jac are so similar that anisotropism must be observed to distinguish them (Fig. 22.1c, d). Rub.sn sample, Praborna, Val d'Aosta, Italy. **c**, **d** (Same scene as Fig. 22.1a: **d+p** and **o+p**, *respectively*). The anisotropism of bra, with dull but visible greenish-grey tones, allows it to be distinguished from jac (isotropic), in extinction. Bright, colourless IR make q very visible

Internal reflections (brownish) very rarely visible, much scarcer than in manganite or jacobsite.

22.2.3 Paragenesis

Paragenesis varied, usually high-temperature, metamorphic or even magmatic (pegmatitic). However, Ramdohr (1982) also cites low-temperature examples, in which he supposes the involvement of surface fluids and the formation of psilomelane or similar gels.

Accompanying minerals: jacobsite, pyrolusite, hausmannite, piemontite.

22.2.4 Identification

It is easily recognized by its low reflectance, its good polish and hypidiomorphic tendency, its weak but visible anisotropism and its association with manganiferous paragenesis. It differs from jacobsite, magnetite, manganite, goethite, hausmannite, as follows:

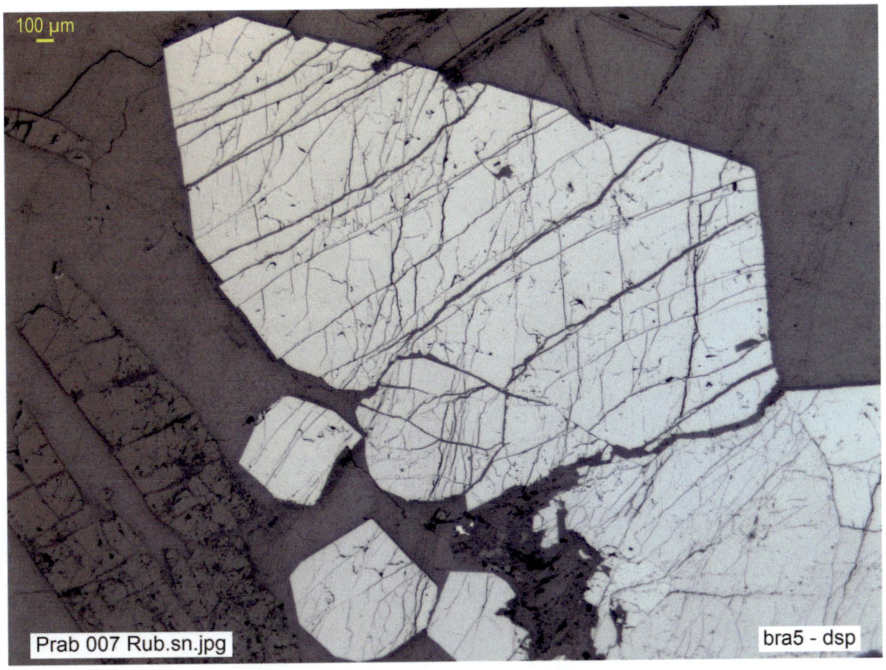

Fig. 22.2 Vein fill of bra (light grey), jac (light grey, similar to bra, but distinctly automorphic), q (grey) and piemontite (**pie**, brownish grey, faintly pinkish shade as compared with q, polishing pits). Figures 22.1a, b and 22.1 c, d offer detailed views of the central zone of Fig. 22.2. Rub-sn, Praborna, Val d'Aosta, Italy (**dsp**)

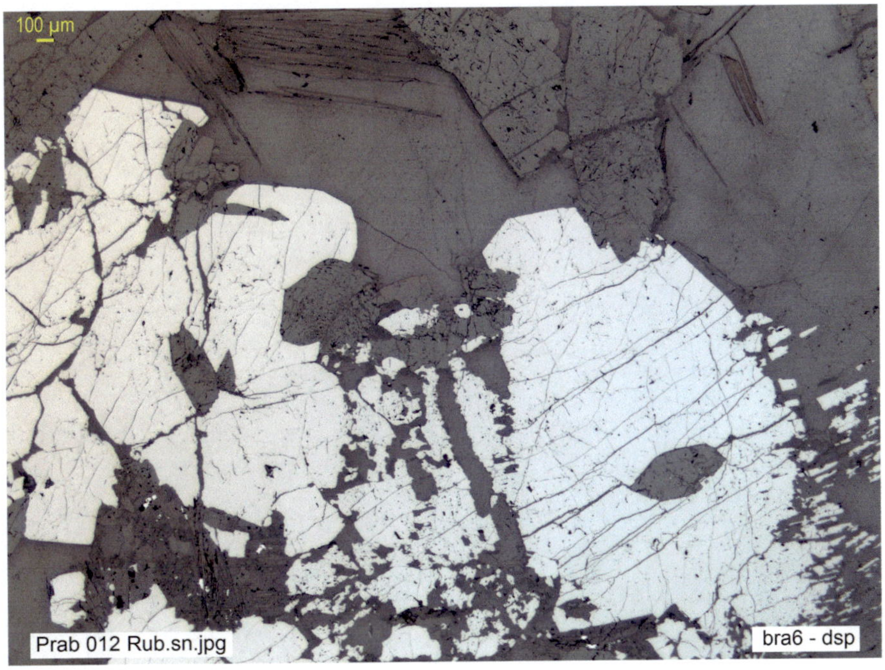

Fig. 22.3 Vein fill of jac (main mass, right and upper left, apparently the same as bra, except for its remarkable automorphic trend, visible despite subsequent partial corrosion), bra, q and piemontite. Figure 22.4a, b offer detailed views of the central zone of the left aggregate (jac, bra, pie). Rub-sn, Praborna, Val d'Aosta, Italy (**dsp**)

- jacobsite and magnetite are isotropic,
- manganite is bireflective, more strongly anisotropic and has many more IR,
- goethite is darker, more strongly anisotropic and shows frequent reddish IR,
- hausmannite is strongly anisotropic and has many twins.

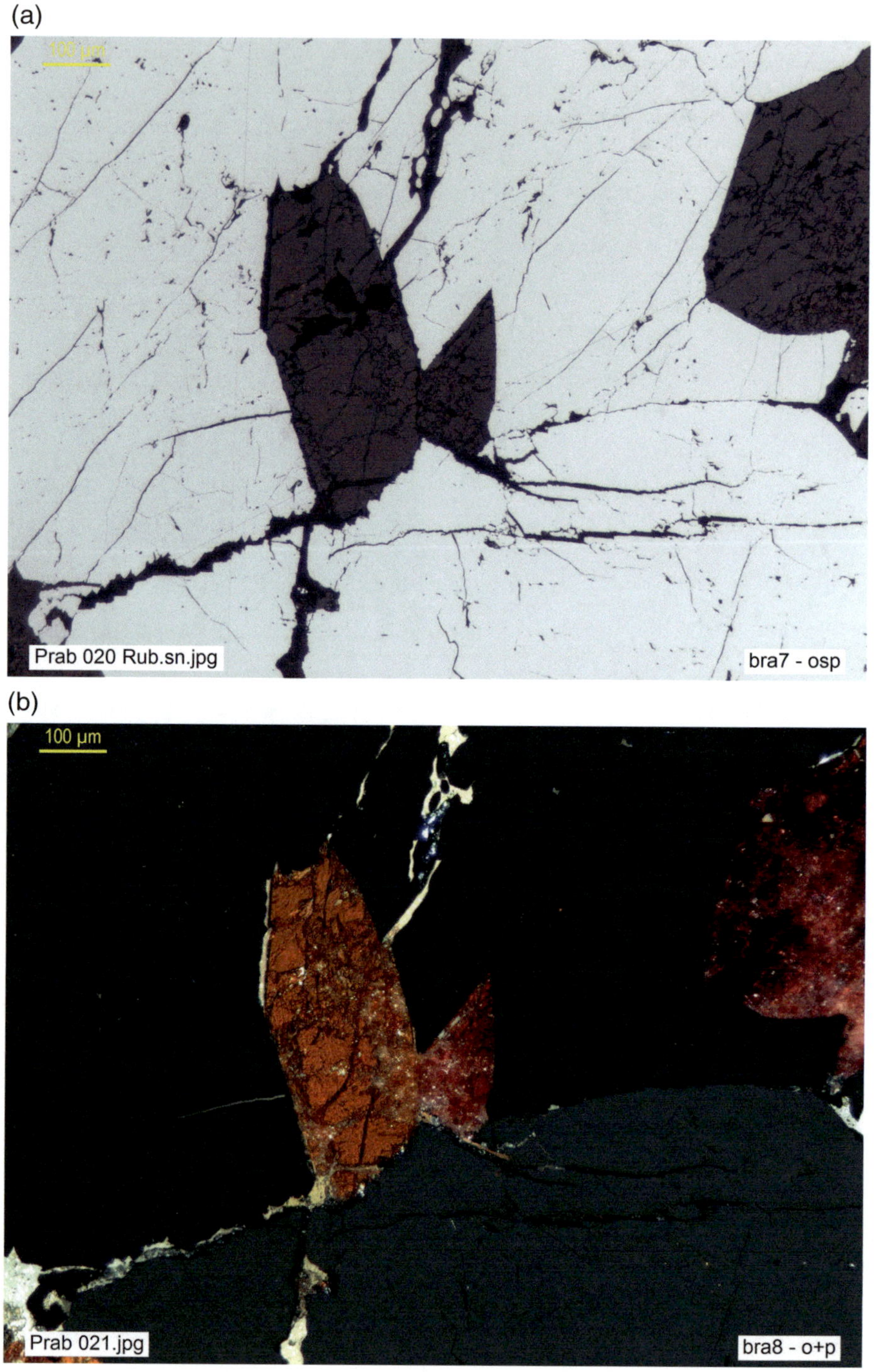

Fig. 22.4 **a**, **b** Detail of Fig. 22.3 (**osp** and **o+p**, *respectively*). While jac remains extinct, bra is distinguished by its greenish gray anisotropism, and piemontite, which forms idiomorphic inclusions in jac, by its anisotropism and by its reddish-brown to purplish IR. Rub-sn, Praborna, Val d'Aosta, Italy (**dsp**)

22.3 Spectral Information

22.3.1 Braunite Reflectance R (in Air, Visible Spectrum) (%)

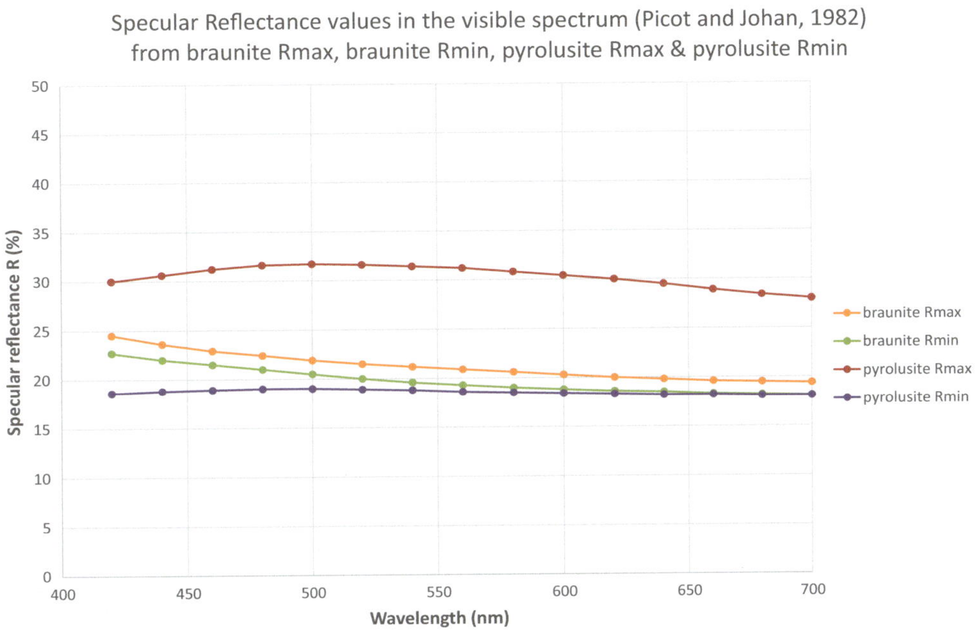

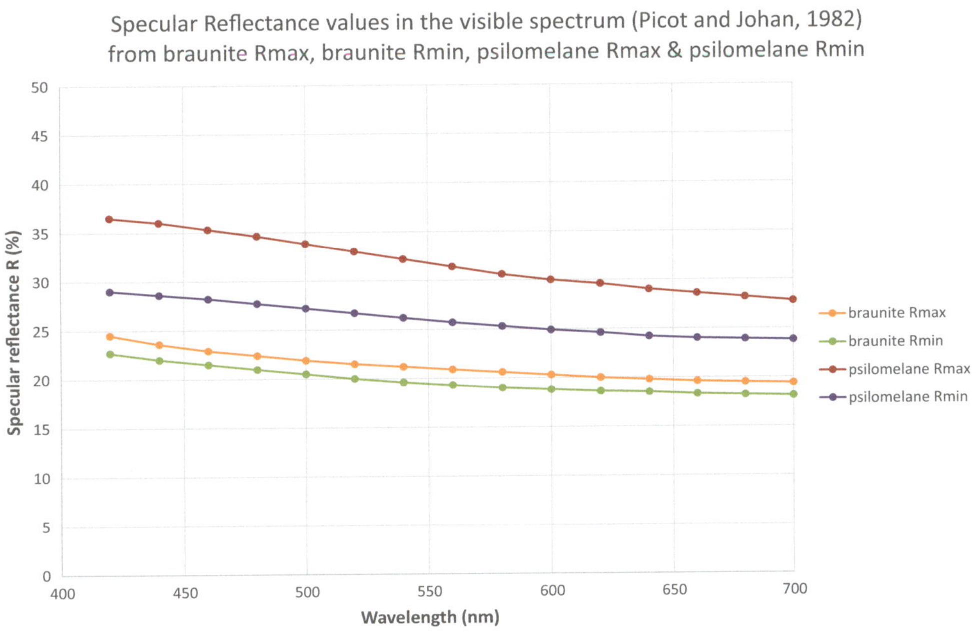

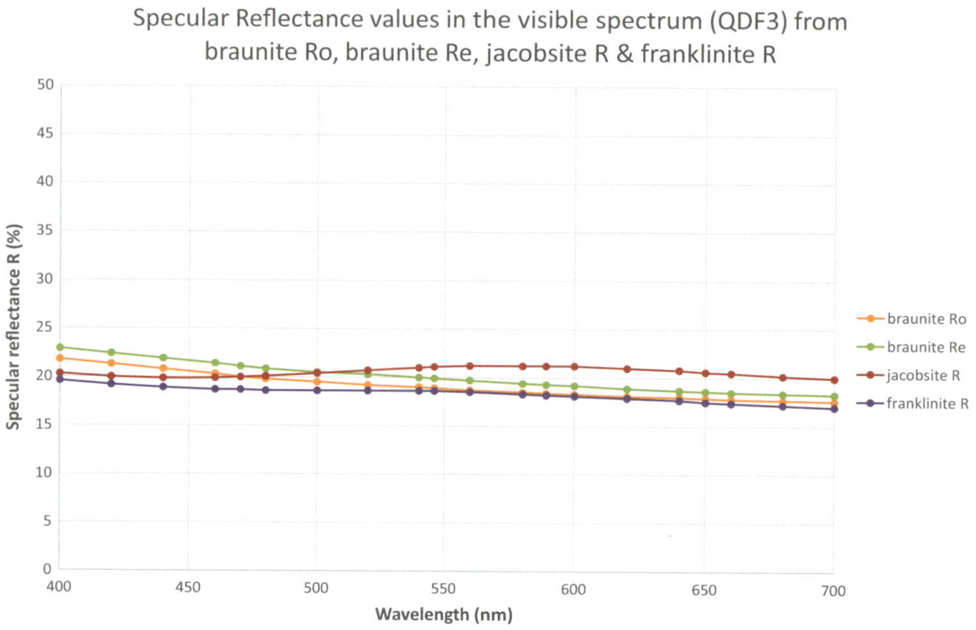

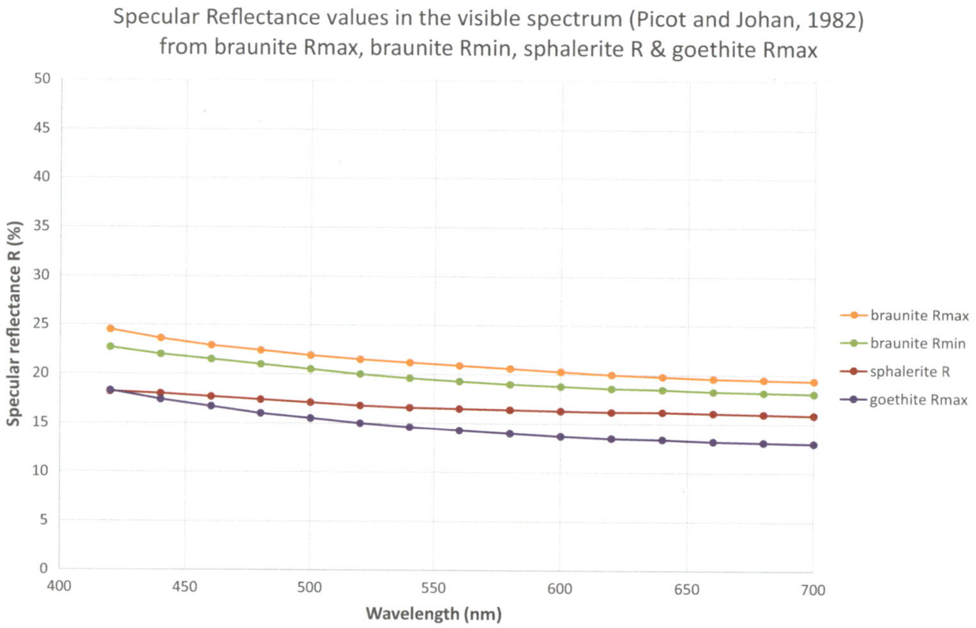

Reference

Picot P, Johan Z (1982) Atlas of ore minerals. B.R.G.M./Elsevier

23 Bravoite (*brv/vae-ctt, vaesite-cattierite series*)

(Fe, Ni, Co)S_2: Cubic

Bravoite	R_{546} (%)	Y (%)	λ_d (nm)	P_e (%)	Colour sensation
Air	39.2	39.7	581	8.2	Variable, yellowish-white or brownish-yellow to purplish brown
Oil	Nd	Nd	Nd	Nd	
Bir: Null (1)			*Pleoc*: No (1)		*Tarn*: Variable with composition
Anis +P: Null (1)			*IR*. No		*Twg*: –
Mor: Euhedral cubic, zoned			*Clv*: –		*PolQ*: Good, variable with composition
Hardness	*PolH*: M-A, > sp, < py (3–4)		*Mohs*: 6–3.5	*Talm*: –	*VHN* $_{(100\ g)}$: –

Obs: Py variety with Ni–Co. The optical data of this table correspond to *Pyrite (nickeloan)* of QDF3, with 31.44% Fe, 13.01% Ni, 1.27% Co and 1.38% Cu

Spectral information: cf. Table and R plots, multispectral data (VNIR, AMCO) compared with visible range (QDF3)

Mineral abbreviations: brv—bravoite, ccp—chalcopyrite, ctt—cattierite, gn—galena, mc—marcasite, ml—millerite, pn—pentlandite, po—pyrrhotite, py—pyrite, sp—sphalerite, vae—vaesite, vio—violarite

23.1 General

Bravoite, in spite of the wide diffusion of the term, is no longer accepted by IMA/COM as a mineral species. By its composition (Fe, Ni, Co)S_2, it could be considered as a variety of pyrite (nickeliferous pyrite, like the one documented with its spectrum in QDF3), characterized by its striking zoned structure, with alternations of colour and hardness that translate compositional changes (*vbgr* Fe/Ni ratio, in the solid solution FeS_2–NiS_2). However, today it is rather understood in the framework of the **vaesite** (*vae* NiS_2), **cattierite** (*ctt* CoS_2) series, the different zones corresponding to different compositions within the series, including possible pyrite intercalations. Indeed, the optical (colour) and physical (polishing hardness) properties of the zones observed in bravoite correspond *roughly* to the terms vaesite-cattierite and pyrite or are transitional between them. Consequently, the usual concept of bravoite is maintained in the following description, with pyrite and the terms vaesite and cattierite or their transitions as reference. For this reason, the descriptive scheme used does not adhere strictly to the systematic model followed in general but deals simultaneously with bravoite, vaesite and cattierite.

23.2 Microscopic Description

23.2.1 SP Observation (*Single Polariser*)

Colour variable according to composition, from yellowish white to purplish brown, roughly ≈ linnaeite or between pyrite and pyrrhotite (Figs. 23.1a, b). The nickeliferous term (vaesite) is much darker than pyrite and shows brownish tones. The cobaltiferous term (cattierite) is white and bright; next to pyrite it shows a slightly bluish tone, while before galena it looks slightly yellowish.

The *luminance* (air) varies from Y = 31% (vaesite) to Y = 37.5% (cattierite), compared to Y = 39.7% (Ni-pyrite) and 53.8% (stoichiometric pyrite), according to QDF3.

Nickeliferous varieties may tarnish (brown to purplish patina).

R. Castroviejo, *A Practical Guide to Ore Microscopy—Volume 1*,
https://doi.org/10.1007/978-3-031-12654-3_23

Fig. 23.1 **a** Concentrate of brv (yellowish-brown to violet-brown shades, zoned), sp (grey), mc (white), gn (light grey, sparse). Sample rusn-Maubach, Germany (**dsp**). **b** Same scene as brv1 (**osp**). **c** Same scene as brv1 (**d+p**), extinction of brv and gn (black), in contrast to the luminosity of sp (internal reflections) and the anisotropism of mc (green). **d**, **e** Variety of morphologies and brv zoning (detail of brv2, **osp**). **f** Ore of en (purplish grey) and td (bluish grey), with interstitial infill of zoned py, ccp (intense yellow) and td. Note the py growth towards the interior of the pore. Sample CR 283, Peru (**dsp**)

Isotropic but usually heterogeneous (*zoned*). Much more heterogeneous than zoned pyrite, which does not usually show large differences in colour and often only reveals its zoned structure by chemical attack or tarnishing (*comp.* Figs. 23.1d, e and f).

Polishing hardness is dependent on composition and therefore variable in different zones of the crystal. Relatively high > pyrrhotite but always ≤ pyrite (minimum hardness in Ni terms, maximum in pyrite). Polishing quality is generally good, better than pyrite.

Reflectance varies according to Fe, Ni, Co contents (cf comparative graphs R). It is higher in cobaltiferous terms than in nickeliferous terms, but given the difficulty of measuring mixed, heterogeneous crystals, Ramdohr's (1980) scepticism about the limited usefulness of reflectance measurements for bravoite is reasonable.

It occurs typically in striking *euhedral* crystals, showing diverse *cubic forms* that can vary in the different zones of the crystal, but it can also be formed without zoning, as a secondary product (e.g., by alteration of pentlandite) in fine aggregates, masses or pseudomorphs that can be confused with violarite.

23.2.2 Observation +P (*Polariser + Analyser at 90°*)

Isotropic (Fig. 23.1c), no IR.

23.2.3 Paragenesis

Parageneses typical of low temperatures, in most cases supergene, *vbgr.* by alteration of pentlandite, which, in the superficial zones, can appear as relicts or ghosts. As a secondary mineral, brv can appear in different types of deposits replacing Ni-Co sulphides, but it may also occur as a primary mineral in sedimentary deposits. It can be found in concentrations of orthomagmatic origin (Sudbury), as well as in stratabound deposits, Lead Belt type, Missouri, USA.

It is accompanied by pyrrhotite, pentlandite, violarite, millerite, galena, sphalerite, etc.

23.2.4 Identification

Its appearance, morphology and zoning (compositional and, consequently, of colour and polishing hardness) are so typical that no confusion is possible when zoned brv is found. Identification is more difficult in bravoite, secondary to pentlandite or non-zoned, often massive or microcrystalline, which can be confused with violarite, sometimes making it necessary to resort to XRD.

23.3 Spectral Information

23.3.1 Bravoite Reflectance Data R (in Air, QDF3; Pyrite, AMCO) (%)

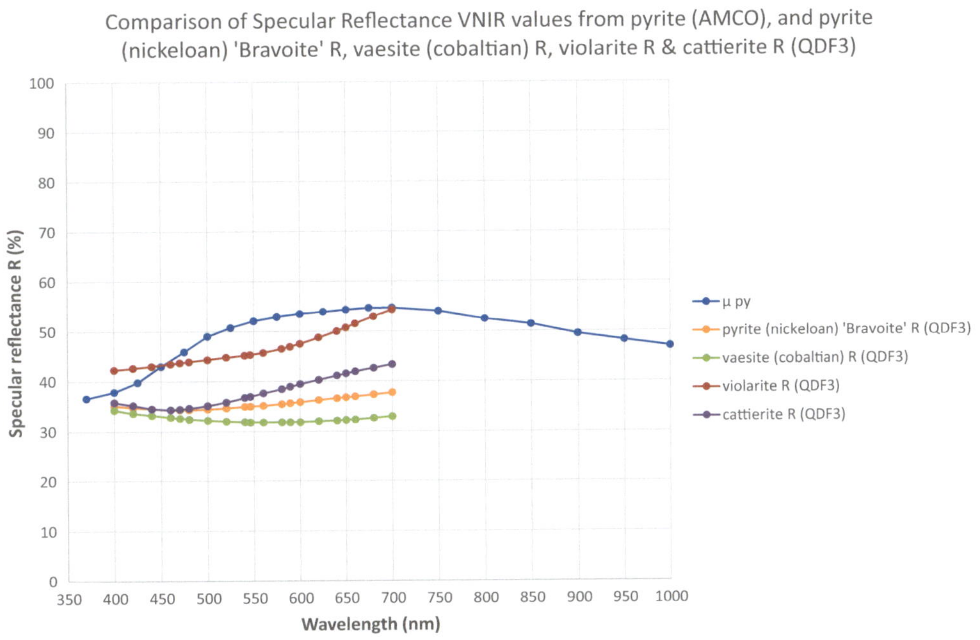

Reference

Ramdohr P (1980) The ore minerals and their intergrowths, vol 2, 2nd edn. Pergamon Press, Oxford, 1205p

24 Breithauptite (*bth*)

NiSb: Hexagonal

Breithauptite		R_{546}(%)	Y(%)	λ_d (nm)	P_e (%)	Colour sensation
Air	R_1	48.0	49.6	598	5.0	Light pink, slightly purplish
	R_2	37.8	40.3	c496	5.4	Pure purplish pink
Oil	R_1	35.2	36.9	595	8.8	Pronounced purplish-pink, light
	R_2	24.5	27.1	C493	7.4	Purple-pink to deep purplish pink
Bir: Medium (4)				*Pleoc*: Strong (4)		*Tarn*: –
Anis + P: Strong (5)				*IR*: No		*Twg*: No
Mor: Aggregates; sometimes euhedral				*Clv*: No		*PolQ*: Good
Hardness	*PolH*: Mean, ≈ nc, < sf (3)			*Mohs*: 5.5	*Talm*: nd	*VHN* $_{(50–100\ g)}$: 412–584

Spectral information: cf. Table and R charts, multispectral data (VNIR, AMCO) and visible range (QDF3)
Mineral abbreviations: Ag—native silver, asp—arsenopyrite, bth—breithauptite, dy—dyscrasite, ln—linnaeite, mau—maucherite, nc—nickeline, po—pyrrhotite, py—pyrite, sf—safflorite, sk—skutterudite

24.1 General

Composition: NiSb, may contain Fe, Co, As, S as impurities.

With higher As content, it changes to *arite* (Ni[As, Sb], cf nickeline).

24.2 Microscopic Description

24.2.1 SP Observation (*Single Polariser*)

Pinkish *colour*, with a strong *pleochroism*, from light pink, slightly purplish to distinctly purplish pink, darker (against Ag an apparently greyish tone is superimposed, effect of the great difference in reflectances, Fig. 24.1a and comparative table of reflectances).

In oil immersion, more saturated pink colours are observed. Light tones of pronounced purplish pink, or darker, intense purplish tones (Fig. 24.1b).

High *reflectance* but < py or asp (cf multispectral comparative graphs). Clearly bireflective, according to QDF3 reflectance curves, parallel in the visible range; however, the dominant visual perception is that of pleochroism, due to the striking colouration of the mineral.

Polishes well. Medium *polishing hardness*, like nickeline: $PolH_{bth}$ > Ag, dy. $PolH_{bth}$ ≤ mau; < sk, sf, po, asp; << py. Does not show cleavage.

It occurs in xenomorphic aggregates, although it occasionally shows idiomorphism, sometimes associated with nickeline in *zoned* grains of alternating composition, which are usually interpreted as derived from the decomposition of *arite* (cf nickeline).

24.2.2 Observation +P (*Polariser + Analyser at 90°*)

Strongly *anisotropic*, with striking polarization colours, in which green tones dominate, as in nickeline, but more luminous and without the blues of the latter (Fig. 24.1c, d). Distinct extinction.

R. Castroviejo, *A Practical Guide to Ore Microscopy—Volume 1*,
https://doi.org/10.1007/978-3-031-12654-3_24

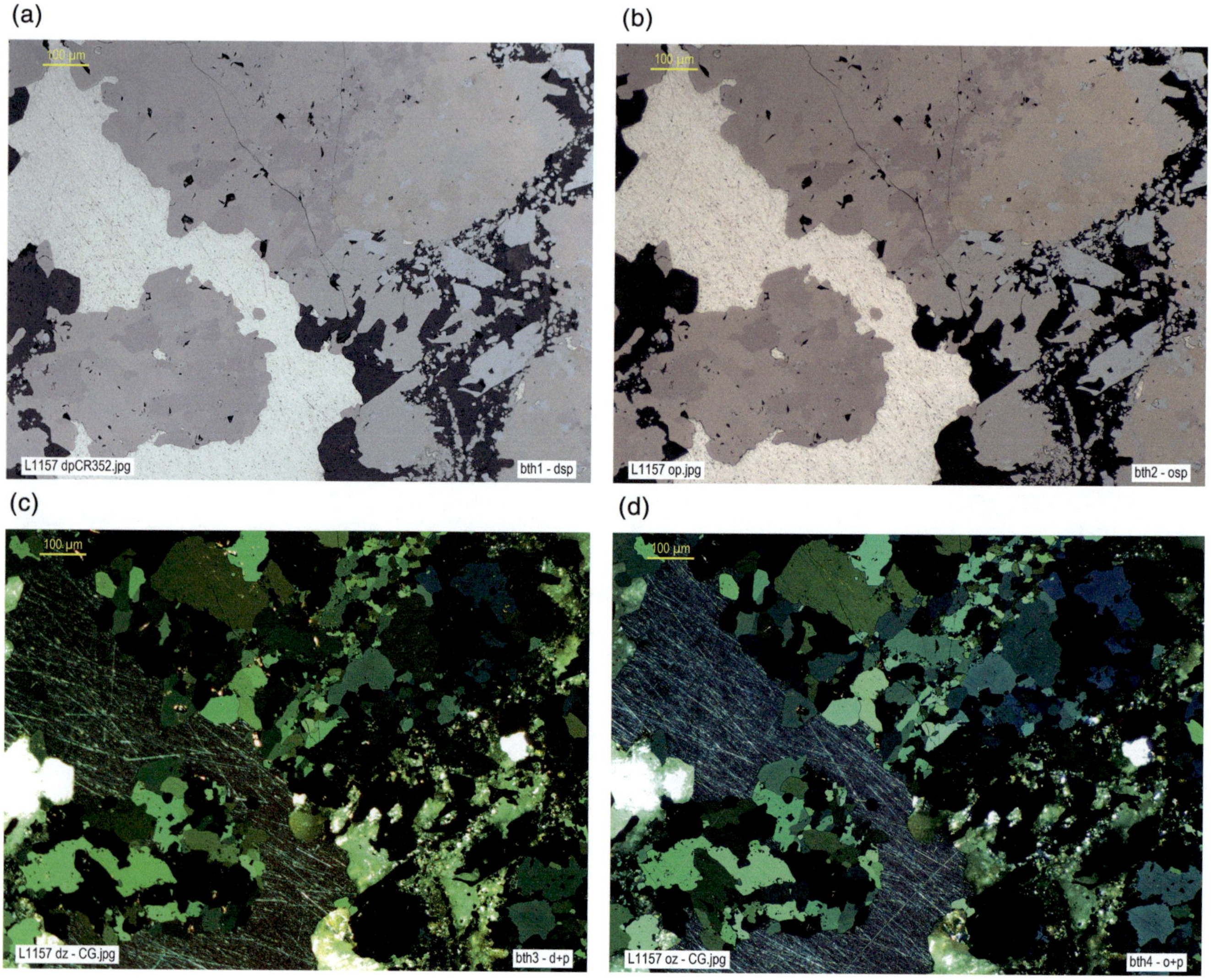

Fig. 24.1 **a** Argentiferous ore with *bth* (purplish pink, strongly pleochroic, forms the main mass, center and lower left zones), *nc* (orange pink, lighter and less pleochroic than bth, aggregate upper right zone), *mau* (pinkish grey), cemented by *Ag* (bright white) and *calcite* (dark grey), sparse *asp* (grey, euhedral—prisms and diamonds) disseminated on calcite. CR352, Cobalt, Ontario, Canada (**dsp**). **b** Same scene as Fig. 24.1 (**osp**)—the higher contrast by immersion allows a much more precise observation of colours, pleochroism and relative reflectance's, which appear lower when in contact with Ag (e.g., asp, normally white, looks grey). **c**, **d** Same scene as Fig. 24.1, (**d +p**) and (**o+p**), respectively. Note the contrast between the strong and bright anisotropism of bth, with dominant green tones and the more moderate anisotropism of nc, with more bluish tones

24.2.3 Paragenesis

Hydrothermal *paragenesis* typical of deposits known as *five-element formations* (Ni–Co–Bi–Ag–U), with native Ag and As/Sb, dyscrasite, Ag sulphosalts, Ni–Co arsenides and sulphoarsenides, linnaeite.

Also in deep, high-temperature *paragenesis* linked to ultramafic or peridotitic formations.

24.2.4 Identification

The pinkish-violet-purplish colour, good polish, pleochroism and anisotropism allow a reliable identification of bth, although as a first impression it could be confused with nickeline, whose appearance is very similar. However, nickeline is lighter, shows orange tones, is much less pleochroic and its anisotropism is weaker (and with less greenish and more bluish polarisation tints) (Figs. 24.1b, c and d)

24.3 Spectral Information

24.3.1 Breithauptite Multispectral Reflectance R (in Air, Without Polariser) (%)

λ (nm)	370	400	425	450	475	500	525	550	575	600	625	650	675	700	750	800	850	900	950	1000
μ	48.4	48.4	45.9	44.9	43.3	43.2	43.6	46.2	49.1	51.9	54.1	56.1	57.5	58.9	61.5	63.3	65.5	66.7	67.1	68.1
P_5	46.8	47.3	44.4	43.9	41.9	41.5	41.2	43.5	46.2	49.1	51.3	53.4	54.8	56.3	58.8	60.7	62.9	64.2	64.4	65.2
P_{20}	47.7	47.8	45.1	44.4	42.6	42.3	42.2	44.5	47.3	50.2	52.4	54.5	55.9	57.4	59.9	61.7	63.8	65.0	65.1	66.3
P_{80}	49.0	48.9	46.5	45.4	43.9	44.0	44.8	48.0	51.0	53.8	55.9	57.7	59.0	60.4	62.9	64.7	67.0	68.2	68.7	69.7
P_{95}	49.5	49.2	46.8	45.7	44.2	44.6	45.8	49.0	52.0	54.9	56.9	58.8	60.2	61.5	64.0	65.8	67.8	69.0	69.7	70.6

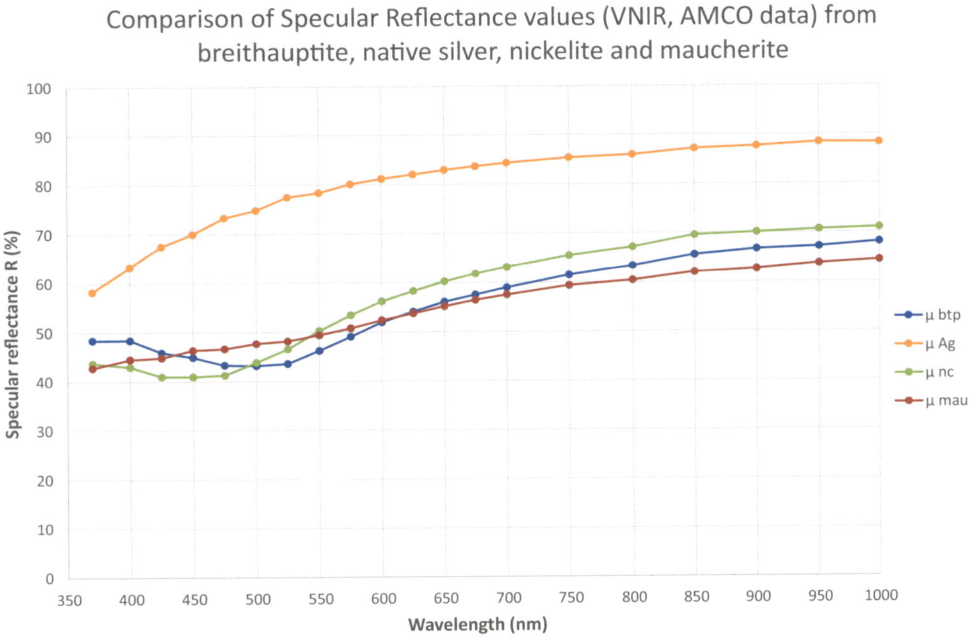

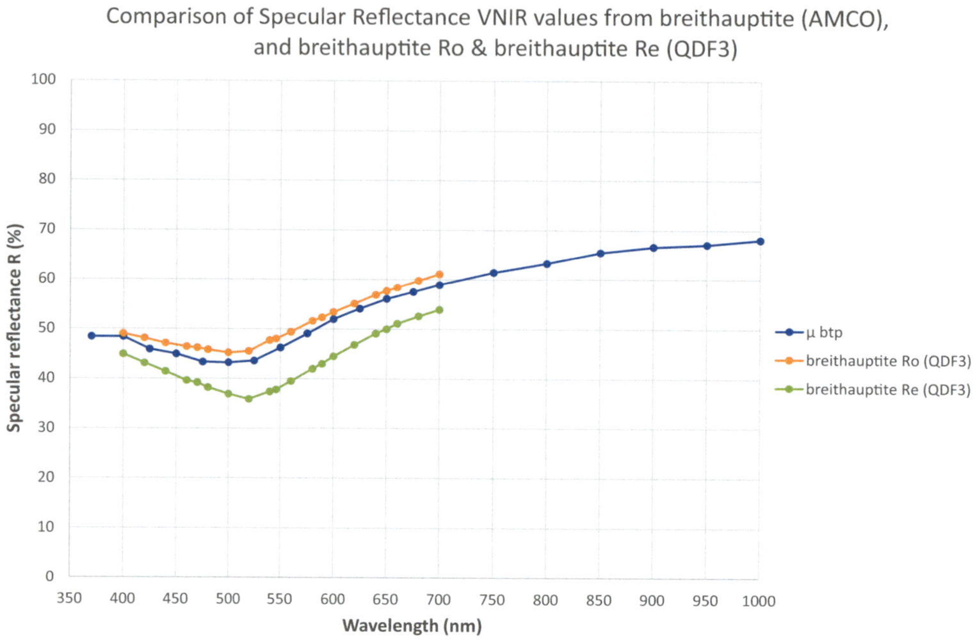

Comparison of Specular Reflectance VNIR values from breithauptite (AMCO), and breithauptite Ro & breithauptite Re (QDF3)
Specular reflectance R (%)
Wavelength (nm)
100
90
80
70
60
50
40
30
20
10
0
350
400
450
500
550
600
650
700
750
800
850
900
950
1000
µ btp
breithauptite Ro (QDF3)
breithauptite Re (QDF3)

Calaverite (*clv*)

25

$AuTe_2$: Monoclinic

Calaverite			R546 (%)	Y (%)	λ_d (nm)	P_e (%)	Colour sensation
Air	R_1		60.8	60.7	574	8.2	White, slightly brownish-pink
	R_2		67.6	67.4	573	5.8	White
Oil	R_1		48.6	48.5	574	10.3	White, slightly yellowish-brown
	R_2		56.6	56.4	573	7.1	White, slightly yellowish
Bir: Weak (3)					*Pleoc*: Faint (3)		*Tarn*: Faint
Anis+P: Weak, but well visible (3)					*IR*: No		*Twg*: Occasional
Mor: Anhedral aggregates. Prismatic xx					*Clv*: No		*PolQ*: Good
Hardness		*PolH*: Low, < gn, << ccp, td (2)			*Mohs*: 2½–3	*Talm*: C	*VHN*$_{(100\ g)}$: 197–213

Obs: cf. Table and R charts, multispectral data (VNIR, AMCO) and visible range (QDF3)

Mineral abbreviations: Ag—native silver, Au—native gold, Bi—native bismuth, bn—bornite, ccp—chalcopyrite, cld—choloradoite, clv—chalaverite, cob—cobaltite, gg—gangue, gn—galena, krn—krennerite, Pt—native platinum, py—pyrite, q—quartz, sp—sphalerite, syl—sylvanite, td—tetrahedrite, Te—native tellurium

25.1 General

Composition: $AuTe_2$, with some Ag instead of Au occasionally.

It has been considered a dimorph of krennerite, but recent studies suggest that compositions are not exactly the same for both species (cf.*krn description*).

25.2 Microscopic Description

25.2.1 SP Observation (*Single Polariser*)

Bright white *colour*, slightly pleochroic towards pinkish-brown (air) or yellowish (immersion) tones, Fig. 25.1a and b.

Good *polishing*. Low *polishing hardness*: $PolH_{clv}$ slightly <gn, << ccp, td.

Very high *reflectance*, $R_{clv} \approx$ Te, Bi. R_{clv} > py, cob. But R_{clv} slightly < Pt, and < < < Au, Ag (cf multispectral comparative charts and Fig. 25.2a).

It occurs in anhedral aggregates (Fig. 25.2a and b), but it has also been found well crystallized in euhedral, tabular or prismatic forms.

25.2.2 Observation +P (*Polariser+ Analyser at 90°*)

Weakly *anisotropic*, with dull greyish to reddish-brown colours (Figs. 25.1c, d and 25.2b). Twinning rarely visible under the microscope.

25.2.3 Paragenesis

Hydrothermal *paragenesis*. Posthumous ore from auriferous mesothermal mineralization. Also found in low-temperature paragenesis, typically epithermal with precious metals (Au, Ag) and tellurides.

Accompanying minerals: gold, silver, galena, chalcopyrite, bornite, tetrahedrite, sphalerite, quartz, krennerite, sylvanite, coloradoite and other tellurides.

R. Castroviejo, *A Practical Guide to Ore Microscopy—Volume 1*,
https://doi.org/10.1007/978-3-031-12654-3_25

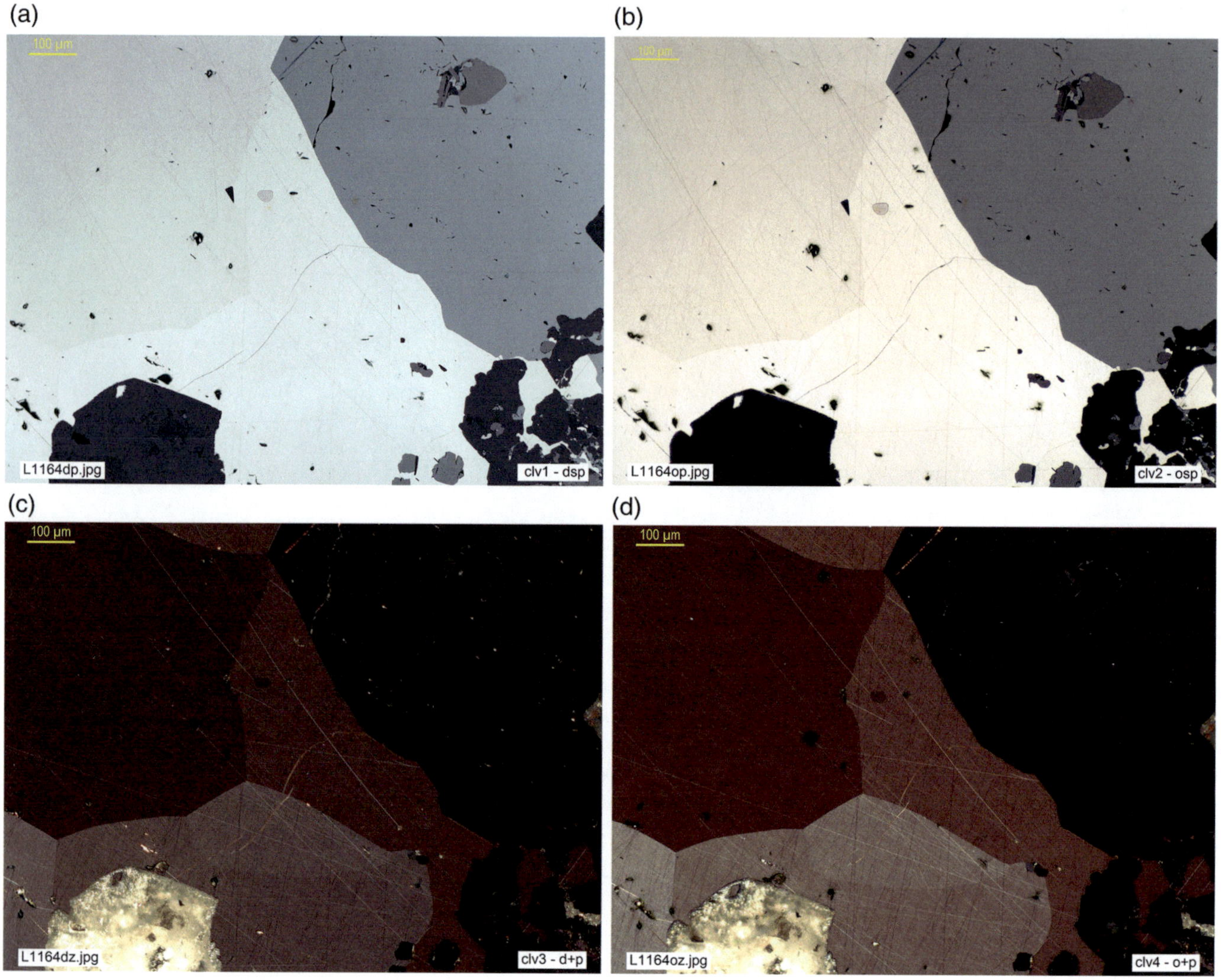

Fig. 25.1 **a**, **b** (Same scene: **dsp** and **osp**, *respectively*). Holocrystalline mass of clv (bright white, slightly pleochroic) and td (grey, upper right), with inclusions of mt (grey, darker than td, relief) and q (very dark grey/black, *osp*). Tiny crystals of cob (pinkish tone slightly darker than clv, relief) included in clv. CR353, Paringa Mine, Kalgoorlie, Australia. **c**, **d** Same scene as Fig. 25.1, (**d+p** and **o+p**, *respectively).* The anisotropism of clv, though weak, distinguishes it from all other components

25.2.4 Identification

The paragenesis is an important key. Calaverite is characterized by its very high reflectance, low hardness, white colour with faint pleochroism and weakly coloured but visible anisotropism.

- clv resembles krennerite, but this one is softer, polishes better, shows net exfoliation and its anisotropism is weaker
- clv is easily distinguished from sylvanite, which is darker (lower reflectance) and softer, has a much more vivid anisotropism and frequent, well visible polysynthetic twins.

Fig. 25.2 **a**, **b** (Same scene: **dsp** and **d+p**, respectively). Wide-field image of the same ore (Fig. 25.1 is a detailed view of the central part): clv (white, dominant in the left half) is notable for its high reflectance and anisotropism; cob (pinkish-white tone, idiomorphic) is abundant in the upper right third of the image and is also seen in small relict inclusions in clv. Incipient corrosion of td (large crystal, center), mt and gg (q and scarce carbonates) by clv

(a)

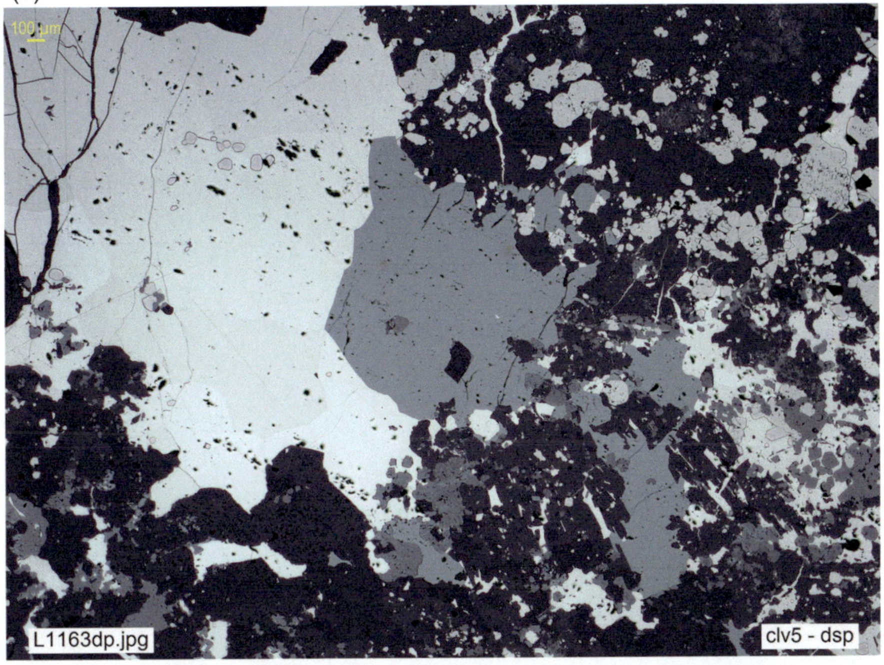

(b)

25.3 Spectral Information

25.3.1 Calaverite Multispectral Reflectance R (in Air, Without Polariser) (%)

λ (nm)	370	400	425	450	475	500	525	550	575	600	625	650	675	700	750	800	850	900	950	1000
μ	44.9	46.9	49.2	52.4	55.2	57.7	59.6	60.9	62.1	63.1	63.7	64.3	64.6	64.9	65.5	65.9	66.9	66.5	67.4	67.0
P_5	40.6	43.0	45.6	49.1	52.1	55.0	57.2	58.5	60.0	61.0	61.7	62.4	62.9	63.3	63.9	64.4	65.4	64.9	65.7	65.4
P_{20}	43.4	45.5	47.7	51.0	53.9	56.5	58.6	59.9	61.2	62.2	62.9	63.6	63.9	64.3	64.9	65.3	66.3	65.7	66.8	66.1
P_{80}	46.6	48.5	50.7	53.8	56.5	58.9	60.8	62.0	63.2	64.1	64.6	65.3	65.4	65.7	66.2	66.6	67.6	67.6	68.2	68.0
P_{95}	47.7	49.6	52.0	55.0	57.6	60.0	61.8	62.9	64.0	64.9	65.3	65.8	66.0	66.3	66.8	67.2	68.2	68.3	68.6	68.6

Comparison of Specular Reflectance values (VNIR, AMCO data) from calaverite, cobaltite, tetrahedrite and native platinum

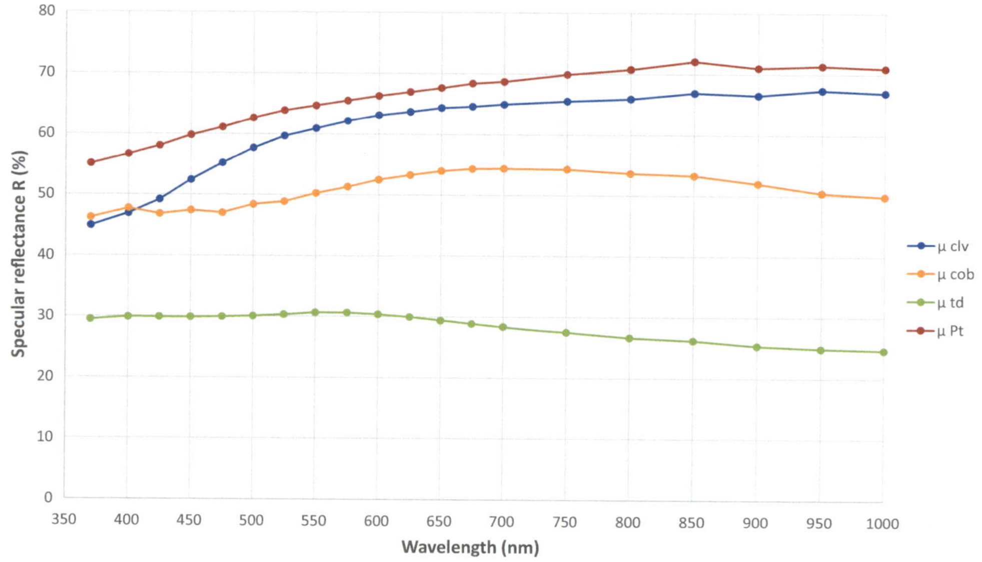

Specular Reflectance values in the visible spectrum (Picot and Johan, 1982) from calaverite Rmax, calaverite Rmin, tellurium Rmax & tellurium Rmin

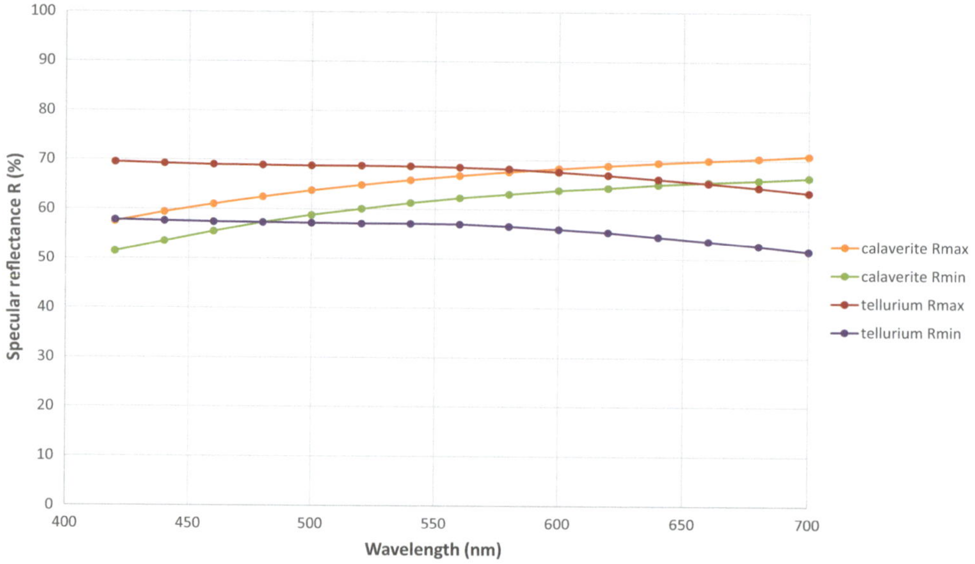

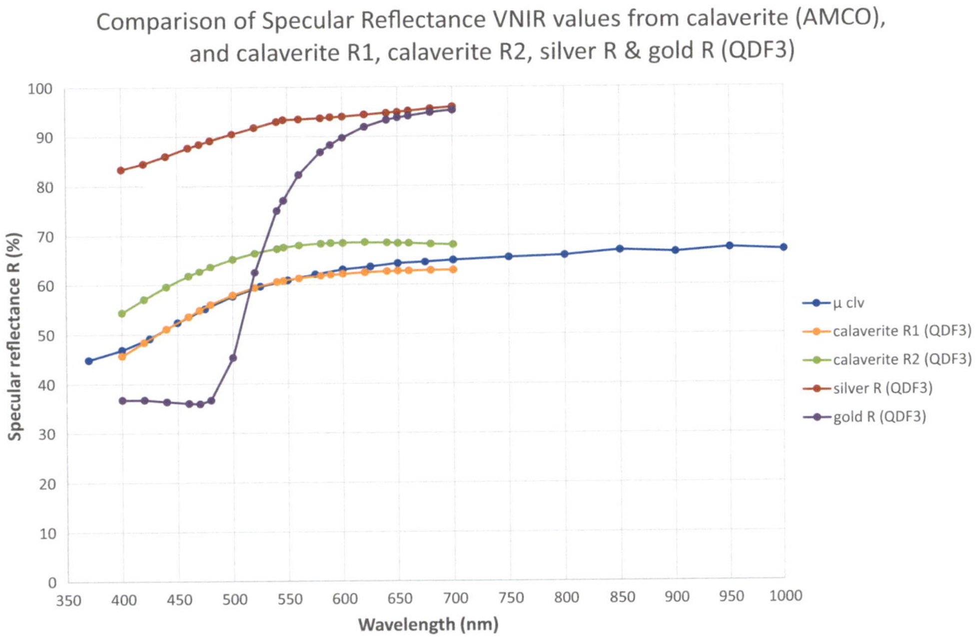
Comparison of Specular Reflectance VNIR values from calaverite (AMCO),
and calaverite R1, calaverite R2, silver R & gold R (QDF3)
Specular reflectance R (%)
Wavelength (nm)
μ clv
calaverite R1 (QDF3)
calaverite R2 (QDF3)
silver R (QDF3)
gold R (QDF3)

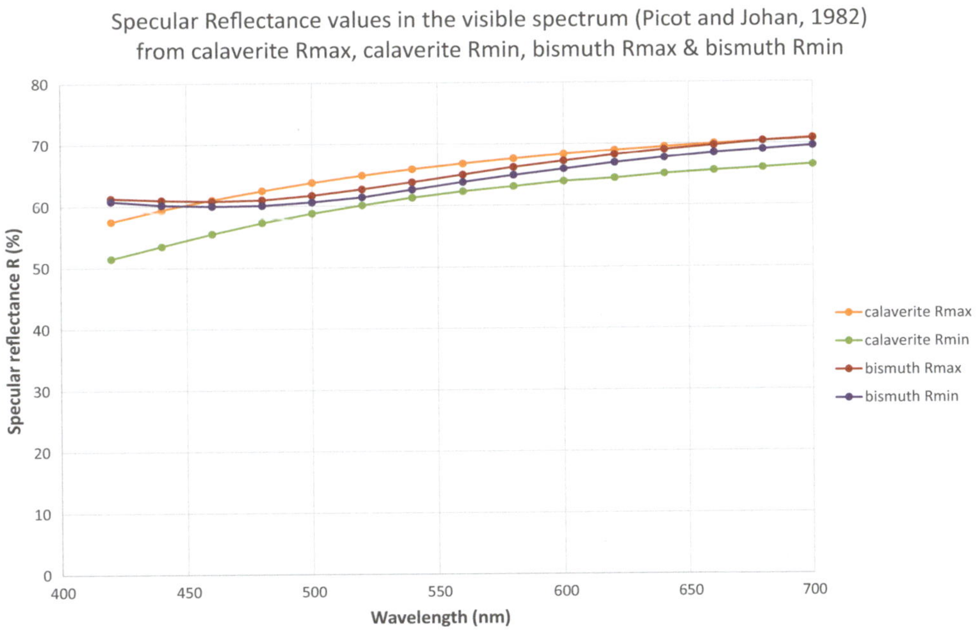
Specular Reflectance values in the visible spectrum (Picot and Johan, 1982)
from calaverite Rmax, calaverite Rmin, bismuth Rmax & bismuth Rmin
Specular reflectance R (%)
Wavelength (nm)
calaverite Rmax
calaverite Rmin
bismuth Rmax
bismuth Rmin

Cassiterite (*cst*)

SnO_2: Tetragonal

26

Casiterite			R546 (%)	Y (%)	λ_d (nm)	P_e (%)	Colour sensation
Air	R_o		10.7	10.7	476	2.4	Grey (dark against sulphides)
	$R_{e'}$		12.5	12.1	476	2.0	
Oil	R_o		1.8	1.8	475	4.1	Very dark grey
	$R_{e'}$		2.4	2.4	475	3.2	
Bir: Medium, but well visible (4)					*Pleoc*: No (always grey)		*Tarn*: No
Anis +P: Distinctly anisotropic (4)					*IR*: Abundant		*Twg*: Very frequent growth twg
Mor: Holocrystalline euh. Massive botr					*Clv*: Weak, // (110)		*PolQ*: Difficult (high hardness)
Hardness		*PolH*: Very high, > py, q (5)			*Mohs*: 6–7	*Talm*: G+	*VHN* $_{(200\ g)}$: 1239–1467

Spectral Information: cf. Table and R charts, multispectral data (VNIR, AMCO) and visible range (QDF3)
Mineral abbreviations: asp—arsenopyrite, bm—bismuthinite, bn—bornite, cc—chalcocite, ccp—chalcopyrite, cst—cassiterite, cv—covellite, dg—digenite, gg—gangue, gn—galena, gth—goethite, ht—hematite, mol—molybdenite, mt—magnetite, py—pyrite, q—quartz, sch—scheelite, sp—sphalerite, ser—sericite, tur—tourmaline, wf—wolframite

26.1 General

Composition: SnO_2, with possible traces of Ta, Nb, W, Fe, Ti, Ge, In, Cu, Zn, Mn. Due to its transparency, it can also be easily identified with transmitted light, **TL**.

Very stable mineral in superficial conditions. Resistant to weathering and fluvial transport, so in addition to the primary deposits mentioned (cf. Paragenesis), it is also found in important detrital concentrations.

26.2 Microscopic Description

26.2.1 SP Observation (*Single Polariser*)

Grey *colour*, slightly lighter than the gangue minerals (quartz and other silicates) that usually accompany it but very dark in comparison with sulphides, especially in immersion (comp. Figs. 26.1a, b, 26.2, 26.3 and 26.4), due to its low reflectance.

Typical low *reflectance*, lower than wolframite, sphalerite, magnetite or goethite, slightly higher than scheelite and higher than quartz and resin. Remarkable bireflectance in shades of grey but without colour changes (not pleochroic in reflected light, as can be deduced from its spectra, R_o and $R_{e'}$, practically flat and parallel. *See* comparative graphs of reflectance values in the visible range). Its transparency makes the **IR**, *internal reflections*, often visible even without an analyser. Sometimes the fluid inclusions can also be seen (Figs. 26.1 and 26.4).

Difficult to *polish*, due to its high hardness but, with care, high quality can be obtained. *Polishing hardness* among the highest, superior to quartz and pyrite. *Vickers microhardness*, VHN, also very high, clearly superior to wolframite. However, its scratch hardness is lower than that of quartz. Cleavage // (110) rarely visible. *Zoning* occasionally visible.

It usually occurs in crystals or crystalline aggregates with a pronounced idiomorphic development (prismatic and acicular habits, (Figs. 26.1a, b and 26.4) but also in cryptocrystalline masses and in collomorphic, botryoidal or fibrous-radiated aggregates (*wood-tin*), typical of volcanic environments.

R. Castroviejo, *A Practical Guide to Ore Microscopy—Volume 1*,
https://doi.org/10.1007/978-3-031-12654-3_26

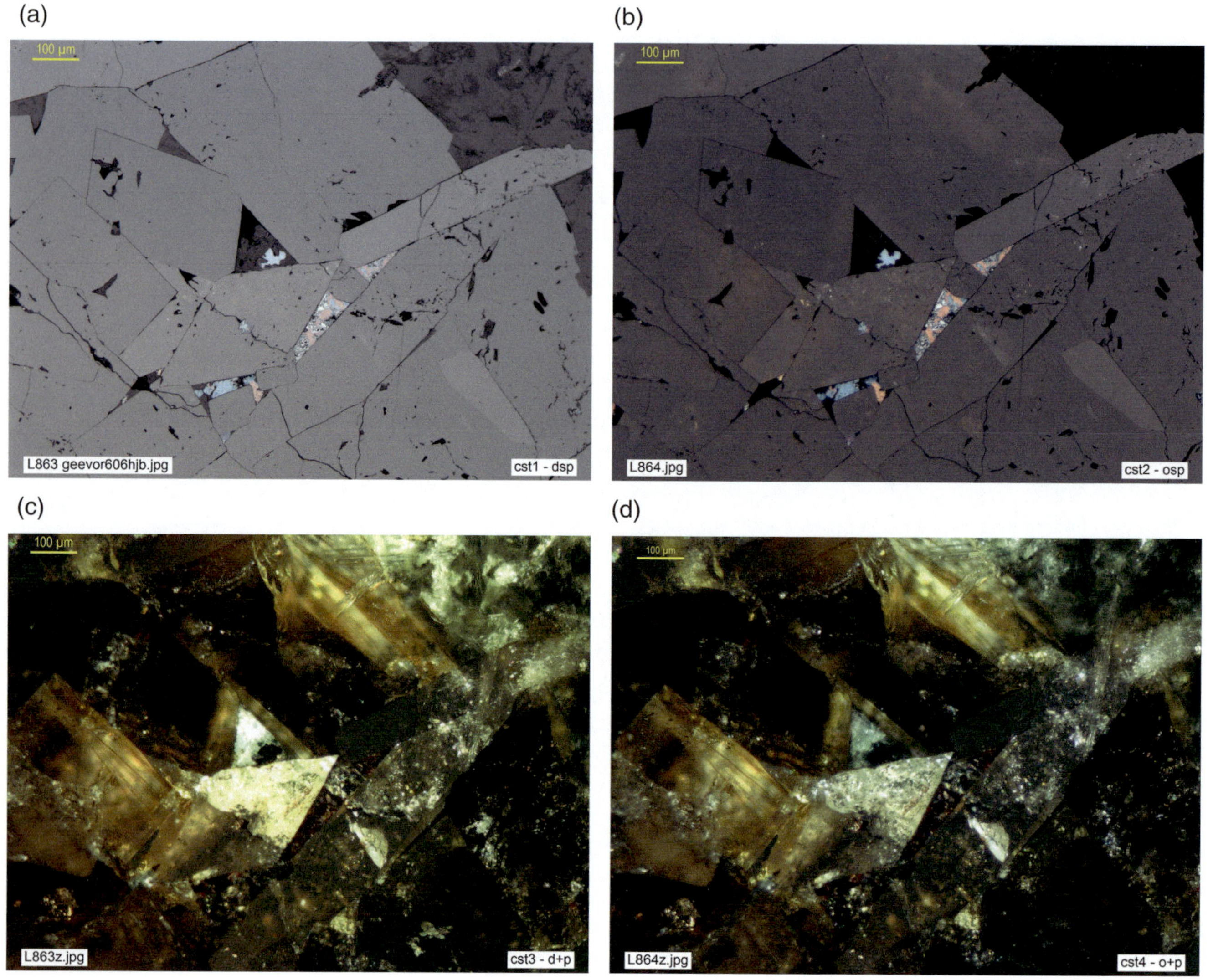

Fig. 26.1 **a** Holocrystalline aggregate of cst (grey, bireflective) with interstitial cc (white), bn (pinkish brown), dg (light blue) and gangue (q, ser: dark grey). M606, Geevor Mine, Cornwall, UK (**dsp**). **b** Same scene as cst1 (**osp**), IR and strong decrease of reflectance are seen in cst, but the contrast increases with gangue, which is practically black. **c, d** Same scene as cst1 (**d+p**) and (**o+p**)

26.2.2 Observation +P (*Polariser + Analyser at 90°*)

Clearly *anisotropic*, although the frequent and luminous IR, *internal reflections*, partly mask the anisotropism (Fig. 26.1c, d). Given the absence of pronounced polarization tints, the visible shades are dependent of the IR colours. These are usually yellowish-brown but can vary to red (cassiterite with Fe_2O_3 contents), reddish-brown, yellow and colourless tones.

Very frequent and characteristic *twinning* (Figs. 26.1c, d and 26.4), usually simple growth twins, sometimes laminar.

26.2.3 Paragenesis

Characteristic pegmatitic, pneumatolitic and hydrothermal *parageneses*, usually in holocrystalline vein fillings or impregnations (*greisen*), related to evolved granites. *Accompanying ores*: arsenopyrite, wolframite, molybdenite, bismuthinite, scheelite, quartz, tourmaline, etc.

It also occurs in volcanic and subvolcanic environments, sometimes as collomorphic aggregates (*wood-tin*).

Although it is not the typical environment, important concentrations of cassiterite have also been found in volcanic-sedimentary massive sulphides, VMS, with pyrite, chalcopyrite, sphalerite, galena, etc. (*Neves-Corvo*, Portugal).

Fig. 26.2 Early cst generation, brecciated and cemented by py, with interstitial q; traces of dg, cv and q in cst microcracks. M606, Geevor Mine, Cornwall, UK (**dsp**)

Fig. 26.3 Same scene as Fig. 26.3 (**osp**)

26.2.4 Identification

Cassiterite is easily identified by its low reflectance (grey colour) with remarkable bireflectance, very high hardness, idiomorphic or colloform development (depending on the plutonic or volcanic environment), anisotropism and internal reflections.

- Cassiterite differs from wolframite by the lower R, more abundant and luminous IR and much higher hardness of cassiterite.
- Compared to scheelite, whose colour is very similar (slightly lower R), cassiterite is much harder and its IRs are more coloured but much scarcer.
- Against quartz (or silicate gangue), cassiterite is brighter and usually shows less IR.

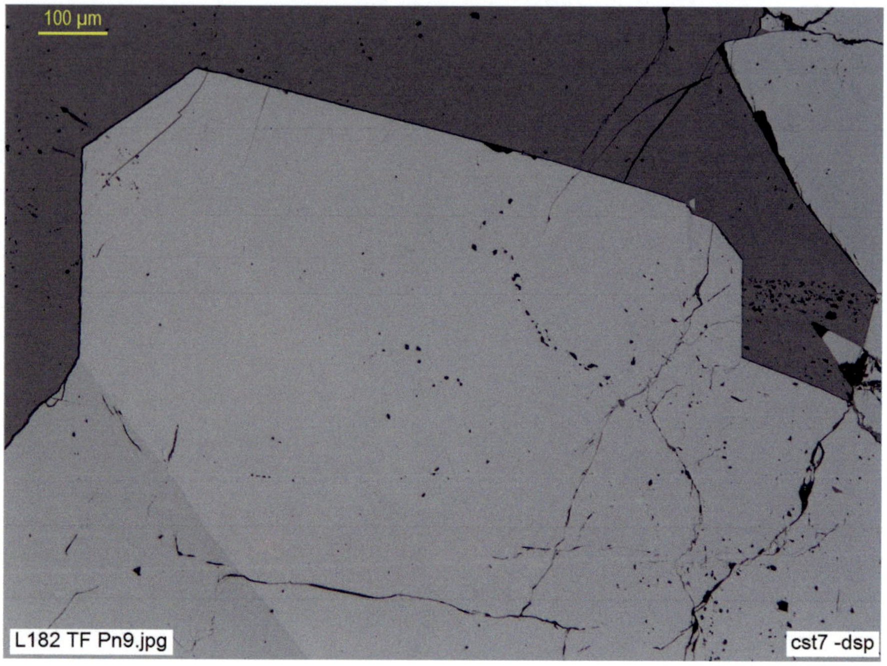

Fig. 26.4 Vein fill of cst (euhedral, grey, bireflective, twinned) and q (anhedral, darker grey), with visible fluid inclusions (black dot trails). The faintly visible brownish-pinkish spots in cst correspond to internal reflections. Sample TF Pn9, Montánchez, Cáceres (**dsp**)

- The influence of adjacent minerals should always be kept in mind when visually estimating the reflectance of cassiterite by brightness. It appears very dark (especially in immersion) against sulphides like pyrite, not so much against hematite or intermediate R sulphides like tetrahedrite but relatively light next to quartz (comp. Figs. 26.1a, b, 26.2, 26.3 and 26.4).

26.3 Spectral Information

26.3.1 Cassiterite Multispectral Reflectance R (in Air, Without Polariser) (%)

λ (nm)	370	400	425	450	475	500	525	550	575	600	625	650	675	700	750	800	850	900	950	1000
μ	13.0	12.8	12.6	12.4	12.2	12.1	12.1	11.9	11.9	11.8	11.7	11.6	11.7	11.6	11.5	11.4	11.4	11.2	11.3	11.2
P_5	12.6	12.5	12.2	12.1	11.8	11.8	11.7	11.6	11.5	11.4	11.3	11.3	11.3	11.2	11.0	11.0	11.0	10.8	10.9	10.7
P_{20}	12.8	12.7	12.5	12.3	12.1	12.0	11.9	11.7	11.7	11.6	11.5	11.4	11.4	11.3	11.3	11.1	11.2	11.0	11.0	11.0
P_{80}	13.1	13.0	12.8	12.6	12.4	12.3	12.2	12.1	12.1	12.0	11.9	11.9	11.9	11.8	11.8	11.6	11.7	11.5	11.6	11.5
P_{95}	13.3	13.2	13.0	12.9	12.6	12.5	12.5	12.3	12.3	12.2	12.2	12.2	12.2	12.2	12.1	12.0	12.0	11.8	12.0	11.9

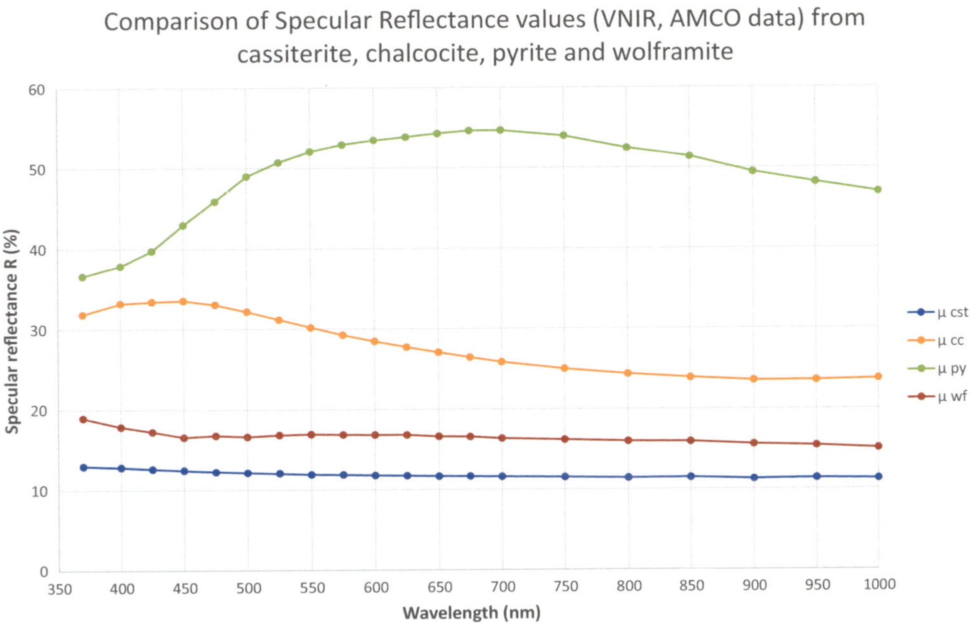
Comparison of Specular Reflectance values (VNIR, AMCO data) from cassiterite, chalcocite, pyrite and wolframite
Specular reflectance R (%)
60
50
40
30
20
10
0
350
400
450
500
550
600
650
700
750
800
850
900
950
1000
Wavelength (nm)
μ cst
μ cc
μ py
μ wf

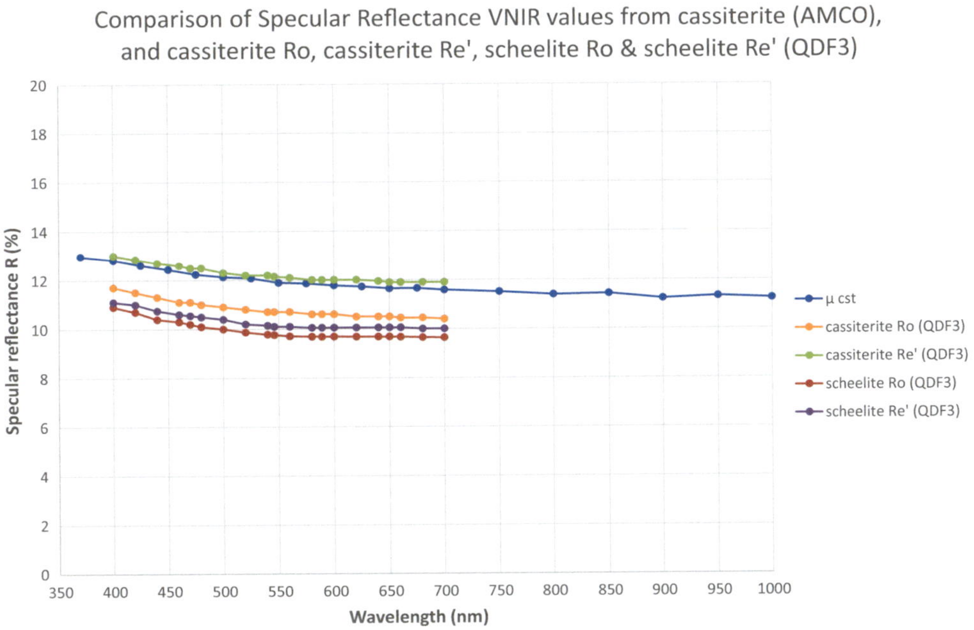
Comparison of Specular Reflectance VNIR values from cassiterite (AMCO), and cassiterite Ro, cassiterite Re', scheelite Ro & scheelite Re' (QDF3)
Specular reflectance R (%)
20
18
16
14
12
10
8
6
4
2
0
350
400
450
500
550
600
650
700
750
800
850
900
950
1000
Wavelength (nm)
μ cst
cassiterite Ro (QDF3)
cassiterite Re' (QDF3)
scheelite Ro (QDF3)
scheelite Re' (QDF3)

Chalcocite (*cc*)

27

Cu_2S: Orthorhombic

Chalcocite		R_{546} (%)	Y (%)	λ_d (nm)	P_e (%)	Colour sensation
Air	R_a	33.4	33.2	480	6.7	Light bluish grey to bluish white
	R_b	33.5	33.1	480	6.8	
	R_c	33.2	32.9	479	7.1	
Oil	R_a	18.3	18.0	479	11.0	
	R_b	18.1	17.9	479	11.1	
	R_c	17.9	17.7	478	11.9	
Bir: Imperceptible (2)				*Pleoc*: Very weak (2)		*Tarn*: Frequent and varied
Anis +P: Weak, but perceptible (3)				*IR*: No		*Twg*: Frequent laminar
Mor: Aggregates, pseudomorphs…				*Clv*: Occasional // (001)		*PolQ*: Good (risk blue polishing tarnish)
Hardness	*PolH*: Low, ≈ gn, < bn (2)			*Mohs*: 3^-	*Talm*: –	*VHN* $_{(100\ g)}$: 84–87, on section (001)

Obs: cf. Table and R charts, multispectral data (VNIR, AMCO) and visible range (QDF3)
Mineral abbreviations: bn—bornite, cc—chalcocite, cv—covellite, dg—digenite, dju—djurleyite, gn—galena, ht—hematite, py—pyrite, stm—stromeyerite, td—tetrahedrite, vio—violarite

27.1 General

Composition: Cu_2S. It may contain impurities of Fe, Mn and Ag, among other cations, while S may be partly replaced by Se or Te. It is a mineral of great economic importance as a Cu ore. The Cu content of cc is ≈80% (79.9%), compared to 34.5% Cu of ccp or 66.5% Cu of cv. Its formation by supergene cementation processes at the expense of other sulphides, such as chalcopyrite or pyrite, may produce a dramatic metallic upgrade of the ore (*supergene enrichment blanket*).

Various compounds are classically understood as modifications of chalcocite, whose complex relationships are briefly sketched, following Ramdohr (1980). Cu_2S formed at low temperature is orthorhombic (β-Cu_2S). It becomes hexagonal by heating. The phase that solidifies from the metallurgical melt is cubic, but it is actually Cu_9S_5, a compound that can also be formed from hydrothermal solutions at moderate temperatures (digenite), and which alters to β-Cu_2S (also known as *paramorphic* or *lamellar chalcocite*).

This author mentions seven "modifications" of chalcocite: orthorhombic (of three possible origins: hypogene, primary supergene and cementative), hexagonal (>103 °C), cubic or α-Cu_2S (digenite), paramorphic (or laminar), djurleite, anilite and grey-pink cc. Their relationships allow some applications as a *geothermometer*. Primary orthorhombic cc indicates T < 103 °C, but if it is paramorphic from hexagonal chalcocite, it indicates genesis at T > 103 °C. Likewise, hexagonal chalcocite can be formed by heating orthorhombic chalcocite above this temperature. Some of these "modifications", *e.g.,* digenite and djurleite, are now recognized as independent species (Roseboom 1962; Pósfai and Buseck 1994).

As far as the ore chalcocite is concerned, it will be characterised here as plainly as possible and, following current criteria (Back and Mandarino 2008), it is described as a single mineral (today cc is considered *monoclinic*). However, this description essentially coincides with that of **djurleite** (dju $Cu_{31}S_{16}$, varying between $Cu^{+}{}_{60}Cu^{2+}{}_{2}S_{32}$ and $Cu^{+}{}_{62}Cu^{2+}{}_{2}S_{32}$, monoclinic, Goble 1985). A reliable distinction of chalcocite and djurleite requires XRD, since their

R. Castroviejo, *A Practical Guide to Ore Microscopy—Volume 1*,
https://doi.org/10.1007/978-3-031-12654-3_27

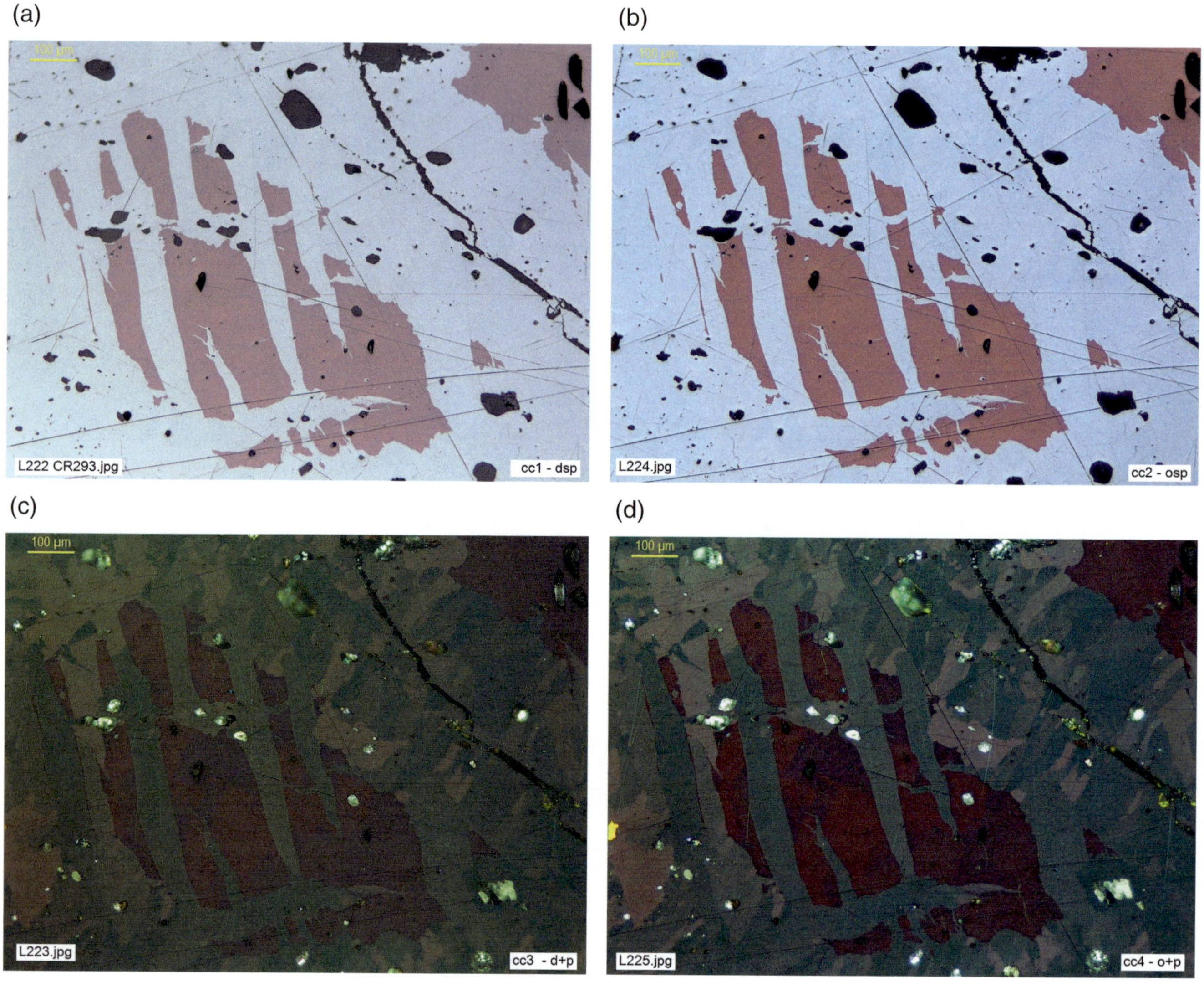

Fig. 27.1 **a** Aggregate of cc (light bluish grey) with corroded, residual bn (brownish-pink) and small q inclusions. CR293, Kipushi, Shaba, Copperbelt RD Congo (**dsp**). **b** Same scene as cc1 (**osp**). **c, d** Same scene as Fig. 27.1, (**d+p**) and (**o+p**) respectively, very weak anisotropism of bn and weak but coloured (not very luminous, greenish and orange tones) of cc

optical properties are similar, although their formula differs slightly. The following description of chalcocite can therefore also be applied, in general lines, for the characterization of djurleite (Fig. 27.5a, b).

Anilite (Cu_7S_4, orthorhombic) has been found finely intergrown with djurleite, but its optical properties could not be properly characterized because it was altered to digenite during the polishing process (Morimoto et al. 1969, in: Uytenbogaardt and Burke 1971).

Digenite is described separately as a different species.

27.2 Microscopic Description

27.2.1 SP Observation (*Single Polariser*)

Bluish-white *colour* (Figs. 27.1a, b, 27.2 and 27.4). The bluish tone is accentuated in contact with galena.

Polishes well, but scratches appear easily. Resins requiring heating >60 °C should be avoided, to prevent low-temperature phase reversal in the Cu_2S-CuS system

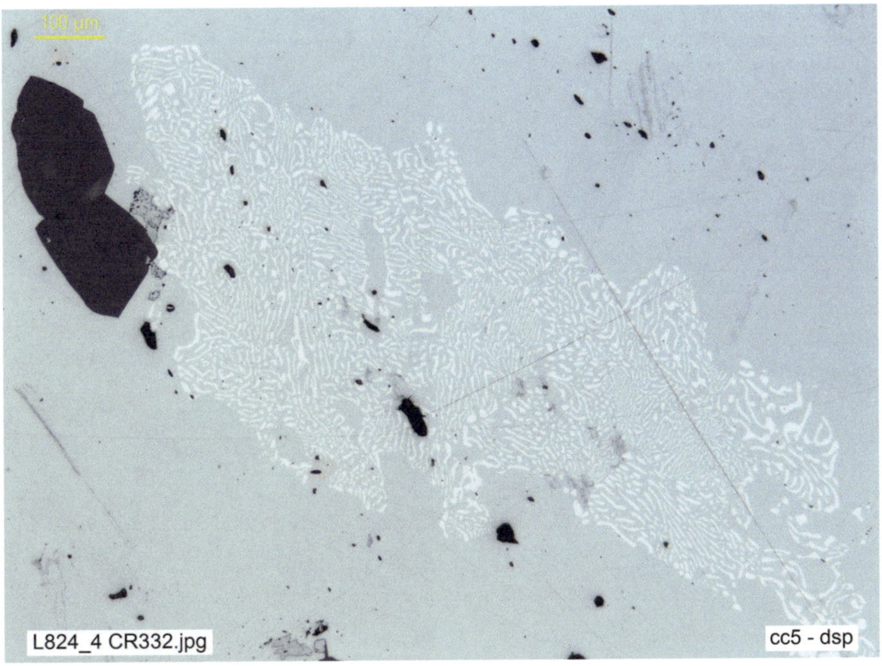

Fig. 27.2 Symplectitic intergrowth of gn (white) in cc (light bluish-grey), small residual inclusions of ht (grey, a little darker than cc, relief) and lamellae (exsolutions?) of stm (grey like ht, but without relief) in cc mass. Sample CR 332, Vrancice, Pribram, Czech Rep. (**dsp**)

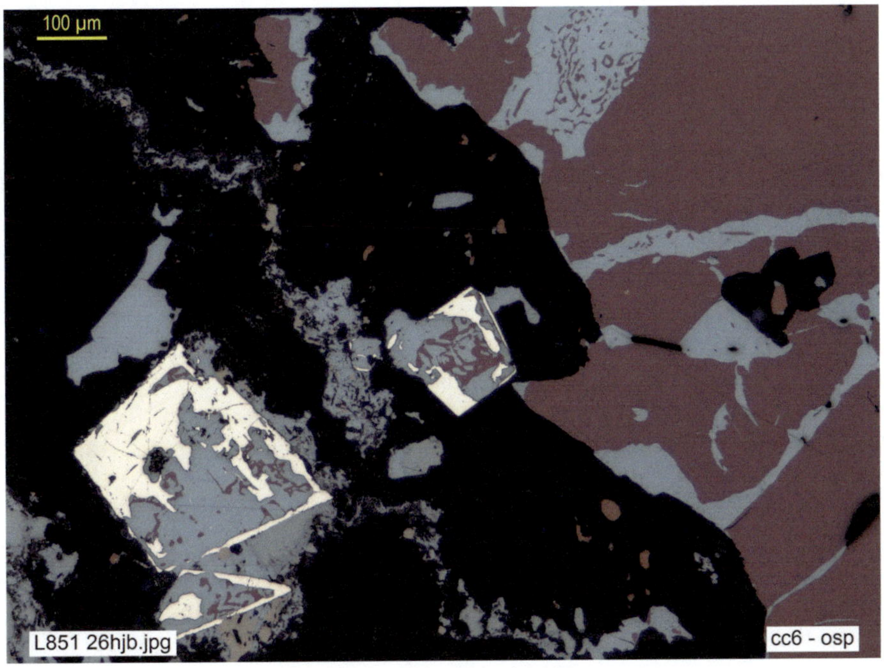

Fig. 27.3 Intergrowth of cc and bn, with partial replacement of py (now reduced to *ghost relicts*) and traces of probable vio (lower left corner, pinkish brown, lighter than bn). Sample 26hjb, Tsumeb, Namibia (**osp**)

(Buerger, in: Ramdohr, 1980). If polishing or cleaning is deficient, sometimes a dark blue surface film or microlayer resembling digenite appears, which masks the calcocite colour and can be cleaned by a quick polishing with 0.25 µm diamond.

Rapid *tarnishing* can present varied tones—vbgr. pinkish in sections//(001)- and mask the true colour of the mineral. If fresh (cc) and tarnished chalcocite (cc_T, *Tarnishing*) are compared, the spectral composition (R) shows a slight decrease in the bluish tone of the latter, cf. graphs R (AMCO).

Moderate *reflectance* of the order of td < gn. Although its structure is anisotropic, its bireflectance is practically nil (cf. comparative multispectral and visible spectrum charts, R_a, R_b and R_c curves, documented in QDF3, overlap) and its pleochroism is rarely perceptible (not to be confused with the differentiated development of tarnishing patina in different grains, depending on orientation).

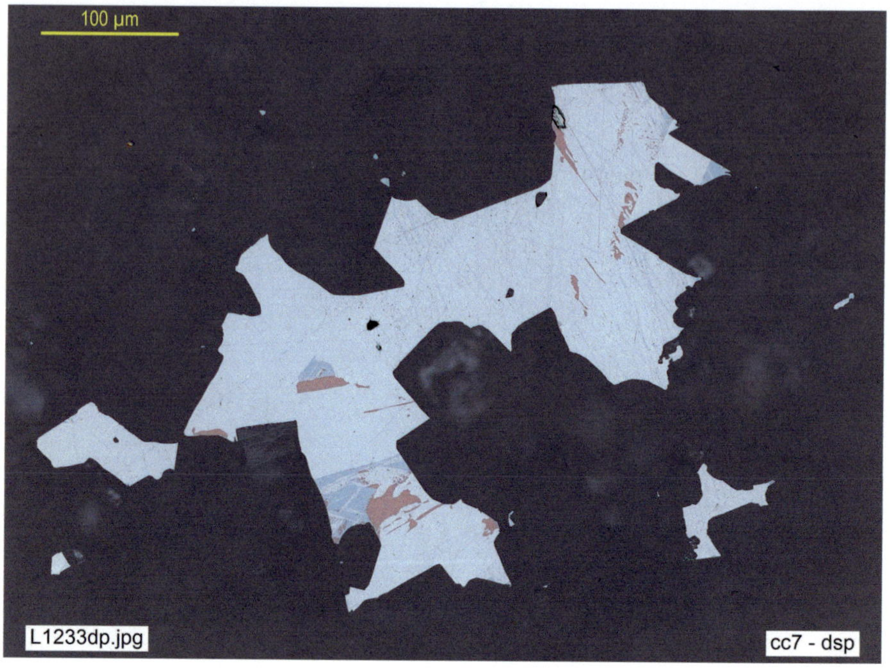

Fig. 27.4 Interstitial aggregate of intergrown cc (bluish white), dg (very light blue) and bn (pinkish brown), filling micro-geode in q. Sample CR 368, Crescent Mine, Olympic Peninsula, Clallan County, Washingon, USA (**dsp**)

Polishing hardness, ***PolH$_{cc}$***, low (transitional between groups low PolH—medium PolH) and orientation dependent, PolH$_{cc}$ $\approx$ gn > sbt. PolH$_{cc}$ < bn, ccp, Ag. Cleavage occasionally visible.

It usually occurs in anhedral aggregates with varied textures, laminar or banded, replacement, filling (Fig. 27.5). Its structure, characterized by the type and morphology of lamellar twins and intergrowths, can be very varied, depending on the primary or secondary genesis, the formation temperature, the evidence of replacement or pseudomorphism, etc. Ramdohr (1982) discusses in detail the morphology of the seven types mentioned and the criteria for distinction. These are sometimes not very obvious, as happens with paramorphic transformation of the high-temperature type (*hexagonal chalcocite*, T $\leq$ 300 °C) into *rhombic chalcocite* (T $\leq$ 103 °C), whose distinction from coarse-grained chalcocite formed at low temperatures is very difficult. It may be manifested by the presence of lanceolate, flamboyant or *oleander leaf-type lamellae*. These are different from the lamellae of the type defined as *lamellar chalcocite*, secondary of digenite, whose octahedral structure, evidenced by the (111) orientations of the lamellae, is inherited by cc (Fig. 27.5a, b).A similar structure is obtained by replacing bornite in the laboratory.

Intergrowths and exsolutions are frequent in hypogene chalcocite, even *myrmekitic* (or, more properly, *symplectitic*, Fig. 27.2) intergrowths, while supergene chalcocite is usually fine-grained or very fine-grained. In all cases, replacement processes are common (Figs. 27.1b, 27.2, 27.3 and 27.5a, b), evidenced by pseudomorphism or by textures and reaction rims, in which chalcocite may be the primary or secondary mineral.

The *artificial patina* obtained with a simple acid etch test helps to identify chalcocite in case of doubt, particularly to distinguish it from the so-called *fahlores* (tetrahedrite *s. l.*).The grain in question is attacked with 1 drop of HNO_3 for 5 s and immediately washed with water; if it is chalcocite, it is covered with a blue patina (of covellite), which does not occur if it is a *fahlore*. This etch test also evidences the lamellar texture of chalcocite, which allows us to distinguish cc that is pseudomorph after digenite, making visible the octahedral orientation of the lamellae.

27.2.2 Observation +P (*Polariser + Analyser at 90°*)

Weakly *anisotropic*, but it shows very characteristic greenish and orange polarization tints, in spite of its low luminosity (Fig. 27.1c, d). They are more visible by slightly uncrossing the analyser or in immersion.

Very frequent and varied lamellar *twins*, as explained (cf. morphology).

27.2.3 Paragenesis

Paragenesis varies in terms of the type of deposit, given its wide formation conditions in the hydrothermal (usually < 200 °C) and supergene domains, occasionally contact metamorphic and, very rarely, orthomagmatic. It is frequent in epithermal deposits and very common in stratabound

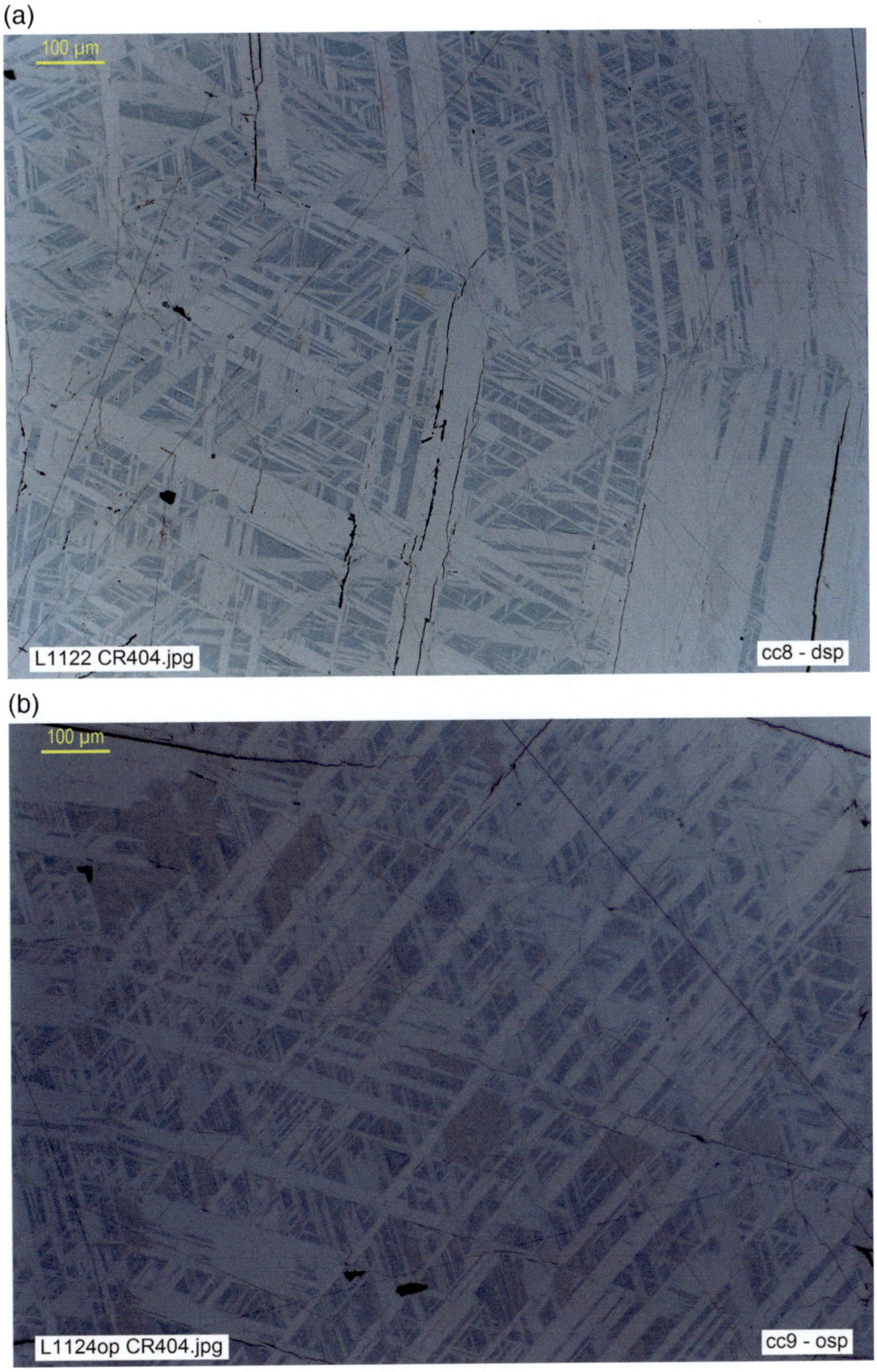

Fig. 27.5 **a, b** Two views, respectively (**dsp**) and (**osp**), of aggregate of dju (blueish-white) and dg (light blue), with *paramorphic* or *lamellar* cc morphology: dju secondary to dg, which is preserved in the interstices between the dju lamellae. The orientation of these lamellae is controlled by the cubic structure of dg and is arranged according to octahedral (111) directions of dg. The identity of dju, whose optical properties are similar to those of cc, has been confirmed by XRD. Sample CR404, Las Cruces Mine, Seville

Copperbelt- or *Kupferschiefer-type* concentrations. Of particular importance are the *porphyry-copper* deposits, which can develop extensive supergene sulphide *enrichment blankets*, as well as the massive sulphide deposits.

Typical companions in hypogene formations are chalcopyrite, bornite, sphalerite, galena, tetredrite, pyrite; and in supergene formations, digenite, covellite, native silver, native copper, cuprite, goethite, among many others.

27.2.4 Identification

Its anisotropism—with dull but unmistakable orange and greenish colour tones—and its lamellar twinning, as well as its colour, moderate reflectance and low polishing hardness are characteristic. The association with cupriferous paragenesis, particularly digenite and covellite, is an additional criterion, which can be confirmed by the artificial blue patina obtained by acid etching (HNO_3).

27.3 Spectral Information

27.3.1 Chalcocite Multispectral Reflectance R (in Air, Without Polariser) (%)

λ (nm)	370	400	425	450	475	500	525	550	575	600	625	650	675	700	750	800	850	900	950	1000
μ	31.9	33.2	33.4	33.5	33.1	32.2	31.2	30.2	29.2	28.5	27.7	27.1	26.4	25.9	25.0	24.3	23.9	23.5	23.5	23.7
P5	27.9	29.6	30.1	30.4	30.2	29.7	29.0	28.2	27.5	26.8	26.2	25.6	25.0	24.5	23.5	22.8	22.3	21.9	21.7	21.6
P20	29.2	31.0	31.6	31.9	31.7	31.0	30.1	29.2	28.4	27.6	26.9	26.3	25.7	25.1	24.1	23.5	23.1	22.7	22.7	22.9
P80	34.9	35.8	35.6	35.4	34.7	33.6	32.4	31.4	30.2	29.4	28.6	27.9	27.3	26.6	25.8	25.2	24.7	24.4	24.4	24.7
P95	35.4	36.2	36.1	35.8	35.1	34.0	32.8	31.7	30.6	29.8	29.0	28.3	27.8	27.2	26.5	25.9	25.4	25.2	25.2	25.4

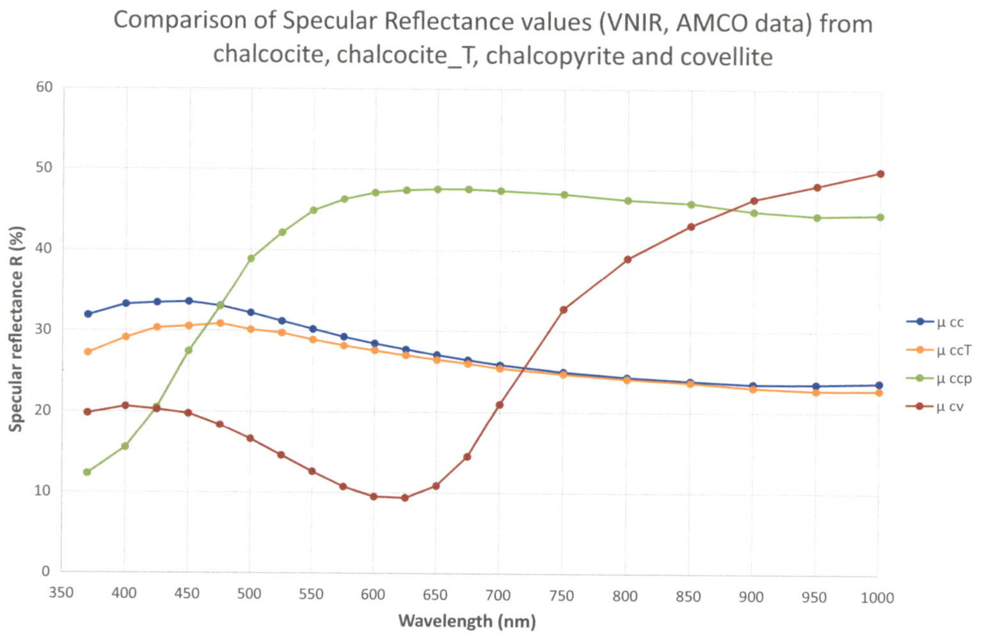

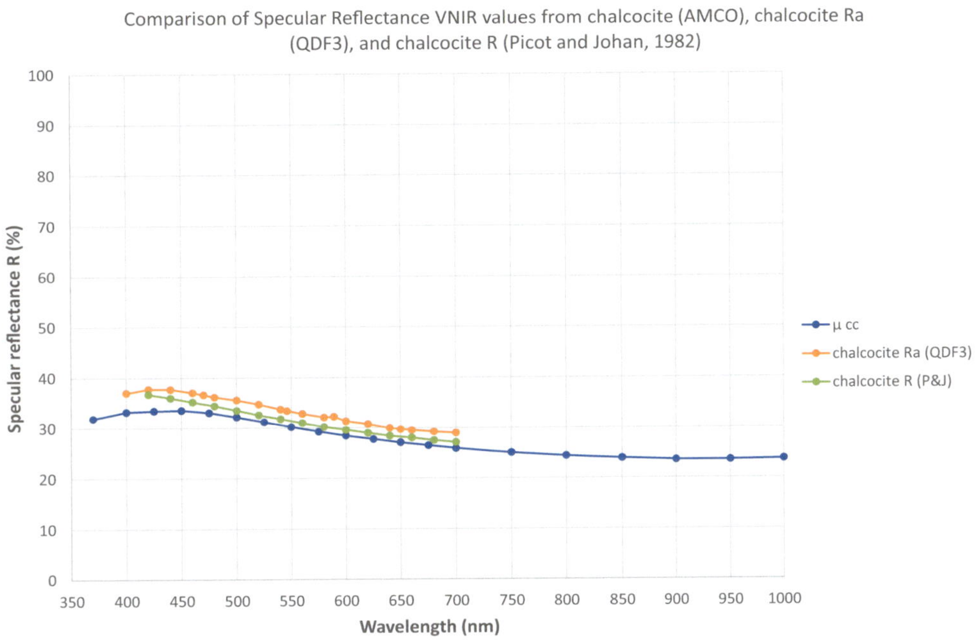

References

Back ME, Mandarino JA (2008) Fleischer's glossary of mineral species 2008. The Mineralogical Record Inc. Tucson, AZ, 345p

Goble RJ (1985) The relationship between crystal structure, bonding and cell dimensions in the copper sulfides. Can Mineral 23:61–76

Pósfai M, Buseck PR (1994) Djurleite, digenite, and chalcocite: intergrowths and transformations. Am Miner 79:308–315

Ramdohr P (1980) The ore minerals and their intergrowths, 2nd edn. Pergamon Press, Oxford. 2 volumes, 1205p

Roseboom EH (1962) Djurleite, Cu1.96S, a new mineral. Am Miner 47:1181–1184

Uytenbogaardt W, Burke EAJ (1971) Tables for microscopic identification of ore minerals, 2nd edn. Elsevier, Amsterdam

28 Chalcopyrite (*ccp*)

$CuFeS_2$: Tetragonal

Chalcopyrite		R_{546} (%)	Y (%)	λ_d (nm)	P_e (%)	Colour sensation
Air	R_o	44.6	44.1	574	24.5	Bright yellow colour
	R_e	45.0	45.1	574	23.6	
Oil	R_o	nd	nd	nd	nd	
	R_e	nd	nd	nd	nd	
Bir: Imperceptible (2)				*Pleoc*: Imperceptible (2)		*Tarn*: Very frequent, dark and coloured
Anis +P: Weak (3)				*IR*: No		*Twg*: Frequent and varied
Mor: Varied, massive, colloidal, etc				*Clv*: Occasional//(111)		*PolQ*: Easy, good
Hardness	*PolH*: Mean: > gn, < sp,po (3)			*Mohs*: 4^-	*Talm*: C	*VHN* $_{(100\ g)}$: 187–203 (b); 181–192 (v)

Obs: cf. Table and R charts, multispectral data (VNIR, AMCO) and visible range (QDF3)
Mineral abbreviations: Ag—native silver, Au—native gold, asp—arsenopyrite, az—azurite, bn—bornite, brv—bravoite, cbn—cubanite, cc—chalcocite, ccp—chalcopyrite, crc—chrysocolla, cv—covellite, dg—digenite, PGE—Pt group elements, gg—gangue, gn—galena, ht—hematite, il—ilmenite, isc—isocubanite, ln—linnaeite, mch—malachite, mc—marcasite, mck—mackinawite, pn—pentlandite, po—pyrrhotite, py—pyrite, sp—sphalerite, stn—stannite, td—tetrahedrite, vae—vaesite, vio—violarite

28.1 General

Chalcopyrite, ***ccp***, composition is $CuFeS_2$, with possible traces of Ag, Au, EGP, Se, Te, In, Tl. It admits variable amounts of Fe, Zn, Sn, Ni, Co in the form of sulphides (FeS, ZnS, etc.), especially at high temperatures. On cooling, ccp must balance its composition, eliminating components in excess in the form of exsolutions—pyrrhotite, cubanite (Fig. 28.2) and/or mackinawite (Figs. 28.7 and 28.8) for FeS, sphalerite (Fig. 28.4) for ZnS, stannite for Sn.

The high-temperature, isometric form used to be called *cubic chalcopyrite* (*chalcopyrrhotite* to other authors), but it is now considered a different species, recognized by IMA as *talnakhite*, IMA formula $Cu_9Fe_8S_{16}$, or $Cu_9(Fe,Ni)_8S_{16}$. The term chalcopyrrhotite has also been discredited by IMA in favor of *isocubanite* (cf description *isc*).

28.2 Microscopic Description

28.2.1 SP Observation (*Single Polariser*)

Light yellow, bright *colour* (Fig. 28.1a, b), although with time it darkens considerably (tarnishing). The colour impression can vary considerably with the composition (e.g., Se impurities give it a brownish colour, cf. Figs. 4.1a, b and 4.2) and with the neighbouring ores (e.g., next to native gold, chalcopyrite appears much darker, with greyish or greenish tones).

Easily *polished,* medium–low *polishing hardness* (Fig. 28.2) between gn and sp. $PolH_{ccp}$ > gn. $PolH_{ccp}$ < Ag, stn, cbn, sp, po; << py. Cleavage//(111) rarely visible under microscope.

R. Castroviejo, *A Practical Guide to Ore Microscopy—Volume 1*,
https://doi.org/10.1007/978-3-031-12654-3_28

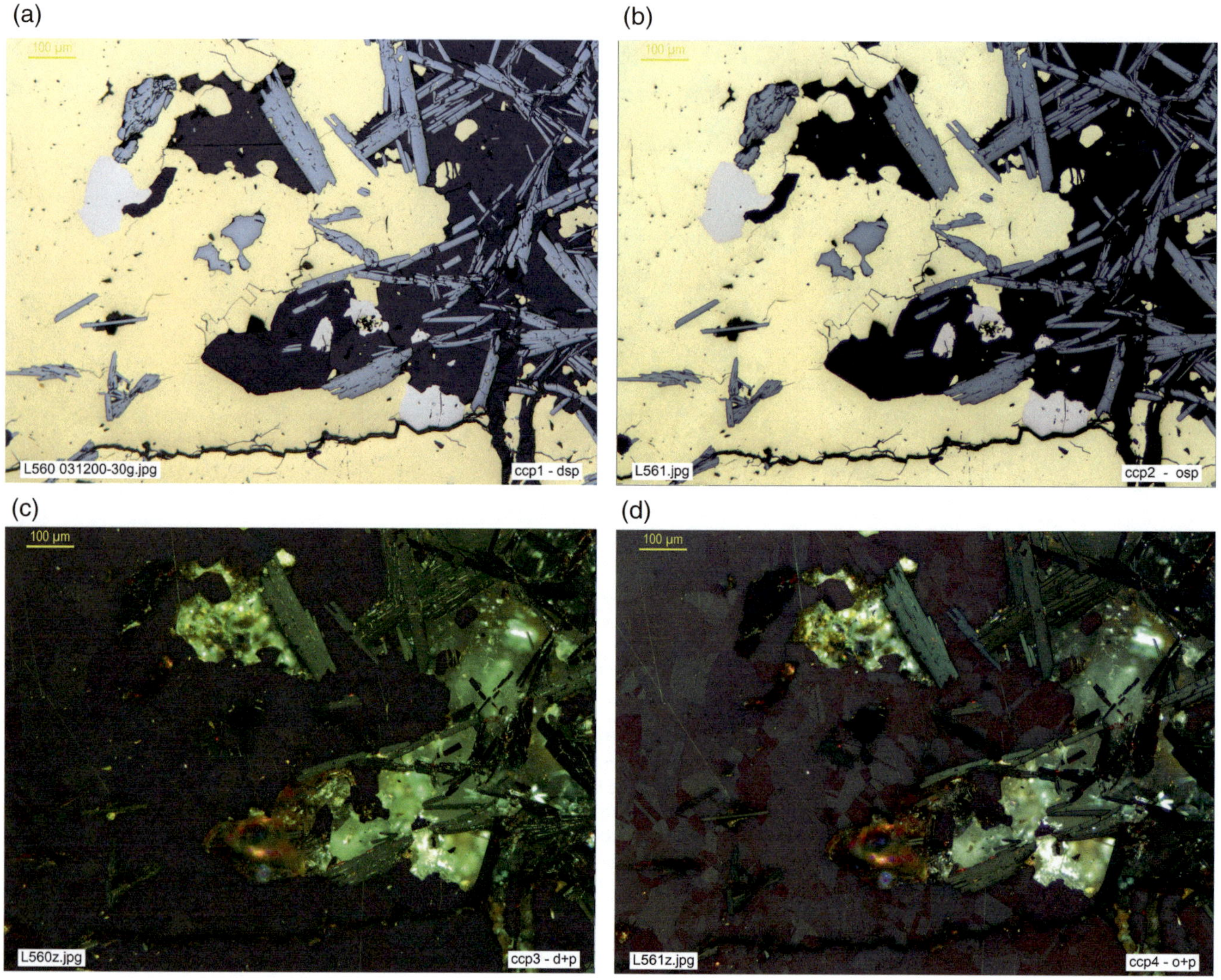

Fig. 28.1 **a**, **b** Metamorphic ore of ccp (yellow) with inclusions of linnaeite (slightly pinkish white; *ln s.str.* composition confirmed by electron microprobe) and relicts of ht (grey), also prismatic, diablastic ht, included in gg (dark grey). Sample 031200_30g, Mina Maruxa, Moeche, Coruña (**dsp**) and (**osp**), respectively. **c**, **d** Same scene as Fig. 28.1 (**d +p**) and (**o+p**), respectively. The weak anisotropism of ccp allows distinguishing the twins and evidences its granoblastic texture, in agreement with the diablastic texture of ht

Relatively high *reflectance* (Figs. 28.2, 28.5 and 28.6), close but slightly higher than galena, lower than pyrite (cf multispectral comparative charts at the end of the description).

Chalcopyrite usually occurs in anhedral masses or aggregates of variable granulometry, frequently as vein filling or disseminated. Also in stratiform concentrations with varied morphologies, laminar, massive, pseudomorph (by replacement) and collomorph. It acquires granoblastic texture in metamorphic contexts. It is sensitive to deformation and shows a more ductile behaviour than sphalerite or pyrite.

28.2.2 Observation +P (*Polariser + Analyser at 90°*)

Weakly *anisotropic* (Fig. 28.1c, d) but, by slightly uncrossing the analyser, varied colours of polarisation (bluish grey, violet, greenish, yellowish) appear, which show the different grain orientations and help to define textures. Twins are also enhanced (comp. Fig. 28.1d with Fig. 28.3). An excess of Fe content can lead to a more intense anisotropism (Uytenbogaardt and Burke 1971).

Twinning is ubiquitous and varied (Figs. 28.1c, d, 28.3, 28.7, 28.8 and 28.9), growth, deformation or transformation.

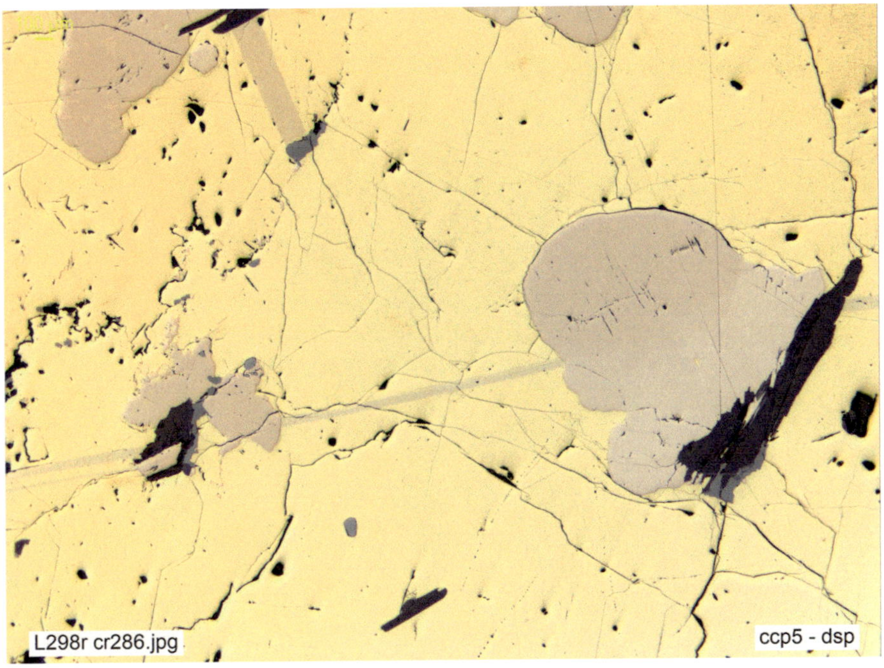

Fig. 28.2 Massive aggregate of ccp, with inclusions of po (brownish-pink) and cbn (like po, but slightly lighter; laminar or prismatic habit), sp (grey), traces of mck (light pink lamellae), rt (grey, slightly lighter than sp) and gg (dark grey).The higher polishing hardness of po makes it appear in relief on ccp (*vbgr* bigger inclusion, right center), which is not the case with cbn, whose hardness is closer to that of ccp. CR 286, Gihen Mine, Sulitjelma, Norway (**dsp**)

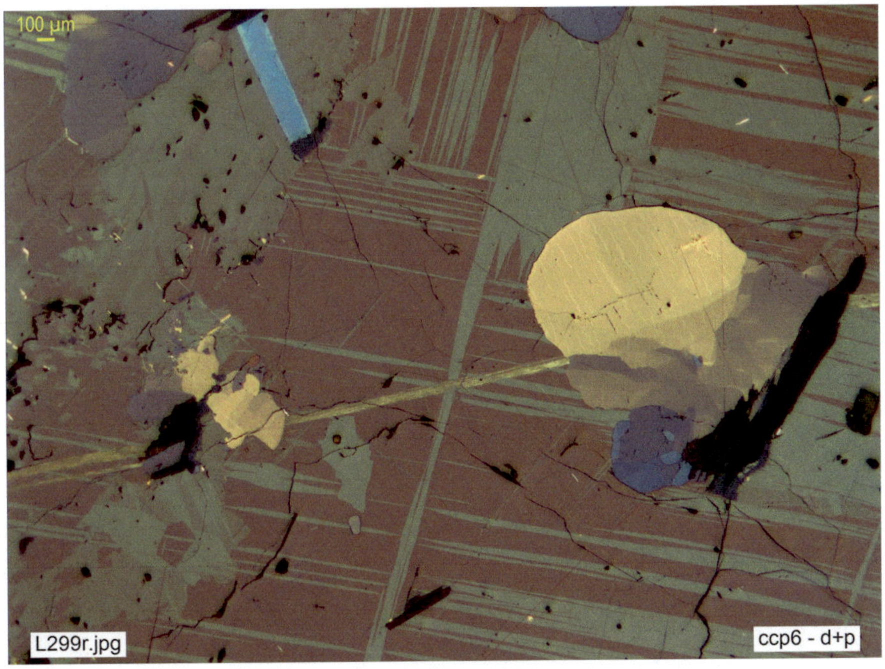

Fig. 28.3 Same scene as ccp5 (**dxp**). Coloured anisotropism and lamellar and lanceolate, grid-like, twins of ccp clearly visible when the analyser is slightly uncrossed. Bluish and yellowish-brown polarisation tints of cbn differ from those of po, mck evidenced by its strong anisotropism (scattered luminous lamellae in ccp)

It may also show several possible combinations, (101), (111), (110) grid. Typical lamellar, lanceolate twins (*oleander leaf*) are traditionally attributed to inversion from high-temperature chalcopyrite.

28.2.3 Paragenesis

Paragenesis varies from high temperature liquid-magmatic concentrations, with Ni sulphides, to volcanic-sedimentary exhalative deposits (Cu-Zn-Pb) and supergene concentrations, passing through magmatic-hydrothermal concentrations (copper porphyries), current deposits in suboceanic *black-smoker* environments and diverse metamorphic concentrations.

It can occur with practically any ore, but it is particularly frequent with sphalerite, pyrite, galena, pyrrhotite, bornite, cubanite, pentlandite, arsenopyrite, chalcocite, covellite, and digenite.

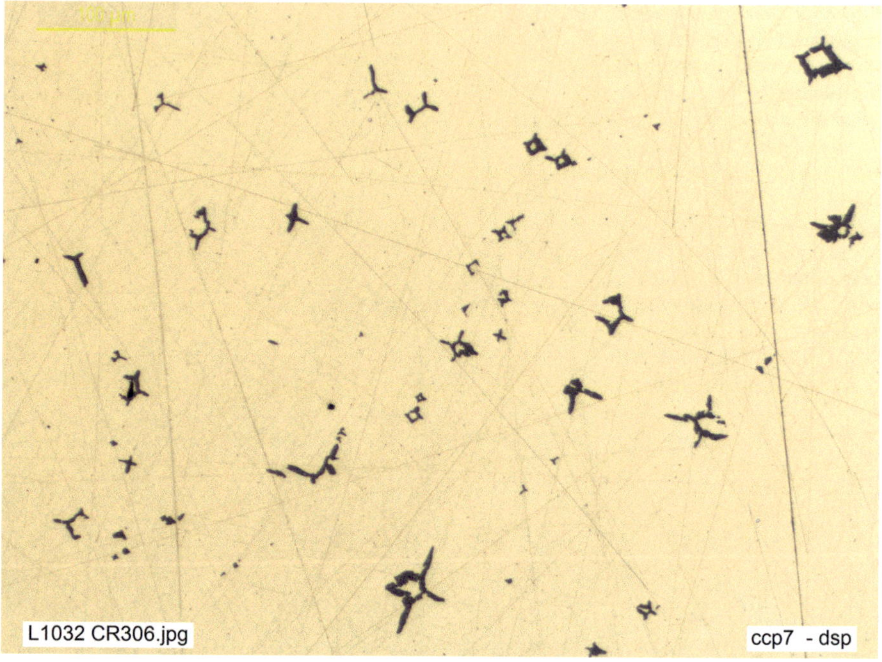

Fig. 28.4 Exsolution of sp (dark grey *stars*) in ccp. CR 306, Mina As Sombras, Lobios, Orense (**dsp**)

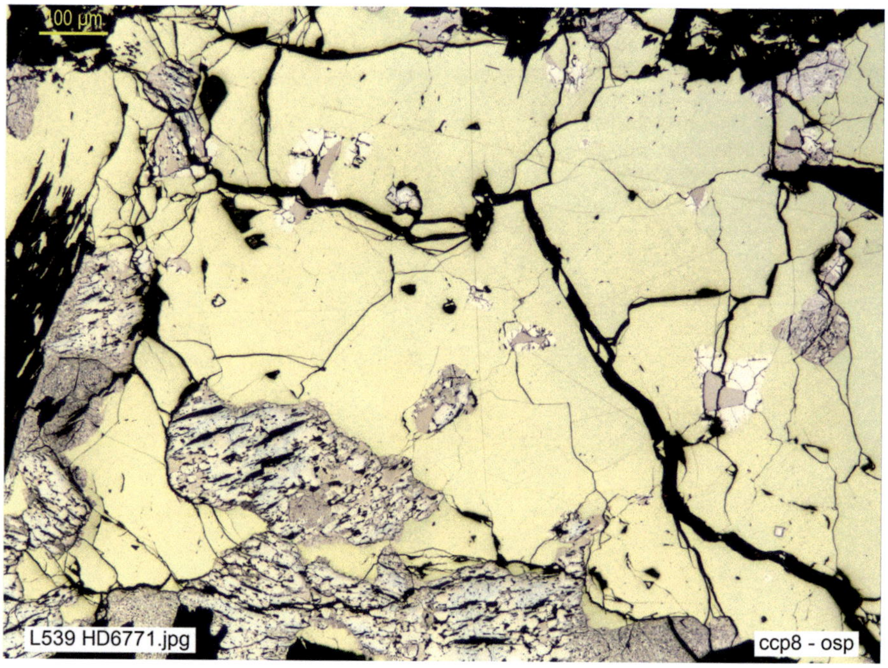

Fig. 28.5 Primary inclusions of po (pinkish-brown) and pn (white) in massive ccp, po largely replaced by secondary aggregates of mc (white, slightly bluish), py (yellowish-white), vio (brownish-pink) and brv/vae (purplish-brown, somewhat darker than po). HD 6771, Horbach, Schwarzwald, Germany (**osp**)

28.2.4 Identification

Its characteristics, particularly its colour and weak anisotropism, make it unmistakable. Although gold also shows intense yellow colour, its reflectance is incomparably higher and its surface appearance is very different, because of the numerous polishing scratches of Au. Millerite shows yellow colour, but paler. It is also pleochroic and strongly anisotropic. The yellow of pyrite is even paler, and pyrite is isotropic and much harder than chalcopyrite. The paragenesis and the frequent association with secondary Cu minerals, such as chalcocite, covellite, digenite, azurite, malachite, chrysocolla, etc. are additional criteria.

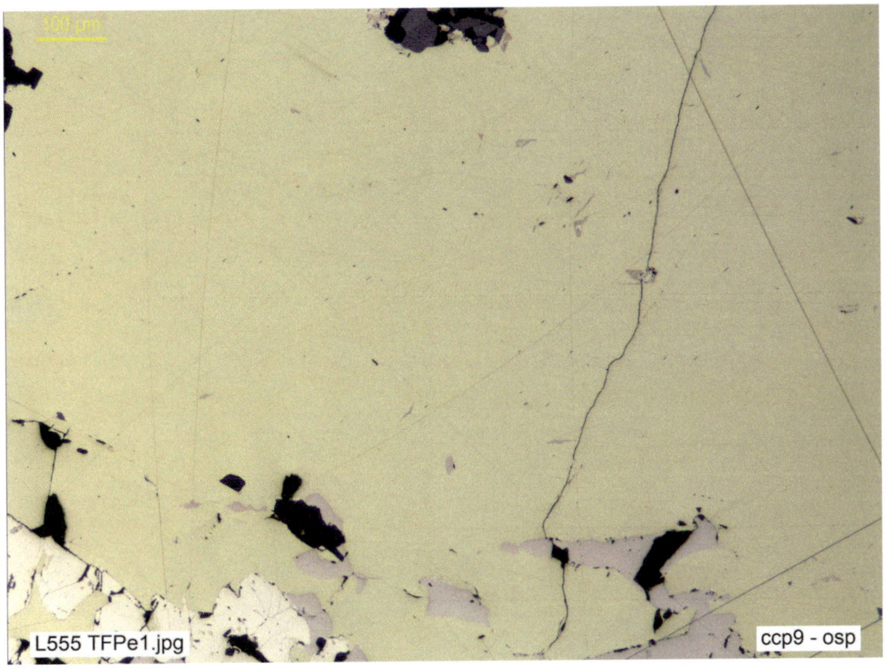

Fig. 28.6 Massive ccp ore with inclusions of pn and po (and traces of mck and il). TF_Pe1, Cobalt Mine, Ontario, Canada (**osp**)

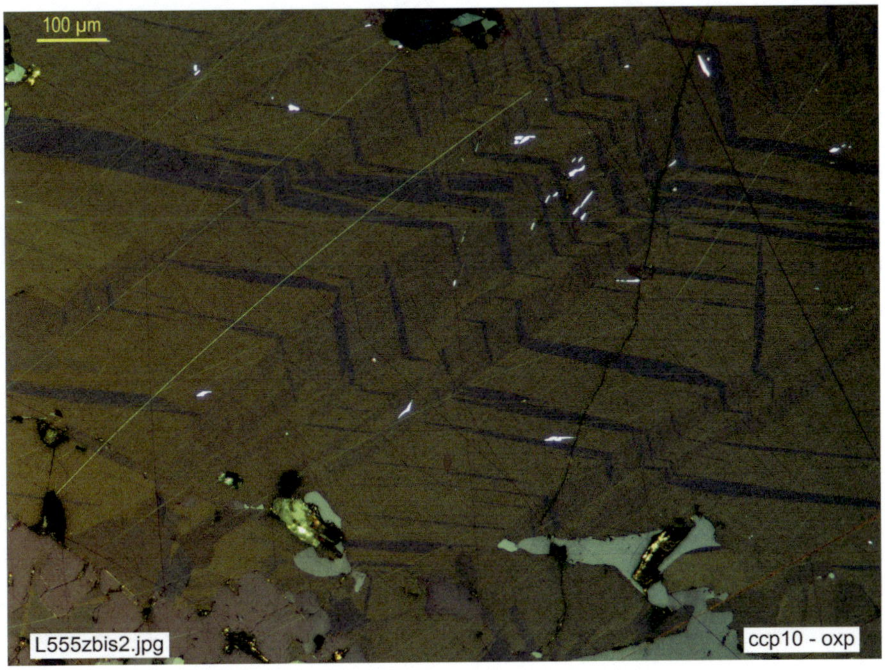

Fig. 28.7 Same scene as ccp9 (**oxp**). Lanceolate lamellar twins of ccp, with *kink-bands* showing deformation. Abundant scattered lamellae of mck (very anisotropic, luminous, white)

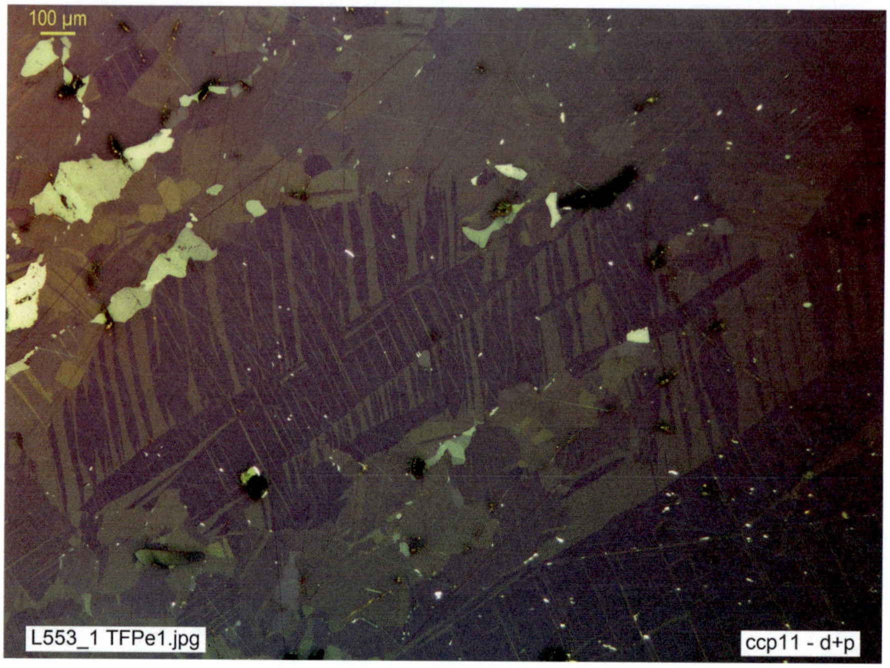

Fig. 28.8 Deformed granoblastic mass of ccp, with po inclusions in microbands (yellowish-white) and dissemination of tiny mck flakes (frequent, very luminous); ccp shows laminar and deformation twins in the central zone. Same sample as ccp9 (**d+p**)

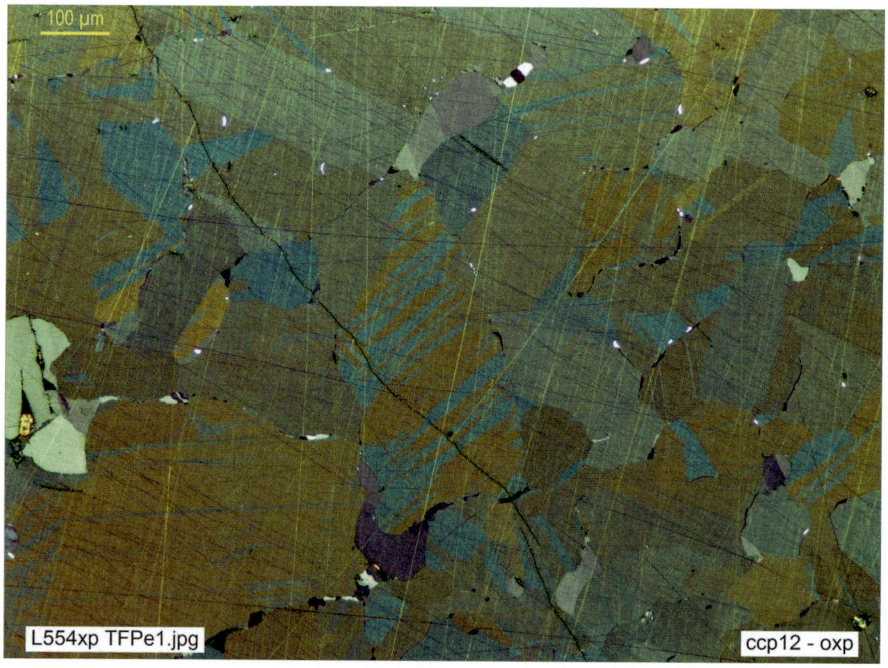

Fig. 28.9 Anisotropism and polarisation colours of chalcopyrite, with partially uncrossed analyser. Twins and granoblastic texture in ccp. Minor inclusions of po and traces of mck. Same sample as ccp9 (**oxp**)

28.3 Spectral Information

28.3.1 Chalcopyrite Multispectral Reflectance R (in Air, Without Polariser) (%)

λ (nm)	370	400	425	450	475	500	525	550	575	600	625	650	675	700	750	800	850	900	950	1000
μ	12.3	15.5	20.5	27.5	33.0	38.9	42.2	44.9	46.3	47.1	47.4	47.6	47.6	47.4	47.0	46.3	45.9	44.9	44.4	44.5
P_5	11.0	14.1	18.9	25.4	31.0	36.9	40.3	43.3	44.9	45.9	46.3	46.6	46.6	46.5	46.2	45.6	45.2	44.1	43.6	43.7
P_{20}	11.4	14.5	19.5	26.3	32.0	37.9	41.2	44.1	45.6	46.5	46.9	47.1	47.1	47.0	46.6	46.0	45.6	44.5	43.9	44.0
P_{80}	13.1	16.5	21.5	28.7	34.1	40.0	43.2	45.8	47.1	47.8	48.0	48.1	48.1	47.9	47.4	46.7	46.3	45.3	44.8	45.0
P_{95}	13.6	17.0	22.1	29.4	34.9	40.7	43.8	46.3	47.6	48.3	48.5	48.5	48.5	48.3	47.7	47.0	46.6	45.6	45.2	45.6

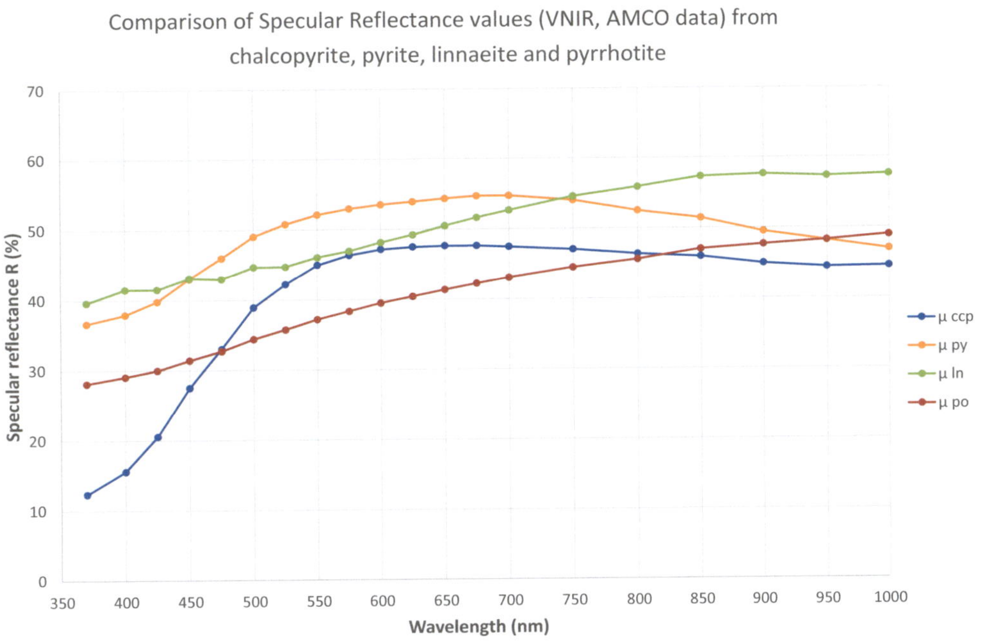

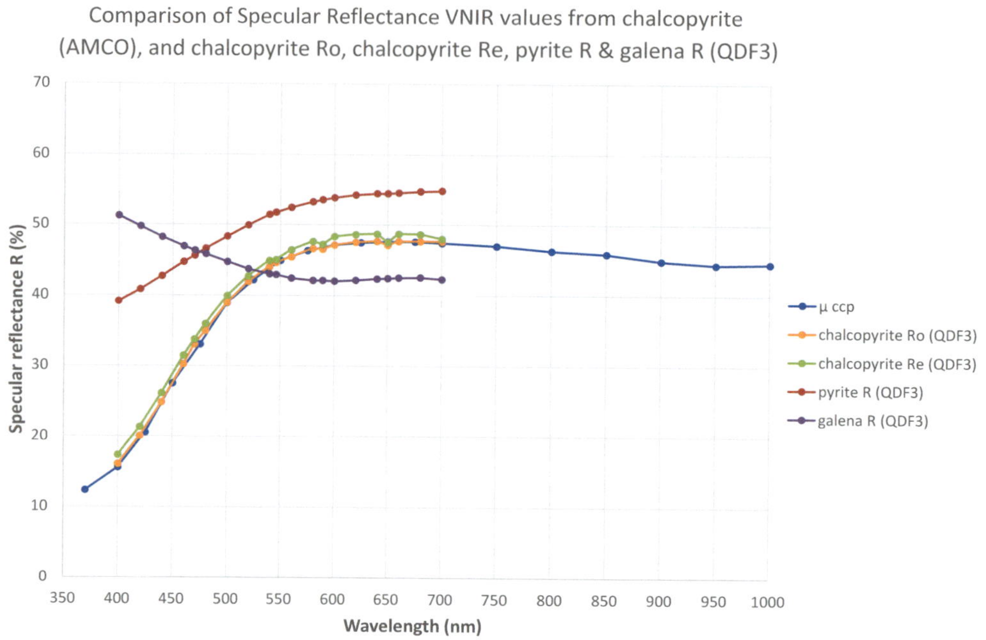

Reference

Uytenbogaardt W, Burke EAJ (1971) Tables for microscopic identification of ore minerals, 2nd edn. Elsevier, Amsterdam

29 Chlorargyrite (*cag/cerargyrite*)

AgCl: Cubic

Chlorargyrite	R_{546} (%)	Y (%)	λ_d (nm)	P_e(%)	Colour sensation
Air	8.8	9.0	589	1.9	Grey
Oil	nd	nd	nd	nd	
Bir: Null (1)			*Pleoc*: Null (1)		*Tarn*: Darkens when exposed to light
Anis +P: Null (1)			*IR*: Colourless		*Twg*: No
Mor: Inlays, cracks. Cub/Octahed			*Clv*: No		*PolQ*: Difficult (very soft, scratches)
Hardness	*PolH*: Very low (1)		*Mohs*: 2	*Talm*: –	*VHN* $_{(100\ g)}$: nd

Obs: Measured reflectance (QDF3 8.8%) < theoretical R (calculated, Ramdohr 1980 11.3%), due to poor polishing?
Mineral abbreviations: ac—acanthite, calc—calcite, gg—gangue, gth—goethite, ht—hematite, mgh—maghemite, mt—magnetite, q—quartz, py—pyrite

29.1 General

Composition: AgCl, miscible with AgBr, may contain I, trace Hg, Na; has been an important silver ore in the past.

29.2 Microscopic Description

29.2.1 SP Observation (*Single Polariser*)

Grey *colour* (Fig. 29.1a, b). *Reflectance*, R, low but higher than usual gangue minerals, even carbonates (compare with calcite Fig. 29.1a, b). Although it is difficult to measure R accurately because of its poor polishing, it can be estimated at ≈ 11%, from its refractive index (n = 2.07), applying Fresnel's formula or directly from the n-R graph (§ 1.2.2). It tends to darken, sometimes rapidly, by exposure to light (*light etching*), tarnishing with a brownish-violet shade. Isotropic.

Difficult to polish, it is usually scratched because it is very soft. Very low *polishing hardness* (≈ ac), lower than all ores considered by Schneiderhöhn (1952), including acanthite.

It usually occurs as an impregnation, in veinlets and in incrustations on silver ores. In open spaces (pores, geodes, druses) it develops cubic and octahedral forms.

29.2.2 Observation +P (*Polariser + Analyser at 90°*)

Isotropic, but its extinction is partly overprinted by clear, generally colourless, *internal reflections* (Fig. 29.1c, d).

29.2.3 Paragenesis

Paragenesis typical of weathering of silver deposits, particularly in oxidation zone (gossan), whenever Cl is available.

29.2.4 Identification

Its inconspicuous appearance means that it can easily go unnoticed. Useful criteria for its recognition are its argentiferous paragenesis, its very low hardness and consequent difficult polishing, its intermediate reflectance between gangue (quartz, calcite) and usual supergene oxides (goethite, hematite, maghemite) and its isotropy, which distinguishes it from the oxidized Pb and Zn minerals present in the gossan.

R. Castroviejo, *A Practical Guide to Ore Microscopy—Volume 1*,
https://doi.org/10.1007/978-3-031-12654-3_29

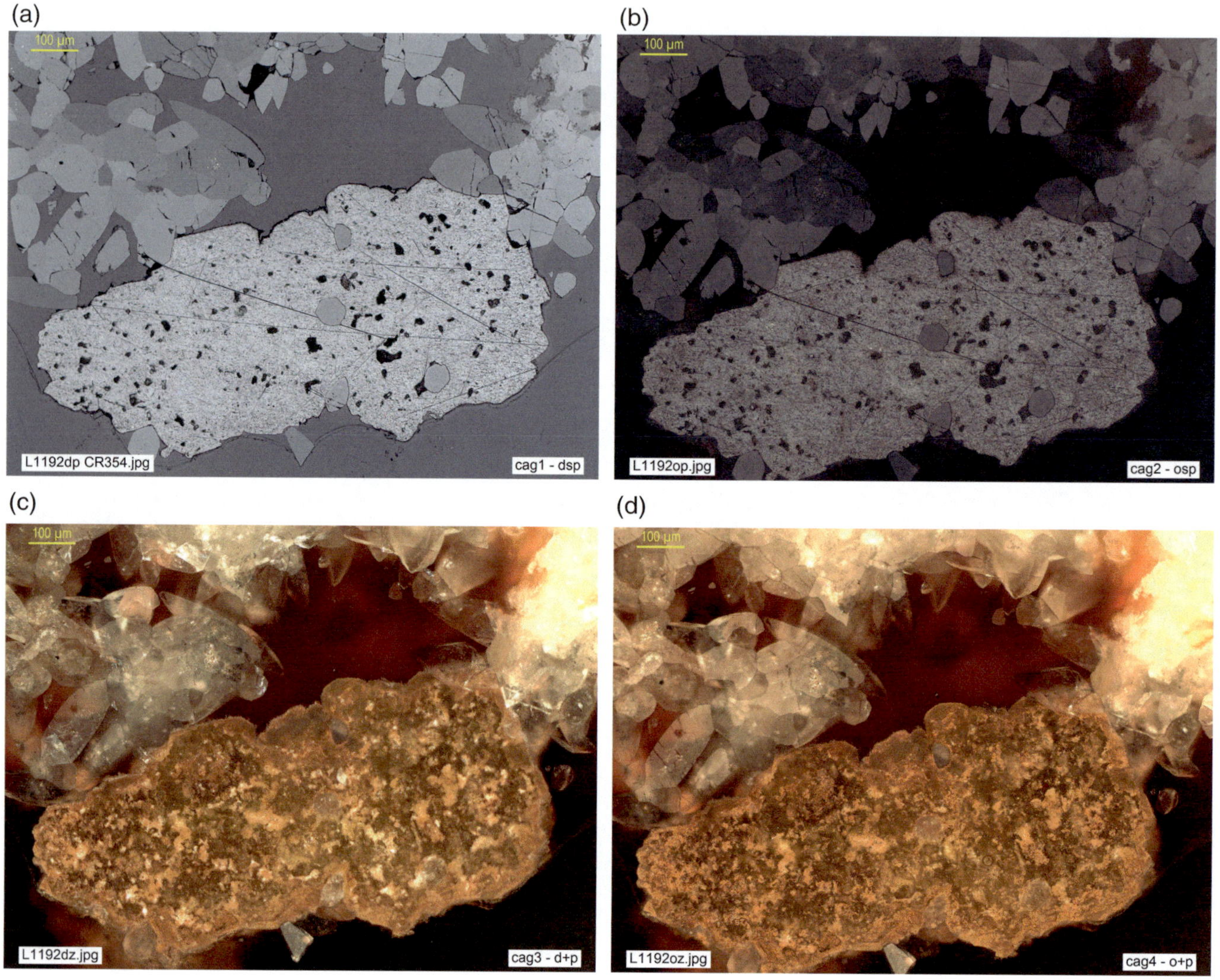

Fig. 29.1 **a**, **b** (Same scene: **dsp** and **osp**, respectively). Impregnation of cag (light grey, scratched and with polishing break-offs, lower half of the image) on microgeode partially sealed by crystalline calcite (darker grey, perfectly polished, strongly bireflective). Note the higher reflectance of cag in the areas where it is well polished, even compared to the brightest calcite position. Compared to the tiny py inclusions in calcite (upper left), cag is much darker. CR354, Broken Hill, New South Wales, Australia. **c**, **d** Same scene as cag1 (**d+p** and **o+p**, respectively), extinction of cag, overprinted by its IR

Reference

Ramdohr P (1980) The ore minerals and their intergrowths, 2nd edn. Pergamon Press, Oxford. 2 volumes, p 1205

Schneiderhöhn H (1952) Erzmikroskopisches Praktikum. E. Schweizerbart'sche Verlagsbuchhandlung, Stuttgart, p 274

Chromite (*chr*)

$FeCr_2O_4$: Cubic

30

Chromite		R_{546} (%)	Y (%)	λ_d (nm)	P_e (%)	Colour sensation
Air		13.5	13.5	478	2.6	Grey
Oil		3.1	3.1	478	4.4	
Bir: Null (1)				*Pleoc*: No (1)		*Tarn*: No
Anis +P: No (1)				*IR*: Generally visible		*Twg*: No
Mor: Euhedral (cubic); granobl, massive				*Clv*: Sporadic//(111)		*PolQ*: Good, but slow (hard)
Hardness	*PolH*: High: > mt, < ht (4)			*Mohs*: 5.5	*Talm*: G+	*VHN* $_{(100\ g)}$: 1248–1456

Obs: cf. Table and R charts, multispectral data (VNIR, AMCO) and visible range (QDF3)
Mineral abbreviations: asp—arsenopyrite, PGE—Pt-group elements, Fe—native iron, gg—gangue, ht—hematite, il—ilmenite, mt—magnetite, nc—nickelite, pn—pentlandite, po—pyrrhotite, py—pyrite, rt—rutile, spn—spinel, ulv—ulvöspinel

30.1 General

Chromite, *chr*, is the main chromium ore, and when Cr-poor, it still finds application as a refractory. The $FeCr_2O_4$ composition ideally attributed to chromite ($Fe^{2+}Cr_2O_4$, according to IMA) is usually not exactly that of most natural chromites, which cover a wide compositional range in the Spinel Subgroup (Oxyspinel Group/Spinel Supergroup, cf. https://www.mindat.org/min-1036.html). That is why Ramdohr (1980), taking into account the possible substitutions between divalent cations (Fe^{2+}, Mg, Mn, Zn) and between trivalent cations (Cr, Al, Fe^{3+}, Ti), explains chromite as "(Fe, Mg) (Cr, Al, Fe)$_2$O$_4$, with small amounts of Mn, Zn, Ti, etc.". For example, in *ferritchromite*, Fe^{3+} replaces Cr and Al. Considered as a species among the spinels, i.e., a Cr-rich spinel, chromite forms several series of solid solution, e.g., chromite-magnetite, chromite-spinel, chromite-hercynite, chromite-magnesiochromite.

The latter is important because of the frequent presence of Mg in natural chromites, which can be favourable for its metallurgical use by improving the Cr/Fe ratio. These peculiarities make it necessary to resort to chemical information to properly characterize the chromite ore. For more information on these aspects, cf. Haggerty (1976) and Bowles et al. (2011).

30.2 Microscopic Description

30.2.1 SP Observation (*Single Polariser*)

Grey *colour* (Fig. 30.1a, b) although, depending on the neighbouring minerals, a brownish shade can be perceived. As well, magnetite next to chromite can appear white (Fig. 30.7). Isotropic, not pleochroic. *Good polishing*, although slow due to its hardness. *High polishing hardness* > mt, asp or nc but < ht, rt or py.

Low reflectance, R > spn and silicates but < mt < ht and << common accompanying sulphides (cf. multispectral comparison graphs). The optical impression varies a lot according to the environment: it looks light grey compared to silicate gangue but dark next to pyrite or nickelite and very dark in immersion (Figs. 30.1a, b, and 30.5).It also depends on the magnification (it looks brighter at lower magnifications, comp. Figs. 30.1a, b and 30.3). R values vary sensitively according to composition (cf. R-graphs, spinel, Cr-spinel, chromite and magnetite) and can be used as an estimator of chr-composition (López et al. 2017). The VHN microhardness also varies. In general, VHN increases and R decreases with increasing Mg & Al contents and decreasing Fe & Cr contents (Bowles et al. 2011).

R. Castroviejo, *A Practical Guide to Ore Microscopy—Volume 1*,
https://doi.org/10.1007/978-3-031-12654-3_30

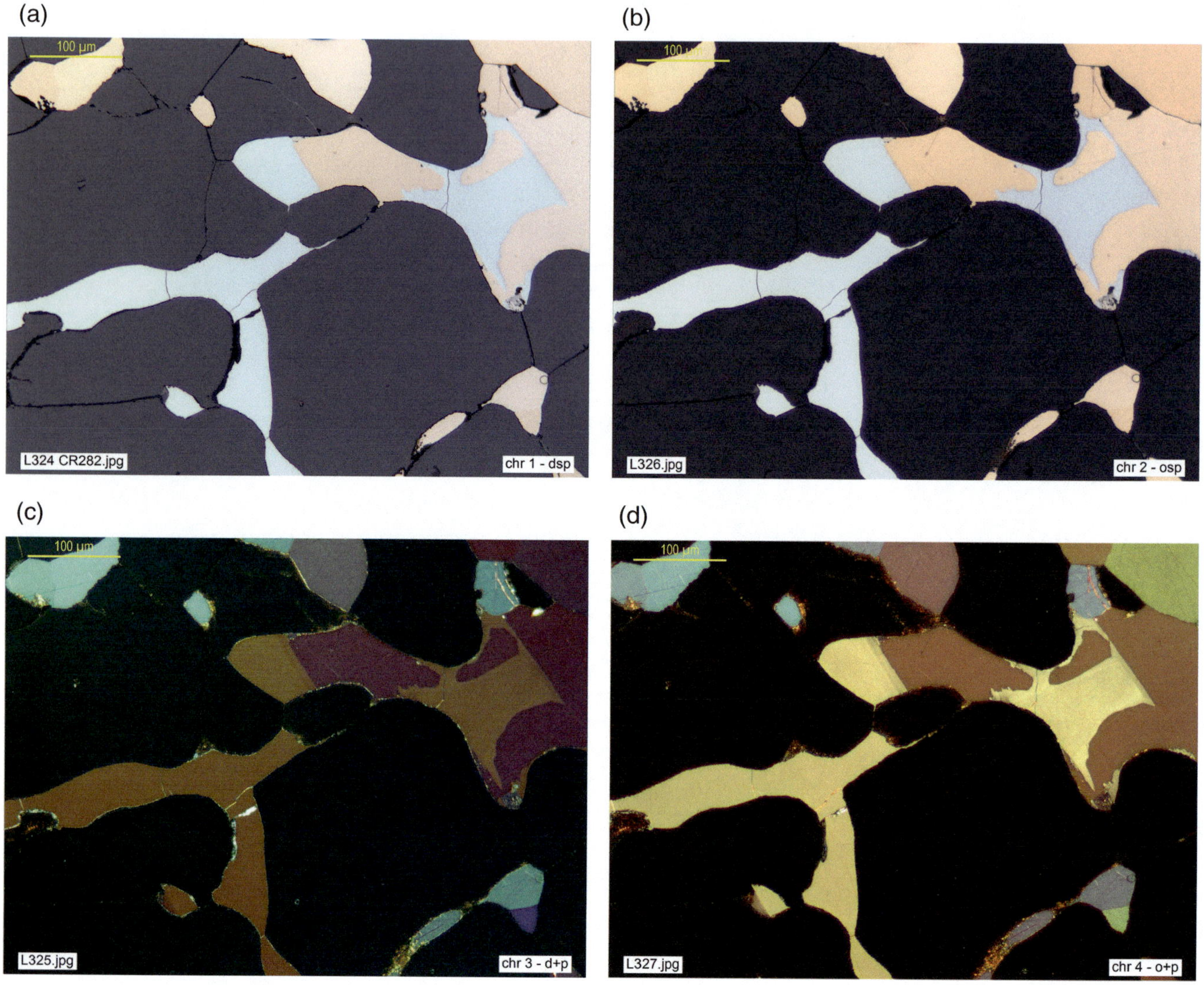

Fig. 30.1 **a, b** (Same scene: **dsp** and **osp**, *respectively*. Figure 30.2 offers a wider view). Massive chr ore (grey) with interstitial nc (pink) and gf (white, partially replacing nc). CR 282, La Gallega Mine (S[a] Ronda, Málaga). **c, d** Same scene as Fig. 30.1 (**d+p** and **o+p**, *respectively*). Complete extinction of chr (isotropic) and internal reflections, visible mainly in immersion and at edges or microcracks; vivid and multicoloured anisotropism, with luminous green tones, of nc and anomalous anisotropism (poorly coloured) of gf

It is usually well crystallized, with cubic or rounded forms, in accumulations, bands or lenticular bodies, or dispersed in gabbros and meteorites (Fig. 30.4). Granoblastic cumulate textures are frequent. As well, in podiform chromitites, brecciated textures show tectonic banding or even mylonitic textures. Deformation, like metamorphism, is usually associated with alteration to more ferriferous varieties, ferritchromite, magnetite, also with neoformation of silicates such as serpentine and chlorite (Figs. 30.6 and 30.7).

Zoning is difficult to observe because of its isotropy but is visible when the compositional differences are sufficient to produce sensitive R variations, which result in different shades of grey (usually darker in the core and lighter in the periphery, due to their greater richness in Cr and Fe, respectively). This primary zoning should not be confused with the later one due to cation mobilization (Fe, Al, Cr) by alteration or metamorphism, which is usually more irregular or linked to microcracks (Figs. 30.5 and 30.7).

30.2.2 Observation +P (*Polariser* + *Analyser at 90°*)

Isotropic, complete extinction. Usually shows, especially in immersion, reddish-brown *internal reflections* (Fig. 30.1c, d), more abundant in the varieties rich in Mg and Al and scarce or absent in the ferriferous terms. Not *twinned.*

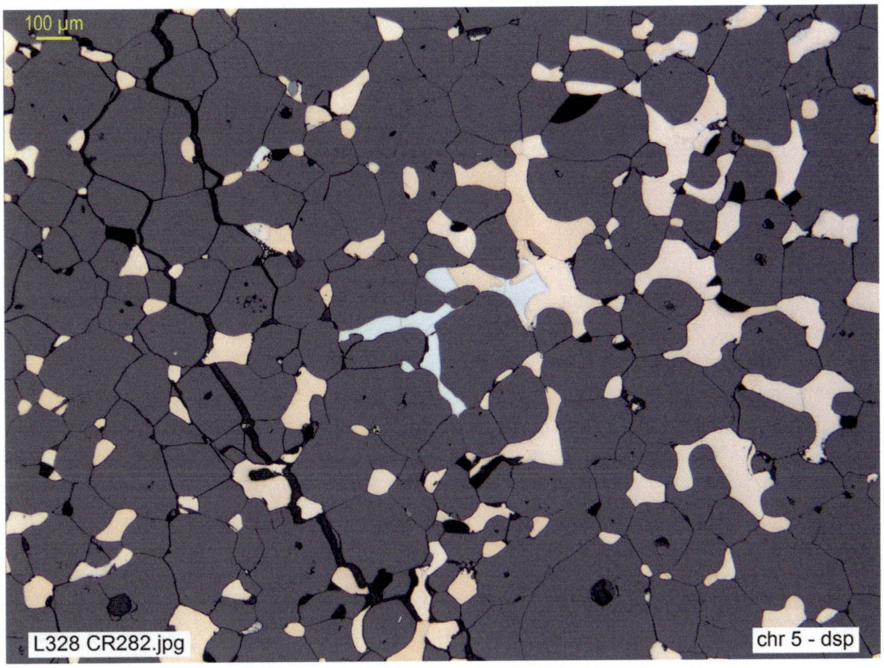

Fig. 30.2 Massive chr ore (grey, granoblastic cumulate) with interstitial nc (pink) and sparse gf (white). Wide-field image of sample CR282, Fig. 30.1 is a detail view of the central zone of Fig. 30.2, (**dsp**)

Fig. 30.3 Cumulate banded chr (light grey) in serpentinized dunite (darker grey). Rub20, Chalkidiki, Greece, (**dsp**)

30.2.3 Paragenesis

Chromite is a typical orthomagmatic mineral, related to ultramafic and mafic rocks such as peridotites, dunites, pyroxenites and, occasionally, gabbros.

As a mineral of early precipitation, it forms by fractional crystallization important concentrations in ultramafic rocks and disseminations in gabbros, which persist in serpentinites and amphibolites and in their detrital derivatives. In these, due to its stability, hardness and high density, it may be upgraded in placers of economic importance. It has also been found, exceptionally, in massive sulphide deposits such as Outukumpu. It also occurs in some types of meteorites and in kimberlites (included in diamond).

Common companions: magnetite, ilmenite, ulvöspinel, nickelite, pyrrhotite, pentlandite, platinum group elements (PGE).

Fig. 30.4 Neuschwanstein Meteorite: Native Fe (white), troilite (light pinkish brown), graphite (brownish grey), chr (dark grey, upper right quarter of the image) and silicates (very dark grey background). Neu 003Rub, Neuschwanstein Meteorit, Bavaria, Germany, (**dsp**)

Fig. 30.5 Podiform chromitite ore, brecciated, with associated incipient alteration, Fe-rich rims (ferritchromite, almost white), on chr microclasts (grey). TP110607.21, Tapo Mine, Tarma (Dept. Junín), Peru, (**osp**)

30.2.4 Identification

Its restricted paragenesis and properties (low reflectance between spinel and magnetite, isotropy, internal reflections, hardness and euhedral tendency, granoblastic overprinting of cumulates and metamorphic ores, and brecciated occurrence in ophiolitic podiform concentrations) allow a reliable diagnosis, but characterization from an economic point of view requires chemical compositional data.

Fig. 30.6 Brecciated chromitite, with almost total chr alteration to more iron-rich varieties (lighter), ferritchromite (light grey) and magnetite (white, microcracks, details in Fig. 30.7) and very scarce relicts of primary chr (grey, less reflective). Secondary serpentine and chlorite (dark grey) included in altered chr. TP 140607.30, Tapo Mine (Tarma, Dept. Junín), Peru (**dsp**)

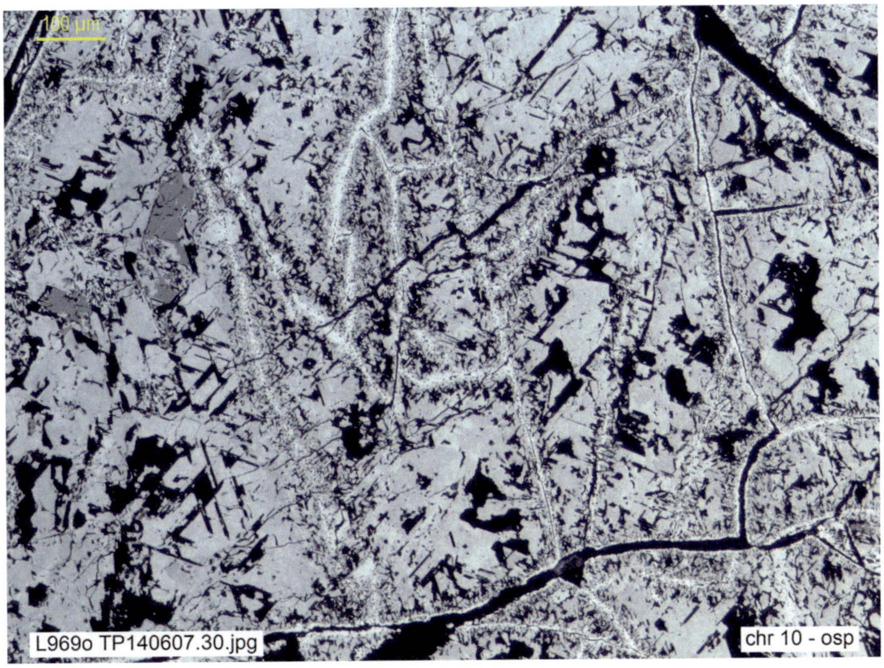

Fig. 30.7 Advanced chr alteration, controlled by fracturing and metamorphism (detail of the central zone of Fig. 30.6), octahedral (111) chr directions enhanced by secondary silicate inclusions (reaction products, serpentine and chlorite, almost black), different Fe contents of secondary chr and ferritchromite manifested by reflectance differences (darker or lighter grey), relicts of primary chr (Fe-poor, darker) clearly visible in centre left of image;,posthumous mt (white) associated with microcracks. TP 140607.30, Tapo Mine (Tarma, Dept. Junín), Peru, (**dsp**)

30.3 Spectral Information

30.3.1 Chromite Multispectral Reflectance R (in Air, Without Polariser) (%)

λ (nm)	370	400	425	450	475	500	525	550	575	600	625	650	675	700	750	800	850	900	950	1000
μ	12.9	12.7	12.3	12.2	12.0	11.9	11.8	11.6	11.5	11.4	11.4	11.3	11.3	11.2	11.1	11.1	11.1	10.9	10.9	10.9
P_5	12.0	11.5	11.2	10.9	10.9	10.8	10.7	10.7	10.5	10.5	10.5	10.3	10.4	10.3	10.4	10.4	10.4	10.2	10.3	10.1
P_{20}	12.1	11.7	11.4	11.1	11.1	10.9	10.8	10.9	10.7	10.6	10.6	10.5	10.5	10.4	10.5	10.5	10.5	10.4	10.4	10.3
P_{80}	13.8	13.8	13.5	13.4	13.1	13.0	12.9	12.6	12.5	12.4	12.3	12.2	12.2	12.1	11.9	11.8	11.8	11.6	11.6	11.6
P_{95}	14.0	13.9	13.7	13.6	13.3	13.2	13.1	12.8	12.7	12.5	12.5	12.3	12.3	12.2	12.1	11.9	11.9	11.7	11.8	11.7

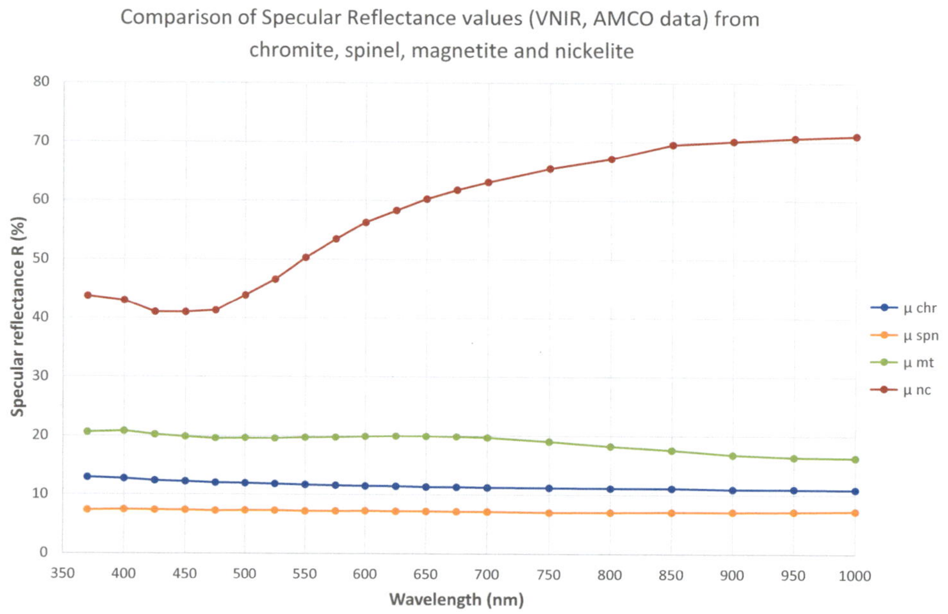

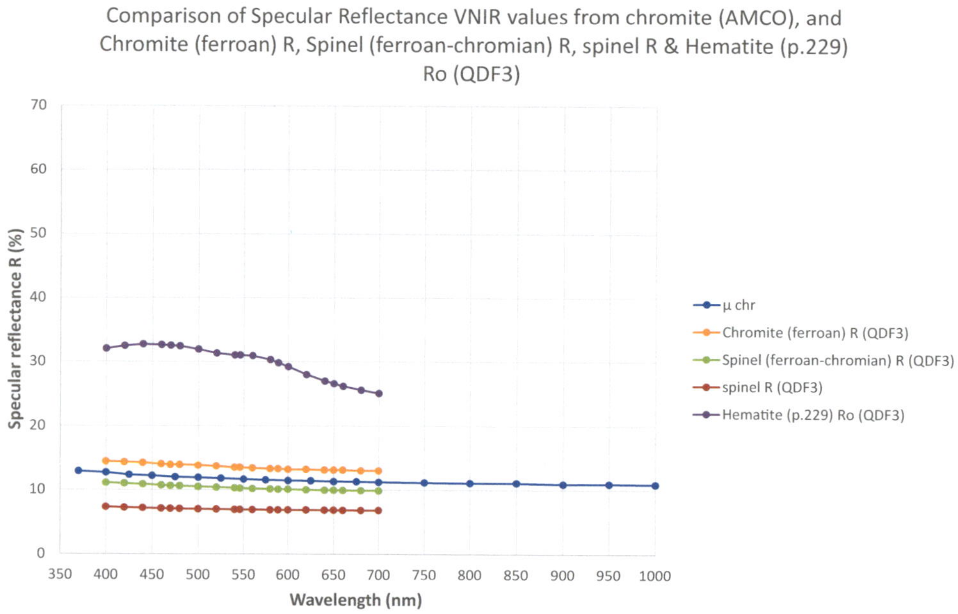

References

Bowles JFW, Howie RA, Vaughan DJ, Zussman J (2011) Rock-forming Minerals, vol. 5A (2nd edn.) Non-silicates: oxides, hydroxides and sulphides. The Geological Society, London, 920p

Haggerty S (1976) Opaque mineral oxides in terrestrial igneous rocks. Ch. 8 (pp. Hg_101–Hg_300), In: Rumble III D (ed) Oxide minerals. Rev. Mineralogy v3. Mineralogical Society of America, Washington, DC

López-Benito A, Gervilla F, Catalina JC, Castroviejo R (2017) Chromite typology and composition characterized through multispectral specular reflectance. Ore Geol Rev 89:132–142. ISSN: 0169-1368. https://doi.org/10.1016/j.oregeorev.2017.06.010

Ramdohr P (1980) The ore minerals and their intergrowths, 2nd edn. Pergamon Press, Oxford. 2 volumes, 1205p

Chrysocolla (*crc*)

31

$(Cu_{2-x}Al_x)H_{2-x}Si_2O_5(OH)_4 \cdot nH_2O$, ($x < 1$): Orthorhombic

Mineral abbreviations: crc—chrysocolla, cup—cuprite, gth—goethite, ht—hematite, lim—limonite, q—quartz.

31.1 General

Silicate mineral, occasionally present as gangue in the oxidation zone of copper deposits, accompanied by other secondary copper products. It may also occur as part of the ore in the oxidized type of Cu deposits, but its economic significance as Cu-ore is small.

It usually occurs in massive aphanitic, cryptocrystalline, botryoidal or fibrous-radiated aggregates; also, as a pseudomorph of cupriferous ores but not in well-developed crystals.

The *crc* optical properties have not been described with reflected light in the available sources but will be briefly characterised below, and they are also documented with photomicrographs (dsp/d+p) to facilitate their recognition.

31.2 Microscopic Description

31.2.1 SP Observation (*Single Polariser*)

As usual in silicate minerals, the crc reflectance is low. Its dull aspect—monotonous, massive, amorphous or cryptocrystalline appearance, with greyish tones (Fig. 31.1a)—attracts little attention.

As befits its refractive indices (n_α = 1.575–1.585, n_β = 1.597, n_γ = 1.598–1.635; birefringence ≈ 0.025–0.050), the specular reflectance of crc is low (R ≤ 6%, little more than quartz or *epoxy* resin). Its bireflectance is moderate, which clearly differentiates it from highly birefringent carbonate species such as malachite or azurite. No pleochroism or euhedral forms are observed, only occasional pseudomorphs.

31.2.2 Observation +P (*Polariser + Analyser at 90°*) Identification

Characteristic of crc are low crystallinity, with botryoidal or fibrous-radiated textures and, above all, the ubiquity of its green or blue-green, luminous, internal reflections (Fig. 31.1b).

It is related to copper-bearing silicate rocks in oxidation environments.

Typical crc companions comprise calc, q, mch, az, ht, gth, cup, etc.

R. Castroviejo, *A Practical Guide to Ore Microscopy—Volume 1*,
https://doi.org/10.1007/978-3-031-12654-3_31

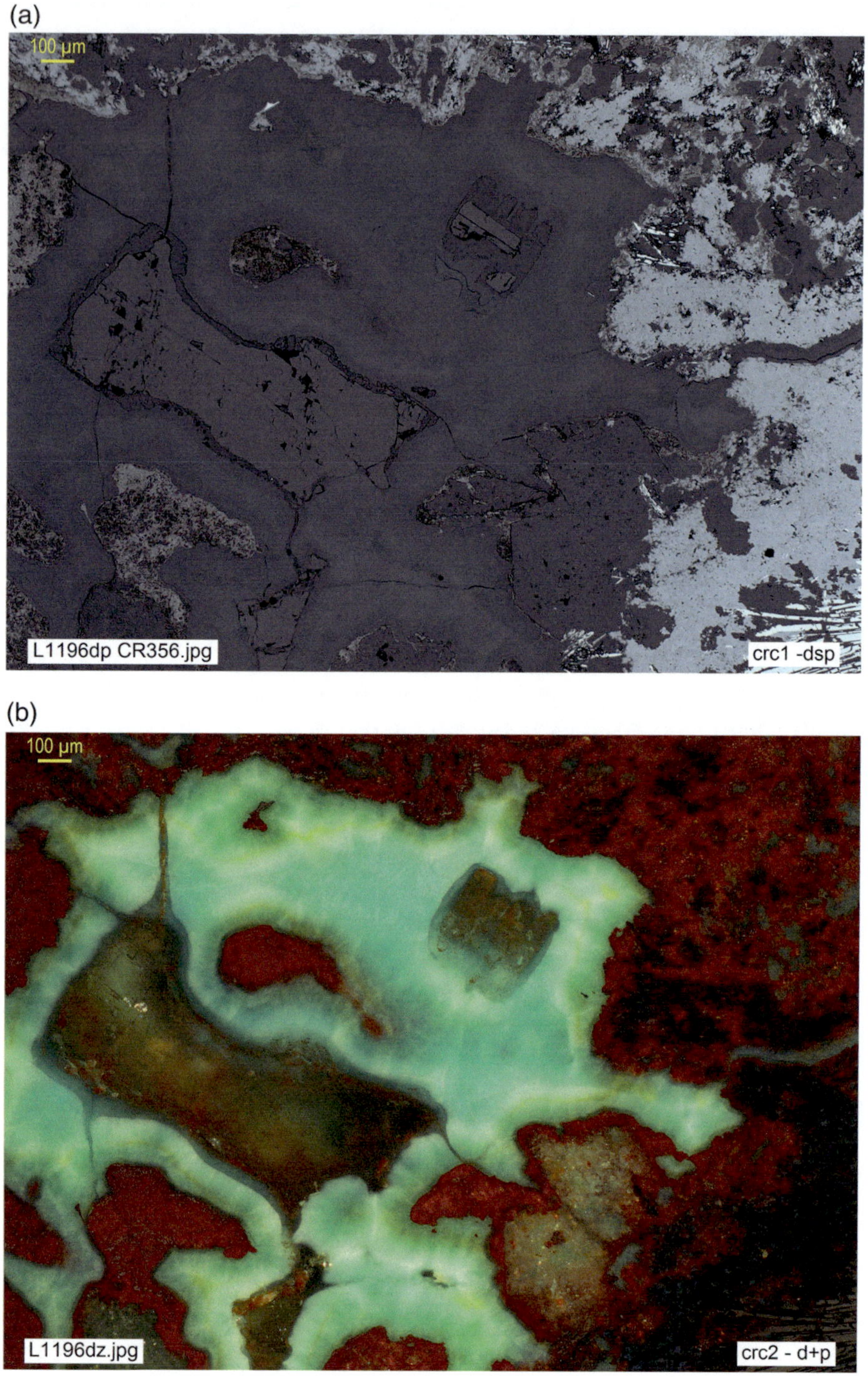

Fig. 31.1 **a** Microgeode filled with crc, in brecciated rock impregnated with ht (acicular bundles, bright bluish grey, lower right corner), cup (bluish grey, main mass, micro- to cryptocrystalline), gth or lim (grey) and q (dark grey). CR356, San Antonio Mine, Sonora, Mexico (**dsp**). **b** Same scene as crc1 (**d+p**), fibrous internal structure of crc, internal reflections are blue-green of crc, red and scarce (ht) or abundant (cup), reddish-brown to yellowish (gth, lim)

32 Cinnabar (*cin*)

HgS: Trigonal

Cinnabar		R_{546} (%)	Y (%)	λ_d (nm)	P_e (%)	Colour sensation
Air	R_o	24.7	24.6	476	602	Greyish white, slightly bluish and slightly pleochroic
	R_e	29.7	29.4	481	6.3	
Oil	R_o	10.6	10.5	475	10.4	
	R_e	14.7	14.4	481	10.2	
Bir: Weak (3)				*Pleoc*: Weak (3)		*Tarn*: Not observed
Anis +P: Distinct (4)				*IR*: Very abundant, red		*Twg*: Rarely visible under the microscope
Mor: Massive, disseminated. Rarely euhedral				*Clv*: Poor		*PolQ*: Unavoidable scratches (inclusions py)
Hardness	*PolH*: Low, < gn (2)			*Mohs*: 2	*Talm*: B	*VHN* $_{(10\ g)}$: 82.4–156

Obs: cf. Table and R charts, multispectral data (VNIR, AMCO) and visible range (QDF3)
Mineral abbreviations: bl—boulangerite, bm—bismuthinite, cup-cuprite, gg—gangue, gn—galena, gth—goethite, Hg—native mercury, mc—marcasite, my—miargyrite, prg—pyrgyrite, pru—proustite, py—pyrite, rl—realgar, sbt—stibnite

32.1 General

Composition: *α-HgS phase*, usually stoichiometric, without impurities (except inclusions). Trimorph with metacinnabar (cubic) and hypercinnabar (hexagonal).

32.2 Microscopic Description

32.2.1 SP Observation (*Single Polariser*)

Greyish white to very light grey (Fig. 32.1a, b), somewhat bluish, weakly bireflective/pleochroic (Fig. 32.2a).

Being very soft, the *polishing quality* depends on the texture and the intergrown minerals. These can produce relief and many scratches (Figs. 32.3a and 32.4) if detached (pyrite microinclusions) but, in favourable cases (pure cinnabar or cin with carbonates), the quality is very good. Low *polishing hardness*: $PolH_{cin}$ < gn, bm, bl; > my, pru-prg.

Average *reflectance*, of the order of td: R_{cin} < py, gn, sbt ($R_{cin} \approx R_{min\ sbt}$ and << $R_{max\ sbt}$); > gth (cf. multispectral comparative plots).

Recognition of the weak bireflectance requires careful observation and, if possible, immersion (Fig. 32.2a). The colour changes very little and sometimes not even contiguous grains of an aggregate are distinguishable, despite their different orientations (Figs. 32.1a, b and 32.4).

It can be found crystallized (rhombohedrons, prisms) in pores and geodes but, as ore, it usually occurs in xenomorphic aggregates, more or less massive or as impregnation, in pores, microcracks, interstitial spaces (Fig. 32.1a, b).

32.2.2 Observation +P (*Polariser + Analyser at 90°*)

Lively and coloured *anisotropism* (shades of green) but frequently masked by the very abundant *internal reflections*, **IR**, so that the crystal may appear as a luminous blood-red mass

R. Castroviejo, *A Practical Guide to Ore Microscopy—Volume 1*,
https://doi.org/10.1007/978-3-031-12654-3_32

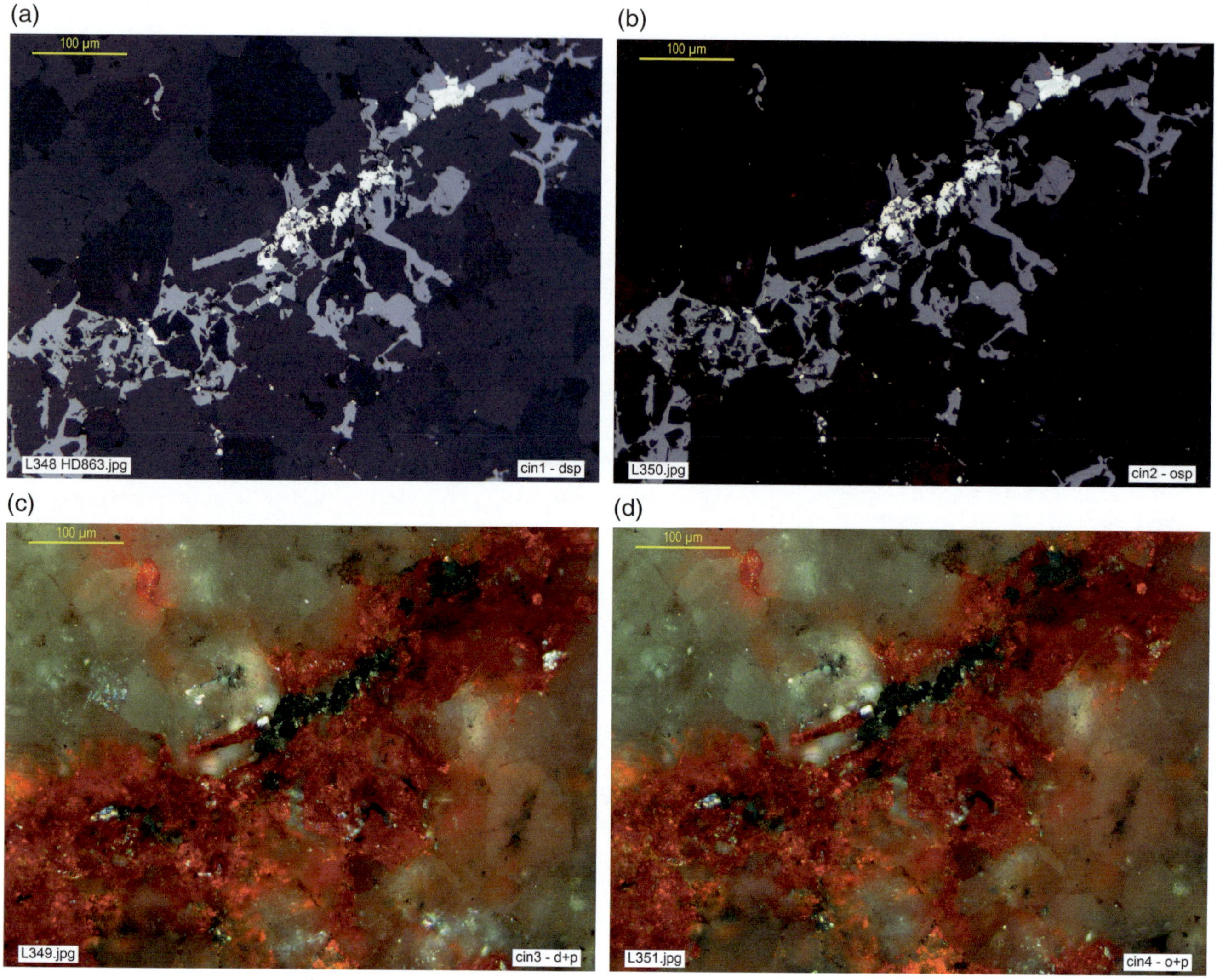

Fig. 32.1 **a, b** (Same scene, **dsp** and **osp**, *re*spectively). Impregnation of cin (grey), py (yellowish-white) and mc (white), replacing dolomitic gg, along a thin microfracture. HD863, Idria, Yugoslavia. **c, d** Same scene as cin1 (**d+p** and **o+p**, resp), massive internal reflections of cin, which overprint its anisotropism

(Figs. 32.1c, d and 32.2b). Sometimes the anisotropism or polarisation tints are easily seen (Figs. 32.3b and 32.5a, b).

Although frequently *twinned*, the twins (contact, {0001}) are rarely visible under the microscope except in deformed ores, in which pressure twins are generated. These are polysynthetic and sinuous, relatively frequent because it is a very soft mineral.

32.2.3 Paragenesis

Typical mineral of low-temperature hydrothermal *paragenesis of* varied origin (epithermal, volcanic-exhalative, diagenetic remobilization, metamorphism). It can appear accompanied by pyrite, stibnite and low-temperature minerals, such as opal, chalcedony, marcasite, or melnikovite pyrite.

32.2.4 Identification

Cinnabar is easy to recognize by its very abundant and luminous red IR, apart from its paragenesis, moderate reflectance, vivid anisotropism with greenish tones (when not overprinted by IR), low hardness and polishing scratches. The presence of native mercury droplets, which appear spontaneously as exudation in some samples (Fig. 32.3a, b) is an indisputable additional criterion.

Metacinnabar is isotropic (cubic), less reflective (slightly darker) and it is more scarce.

Fig. 32.2 **a, b** (Same scene, **dsp** and **osp**, *respectively*). Bireflectance of cin and massive IR, seen in immersion. IR not only overprint anisotropism (Fig. 32.2b) but are even clearly perceived without analyser (*osp*, Fig. 32.2a). Impregnation of cin and trace py in dolomitic gg. TFPn17, Sierra de Gádor, Almería

(a)

(b)

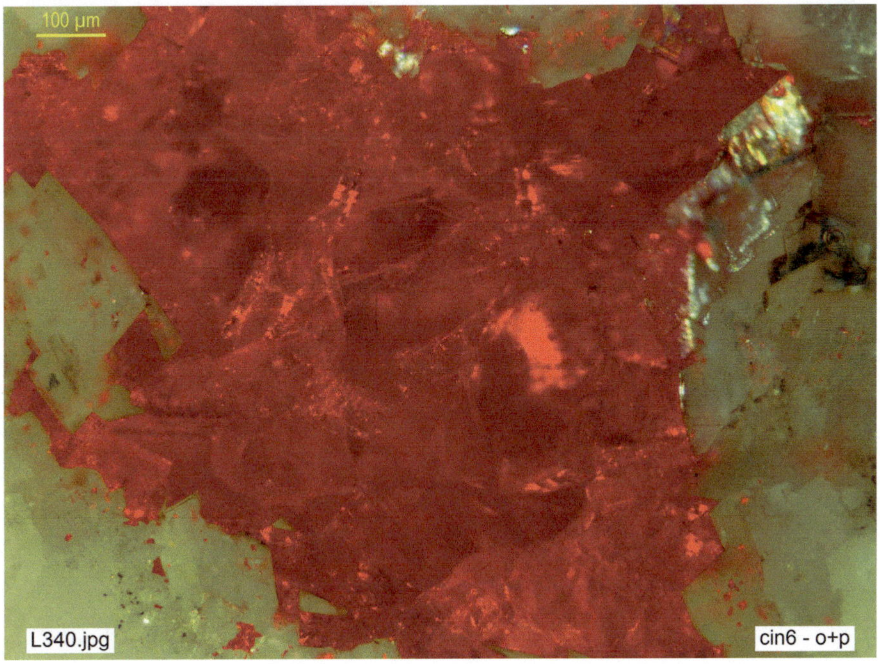

Cinnabar resembles *proustite* by IR, but the anisotropism of the latter lacks green tones.

Realgar is less reflective, polishes worse and its IR are lighter.

Cuprite is much harder, less reflective, more bluish and is usually accompanied by native copper.

Fig. 32.3 **a, b**(Same scene, **dsp** and **d+p**, respectively). Ore of cin, with three drops of spontaneously exuded liquid Hg (dark grey and very bright white with dark interference halos, without and with analyser, *respectively*). Note the greenish anisotropism of cin. Sample C12, Almadén, Ciudad Real

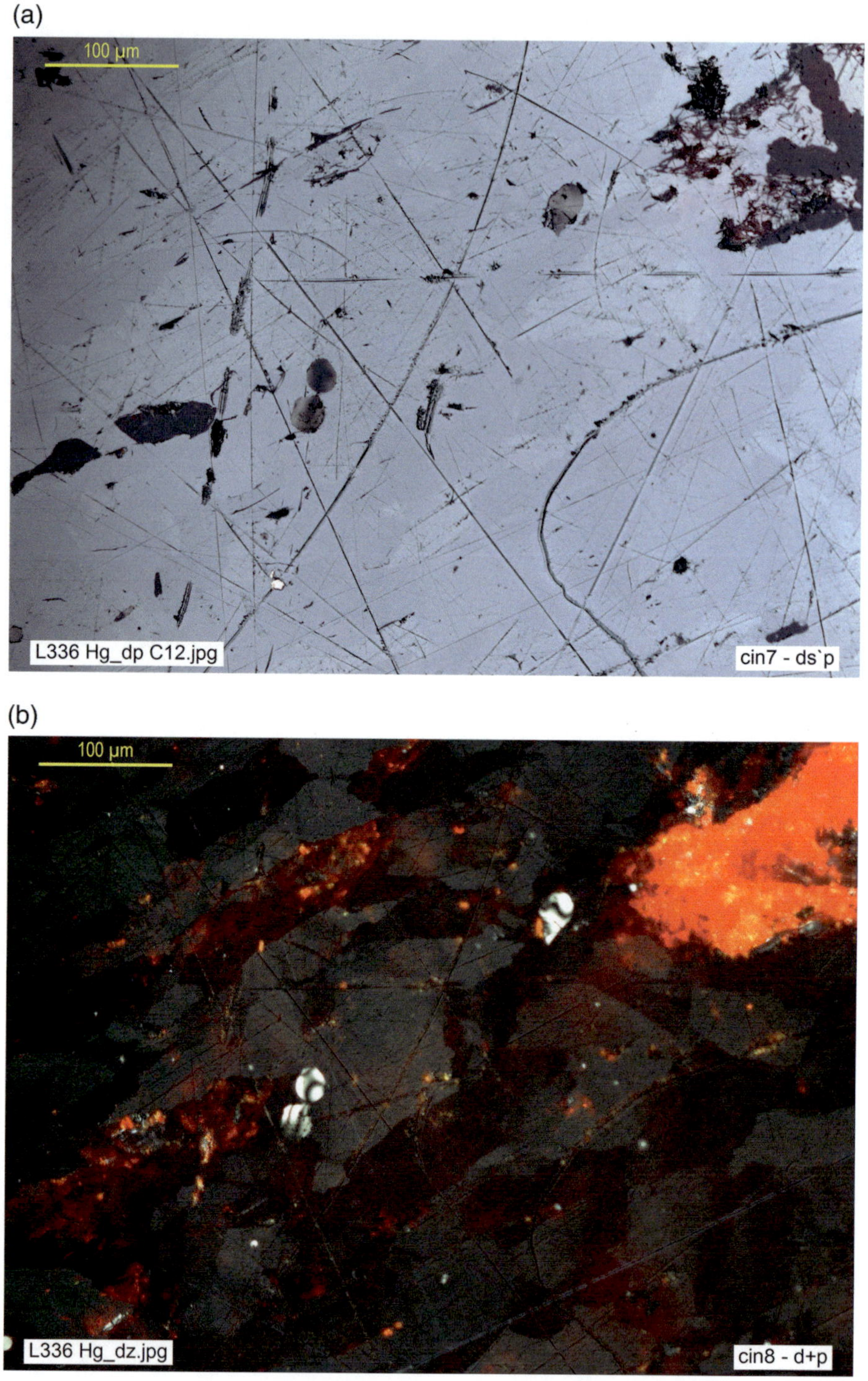

Fig. 32.4 Impregnation of cin and (trace) py, in dolomite. Sample TF Pn 17, Sierra de Gádor, Almería (**dsp**)

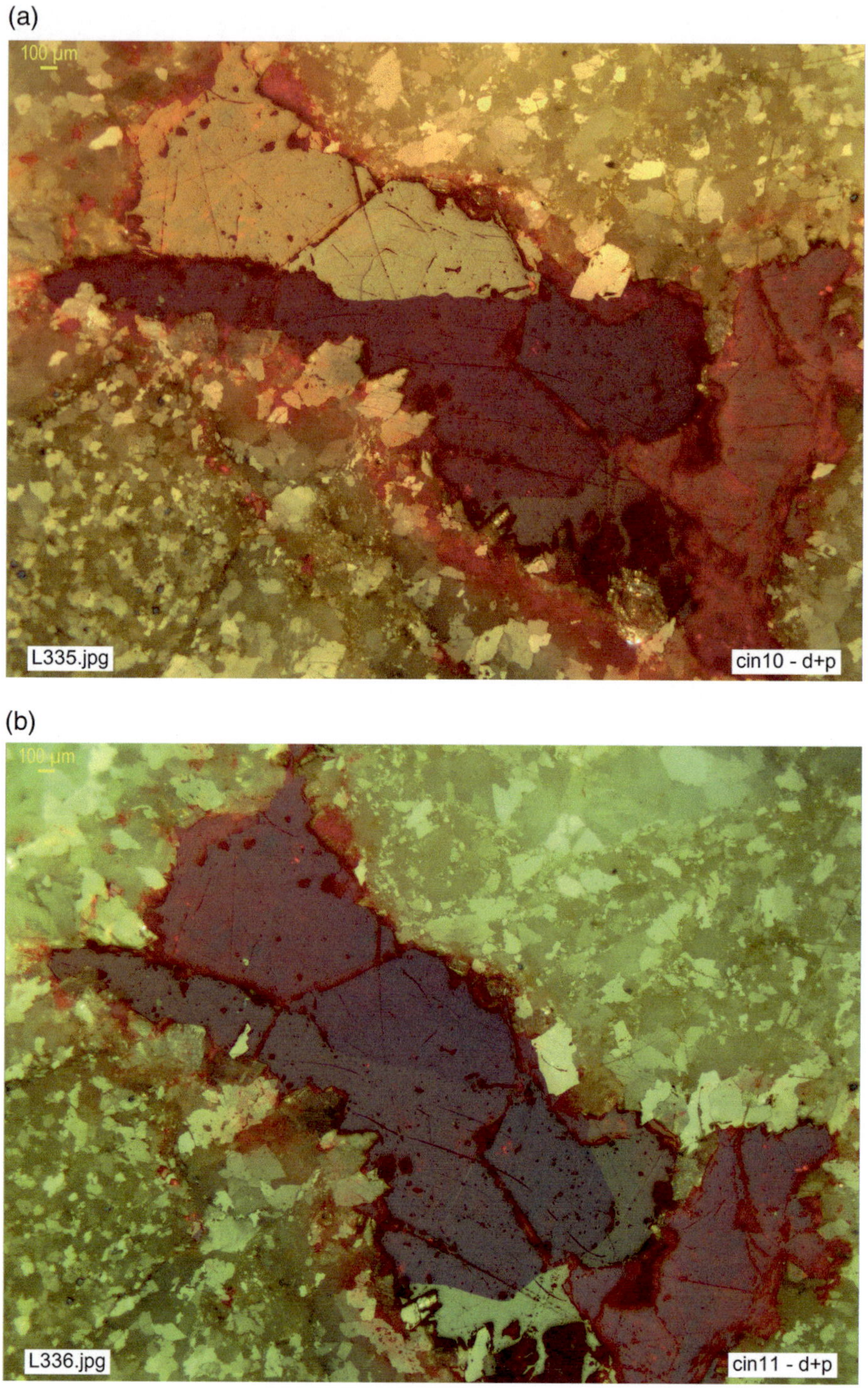

Fig. 32.5 **a, b** Same scene as Fig. 32.4, but now (**d+p**) in two different orientations (rotating the stage) to observe variation of cin polarization tints

32.3 Spectral Information

32.3.1 Cinnabar Multispectral Reflectance R (in Air, Without Polariser) (%)

ortional``>λ (nm)	370	400	425	450	475	500	525	550	575	600	625	650	675	700	750	800	850	900	950	1000
μ	32.1	31.1	29.0	27.8	26.9	26.5	26.0	25.4	25.1	24.8	25.7	26.3	26.2	26.2	25.9	25.7	25.5	25.5	25.3	25.4
P_5	30.9	30.0	28.0	26.8	25.9	25.6	25.1	24.6	24.3	24.1	24.5	24.5	24.3	24.2	23.9	23.5	23.3	23.2	23.0	23.0
P_{20}	31.6	30.7	28.6	27.5	26.5	26.1	25.6	25.0	24.7	24.5	25.1	25.6	25.5	25.4	25.1	24.8	24.6	24.6	24.4	24.5
P_{80}	32.7	31.6	29.5	28.3	27.3	26.9	26.5	25.9	25.5	25.2	26.3	27.1	27.0	27.1	26.9	26.6	26.4	26.5	26.2	26.3
P_{95}	33.4	32.0	29.8	28.7	27.8	27.6	27.2	26.5	26.0	25.6	26.9	27.9	27.8	27.8	27.6	27.4	27.2	27.3	27.1	27.2

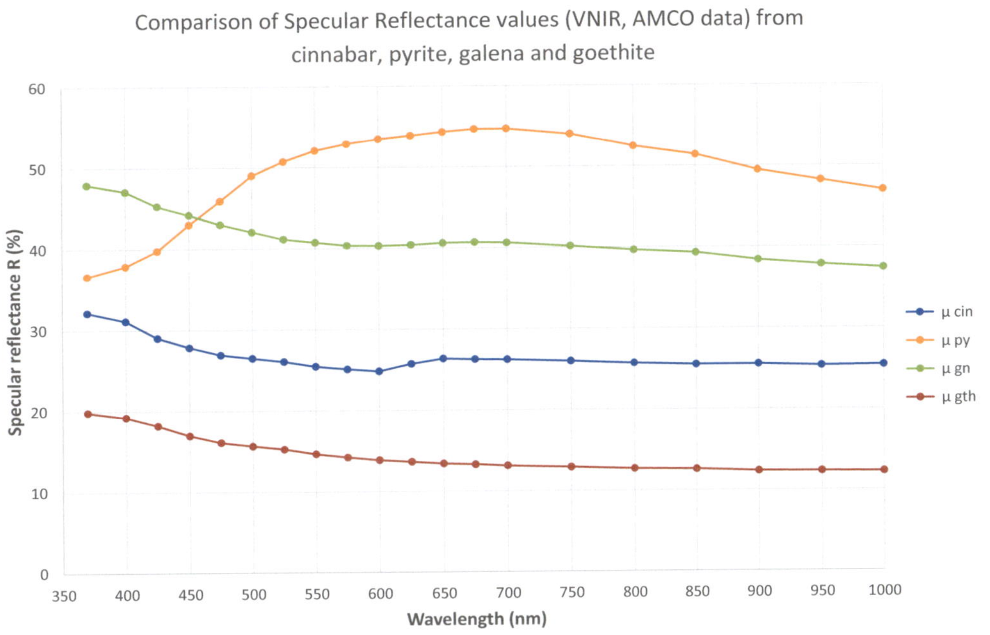

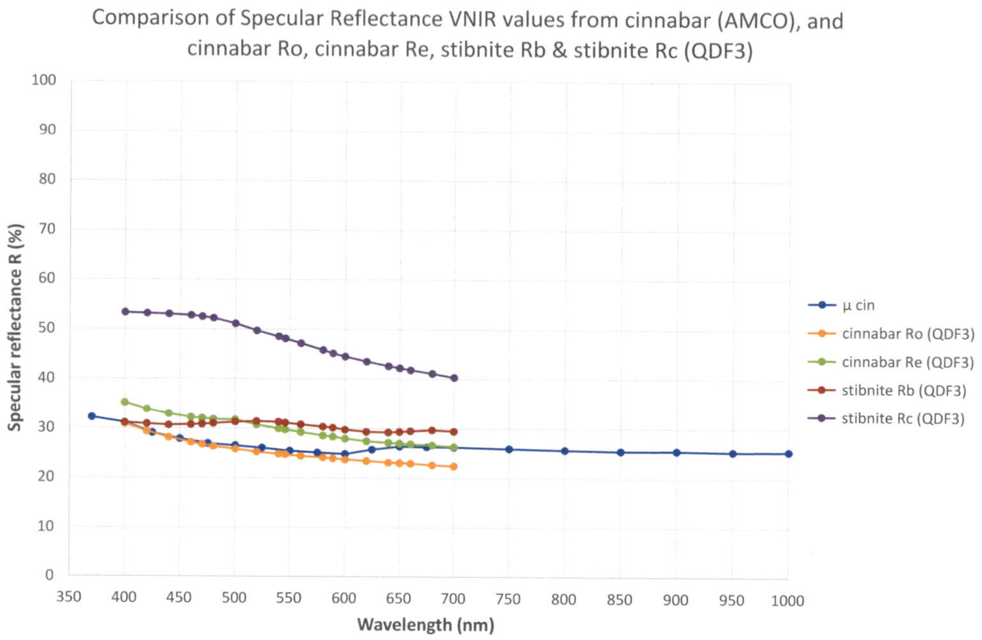
Comparison of Specular Reflectance VNIR values from cinnabar (AMCO), and cinnabar Ro, cinnabar Re, stibnite Rb & stibnite Rc (QDF3)
Specular reflectance R (%)
100
90
80
70
60
50
40
30
20
10
0
350
400
450
500
550
600
650
700
750
800
850
900
950
1000
Wavelength (nm)
µ cin
cinnabar Ro (QDF3)
cinnabar Re (QDF3)
stibnite Rb (QDF3)
stibnite Rc (QDF3)

33 Clausthalite (*clt*)

PbSe: Cubic

Clausthalite	R_{546} (%)	Y (%)	λ_d (nm)	P_e (%)	Colour sensation
Air	53.6	53.2	478	7.3	White
Oil	38.7	38.4	477	11.6	
Bir: Null (1)			*Pleoc*: Null (1)		*Tarn*: No
Anis +P: Isotropic (1)			*IR*: No		*Twg*: No
Mor: Xenomorphic masses and aggregates			*Clv*: (100), triangular pits		*PolQ*: Good
Hardness	*PolH*: Low: ≈ um (2)		*Mohs*: 2½–3	*Talm*: A	*VHN* (100 g): 44–49

Note: cf. Table and R charts, visible range data (QDF3)
Mineral abbreviations: ac—acanthite, Ag—native silver, agl—aguilarite, Au—native gold, bn—bornite, ccp—chalcopyrite, clt—clausthalite, el—electrum, fse—ferroselite, gg—gangue, gn—galena, my—miargyrite, nau—naumannite, py—pyrite, Sb—native antimony, sbt—stibnite, td—tetrahedrite, um—umangite

33.1 General

It is the most common Se mineral. Composition: PbSe, with possible minor to trace contents of Cu, Co, Hg. It forms, with galena, the PbSe-PbS series (clausthalite-galena).

Low *polishing hardness*: $PolH_{clt}$ < nau, gn; $PolH_{clt} \geq$ um; $PolH_{clt}$ > my, Sb (according to Uytenbogaardt and Burke 1971, clausthalite shares with naumannite and, in general with selenides, a PolH < sbt, very low).

It is found in masses or microgranular aggregates, generally anhedral.

33.2 Microscopic Description

33.2.1 SP Observation (*Single Polariser*)

White *colour*, like galena, but noticeably brighter. Its brightness (luminance) is ≈ py, although its reflectance spectrum is very different because of the yellow colour of pyrite. High *Reflectance*, ***R***: R_{clt} > ***gn***, ***ccp***; ≫ um; ≪ Au, Ag, el (Fig. 33.1a, b; cf. comparative reflectance plots). Isotropic.

Good *polish*, cleavage (100) occasionally gives rise to *triangular pits*, as in galena.

33.2.2 Observation +P (*Polariser + Analyser at 90°*)

Isotropic, but with incomplete extinction (Fig. 33.1c, d).

33.2.3 Paragenesis

Varied *parageneses*: hydrothermal veins or metamorphic concentrations, in different lithologies (limestones, greywackes, diabases, carbonaceous shales, often with carbonate gangue).

R. Castroviejo, *A Practical Guide to Ore Microscopy—Volume 1*,
https://doi.org/10.1007/978-3-031-12654-3_33

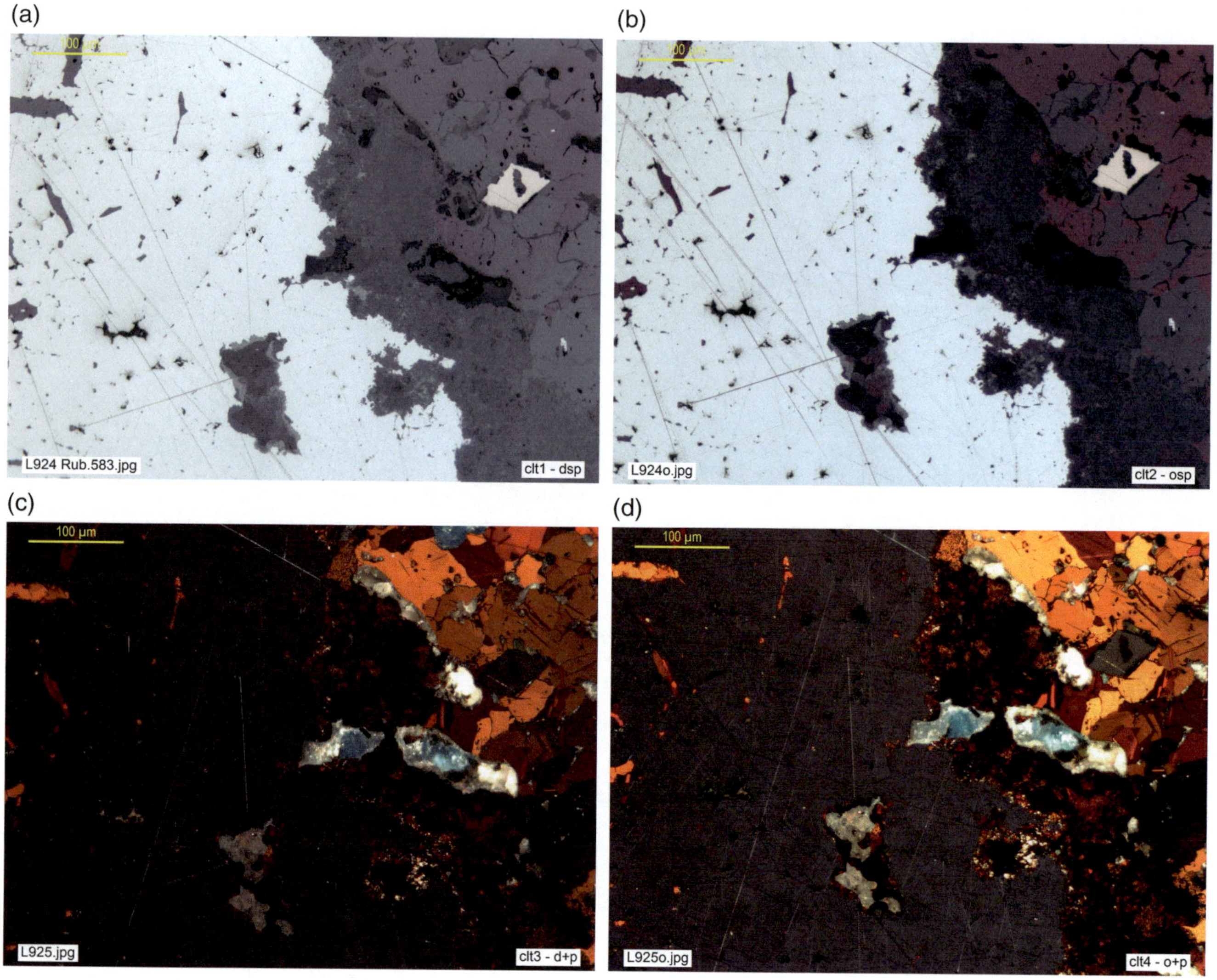

Fig. 33.1 **a**, **b** (Same scene **dsp** and **osp**, *respectively*). Ore of clt (white) and um (purplish-grey to purplish-brown, up. right), with ferroselite (*fse*, yellowish-white, bright) included in um. Marginal replacement of clt by secondary carbonates (cerussite and azurite, gray). Rub.583, San Francisco Mine, La Rioja, Argentina. **c**, **d** (Same scene as *clt1*, d+p and o+p, *respectively*). The isotropy of clt clearly distinguishes it from um, strongly anisotropic; fse is anisotropic, but here it is almost in the extinction position

The companions can be, consequently, varied—umangite and other selenides, gold, chalcopyrite, bornite and other sulphides, urn (pitchblende). By weathering, it is typically altered to cerussite accompanied by native Se.

33.2.4 Identification

Its resemblance to galena is typical, from which it is distinguished, however, by its higher reflectance and by its association with other selenides, such as umangite. In a selenide paragenesis, moreover, alteration to cerussite is a useful clue.

33.3 Spectral Information

33.3.1 Clausthalite Reflectance R (in Air, Visible Spectrum) (%)

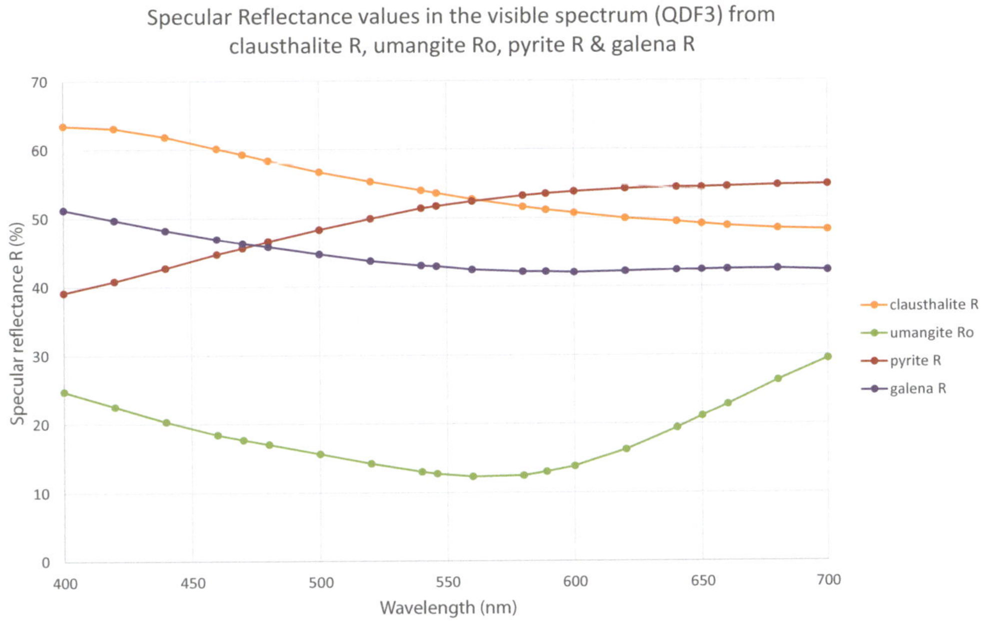

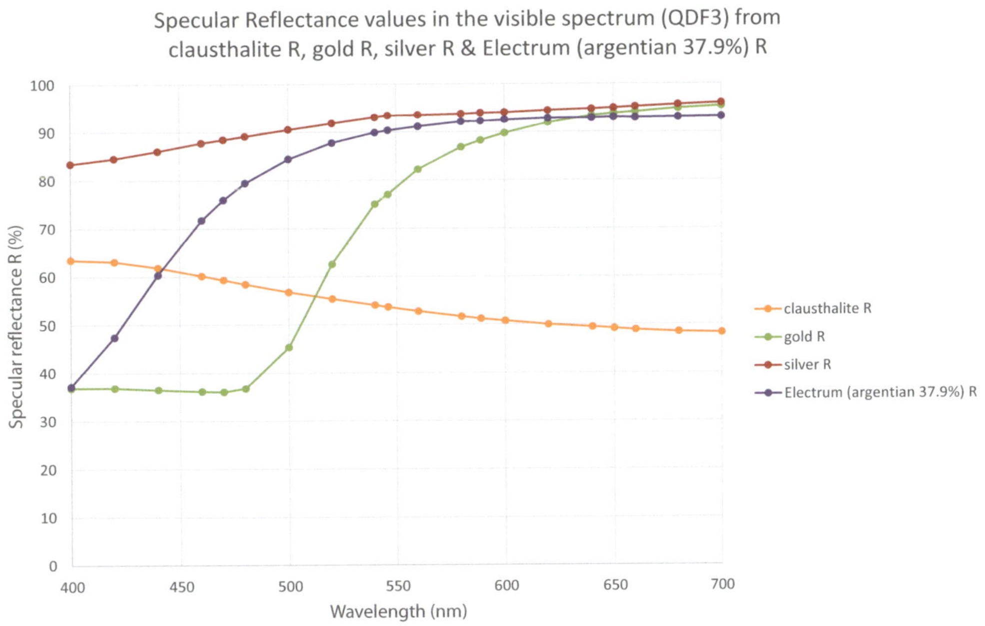

Reference

Uytenbogaardt W, Burke EAJ (1971) Tables for microscopic identification of ore minerals, 2nd edn. Elsevier, Amsterdam

34 Cobaltite (*cob*)

CoAsS: Orthorhombic (Pseudocubic)

Cobaltite	R'_{546} (%)	Y (%)	λ_d (nm)	P_e (%)	Colour sensation
Air	50.6	51.0	581	4.1	Bright white, slightly pinkish
Oil	36.2	36.7	581	6.1	White with a clear pinkish tinge
Bir: Very weak (2)			*Pleoc*: Very weak (2)		*Tarn*: Faint, very slow
Anis +P: Weak but visible (3)			*IR*: –		*Twg*: Almost always twinned
Mor: Euhedral (frequent cubes)			*Clv*: {001}		*PolQ*: Difficult (very hard)
Hardness	*PolH*: High, almost ≈ py (4)		*Mohs*: 5.5	*Talm*: G	*VHN* $_{(200\ g)}$: 1171–1346

Obs: cf. Table and R charts, multispectral data (VNIR, AMCO) and visible range (QDF3)
Mineral abbreviations: Ag—native silver, asp—arsenopyrite, Bi—native bismuth, cbn—cubanite, ccp—chalcopyrite, ln—linnaeite, lo—löllingite, mt—magnetite, po—pyrrhotite, py—pyrite, rm—rammelsbergite, sf—safflorite, sk—skutterudite

34.1 General

Composition: CoAsS, may contain, as impurities, Fe, Ni, Pb, Sb.

34.2 Microscopic Description

34.2.1 SP Observation (*Single Polariser*)

White *colour* with an appreciable pinkish tone next to minerals such as skutterudite or arsenopyrite and especially distinct in immersion (Fig. 34.1a, b). Its hardness can cause difficulties in polishing, but the quality can be very good with careful work.

Reflectance high, ≈ py > ccp (cf. comparative multispectral charts). Anisotropic (Fig. 34.1c, d), but birefringence and pleochroism are practically imperceptible.

High *polishing hardness*, almost like py > asp, mt, lo.

Usually *euhedral* (pseudocubic), with predominance of (100) and (210), cob typically occurs in cubic forms and sometimes in large crystals (Fig. 34.2a). Cleavage {001} perfect, but not always visible under the microscope.

34.2.2 Observation +P (*Polariser + Analyser at 90°*)

Distinct but weakly *anisotropic*, with not very striking polarization tints (more saturated in immersion, Figs. 34.1c, d and 34.3), which are better perceived in the larger crystals but can go unnoticed if the grain size is fine. Slightly uncrossing the analyser (*xp*), a greater variety of tones can be perceived (yellowish-brown, greenish-grey, bluish, greenish, purplish).

Almost always *twinned*, with varied and complex, lamellar, intergrown twins (Fig. 34.3).

34.2.3 Paragenesis

Various types of deposits comprise cob as a minor (accessory) component, but cob is characteristic of hydrothermal veins of the *five-element type* (Ni–Co–Bi–Ag–U), like those of Saxony (Germany) or Cobalt (Ontario, Canada), or of the *fahlband* type (e.g., Norway).

Commonly accompanied by pyrrhotite, chalcopyrite, arsenopyrite, cubanite, native bismuth, native silver, Ni–Co arsenides.

R. Castroviejo, *A Practical Guide to Ore Microscopy—Volume 1*,
https://doi.org/10.1007/978-3-031-12654-3_34

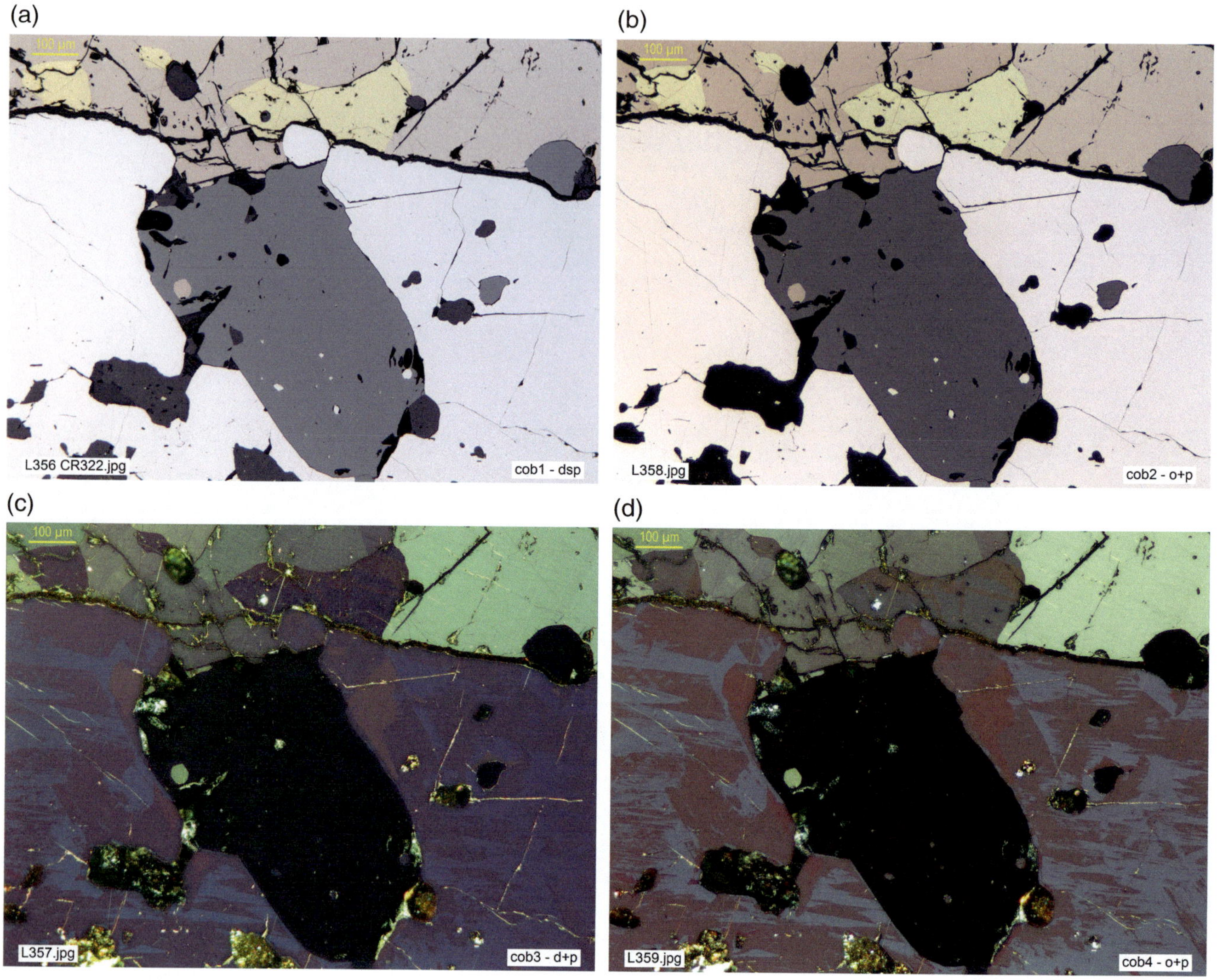

Fig. 34.1 **a** Aggregate of cob (slightly pinkish white), mt (grey), po (pinkish brown), ccp (yellow) and silicate gg (dark grey). CR322, Hakanibooda, Sweden (**dsp**). **b** Same scene as cob1 (**osp**), increased contrast and intensity of the pinkish hue of cob. **c**, **d** Same scene as cob1 (**d+p**) and (**o+p**), respectively, anisotropism of cob, po and ccp, cob and ccp twins

34.2.4 Identification

Typical are its pinkish-white colour (unlike other arsenides, such as safflorite or skutterudite, which are pure white), as well as its idiomorphism, high hardness, weak anisotropism and complex twinning. Pyrite is isotropic and shows a yellowish tone, very different from the pinkish colour of cobaltite. Arsenopyrite is more strongly anisotropic. Its colour resembles that of some minerals of the linnaeite group, but these are softer (less relief), polish more easily and are isotropic.

(a)

(b)

Fig. 34.2 **a** Cob idioblast with poikilitic inclusions of gg, mt, (po), surrounded by po (mt, ccp). CR322, Hakanibooda, Sweden (**dsp**). **b** Same scene as cob5 (**d +p**), anomalous anisotropism due to unavoidable chromatic aberrations with wide field objective (2.5x)

Fig. 34.3 Same sample as cob5 (**d+p**). Large cobaltite phenocryst (idioblast, center, and most of the image), included in massive po + ccp crystal aggregate (upper margin). Twinning in cob, with gg and mt inclusions in idioblast. Typical cob polarization colours, brighter than cob3-4 because of the wide-field objective (imperfect extinction)

34.3 Spectral Information

34.3.1 Cobaltite Multispectral Reflectance R (in Air, Without Polariser) (%)

λ (nm)	370	400	425	450	475	500	525	550	575	600	625	650	675	700	750	800	850	900	950	1000
μ	46.2	47.7	46.8	47.4	47.0	48.4	48.9	50.2	51.3	52.5	53.3	53.9	54.3	54.4	54.3	53.7	53.3	52.0	50.5	49.9
P_5	45.1	46.6	45.4	46.5	45.7	47.5	47.9	49.5	50.5	51.8	52.6	53.3	53.7	53.7	53.7	53.0	52.7	51.5	49.9	49.1
P_{20}	45.6	47.2	46.2	47.0	46.5	48.0	48.4	49.9	50.9	52.1	52.9	53.6	54.0	54.0	54.0	53.4	53.1	51.8	50.2	49.7
P_{80}	46.8	48.3	47.4	47.9	47.6	48.8	49.4	50.6	51.7	52.9	53.6	54.3	54.6	54.7	54.6	53.9	53.5	52.2	50.7	50.1
P_{95}	47.5	48.7	47.9	48.5	47.9	49.2	49.6	50.9	51.9	53.1	53.9	54.5	54.9	54.9	54.8	54.1	53.7	52.4	51.0	50.4

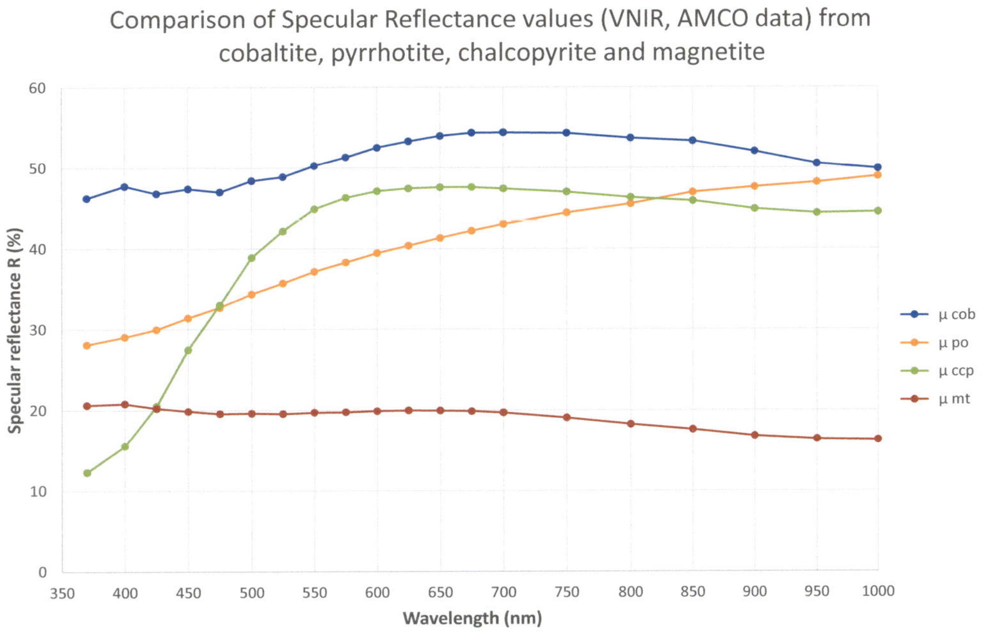
Comparison of Specular Reflectance values (VNIR, AMCO data) from cobaltite, pyrrhotite, chalcopyrite and magnetite
Specular reflectance R (%)
Wavelength (nm)
µ cob
µ po
µ ccp
µ mt

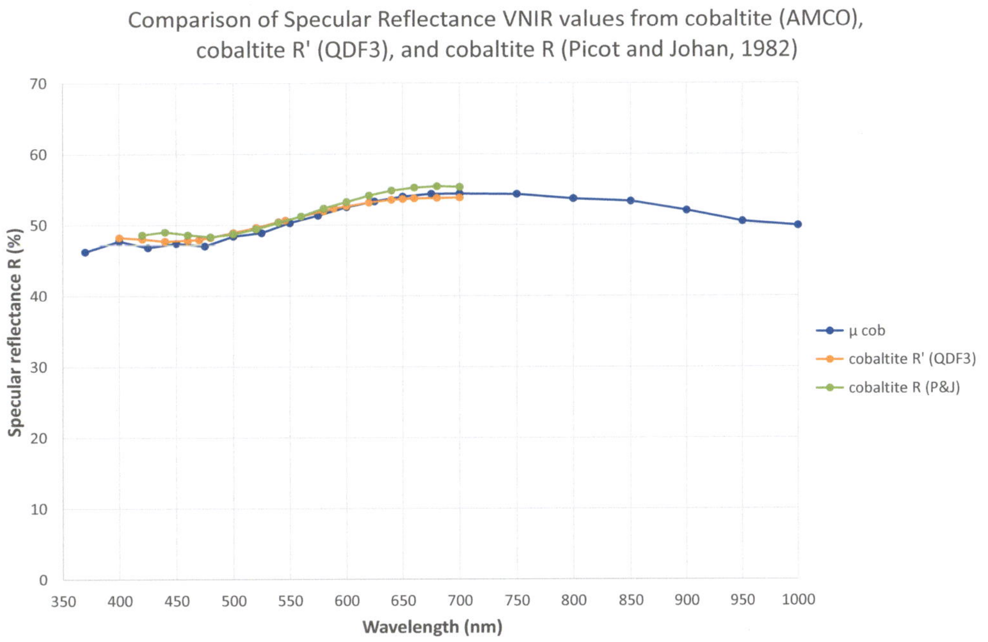
Comparison of Specular Reflectance VNIR values from cobaltite (AMCO), cobaltite R' (QDF3), and cobaltite R (Picot and Johan, 1982)
Specular reflectance R (%)
Wavelength (nm)
µ cob
cobaltite R' (QDF3)
cobaltite R (P&J)

35 Coloradoite (*colo*)

HgTe: Cubic

Coloradoita	R_{546} (%)	Y (%)	λ_d (nm)	P_e (%)		Colour sensation
Air	36.4	37.1	598	1.5		Grey
Oil	21.7	22.3	c494	1.2		
Bir: Null (1)			*Pleoc*: Null (1)			*Tarn*: Fast, dark reddish-brown tone
Anis +P: Null (1)			*IR*: No			*Twg*: No
Mor: Disseminated, granular massive			*Clv*: No			*PolQ*: Good, but sometimes scratchy
Hardness	*PolH*: Low, < clv, gn (2)		*Mohs*: 2½		*Talm*: C–	*VHN* $_{(100\ g)}$: 25–28

Note, spectral information: cf. R charts, visible range (QDF3)
Mineral abbreviations: Au—native gold, bn—bornite, colo—coloradoite, clv—calaverite, gg—gangue, gn—galena, il—ilmenite, krn—krennerite, py—pyrite, q—quartz, syl—silvanite, td—tetrahedrite

35.1 General

Composition: HgTe, occasional Pb impurities.

35.2 Microscopic Description

35.2.1 SP Observation (*Single Polariser*)

Grey *colour*, occasionally with a slight brownish tinge. This is enhanced over time by tarnishing. Soft mineral, polishes well, although with scratches. *Polishing hardness*: PolH-colo < clv, syl, gn.

Moderate *reflectance*, R colo > il, bn, but < gn, py; ≪ krn, clv, Au (Fig. 35.1a and b, cf. comparative graphs of reflectance spectra).

It occurs disseminated in grains or xenomorphic aggregates.

35.2.2 Observation +P (*Polariser + Analyser at 90°*)

Isotropic (Figs. 35.1a–d).

35.2.3 Paragenesis

Hydrothermal *parageneses* of precious metals and tellurides in volcanic (epithermal) and intrusive environments.

Companion ores: gold, pyrite, tetrahedrite, bornite, galena, telluride, sylvanite, krennerite, calaverite, petzite and other tellurides.

35.2.4 Identification

Paragenesis is an important datum. Even so, it can be difficult to identify.

Coloradoite has few distinctive properties, and these are also partly shared by petzite (Ag_3AuTe_2), which it closely resembles, although petzite is slightly anisotropic. Two specific features may support the diagnosis:

1. the peculiar shape of the reflectance spectrum of coloradoite in the visible range, whose curve, if observed in detail, shows two maxima. These do not exist in petzite, which is somewhat more reflective (cf. second comparative graph of reflectance spectra); and
2. the attack with concentrated HNO_3, to which petzite is very sensitive, while coloradoite is more resistant.

R. Castroviejo, *A Practical Guide to Ore Microscopy—Volume 1*,
https://doi.org/10.1007/978-3-031-12654-3_35

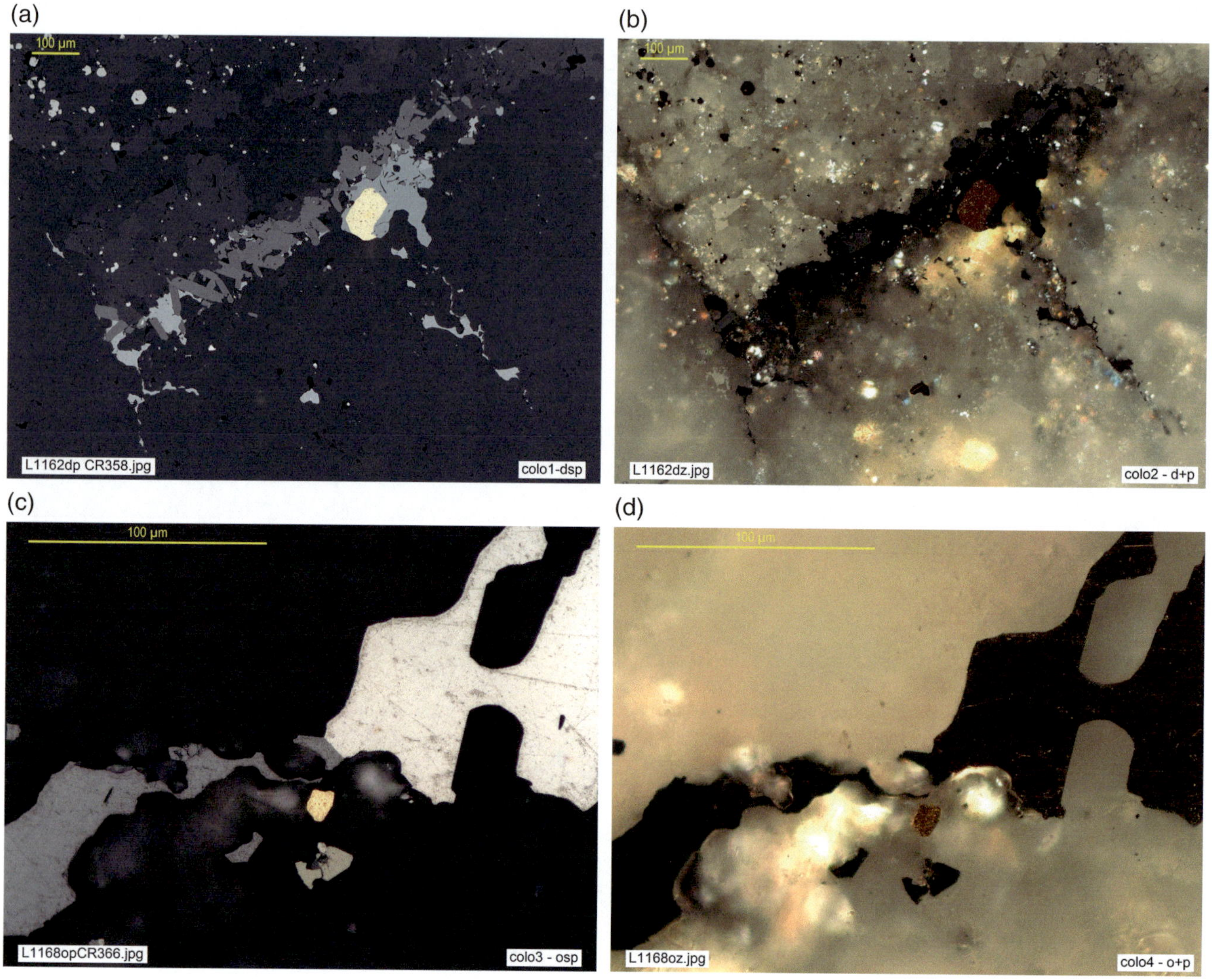

Fig. 35.1 **a**, **b** (Same scene: **dsp** and **d+p**, respectively). Brecciated quartz-carbonate vein, with Au and colo. Fine dissemination of euhedral py (cubic crystals, upper left third of image) in the carbonate microclast. The contact between it and the siliceous microclasts (lower right) is welded by il (dark brownish grey, the only anisotropic species of this ore), td (bluish grey), Au and posthumous colo (light grey compared to il and td, isotropic), which seal a thin sheared band. Colo also visible as breccia impregnation, infiltrating into microcracks and interstices throughout the image. CR358, Oroya-Brownhill Mine, Kalgoorlie, Australia. **c**, **d** (Same scene: **osp** and **o+p**, respectively). Interstitial deposition of colo (grey), krn (white) and Au, with scarce py scattered in gg (lower center, py looks yellowish-brownish-grey next to Au and krn but always lighter than colo). With analyser, the weak anisotropism of krn is barely noticeable, given the luminous oversaturation by the massive internal reflections of gg (q). CR366, Oroya-Brownhill Mine, Kalgoorlie, Australia

35.3 Spectral Information

35.3.1 Coloradoite Reflectance R (in Air, Visible Spectrum) (%)

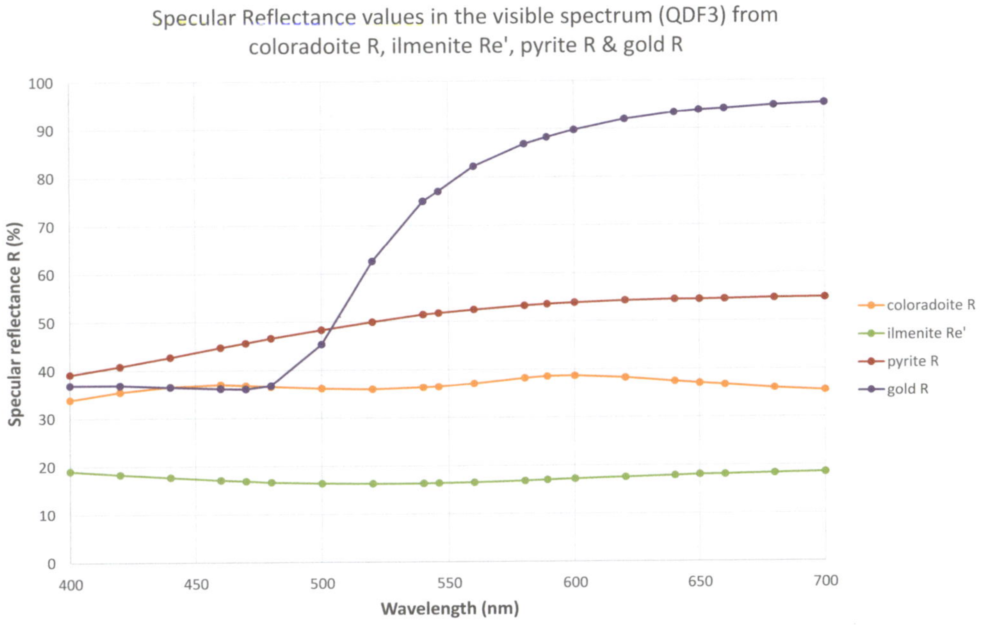

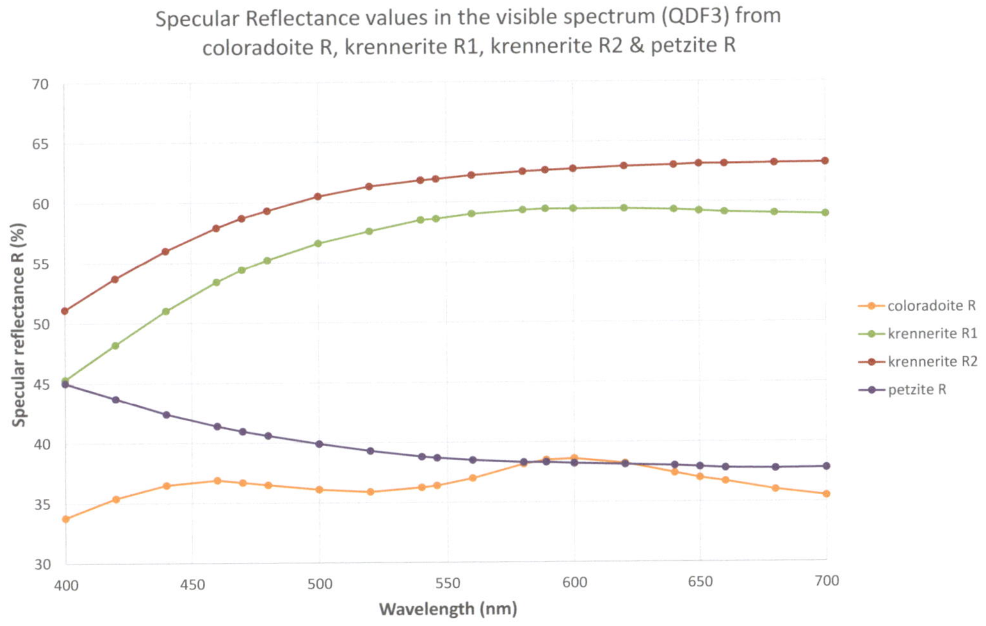

Columbotantalite (*cbt/coltan*)

36

(Fe, Mn)(Nb, Ta)$_2$0$_6$: Orthorhombic

Columbotantalite		R_{546} (%)	Y (%)	λ_d (nm)	P_e (%)	Colour sensation
Air	R_1	15.9	Nd	Nd	Nd	Grey
	R_2	16.8	Nd	Nd	Nd	
Oil	R_1	Nd	Nd	Nd	Nd	
	R_2	Nd	Nd	Nd	Nd	
Bir: Very weak (2)				*Pleoc*: Very weak (2)		*Tarn*: No
Anis + P: Weak (3)				*IR*: Reddish, varied		*Twg*: Occasionally
Mor: Euhedral, tabular, radial aggregates				*Clv*: Visible//(100), (010)		*PolQ*: Good, but slow
Hardness	*PolH*: Very high, > py (5)			*Mohs*: 6	*Talm*: nd	*VHN* (nd on weight): ≈ 480–850

Obs: R according to Picot and Johan (1983), data and charts *cf ibid.* and VNIR Spectral Information (AMCO). VHN according to Tarkian and Liessmann (1991). No QDF3 data

Mineral abbreviations: cb—columbite, cbt—columbotantalite, cst—cassiterite, gg-gangue, hub—hübnerite, il—ilmenite, mt—magnetite, mlt—microlite, py—pyrite, pyc—pyrochlore, q—quartz, rt—rutile, sch—scheelite, sp—sphalerite, tan—tantalite, tp—tapiolite, wf—wolframite

36.1 General

The generic term columbotantalite, ***cbt***, comprises the different compositional varieties of columbite and tantalite, classically treated (e.g. Ramdohr 1980) as an isomorphic, continuous solid solution series between the members (Fe, Mn)Nb_2O_6 (*niobite*, now columbite, ***cb***) and (Fe, Mn)Ta_2O_6 (*tantalite*, ***tan***), a series also known as *columbite-tantalite* or *niobite-tantalite* (or *coltan* in mining parlance). From this point of view, in general, it is practical for this series to apply the formula used by Picot and Johan (1982): (Fe, Mn)(Nb, Ta)$_2O_6$, as well as (Fe, Mn)(Ta, Nb)$_2O_6$ for tapiolite, ***tp***.

Recent investigations have led to important modifications in the understanding of these mineral species, which are now comprised in different series. While the dimorphism of the different tantalite (orthorhombic) species with the corresponding tapiolite (tetragonal) species is established, according to IMA/COM, the generic term columbotantalite should be avoided in favour of the new definitions of mineral species (Box 36.1), but the reality is complex. As Ramdohr (1980) concludes, noting that diagnostic optical observations are not yet sufficient, a precise optical differentiation of columbite and tantalite is not possible. The same is true for the currently recognized species (Box 36.1), based on accurate EMP compositions and on crystal-chemical criteria.

As the descriptive strategy of this text is based on the properties observable with optical microscopy, which do not allow a reliable distinction of the newly defined species, the optical description presented is necessarily undifferentiated, relying on a general description of columbotantalite (***cbt***): Figs. 36.1 to 36.2. The discussion of some specific examples follows (Figs. 36.3a–d, 36.4a–d, 36.5, 36.6, 36.7, 36.8, 36.9 and 36.10), to compare some of the most common species of columbite (*cb*: Figs. 36.3a–d) and tantalite (*tan*: Figs. 36.4a–d), as well as tapiolite (*tp*: Figs. 36.4 to 36.10). These have been provided for this purpose by Prof. H. Beurlen (Federal Univ. Pernambuco, Brazil), with precise EMPA and XRD (Beurlen et al. 2007, 2008; Thomas et al. 2011) analyses.

It is essential to point out that columbotantalite, **coltan** for the general public and raw material of great value for the electronics industry, for years has also been the support of criminal activity of violent militarized mafias of foreign origin, which control its extraction and distribution in the

R. Castroviejo, *A Practical Guide to Ore Microscopy—Volume 1*,
https://doi.org/10.1007/978-3-031-12654-3_36

Box 36.1 Updated terminology for Columbite and Tantalite

The terminology accepted by IMA/COM includes: Columbite-(Fe), Columbite-(Mg) Columbite-(Mn), Stibiocolumbite, Stibiotantalite, Tantalite-(Fe), Tantalite-(Mg) and Tantalite-(Mn).
These are defined by Back and Mandarino (2008), and https://www.mindat.org/min-10303.html, *visit. 2020/01/24-* as follows:

- Columbite-(Fe): $Fe^{2+}Nb_2O_6$, orthorhombic, formerly known as columbite or ferrocolumbite and now as term (Fe) of two series, Columbite-(Fe)-Tantalite-(Fe) and Columbite-(Fe)-Columbite-(Mn);
- Columbite-(Mn): $(Mn^{2+},Fe^{2+})(Nb,Ta)_2O_6$, orthorhombic, formerly manganocolumbite, term of two series, Columbite-(Fe)-Columbite-(Mn) and Columbite-(Mn)-Tantalite-(Mn);
- Columbite-(Mg): $(Mg,Fe^{2+},Mn^{2+})(Nb,Ta)_2O_6$, orthorhombic, formerly magnesiocolumbite;
- Tantalite-(Fe): $Fe^{2+}Ta_2O_6$, formerly ferrotantalite, orthorhombic, dimorph with tapiolite-(Fe); it forms two series with Tantalite-(Mg) and Tantalite-(Mn) and with Columbite-(Fe);
- Tantalite-(Mn): $Mn^{2+}Ta_2O_6$, formerly manganotantalite, orthorhombic, dimorphic with tapiolite-(Mn); forms series with Tantalite-(Fe), Tantalite-(Mg) and Columbite-(Mn);
- Tantalite-(Mg): $(Mg, Fe)(Ta,Nb)_2O_6$, formerly magnesiotantalite, orthorhombic, dimorphic with tapiolite-(Mg); forms series with Tantalite-(Fe) and Tantalite-(Mn) and with Columbite-(Mg)

Democratic Republic of the Congo. It is a planned aggression, sustained over time and intolerable from an ethical point of view because of its cruelty against the local population and the genocidal dimension it has reached, given the ineffectiveness of the supposed control by the mediators of international organizations. The attempts of traceability in the mineral market, to identify and block the *pirate traffic* of coltan, have not given the desired results for the moment. When talking about raw materials and sustainable development, these aspects cannot be forgotten.

As for the cbt metallurgy, it should be noted that its density is close to that of cassiterite, with which it is often associated, and this is problematic for gravimetric concentration.

Stibiocolumbite: $SbNbO_4$ forms a series with Stibiotantalite, $SbTaO_4$, both orthorhombic.

36.2 Microscopic Description

36.2.1 SP Observation (*Single Polariser*)

Colour grey (Fig. 36.1a, b), slightly brownish (reminiscent of wolframite or magnetite). Low *reflectance*, ≈wf or sp; > cst, sch, pyc-mlt; slightly < tp, mt (cf. Figs. 36.4a, b and 36.8 and spectral information). The first R graph, multispectral VNIR, compares columbotantalite with other common ores. The second compares different species of the columbotantalite group, evidencing the coincidence or proximity of their VNIR values. The third shows the difference, small but perceptible, between columbotantalite and tapiolite in the visible range (Picot and Johan 1982).

Weakly pleochroic. Oscillatory or growth zoning (EMP), rarely visible with an optical microscope, given the small difference in appearance between the different compositional terms (Fig. 36.2a, b).

Good *polishing* but slow because of its hardness. Very high *polishing hardness*: $PolH_{cbt}$ > py, but < q, cst. Distinct cleavage // (100), occasionally visible (Fig. 36.6), and // (010), less frequent. No tarnishing has been observed on polished sections preserved in dry environment (in laboratory) but has been cited on hand samples.

Vickers Microhardness, VHN, varies with composition. For example, for columbite-(Fe), $VHN_{100} = 652 - 739$ kg/mm^2 (https://www.mindat.org/min-1514.html, *consult. 2020/01/24*), while for tantalite-(Mn) $VHN_{100} = 488 - 681$ kg/mm^2 (https://www.mindat.org/min-2522.html, *consult. 2020/01/24*). According to Uytenbogaardt and Burke (1971), for the cbt-(Fe,Mn) group, VHN varies in a wide range, between 240 and 1021, with no relation between the Nb/Ta ratio and the VHN value. The VHN range ≈ 480–850 (Tarkian and Liessmann 1991) has been taken as the most representative in general.

In general, cbt occurs in aggregates with a marked idiomorphic tendency, in tabular or prismatic crystals (Figs. 36.2a and 36.6) and sometimes in radiating aggregates. It is also included in *cassiterite*. Tantalite can also be found associated with *tapiolite*, with or without substitution, between both (Figs. 36.8, 36.9 and 36.10).

The variations in appearance, colour and reflectance between columbite and tantalite are almost imperceptible, as can be seen in the zoned crystal of Fig. 36.2a. However, other images of different samples of columbite and tantalite, as well as tapiolite (tp) associated with tantalite (Figs. 36.3 to 36.10), are also presented for comparison.

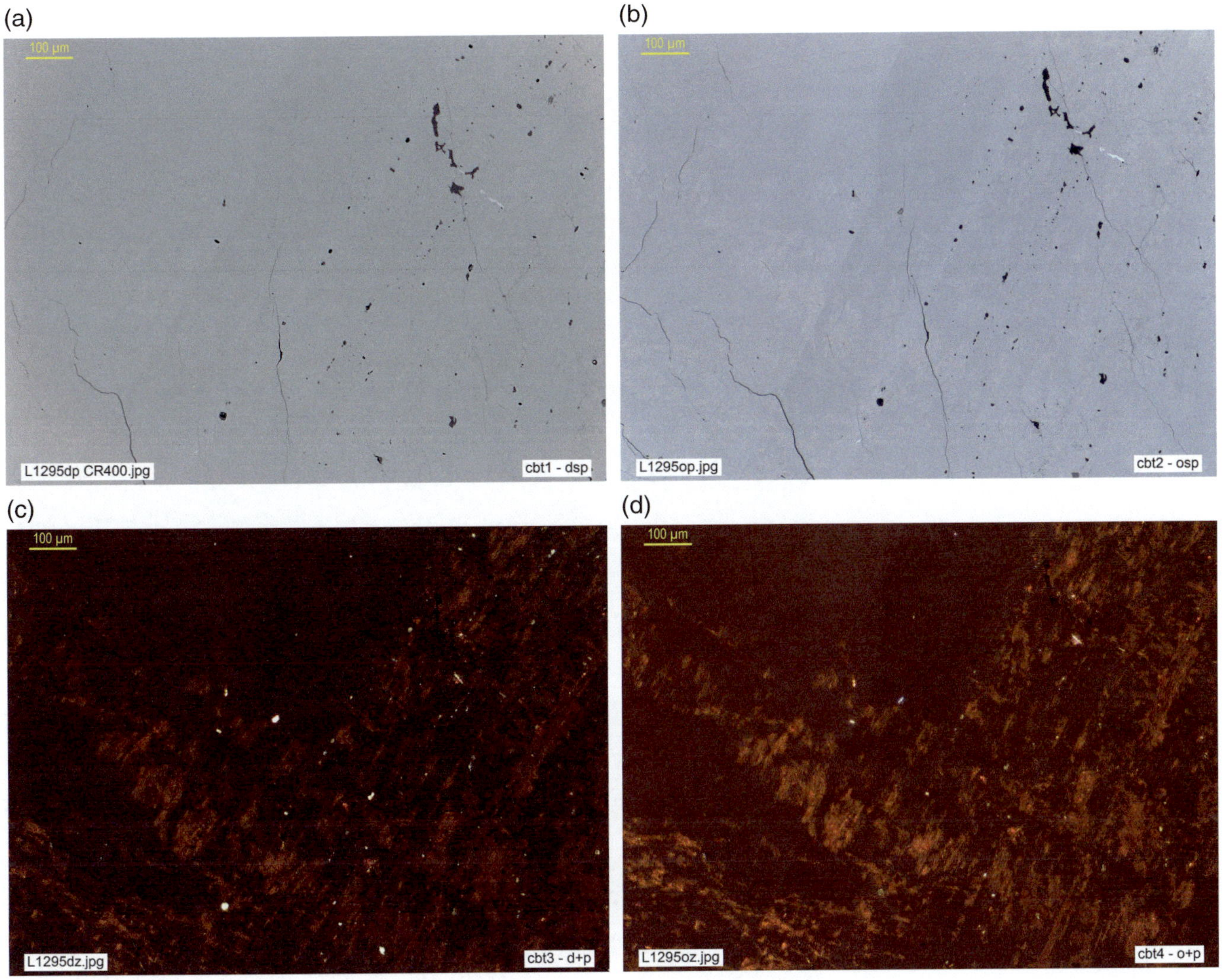

Fig. 36.1 **a, b** (Same scene: **dsp** and **osp**, respectively). Single crystal of cbt with compositional zoning, visible by the fine silicate inclusions (dark grey) delimiting successive growth zones, starting from a more homogeneous core (upper left). Microzones composed alternatively of cb and tan, difficult to distinguish optically (although they show a better contrast in immersion, *Fig.* 36.1b) but evidenced by micro-analysis (**EMP**). CR 400, Borborema, Brazil. **c, d** (Same scene as cbt-1: **d+p** and **o+p**, respectively). Abundant internal reflections, especially where tantalite dominates. Weak anisotropism, although enough to evidence twinning in the core (upper center), also evidenced by slight bireflectance (in Fig. 36.1b)

36.2.2 Observation +P (*Polariser + Analyser at 90°*)

Weakly *anisotropic*, polarisation tints are even less coloured than wolframite (Figs. 36.1c, d and 36.2b, 36.9 and 36.10). Straight extinction. Occasionally *twinned* (Figs. 36.9 and 36.10).

It generally shows, especially in immersion, reddish *internal reflections*, more abundant than in wolframite (except hübnerite). IR are scarcer in Nb-rich columbotantalites (columbite) and more frequent and luminous in Ta-rich (tantalite) or Mn-rich terms, in which they can largely overprint the anisotropism of the mineral (comp. Figs. 36.1c, d, 36.2b with 36.5, 36.9 and 36.10).

Fig. 36.2 **a, b** (Same scene: **dsp** and **dxp**). Wide field image (obj. 2.5x) of the single crystal of cbt: Fig. 36.1. Alternating cb and tan microzones marking oscillatory zoning (EMP), barely visible by optical microscopy (and only with partially uncrossed analyser, cbt6)

(a)

(b)

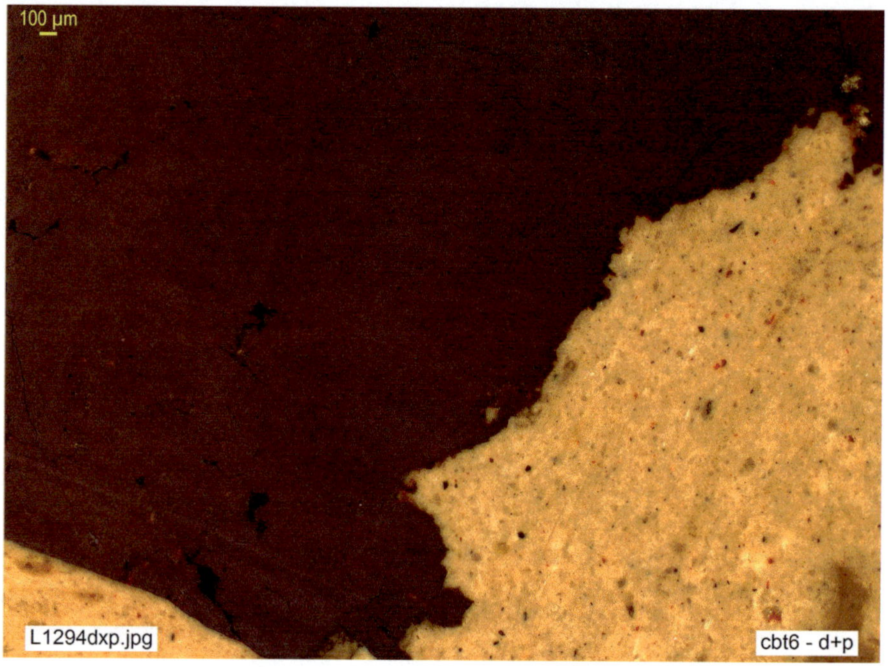

36.2.3 Paragenesis

Paragenesis typically associated with plutonic environments, particularly leucogranites or S-type evolved granites and pegmatites, with Be, Li, TR, accompanied by cassiterite, wolframite, pyrochlore, ilmenite, rutile, tapiolite.

36.2.4 Identification

Characterized by its strong tendency to idiomorphism, high polishing hardness, low reflectance, weak pleochroism, weak anisotropism and abundant IR. A typical criterion is also the paragenesis and the association with cassiterite, in which it forms exsolutions, or with tapiolite (tantalite).

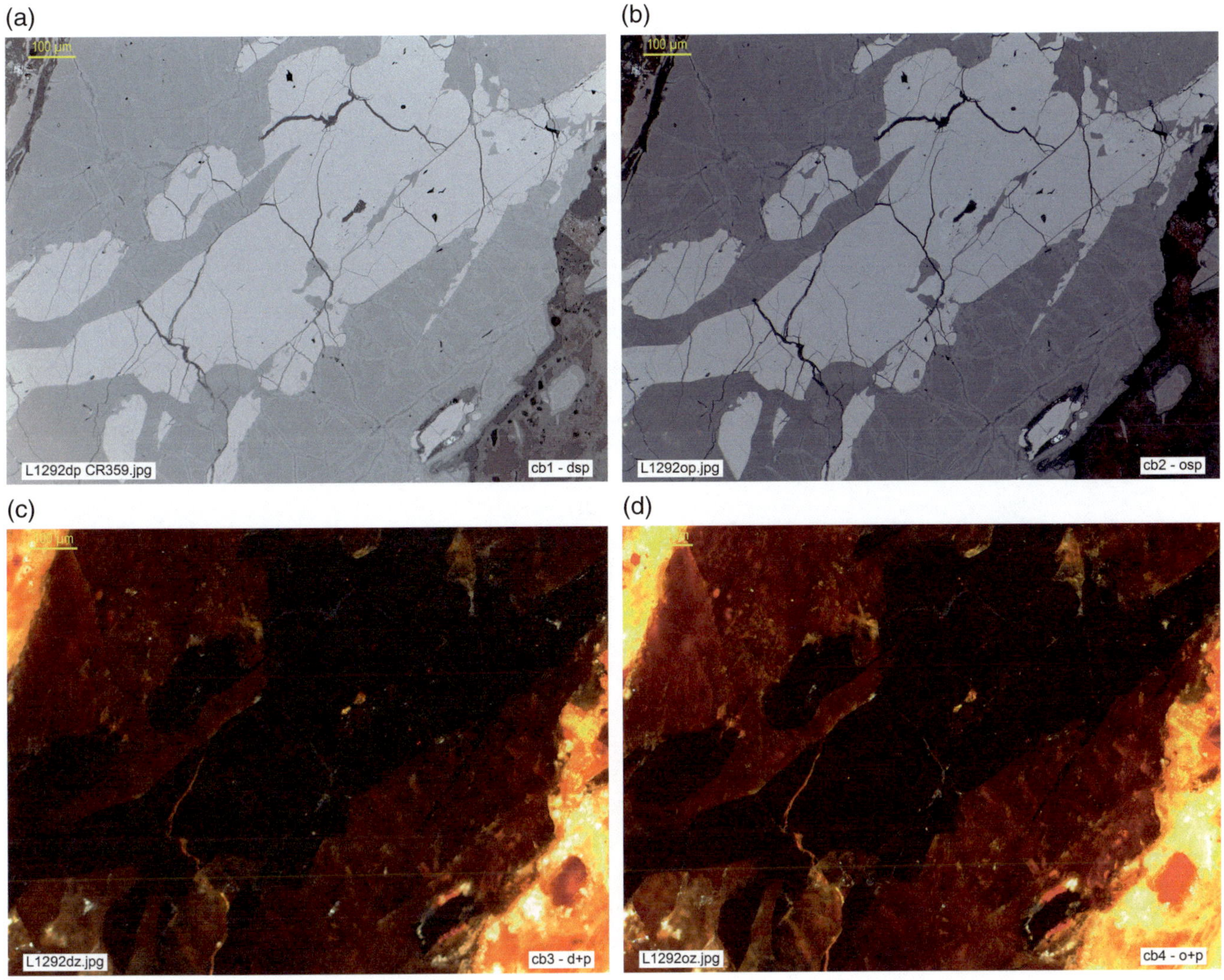

Fig. 36.3 **a** Ferrocolumbite (columbite-(Fe), after updated terms, light grey, center image) brecciated, included in a mass of pyc-mlt (darker grey) that cements and partly replaces it. The whole is cemented by veinlets of siliceous gg impregnated with limonite (upper left and lower right). CR-359, Picahy, Paraiba do Norte, Brazil (**dsp**). **b** Same scene as Fig. 36.3a (**osp**). **c** Same scene as cb1 (**d+p**). Reddish RI of cb, not very visible and scarce as corresponds to its composition (Fe, Nb); more abundant and lighter in pyc-mlt (isotropic). **d** Same scene as Fig. 36.3a (**o+p**)

Among the minerals that may occur in association, confusion with cassiterite, tapiolite, wolframite and magnetite is possible, but careful observation should avoid this:

- cbt is distinguished from *cassiterite* by the lower reflectance and lighter internal reflections of the latter,
- *tapiolite* shows higher reflectance and a more vivid and coloured anisotropism.

- cbt resembles *wolframite*, but the latter is pleochroic and more strongly anisotropic and, in addition, shows oblique extinction,
- *magnetite* is isotropic and has no IR.

As explained, the precise determination of columbite and tantalite species is beyond the capabilities of the optical microscope (Figs. 36.1a–d and 36.2a, b).

However, some general characterizations of columbite and tantalite are possible, illustrated by the additional photomicrographs Figs. 36.3 to 36.10:

- **COLUMBITE** (Fe,Mn)Nb_2O_6, (Figs. 36.3a–d)—Its higher Fe and Nb content enhances the scarcity of IR (Figs. 36.3c, d), while the single polariser setting shows its similarity to columbotantalite (no appreciable difference in aspect). The Vickers microhardness of columbite-(Fe) is higher than that of tantalite-(Mn).

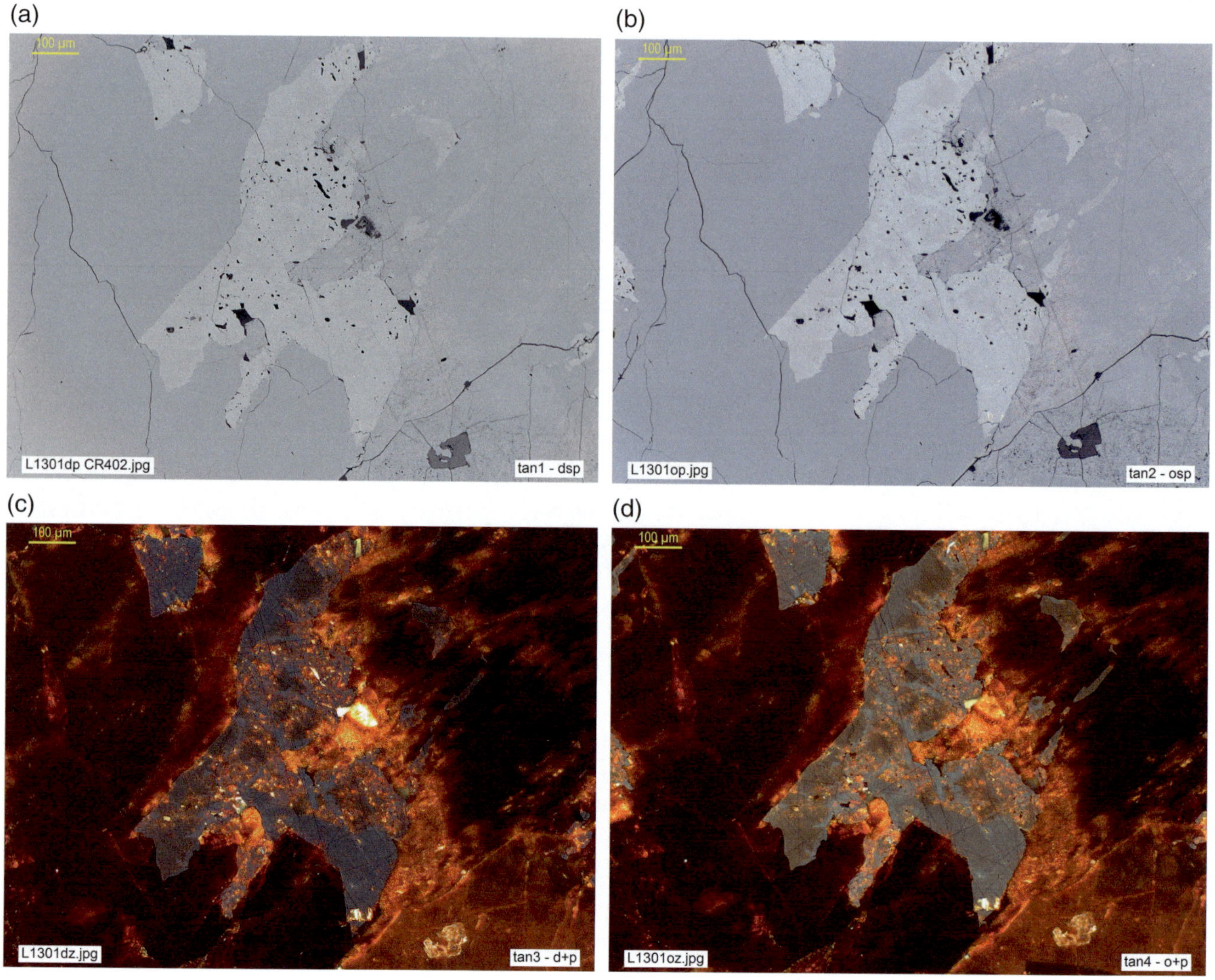

Fig. 36.4 **a** Manganotantalite (or tan-(Mn), grey) with inclusions of tp (lighter, center of image) and quartz (dark grey, bottom right-hand). CR 402, Borborema, Brazil (**dsp**). **b** Same scene as tan1 (**osp**). **c** Same scene as tan1 (**d+p**). Effects of anisotropism in tp more intense and coloured (blue-green) than in tan. Abundant IR in both. **d** Same scene as tan1 (**o+p**). Effects of anisotropism in tp more intense and coloured (blue-green) than in tan (*comp.* Fig. 36.5). Abundant IR in both

- **TANTALITE** (Mn,Fe)Ta_2O_6 (Figs. 36.4a–d, 36.5, 36.6, 36.7, 36.8, 36.9, and 36.10)—It is typically composed of manganiferous members, ferriferous tantalites being very rare, unlike its dimorph tapiolite, which is usually ferriferous (Fe,Mn)$(Ta,Nb)_2O_6$; *tan* and *tp* sometimes appear intergrown (Figs. 36.4a–d, 36.8, 36.9 and 36.10) and may show replacement textures. The IR of tan are lighter and more abundant than those of columbite. The high content of Ta and also that of Mn (Fig. 36.5) contribute to this. The Vickers microhardness of tantalite-(Mn) is lower than that of columbite-(Fe).

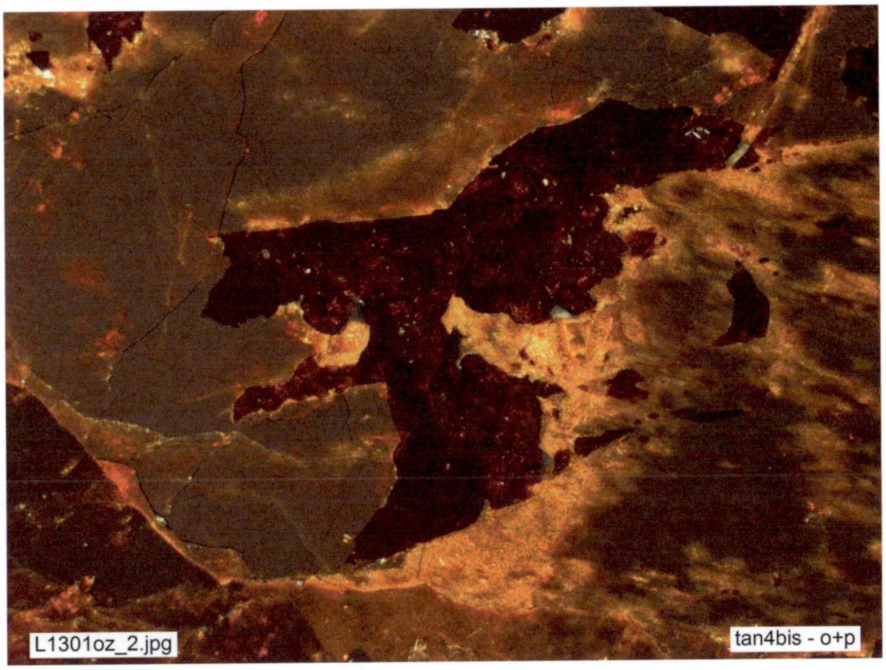

Fig. 36.5 Image of Fig. 36.4d, *rotated* to tp extinction, to better observe the weak anisotropism of tan, as well as its abundant IR, consistent with its high Mn (**o+p**) content

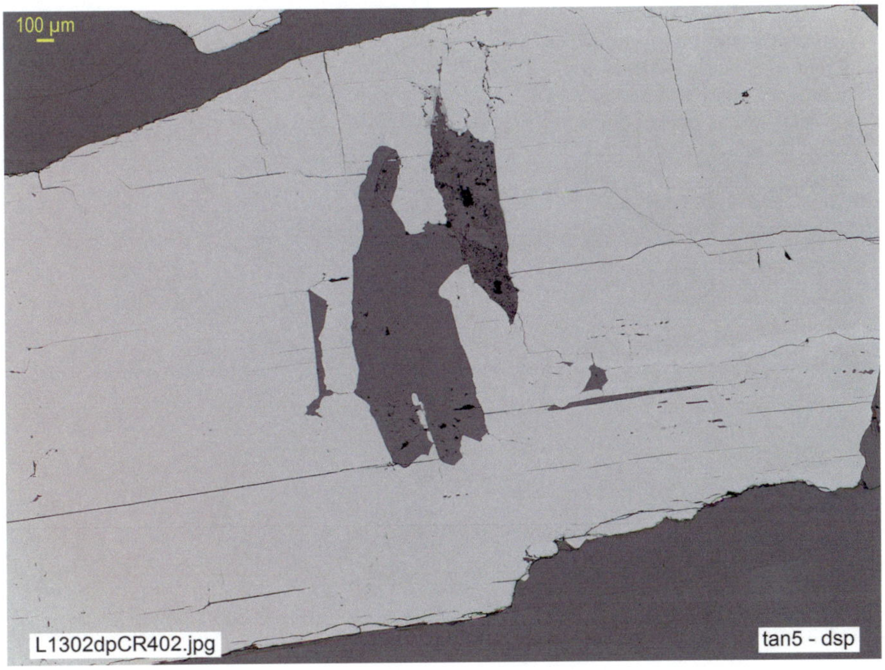

Fig. 36.6 Concentrate of tan (wide field image, obj. 2.5x), euhedral tan grain, light grey, with prismatic habit and cleavage // (100). Sample CR 402, Alto do Giz Pegmatite, Borborema, Brazil (**dsp**)

Fig. 36.7 Same scene as tan5 (**d+p**). The abundant reddish IR almost completely mask the anisotropism of tan

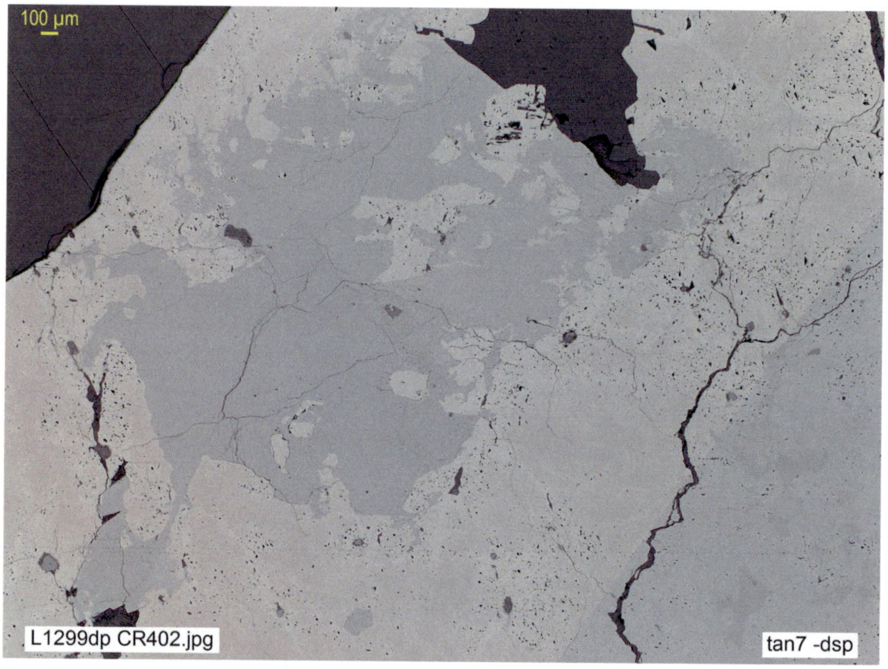

Fig. 36.8 Relict core of tan-(Mn), grey, included in large prismatic and twinned crystal of tp, lighter and pleochroic between light brownish grey and darker bluish grey shades (wide field image of the tan-tp concentrate, of which images Figs. 36.4a–d show details). CR 402, Borborema, Brazil (**dsp**)

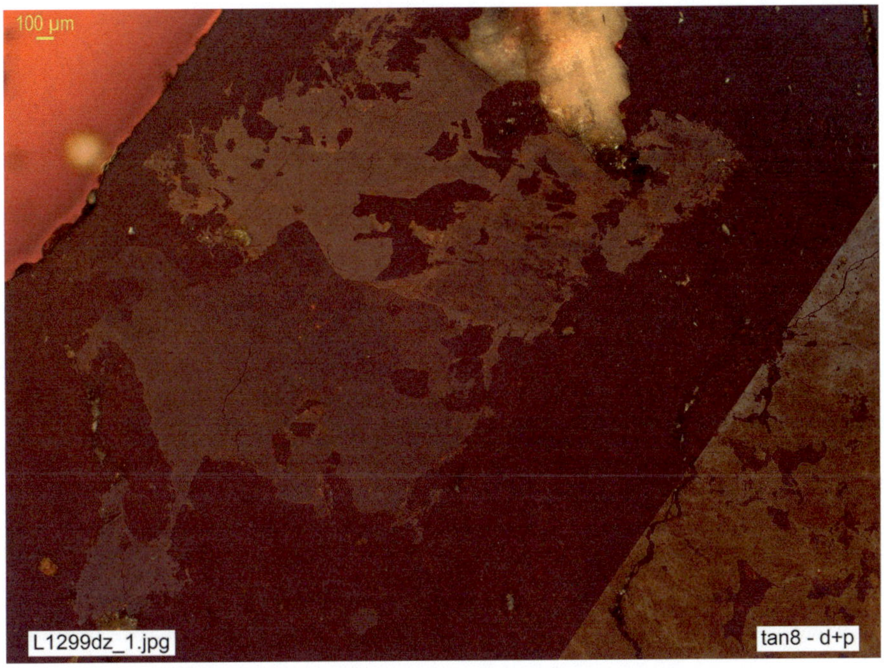

Fig. 36.9 Same scene as tan7 (**d +p**). The extinction position of one of the twin elements of tp allows appreciation of the weak anisotropism of tan, also twinned, and the abundant IR of both. (The red colour on the upper left corner is that of the embedding resin)

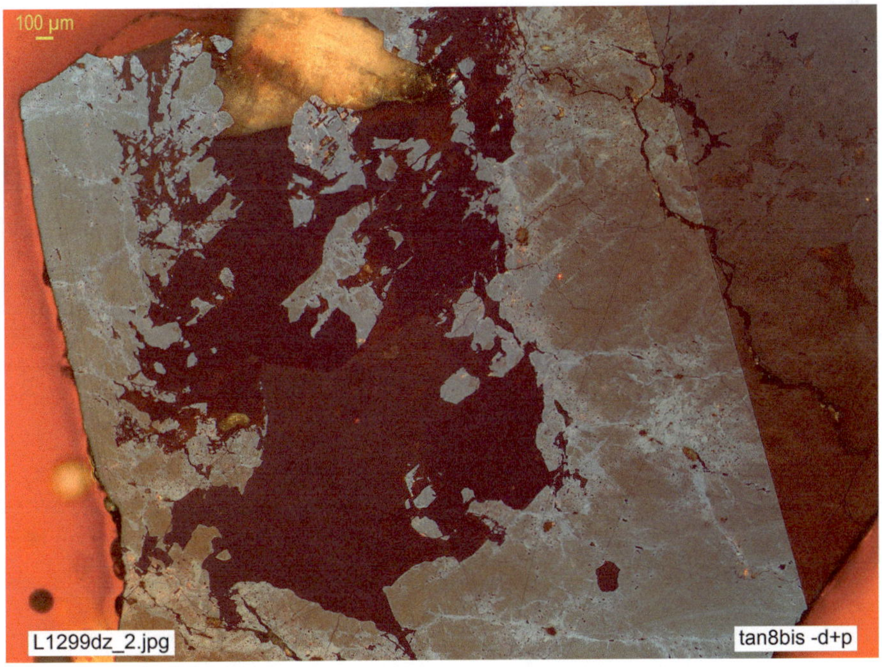

Fig. 36.10 Same scene as tan8 (**d+p**) but rotated to enhance the different anisotropism of tan and tp, as well as the replacement texture of tan by tp

36.3 Spectral Information

36.3.1 Columbotantalite Multispectral Reflectance R (in Air, Without Polariser) (%)

λ (nm)	370	400	425	450	475	500	525	550	575	600	625	650	675	700	750	800	850	900	950	1000
μ	18.5	18.0	17.7	17.3	17.1	16.8	16.6	16.5	16.3	16.2	16.1	16.0	15.9	15.8	15.7	15.5	15.5	15.3	15.2	15.0
P_5	17.8	17.5	17.2	16.8	16.6	16.3	16.2	16.0	15.9	15.7	15.7	15.6	15.5	15.4	15.3	15.1	15.0	14.9	14.8	14.6
P_{20}	18.1	17.6	17.3	16.9	16.7	16.4	16.3	16.1	16.0	15.8	15.8	15.6	15.6	15.5	15.4	15.2	15.1	14.9	14.9	14.7
P_{80}	19.0	18.4	18.1	17.7	17.5	17.1	17.0	16.8	16.6	16.5	16.4	16.3	16.2	16.1	16.0	15.8	15.7	15.5	15.4	15.2
P_{95}	19.2	18.5	18.2	17.8	17.6	17.3	17.1	17.0	16.8	16.7	16.6	16.5	16.4	16.3	16.2	16.0	15.9	15.7	15.6	15.4

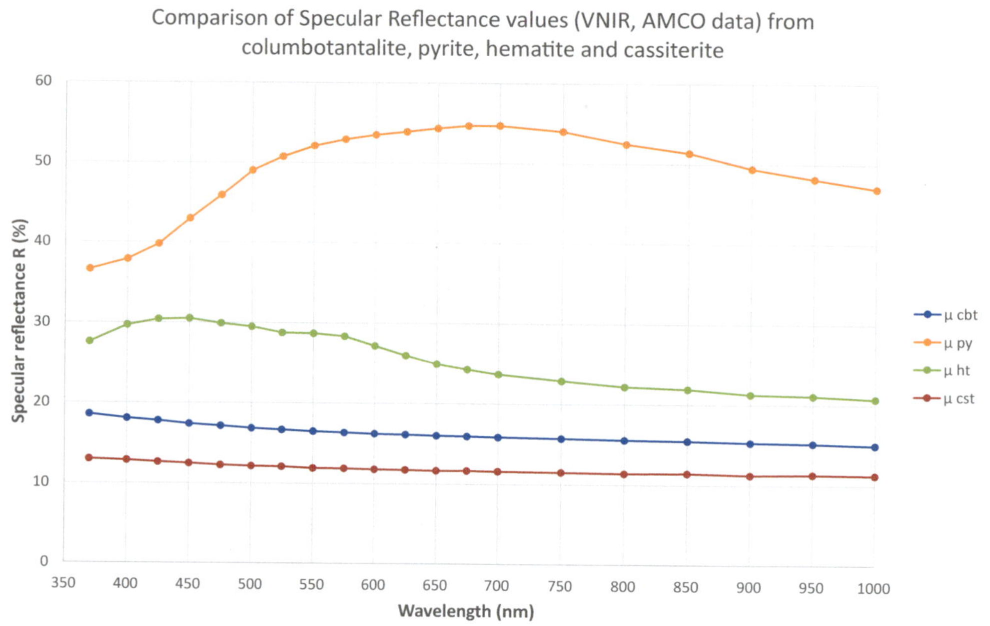

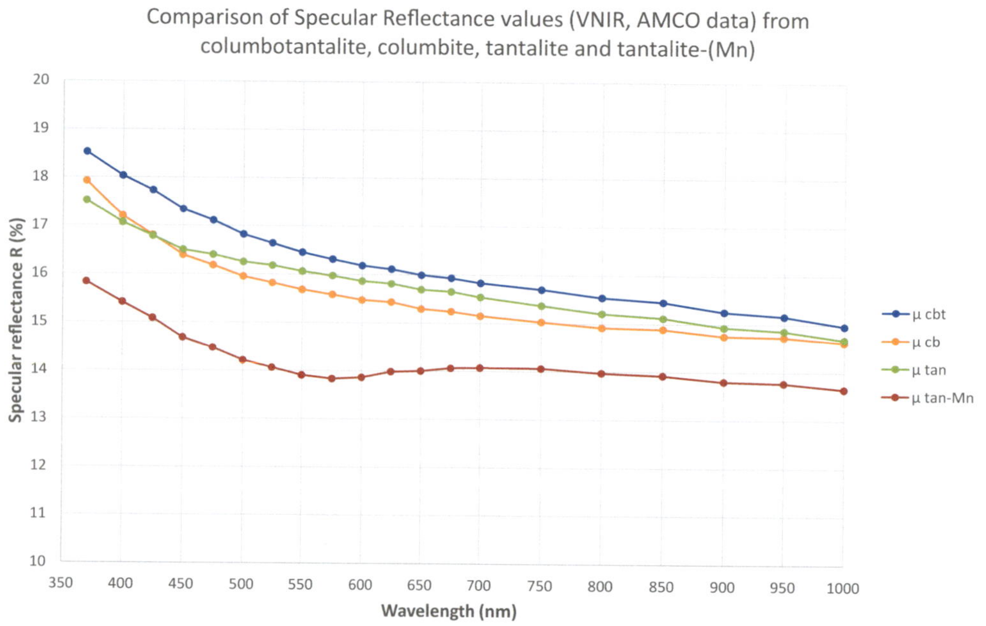

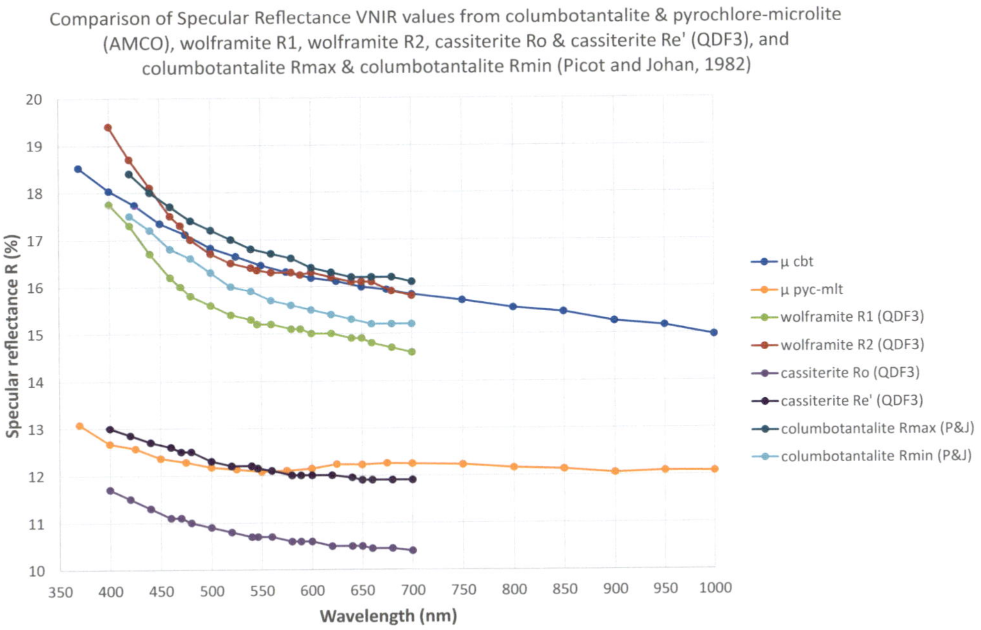

Comparison of Specular Reflectance VNIR values from columbotantalite & pyrochlore-microlite (AMCO), wolframite R1, wolframite R2, cassiterite Ro & cassiterite Re' (QDF3), and columbotantalite Rmax & columbotantalite Rmin (Picot and Johan, 1982)
Specular reflectance R (%)
Wavelength (nm)
µ cbt
µ pyc-mlt
wolframite R1 (QDF3)
wolframite R2 (QDF3)
cassiterite Ro (QDF3)
cassiterite Re' (QDF3)
columbotantalite Rmax (P&J)
columbotantalite Rmin (P&J)

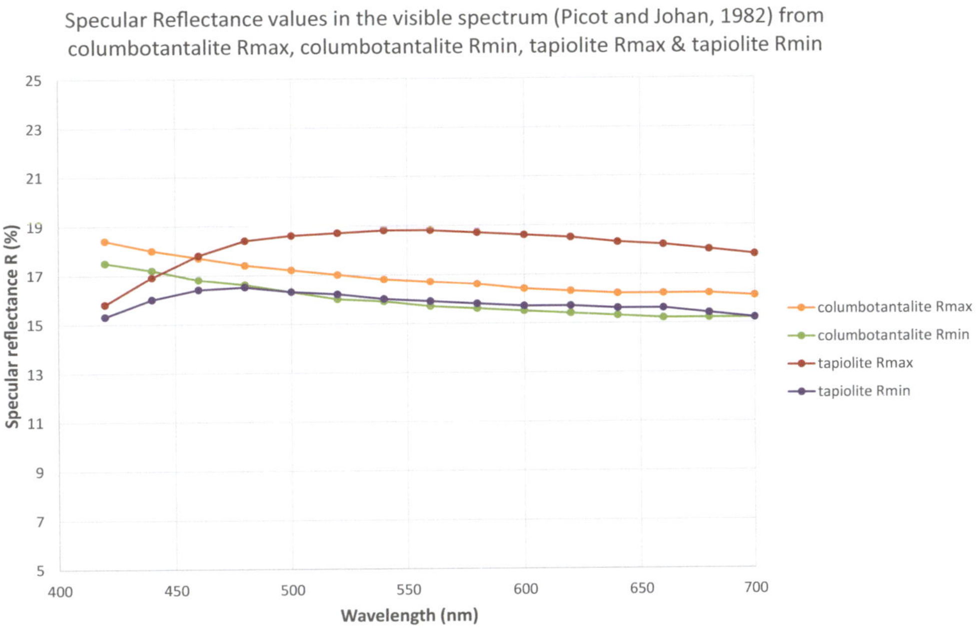

Specular Reflectance values in the visible spectrum (Picot and Johan, 1982) from columbotantalite Rmax, columbotantalite Rmin, tapiolite Rmax & tapiolite Rmin
Specular reflectance R (%)
Wavelength (nm)
columbotantalite Rmax
columbotantalite Rmin
tapiolite Rmax
tapiolite Rmin

References

Back ME y Mandarino JA (2008) Fleischer's glossary of mineral species 2008. The Mineralogical Record Inc. Tucson, AZ, 345p

Beurlen H, Barreto SB, Silva D, Wirth R, Olivier P (2007) Titanian ixiolite-niobian rutile intergrowths from the Borborema Pegmatite Province. NE Brazil Canad Mineralogist 45:1367–1387

Beurlen H, Da Silva M, Thomas R, Soares D, Olivier P (2008) Na-Ta-(Ti-Sn) oxide mineral chemistry as tracer of rare-element granitic pegmatite fractionation in the Borborema Province, NE Brazil. Miner Deposita 43:207–228

Picot P, Johan Z (1982) Atlas of ore minerals. B.R.G.M./Elsevier

Ramdohr P (1980) The ore minerals and their intergrowths, 2nd edn, vol 2. Pergamon Press, Oxford, 1205p

Tarkian M, Liessmann W (1991) A Guide for optical and analytical identification of ore minerals. Clausthaler Geol Abh 47, Verlag Sven von Loga, Köln, 78p

Thomas R, Davidson P, Beurlen H (2011) Tantalite-(Mn) from the Borborema Pegmatite Province, NE Brazil: conditions of formation and melt- and fluid-inclusion constraints on experimental studies. Miner Deposita 46:749–759

Uytenbogaardt W, Burke EAJ (1971) Tables for microscopic identification of ore minerals, 2nd edn. Elsevier, Amsterdam

Copper(*Cu/Native Copper*)

37

Cu: Cubic

Copper	R_{546} (%)	Y (%)	λ_d (nm)	P_e (%)	Colour sensation
Air	64.6	74.4	588	22.3	Intense, bright reddish pink colour
Oil	56.9	68.7	588	28.5	
Bir: Null (1)			*Pleoc*: Null (1)		*Tarn*: Fast, reddish-brown
Anis +P: Isotropic, imperfect extinction (2)			*IR*: No		*Twg*: (111) visible by struct. etching
Mor: Impregnation, microcracks			*Clv*: No		*PolQ*: Easy, but frequent scratches
Hardness	*PolH*: Medium, > gn, cc (3)		*Mohs*: 2.5–3	*Talm*: B	*VHN* $_{(100\ g)}$: 79–99

Obs: cf. Table and R charts, multispectral data (VNIR, AMCO) and visible range (QDF3)
Mineral abbreviations: Ag—native silver, Au—native gold, bn—bornite, cc—chalcocite, ccp—chalcopyrite, Cu—native copper, cup—cuprite, gn—galena, ht—hematite, lim—limonite
Colour abbreviations

37.1 General

Composition: Cu. May contain traces of Au and Ag. May incorporate As in solid solution, transition to the *whitneyite* variety (Cu, As) or to the species recognized by IMA/COM algodonite ($Cu_{1-x}As_x$; x ≈ 0.15). The incorporation of As reduces reflectance and increases hardness and brittleness.

37.2 Microscopic Description

37.2.1 SP Observation (*Single Polariser*)

Intense pinkish *colour* (copper red) and bright when freshly polished (Fig. 37.1a, b), but it tarnishes quickly after polishing, if the surface is not protected, developing a reddish-brown *patina* (Fig. 37.2). Polishes well, but it is difficult to avoid scratches. *Polishing hardness* medium–low, ≈ bn, ccp; > gn, cc; < Ag, Au, cup, ht, lim.

High *reflectance*, with an unmistakable spectral response, characterized by a sharp rise in R towards red, which places it clearly above Ag from 600 nm on and throughout the near IR range.[1] Similar to, though less extreme (and white) than Fe (Fig. 37.6).

It occurs in intergrown crystalline aggregates, filling microcracks or as impregnation (Fig. 37.4), interstitial, in intergranular films, occasionally botryoidal. Frequent association with cuprite, in small inclusions (Fig. 37.5). It can also be found in meteorites, with troilite and native Fe (Fig. 37.6).

37.2.2 Observation +P (*Polariser + Analyser at 90°*)

Isotropic but with imperfect extinction (Figs. 37.1c, d and 37.3), which may be accentuated by polishing scratches.

Lamellar *twinning* (111) visible after structural etching.

[1] Ag is currently the mineral substance with the highest known reflectance, but this must be understood as referring to white light or for λ < 600 nm (cf. multispectral comparative graphs).

R. Castroviejo, *A Practical Guide to Ore Microscopy—Volume 1*,
https://doi.org/10.1007/978-3-031-12654-3_37

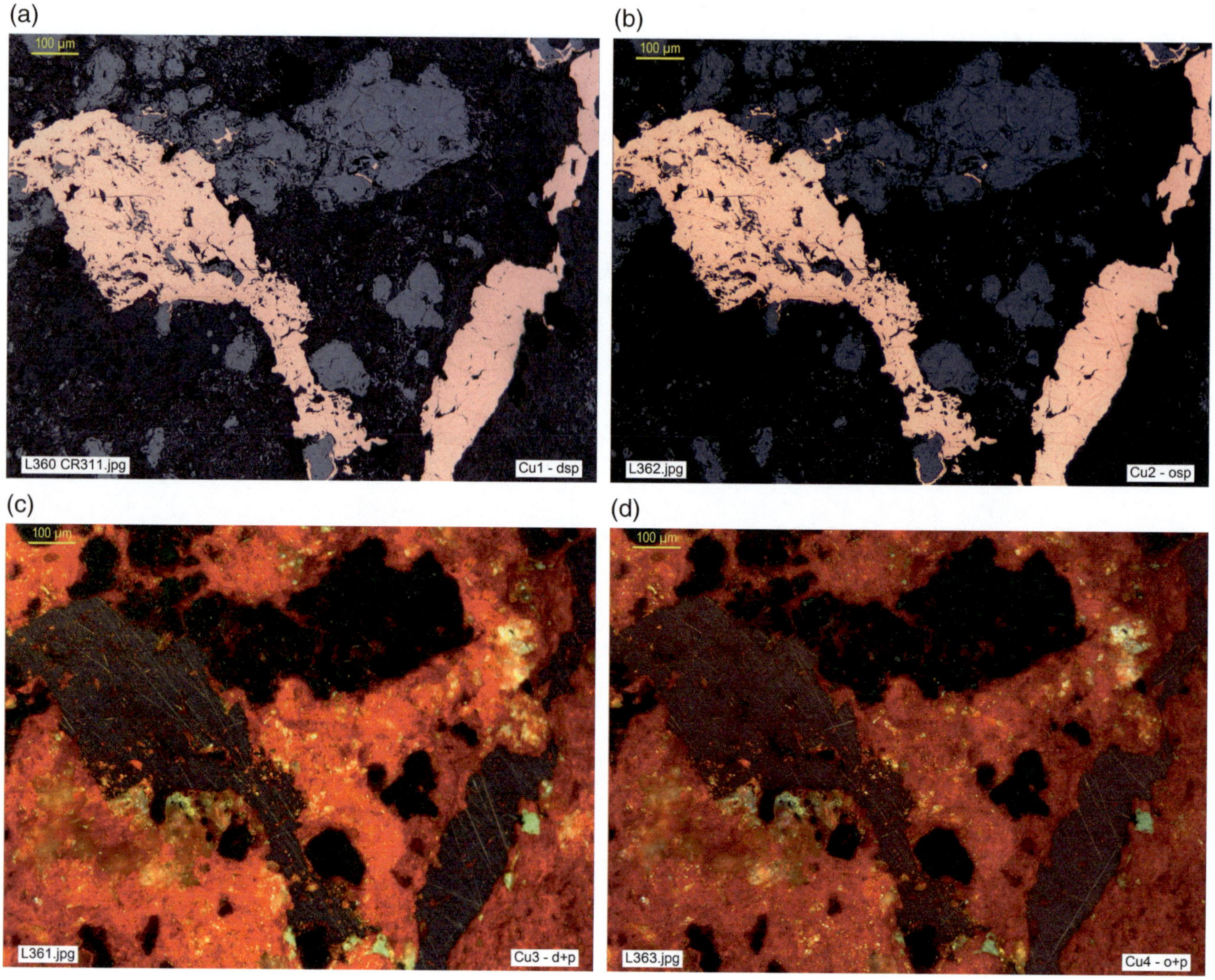

Fig. 37.1 **a** Supergene Cu impregnation (intense pink, bright), massive, fine-grained ht-lim (grey) and scarce cup (bluish-grey) in porphyry copper. CR311, Milpillas, Mexico (**dsp**). **b** Same scene as Cu1 (**osp**). **c**, **d** Same scene as Cu1 (**d+p**) and (**o+p**), respectively. Imperfect Cu extinction. Anisotropism of ht, lim and cup masked by internal reflections (comp. Cu4bis)

37.2.3 Paragenesis

The parageneses of native copper are varied, although it is abundant in few places. It occurs as supergene product, mainly secondary from chalcocite and associated with cuprite, at the bottom of the oxidation zone of cupriferous sulphide masses.

It is also found associated with basaltic rocks in Keewenaw-type deposits, Lake Superior, USA, forming disseminations of disputed genesis.

As a primary magmatic product it is seen in some meteorites. Some occurrences, understood as primary magmatic as well, have been cited accompanying PGM (Bushveld, South Africa).

37.2.4 Identification

It is practically unmistakable, due to its colour and high reflectance. The association with cupriferous paragenesis, and particularly with cuprite, are additional criteria.

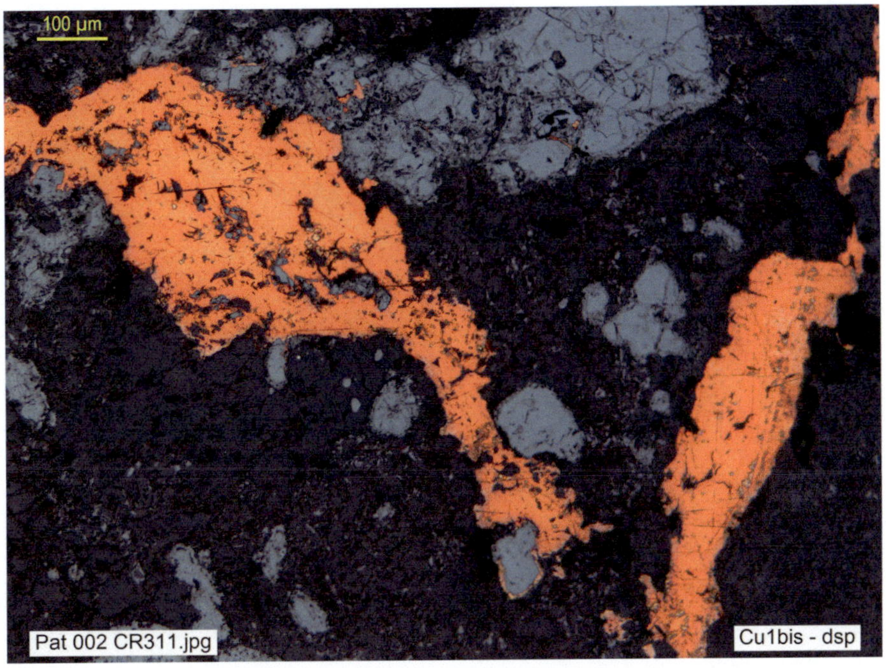

Fig. 37.2 Same scene as Cu1, effect of tarnishing (darker surface of Cu) in two months (**dsp**)

Fig. 37.3 Same scene as Cu4 (**oxp**), incomplete Cu extinction and oxides anisotropism (cup gives green tones)

Fig. 37.4 Quartz vein, slightly brecciated, with Cu in intergranular films and interstices and slight cup and lim impregnation. CR311, Milpillas, Mexico (**dsp**)

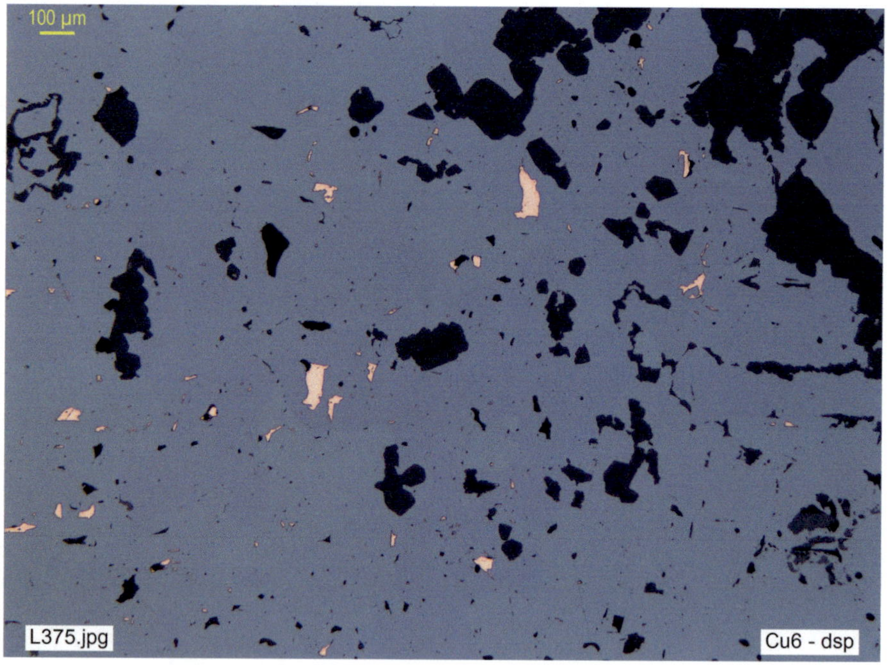

Fig. 37.5 Dissemination of Cu in massive cup. CR287, Onganja, Namibia (**dsp**)

Fig. 37.6 Native Cu (pink) and Fe (white), troilite (brown), scarce mt (grey) and silicates (grey, almost black). Note the dark appearance of troilite (stoichiometric po) against native metals and the disappearance of its pinkish tones next to Cu. Sample hjb/sn, El Desaguadero Meteorite, fallen October 2007, Puno, Peru (**osp**)

37.3 Spectral Information

37.3.1 Copper Multispectral Reflectance R (In Air, Without Polariser) (%)

λ (nm)	370	400	425	450	475	500	525	550	575	600	625	650	675	700	750	800	850	900	950	1000
μ	42.4	46.2	49.5	52.6	54.8	56.6	58.1	62.4	76.2	87.7	90.6	92.1	93.0	93.7	94.7	95.4	96.6	97.3	98.2	98.3
P_5	40.7	44.3	47.4	50.5	52.7	54.6	56.1	60.5	74.0	85.6	88.6	90.2	91.2	92.1	93.3	94.2	95.5	96.3	97.3	97.5
P_{20}	41.5	45.3	48.6	51.7	53.9	55.8	57.3	61.6	75.3	86.9	89.9	91.4	92.3	93.2	94.2	95.0	96.2	97.0	97.9	98.0
P_{80}	43.3	47.2	50.4	53.6	55.8	57.6	59.0	63.3	77.2	88.7	91.5	93.0	93.7	94.5	95.4	96.0	97.1	97.8	98.6	98.7
P_{95}	44.0	48.0	51.2	54.3	56.5	58.2	59.6	64.0	78.0	89.3	92.1	93.6	94.3	94.9	95.8	96.4	97.5	98.1	99.0	99.1

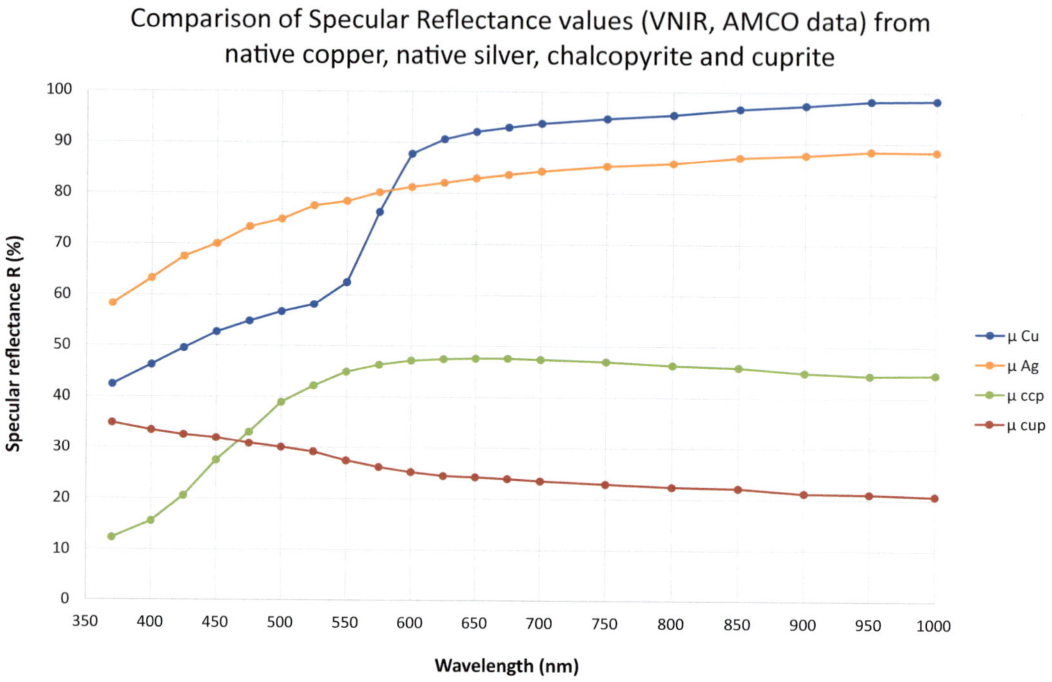
Comparison of Specular Reflectance values (VNIR, AMCO data) from native copper, native silver, chalcopyrite and cuprite
Specular reflectance R (%)
Wavelength (nm)
µ Cu
µ Ag
µ ccp
µ cup

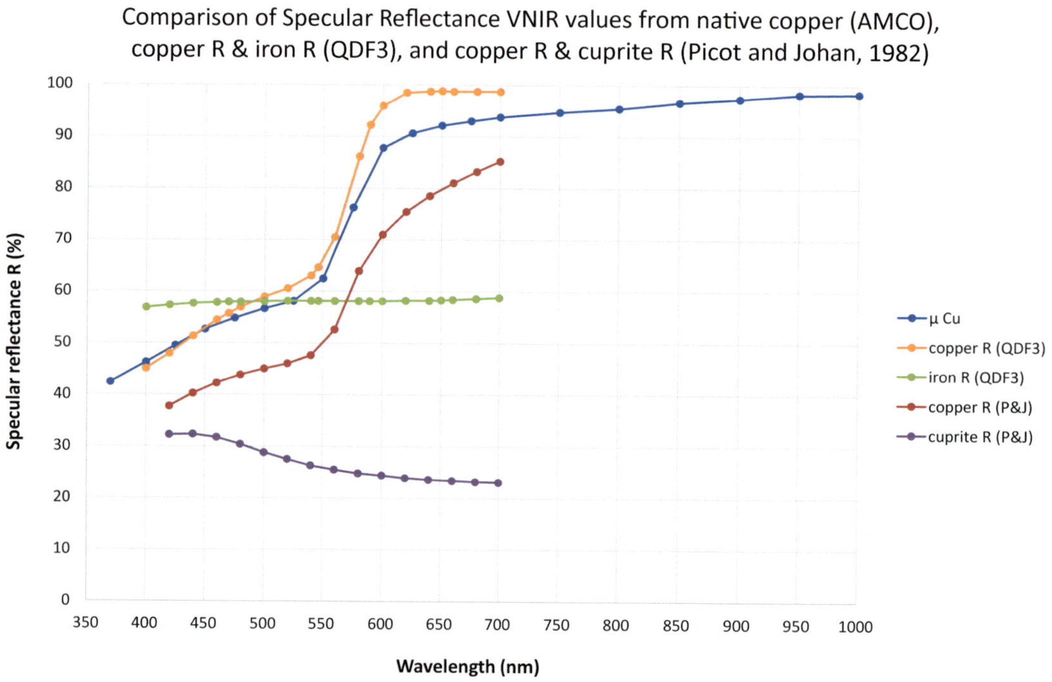
Comparison of Specular Reflectance VNIR values from native copper (AMCO), copper R & iron R (QDF3), and copper R & cuprite R (Picot and Johan, 1982)
Specular reflectance R (%)
Wavelength (nm)
µ Cu
copper R (QDF3)
iron R (QDF3)
copper R (P&J)
cuprite R (P&J)

38 Cosalite (*cos*)

$Pb_2Bi_2S_5$: Orthorhombic

Cosalite		R_{546} (%)	Y (%)	λ_d (nm)	P_e (%)	Colour sensation
Air	R_1	41.4	41.4	474	4.6	Greyish white to light grey
	R_2	45.7	45.9	473	3.4	
Oil	R_1	26.9	27.0	475	7.5	
	R_2	31.4	31.5	474	5.7	
Bir: Weak (3)				*Pleoc*: Weak (3)		*Tarn*: –
Anis +P: Visible and coloured YlBr (3)				*IR*: No		*Twg*: No
Mor: Usual fibrous aggregates				*Clv*: Not frequent		*PolQ*: Good
Hardness	*PolH*: Medium–low (3)			*Mohs*: 2½- 3	*Talm*: B	*VHN* $_{(50\ g)}$: 83–161

Obs: cf. Table and R charts, multispectral data (VNIR, AMCO) and visible range (QDF3). VHN Uytenbogaardt-Burke (1971)
Mineral abbreviations: Bi—native bismuth, bm—bismuthinite, ccp—chalcopyrite, cos—cosalite, cst—cassiterite, gn—galena, py—pyrite, wf—wolframite
Colour abbreviations: Yl—yellow, YlWh—yellowish white, GrWh—greyish white, YlBr—yellowish brown

38.1 General

The traditionally accepted composition of cosalite (*cos*), $Pb_2Bi_2S_5$, has been questioned, given the usual presence of Cu and Ag. According to Topa and Makovicky (2010), these two metals would enter into the cosalite formula, defining a unique mode of solid solution. It may also contain Fe impurities and, exceptionally, Ramdohr (1980) cites also notable Se contents.

38.2 Microscopic Description

38.2.1 SP Observation (*Single Polariser*)

White to light grey *colour*, resembling galena, *gn*, and bismuthinite, *bm*. The colour is intermediate between those of both sulphides (Figs. 38.1a, b and 38.2). As suggested by their reflectance spectra, ***R***, cos looks slightly more yellowish than galena and more bluish than bm (cf multispectral comparative plots). The pleochroism is weak (Fig. 38.3).

It polishes well and its polishing hardness, slightly above the medium–low limit, is very similar to that of gn, although somewhat higher. Clearly lower than py (Fig. 38.3).

It usually occurs in fibrous aggregates, at times somewhat recrystallised. Its decomposition is frequent, giving rise to fine intergrowths of native bismuth, bismuthinite, galena and chalcopyrite.

38.2.2 Observation +P (*Polariser + Analyser at 90°*)

Anisotropic, with not very showy but peculiar colours, varying from dark brown to more luminous yellowish-brown tones (Fig. 38.1c, d). It does not show *internal reflections* or *twins*.

R. Castroviejo, *A Practical Guide to Ore Microscopy—Volume 1*,
https://doi.org/10.1007/978-3-031-12654-3_38

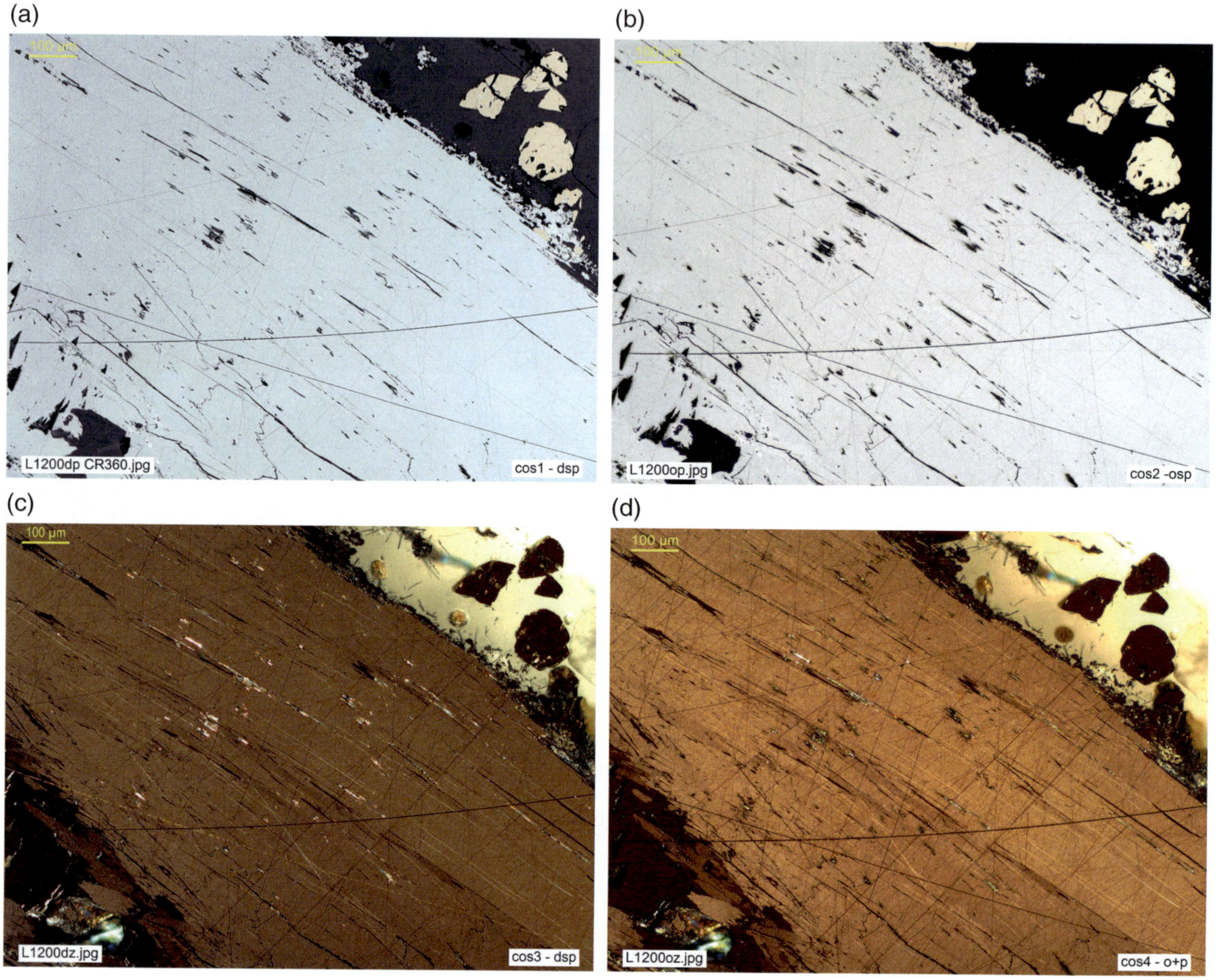

Fig. 38.1 **a**, **b** (Same scene: **dsp** and **osp**, *respectively*). Aggregate of cosalite, *cos* (*GrWh*, greyish white, slightly yellowish in oil, *osp*; fibrous, filling almost the whole image) and minor galena, *gn* (*GrWh*, slightly bluish, *triangular pits*; visible at bottom left corner). Decay of cos along narrow marginal band (upper right edge) to fine-grained *Bi* (white), bismuthinite, *bm* (*GrWh*) and chalcopyrite, *ccp* (yellow). CR360, Cariboo Au Mine, Wells, BC, Canada. **c**, **d** (Same scene as Fig. 38.1 a: **d+p** and **o+p**, *respectively*). The isotropic character of gn (always in extinction) contrasts with the distinct and coloured anisotropism of cos, which allows an immediate distinction of the two

38.2.3 Paragenesis

Parageneses varied, essentially hydrothermal or pneumatolitic-hydrothermal and contact metasomatic, but the range is wide. It has also been cited in pegmatites and even in epithermal veins. It is a relatively ubiquitous mineral, although not abundant.

It is usually accompanied by cassiterite, wolframite, bismuthinite, bismuth and sulphosalts of Bi, galena, chalcopyrite.

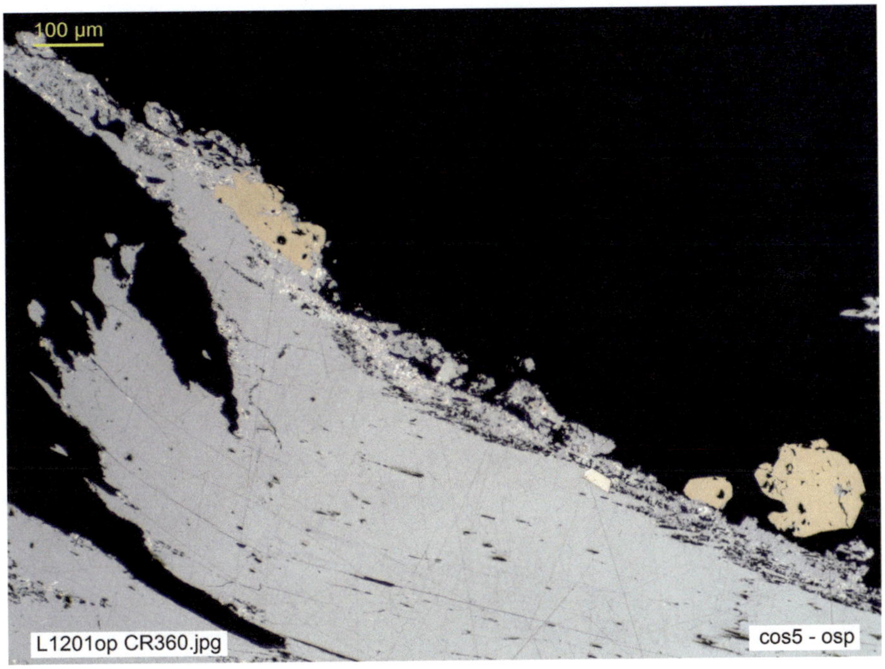

Fig. 38.2 Decay of cos, marginal band of secondary native bismuth, Bi (white, bright), bm and gn (grey), with associated ccp (yellow). Small marginal inclusion of py (YlWh, relief) in cos. Sample CR360, Cariboo Au Mine, Wells, Brit. Columbia, Canada (**osp**)

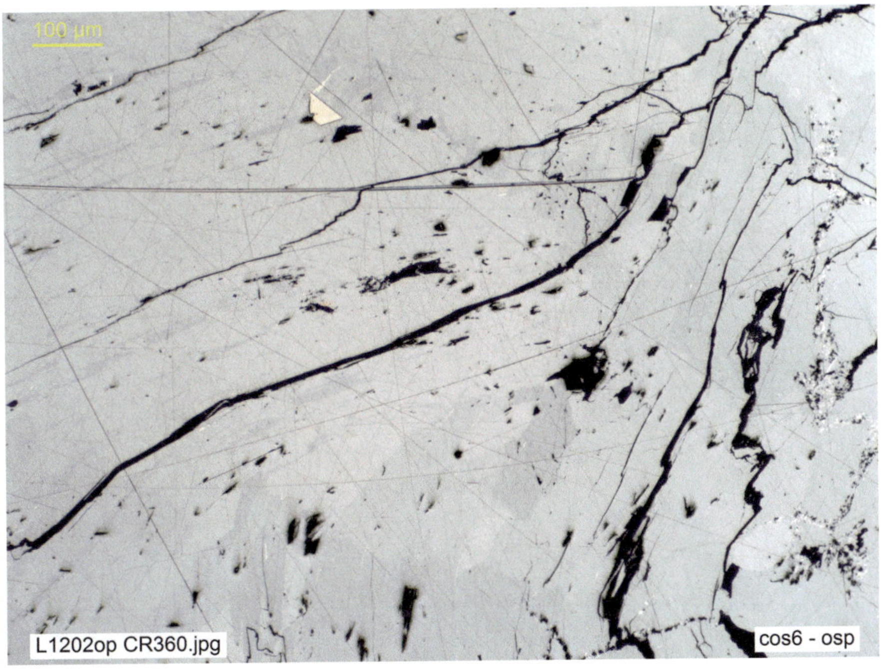

Fig. 38.3 Properties of cos (oil immersion): slight bireflectance and weak pleochroism, evidenced by comparison of sectors of different orientation. The small inclusion of py (YlWh, relief) provides a reference for cos reflectance, clearly lower than py. CR360, Cariboo Au Mine, Wells, Brit. Columbia, Canada (**osp**)

38.2.4 Identification

Although it resembles other Pb sulphosalts, it can be recognized by its higher reflectance and its typical polarisation colours, in addition to its fibrous morphology and the total absence of twins. Another fact to consider is its occasional substitution by bismuthinite or the typical alteration to fine aggregates of Bi, bm, gn, secondary to cos.

38.3 Spectral Information

38.3.1 Cosalite Multi-spectral Reflectance R (in Air, Without Polariser) (%)

λ (nm)	370	400	425	450	475	500	525	550	575	600	625	650	675	700	750	800	850	900	950	1000
μ	43.5	44.5	44.9	44.5	44.2	43.5	43.0	42.3	42.0	41.7	41.6	41.4	41.1	41.0	40.6	40.1	39.4	38.7	37.9	37.4
P_5	42.1	43.3	43.4	43.1	42.8	42.2	41.8	41.2	41.0	40.8	40.6	40.5	40.2	40.2	39.7	39.1	38.6	37.8	37.3	36.5
P_{20}	42.8	44.0	44.3	43.9	43.7	42.9	42.5	41.8	41.6	41.3	41.1	41.0	40.7	40.6	40.1	39.6	39.0	38.2	37.5	36.9
P_{80}	44.2	45.1	45.5	45.2	44.8	44.1	43.5	42.8	42.5	42.2	42.1	41.9	41.6	41.5	41.0	40.6	39.8	39.3	38.5	38.0
P_{95}	44.9	45.8	46.2	45.8	45.3	44.7	44.0	43.3	43.0	42.7	42.5	42.4	42.1	42.0	41.5	41.1	40.2	39.6	38.9	38.4

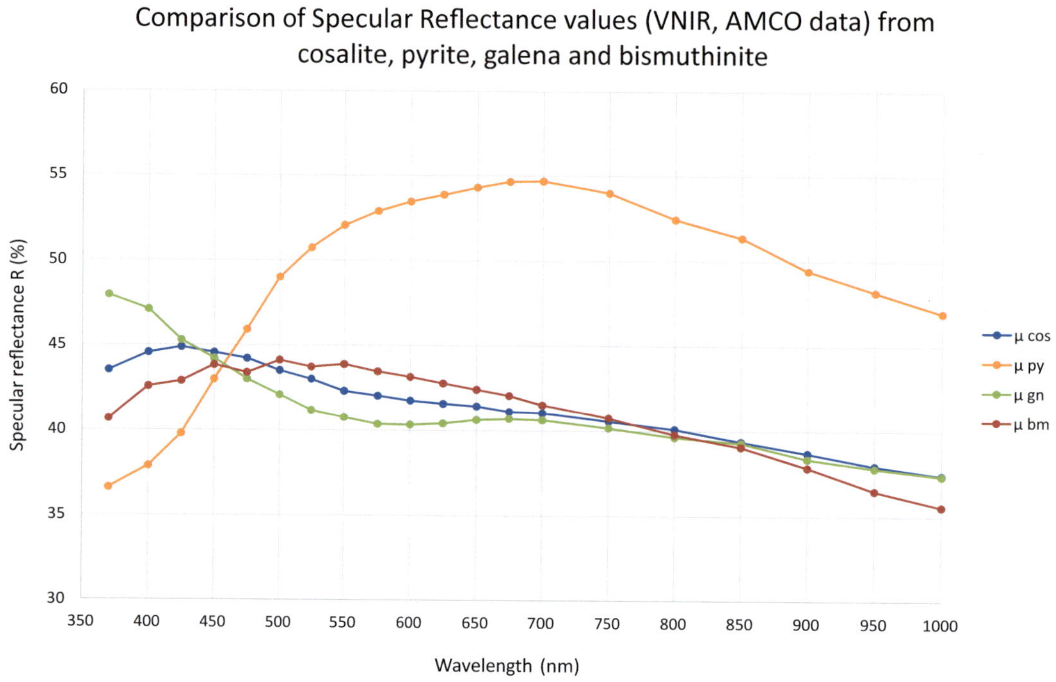

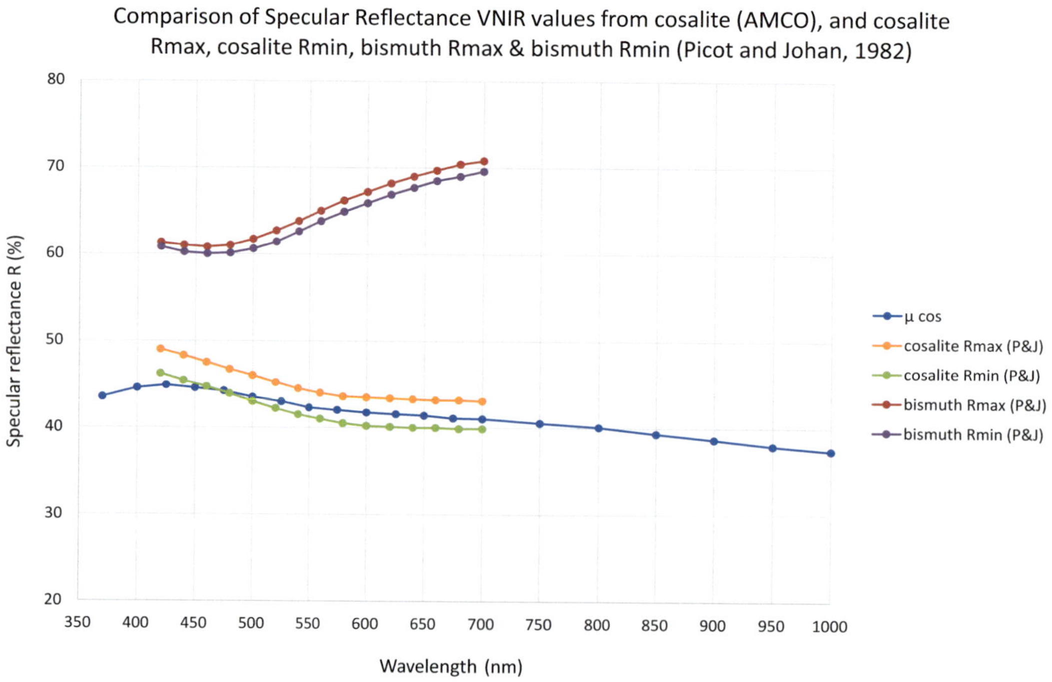

References

Ramdohr P (1980) The ore minerals and their intergrowths, 2nd edn, 2 volumes. Pergamon Press, Oxford, 1205p

Topa D, Makovicky E (2010) The crystal chemistry of cosalite based on new electron-microprobe data and single-crystal determinations of the structure. Can Mineral 48:1081–1107

Uytenbogaardt W, Burke EAJ (1971) Tables for microscopic identification of ore minerals, 2nd edn. Elsevier, Amsterdam

Covellite (*cv, covelline*)

39

CuS: Hexagonal

Covellite		R_{546} (%)	Y (%)	λ_d (nm)	P_e (%)	Colour sensation
Air	R_o	6.7	6.5	476	43.1	Intense blue, very saturated
	R_e	23.7	23.6	476	14.9	Light blue to bluish white
Oil	R_o	0.9	1.8	c546	50.8	Colour change: purple to RVi
	R_e	9.9	10.0	474	25.0	Intense blue, with ViBlu shades
Bir: Strong (5)				*Pleoc*: Strong (5)		*Tarn*: No
Anis+P: Strong, striking, very bright (5)				*IR*: Not usual		*Twg*: Frequent deformation twins
Mor: Euhedral, lamellar; aggregates				*Clv*: // (001) occasional		*PolQ*: Polishes well but scratches
Hardness	*PolH*: Low, ≈cc, <gn; >>ac			*Mohs*: 1.5+	*Talm*: B+	*VHN* $_{(50\ g)}$: 92–110

Note Frequent in Cu deposits, typical in the secondary sulphide enrichment blanket. *Pleoc*. and *Anis*. very intense and unmistakable, despite the low reflectance
Spectral information: cf. Table and R charts, data from multispectral (VNIR, AMCO) and visible (QDF3) ranges
Mineral abbreviations: ac—acanthite, cc—chalcocite, cv—covellite, gn—galena
Colour abbreviations: Blu—blue, RVi—reddish violet, ViBlu—violet-blue

39.1 General

Copper sulphide (66.48% Cu, 33.52% S, often with traces of Fe); typical but not exclusive of the supergene enrichment zone (or secondary sulphide blanket) of copper-bearing deposits. Possible confusion with *spionkopite* ($Cu_{39}S_{28}$, formerly *blaubleibender Covellin* or "everblue covellite" of Ramdohr 1980) and *yarrowite* (Cu_9S_8), as explained below (cf Identification).

39.2 Microscopic Description

39.2.1 SP Observation (*Single Polariser*)

Covellite, ***cv***, is easily recognised by its striking optical properties. It is blue-colored, and strongly bireflectant and pleochroic. The *reflectance* spectra of cv are very characteristic, both in the visible range and in the near IR (*cf.* Table and R graphs).

As expected from the R spectra, cv *bireflectance* is very pronounced. In the darkest position, R_o < sphalerite; in the lightest, R_e slightly > digenite and enargite, although < chalcocite and tetrahedrite-tennantite.

The comparison of R_o and R_e values in the visible range explains the strong and unmistakable *pleochroism* of cv, characterized by the passage from an intense blue colour (R_o) to a very light blue or bluish-white tone (R_e). Additionally, cv shows a noteworthy peculiarity in oil immersion (*osp setting*). A pronounced colour shift is observed: purple to reddish-violet (or purple-red) hues appear (*osp*) instead of the dark blue seen with dry objective (*dsp setting*, Figs. 39.1a and b), as corresponds to the cv reflectance spectra in immersion (QDF3).

Low *polishing hardness*: $PolH_{cv}$ ≈ chalcocite (> or < , depending on orientation), < galena, >>a (acanthite). However, cv polishes well, despite being soft (polishing scratches or minor pull-outs, almost unavoidable). Occasional (001) cleavage.

Frequently euhedral, as disseminated flakes or lamellar crystallites, although it also occurs as massive aggregates.

R. Castroviejo, *A Practical Guide to Ore Microscopy—Volume 1*,
https://doi.org/10.1007/978-3-031-12654-3_39

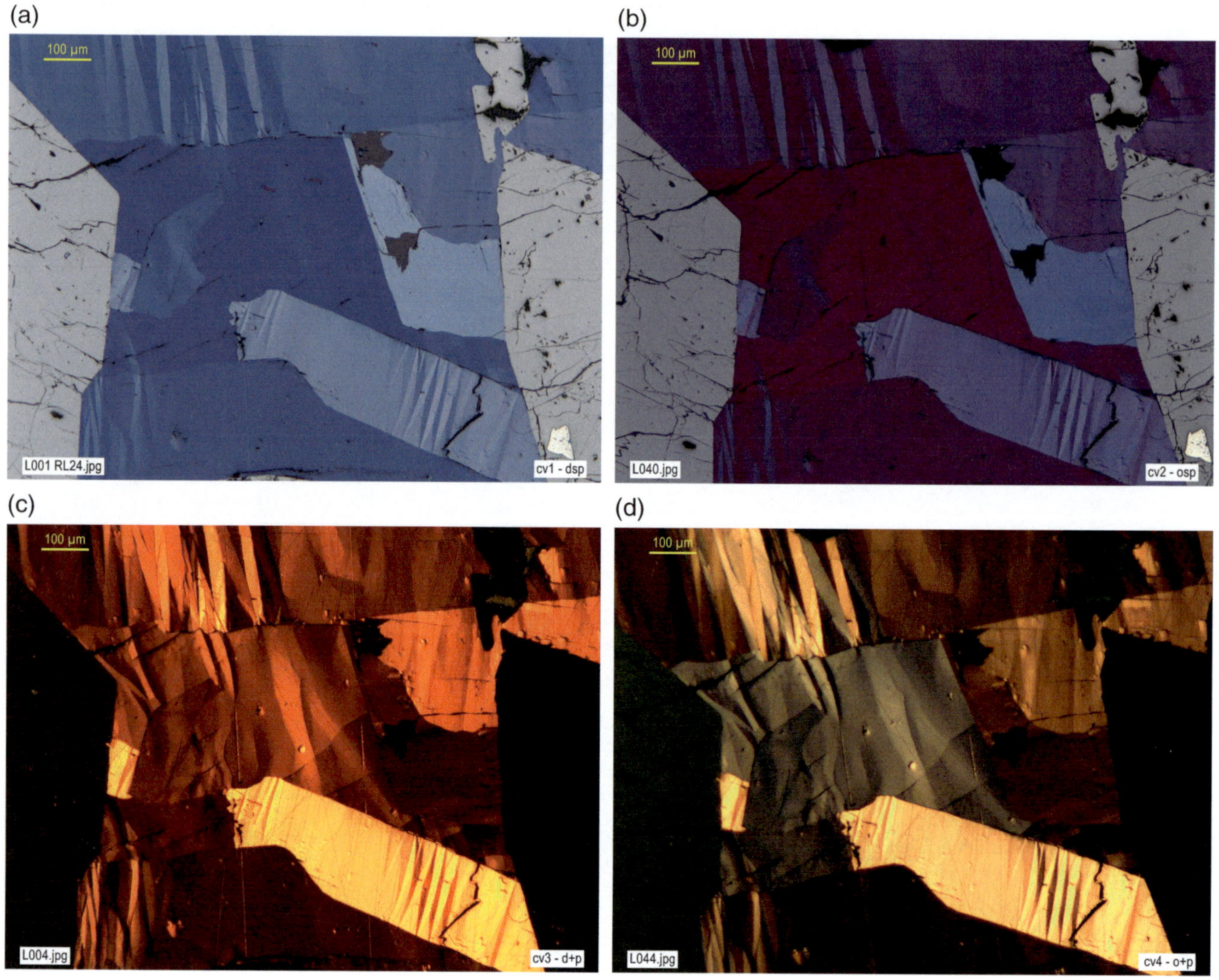

Fig. 39.1 **a** Crystalline aggregate of covellite (blue, strongly pleochroic), enargite (grey) and pyrite (brightest, pale yellow, small inclusion in enargite) (**dsp**). Sample RL24, Bor (Serbia). **b** Same scene (**osp**). Note the new purplish colours of cv in oil. **c** Same scene (**d+p**). **d** Same scene (o+p)

39.2.2 Observation +P (*Polariser + Analyser at 90°*)

Very strong and striking anisotropism, changing from very bright yellow-orange tones to much duller reddish-brown to reddish-orange tones that darken even more with immersion (*o+p* setting) and take on a greyish hue (Figs. 39.1c and d). Internal reflections are usually not observed, but have been cited (Ramdohr 1980).

Although not considered true twins, deformation frequently produces subparallel lamellae similar to polysynthetic twins, very apparent under the microscope (Fig. 39.1c and d).

39.2.3 Paragenesis

Covellite is usually present, at least in traces, in most of the cupriferous parageneses. It is typically present as a neoformed mineral, replacing primary sulphides such as chalcopyrite or pyrite, which disappeared or appear as relicts in the secondary sulphide blanket. This process is often responsible for an important supergene enrichment in copper deposits.

Frequent companions of cv are digenite, chalcocite, bornite, chalcopyrite, fahlores (tetrahedrite-tennantite series), occasionally djurleite, enargite, luzonite, pyrite, native copper and, more rarely, oxides, carbonates or silicates like cuprite, malachite, azurite, chrysocolla, etc.

39.2.4 Identification

The optical properties of covellite are so clearly diagnostic that they make its microscopic identification unequivocal. Even in very small quantities and in fine grain sizes, it is not difficult for a careful observer to notice it. Rotating the stage (with+*p setting*), its presence is easily evidenced by its luminous polarization tints, with very bright golden yellow-orange tones, which stand out against the dark surroundings.

It might be confused with the new species *spionkopite* ($Cu_{39}S_{28}$, trigonal, *blaubleibender Covellin* of Ramdohr), which is less common and is distinguished by its higher reflectance and by the permanence of the blue colour in immersion (without analyser, *osp setting*), instead of the change to purple to purplish-red tones typical of covellite (cf. images in description dg, Figs. 43.1a–d). A similar case is that of *yarrowite* (Cu_9S_8, trigonal), which acquires in immersion a light faintly purplish tone, different from that of cv.

39.3 Spectral Information

39.3.1 Covellite Multispectral Reflectance R (in Air, Without Polariser) (%)

λ (nm)	370	400	425	450	475	500	525	550	575	600	625	650	675	700	750	800	850	900	950	1000
μ	19.8	20.6	20.3	19.8	18.3	16.6	14.6	12.6	10.7	9.5	9.4	10.9	14.5	21.0	32.8	39.1	43.2	46.4	48.1	49.9
P_5	14.7	15.7	15.7	15.2	13.9	12.0	9.9	7.8	5.8	4.2	3.5	5.0	10.1	18.7	27.9	31.4	33.7	36.0	36.9	38.2
P_{20}	16.5	17.7	17.4	16.9	15.3	13.3	11.1	8.9	6.8	5.2	4.5	5.9	11.1	20.1	28.7	32.4	35.0	36.9	38.1	39.2
P_{80}	23.4	23.9	23.2	22.6	21.1	19.5	17.5	15.6	13.9	13.0	13.3	15.0	17.6	22.0	37.7	46.9	52.3	56.3	57.8	59.7
P_{95}	24.4	24.8	24.0	23.3	21.8	20.0	18.1	16.2	14.5	13.7	14.2	15.9	18.6	22.9	39.3	49.2	55.4	59.5	61.7	63.6

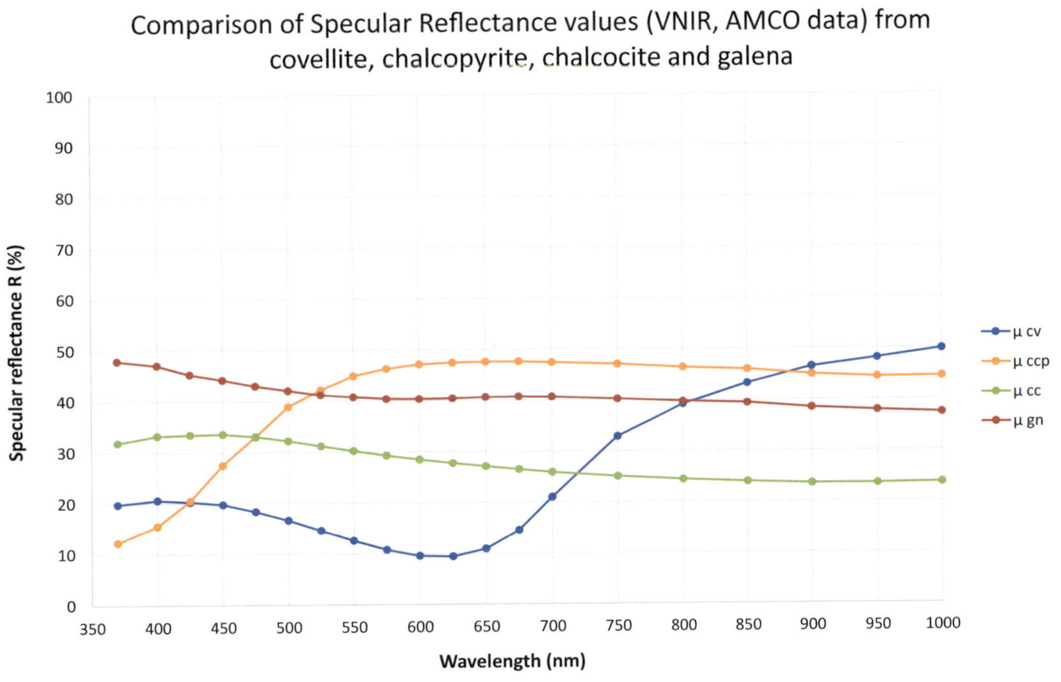

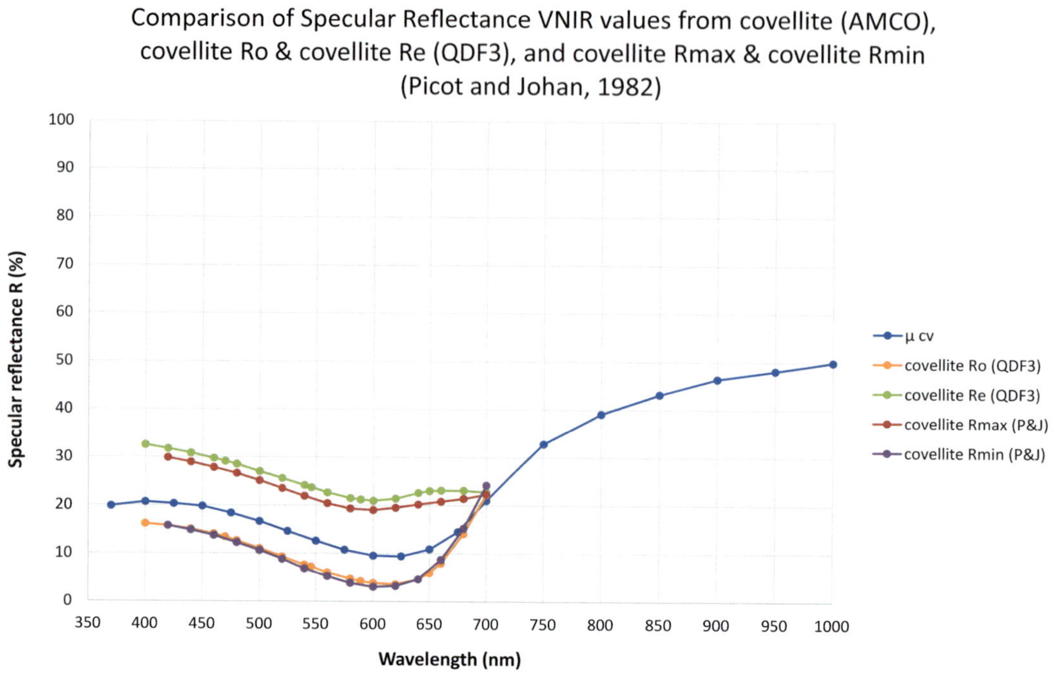

Reference

Ramdohr P (1980) In: The ore minerals and their intergrowths. vol 2, 2nd edn. Pergamon Press, Oxford, pp 1205

Cryptomelane (*cry*)

40

$K_{1\text{-}1.5}$ $(Mn^{4+}, Mn^{2+})_8$ O_{16}: Monoclinic

CRYPTOMELANE		R_{550} (%)	Y (%)	λ_d (nm)	P_e (%)	Colour sensation
Air	$R_{1?}$	26.7	nd	nd	nd	Greyish white
	$R_{2?}$	26.7	nd	nd	nd	Grey
Oil	R_1	nd	nd	nd	nd	Light grey
	R_2	nd	nd	nd	nd	Dark grey
Bir: Variable (2–4)				*Pleoc*: Very weak (2)		*Tarn*: –
Anis +P: Distinct, not coloured (4)				*IR*: No		*Twg*: –
Mor: Collomorf, banded, botryoidal				*Clv*: No		*PolQ*: Variable
Hardness	*PolH: Variable (4)*			*Mohs*: 5–6½		*VHN* $_{(nd\ g)}$: 525–1048

Obs: R and VHN data: Uytenbogaardt and Burke, 1971

Mineral abbreviations: bra—braunite, cry—cryptomelane, gg—gangue, gth—goethite, lim—limonite, mng—manganite, prl—pyrolusite, ps—psilomelane, q—quartz, rom—romanechite

40.1 General

A frequent component of oxidized Mn ores, cryptomelane (*cry*) is presented by Uytenbogaardt and Burke (1971) as the most common of the Mn minerals, together with pyrolusite (*prl*). This statement is difficult to prove. It is supposed to be a common component of the very common psilomelane, *psi*; however, psi usually contains also other cryptocrystalline or submicroscopic species, such as romanechite, hollandite, etc., which cannot be visually distinguished from each other or from cryptomelane in most of the samples. Nevertheless, cryptomelane is, in any case, an important Mn ore.

Composition: the formula of cryptomelane varies according to the sources; the formula published by Back and Mandarino (2008) is well accepted—$K_{1\text{–}1.5}(Mn^{4+}, Mn^{2+})_8O_{16}$. However, possible contents of adsorbed metals (e.g., Co, Ni, Ag, Cu, W) must be considered.

40.2 Microscopic Description

It is a practically unanimous opinion that cryptomelane, like the other usual components of psilomelane, cannot generally be reliably identified with the optical microscope. Indeed, its extremely fine grain size defies the resolving power of this microscope and makes it necessary to resort to auxiliary techniques (X-Ray, electron microprobe). For this reason, no detailed description of the optical properties of cryptomelane comparable to those usually published for common ores (or to those presented in this text for manganite, pyrolusite, romanechite, etc.) is presented here (nor is it known in the current literature, except in very general treatments, e.g., Uytenbogaardt and Burke 1971).

However, since it is a common species, its microscopic appearance is documented graphically, with the usual series of four basic photomicrographs (Figs. 40.1a–d). These can be

R. Castroviejo, *A Practical Guide to Ore Microscopy—Volume 1*,
https://doi.org/10.1007/978-3-031-12654-3_40

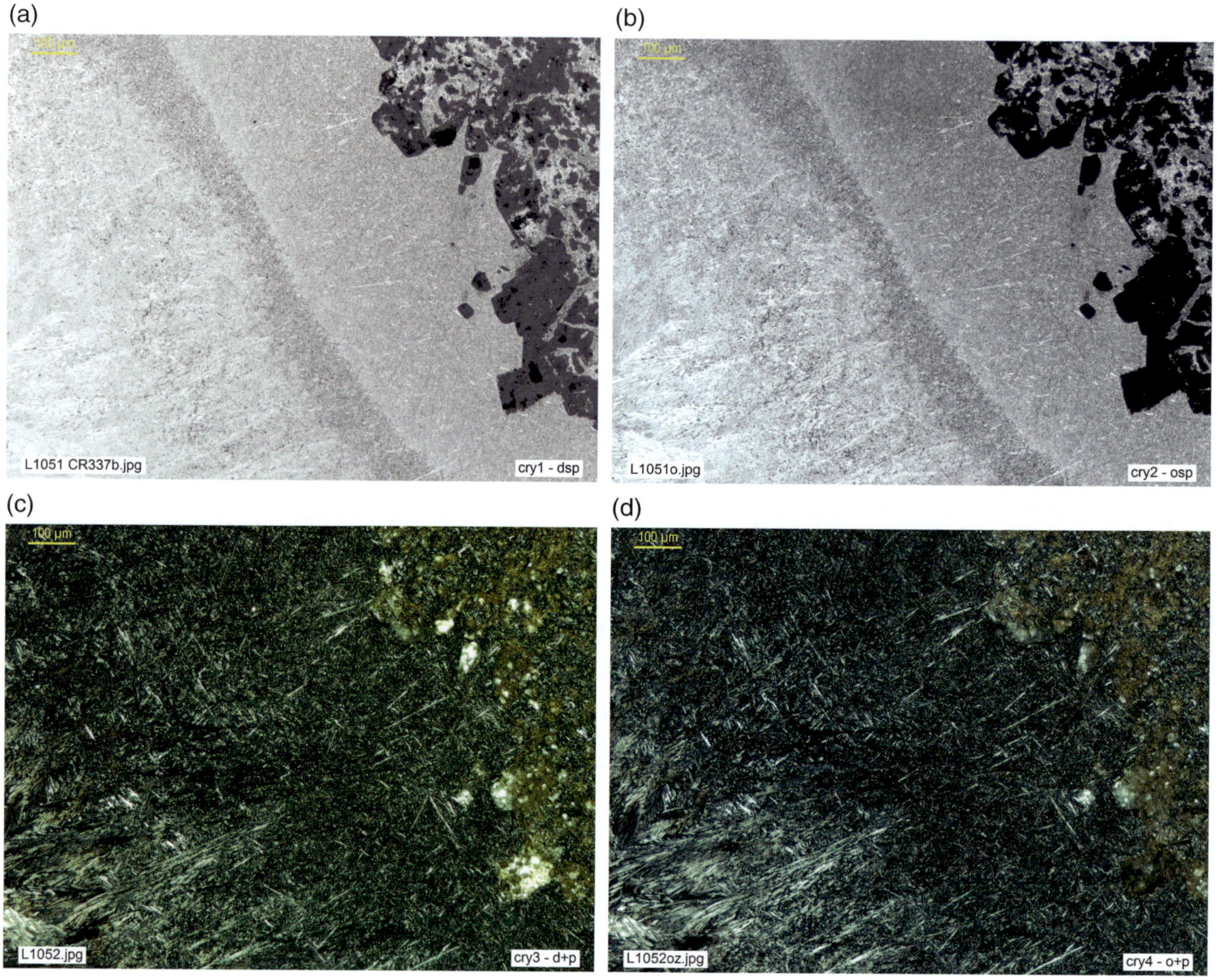

Fig. 40.1 **a, b**(**dsp** and **osp**, respectively)—Banded manganiferous ore (greyish white), deposited on probable barite gangue (prismatic, dark grey, almost black in oil; corroded and partially replaced by the ore). As confirmed by XRD, the first and second bands in the ore, darker and cryptocrystalline, are composed of dominant cry, with scattered acicular inclusions of rom (more visible with analyser, Fig. 40.1c, d), while the third (bottom left) is mostly composed of acicular and fibrous-radiating rom, with subordinate cryptocrystalline cry. CR337b, Clemira Vein, El Remanso, Santiago del Estero, Argentina. **c, d**(same scene as Fig. 40.1a: **d+p** and **o+p**, respectively)—The anisotropism in whites and greys of the ore evidences the acicular crystals of rom, but it is hardly noticeable in cry because of its very fine grain size, which even makes some zones look isotropic. In this case, both minerals can be distinguished under the optical microscope only by their textures

completed with the *Summary Table* above, in which the scarce information found is presented. For general characterization, we refer to what has been said for psilomelane (cf. psi description), taking into account that psilomelane is a mixture with several species of micro- to cryptocrystalline Mn oxides, cryptomelane being one of the main components (Ramdohr 1980).

40.2.1 Paragenesis

Like psilomelane, cryptomelane is a frequent component of most oxidized Mn deposits, either as a primary or as a secondary mineral, replacing other Mn minerals (cf psi).

Companion minerals: psilomelane, romanechite, pyrolusite, braunite, manganite, goethite.

40.2.2 Identification

It is typically associated with parageneses of Mn oxides and, in particular, with psilomelane, in which it usually occurs in cryptocrystalline or submicroscopic aggregates. Its optical properties are similar to those of the other psilomelane components.

References

Back ME and Mandarino JA (2008) Fleischer's glossary of mineral species 2008. The Mineralogical Record Inc. Tucson, AZ, 345p

Correa MJ (2003) La Mineralización de Manganeso en el Distrito El Remanso y sus Relaciones Metalogenéticas, Sierras Pampeanas Orientales, Provincia de Santiago del Estero. Tesis Doctoral, Universidad de La Plata (Argentina), 179 pp + Anexos

Ramdohr P (1980) The ore minerals and their intergrowths, vol 2, 2nd edn. Pergamon Press, Oxford. 1205p

Uytenbogaardt W, Burke EAJ (1971) Tables for microscopic identification of ore minerals, 2nd edn. Elsevier, Amsterdam

Cubanite (*cbn*)

41

$CuFe_2\ S_3$: Orthorhombic

Cubanite		R_{546} (%)	Y (%)	λ_d (nm)	P_e (%)	Colour sensation
Air	R_1	35.4	35.5	577	17.0	Pinkish brown, pleochroic
	R_2	39.4	39.4	575	12.3	
Oil	R_1	23.9	24.1	577	17.9	
	R_2	28.2	28.1	574	10.4	
Bir: Medium (4)				*Pleoc*: Visible (4)		*Tarn*: Not observed
Anis + P: Distinct, blue tones (4)				*IR*: No		*Twg*: (110)
Mor: Tabular exsolucs // (111) in ccp				*Clv*: // (001)		*PolQ*: Good
Hardness	*PolH*: Med.: > ccp; < po (3)			*Mohs*: 31/2	*Talm*: C	*VHN* $_{(nd\ g)}$: 150–264

Obs: cf. *R data:* Table, charts, measures VNIR and visible ranges (AMCO & QDF3, *resp.*). VHN: Uytenbogaardt and Burke (1971)
Mineral abbreviations: cbn—cubanite, ccp—chalcopyrite, il—ilmenite, isc—isocubanite, mck—mackinawite, mt—magnetite, po—pyrrhotite, rt—rutile, sp—sphalerite

41.1 General

Composition: $CuFe_2S_3$, dimorph with isocubanite (*isc*). Undergoes irreversible transformation to isocubanite by heating to >200 °C (transformation to **iss**, *intermediate solid solution* according to Cabri et al. 1973; iss, dominant high-temperature cubic phase above 500 °C, has a broader composition than strict isocubanite, cf. Craig and Scott 1976, and discussion in Bowles et al. 2011, also *isocubanite* description).

Can be found in exsolutions in chalcopyrite but is, by far, less abundant than ccp. Magnetic.

41.2 Microscopic Description

41.2.1 SP Observation (*Single Polariser*)

Pinkish-brown, moderately *pleochroic* (light pink to pinkish-grey), pyrrhotite-like (Figs. 41.1a, b and 41.2).

Moderate *reflectance*: R cbn <ccp and >sp or mt. Very close to pyrrhotite (and replicating its pleochroism) in the visible spectrum, although lower for $\lambda < 500$ nm and for $\lambda > 750$ nm (cf. multispectral comparison graphs Figs. 41.1a, b and 41.2).

Good *polish*. Cleavage // (001), though rarely visible under microscope. Medium *polishing hardness*, close to but slightly > ccp; < sp and distinctly < po.

It occurs in tabular crystals, actually exsolution lamellae // (111), in high-temperature chalcopyrite, which frequently includes pyrrhotite or sphalerite and also contains minute inclusions of mackinawite (Fig. 41.2a, b). Sometimes cubanite is also unmixed from high-temperature pyrrhotite.

41.2.2 Observation +P (*Polariser + Analyser at 90°*)

Anisotropic, with typical polarization tints (light, bluish shades), Figs. 41.1c, d and 41.2b (also Fig. 28.3).

R. Castroviejo, *A Practical Guide to Ore Microscopy—Volume 1*,
https://doi.org/10.1007/978-3-031-12654-3_41

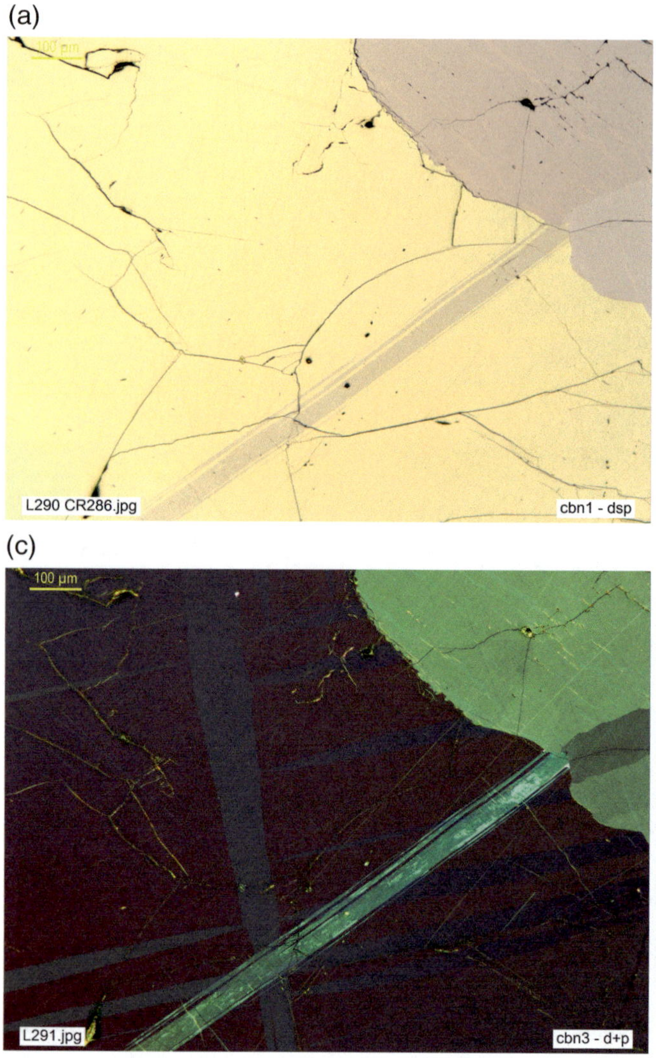

Fig. 41.1 **a**, **b** (Same scene: **dsp** and **osp**, *respectively*). Lamellar exsolution of cbn (light pinkish brown) in ccp (yellow), with large inclusion of po (pleochroic, upper right corner). The colour of cbn is very similar to that of po in one of its orientations (R_2). CR 286, Gihen Mine (Sulitjelma district, Norway). **c**, **d** Same scene as cbn1 (**d+p** and **o+p**, *respectively*). Lanceolate twins and weak anisotropism of ccp, stronger anisotropism and bluish of cbn. More greenish polarization tints of po

41.2.3 Paragenesis

Cubanite is frequently found in the sulphide phase of liquid-magmatic deposits (with Cu-Ni sulphides), as well as in pegmatitic to high-temperature hydrothermal deposits, usually in lamellar exsolutions in chalcopyrite. It can also be formed by metamorphic modification of pre-existing copper sulphide deposits, if the required temperature conditions have been reached.

Companions: chalcopyrite, pyrrhotite, sphalerite, mackinawite, pentlandite, magnetite, ilmenite, rutile.

41.2.4 Identification

Identification of cbn is not problematic. Its common paragenesis, mainly in liquid-magmatic deposits, and its association with chalcopyrite in laminar exsolutions are typical, as are the colour, reflectance and pleochroism similar to pyrrhotite. However, it is easily distinguished from po because cbn is softer (polishing hardness close to that of ccp; < sp, po), to the point that pyrrhotite usually appears in relief in contact with chalcopyrite and cubanite (Fig. 41.2a) Compared with chalcopyrite, cbn anisotropism is distinctly more striking and lively, with typical bluish tones.

Fig. 41.2 **a, b** (**dsp** and **d+p**, respectively). Lamellar exsolutions of cbn (light pinkish brown) in massive ccp. As inclusions in ccp, appear: sp (grey, coarse grain to the right), po (colour like cbn, small, upper edge of sp) and mck (white–grey, tiny flakes, scattered).In ccp and sp, rt (the largest ≈ 20 μ in sp, grey slightly lighter than sp). The different polishing hardness of po (harder than ccp it appears in relief) and cbn (softer, ≈ ccp) can be appreciated. The analyser (*d +p*, F cbn6) discloses the orthogonal twins of ccp, as well as the typical anisotropism of cbn (lively, bluish shades), mck (very strong but colourless) and rt (weaker, yet distinct, comp sp in extinction). CR 286, Gihen Mine (Sulitjelma district, Norway)

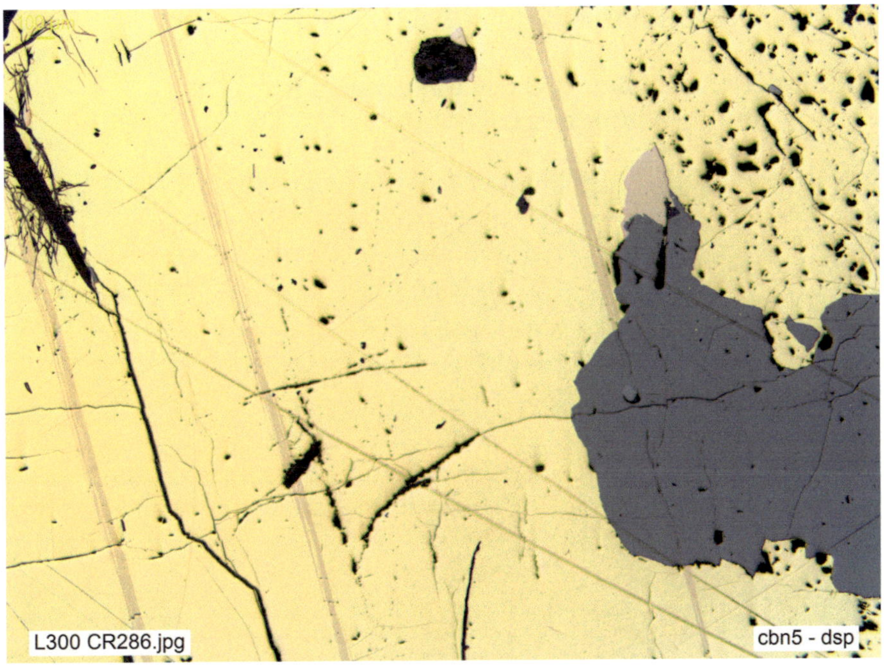

(a)

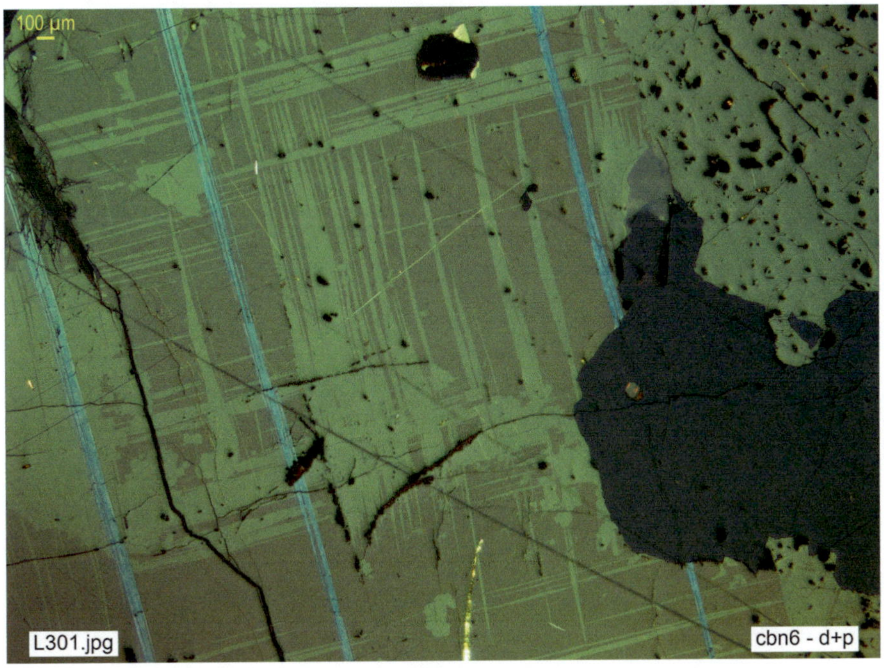

(b)

41.3 Spectral Information

41.3.1 Cubanite Multispectral Reflectance R (in Air, Without Polariser) (%)

λ (nm)	370	400	425	450	475	500	525	550	575	600	625	650	675	700	750	800	850	900	950	1000
μ	16.7	19.4	23.1	26.9	30.5	33.6	35.9	37.6	39.0	40.0	40.9	41.8	42.6	43.3	44.2	44.7	45.2	44.8	44.5	44.3
P_5	16.0	18.9	22.5	26.3	29.9	32.9	35.2	36.9	38.2	39.2	40.0	40.9	41.6	42.3	43.2	43.6	44.0	43.6	43.1	42.8
P_{20}	16.5	19.2	22.9	26.7	30.3	33.4	35.6	37.3	38.6	39.6	40.5	41.4	42.1	42.8	43.7	44.1	44.6	44.2	43.8	43.6
P_{80}	17.0	19.7	23.4	27.2	30.9	33.9	36.2	38.0	39.3	40.4	41.4	42.3	43.1	43.8	44.8	45.2	45.7	45.5	45.2	45.0
P_{95}	17.3	19.9	23.6	27.4	31.1	34.1	36.5	38.2	39.6	40.7	41.6	42.6	43.4	44.1	45.1	45.6	46.1	45.9	45.6	45.4

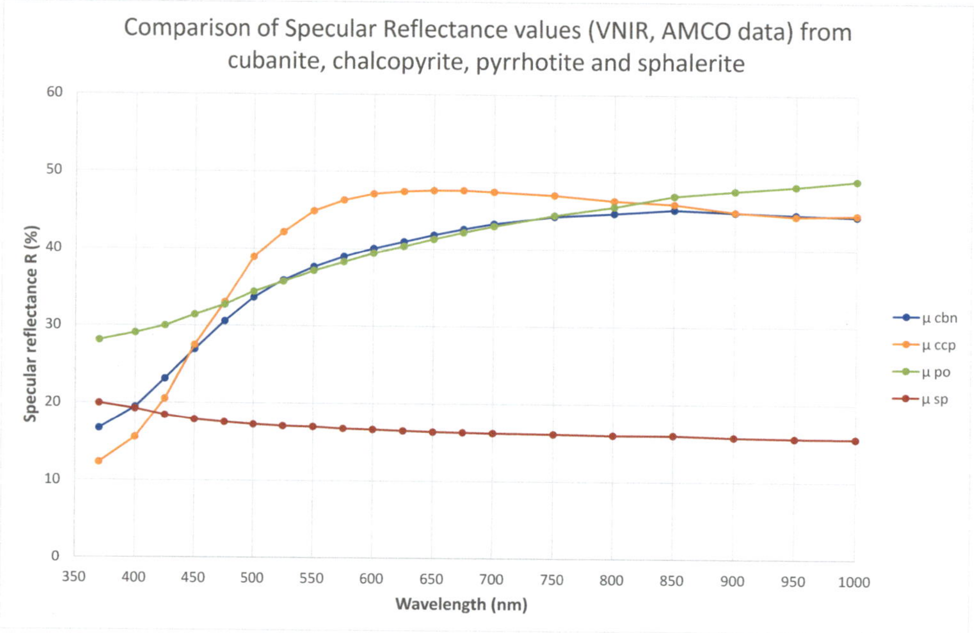

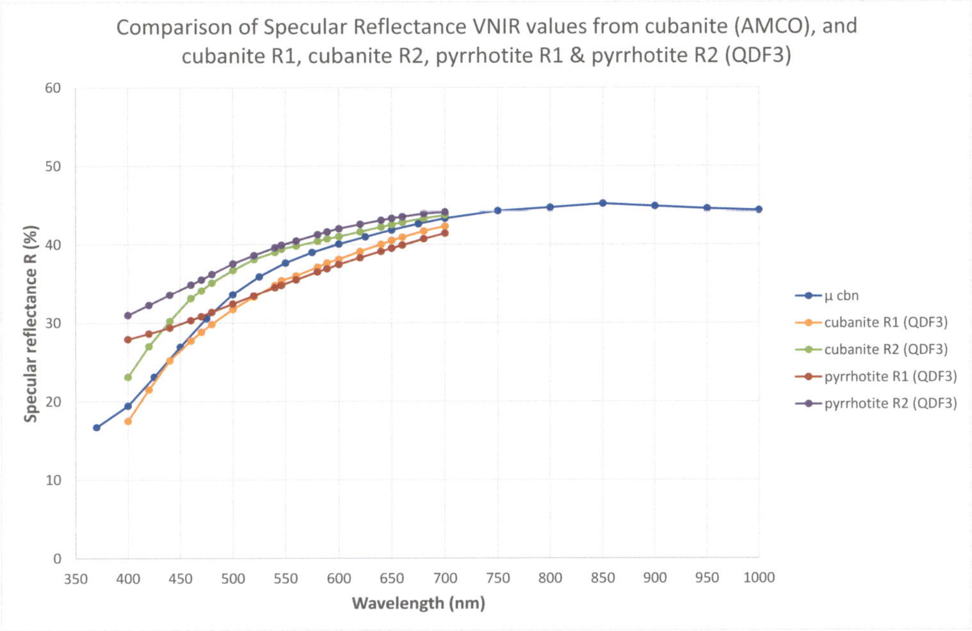

References

Bowles JFW, Vaughan DJ, Howie RA, Zussman J (2011) Rock-forming minerals, vol 5A: Non-Silicates: Oxides, Hydroxides and Sulphides. 2nd Edn. Chalcopyrite: Phase Relations in the ternary system Cu-Fe-S (p 780 ff). Geological Soc London, 920p (ISBN-13: 978–1862393158)

Cabri JL (1973) New data on phase relations in the Cu-Fe-S system. Econ Geol 68:443–454

Cabri JL, Hall R, Szymanski JT, Stewart JM (1973) On the transformation of cubanite. Can Mineral 12:33–38

Craig JR, Scott SD (1976) Sulphide phase equilibria. In: Ribbe PH (ed) Sulfide Mineralogy. Mineralogical Soc America, Rev Mineralogy 1(5), CS1–110

Uytenbogaardt W, Burke EAJ (1971) Tables for microscopic identification of ore minerals, 2nd edn. Elsevier, Amsterdam

Cuprite (*cup*) 42

Cu_2O: Cubic

Cuprite		R_{546} (%)	Y (%)	λ_d (nm)	P_e(%)	Colour sensation
Air		26.6	26.3	480	10.5	Bluish grey
Oil		12.3	12.1	481	14.4	Darker bluish grey
Bir: Imperceptible (1)				*Pleoc*: Weak, anomalous (3)		*Tarn*:–
Anis +P: Remarkable, anomalous (3)				*IR*: Abundant, red		*Twg*: Not visible
Mor: Aggregates, euhedral xx, acicular				*Clv*: (111) not very visible		*PolQ*: Variable
Hardness	*PolH*: Medium, ≈ gth, Lim (3)			*Mohs*: 4^-	*Talm*: D–	*VHN* $_{(100\ g)}$: 193–207

Obs: cf. Table and R charts, multispectral data (VNIR, AMCO) and visible range (QDF3)
Mineral abbreviations: Ag—native silver, Au—native gold, az—azurite, cbn—cubanite, cc—chalcocite, ccp—calcopyrite, Cu—native copper, cup—cuprite, cv—covellite, gn—galena, gth—goethite, ht—hematite, lim—limonite, mch—malachite, prg—pyrargyrite, pru—proustite, sp—sphalerite, td—tetrahedrite

42.1 General

Composition: Cu_2O. Frequent in the oxidation zone, as a secondary, weathering product of copper sulphide ores. *Chalcotrichite* is a fibrous variety.

42.2 Microscopic Description

42.2.1 SP Observation (*Single Polariser*)

Bluish-grey *colour*, especially next to Cu and in immersion (Figs. 42.1a, b). Its reflectance is moderate and shows a VNIR spectral response very similar to chalcocite and hematite (cf multispectral comparison charts; in the visible range, it is between R_o and R_e of hematite, as can be seen with QDF3 data). R higher than goethite. As a cubic, optically isotropic mineral, its *bireflectance* is nil, but it may show a slight anomalous pleochroism.

Occurs in intergrown crystal aggregates of variable grain size; euhedral (octahedron, cube) or acicular (chalcotrichite) development, if growing in free spaces. Cleavage // (111) rarely visible.

Difficult *polishing*, not easy to avoid pull-outs. Medium *polishing hardness* ($PolH_{cup}$ ≈ mch, az or Lim) between sp and ccp ($PolH_{cup}$ < sp & $PolH_{cup}$ > Cu, ccp, cc, gn); cup remains in relief against chalcocite.

42.2.2 Observation +P (*Polariser + Analyser at 90°*)

Clearly *anisotropic*, despite its cubic symmetry, an anomaly not yet explained. It shows characteristic blue-green polarisation tints, visible especially if the analyser is slightly uncrossed (Fig. 42.1c, d *o+p*, comp. Fig. 42.2 oxp).

Very abundant deep red *internal reflections*, so much so that, surprisingly, pleochroism and anisotropism can be

R. Castroviejo, *A Practical Guide to Ore Microscopy—Volume 1*,
https://doi.org/10.1007/978-3-031-12654-3_42

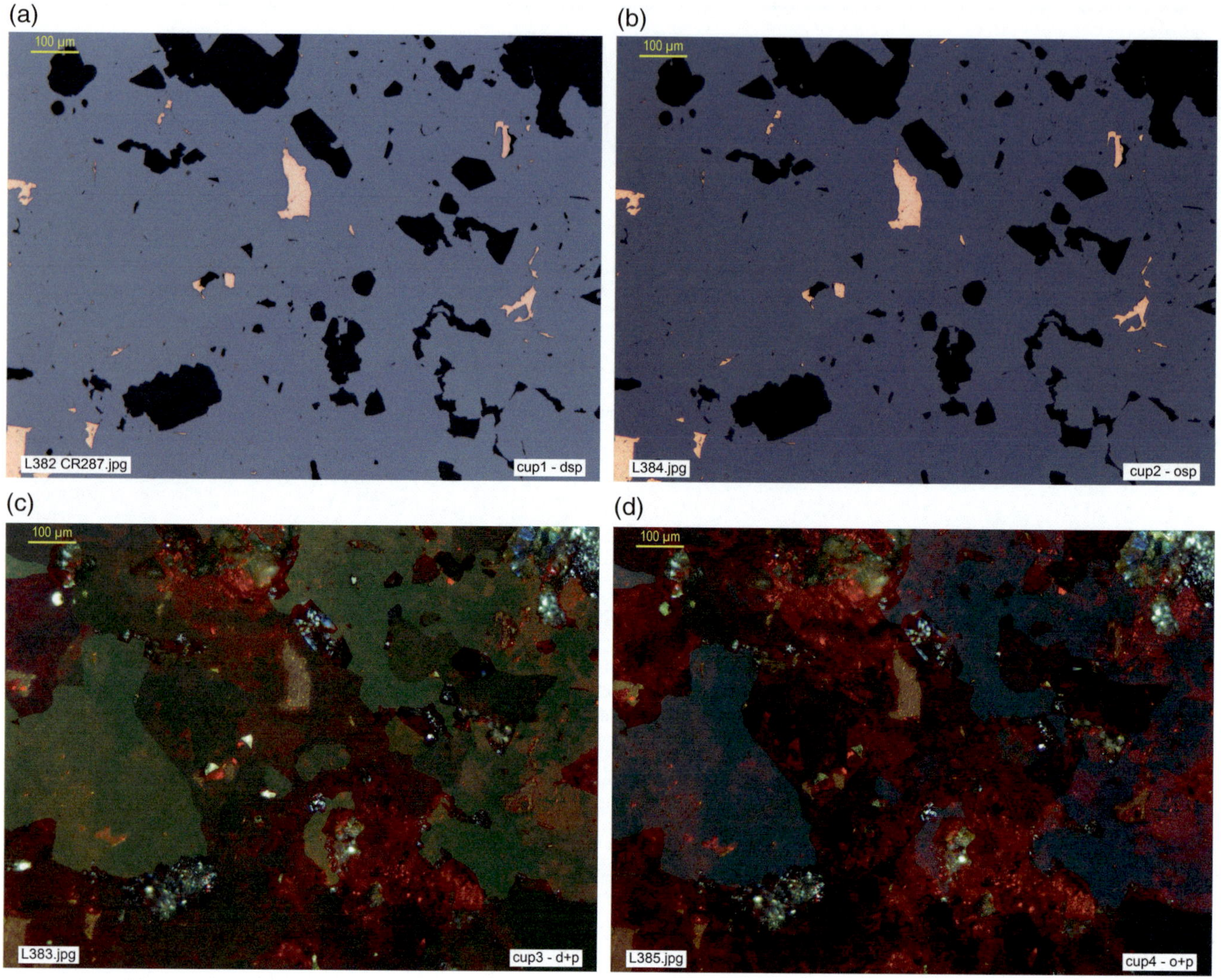

Fig. 42.1 **a** Massive cuprite (bluish grey) with Cu inclusions (pink). CR 287, Onganja, Namibia (**dsp**). **b** Same scene as cup1 (**osp**). **c**, **d** (Same scene as cup1: ***d* + *p*** and ***o* + *p***, *respectively*) Effect of anisotropism and internal reflections of cup. The greenish tones are best seen by slightly uncrossing the analyser (cf. Fig. 42.2)

observed better in air than in immersion (comp. Fig. 42.1c, d, *d+p* & *o+p*).

42.2.3 Paragenesis

Typical of the oxidation zone of cupriferous deposits, with native copper, malachite, azurite, goethite, limonite and eventual residual sulphides (chalcocite, covellite, chalcopyrite). It replaces chalcocite (Figs. 42.3 and 42.4) and sometimes copper. More abundant if carbonates are present. With limonite, by alteration of chalcopyrite, it forms cryptocrystalline powdery masses (*brickore*). Incompatible with hematite and magnetite (after Uytenbogaardt and Burke 1971).

42.2.4 Identification

Its colour, hardness, moderate reflectance, coloured anisotropism and abundant internal reflections, together with the cupriferous paragenesis, allow a safe identification. Other minerals with similar colour and internal reflections, such as the *red silver ores* or *ruby silvers* (e.g., proustite-pyrgyrite series) are much softer.

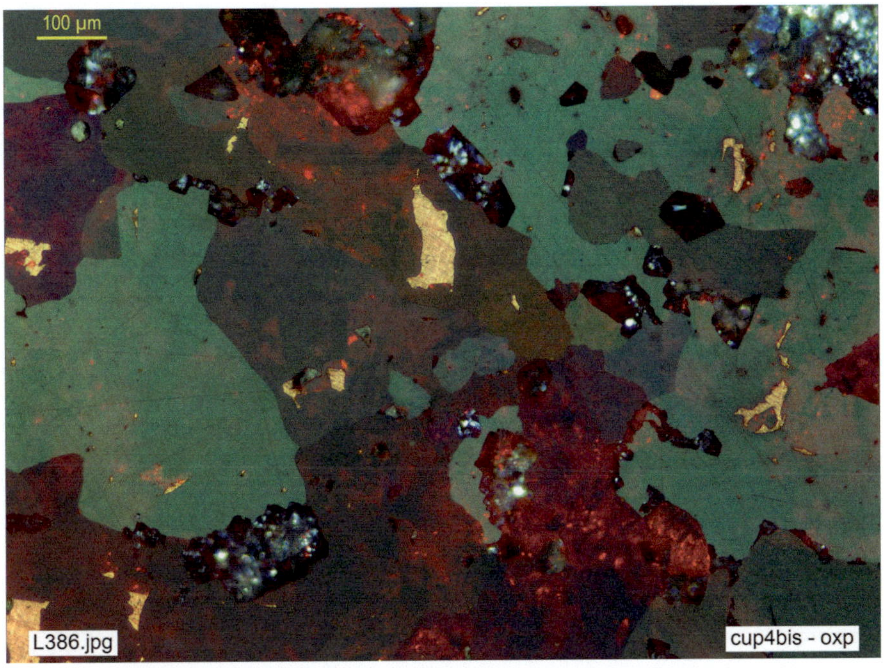

Fig. 42.2 Same scene as Fig. 42.1d (**oxp**). The anisotropism of cup, partly overprinted by the internal reflections, IR, is better seen with slightly uncrossed analyser (≈88°). Distinct bluish-green polarisation tints appear

Fig. 42.3 Mixed copper ore: chalcocite, cc (slightly bluish white), and mch (much darker, grey, bireflective), with minor cup (grey) and gth (darker grey). Two generations of sulphide (cc). At the core, a likely hypogene first generation, *cc1* (ovalshaped, enclosed by gently lobed grey reaction rim). In the outer zone, *cc2*, a concentric band or supergene *corona* on cc1, encrusted along the mch contact by a very thin rim of cryptocrystalline cup (very tightly intergrown with cc). Inclusions of q (very dark) and faint dissemination of gth. CR376, Tsumeb, Namibia (**dsp**)

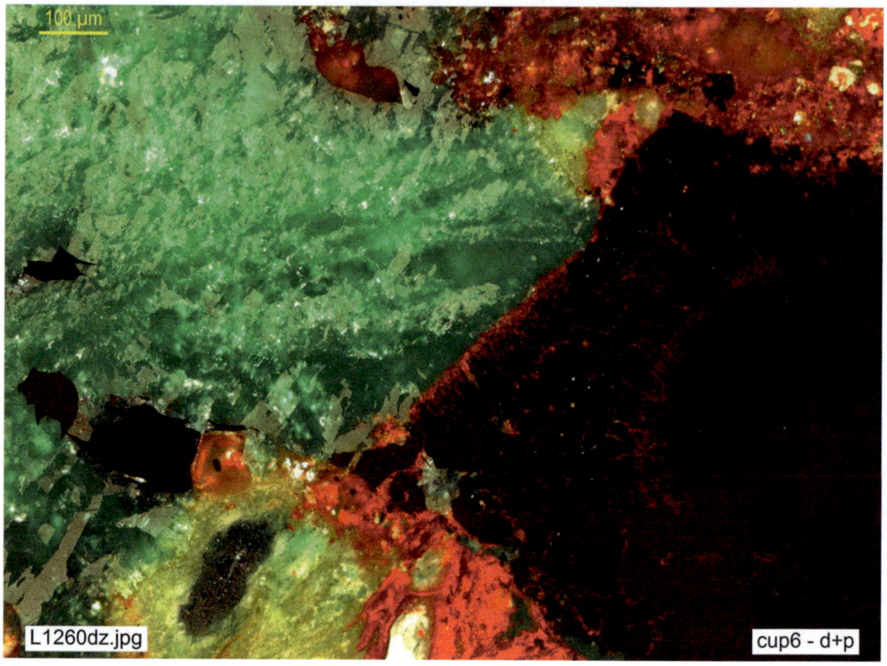

Fig. 42.4 (Same scene as Fig. 42.3: **d+p**). Red internal reflections identify the cup of the two generations, marking first a slight oxidative interval between *cc1* and *cc2*; then, a change from sulphidic to oxidized copper ore, suggested by the transition from cc2 (dark, no IR) to mch (massive green IR)

42.3 Spectral Information

42.3.1 Cuprite Multispectral Reflectance R (in Air, Without Polariser) (%)

λ (nm)	370	400	425	450	475	500	525	550	575	600	625	650	675	700	750	800	850	900	950	1000
μ	34.8	33.4	32.5	31.9	30.8	30.1	29.1	27.4	26.2	25.2	24.5	24.3	23.9	23.5	22.9	22.4	22.2	21.2	21.1	20.7
P_5	34.3	32.9	32.0	31.1	30.2	29.4	28.7	27.1	25.9	24.9	24.2	23.9	23.6	23.2	22.6	22.1	21.9	21.0	20.7	20.4
P_{20}	34.6	33.1	32.3	31.4	30.5	29.7	28.9	27.2	26.0	25.1	24.4	24.1	23.7	23.3	22.7	22.2	22.0	21.1	20.9	20.5
P_{80}	35.1	33.7	32.7	32.4	31.2	30.5	29.4	27.6	26.3	25.3	24.6	24.4	24.1	23.7	23.1	22.6	22.3	21.3	21.3	20.8
P_{95}	35.3	33.9	32.9	33.0	31.5	30.9	29.6	27.8	26.5	25.5	24.7	24.7	24.4	24.1	23.5	23.0	22.8	21.7	21.8	21.3

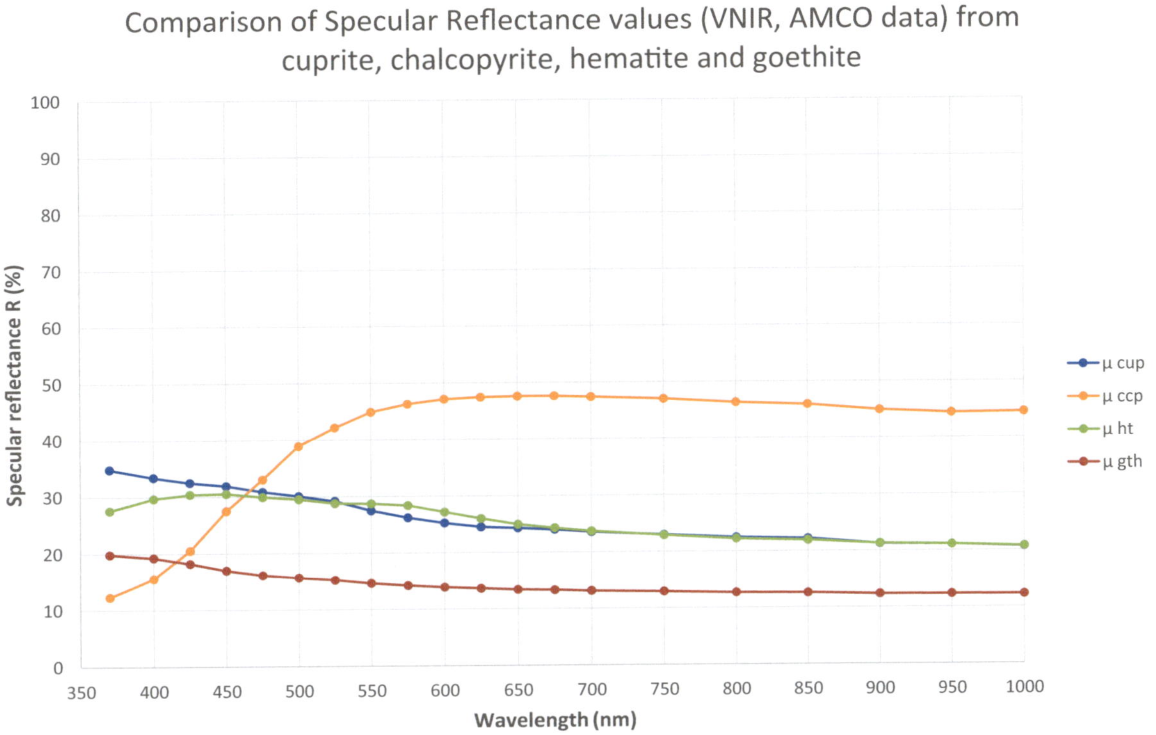

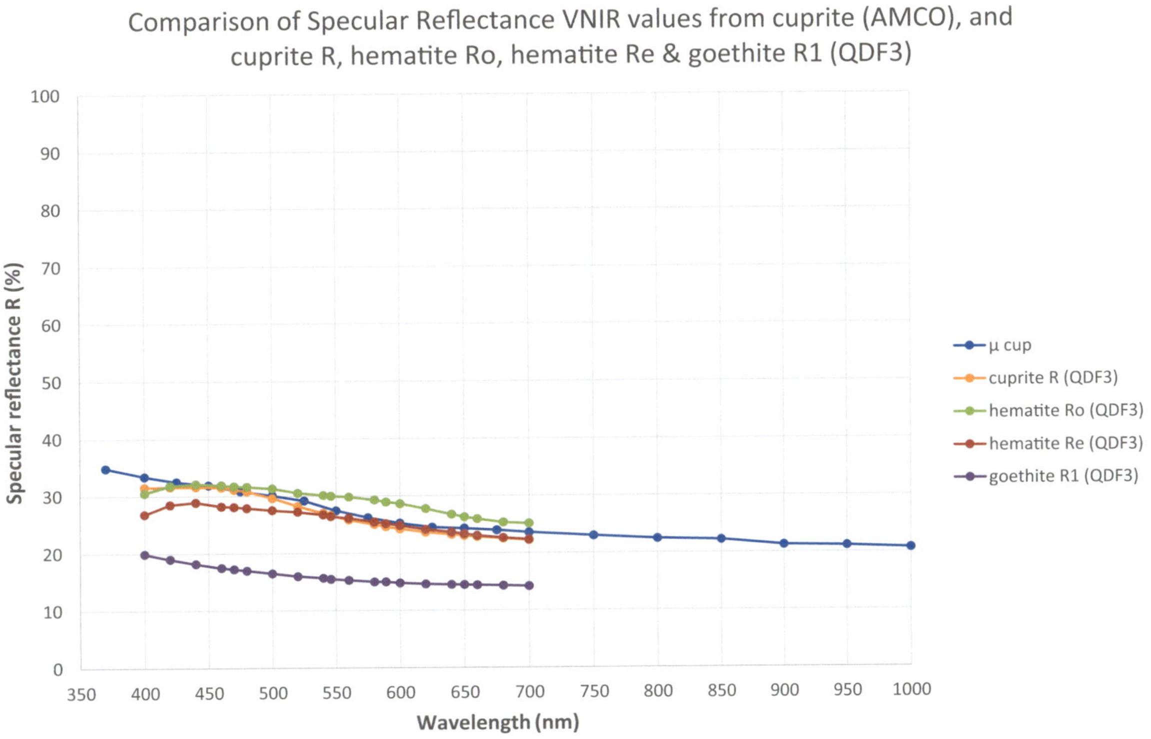

Reference

Uytenbogaardt W, Burke EAJ (1971) Tables for microscopic identification of ore minerals, 2nd edn. Elsevier, Amsterdam

43 Digenite (*dg/neodigenite*)

Cu_9S_5: Cubic

Digenite		R_{546} (%)	Y (%)	λ_d (nm)	P_e (%)	Colour sensation
Air		21.9	21.3	480	14	Light bluish grey, low gloss
Oil		8.61	8.3	479	25.9	Idem darker
Bir: Null (1)				*Pleoc*: Null (1)		*Tarn*:–
Anis +P: Isotropic (1)				*IR*: No		*Twg*: Seldom
Mor: Anhedral aggregates, dissem, pseudomf				*Clv*: (111) distinct		*PolQ*: Good
Hardness	*PolH*: Low ≈ gn (3)			*Mohs*: 2.5–3	*Talm*:	*VHN* $_{(100\ g)}$: 86–106

Obs: cf. Table and R charts, multispectral data (VNIR, AMCO) and visible range (QDF3)
Mineral abbreviations: Ag—native silver, Au—native gold, anil—anilite, bn—bornite, bzl—berzelianite, cc—chalcocite, cv—covellite, dg—digenite, dju—djurleite, gn—galena, py—pyrite, sp—sphalerite, td—tetrahedrite

43.1 General

Formerly understood as a variety or modification of chalcocite (*blue isotropic chalcocite*) and later also called neodigenite, today it is a credited species, of composition Cu_9S_5 (formula $Cu_{1.8}S$ according to IMA/COM), with possible impurities of Fe or Ag.

Digenite formed above 83 °C can generate on cooling varied exsolutions, drops, spindles or lamellae of chalcocite, covellite, chalcopyrite. If formed below 83 °C, it usually includes bornite.

43.2 Microscopic Description

43.2.1 SP Observation (*Single Polariser*)

Light blue *colour* (Fig. 43.1a, b), not bright but striking despite its surprisingly low luminance (Y: 21.3% in air and 8.3% in oil) and *reflectance*, (R, *cf* table and multispectral comparative charts):

- in air, and for white light (λ = 546 nm), R_{dg} < cc, td, gn
- as compared with bn and cv, it depends on wavelength, R_{dg} > bn, cv, for lower λ, but this changes for the larger λ, towards red and beyond (*i.e.*, throughout the NIR or near-infra-red range, up to 1000 nm), R_{dg} < bn for λ 590 nm, and R_{dg} < cv for $\lambda \geq$ 680 nm.

The digenite blue colour gives variable shades according to the Cu content but always less intense than the covellite blue. The Fe content attenuates or suppresses the blue colour.

Good *polishing*. Isotropic. Distinct octahedral cleavage (111). *Polishing hardness* low, ≈ gn (low-medium limit); > cv, < bn, ccp, Ag, Au, sp.

It occurs frequently intergrown with chalcocite, covellite, bornite. Stability diagrams in the Cu_2S-CuS system (Buerger, 1941, and Morimoto and Kullerud, 1963, Ramdohr, 1980) explain the decomposition of high-temperature digenite on cooling below 78 °C, either to cc+dg, or to dg +cv, depending on the composition (% Cu_2S versus % CuS) of the system. Low temperature transformation to dg+dju or

R. Castroviejo, *A Practical Guide to Ore Microscopy—Volume 1*,
https://doi.org/10.1007/978-3-031-12654-3_43

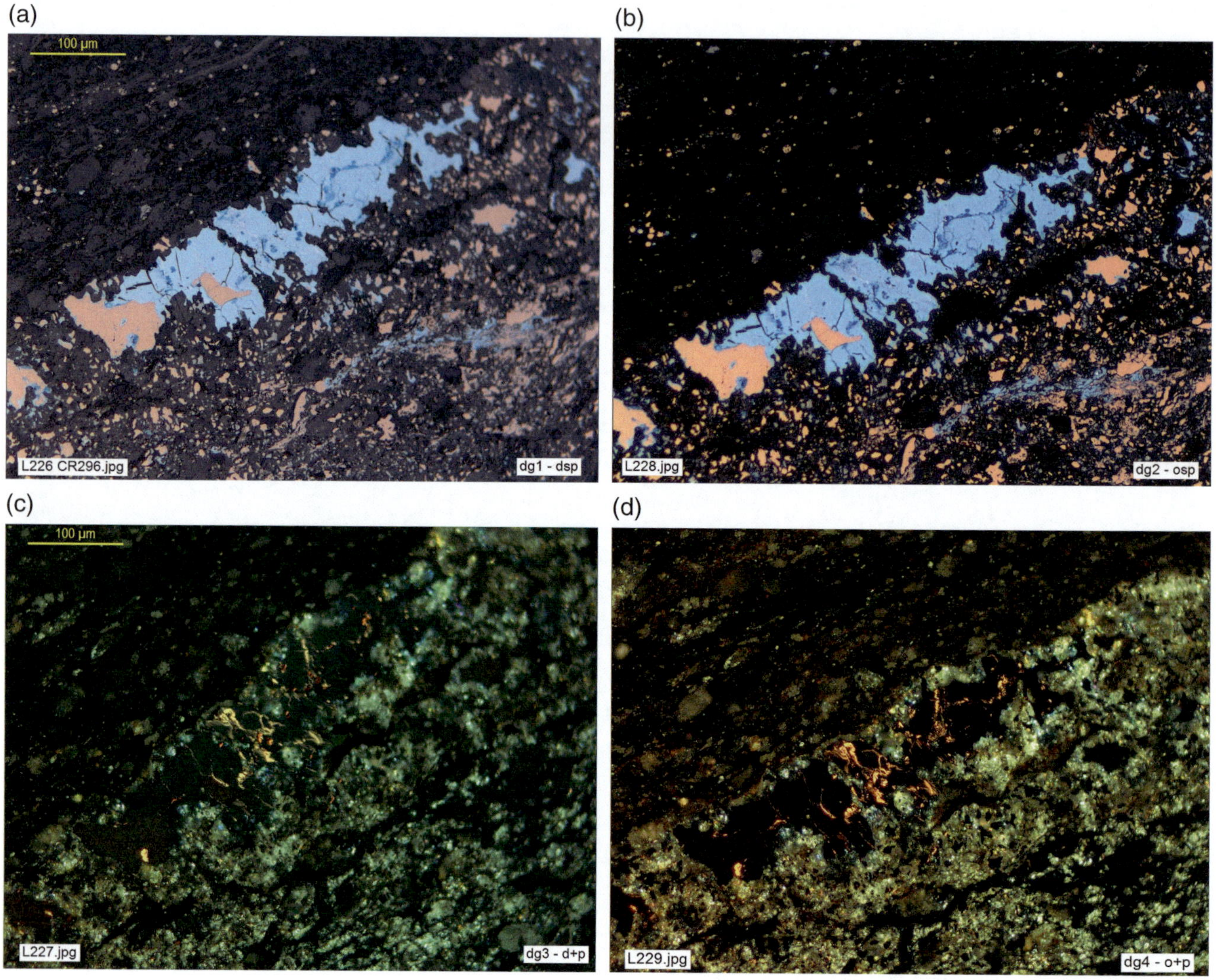

Fig. 43.1 **a**, **b** (Same scene: **dsp** and **osp**, *respectively*). *Kupferschiefer* ore: dg (light blue), bn (pink) and *spionkopite* (dark blue), disseminated and in microbands. The permanent blue colour identifies *spionkopite* (former "*blaubleibender Covellin*" or *everblue cv* variety—Ramdohr, 1980-, lacking the purple or violet hues typical of cv in oil immersion, now credited as an independent species). CR 296, Lubin, Poland. **c**, **d** (Same scene as dg1: **d+p** and **o+p**, *respectively*). Extinction of dg, unlike bn (very weakly) and spionkopite (strongly) anisotropic (≈cv)

dg+anilite is also possible. Likewise, cc+cv and cc+bn are frequent. The result may be a spectacular pseudo-octahedral lamellar lattice of cc (or dju) lamellae, with (001) // (111) of dg (cf Fig. 25.7a, b).

43.2.2 Observation +P (*Polariser + Analyser at 90°*)

Isotropic, perfect extinction (Fig. 43.1c, d). Occasional twinning.

43.2.3 Paragenesis

Varied *parageneses*, hypo- or supergene, primary or secondary. Different types of copper-bearing deposits, as *Kupferschiefer*/Copperbelt, VMS (massive sulphides), *copper-porphyries* especially in the supergene sulphide enrichment blanket (or *cementation* zone).

Most frequent companions: bornite, covellite, chalcocite, chalcopyrite, enargite, tetrahedrite.

43.2.4 Identification

It is easily recognized by its typical colour, with moderate to low reflectance and hardness, octahedral cleavage and isotropy. The VNIR spectra are a useful additional tool.

The association with cupriferous paragenesis helps to distinguish it from berzelianite, which resembles it but is much rarer, a bit lighter, occurs with selenides and shows no cleavage (Figs. 43.1a–d)..

43.3 Spectral Information

43.3.1 Digenite Multispectral Reflectance R (in Air, Without Polariser) (%)

λ (nm)	370	400	425	450	475	500	525	550	575	600	625	650	675	700	750	800	850	900	950	1000
μ	28.8	28.5	27.7	27.1	26.4	25.3	24.2	22.7	21.6	20.3	19.2	18.0	17.1	16.2	14.6	13.4	12.7	12.7	13.4	15.0
P_5	26.1	25.7	24.9	24.3	23.6	22.6	21.5	20.0	18.9	17.6	16.4	15.2	14.4	13.6	12.2	11.2	10.8	11.4	12.1	12.9
P_{20}	27.4	27.1	26.3	25.7	25.1	24.0	22.8	21.3	20.1	18.7	17.4	16.0	15.1	14.2	12.6	11.6	11.3	12.1	12.5	13.3
P_{80}	30.0	29.8	29.1	28.4	27.6	26.6	25.4	24.0	22.8	21.6	20.6	19.5	18.6	17.6	16.0	14.6	13.6	13.0	14.4	17.2
P_{95}	30.3	30.1	29.4	28.8	27.9	27.0	25.7	24.3	23.1	22.0	21.0	19.9	19.1	18.1	16.5	15.1	14.1	14.3	16.1	18.8

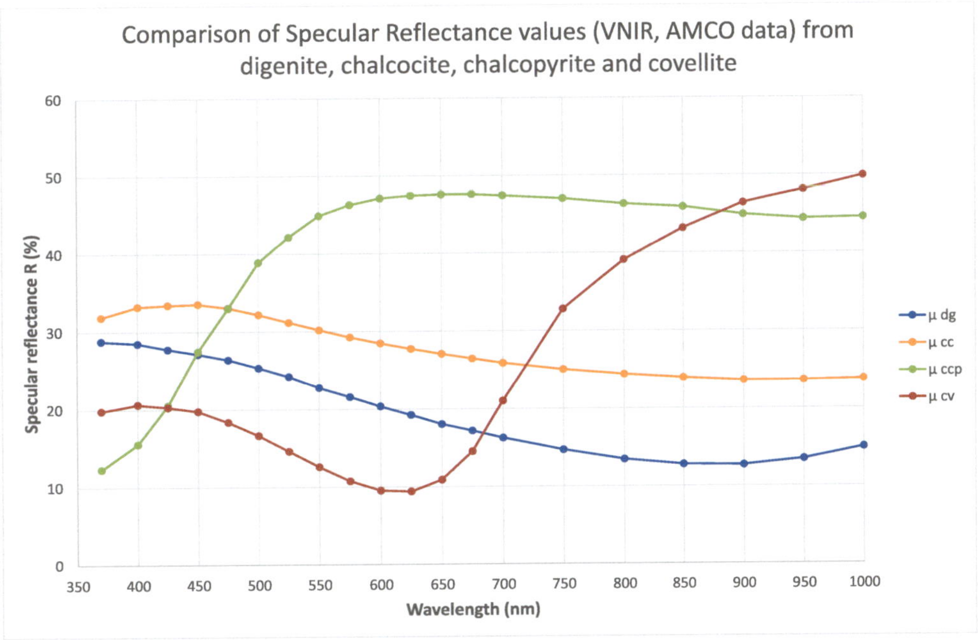

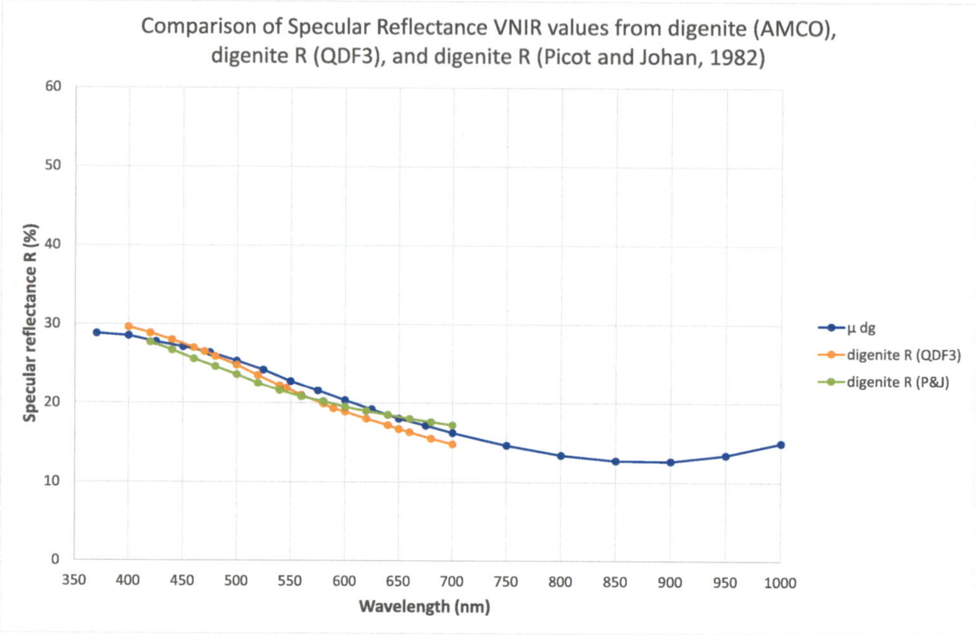
Comparison of Specular Reflectance VNIR values from digenite (AMCO), digenite R (QDF3), and digenite R (Picot and Johan, 1982)
Specular reflectance R (%)
0
10
20
30
40
50
60
350
400
450
500
550
600
650
700
750
800
850
900
950
1000
Wavelength (nm)
μ dg
digenite R (QDF3)
digenite R (P&J)

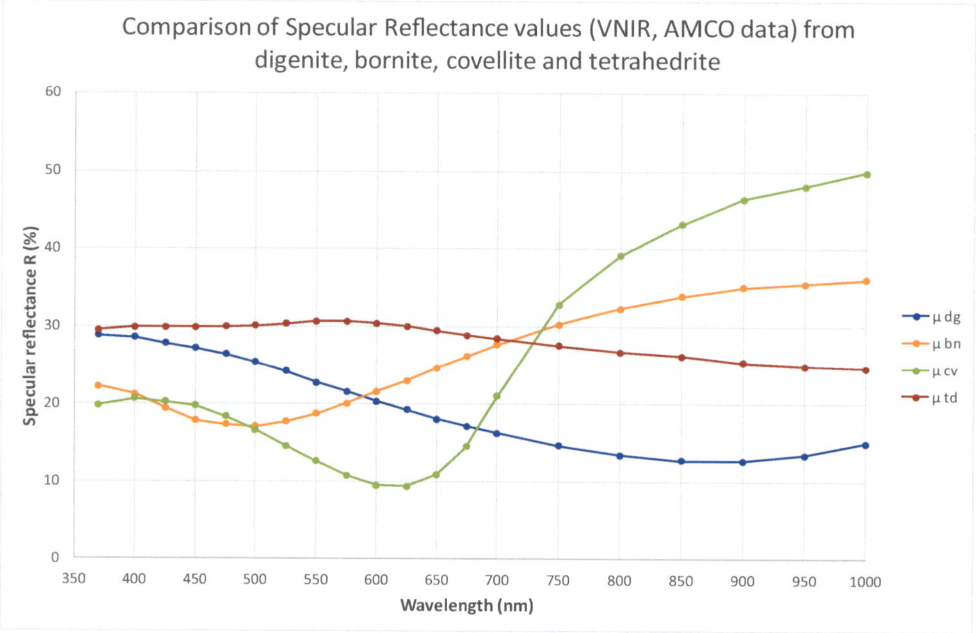
Comparison of Specular Reflectance values (VNIR, AMCO data) from digenite, bornite, covellite and tetrahedrite
Specular reflectance R (%)
0
10
20
30
40
50
60
350
400
450
500
550
600
650
700
750
800
850
900
950
1000
Wavelength (nm)
μ dg
μ bn
μ cv
μ td

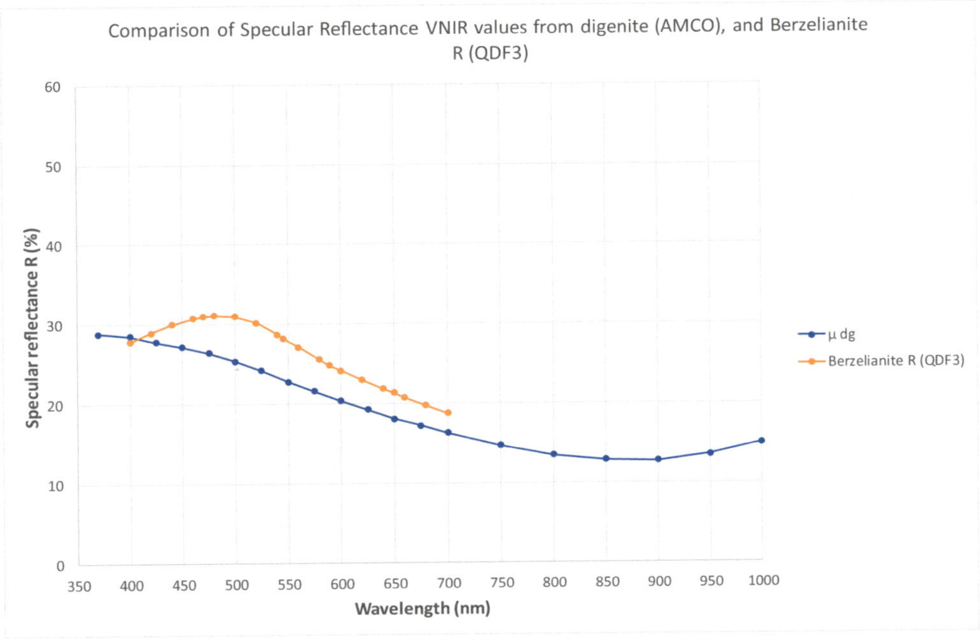
Comparison of Specular Reflectance VNIR values from digenite (AMCO), and Berzelianite R (QDF3)
Specular reflectance R (%)
Wavelength (nm)
μ dg
Berzelianite R (QDF3)
0
10
20
30
40
50
60
350
400
450
500
550
600
650
700
750
800
850
900
950
1000

Dyscrasite (*dy, dyscrasite, stibiotriargentite*) 44

Ag_3Sb: Orthorhombic

Dyscrasite		R_{546} (%)	Y (%)	λ_d (nm)	P_e (%)	Colour sensation
Air	R_1	60.1	59.9	561	1.0	White to slightly yellowish-white, glossy
	R_2	62.8	62.7	572	2.0	
Oil	R_1	47.9	47.6	531	0.4	
	R_2	51.3	51.2	571	1.9	
Bir: Very weak (2)				*Pleoc*: Very weak (2)		*Tarn*: Slow tarnishing
Anis +P: Weak (3)				*IR*: No		*Twg*: Occasional
Mor: Euhedral Ps-hexag.; massive mx				*Clv*: Rarely visible		*PolQ*: Good, with occasional scratches
Hardness	*PolH*: Low (>Sb, As; < gn) (2)			*Mohs*: 4^-	*Talm*: C	*VHN* $_{(100\ g)}$: 153–179

Note Fine exsolutions in gn or Ag, often intergrown with allargentum Cf. reflectance spectra in the visible range (QDF3 data and Picot and Johan 1982)

Mineral abbreviations: Ag—native silver, As—native arsenic, Bi—native bismuth, btp—breithauptite, cerv—cervantite, dy—dyscrasite, gn—galena, lo—löllingite, my—miargyrite, pru—proustite, prg—pyrargyrite, Sb—native antimony, sf—safflorite

44.1 General

Composition: Ag_3Sb or, according to the IMA formula, $Ag_{3+x}Sb_{1-x}$ ($x \approx 0.2$). At Broken Hill, Australia, $Ag_{3.15}Sb_{0.85}$ has been cited. The metal contents (wt.%) corresponding to the original Ag_3Sb formula are ≈ 72.7% Ag and 27.3% Sb, but in dyscrasite synthesized at 300 °C, Sb contents of 22.9 to 26.9% have been found (Somanchi 1966). It shows complex relationships in the Ag-Sb binary system. Ag_3Sb is stable as dyscrasite up to 400 °C but transforms at 440 °C into a high-temperature cubic phase (*animikite*, Schroecke and Weiner 1981). For richer Ag compositions, other intermetallic phases appear, such as *allargentum*, a mineral defined by Ramdohr (1980), who describes, in the Cobalt, Ontario ores, three finely intergrown phases, difficult to characterize and with complex relationships, identifying phase III as allargentum (cf. description allargentum). At the Sb end, native Sb is formed, without Ag (Keighin and Honea 1969). Often dyscrasite contains some Hg and, occasionally, traces of As.

44.2 Microscopic Description

44.2.1 SP Observation (*Single Polariser*)

Colour white to yellowish white ("cream" for some authors). It looks very faintly brownish in direct contact with native Sb (Fig. 44.1a, b). Good *polishing* but not usually scratch-free. It tarnishes in the open air but, by far, not as fast as native silver.

High *reflectance*, higher than common sulphides (comp. galena, Fig. 44.1a, b) but less than Ag or even Sb. Very weak bireflectance and pleochroism, although perceptible in oil (cf. reflectance spectra in the visible range, with QDF3 data and Picot and Johan 1982).

Rarely visible cleavage.

It occurs in intergrowths or allotriomorphic aggregates of very fine grain size, often as exsolutions (in galena, in Sb or other phases of the Ag-Sb system) or intergrown with native As (formerly known as *arsensilber*, Ramdohr 1980).

R. Castroviejo, *A Practical Guide to Ore Microscopy—Volume 1*,
https://doi.org/10.1007/978-3-031-12654-3_44

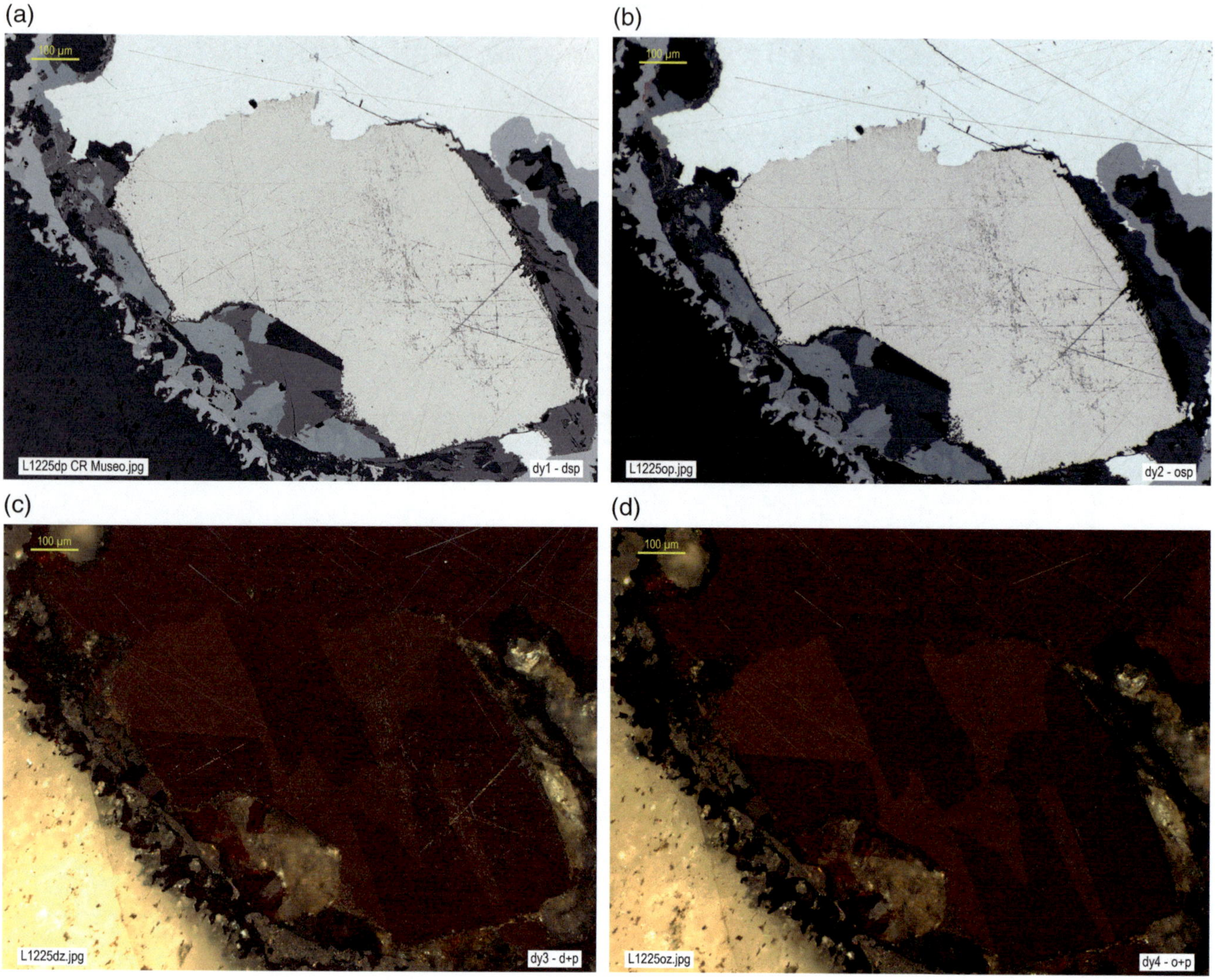

Fig. 44.1 **a**, **b** (Same scene: **dsp** and **osp**, respectively). Silver ore, in quartz gangue: Sb (white, top) and dy (somewhat darker and with a faint pinkish-brownish tinge against Sb, center), surrounded by a thin, inner band of pyrargyrite (bluish grey), and an outer crown of galena (lighter grey, left-hand side). Sb ochre (cervantite, ***cerv***, dark grey) intergrown with prg and encrusting dy (inner band with prg, and much darker). In oil (Fig. 44.1b, **dsp**) the higher contrast of the ore components, the near disappearance of cerv and the total disappearance of quartz (R ≈ 0%: black) can be observed. Sample CR398: Hiendelaencina, Guadalajara. **c**, **d** (Same scene as dy1: **d+p** and **o+p**, respectively). Note the anisotropism and twinning of dy, as well as the anisotropism and internal reflections of prg, more distinct in immersion F dy4, and of cerv

Occasionally, in well-developed, idiomorphic crystals, typically pseudohexagonal (by twinning of three individuals, like aragonite).

Low *polishing hardness* (≈ cinnabar), slightly > Sb, As, my, but < gn. *Vickers microhardness* (load 100 g): 153–179. *Mohs hardness*: 3.5–4, *Talmage H*: C.

44.2.2 Observation +P (*Polariser + Analyser at 90°*)

Weakly *anisotropic*, with dull brownish to brownish-pinkish tones, clearer in oil. No *internal reflections*. Occasional *twinning*, in well-developed crystals (Fig. 44.1c, d).

44.2.3 Paragenesis

Typically hypogene hydrothermal, e.g., in veins of the *five-element formation* (Ni-Co-Bi-Ag-U) type, as in the Erzgebirge, Saxony (Germany), while in Cobalt, Ontario, it seems to be rather scarce, allargentum appearing instead (with which it had been confused). However, in trace amounts it can be found in various types of deposits and environments, even as a possible supergene mineral. As companion ores, galena, As, Sb, Ag, safflorite, breithauptite, löllingite, Ag sulphosalts (proustite, pyrargyrite, miargyrite), etc. are mentioned.

44.2.4 Identification

Because of its colour and high reflectance, dyscrasite can be confused with Sb, Ag or Bi, but careful observation avoids confusion.

- Against Ag, dyscrasite is much less reflective, shows a yellowish to brownish hue, tarnishes much more slowly and is anisotropic.
- Against Sb, dyscrasite is less reflective, shows a brownish to brownish-pinkish tone, tarnishes (in a few days) and is less anisotropic.
- Against As, dyscrasite is slightly more reflective, tarnishes much more slowly and lacks polysynthetic twins.

44.3 Spectral Information

44.3.1 Dyscrasite Multispectral Reflectance R (in Air) (%)

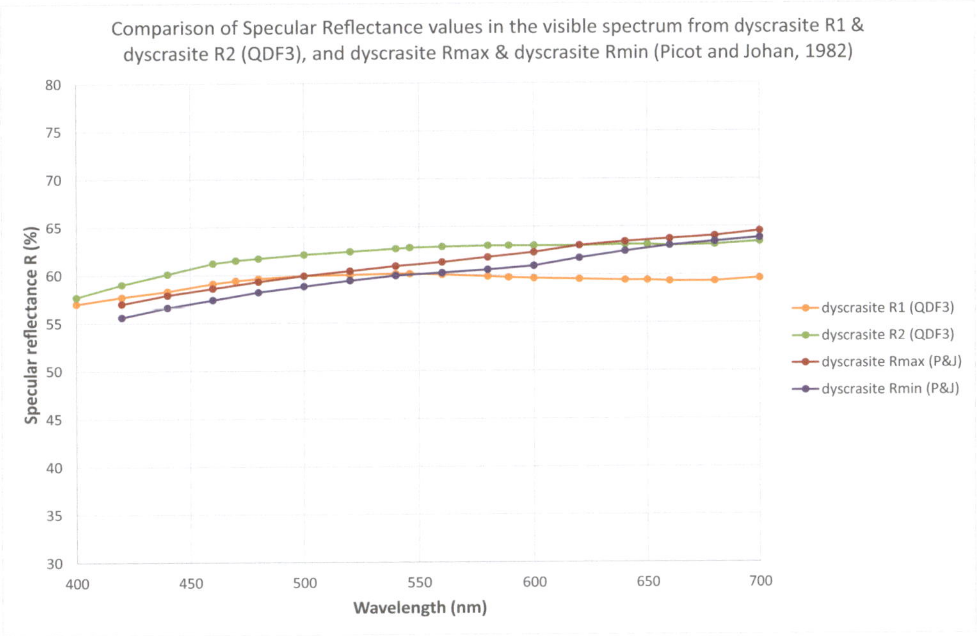

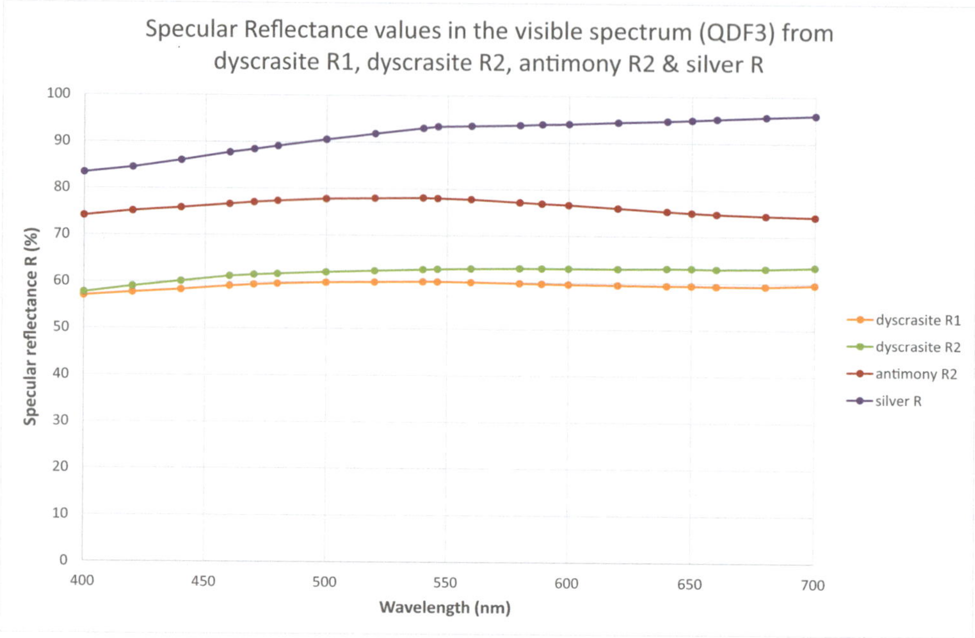

References

Keighin CW, Honea RM (1969) The system Ag-Sb-S from 600 °C tp 200 °C. Mineral Deposita 4:153–171

Picot P, Johan Z (1982) In: Atlas of ore minerals. B.R.G.M./Elsevier

Ramdohr P (1980) In: The ore minerals and their intergrowths. vol 2. 2nd edn. Pergamon Press, Oxford, pp 1205

Schröcke H, Weiner KL (1981) Mineralogie. Walter de Gruyter, Berlin, p 952

Somanchi S (1966) Subsolidus phase relations in the system Ag-Sb. Can J Earth Scii 3:211–222

Electrum (*el/Gold-Silver Alloy, Aurian Silver Ag–Au*) 45

AgAu: Cubic

Electrum	R_{546} (%)	Y (%)	λ_d (nm)	P_e (%)	Colour sensation
Air	90.3	89.6	572	16.4	Pale yellow to YlWh, very shiny
Oil	86.7	86.1	572	18.9	
Bir: No (1)			*Pleoc*: No (1)		*Tarn*: No
Anis +P: Isotropic, imperf. extinction (2)			*IR*: No		*Twg*: –
Mor: Fine aggregates, allotriomorphic			*Clv*: No		*PolQ*: Easy, but many scratches
Hardness	*PolH*: Medium (3)		*Mohs*: 3–	*Talm*: B	*VHN* $_{(50\ g)}$: 59–67

Spectral information: cf. Table and R-charts, multispectral (VNIR, AMCO) and visible (QDF3; Picot and Johan 1982) ranges
Mineral abbreviations: Ag—native silver, As—native arsenic, Au—native gold, Bi—native bismuth, ccp—chalcopyrite, el—electrum, gn—galena, Sb—native antimony, td—tetrahedrite

45.1 General

There is no unanimity about its precise composition. It is a solid solution AgAu, as major elements, in varying proportions. An acceptable range is between $Ag_{0.7}Au_{0.3}$ and $Ag_{0.3}Au_{0.7}$ (Spiridonov and Yanakieva 2009), but criteria vary according to authors, *vbgr*, Ag $\approx$ 30–45% (Ramdohr 1980), Ag $\geq$ 25% (Uytenbogaardt and Burke 1971) or $\approx$ $Ag_{0.4}Au_{0.6}$ and $Ag_{0.24}Au_{0.76}$ (the two compositions analysed in QDF3). Consequently, the term *electrum* (or its equivalent *gold-silver alloy*) is losing scientific acceptance as a mineralogical species in favor of *aurian silver Ag–Au*, preferred by IMA/COM. However, the chronological priority, the practical significance and the wide diffusion in the mining literature of the traditional term, electrum, advise not to ignore it.

As minor or trace elements can be found: Hg, Cu, Sb, etc. (The same as in native Ag and Au.)

Its properties are intermediate between those of the noble metals Au and Ag.

45.2 Microscopic Description

45.2.1 SP Observation (*Single Polariser*)

Bright yellowish white *colour* (Fig. 45.1), yellower as the Au content increases, resulting in comparison more yellow than silver and paler, although brighter than pure gold. It is easily polished but, due to its low hardness (VHN 59–67), polishing scratches are unavoidable and very numerous (Figs. 45.1 and 45.2a, b). Unlike native silver, in polished section, it is not usually darkened by surface alteration patina.

Its reflectance (90.3%, 546 nm) is very high, although not as high as that of pure silver. In any case, its brightness is very intense (Y%, air: 89.6) and, as with native silver, next to it the other minerals appear darker, increasing this sensation in oil because of the greater contrast (comp. Figs. 45.1 and 45.2a with Fig. 45.3). There is a relationship between the magnitude and spectral composition of the specular reflectance and the Au:Ag ratio which, completing the data,

R. Castroviejo, *A Practical Guide to Ore Microscopy—Volume 1*,
https://doi.org/10.1007/978-3-031-12654-3_45

might be used as an indicator of its *fineness* or Au content (cf. QDF3 spectra).

Isotropic, although reflection in the different polishing scratches may give a false idea of anisotropism when the stage is rotated.

Medium *polishing hardness* (≈ ccp; > gn and < td), chalcopyrite-like, between that of galena and tetrahedrite. Vickers microhardness: 59–67 (charge 50 g). Mohs hardness: 2.5–3, Talmage: B.

It usually occurs disseminated or as hydrothermal filling, often cementing microcracks or included in sulphides (Figs. 45.1, 45.2a and 45.3). It commonly forms allotriomorphic grains or aggregates, occasionally developing peculiar dendritic or filiform textures, but its morphology can be varied. Nuggets in placers.

45.2.2 Observation +P (*Polariser + Analyser at 90°*)

Isotropic as a cubic mineral but with *incomplete or imperfect extinction* and anomalous luminosity, reinforced by the polishing scratches and the false polarization effects associated with them (Fig. 45.2b). It shows no internal reflections.

45.2.3 Paragenesis

Great economic interest. Paragenesis generally complex, typical of hydrothermal and especially epithermal deposits. The weathering of these primary ores, with the consequent surface mobilization and/or redistribution of Au and Ag, gives rise to notable secondary concentrations or supergene enrichment. In the detrital concentrations (placers), there is a characteristic Au–Ag zoning of the auriferous nuggets. These acquire a *corona* (crown) or external rim of almost pure gold, due to the preferential leaching and extraction of the (more soluble) Ag, leaving a remnant with increased *fineness*.

45.2.4 Identification

Electrum is generally easily recognized. Its properties are intermediate between Ag and Au: very high reflectance, yellowish-white colour (more yellow with increasing gold content), frequent polishing scratches, medium polishing hardness, isotropic character but with incomplete extinction and environment or paragenesis. Unlike native Ag, it does not tarnish.

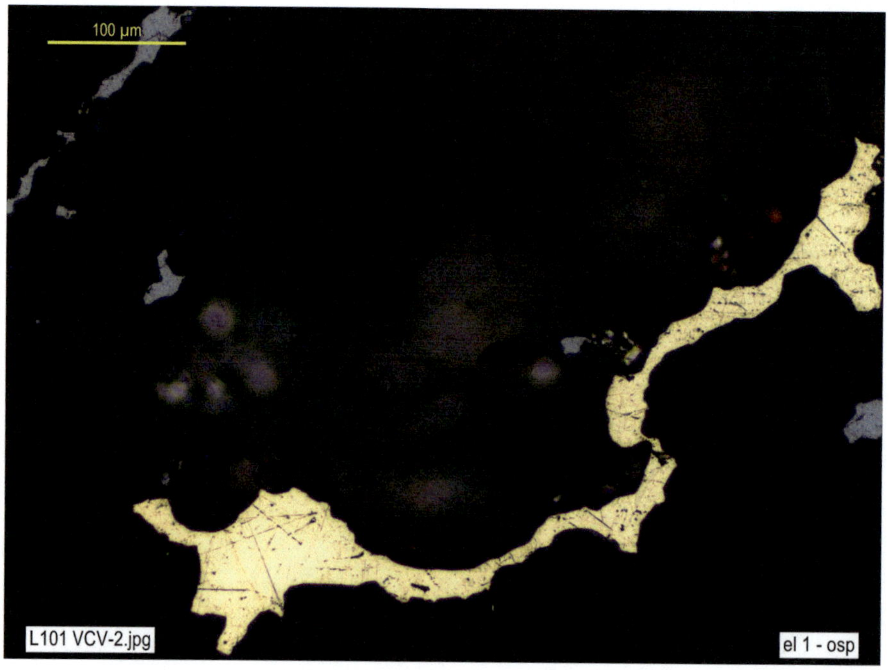

Fig. 45.1 Electrum-filled microcrack (69.9% Au, 30.1% Ag, EMP analysis) in quartz gangue (black) with disseminated galena (grey). The high contrast of reflectances is enhanced by immersion, which makes the galena look like a relatively dark mineral. Sample VCV2, Vilarcovo, Santiago de Compostela (**osp**)

Fig. 45.2 **a** Electrum included in galena (grey, triangular pits), disseminated in quartz (black). Electrum analysed by electron microprobe: 70% Au, 30% Ag. Sample VCV2, Vilarcovo, Santiago de Compostela (**osp**). **b** Same scene as *el-2* (**o+p**): note the incomplete extinction of el

(a)

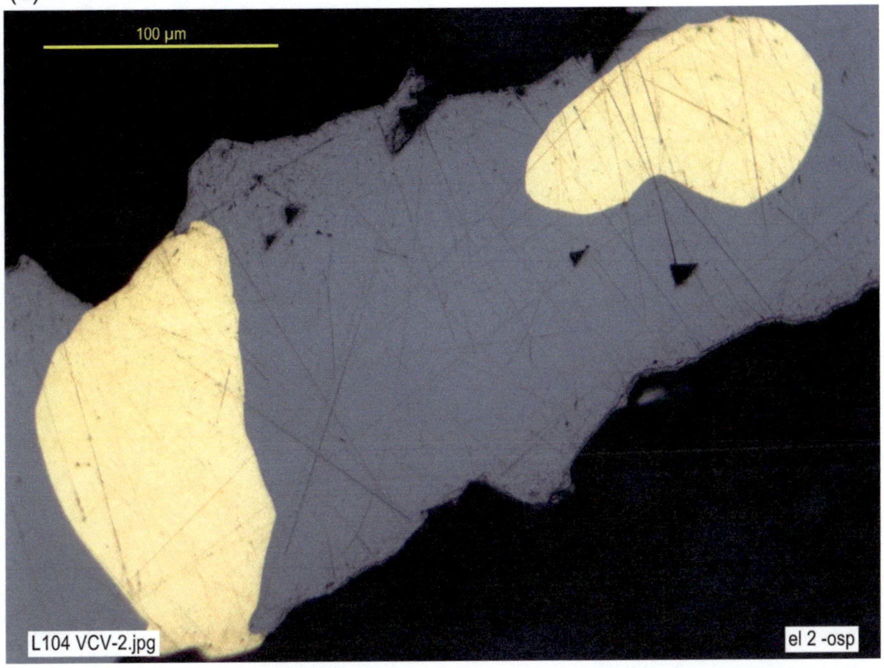

(b)

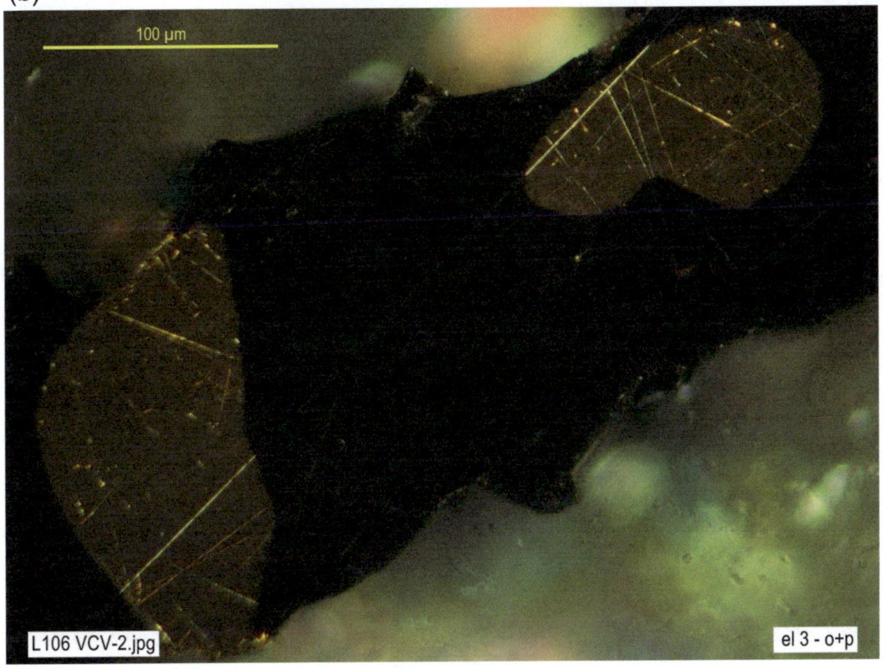

Among the noble metals, it is distinguished from platinum because Pt is whiter, with hardly any polishing scratches, is less reflective and shows clearly higher polishing hardness. Native Bi, Sb and As are softer, less reflective (darker) and anisotropic.

Fig. 45.3 Pyrite aggregate (yellowish white, right-hand half of the image) with inclusions of electrum (light yellow, very bright), marcasite (white, with a slight greenish tint), chalcopyrite (yellow), galena (grey), sphalerite (dark grey) and gangue (very dark grey, almost black). The same sulphides appear intergrown with gangue in the upper left corner. Note the lower contrast, compared to Figs. 45.1 and 45.2a, due to observation in air. Sample from Rammelsberg (Banderz), Niedersachsen, Germany (**d+p**)

45.3 Spectral Information

45.3.1 Electrum Multispectral Reflectance R (in Air, Without Polariser) (%)

λ (nm)	370	400	425	450	475	500	525	550	575	600	625	650	675	700	750	800	850	900	950	1000
μ	36.4	38.6	40.7	43.0	45.0	48.0	52.6	61.3	70.8	77.0	80.3	82.8	84.4	85.3	87.0	87.5	88.7	89.2	89.8	89.8
P_5	34.5	36.8	38.6	40.7	42.5	45.6	50.0	58.5	67.7	73.9	77.3	79.9	81.6	82.6	84.5	85.2	86.5	87.1	88.0	88.1
P_{20}	35.6	37.8	39.8	42.1	44.0	47.1	51.5	60.1	69.5	75.7	79.1	81.5	83.2	84.1	85.9	86.5	87.7	88.2	89.0	89.1
P_{80}	37.2	39.5	41.6	44.1	46.0	49.0	53.7	62.5	72.1	78.3	81.7	84.1	85.8	86.6	88.3	88.6	89.8	90.2	90.6	90.7
P_{95}	37.9	40.2	42.3	45.0	46.7	49.9	54.4	63.5	73.1	79.4	82.8	85.2	86.9	87.8	89.6	89.8	91.0	90.7	91.8	91.3

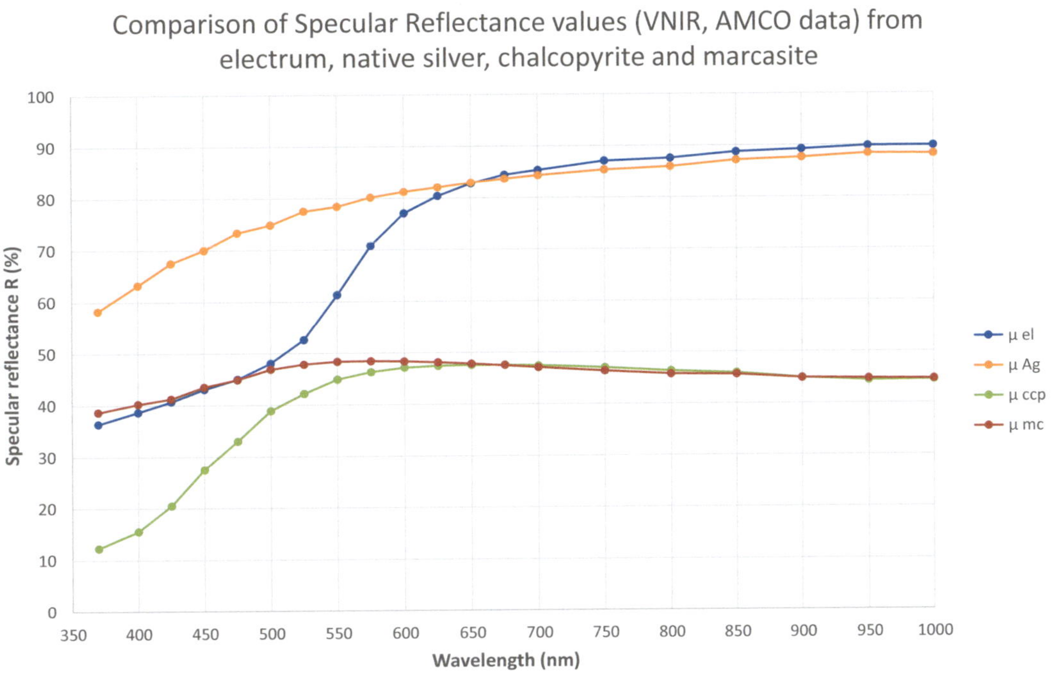

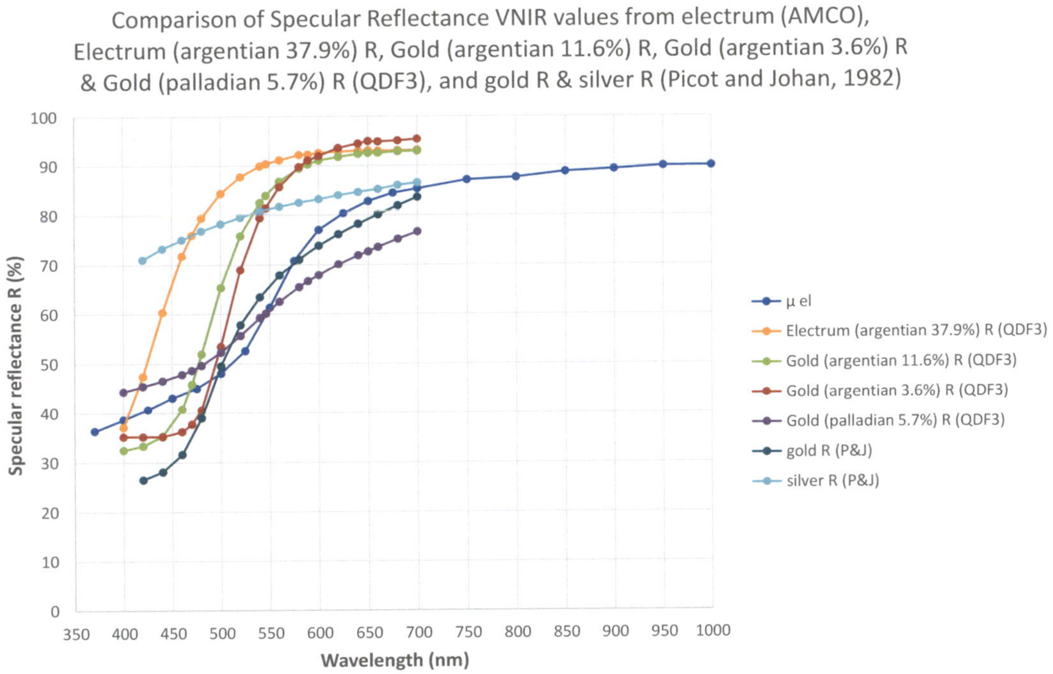

References

Picot P, Johan Z (1982) Atlas of ore minerals. B.R.G.M./Elsevier

Ramdohr P (1980) The ore minerals and their intergrowths, 2nd edn, 2 vols. Pergamon Press, Oxford, 1205 p

Spiridonov E, Yanakieva D (2009) Modern mineralogy of gold: overview and new data. ArcheoSci Rev D'archéométrie 33:67–93

Uytenbogaardt W, Burke EAJ (1971) Tables for microscopic identification of ore minerals, 2nd edn. Elsevier, Amsterdam

Enargite (*en*)

46

Cu_3 AsS_4: Orthorhombic

ENARGITE		R_{546} (%)	Y (%)	λ_d (nm)	P_e (%)	Colour sensation
Air	R_1	24.2	24.4	474	3.8	Pinkish grey, variable shades
	R_2	25.2	25.5	c508	1.2	Light pinkish grey, variable shades
Oil	R_1	10.9	11.1	468	6.6	Pinkish grey, variable shades
	R_2	11.6	12.0	c518	2.7	Light pinkish grey, variable shades
Bir: Weak (3)				*Pleoc*: Appreciable (3)		*Tarn*: Not usual
Anis+P: Distinct and coloured (4)				*IR*: Infrequent		*Twg*: Occasional
Mor: Massive, tabular, prismatic				*Clv*: Frequent // (110)		*PolQ*: Good
Hardness	*PolH*: Medium, > ccp, < sp (3)			*Mohs*: 3	*Talm*: D	*VHN* $_{(100\ g)}$: ≈300

Obs: cf. Table and R charts, multispectral data (VNIR, AMCO) and visible range (QDF3)
Mineral abbreviations: Ag—native silver, bn—bornite, cc-chalcocite, ccp—chalcopyrite, cup-cuprite, cv—covellite, en—enargite, gg—gangue, gn—galena, lz—luzonite, po—pyrrhotite, py—pyrite, sp—sphalerite, tn—tennantite, td—tetrahedrites

46.1 General

Dimorph of luzonite, lz (tetragonal), composition Cu_3AsS_4, may contain notable amounts of Sn or Sb, trace amounts of Pb, Zn, Fe, Ag, Ge. It is easily altered to tennantite. The process requires few chemical or volume changes, the result being known as "mottled enargite", an aggregate of tennantite with residual enargite. This secondary tennantite (±en) was also named "green enargite", to differentiate it from the real enargite or "pink enargite".

Important as a Cu ore only in a few deposits, e.g., Tsumeb, Namibia; Chuquicamata, Chile; Cerro de Pasco, Peru. The As content of en poses a risk due to its environmental impact.

46.2 Microscopic Description

46.2.1 SP Observation (*Single Polariser*)

Light pinkish-brown to brownish-grey colour, slightly pleochroic. The subjective impression of colour, brightness and pleochroism can vary with the composition but especially with the neighbouring ores. It changes according to the accompanying minerals, as can be seen by comparing

R. Castroviejo, *A Practical Guide to Ore Microscopy—Volume 1*,
https://doi.org/10.1007/978-3-031-12654-3_46

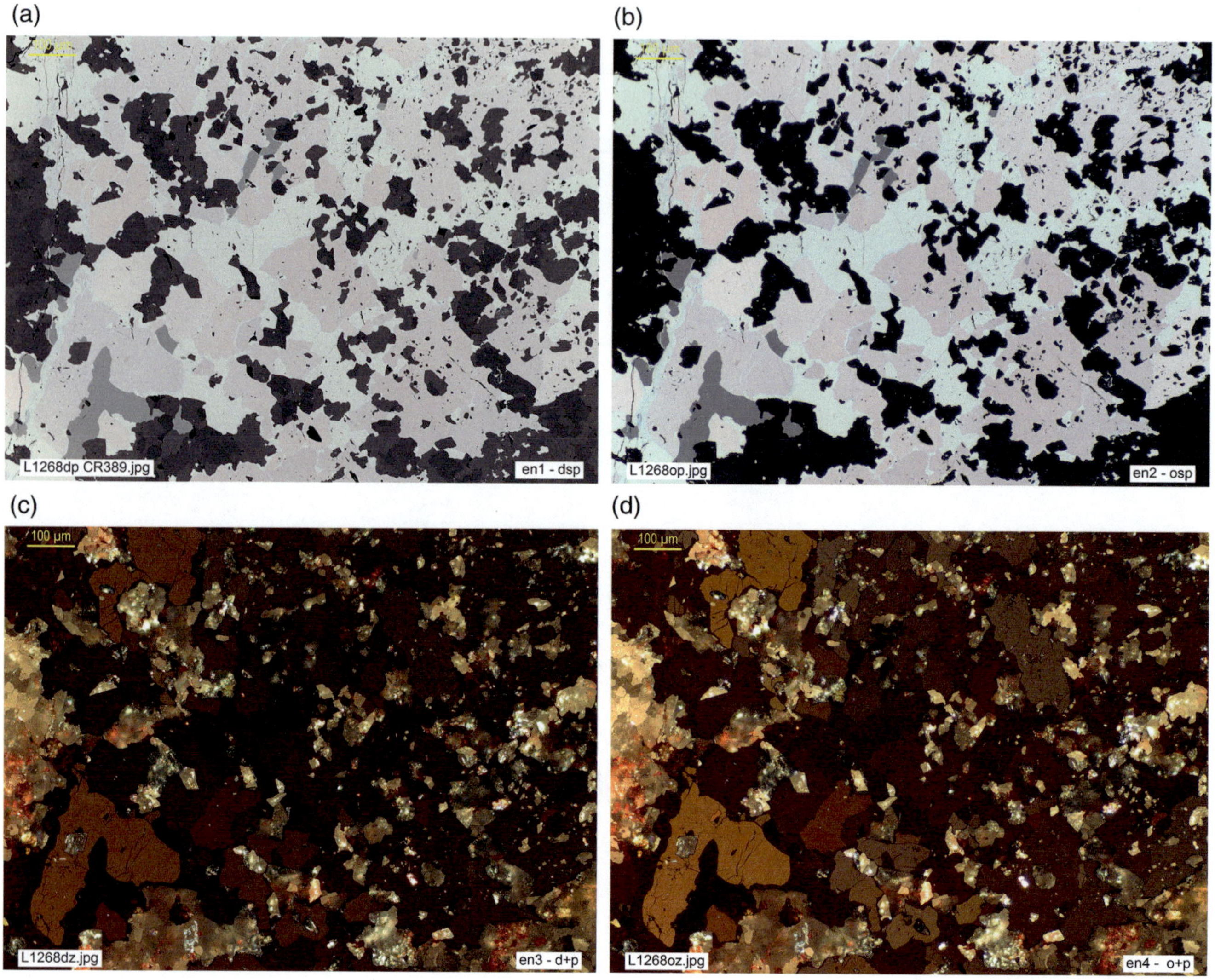

Fig. 46.1 **a, b** (Same scene: **dsp** and **osp**, respectively). Intergrowth of en (pleochroic, light pinkish-brown to brownish grey shades), tn (very light grey with a light bluish-green hue) and quartz-carbonate gg (very dark grey to black), with minor sp (grey).Partial replacement of en by tn. CR 389, Tsumeb Mine, Namibia (**dsp**) and (**osp**), respectively. **c, d** (Same scene as en1: **d+p** and **o+p**, respectively) Unlike the cubic tn and sp (in extinction), en shows distinctly coloured anisotropism. Red internal reflections, **IR**, of tn and sp. Massive and colourless IR of gg

Figs. 46.1a, b, 46.2, 46.4, and 46.5. In immersion, it darkens and gives more saturated colours (Figs. 46.1b and 46.2).

Reflectance is close to that of fahlores (td-tnn) but slightly lower in the visible range. The pinkish-brown tone of en makes a difference with the bluish-greenish olive grey shades of td-tnn (cf. multispectral comparative graphs and Fig. 46.2).

Good *polishing*, although *cleavage* // (110) appears easily if the grain size is coarse. Medium *polishing hardness*, between td and sp ($PolH_{en} \geq$ td, > cup, bn, gn, cc; $PolH_{en} \leq$ sp, < po, py).

It usually occurs in massive aggregates, frequently coarse grained and anhedral, although crystals with dominant prismatic or tabular habits can occasionally be found. Relict prismatic outlines may also be seen. Zoning frequent but observable only by structural etching.

46.2.2 Observation +P (*Polariser + Analyser at 90°*)

Clearly *anisotropic*, with intense, vivid and varied polarisation colours: orange-brown, bluish-green or multicoloured,

Fig. 46.2 Partial replacement of en (grey, with slightly brownish-violet-pink tinge, top left and right and bottom left corners) by tn (greenish olive grey), py (yellowish white, zoned) and ccp (yellow). CR 283, 251102_1 Peru (**osp**)

Fig. 46.3 Same scene as en5 (**oxp**). Lively greenish and purplish polarisation tints of en, moderate anisotropism of ccp and anomalous anisotropism of py, distinctly perceived in oil with slightly uncrossed polars

even violet, especially if the analyser is slightly uncrossed (Figs. 46.1, 46.3 and 46.6). Extinction straight but not always complete.

No *internal reflections,* but sometimes cited.

Unlike in luzonite, the observable *twinning* at hand specimen scale is not usually seen under the microscope in en.

46.2.3 Paragenesis

Mineral usually of hypogene hydrothermal genesis, for example in *copper porphyries*. As main mineral, Ramdohr (1980) considers en restricted to Fe-poor Cu–As hydrothermal *paragenesis.*

Fig. 46.4 Aggregate of en, cv and py. RL24 ETSIME, Bor, Serbia (**dsp**). Grey coloured en next to blue cv

Fig. 46.5 Holocrystalline mass of en, slightly pleochroic (grey to brownish-grey), with traces of py +td (intergranular films or interstitial) and of ccp (microcracks and pores). CR 283, Sample 251102_1, Peru (**dsp**)

Typical of epithermal Au (Ag) and Au–Cu (Ag) deposits, of the high sulphidation type, accompanied by pyrite and bornite or pyrite and covellite. Its presence makes it possible to distinguish this type from those of low (or intermediate) sulphidation.

It is usually an early mineral, often brecciated and/or altered to tennantite (Figs. 46.1a, b and 46.2). However, it has also been cited as supergene.

Accompanying ores: tennantite, luzonite, pyrite, bornite, covellite, digenite, chalcocite.

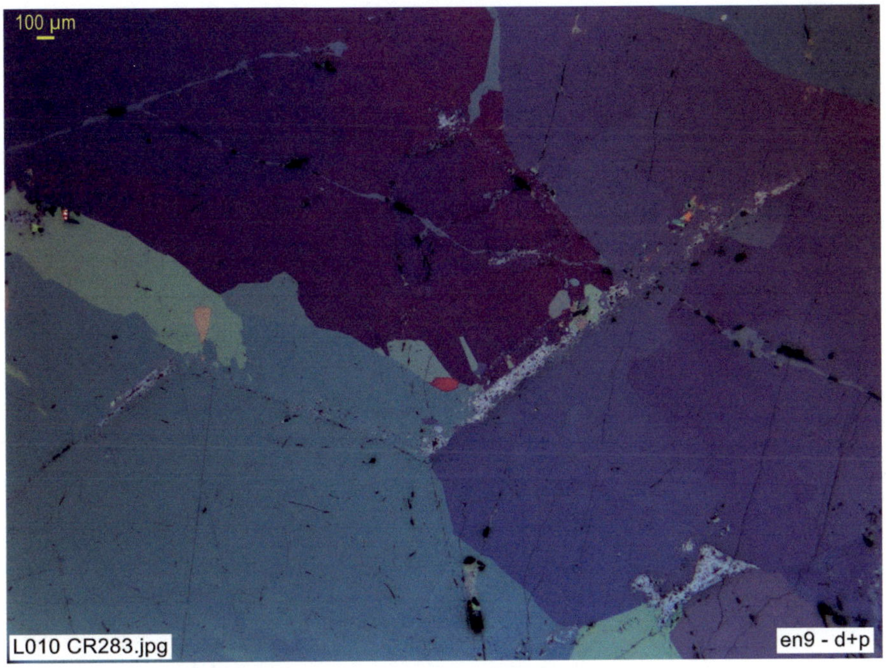

Fig. 46.6 Same scene as en8 (**d+p**): strong and coloured anisotropism of en

46.2.4 Identification

Enargite is recognized by its colour, by its moderate pleochroism and low reflectance and its distinct and coloured anisotropism. Paragenesis is an additional criterion.

It is distinguished from luzonite by the constant presence of polysynthetic twins in lz.

46.3 Spectral Information

46.3.1 Enargite Multispectral Reflectance R (in Air, Without Polariser) (%)

λ (nm)	370	400	425	450	475	500	525	550	575	600	625	650	675	700	750	800	850	900	950	1000
μ	27.6	27.4	27.0	26.9	26.8	26.5	26.3	26.0	26.0	26.2	26.4	26.7	26.9	27.0	26.8	26.8	27.2	27.1	26.1	25.4
P_5	25.0	25.6	25.4	25.7	25.5	25.5	25.1	25.0	24.9	25.1	25.4	25.7	26.1	26.2	26.2	26.1	26.5	26.5	25.6	24.8
P_{20}	26.7	26.5	26.3	26.3	26.2	26.0	25.7	25.5	25.4	25.6	25.9	26.2	26.5	26.6	26.5	26.5	26.8	26.9	25.9	25.1
P_{80}	28.7	28.4	27.9	27.5	27.4	27.1	26.9	26.6	26.7	26.7	26.9	27.1	27.3	27.3	27.0	27.1	27.6	27.4	26.4	25.6
P_{95}	29.5	29.2	28.5	28.1	27.8	27.5	27.4	27.0	27.1	27.1	27.3	27.5	27.5	27.6	27.2	27.5	27.9	27.6	26.6	25.8

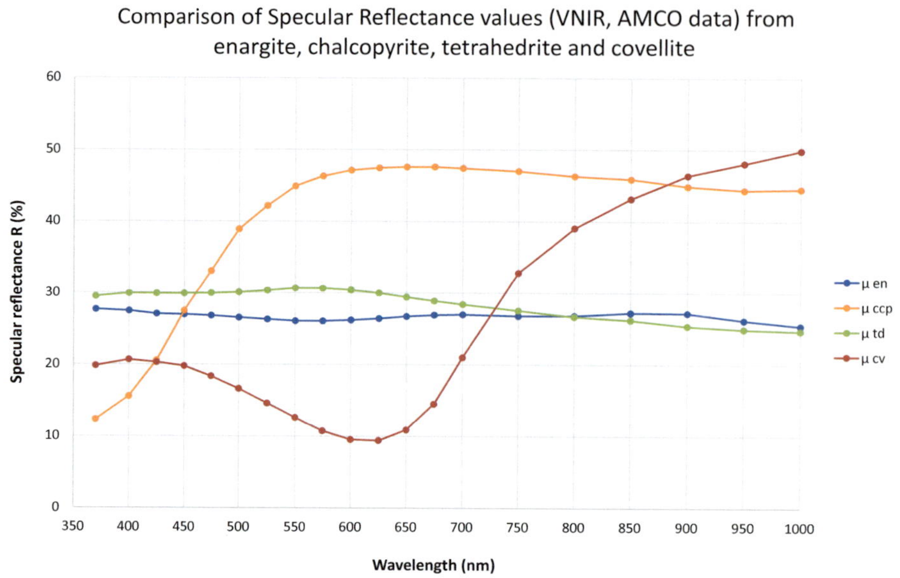

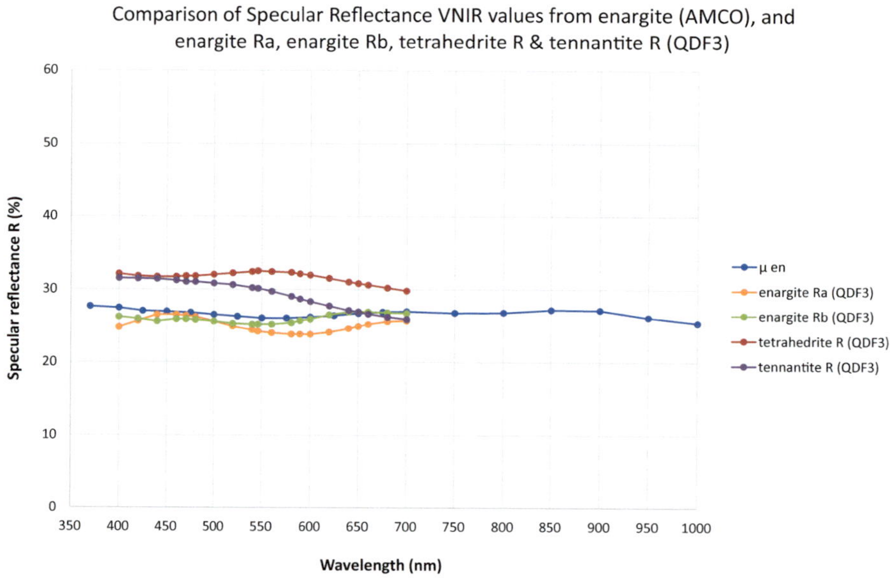

Reference

Ramdohr P (1980) The ore minerals and their intergrowths, 2nd edn, vol 2. Pergamon Press, Oxford, 1205p

Erlichmanite (*erl*); Laurite (*lrt*)

47

OsS_2: Cubic

Erlichmanite	R_{546} (%)	Y (%)	λ_d (nm)	P_e (%)	Colour sensation
Air	42.1	41.8	482	2.9	White–grey to bluish grey
Oil	25.1	25.8	481	4.7	
Bir: Null (1)			*Pleoc*: No (1)		*Tarn*: No
Anis +P: Isotropic (1)			*IR*: No		*Twg*: Not observed
Mor: Subhedral tendency (cub, octdr …)			*Clv*: nd		*PolQ*: Difficult (high hardness)
Hardness	PolH: Very high (5)		Mohs: nd	Talm: nd	VHN $_{(50\ g)}$: 1358

Note: Optical parameters according to QDF3. VHN according to Cabri (2002)
Mineral abbreviations: brg—braggite, chr—chromite, cpr—cooperite, erl—erlichmanite, gg—gangue, Ir—native iridium, lrt—laurite, pn—pentlandite, po—pyrrhotite, Os—native osmium, Pt—native platinum, py—pyrite, spy—sperrylite

Laurite (*lrt*)			RuS_2		Cubic
Laurite	R_{546} (%)	Y (%)	λ_d (nm)	P_e (%)	Colour sensation
Air	42.6	42.3	477	6.5	White to slightly bluish grey
Oil	26.6	26.3	477	9.9	
Bir: Null (1)			*Pleoc*: No (1)		*Tarn*: No
Anis +P: Isotropic (1)			*IR*: No		*Twg*: Not observed
Mor: Subhedral tendency (cub, octdr …)			*Clv*: (001)		*PolQ*: Difficult (very high hardness)
Hardness	*PolH*: Extremely high (5)		*Mohs*: 7	*Talm*: nd	*VHN* $_{(100\ g)}$: 1650^+, *VHN*$_{(25\ g)}$: 2870

Note: Optical parameters according to QDF3. VHN according to Cabri (2002)
Mineral abbreviations: chr—chromite, cpr—cooperite, erl—erlichmanite, gg—gangue, Ir—native iridium, lrt—laurite, PGM—platinum group minerals, pn—pentlandite, po—pyrrhotite, Os—native osmium, Pt—native platinum, py—pyrite, spy—sperrylite

47.1 General

Erlichmanite, *erl*, and laurite, *lrt*, are two isostructural species belonging to the pyrite group, defining the isomorphic cubic $(Os,Ru)S_2$ series. As typically occurs in the 109 **PGM** (*Platinum Group Minerals*) credited by IMA (*cf.* comments on PGE and PGM in *Pt description* and more data in Cabri, ed. 2002), several of the remaining four **PGE** (Platinum Group Elements) may be present in their lattice, in addition to Os and Ru.

Erlichmanite, *erl*, and laurite, *lrt*, form a complete solid-solution series between OsS_2 (erl) and RuS_2 (lrt). As minor or trace elements Rh, Ir, Pt, Pd, Ni, Cu, As, Te, Bi, Sb, Pb have been cited.

R. Castroviejo, *A Practical Guide to Ore Microscopy—Volume 1*,
https://doi.org/10.1007/978-3-031-12654-3_47

Fig. 47.1 **a, b** (*detail of bottom right-hand corner of Fig.* 47.2: **osp** and **o+p**, *respectively*). PGM-chr ore, mainly Pt, with microcrack infill of erl-lrt (bluish grey, isotropic) and another mineral also grey and of similar reflectance, but slightly pleochroic (bluish to very faint violet-brown shades) and distinctly anisotropic. Possibly *cpr* (cooperite, PtS) or *brg* (braggite, ‹Pt, Pd›S) Tabular (well visible) and thin, lamellar (very fine to submicroscopic size) inclusions of Os (bluish-white, relief, clearly anisotropic with characteristic orange tones) distributed throughout the section, one of them as part of the microcrack infill, very thin but distinctly visible. EMP analyses confirm erl-lrt as (**Os**,Ru)S_2 near the erl term. **c, d** (*detail of Fig.* 47.2, *center*: **osp** and **o+p**, *respectively*). Pt (white) with chr (dark grey, isotropic) and silicate gg (black, massive, light green IR). Fine-sized erl (+cpr/brg?) and Os inclusions in Pt-chr contact. Same ore and description as Figs. 47.1a, b and 47.2

Although erlichmanite and laurite are each recognized as a species, their optical properties are very similar and, given their frequent association in intermediate terms of the isomorphic series, a joint description has been chosen since, in order to determine the precise composition, specific analytical techniques (electron microprobe) must be used.

47.2 Microscopic Description

47.2.1 SP Observation (*Single Polariser*)

White *colour*; greyish to bluish-grey next to Pt (Figs. 47.1a and c). Moderate *reflectance compared* to usual

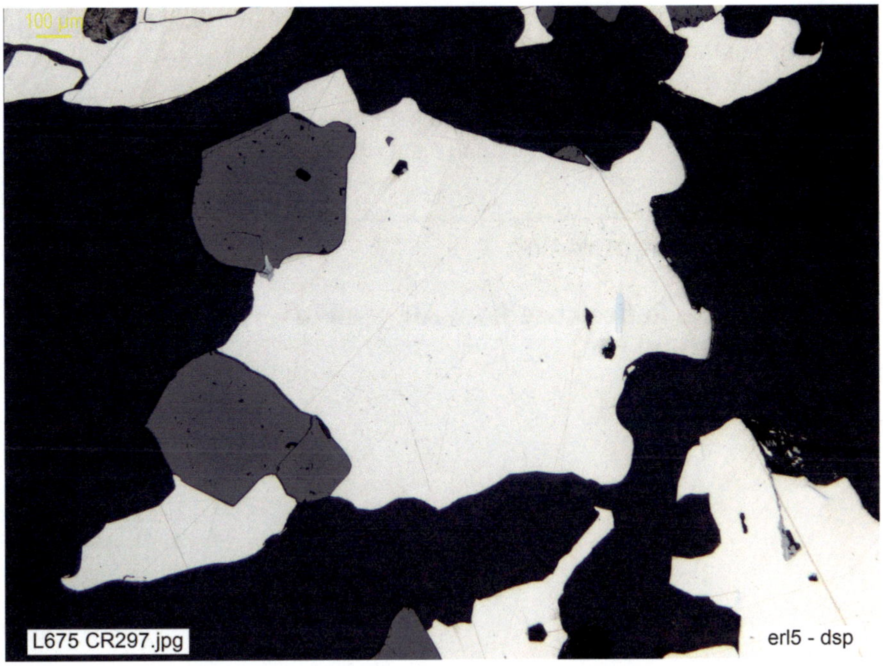

Fig. 47.2 PGE-Cr ore: Pt (white, main ore) intergrown with chr (grey) and gg (very dark grey). Inclusions in Pt: Os (bluish white, tabular or lamellar) and erl-cpr-brg? (light grey), discussed in detail in Fig. 47.1a, b (lower right field in Fig. 47.2, *inclusions associated with microcracks in Pt grain*) and in Fig. 47.1c, d (left edge of central Pt grain in Fig. 47.2 *contact with chr*). CR 297, Fox Gulch, Goodnews Bay, Alaska (**dsp**)

companions, lower than sperrylite or pyrite, distinctly lower than Pt, Os or Ir, making the colour of erlichmanite or laurite greyish by contrast, occasionally with a slight bluish tinge, but much lighter than chromite (Figs. 47.1c and 47.2). Isotropic.

Difficult to *polish* because of its hardness. Very high *polishing hardness*. Laurite is, together with eskolaite, the hardest of the metallic minerals (Picot and Johan 1982); therefore, compared to Pt it shows fewer polishing scratches, as can be seen in Figs. 47.1a–d (many scratches disappear when passing from Pt to erl-lrt). Very high *Vickers microhardness*, varying between the values $VHN_{50} = 1358$ kg/mm^2 (erl) and $VHN_{100} = 1650$ to $VHN_{25} = 2870$ kg/mm^2 (lrt), of the terms erlichmanite and laurite, respectively (Cabri 2002).

They occur in tiny inclusions with subhedral tendency (cubic and octahedral forms, among others) in Pt or Fe–Pt alloys, sometimes also in chromite and others.

47.2.2 Observation +P (*Polariser + Analyser at 90°*)

Isotropic (Figs. 47.1b, c).

47.2.3 Paragenesis

Chromium-nickel and PGE deposits in ultramafic intrusive rocks of the Alaskan type, stratified complexes, ophiolites and in derived detrital deposits.

47.2.4 Identification

White to greyish colour (depending on the neighbouring minerals), the relatively low reflectance among PGMs, the isotropic character, the extremely high hardness (especially of laurite) and the paragenesis are the most useful properties for laurite-erlichmanite recognition.

Companion ores: sperrylite, Pt or Fe–Pt alloys, Os, cooperite, braggite, pentlandite, chromite and other PGMs.

47.3 Spectral Information

47.3.1 Laurite Reflectance R (in Air, Visible Spectrum) (%)

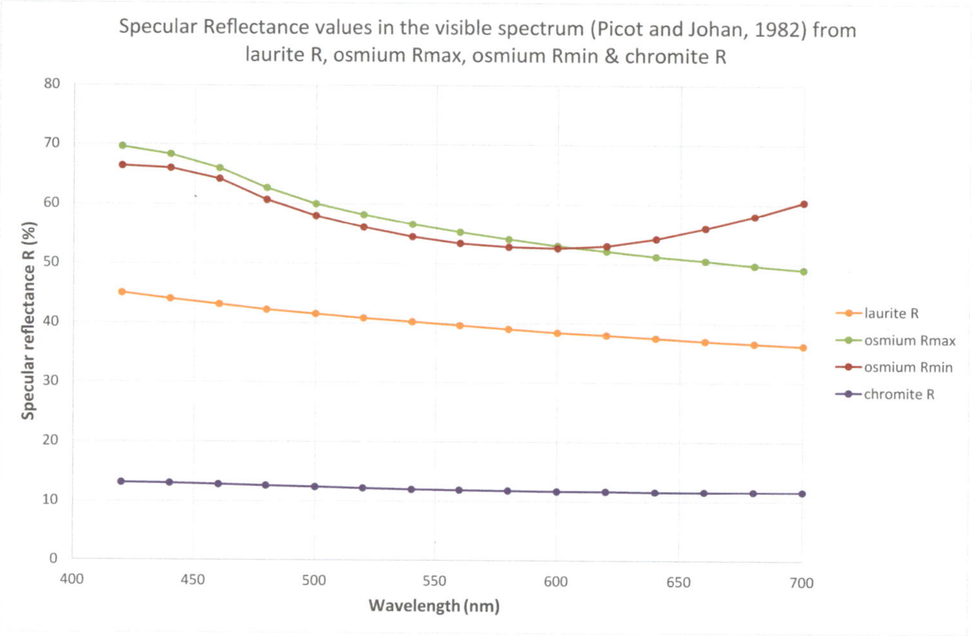

References

Cabri L (2002) The platinum-group minerals, pp 13–129. In: Cabri L (ed) CIM special, vol 54. Canadian Institute of Mining, Metallurgy and Petroleum, Calgary, Alberta, Canada, p 852

Cabri LJ (ed) (2002) The geology, geochemistry, mineralogy and mineral beneficiation of platinum group minerals. Canadian Institute of Mining, Metallurgy and Petroleum, Calgary, Alberta, Canada, p 852. ISBN 1-894495-27-5

Picot P, Johan Z (1982) Atlas of ore minerals. BRGM/Elsevier

Famatinite (*fm/stibioluzonite*) 48

Cu_3SbS_4: Tetragonal

Famatinite		R_{546} (%)	Y (%)	λ_d (nm)	P_e(%)	Colour sensation
Air	R_1	23.8	nd	nd	nd	GyBr with tones RdBr to ViBr
	R_2	25.0	nd	nd	nd	
Oil	R_1	nd	nd	nd	nd	GyBr with tones *ViBr* to RdBr
	R_2	nd	nd	nd	nd	
Bir: Weak (3)				*Pleoc*: Visible (3)		*Tarn*: nd
Anis+P: Strong (4)				*IR*: No		*Twg*: Very frequent, polysynthetic
Mor: Massive aggregates				*Clv*: Occasional		*PolQ*: Good
Hardness	*PolH*: Mean: ≈ en, > ccp (3)			*Mohs*: 3.5	*Talm*: D	*VHN* $_{(100\ g)}$: 205–397

Note No QDF3 multispectral data. R1 and R2 are Rmin and Rmax from Picot and Johan (1982) and the values used for the R-λ plot are also those of these authors. VHN according to Uytenbogaardt and Burke (1971) data for *lz-stibioluzonite*

Mineral abbreviations: Au—native gold, ccp—chalcopyrite, en—enargite, fm—famatinite, lz—luzonite, po—pyrrhotite, py—pyrite, sp—sphalerite, tn—tennantite, td—tetrahedrite

Colour abbreviations: Br—brown, Gn—green, GyBr—greyish brown, RdBr—reddish brown, ViBr—violetbrown, Yl—yellow, YlBr—yellowish brown

48.1 General

Mineral recognized as a species by IMA/COM, with the formula Cu_2CuSbS_4. Its definition had been complicated in the past by published confusing or contradictory information.

Composition Cu_3SbS_4. Occasional minor contents of Ag, Bi, Fe, Te, As, Sn, Se. May contain Se, up to ≈ 24%, in the *seleniferous famatinite* variety.

Higher amounts of As correspond to Ramdohr (1982) *stibioluzonite,* although this would now be considered rather an As-rich variety of famatinite, while Strunz's stibioluzonite is equivalent to famatinite.

Variable As/Sb ratios, $Cu_3(As, Sb)S_4$, define the *fm-lz series*, formerly luzonite-stibioluzonite of Uytenbogaardt and Burke (1971) and Ramdohr (1982). More information in Paar et al (2016) and *mindat* (https://www.mindat.org/min-1451.html).

48.2 Microscopic Description

48.2.1 SP Observation (*Single Polariser*)

Brownish-grey *colour*, with pinkish to violet tones, more accentuated in immersion (Figs. 48.1a, b). Slightly pleochroic. Good *polishing*. Medium *polishing hardness*, $PolH_{fm}$ ≈ en: ≥ td-tn, Au, ccp; <sp, po, py.

Moderate *reflectance*, relatively low compared to other sulphides and lower than fahlores (or tetrahedrite-tennantite, Fig. 48.1). Weak bireflectance and perceptible pleochroism,

R. Castroviejo, *A Practical Guide to Ore Microscopy—Volume 1*,
https://doi.org/10.1007/978-3-031-12654-3_48

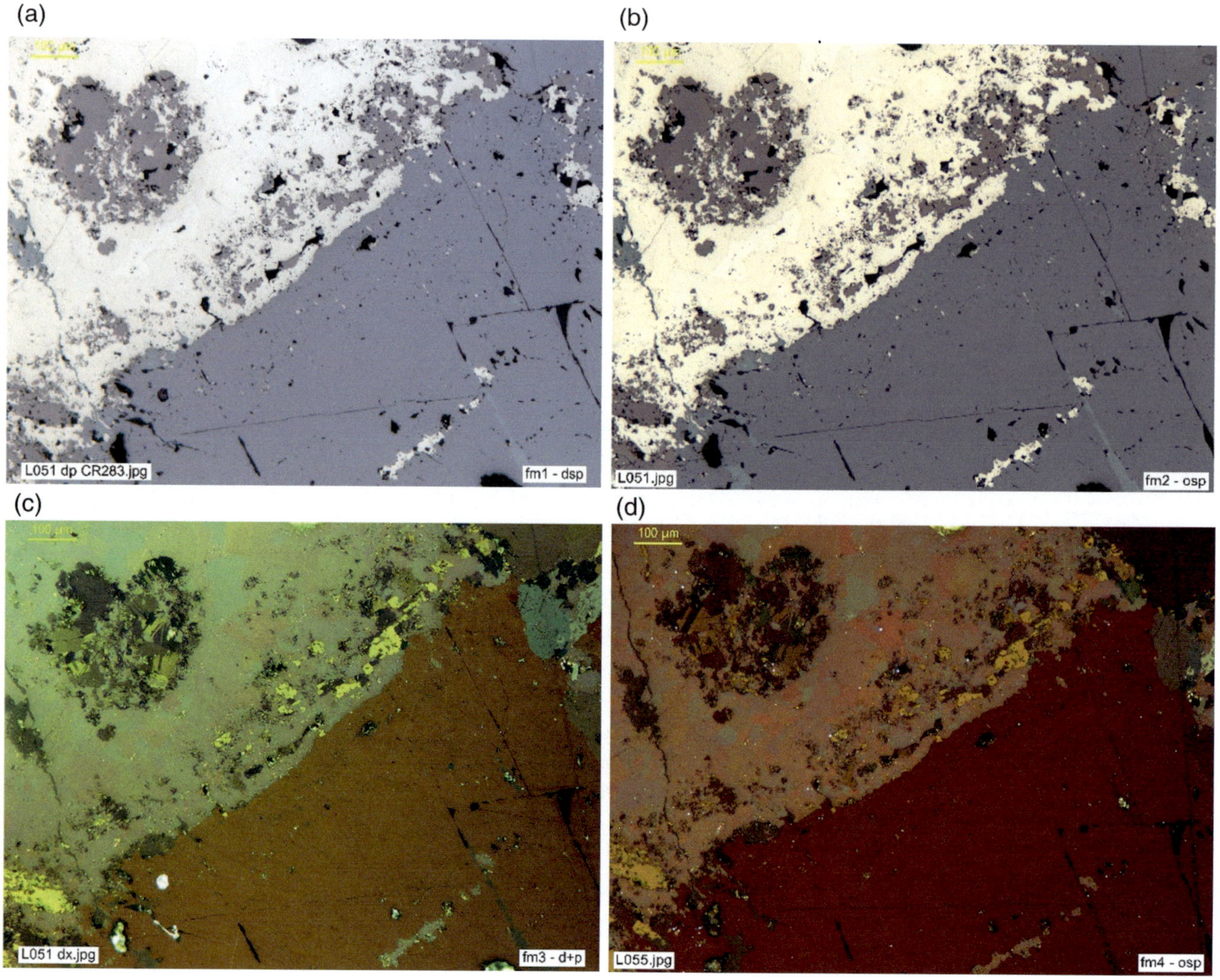

Fig. 48.1 **a** Aggregate of *py* (yellow, faintly zoned, upper left half of image) with inclusions of *fm* (pinkish brown to purplish brown), partially replaced by massive *en* (purplish grey, lower right half of image) and by scarce late *tn* (filling microcracks, greenish grey, lighter than *en*). Sample CR 283, Peru 251102_1, (**dsp**). **b** (Same scene as Fig. 48.1a, osp). Immersion enhances py zoning and reflectance differences (R py > tn > fm, en). **c**, **d** (Same scene as Fig. 48.1a, **d+p** and **o+p**, respectively). Frequent polysynthetic twins and strong anisotropism of *fm*, in YlBr, Yl and Gn tones, unlike the duller tones of *en* and the (exceptional) weak anisotropism of *py*

although moderate, as can be expected from its spectral curves (cf graph of reflectance values).

It is often found included in tetrahedrite-tennantite, disseminated or in aggregates of varied morphology and grain size, massive, granular, concentric, botryoidal.

48.2.2 Observation +P (*Polariser + Analyser at 90°*)

Strongly *anisotropic*, with bright and coloured tones—yellowish-brown, yellowish and greenish (Figs. 48.1c, d). Very frequent and characteristic *polysynthetic twinning*, with orthogonal directions.

48.2.3 Paragenesis

Paragenesis more restricted than enargite and luzonite, typical of epithermal deposits, with pyrite, enargite, chalcopyrite, tetrahedrite-tennantite, native gold, tellurides.

48.2.4 Identification

Typical for famatinite are its colour, anisotropism and twinning, as well as the paragenesis.

Famatinite resembles luzonite but famatinite is more purplish, not orange, shows more yellowish, less greenish anisotropism colours and *two* sets of intersecting, orthogonal polysynthetic twinsThe twins and the vividly coloured anisotropism of famatinite distinguish it from enargite.

48.3 Spectral Information

48.3.1 Famatinite Multispectral Reflectance R (in Air) (%)

References

Picot P, Johan Z (1982) In: Atlas of ore minerals. B.R.G.M./Elsevier

Uytenbogaardt W, Burke EAJ (1971) Tables for microscopic identification of ore minerals, 2nd edn. Elsevier, Amsterdam

Paar WH, Brodtkorb MK, Putz H, Martin RF (2016) In: Atlas of ore minerals: focus on epithermal deposits of Argentina. Canadian Mineralogist, Spec. Pub. 11, Quebec, pp 402

Franklinite (*fk, [fkl]*)

49

$Zn^{2+}Fe^{3+}{}_2O_4$: Cubic

Franklinite		R_{546} (%)	Y (%)	λ_d (nm)	P_e (%)	Colour sensation
Air		18.6	18.4	485	1.7	Light grey
Oil		6.2	6.1	485	3.1	
Bir: Null (1)				*Pleoc*: Null (1)		*Tarn*: No
Anis +P: Isotropic (1)				*IR*: Scarce, red		*Twg*: (111)
Mor: Euhedral tendency (octahedra)				*Clv*: Distinct, octahedral		*PolQ*: Good
Hardness	*PolH*: High, ≈ mt (4)			*Mohs*: 5½–6	*Talm*: F^+	*VHN* $_{(100\ g)}$: 852–882

Note: cf. R charts (% specular reflectance, in the visible range), according to QDF3 data
Mineral abbreviations: bra—braunite, fk (*fkl*)—franklinite, gg—gangue, gth—goethite, ht—hematite, jac—jacobsite, lim—limonite, mt—magnetite, prl—pyrolusite, psi—psilomelane, sp—sphalerite, spn—spinel

49.1 General

Franklinite, *fk*, is a typical ferritic spinel, like magnetite or jacobsite among others. All of them are characterized by forming solid solution at high temperature and, in addition, by very similar physical properties. Franklinite's description could be replaced in many aspects by that of the most common species, magnetite, to which reference will be made in due course to avoid repetition.

Composition: $Zn^{2+}Fe^{3+}{}_2O_4$; taking into account the possibilities of solid dissolution, it could also be formulated as (Zn, Mn^{2+}, Fe^{2+}) (Fe^{3+}, Mn^{3+})$_2O_4$ (Back and Mandarino 2008). Possible minor contents of Ti, Al, Ca, Mg. As explained when treating spinel and magnetite (cf. both descriptions), it belongs to the Spinel Subgroup, which is in the Oxyspinel Group within the Spinel Supergroup. Isostructural with magnetite.

49.2 Microscopic Description

49.2.1 SP Observation (*Single Polariser*)

Colour light grey (Figs. 49.1a, b and 49.2a), good *polish*. Moderate *reflectance* (R_{fk} < ht; R_{fk} slightly < mt, jac; $R_{fk} \geq$ sp; R_{fk} distinctly > gth, spn, gg; cf. comparative charts). Isotropic. Occasional octahedral cleavage visible.

It occurs in idiomorphic crystals, often with rounded octahedral forms and in granoblastic aggregates. Sometimes with magnetite // (111) exsolutions which, according to Ramdohr (1980), are almost always found in the so-called "magnetic franklinite".

49.2.2 Observation +P (*Polariser + Analyser at 90°*)

Isotropic, red *internal reflections,* not abundant, but usually visible, at least in immersion (Figs. 49.1c, d and 49.2b). *Twinning* (111) sometimes visible.

49.2.3 Paragenesis

Metamorphic *parageneses*, whose protoliths are supposed to be Fe−, Mn− and Zn-rich carbonate sedimentary pre-concentrations. It has been found only in the Franklin Mining District (Franklin, Trotter, and Sterling mines), Sussex County, N. Jersey, USA.

R. Castroviejo, *A Practical Guide to Ore Microscopy—Volume 1*,
https://doi.org/10.1007/978-3-031-12654-3_49

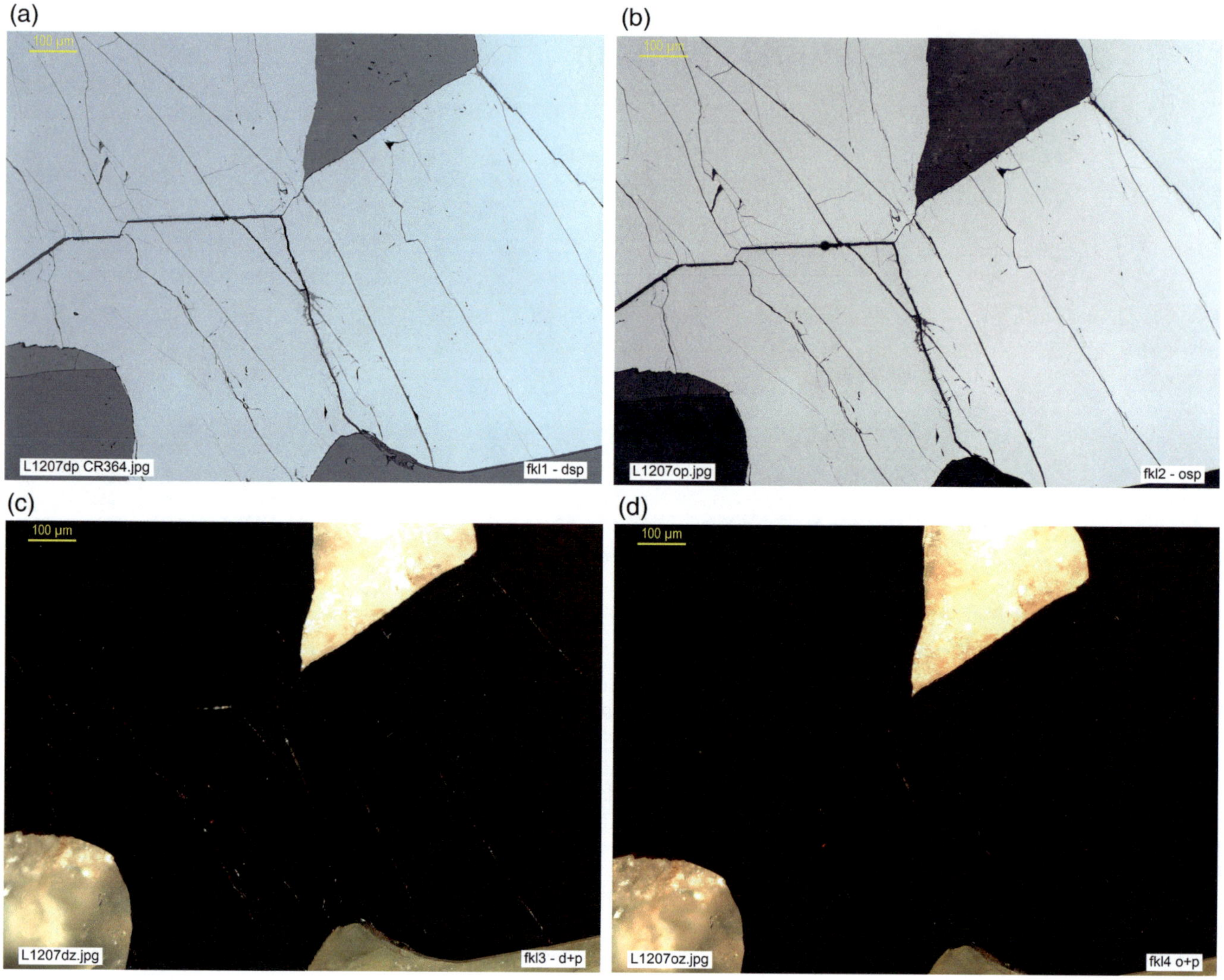

Fig. 49.1 **a**, **b** (**dsp** and **osp**, respectively)—Granoblastic aggregate of fk (light gray) and silicates. CR364, Franklin Furnace, New Jersey, USA. **c**, **d** (Same scene as fkl: **d+p** and **o+p**, respectively). The extinction of fk (isotropic), facilitates the observation of its red IR, visible especially in immersion (Fig. 49.1b)

Companions: zincite, andradite, calcite, willemite, magnetite, spinel, jacobsite, sphalerite.

49.2.4 Identification

Paragenesis is an important criterion, in particular the *association with zincite* (ZnO, dark gray, striking for its luminous and massive red–orange IR, visible even with no analyser), as well as metamorphic textures. Possible confusions:

- fk is very similar to mt, from which it is distinguished by lacking the brownish hue usually seen in mt and by having IR, which mt lacks;
- also looks very much like jac, which is slightly lighter and yellowish;
- confusion with bra can be ruled out, because bra is anisotropic.

Fig. 49.2 **a**, **b** (**dsp** and **d+p**, respectively). Metamorphic texture of fk ore with silicate gg. Smooth contacts and polygonal contours, with triple junctions. Weak anomalous anisotropism of some of the fk grains (Fig. 49.2b, center left). CR364, Franklin Furnace, New Jersey, USA

(a)

(b)

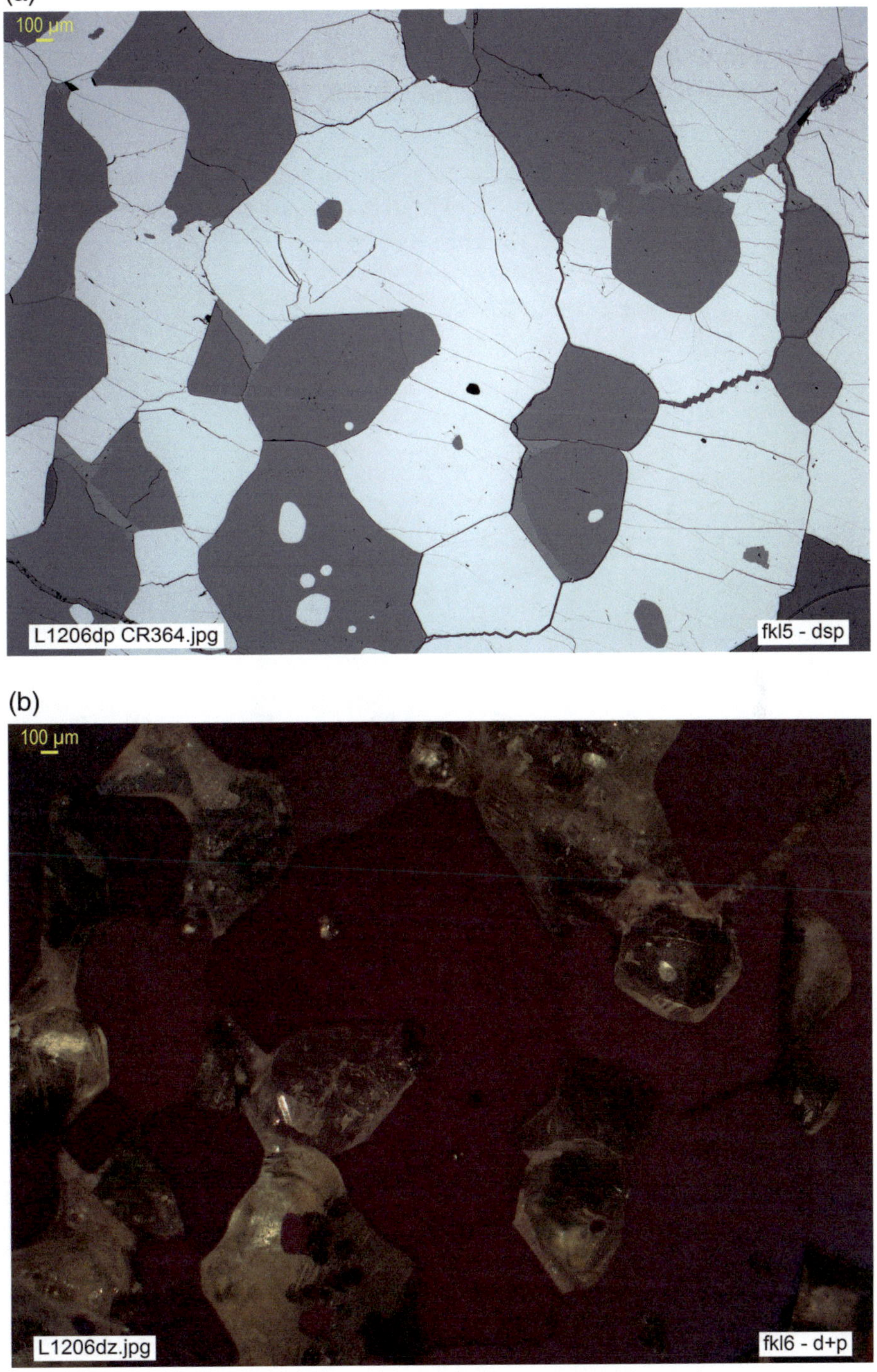

49.3 Spectral Information

49.3.1 Franklinite Reflectance R According to QDF3 (in Air, Visible Spectrum) (%)

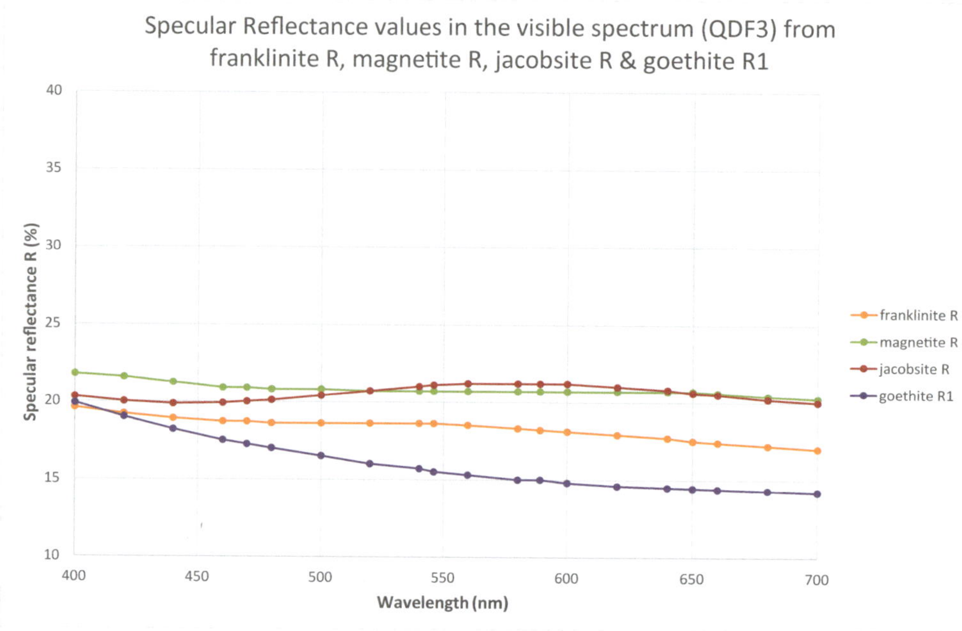

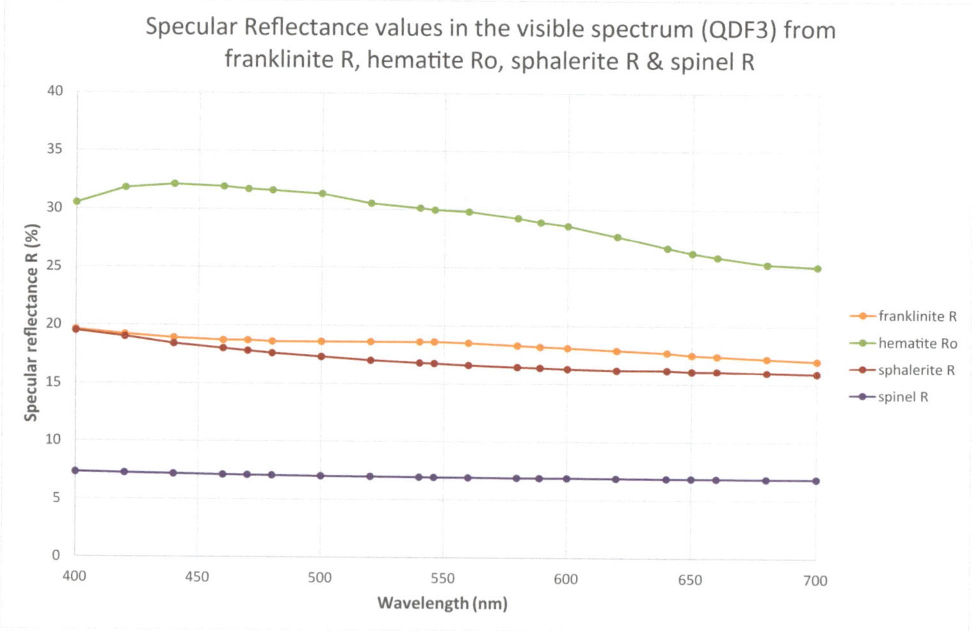

References

Back ME, Mandarino JA (2008) Fleischer's glossary of mineral species 2008. The Mineralogical Record Inc. Tucson, AZ, p 345

Ramdohr P (1980) The ore minerals and their intergrowths, vol 2, 2nd ed. Pergamon Press, Oxford, p 1205

Freieslebenite (*flb*)

50

$PbAgSbS_3$: Monoclinic

Freieslebenite		R_{546} (%)	Y (%)	λ_d (nm)	P_e (%)	Colour sensation
Air	R_1	35.6	34.5	480	3.4	Greyish white
	R_2	39.5	39.4	480	3.1	
Oil	R_1	20.8	20.7	478	6.5	
	R_2	24.4	24.3	478	6.1	
Bir: Weak/very weak (3)				*Pleoc*: Weak/very weak (3)		*Tarn*: Not observed
Anis + *P*: Distinct and coloured (4)				*IR*: No		*Twg*: Polysynthetic (not always)
Mor: Prismatic crystals and aggregates				*Clv*: (001), (110)		*PolQ*: Good, but scratched
Hardness	*PolH*: Low, << gn (2)			*Mohs*: $2^1/_2$	*Talm*: nd	*VHN* $_{(20\ g)}$: 85–140

Obs: cf. Table and R charts, multispectral data (VNIR, AMCO) and visible range (QDF3)
Mineral abbreviations: bnn—bournonite, Carb—carbonates, fbg—freibergite, flb—freieslebenite, gg—gangue, gn—galena, my—miargyrite, pc—pearceite, plb—polybasite, prg—pyrargyrite, pru—proustite, q—quartz, sbt—stibnite, td—tetrahedrite

50.1 General

Rare Ag ore. Composition: $PbAgSbS_3$, with possible traces of Cu and Fe.

50.2 Microscopic Description

50.2.1 SP Observation (*Single Polariser*)

Colour greyish white (Fig. 50.1), very weakly pleochroic towards greenish-brownish shades (pleochroism can be best seen by comparing contiguous grains with different orientations, especially when immersed).

Medium *reflectance* (higher range of the group), between galena and tetrahedrite-freibergite, >prg or my (Figs. 50.1 and 50.2a, cf multispectral comparison charts).

It occurs in aggregates and, under favourable conditions, in prismatic or tabular crystals.

50.2.2 Observation +P (*Polariser + Analyser at 90°*)

Distinct and coloured *anisotropism*, with varied colour effects, the reddish to bluish purple tints being typical but not exclusive (comp. Figures 50.1c, d, 50.2b and 50.3. The tones vary according to the setting—observation in air or in immersion, also according to the magnification of the objective).

No *internal reflections*. Polysynthetic *twinning* visible on some grains (not all).

50.2.3 Paragenesis

Rare mineral, found as a late phase in some hydrothermal silver deposits, such as Hiendelaencina (Spain) and Freiberg (Germany), accompanied by *ruby silvers* (proustite, pyrargyrite), miargyrite, tetrahedrite-freibergite and other silver sulphosalts, as well as galena.

R. Castroviejo, *A Practical Guide to Ore Microscopy—Volume 1*,
https://doi.org/10.1007/978-3-031-12654-3_50

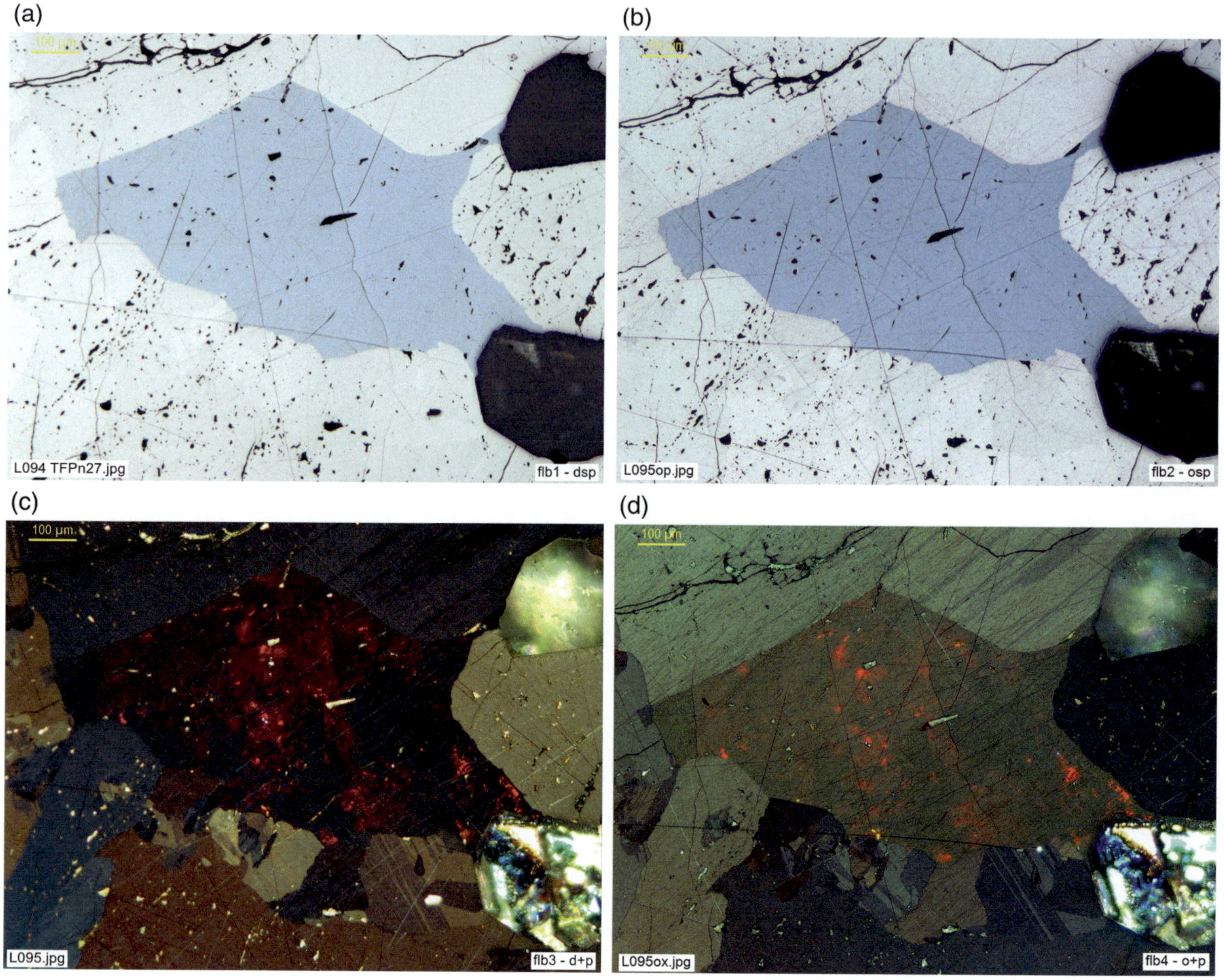

Fig. 50.1 **a**, **b** (**dsp** and **osp**, respectively)—Aggregate of flb (greyish white) with inclusions of prg (bluish grey) and quartz (dark grey). TFPn27, Hiendelaencina, Guadalajara. **c**, **d** (Same scene as flb1: **d+p** and **o+p**, respectively). Coloured anisotropism of flb, which in some grains shows polysynthetic twins; prg, also anisotropic, is characterized by very abundant, red IR

50.2.4 Identification

From the point of view of reflected light optical microscopy, freieslebenite is poorly documented in the literature and the published data are not always compatible (e.g., according to Picot and Johan 1982, freieslebenite is never twinned, but according to all other sources consulted it is usually twinned and, in the available material from Hiendelaencina, polysynthetic twins are well visible, Figs. 50.1c, d and 50.2).

Freieslebenite can be characterized by its low hardness (many polishing scratches), weak pleochroism, occasional polysynthetic twinning and, above all, its distinctly coloured anisotropism.

Its anisotropism clearly distinguishes flb from minerals like antimonite or bournonite (anisotropic) and fahlores (tetrahedrite, freibergite, isotropic).

The paragenesis is an important data and the absence of internal reflections differentiates it from other silver sulphosalts such as proustite, pyrargyrite, miargyrite, pearceite, polybasite.

Fig. 50.2 **a**, **b** (**dsp** and **d+p**, respectively)—Hydrothermal infill of flb (light grey, clearly anisotropic in bluish and purplish-red tones), q and Carb (dark grey, massive light IR), with traces of td (grey, upper centr border of flb, in contact with q) and my (slightly bluish grey, small inclusions in flb and in q). TFPn27, Hiendelaencina, Guadalajara

(a)

(b)

Fig. 50.3 Detail of flb6 (top center, **o+p**): flb anisotropic and twinned, td in extinction (cubic)

50.3 Spectral Information

50.3.1 Freieslebenite Reflectance R (in Air, Visible Spectrum) (%)

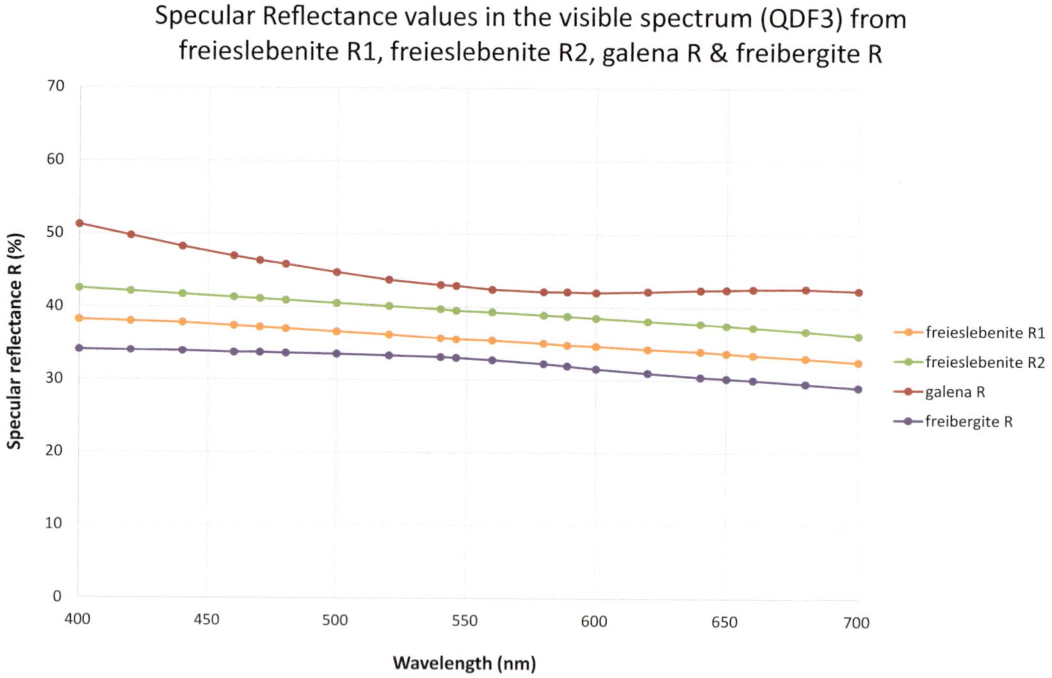

Reference

Picot P, Johan Z (1982) Atlas of ore minerals. B.R.G.M./Elsevier

Galena (*gn*)

PbS: Cubic

51

Galena		R_{546} (%)	Y (%)	λ_d (nm)	P_e (%)	Colour sensation
Air		42.9	43.0	472	4.9	White or light grey
Oil		27.9	28.0	471	8.0	
Bir: Null (1)				*Pleoc*: No (1)		*Tarn*: Darkens in hand sample
Anis +P: Null (1)				*IR*: No		*Twg*: Rarely visible
Mor: Allotriomorphic, massive, rare xx				*Clv*: // (100) perfect		*PolQ*: Imperfect (scratches, triang pits)
Hardness	*PolH*: Low-Middle limit (2)			*Mohs*: 2 1/2	*Talm*: B	*VHN* $_{(50\ g)}$: 61–74

Obs **:** cf. Table and R charts, multispectral data (VNIR, AMCO) and visible range (QDF3)
Mineral abbreviations: ac—acanthite, Ag—native silver, asp—arsenopyrite, Au—native gold, bm—bismuthinite, bnn—bournonite, cc—chalcocite, ccp—chalcopyrite, cin—cinnabar, cv—covellite, dg—digenite, gg—gangue, gf—gersdorffite, gn—galena, lo—löllingite, mc—marcasite, pc—pearceite, po—pyrrhotite, prg—pyrargyrite, pru—proustite, py—pyrite, q—quartz, Sb—native antimony, sbt—stibnite, sf—safflorite, sk—skutterudite, sp—sphalerite, td—tetrahedrite

51.1 General

Galena, the most important lead ore, is composed of PbS with possible impurities of Se, Te, Cu, Bi, Zn, Hg, Cd, among others, with the usual presence, as traces, of silver in solid solution (Ag_2S). The argentiferous or silver-rich galena is an ore of lead and silver.

In this case, the precious metal—for Ag contents > 0.1% or even lower, according to Ramdohr, 1980—is no longer found in solid solution but in small inclusions of native silver and/or argentiferous minerals.

The same is true for the high contents of other metals (Bi, Sb, As, Ni, etc.) cited in the literature which, according to Bowles et al. (2011), are explained by the presence of tiny inclusions of the corresponding sulphosalts.

51.2 Microscopic Description

51.2.1 SP Observation (*Single Polariser*)

Colour bright white to greyish white, with slightly variable colour shades depending on the neighbouring ores (comp. Figures 51.1a, b and 51.3).The perception of brightness in gn is greatly reduced facing highly reflective ores (native silver, native antimony, native gold, etc.), galena appearing in these cases with a very dull grey hue (Fig. 51.5).

Reflectance, R, relatively high, > td and ≫ sp, although < py (cf. multispectral comparative plots and Figs. 51.1a, b, 51.2, 51.3 and 51.4). Not pleochroic.

Due to its low *polishing hardness* (at the limit between the low and medium PolH groups, > cin, cv, sbt, ac; ≈ bm, cc, dg; < ccp, Ag, td, sp, po), it is not easy to achieve a perfect *polish*, resulting almost always in a scratched surface, lined with typical pull-outs (*triangular pits*, Fig. 51.2), which mirror the cubic structure of galena and its perfect // (100) cleavage.

Given its abundance and presence in diverse associations, it is often used as a reference or term of comparison of colour, reflectance, and hardness to describe other species. For example, the very varied argentiferous ores (Figs. 51.2 and 51.3) or the sulphosalts with As, Sb, Bi, Ag, Zn, Pb, etc. are frequently found as inclusions in galena.

It usually occurs in masses or allotriomorphic aggregates, although under favourable conditions it shows idiomorphic crystalline forms such as cubes and octahedrons. Zoned crystals are frequent, but the *zoning* is rarely visible under the microscope except with structural etching; however, it is sometimes revealed by weathering. Its low hardness and

R. Castroviejo, *A Practical Guide to Ore Microscopy—Volume 1*,
https://doi.org/10.1007/978-3-031-12654-3_51

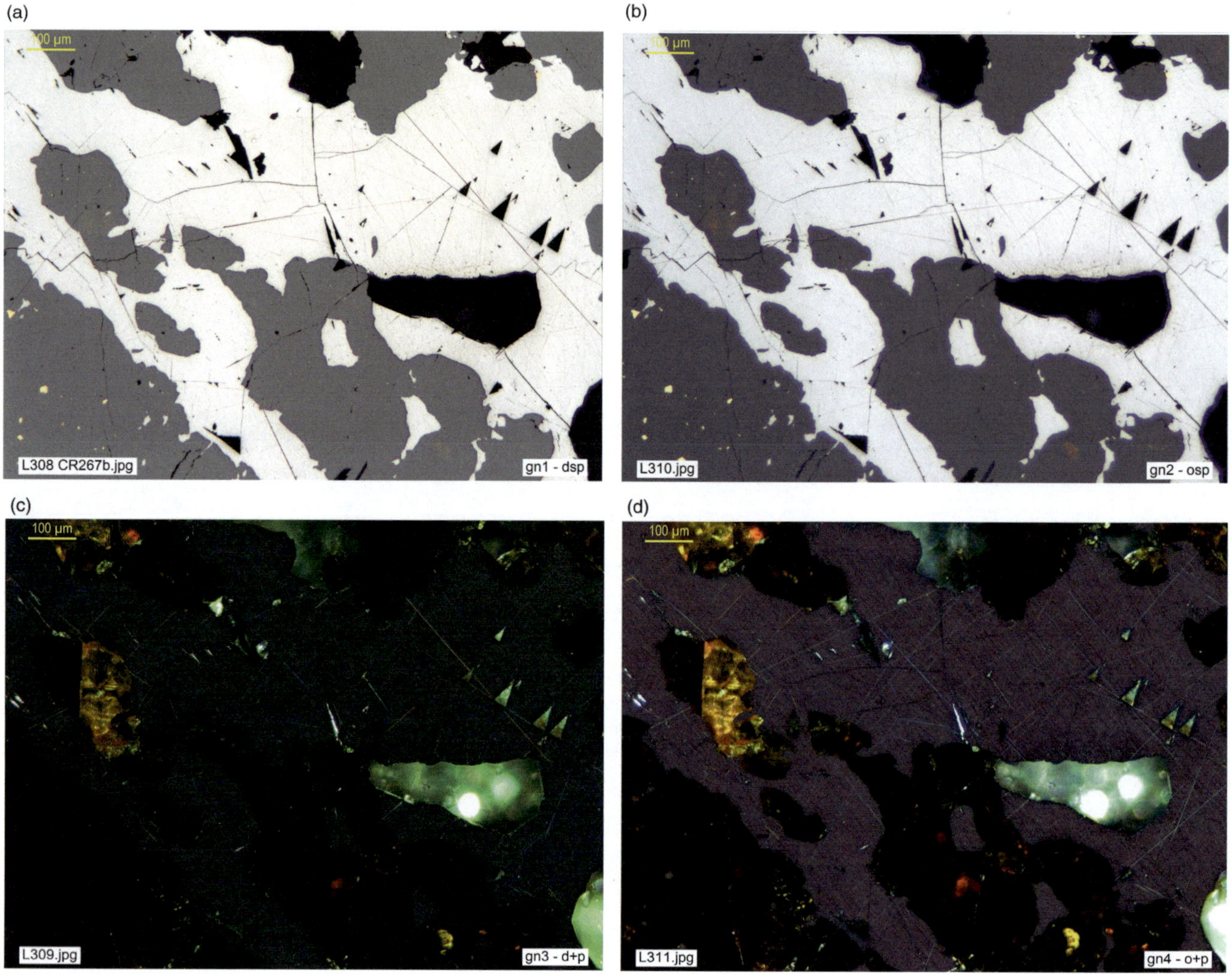

Fig. 51.1 **a**, **b** (**dsp** and **osp**, respectively)—Intergrowth of gn (white), sp (grey) and q (dark grey); polishing scratches and triangular pits in gn; small inclusions of ccp (yellow) and one of py (yellowish-white, relief) in sp and gn, respectively. CR 267 b, Bunker Hill Mine, Coeur d'Alene, Idaho, USA. **c**, **d** Same scene as gn1 (**d+p** and **o+p**, respectively). Extinction of gn and sp. IR of sp (yellowish-brown) and q (colourless to greenish), most abundant in immersion

plasticity facilitate its *deformation*, generally showing a typical *ductile* behaviour (Fig. 51.6).

51.2.2 Observation +P (*Polariser + Analyser at 90°*)

Isotropic (Fig. 51.1c, d), although cases of exceptional anisotropy have been reported (e.g., due to deformation) and, in extreme cases, very abundant polishing pits may give a false sense of anisotropy. Twinning is varied, though rarely visible, unless by etching or electrolytic attack.

51.2.3 Paragenesis

Galena occurs in a large number of deposits, and it could be considered almost ubiquitous as an accessory mineral. However, as an economic lead ore it is extracted from a restricted number of deposit types, essentially large tonnage hydrothermal orebodies. Typical are hydrothermal Pb–Zn–Ag seams (Coeur d'Alene, Idaho), various stratabound deposits in sedimentary (*Sedex* types, *MississippiValley*), volcano-sedimentary (massive sulphide) and metamorphic rocks (Broken Hill, Australia). It also occurs in some skarn formations.

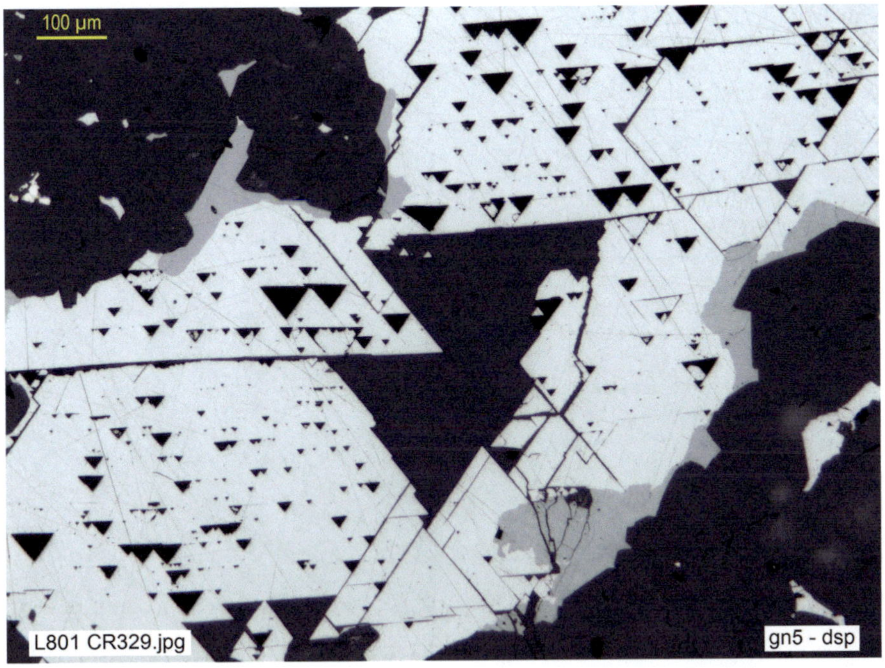

Fig. 51.2 Argentiferous gn ore: gn (greyish white) with thin rim of pru (light grey) and traces of py (in gg); numerous polishing pull-outs (*triangular pits*) evidence the cubic structure of gn. CR 329, Imiter, Morocco (**dsp**)

Fig. 51.3 Ore of the *Five Element Formation* type (Ni, Co, Bi, Ag, U): dominant gn (center), with inclusions of ccp (yellow); marginal coating and inclusions of prismatic pc (grey) intergrown with white arsenides (sf, lo, gf, sk). CR 328, Freiberg, Saxen, Germany (**osp**)

It is usually accompanied by sphalerite, pyrite, marcasite, chalcopyrite, native silver, tetrahedrite, bournonite, pyrargyrite and other sulphosalts, arsenopyrite, and pyrrhotite, among many other possible minerals, due to its ubiquity.

In the supergene oxidation zone, gn may be replaced by anglesite and cerussite, more rarely also pyromorphite or wulfenite, which sometimes form an insoluble layer over the original galena grains, thus protect them from weathering.

The total absence of gn in liquid-magmatic deposits is remarkable.

51.2.4 Identification

Galena is easily identified in a hand sample, as well as under the microscope, by its low hardness, abundant polishing

Fig. 51.4 Intergrowth of gn and sp, the latter with some inclusions of po (pinkish brown), ccp (yellow) and asp (associated with po inclusion, upper right) and scarce prg (lighter grey than sp, associated with po inclusion). Simple, smooth contacts, even with gg, suggesting equilibrium (metamorphic) paragenesis. The slight relief observed in the contacts indicates that the polishing hardnesses vary in the order asp > po > sp > ccp > gn. CR365, Zn Corp Mine, Broken Hill, NSW, Australia (**dsp**)

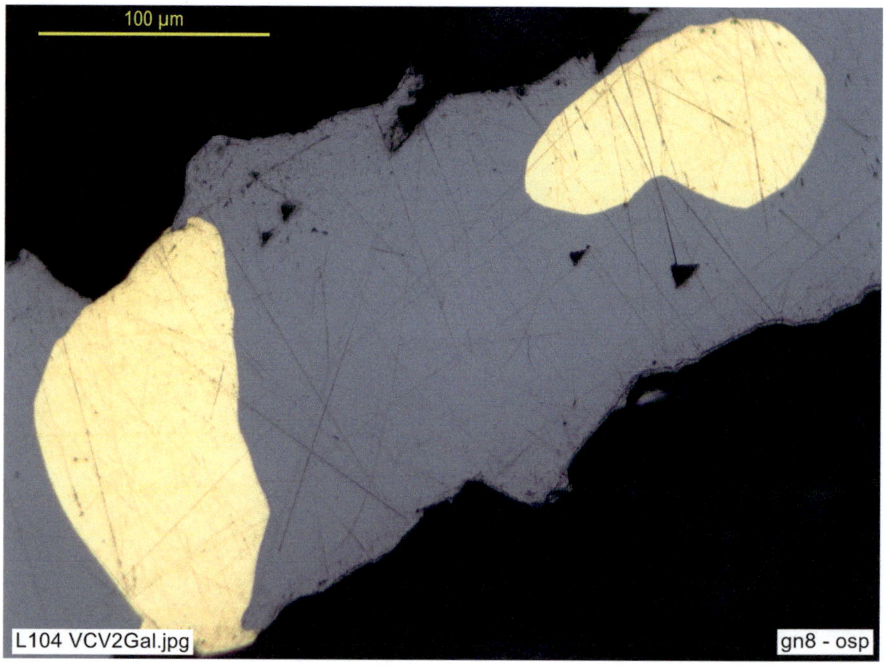

Fig. 51.5 Microcracks cemented by gn (dark grey) and Au (bright yellow *electrum*), in quartz vein (black). VCV2Gal, Vilarcovo, Santiago de Compostela (**osp**)

scratches and *triangular pits*, bright white colour, relatively high reflectance (though lower than pyrite and slightly < ccp, with $R_{546}\%$ = 42.9-gn versus 45.8-ccp or in white light, with Y%: 43.0-gn, versus 45.3-ccp, always in air), isotropic character and presence of the so-called "B-G-P mineralisation" (Blende-Galena-Pyrite), which reminds us of its common paragenetic association with pyrite and sphalerite (± chalcopyrite, tetrahedrite).

The minerals with which it might be confused are very scarce and are usually anisotropic (sulphosalts) and do not have *triangular pits*, or if they do (altaite, clausthalite), they differ in their reflectance.

Common accompanying ores are sphalerite, pyrite, marcasite, chalcopyrite, native silver, tetrahedrite, bournonite, pyrargyrite and other sulphosalts, arsenopyrite, and pyrrhotite.

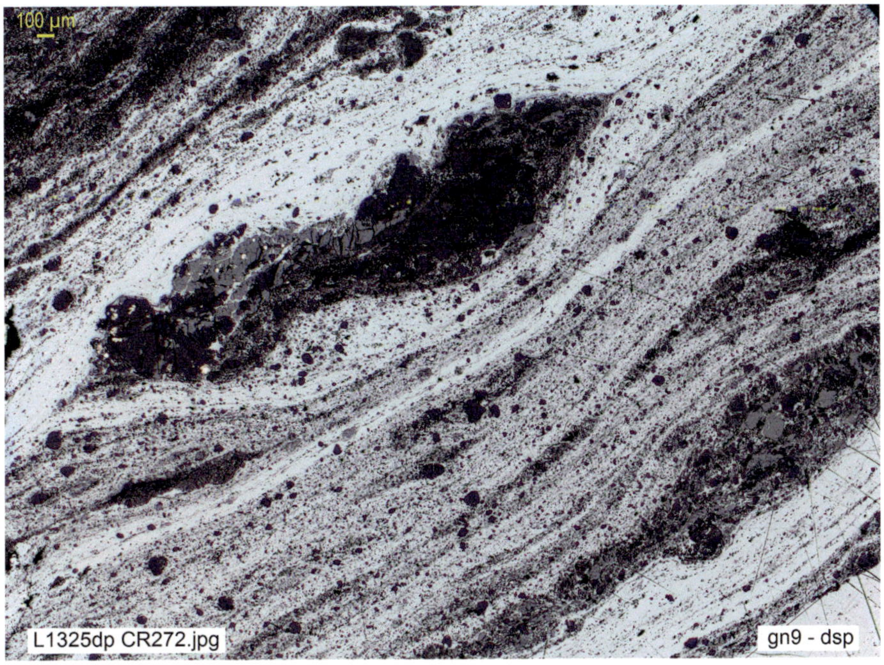

Fig. 51.6 Mylonitic ore with ductile gn matrix and mylonitic schistosity, marked by alignments of q (dark grey) and sp (grey) microclasts, some of which contain py microinclusions. CR-272, Cierco Mine, Lérida

51.3 Spectral Information

51.3.1 Galena Multispectral Reflectance R (in Air, Without Polariser) (%)

λ (nm)	370	400	425	450	475	500	525	550	575	600	625	650	675	700	750	800	850	900	950	1000
μ	47.9	47.1	45.3	44.2	43.0	42.1	41.2	40.8	40.4	40.3	40.4	40.6	40.7	40.6	40.2	39.6	39.3	38.4	37.8	37.4
P_5	44.2	43.4	41.9	40.9	40.0	39.2	38.5	38.4	38.2	38.3	38.5	38.9	39.0	39.1	38.8	38.4	38.1	37.2	36.9	36.3
P_{20}	46.4	45.7	44.1	43.0	41.9	41.0	40.2	39.9	39.6	39.5	39.7	39.9	40.0	40.0	39.6	39.2	38.8	38.0	37.4	36.9
P_{80}	49.6	48.7	46.7	45.7	44.3	43.3	42.3	41.8	41.3	41.3	41.3	41.5	41.5	41.3	40.8	40.2	39.8	38.8	38.2	37.8
P_{95}	50.3	49.4	47.4	46.4	45.0	44.0	42.8	42.3	41.8	41.7	41.7	41.9	41.9	41.7	41.2	40.5	40.1	39.1	38.6	38.1

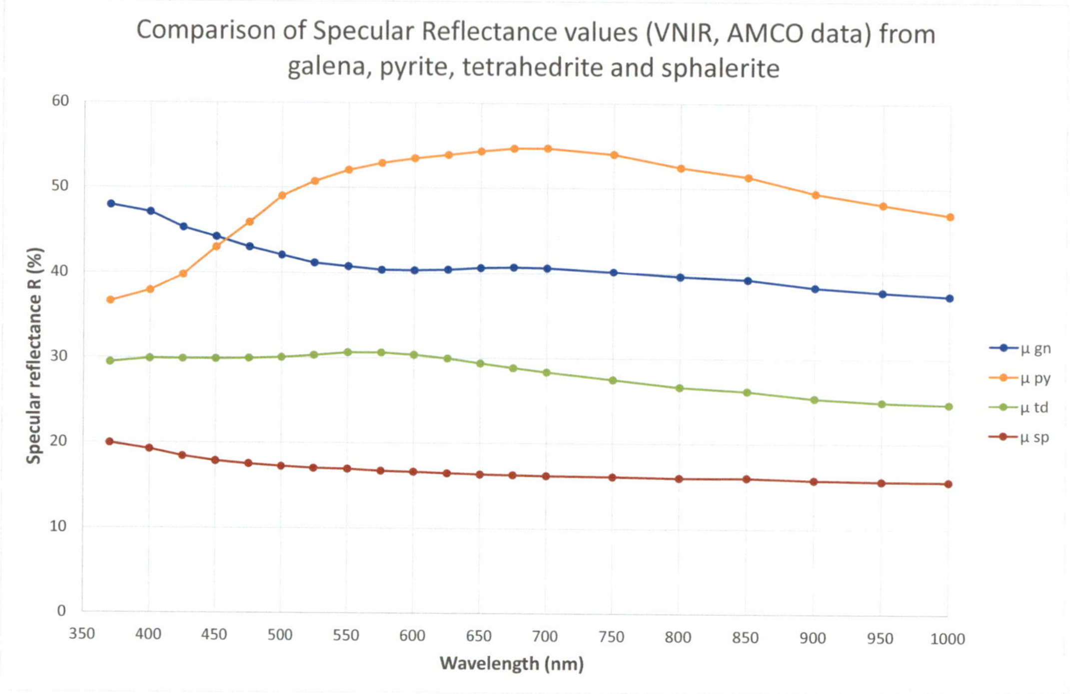
Comparison of Specular Reflectance values (VNIR, AMCO data) from galena, pyrite, tetrahedrite and sphalerite
Specular reflectance R (%)
Wavelength (nm)
µ gn
µ py
µ td
µ sp

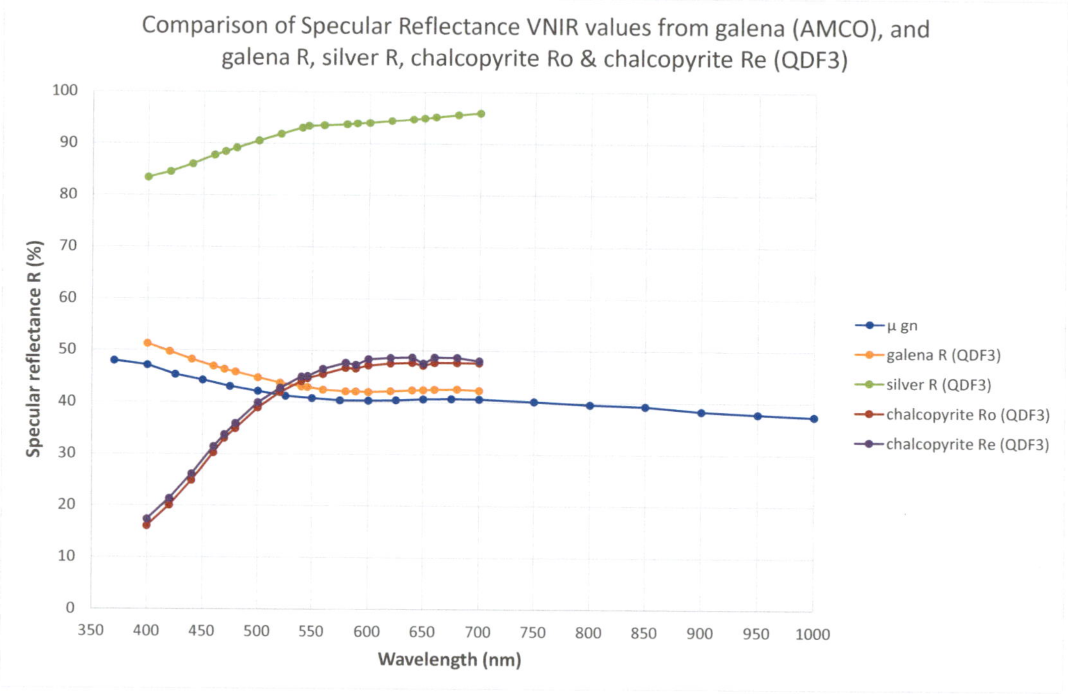
Comparison of Specular Reflectance VNIR values from galena (AMCO), and galena R, silver R, chalcopyrite Ro & chalcopyrite Re (QDF3)
Specular reflectance R (%)
Wavelength (nm)
µ gn
galena R (QDF3)
silver R (QDF3)
chalcopyrite Ro (QDF3)
chalcopyrite Re (QDF3)

Geocronite (*gc*)

52

Pb_{14} $(Sb,As)_6$ S_{23}: Monoclinic

Geocronite			R_{546} (%)	Y (%)	λ_d (nm)	P_e (%)	Colour sensation
Air	R_1		38.8	38.6	482	2.1	Greyish white
	R_2		42.2	42.0	481	3.0	
Oil	R_1		23.1	22.9	480	4.1	
	R_2		26.5	26.3	479	5.4	
Bir: Weak (3)					*Pleoc*: Weak (3)		*Tarn*: -
Anis +P: Distinct, faintly coloured (4)					*IR*: No		*Twg*: Very frequent
Mor: Anhedral aggregates. Tabular					*Clv*: No		*PolQ*: Good
Hardness		*PolH*: Low, < gn (2)			*Mohs*: 2½-3	*Talm*: B	*VHN* $_{(100\ g)}$: 144–160

Obs: cf. Table and R charts, multispectral data (VNIR, AMCO) and visible range (QDF3)
Mineral abbreviations: bl—boulangerite, bnn—bournonite, ccp—chalcopyrite, gc—geocronite, gn—galena, gtt—gratonite, jm—jamesonite, jd—jordanite, py—pyrite, Sb—native antimony, sbt—stibnite, sp—sphalerite, Ss—sulphosalts, td—tetrahedrite

52.1 General

Geochronite, ***gc***, is isostructural with jordanite, ***jd***. Its composition, $Pb_{14}(Sb,As)_6S_{23}$, is variable by As-Sb substitution in the geocronite-jordanite series, $Pb_{14}(Sb,As)_6S_{23}$—$Pb_{14}(As,Sb)_6S_{23}$.

52.2 Microscopic Description

52.2.1 SP Observation (*Single Polariser*)

Greyish white *colour* (Fig. 52.1a, b), similar to galena, ***gn***, although its *Reflectance*, ***R***, is slightly lower and shows a weak ***bireflectance***, while gn is isotropic.

However, the bireflectance of gc is very low, if compared to typically bireflective ores such as stibnite, ***sbt***. Its spectral curves are very similar to those of jordanite, **jd**, or boulangerite, bl, and they are also close to those of other sulphosalts of similar appearance (cf. multispectral comparison plots).

Slightly pleochroic, especially in immersion (Fig. 52.3a). A subtle change of shades from light grey to slightly greenish white can be seen.

Polishes well. *Polishing hardness*, *PolH*, low, although variable with composition, *PolH gc* is lower than that of galena and much lower than that of chalcopyrite.

It occurs in anhedral aggregates, with rounded or irregularly intergrown grains, whose contours are more clearly distinguishable with analyser (+p).

52.2.2 Observation +P (*Polariser + Analyser at 90°*)

Distinctly *anisotropic*, although the polarization tints are dull, not very striking (Figs. 52.1c, d, 52.2 and 52.3), ranging from greyish-brown to bluish-green or bluish-grey tones.

No *internal reflections*. Frequent *twinning*, usually polysynthetic (Fig. 52.2).

52.2.3 Paragenesis

Typical hydrothermal *parageneses*, either in hypothermal formations with late Sb input or in shallow, low-temperature veins.

R. Castroviejo, *A Practical Guide to Ore Microscopy—Volume 1*,
https://doi.org/10.1007/978-3-031-12654-3_52

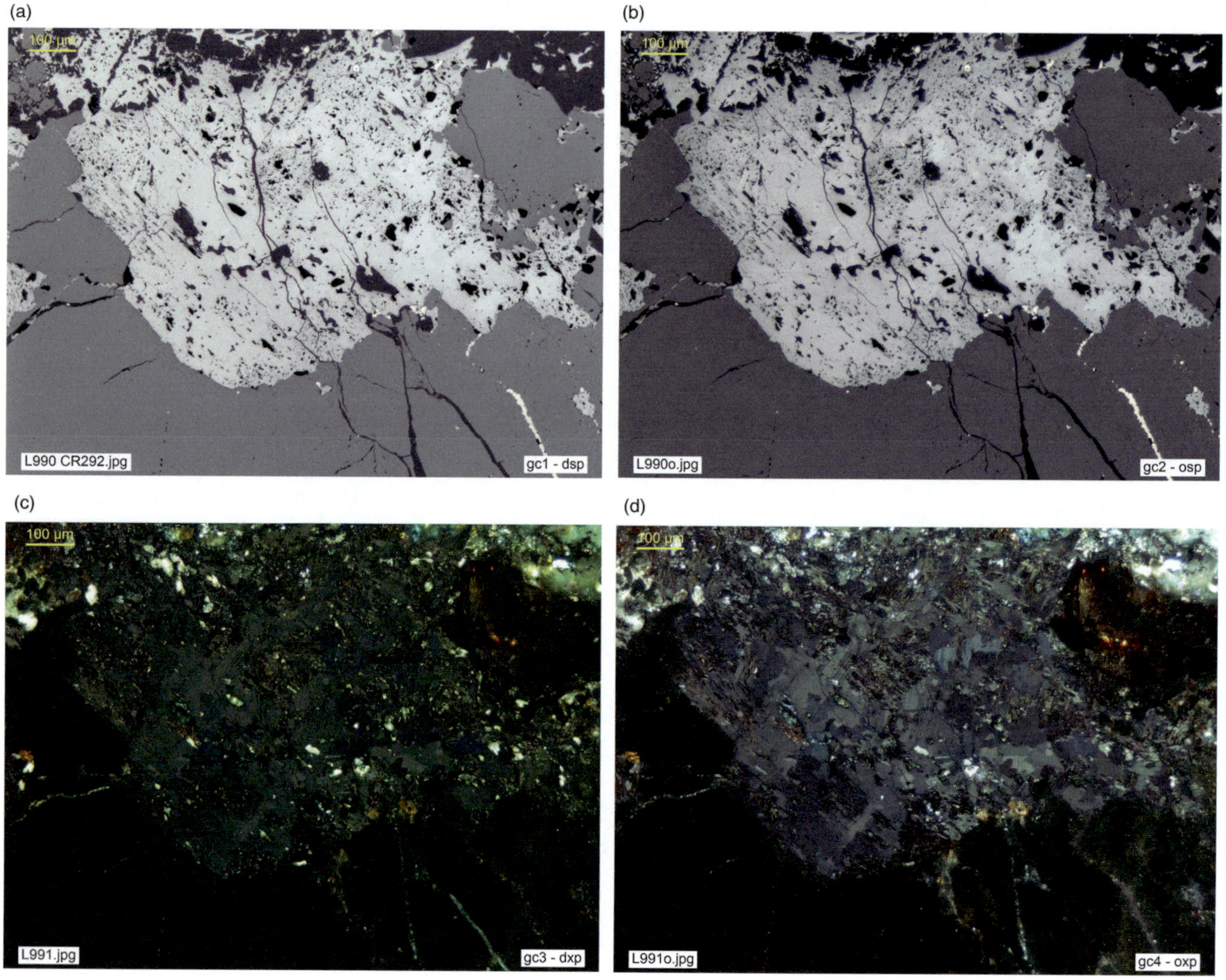

Fig. 52.1 **a, b** (Same scene: **dsp** and **osp**, *respectively*). Ore of geochronite, *gc* (greyish white, porous, intergrown with gg) and sphalerite, sp (grey), with microcracks sealed by pyrite, *py*. Gangue, *gg* (black), top border of image. The higher polishing hardness of sp, py and gg makes them show up in relief on gc. CR 292, Angelica Mine, Huachocolpa, Huancavelica (Peru). **c, d** (Same scene as Fig. 52.1a: **dxp** and **oxp**, *respectively*). Twinning and anisotropism of gc, whose polarisation tints are more clearly perceived by slightly rotating the analyser and in immersion (**oxp**). Total extinction of sp, whose internal reflections are, in this case, rather weak (ferriferous variety)

It is usually accompanied by jordanite, tetrahedrite and other sulphosalts of Pb, Sb, As but also by common sulphides such as galena (***gn***), pyrite (***py***), or sphalerite (***sp***). It occasionally forms intergrowths, crystals or mixed aggregates with jordanite.

52.2.4 Identification

Its white colour and reflectance values give it a similar appearance to galena and other Pb-Sb sulphosalts, ***Ss***. However, twins are a good distinguishing feature: they characterize gc and are not frequent in jd, the arsenical term of the gc-jd series. In addition:

- gc differs from gn by its anisotropism.
- against the mentioned Ss, in general,
  - gc shows weaker pleochroism
  - and it occurs in fine anhedral aggregates, instead of the tabular or acicular crystals typical of most of these sulphosalts.
- in front of the Ss of the tetrahedrite group, there is no room for confusion because td is much darker (less reflective) and isotropic.

Fig. 52.2 Detail of anhedral aggregate of geocronite, with anisotropism and visible twins (**oxp**). Sample CR 292, Mina Angelica, Huachocolpa, Huancavelica, Peru

Fig. 52.3 **a, b** (Same scene: **osp/o+p**, *respectively*). Geocronite, *gc* (grey-white bottom left), jordanite, *jd* (*GrWh*, fills top right image half, including many tiny quartz crystals, black) and sphalerite, *sp* (grey, center and left), with scarce pyrite, *py*. Comparing contiguous grains in the aggregates of gc and especially of jd, the bireflectance and the slight pleochroism of both minerals (**osp**) are perceived. With crossed polars (**o+p**), gc and jd are distinguished by their anisotropism (dull and weakly coloured in gc, vivid and striking in jd) and by twinning, distinct in gc, absent in jd. Cubic sp shows extinction and IR. Sample CR 292, Mina Angelica, Huachocolpa, Huancavelica, Peru

(a)

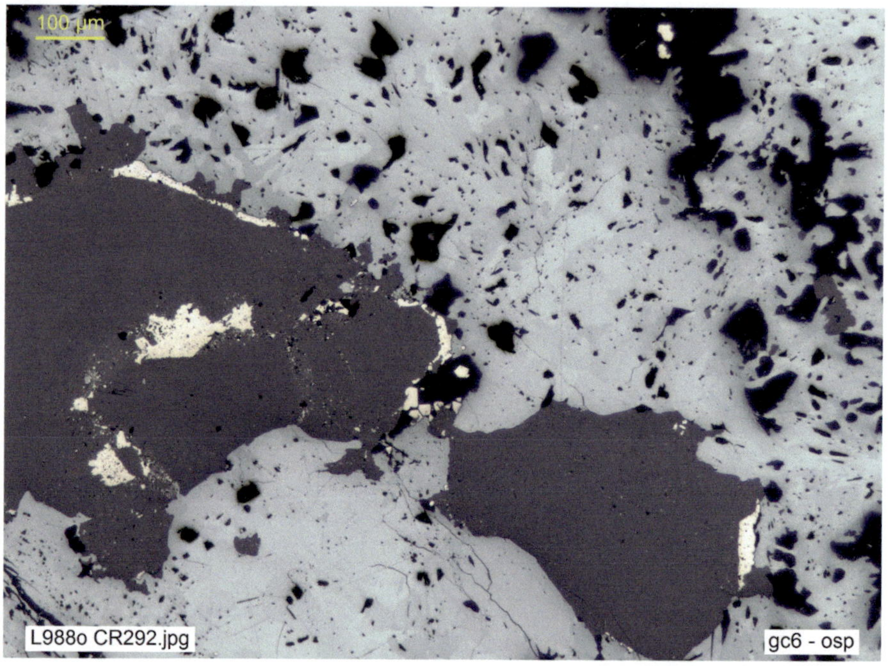

(b)

52.3 Spectral Information

52.3.1 Geocronite Multispectral Reflectance R (in Air, Without Polariser) (%)

λ (nm)	370	400	425	450	475	500	525	550	575	600	625	650	675	700	750	800	850	900	950	1000
μ	41.7	41.2	40.7	40.3	39.9	39.6	39.3	39.0	38.7	38.3	37.9	37.4	36.8	36.2	35.0	33.8	32.8	33.3	33.5	33.5
P_5	39.7	39.2	38.8	38.4	38.2	38.0	37.6	37.2	36.8	36.4	35.8	35.2	34.6	34.0	32.8	32.0	30.8	31.6	31.6	31.6
P_{20}	40.9	40.4	40.0	39.6	39.2	38.8	38.6	38.4	38.2	38.0	37.4	36.8	36.2	35.6	34.4	33.2	32.4	32.4	32.4	32.4
P_{80}	42.7	42.0	41.4	40.8	40.6	40.4	40.0	39.6	39.2	38.8	38.4	38.0	37.4	36.8	35.6	34.4	33.6	34.0	34.4	34.4
P_{95}	43.5	42.8	42.2	41.6	41.4	41.2	40.6	40.0	39.8	39.6	39.0	38.4	37.8	37.2	36.0	34.8	34.0	34.8	35.6	35.6

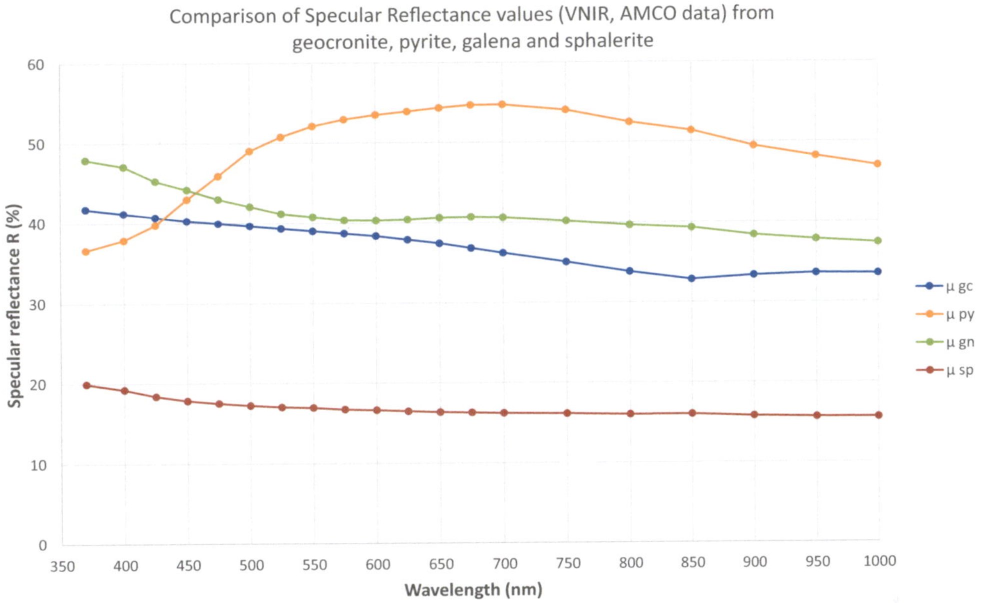

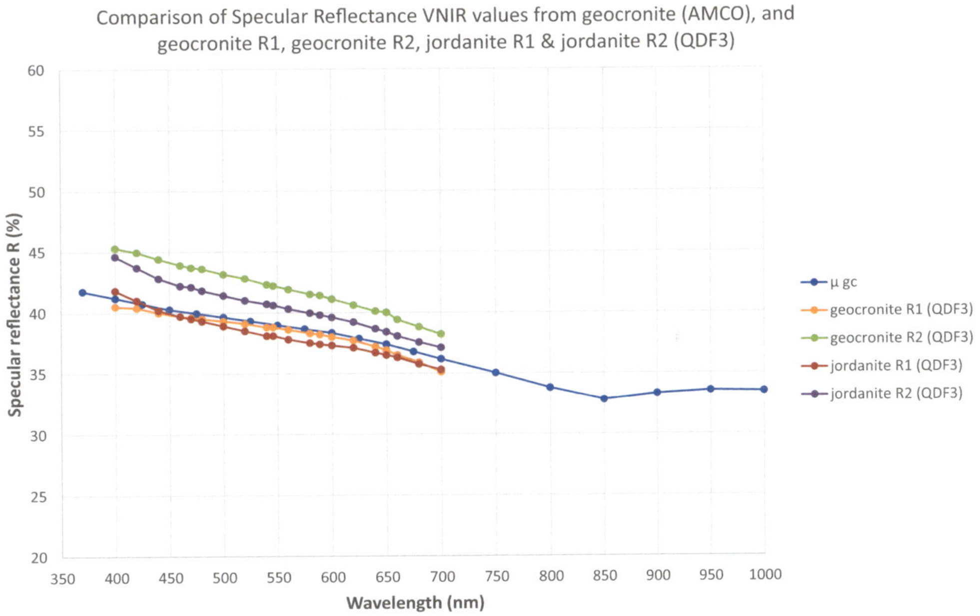

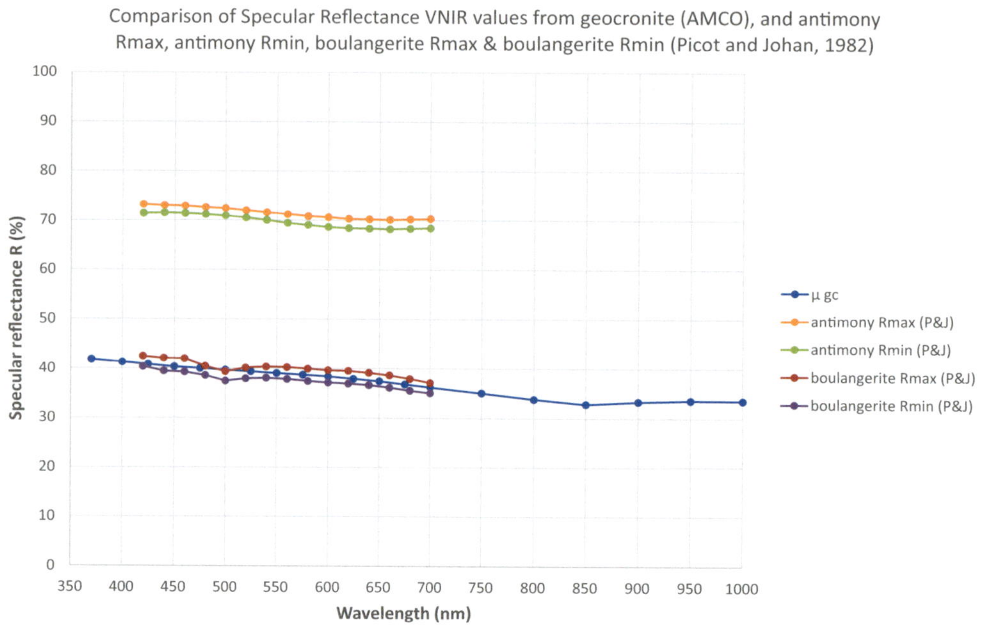
Comparison of Specular Reflectance VNIR values from geocronite (AMCO), and antimony Rmax, antimony Rmin, boulangerite Rmax & boulangerite Rmin (Picot and Johan, 1982)
Specular reflectance R (%)
Wavelength (nm)
µ gc
antimony Rmax (P&J)
antimony Rmin (P&J)
boulangerite Rmax (P&J)
boulangerite Rmin (P&J)

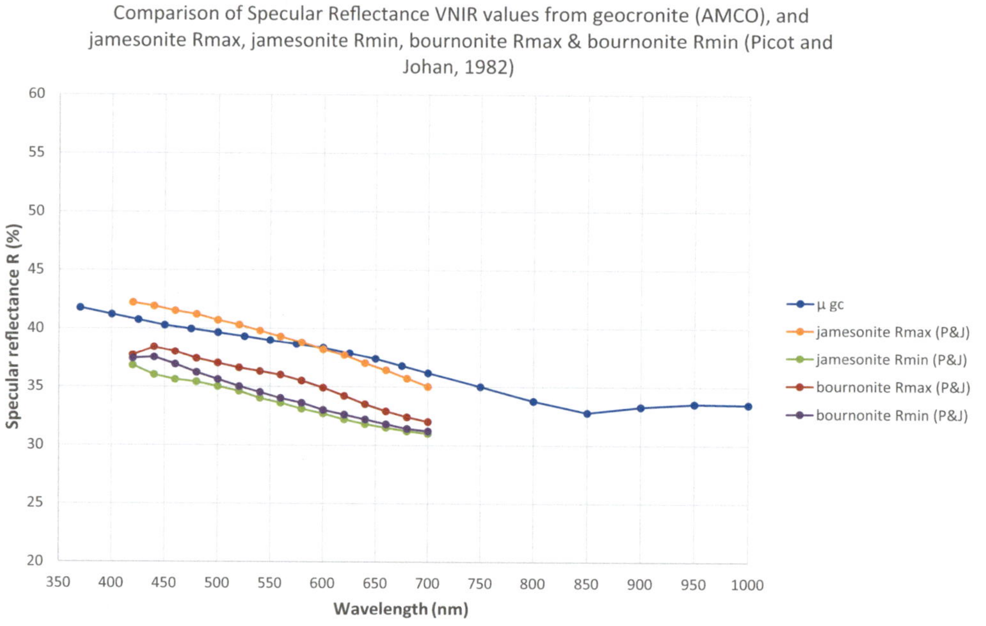
Comparison of Specular Reflectance VNIR values from geocronite (AMCO), and jamesonite Rmax, jamesonite Rmin, bournonite Rmax & bournonite Rmin (Picot and Johan, 1982)
Specular reflectance R (%)
Wavelength (nm)
µ gc
jamesonite Rmax (P&J)
jamesonite Rmin (P&J)
bournonite Rmax (P&J)
bournonite Rmin (P&J)

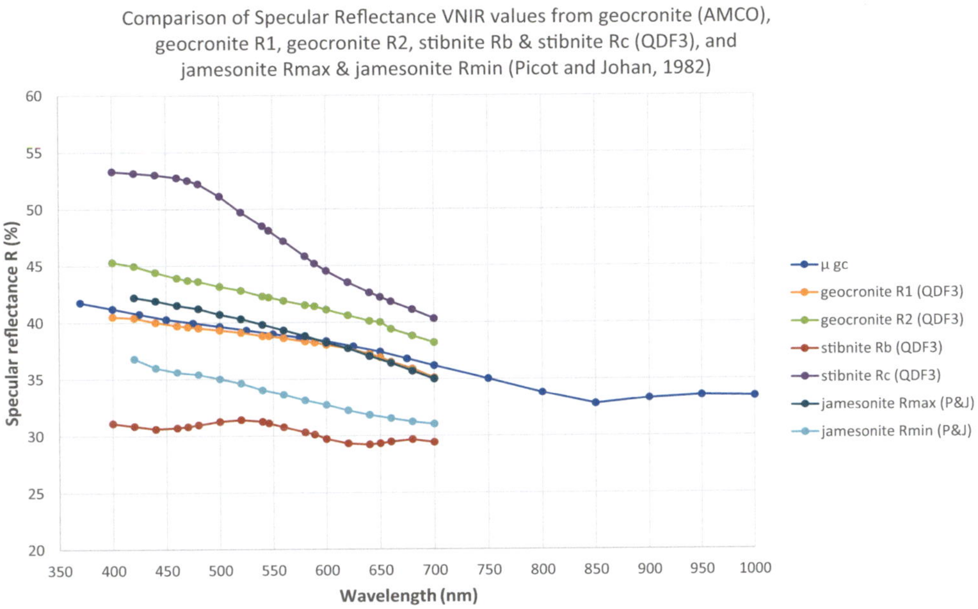
Comparison of Specular Reflectance VNIR values from geocronite (AMCO), geocronite R1, geocronite R2, stibnite Rb & stibnite Rc (QDF3), and jamesonite Rmax & jamesonite Rmin (Picot and Johan, 1982)
Specular reflectance R (%)
Wavelength (nm)
µ gc
geocronite R1 (QDF3)
geocronite R2 (QDF3)
stibnite Rb (QDF3)
stibnite Rc (QDF3)
jamesonite Rmax (P&J)
jamesonite Rmin (P&J)

53 Gersdorffite (*gf*)

NiAsS: Cubic

Gersdorffite		R_{546} (%)	Y (%)	λ_d (nm)	P_e (%)	Colour sensation
Air		45.6	45.7	472	3.1	White
Oil		31.0	31.2	470	4.7	
Bir: Null (1)				*Pleoc*: Null (1)		*Tarn*: Darkens in a few weeks
Anis +P: Isotropic (not always) (2)				*IR*: No		*Twg*: Occasional
Mor: Cubic idiomorphic or interstitial				*Clv*: // (100), *triangul pits*		*PolQ*: Good
Hardness	*PolH*: High, > po, ≪ py (4)			*Mohs*: 5.5	*Talm*: F^{-}	*VHN* $_{(100\ g)}$: 657–767

Obs: cf. Table and R charts, multispectral data (VNIR, AMCO) and visible range (QDF3; Picot and Johan 1982). There can be remarkable variations of R, depending on the composition (Sb, Fe, Co, S contents). QDF3 presents up to five different spectra, of which that of *Gersdorffite*coincides in the visible range with the AMCO spectrum and is considered the most representative for being the closest to the ideal formula of the ore (Ni 35.42%; As 45.23%; S 19.35%)
Mineral abbreviations: asp—arsenopyrite, bth—breithauptite, ccp—chalcopyrite, chr—chromite, cob—cobaltite, gf—gersdorffite, gg—gangue, gn—galena, ln—linnaeite, lo—löllingite, nc—nickelite, po—pyrrhotite, py—pyrite, sk—skutterudite, sp—sphalerite, ul—ullmannite

53.1 General

Composition: NiAsS, can incorporate variable contents of Sb, Fe, Co, S. Ramdohr (1980) points out notable variations in the ratios (Ni, Co, Fe):(S, As, Sb), between 1:2 and 2:3, as well as in the proportion As:S, with frequent excess of As over S. Higher contents of Sb (*korynite*) mark a gradual transition to ullmannite.

53.2 Microscopic Description

53.2.1 SP Observation (*Single Polariser*)

White *colour* (Fig. 53.1a, b), very well *polished*. Tarnishes (dark) in a few weeks. Isotropic.

High *reflectance* (cf. multispectral comparative charts) but with remarkable variations according to composition (QDF3 shows luminance Y% values ranging, in air, between 45.7 and 57.1). In white light, $R_{gf} \approx R_{ul}$, > gn, < py, lo, sk.

Polishing hardness high but among the lowest of the high-PolH group (4, near transition to the medium-PolH group, 3, cf. Ax. 4.2), ≈ ul; > ln, po, ccp, sp; < lo, asp, ≪ py.

Distinct euhedral (cubic) tendency, with good development of (100) and (111). Frequent zoning. Very marked *cleavage* // (100), gives rise to frequent pull-outs during the polishing process, triangular (*triangular pits*) or longitudinal, which does not prevent excellent polishing.

53.2.2 Observation +P (*Polariser + Analyser at 90°*)

Isotropic but sometimes shows weak anisotropism or imperfect extinction (Fig. 53.1c, d).

Complex *twinning*, like cobaltite, may be seen (common according to Picot and Johan 1982).

R. Castroviejo, *A Practical Guide to Ore Microscopy—Volume 1*,
https://doi.org/10.1007/978-3-031-12654-3_53

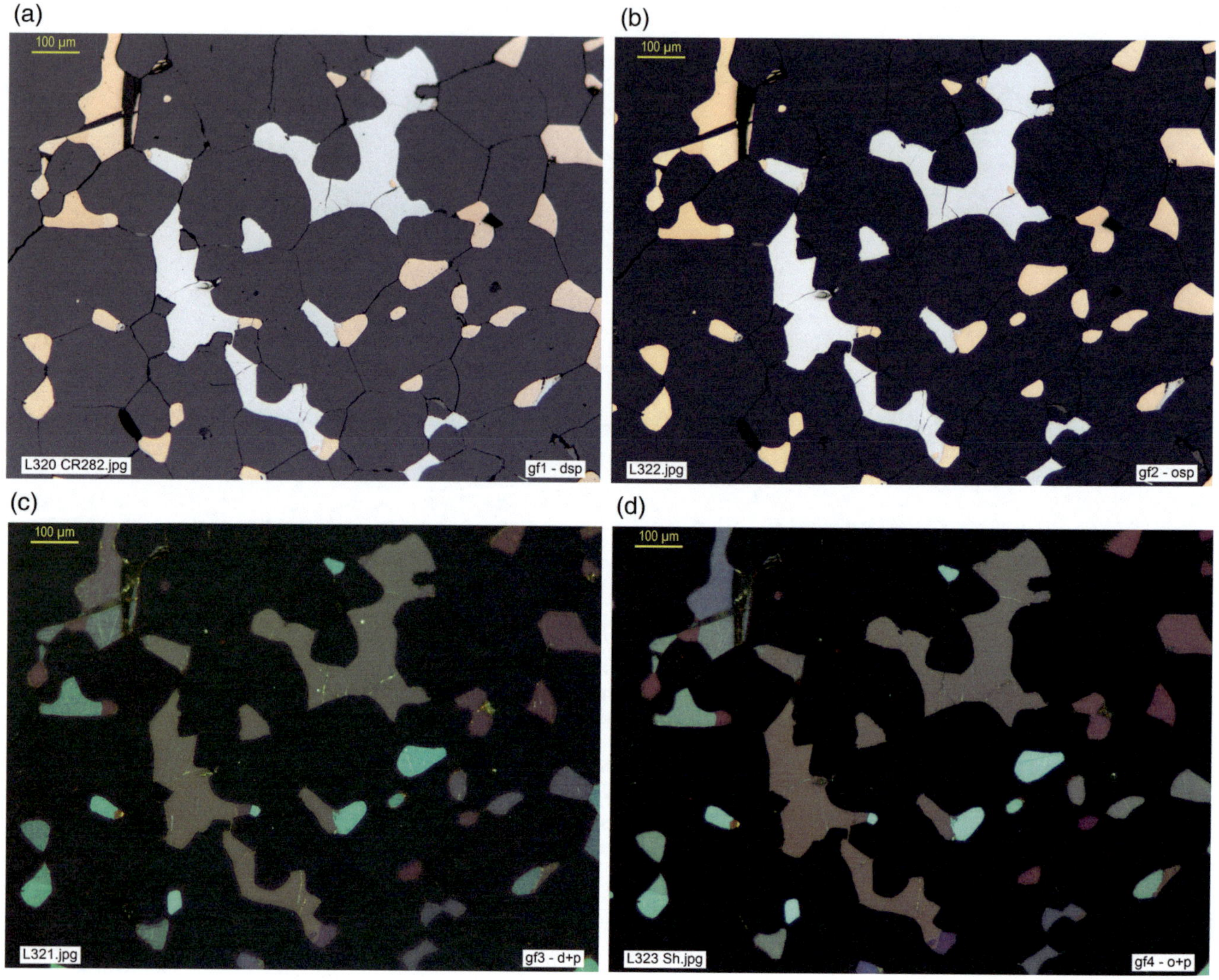

Fig. 53.1 **a, b** (Same scene as gf1: **dsp** and **osp**, respectively). Late, interstitial infill of gf (white) and nc (pink) in cumulus aggregate of chr (grey). The idiomorphic tendency or the typical cleavage usually shown by gf are not visible in this ore, due to its interstitial character and fine grain size. CR 282, Mina La Gallega, S^{a} Ronda, Málaga. **c, d** (Same scene as gf1: **d+p** and **o+p**, respectively). The imperfect extinction of gf contrasts (especially in immersion) with total extinction of chr and vivid anisotropism of nc

53.2.3 Paragenesis

Uncommon mineral, although perhaps because of its unremarkable properties it may have gone unnoticed in many studies. Cited as abundant only in a few deposits, as in Cobalt, ON. Also found in Sudbury, Ontario, where gf is typically found as a late hydrothermal ore, linked to the main *paragenesis* of nickeliferous pyrrhotite. Found in some peculiar chromite-nickelite deposits in peridotites, as well.

53.2.4 Identification

In a quick examination, gersdorffite can be confused, by its appearance, with skutterudite or ullmannite. However, gersdorffite exhibits some characteristic features that should be noted, such as its generally very developed cleavage and its tendency to tarnish.

- It differs from skutterudite by the lower reflectance and the frequent cleavage of gersdorffite, as well as by the yellowish tone that gersdorffite acquires in direct contact with skutterudite.
- It is distinguished from ullmannite by the distinct extinction of the latter (always perfectly isotropic), as opposed to the weak (anomalous) anisotropism or imperfect extinction of gersdorffite, in many cases, as well as by the presence of complex twins (of the cobaltite type) in gersdorffite and the scarcer appearance of zoning and cleavage in ullmannite.
- Finally, the minerals of the linnaeite group are softer and show a pinkish tone, absent in gersdorffite.

53.3 Spectral Information

53.3.1 Gersdorffite Multispectral Reflectance R (in Air, Without Polariser) (%)

λ (nm)	370	400	425	450	475	500	525	550	575	600	625	650	675	700	750	800	850	900	950	1000
μ	44.9	45.7	46.2	46.6	46.5	46.3	46.3	45.9	45.9	45.9	46.0	46.4	46.8	47.0	47.8	48.0	48.9	48.4	48.7	48.4
P_5	42.2	43.4	44.2	44.4	44.6	44.3	44.5	44.0	44.0	44.0	44.2	44.7	45.2	45.6	46.5	46.7	47.4	46.5	46.3	45.9
P_{20}	43.7	44.7	45.6	45.9	46.0	45.6	45.7	45.1	45.1	45.0	45.2	45.6	46.0	46.4	47.3	47.6	48.1	47.4	47.1	46.8
P_{80}	46.7	47.1	47.0	47.5	47.2	47.1	47.2	47.0	47.1	47.2	47.3	47.5	48.0	47.9	48.2	48.5	49.6	49.4	50.2	49.9
P_{95}	47.3	47.9	47.7	48.2	47.8	48.1	47.9	47.9	48.0	48.2	48.2	48.5	48.9	48.9	49.1	49.0	49.8	49.6	50.4	50.2

Comparison of Specular Reflectance values (VNIR, AMCO data) from gersdorffite, nickelite, chromite and pyrite

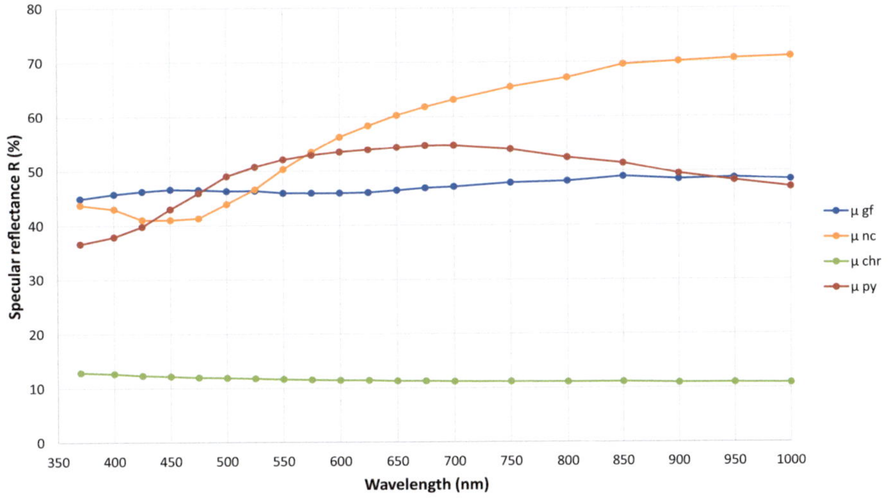

Comparison of Specular Reflectance VNIR values from gersdorffite (AMCO), gersdorffite I R, löllingite R1, löllingite R2 & skutterudite R (QDF3), and arsenopyrite Rmax & arsenopyrite Rmin (Picot and Johan, 1982)

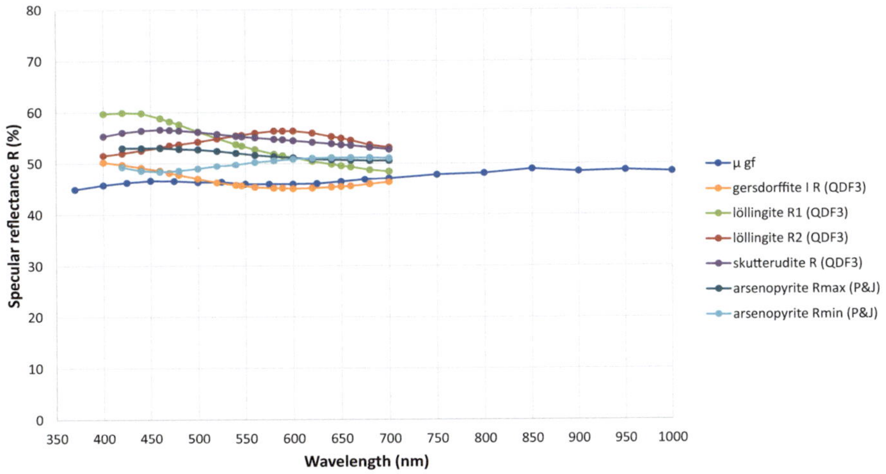

References

Picot P, Johan Z (1982) Atlas of ore minerals. B.R.G.M./Elsevier

Ramdohr P (1980) The ore minerals and their intergrowths, vol 2, 2nd ed. Pergamon Press, Oxford, p 1205

54 Glaucodot (*gld*)

$(Co_{0.5}Fe_{0.5})AsSAsS$: Orthorhombic

Glaucodot			R_{546} (%)	Y (%)	λ_d (nm)	P_e (%)	Colour sensation
Air	*R_1*		50.2	–	–	–	White
	R_2		51.1	–	–	–	
Oil	*R_1*		–	–	–	–	
	R_2		–	–	–	–	
Bir: Imperceptible (2)					*Pleoc*: Imperceptible (2)		*Tarn*: –
Anis +P: Weak (3)					*IR*: –		*Twg*: Occasional
Mor: Prismatic, massive					*Clv*: (010)		*PolQ*: Easier than asp or cob
Hardness		*PolH*: High, but < asp (4)			*Mohs*: 5	*Talm*: F	*VHN* $_{(100–200\ g)}$: 841–1277

Obs: cf. R charts, visible range (Picot and Johan 1982; no gld data in QDF3). VHN: Uytenbogaardt and Burke (1971)
Mineral abbreviations: asp—arsenopyrite, ccp—chalcopyrite, cob—cobaltite, gg—gangue, gld—glaucodot, il—ilmenite, lo—löllingite, sf—safflorite, po—pyrrhotite, py—pyrite

54.1 General

Although the currently accepted formula (IMA/COM) is $(Co_{0.5}Fe_{0.5})$ AsS, Ramdohr (1980) calls *«"glaucodot" the mixed crystals of the arsenopyrite series with Co > ½ Fe content»*. He also points out that in *«single cases the ratio CoAsS: FeAsS may attain 6:1, i.e., nearly the composition of cobaltite»*.

In addition, it may contain Ni as a minor element. According to Back and Mandarino (2008), glaucodot is (Co, Fe)AsS, dimorphous with alloclasite.

54.2 Microscopic Description

54.2.1 SP Observation (*Single Polariser*)

White *colour*, ≈ asp, showing a slightly bluish tinge against arsenopyrite, and a yellowish tinge facing safflorite or löllingite (Figs. 54.1a, b and 54.2). Good *polish*.

High *reflectance*, **R**, between pyrite and gersdorffite (cf. Reflectance charts); R gld ≈ asp (arguably).[1] Always, R gld > ccp and gn.

Bireflectance/pleochroism hardly noticeable, although it is anisotropic.

Marked idioblastic tendency, also poikiloblastic (Figs. 54.1b and 54.2). Occasional cleavage // (010). *Polishing hardness* high (> po) but < asp, cob, py.

54.2.2 Observation +P (*Polariser + Analyser at 90°*)

Distinctly but moderately *anisotropic*, with colours (bluish and brownish) not very pronounced (< asp) and much less

[1] Actually, R gld > R asp according to Ramdohr (1980), but R gld < R asp according to Uytenbogaardt and Burke (1971).This discrepancy might be attributed to compositional differences. However, a simple alternative explanation is deduced by comparing the R_1 and R_2 reflectance values measured on oriented sections of both ores by Picot and Johan (1982). Their results (λ = 540 nm) for gld/asp (respectively) are: R_1 = 50.2/49.7 and R_2 = 51.1/52.0, i.e., the reflectance (or brightness, as perceived by the human eye) will be R gld > R asp or the opposite depending on whether the section is observed in the R_1 or R_2 orientation.

R. Castroviejo, *A Practical Guide to Ore Microscopy—Volume 1*,
https://doi.org/10.1007/978-3-031-12654-3_54

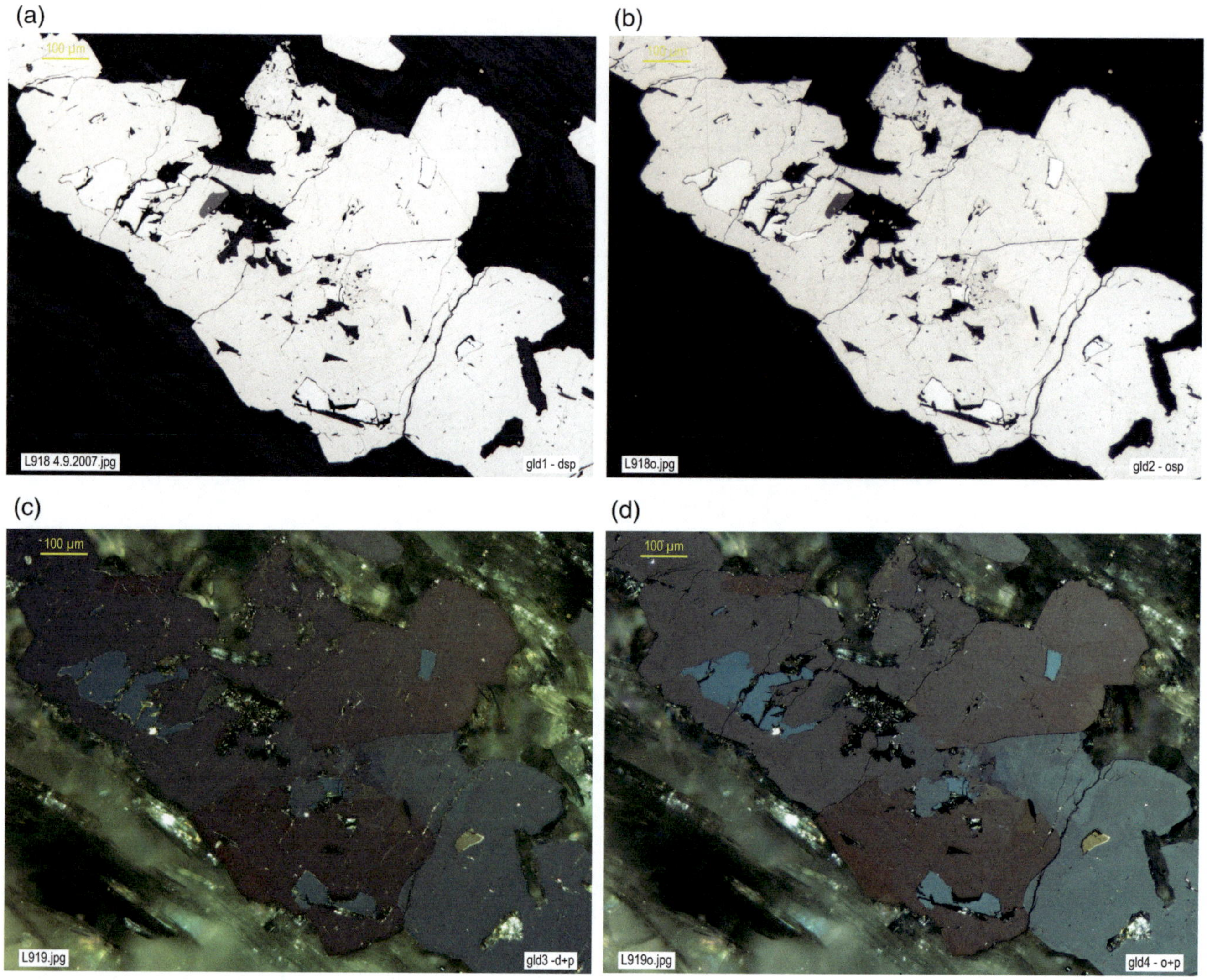

Fig. 54.1 **a, b** (Same scene: **dsp** and **osp**, respectively). Poikiloblastic aggregate of gld (white) with relicts of sf (brighter white) and small inclusions of il (grey) and gg (dark grey). Tiny ccp (yellow) particles scattered in gg (micaceous schist). In oil (osp), the slight yellowish tinge of gld, as opposed to pure white of sf, is enhanced. Identities of gld and sf confirmed by EMPA. 492007, St. Maria Anna Mine, Przecznica, Isergebirge, Lower Silesia (Poland). **c** Same scene as gld1 (**d+p**): the stronger anisotropism, with more vivid polarization colours, makes it easier to distinguish the sf inclusions; il shows greenish tones. **d** Same scene as gld1 (**o+p**): reinforcement of the anisotropism contrast between gld (twinned) and sf

luminous than safflorite (Figs. 54.1c, d and 54.3). It may show irregular twins, like arsenopyrite.

54.2.3 Paragenesis

Hydrothermal high temperature *parageneses* occurring in deep environments, rarely also in contact metamorphic deposits.

Accompanying ores: arsenopyrite, cobaltite, Ni–Co arsenides, linnaeite, etc. Also, in weathered outcrops, erythrite encrustations.

54.2.4 Identification

It resembles arsenopyrite, but its anisotropism is weaker and its hardness is < asp.

The alteration visible in the field on weathered deposits can provide a useful additional criterion. The presence of *cobaltbloom* (erythrite) in hand sample on S-As ores, even on apparent asp, is a good hint to suspect the likely presence of gld, since it highlights Co, a typical component of glaucodot, but not of arsenopyrite.

Fig. 54.2 Idioblasts of gld, with residual inclusions of sf; il, ccp and small relicts of altered po (*Zwischenprodukt* of Ramdohr 1980), scattered in micaceous schist. Same provenance as gld1 (**osp**)

Fig. 54.3 Same scene as gld5 (**o +p**): immersion highlights the more vivid and luminous anisotropism of sf

54.3 Spectral Information

54.3.1 Glaucodot Reflectance R (in Air, Visible Range) (%)

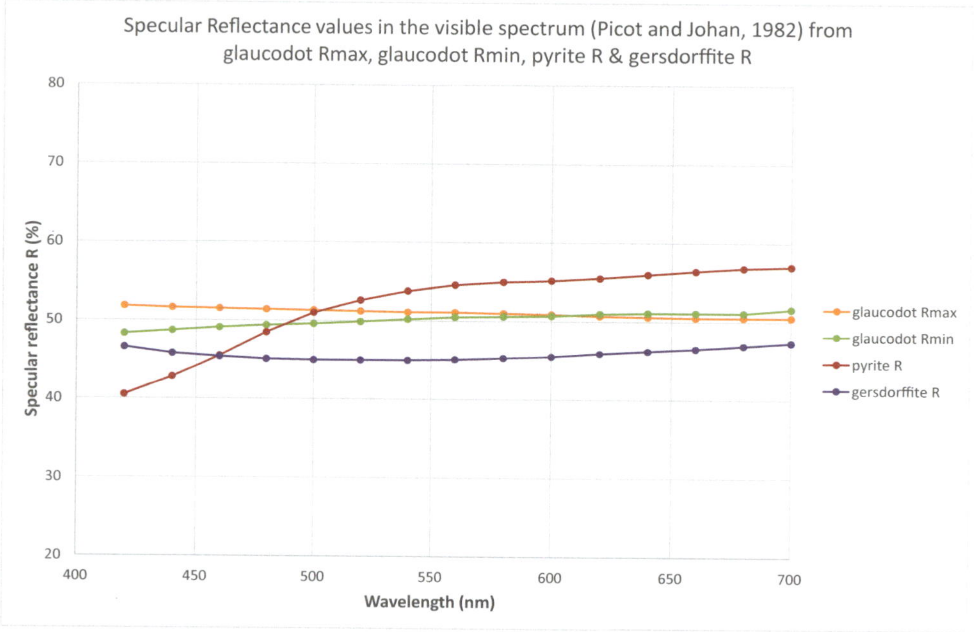

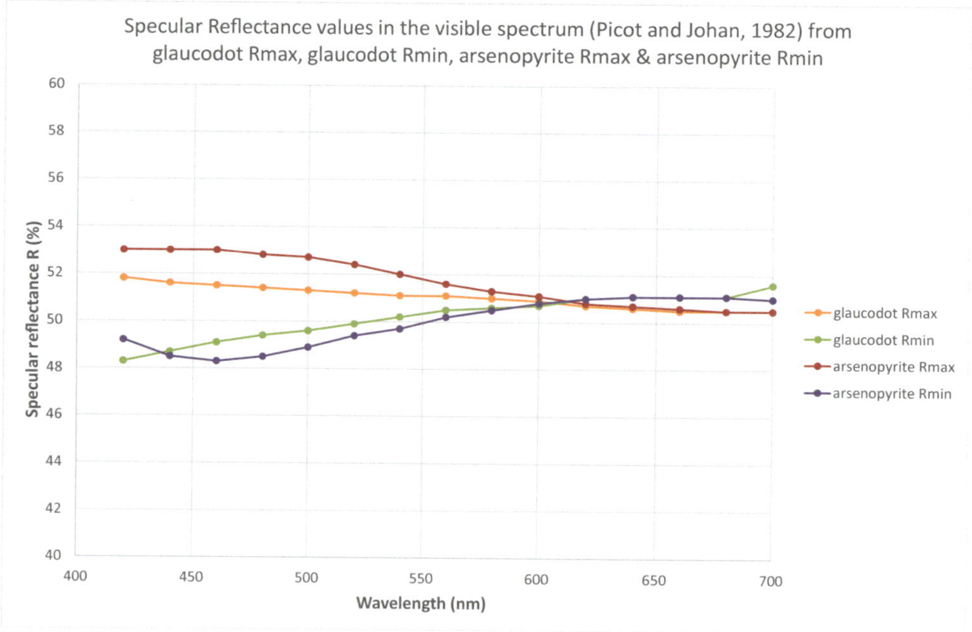

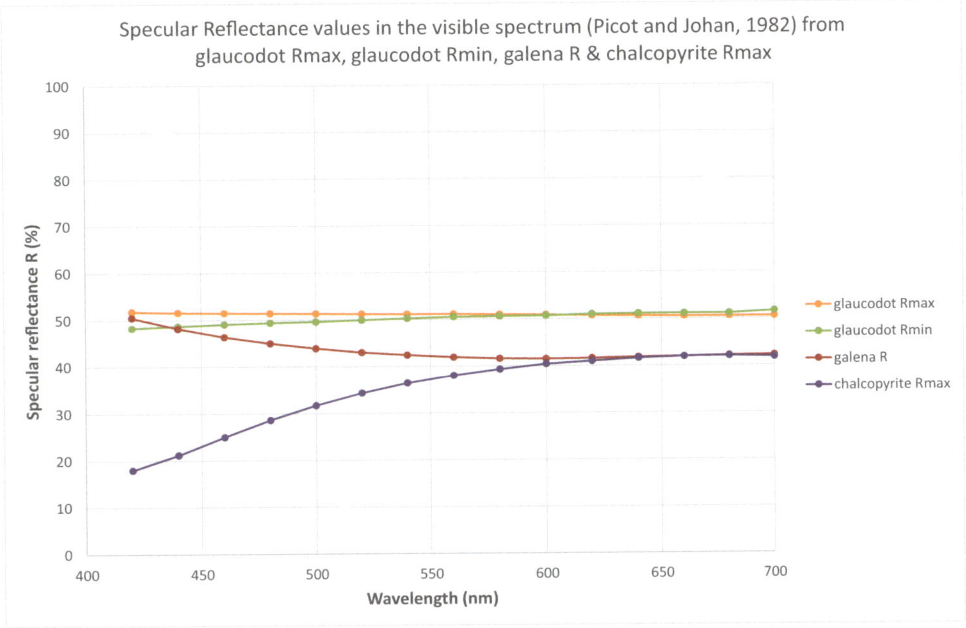

References

Back ME, and Mandarino JA (2008) Fleischer's glossary of mineral species 2008. The Mineralogical Record Inc., Tucson, AZ, p 345

Picot P, Johan Z (1982) Atlas of ore minerals. B.R.G.M./Elsevier

Ramdohr P (1980) The ore minerals and their intergrowths, vol 2, 2nd ed. Pergamon Press, Oxford, p 1205

Uytenbogaardt W, Burke EAJ (1971) Tables for microscopic identification of ore minerals, 2nd edn. Elsevier, Amsterdam

Goethite (*gth/Nadeleisenerz*) 55

FeO(OH): Orthorhombic

Goethite		R_{546} (%)	Y (%)	λ_d (nm)	P_e (%)	Colour sensation
Air	R_1	15.5	15.5	476	7.6	Grey, slightly bluish (variable according to porosity, crystallinity, polishing quality)
	R_2	17.5	17.5	476	9.2	
Oil	R_1	4.3	4.3	476	15.3	
	R_2	5.5	5.5	474	18.2	
Bir: Medium (4)				*Pleoc*: Weak (3)		*Tarn*: No
Anis +P: Distinct (4)				*IR*: Abundant		*Twg*: Not observed
Mor: Botryoidal, acicular, massive, etc				*Clv*: Occasional //(010)		*PolQ*: Variable
Hardness	*PolH*: Medium (3)			*Mohs*: 5	*Talm*: E	*VHN* $_{(100\ g)}$: 609–847

Obs: cf. Table and R charts, multispectral data (VNIR, AMCO) and visible range (QDF3)
Mineral abbreviations: ccp—chalcopyrite, gg—gangue, gth—goethite, ht—hematite, lpc—lepidocrocite, mt—magnetite, psi—psilomelane, py—pyrite, q—quartz, sp—sphalerite

55.1 General

A very common mineral in the surface environment and an important Fe ore. FeO(OH) or α-$Fe^{3+}O(OH)$, polymorph of lepidocrocite. It can contain variable amounts of Mn. It can also adsorb H_2O, Al_2O_3, SiO_2, CaO, precious metals, etc. (cf. *limonite*).

55.2 Microscopic Description

55.2.1 SP Observation (*Single Polariser*)

Grey *colour*, slightly bireflective and pleochroic. The optical perception is variable according to the grain size, degree of hydration, adsorbed substances, porosity, etc. It may show slightly brownish and, especially in immersion, bluish shades of grey (Fig. 55.1a, b).

Moderate to low *reflectance*: R_{gth} < mt < ht; between R_1 and R_2 of lpc, > gg (cf. multispectral comparison charts).

Variable *polishing* quality and hardness. Good polishing on pure, crystalline materials, whose polishing hardness is medium: $PolH_{gth}$ > ccp, < sp, mt. Cleavage // (010), perfect in hand specimen, is rarely visible under the microscope.

It occurs in prismatic, tabular, acicular crystals and radiating aggregates, frequently spherulitic, botryoidal or reniform (Figs. 55.1a, b and 55.2a). Also in earthy masses, oolitic deposits, secondary supergene concentrations of sulphides (oxidation zone, *gossan*), laterites. In trace amounts, very widespread, practically ubiquitous as supergene limonitic impregnation in soils.

55.2.2 Observation +P (*Polariser + Analyser at 90°*)

Distinct anisotropism with bluish-grey tones (Fig. 55.1c, d), although often—and especially in immersion—masked by abundant reddish-brown *internal reflections*, brighter and lighter (orange to yellowish) in the colloform aggregates as hydration increases and reflectance decreases (**limonites**, often *isotropic*, Fig. 55.2b; cf. *lim description*).

R. Castroviejo, *A Practical Guide to Ore Microscopy—Volume 1*,
https://doi.org/10.1007/978-3-031-12654-3_55

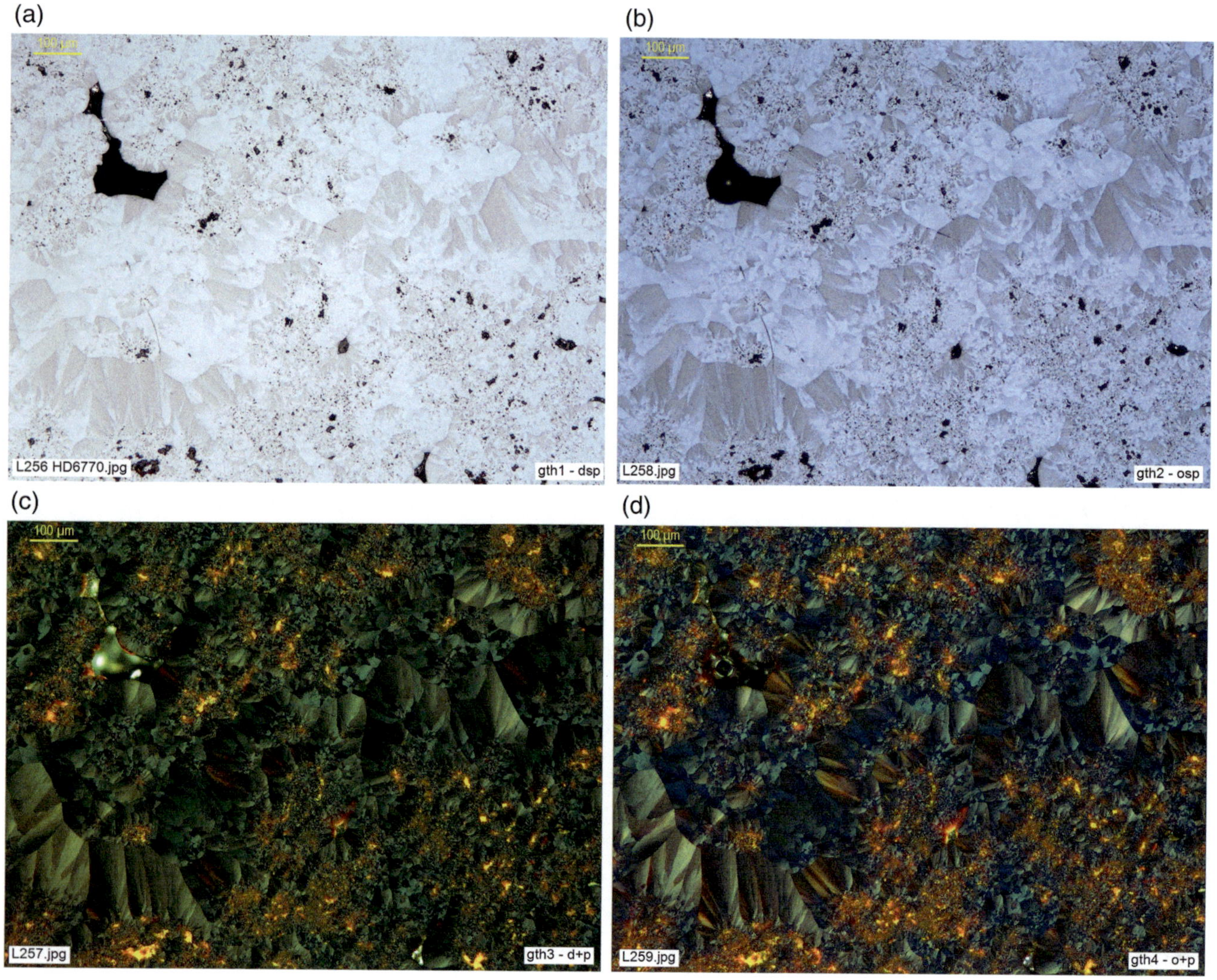

Fig. 55.1 **a, b** (**dsp** and **osp**, respectively). Goethite, gth, aggregate, with transitions from micro-/crypto-crystalline porous zones to radiating recrystallized zones. HD 6770, Friedberg, Hungen, Germany. **c, d** (Same scene as gth1: **d+p** and **o+p**, respectively). Anisotropism and IR of gth. Central band of gth with incipient spherulitic development corresponds to the recrystallized zone of gth1-2; porous, more limonitic zones are less reflective and show lighter, yellowish IR

55.2.3 Paragenesis

Typical weathering product, particularly in secondary supergene concentrations of sulphides (oxidation zone, *gossan*), laterites and limonites but also as a primary mineral in hydrothermal and sedimentary formations.

Goethite is a mineral that continues to form on the earth's surface on a daily basis, due to the poor solubility of Fe^{3+} under natural conditions (oxidizing, pH $\approx$ neutral).

Associated with hematite, lepidocrocite, psilomelane, romanechite, cuprite, sulphide ghosts and additionally (in limonites) with jarosite, opal, ferruginous clays, etc. in superficial deposits that can also retain precious metals and native copper.

55.2.4 Identification

Its colour, low reflectance, distinct anisotropism with bluish tones, abundant internal reflections and colomorphic textures, frequently reniform, botryoidal or fibrous-radial, allow easy recognition. Its association with oxidation processes and/or sulphide relicts or ghosts, as well as its paragenesis are important data.

- Facing *hematite*, goethite is darker and shows more abundant and clearer IR.
- Against *magnetite, it is* darker and anisotropic.
- Beside *lepidocrocite*, it is more frequently botryoidal, less anisotropic, shows lighter IR and is less bireflective, with

Fig. 55.2 **a, b** Botryoidal aggregate of limonite, showing alternation of bluish-grey bands, more brilliant, crystalline and clearly anisotropic (goethite dominant), with brownish-grey limonite bands, duller and less anisotropic but with more abundant and luminous IR (yellowish and orange tones). TF PN6, Mina Sar, Santiago de Compostela (**osp** and **o+p**, respectively)

a

b

a reflectance always lower than R_2 of lepidocrocite. Between R_2 and R_1 of lepidocrocite, in the case of goethite (s.str.), more variable, even < R_1 of lepidocrocite, in the case of *limonites*, especially the most hydrated and rich in jarosite or silicates.

- It is less reflective than *psilomelane*, which lacks IR and is usually cryptocrystalline. However, their natural association in mixed (lim+psi) oxides, mostly massive, submicroscopic or amorphous, may make their distinction impossible in some cases.

55.3 Spectral Information

55.3.1 Goethite Multispectral Reflectance R (in Air, Without Polariser) (%)

λ (nm)	370	400	425	450	475	500	525	550	575	600	625	650	675	700	750	800	850	900	950	1000
μ	19.8	19.2	18.2	17.0	16.1	15.6	15.2	14.6	14.2	13.8	13.6	13.4	13.3	13.1	12.9	12.6	12.6	12.3	12.3	12.2
P_5	18.1	17.6	16.6	15.5	14.8	14.4	14.0	13.5	13.2	12.8	12.6	12.4	12.3	12.2	12.0	11.8	11.7	11.4	11.3	11.2
P_{20}	19.0	18.5	17.4	16.3	15.5	15.1	14.7	14.1	13.8	13.4	13.2	13.0	12.9	12.7	12.5	12.3	12.3	12.0	11.9	11.8
P_{80}	20.7	20.0	19.0	17.6	16.8	16.3	15.8	15.1	14.7	14.3	14.1	13.8	13.7	13.5	13.3	13.0	12.9	12.7	12.7	12.7
P_{95}	21.1	20.5	19.5	18.1	17.2	16.7	16.2	15.5	15.1	14.7	14.4	14.1	14.0	13.8	13.5	13.3	13.2	13.0	13.0	13.0

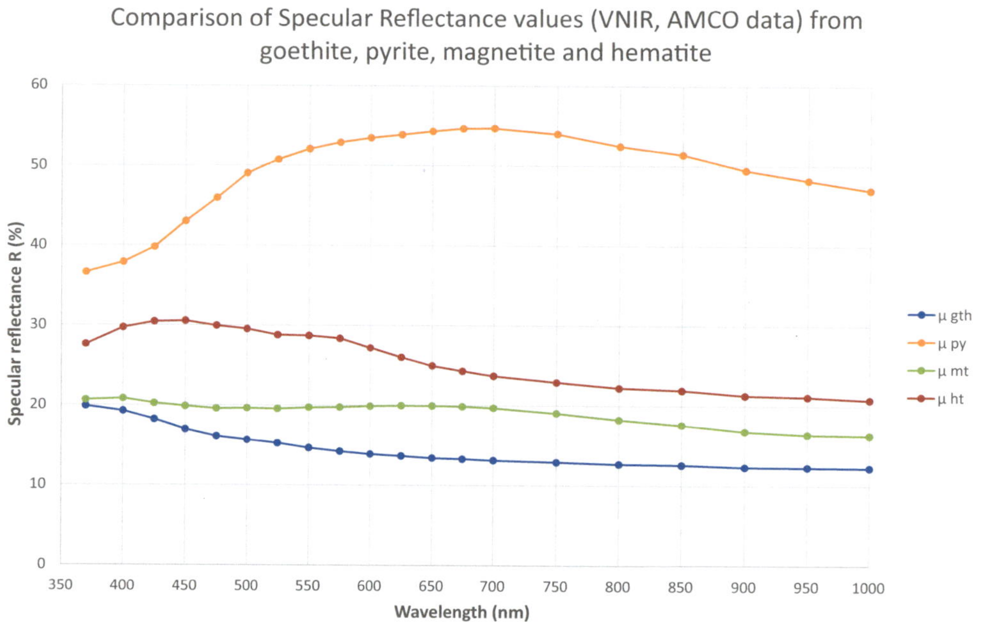

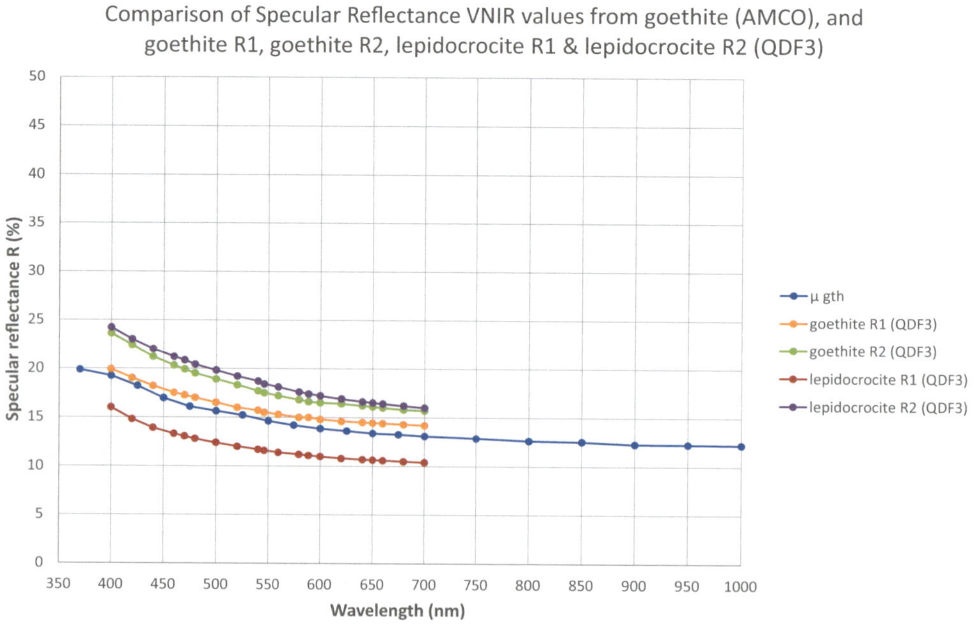

Gold (*Au, Native Gold*) 56

Au: Cubic

Gold	R_{546} (%)		Y (%)	λ_d (nm)	P_e (%)	Colour sensation
Air	77.0		76.1	578	39.8	Intense, very bright yellow
Oil	71.5		70.7	578	50.3	Bright, intense yellow
Bir: Null (0)				*Pleoc*: No (0)		*Tarn*: -
Anis+P: Imperfect Extinction Isotr. (2)				*IR*: No		*Twg*: Occasional
Mor: Allotriomorphic (incl, diss, detr.)				*Clv*: No		*PolQ*: Easy, but frequent scratches
Hardness	*PolH*: Medium, > gn (3)			*Mohs*: 3^-	*Talm*: B	*VHN* $_{(10\ g)}$: 30–34

Obs: Variable contents of Ag, Cu, Bi, Hg, Pd, Pt, which influence colour and optical parameters (QDF3)
Spectral information: cf. Table and R charts, multispectral data (VNIR, AMCO) and visible range (QDF3)
Mineral abbreviations: Au—native gold, ccp—chalcopyrite, el—electrum, gn—galena, mt—magnetite, po—pyrrhotite, Pt—native platinum, sp—sphalerite, td—tetrahedrite

56.1 General

Native gold generally contains some Ag, with which it can form solid solution in any proportion, usually defined by its *fineness*, i.e., the gold content (in ‰). For significant Ag contents, in the range of 30–70%, the term *electrum* is used, although it is not credited as mineral species (cf. discussion on electrum). Likewise, although less commonly, Au admits unlimited solid solution with Pt. Au may also contain other metals such as Cu, Bi, Hg, Pd, Rh, which may condition the colour tone and possible tarnishing. Current research (Chapman et al. 2021) aiming to understand their distribution points to some of them (e.g., Ag, Cu, Hg, precisely determined by EPMA) as potential tools to characterize the source.

Alloying with Hg (Au, Hg) is known as *gold amalgam* or *mercurial gold* for a Hg content ≈ 60% (Hey 1975), but the use of amalgam as a mineralogical species has been rejected (discussion, cf. amalgam, *am*), in favor of other phases whose definition satisfies IMA-COM requirements such as *aurihydrargirumite*, Au_6Hg_5, *yiyangite*, Au_3Hg, and *weishanite* $(Au, Ag)_{1.2}Hg_{0.8}$. However, since gold amalgam has acquired some relevance as an industrial product, generated in gold concentration processes by amalgamation that can produce a hazardous environmental impact, it has been considered appropriate to document its appearance (Figs. 7.3−7.6, cf. amalgam description).

56.2 Microscopic Description

56.2.1 SP Observation (*Single Polariser*)

Intense, luminous yellow *colour*, so much so that other minerals, even sulphides, often appear pale and dark by comparison; the effect is reinforced by immersion (Figs. 56.1a, b).

However, appreciable contents of other metals such as Ag, Cu, Pd or Hg may influence the colour tone and brightness: whiter and more reflective with Ag, more pinkish or reddish (Cu), paler (Hg), whiter and slightly brownish (Pd).

Although *polishing* is not difficult, it usually appears very scratched (Figs. 56.2, 56.3 and 56.5), given its low hardness (VHN). Native gold is very stable, so it does not tarnish but, exceptionally, it can darken due to impurities or, simply, due to dirt retained in the scratches or polishing pull-outs.

Very high *reflectance*, much more than pyrite and more than native bismuth but, in the visible range, much less than

R. Castroviejo, *A Practical Guide to Ore Microscopy—Volume 1*,
https://doi.org/10.1007/978-3-031-12654-3_56

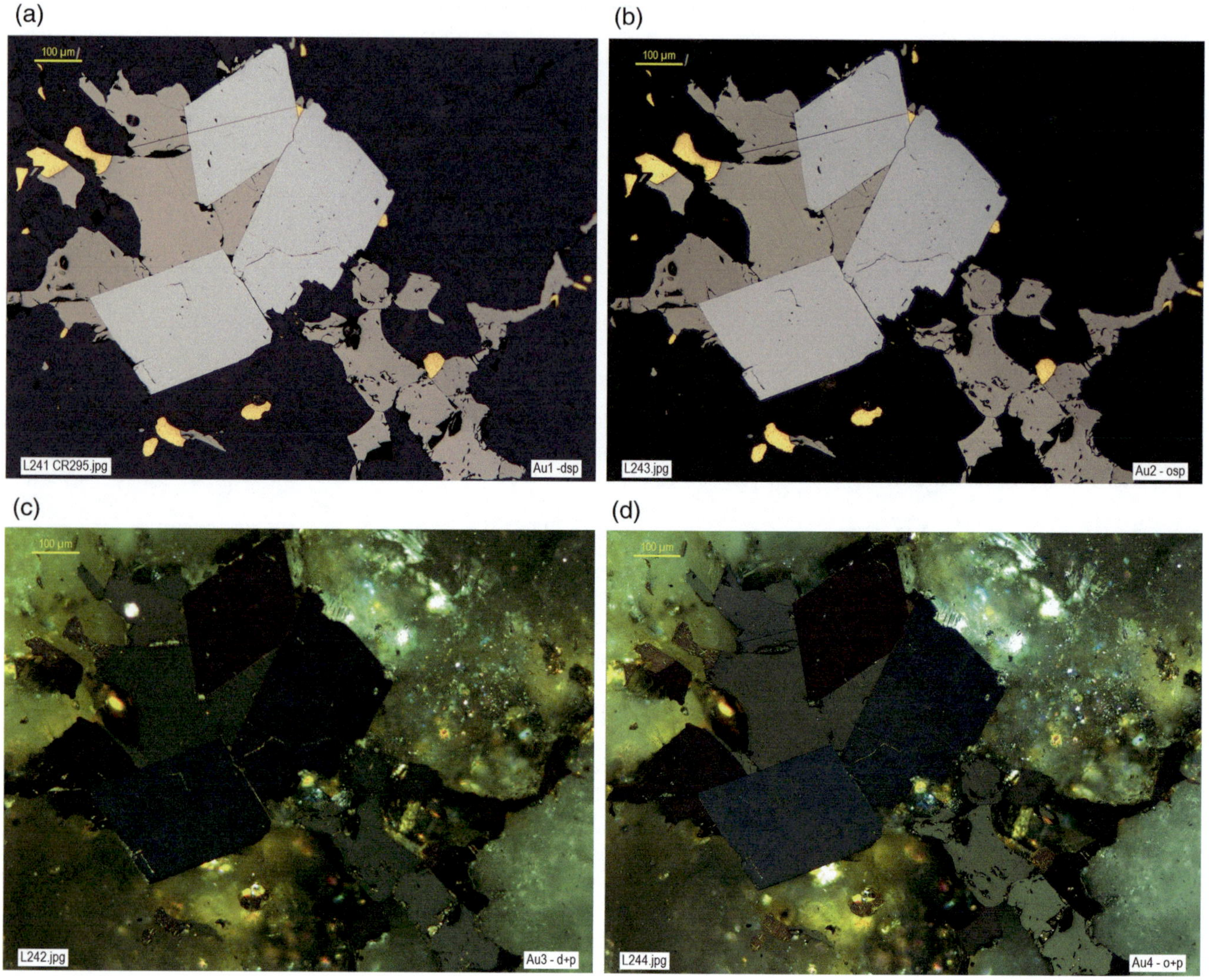

Fig. 56.1 **a** Gold (yellow) disseminated in asp aggregate (white) + po (brownish grey) and in quartz gangue (dark grey). Note the intense yellow colour of the gold (*fineness* 950, composition analyzed by electron microprobe: 95% Au, 5% Ag). CR295: La Rinconada Mine, Puno, Peru (**dsp**). **b** Same scene as Fig. 56.1 (**osp**). Note the higher contrast and brighter appearance of the immersed gold. **c** Same scene as Fig. 56.1 (**d+p**). **d** Same scene as Fig. 56.1 (**o+p**). Note the incomplete extinction of the gold and the better definition of the asp twins in immersion

native silver, cf. Table and R charts, multispectral data (VNIR, AMCO) and visible range (QDF3).

Isotropic, not pleochroic. Shows no exfoliation.

Medium *polishing hardness*, $PolH_{Au}$ ≈ chalcopyrite,[1] > gn, < sp or td. *Vickers microhardness*: 30–34 (10 g load, QDF3) to 41–94 (Uytenbogaardt and Burke 1971). *Mohs hardness*: 2.5–3. *Talmage H*: B.

Usually allotriomorphic, in tiny inclusions (e.g., in sulphides, Fe oxides, quartz) or filling microcracks, interstitial spaces, intergranular films and pores (Fig. 56.4). In some cases (*invisible gold* deposits), Au has no mineralogical expression because it is dispersed at the atomic scale in the crystal lattice of other minerals (e.g., pyrite or arsenopyrite), but in some cases it has been shown that it is only the very fine grain size that makes the gold "invisible".

Compositional zoning[2] is occasionally noticeable by colour changes, especially in detrital deposits, where it can

[1] Although for Ramdohr (1980) $PolH_{Au}$ is similar to that of *ccp*, after the compilation by Schneiderhöhn (1952, cf *Annex 4.2.1.1*) *Au* is harder than *ccp* (and also than *en, mol, cub*), but slightly less than *el* and *Ag*. Possibly, the elemental compositions of the gold grains compared by both authors was different.

[2] Overview with recent analytical data, cf Spiridonov and Yanakieva (2009)

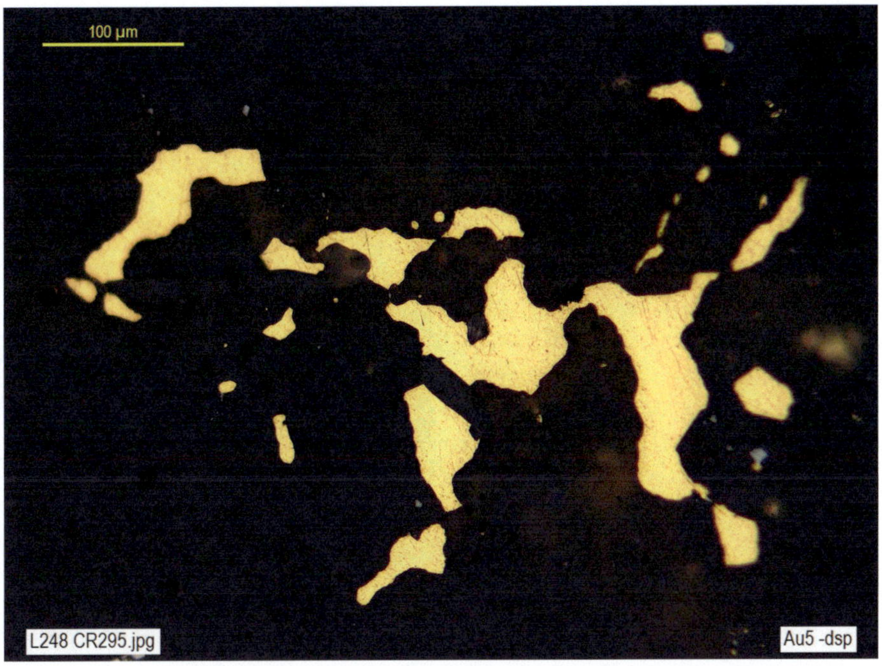

Fig. 56.2 Gold (yellow) disseminated in quartz gangue (dark grey). CR295: La Rinconada Mine, Puno, Peru (**dsp**)

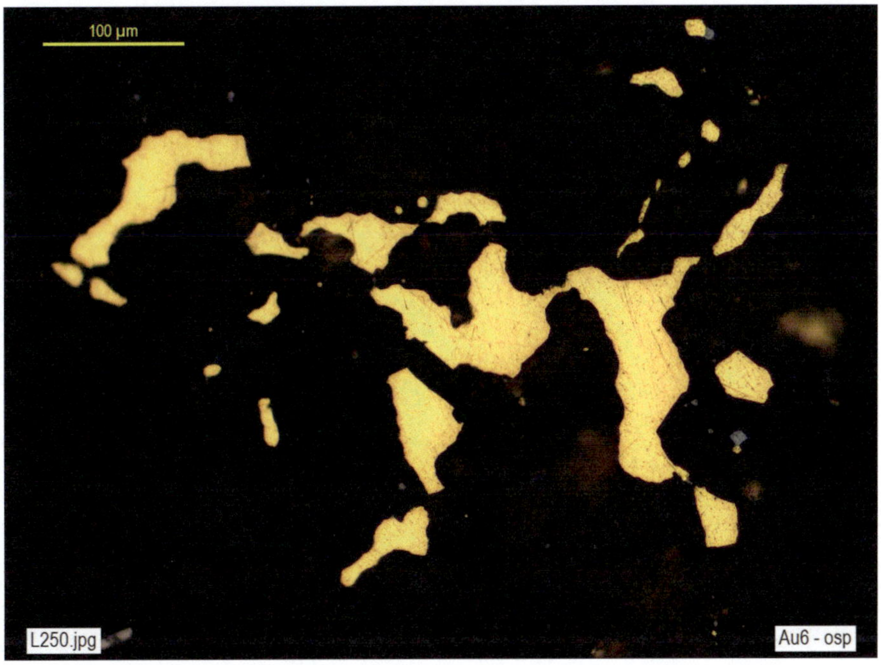

Fig. 56.3 Same scene as Fig. 56.2 (**osp**). More intense yellow colour of gold (*fineness* 950) in immersion

be observed at present as an on-going, actualistic process that can be precisely analyzed with MPA. This is a natural refining process by peripheral leaching of Ag from the mixed AuAg nuggets, leaving a distinctly yellower *corona*, i.e., a very gold-rich rim around the nugget (Fig. 56.5).

56.2.2 Observation +P (*Polariser* + *Analyser at 90°*)

Isotropic, but incomplete extinction (Figs. 56.1c, d), an effect accentuated by the numerous polishing scratches.

Fig. 56.4 Gold (yellow) disseminated: interstitial, in intergranular spaces and in microcracks and inclusions in quartz (dark grey) or associated with pyrrhotite (greyish brown) and arsenopyrite (white). The intense yellow colour of the gold corresponds to a high purity (*fineness* 950). Same sample as Fig. 56.1 (wider field, **dsp**)

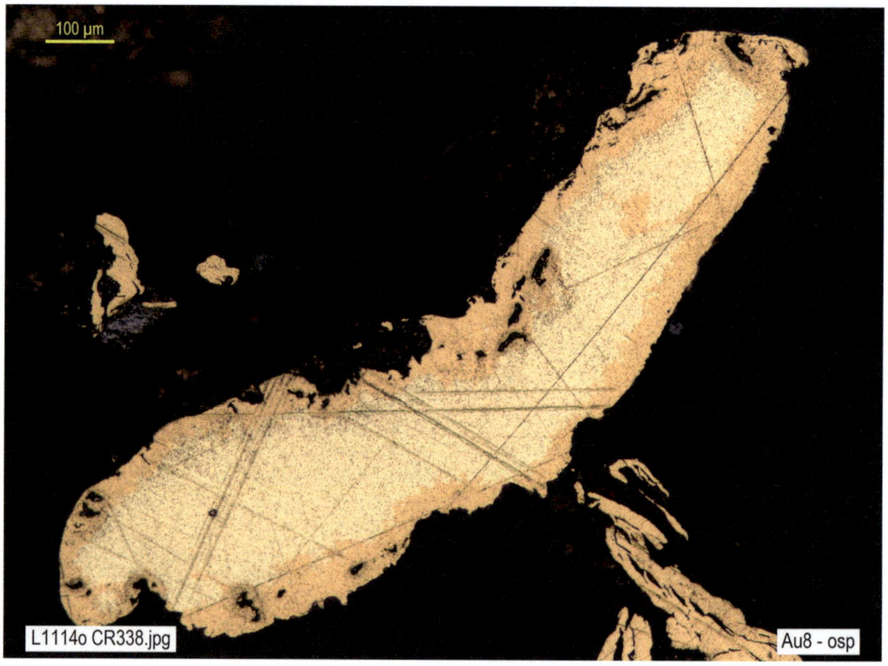

Fig. 56.5 Gold nugget, composed of an electrum core (pale yellow) and a crust enriched in Au (more pronounced yellow colour, in this case with a slight pinkish tinge, probably due to impurities) by selective dissolution of the silver from the rims. CR338: pan concentr. Río Alagón, Rincón del Obispo, Cáceres; **osp**

No *internal reflections*. Occasionally, laminar, // (111), *growth twins* are observed, although the fine grain size does not make it easy to distinguish them.

56.2.3 Paragenesis

Although it is a scarce mineral, its *parageneses* are varied. Hypogenic: metamorphic hydrothermal concentrations or orogenic gold deposits, epithermal, skarn, ophiolites. Supergene: detrital (placers and paleoplacers, vbgr. Witwatersrand), surficial mobilizations and impregnations (e.g., in laterites, cf. Figs. 1.51g and 1.51h, vol. 2), oxidation blankets of sulphide deposits, iron hats or gosssan formations (with Fe oxides). Mixed: associated with volcano-sedimentary massive sulphides.

The list of *associated minerals* in *gold-bearing assemblages* can be, therefore, wide and varied: arsenopyrite,

pyrite, galena, pyrrhotite, stibnite, cobaltite, hematite, goethite, uraninite, etc.

56.2.4 Identification

Its intense yellow colour, its low hardness and its high reflectance are typical, in spite of usually imperfect polishing due to the frequent scratches. All this makes its presence very conspicuous, while the other minerals next to it appear pale and—except Ag—dark (Fig. 56.4), an effect even more evident in immersion. Therefore, its recognition should not be a problem.

- However, it can be confused with sulphides such as pyrite or chalcopyrite in very fine inclusions (< 5 μm) isolated in gangue, given the difficulty for the human eye to quantify the brightness without an adequate reference, since these sulphides also appear as very bright against the dark background of the gangue.In these cases, it is convenient to resort to 20 × or 50 × immersion objectives, which will dispel any doubt.
- Chalcopyrite is darker.
- Native silver is white and brighter. Although tarnishing may give it a gold-like appearance, careful cleaning or polishing will eliminate any confusion.
- Electrum is brighter, but paler and its colour and reflectance vary with Ag content (generally whiter and > R with increasing Ag content).
- Native copper, reddish.
- Native bismuth, paler, less reflective and anisotropic.
- Native platinum, white and harder, better polished.

56.3 Spectral Information

56.3.1 Gold Multispectral Reflectance R (in Air, Without Polariser) (%)

λ (nm)	370	400	425	450	475	500	525	550	575	600	625	650	675	700	750	800	850	900	950	1000
μ	30.8	32.3	32.7	33.0	33.8	40.5	54.5	66.3	73.7	78.3	81.0	83.3	84.9	85.9	87.5	87.9	89.0	89.8	90.3	90.4
P_5	28.2	29.5	29.9	29.9	30.4	36.2	48.3	60.0	67.5	72.2	74.9	77.1	78.7	79.8	81.5	82.2	83.4	84.1	84.7	84.7
P_{20}	29.4	30.8	31.3	31.3	31.8	38.7	52.1	63.9	71.5	76.0	78.8	81.1	82.7	83.7	85.3	85.7	86.7	87.6	88.0	88.0
P_{80}	32.3	33.8	34.2	34.7	35.9	42.3	57.0	68.9	76.3	80.8	83.6	85.9	87.4	88.5	90.0	90.3	91.5	92.3	92.8	93.0
P_{95}	33.5	35.2	35.9	36.8	38.2	44.5	59.4	70.7	78.1	82.5	85.3	87.6	89.0	90.1	91.5	91.9	93.6	93.6	94.5	94.1

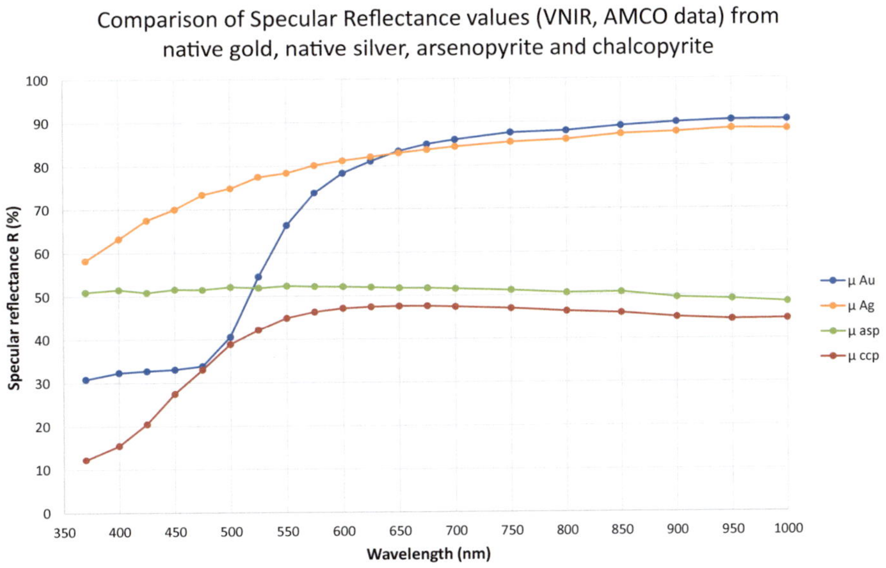

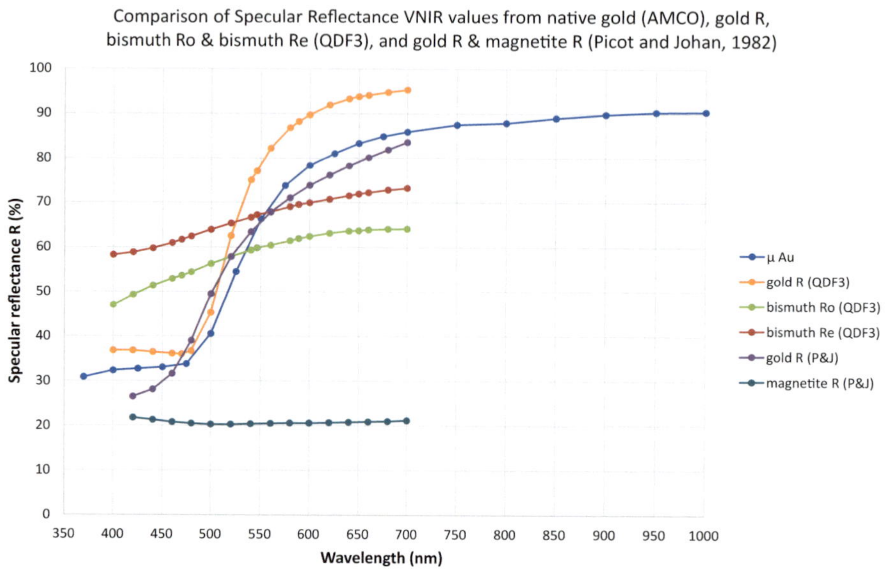

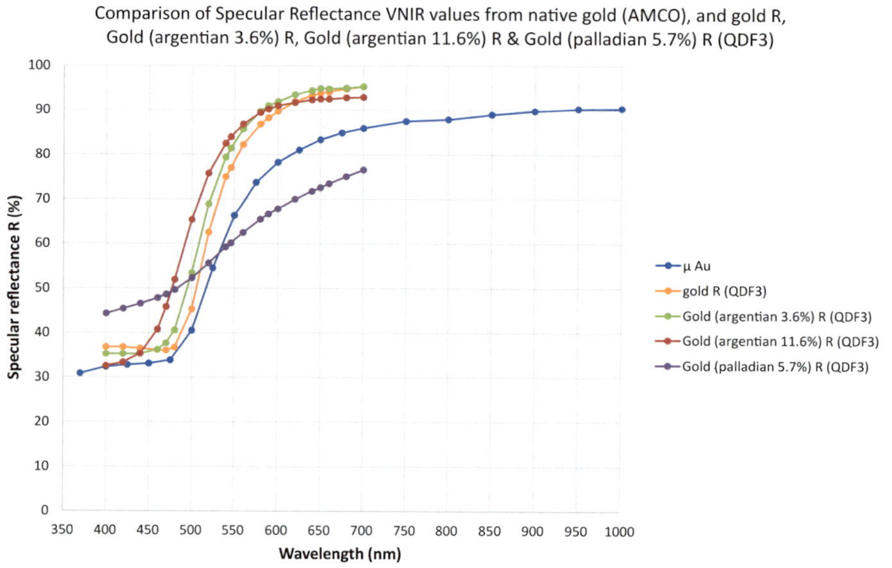

References

Chapman RJ, Banks DA, Styles MT, Walshaw RD, Piazolo S, Morgan DJ, Grimshaw MR, Spence-Jones CP, Matthews TJ, Borovinskaya O (2021) Chemical and physical heterogeneity within native gold: implications for the design of gold particle studies. Miner Deposita 56:1563–1588. https://doi.org/10.1007/s00126-020-01036-x

Hey MH (1975) An index of mineral species and varieties arranged chemically (Chemical Index of Minerals). British Museum (Nat. History), vol 3. London

Ramdohr P (1980) The ore minerals and their intergrowths, vol 2. 2nd edn, Pergamon Press, Oxford, pp 1205

Schneiderhöhn H (1952) Erzmikroskopisches Praktikum. E. Schweizerbart'sche Verlagsbuchhandlung, Stuttgart, p 274

Spiridonov E, Yanakieva D (2009) Modern mineralogy of gold: overview and new data. ArcheoSciences 33:67–73

Uytenbogaardt W, Burke EAJ (1971) Tables for microscopic identification of ore minerals, 2nd edn. Elsevier, Amsterdam

Graphite (*gra*)

C: Hexagonal

57

Graphite		R_{546} (%)	Y (%)	λ_d (nm)	P_e (%)	Colour sensation
Air	R_o	26.4	26.6	581	4.9	Grey to yellowish brown, very strongly bireflective and pleochroic
	$R_{e'}$	6.2	6.2	c497	0.6	
Oil	R_o	14.9	15.1	581	6.2	
	$R_{e'}$	0.5	0.5	587	2.7	
Bir: Extremely strong (5)				*Pleoc*: Very strong (5)		*Tarn*: No
Anis +P: Strong (5)				*IR*: No		*Twg*: No
Mor: Markedly lamellar. Spherulitic				*Clv*: Perfect {0001}		*PolQ*: Difficult
Hardness	*PolH*: Medium, > ccp, < sp (3)			*Mohs*: 1–2	*Talm*: nd	*VHN* $_{(10\ g)}$: 7–11

Obs: cf. Table and R charts, multispectral data (VNIR, AMCO) and visible range (QDF3)
Mineral abbreviations: Ag—native silver, Au—native gold, ccp—chalcopyrite, gg—gangue, gn—galena, gth—goethite, ht—hematite, mgh—maghemite, mol—molybdenite, mt—magnetite, po—pyrrhotite, py—pyrite, q—quartz, rt—rutile, sp—sphalerite, td—tetrahedrite

57.1 General

Composition: C. Diamond polymorph.

Graphite properties may vary slightly with increasing interlayer distance (semi-graphites, pre-graphites, pre-graphitized organic matter). Its crystallinity varies as a function of temperature and metamorphic grade.

57.2 Microscopic Description

57.2.1 SP Observation (*Single Polariser*)

Grey to bright yellowish-brown *colour*, with a remarkable peculiarity, which is obvious when rotating the stage (Figs. 57.1a, b and 57.2a).The strong contrast between the scarce brightness (dark grey colour, comparable to gangue) in the $R_{e'}$ orientation and the noticeably more vivid gloss (yellowish-brown colour, metallic appearance) in the R_o orientation.

In general, the *Reflectance, R,* can be characterized as moderate or low to very low (between R of quartz or gangue and R of hematite, tetrahedrite or rutile). Although in absolute value its maximum reflectance (R_o) is no more than moderate (< td, for example), in relative terms it is enormous, higher than the minimum ($R_{e'}$) by a factor of x4 (air) and x30 (oil).

This strong bireflectance is reflected in the variation of the multispectral R values, even without polariser, in randomly oriented grains, as shown by the scatter of the curves corresponding to the P_5–P_{95} percentiles of graphite in the AMCO R plot (cf. summary table above and AMCO and QDF3 comparative charts, below).

The different slopes of both curves (R_o and $R_{e'}$), evidenced in Graph QDF3, show that graphite is not only *strongly bireflective*, but also *strongly pleochroic*.

Polishing is difficult (soft, unctuous ore), easier in fine-grained aggregates. *Polishing hardness is* medium, higher than might be expected from scratch hardness: PolH gra > mol, ccp, gn; < Ag, Au, td, sp, po. *Cleavage* (0001) is usually visible, often bent by deformation, given the ductility of the mineral.

It usually occurs in the form of lamellae or leafy aggregates, usually in metamorphic rocks. Exceptionally, it has a spherulitic structure (cf. Fig. 1.25b, c, vol. 2).

R. Castroviejo, *A Practical Guide to Ore Microscopy—Volume 1*,
https://doi.org/10.1007/978-3-031-12654-3_57

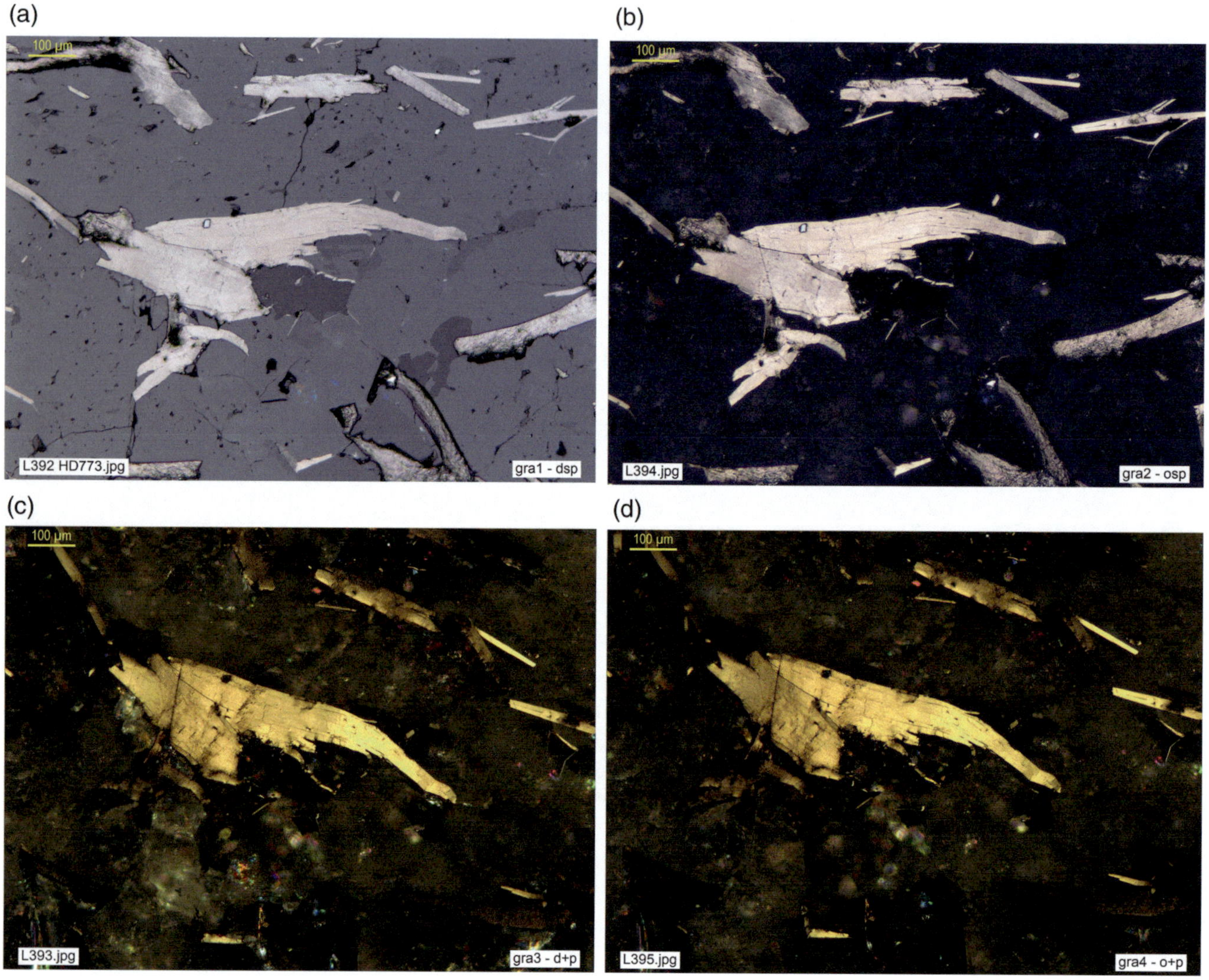

Fig. 57.1 **a, b** (**dsp** and **osp**, respectively)—Quartz-schist with gra and small inclusions of po (in gra) and py (in q). Note the strong pleochroism of gra, although the metamorphic orientation of the rock causes the gra lamellae to be almost all subparallel and, therefore, not to display in a single image the extreme bireflectance of the mineral to its full extent, as can be seen in Fig. 57.2a. The tiny particles of py and po serve as a reference to appreciate the reflectance. HD 773, Kropfmuehl, Passau, Germany. **c, d** Same scene as gra1 (**d+p** and **o+p**, respectively). Strong, striking anisotropism of gra

57.2.2 Observation +P (*Polariser + Analyser at 90°*)

Strongly anisotropic, very striking for its brightness in contrast to the extinction position, but the colour effects are not very varied (yellowish tones, Fig. 57.1c, d, 57.2b and 57.3b).

57.2.3 Paragenesis

It is found in many metamorphic rocks (especially if they come from protoliths rich in organic matter), but it is also found in seams and in magmatic rocks, carbonatite complexes (cf. spherulitic texture and Fig. 1.25b, c, vol. 2), detrital deposits (Fig. 57.3a, b), etc. Being such a widespread mineral, the possible accompanying minerals are countless.

57.2.4 Identification

It is easily recognized by its leafy morphology, with clearly visible cleavage and grey to yellowish-brown tones but, above all, by its extreme bireflectance, pleochroism and anisotropism, only comparable to those observed in minerals

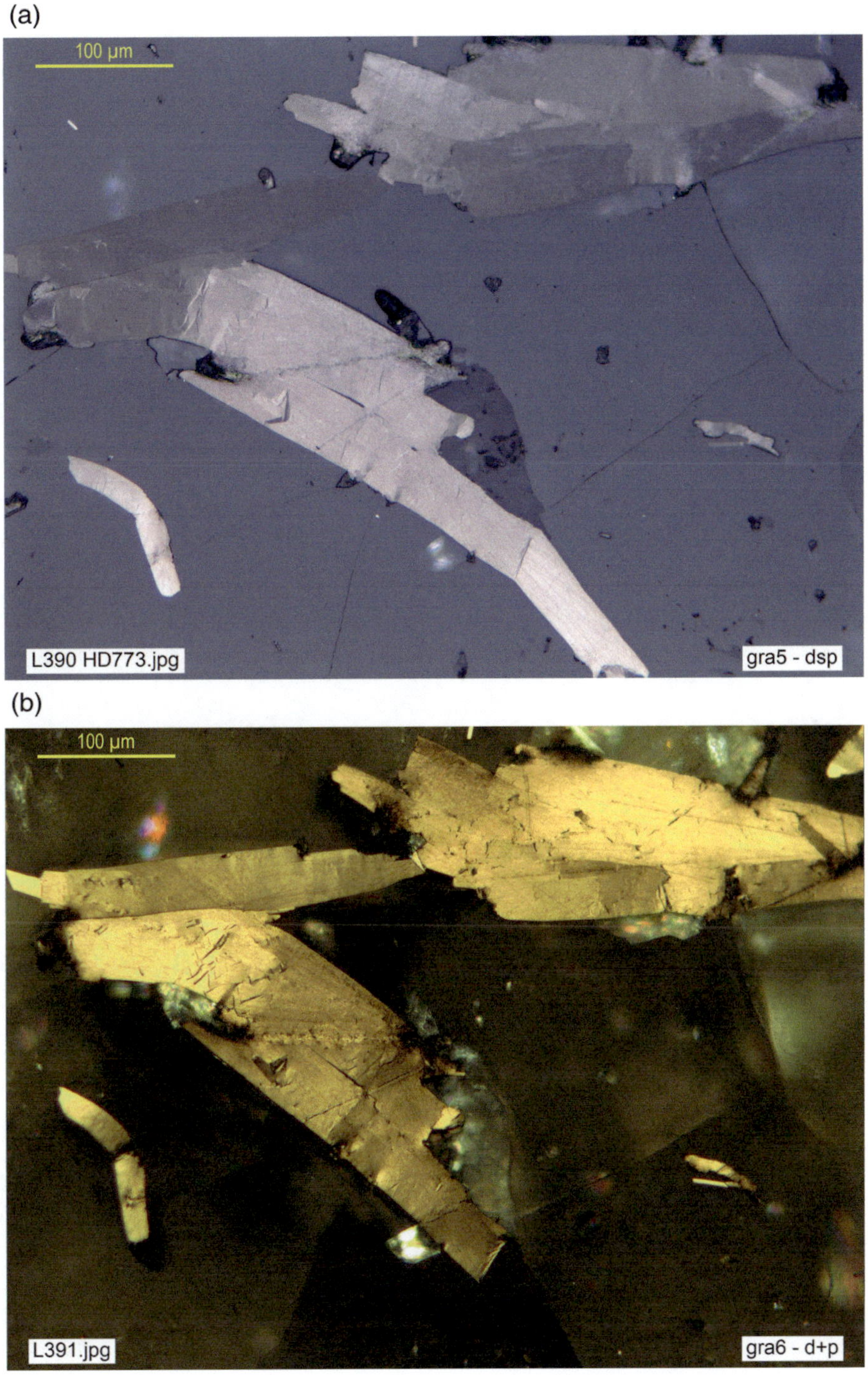

Fig. 57.2 **a, b** (Same image: **dsp** and **d+p**, respectively). Extreme bireflectance of graphite, which is similar to gg in $R_{e'}$ orientation (≈ 5%, horizontal lamella, upper left, distinct only with analyser: Fig. 57.2b) and much brighter, exceeding 25% in R_o (subvertical lamella, lower left). Comparing the **dsp** and **d+p** images, it is obvious that, thanks to the strong gra anisotropism, some of the grains that in Fig. 57.2a might go unnoticed, mistaken for gg, stand out in Fig. 57.2b for their luminosity. HD 773, Kropfmuehl, Passau, Germany

such as molybdenite, mackinawite or valleriite. These can be distinguished without difficulty:

- *molybdenite* is much more reflective than graphite and polarizes in pinkish-white tints.
- *mackinawite* is white and shows much higher R-values than graphite: R_{max} up to 49%, $R_{min} \approx$ 21–25%.
- *valleriite*, which is the most similar to graphite, has R_{min} (≈ 10%) > R_{min} of graphite and shows, when uncrossing the analyser, bluish anisotropism effects, absent in graphite.

Fig. 57.3 **a**, **b** (Same image: osp and o+p, respectively)—Oolitic sandstone with gra (greyish-brown folded lamellae), maghemite (white), magnetite (brownish-grey), goethite (grey) and quartz (dark grey-black). The strong anisotropism of gra, with very luminous yellowish tones, is immediately noticeable (+p). CR 325, Gora Djebilet, Tindouf, Algeria

(a)

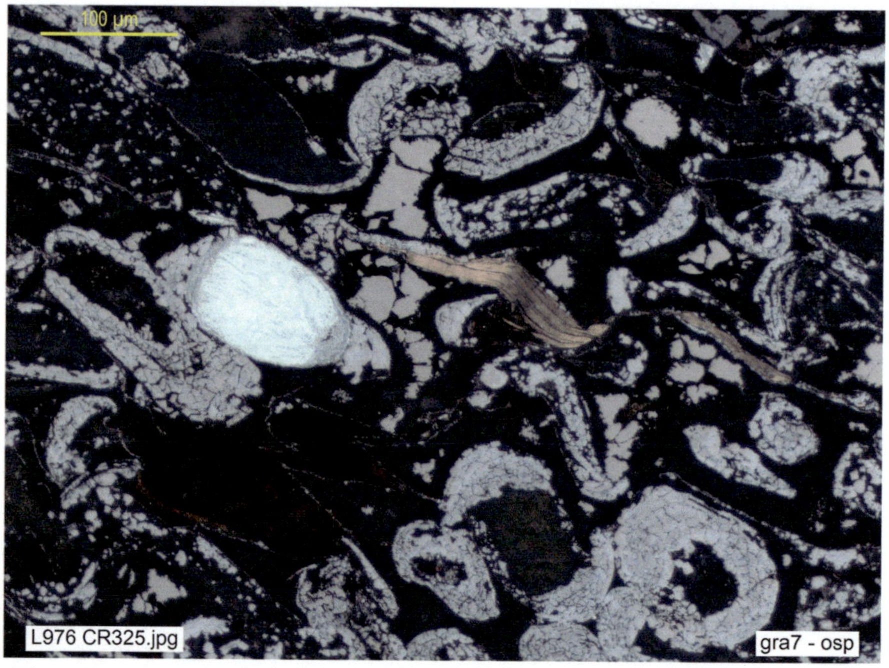

(b)

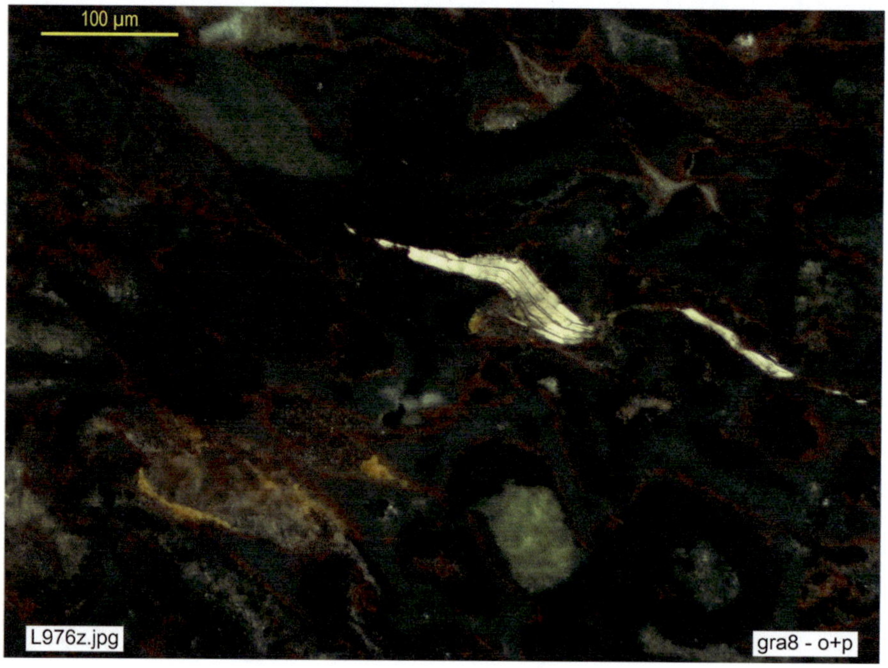

57.3 Spectral Information

57.3.1 Graphite Multispectral Reflectance R (in Air, Without Polariser) (%)

λ (nm)	370	400	425	450	475	500	525	550	575	600	625	650	675	700	750	800	850	900	950	1000
μ	10.9	11.3	11.5	11.7	11.9	12.2	12.6	12.8	13.1	13.4	13.6	13.9	14.1	14.4	14.9	15.4	15.9	16.5	17.0	17.5
P_5	7.0	7.8	8.3	8.7	8.9	9.4	9.8	10.1	10.4	10.7	10.9	11.3	11.4	11.7	12.1	12.5	12.9	13.4	13.6	14.1
P_{20}	8.7	9.3	9.7	10.1	10.3	10.7	11.2	11.4	11.7	12.0	12.2	12.6	12.7	13.0	13.3	13.6	14.0	14.5	14.8	15.2
P_{80}	12.8	13.0	13.1	13.2	13.3	13.6	13.9	14.0	14.3	14.6	14.8	15.2	15.4	15.8	16.4	17.0	17.7	18.5	19.1	19.8
P_{95}	14.9	15.0	15.1	15.2	15.2	15.4	15.6	15.7	15.9	16.1	16.2	16.5	16.6	16.9	17.5	18.1	18.9	19.7	20.5	21.3

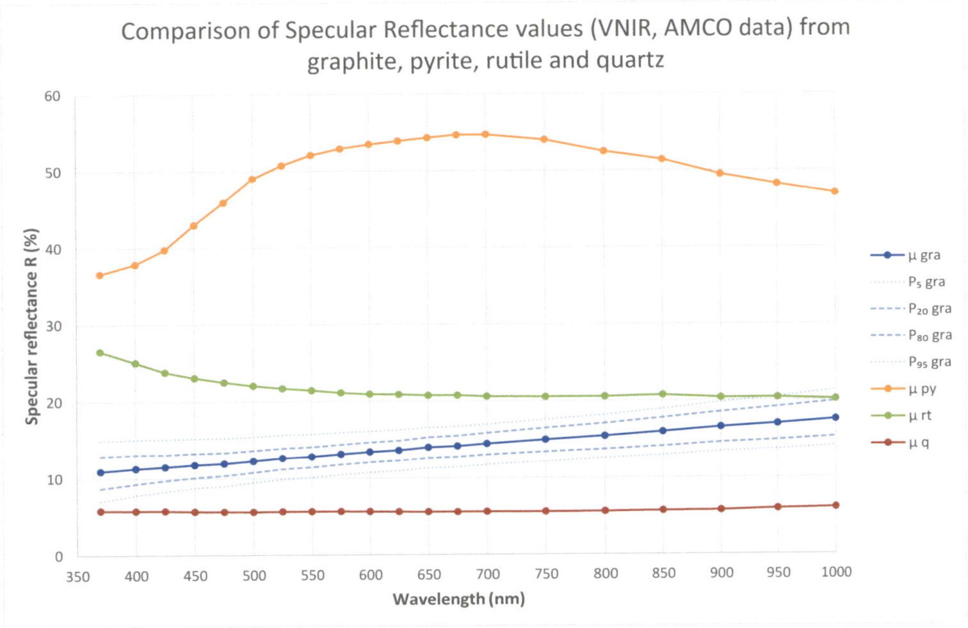

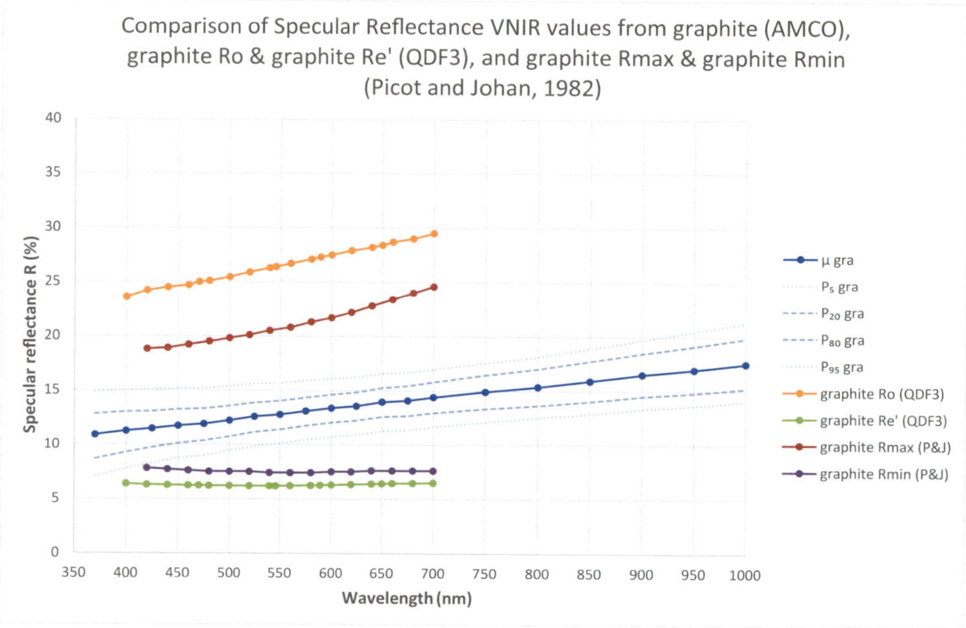
Comparison of Specular Reflectance VNIR values from graphite (AMCO), graphite Ro & graphite Re' (QDF3), and graphite Rmax & graphite Rmin (Picot and Johan, 1982)
Specular reflectance R (%)
Wavelength (nm)
0
5
10
15
20
25
30
35
40
350
400
450
500
550
600
650
700
750
800
850
900
950
1000
µ gra
P5 gra
P20 gra
P80 gra
P95 gra
graphite Ro (QDF3)
graphite Re' (QDF3)
graphite Rmax (P&J)
graphite Rmin (P&J)

58 Gratonite (*gtt*)

$Pb_9As_4S_{15}$:Trigonal

Gratonite		R_{546} (%)	Y (%)	λ_d (nm)	P_e (%)	Colour sensation
Air	R_o	36.2	36.4	456	1.8	White to greyish-white, slightly yellowish against gn
	R_e	36.2	36.3	439	1.1	
Oil	R_o	20.6	20.8	c567	2.9	
	R_e	20.6	20.8	c564	2.2	
Bir: Very weak (2)				*Pleoc*: Very weak (2)		*Tarn*: nd
Anis +P: Very weak (3)				*IR*: Reds, very rare		*Twg*: No
Mor: Usually euhedral				*Clv*: nd		*PolQ*: Good
Hardness	*PolH*: Low (2)			*Mohs*: 2½	*Talm*: B	*VHN* $_{(100\ g)}$: 130–146

Obs: cf. Table and R charts, multispectral data (VNIR, AMCO) and visible range (QDF3). VHN, QDF3
Mineral abbreviations: ang—anglesite, crs—cerussite, en—enargite, gc—geocronite, gn—galena, gtt—gratonite, jd—jordanite, py—pyrite, rl—realgar, sp—sphalerite, Ss—sulphosalts, td—tetrahedrite (*s.l.*), wz—wurtzite
Colour abbreviations: GrWh—greyish white

58.1 General

The composition of gratonite, ***gtt***, is that of a lead sulphoarsenide ($Pb_9As_4S_{15}$) close to jordanite, ***jd***, but purely arsenical. It may contain impurities of Fe and Sb. It appears frequently associated with secondary jd which, after replacing gtt, usually decomposes, forming galena, **gn**, among other secondary phases, which finally alters to anglesite and cerussite (**ang**, ***crs***).

According to experimental data (Roland 1968), gtt would be a low-temperature dimorph of jd, although textural observations suggest that jd can also form at low temperatures (cf. description jordanite, ***jd***, where these aspects are discussed in more detail).

58.2 Microscopic Description

58.2.1 SP Observation (*Single Polariser*)

White *colour*, similar to galena, but slightly yellowish in direct comparison. Very similar to jd, in colour and *reflectance* values (cf. multispectral comparative plots, Figs. 58.1a, b and 58.2a). Very weak pleochroism, barely perceptible on contiguous grain boundaries.

Good *polish*. Low *polishing hardness* (towards the medium–low limit, slightly < jd).

Typically occurs in perfect idiomorphic crystallites (hexagonal or trigonal sections), disseminated in *schalenblende* (cf. ***sp-wz***), and also in aggregates of tabular outlines, often pseudomorphosed by secondary jordanite (Figs. 58.1 a–d and 58.2b).

58.2.2 Observation +P (*Polariser + Analyser at 90°*)

Anisotropic but with very weak anisotropism effects (Figs. 58.1c, d and 58.2b), hardly recognizable (except on grain boundaries, ***o + p*** or ***oxp***). The anisotropism of gtt is the weakest of all non-cubic Pb-As sulphosalts, according to Ramdohr (1980). Straight extinction.

IR, *internal reflections*, only rarely visible (red, very scarce, Fig. 58.1c, d). *No twinning* observed.

R. Castroviejo, *A Practical Guide to Ore Microscopy—Volume 1*,
https://doi.org/10.1007/978-3-031-12654-3_58

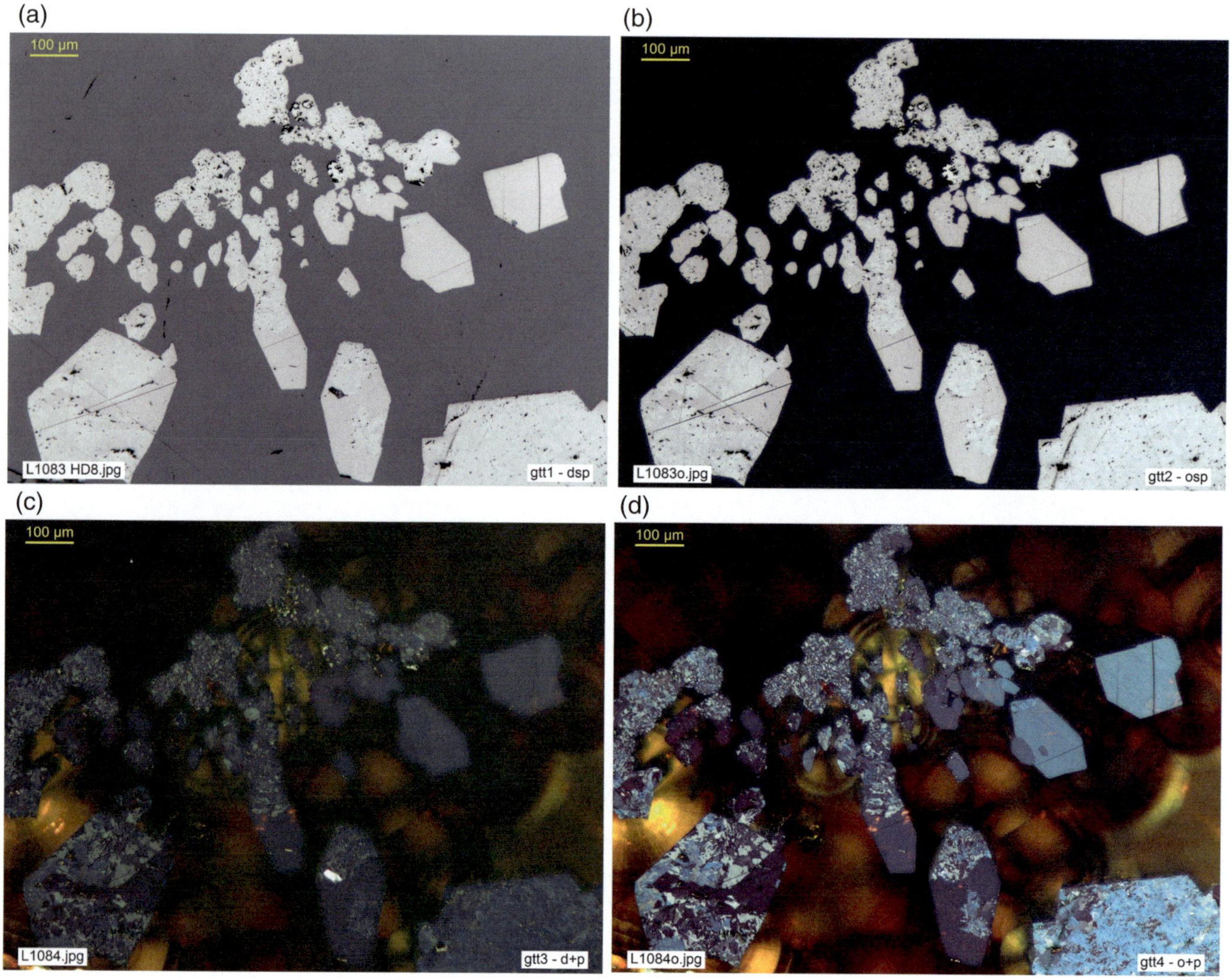

Fig. 58.1 **a, b** (Same scene: **dsp** and **osp**, respectively). Gratonite, ***gtt*** (light grey-white), in disseminated, idiomorphic crystals in a massive background of sphalerite-wurtzite (*schalenblende*, grey). Most of the gtt crystals are pseudomorphosed by jordanite, ***jd*** (colour like gtt), which is the most abundant phase after *schalenblende*. HD8, Wiesloch, Baden-Wuerttenberg, Germany. **c, d** (Same scene as Figs. 58.1a: **d+p** and **o+p**, respectively). Vivid anisotropism of jd, especially in immersion (Fig. 58.1d), which contrasts with the dull tones of gtt. Internal reflections, I*R*, abundant in *schalenblende* and much scarcer in gtt

58.2.3 Paragenesis

Hydrothermal *paragenesis*. Infrequent (Wiesloch, Germany; Silesia, Poland; Cerro de Pasco, Peru).

Typically associated, as disseminations, with *schalenblende* (cf. ***sp-wz***) in Wiesloch (Germany). Other associates: gn, jd, py, en, td (*s. l.*), rl, crs, ang.

58.2.4 Identification

Its colour and reflectance differentiate, without possible confusion, gtt from sphalerite and wurtzite (*schalenblende:* cf. ***sp-wz***). Although it resembles *gn*, its colour and R distinguish them: *gtt* is more yellowish and less reflective, apart from its anisotropism, morphology, etc.

The distinction is more delicate with respect to other sulphosalts that have similar colour and spectral curves, such as *jd*, but is still possible because of the extremely weak anisotropism and the morphology—typical euhedral, hexagonal habit—of gtt. Paragenesis, especially when it occurs in its typical association with sphalerite and wurtzite (*schalenblende:* cf. ***sp-wz***), also provides a useful criterion.

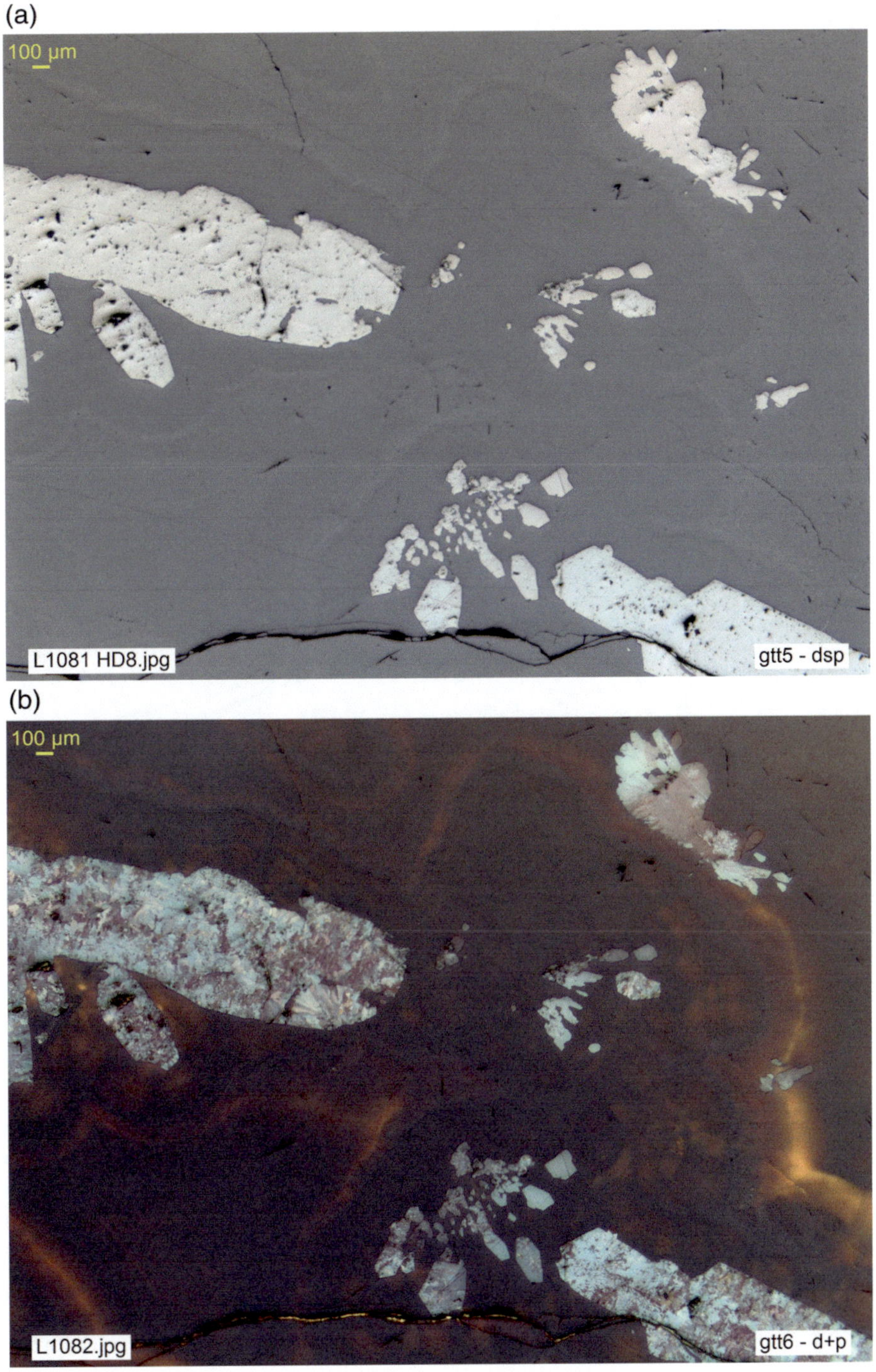

Fig. 58.2 **a, b** (Same scene: **dsp** and **d+p**, *respectively*). Wide-field image: *schalenblende* deposition (grey, massive, IR and concentric, botryoidal-spherulitic structure) on microcrystalline aggregates of *jd* (GrWh, clearly anisotropic) pseudomorphic after gratonite, *gtt*. Scarce fresh, monocrystalline, idiomorphic relicts of gtt (hexagonal sections, GrWh like jd, but very weakly anisotropic), preserved as inclusions in *schalenblende* (well visible in centre-lower zone of the image, enlarged in Fig. 58.1a). HD8, Wiesloch, Baden-Wuerttenberg, Germany

58.3 Spectral Information

58.3.1 Gratonite Multispectral Reflectance R (in Air, No Polariser) (%)

λ (nm)	370	400	425	450	475	500	525	550	575	600	625	650	675	700	750	800	850	900	950	1000
μ	38.9	38.1	37.5	36.8	36.4	36.0	36.2	36.3	36.5	36.6	36.0	35.4	34.2	33.1	31.8	30.9	30.3	30.3	29.9	29.9
P_5	37.9	37.2	36.6	36.0	35.6	35.2	35.4	35.6	35.8	36.0	35.2	34.4	33.2	32.0	31.2	30.0	29.6	29.6	29.2	29.2
P_{20}	38.3	37.6	37.0	36.4	36.0	35.6	35.8	36.0	36.2	36.4	35.6	34.8	33.6	32.4	31.2	30.4	30.0	30.0	29.6	29.6
P_{80}	39.1	38.4	37.8	37.2	36.8	36.4	36.6	36.8	37.0	37.2	36.6	36.0	34.8	33.6	32.0	31.2	30.8	30.8	30.4	30.4
P_{95}	39.5	38.8	38.2	37.6	37.2	36.8	37.0	37.2	37.4	37.6	37.0	36.4	35.2	34.0	32.4	31.6	31.2	31.2	30.8	30.8

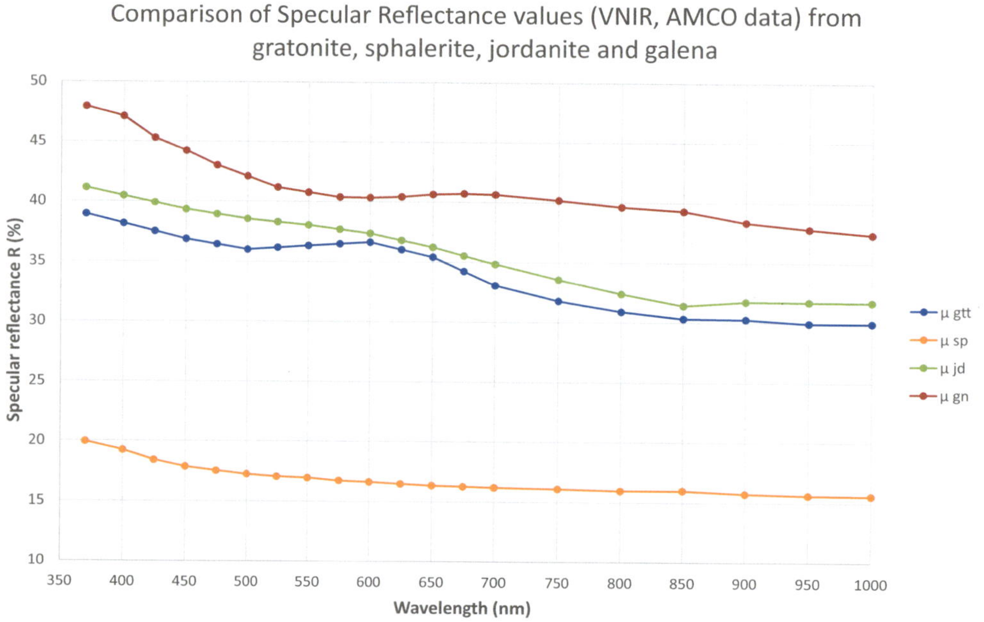

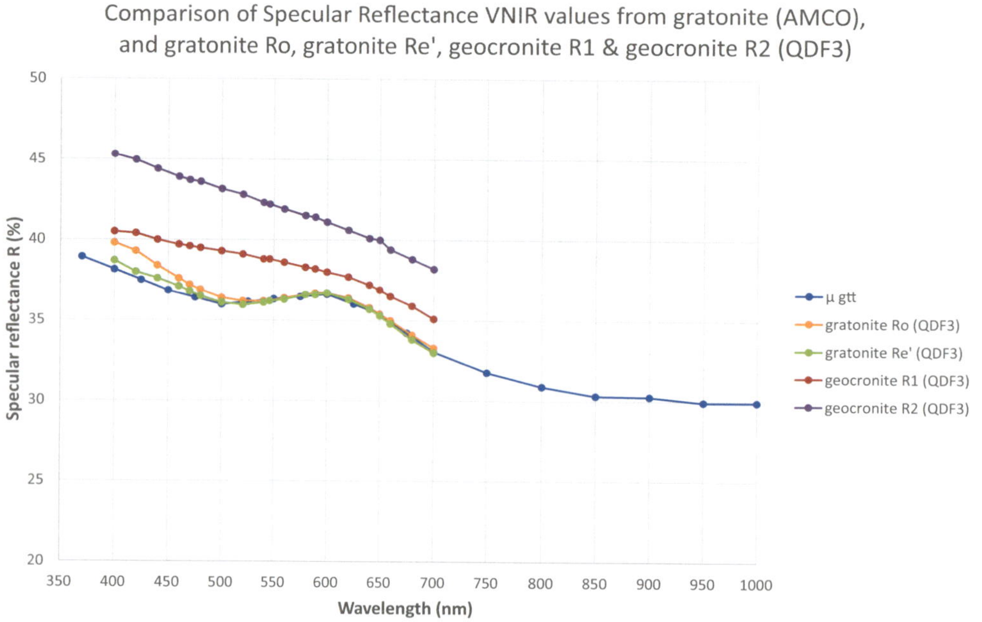

References

Roland GW (1968) The system Pb–As–S. Composition and stability of jordanite. Miner Deposita 3:249–260

Ramdohr P (1980) The ore minerals and their intergrowths, 2 vol, 2nd ed. Pergamon Press, Oxford, p 1205

Gudmundite (*gud*)

59

FeSbS: Monoclinic (Pseudorhombic)

Gundmundite		R_{546} (%)	Y (%)	λ_d (nm)	P_e (%)	Colour sensation
Air	R_1	51.5	51.9	583	3.0	Bright white
	R_2	56.6	56.1	495	1.0	
Oil	R_1	37.8	38.2	584	3.8	
	R_2	43.3	42.7	489	2.3	
Bir:Weak (3)				*Pleoc*: Weak (3)		*Tarn*: –
Anis +P: Strong (4)				*IR*: –		*Twg*: Twinning frequent, varied
Mor: Euhedral tendency < asp				*Clv*: –		*PolQ*: Very good, > asp
Hardness	*PolH*: High, > po, < asp (4)			*Mohs*: 5.5	*Talm*: –	*VHN* $_{(100\ g)}$: 572–673

Obs: cf. Table and R charts, multispectral data (VNIR, AMCO) and visible range (QDF3; Picot and Johan 1982)
Mineral abbreviations: asp—arsenopyrite, bth—breithauptite, ccp—chalcopyrite, gg—gangue, gn—galena, gud—gudmundite, lo—löllingite, mc—marcasite, mg—magnetite, po—pyrrhotite, prm—parammelsbergite, py—pyrite, rm—rammelsbergite, sbt—stibnite, sf—safflorite, sp—sphalerite, td—tetrahedrite

59.1 General

Composition: FeSbS, with occasional traces of Ni.

59.2 Microscopic Description

59.2.1 SP Observation (*Single Polariser*)

Bright white, slightly pinkish *colour* (very faint shade, perceptible in immersion, Fig. 59.1a, b). Very good *polish*. High *reflectance* (> lo, sf, asp, mc, but < rm, prm; *cf* multispectral comparison charts). Anisotropic, bireflectance and pleochroism of gud are > arsenopyrite but still weak, visible mostly in immersion and comparing adjacent grains (white and pinkish tones).

High *polishing hardness* but close to pyrrhotite, i.e., it is among the lowest of the *high PolH group*: > po, sf, rm but < mt, asp.

It occurs frequently in crystals with elongated prismatic to acicular habit, but its euhedral tendency is less pronounced than that of arsenopyrite.

59.2.2 Observation +P (*Polariser + Analyser at 90°*)

Strong *anisotropism* (> arsenopyrite), with pronounced and characteristic colours (orange, red-violet and bluish tones, reinforced in immersion, Fig. 59.1c, d), which become more luminous and enriched with light orange-yellowish and greenish tones when the analyser is slightly uncrossed (2–3°) (Fig. 59.2a, b).

R. Castroviejo, *A Practical Guide to Ore Microscopy—Volume 1*,
https://doi.org/10.1007/978-3-031-12654-3_59

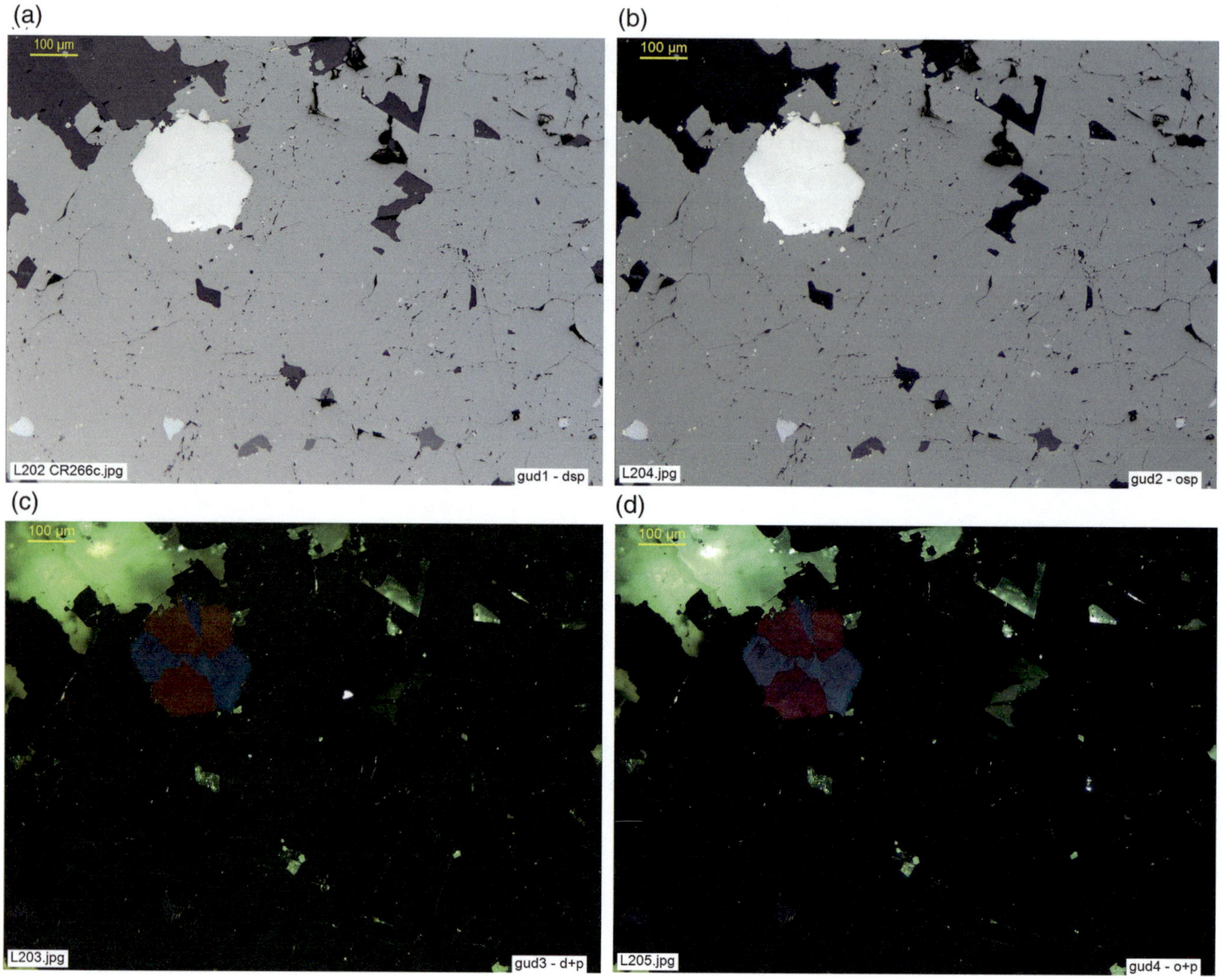

Fig. 59.1 **a** Grain of gud (white) included in massive td (grey), which also includes gn, sp, quartz-carbonate gg and tiny particles of ccp and py. Incipient corrosion of gud by td, resulting in an indented outline that partially masks the euhedral morphology of gud. CR266c: Lucky Friday Mine, Coeur d'Alene, Idaho, USA (**dsp**). **b** Same scene as gud1 (**osp**). Note the slight pink pleochroism of gud and the darkening of gg, which appears black in immersion, while sp remains a dark grey. **c, d** Same scene as gud1 (**d+p**) and (**o+p**), respectively, twinning and strong and coloured anisotropism of gud, accentuated in immersion

Frequent, varied, laminar or irregular twinning (Figs. 59.1c, d and 59.2a, b).

59.2.3 Paragenesis

Hydrothermal *paragenesis* of posthumous phases in high-temperature sulphide deposits, associated with late Sb inputs or decay of early Sb minerals. Under these conditions, it is a relatively common minor component and may occur in the company of numerous species: tetrahedrite, stibnite, pyrrhotite, arsenopyrite, breithauptite, pyrite, galena, sphalerite, etc.

59.2.4 Identification

Its recognition is not difficult, except in very fine aggregates (frequent). Its strong anisotropism is typical, with *orange* to bluish and greenish polarization tints, different from those of arsenopyrite or löllingite. Its pinkish-white pleochroism (> arsenopyrite) and its relatively low polishing hardness ($\approx$ pyrrhotite, $\ll$ arsenopyrite), as well as the association with antimoniferous species (*vbgr.* tetrahedrite) are also useful criteria.

Fig. 59.2 **a, b** Same scene as gud1 (**dxp**) and (**oxp**), respectively. When the analyser is slightly uncrossed, gud shows brighter and livelier polarisation tints, appearing orange-yellow and bluish-green tones

(a)

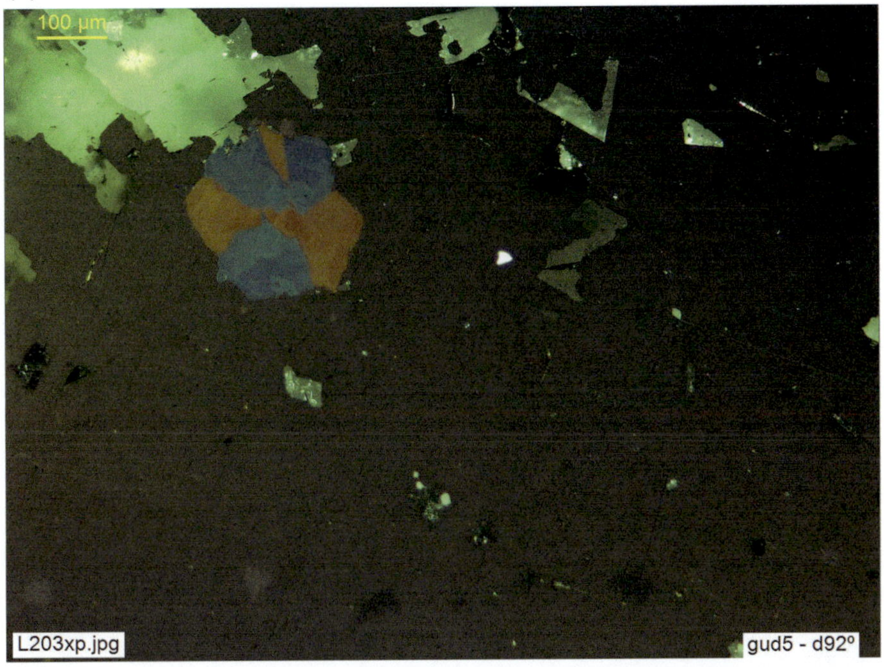

(b)

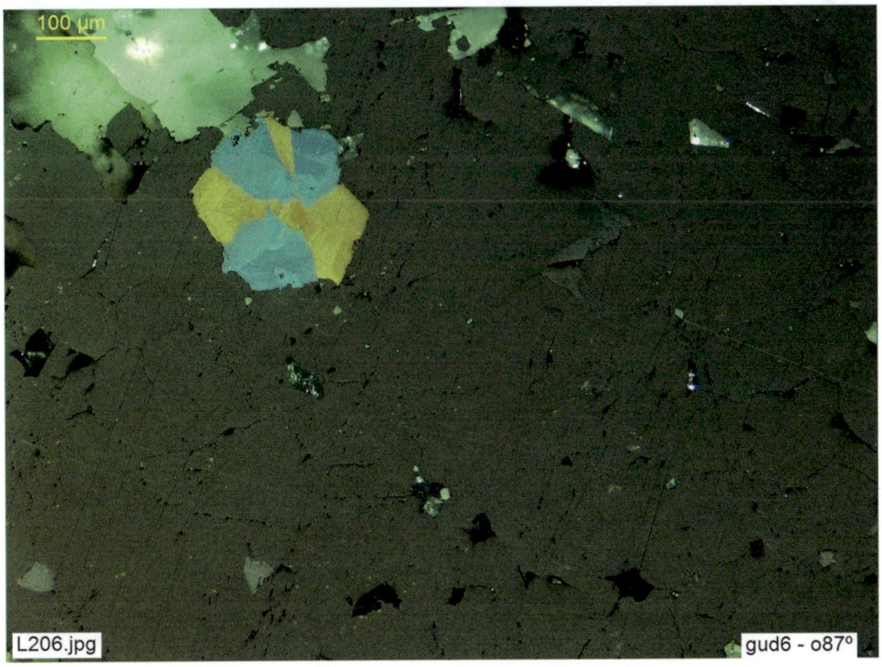

59.3 Spectral Information

59.3.1 Gudmundite Multispectral Reflectance R (in Air, Without Polariser) (%)

λ (nm)	370	400	425	450	475	500	525	550	575	600	625	650	675	700	750	800	850	900	950	1000
μ	46.4	47.5	47.7	48.4	48.9	49.5	50.1	50.2	50.5	50.5	50.4	50.4	50.4	50.3	50.0	49.6	49.7	49.1	48.9	48.5
P_5	44.9	46.2	46.4	47.3	48.0	48.8	49.2	49.5	49.8	49.9	49.8	49.8	49.8	49.7	49.4	49.0	48.8	48.6	48.4	48.1
P_{20}	45.9	47.2	47.4	48.2	48.8	49.4	49.9	50.1	50.4	50.5	50.5	50.4	50.4	50.3	49.9	49.5	49.3	49.0	48.8	48.4
P_{80}	48.2	49.3	49.3	49.9	50.4	51.0	51.6	51.8	52.1	52.1	52.0	51.9	52.0	51.8	51.6	51.2	51.4	50.7	50.5	50.1
P_{95}	49.6	50.3	50.3	50.7	51.3	51.6	52.0	52.1	52.3	52.4	52.4	52.3	52.4	52.2	52.1	51.5	51.7	50.9	50.7	50.4

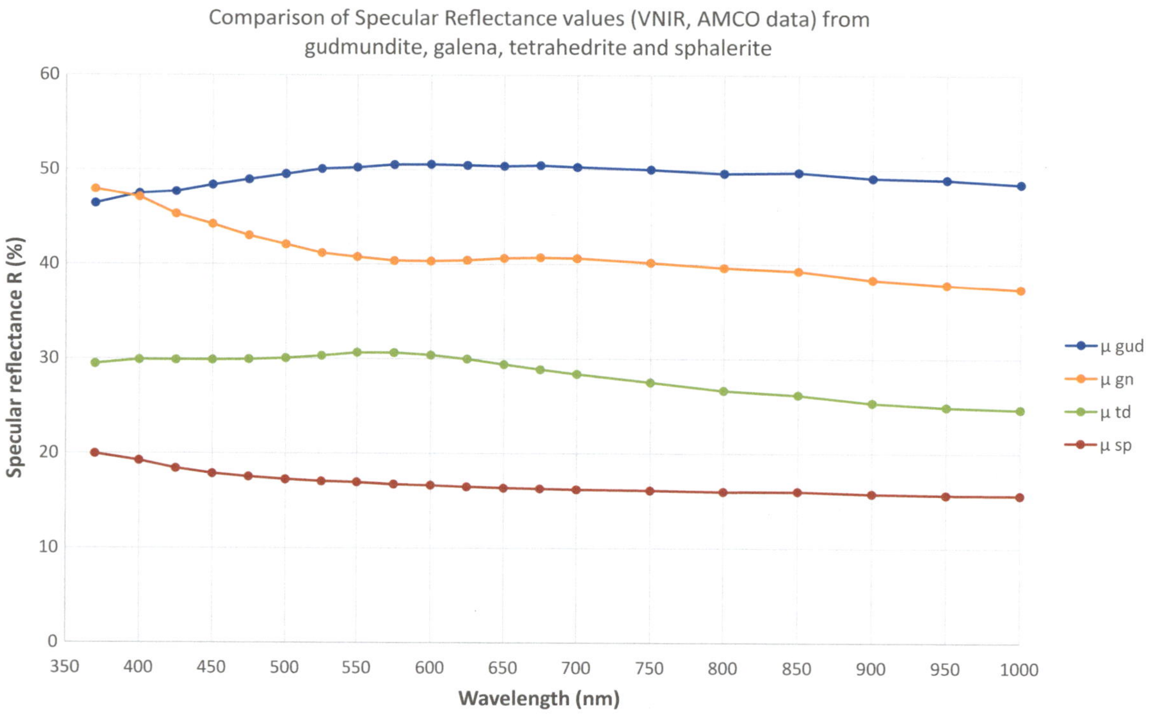

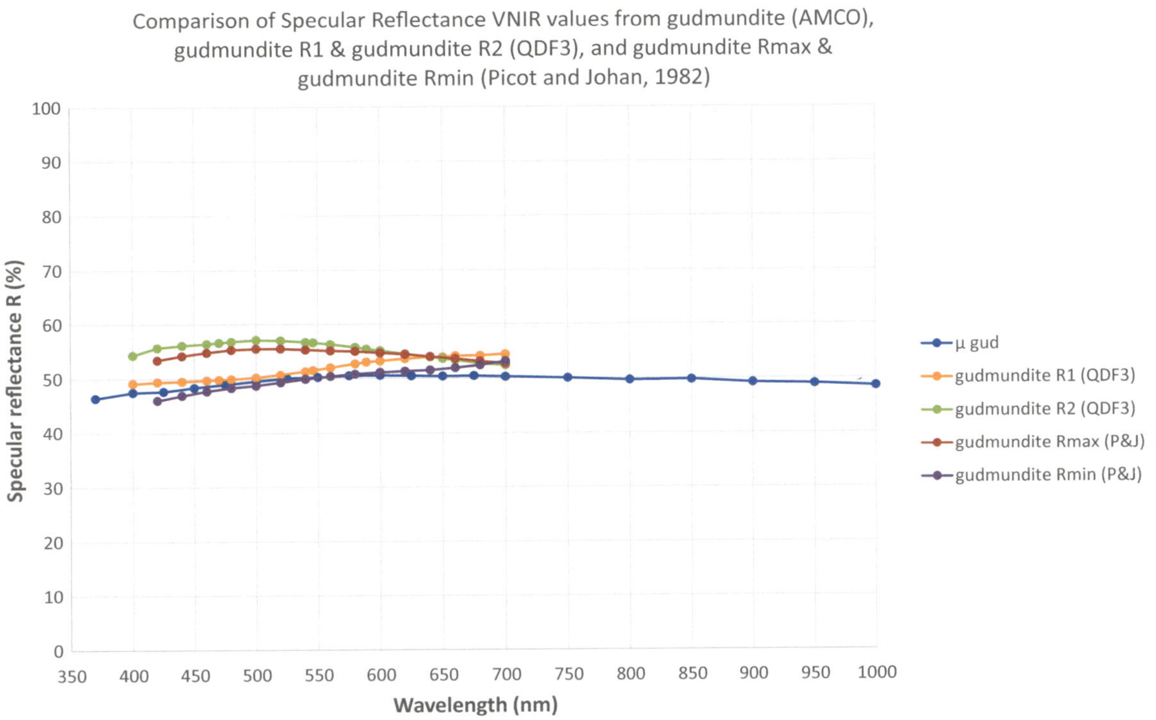

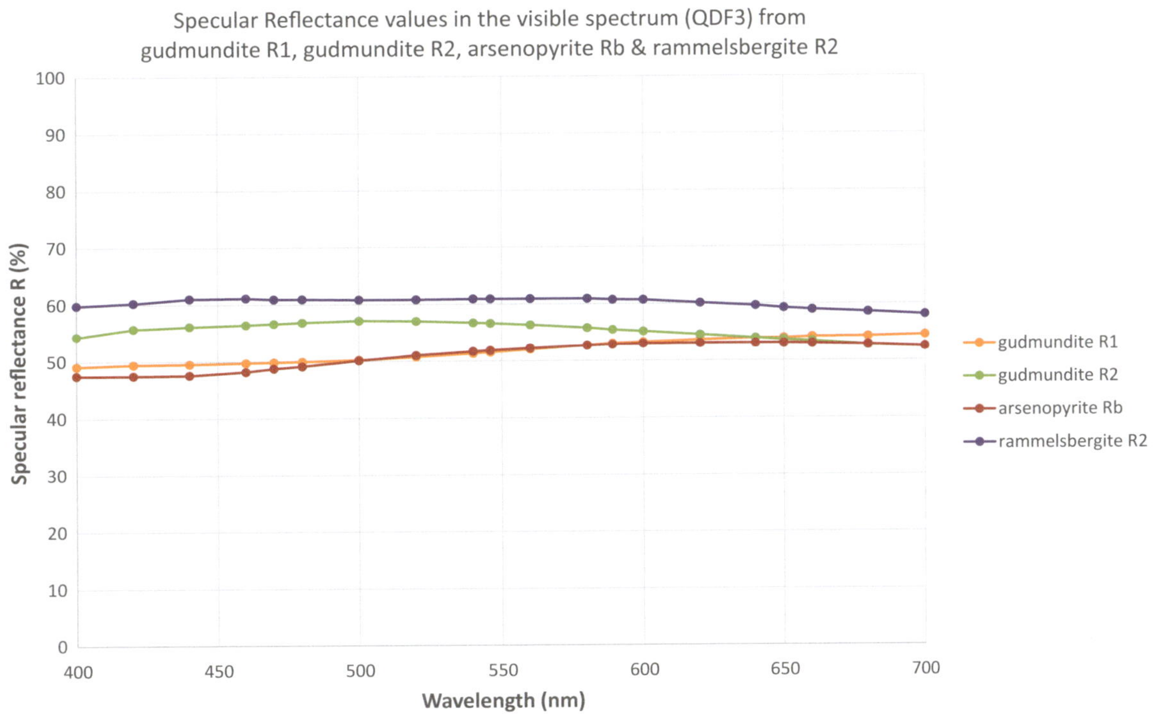

Reference

Picot P, Johan Z (1982) Atlas of ore minerals. B.R.G.M./Elsevier

Hematite (*ht/red iron ore, specularite*)

60

Fe_2O_3: Trigonal

Hematite		R_{546} (%)	Y (%)	λ_d (nm)	P_e (%)	Colour sensation
Air	R_1	30.0	29.7	483	4.7	White to bluish-grey
	R_2	26.4	26.1	483	5.6	
Oil	R_1	15.1	14.9	482	8.6	
	R_2	12.0	11.8	482	10.0	
Bir: Medium (4)				*Pleoc*: Weak (3)		*Tarn*: No
Anis +P: Distinct (4)				*IR*: Red		*Twg*: Frequent laminar
Mor: Tabular, fibrous, massive				*Clv*: Occasional (apparent)		*PolQ*: Difficult (very hard)
Hardness	*PolH*: High (4)			*Mohs*: 5–6	*Talm*: G	*VHN* $_{(100\ g)}$: 1000–1100

Obs: cf. Table and R charts, multispectral data (VNIR, AMCO) and visible range (QDF3)
Mineral abbreviations: asp—arsenopyrite, Au—native gold, ccp—chalcopyrite, chr—chromite, cst—cassiterite, gg—gangue, gth—goethite, il—ilmenite, lpc—lepidocrocite, mt—magnetite, py—pyrite, q—quartz, rt—rutile, tn—tennantite, td—tetrahedrite

60.1 General

Important iron ore and, therefore, one of the most indispensable minerals for industry. Composition: Fe_2O_3, with frequent impurities of Mg, Mn, Al. It can incorporate Ti, showing unlimited miscibility with ilmenite ($FeTiO_3$) at high temperature. Maghemite dimorph.

60.2 Microscopic Description

60.2.1 SP Observation (*Single Polariser*)

Bright white *colour* if isolated but usually looks bluish grey (next to pyrite, gold, chalcopyrite) and appears much darker, especially in immersion (Figs. 60.1a, b).

Moderate *reflectance*, ≈ td-tnn, but relatively high for an oxide (R > mt, il, rt; cf comparative multispectral charts), although it decreases with high Ti contents. Moderately *bireflective*, but not pleochroic. Turning the stage changes only from lighter to darker, so the change may go unnoticed to the careless observer but becomes evident comparing contiguous sections with different orientations (Fig. 60.2).

High *polishing hardness*: $PolH_{ht}$ > mt, il, asp, but < rt, py, q, cs. With perseverance, despite its hardness, a *good polish* can be achieved. It lacks cleavage, but sometimes pseudo-exfoliations appear during polishing of deformed materials.

It occurs crystallized (tabular, prismatic, lamellar, even acicular or fibrous habit) or in earthy, reniform, botryoidal masses, in oolitic deposits and impregnations. It can develop diablastic and nematoblastic textures in metamorphic ores (Figs. 60.4, 60.5a and 60.6) and, in magmatic Fe–Ti ores, oriented lamellae by unmixing in ilmenite (Fig. 60.3).

It replaces sulfides in oxidizing environments, but in other conditions the reverse may be the case (Fig. 60.5a, b). By oxidation of magnetite, it forms *martite*, a lattice of lamellae oriented according to reticular directions (111) of mt (Figs. 60.4 and 60.6).

Fequently intergrown with goethite, lepidocrocite, jarosite, opal, ferruginous clays, etc. in cryptocrystalline or submicroscopic aggregates, often colomorphic, known as **limonite** (cf. *lim description*).

R. Castroviejo, *A Practical Guide to Ore Microscopy—Volume 1*,
https://doi.org/10.1007/978-3-031-12654-3_60

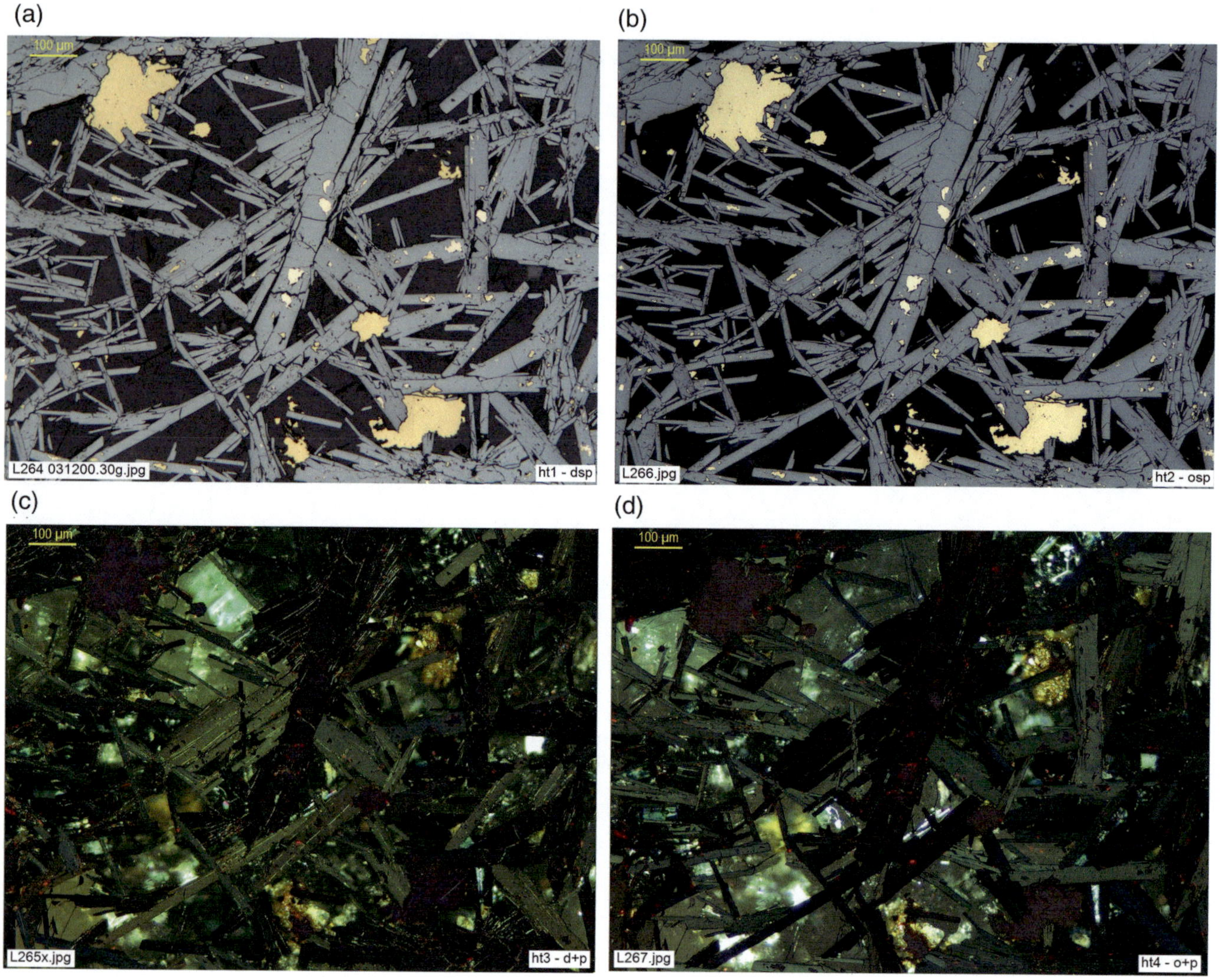

Fig. 60.1 **a, b** (**dsp** and **osp**, respectively). Diablastic intergrowth of ht (grey) and q (dark grey, interstitial) with ccp (yellow). 031200.30g, Mina Maruxa, Moeche (Coruña). **c, d** Same scene as ht1 (**d+p** and **o+p**, respectively). Anisotropism and internal reflections of ht

60.2.2 Observation +P (*Polariser + Analyser at 90°*)

Clearly *anisotropic*, although little coloured (Figs. 60.1c, d and 60.5b), grey and greenish tones, more varied in diagonal position. Usually shows red *internal reflections.* Lamellar *twinning* (1011) frequent.

60.2.3 Paragenesis

Very important Precambrian sedimentary concentrations, such as BIF or iron banded formations, itabirites, taconites, frequently metamorphosed, and also in metamorphic skarn concentrations. All of them are of great economic importance.

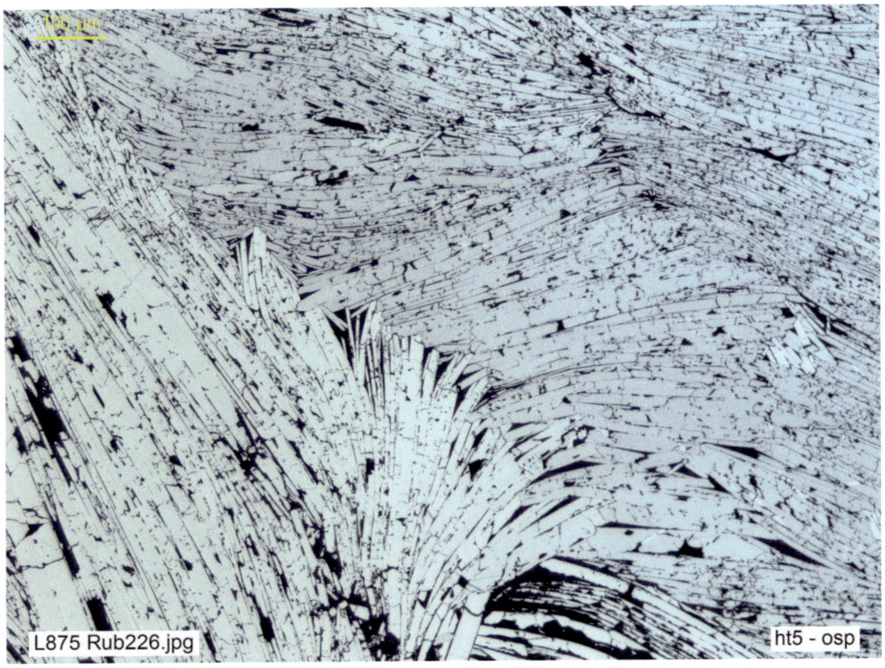

Fig. 60.2 Acicular aggregates of ht. Comparing beams of different orientation shows bireflectance. Sample Rub226, St. Martin am Silberberg, bei Huettenberg, Kaernten, Austria (**osp**)

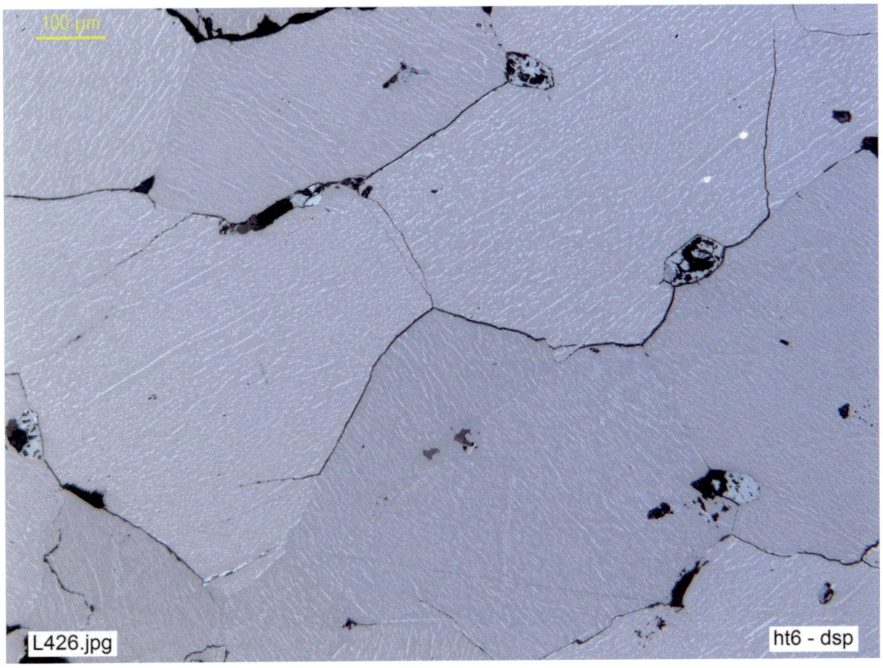

Fig. 60.3 Exsolution of ht (white lamellae) in il (grey to brownish-pink). HD 6764, Abu Galqua, Egypt (**dsp**)

In small quantities, it is a common component of soils and weathered rocks, particularly abundant in lateritic profiles. It is an essential component, with goethite, of *gossan*, a limonitic ferruginous deposit formed by weathering of sulphide masses, which may contain interesting concentrations of precious metals (Au, Ag).

Usual companions: magnetite, ilmenite, rutile, chromite, goethite, lepidocrocite. Frequently in exsolution lamellae in

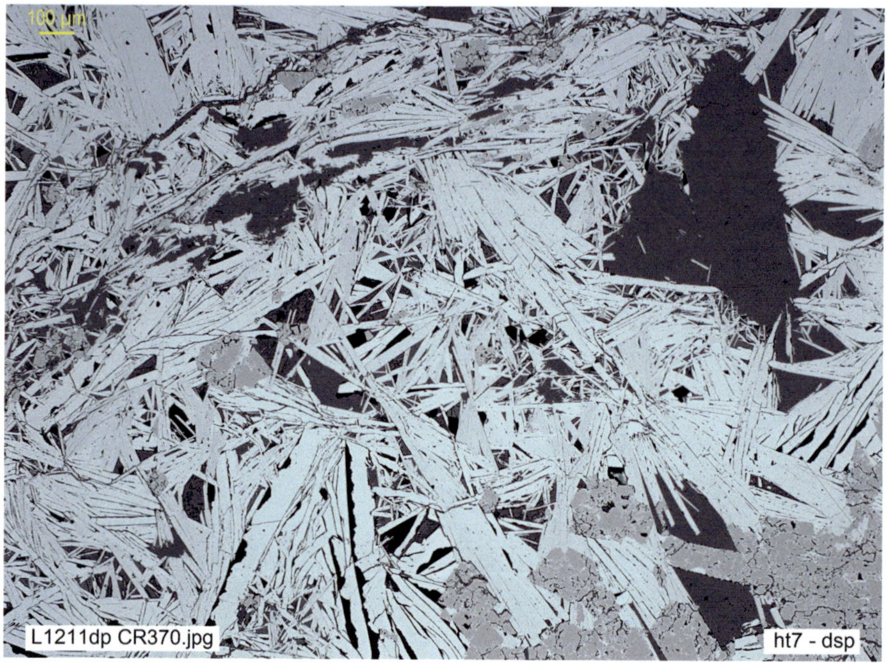

Fig. 60.4 Intergrowth of ht (grey, acicular) with mt (brownish grey, partially altered to martite). CR 370, Mina Recanzoni, Alidos, Mexico (**dsp**)

ilmenite from magmatic deposits, as well as replacing magnetite following exfoliation directions (*martite*).

60.2.4 Identification

Its colour, hardness, reflectance, anisotropism and internal reflections, together with paragenesis provide reliable criteria.

Its anisotropism distinguishes it from maghemite. It resembles lepidocrocite, but the latter is somewhat darker (more transparent, with more abundant internal reflections) and softer. Both (maghemite and lepidocrocite) are much less abundant than hematite, with which they are often associated. It may be impossible to distinguish them visually in earthy masses and in cryptocrystalline aggregates (limonite), in which they are associated with goethite, lepidocrocite, ferruginous clays, etc.

These aggregates sometimes appear isotropic, because of their submicroscopic grain size or their content in hydrated ferruginous gels, but the abundant reddish to yellowish-brown internal reflections denote the presence of Fe^{3+}. Otherwise, their scarcity or absence may suggest the presence of Mn oxides, also abundant in the surficial environment.

Fig. 60.5 **a, b** (**dsp** and **d+p**, resp)—Aggregate of diablastic ht, partially corroded by massive ccp; small inclusions of py and ln in ccp. M031200.30g, Maruxa Mine, Moeche, Coruña

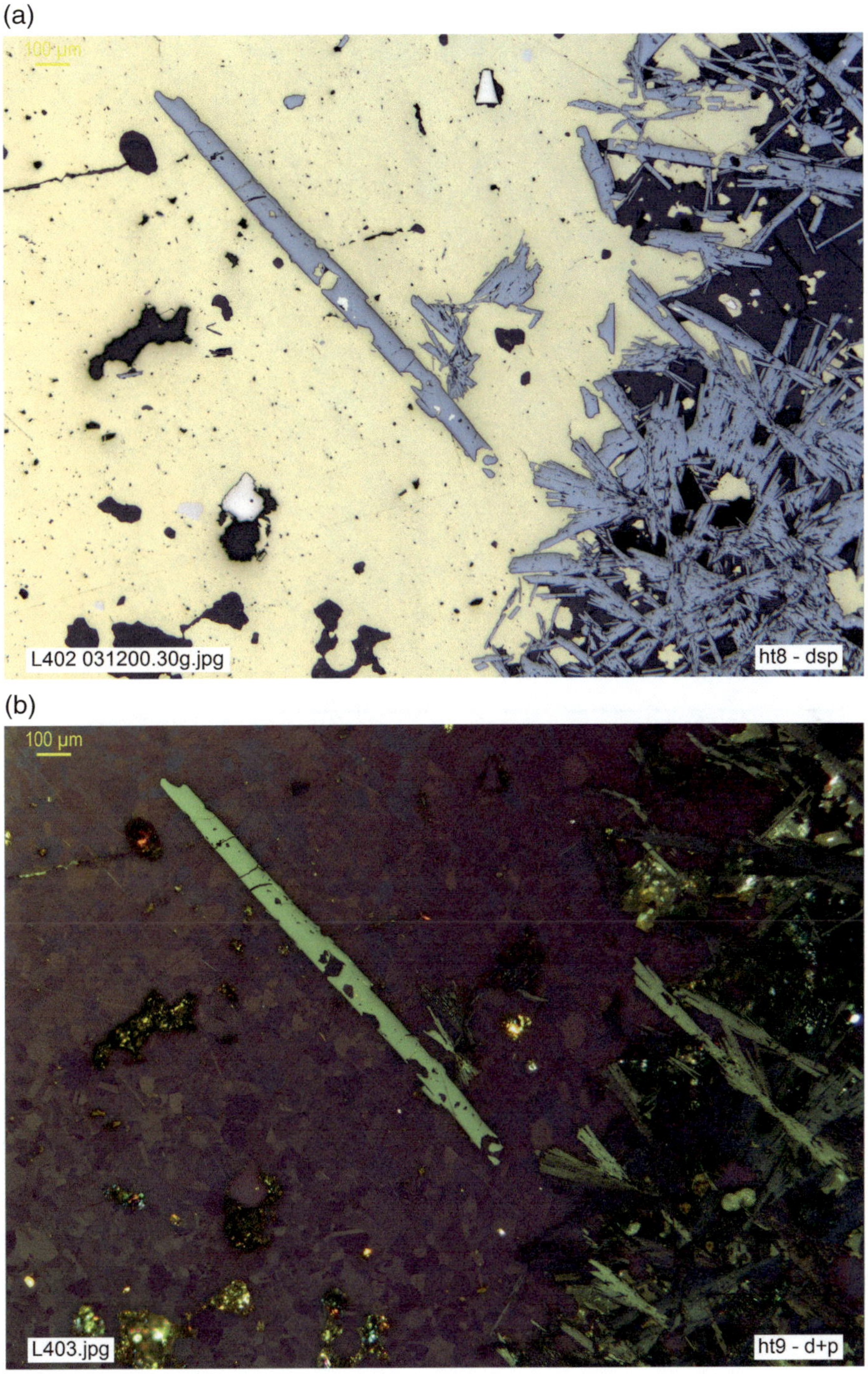

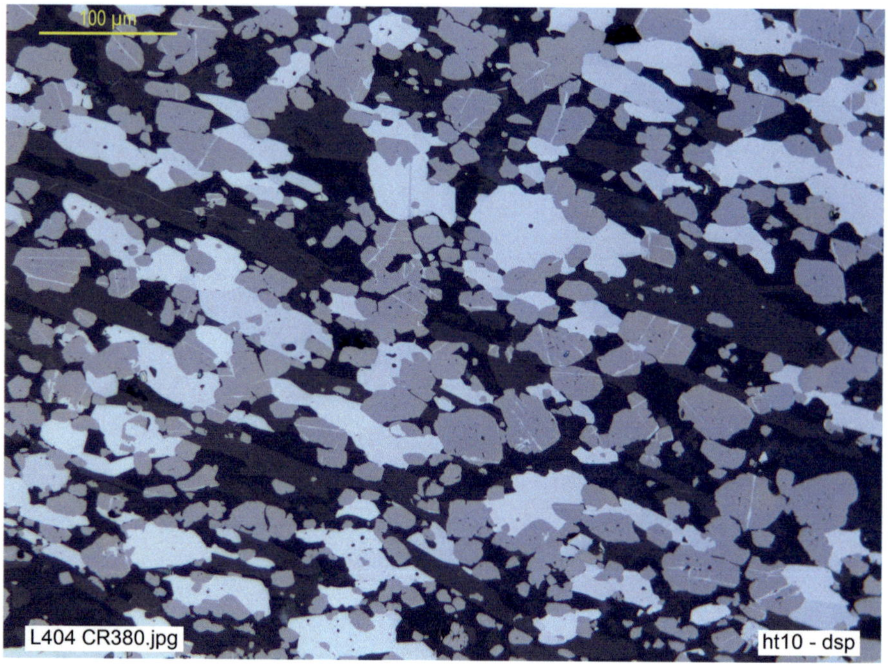

Fig. 60.6 BIF ore: nematoblastic ht intergrown with mt (with incipient martitization) and silicates. CR 280, Harare, Zimbabwe (**dsp**)

60.3 Spectral Information

60.3.1 Hematite Multispectral Reflectance R (in Air, Without Polariser) (%)

λ (nm)	370	400	425	450	475	500	525	550	575	600	625	650	675	700	750	800	850	900	950	1000
μ	27.6	29.7	30.4	30.5	29.9	29.5	28.8	28.7	28.3	27.2	26.0	24.9	24.3	23.7	22.9	22.2	21.9	21.3	21.1	20.7
P_5	26.1	28.5	29.4	29.3	28.8	28.3	27.8	27.6	27.1	25.9	24.8	23.8	23.2	22.7	22.0	21.4	21.1	20.5	20.2	19.9
P_{20}	26.9	29.0	29.8	29.7	29.2	28.7	28.1	28.0	27.5	26.4	25.3	24.3	23.6	23.1	22.3	21.7	21.4	20.8	20.6	20.3
P_{80}	28.4	30.5	31.1	31.5	30.9	30.6	29.5	29.5	29.5	28.2	26.9	25.7	25.0	24.3	23.5	22.8	22.5	21.8	21.6	21.3
P_{95}	28.8	30.9	31.5	32.0	31.3	30.9	29.9	30.2	30.0	28.7	27.4	26.2	25.5	24.8	23.9	23.1	22.8	22.2	21.9	21.6

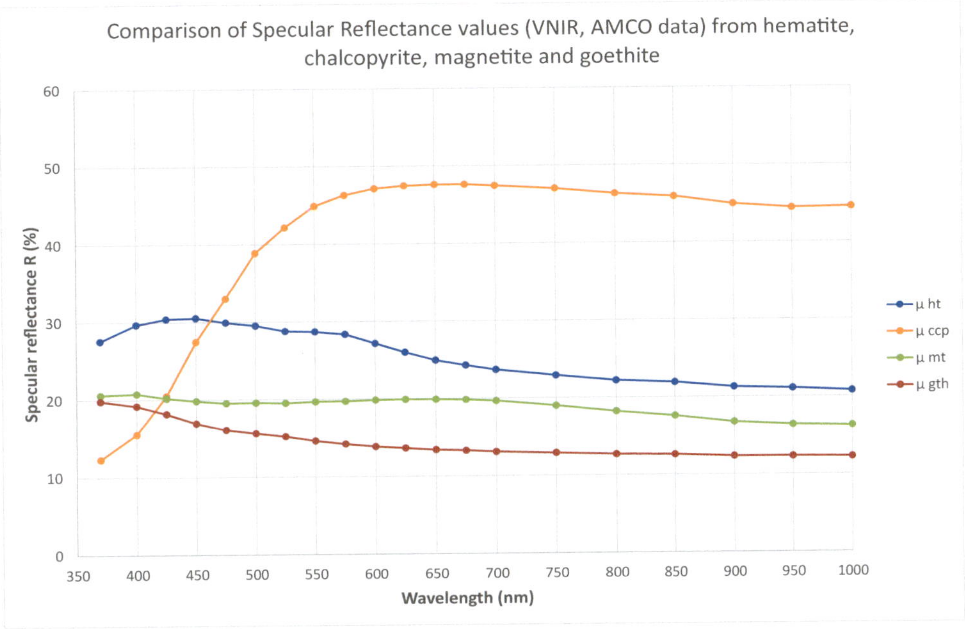
Comparison of Specular Reflectance values (VNIR, AMCO data) from hematite, chalcopyrite, magnetite and goethite
Specular reflectance R (%)
Wavelength (nm)
µ ht
µ ccp
µ mt
µ gth
0
10
20
30
40
50
60
350
400
450
500
550
600
650
700
750
800
850
900
950
1000

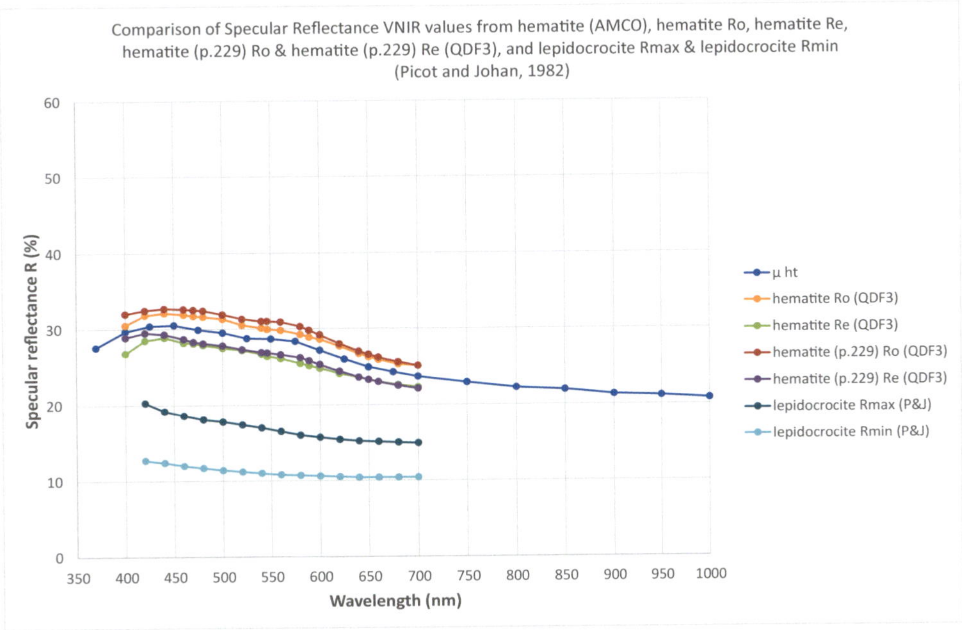
Comparison of Specular Reflectance VNIR values from hematite (AMCO), hematite Ro, hematite Re, hematite (p.229) Ro & hematite (p.229) Re (QDF3), and lepidocrocite Rmax & lepidocrocite Rmin (Picot and Johan, 1982)
Specular reflectance R (%)
Wavelength (nm)
µ ht
hematite Ro (QDF3)
hematite Re (QDF3)
hematite (p.229) Ro (QDF3)
hematite (p.229) Re (QDF3)
lepidocrocite Rmax (P&J)
lepidocrocite Rmin (P&J)
0
10
20
30
40
50
60
350
400
450
500
550
600
650
700
750
800
850
900
950
1000

Ilmenite (*il*)

61

$FeTiO_3$: Trigonal

Ilmenite		R_{546} (%)	Y (%)	λ_d (nm)	P_e (%)	Colour sensation
Air	R_o	19.2	19.5	c553	2.2	Grey/pinkish brown, pleochroic
	$R_{e'}$	16.4	16.7	c537	3.1	
Oil	R_o	6.7	6.8	c535	3.6	
	$R_{e'}$	4.9	5.1	c509	5.3	
Bir: Medium (4)				*Pleoc*: Distinct (4)		*Tarn*: No
Anis +P: Distinct (4)				*IR*: Exceptionally		*Twg*: Frequent, polysynthetic
Mor: Euhedral, tabular, massive				*Clv*: –		*PolQ*: Good
Hardness	*PolH*: High (4)			*Mohs*: 5–6	*Talm*: G	*VHN* $_{(100g)}$: 566–698

Obs: cf. Table and R charts, multispectral data (VNIR, AMCO) and visible range (QDF3)
Mineral abbreviations: ana—anatase, cbt—columbotantalite, ccp—chalcopyrite, chr—chromite, ht—hematite, il—ilmenite, ilr—ilmenorutile, isc—isocubanite, lim—limonite, mt—magnetite, pn—pentlandite, po—pyrrhotite, py—pyrite, rt—rutile, spn—spinel, tit—titanite, tp—tapiolite, ulv—ulvöspinel (ulvite)
Colour abbreviations: BluGy—bluish grey, BrGy—brownish grey, PiBr—pinkish brown, PiGy—pinkish grey, ViGy—violet grey

61.1 General

Ilmenite is one of the main titaniferous ores. Its formula according to IMA is $Fe^{2+}Ti^{4+}O_3$, but it can contain variable amounts of Mg or Mn and traces of V, Nb, Ta. With the terms Mg and Mn, it forms the series *geikielite-ilmenite* ($MgTiO_3$–$FeTiO_3$) and *ilmenite-pyrophanite* ($FeTiO_3$–$MnTiO_3$), respectively. At high temperatures, there is remarkable miscibility with Fe_3O_4 (magnetite).

61.2 Microscopic Description

61.2.1 SP Observation (*Single Polariser*)

Grey *colour*, clearly pleochroic, with a remarkable pinkish-brownish tone (Fig. 61.1a, b) that varies with composition (it is less pronounced in varieties rich in Mg or Mn) and whose perception depends also on the adjacent ores. The pinkish tone is well appreciated in front of hematite or goethite, but it is much duller in front of pyrrhotite or isocubanite (Fig. 61.2) and even next to magnetite (Fig. 61.4a).

Moderate *reflectance*: $R_{il} \leq R$ mt; < rt and distinctly < ht. R varies with composition, being lower in Mg or geikielite varieties.

Distinct bireflectance and pleochroism, visible even within the same grain, comparing twins (greyish-pink, with light brownish to purplish tones, darker, Fig. 61.5a).

Good *polish,* high *polishing hardness*, $PolH_{il}$ > mt; < chr, ht, py.

Usually well crystallized, with *euhedral* tendency and tabular habit, sometimes laminar or rhombohedral, sometimes massive. It frequently presents hematite exsolutions (Figs. 61.1a, b). It can also be found unmixed in titaniferous magnetite (Fig. 61.4a, b).

R. Castroviejo, *A Practical Guide to Ore Microscopy—Volume 1*,
https://doi.org/10.1007/978-3-031-12654-3_61

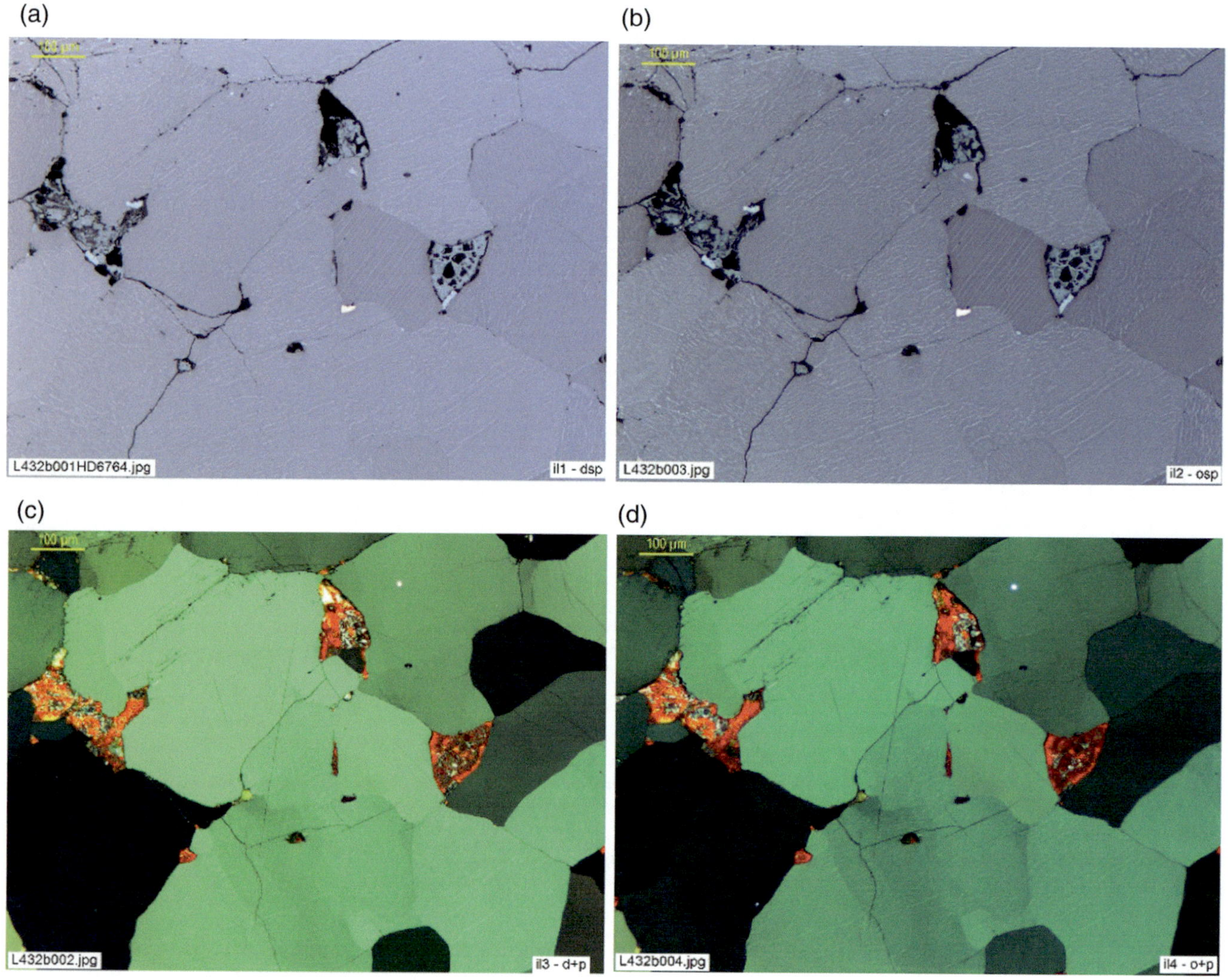

Fig. 61.1 **a** Crystalline aggregate of il (brownish grey and ViGy to PiBr) with fine exsolutions of ht in subparallel lamellae and interstitial spaces filled with gth (grey to BluGy, somewhat darker). Tiny inclusion of po (pinkish brown, bright) toward the center, provides reference for value of R. HD-6764, Abu Galqua, Egypt (**dsp**). **b** Same scene as ilr1 (**osp**). **c, d** Same scene as Fig. 61.1 (**d+p**) and (**o+p**), respectively. Distinct and coloured anisotropism (greenish tones) of il, reddish-brown RI of gth

61.2.2 Observation +P (*Polariser + Analyser at 90°*)

Clearly *anisotropic*, with striking greenish polarization colours, particularly in immersion (Figs. 61.1c, d, 61.3, 61.4b and 61.5b). Frequent polysynthetic twins (Fig. 61.5b).

Usually, it does not show *internal reflections* or they are minuscule and scarce IR, but in varieties rich in Mg and, above all, Mn (geikielite and pyrophanite), reddish IR are usually observed.

61.2.3 Paragenesis

Mineral frequent in nature, although less frequent than magnetite. Formed at high temperature, usually > 500 °C and therefore present in a variety of magmatic rocks, intrusive, extrusive, pegmatites, etc.

Typical companion of magnetite in Ti–Fe concentrations (titano-magnetites) associated with mafic rocks. It can also be found in metamorphic rocks and in alpine-type seams.

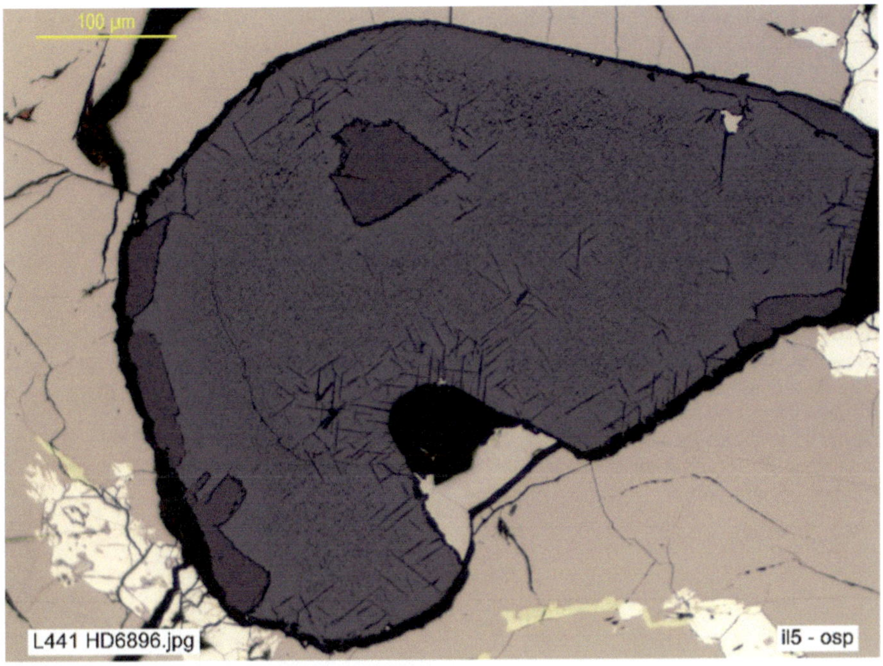

Fig. 61.2 Aggregate of po-isc (pinkish brown) with interstitial pn-ccp and, in the centre, large inclusion of mt (grey) with subordinate il (purplish grey or ViGy) and oriented exsolutions of spn/ulv (grey/BrGy, dark). Compare the colour of il in Fig. 61.1a, b (thc typical onc) with that of Fig. 61.2 (next to po/isc the pinkish hue is no longer perceived; instead, ViGy is seen). HD 6896, Sudbury (**osp**)

Fig. 61.3 Same scene as Fig. 61.2 (**o+p**): isotropy of mt, icb, spn and ulv (dark), unlike the distinctly anisotropic il and po

Placer deposits, as detrital derivatives of all the mentioned formations. In spite of its resistance to weathering, it is not uncommon for il to show different degrees of alteration to Ti oxides and leucoxene (cf. anatase, titanite, rutile), especially in tropical climates.

Companions: magnetite, hematite, rutile, spinel, ulvöespinel.

61.2.4 Identification

It is easily recognized by its pink colour, pleochroism (grey to brownish-pink), low reflectance, hardness, distinct anisotropism and frequent twinning, with the added criterion of its paragenesis (magnetite, rutile, hematite, etc.). Nevertheless, its variable composition (Mg, Mn) and the influence of neighbouring ores on the subjective perception can lead to confusion as to its colour.

(a)

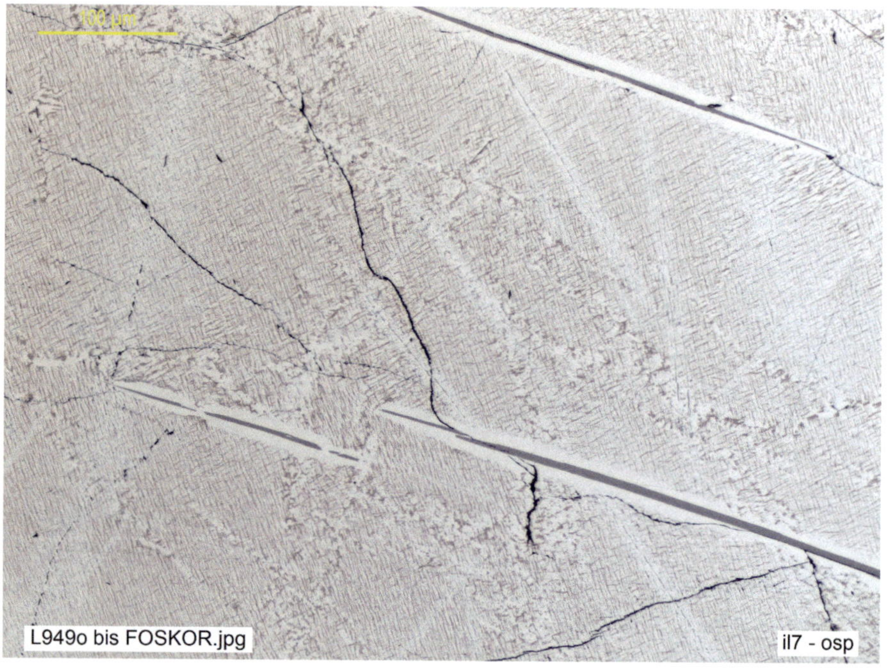

(b)

Fig. 61.4 **a, b** (**osp** and **o+p** , *resp*ectively) Lattice-controlled exsolution of lamellar il (grey to BrGy) in mt (lighter PiGy). Two sets of lamellae are visible: (i) an extremely fine network of tiny lenticular lamellae, and (ii) a few long, isolated, spindle-shaped lamellae, cutting through (i). They follow cubic (100) and octahedral (111) lattice planes of mt, respectively. The set (i) disappears and leaves an elongated, light-coloured halo of pure mt around each of the (ii) lamellae, as these have taken all the Ti available in the immediate vicinity. Foskor Mine, Palabora, South Africa

- Sometimes it can be difficult to distinguish il from titanomagnetite. In these cases, the polishing hardness is a good criterion—magnetite is softer.
- Geikielite is more greyish and darker and usually shows IR.
- Pyrophanite, also darker, typically shows abundant reddish IR and is often accompanied by other manganiferous species.
- The colour, reflectance and pleochroism, as well as the hardness, the anisotropism and the absence of IR avoid any confusion with hematite.
- In isolated and small grains, it might be confused with rutile, but the latter has abundant IR. However, in some cases and in peculiar paragenesis (*e.g.,* pegmatites with Nb, Ta, TR) the distinction can be delicate (cf. ilmenorutile, columbotantalite, tapiolite).

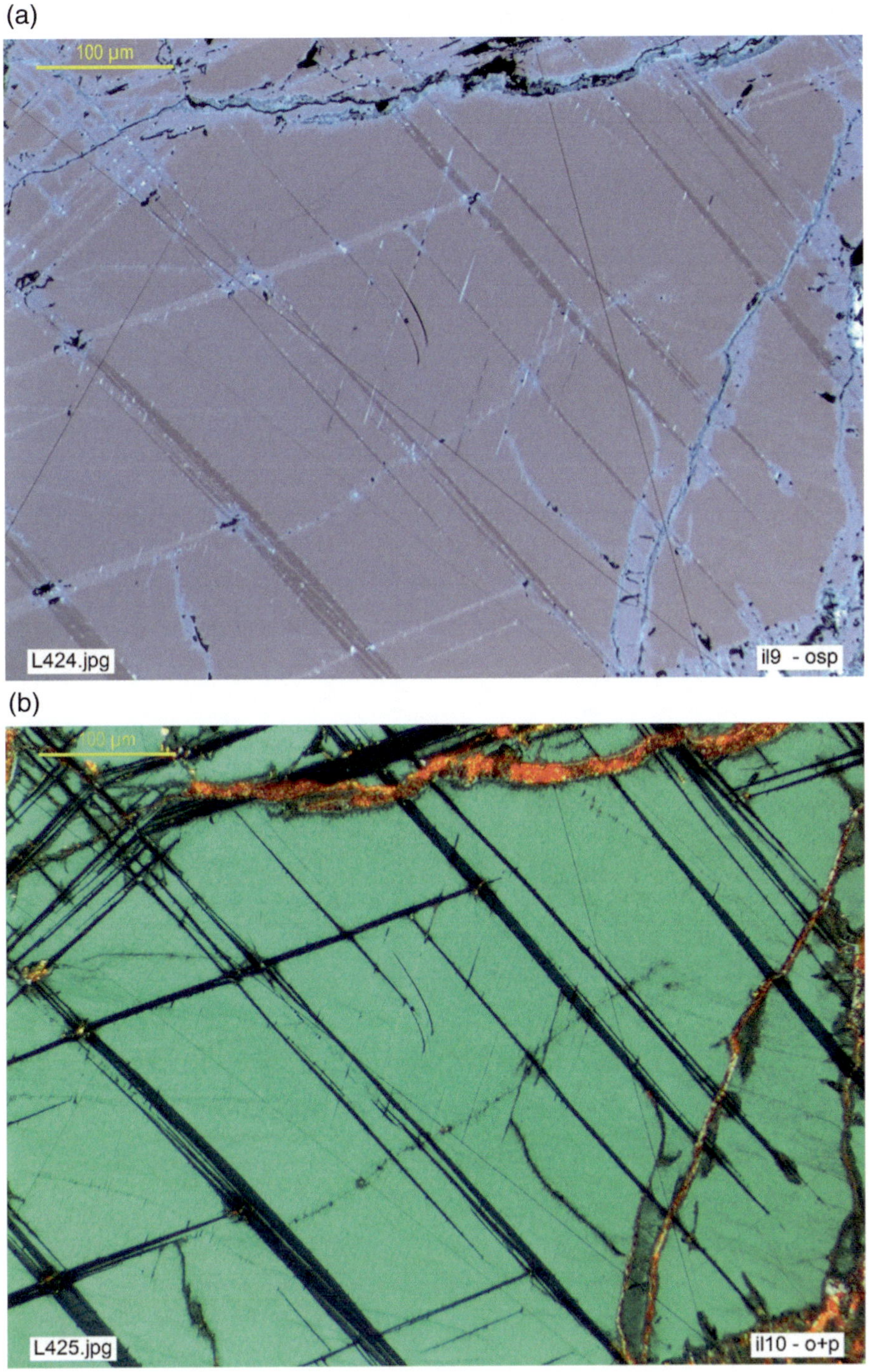

Fig. 61.5 **a, b** Single crystal of il, twinned and with two generations of ht inclusions: (1) exsolutions (lamellae oriented after the trigonal lattice of il); (2) microcrack infill, with gth/lim. Additionally, an extremely fine, hardly visible, mt (?) network exsolved from il. HD 6858, Agracha, Western Sahara (**osp**) and (**o+p**), respectively

61.3 Spectral Information

61.3.1 Ilmenite Multispectral Reflectance R (in Air, Without Polariser) (%)

λ (nm)	370	400	425	450	475	500	525	550	575	600	625	650	675	700	750	800	850	900	950	1000
μ	20.6	20.2	19.0	18.4	17.8	17.8	17.6	18.0	18.1	18.3	18.5	18.6	18.7	18.8	18.9	18.6	18.6	18.0	17.6	17.4
P_5	19.1	19.0	18.0	17.3	16.9	16.8	16.8	17.2	17.4	17.7	17.9	18.0	18.1	18.2	18.2	18.0	17.8	17.3	17.0	16.8
P_{20}	19.6	19.4	18.2	17.6	17.1	17.1	17.0	17.5	17.6	17.8	18.1	18.1	18.2	18.3	18.4	18.1	18.0	17.5	17.2	16.9
P_{80}	21.6	21.2	20.0	19.3	18.7	18.6	18.4	18.6	18.7	18.9	19.0	19.1	19.4	19.5	19.6	19.5	19.4	18.7	18.2	17.8
P_{95}	22.2	21.8	20.4	19.9	19.2	19.1	18.9	19.0	19.1	19.3	19.5	19.8	20.1	20.2	20.3	20.1	20.1	19.3	18.8	18.5

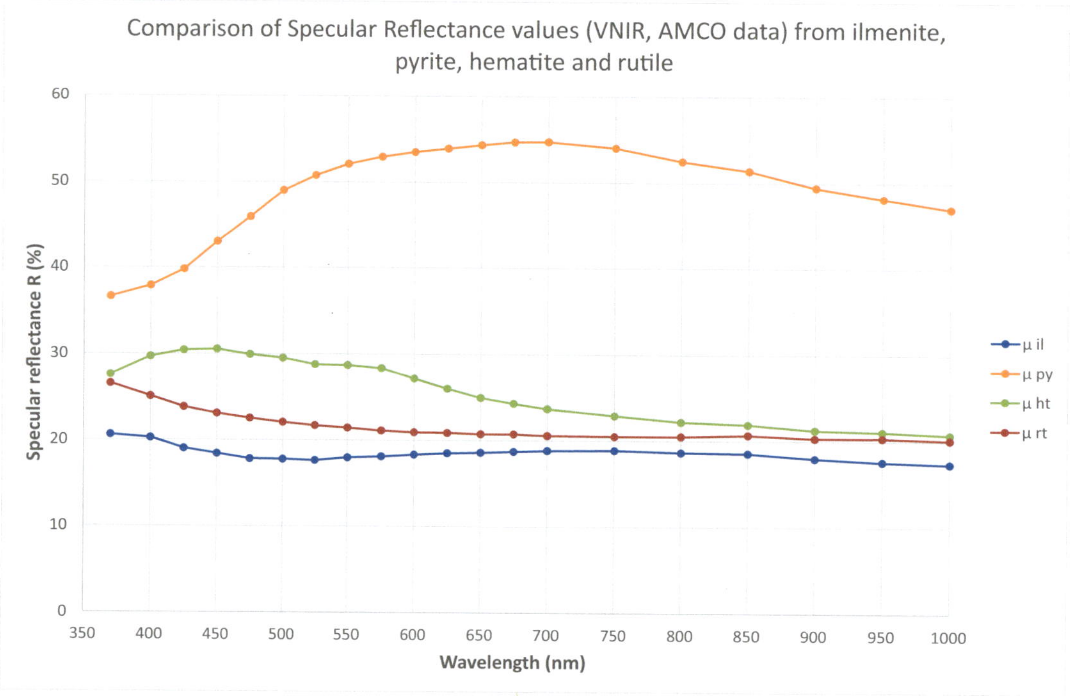

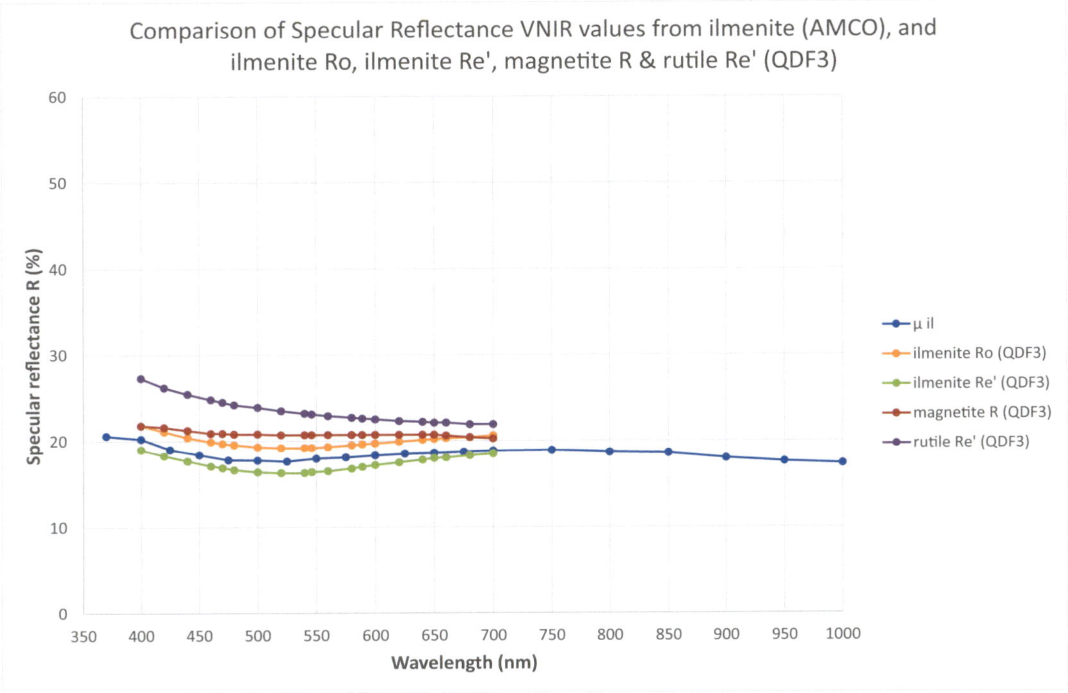
Comparison of Specular Reflectance VNIR values from ilmenite (AMCO), and ilmenite Ro, ilmenite Re', magnetite R & rutile Re' (QDF3)
Specular reflectance R (%)
60
50
40
30
20
10
0
350
400
450
500
550
600
650
700
750
800
850
900
950
1000
Wavelength (nm)
μ il
ilmenite Ro (QDF3)
ilmenite Re' (QDF3)
magnetite R (QDF3)
rutile Re' (QDF3)

62 Ilmenorutile (*ilr*)

(Ti, Nb, Fe)O_2 : Tetragonal

Ilmenorutile			R_{546} (%)	Y (%)	λ_d (nm)	P_e(%)	Colour sensation
Air	R_o		nd	nd	nd	nd	Slightly brownish grey
	$R_{e'}$		nd	nd	nd	nd	
Oil	R_o		nd	nd	nd	nd	
	$R_{e'}$		nd	nd	nd	nd	
Bir: Weak (3)					*Pleoc*: Weak (3)		*Tarn*: No
Anis +P: Notable (3)					*IR*: Present (rare)		*Twg*: Frequent
Mor: Euhedral, prismatic					*Clv*: Occasional (110)		*PolQ*: Good
Hardness		*PolH*: High (4)			*Mohs*: 6	*Talm*: G?	*VHN* $_{(?)}$: 803–1180 (Nb), 911–1239 (Ta)

Remark: Rutile variety (*cf. rt)*. No data in QDF3. VHN according to Uytenbogaardt and Burke (1971), seems higher than expected (standard weight 100 gf), suggesting that it was possibly measured with a lighter weight

Mineral abbreviations: cb—columbite, cbt—columbotantalite, gg—gangue, ht—hematites, il—ilmenite, ilr—ilmenorutile, mt—magnetite, py—pyrite, q—quartz, rt—rutile, tan—tantalite, tp—tapiolite

62.1 General

Today, *Ilmenorutile*, **ilr**, is considered by *IMA/COM* as a Nb-rich variety of rutile rt (*Nb-rt*; after *Mindat*[1] with Formula [Ti, Nb]O_2), while the Ta-rich variety (*Ta-rt*) is called *strüverite* (*íbid.*). However, both were traditionally considered as distinct species. For Strunz and Nickel (2001), *ilr* is a mineral of the Rutile Group, with formula (Ti, Nb, Fe)O_2. The compositional and equilibrium relationships between these phases are very complex (cf. Beurlen et al. 2007, 2008).

Ilmenorutile, treated as a mineral species and defined as a solid solution of (Fe, Mn) $(Nb, Ta)_2$ 0_6 and Ti_3O_6 ($\approx$50%) by Ramdohr (1980) is also, for other authors, a mineral of formula (Ti, Nb, Fe^{3+})O_2, isostructural with rutile. As a Nb term, it is part of a series whose Ta term is strüverite [(Ti, Ta, Fe^{3+})O_2]. However, ilmenorutile is considered as a variety of rutile by Hey (1975, 18.1.4) and discredited, although mentioned, by Back and Mandarino (2008).

Cerny et al (1964), make a systematic study of the known ilmenorutiles and conclude that there are two types: homogeneous (stable, containing more Ta than Nb) and inhomogeneous (containing more Nb than Ta and also columbite exsolutions), but that both should be considered as varieties of rutile and called Ta-rutile and Nb-rutile, respectively.

In order to better understand the complex reality of this substance, it will be described in two steps:

1. First, a brief optical characterization of ilmenorutile by reflected light, **LR**, is made according to the scarce data published by Cerny et al. (1964), Uytenbogaardt and Burke (1971), Ramdohr (1980), Bowles et al (2011).
2. Then, the study of an old collection sample, supposed to be *ilmenorutile* from Ljosland (Iveland, Norway) is discussed, comparing LR and SEM observations, which will show the necessity of both complementary techniques for the study of ilmenorutile.

1. **Ilmenorutile Optical Characterization by RL Microscopy**

[1] cf. https://www.mindat.org/min-2014.html (*last visit 2021.12.21*).

R. Castroviejo, *A Practical Guide to Ore Microscopy—Volume 1*,
https://doi.org/10.1007/978-3-031-12654-3_62

62.1.1 SP Observation (*Single Polariser*)

Grey, slightly brownish or pinkish *colour* compared to rutile, **rt**, less pleochroic than ilmenite. Relatively high brightness for an oxide; less than hematite but distinctly more than gangue.

Moderate *reflectance*. There are no QDF3 data. According to the values published by Uytenbogaardt and Burke (1971) and those obtained in Ta-rt (cf. rutile description, AMCO measurements), R of ilr is $\leq$ rt and certainly < ht.

Weakly bireflective (bireflectance can be observed only by comparing neighbouring grains with different orientation). Good *polish* and high *polishing hardness* ($\geq$ rutile, Uytenbogaardt and Burke, 1971).

62.1.2 Observation +P (*Polariser + Analyser at 90°*)

Moderate *anisotropism*, its effects being often more visible than in rutile, not overprinted by *internal reflections* (scarce and darker in ilmenorutile, compared to rt). Frequently *twinned*.

62.1.3 Paragenesis

As a primary mineral, it can be found mainly in pegmatites, also in alkaline complexes, and, naturally, in its detrital derivatives. Companions: columbotantalite, tantalite, tapiolite, hematite, magnetite, ilmenite, etc.

62.1.4 Identification

As a variety of rutile, the same criteria identifying rt apply, distinguishing ilr by the following: its lower reflectance; grey colour, sometimes slightly more pinkish or brownish, which allows appreciation of a weak pleochroism; scarcer and darker internal reflections; anisotropism distinct (commonly less than rt, but not masked by IR). In short, its optical properties are intermediate between those of tapiolite and rutile (Meixner 1951, in: Cerny et al, 1964; Ramdohr, 1980).

However, as happens with columbotantalite, the optical properties alone do not ensure the reliable determination of the composition and, given its potential *significance* as Nb/Ta ore, *it is advisable to check* by other techniques, particularly EMP and XRD, should the presence of ilmenorutile be suspected and the paragenesis be appropriate. It should be remembered that contents of up to 46.9% Nb_2O_5 and 34.5% Ta_2O_5 have been found in these rutile varieties (Cerny et al. 1981, 1999 and Groat et al. 2003, in: Bowles et al., 2011, p. 97).

The study of the cited sample of supposed ilmenorutile from Ljosland is presented below.

2. **Sample CR369 (LJOSLAND)**

62.2 Hand-Sample Description

Fragment of pegmatite, comprising large idiomorphic crystals of a black, hard, mineral with long prismatic habit (centimetric) and metallic luster, whose polished section is described below.

62.3 Microscopic Study (Reflected Light)

- **Mineralogical Composition**
  - Main components: ilmenite, hematite, silicates
  - Minor components: rutile, carbonates.
- **Description**

The main ore mineral occupies 30% of the section and shows the optical properties of an atypical ilmenite, because of its relatively weak pleochroism and anisotropism (Figs. 62.1a–d).

Well crystallized, *euhedral* and *prismatic* in habit (Figs. 62.2 and 62.3), it has numerous minute exsolutions of oriented hematite and occasional much larger secondary inclusions of rutile and carbonate (Figs. 62.1a, b).

Ilmenite shows good *polish. Colour* brownish grey, slightly pleochroic (light brownish grey to darker brownish-violet grey). Moderate *reflectance*: < rt, ht (Fig. 62.1a, b). Moderately *anisotropic*, with greenish polarization tints, similar to those typical of ilmenite, although much less luminous (Fig. 62.1c, d).

62.4 Microscopic Study (SEM)

Carried out at CSIC Laboratories, National Museum of Natural Sciences, Madrid (MNCN Service: Non- Destructive Techniques, Dr. Laura Tormo), on the same section studied by RL.

62.4.1 Results

The main mineral, which forms the essential mass of the prismatic crystals, is an ilmenite enriched in Mn (14.36 to 14.82% MnO_2) and in Nb–Ta (0.67 to 0.91% Nb_2O_5 + Ta_2O_5, Nb clearly dominating over Ta), with numerous inclusions and microcracks (Fig. 62.4a).

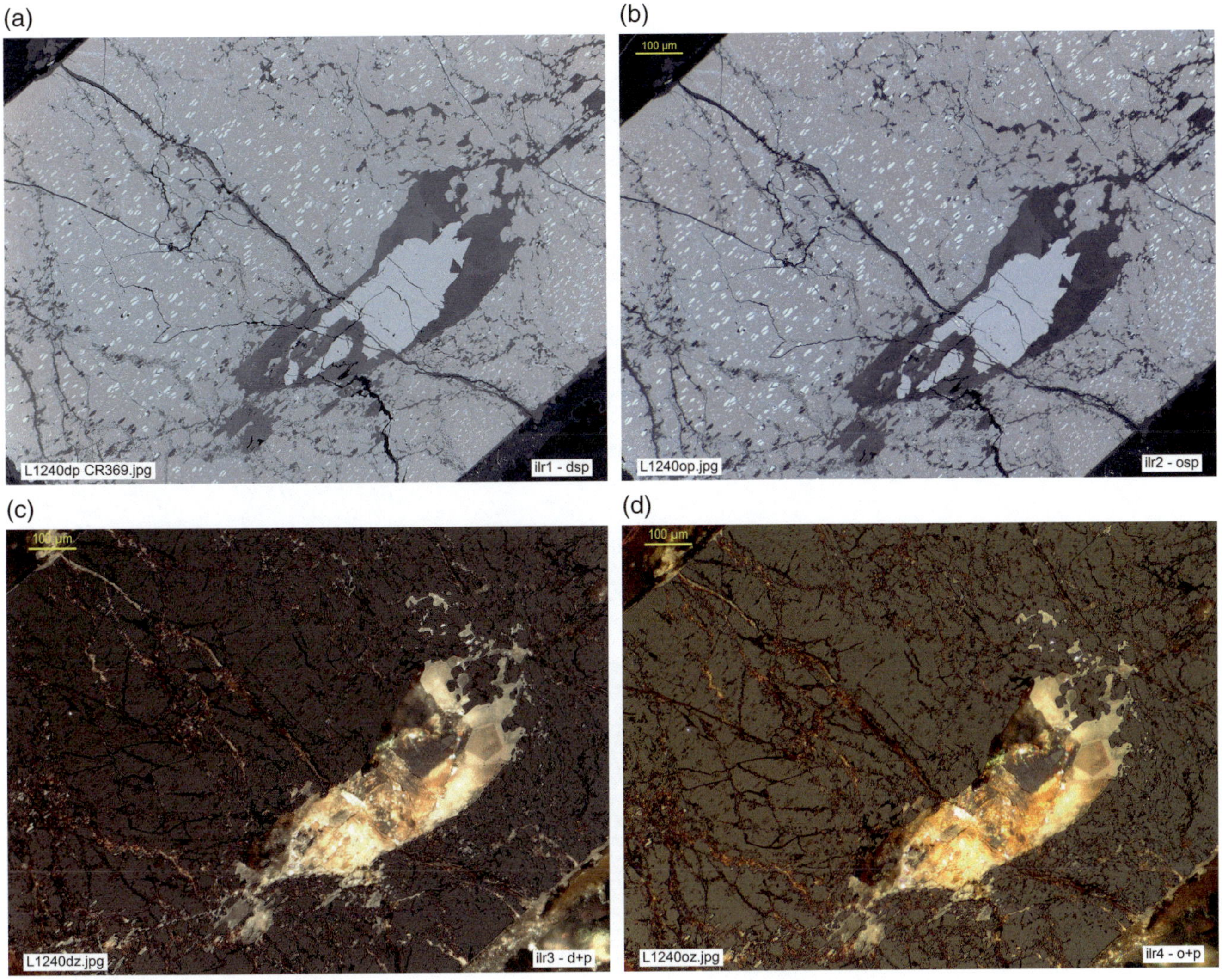

Fig. 62.1 **a** Single crystal of il (brownish grey), microcracked and partially altered, with fine exsolution of ht (white microlenticles, oriented //c axis) and major inclusion of rt (light grey) and carbonate (darker grey and bireflective). Detail of Fig. 62.2. CR 369, Ljosland, Iveland, Norway (**dsp**). **b** Same scene as ilr1 (**osp**). Note R_{ht} > rt > il > gg (carbonate). **c, d** (Same scene as Fig. 62.1: d+p and o+p, *respectively*). Few IR and moderate but distinct anisotropism of il (especially in oil), unlike rt and Carb whose anisotropism is overprinted by massive, bright internal reflections, IR

The ilmenite inclusions comprise mainly hematite, with high Ti (18.7% TiO_2) and Mn (4.5% MnO_2) contents. Small bright elongate corpuscles ($\approx$0.5 to 1 µm × 2 to 10 µm), mostly oriented in a direction cross cutting the hematite exsolutions (Fig. 62.4b, enlarged view, top central right area of Fig. 62.4a), appear frequently scattered in a very subordinate quantity. These stand out for their brightness, which denotes a high **MAN** (*Mean Atomic Number*), the highest of all the observed components. Their composition is similar to that traditionally attributed to ilmenorutile, $\approx$ (Ti, Nb, Ta, Mn, Fe)O_2. They contain ***17.01% Nb_2O_5, 1.7% Ta_2O_5***, 10.58% MnO_2, 31.92% Fe_2O_3, 32.33% TiO_2, with 1.19% Y_2O_3 and 1.6% Sc_2O_3, among other impurities.

Microcrack infill, besides limonite and carbonates, comprises an oxidized, amorphous or cryptocrystalline material of high, striking brightness (high MAN), which is rich in REE: **Ce** (14.49% Ce_2O_3), **Nd** (10.09% Nd_2O_3), La (5.08% La_2O_3), Sm (2.49% Sm_2O_3), Pr (2.12% Pr_2O_3), with 18.86% TiO_2, 12.16% Fe_2O_3, 4.38% CaO, 4.20% MnO_2 and 22.61% CO_2, a.o. (Fig. 62.4c).

62.4.2 Discussion: Comparison of RL and SEM Results

As suggested by the reflected light study, SEM microanalysis confirms that the assumed ilmenorutile (the large

Fig. 62.2 Prismatic single crystal of il, with oriented ht unmixing and major, possibly secondary, rt and carbonate inclusions (general view, Fig. 62.1 shows detailed view of top right-hand corner). CR 369, Ljosland, Iveland, Norway (**dsp**)

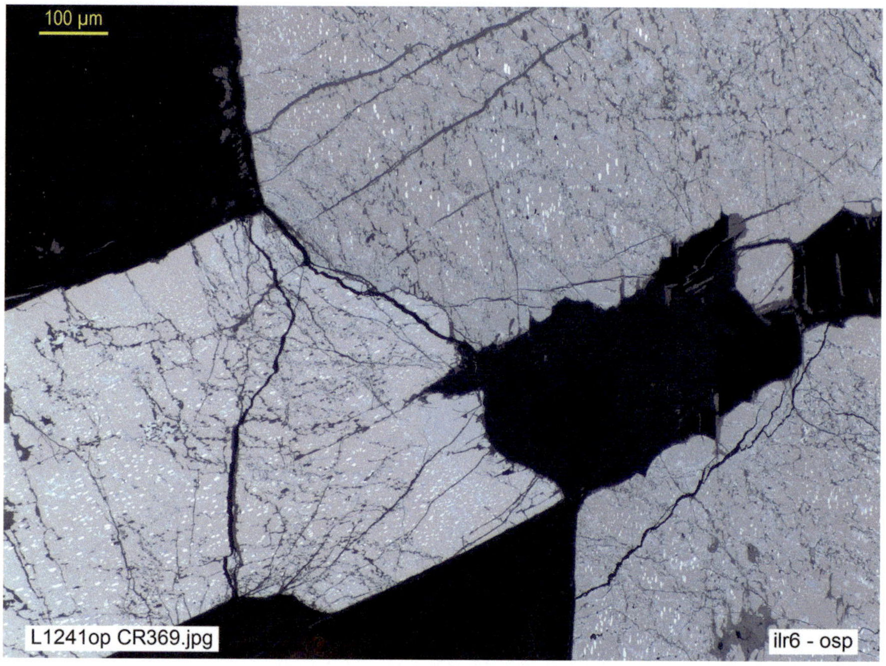

Fig. 62.3 Prismatic crystals of il (with ht exsolution, as in Fig. 62.1) whose different orientations make visible, with immersion, their weak pleochroism. CR 369, Ljosland, Iveland, Norway (**osp**)

prismatic crystals) is actually a peculiar ilmenite, with anomalous composition and optical properties. Its Nb and Ta content may be of economic interest.

However, the highest Nb and Ta grades (17.01% Nb_2O_5, 1.7% Ta_2O_5,) are found in another, minor species, which occurs as tiny lamellae (≤ 10 µm) exsolved in ilmenite, broadly identified by their composition as **ilmenorutile** and mostly associated with the exsolved hematite. Their optical properties differ in some aspects from those of common rutile. Their low reflectance makes them hardly visible under reflected light (Fig. 62.5a). In fact, they are clearly visible only with 50 × oil-immersion objective. They are seen as dark lamellae in Fig. 62.5b, Fig. 62.6a, b and correspond closely to those standing out for their brightness in the SEM image (*BSED*, Fig. 62.4b), but are "invisible" (seem to disappear) in Fig. 62.5a (20x, air).

Fig. 62.4 **a, b, c** BSED images of the mineral phases analyzed by SEM (description in text)

(a)

(b)

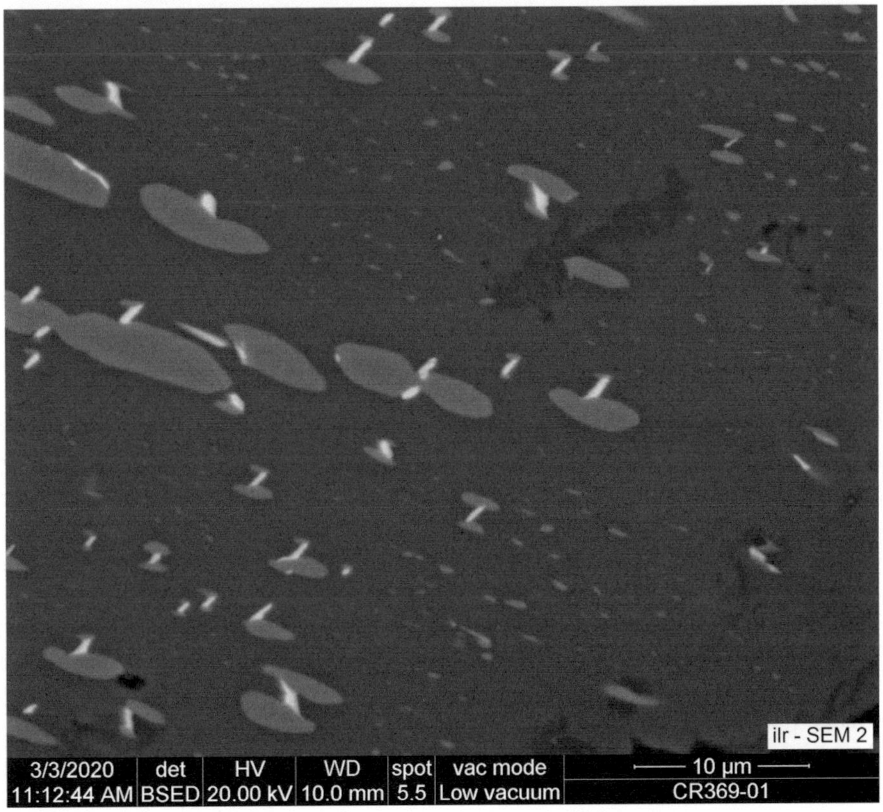

Fig. 62.4 (continued)

(c)

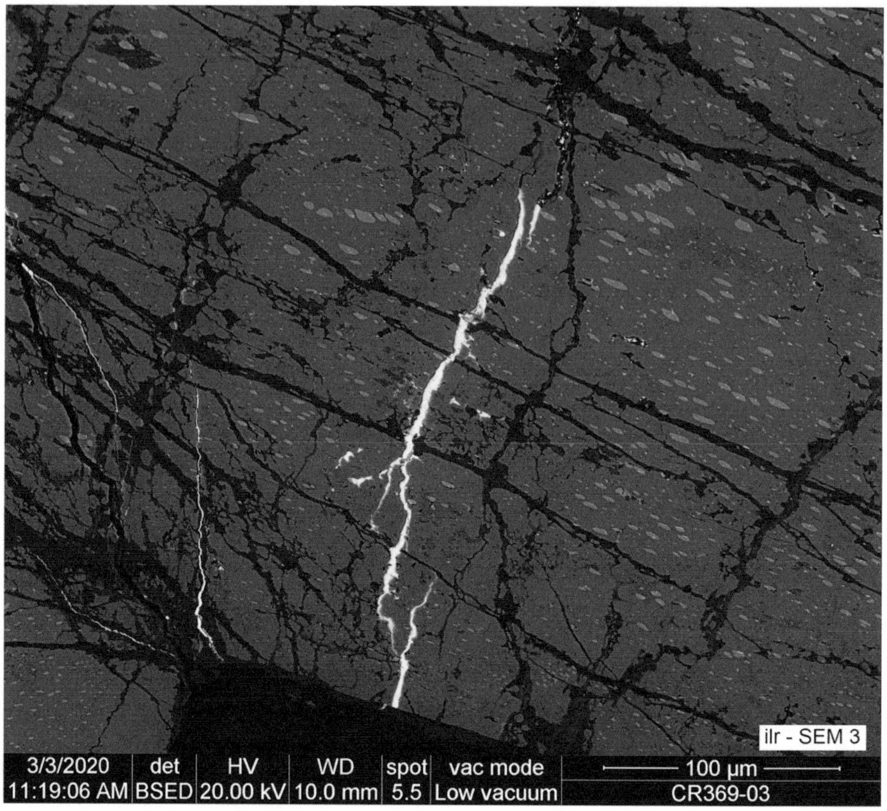

Their spatial relationship with hematite suggests that they have segregated as a consequence of Nb-(Ta) supersaturation of ilmenite, due to Fe-depletion in il by previous exsolution of hematite. These data are consistent with the characterization of these dark lamellae as *inhomogeneous ilmenorutile* (Cerny et al. 1964), since the Nb: Ta ratio is ≈10:1. According to these authors, it is a variety of rutile (*niobian rutile* or Nb-rt). In this sample, however, the main species is not rutile but ilmenite and there is no associated columbite.

The SEM also suggests the circulation of late fluids, channelled by a network of microfissures and responsible for a posthumous REE-rich paragenesis, present now as microcrack infill. It has not been possible to determine with reflected light the mineralogical composition of these "invisible" hydrothermal filling phases, but the presence of carbonates and the abundant Ce detected suggest the presence of bastnäsite, among other possible species.

62.4.3 Conclusions

Ilmenorutile is not a mineral species but a variety of rutile that can usually be characterized according to the criteria of Cerny et al. (1964) as:

- *tantalian rutile* (Ta-rt), also designated as "homogeneous ilr" (stable), or
- *niobian rutile* (Nb-rt) or "inhomogeneous ilr" (unstable).

Its rigorous determination cannot be done only with optical microscopy. It requires complementary techniques such as XRD, EPM or SEM. However, as noted by these authors, there are still problems to be solved to properly understand the nature and complex relationships of the possible minerals involved (rutile, mossite, tapiolite, columbite, ilmenite, hematite, etc.).

The previously assumed ilmenorutile in this sample is actually ilmenite (Nb-il), with exsolutions of hematite and, subsequently, of a mineral rich in Nb and Ta (Nb: Ta ≈10:1) that resembles the composition traditionally attributed to ilmenorutile but whose reflectance is distinctly lower than that of common rutile.

Subsequently, a posthumous generation of scarce species, not identified with optical microscopy, and rich in REE (Ce, Nd, La, Sm, Pr) is deposited as microcrack infill, accompanied by carbonates, oxides and scarce silicates, suggesting the possible presence of bastnäsite as the main REE carrier.

The observations made prove that, in samples of this type, optical microscopy and SEM microanalysis provide

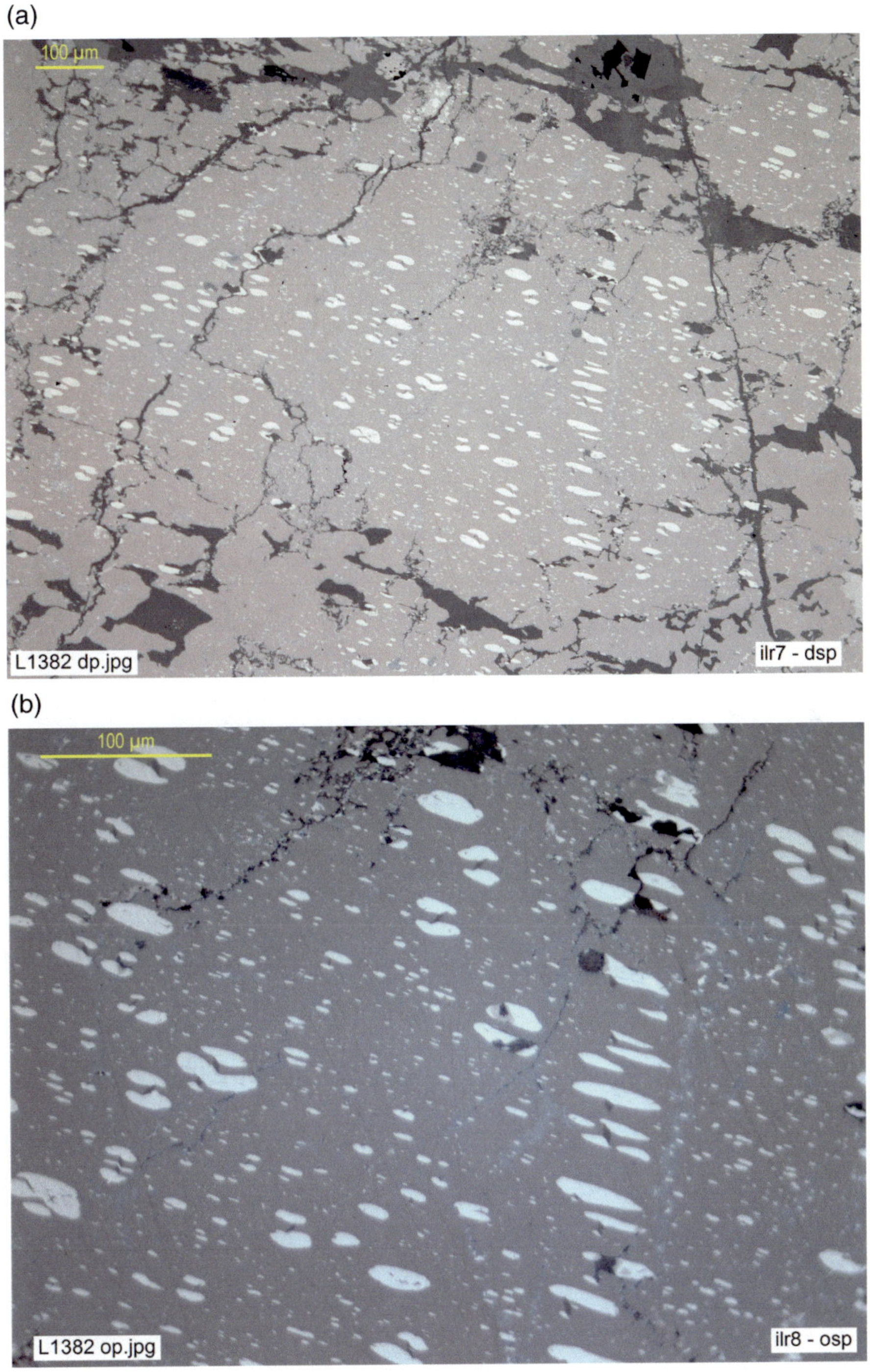

Fig. 62.5 **a, b** Unmixed ht (white) in coarse, crackled il crystal, with microcrack filling by late gth/lim (bluish grey) and gg (dark grey). The bright ilr corpuscles of Fig. 62.4b are seen as tiny, dark unmixed lamellae (ilr) in Fig. 62.5b. CR369, Ljosland ore, observed with RL, objective 20x (**dsp**) and 50x (**osp**, detail of right hand central area of Fig. 62.5a), *respectively*

complementary information, both of which are necessary. As Cerny et al. (1964, p 169) conclude: "For distinguishing niobian and tantalian rutile the study of polished sections is of first importance…/… present (chemical) data are scarce. Therefore we recommend to investigate all new niobian and tantalian rutiles in full details, to obtain new quantitative data which will supplement our study".

Fig. 62.6 **a, b** Tiny exsolutions of ilr (tiny, thin, dark lamellae) in il (main ore, greyish brown), associated with previous ht unmixing (bright, coarser lenticles) and oriented in directions cross cutting the ht alignment. Microcrack cement by oxides and silicates (Fig. 62.6b). CR369, Ljosland ore, observed with RL, 50 × obj. (**osp**)

(a)

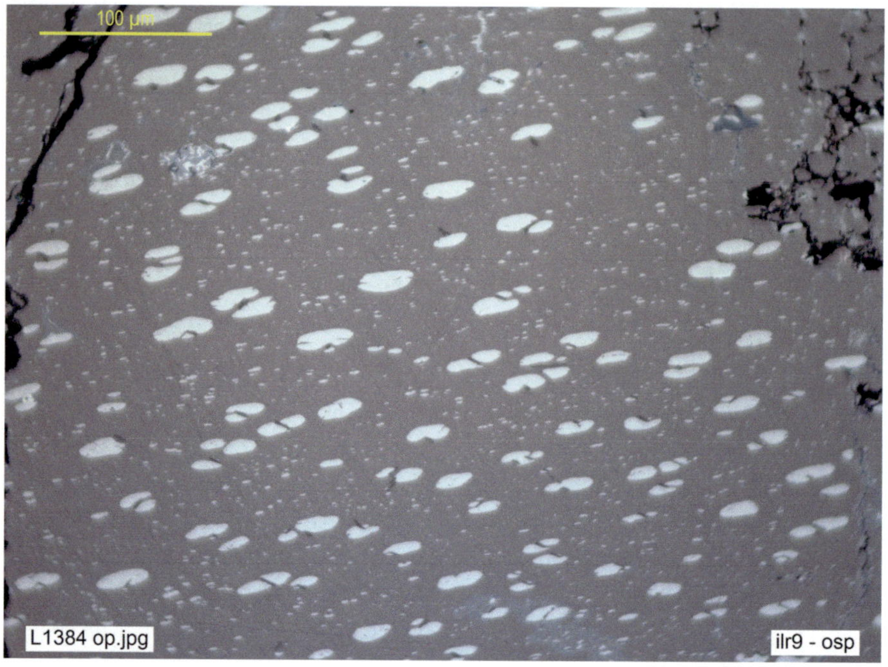

(b)

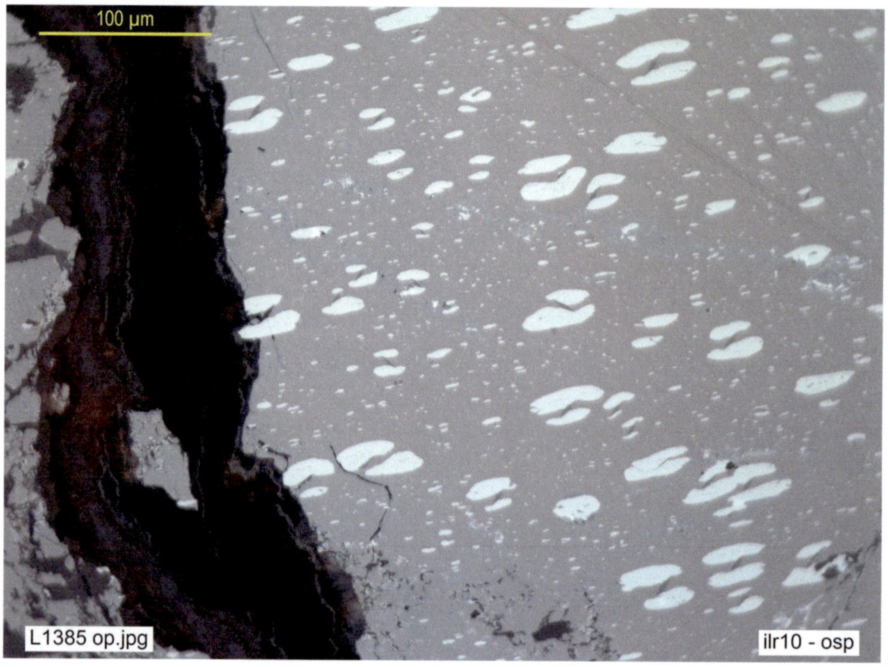

References

Back ME y Mandarino JA (2008) Fleischer's glossary of mineral species 2008. The Mineralogical Record Inc. Tucson, AZ, 345p

Beurlen H, Barreto SB, Silva D, Wirth R, Olivier P (2007) Titanian ixiolite-niobian rutile intergrowths from the Borborema Pegmatite Province. NE Brazil. Canad Mineralogist 45:1367–1387

Beurlen H, Da Silva M, Thomas R, Soares D, Olivier P (2008) Na-Ta-(Ti-Sn) oxide mineral chemistry as tracer of rare-element granitic pegmatite fractionation in the Borborema Province, NE Brazil. Miner Deposita 43:207–228

Bowles JFW, Howie RA, Vaughan DJ, Zussman J (2011) Rock-forming Minerals, vol 5A (2nd ed) Non-silicates: Oxides, Hydroxides and Sulphides. The Geological Society, London 920p

Cerny P, Cech F, Povondra P (1964) Review of ilmenorutile-strüverite minerals. Neuew Jahrb Miner Abh 101:142–172

Hey MH (1975) An index of mineral species and varieties arranged chemically (Chemical Index of Minerals), vol 3. British Museum (Nat. History), London

Ramdohr P (1980) The ore minerals and their intergrowths, vol 2, 2nd edn. Pergamon Press, Oxford, 1205p

Strunz H, Nickel EH (2001) Strunz Mineralogical Tables, 9th edn. E. Schweizrbart'sche Verlagsbuchhandlung, Stuttgart, p 870

Uytenbogaardt W, Burke EAJ (1971) Tables for microscopic identification of ore minerals, 2nd edn. Elsevier, Amsterdam

Iridium (*ir*)

Ir: Cubic

63

Iridium		R_{546} (%)	Y (%)	λ_d (nm)	P_e (%)	Colour sensation
Air		71.0	71.4	582	2.2	White
Oil		nd	nd	nd	nd	
Bir: Null (1)				*Pleoc*: No (1)		*Tarn*: No
Anis + P: Null (1)				*IR*: No		*Twg*: No
Mor: Euhedral (cubic) or anhedral inclusions				*Clv*: No		*PolQ*: Easy
Hardness	*PolH*: High, > Pt (4)			*Mohs*: 6 1/2	*Talm*: nd	*VHN* $_{(100\ g)}$: 946–988

Obs: cf. Table and R charts, multispectral data (VNIR, AMCO) and visible range (QDF3)
Mineral abbreviations: Ag—native silver, chr—chromite, Fe—native iron, gg—gangue, Ir—native iridium, Os—native osmium, Pd—native palladium, Pt—native platinum, spy—sperrylite

63.1 General

Ir is, together with Os, Ru, Pt, Pd and Rh, one of the six **PGE** (*Platinum Group Elements*). Isomorphic with Pt and Rh, it is also, as a native metal, one of 109 **PGM** (Platinum *Group Minerals)* recognized by IMA (cf. comments on PGE and PGM in *Pt description* and more data in Cabri, ed. 2002).

Composition: Ir. However, some authors prefer the formula (Ir, Os, Ru), which recalls that the former *osmyridium* and *rutenosmyridium* species, alloys with Ir as the main component, have been incorporated into the current, broader re-definition of Ir by IMA, while *rutheniridosmin* is credited (cf Osmium). The native iridium mineral, **Ir**, which is usually found associated with Pt or Pt-Fe in inclusions, usually contains Os, Ru, Pt and occasionally minor to trace amounts of Rh, Pd, Fe, Cu and Ni (Cabri 2002).

63.2 Microscopic Description

63.2.1 SP Observation (*Single Polariser*)

Pure white *colour* (Fig. 63.1a, b), very similar to native platinum but, in direct comparison, purer (without the faint yellowish tinge of platinum) and somewhat brighter. Also better polished (less scratches) than platinum and in relief on platinum (Figs. 63.1a, b, 63.2 and 63.3).

Very high *reflectance*, slightly higher than Pd but lower than Ag: R_{Ir} > Pt, Os, Fe, spy; R_{Ir}>> chr (cf multispectral comparison charts). Isotropic. High *polish hardness* (>Pt).

It occurs included in native platinum, in exsolutions (Figs. 63.1a, b, 63.2 and 63.3).

63.2.2 Observation +P (*Polariser + Analyser at 90°*)

Isotropic (Fig. 63.1c, d).

63.2.3 Paragenesis

Very limited *paragenesis*, usually included in platinum (or PtFe alloys), in connection with ultramafic intrusions and in ophiolitic series, particularly in chromitites. Also, in placers derived from these formations.

Companion ores: native platinum and MGP.

R. Castroviejo, *A Practical Guide to Ore Microscopy—Volume 1*,
https://doi.org/10.1007/978-3-031-12654-3_63

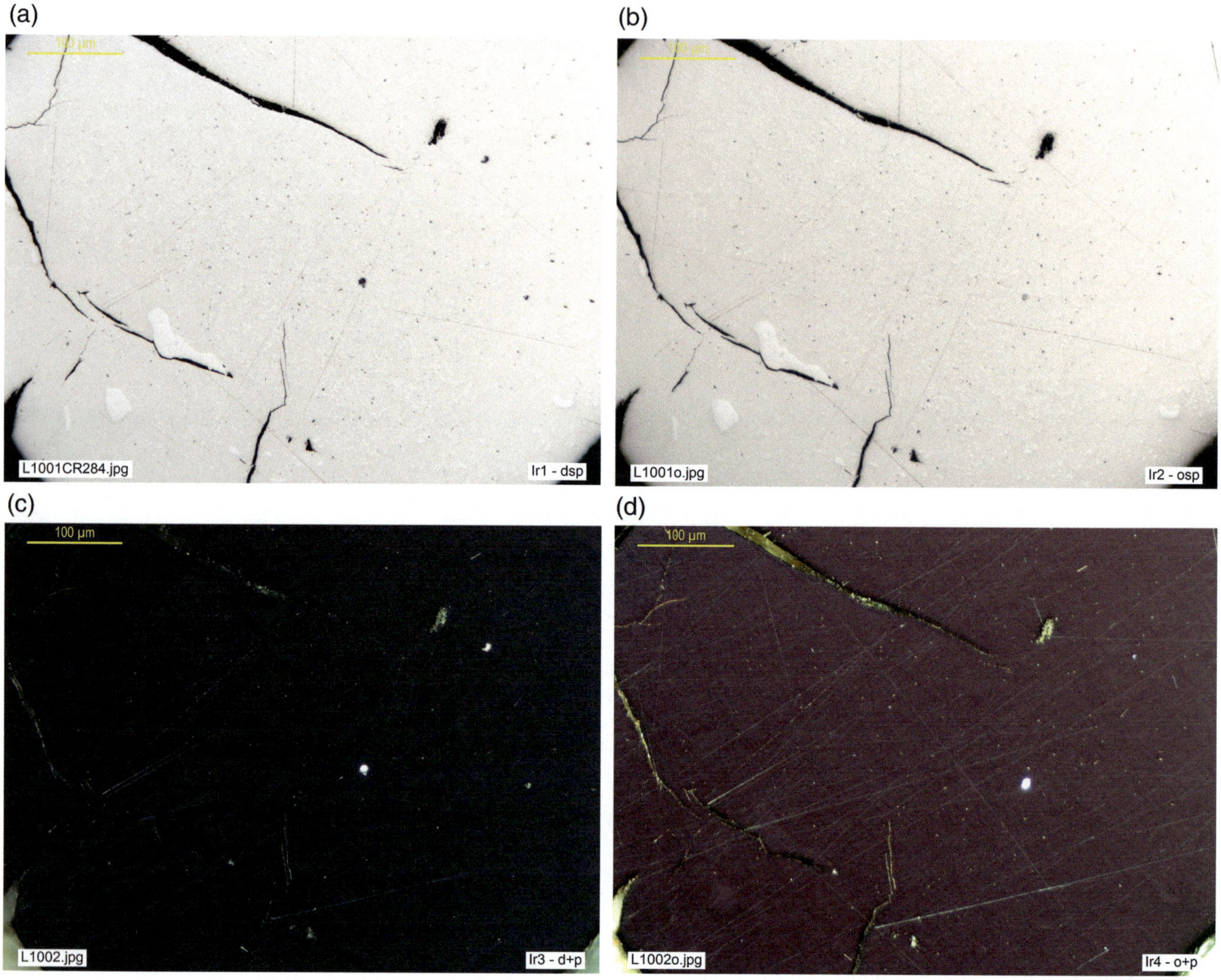

Fig. 63.1 **a, b** (**dsp** and **osp**, respectively)—Ir inclusions (white, relief) in Pt nugget. Reflectance and polishing hardness of Ir (>Pt) allow observation of a thin emulsion of Ir in the whole Pt mass, which disappears in the immediate vicinity of the major inclusions, thus evidencing its origin by exsolution. The compositions of Pt, which contains some Fe, and Ir have been confirmed by electron microprobe. CR 284, Urals, Russia. **c, d** Same scene as Ir1 (**d+p** and **o+p**, respectively). The cubic symmetry of Pt and Ir, both in extinction, makes their distinction difficult (**+p**), enhancing instead, due to their anisotropism, thin Os lamellae included in Pt (top right-hand corner), hardly visible without analyzer (**dsp** and **osp**: Fig. 63.1a and b). However, Ir is easily distinguished with i.c. by its relief (**d+p.w**: cf. Fig. 63.3)

63.2.4 Identification

The white colour, slightly brighter than platinum and palladium, and its greater hardness, which makes it appear in relief on platinum, together with its isotropy that differentiates it from osmium, allow easy recognition, since it is only known in nature as inclusions (unmixed) in platinum linked to ultramafic formations or placers derived from these.

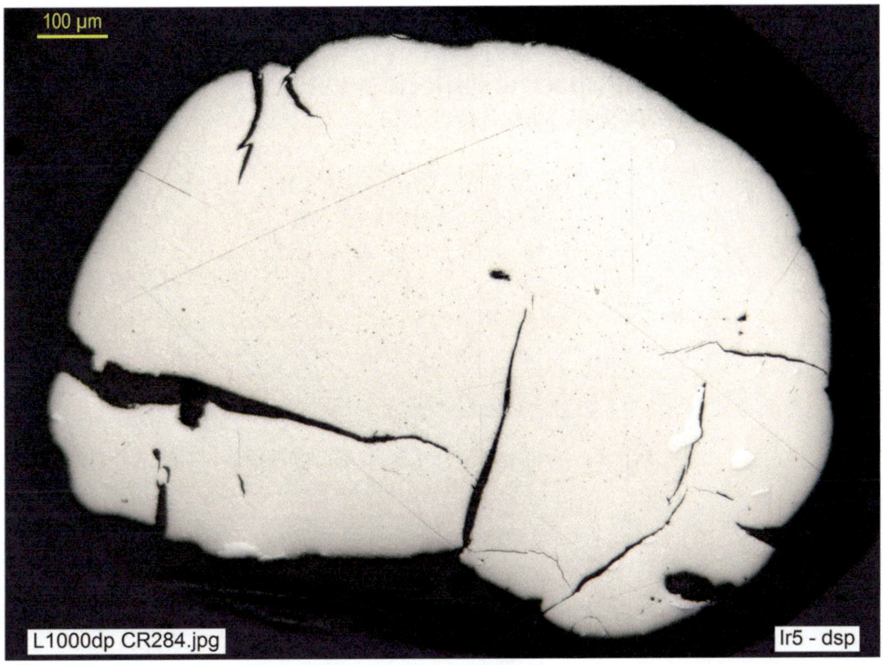

Fig. 63.2 Pt nugget (full grain of Fig. 63.1a and also of Fig. 63.3, which are detailed views of the right hand side of the nugget, rotated 90º clockwise), with Ir in discrete inclusions (10–80 µm, white, in relief, R and DPul > Pt) and in fine emulsion throughout the Pt volume. Fine Os lamellae, inconspicuous at this scale, included in Pt. CR 284, Urals, Russia (**dsp**)

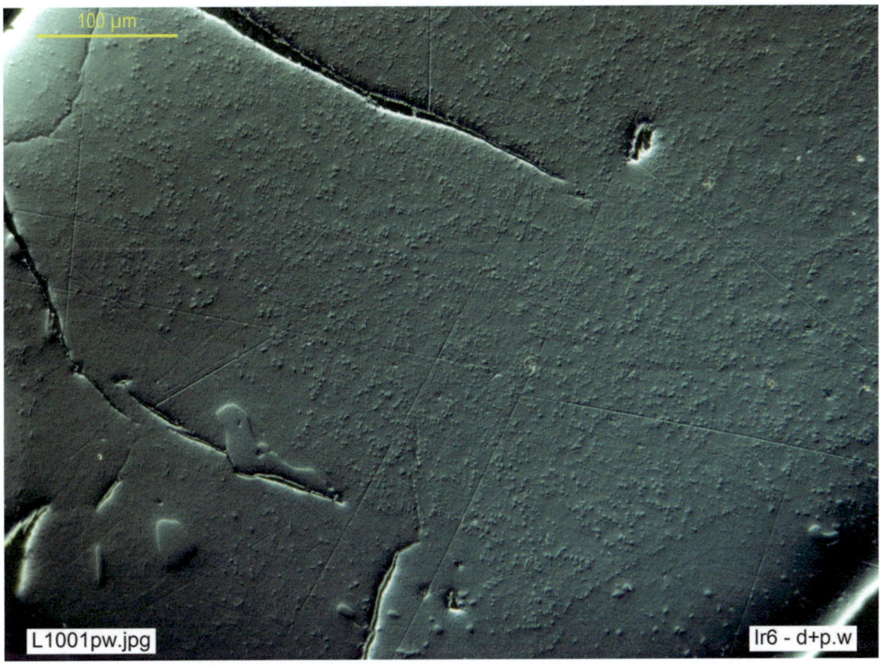

Fig. 63.3 Same scene as Fig. 63.1a (**d+p.w**). View in interference contrast, highlighting in relief the exsolution of Ir, distributed in a thin emulsion in the whole mass, except in the surroundings of the major inclusions (left). Also visible in relief, although scarce, are the thin lamellae of Os included in Pt

63.3 Spectral Information

63.3.1 Iridium Multispectral Reflectance R (in Air, Without Polariser) (%)

λ (nm)	370	400	425	450	475	500	525	550	575	600	625	650	675	700	750	800	850	900	950	1000
μ	62.8	64.1	66.0	66.8	68.8	69.2	70.7	70.6	71.2	71.6	72.0	72.3	72.6	72.9	73.6	74.3	75.6	76.6	77.5	77.4
P_5	58.6	60.3	62.2	63.4	65.3	66.1	67.6	67.8	68.5	69.0	69.4	69.9	70.1	70.6	71.5	72.3	73.8	75.1	76.0	76.0
P_{20}	60.2	61.9	63.6	64.7	66.6	67.3	68.7	68.8	69.5	69.9	70.3	70.8	71.0	71.4	72.2	72.9	74.4	75.6	76.5	76.5
P_{80}	65.2	66.0	68.0	68.7	70.8	71.0	72.4	72.2	72.8	73.1	73.4	73.7	74.0	74.3	74.9	75.5	76.8	77.7	78.5	78.3
P_{95}	66.4	67.1	69.1	69.4	71.7	71.6	73.2	72.9	73.5	73.7	74.0	74.3	74.6	74.8	75.5	76.0	77.3	78.3	79.2	79.0

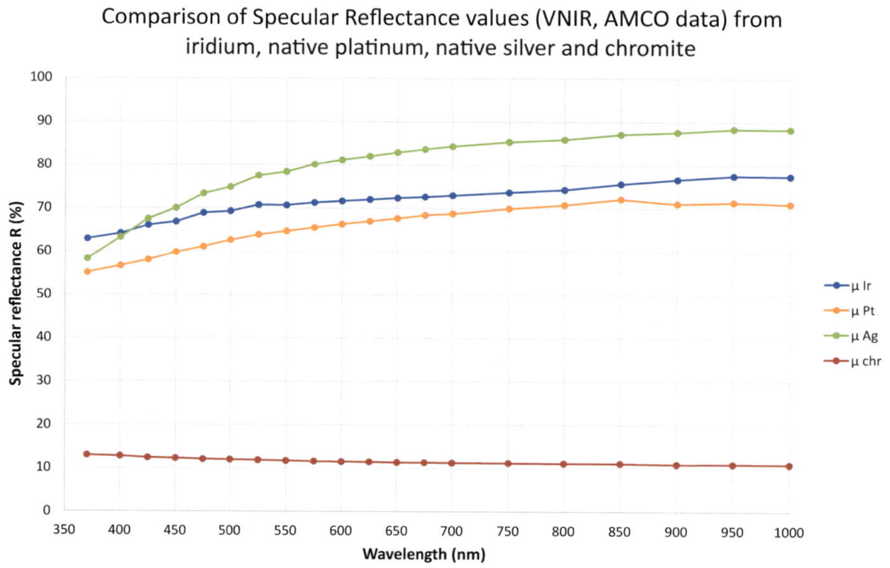

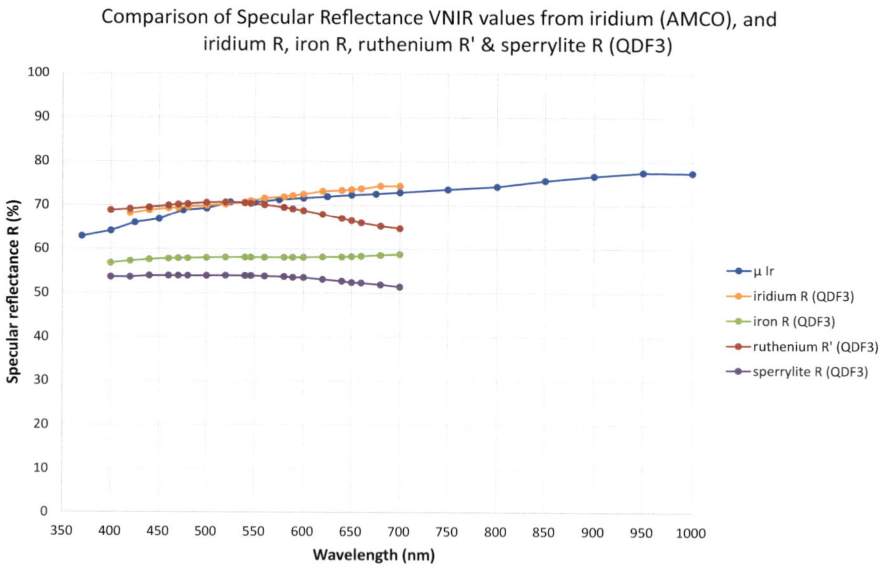

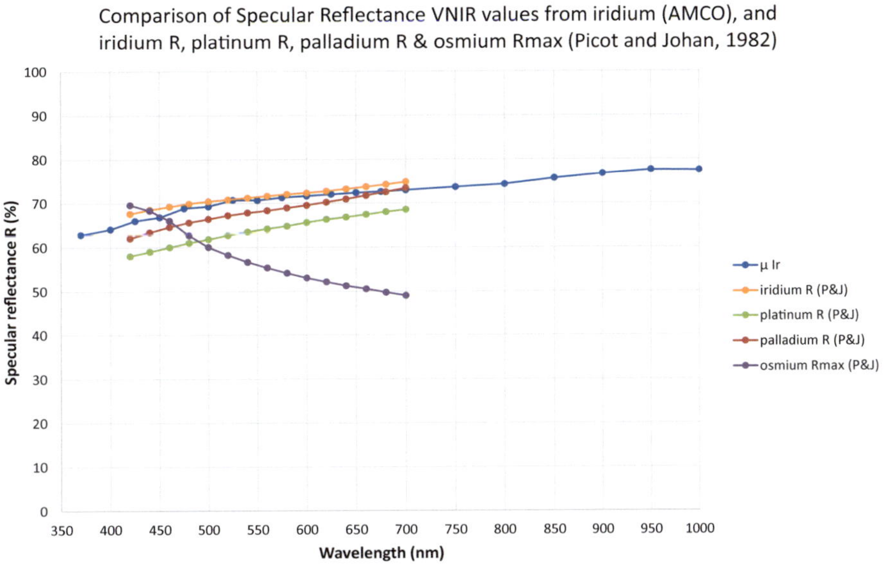

References

Cabri LJ (2002) The platinum-group minerals. In: Cabri LJ (ed) CIM Special, vol 54. Canad. Inst Mining, Metall. and Petroleum, Calgary, Alberta, Canada, pp. 13–129, 852

Cabri LJ (2002, ed) The geology, geochemistry, mineralogy and mineral beneficiation of platinum group minerals. Canadian Institute of Mining, Metallurgy and Petroleum, Calgary, Alberta, Canada, pp 852. (ISBN 1–894495–27–5)

Iron (*Fe/Native Iron, cohenite, troilite, steel*) 64

Fe: Cubic

Iron	R_{546} (%)		Y (%)	λ_d (nm)	P_e (%)	Colour sensation
Air	58.1		58.1	571	0.4	White
Oil	45.2		45.2	577	0.1	
Bir: Null (1)				*Pleoc*: No (1)		*Tarn*: Avoidable with lacquer
Anis +P: Isotropic (1)				*IR*: No		*Twg*: No (terrestrial). Lamellar (meteor, siderurg)
Mor: Disseminated interstitial particles, droplets				*Clv*: //(100), not visible microsc		*PolQ*: Good
Hardness	*PolH*: Mean: < mt, < < coh (3)			*Mohs*: 4 1/2	*Talm*: nd	*VHN* $_{(100\ g)}$: 110–117

Note: cf. Specular reflectance graphs, R%, visible range (QDF3)
Mineral abbreviations: brk—brookite, chr—chromite, coh—cohenite, Cu—native copper, Fe—native iron, gg—gangue, gra—graphite, hub—hübnerite, il—ilmenite, Ir—native iridium, mt—magnetite, po—pyrrhotite, Pd—native palladium, Pt—native platinum, py—pyrite, sch—scheelite, sp—sphalerite, tr—troilite

64.1 General

Given its marked tendency to oxidation, iron in its native state, *Fe*, is only exceptionally found in the terrestrial environment (*vbgr.* in basalts intruding coal layers—Uivfaq, Disko Island, Qeqertalik, Greenland or in some serpentinites generated in peculiar *redox* conditions), being instead a normal component in meteorites (*cosmic* Fe, which is supposed to be analogous to that of the terrestrial nucleus) and, of course, in metallurgical production. The steel industry, including both iron and its various alloys (particularly steel), as well as its industrial derivatives and waste (scrap), has been for ages and is still of utmost importance for human development. The first two types, identified as *native iron* or natural Fe (Figs. 64.1a, b, 64.2a, and 64.3a) represent an insignificant proportion of the metal accessible on the earth's surface, and their incidence in mining is nil.

The third type, man-made metal (*iron and steel*), is obviously not an ore but an industrial product and, as such, belongs rather to the field of metallurgy. However, it is considered of interest to mineralogists and plant engineers at least to be able to recognize it, since they will find it sooner or later. In the plant process, sometimes also in drill holes, small steel chips or shards (from crushers, mills, screens, etc.) are often released, which may then appear mixed with the ore (Figs. 64.4 and 64.5) and at first may be confusing for unaware observers. This is reason enough to include Fe among the minerals of interest, although natural Fe itself is not the object of mining. For this reason, the aim of what follows is just a brief characterization of iron, *Fe*, by optical methods. The description comprises steel also, without going into the different alloys and metallurgical products which are the object of metallography, nor into the details of the study of meteorites. Nevertheless, the recognition of cohenite and troilite, two extraterrestrial phases found in meteorites, is briefly sketched, for their possible relationships with iron.

Composition: Fe, sometimes intergrown with cohenite (Fe_3C). In meteorites, notable Ni contents ($\geq$ 5%, may be > 50% Ni). It may also contain minor amounts of other elements: P, Co, S, etc.

R. Castroviejo, *A Practical Guide to Ore Microscopy—Volume 1*,
https://doi.org/10.1007/978-3-031-12654-3_64

64.2 Microscopic Description

64.2.1 SP Observation (*Single Polariser*)

White *colour*. Good *polish*. With time, it can acquire a *patina* of oxidation, avoidable by lacquering. *Polishing hardness* lower than that of magnetite, ***mt***, and even lower than that of cohenite, ***coh***, which is an interesting fact for the recognition of cohenite without etching.

Moderately high *reflectance*: R similar to that of cohenite, higher than troilite (stoichiometric pyrrhotite) but clearly lower than Pt and EGP in general (cf. comparative graphs of reflectance in the visible spectrum). Isotropic.

Terrestrial Fe, very rare, occurs as disseminated particles or droplets. It fills interstices between silicates in basalt contaminated by reducing sediments and, not infrequently, occurs intergrown with **cohenite**, *coh* (cf. characterization below). Very rarely, it also occurs in serpentinites (with awaruite, heazlewoodite, magnetite, etc.) and in very bituminous sediments. Cosmic Fe is also found disseminated in meteorites, accompanied by **troilite**, *tr* (cf. characterization below and Figs. 64.1a–d, 64.2a, b and 64.3a, b), graphite, native copper, etc.

In plant concentrates and, more rarely, also in drill cores, small **steel** fragments, splinters, and balls are sometimes found, torn from the balls or bars of the mills in the comminution process. These fragments are totally unrelated to the ore and the microscopist must recognize them to avoid serious errors, such as their possible confusion with Pt or precious metals. Fortunately, they are easily identified because their optical characteristics are similar to those of *natural* Fe and, in addition, unlike the latter, they do not appear intergrown but isolated and their morphology is characteristic—clastic (angular fragments or splinters) or rounded.

64.2.2 Observation +P (*Polariser + Analyser at 90°*)

Isotropic (Figs. 64.1c, d, 64.2b and 64.3b). *Lamellar twins* have been found in meteoritic Fe and in iron and steel Fe but not in *natural* terrestrial Fe.

64.2.3 Paragenesis

Natural Fe has been found only in basalts, serpentinites and exceptional sediments (terrestrial domain) and as a common component in meteorites (cosmic domain).

Accompanying minerals: cohenite, troilite, graphite, copper, silicates, etc. None of them are currently being mined.

Iron and steel Fe is found in the mining environment and with some frequency in plant concentrates, as contamination generally derived from the comminution process (mill liner and ball or bar pull-outs).

64.2.4 Identification

Fe is easily recognized by its color, reflectance, hardness, isotropy and paragenesis. I It occurs only exceptionally in terrestrial domains (*e.g.,* basalts from Disko Island, Greenland) or is common in such exceptional rocks as meteorites. *Cohenite* has a similar appearance, but the usual Fe-coh association in intergrown aggregates facilitates the distinction—cohenite is clearly harder and anisotropic.

When Fe occurs in a concentrate, it is generally **steel**, whose optical properties coincide, for recognition purposes, with those described for Fe. It should not be confused with precious metals, particularly Pt, Pd and Ir, which resemble it. Steel (Figs. 64.4 and 64.5) is easily recognized because it occurs in isolated particles, not intergrown with the ore, and with characteristic shapes—rounded (balls) or clastic, due to impact (irregular or angular fragments, shards, chips, shavings).

64.2.4.1 Cohenite Brief Microscopic Characterization

The mineral cohenite, *cohenite*, also known as cementite in metallurgy, has Fe_3C composition and orthorhombic symmetry.

It forms intergrowths with Fe, either with natural Fe (terrestrial or cosmic) or with steel Fe. The morphology of these intergrowths can vary; myrmekitic-like morphologies are frequent. In meteorites, cohenite is found especially in iron meteorites, in which it is often abundant, associated with graphite and troilite. For some authors, its formula is $(Fe, Ni)_3C$, because it usually contains also minor amounts of Ni (and sometimes Co), replacing Fe.

It polishes well and with fewer polishing scratches than Fe, because of its greater hardness. Its color is bright white, like Fe but, next to Fe, cohenite shows a slightly yellowish tinge (Fig. 64.3a). Moderately high reflectance, like Fe. Bireflectance is hardly noticeable but is visible in immersion, comparing contiguous grains with different orientation. Weak anisotropism but well visible in immersion (Fig. 64.2b).

Identification: Although due to its color and reflectance it can be confused with Fe, the anisotropism of cohenite and its higher polishing hardness easily distinguish it. In routine work for the metallurgical industry, with metallographic microscopy (without polarization), chemical etching is often used as an alternative criterion. This is superfluous when a

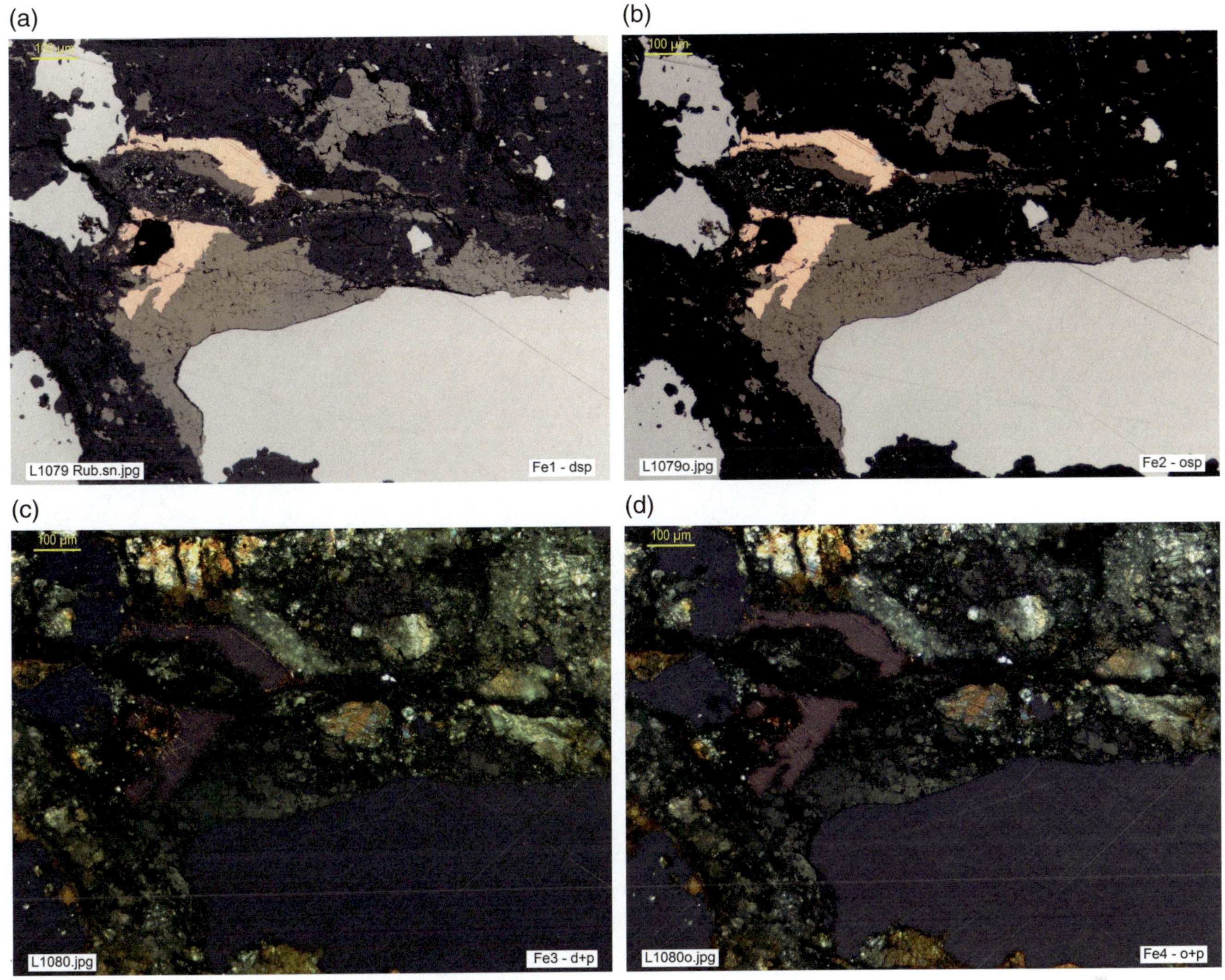

Fig. 64.1 a, b (**dsp** and **osp**, respectively)—Fe (white), troilite (brownish), Cu (pinkish, shiny) and silicates (dark grey). Note the subjective perception of the shade of tr, dark here, next to Fe and Cu, in contrast to the much lighter appearance next to gra, chr and gg, as seen in Fig. 64.3a. Rub.sn, El Desaguadero Meteorite, Puno, Peru. **c, d** Same scene as Fe1 (**d+p** and **o+p**, respectively). Fe and Cu in extinction. Troilite and silicates, anisotropic

polarizing microscope is available, such as those used for ore microscopy.

64.2.4.2 Troilite Brief Microscopic Characterization

Troilite, *tr*, is stoichiometric FeS, of hexagonal symmetry, without the Fe deficit (vacancies) of pyrrhotite ($Fe_{1-x}S$). Troilite is paramagnetic and pyrrhotite, ferromagnetic, varying according to the different polytypes (monoclinic or hexagonal symmetry). Although troilite is considered by some to be a pyrrhotite polytype, it maintains its status as an IMA *grandfathered* species.

The optical properties and parameters of troilite are (according to QDF3 data) similar to those of pyrrhotite, but its reflectance is slightly lower and it looks a little darker:

- *Y % pyrrhotite (p. 467)* oriented after R1 and R2: 36.9 and 40.5 'air', *respectively*; 23.6 and 27.1 'oil', *respectively*
- *Y % troilite* oriented after R1 and R2: 33.9 and 38.7 'air', *respectively*; 22.1 and 26.3 'oil', *respectively*), which is accentuated with time by its fast tarnishing.

Troilite colour and pleochroism (Figs. 64.1a, b and 64.3a), in pinkish-brown to yellowish-brown shades as pyrrhotite, are characterized in both minerals by:

- identical values λd (λd for R1: 580 nm and λd for R2: 578 nm, both in air and oil for both minerals),
- but higher troilite P_e (*P_e % pyrrhotite (p. 467)*, according to R1 and R2: 9.7 and 8.9 'air', *respectively*; 14.1 and

Fig. 64.2 **a, b** (same scene: **osp** and **o+p**, respectively). Aggregate of intergrown Fe (white, isotropic) and coh (yellowish white, weakly anisotropic) in basaltic groundmass, containing also a fine dissemination of Fe, troilite (pinkish brown) and acicular il (greyish brown, darker than troilite). The higher hardness of coh ($PolH_{coh}$ > Fe) is evidenced by the disappearance of many polish scratches when switching from Fe to coh. The anisotropism of tr and il is barely noticeable here, hidden by the abundant IR of the silicates. Rub. DiskIs sample, Disko Island, Greenland

12.3 'oil', *respectively*; P_e *% troilite*, according to R1 and R2: 14.6 and 11.2 'air', *respectively*; 18.1 and 13.2 'oil', *respectively*).

Its distinct anisotropism is, as in pyrrhotite, characteristic (Fig. 64.1c, d, although here it is hardly visible because of the luminous supersaturation of the immediate surroundings due to the abundant internal reflections of the gangue).

Given the similarity of both minerals, interested readers are referred to the description of pyrrhotite for other properties of troilite and for its microscopic identification.[1]

Troilite is typically found in meteorites, hence its synonym *cosmic pyrrhotite,* and its presence in terrestrial rocks such as serpentinites and ultramafic complexes is

[1] Further information: https://www.mindat.org/show.php?id=4029&ld=2&pho.

Fig. 64.3 **a, b** (same scene: **dsp** and **d+p**, respectively) Fe (white, isotropic), gra (brownish grey, strongly anisotropic), tr (pinkish brown), chr (grey, isotropic, yellowish-brown internal reflections) and silicates (dark grey, colourless or greenish IR, abundant). Rub.NMet sample, Neuschwanstein Meteorite, Bavaria, Germany

exceptional. The opposite occurs with pyrrhotite. The presence of troilite has also been observed in metallurgical slags and coal ashes. According to modern data from space probes, troilite could be a frequent mineral on some planets or satellites of the solar system, such as Mars or Ganymede.

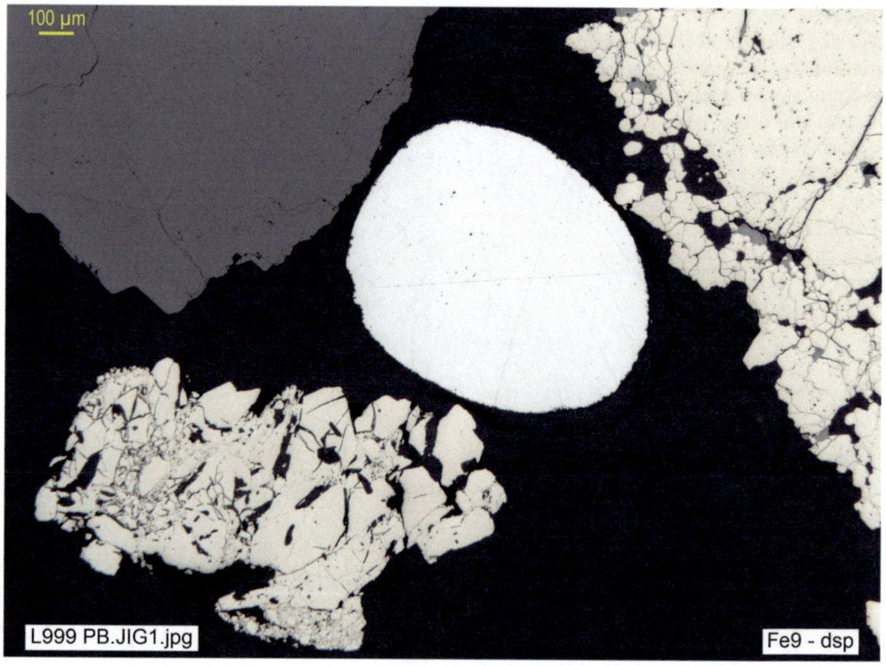

Fig. 64.4 W mine plant concentrate: hübnerite, ***hub*** (grey, top left), py with sparse interstitial sphalerite and, in the center, steel ball (white). PB-Jig1, Pasto Bueno Mine, Ancash, Peru (**dsp**)

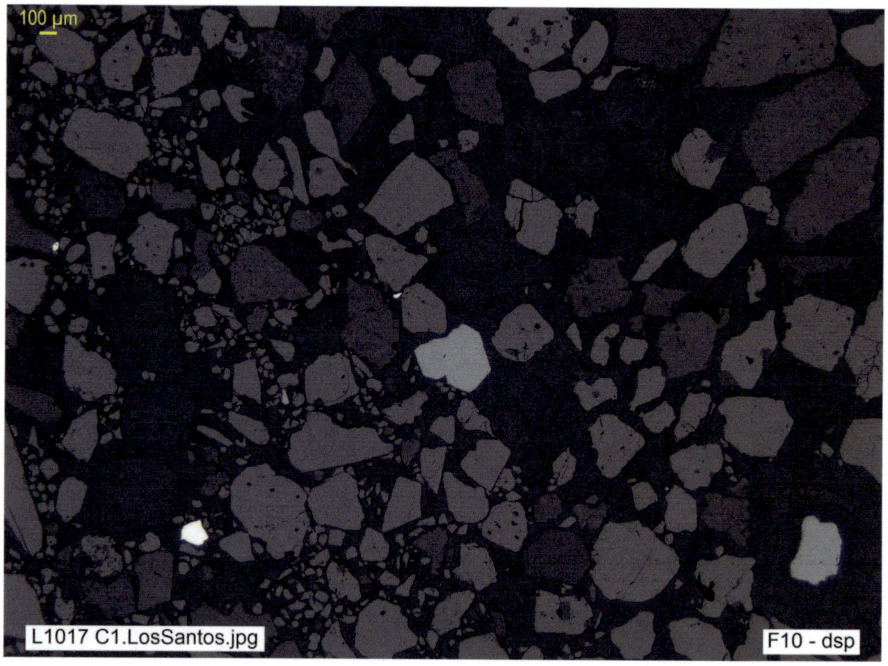

Fig. 64.5 Scheelite concentrate, with: sch (main mineral, grey); gg (darker gray, various shades); rutile, *rt* (two very visible grains, center and bottom right, much lighter grey than sch) and steel (white, very bright, one small fragment lower left and several tiny shards scattered), probable contamination from milling in the plant. Sample C1.LMA, Los Santos Mine, Cáceres (dsp)

64.3 Spectral Information

64.3.1 Iron Multispectral Reflectance R (in Air, Without Polariser) (%)

λ (nm)	370	400	425	450	475	500	525	550	575	600	625	650	675	700	750	800	850	900	950	1000
μ	48.6	50.0	51.1	51.9	52.4	52.8	52.9	53.0	53.2	53.5	53.7	53.9	54.2	54.5	55.1	55.5	56.2	56.7	56.9	56.8
P_5	40.7	42.2	43.8	45.1	45.8	46.2	46.5	46.6	46.6	47.0	47.2	47.4	47.6	47.9	48.2	48.5	49.0	49.3	49.4	49.1
P_{20}	46.4	48.1	49.4	50.4	51.2	51.6	51.8	51.9	52.1	52.3	52.4	52.6	52.9	53.1	53.6	53.9	54.4	54.8	54.8	54.4
P_{80}	51.4	52.7	53.6	54.3	54.7	55.0	55.1	55.2	55.4	55.7	55.9	56.2	56.5	56.8	57.5	58.0	58.9	59.5	60.0	60.2
P_{95}	52.5	53.7	54.5	55.1	55.5	55.8	55.9	55.9	56.1	56.4	56.6	56.9	57.3	57.6	58.2	58.6	59.5	60.0	60.5	60.8

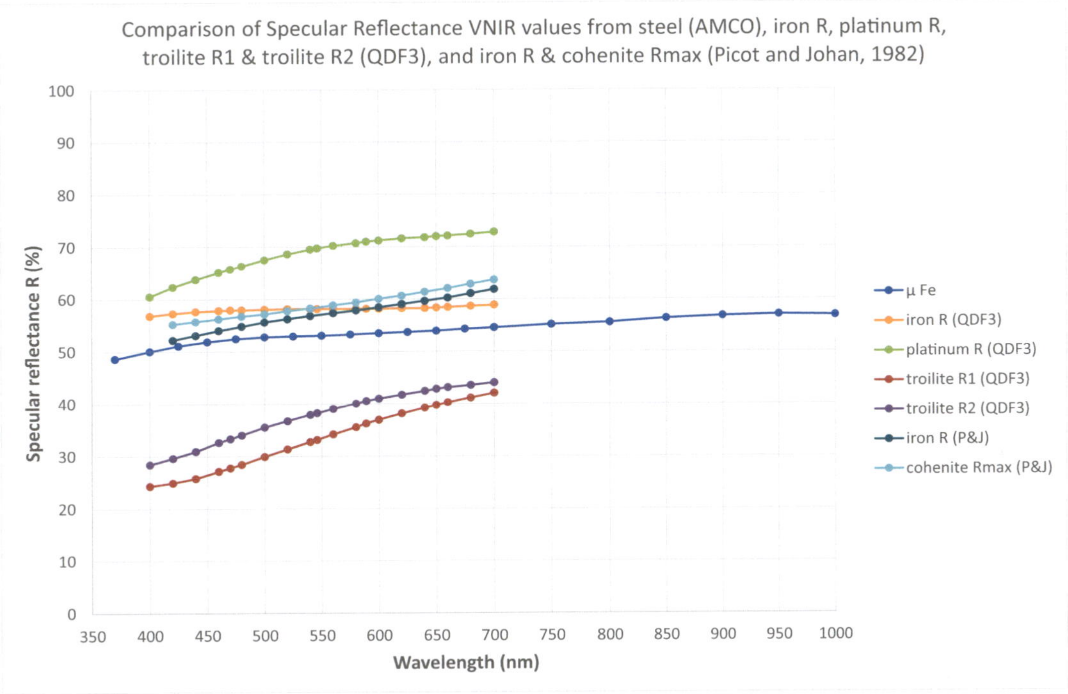

Isocubanite (*isc/chalcopyrrhotite*) 65

$CuFe_2S_3$: Cubic

Isocubanite		R_{546} (%)	Y (%)	λ_d (nm)	P_e (%)	Colour sensation
Air		35.2	35.5	576	17.8	Pinkish brown
Oil		23.1	23.2	576	22.0	
Bir: Null (1)				*Pleoc*: No (1)		*Tarn*: Variable, sometimes very fast
Anis +P: Isotropic (1)				*IR*: No		*Twg*: // (111)
Mor: Anhedral aggr. disseminated				*Clv*: Not observed		*PolQ*: Good
Hardness	*PolH*: Medium (3)			*Mohs*: 31/2	*Talm*: nd	*VHN* $_{(100\ g)}$: 175

Note: cf. R charts, visible range (QDF3)
Mineral abbreviations: cbn—cubanite, ccp—chalcopyrite, isc—isocubanite, mck—mackinawite, mt—magnetite, pn—pentlandite, po—pyrrhotite, py—pyrite, sp—sphalerite

65.1 General

Traditionally known as chalcopyrrhotite or cubic cubanite, isocubanite (***isc***) is the high-temperature phase of $CuFe_2S_3$. Its dimorph cubanite (*cbn*) is transformed into isc above 200 °C. As a laboratory synthetic phase, isocubanite is called **iss**, *intermediate solid solution* (sphalerite-type disordered structure phase, cubic *fcc*, dominant above 500 °C) although, in reality, the compositional field of iss is wider than the stoichiometric field of isocubanite. It also includes that of other species, as demonstrated by numerous experimental investigations carried out on the Cu–Fe–S ternary system (talnakhite, mooihoekite, haycockite, etc. For more information and discussion of experimental results, cf. Craig and Scott 1976, and Bowles et al 2011).

As a high-temperature mineral, isocubanite is unstable upon cooling and, therefore, its occurrence is restricted. It is currently deposited with submarine exhalative massive sulfides, but in fossil deposits it is preserved only under certain conditions (e.g., very rapid cooling or particular compositions). Regarding its use as a geological thermometer, Ramdohr (1980) proposes a formation temperature above 240 °C (and probably much higher). Despite its instability, this author argues that isocubanite has gone unnoticed in many cases (by confusion with pyrrhotite, cubanite or even chalcopyrite) and, therefore, it would be much more frequent in nature than it seems. It may contain Zn or Ni as impurities.

65.2 Microscopic Description

65.2.1 SP Observation (*Single Polariser*)

Pinkish-brown *colour*, similar to pyrrhotite but without its pleochroism (Fig. 65.1a, b). In meteorites, possibly because they contain Ni, the colour of isocubanite is pinker and brighter (Fig. 65.2).

Reflectance similar to those of cubanite and pyrrhotite, practically coincident with R_1 of pyrrhotite and cubanite. Therefore, $R_{isc} < R_2$ of po/cbn and clearly < pn, py, ccp (Fig. 65.1a, b; *cf.* reflectance comparative graphs). As a cubic mineral, it is neither bireflective nor pleochroic. *Medium polishing hardness*: > ccp; < po.

R. Castroviejo, *A Practical Guide to Ore Microscopy—Volume 1*,
https://doi.org/10.1007/978-3-031-12654-3_65

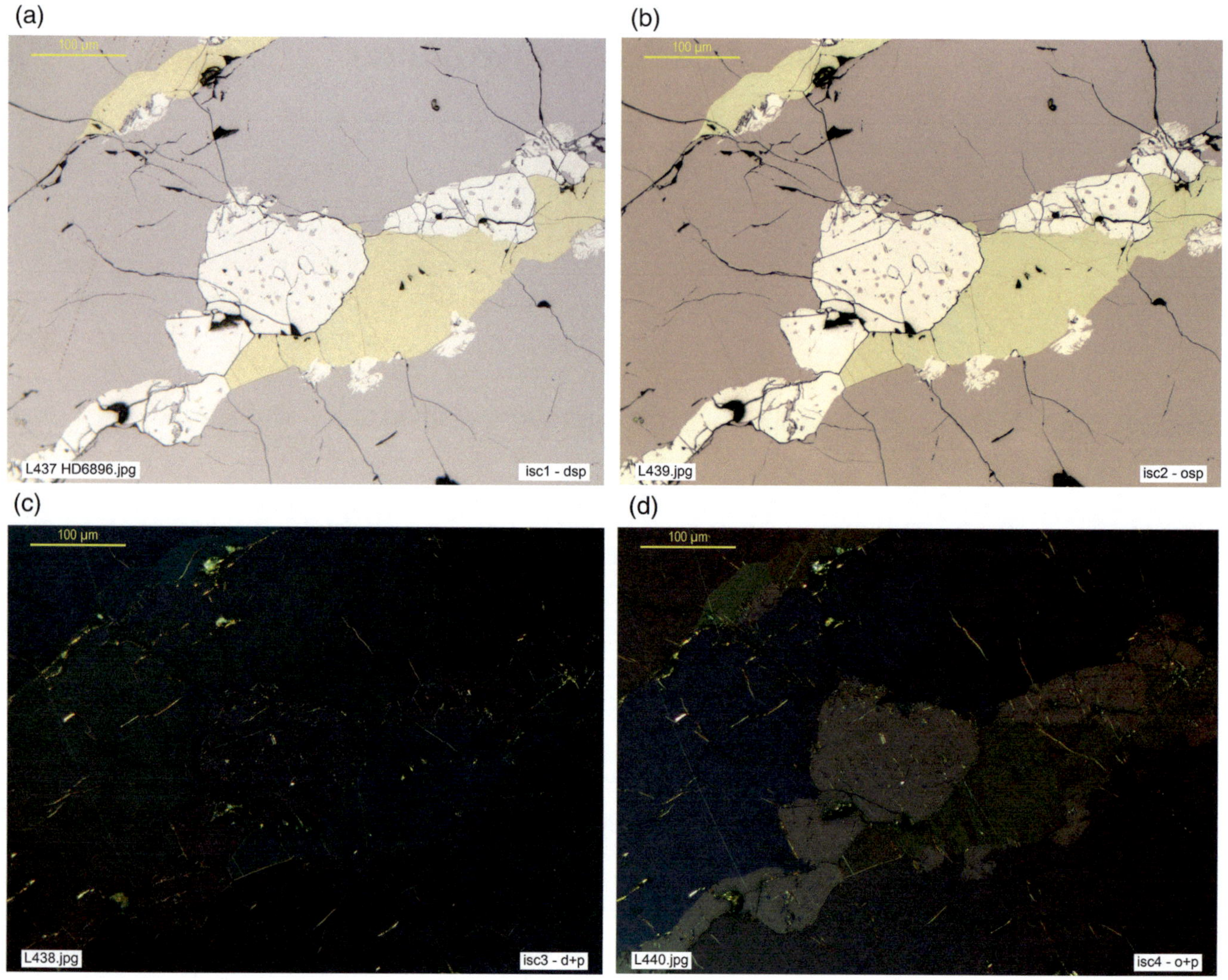

Fig. 65.1 **a, b** (**dsp** and **osp**, respectively)—copper-nickel sulfide ore: massive isc (pinkish-brown), with fine banded intercalations of ccp (yellow) and pn (white). HD 6896, Sudbury, Ontario, Canada. **c, d** Same scene as isc1 (**d+p** and **o+p**, respectively). With crossed polars, the cubic phases, isc and pn, remain dark (extinction), unlike the anisotropic (and twinned) ccp

It occurs in small masses or aggregates included in sulphides or disseminated (e.g., in meteorites). Because of its instability in shallow conditions, it rarely shows tiny automorphic crystals preserved as inclusions, e.g., octahedrons embedded in chalcopyrite. It may remain metastable at room temperature, though it more often decays to lower-temperature phases, forming varied assemblages (as ccp+cbn, ccp+po, ccp+mck, ccp+isc… Ramdohr 1980).

65.2.2 Observation +P (*Polariser* + *Analyser at 90°*)

Isotropic, always in extinction (Fig. 65.1c, d).

65.2.3 Paragenesis

Typical of *high-temperature sulphide paragenesis*, magmatic or liquid-magmatic (e.g., Cu-Ni sulphides), pegmatitic, metamorphic. Also in meteorites.

It may occur with chalcopyrite, pyrrhotite, cubanite, mackinawite, pentlandite, and it also occurs with native iron and troilite (in meteorites).

65.2.4 Identification

Its identification is not difficult for an attentive observer, taking into account the paragenesis. Its colour and

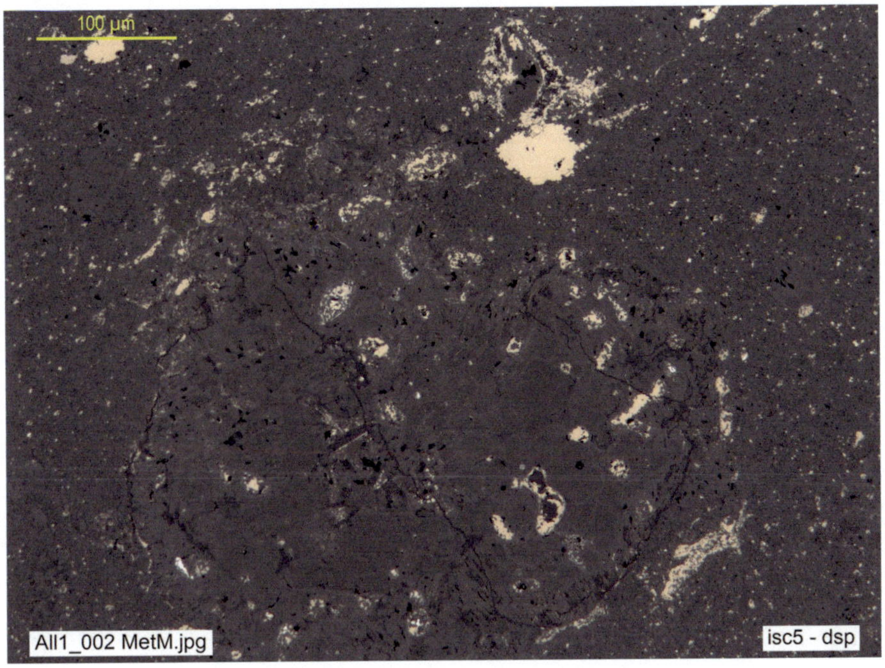

Fig. 65.2 Dissemination of isc (pink). Allende1 Meteorite, Mexico (dsp)

reflectance, similar to pyrrhotite or cubanite, which can be misleading, suggest isocubanite as soon as its isotropy is verified. It is also distinguished from pyrrhotite by its lower hardness. However, it is probable that it has often been confused with pyrrhotite, cubanite or chalcopyrite. This is the only explanation, according to Ramdohr (1980), for the frequency with which "particularly soft" pyrrhotite or "brown or isotropic" chalcopyrite are cited in the literature.

65.3 Spectral Information

65.3.1 Isocubanite Reflectance R (%) in the Visible Spectrum (in Air)

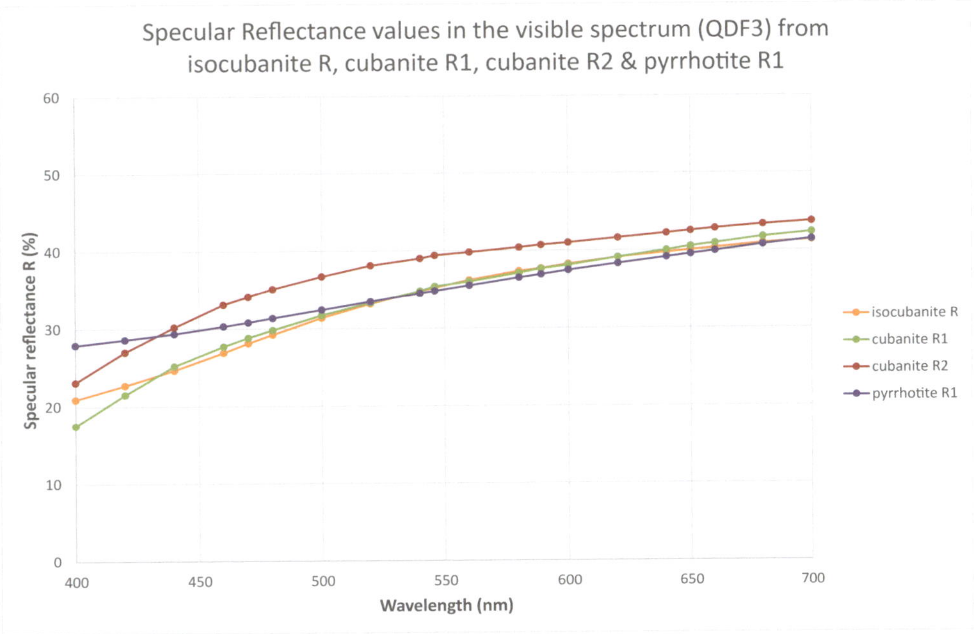

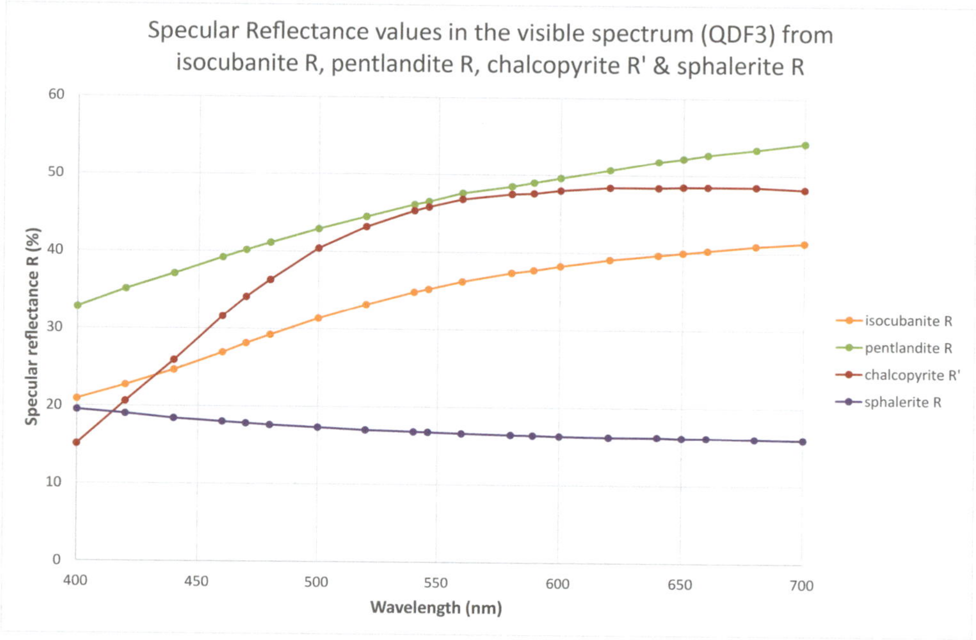

References

Bowles JFW, Vaughan DJ, Howie RA, Zussman J (2011) Rock-forming minerals Vol. 5A: non-silicates: oxides, hydroxides and sulphides, 2nd Edn. Chalcopyrite: phase relations in the ternary system Cu-Fe-S (p 780 ff). Geological Soc. London, 920 p. ISBN-13: 978-1862393158

Craig JR, Scott SD (1976) Sulphide phase equilibria. In: Ribbe PH (ed) Sulfide mineralogy. Mineralogical society of America, reviews in mineralogy 1, Ch 5, CS1–110

Ramdohr P (1980) The ore minerals and their intergrowths, 2nd edn, vol 2. Pergamon Press, Oxford, 1205 p

66 Jacobsite (*jac*)

$Mn^{2+}Fe^{3+}{}_2O_4$: Cubic

Jacobsite		R_{546} (%)	Y (%)	λ_d (nm)	P_e (%)	Colour sensation
Air		21.1	21.0	572	2.8	Grey
Oil		8.2	8.0	570	4.9	
Bir: Null (1)				*Pleoc*: Null (1)		*Tarn*: No
Anis +P: Isotropic (1)				*IR*: Scarce, red		*Twg*: No
Mor: Euhedral tendency (octahedra)				*Clv*: (111) occasional		*PolQ*: Good
Hardness	*PolH*: High, ≈ mt (4)			*Mohs*: 6	*Talm*: nd	*VHN* $_{(100\ g)}$: 720–813

Note: cf. Table and R charts, visible range data (QDF3; Picot and Johan 1982)
Mineral abbreviations: bra—braunite, fkl—franklinite, gg—gangue, gth—goethite, ht—hematite, jac—jacobsite, lim—limonite, mt—magnetite, pie—piemontite, prl—pyrolusite, psi—psilomelane, q—quartz

66.1 General

Jacobsite, *jac*, is one of the ferric spinels, like magnetite or franklinite among others. They are characterized by forming solid solution at high temperature and, in addition, by very similar physical properties. Its description could be replaced in many respects by that of the most common species, magnetite, to which reference will be made in due course to avoid repetition.

Composition: $Mn^{2+}Fe^{3+}{}_2O_4$, although actually it might be more realistic to express it as QDF3: (Mn^{2+}, Fe^{2+}, Mg)(Fe, Mn)$^{3+}{}_2O_4$, taking into account the possibilities of substitution of Mn, Mg and Fe. In addition, it can have Zn as a minor element.

With magnetite, it forms the *isomorphic series jacobsite—magnetite*: $Mn^{2+}Fe^{3+}{}_2O_4$–$Fe^{2+}Fe^{3+}{}_2O_4$. As explained in the discussion of spinel and magnetite (cf. both descriptions), it belongs to the Spinel Subgroup, which is in the Oxyspinel Group, within the Spinel Supergroup.

66.2 Microscopic Description

66.2.1 SP Observation (*Single Polariser*)

Grey *colour*, very similar to braunite. Good *polish*. High *polishing hardness*, like magnetite.

Low *reflectance*, as magnetite or braunite (cf. comparative graphs of reflectances, R. Note the similarity of values with magnetite in QDF3 and the slight differences between QDF3 and Picot and Johan 1982, explainable by the recognized incidence in the values of R of possible compositional variations between the measured samples).

Isotropic.

Occasional visible octahedral exfoliation.

It occurs in euhedral crystals (octahedral) and in fine aggregates, often partly altered to secondary oxides such as goethite, pyrolusite, hematite.

R. Castroviejo, *A Practical Guide to Ore Microscopy—Volume 1*,
https://doi.org/10.1007/978-3-031-12654-3_66

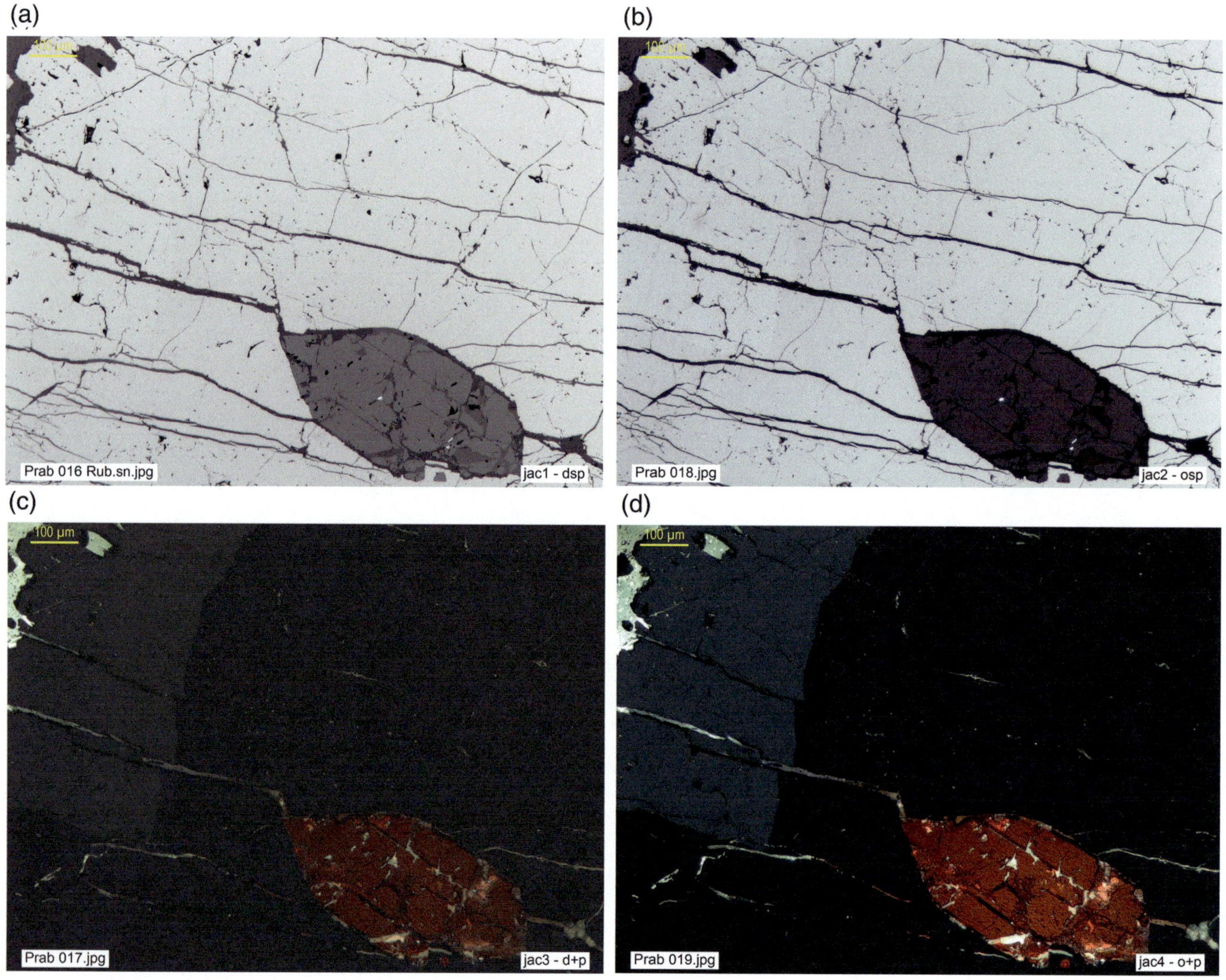

Fig. 66.1 **a, b** (**dsp** and **osp**, respectively)—Aggregate of jac and bra (light grey, apparently identical) with idiomorphic inclusion of silicate piemontite, *pie* (grey), and quartz-cemented microcracks (dark grey, black in oil). Rub.sn, Praborna, Val d'Aosta, Italy. **c, d** Same scene as jac1 (**d+p** and **o+p**, respectively). The anisotropism of *bra* (top left) distinguishes it from *jac* (isotropic, dark, main ore), *pie* characterized by its anisotropism and red IR

66.2.2 Observation +P (*Polariser + Analyser at 90°*)

Isotropic. Sometimes red *internal reflections*, more abundant with increasing Mn content.

66.2.3 Paragenesis

Metamorphic *paragenesis*, presumably derived from sedimentary pre-concentrations of Mn.

Companions: braunite, pyrolusite, franklinite, psilomelaneand, as secondary products of weathering, goethite, hematite, limonite.

66.2.4 Identification

Jacobsite is strongly magnetic and very similar to magnetite, which lacks IR. Additionally, paragenesis is an important criterion.

It also resembles braunite, but this one is anisotropic, although weakly (Figs. 66.1a–d and 66.2a, b).

Fig. 66.2 **a, b** (same scene: **dsp** and **d+p**, respectively)—Ore of *jac* (light grey, idiomorphic, octahedral, always in extinction, *d+p*) and *pie* (***dsp***: dark grey, idiomorphic, prismatic habit; ***d + p***: twinned, reddish-brown colours due to anisotropism and IR) with quartz (*q*: grey, anisotropic, colourless IR). Rub.sn sample, Praborna, Val d'Aosta, Italy

(a)

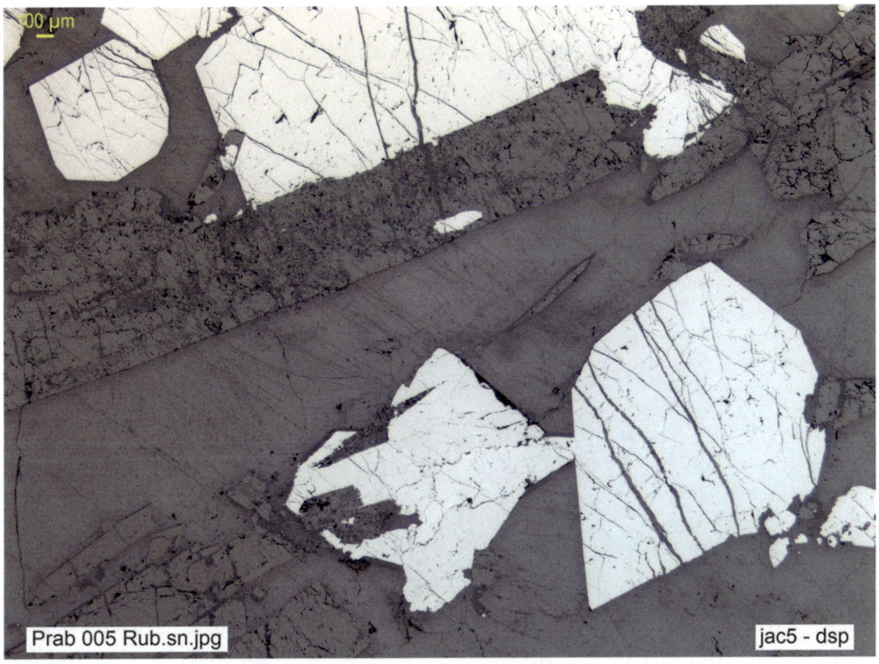

(b)

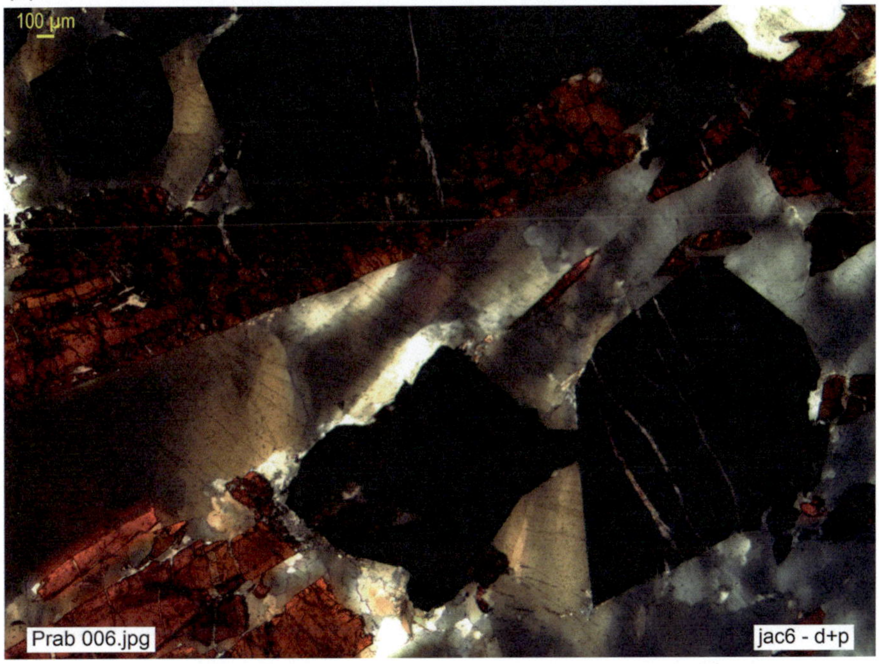

66.3 Spectral Information

66.3.1 Jacobsite Reflectance R (in Air, Visible Spectrum) (%)

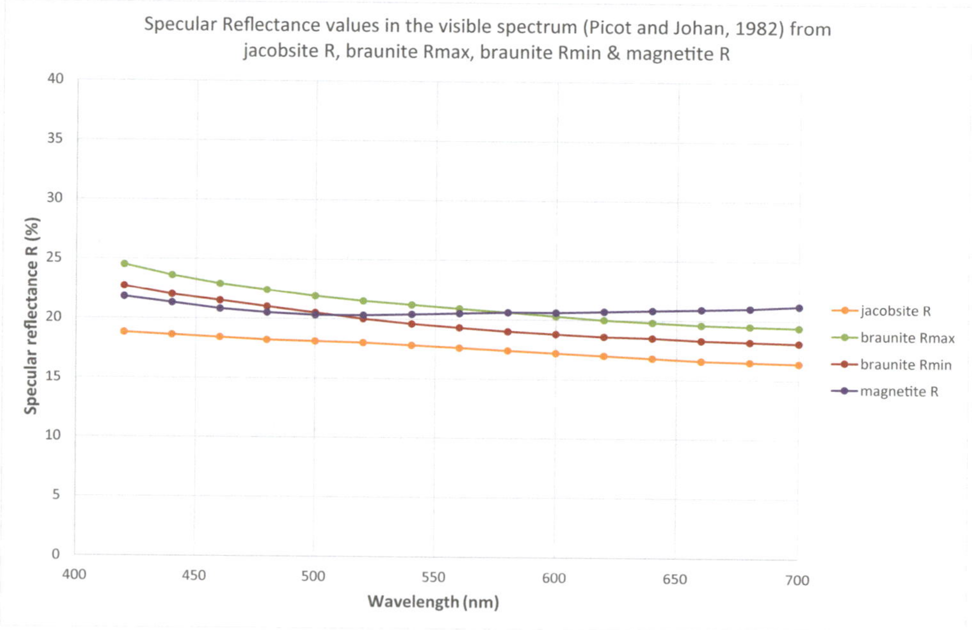

Reference

Picot P, Johan Z (1982) Atlas of ore minerals. B.R.G.M./Elsevier

67 Jamesonite (*jm*)

$Pb_4FeSb_6S_{14}$: Monoclinic

Jamesonite			R_{546} (%)	Y (%)	λ_d (nm)	P_e (%)	Colour sensation
Air	R_1		36.4	36.2	483	2.8	Greyish white
	R_2		44. 2	43.7	487	2.4	
Oil	R_1		20.9	20.6	483	4.9	
	R_2		28.8	28.3	485	4.5	
Bir: Medium (3–4)					*Pleoc*: Medium (3–4)		*Tarn*: –
Anis +P: Distinct and coloured (4)					*IR*: No		*Twg*: Ubiquitous, polysynthetic
Mor: Prismatic, fibrous or acicular					*Clv*: Visible		*PolQ*: Good
Hardness		*PolH*: Low (2)			*Mohs*: 2½	*Talm*: B+	*VHN* $_{(100\ g)}$: 66–86

Obs **:** cf. Table and R charts, multispectral data (VNIR, AMCO) and visible range (QDF3)
Mineral abbreviations: asp—arsenopyrite;,bar—barite, bl—boulangerite, bnn—bournonite, ccp—chalcopyrite, gc—geocronite, gg—gangue, gn—galena, jm—jamesonite, py—pyrite, q—quartz, sbt—stibnite, sid—siderite, sp—sphalerite, Ss—Sulphosalt, stn—stannite, td—tetrahedrite

67.1 General

For jamesonite, ***jm***, the composition $Pb_4FeSb_6S_{14}$ is accepted, although it seems that the Fe content can vary and it usually contains traces of Cu, Zn, Bi (±Ag).

67.2 Microscopic Description

67.2.1 SP Observation (*Single Polariser*)

Greyish-white *colour* (Fig. 67.1a, b), similar to that of galena, ***gn***, but with a slight greenish tinge, as opposed to the whiter or bluish tone of gn, which is somewhat more reflective (lighter, Fig. 67.3). In addition, jm shows a variable pleochroism, while gn has none.

Reflectance, ***R***, moderate, similar to boulangerite, ***bl***, or bournonite, ***bnn***, but lower than gn (cf. multispectral comparative graphs, R_{jm} is between those of tetrahedrite, ***td***, and gn, i.e., $R_{td} \leq R_{jm} < R_{gn}$). Against pyrite, ***py***, or chalcopyrite, ***ccp***, jm is perceived with a grey, relatively dark, tone, especially in oil (Fig. 67.1a, b). Good *polishing*. Low *polishing hardness*, < gn.

Generally idiomorphic, commonly with a fibrous or acicular habit, although sometimes it also forms thicker prismatic crystals or irregular aggregates. Cleavage {001} usually visible.

67.2.2 Observation +P (*Polariser + Analyser at 90°*)

Distinctly *anisotropic* and coloured in brownish and greenish or bluish-green tones, very variable with orientation (Fig. 67.1a, b, compare Fig. 67.2) and more saturated and darker when partially uncrossing the analyser.

Usually *twinned*. It almost always shows striking lamellar or polysynthetic twins (Fig. 67.1c, d), parallel to the elongation of the crystal, unless it occurs in very fine acicular crystals.

R. Castroviejo, *A Practical Guide to Ore Microscopy—Volume 1*,
https://doi.org/10.1007/978-3-031-12654-3_67

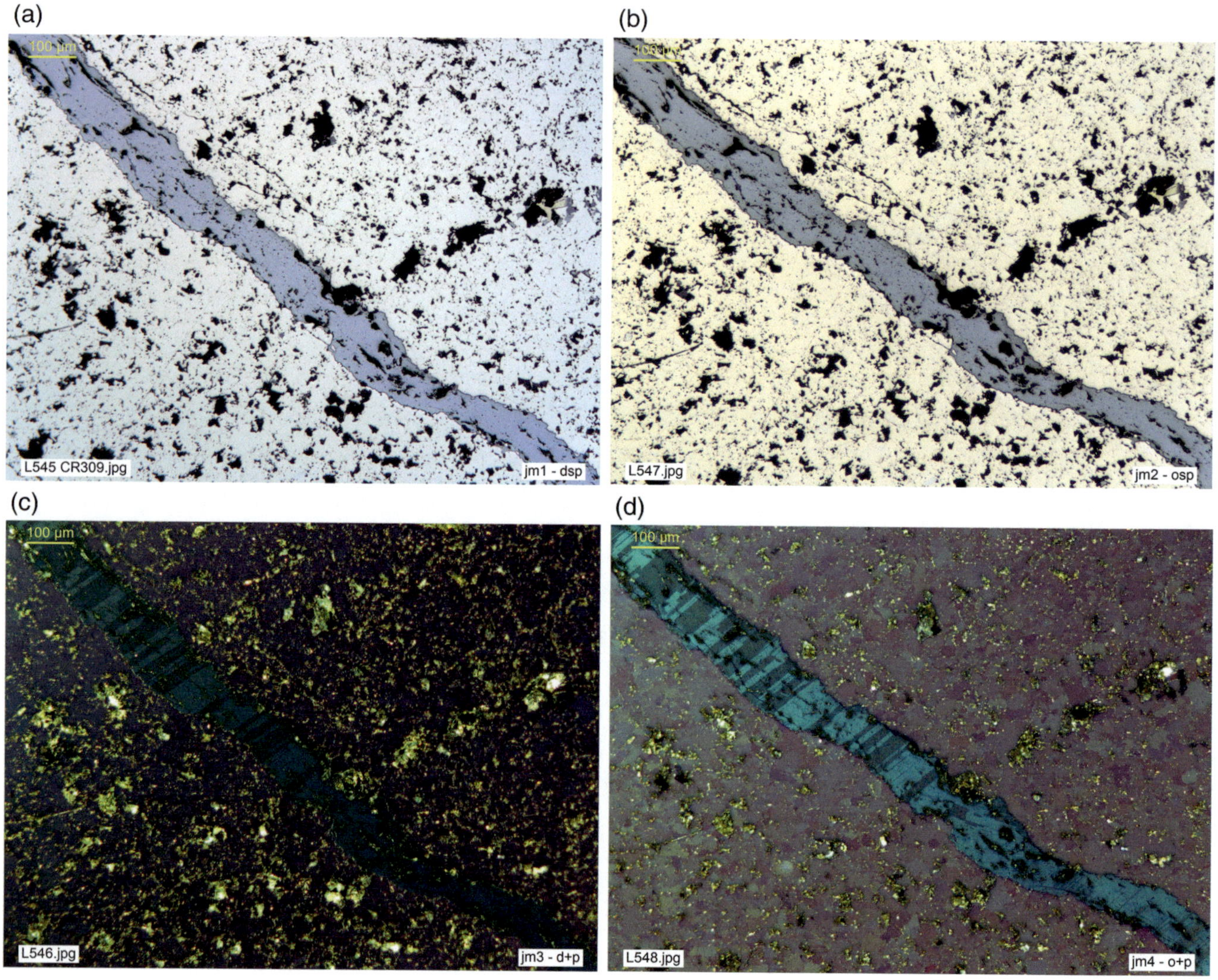

Fig. 67.1 **a, b** (Same scene: **dsp** and **osp**, *respectively*). Microvein of jamesonite, *jm* (grey), in microcrystalline pyritic mass with traces of interstitial sphalerite, chalcopyrite and gangue. CR-309, massive sulphides from the IPB, Aznalcóllar, Sevilla. **c, d** (Same scene as Fig. 67.1a: **d+p** and **o+p**, *respectively*). Twinning and anisotropism of jm

67.2.3 Paragenesis

Hydrothermal *paragenesis* (argentiferous or Pb–Zn–Ag seams, Au–Sb seams, etc.), in which it generally appears as a late, moderate-temperature mineral.

It can be accompanied by *gn* , *sp* , *ccp, asp, py* , *stn, q, sid, bar*, apart from other Pb sulfoantimonides.

67.2.4 Identification

At first glance, jamesonite resembles several minerals, such as galena, **gn**, stibnite, **sbt**, and several Pb–Sb *Ss*, in particular, boulangerite, **bl**. Confusion with the first two is easily resolved by noting the isotropic character of gn and the much stronger bireflection and anisotropism of sbt, as well as the higher reflectance of both. The ubiquity of polysynthetic twins in jm and their orientation, parallel to the prism axis or crystal elongation, distinguish it from bl.

Bournonite, ***bnn***, geochronite, ***gc***, and tetrahedrite, ***td***, are also Ss that, at first impression, could resemble jm. They can be distinguished by the typical parquetry of bnn, as well as by the isotropy and lower reflectance of td, while the usual morphology (anhedral aggregates, not fibrous) and the different twinning of gc facilitate the distinction.

Fig. 67.2 Veinlet of jm in py aggregate (same scene as Fig. 67.1d, after rotating the stage 90°), remarkable variation of jm polarisation colours with orientation. CR-309, IPB massive sulphides, Aznalcóllar, Sevilla (**o +p**)

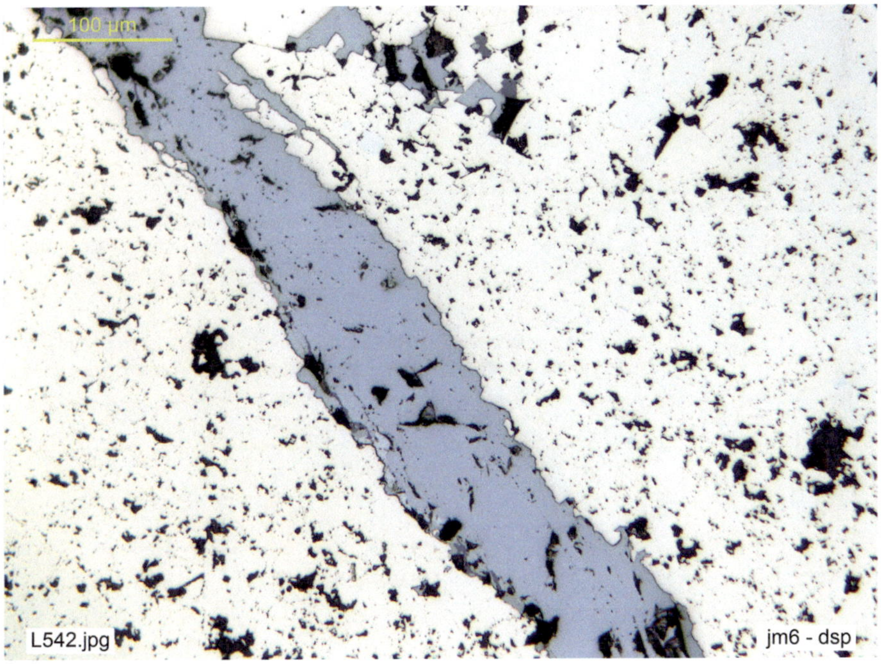

Fig. 67.3 Microcrystalline *polymetallic sulphide* ore (massive py ± ccp, interstitial gn and sp) with jm microvein (± sp, gn, bm). Comparison of jm with gn (interstitial in py, visible top center of image) evidences the resemblance of both minerals and the slightly higher brightness of gn ($R_{gn} > R_{jm}$). CR-309, IPB, Aznalcóllar, Sevilla (**dsp**)

67.3 Spectral Information

67.3.1 Jamesonite Multispectral Reflectance R (in Air, Without Polariser) (%)

λ (nm)	370	400	425	450	475	500	525	550	575	600	625	650	675	700	750	800	850	900	950	1000
μ	34.1	34.8	34.6	35.3	34.8	35.2	34.7	34.7	34.4	34.1	33.7	33.2	32.8	32.1	31.2	30.3	29.7	28.7	27.9	27.4
P_5	32.5	33.3	33.3	33.9	33.6	34.0	33.7	33.7	33.5	33.1	32.8	32.3	31.9	31.2	30.3	29.5	29.0	28.0	27.3	26.7
P_{20}	33.5	34.2	34.2	34.8	34.5	34.9	34.4	34.4	34.0	33.6	33.2	32.7	32.3	31.6	30.6	29.8	29.3	28.3	27.5	27.0
P_{80}	34.9	35.5	35.3	35.9	35.3	35.6	35.1	35.2	34.9	34.6	34.2	33.7	33.3	32.7	31.8	30.9	30.2	29.1	28.3	27.8
P_{95}	35.5	36.0	35.7	36.2	35.5	35.9	35.4	35.5	35.3	35.1	34.7	34.3	33.8	33.1	32.2	31.2	30.5	29.4	28.5	28.1

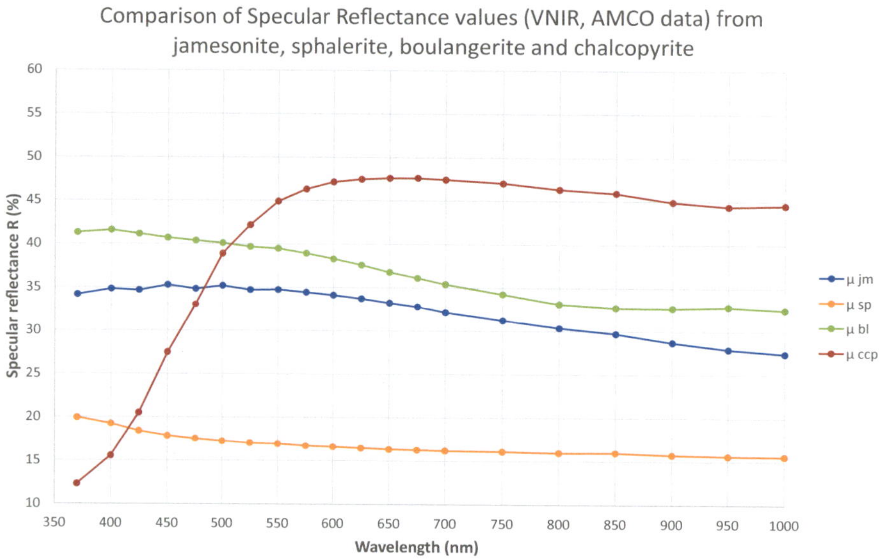

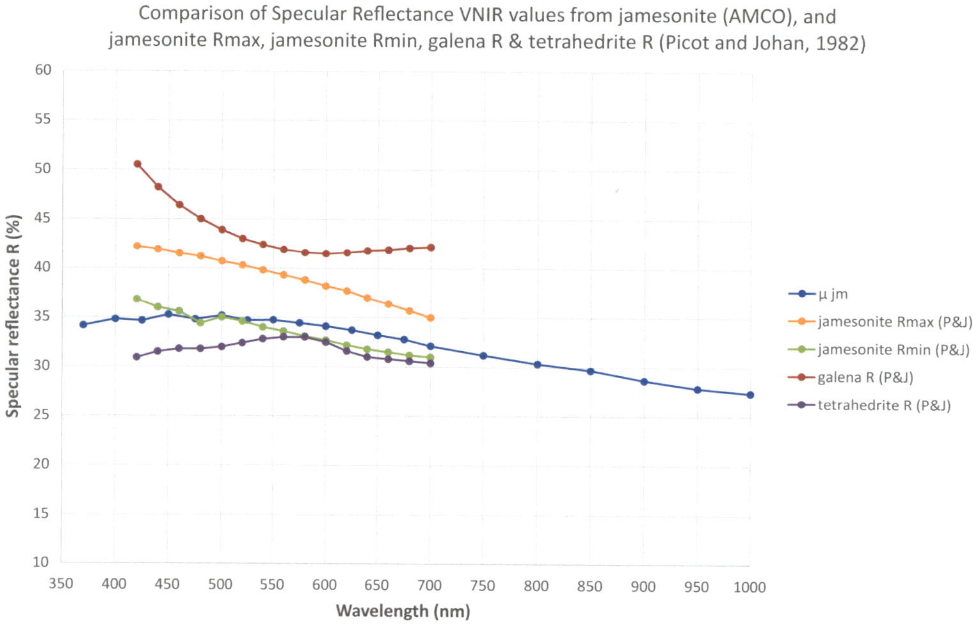

Jordanite (*jd*)

68

$Pb_{14}(As,Sb)_6S_{23}$: Monoclinic

Jordanite			R_{546} (%)	Y (%)	λ_d (nm)	P_e (%)	Colour sensation
Air	R_1		38.1	37.9	478	2.9	Greyish white
	R_2		40.6	40.3	479	3.0	
Oil	R_1		22.6	22.5	477	5.1	
	R_2		25.1	24.9	478	5.3	
Bir: Weak (3)					*Pleoc*: Weak (3)		*Tarn*: –
Anis +P: Net (3–4)					*IR*: No		*Twg*: Occasionally
Mor: Fibrous-radial, reniform					*Clv*: –		*PolQ*: Good
Hardness		*PolH*: Medium–low (3)			*Mohs*: 3	*Talm*: C	*VHN* $_{(50\ g)}$: 106–141

Obs: cf. Table and R charts, multispectral data (VNIR, AMCO) and visible range (QDF3)
Mineral abbreviations: gc—geocronite, gn—galena, gtt—gratonite, jd—jordanite, py—pyrite, Sb—native antimony, sp—sphalerite, td—tetrahedrite, wz—wurtzite

68.1 General

Jordanite, ***jd***, was traditionally considered a lead sulphoarsenide ($Pb_{14}As_6S_{23}$) but now it is understood as $Pb_{14}(As,Sb)_6S_{23}$, isostructural with geochronite and a member of the geochronite-jordanite series: $Pb_{14}(Sb,As)_6S_{23}$–$Pb_{14}(As,Sb)_6S_{23}$ (cf. geochronite, ***gc***).

Jordanite is dimorphous of gratonite, **gtt**, which, according to experimental data (Roland 1968), appears to be the low-temperature form, since it transforms into jd on heating to 250 °C. If heating is continued, jd remains stable until 549 ± 3 °C, at which temperature it melts, decomposing into galena, liquid and S-rich vapour.

However, before defining jd as a high-temperature mineral, attention should be paid to the textural data, from whose observation Ramdohr (1980) deduces that a common feature of all jd deposits, besides a high arsenic concentration (or high As:Sb ratio), is precisely *a very low formation temperature* (p. 756).

This conclusion is in agreement with the textures described in jd, which suggest its frequent formation as fine grained aggregates pseudomorphous after gtt (a process usually related to posthumous evolution and cooling). When jd shows signs of incipient crystallization, *etching* shows evidence of its evolution from a primary deposition of jd as a gel.

The concordance of this seemingly contradictory data set suggests that jd may be stable over a wide temperature range. Unlike gtt, it does not decompose upon heating above 250 °C (up to ≈ 550 °C) and, unlike other high- or intermediate-temperature hydrothermal phases (such as argentite, which is always replaced by acanthite by cooling below ≈ 177 °C), it does not decompose upon cooling.

68.2 Microscopic Description

68.2.1 SP Observation (*Single Polariser*)

Greyish-white *colour*, comparable to galena, ***gn***, but a little less reflective than gn (Fig. 68.1a, b) and closer to geochronite, jd is situated by its *reflectance*, ***R***, between geochronite and gratonite (cf. multispectral comparative graphs). Weakly bireflective (varies only in shade of grey).

Polishes well. *Polishing hardness* (3) in the medium–low limit, slightly higher than galena.

R. Castroviejo, *A Practical Guide to Ore Microscopy—Volume 1*,
https://doi.org/10.1007/978-3-031-12654-3_68

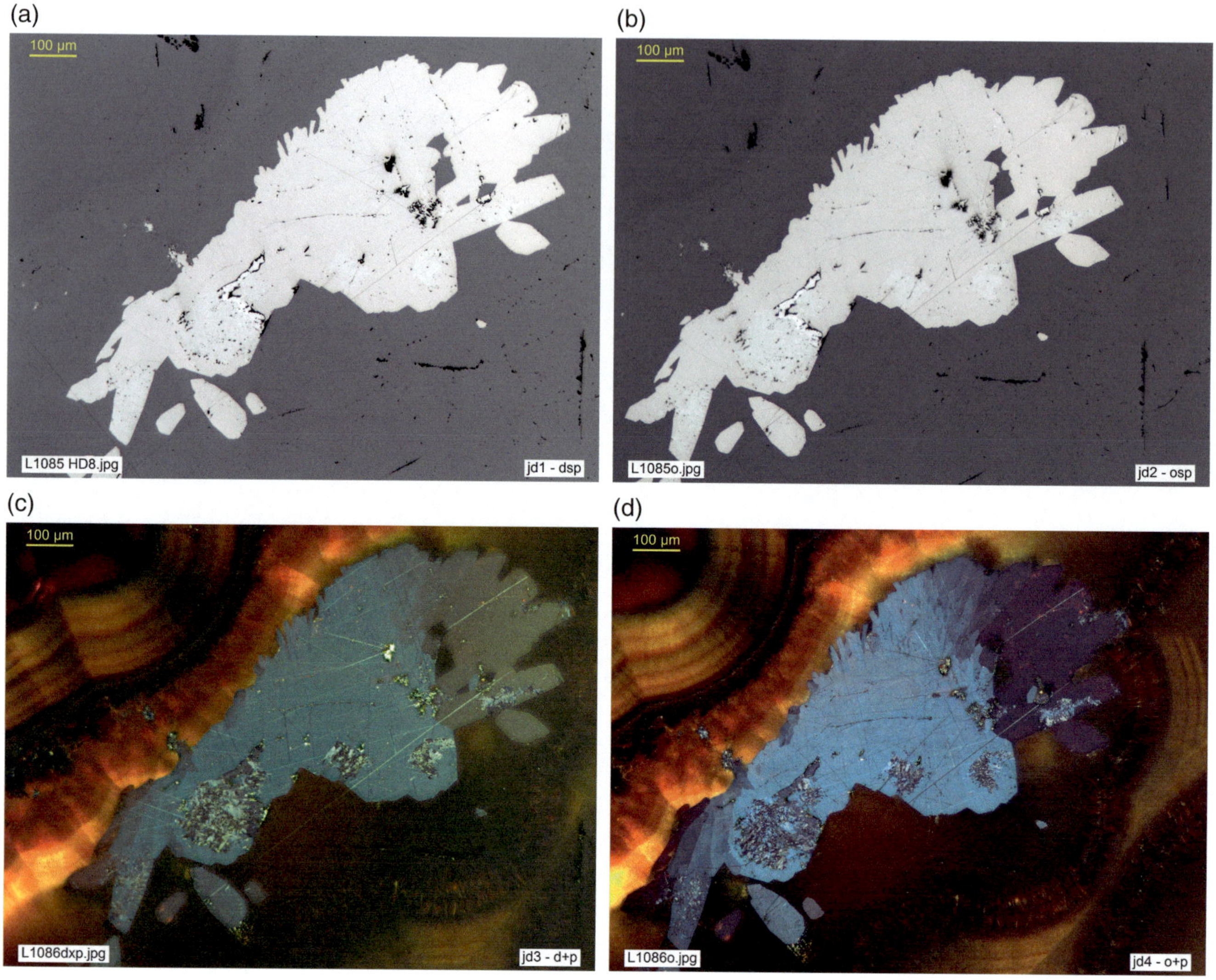

Fig. 68.1 a, b (Same scene: **dsp** and **osp**, respectively). Sphalerite-wurtzite (*schalenblende*, grey) mass including fibrous-radiated crystalline aggregate of jordanite, *jd* (greyish-white, weakly bireflective), which grows on relict cores composed of microcrystalline jd and *gn* (galena, white, isotropic), possibly ghosts of substituted *gtt*. Small scattered idiomorph crystals of *gtt* (gratonite, colour like jd). HD8, Wiesloch, Baden-Wuerttenberg, Germany. **c, d** (Same scene as Fig. 68.1a: **d+p** and **o+p**, respectively). Spherulitic structure and IR of *schalenblende*. Coloured anisotropism of jd and much duller of gtt, which shows rare red IR. Extinction of gn (cubic)

It tends to occur in concentric, fibrous-radial or spherulitic aggregates, sometimes associated with *schalenblende* (cf. sphalerite, ***sp***). Other times, it is more or less evolved towards crystalline, fibrous or tabular forms, occasionally euhedral (Fig. 68.1a, b).

68.2.2 Observation +P (*Polariser* + *Analyser at 90°*)

Anisotropic, with anisotropism effects coloured (bluish-green) but not very luminous, more distinct than those of gc, more vivid in immersion, with more defined polarisation tints in bluish and violet tones (Fig. 68.1c, d). No *internal reflections, **IR***. *Twinning* occasionally observed (Ramdohr 1980) in euhedral forms and in coarser aggregates but not common.

68.2.3 Paragenesis

Varied hydrothermal *paragenesis*, rich in As. It is usually accompanied by sphalerite-wurtzite (*schalenblende*), geochronite, gratonite, galena and arsenical lead sulfosalts.

68.2.4 Identification

Although it has features in common with other sulphosalts, it is distinguished by:

- its relatively high reflectance, close to that of gn,
- its coloured anisotropism (bluish and violet tones),
- its frequent spherulitic, reniform or recrystallized colomorphic textures,
- the scarcity of twins and
- the total absence of IR, *internal reflections*.

68.3 Spectral Information

68.3.1 Jordanite Multispectral Reflectance R (in Air, Without Polariser) (%)

λ (nm)	370	400	425	450	475	500	525	550	575	600	625	650	675	700	750	800	850	900	950	1000
μ	41.1	40.4	39.9	39.3	38.9	38.5	38.3	38.0	37.7	37.4	36.8	36.2	35.5	34.8	33.6	32.4	31.4	31.7	31.7	31.7
P_5	39.3	38.8	38.4	38.0	37.4	36.8	36.8	36.8	36.4	36.0	35.6	35.2	34.4	33.6	32.4	31.2	30.0	30.4	30.4	30.4
P_{20}	39.7	39.2	38.8	38.4	38.0	37.6	37.4	37.2	37.0	36.8	36.2	35.6	34.8	34.0	32.8	31.6	30.8	31.2	30.8	30.8
P_{80}	42.6	41.6	40.8	40.0	39.8	39.6	39.2	38.8	38.4	38.0	37.4	36.8	36.2	35.6	34.4	33.2	32.0	32.4	32.4	32.4
P_{95}	43.4	42.4	41.6	40.8	40.4	40.0	39.8	39.6	39.2	38.8	38.2	37.6	37.0	36.4	35.2	34.0	32.8	33.2	33.2	33.2

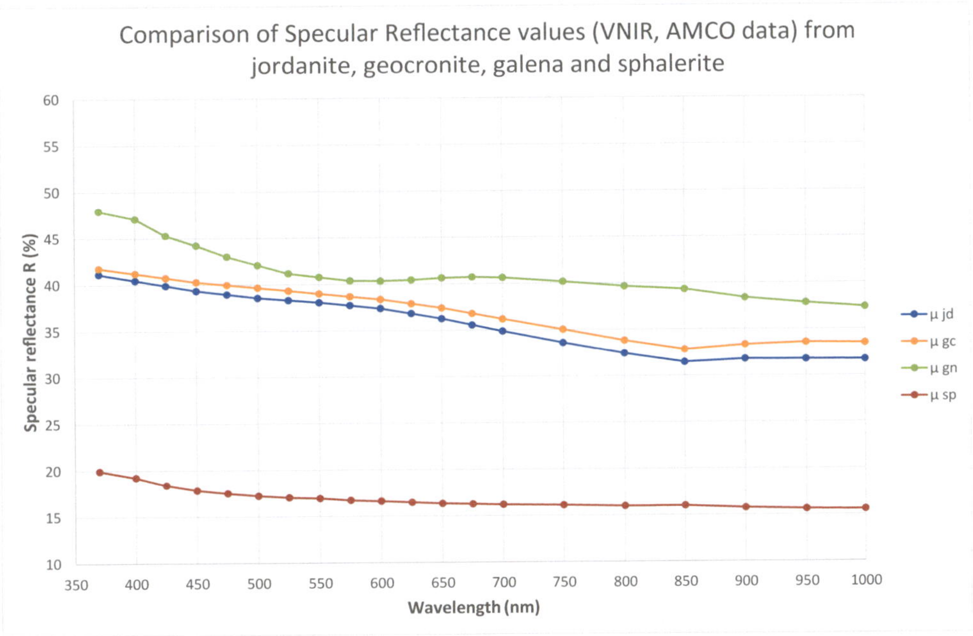

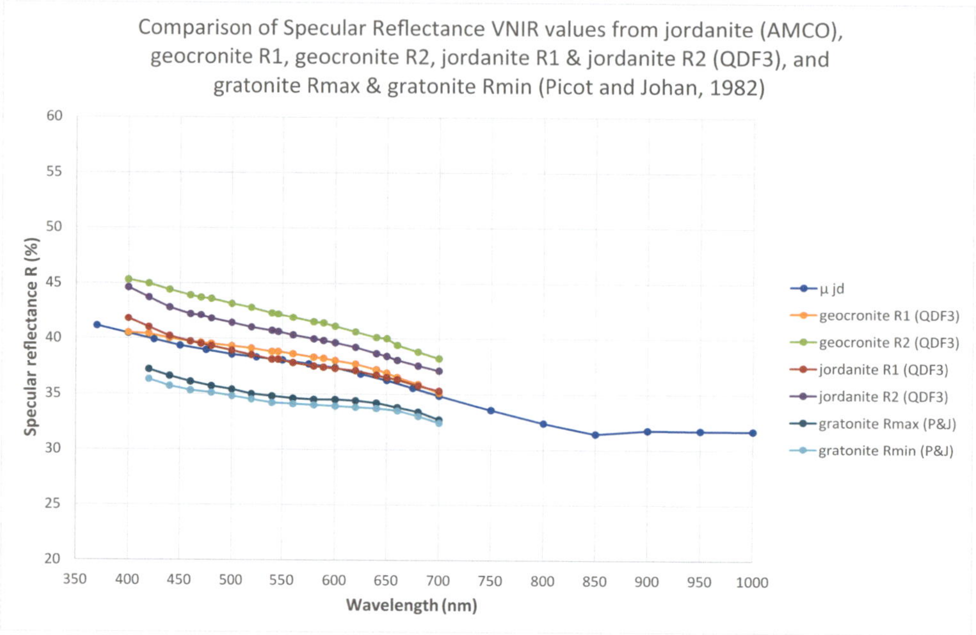

References

Roland GW (1968) The system Pb–As–S. Composition and stability of jordanite. Miner Deposita 3:249–260

Ramdohr P (1980) The ore minerals and their intergrowths, vol 2, 2nd ed. Pergamon Press, Oxford, p 1205

69 Kermesite (*ker*)

Sb_2S_2O: Triclinic

Kermesite		R_{546} (%)	Y (%)	λ_d (nm)	P_e (%)	Colour sensation
Air	R_a	25.1	25.0	476	6.9	Grey, pleochroic (GrBlu–BrGr)
	R_b	30.5	30.3	477	7.1	
	R_b	25.9	25.8	476	7.0	
Oil	R_a	10.7	10.7	475	11.2	Grey, darker but with more pronounced pleochroism (GrBlu–BrGr–GrOl)
	R_b	15.1	14.9	476	11.6	
	R_c	11.3	11.2	475	11.5	
Bir: Medium (4)				*Pleoc*: BluGy–BrGy (4)		*Tarn*: No
Anis +P: Strong and coloured (5)				*IR*: Abundant		*Twg*: -
Mor: Fine-g aggregates, fib-rad bundles				*Clv*: {100}		*PolQ*: Good, despite the low hardness
Hardness	*PolH*: Very soft (1)			*Mohs*: 1–1½	*Talm*: B	*VHN* $_{(5\ g)}$: 30–90

Obs **:** cf. Table and graphs R, Specular Reflectance: visible range data (QDF3; Picot and Johan 1982)
Mineral abbreviations: gg—gangue, gth—goethite, ker—kermesite, pru—proustite, prg—pyrargyrite, Sb—native antimony, sbt—stibnite
Colour abbreviations: Blu—blue, Gy—grey, Ol—olive, Br—brown, Vi—violet, BluGy—bluish—grey, YlGn—Yellowish—green, etc.

69.1 General

It is a typical companion of stibnite, *sbt*, as a secondary product of stibnite oxidation at surface conditions, although it has also been cited as a primary mineral. Its peculiar composition, incorporating S and O simultaneously, makes it a mineral of restricted occurrence, although it is very frequent in altered stibnite ores.

69.2 Microscopic Description

69.2.1 SP Observation (*Single Polariser*)

Grey *colour*, bireflective and with marked pleochroism (BluGy–BrGy–OlGy), more defined in immersion despite looking darker (Fig. 69.1a, b). *Polishing* of reasonable quality in spite of its low hardness, which can make relief inevitable (Fig. 69.2a).

Reflectance, R, relatively high for an oxidized ore, although always lower than stibnite and dropping markedly in immersion (Figs. 69.1a, b and 69.2a), $R_{ker} \ll R_{sbt}$ but $R_{ker} > R_{gth}$ and $\gg$ R of other associated oxidation products, such as antimony ochres (cf. comparative R graphs, which delimit R_{ker} between a maximum value $R_b \approx$ R minimum of stibnite and a minimum $R_a > R_{gth}$).

Very low *polishing hardness* (≈ between acanthite and stephanite). *Vickers microhardness* (30–90); *Mohs* (1–1½) and *Talmage* (B) *hardnesses* are also very low.

It occurs in thin, brittle encrustations, often with a fibrous-radiated structure.

69.2.2 Observation +P (*Polariser + Analyser at 90°*)

Strongly anisotropic, with vivid and varied colours (BluGn, Vio, Gn, YlGn), well perceptible in spite of the *very abundant, red,* ***IR***, internal reflections (Figs. 69.1c, d and 69.2b).

R. Castroviejo, *A Practical Guide to Ore Microscopy—Volume 1*,
https://doi.org/10.1007/978-3-031-12654-3_69

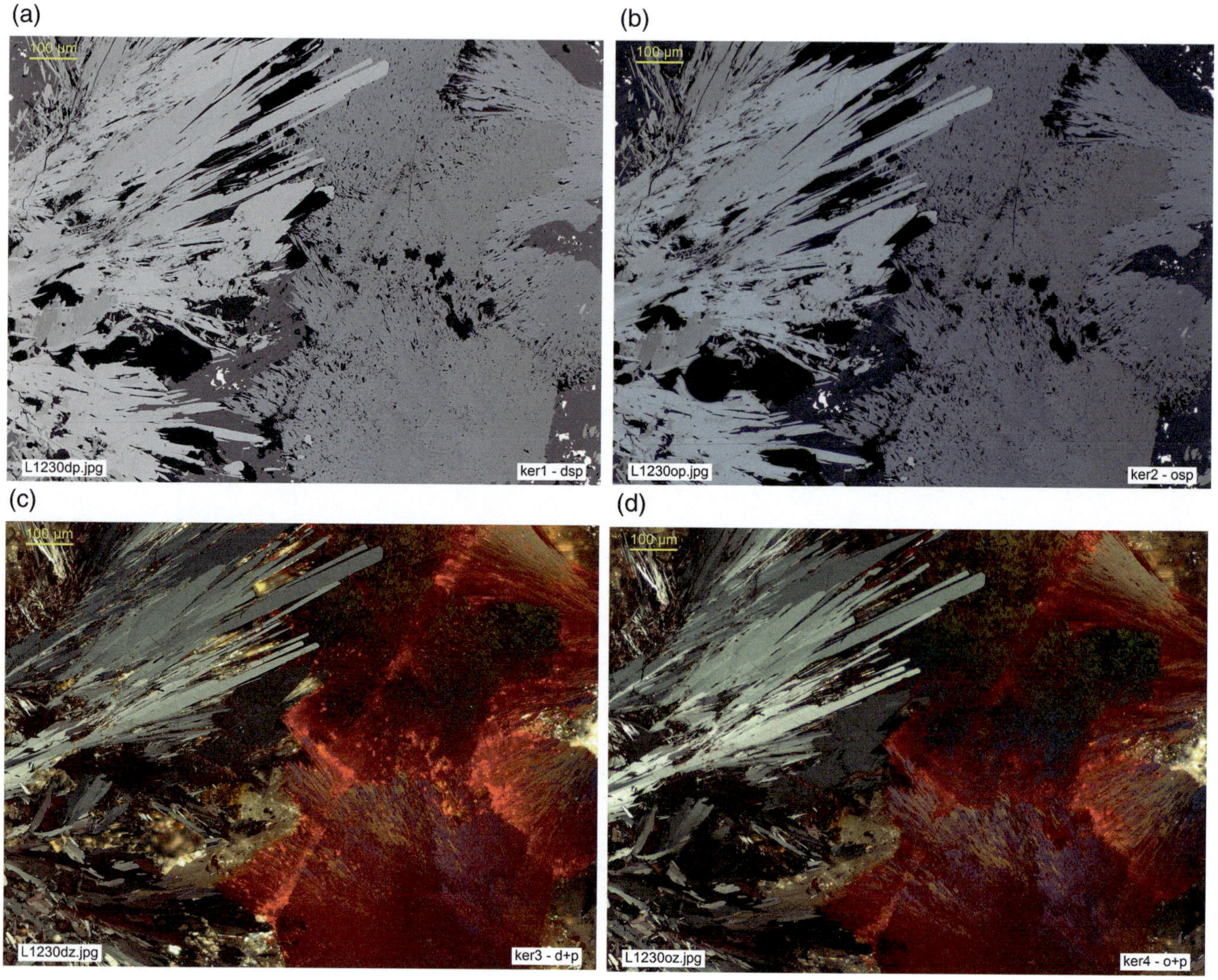

Fig. 69.1 **a, b** (**dsp** and **osp**, respectively). Fibrous-radiating aggregate of relatively fresh sbt (light grey, distinctly bireflective) on left and, on right, completely replaced by ker (darker grey, bireflective and pleochroic from BluGy to BrGy). Scarce interstitial gg (very dark, black in oil) and abundant impregnation of *antimony ochre* (very dark grey, especially in oil, but always lighter than gg), secondary to Sb (white, very bright), tiny relicts of which occur scattered throughout the field. CR 405, West Gore Mines, Hants, Nova Scotia, Canada. **c, d** (Same scene as Fig. 69.1a: **d+p** and **o+p**, respectively). The abundant red IR and the strong and coloured anisotropism of ker allow its easy identification. The anisotropism of sbt is also strong, but less coloured

69.2.3 Paragenesis

Paragenesis typical of oxidation zone of antimony ores, mostly associated with stibnite.

69.2.4 Identification

The paragenesis is an important factor which, together with the fibrous structure and the strong anisotropism, with bright and varied colours, well visible in spite of the very abundant IR, assure its recognition in most cases. However, in Sb-bearing Ag deposits, kermesite can occur together with some argentiferous ores which display similar properties of anisotropism and IR, such as proustite and pyrargyrite. In these cases, the inclusions of antimony-bearing relicts (stibnite, Sb, antimony ochres), the brittle character, and the fibrous structure of kermesite, as well as its lower R in immersion, help to distinguish them.

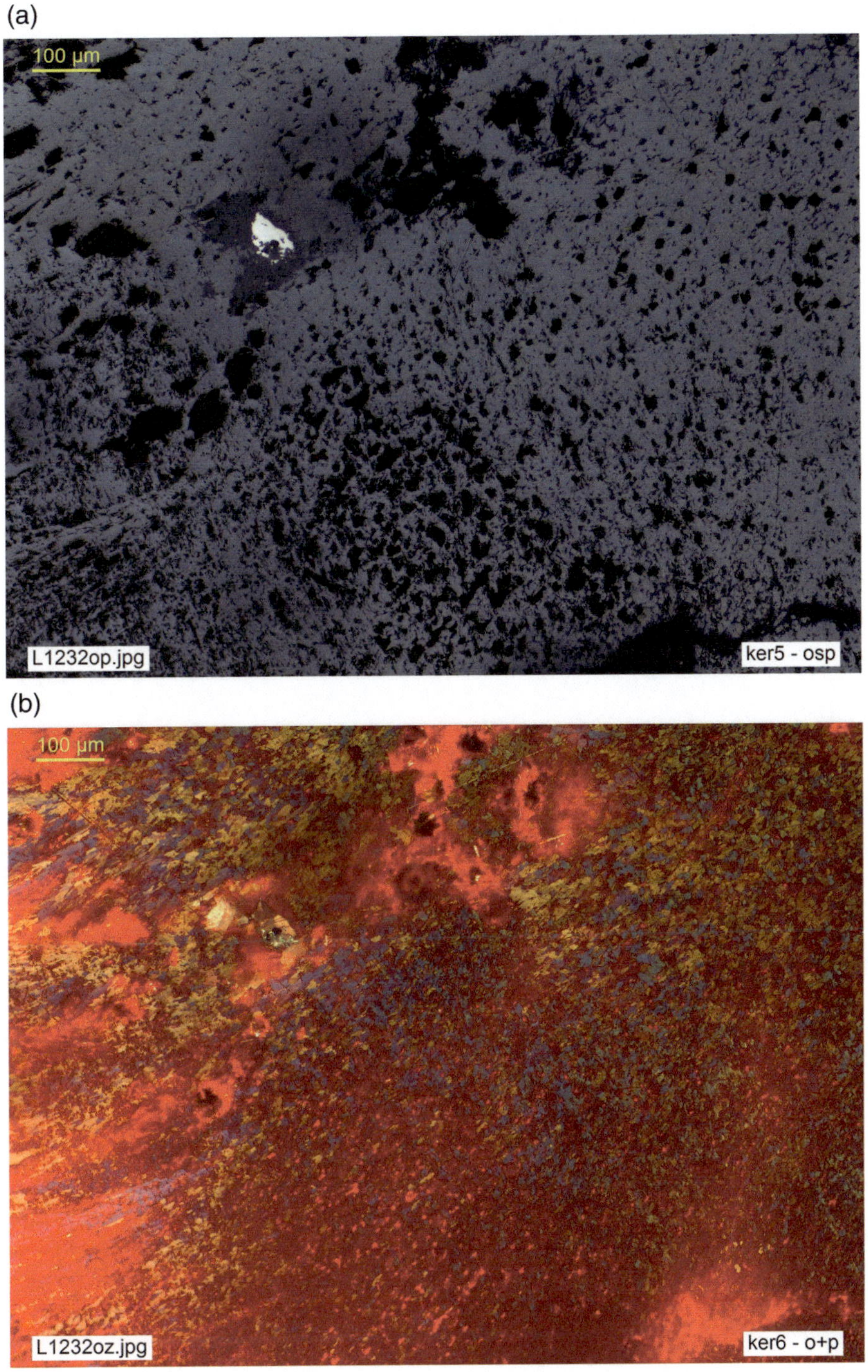

Fig. 69.2 **a, b** (Same scene, **osp** and **o+p**, respectively). Microcrystalline aggregate of ker, including antimony ochers and Sb relicts. The pleochroism of ker, less evident here than in Fig. 69.1b due to the finer grainsize, is still noticeable comparing contiguous grains. The characteristic IR and the intense and multicoloured (Blu, Vi, BluGn, ViBlu, BrGn, YlGn) anisotropism of ker are well displayed

69.3 Spectral Information

69.3.1 Kermesite Multispectral Reflectance R (in Air, Visible Range) (%)

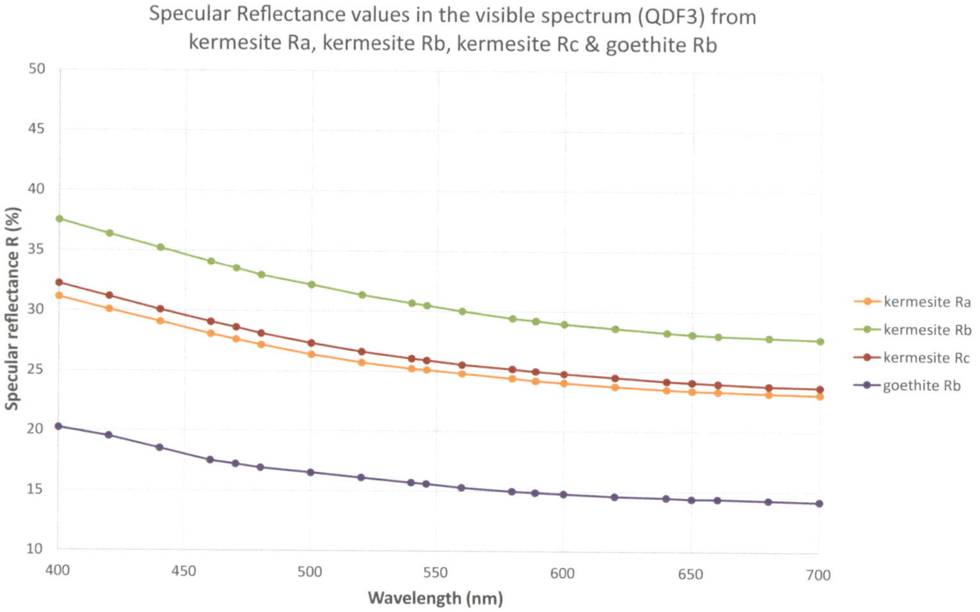

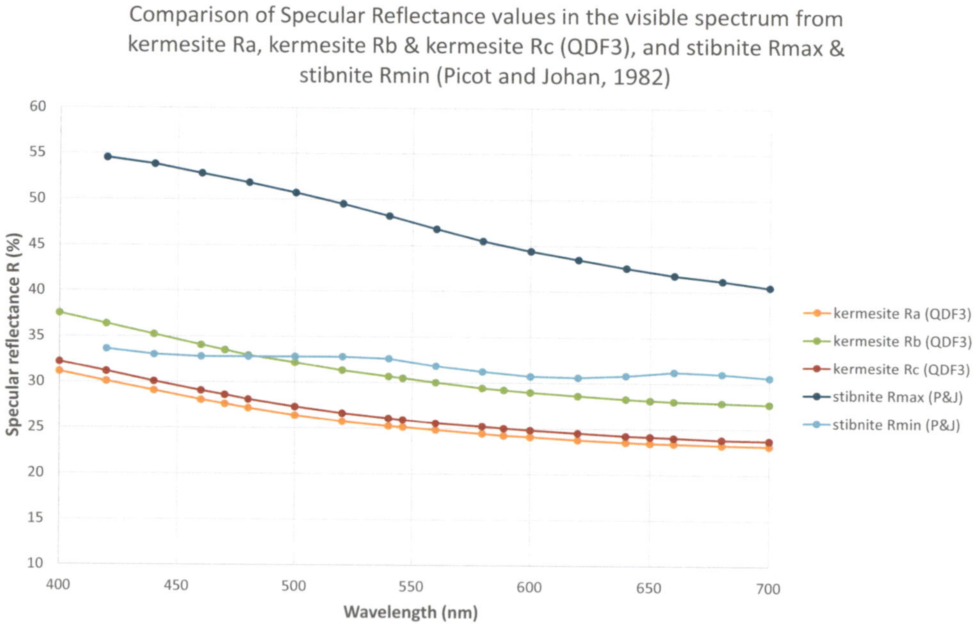

Reference

Picot P, Johan Z (1982) Atlas of ore minerals. B.R.G.M./Elsevier

Krennerite (*krn*)

70

Au_3AgTe_8: Orthorhombic

Krennerite		R_{546} (%)	Y (%)	λ_d (nm)	P_e (%)	Colour sensation
Air	R_1	58.6	58.4	573	6.3	White
	R_2	61.9	61.9	573	4.7	
Oil	R_1	46.4	46.1	572	7.1	White, very faint yellowish tone
	R_2	50.1	50.1	573	5.5	
Bir: Very weak (2)				*Pleoc*: Very weak (2)		*Tarn*: nd
Anis +P: Weak (3)				*IR*: No		*Twg*: Not usual
Mor: Anhedral aggregates				*Clv*: //(001)		*PolQ*: Good
Hardness	*PolH*: Low (2)			*Mohs*: 2–3	*Talm*: C	*VHN* $_{(100\ g)}$: 166–186

Note: cf. Table and R charts, data and visible range (QDF3)
Mineral abbreviations: Ag—native silver, Au—native gold, colo—coloradoite, clv—calaverite, gg—gangue, gn—galena, il—ilmenite, krn—krennerite, Pt—native platinum, py—pyrite, q—quartz, syl—silvanite, td—tetrahedrite, Te—native tellurium

70.1 General

Krennerite, ***krn***, with formula $AuTe_2$ or, for argentiferous varieties (Au, Ag) Te_2, was considered as a dimorph of calaverite, ***clv***. However, recent work suggests that the composition of both species is not exactly the same and the formula Au_3AgTe_8 is proposed for krennerite.

After work from Dye and Smyth (2012), comparing experimental data, equilibrium ratios, textures and unmixings observed in calaverite, krennerite and sylvanite (***syl***, ‹Au, Ag›$_2$Te$_4$), it is deduced that the presence of Ag is indispensable for the structure of krennerite, whose stability would restrict the Au:Ag ratio to ≈3:1.

Thus, because of its silver content (≈3.5–6.5% Ag), krennerite occupies an intermediate place between calaverite (0 to ≈3% Ag) and sylvanite (≈7.5–13.5% Ag). Although at high temperature there exist solid solutions that overflow these limits, upon cooling below ≈350 °C krn_{ss} + liquid ceases to coexist, krn_{ss} gradually becomes impoverished in Ag and the excess Ag passes to syl_{ss}, which now coexists with krn_{ss}. This would explain the exsolution of sylvanite to krennerite in natural systems.

70.2 Microscopic Description

70.2.1 SP Observation (*Single Polariser*)

White *colour*, very similar to calaverite but somewhat darker and less coloured, very weakly pleochroic (Fig. 70.1a, b).

Good *polishing* quality. Low *polishing hardness*, lower than calaverite, may appear scratched.

Reflectance very high, higher than pyrite but distinctly lower than Ag, Au; R krn > colo (Fig. 70.1a, b) and slightly < clv. *Bireflectance* is also somewhat lower than calaverite (cf. comparative graphs of reflectances).

It occurs disseminated, usually in anhedral aggregates; it rarely displays prismatic forms, in isolated grains. Cleavage//(001).

R. Castroviejo, *A Practical Guide to Ore Microscopy—Volume 1*,
https://doi.org/10.1007/978-3-031-12654-3_70

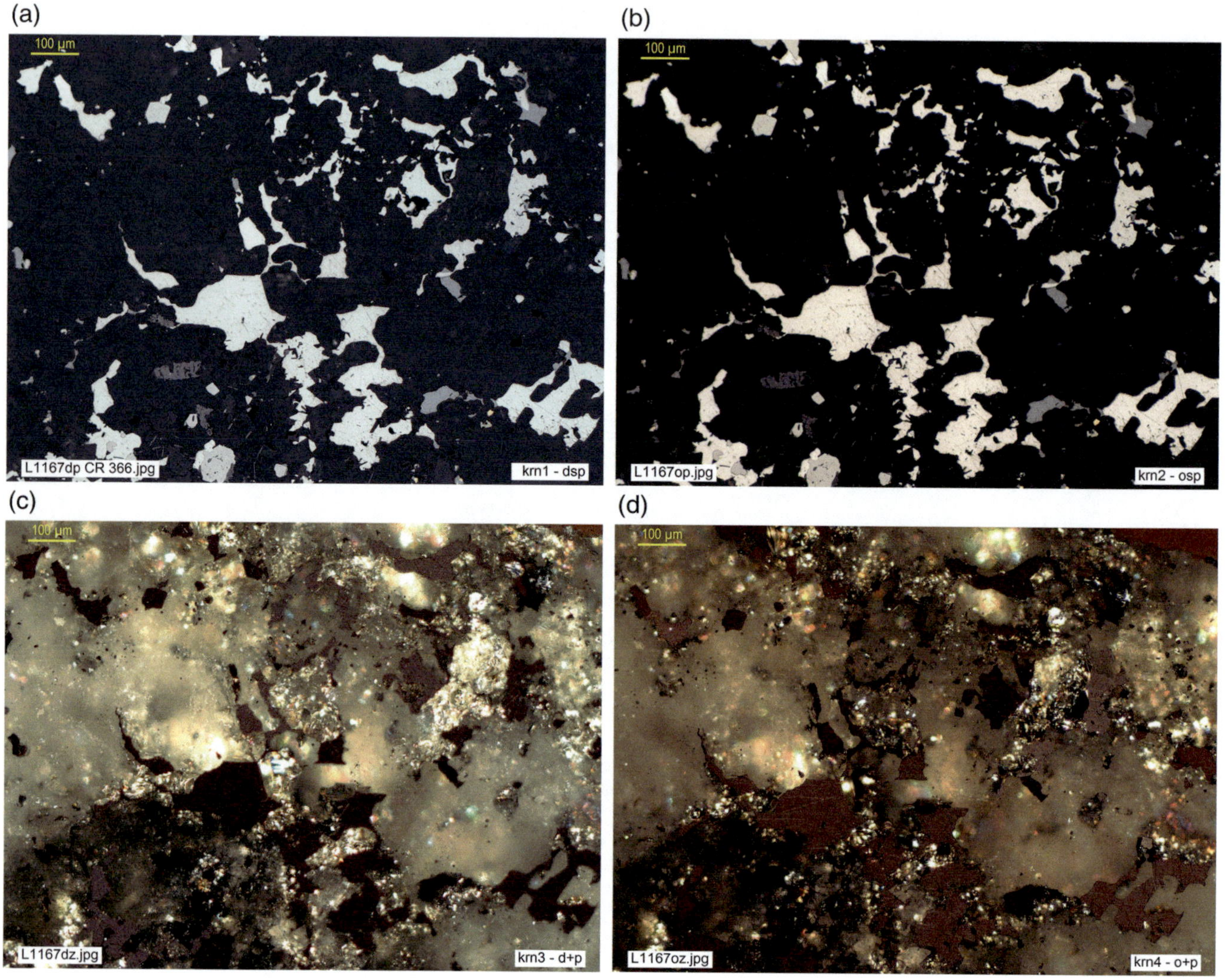

Fig. 70.1 **a, b** (**dsp** and **osp**, respectively)—impregnation of krn (white) in quartz-carbonate gg (grey/black) with scarce disseminated py (bottom left, relief, included in krn, against which py looks darker and slightly pinkish), coloradoite (grey, three main grains, to the right), il (dark greyish brown, tiny prisms), and Au (a ≈ 15 µm particle, bright yellow, bottom right). CR366, Oroya-Brownhill Mine, Kalgoorlie, Australia. **c, d** Same scene as krn1 (d+p and o+p, respectively): Weak anisotropism of krn; massive IR in gg

70.2.2 Observation +P (*Polariser + Analyser at 90°*)

Weak and faintly coloured, though visible, *anisotropism* (Fig. 70.1c, d). Not usually *twinned.*

70.2.3 Paragenesis

Hydrothermal *parageneses*, particularly in mesothermal or epithermal veins, with precious metals and tellurides.

Companion ores: pyrite, gold, silver, galena, coloradoite, calaverite, sylvanite and other tellurides.

70.2.4 Identification

Krennerite is characterized by very high reflectance, low hardness, white colour (slightly yellowish in immersion), weak anisotropism, faintly coloured but visible.

- The paragenesis is an important key.
- It is very similar to calaverite, but krennerite is somewhat less reflective and its anisotropism is weaker.
- It is more easily distinguished from sylvanite, which has a much more vivid anisotropism and frequent, well visible, polysynthetic twins.

70.3 Spectral Information

70.3.1 Krennerite Reflectance R (in Air, Visible Spectrum) (%)

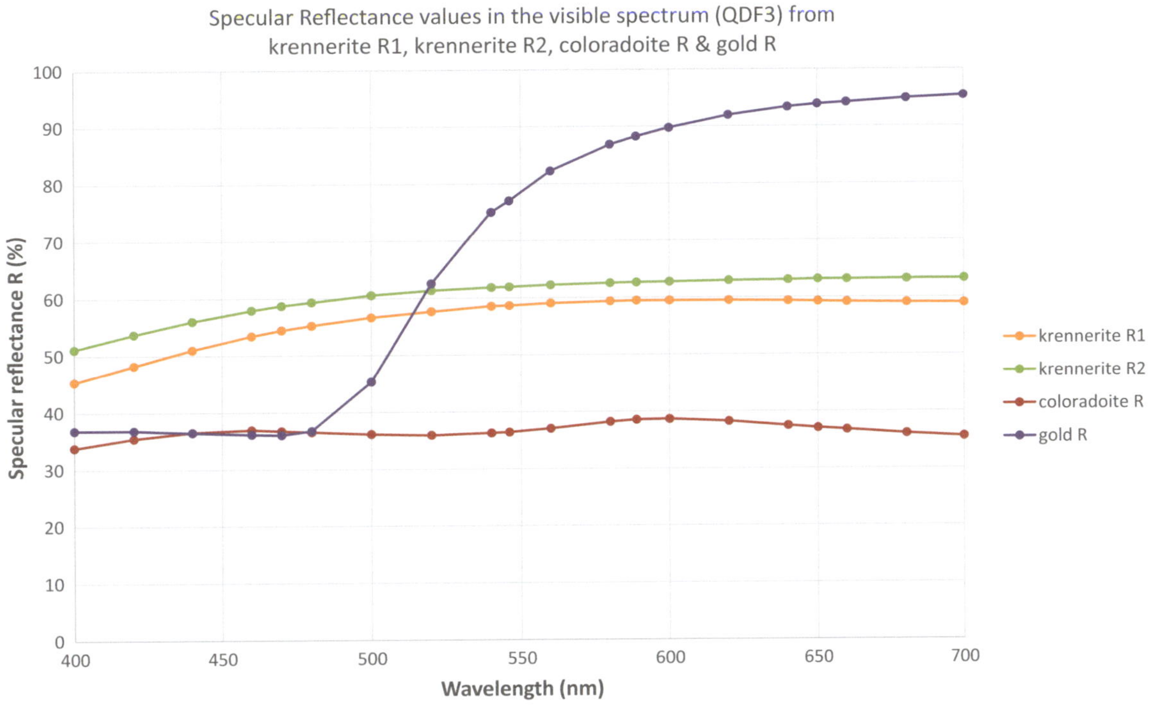

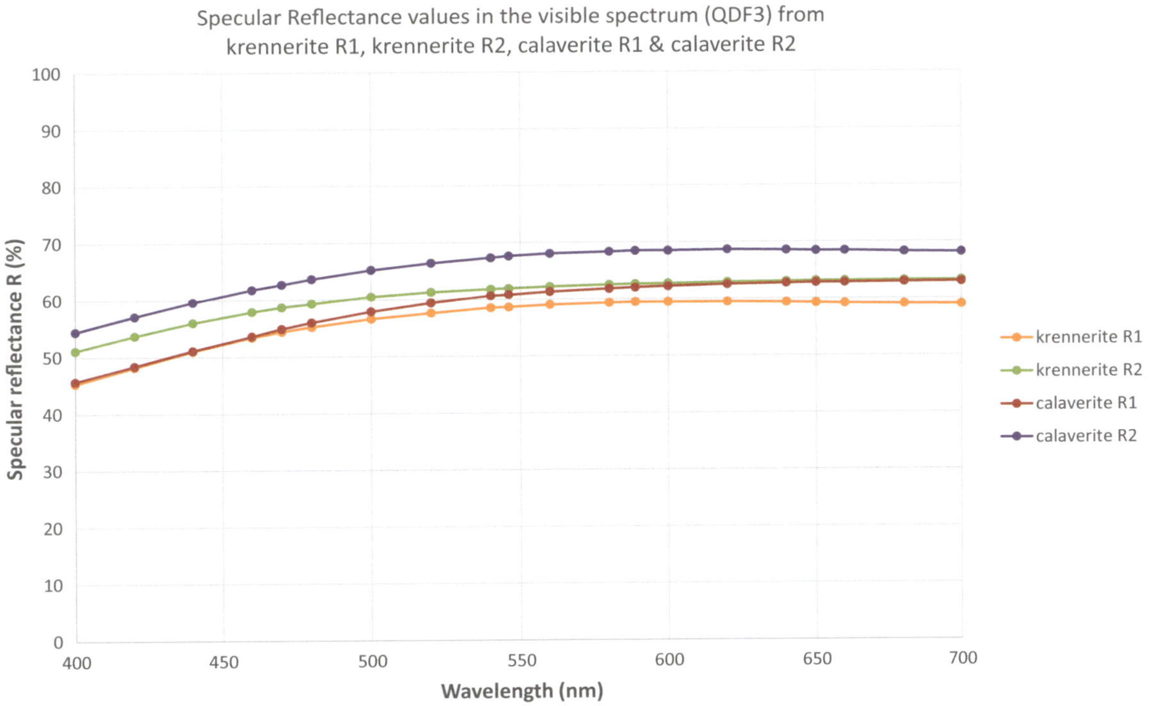

Reference

Dye MD, Smyth JR (2012) The crystal structure and genesis of krennerite, Au_3AgTe_8. Can Mineral 50:119–127. https://doi.org/10.3749/canmin.50.1.119

Lepidocrocite (*lpc/Rubinglimmer*)

71

γ-$Fe^{3+}O(OH)$: Orthorhombic

Lepidocrocite		R_{546}(%)	Y (%)	λ_d (nm)	P_e(%)	Colour sensation
Air	R_1	11.6	11.5	476	8.7	Greyish white, highly bireflective
	R_2	18.4	18.3	478	9.1	
Oil	R_1	2.1	2.1	472	19.4	
	R_2	6.2	6.1	477	18.3	
Bir: Strong (5)				*Pleoc*: <<Birr (2)		*Tarn*: No
Anis +P: Strong (5)				*IR*: Abundant, red		*Twg*: Not observed
Mor: Tabular, fibrous, (massive)				*Clv*: // (010) occasional		*PolQ*: Variable
Hardness	*PolH*: Medium (3)			*Mohs*: 5	*Talm*: nd	*VHN* $_{(mean\ 100–200–300\ g)}$: 402

Obs : cf. R charts (visible spectrum: QDF3 1993; Picot and Johan 1982).
Mineral abbreviations: gg—gangue, gth—goethite, ht—hematite, lpc—lepidocrocite, py—pyrite

71.1 General

Composition: $Fe^{3+}O(OH)$, with possible substitution of Fe^{3+} by Mn^{3+} or Co^{3+}. Formula also expressed as γ-$Fe^{3+}O(OH)$, dimorph of goethite. Mineral much scarcer than goethite or hematite.

71.2 Microscopic Description

71.2.1 SP Observation (*Single Polariser*)

Greyish-white colour. Moderate reflectance, relatively high for an oxide but always lower than hematite. *Strongly bireflective*. R_1 and R_2 of goethite are between R_1 and R_2 of lepidocrocite (cf comparative graphs in the visible spectrum). Although by its bireflectance it is usually described as clearly pleochroic, actually its pleochroism is scarce, since the colour practically does not change.It remains between white and grey, with parallel spectral curves R_1 and R_2 (Figs. 71.1a, b and 71.2). Cleavage //(010) is rarely visible under the microscope.

It occurs, usually associated with goethite, in tabular or lamellar crystallites, in microcracks, covering geodes, etc. (Fig. 71.4a). Sometimes it can be, with goethite, a main component of limonite (cf Fig. 72.2a–d), in cryptocrystalline or earthy masses. Frequently secondary to pyrite, it is sometimes also pseudomorphized by goethite.

71.2.2 Observation +P (*Polariser + Analyser at 90°*)

Strongly anisotropic (Figs. 71.1c, d and 71.4b), in poorly coloured but striking, mainly white and grey tones (turning to bluish, when slightly uncrossing the analyzer, Fig. 71.3).

Unlike goethite, the anisotropism is not masked by the abundant red or orange *internal reflections.*

Twinning not observed.

R. Castroviejo, *A Practical Guide to Ore Microscopy—Volume 1*,
https://doi.org/10.1007/978-3-031-12654-3_71

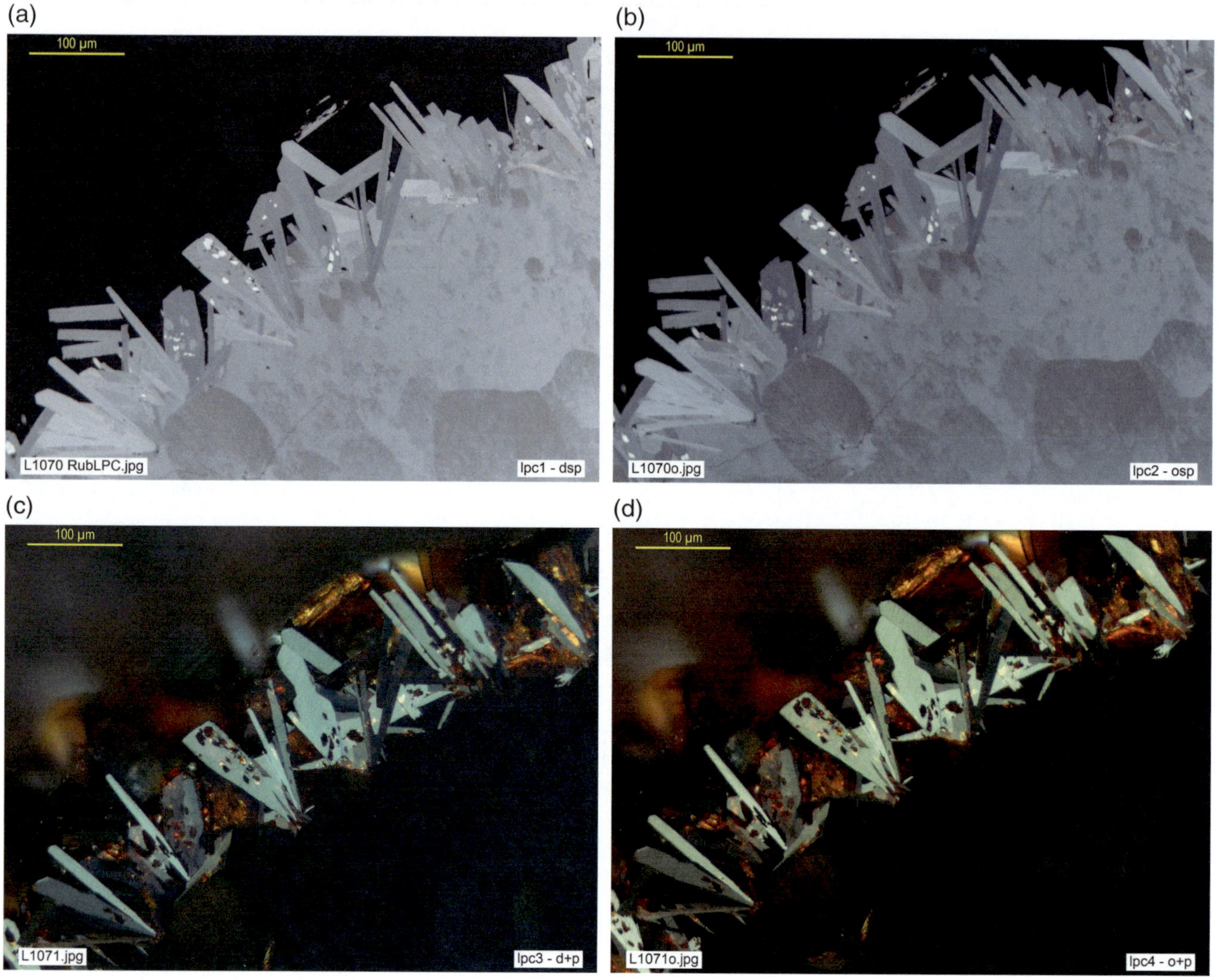

Fig. 71.1 **a, b** Tabular crystals of lpc (greyish-white, bireflective, with small inclusions of ht, white) on botryoidal aggregate of gth (grey, lower right half of image). To better appreciate the strong bireflectance of lpc, the same scene is reproduced, rotated, in Fig. 71.2. MRubLPC, Kahlberger Zug, Olpe, Sauerland, Germany (**dsp** and **osp**, respectively). **c, d** Same scene as Fig. 71.1a (**d+p** and **o+p**, respectively). Poorly coloured but strong anisotropism, despite abundant internal reflections (red or orange) of lepidocrocite

71.2.3 Paragenesis

Frequent, though not abundant, in oxidation zone *parageneses*, typically secondary to pyrite. Given its similarity to hematite, it has probably gone unnoticed on more than one occasion.

Companions: pyrite, goethite, hematite and other species of the oxidation zone.

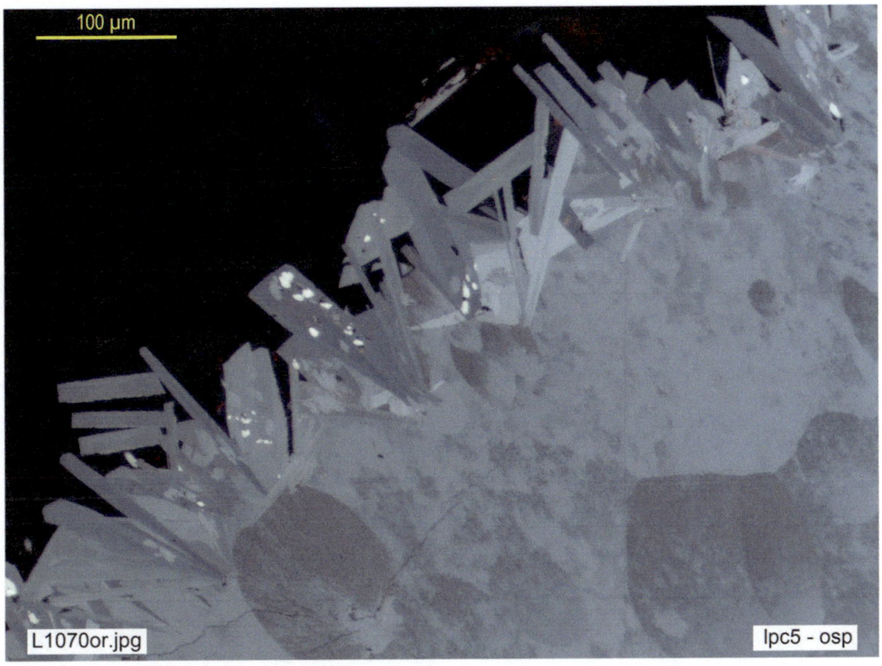

Fig. 71.2 Same scene as 71.1b, with rotation of the polariser, to show the strong bireflectance of lpc (**osp**)

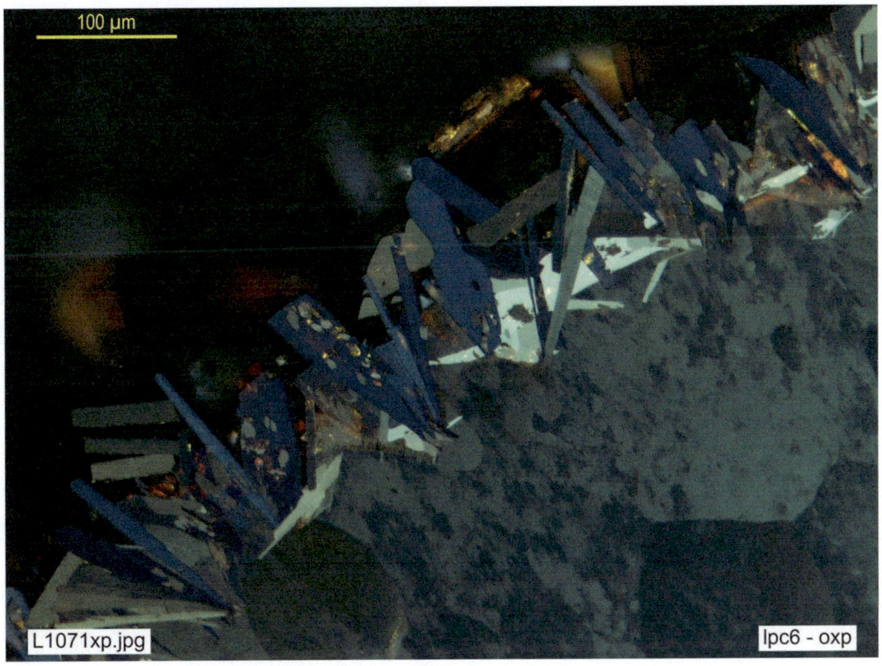

Fig. 71.3 Same scene as 71.1d, slightly uncrossing the analyser to observe the appearance of bluish tones of anisotropism in lpc (**o+p**)

71.2.4 Identification

Its moderate reflectance (<ht, but $\geq$ gth), IR, colour and anisotropism (reminiscent of hematite, but hematite is harder and brighter), together with its paragenesis, allow a reliable identification.

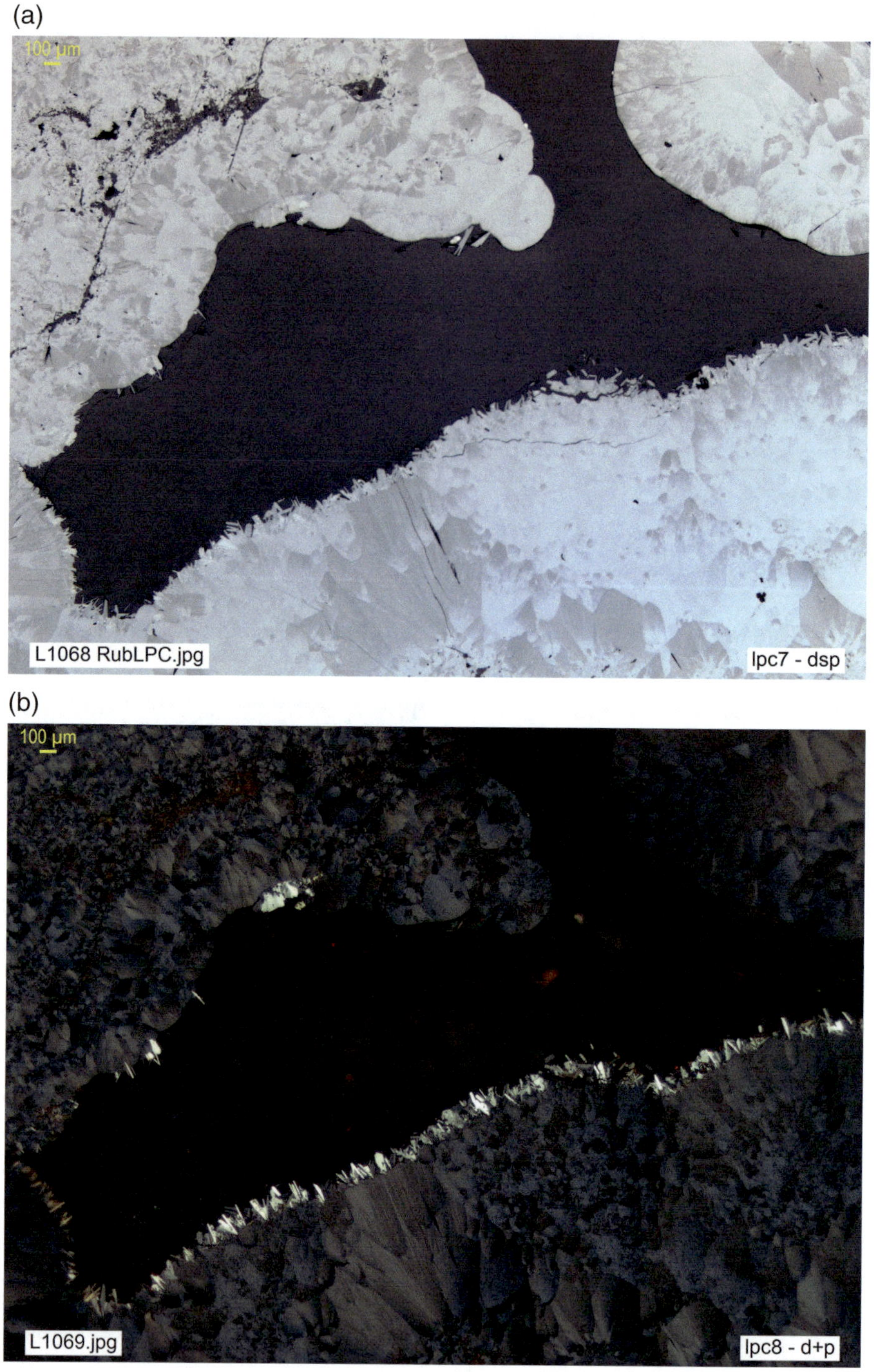

Fig. 71.4 **a, b** (**dsp** and **d+p**, respectively)—Botryoidal aggregate of gth (brownish grey-bluish grey), coated with fine lpc (strong greyish-white anisotropism, Fig. 71.4b), growing towards the interior of the geode. (Fig. 71.1a shows in detail the lower left rim of the cavity). MRubLPC, Kahlberger Zug, Olpe, Sauerland, Germany

71.3 Spectral Information

71.3.1 Lepidocrocite Reflectance in the Visible Range (in Air) (%)

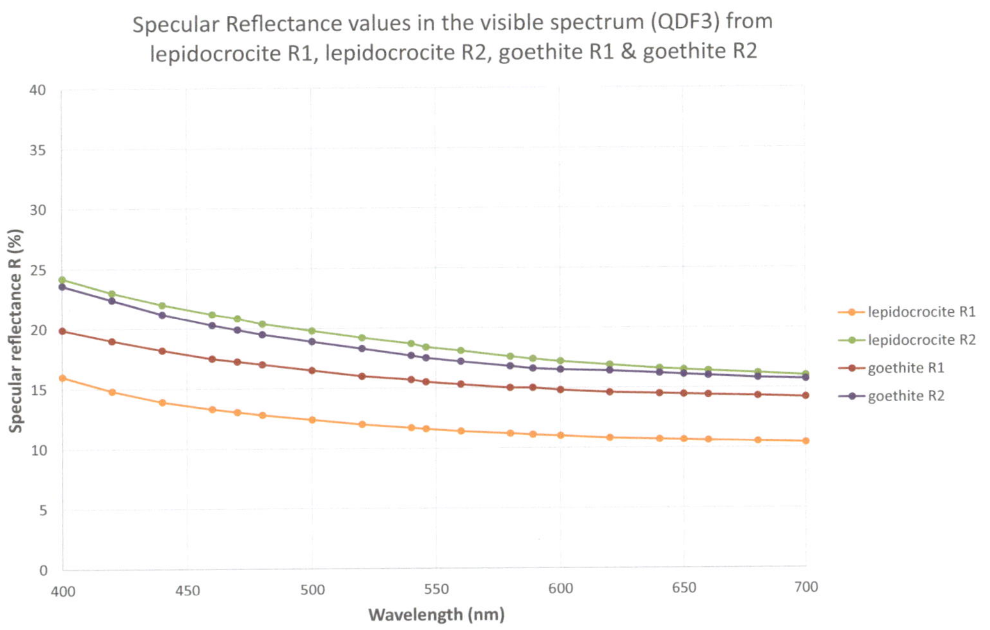

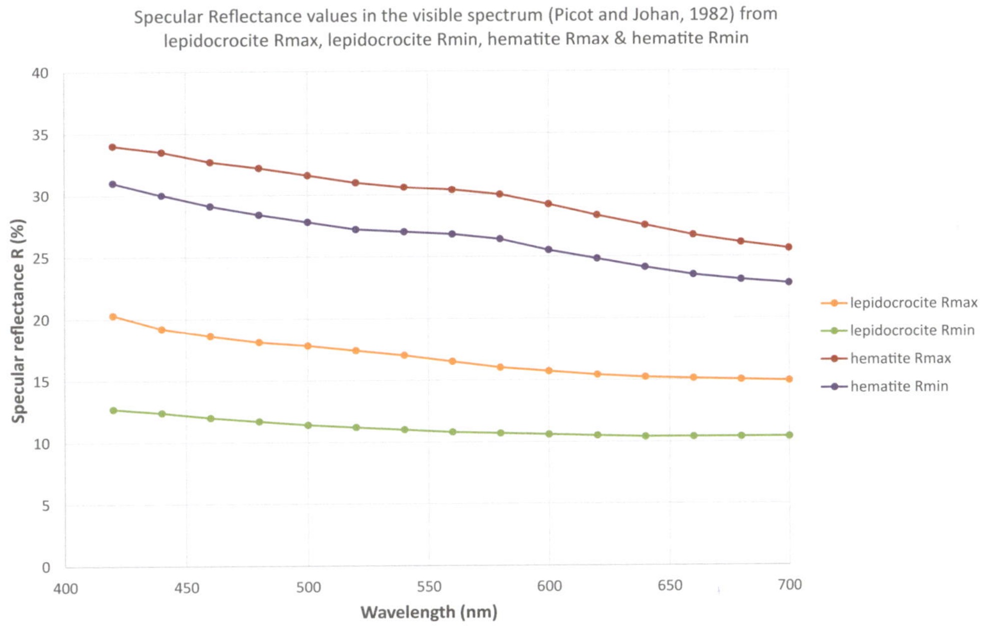

References

Picot P, Johan Z (1982) Atlas of ore minerals. B.R.G.M./Elsevier

QDF3 (1993) Quantitative data file… cf. In: Criddle AJ, Stanley CJ (eds)

Limonite (*lim*)

72

Hydrated Fe oxide: –

«Limonite is nowadays used as a rock name for hydrous masses of iron oxides, when one cannot or does not wish to make any statements regarding their mineralogical nature and homogeneity.»

(Ramdohr 1980, p. 1071).

72.1 General

The definition in the heading invites us to dispense with the classical concept of limonite as *brown iron ore* = $Fe_2O_3 \cdot 1\frac{1}{2}\ H_2O$ (i.e. $2Fe_2O_3 \cdot 3H_2O$) and, by referring to a *rock of hydrated iron oxides*, hints at the great variety that can be found in limonites, a term discredited by IMA as a mineral species. However, since it can be both an ore and an important exploration guide (*gossan*), its use is unavoidable for miners and exploration geologists and should not be ignored. Therefore, such a general definition as a *rock of hydrated iron oxides* is not sufficient.

In practice and following Blain and Andrew (1977) and Blanchard (1968, 1984), we refer to limonite as a field term for a mixture or aggregate of *amorphous, colloidal iron oxides or microcrystalline goethite, with subordinate amounts of silica (jasper), hematite, jarosite, lepidocrocite or manganese oxides in varying proportions and states of hydration*. The mineralogy can be even more varied, taking into account the primary associations from which the limonite is derived (massive sulphides, Cu-Ni deposits, Pb–Zn ores, carbonates, etc.).In addition to the oxides goethite, lepidocrocite, hematite, maghemite, psilomelanes (s.l.), silica and its possible hydrated and/or amorphous derivatives, it may contain occasional clay minerals, relics of sulphides or carbonates such as siderite, and various oxidation products of these, such as jarosite, anglesite, gypsum, plumbojarosite, cerussite, smithsonite, hemimorphite, pyromorphite, scorodite, etc.

In addition to mineralogy, texture can provide, along with field data, important information for exploration, such as the distinction of:

- *indigenous* limonite (formed directly on the sulphide it replaces),
- *transported* limonite, which implies a reduced mobility of Fe^{3+} and precipitation close to the primary sulphide by a few centimetres, and
- *exotic* limonite, deposited outside the limits of the orebody.

These types refer, evidently, to *secondary* limonites (usually derived from sulphides and, to a lesser extent, from carbonates). Although not so frequent, there are also *primary* limonites, by direct precipitation of hydrothermal solutions or in sedimentary environments. Their distinction is not difficult, taking into account field data and microscopic textures.

According to Blanchard (1984), there is no mineral species corresponding to the formula, $2Fe_2O_3 \cdot 3H_2O$, attributed to limonite in traditional texts. For the description, therefore, a specific format must be followed, since its polymineral or amorphous character cannot fit into the usual model. Its composition, expressed in general terms as "hydrated iron oxides", varies according to local conditions, adsorption of Mn, Si, precious metals or other cations derived from the primary ore. Their textures are also very varied, dominating colomorphic, spherulitic, botryoidal, reniform, submicroscopic, sometimes amorphous textures, with specific features such as relicts, ghosts or replicas of sulphides, boxwork, etc.

R. Castroviejo, *A Practical Guide to Ore Microscopy—Volume 1*,
https://doi.org/10.1007/978-3-031-12654-3_72

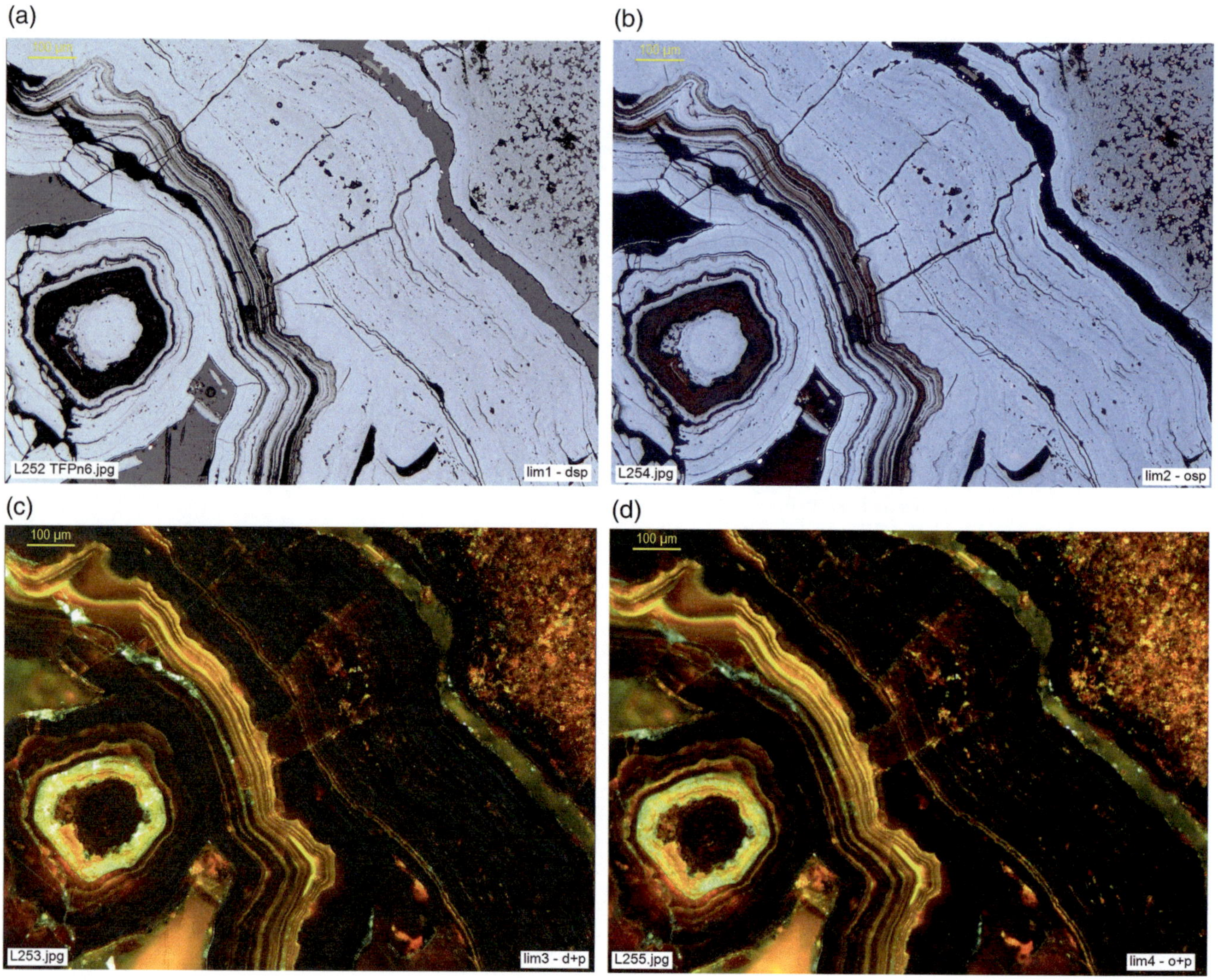

Fig. 72.1 **a, b, c, d** (**dsp**, **osp**, **d+p**, **o+p**, respectively)—Colomorphic limonite, with alternating bands dominated by gth (bluish grey-grey) and thin clay-rich bands of lower reflectance and internal reflections, much more abundant and clearer, luminous. TF Pn6, Sar Mine, Santiago de Compostela

72.2 Microscopic Description

72.2.1 SP Observation (*Single Polariser*)

Grey to greyish-brown, with variable but low reflectance (R $\leq$ goethite), heavily hydrated varieties may show yellowish coloration (Figs. 72.1a, b and 72.2a, b).

Scarce to nil crystallinity, in general. In many cases it forms masses of submicroscopic grains. Variable morphology, frequently colomorphic, also in aggregates formed as pseudomorphs after sulphides or siderite, as suggested by the textures displayed (*replica*, *phantom*, *boxwork* Figs. 72.3, 72.4, 72.5, 72.6 and 72.7a, b). The heterogeneous nature of the mineral is shown by the coexistence of several species, finely intergrown or in aggregates (Figs. 72.1a, b and 72.2a, b), such as goethite, jarosite, hematite, jasper, although they are not always visible (submicroscopic or amorphous aggregates).

72.2.2 Observation +P (*Polariser + Analyser at 90°*)

The anisotropism of components such as goethite, hematite, lepidocrocite, etc. is not always visible, because of the fine grain size or because of their low/nil crystallinity. If the grain size is sufficient, the anisotropism of species such as

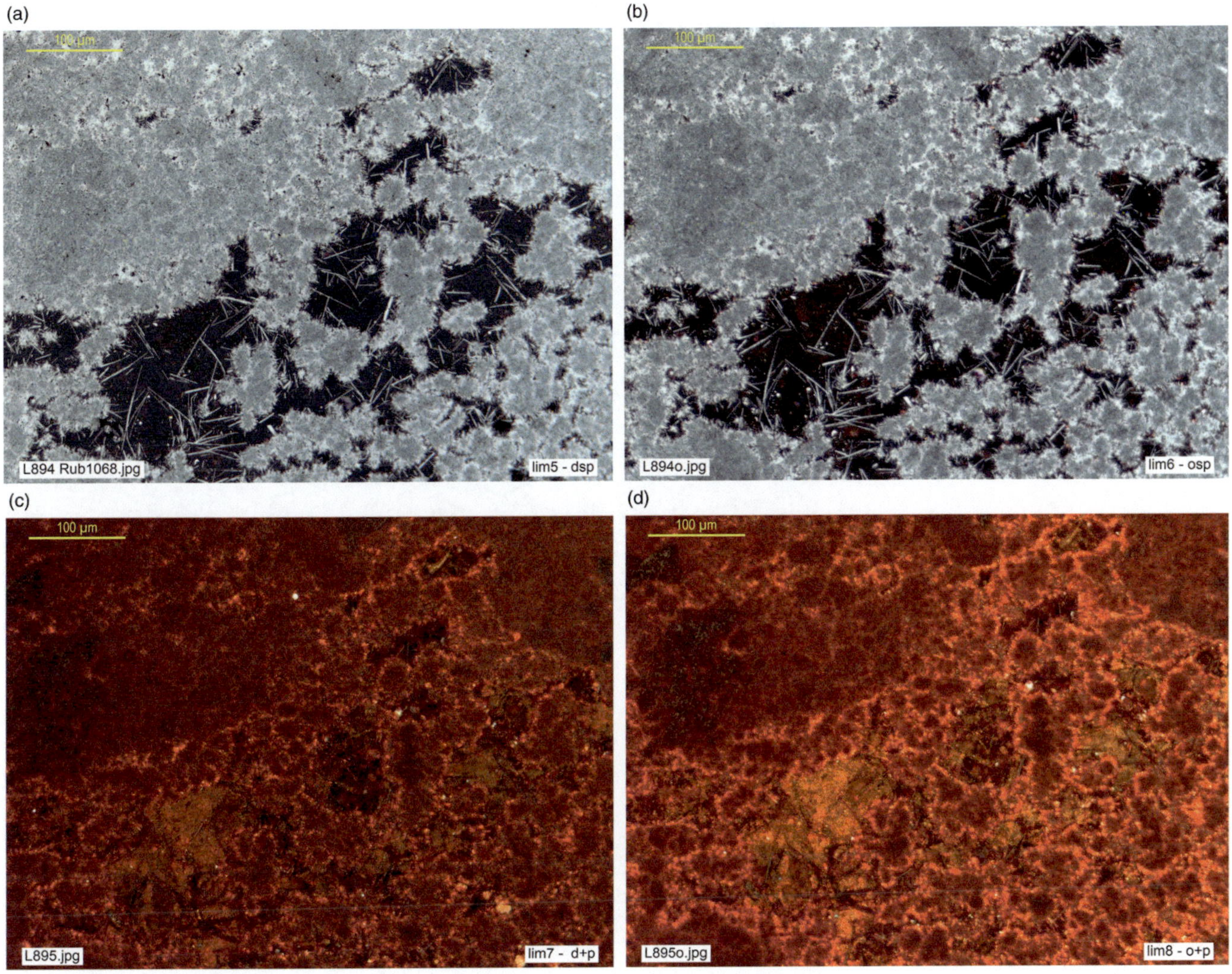

Fig. 72.2 **a, b, c, d** (**dsp**, **osp** , **d+p**, **o+p**, respectively)—Micro- to cryptocrystalline limonite, fibrous-radiated texture: lepidocrocite (light grey, main component) intergrown with gth (grey) and silicates (dark grey-brown) and acicular ht (white) rim. Rub1068, Ober-Neissen, Nassau, Germany

goethite, hematite, lepidocrocite, jarosite, etc. can be recognized in favourable cases. In all cases, *internal reflections* (Figs. 72.1c, d and 72.2c, d) are abundant, reddish-brown to yellowish-brown, more yellow and luminous the more hydrated or jarositic the ferruginous gels are (except if they contain psilomelane or manganese oxides).

72.2.3 Paragenesis

Primary limonites typically occur in association with goethite, hematite, lepidocrocite, maghemite, jasper in materials deposited in oxidizing environments, including also the shallow levels of high sulphidation epithermal formations.

Secondary limonites, common in the oxidation zone of sulphide or ferriferous deposits, apart from their own components, may contain various materials, particularly relicts of pyrite, chalcopyrite, marcasite, siderite and any of the primary minerals from which they are derived.

Its association with sulphides and precious metals (Au, Ag, PGM) in the oxidation zone of some deposits, e.g. volcanogenic massive sulphides and epithermal concentrations, gives limonite a remarkable exploration potential (cf. F 43.10.a–b and g–h, F43.94.f–h).

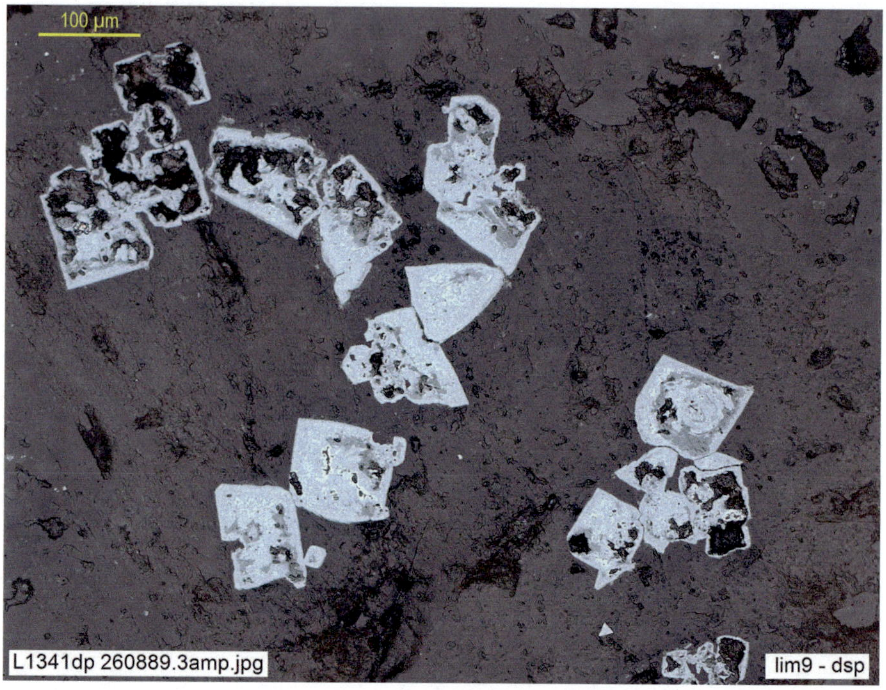

Fig. 72.3 Py pseudomorphic lim cubes. Sample 260889.3, Rodalquilar, Almería (**dsp**)

Fig. 72.4 Limonitic *boxwork* on py ghost. Sample 260889.3, Rodalquilar, Almería (**dsp**)

72.2.4 Identification

In spite of the notable variations pointed out, the paragenesis, together with its colour, low reflectance, usual colomorphic or pseudomorphic textures, with occasional relict sulphides and abundant internal reflections, in addition to the field relations, make its identification unproblematic.

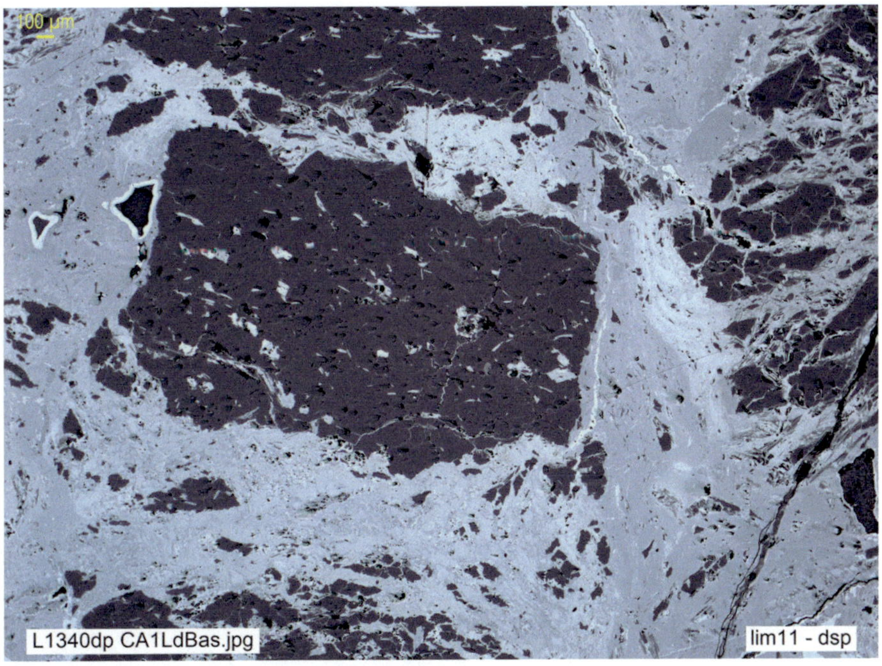

Fig. 72.5 Vein infill of quartz (dark grey), hosting pyritic stockwork totally pseudomorphosed by limonite (grey). CA.1LdB, Lomo de Bas, Cantal, Murcia (**dsp**)

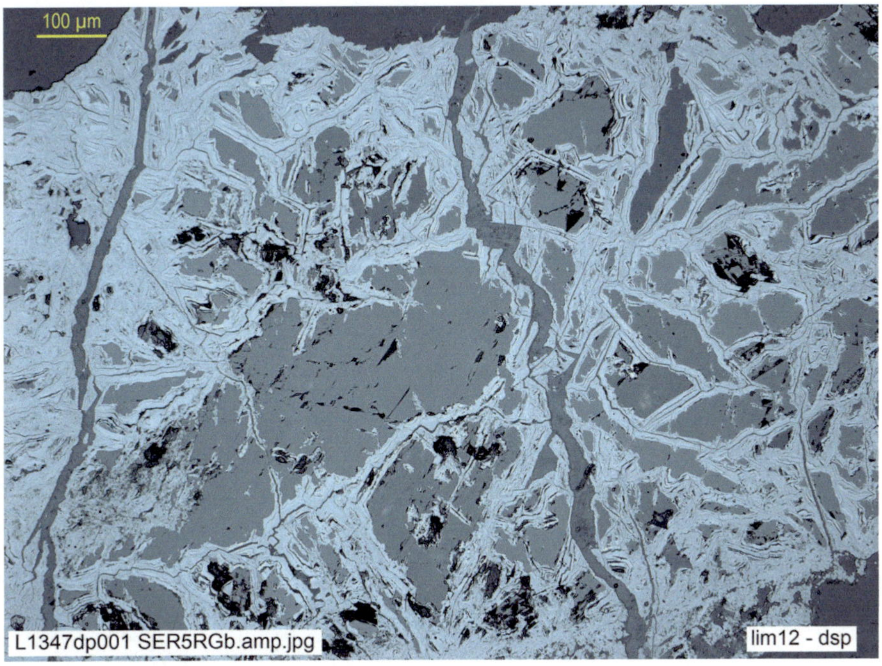

Fig. 72.6 Limonitisation of siderite, preserving its rhombohedral cleavage, inherited by lim (incipient *boxwork*). SER. 5RGb, Rio Gibranzos, Cáceres (**dsp**)

(a)

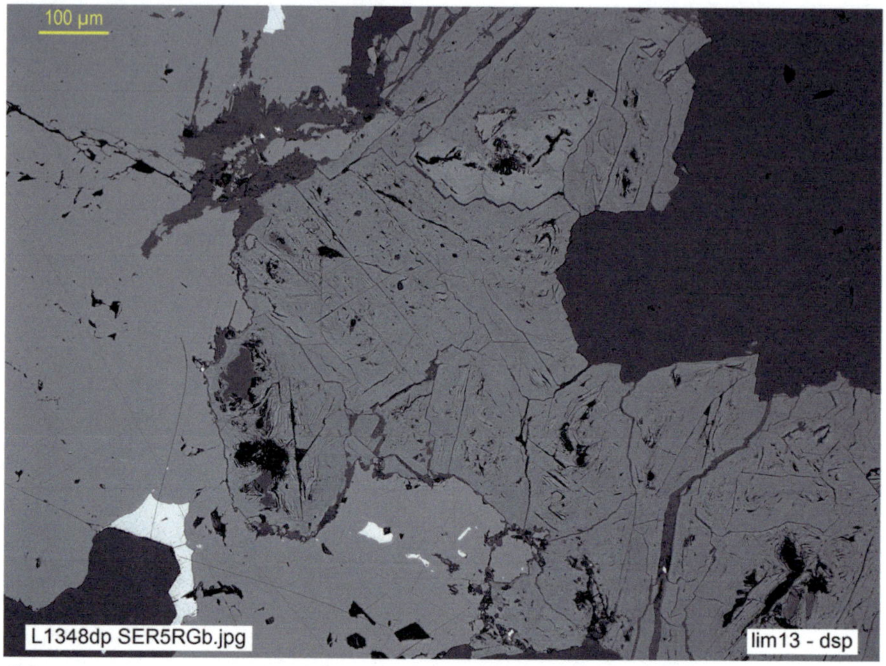

(b)

Fig. 72.7 **a, b** (**dsp** and **d+p**, resp)—Partial replacement of sphalerite (grey, with bright yellowish, massive internal reflections, left half of the image), pseudomorphosed by limonite, which preserves some relicts and the cubic cleavage of sp, accessory gn (white) included in sp. SER.5RGb, Río Gibranzos, Cáceres

References

Blain CF, Andrew RL (1977) Sulphide weathering and the evaluation of gossans in mineral exploration. Minerals Sci Engn 2:219–150

Blanchard R (1968) Interpretation of leached outocrops. Nevada Bur Mines, Reno, NEV, USA, Bull 66:166 p

Blanchard R (1984) Interpretation of leached outcrops. Nevada Bureau of Mines, Bull 66 (3rd Printing) 196 p

Ramdohr P (1980) The ore minerals and their intergrowths, 2nd edn, vol 2. Pergamon Press, Oxford, 1205 p

Linnaeite, Group (*ln/Various Names According to Compositions*) 73

$A^{[4]}B_2^{[6]}S_4$: Cubic

Linnaeite		R_{546} (%)	Y (%)	λ_d (nm)	P_e (%)	Colour sensation
Air		47.3	47.8	580	7.0	Pinkish white, bright
Oil		34.2	34.7	580	9.1	
Bir: Nil (1)				*Pleoc*: Nil (1)		*Tarn*: Occasional
Anis +P: Isotropic (1)				*IR*: No		*Twg*: //(111) not visible in polished section
Mor: Idiomorphic (octahedral forms)				*Clv*: // (100) and (111)		*PolQ*: Excellent
Hardness	*PolH*: Medium, ≈ po (3)			*Mohs*: 5	*Talm*:–	*VHN* $_{(50\ g)}$: 435–558

Obs: The optical parameters documented in this table correspond to a ln with 41.6% S, 34.3% Co, 20.7% Ni and 3.9% Fe (QDF3); cf. Table and R plots, multispectral data (VNIR, AMCO) and visible range (QDF3)

Mineral abbreviations: Ag—native silver, asp—arsenopyrite, bn—-bornite, ccp—chalcopyrite, car—carrollite, cob—cobaltite, dg—digenite, gf—gersdorffite, gg—gangue, gn—galena, grg—greigite, ht—hematite, ln—linnaeite, lo—löllingite, mc—marcasite, ml—millerite, nc—nickelite, pdm—polidymite, po—pyrrhotite, py—pyrite, sg—siegenite, sk—skutterudite, sp—sphalerite, td—tetrahedrite, ul—ullmannite, vio—violarite

73.1 General

There is a large compositional variation within the group. This is reflected in the formula proposed by Strunz and Nickel (2001): $A^{[4]}B_2^{[6]}S_4$, referring to a spinel-like structure, where $\boldsymbol{A}^{2+}$ is usually Fe, Ni, Co, Cu and $\boldsymbol{B}^{3+}$, Co or Ni (and Cr in *daubréelite*), with compact cubic packing of the S atoms and tetrahedral coordination [4] for A and octahedral [6] for B.

By custom, the group is designated as Linnaeite (*s.l.*) but to speak with precision, reference should be made specifically to the credited species. Some of these, found in nature, are:

- *Linnaeite* (*ln s.str*: $CoCo_2S_4$)
- *Siegenite* (*sg*: $CoNi_2S_4$)
- *Polydimite* (*pdm*: $NiNi_2S_4$)
- *Violarite* (*vio*: $FeNi_2S_4$)
- *Greigite* (*grg*: $FeFe_2S_4$)
- *Carrollite* (*car*: $CuCo_2S_4$)
- *Fletcherite* (*fle*: $Cu[Ni,Co]_2S_4$)

However, the existence of intermediate compositions, which define several series of solid solutions and defy any simplistic scheme (cf. Wagner and Cook 1999) should never be forgotten.

Optical properties, and among them particularly colour and spectral response (expressed in multispectral specular reflectance curves), vary according to composition, but it has not yet been possible to establish a bi-univocal relationship between colour and composition, as summarized by Ramdohr (1980, p. 702): «*The writer has not yet succeeded in using the doubtlessly existing differences in colour for the distinction of the individual minerals (obviously besides violarite and greigite!)*».

73.2 Microscopic Description

73.2.1 SP Observation (*Single Polariser*)

Bright pinkish white *colour*, with varied shades (Figs. 73.1a, b, 73.2, 73.3, 73.4 and 73.5), depending on the composition and, for its subjective appreciation, also on the accompanying minerals (very similar to cobaltite, more pinkish against chalcopyrite, purplish to pinkish brown against millerite, pinkish to yellowish brown against gersdorffite or ullmannite). Very good *polishing* is achieved. Medium

R. Castroviejo, *A Practical Guide to Ore Microscopy—Volume 1*,
https://doi.org/10.1007/978-3-031-12654-3_73

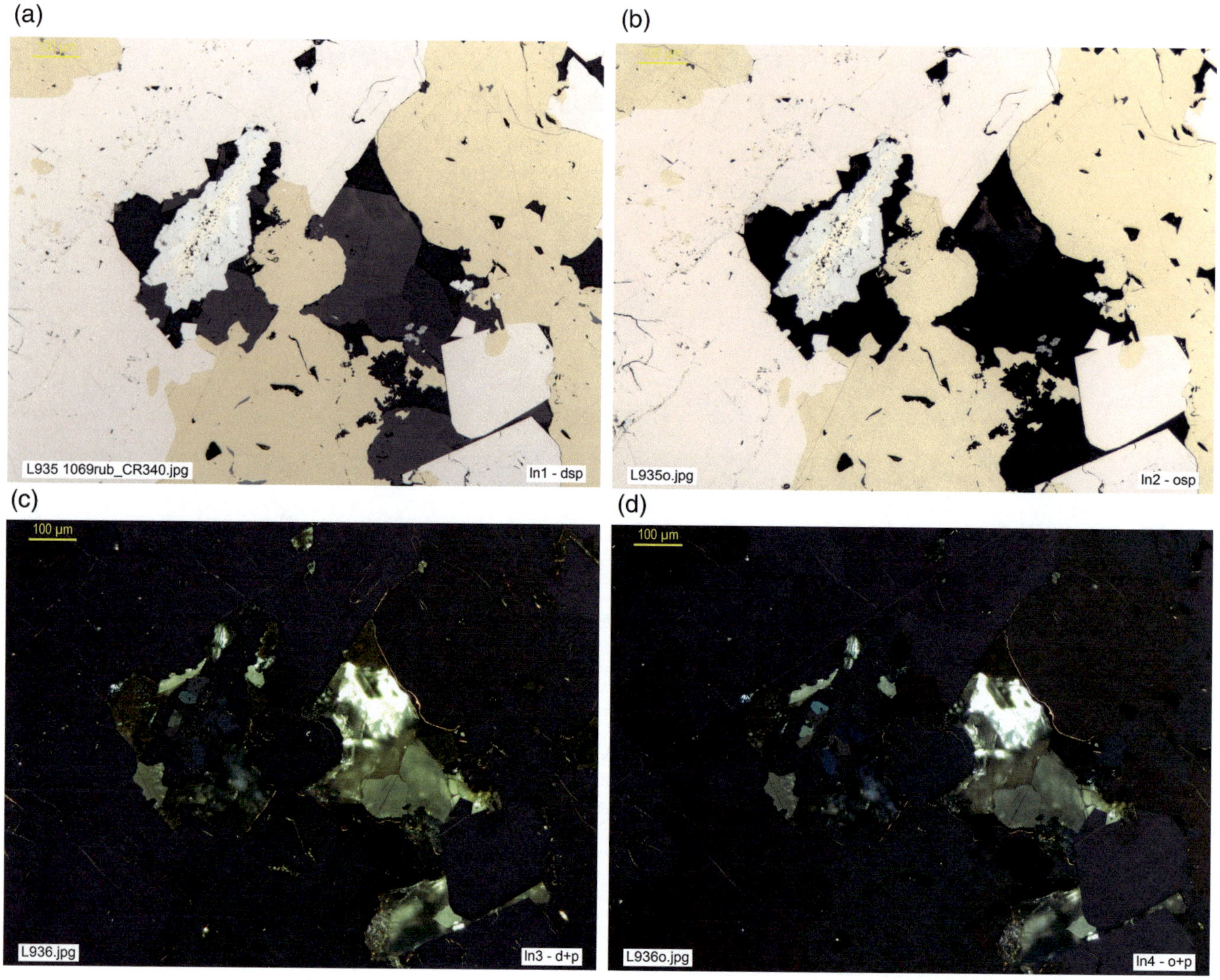

Fig. 73.1 **a** Ore of sg (pinkish-white) and ccp (yellow), with carbonate gg and, in the centre, aggregate of intergrown py (yellowish-white) and mc (white, with a faint bluish-green tint). Sample 1069/CR 340-RCB, Müsen, Siegerland, Germany (**dsp**). **b** Same scene as ln1 (**osp**). **c, d** Same scene as ln1, (**d+p**) and (**o+p**), respectively. Strong anisotropism of mc (with its characteristic green tints) and weak anisotropism of ccp

polish hardness, >td, Ag, ccp; <sk, po, gf; << asp, py. Cleavage//(100) common, //(111) occasionally visible.

Reflectance is high but variable, according to composition. In general, R ln < R py, ml, sk (*cf* comparative R graphs) but > gn, ccp, po. The optical impression of brightness (R in white light) is close to that of chalcopyrite but noticeably lower in some species, *vbgr* greigite (≈ pyrrhotite), or violarite (secondary to pentlandite, cf Figs. 93.4a, b). Isotropic.

It is usually well crystallized in cubic forms, with preferential idiomorphic development according to (111), which makes the *octahedron* the most common form, although alteration or frequent replacement reactions may mask it. Minerals of the group also appear as secondary phases in substitution of other sulphides (even of other species of the group), in which case pseudomorphic linnaeites (Figs. 73.4, 73.5) usually result.

73.2.2 Observation +P (*Polariser + Analyser at 90°*)

Isotropic, but in some cases shows weak anisotropism. Twinning // (111), frequent in some species, is not observed under the microscope. Zonation is also absent.

Fig. 73.2 Crystalline aggregate of ln (*s.l.*, probably sg, white–pink) included and partially replaced by ccp (yellow). Corrosion of the ln borders partly hides its euhedral morphology. Inclusions of gn (grey) and gg in ccp. Sample 1069rub-CR340, Müsen, Siegerland, Germany (**osp**)

Fig. 73.3 Inclusions of ln (*s.str.* -confirmed EMPA), white with a very faint pinkish tinge, in ccp, with diablastic ht (bluish grey) included in ccp and in gg (black). C. Sample 031,200-30 g, Mina Maruxa, Moeche, Coruña (**osp**)

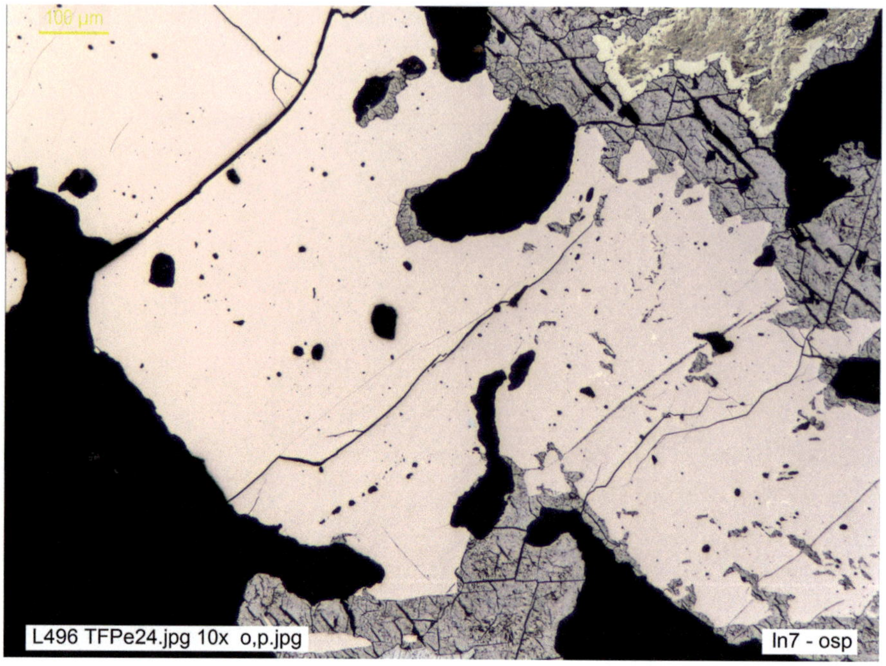

Fig. 73.4 Aggregate of vio (*s.str.*, brownish-pink) and py (yellow), partially replaced, respectively, by a secondary paragenesis of Ni-vio (Ni-rich vio or Fe-pdm, purplish, darker and with prominent octahedral cleavage) and melnicovite (yellowish-grey cryptocrystalline aggregates on py). Compositions analysed by EMP. Sample TFPe24, Dry Nickel Mine, Bindura, Zimbabwe (**osp**)

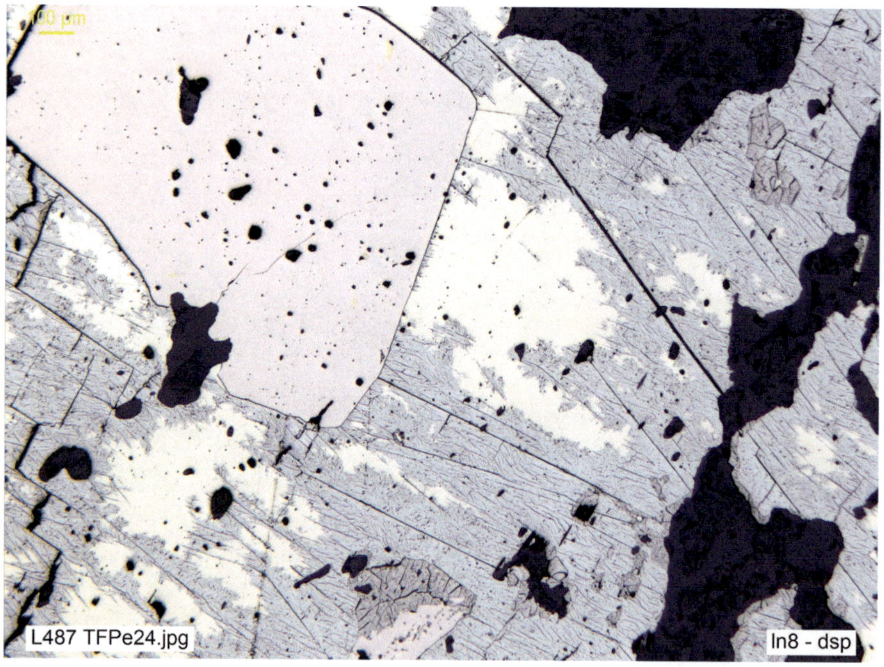

Fig. 73.5 Mass of ml (yellowish-white) with partially corroded euhedral inclusions of vio (pinkish). Both altered to a secondary paragenesis of linnaeites of differentiated compositions, ml is replaced by the nickeliferous term pdm (purplish colour, light) and vio, by Ni-vio (or Fe-pdm, darker, crystal, bottom center, and pseudomorphs, top right), which is similar to that of Fig. 73.4. Same sample as In7 (**dsp**)

73.2.3 Paragenesis

Parageneses varied, as corresponds to their compositional variety—from magmatic (Sudbury, Ontario, polydimite, violarite, contested), through pneumatolitic (replacement), to sedimentary (greigite). The most typical are the hydrothermal vein paragenesis (e.g., the sideritic veins with siegenite from Siegerland, Westphalia, Germany, or the *five element* type deposits) and the stratabound *Copperbelt* type (Zambia and Katanga, DR Congo, siegenite, carrollite).

Accompanying minerals: chalcopyrite, millerite, pyrite, Ni-Co arsenides, pyrrhotite, pentlandite, hematite, galena, bornite, digenite, etc.

73.2.4 Identification

Pink colour, euhedral tendency (octahedra) and relatively high (but < pyrite) reflectance characterize, in general, the minerals of the group. Nevertheless, it is not easy to distinguish among them, particularly linnaeite, siegenite and polydimite. The electron microprobe can be a great help in safe identification. Good polish is another additional criterion, but in secondary species it can be ambiguous (Figs. 73.4 and 73.5) and in sedimentary species, it is difficult when dealing with friable materials. The distinction between violarite and *bravoite* secondary to pentlandite can be difficult, but the recognition of cubic forms (not octahedral) and the bravoite zoning, when they can be observed, may help. Also the morphology can be ambiguous in cases of alteration or in linnetites of secondary origin (pseudomorphs). In the case of greigite, its strong magnetic character is a useful hint.

Because of the similarity to other white sulphides/arsenides and the variation of properties within the linnaeite group (*s.l.*), misidentifications are possible.

- Linnaeites (*s.l.*) are distinguished from cobaltite, whose colour may look similar, by the greater hardness and (weak) anisotropism of the latter.
- Arsenopyrite is whiter, distinctly harder and shows a distinct, coloured anisotropism.
- In the case of greigite (and to a lesser extent violarite), its pyrrhotite-like colour can be confusing, but po is pleochroic and anisotropic.
- In the case of linnaeite or polydymite, their coloured tones (pinkish to yellowish-brown, in direct comparison) help to distinguish them from ullmannite, gersdorffite and skutterudite, which are whiter and more reflective.
- In all these cases, specular reflectance spectra can be very useful (cf graphs at the end).

73.3 Spectral Information

73.3.1 Linnaeite Multispectral Reflectance R (in Air, Without Polariser) (%)

λ (nm)	370	400	425	450	475	500	525	550	575	600	625	650	675	700	750	800	850	900	950	1000
μ	39.6	41.5	41.5	43.1	42.9	44.6	44.7	46.0	46.9	48.1	49.2	50.4	51.6	52.6	54.5	55.8	57.3	57.6	57.3	57.6
P_5	38.8	40.8	40.4	42.3	42.0	43.9	43.8	45.2	46.1	47.4	48.5	49.8	51.0	52.0	54.0	55.3	56.7	57.0	56.7	57.0
P_{20}	39.3	41.3	41.1	42.8	42.5	44.3	44.3	45.6	46.5	47.8	48.9	50.2	51.3	52.4	54.3	55.6	57.0	57.3	57.0	57.2
P_{80}	40.0	41.7	42.0	43.4	43.4	45.0	45.0	46.4	47.2	48.4	49.5	50.7	51.9	52.9	54.8	56.1	57.5	57.8	57.5	57.8
P_{95}	40.2	41.9	42.3	43.6	43.7	45.2	45.3	46.6	47.4	48.6	49.7	50.9	52.0	53.0	55.0	56.3	58.0	58.4	58.3	58.7

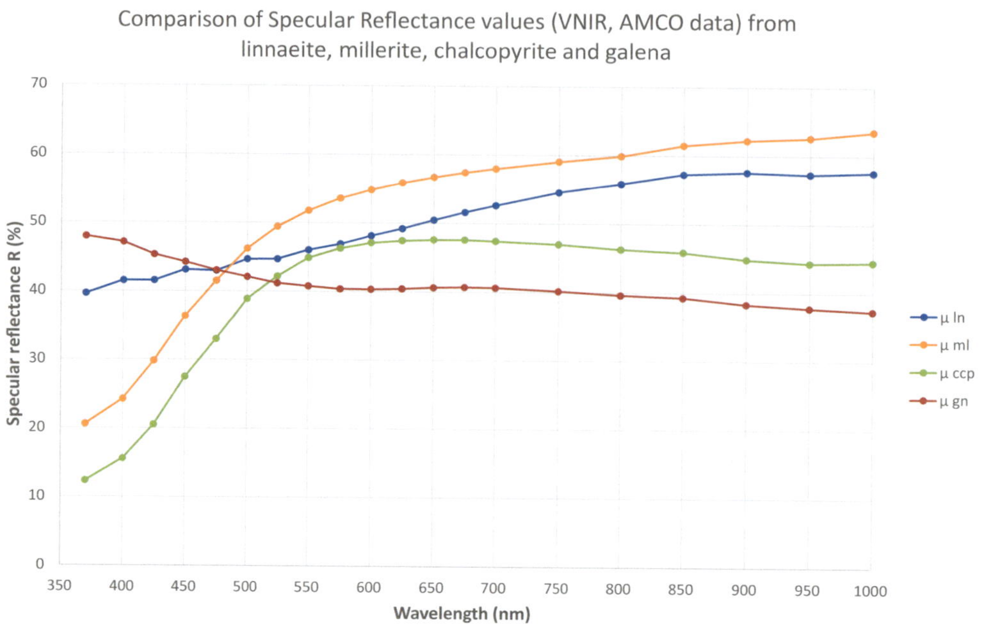

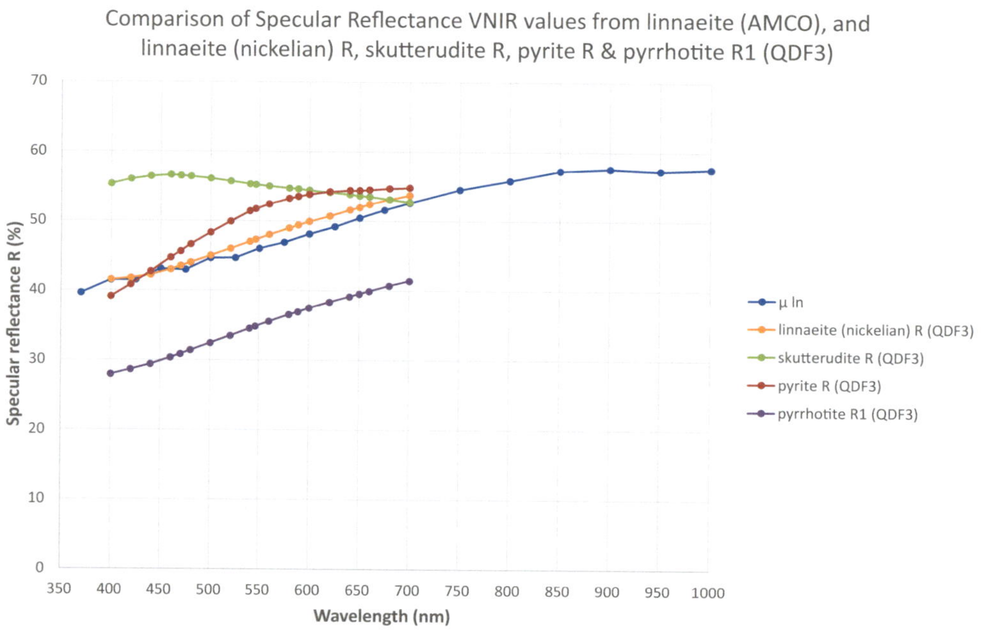

References

Ramdohr P (1980) In: The ore minerals and their intergrowths, vol 2. 2nd edn. Pergamon Press, Oxford, pp 1205

Strunz H, Nickel EH (2001) Strunz mineralogical tables, 9th edn. E. Schweizrbart'sche Verlagsbuchhandlung, Stuttgart, p 870

Wagner T, Cook NJ (1999) Carrollite and related minerals of the linnaeite group: solid solutions and nomenclature in the light of new data from the Siegerland District Germany. Canad Mineralogist 37 (3):545–558

Löllingite (*lo*)

74

$FeAs_2$: Orthorhombic

Löllingite			R_{546} (%)	Y (%)	λ_d (nm)	P_e (%)	Colour sensation
Air	R_1		53.4	53.1	478	5.9	Bright white, slightly pleochroic
	R_2		55.5	55.5	573	2.8	
Oil	R_1		38.4	38.1	478	9.3	
	R_2		41.6	41.5	573	4.0	
Bir: Weak (3)					*Pleoc*: Weak (3)		*Tarn*: -
Anis +P: Strong and coloured (5)					*IR*:–		*Twg*: Occasional, varied
Mor: Idiomorphic (prismatic), massive					*Clv*:–		*PolQ*: Good
Hardness		*PolH*: High but < asp (4)			*Mohs*: 5.5	*Talm*: F-	*VHN* $_{(100\ g)}$: 859–920

Obs: Mineral typically ignored (often confused with asp); cf. Table and R charts, multispectral data (VNIR, AMCO) and visible range (QDF3)
Mineral abbreviations: asp—arsenopyrite, ccp—chalcopyrite, lo—löllingite, mt—magnetite, po—pyrrhotite, py—pyrite, rm—rammelsbergite sf—safflorite sk—skutterudite

74.1 General

Composition: $FeAs_2$, often with some Co and Ni and traces of S. Intergrowths and possible miscibility with safflorite. Possible partial substitution of As by Sb.

74.2 Microscopic Description

74.2.1 SP Observation (*Single Polariser*)

White *colour*, very similar to arsenopyrite, but slightly bluer and brighter than arsenopyrite. *Pleochroism*: weak, but visible even in air and more distinct than that of arsenopyrite, with faint bluish and yellowish tones (Figs. 74.1a, b, 74.2 and 74.6).

Polishes well, better than arsenopyrite. High *polishing hardness*, > po, sp, ccp, sf but < mt, asp, py.

High *reflectance*, R: slightly > asp, sf, but < sk, rm; distinctly < native Ag, Au or even Bi, against which löllingite looks grey, relatively dark, especially in oil (Figs. 74.2 and 74.6; cf multispectral comparison plots at the end of the description).

Tendency to idiomorphic development (prismatic crystals, sometimes in radial aggregates), idioblastic or massive granoblastic (Figs. 74.4, 74.5, 74.6, 74.1a–d).

74.2.2 Observation +P (*Polariser + Analyser at 90°*)

Strongly *anisotropic*, with vivid colours in bluish, brownish and, especially when the analyzer is partially uncrossed, slightly yellowish tones (Figs. 74.1c, d, 74.3, 74.5 and 74.7). According to Ramdohr (1980), löllingite has the peculiarity that its anisotropism is more pronounced in air than in immersion. The anisotropism sometimes shows a characteristic *zoning* (Fig. 74.5). No *internal reflections*. Occasional twinning.

74.2.3 Paragenesis

Usually lo is an accessory mineral, but it is more frequent than commonly assumed, as it is often misidentified as arsenopyrite (confusion with asp). It has been found in pegmatitic, hydrothermal and metamorphic, skarn or contact

R. Castroviejo, *A Practical Guide to Ore Microscopy—Volume 1*,
https://doi.org/10.1007/978-3-031-12654-3_74

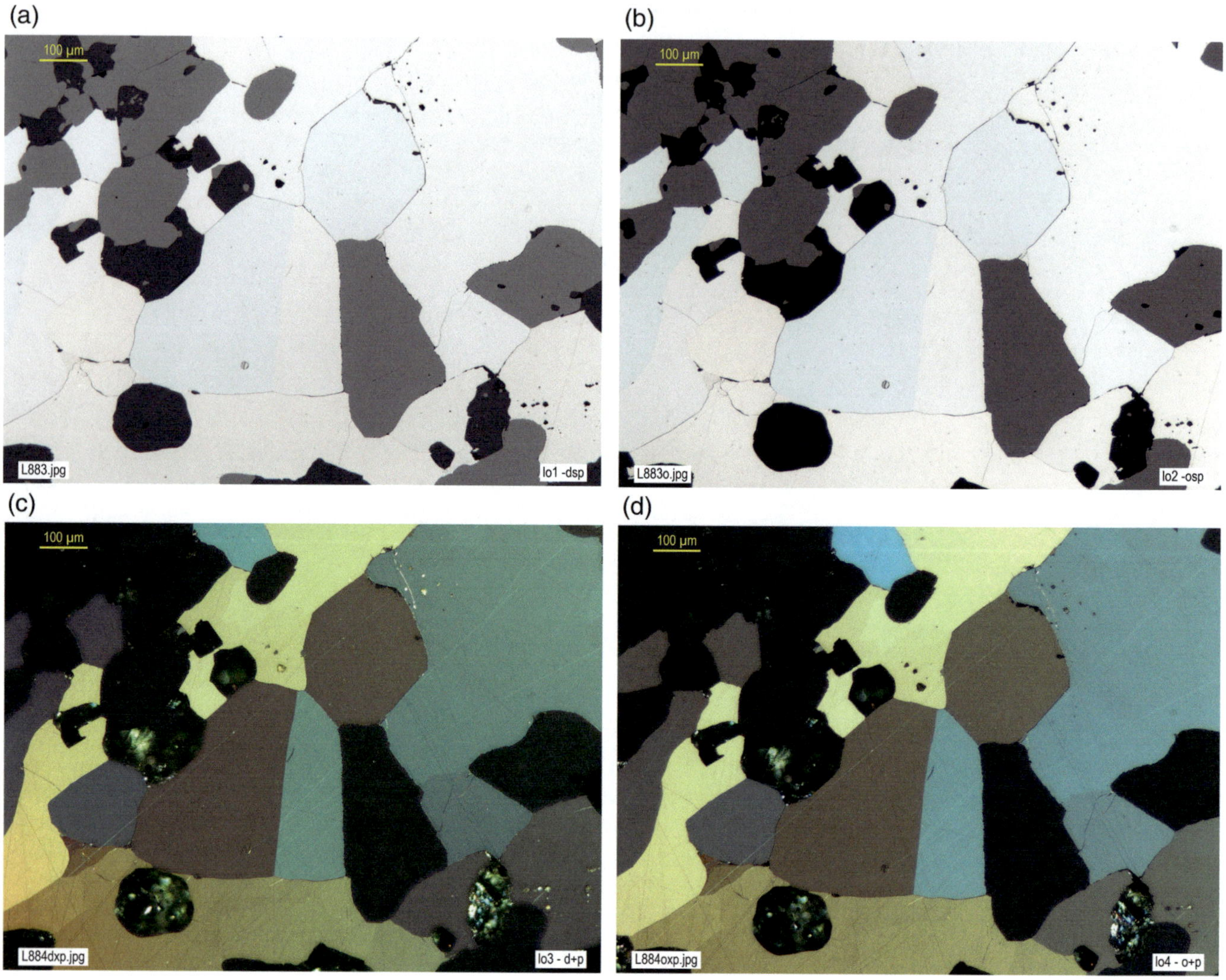

Fig. 74.1 **a** Granoblastic aggregate of löllingite (white), magnetite (grey) and quartz (dark grey). HJB_RUB81-15_ Burguillos del Cerro, Badajoz (**dsp**) **b** Same scene as lo1 (**osp**): slight pleochroism of lo. **c** Same scene as lo1 (**d+p**) **d** Same scene as lo1 (**o+p**)

pneumatolytic deposits. It can be found together with arsenopyrite, sphalerite, galena, Bi/Ag minerals, Ni/Co arsenides, magnetite (in skarn formations), etc.

74.2.4 Identification

Typical characteristics for its diagnosis are: its tendency to idiomorphism, white colour, slight pleochroism (but visible both in air and in immersion), high reflectance, good polish and strong anisotropism, with vivid polarization colours. However, it can be confused with arsenopyrite and with the minerals of the safflorite and rammelsbergite group.Arsenopyrite and safflorite are less anisotropic and rammelsbergite is more reflective but polarizes with less coloured tones and shows abundant polysynthetic twins.

Fig. 74.2 Aggregate of lo, mt, Bi, (bm). Metamorphic ore with granoblastic texture of lo and mg; sparse interstitial Bi in lo (centre). HJB. RUB81-15, Burguillos del Cerro, Badajoz (**dsp**)

Fig. 74.3 Same scene as lo5 (**d +p**). Strong anisotropism of lo and Bi, faint (anomalous) anisotropism of mg

Fig. 74.4 Aggregate of lo, sf, rm, sk. The reflectance differences (rm > sk > lo > sf) allow us to distinguish these species, all of them "white" in colour, by the decreasing brightness from rm to sf; lo is also distinguished by its pleochroism in bluish and yellowish tones and occurs in roughly radiating prismatic aggregates in the centre of the image (included in a diagonal band of rm, brighter), which are zoned, corresponding to lo composition in the periphery and sf (somewhat darker) in the core; sk occupies the bottom right corner. Sample HJB-rub404, Bou Azzer, Morocco (**dsp**)

Fig. 74.5 Same scene as lo7 (**d +p**). Compositional zoning of lo, enhanced by the progressive decrease of anisotropism towards the centre; duller polarization colours, corresponding to higher Co contents (reaching in the core sf compositions, confirmed by EMPA); rm anisotropic and with abundant polysynthetic twins; sk isotropic (dark)

Fig. 74.6 Aggregate of lo, Bi, bm, q. In contact with Bi (highly reflective), lo appears less bright (grey tones), unlike the lo of Fig. 74.1a and b, which appears bright white in contrast to mg, dark. Scarce bm (grey, R < lo, anisotropic) in the centre. HJBrub1070, Hüttenberger Erzrevier, Lölling, Schacht Hauptlager, Kärnten, Austria (**osp**)

Fig. 74.7 Same scene as lo9 (**o +p**)

74.3 Spectral Information

74.3.1 Löllingite Multispectral Reflectance R (in Air, Without Polariser) (%)

λ (nm)	370	400	425	450	475	500	525	550	575	600	625	650	675	700	750	800	850	900	950	1000
μ	51.8	52.6	53.1	53.1	53.6	53.4	53.6	53.3	53.3	53.0	52.7	52.4	51.9	51.7	51.0	50.6	50.3	50.3	50.3	50.1
P_5	50.3	51.0	51.8	51.8	52.5	52.2	52.7	52.3	52.3	52.1	51.8	51.6	51.2	51.0	50.3	49.9	49.5	49.6	49.5	49.4
P_{20}	51.0	51.8	52.5	52.4	53.2	52.8	53.2	52.8	52.8	52.6	52.3	52.1	51.6	51.3	50.6	50.3	49.9	49.9	49.8	49.6
P_{80}	52.7	53.6	53.9	54.1	54.3	54.3	54.3	54.0	53.9	53.7	53.3	53.0	52.4	52.1	51.4	51.2	50.9	50.9	50.8	50.7
P_{95}	53.8	54.5	54.7	54.7	54.9	54.7	54.6	54.3	54.3	54.1	53.8	53.5	52.9	52.6	51.9	51.5	51.4	51.3	51.4	51.1

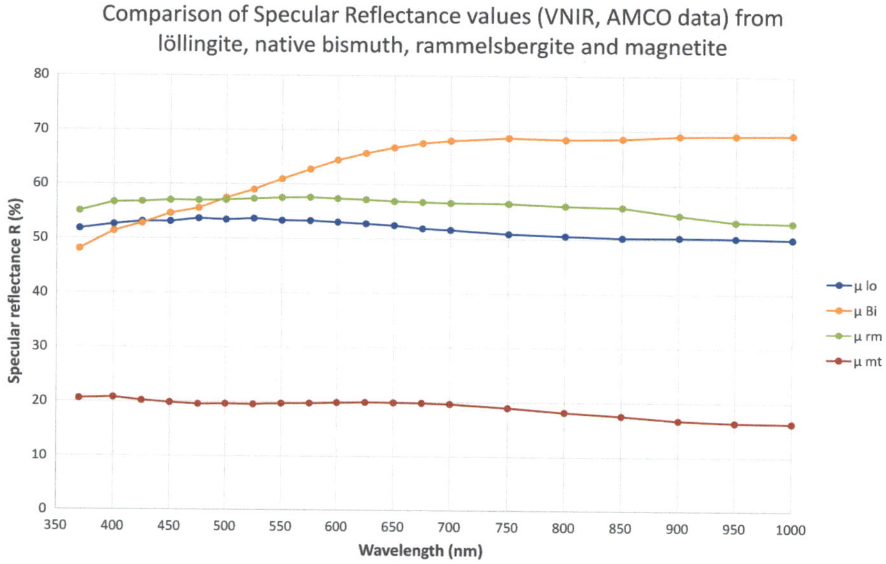

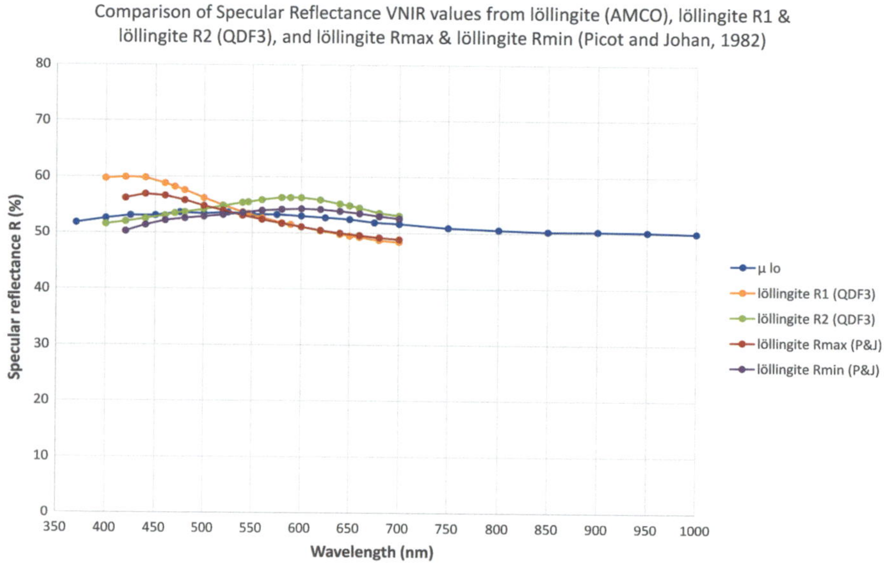

Reference

Ramdohr P (1980) In: The ore minerals and their intergrowths. vol 2, 2nd edn. Pergamon Press, Oxford, pp 1205

Luzonite (*lz*)

75

Cu_3AsS_4: Tetragonal

Luzonite		R_{546} (%)		Y (%)	λ_d (nm)	P_e (%)	Colour sensation
Air	R_1	24.6		25.1	598	3.7	Pinkish brown
	R_2	27.4		27.6	586	5.7	Brown Yl-Or
Oil	R_1	12.1		12.7	595	6.4	Purple-brown
	R_2	14.0		14.1	583	9.4	Brown Pi-Or clear
Bir: Weak (3)					*Pleoc*: Weak/net (3)		*Tarn*: No
Anis +P: Net and coloured Gn (4)					*IR*: No		*Twg*: Polysynthetic constant laminar
Mor: Massive, allotriomorphic					*Clv*: Not visible		*PolQ*: Good
Hardness	*PolH*: Medium, ≈ en (enargite) (3)				*Mohs*: 3.5	*Talm*: D	*VHN* $_{(100\ p)}$: ≈350–400

Obs: cf. Table and R charts, multispectral data (VNIR, AMCO) and visible range (QDF3; Picot and Johan 1982)
Mineral abbreviations: cc—chalcocite, ccp—chalcopyrite, cup—cuprite, cv—covellite, en—enargite, gg—gangue, gn—galena, lz—luzonite, po—pyrrhotite, py—pyrite, sp—sphalerite, tn—tennantite, td—tetrahedrite

75.1 General

Dimorph of enargite (*en*, orthorhombic), of composition Cu_3AsS_4, ***lz*** may contain variable amounts of Sb and minor amounts of Sn. Frequently associated with enargite, which tends to replace it by paramorphic alteration.

As in the case of enargite, it can also alter to tennantite, chalcocite, covellite, etc.

75.2 Microscopic Description

75.2.1 SP Observation (*Single Polariser*)

Light *pinkish-brown colour*, with yellowish, orange or purplish shades, depending on the orientation. The colour perception may vary according to the contiguous ores (e.g., it looks darker and browner next to pyrite, Fig. 75.1b). Next to enargite, the enargite is perceived as grey or bluish grey (Fig. 75.1a, b).

Distinctly *pleochroic* with visible colour contrast, especially in oil, but weakly bireflective.

Moderate *reflectance*, close to that of enargite, although with a spectrum not so flat, as it corresponds to its more intense and varied colouration (cf. comparative R graphs).

Very good *polishing* quality, no noticeable cleavage. Medium *polishing hardness*, very similar to enargite, between td and sp ($\geq$ td, > cup, bn, gn, cc; $\leq$ sp, < po, py).

It is often intergrown with enargite, which tends to replace it.

75.2.2 Observation +P (*Polariser + Analyser at 90°*)

Clearly *anisotropic*, with characteristic luminous, greenish tones (*e.g.*, green or bluish-green to yellowish-green hues). The anisotropism is even more remarkable than that of en, with higher contrast, due to the effect of the twinning (Fig. 75.1c, d).

The extinction is usually oblique to the twin plane.

Lamellar, polysynthetic, omnipresent and eye-catching twinning.

R. Castroviejo, *A Practical Guide to Ore Microscopy—Volume 1*,
https://doi.org/10.1007/978-3-031-12654-3_75

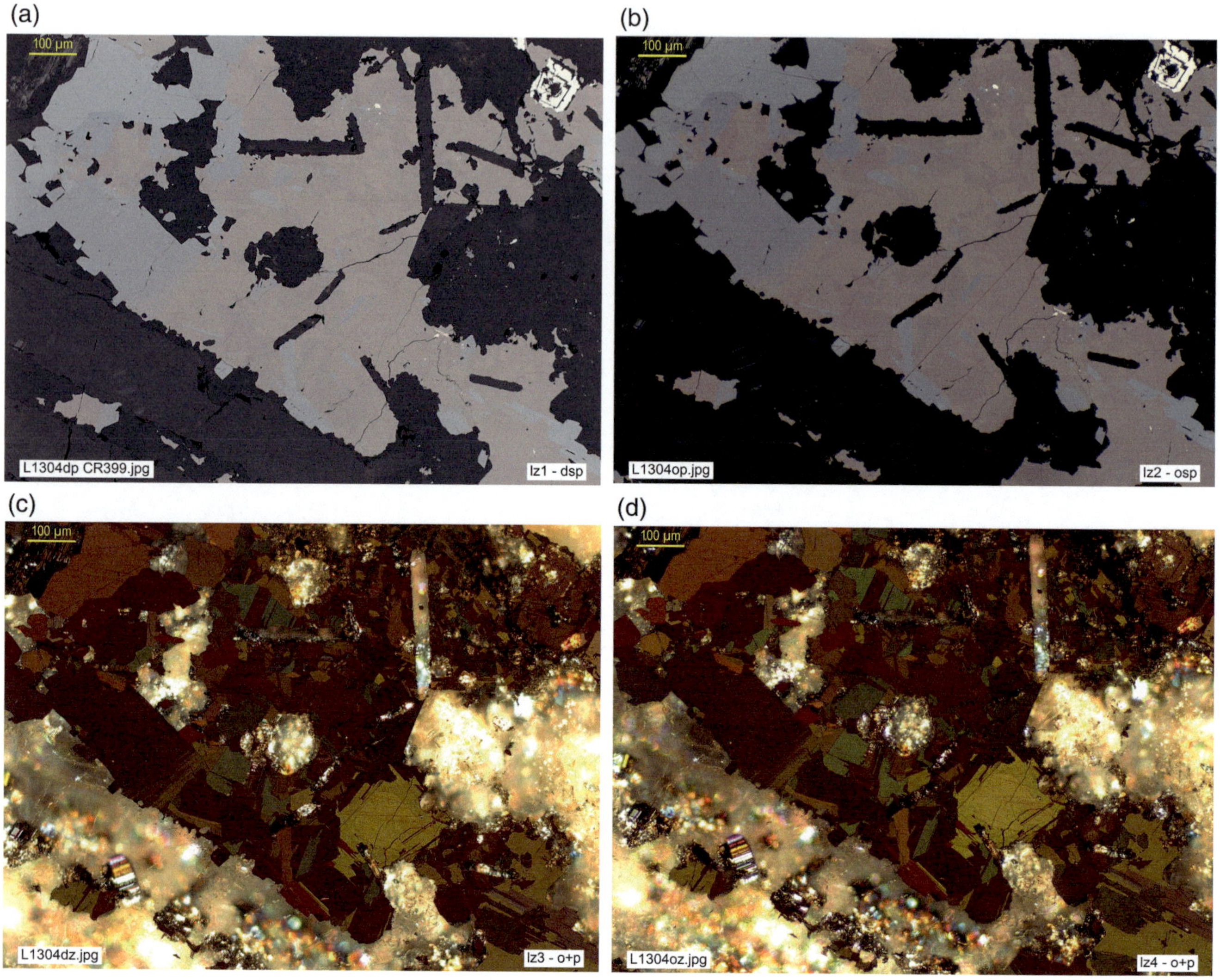

Fig. 75.1 **a, b** (Same scene: **dsp** and **osp**, respectively). Intergrowth of **lz** (brownishpink/orange) and **en** (greyish), both pleochroic, with **q** (dark grey/black) and sparse **py** (yellowishwhite). Note the different colour impression (especially in immersion, **osp**) of lz *vs.* en and lz *vs.* py, orange in the first case and brownish (right upper corner) in the second. CR 399, Lepanto (Luzon, Philippines). **c, d** (Same scene as lz1: **d+p** and **o+p**, respectively). Note the greenish polarisation tints and the constant polysynthetic twinning, both characteristic and clearly differentiating lz from en

75.2.3 Paragenesis

Paragenesis is usually hypogene hydrothermal like enargite, but luzonite is less frequent. It can occur in porphyry copper and epithermal deposits, and it may be altered to tennantite.

Accompanying ores: tennantite, enargite, pyrite, bornite, covellite, digenite, chalcocite.

75.2.4 Identification

Given their association, the greatest risk of confusion is with enargite, which is much more common. They are easily distinguished by the more intense colouration in pink-orange tones and the more striking anisotropism in green shades of luzonite, as well as by the constant presence of polysynthetic twins in lz.

75.3 Spectral Information

75.3.1 Luzonite Multispectral Reflectance R (in Air, Without Polariser) (%)

λ (nm)	370	400	425	450	475	500	525	550	575	600	625	650	675	700	750	800	850	900	950	1000
μ	27.8	27.5	25.9	24.7	24.2	24.8	25.7	26.1	27.1	27.9	28.6	29.2	29.5	29.7	29.2	28.7	28.9	28.6	29.0	27.9
P_5	26.8	26.6	25.3	24.2	23.8	24.4	25.2	25.5	26.2	26.9	27.6	28.4	29.0	29.2	28.8	28.3	28.5	28.0	28.4	27.3
P_{20}	27.1	26.9	25.4	24.4	23.9	24.6	25.4	25.6	26.5	27.2	27.9	28.7	29.2	29.4	29.0	28.4	28.6	28.2	28.7	27.5
P_{80}	28.6	28.2	26.3	24.9	24.4	25.0	26.0	26.6	27.7	28.6	29.3	29.7	29.9	30.0	29.5	28.9	29.3	29.0	29.4	28.3
P_{95}	28.8	28.3	26.5	25.2	24.7	25.4	26.4	26.9	27.9	28.8	29.5	29.9	30.1	30.1	29.7	29.1	29.4	29.2	29.6	28.7

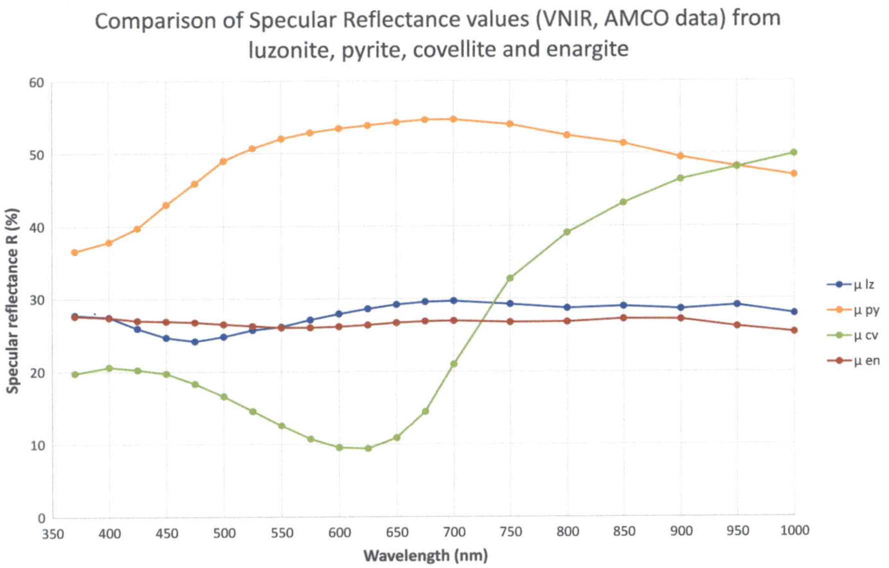

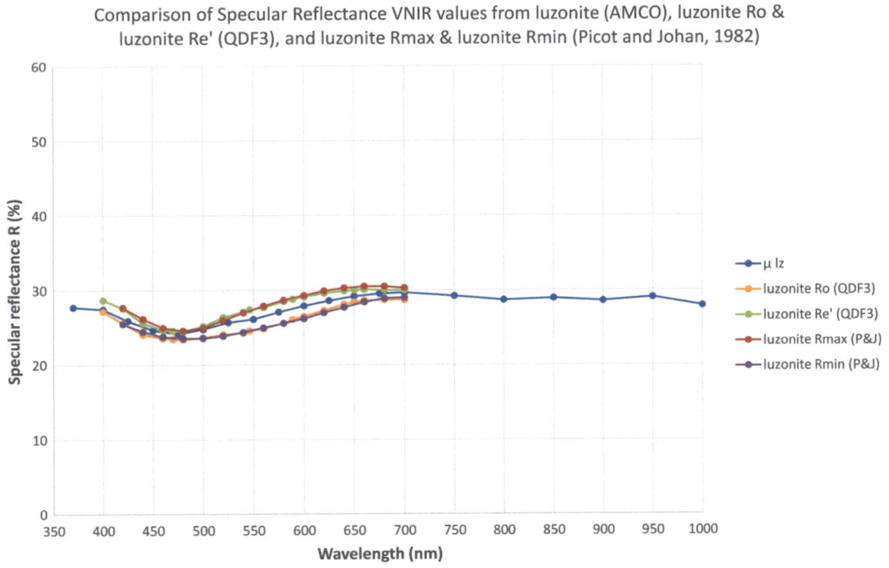

Reference

Picot P, Johan Z (1982) Atlas of ore minerals. B.R.G.M./Elsevier

Mackinawite (*mck*)

76

(Fe, Ni)$_9$S$_8$: Tetragonal

Mackinawita		R_{546} (%)	Y (%)	λ_d (nm)		P_e (%)	Colour sensation
Air	R_o	44.6	44.8	577		10.6	Yellowish white
	$R_{e'}$	24.0	24.3	584		4.1	Grey
Oil	R_o	31.8	32.0	576		12.9	Pinkish-brown to light yellowish
	$R_{e'}$	11.1	11.4	583		8.2	Dark grey
Bir: Strong (5)				*Pleoc*: Strong (5)			*Tarn*: –
Anis +P: Very strong (5)				*IR*: No			*Twg*: Occasional, lamellar
Mor: Tabular. Unmix. lam, flames, spindles				*Clv*: (001)			*PolQ*: Variable
Hardness	*PolH*: Variable (≈3)			*Mohs*: 2½	*Talm*: –		*VHN* $_{(15\ p)}$: 63–89

Note: cf. Table and graphs R: visible range (QDF3 and Picot and Johan 1982)
Mineral abbreviations: asp—arsenopyrite, cbn—cubanite, ccp—chalcopyrite, gg—gangue, gra—graphite, grg—greigite, il—ilmenite, isc—isocubanite, mck—mackinawite, mol—molybdenite, mt—magnetite, pn—pentlandite, po—pyrrhotite, py—pyrite, sp—sphalerite, vll—valleriite

76.1 General

The recognition of the mineral mackinawite, ***mck***, was late and laborious. It was a result of the revision of the older species *valleriite* , *vll*, with which it was initially confused because they were tightly intergrown. Although defined in serpentinites from the *Mackinaw* Mine (*locus typicus*: *l.t.*, Monte Christo district, Washington, USA), its presence has been later confirmed in different environments—low temperature aqueous systems, including sedimentary, hydrothermal and magmatic environments.

Although some authors simplify it as FeS, its composition is variable. Bowles et al. (2011) characterize mackinawite as $Fe_{1+x}S$: *a tetragonal ferrous sulphide deficient in* S. Ramdohr (1980) expresses it as (Fe, Ni, Co)S, with possible Ni contents up to 8% and Co up to 10%, while for other authors[1] it is formulated as $(Fe, Ni)_9S_8$, or $(Fe, Ni)_{1+x}S$ (x = 0–0.11) and, for IMA, $(Fe, Ni)_{1+x}S$ (x = 0–0.07), with Co or Cu as frequent impurities.

76.2 Microscopic Description

76.2.1 SP Observation (*Single Polariser*)

White *colour* (pinkish white or yellowish to grey), with strong *bireflectance* and intense and characteristic *pleochroism*, ranging from a light yellowish or pinkish brown (R_o) reminiscent of pyrrhotite, to a dark grey ($R_{e'}$) close to sphalerite (Figs. 76.1a, b, 76.2 and 76.3).

Reflectance, R, extremely variable according to orientation (comp. Figs. 76.2 and 76.3) and dependent also on composition, although data vary according to sources. R seems to increase with Ni content (*comp.* graphs QDF3, below), but according to Picot and Johan (1982) it is higher in cobaltiferous varieties. In general, R can be said to lie between pentlandite and sphalerite ($R_{mck} < R_{pn}$ and $>R_{sp}$). In the lightest position (R_o), R_{mck} approaches pentlandite and is $>R_{ccp\ \&\ po}$, while in the darkest one ($R_{e'}$), $R_{mck} \leq R_{mt}$ but $\geq R_{sp}$ (cf. graphs R, QDF3 and Picot and Johan 1982, visible range). Finally, R can still be strongly lowered in particular sections by poor polishing.

[1] cf. https://www.mindat.org/min-2512.html, *visit. 2021/03/05.*

R. Castroviejo, *A Practical Guide to Ore Microscopy—Volume 1*,
https://doi.org/10.1007/978-3-031-12654-3_76

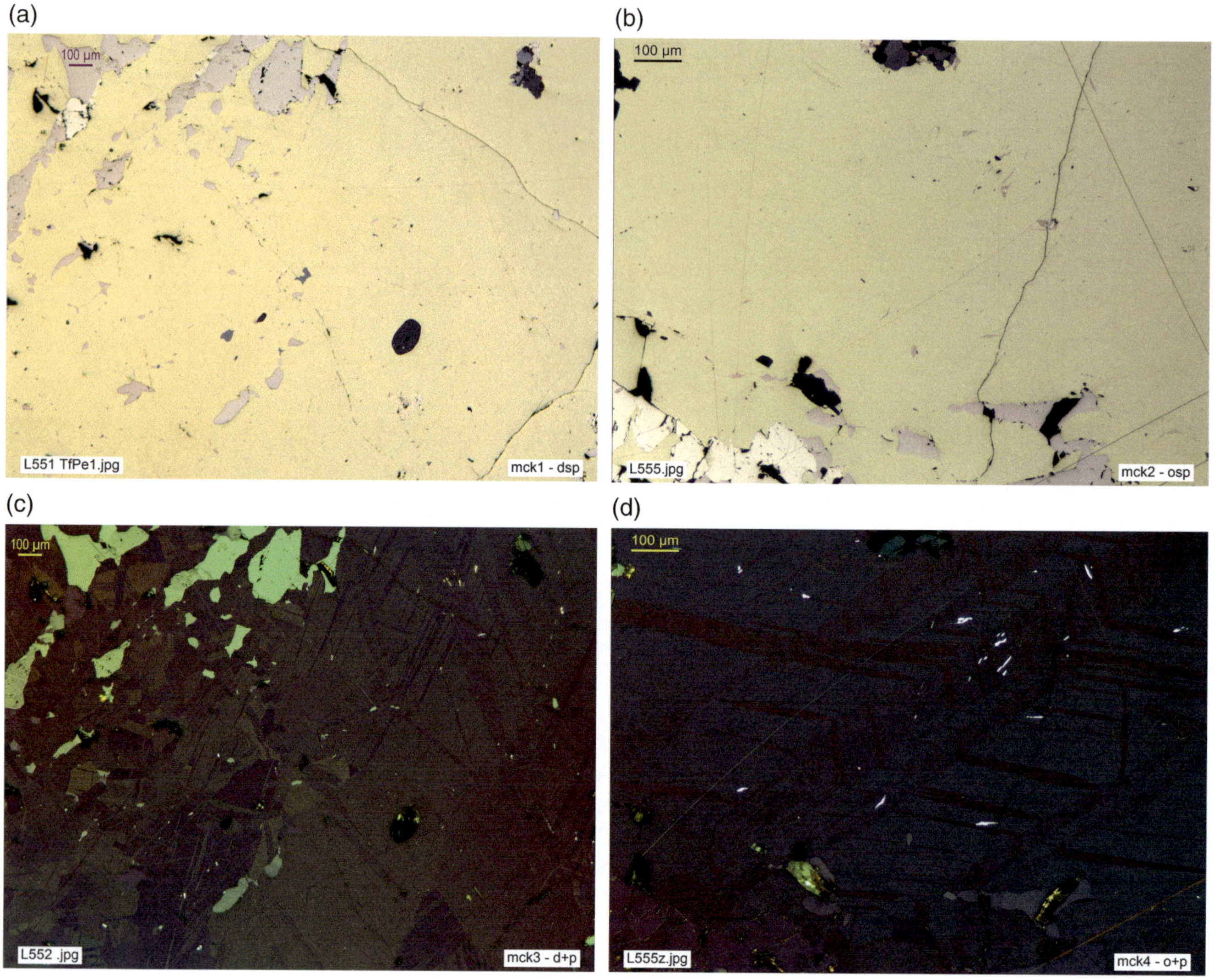

Fig. 76.1 **a** Massive ccp ore (yellow), including po (pinkish brown) with associated pn (white), as well as minute particles of sp (grey, sparse, visible ≈ center image), il (brownish grey, top right corner) and mck. In this ore, mck occurs in numerous lamellae or very fine particles (≤ 10 µm) scattered in the ccp mass, which, depending on orientation, can be confused with ccp or po (comp. Fig. 76.1c). Sample TF_Pe1, Cobalt Mine, Ontario, Canada (**dsp**). **b** Massive ccp ore with accessory pn, po, gg and il (top center) in inclusions. Tiny lamellae of mck, scattered in ccp which, by their strong pleochroism, might be confused (if // one of the main R directions) with ccp or po (R_o), and even with mt or sp ($R_{e'}$), but they are immediately identified with analyser by their unmistakable anisotropism (comp. Fig. 76.1d, **o+p**). Sample TF_Pe1, Cobalt Mine, Ontario, Canada (**osp**). **c** Same scene as Fig. 76.1a (**d+p**). The anisotropism, visible in ccp (twinned), po and il, is extremely intense in mck, which now appears as numerous very bright white (or faintly coloured) lamellae, evident against the dark background of ccp (comp. Fig. 76.1a). **d** Same scene as Fig. 76.1c (**o+p**). Striking anisotropism of mck (very bright white lamellae), contrasting with ccp (twinned), po, il and gg

Polishing quality depends on orientation and grain size. Good in minute inclusions in chalcopyrite, but it can be very bad on basal sections of coarse aggregates.

It is usually finegrained, e.g., as platelets or spindles included in chalcopyrite, pentlandite or isocubanite, possibly due to unmixing or decay of non-stoichiometric sulphides. It may show tabular or lamellar habit but seldom massive aggregates. These are difficult to polish.

76.2.2 Observation +P (*Polariser* + *Analyser at 90°*)

Its *extreme anisotropism* (Figs. 76.1c, d and 76.4) is very characteristic, comparable in its intensity to that of graphite or molybdenite, although, unlike these, it has little or no colouring (bright white tones dominate, striking for their luminosity).

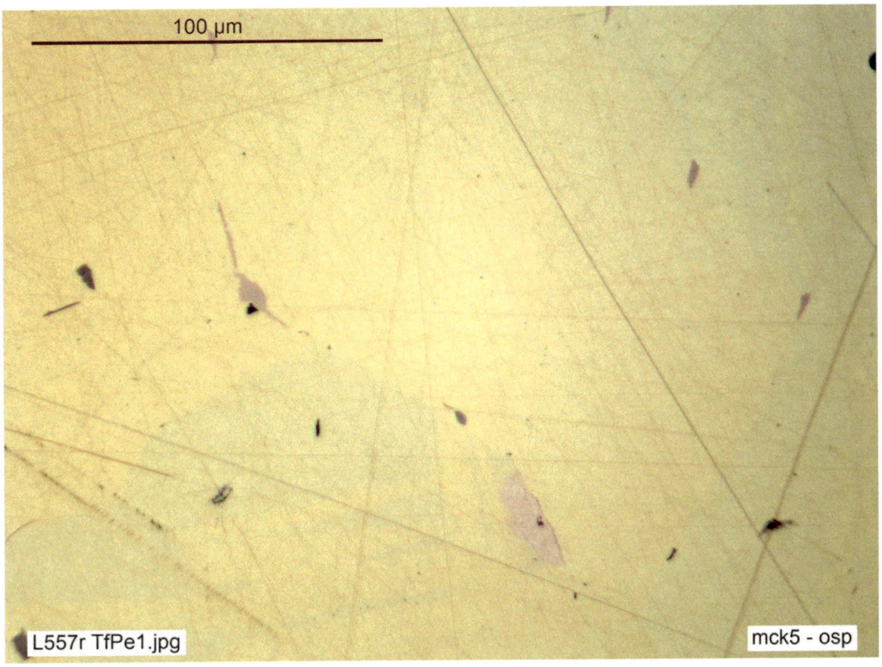

Fig. 76.2 Massive ccp vein fill, with small inclusions of mck and po. Towards bottom center-right, well visible inclusion of mck, oriented after (R_o). It looks like po, although mck is somewhat lighter. Sample TF_Pe1, Cobalt Mine, Ontario, Canada (**osp**)

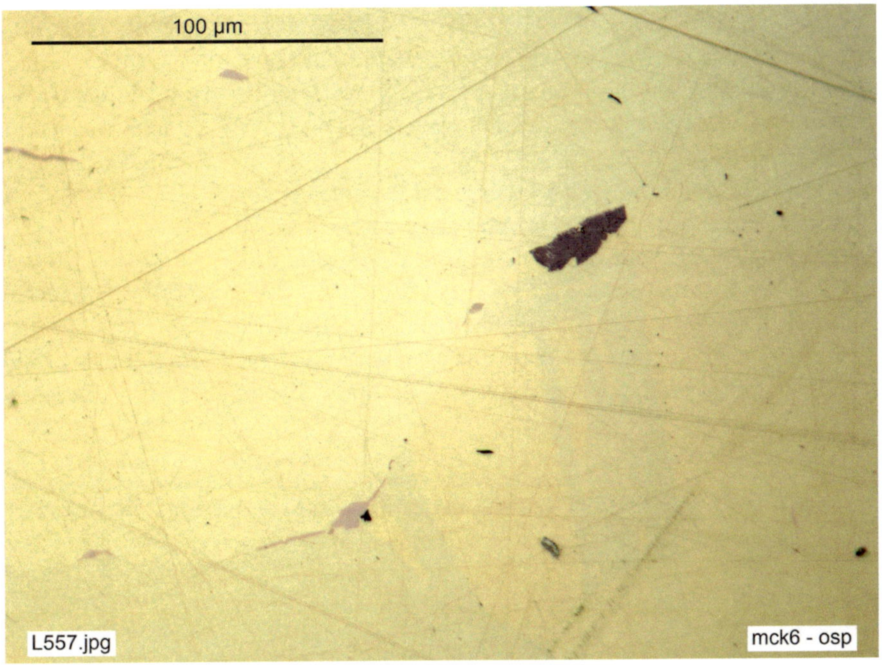

Fig. 76.3 Same scene as Fig. 76.2 (**osp**), but with the stage rotated 90° counterclockwise, so that mck is now oriented after ($R_{e'}$).It looks dark grey and resembles sp, although sp is somewhat darker

76.2.3 Paragenesis

Paragenesis typical of altered serpentinites (*l.t.*, *Mackinaw* Mine, *Monte Cristo* district, Washington, USA). However, mackinawite occurs most often in high-temperature or magmatic Cu–Ni deposits, associated with chalcopyrite and pentlandite as a secondary product, according to Ramdohr (1982), due to the decomposition of high-temperature forms of these.

However, its formation has also been demonstrated in sediments and in low-temperature aqueous systems as a precursor, together with greigite, of pyrite or pyrrhotite. Under these conditions, the transformation of mackinawite to pyrite does not occur purely in the solid state; it is apparently activated by diagenesis, through preliminary dissolution of mackinawite by diagenetic aqueous fluids (cf. Bowles et al. 2011). It is also found in sulphides in volcanic rocks and in hydrothermal vents in marine sediments. It occurs as well in some meteorites.

Usual companions: pyrrhotite, chalcopyrite, pentlandite, cubanite, cobaltite, isocubanite, magnetite, arsenopyrite.

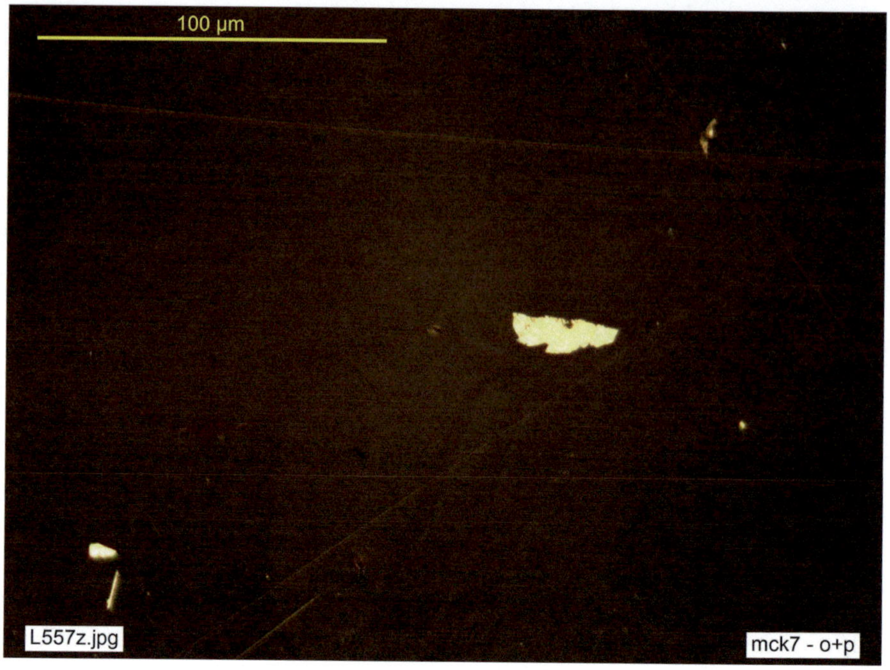

Fig. 76.4 Same scene as Fig. 76.2, rotated 45° from the extinction position of mck (**o+p**). The extreme anisotropism of mck results in such a brightness at this position that it oversaturates the camera

76.2.4 Identification

The characteristic optical properties (extreme bireflectance, pleochroism and anisotropism) make mackinawite easily recognizable, also taking into account paragenesis.

- It might be confused with valleriite, but the latter is distinctly darker (less reflective) and its pleochroism and anisotropism are also lower; moreover, the anisotropism of valleriite is more coloured.
- It differs from graphite because the minimum reflectance of graphite (R'_e) is much lower, similar to that of gangue (nil in immersion: gra looks black//R'_e), while that of mackinawite is notably higher (between those of sphalerite and magnetite); moreover, the anisotropism of graphite is more coloured.
- It differs from molybdenite because, apart from the paragenesis, the maximum reflectance of the latter is slightly higher and its anisotropism is slightly more coloured.

76.3 Spectral Information

76.3.1 Mackinawite Reflectance R (in Air, Visible Spectrum) (%)

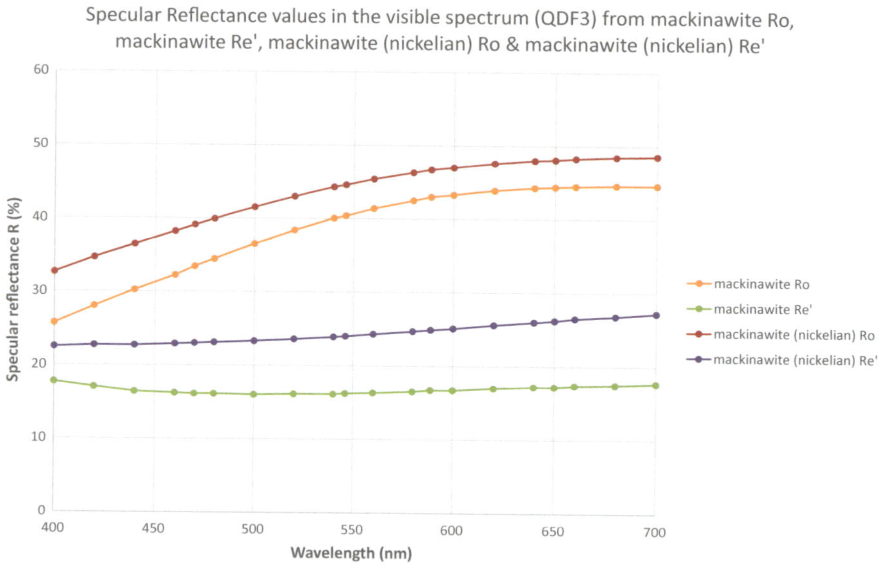

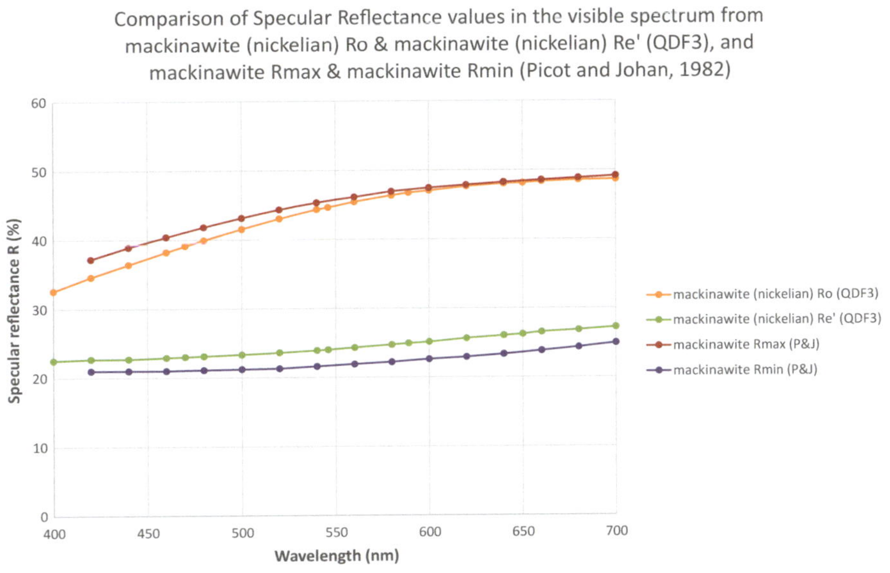
Comparison of Specular Reflectance values in the visible spectrum from mackinawite (nickelian) Ro & mackinawite (nickelian) Re' (QDF3), and mackinawite Rmax & mackinawite Rmin (Picot and Johan, 1982)
Specular reflectance R (%)
Wavelength (nm)
mackinawite (nickelian) Ro (QDF3)
mackinawite (nickelian) Re' (QDF3)
mackinawite Rmax (P&J)
mackinawite Rmin (P&J)

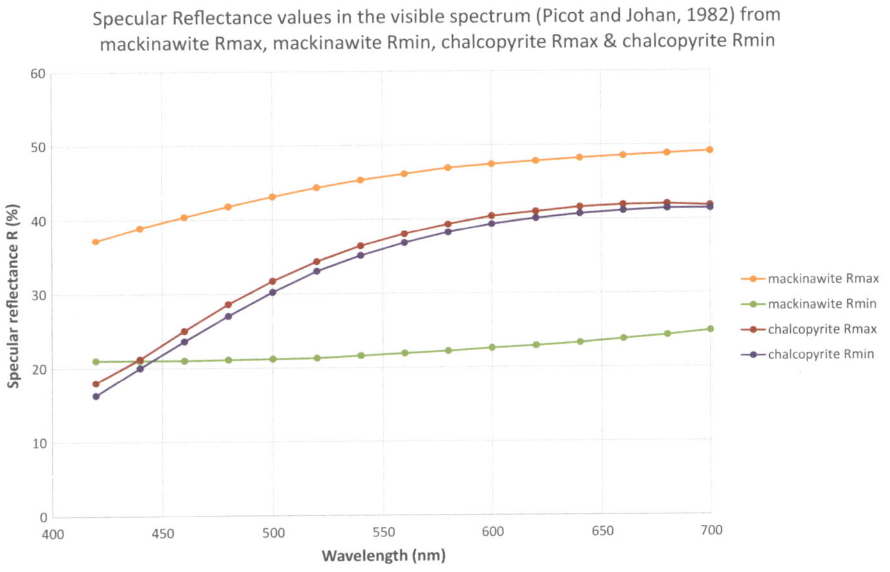
Specular Reflectance values in the visible spectrum (Picot and Johan, 1982) from mackinawite Rmax, mackinawite Rmin, chalcopyrite Rmax & chalcopyrite Rmin
Specular reflectance R (%)
Wavelength (nm)
mackinawite Rmax
mackinawite Rmin
chalcopyrite Rmax
chalcopyrite Rmin

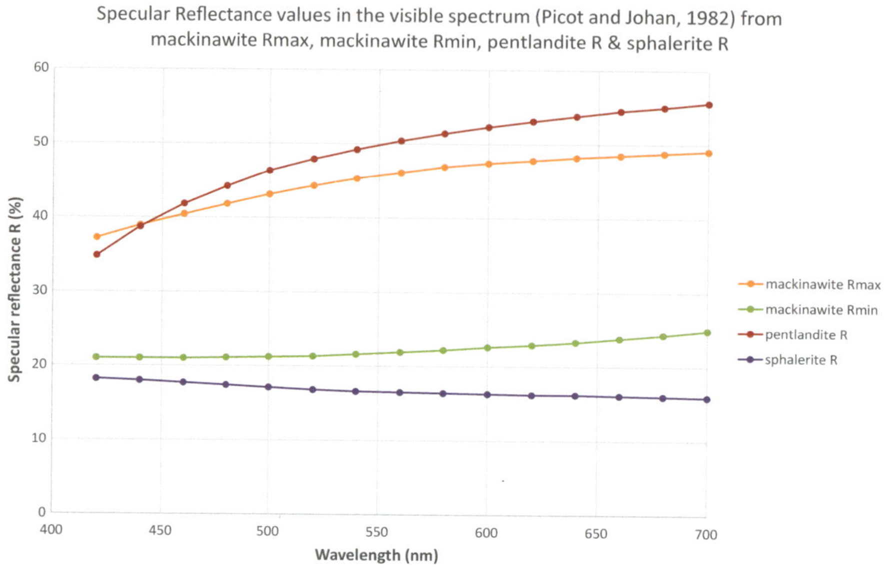
Specular Reflectance values in the visible spectrum (Picot and Johan, 1982) from mackinawite Rmax, mackinawite Rmin, pentlandite R & sphalerite R
Specular reflectance R (%)
Wavelength (nm)
mackinawite Rmax
mackinawite Rmin
pentlandite R
sphalerite R

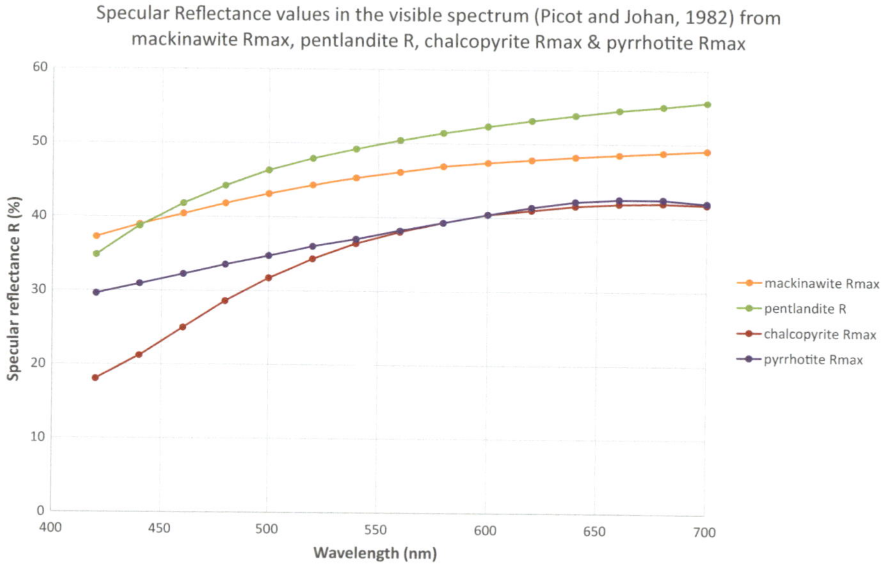
Specular Reflectance values in the visible spectrum (Picot and Johan, 1982) from mackinawite Rmax, pentlandite R, chalcopyrite Rmax & pyrrhotite Rmax
Specular reflectance R (%)
Wavelength (nm)
mackinawite Rmax
pentlandite R
chalcopyrite Rmax
pyrrhotite Rmax

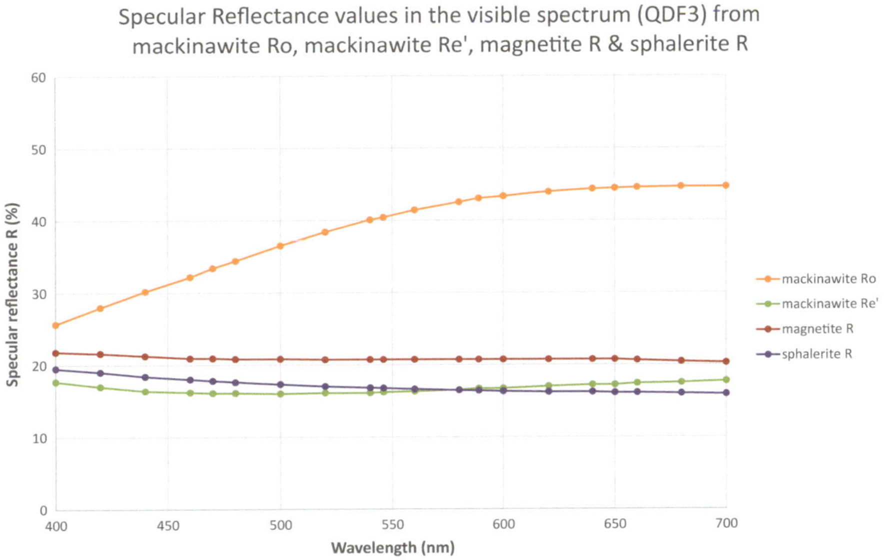

References

Bowles JFW, Howie RA, Vaughan DJ, Zussman J (2011) Rock-forming minerals, vol. 5A, 2nd edn., Non-silicates: oxides, hydroxides and sulphides. The Geological Society, London, 920p

Picot P, Johan Z (1982) Atlas of ore minerals. B.R.G.M./Elsevier

Ramdohr P (1980) The ore minerals and their intergrowths, 2nd edn. Pergamon Press, Oxford, 2 volumes, 1205p

Maghemite (*mgh*)

77

γ-Fe_2O_3: Cubic

Maghemite		R_{546} (%)	Y (%)	λ_d (nm)	P_e (%)	Colour sensation
Air		24.4	24.1	482	7.6	White to bluish grey
Oil		10.3	10.0	478	16.3	
Bir: Null (1)				*Pleoc*: No (1)		*Tarn*: No
Anis +P: Null (1)				*IR*: Scarce		*Twg*: Not observed
Mor: Pseudomorph of mt				*Clv*: Not observed		*PolQ*: Variable
Hardness	*PolH*: High, >mt, <ht (4)			*Mohs*: 6	*Talm*: nd	*VHN* $_{(50\ g)}$: 412

Note: cf. R charts, visible range data (QDF3)
Mineral abbreviations: gth—goethite, ht—hematite, il—ilmenite, lim—limonite, lpc—lepidocrocite, mgh—maghemite, mt—magnetite, ulv—ulvöspinel

77.1 General

A generally scarce mineral, also known as ferromagnetic ferrioxide (γ-Fe_2O_3), maghemite, ***mgh***, has been redefined by IMA (2018) as $(Fe^{3+}_{0.67}\square_{0.33})Fe^{3+}_2O_4$ and included in the Spinel Subgroup, inside the Oxyspinel Group, within the Spinel Supergroup.[1]

Its magnetic properties make mgh important for science —paleomagnetism—and for industry (e.g., synthetic maghemite for magnetic tapes).

77.2 Microscopic Description

77.2.1 SP Observation (*Single Polariser*)

White to bluish grey, the bluish hue is more visible in oil. Brighter (R>) than magnetite and goethite, but <hematite (Fig. 77.1a, b).

Medium–low *reflectance*, between magnetite and hematite, superior to goethite (cf. comparative R charts) and limonite. Isotropic. *Polishing hardness* slightly higher than magnetite, distinctly lower than hematite.

It usually appears as an oxidation patina on magnetite, which it sometimes completely pseudomorphizes; occasionally massive, rarely in octahedral crystals or acicular incrustations.

77.2.2 Observation +P (*Polariser + Analyser at 90°*)

Isotropic. Very few *internal reflections* (brownish or yellowish-red, Figs. 77.1c, d and 77.2b). These two properties clearly distinguish it from hematite and goethite (both anisotropic and, especially goethite, with abundant IR).

77.2.3 Paragenesis

Maghemite has been described as a metastable form of Fe_2O_3, or, rather, as an Fe-deficient magnetite. By heating, it is transformed into hematite. It is typically formed in superficial environments, associated with limonite, goethite or lepidocrocite, as a secondary mineral of magnetite (Fig. 77.1a, b). It is also found as a supergene or diagenetic

[1] cf. Spinel Description and https://www.mindat.org/min-2533.html.

R. Castroviejo, *A Practical Guide to Ore Microscopy—Volume 1*,
https://doi.org/10.1007/978-3-031-12654-3_77

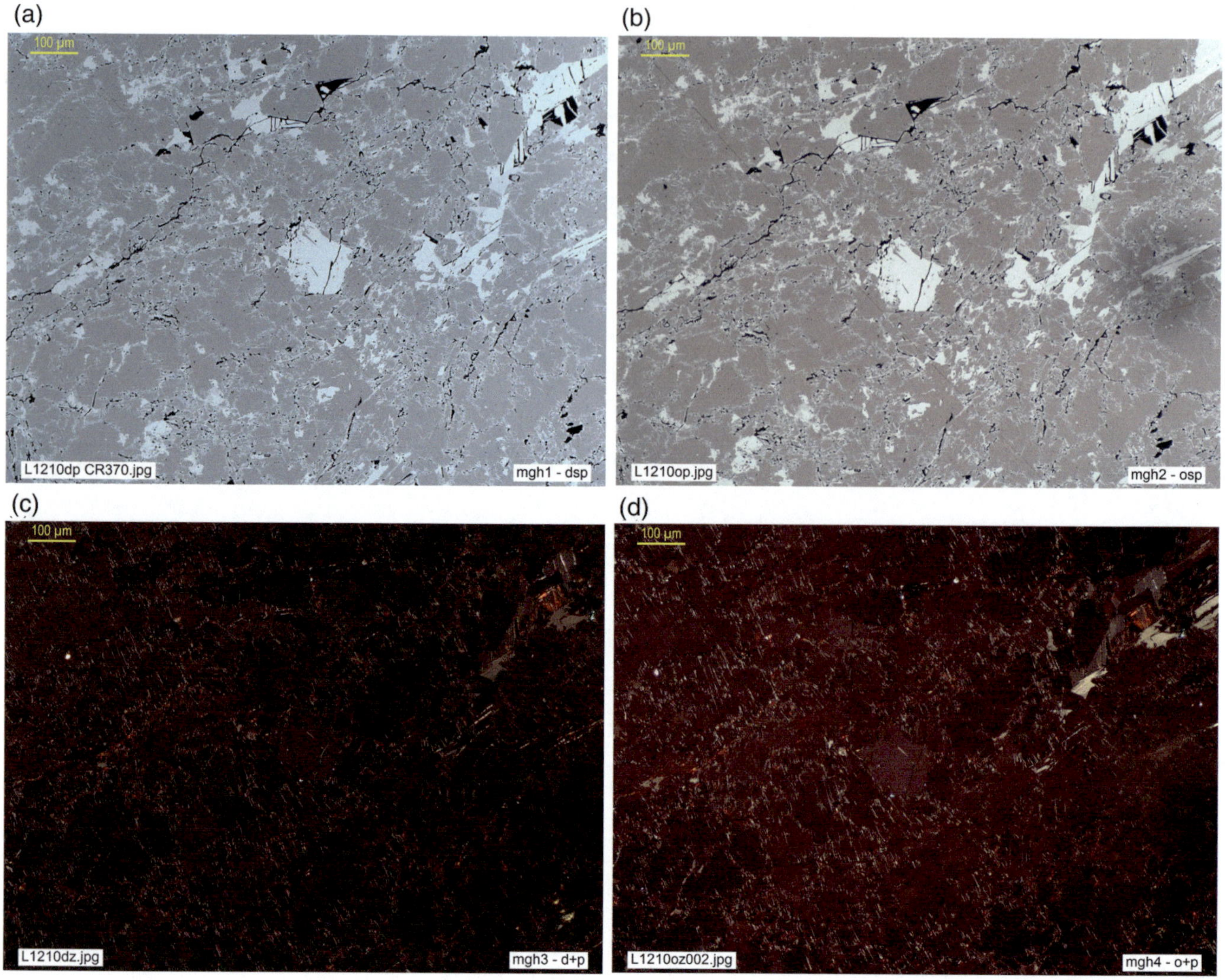

Fig. 77.1 **a, b** (Same scene: **dsp** and **osp**, *respectively*). Inclusions of mgh (greyish white, isotropic, e.g., center of image) in massive aggregate of mt (brownish grey). Incipient mt alteration to ht (white, bireflective, anisotropic, IR): platelets, specks, intergranular films or tabular crystals (top right). CR370, Recanzoni Mine, Alidos, Mexico. **c, d** Same scene as mgh1 (**d+p** and **o+p**, respectively), isotropy and very scarce internal reflections of mgh, unlike ht, which is easily distinguished by its bireflectance, anisotropism and frequent IR

mineral in laterites, in oolitic Fe deposits (Fig. 77.2a, b) and in the cement of sandstones.

The mineral mgh, generally scarce, is relatively frequent as a product of magnetite alteration (from encrustations or thin coatings through pseudomorphs to massive substitution), especially of Ti-magnetite, which is transformed into *titanomaghemite*, a process alternative to martitization and of more irregular morphology than this one.

Associated minerals: magnetite, hematite, goethite, ulvöspinel (from relict inclusions in Ti-magnetite), ilmenite, limonite (advanced oxidation process).

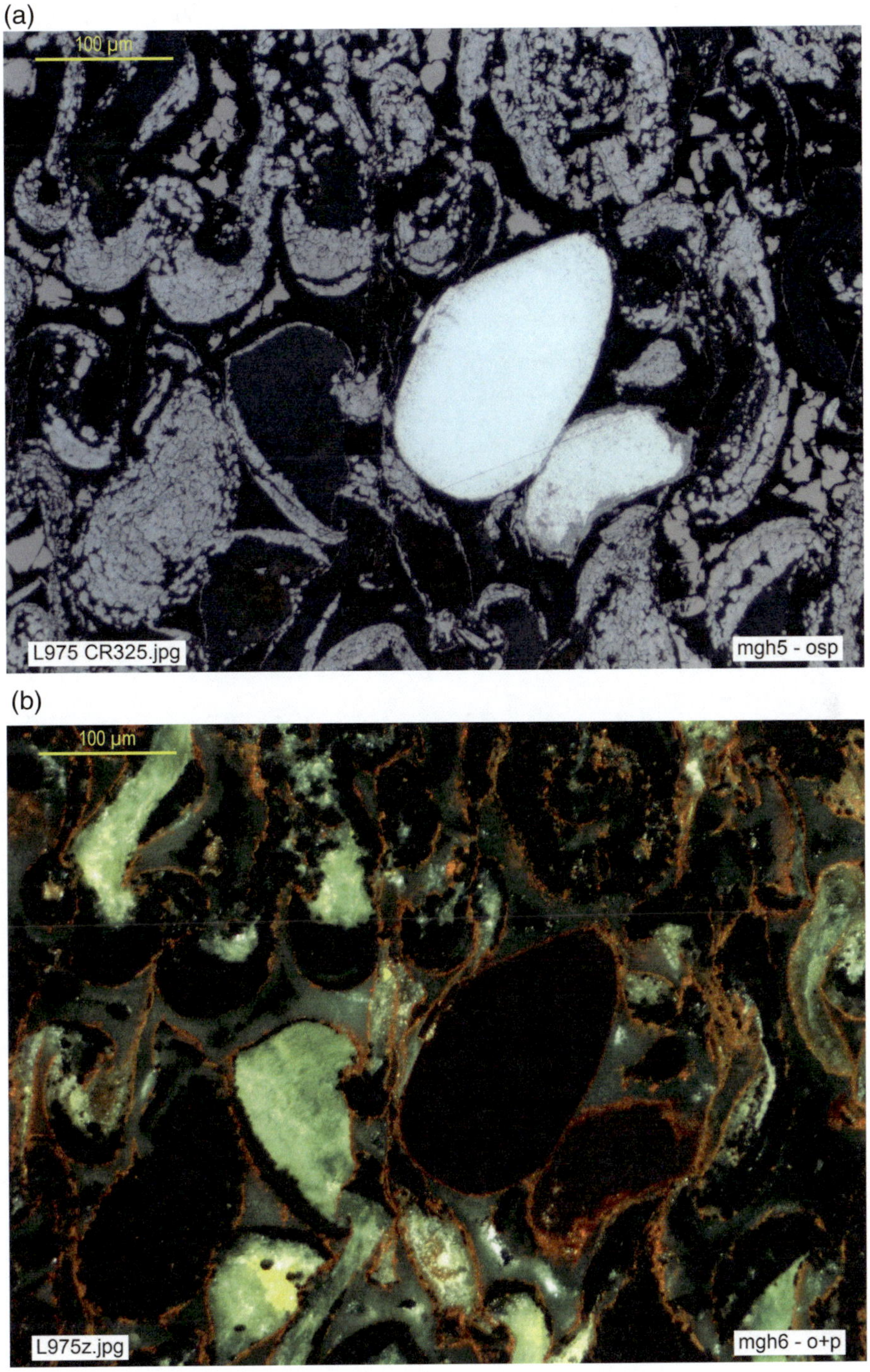

Fig. 77.2 **a, b** (**osp** and **o+p**, *respectively*)—Nodule of mgh (white, center), in oolitic deposit of gth (bluish grey, darker than mgh) intergrown with mt (brownish grey, darker than mgh), interstitial mt, in matrix; marginal alteration of mgh to limonite (gth). Image (**o+p**) shows the isotropic character of mgh (which excludes ht) and the anisotropy and IR of gth. CR325, Gora Djebilet (Tindouf, Algeria)

77.2.4 Identification

Its colour, reflectance, isotropy and scarcity (or absence) of internal reflections, together with its paragenesis and textural relationships as a secondary mineral to magnetite facilitate its identification.

77.3 Spectral Information

77.3.1 Maghemite Reflectance R (in Air, Visible Spectrum QDF3) (%)

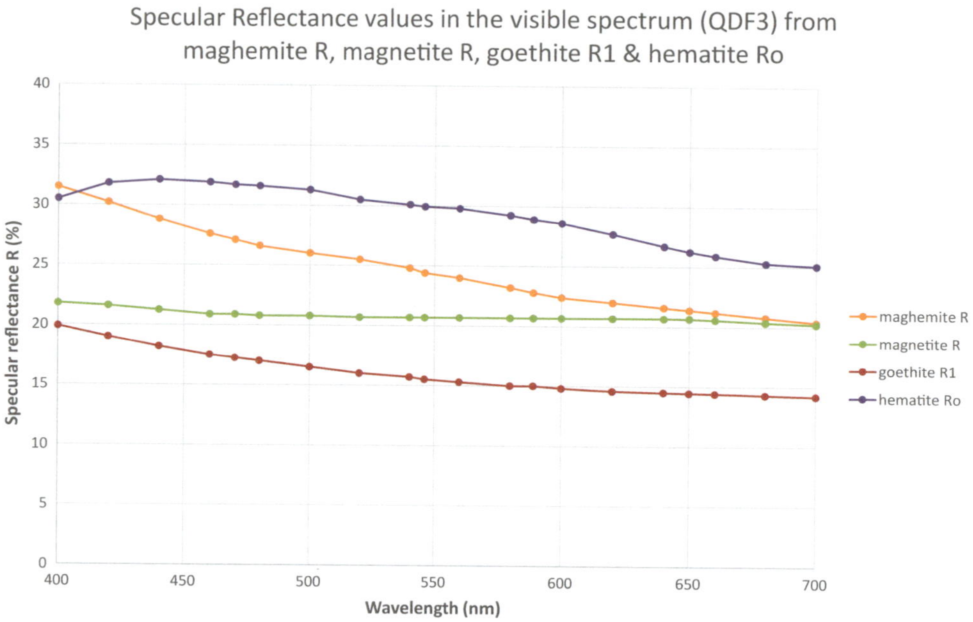

Magnetite (*mt*) 78

$Fe^{2+}Fe^{3+}{}_2O_4$: Cubic

Magnetite		R_{546} (%)	Y (%)	λ_d (nm)	P_e(%)	Colour sensation
Air		19.9	19.9	c560	0.6	Grey to brownish grey
Oil		8.6	8.6	c558	0.7	
Bir: Null (1)				*Pleoc*: No (1)		*Tarn*: No, except for sloppy polishing
Anis +P: Isotropic (or anom. anisotr) (1)				*IR*: No		*Twg*: {111}
Mor: Octahedral, massive, granoblastic				*Clv*: {111}		*PolQ*: Good
Hardness	*PolH*: High (4)			*Mohs*: 6	*Talm*: F	$VHN_{(100\ g)}$: 681−792

Obs: cf. Table and R charts, multispectral data (VNIR, AMCO) and visible range (QDF3)
Mineral abbreviations: ccp—chalcopyrite, chr—chromite, cst—cassiterite, gg—gangue, gn—galena, ht—hematite, il—ilmenite, mgh—maghemite, mt—magnetite, po—pyrrhotite, py—pyrite, q—quartz, sp—sphalerite, spn—spinel, ulv—ulvöspinel, wf—wolframite

78.1 General

Important iron ore. Composition $Fe^{2+}Fe^{3+}{}_2O_4$, with possible substitutions of Fe^{2+} by Mg, Mn, Ni, Zn, Ti and of Fe^{3+} by Cr, Al, Ti, V. With jacobsite, it forms the series $Mn^{2+}Fe^{3+}{}_2O_4$–$Fe^{2+}Fe^{3+}{}_2O_4$.With magnesioferrite, it forms the series $MgFe^{3+}{}_2O_4$–$Fe^{2+}Fe^{3+}{}_2O_4$. Within the Spinel Supergroup and the Oxyspinel Group, it is found in the Spinel Subgroup.[1]

78.2 Microscopic Description

78.2.1 SP Observation (*Single Polariser*)

Grey *colour*, with a slight *brownish* tinge that tends to be more accentuated in samples rich in Ti (*titanomagnetites*). Not pleochroic or bireflective (Figs. 78.1a, b).

Reflectance **R** typical of oxides—low. However, R mt is slightly > R il (Figs. 78.3, 78.4a and 78.7), clearly >R chr, cs, spn and q or gangue (Figs. 78.3, 78.4a, 78.5 and 78.6). R mt <R ht (Figs. 78.2 and 78.7) and distinctly <R sulphides (Figs. 78.1a, b and 78.5), with exceptions such as sphalerite, whose R is ≈ R mt, but slightly lower (cf multispectral comparative plots). Immersion markedly accentuates the contrast and is an effective aid in distinguishing sphalerite and magnetite.

Magnetite polishes well, except in altered samples or samples with many inclusions. Although it does not usually tarnish, if the cloth has become too dry during polishing, the process will be defective and a bluish film or oxidation patina will form in the surface. It does not usually show *cleavage*, but sometimes polishing reveals it, especially {111}. High *polishing hardness*: PolH mt >po, >sp; but < wf, il, chr, ht and, of course, <<py, q, cs, spn.

Magnetite occurs disseminated or, more typically, in massive aggregates, sometimes banded (BIF). It shows a marked tendency to automorphic development (typically but not exclusively octahedral forms) and, especially in skarn formations, growth zoning (Figs. 78.1a, b, sometimes enhanced by alteration. Disseminated in gabbros or basalts, magnetite develops peculiar morphologies, with skeletal or graphic textures and Widmanstätten-type structures (cf. Figs. 1.1 and 1.2, in Chapter 1, vol. 2). As an early mineral, it frequently appears deformed and it is also often recrystallized.

[1] *cf.* Spinel description, and https://www.mindat.org/min-2538.html.

R. Castroviejo, *A Practical Guide to Ore Microscopy—Volume 1*,
https://doi.org/10.1007/978-3-031-12654-3_78

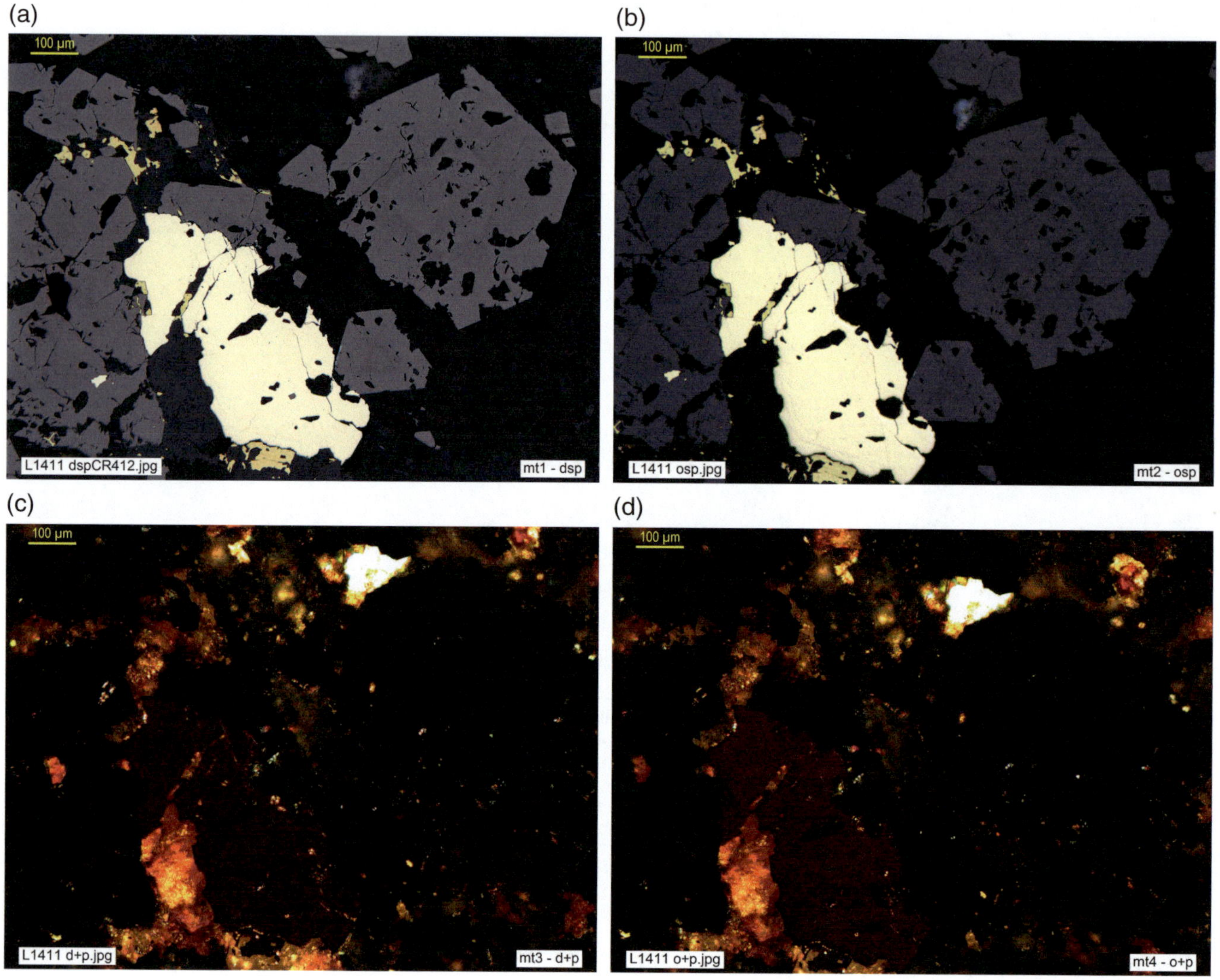

Fig. 78.1 **a** Ore of mt (grey), py and ccp, in *skarn* rock (dark background). Automorphic tendency shows up in the idio-poikiloblastic texture of mt, showing faint zoning (top right). CR412, Mina Cala, Badajoz (**dsp**). **b** Same scene as Fig. 78.1 (**osp**): Note the enhanced contrast between mt and sulphides in immersion. **c**, **d** Same scene as Fig. 78.1 (**d+p** and **o+p**, respectively)

78.2.2 Observation +P (*Polariser + Analyser at 90°*)

Although, as a cubic mineral, magnetite is optically *isotropic* (Figs. 78.1c, d, 78.2b and 78.4b), it is not uncommon for it to show anomalous anisotropism; Ramdohr (1982) chooses to describe it as "approximately isotropic".

It does not show *internal reflections*; however, according to Ramdohr (1982), they could appear in transitional terms to jacobsite or magnesioferrite. *Twinning* (111) may be visible in polished section.

78.2.3 Paragenesis

Varied *parageneses*:

- in intrusive and extrusive magmatic environments (deposits of titanomagnetite, Ti-mt, in gabbros, deposits of the Kiruna type, Sweden, or of the El Laco type, Chile, carbonatites, meteorites),
- sedimentary (*vbgr*. oolitic iron ores, *BIFs* or Banded Iron Formations) and

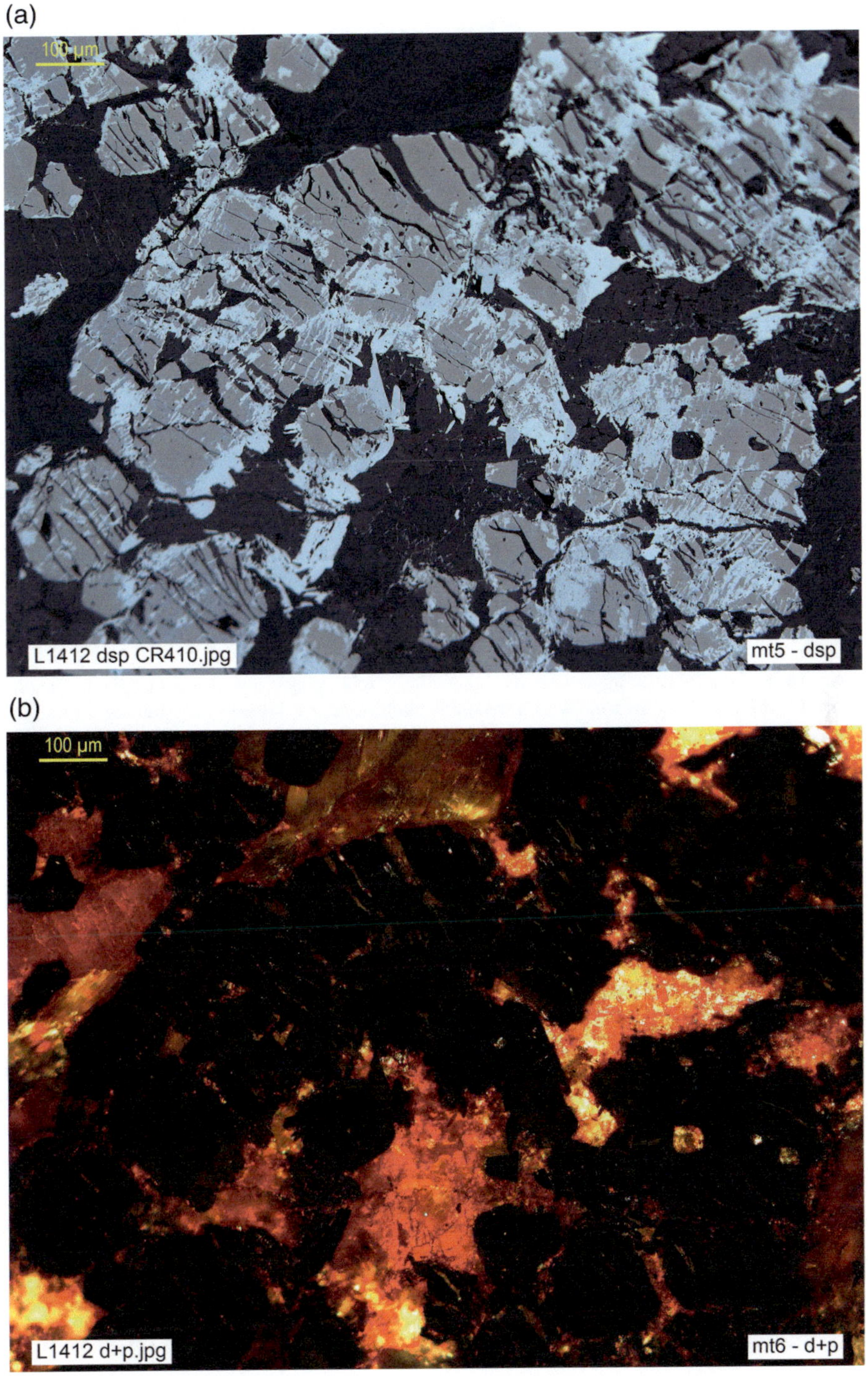

Fig. 78.2 **a**, **b** (Same scene: **dsp** and **d+p**, respectively)—Granoblastic aggregate of mt (brownish grey, isotropic) with incipient alteration to ht (bluish grey, lighter; red internal reflections; its anisotropism is, in this case, masked by the luminosity of the gangue) in calc-silicate hornfels (*skarn*). CR410, Cala Mine, Badajoz

- metamorphic (BIF and contact metasomatic, among others). In some deposits (magmatic Ti-mt, Palabora-type carbonatites), associated with ilmenite, spinel and/or ulvöspinel, with characteristic exsolution textures (Figs. 78.3 and 78.4a, b and 78.7).
- Magnetite is typically altered to *martite* (hematite) by oxidation, initially favored by cleavage, following planes (111) but tending to extend later to the whole crystal (Fig. 78.7). It is also altered to maghemite. It can also be formed as a secondary product, for example by alteration of pyrrhotite, resulting in an aggregate of pyrite +magnetite.
- All this makes possible accompanying minerals numerous, typically being hematite, goethite, maghemite, ilmenite, ulvöspinel, spinel, chromite, pyrite, chalcopyrite.

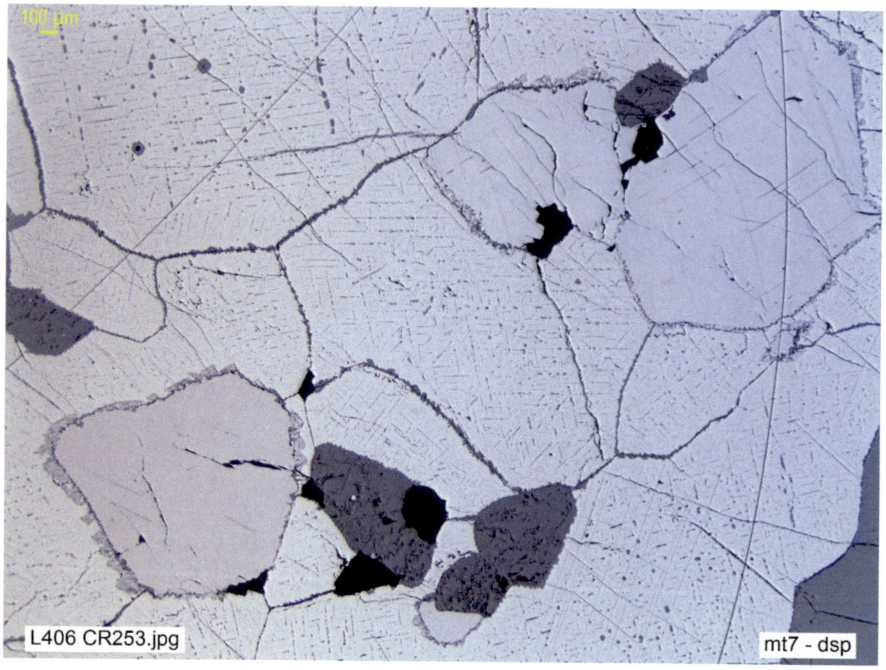

Fig. 78.3 Granoblastic aggregate of magnetite (light grey) and ilmenite (grey/pinkish-brown) with interstitial inclusions of spinel (dark grey). Reaction edges mt-il; mt slightly pinkish due to its Ti content (titano-magnetite), which makes it similar to il, though easily distinguishable (*e.g.*, bottom left) by the il colour, lower reflectance and anisotropism. Exsolutions in mt of 1) spn in isolated idiomorphic crystallites, in intergranular films and in aligned particles according to (100) of mt; 2) il in acicular crystallites according to (111) of mt. CR253, L. Berta, Quebec, Canada (**dsp**)

Along with hematite, magnetite is a very important iron ore.

78.2.4 Identification

Its optical and magnetic properties, together with its high hardness, make magnetite easy to identify. A simple magnet can easily save a lot of time.

However, its inconspicuous appearance under the microscope (low reflectance, grey colour with no or little anisotropism, absence of **IR**-internal reflections) can make it go unnoticed if it is not abundant. In addition, it can be confused with other minerals:

- particularly confused with sphalerite (grey, isotropic, R $\approx$ magnetite), but observation with an immersion lens allows us to clear any doubts, by increasing the contrast of R (sphalerite looks noticeably darker in oil) and making the sphalerite IR more visible and luminous.
- The same happens with chromite (darker than magnetite, IR).
- Paragenesis helps to avoid confusion with braunite (anisotropic) or jacobsite (more yellowish-brown, and red IR).
- Confusion with ilmenite, whose colour and R resemble those of magnetite if the latter is titaniferous (Ti-mt), is easily avoided by observing the clear anisotropism of ilmenite, as well as its pleochroism, well visible in immersion.

Fig. 78.4 **a**, **b** (Same scene: **dsp** and **d+p**, respectively)—Magnetite with two types of ilmenite exsolutions: 1) lamellar, showing as very thin but well visible spindles or acicular crystals (>200 µm long), grey (due to contained Mg/Mn) and distinctly anisotropic (F mt9), preferentially unmixed according to (111) of mt; 2) tiny specks with the usual brownish-pinkish shades of il, distributed throughout the mass of mt, except in the immediate vicinity of type 1 spindles, depleted in Ti by the segregation of the latter. Rub.L. Erz-8.11, Foskor-Mine, Palabora, South Africa

(a)

(b)

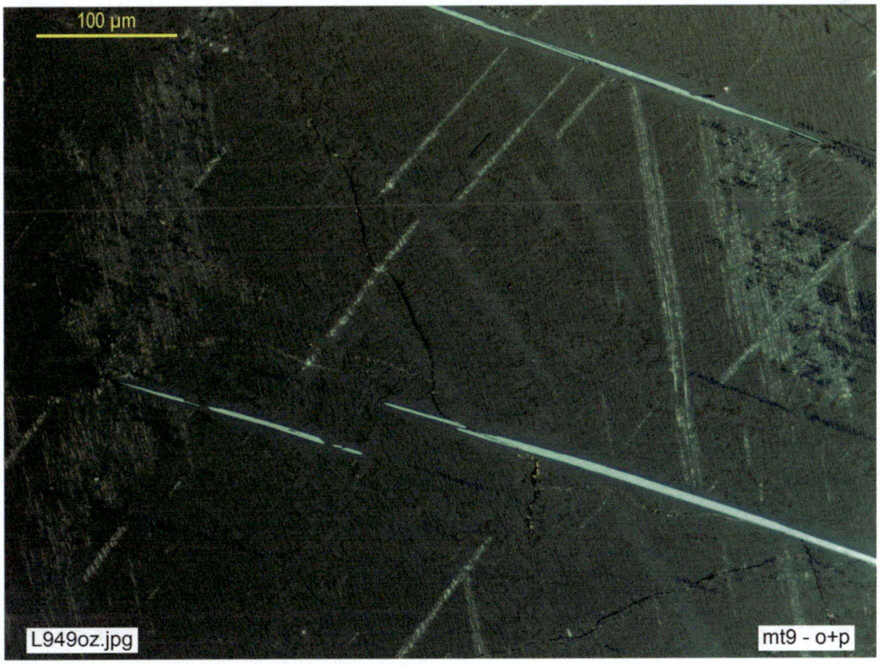

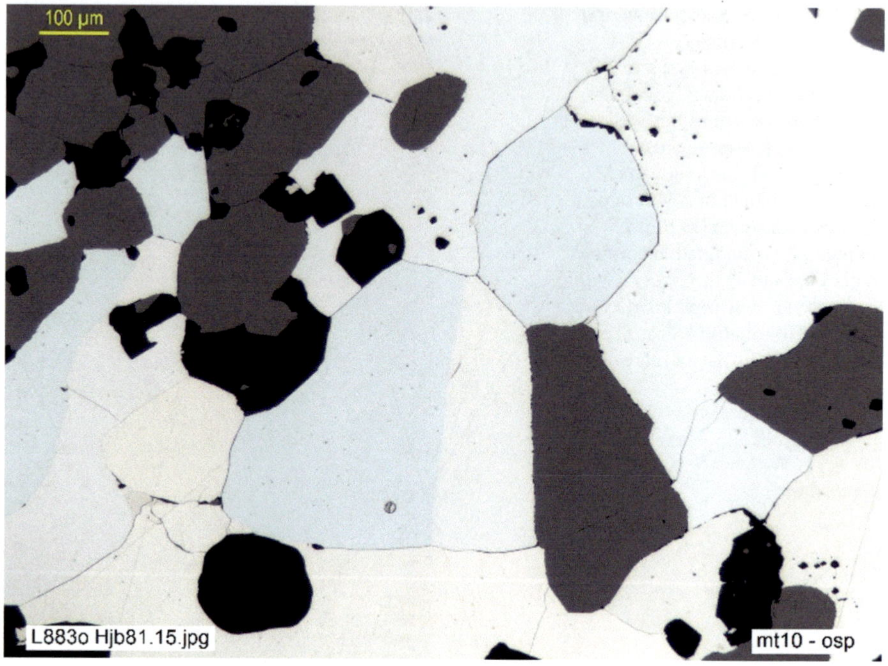

Fig. 78.5 Granoblastic aggregate of löllingite (greyishwhite, pleochroic), magnetite (grey) and quartz (black) Hjb81.15, Burguillos del Cerro, Badajoz (**osp**)

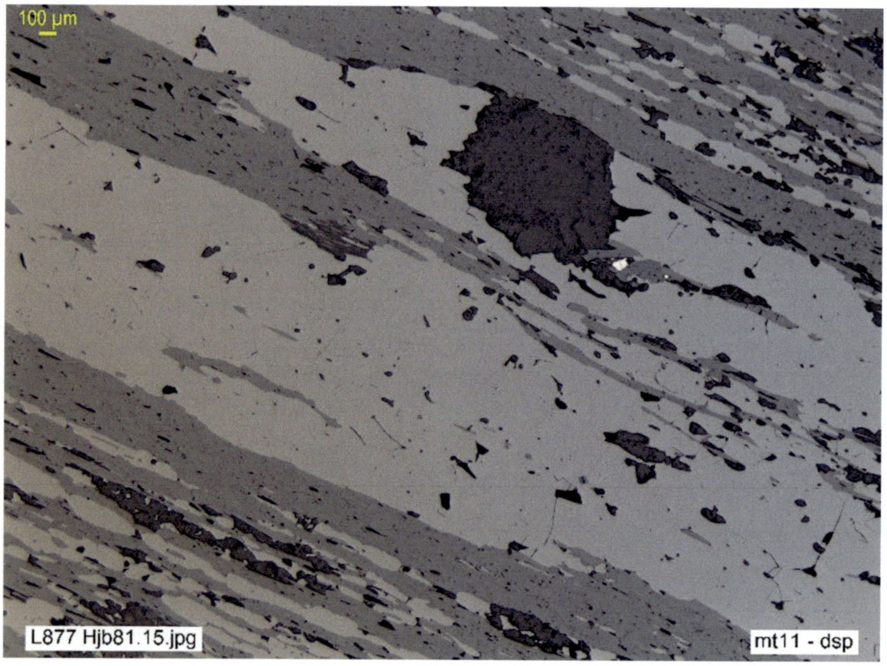

Fig. 78.6 Banded ore of magnetite (brownish grey), vonsenite (grey) and quartz (dark grey), with small inclusions of pyrite (yellowish white). Hjb81.15, Burguillos del Cerro, Badajoz (**dsp**)

Fig. 78.7 Granoblastic ilmenite (pinkish brown, twinned) and magnetite (brownish grey, center) ore, with interstitial spaces occupied by a fine dark grey aggregate (probable secondary product of pyrrhotite). Alteration to ht (light bluish grey), scarce in il, almost complete in mt (*martite*). HD 6858, Agracha, Sahara (**dsp**)

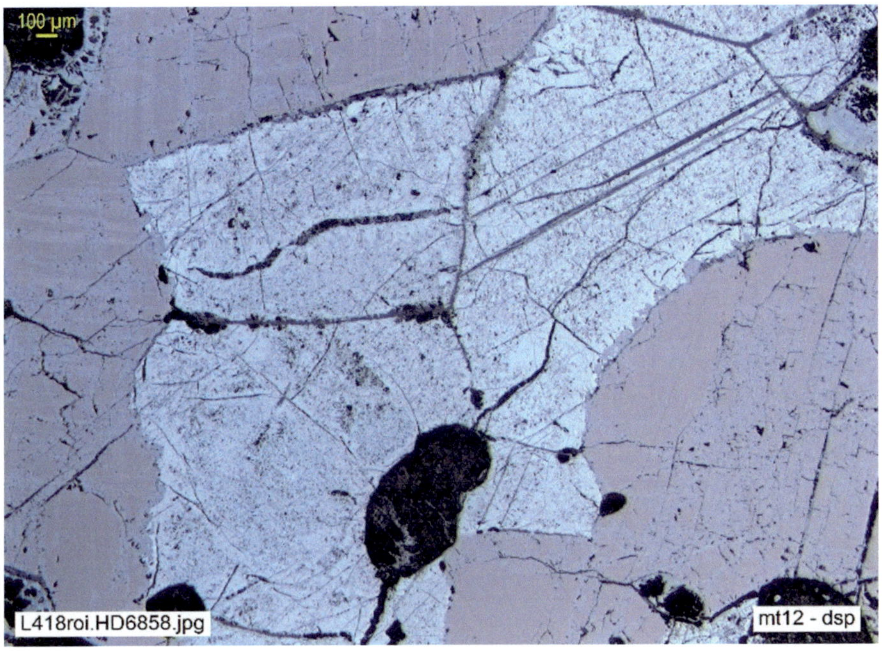

78.3 Spectral Information

78.3.1 Magnetite Multispectral Reflectance R (in Air, Without Polariser) (%)

λ (nm)	370	400	425	450	475	500	525	550	575	600	625	650	675	700	750	800	850	900	950	1000
μ	20.6	20.8	20.2	19.8	19.5	19.6	19.5	19.7	19.7	19.9	19.9	19.9	19.8	19.7	19.0	18.2	17.6	16.8	16.4	16.3
P_5	19.2	19.4	18.9	18.6	18.3	18.4	18.3	18.6	18.5	18.6	18.8	18.7	18.7	18.6	18.0	17.3	16.7	15.9	15.5	15.4
P_{20}	19.9	20.2	19.5	19.2	18.9	19.0	18.9	19.1	19.1	19.3	19.4	19.3	19.2	19.1	18.5	17.7	17.2	16.4	16.1	15.9
P_{80}	21.6	21.6	20.9	20.4	20.1	20.1	20.0	20.2	20.3	20.4	20.4	20.4	20.3	20.1	19.5	18.6	18.0	17.1	16.7	16.7
P_{95}	21.9	21.9	21.2	20.6	20.3	20.2	20.2	20.4	20.4	20.5	20.6	20.6	20.5	20.3	19.6	18.9	18.2	17.4	17.0	17.1

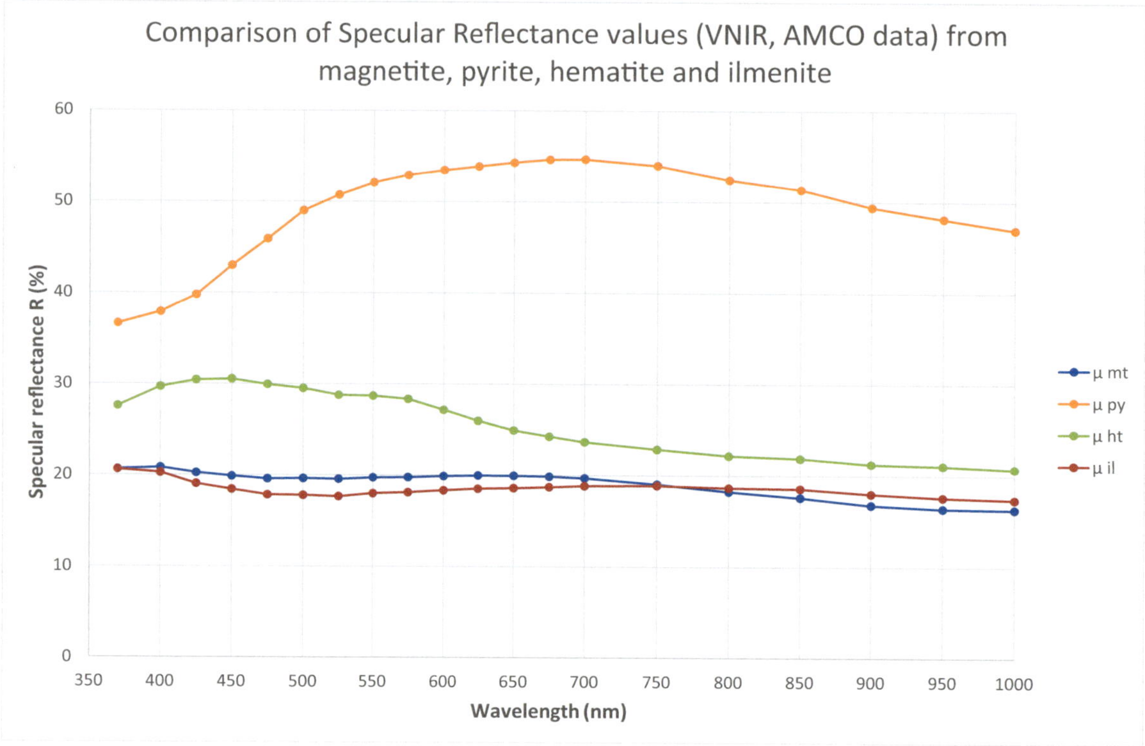
Comparison of Specular Reflectance values (VNIR, AMCO data) from magnetite, pyrite, hematite and ilmenite
Specular reflectance R (%)
Wavelength (nm)
0
10
20
30
40
50
60
350
400
450
500
550
600
650
700
750
800
850
900
950
1000
µ mt
µ py
µ ht
µ il

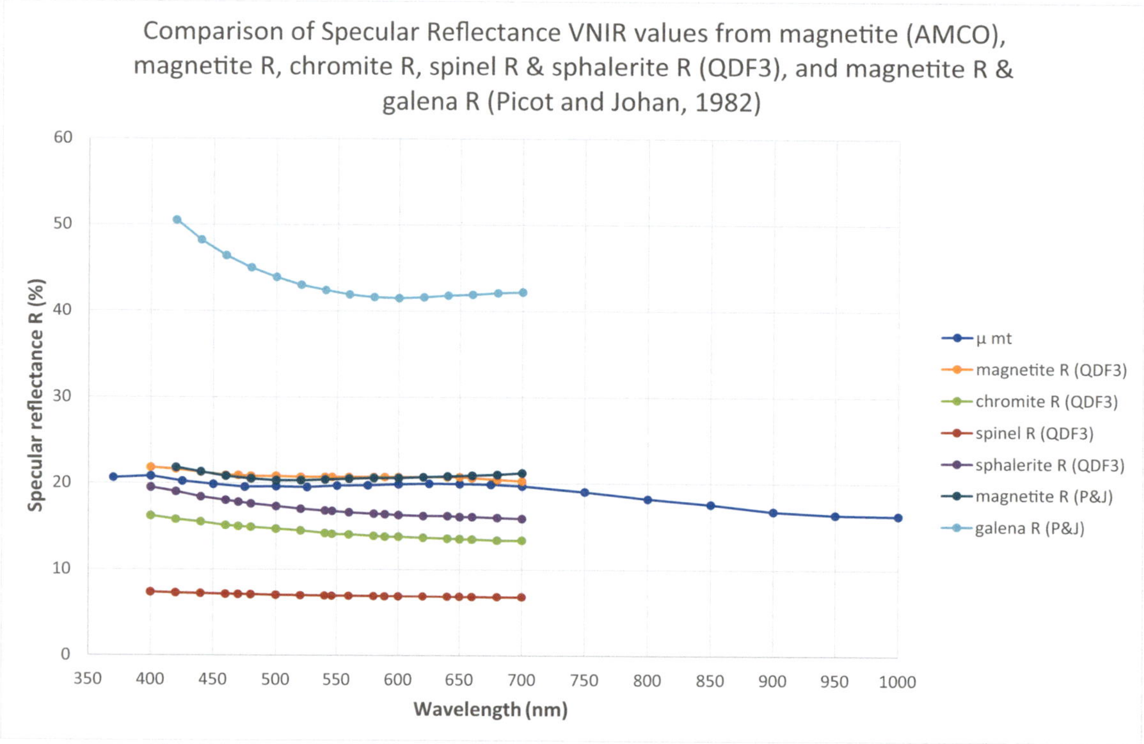
Comparison of Specular Reflectance VNIR values from magnetite (AMCO), magnetite R, chromite R, spinel R & sphalerite R (QDF3), and magnetite R & galena R (Picot and Johan, 1982)
Specular reflectance R (%)
Wavelength (nm)
0
10
20
30
40
50
60
350
400
450
500
550
600
650
700
750
800
850
900
950
1000
µ mt
magnetite R (QDF3)
chromite R (QDF3)
spinel R (QDF3)
sphalerite R (QDF3)
magnetite R (P&J)
galena R (P&J)

79 Malachite (*mch*)

$Cu_2(CO_3)(OH)_2$: Monoclinic

Malachite		R_{546} (%)	Y (%)	λ_d (nm)	P_e (%)	Colour sensation
Air	R_1	6.3	6.3	463	10.1	Grey
	R_2	9.2	9.1	483	3.7	Light grey
Oil	R_1	nd	nd	nd	nd	Dark grey
	R_2	nd	nd	nd	nd	Grey
Bir: Notable (4)				*Pleoc*: Weak/very weak (2)		*Tarn*: No
Anis +P: Distinct to strong (4)				*IR*: Green, luminous		*Twg*: (100)
Mor: Prismat/acicular aggregates				*Clv*: Perfect (001)		*PolQ*: Good
Hardness	*PolH*: Medium, <td (3)			*Mohs*: 3.5+	*Talm*: –	*VHN* $_{(100\ p)}$: nd

Obs: The grey chromatic sensation (single polarizer, *sp setting*) changes dramatically with the analyzer crossed at 90° (+*p*). The colour of the mineral is then perceived in the form of abundant and luminous green internal reflections, very striking. The low reflectance of mch, which contrasts with that of the accompanying sulphides, oxides and native Cu, does not totally hide the green colour of the mineral (R maximum at ≈550 nm, cf. R data and multispectral plots)
Mineral abbreviations: Ag—native silver, Au—native gold, az—azurite, cc—chalcocite, ccp—chalcopyrite, Cu—native copper, cv—covellite, dg—digenite, gn—galena, mch—malachite, sp—sphalerite, td—tetrahedrite

79.1 General

Composition: $Cu_2(CO_3)(OH)_2$, with possible traces of Zn, Co, Ni.

Very frequent mineral in the *oxidation zone* of copper deposits. A copper ore in the *copper oxide deposit* type.

79.2 Microscopic Description

79.2.1 SP Observation (*Single Polariser*)

Grey *colour*, relatively dark, but in polycrystalline aggregates shows a variable level of grey according to the orientation of the grains (Figs. 79.1a, b). The green colour of the mineral in hand specimen is scarcely, if at all, perceived with reflected light, but, being a transparent mineral, it is evident with transmitted light. The same happens with the pleochroism.

Low *reflectance compared* to sulphide ores but clearly higher than common gangue minerals such as quartz.

Clearly *bireflective*. If grains with different orientations are compared, reflectance differences of up to almost 50% can be observed, due to the high birefringence typical of carbonates, as can be seen in graph R (QDF3). Therefore, the perception of bireflectance predominates over pleochroism, which is not very noticeable with reflected light. What is seen is the change in the light/dark shade of grey.

Mch polishes well. Medium *polishing hardness*: > Au, Ag, calcite, ccp, Cu, gn; < dolomite, td, sp, siderite.

It occurs in *fibrous* and fibrous-radial aggregates. Perfect *cleavage*//(001).

79.2.2 Observation +P (*Polariser + Analyser at 90°*)

Clearly *anisotropic*, but the anisotropism is masked by the abundant and luminous green *internal reflections* (Fig. 79.1c, d). Oblique extinction.

Penetration or polysynthetic *twinning*//(100).

R. Castroviejo, *A Practical Guide to Ore Microscopy—Volume 1*,
https://doi.org/10.1007/978-3-031-12654-3_79

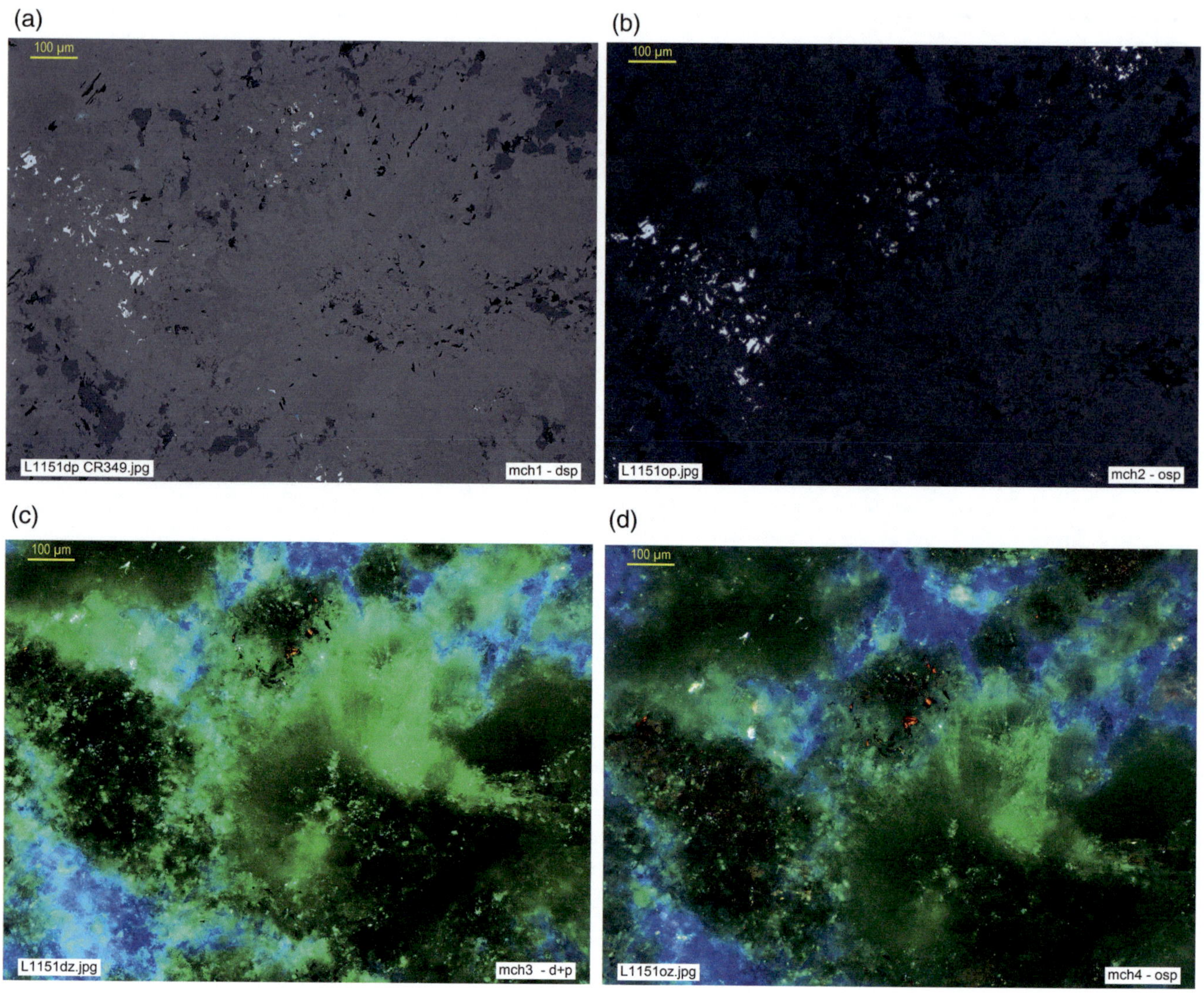

Fig. 79.1 **a** Aggregate of mch and az (grey, bireflective), with disseminated relicts of cc (white), dg (light blue), cv (deep blue) and possible dolomite (darker grey). CR349, Grand Gulch, Coconino County, Ariz. USA (**dsp**). **b** Same scene as mch1 (**osp**): Note the sharp decrease in reflectance (darkening) of az and mch and the greater contrast with the relict sulphides, which now appear brighter. **c, d** Same scene as mch1 (**d+p**) and (**o+p**), respectively. The anisotropism of the carbonates, mch and az, is now masked by the massive internal reflections, green and blue respectively, while only cv can be distinguished from the sulphides, due to its strong golden anisotropism

79.2.3 Paragenesis

It appears as a by-product of supergene alteration, typically in the oxidation zone, in virtually all copper deposits with carbonates. Its presence in the field, conspicuous by its colour even in small quantities, can serve as an exploration guide to detect even low Cu contents.

79.2.4 Identification

Its optical properties, particularly its bireflectance and internal reflections, are so typical that they make it easy to identify mch in copper deposits. However, in case of doubt, it can be confirmed by effervescence with CO_2 release, by attack with dilute acid.

79.3 Spectral Information

79.3.1 Malachite Multispectral Reflectance R (in Air, Without Polariser) (%)

λ (nm)	370	400	425	450	475	500	525	550	575	600	625	650	675	700	750	800	850	900	950	1000
μ	10.1	9.7	9.7	9.7	9.9	10.3	10.4	10.0	9.3	8.7	8.4	8.3	8.2	8.2	8.2	8.2	8.2	8.1	8.2	8.2
P_5	9.6	9.2	9.2	9.1	9.3	9.5	9.5	9.2	8.6	8.2	8.0	7.8	7.8	7.7	7.7	7.7	7.7	7.6	7.6	7.6
P_{20}	9.9	9.5	9.4	9.3	9.5	9.8	9.9	9.5	8.9	8.4	8.2	8.1	8.0	8.0	8.0	7.9	8.0	7.9	7.9	7.9
P_{80}	10.3	10.0	10.0	10.0	10.4	10.8	11.0	10.5	9.7	8.9	8.7	8.5	8.4	8.4	8.4	8.4	8.4	8.4	8.4	8.4
P_{95}	10.6	10.2	10.2	10.3	10.7	11.3	11.5	10.9	10.0	9.2	8.9	8.7	8.7	8.7	8.6	8.6	8.7	8.6	8.6	8.6

Comparison of Specular Reflectance values (VNIR, AMCO data) from malachite, quartz, covellite and azurite

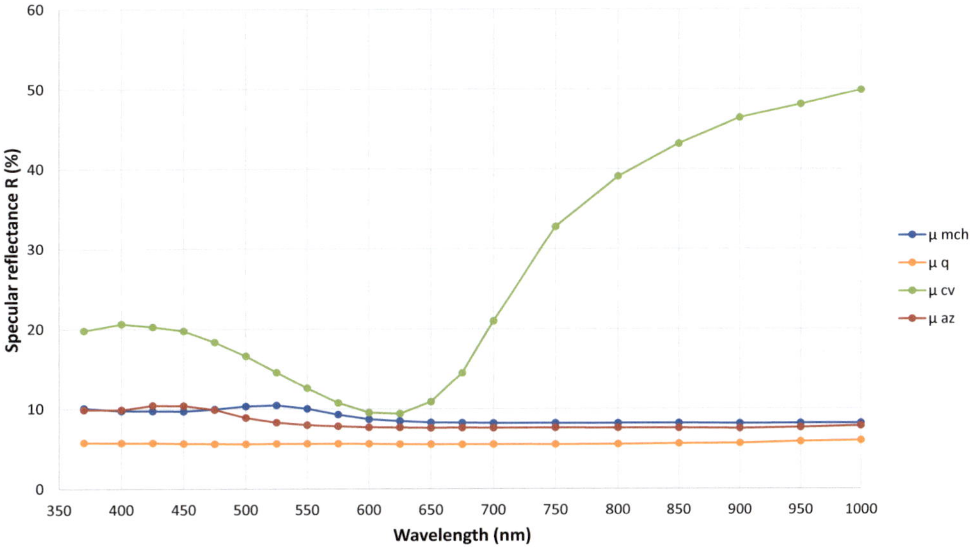

Specular Reflectance values in the visible spectrum (QDF3) from malachite R1, malachite R2, copper R & cuprite R

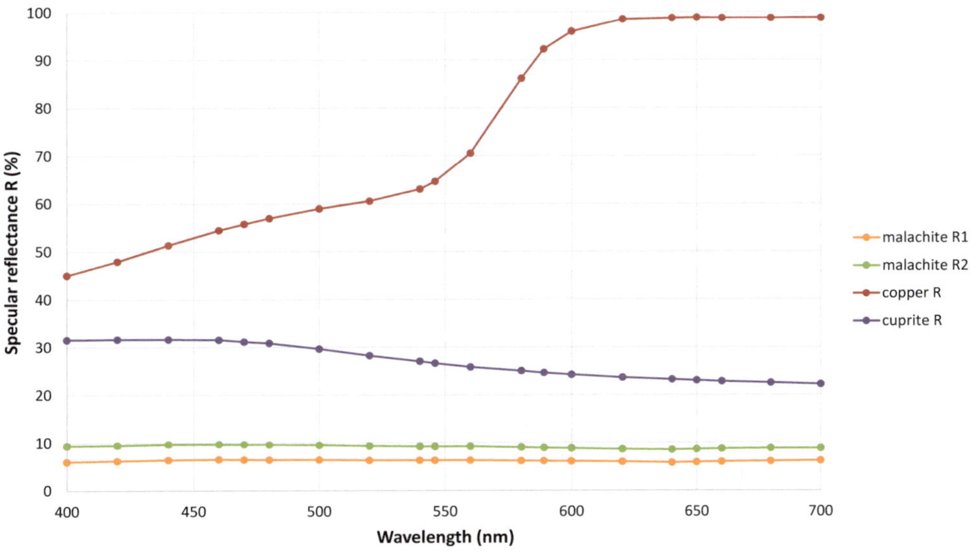

Manganite (*mng*)

80

$Mn^{3+}O(OH)$: Monoclinic

Manganite		R_{546} (%)	Y (%)	λ_d (nm)	P_e (%)	Colour sensation
Air	R_1	14.1	14.0	485	2.9	Dark grey
	R_2	20.5	20.3	484	3.9	Grey
Oil	R_1	3.5	3.4	485	5.7	Dark grey
	R_2	7.5	7.3	483	7.3	Grey
Bir: Medium (4)				*Pleoc*: Very weak (2)		*Tarn*: No
Anis +P: Distinct, extinc // elongation (4)				*IR*: Red, scattered		*Twg*: {100} lamellar, sometimes visible
Mor: Euhedral prismatic, acicular, botr				*Clv*: {010} distinct		*PolQ*:
Hardness	*PolH*: Medium–high, < mt (4			*Mohs*: 4	*Talm*: E	*VHN* $_{(100\ g)}$: 630–743

Obs: cf. Table and R charts, multispectral data (VNIR, AMCO) and visible range (QDF3)
Mineral abbreviations: gth—goethite, mng—manganite, prl—pyrolusite, psi—psilomelane, rom—romanechite

80.1 General

Composition: MnOOH, minor or trace contents of Fe, Cu, Pb, Al, Ba, Ca. Seldom found in nature, due to its instability in oxidizing conditions and its tendency to alter to pyrolusite by weathering.

80.2 Microscopic Description

80.2.1 SP Observation (*Single Polariser*)

Grey *colour*, distinctly bireflective but scarcely pleochroic (colour varies only in shades of grey), good *polish*. *Polishing hardness* much lower than pyrolusite.

Relatively low *reflectance*, R mng > gth, but < psi or rom and distinctly < prl, which usually includes it and, next to which, mng appears distinctly darker (Figs. 80.1a, b and 80.2a; cf multispectral comparison plots).

Distinct cleavage, almost always marked by alteration to pyrolusite, which propagates preferentially along the cleavage planes and from the edges. This process involves the dehydration of MnOOH and the appearance in pyrolusite of shrinkage cracks parallel to these planes (*pyrolusite pits*), very characteristic of the pyrolusite secondary after manganite (Figs. 80.1a, b and 80.2a).

It occurs in crystals with prismatic to acicular idiomorphic tendency and in microcrystalline and botryoidal aggregates, but it is almost always altered in different grades to pyrolusite, which may pseudomorphize it totally, sometimes preserving scarce relict inclusions (Fig. 80.2a).

80.2.2 Observation +P (*Polariser + Analyser at 90°*)

Distinctly *anisotropic* but in shades not very coloured. Extinction parallel to elongation. *Internal reflections*

R. Castroviejo, *A Practical Guide to Ore Microscopy—Volume 1*,
https://doi.org/10.1007/978-3-031-12654-3_80

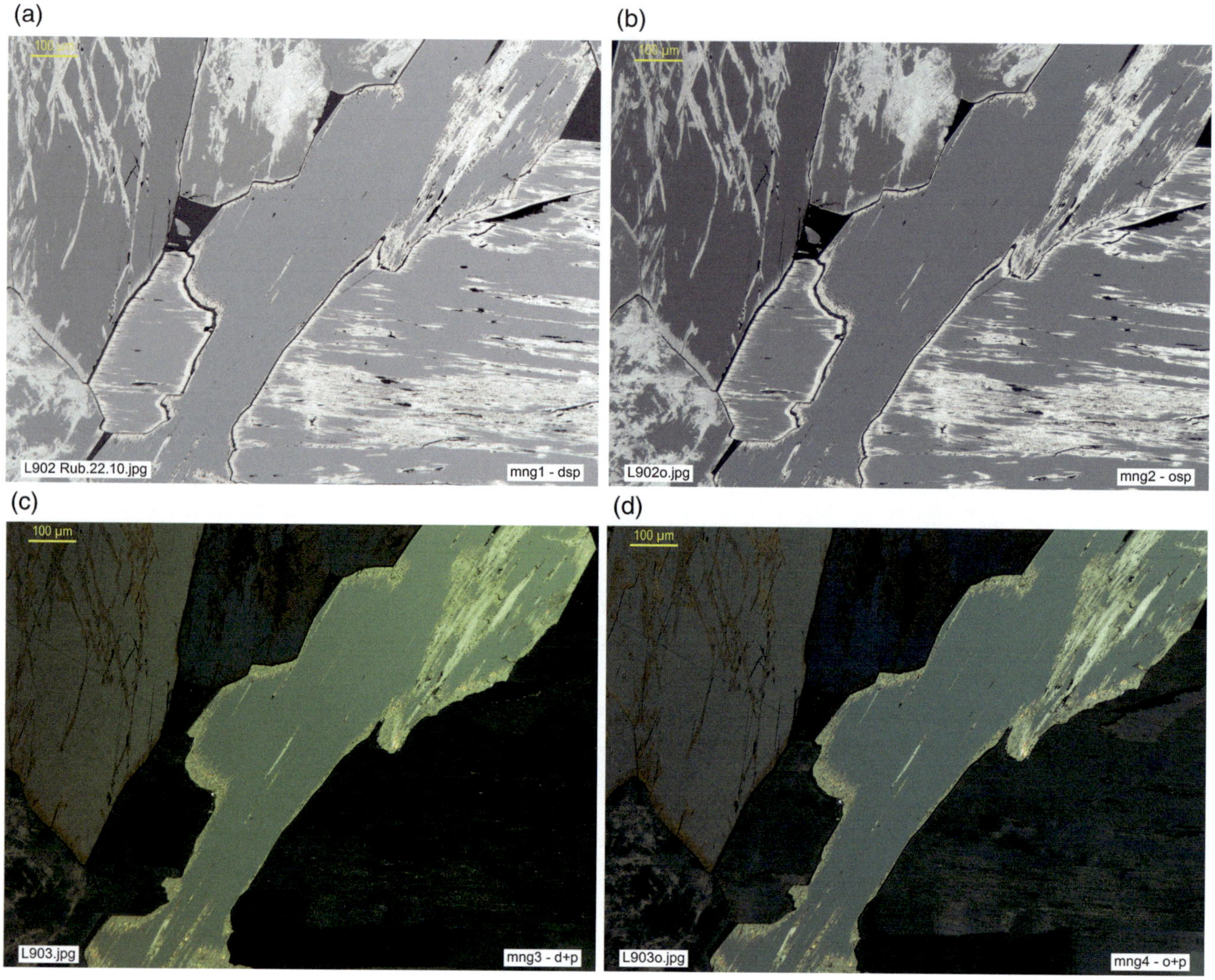

Fig. 80.1 **a**, **b** (Same scene: **dsp** and **osp**, respectively). Aggregate of mng (grey, good polish) partially altered to prl (white, rough surface due to shrinkage cracks) on the edges and along cleavage planes. Comparing grains of different orientation, the bireflectance of mng is apparent. Sample 1073 Rub.22.10, Ilfeld, Germany. **c**, **d** Same scene as Fig. 80.1a (**d+p** and **o+p**, respectively). Scattered red IR in mng (more visible on cleavage planes). Distinct anisotropism of mng (poorly coloured) and prl (bright yellow tints)

scattered and visible mainly as red dots on cleavage planes, not always abundant (Figs. 80.1c, d and 80.2b). Polysynthetic {100} *twinning* occasionally visible (Fig. 80.2b).

80.2.3 Paragenesis

Shallow Mn oxide, low temperature, mainly sedimentary *parageneses*. Usually replaced by pyrolusite, so it is a generally scarce mineral, except in cases where it is preserved fresh, being protected from weathering.

Companions: mainly pyrolusite (secondary to manganite) and also the remaining Mn oxides.

80.2.4 Identification

Given its association with pyrolusite and with oxidized manganiferous parageneses, it is easily distinguished by its grey colour, with low reflectance but clearly bireflective, its exfoliation, its distinct but poorly coloured anisotropism, its red internal reflections and, very typically, by its almost constant alteration to pyrolusite.

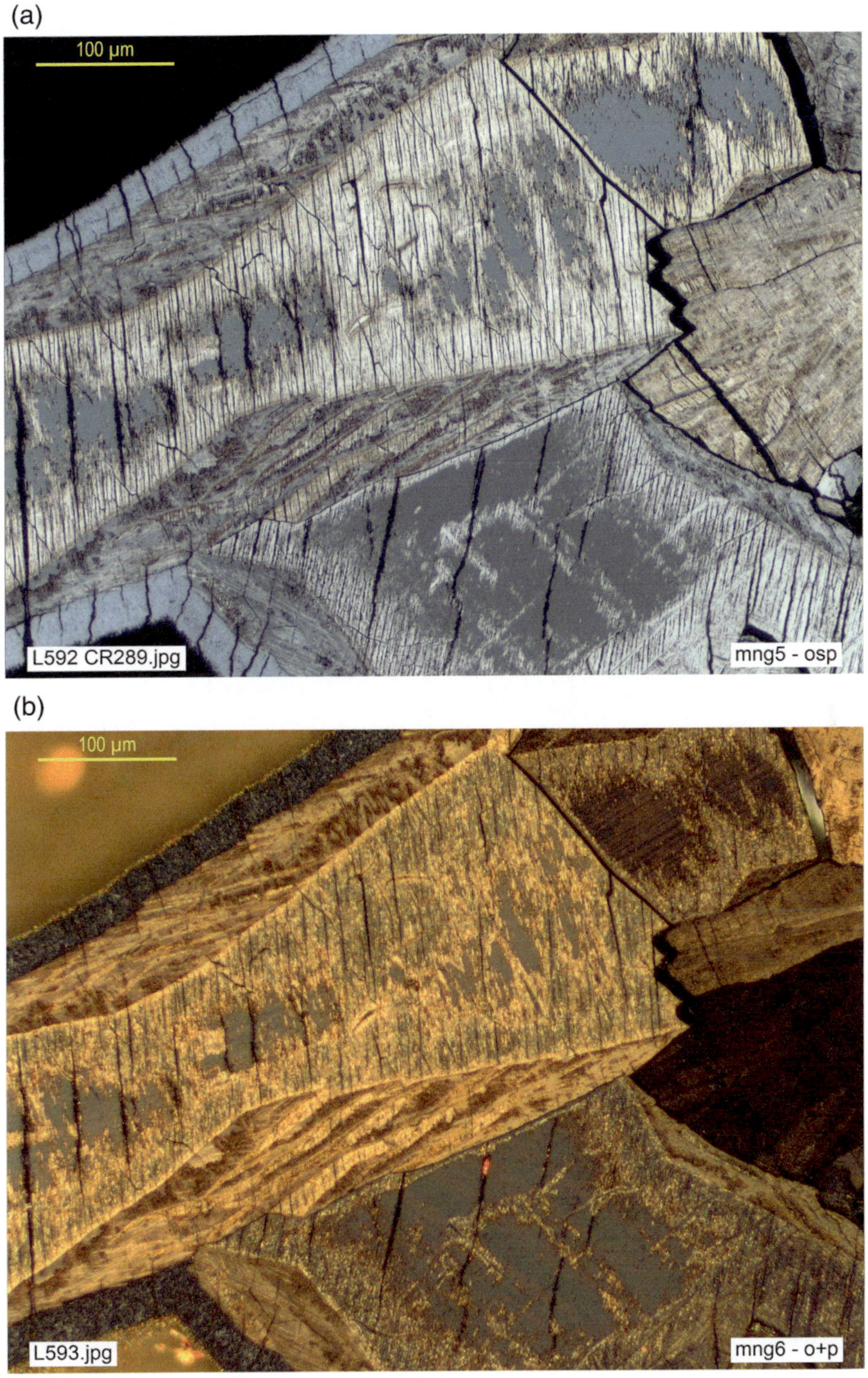

Fig. 80.2 **a**, **b** (Same scene: **osp** and **o+p**, respectively). Relicts of mng (dark grey, bireflective; anisotropic, red IR; twinned: Fig. 80.2b, top right) included in secondary prl (white, cracks// exfoliation of mng, anisotropism in yellowish tones), in turn, partly replaced by psi (grey lighter than mng, weak anisotropism, poorly coloured), which forms a cryptocrystalline encrustation on the outer edges of prl. CR 289, Amalienhohe Mine, Waldalgesheim, Bingerbrueck, Germany

80.3 Spectral Information

80.3.1 Manganite Multispectral Reflectance R (in Air, Without Polariser) (%)

λ (nm)	370	400	425	450	475	500	525	550	575	600	625	650	675	700	750	800	850	900	950	1000
μ	19.3	19.8	20.1	20.1	19.9	19.7	19.4	19.1	18.7	18.4	18.2	18.0	18.0	17.8	17.7	17.6	17.7	17.4	17.4	17.3
P_5	16.7	17.2	17.2	17.1	16.9	16.8	16.7	16.6	16.4	16.2	16.1	16.0	16.1	15.9	16.0	16.0	16.2	16.0	16.0	16.0
P_{20}	18.3	18.8	19.0	19.0	18.8	18.6	18.3	18.1	17.8	17.5	17.4	17.2	17.2	17.0	16.9	16.9	17.1	16.8	16.8	16.7
P_{80}	20.4	21.0	21.3	21.4	21.2	20.9	20.5	20.1	19.7	19.4	19.1	18.9	18.8	18.6	18.5	18.4	18.5	18.2	18.1	18.0
P_{95}	21.3	21.9	22.3	22.4	22.2	21.8	21.5	21.0	20.6	20.2	20.0	19.7	19.6	19.4	19.2	19.1	19.2	18.9	18.8	18.7

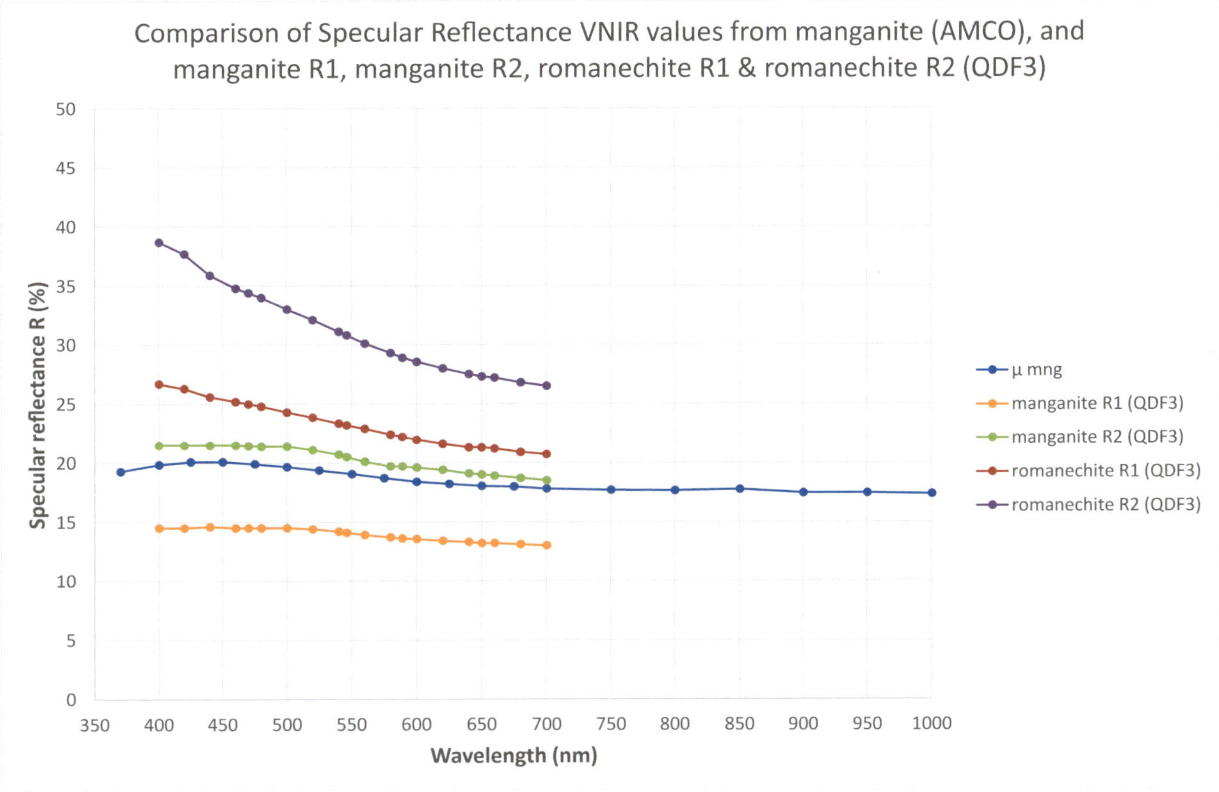

Comparison of Specular Reflectance VNIR values from manganite (AMCO), and manganite R1, manganite R2, romanechite R1 & romanechite R2 (QDF3)
Specular reflectance R (%)
Wavelength (nm)
0
5
10
15
20
25
30
35
40
45
50
350
400
450
500
550
600
650
700
750
800
850
900
950
1000
µ mng
manganite R1 (QDF3)
manganite R2 (QDF3)
romanechite R1 (QDF3)
romanechite R2 (QDF3)

81 Marcasite (*mc*)

FeS_2: Orthorhombic

Marcasite		R_{546} (%)	Y (%)	λ_d (nm)	P_e (%)	Colour sensation
Air	R_1	49.1	48.6	573	5.8	YlWh to very faintly BrWh, bright, slightly pleochroic
	R_2	56.2	55.3	566	6.2	
Oil	R_1	34.3	33.7	572	7.9	YlWh to more saturated BrWh, bright, pleochroic
	R_2	42.3	41.2	564	7.6	
Bir: Weak to Medium (3–4)				*Pleoc*: Moderate (3–4)		*Tarn*: Variable, sometimes fast and destructive
Anis +P: Very strong (5)				*IR*: No		*Twg*: Very frequent and varied
Mor: Prismatic aggreg., fibr-radial; m'x / x'x				*Clv*: Occasional // (101)		*PolQ*: Difficult but good
Hardness	*PolH*: High, almost ≈ py (4)			*Mohs*: 6.5-	*Talm*: E+	*VHN* $_{(200\ g)}$: 915–1099

Obs **:** cf. Table and R charts, multispectral data (VNIR, AMCO) and visible range (QDF3)
Mineral abbreviations: Ag—native silver, asp—arsenopyrite, cb—columbite, ccp—chalcopyrite, cst—cassiterite, gg—gangue, gn—galena, mt—magnetite, mc—marcasite, po—pyrrhotite, py—pyrite, sp—sphalerite, spn—spinel
Colour abbreviations: BluWh, BrWh, GnWh, YlWh, respectively: bluish white, brownish white, greenish white, yellowishwhite

81.1 General

A polymorph of pyrite, marcasite is a metastable form of FeS_2 (cf pyrite). Unstable above 350 °C, it is relatively common in shallow environments and as a late mineral. Frequently pseudomorphic, replacing various Fe-rich sulphides, particularly pyrrhotite.

Very unstable by weathering and in oxidizing environments, to the point of samples from study collections being spontaneously destroyed with time, left alone at atmospheric conditions.

Marcasites with possible minor contents of Cu, Co, Ni, As have been reported.

81.2 Microscopic Description

81.2.1 SP Observation (*Single Polariser*)

White *colour* with a slight yellowish to brownish (sometimes pinkish) tone if isolated.In this case, *pleochroism* is clearly perceived by comparing contiguous grains (Fig. 81.1a, b). However, the perception of the colour depends strongly on the neighbouring ores. In the frequent case of association with pyrite, marcasite is seen with a bluish-green tone, reinforced in immersion, which allows it to be easily distinguished from pyrite (Fig. 81.2); this also occurs next to arsenopyrite or native silver.

High *reflectance*, very close but slightly lower than pyrite, coincident with chalcopyrite in the near infrared range and clearly superior to pyrrhotite in the visible range (cf multispectral comparative plots).

Polishing is not easy, because of its hardness, but good quality can be obtained. High *polishing hardness*, almost as high as py, >almost all ores but <cb, cst, spn. Its instability in oxidizing and humid environments sometimes causes it to rapidly form a surface alteration *patina*, which can become totally destructive of the polished section by penetrating into the interior of the sample and decomposing the mineral. This is accentuated in very fine grained aggregates, but it is not observed in all samples. In crystalline

R. Castroviejo, *A Practical Guide to Ore Microscopy—Volume 1*,
https://doi.org/10.1007/978-3-031-12654-3_81

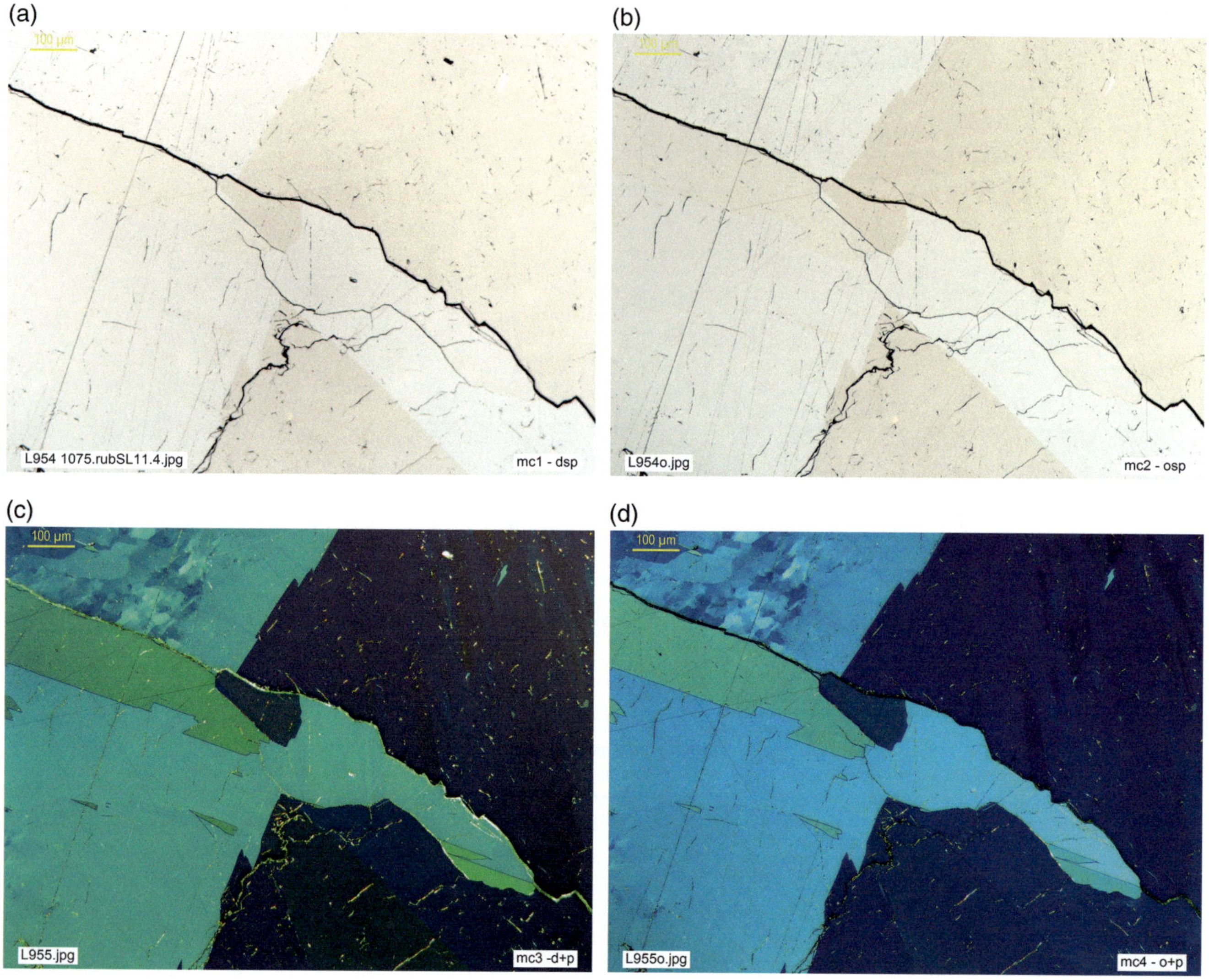

Fig. 81.1 a, b, c, d Single-mineral aggregate of holocrystalline, pleochroic and strongly anisotropic mc. Same scene seen, respectively: (**dsp**), (**osp**), (**d+p**) and (**o+p**). Sample 1075rubSL11.4, Friedrich der Grosse Mine (Germany)

aggregates, there may be occasional exfoliation // (101) and zoning.

Marcasite occurs in crystalline aggregates of various morphologies (e.g., tabular or fibrous-radial) but also in micro to cryptocrystalline masses and often as a pseudomorph of other Fe-rich sulphides, from which it forms as a secondary mineral under the effect of acidic solutions at moderate temperatures.

Alteration of mc has also been noted; pyrite inevitably replaces it at temperatures above 350 °C, at which marcasite becomes unstable.

Associated with pyrite ± magnetite, mc is a typical component of pyrrhotite alteration, forming so-called "*bird's eye*" aggregates (cf Fig. 102.7). It also forms cryptocrystalline masses that have received the name "*melnicovite marcasite*" (*cf* pyrite).

81.2.2 Observation +P (*Polariser + Analyser at 90°*)

Strongly *anisotropic*, with bright and luminous colours, typically green and blue-green (Figs. 81.1c, d, 81.3 and 81.4a, b, c), varied and dependent also on orientation, sometimes even with yellowish tones.

Twinning practically ubiquitous, with varied forms: laminar, arrowheaded, sectored (Fig. 81.4a, b, c).

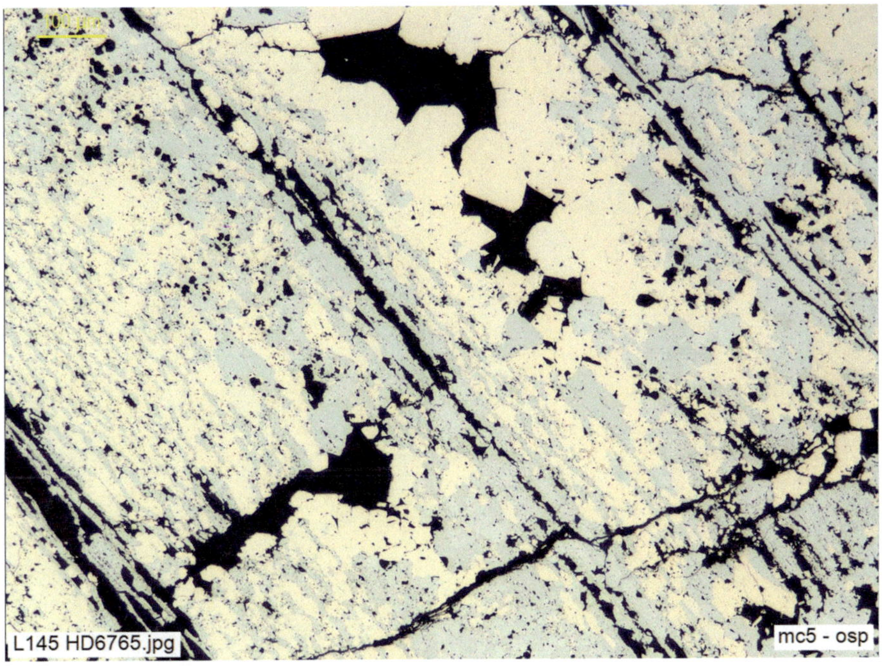

Fig. 81.2 Aggregate of mc (GnWh to BluWh) and py (pale yellow), with partial replacement of py by mc. Sample HD6765, Teufelsgrund, Schwarzwald, Germany (**osp**)

Fig. 81.3 Same scene as mc5 (**o+p**)

81.2.3 Paragenesis

Paragenesis varied, but usually as a late mineral. Sometimes primary and very well crystallized. Often formed in shallow environments, as a secondary product by decomposition of Fe-rich sulphides, which it tends to pseudomorphize.

Most common companions: pyrite, galena, sphalerite, pyrrhotite, melnicovite, magnetite.

81.2.4 Identification

It is unmistakable for its strong anisotropism with greenish to bluish-green, intense and luminous hues, in addition to its colour, hardness, twinning and low temperature or secondary parageneses.

Fig. 81.4 a, b, c Twinning in monomineralic aggregate of mc. Microscope settings respectively (**d+p**), (**d+p**) and (**o+p**). Sample 1075rubSL11.4, Friedrich der Grosse Mine (Germany)

(a)

(b)

Fig. 81.4 (continued)

(c)

81.3 Spectral Information

81.3.1 Marcasite Multispectral Reflectance R (in Air, Without Polariser) (%)

λ (nm)	370	400	425	450	475	500	525	550	575	600	625	650	675	700	750	800	850	900	950	1000
μ	38.7	40.2	41.2	43.5	44.9	46.9	47.8	48.3	48.4	48.3	48.1	47.8	47.5	47.0	46.3	45.7	45.6	44.9	44.7	44.6
P_5	35.1	37.2	38.7	41.1	42.8	44.8	46.1	46.8	47.0	47.0	46.8	46.6	46.3	45.9	45.1	44.6	44.5	43.8	43.6	43.4
P_{20}	37.3	39.0	40.1	42.6	43.9	45.9	47.0	47.6	47.7	47.7	47.5	47.2	46.9	46.4	45.7	45.1	45.1	44.4	44.2	44.1
P_{80}	40.2	41.6	42.3	44.6	46.0	47.9	48.7	49.1	49.2	49.1	48.8	48.5	48.2	47.7	47.0	46.3	46.1	45.4	45.2	45.3
P_{95}	41.5	42.5	43.4	45.3	46.8	48.6	49.5	49.7	49.7	49.6	49.3	49.0	48.7	48.2	47.5	46.8	46.7	45.7	45.9	45.8

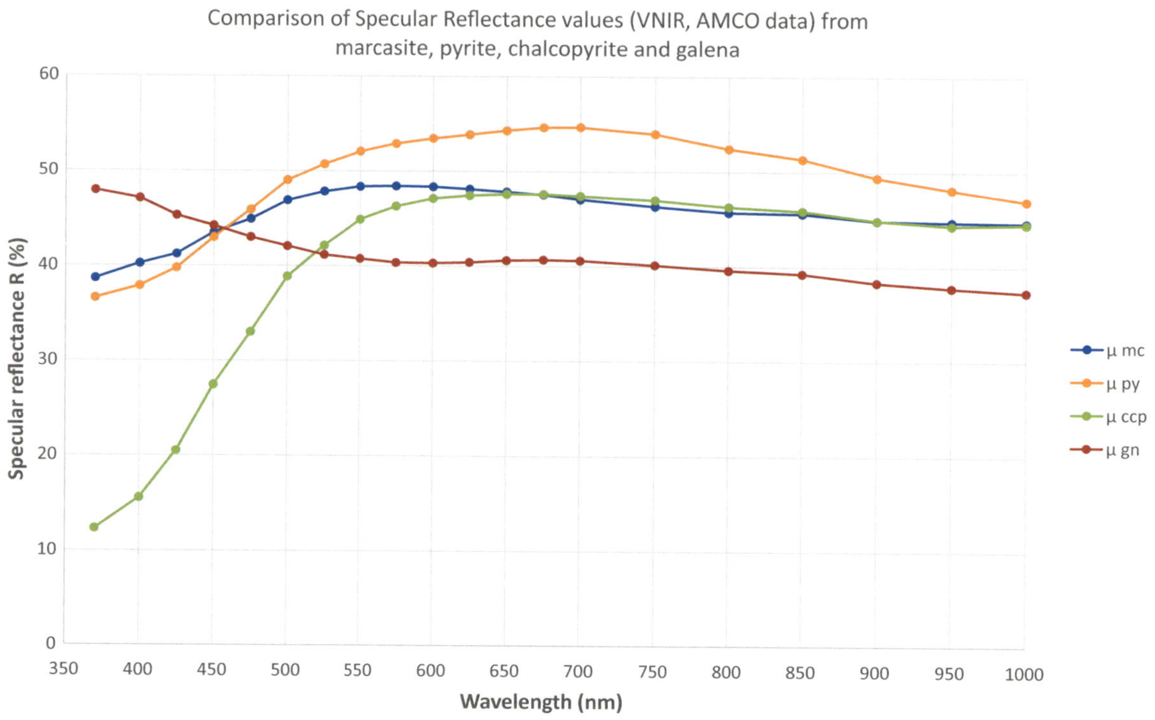
Comparison of Specular Reflectance values (VNIR, AMCO data) from
marcasite, pyrite, chalcopyrite and galena
Specular reflectance R (%)
Wavelength (nm)
0
10
20
30
40
50
60
350
400
450
500
550
600
650
700
750
800
850
900
950
1000
µ mc
µ py
µ ccp
µ gn

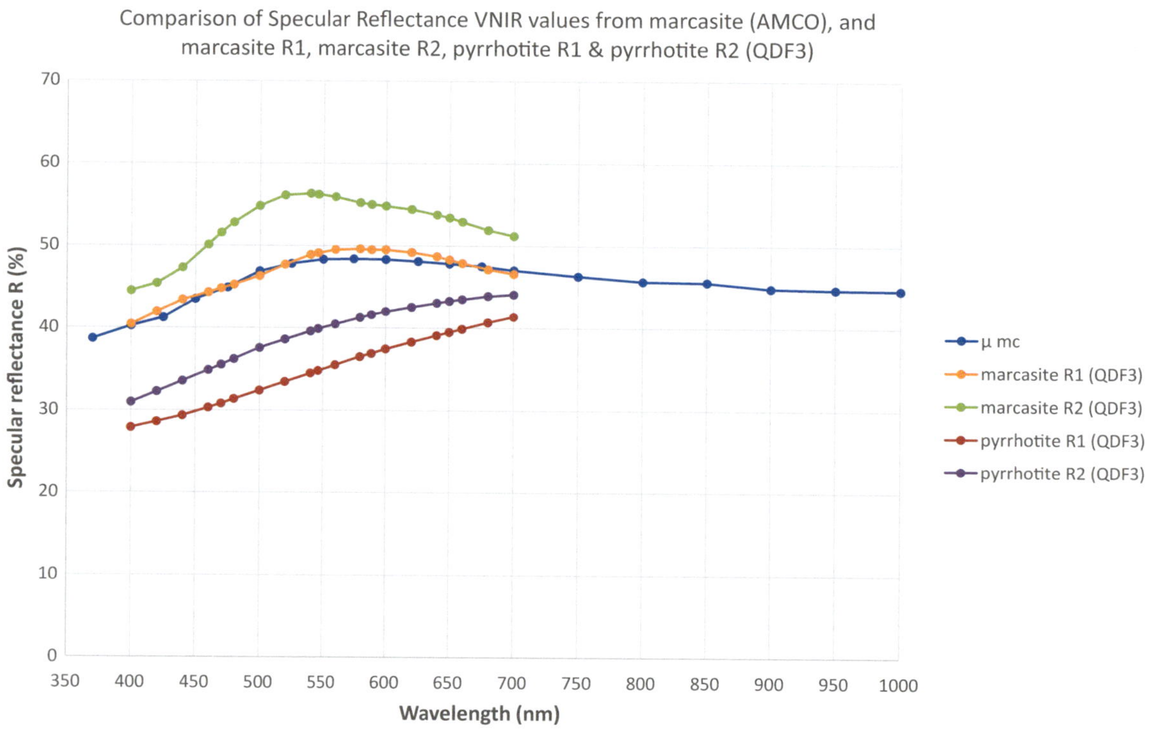
Comparison of Specular Reflectance VNIR values from marcasite (AMCO), and
marcasite R1, marcasite R2, pyrrhotite R1 & pyrrhotite R2 (QDF3)
Specular reflectance R (%)
Wavelength (nm)
0
10
20
30
40
50
60
70
350
400
450
500
550
600
650
700
750
800
850
900
950
1000
µ mc
marcasite R1 (QDF3)
marcasite R2 (QDF3)
pyrrhotite R1 (QDF3)
pyrrhotite R2 (QDF3)

Maucherite (*mau*)

82

$Ni_{11}As_8$: Tetragonal

Maucherite		R_{546} (%)	Y (%)	λ_d (nm)	P_e (%)	Colour sensation
Air	R_1	48.5	49.4	586	4.7	Bright white, slightly pinkish
	R_2	49.6	50.5	583	5.5	
Oil	R_1	35.0	36.0	587	6.9	
	R_2	36.0	36.9	584	7.9	
Bir: Imperceptible (2)				*Pleoc*: Imperceptible (2)		*Tarn*: Rare in polished section
Anis +P: Weak and faintly coloured (3)				*IR*: No		*Twg*: Lamellar
Mor: Anhedral aggregates; prisms				*Clv*: No		*PolQ*: Very good
Hardness	*PolH*: Medium, ≥ nc, <po (3)			*Mohs*: 5	*Talm*:	*VHN* $_{(100\ g)}$: 623–724

Obs: cf. Table and R charts, multispectral data (VNIR, AMCO) and visible range (QDF3)
Mineral abbreviations Ag—native silver, asp—arsenopyrite, bth—breithauptite, ccp—chalcopyrite, cob—cobaltite, dy—dyscrasite, gf—gersdorffite, gg—gangue, ln—linnaeite, lo—löllingite, mau—maucherite, nc—nickeline, pn—pentlandite, prm—pararammelsbergite, po—pyrrhotite, py—pyrite, rm—rammelsbergite, sf—safflorite, sk—skutterudite, sp—sphalerite, ul—ullmannite

82.1 General

Composition: $Ni_{11}As_8$, minor Co, Fe, Cu, Sb, S contents.

82.2 Microscopic Description

82.2.1 SP Observation (*Single Polariser*)

White *colour* when isolated, but mau can appear with faint colour shades, especially noticeable in immersion, when accompanied, *vbgr*: pinkish white to yellowish white against purely white minerals; greyish with very faint purplish to pinkish brown shades next to nickeline or breithauptite (Figs. 82.1a, b and 82.2); brownish grey against löllingite; grey next to native silver.

Although anisotropic, it shows no appreciable bireflectance or pleochroism.

Relatively high *reflectance*, between linnaeite and nickeline, very close to pyrite, but lower than skutterudite. In white light, R mau ≥ bth but < cob, nc, py, according to Picot and Johan (1982). Multispectral values, cf comparative R table and graphs.

Medium *polishing hardness*, PolH mau ≥ nc, ≤ prm; > ln, sp, ccp; <sf, po, gf, lo.

It does not show any cleavage.

It occurs, frequently associated with nickeline, in xenomorphic aggregates, sometimes with a prismatic or tabular habit, which can give rise to a fibrous texture.

82.2.2 Observation +P (*Polariser + Analyser at 90°*)

Anisotropism weak, faintly coloured, but sufficient to distinguish lamellar *twinning* in adequate sections (Figs. 82.3, comp. Fig. 82.1c, d).

82.2.3 Paragenesis

Hydrothermal *parageneses* typical of deposits of the *five-element* type (Ni–Co–Bi–Ag–U), with nickeline, breithauptite, native silver, dyscrasite, Ag sulphosalts, arsenides

R. Castroviejo, *A Practical Guide to Ore Microscopy—Volume 1*,
https://doi.org/10.1007/978-3-031-12654-3_82

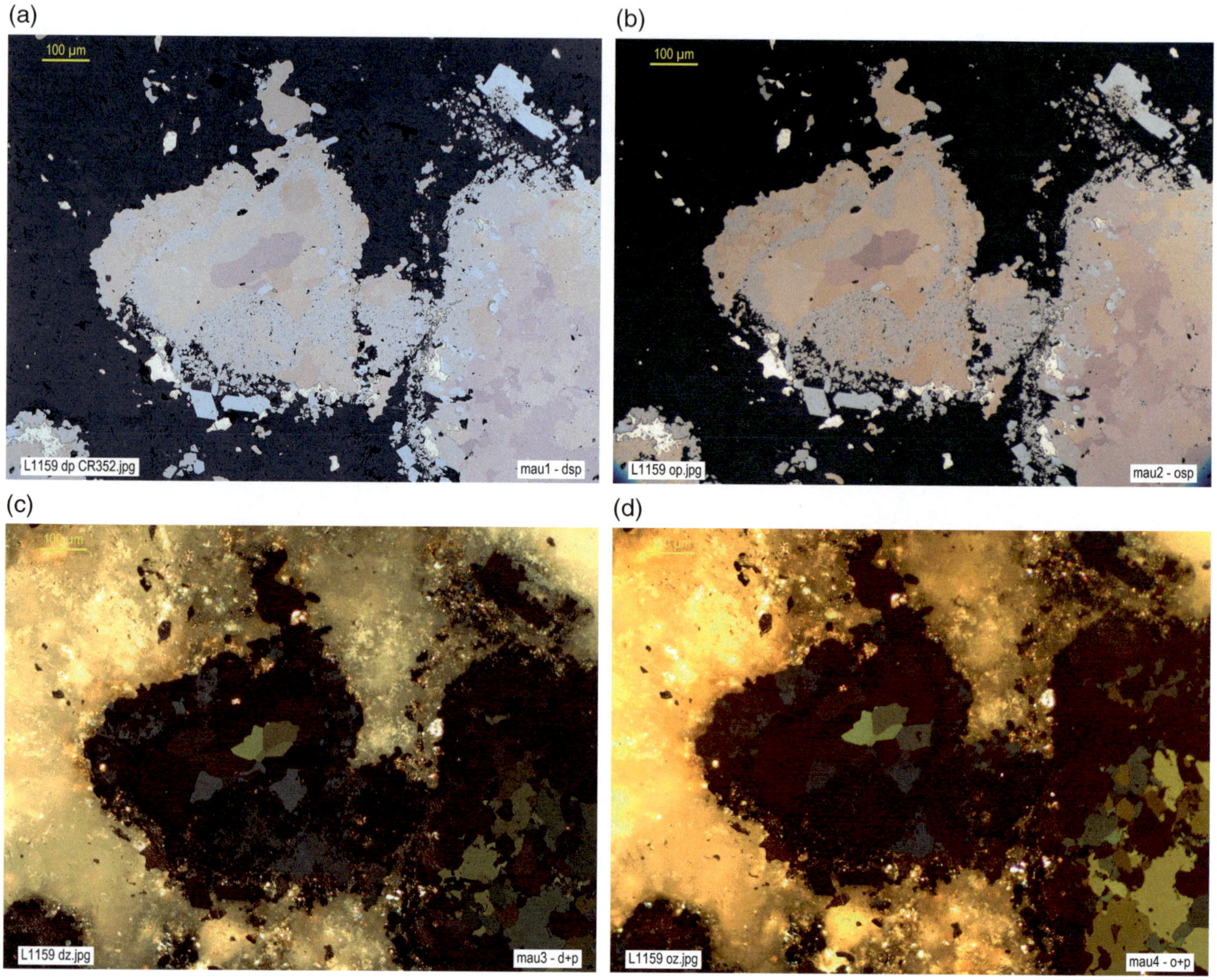

Fig. 82.1 **a, b**(Same scene, **dsp** and **osp**, respectively). Concentric aggregates of intergrown mau, nc and bth, with interstitial Ag and disseminated idiomorphic asp in calcite gg. The center of the aggregates is occupied by bth, easily recognizable by its peculiar colour and pleochroism (pale purplish pink to darker purplish pink). Around it, alternate concentric bands or *coronas* of nc (light orange-pink, less strongly pleochroic than bth) and mau (grey, with a very faint brownish-pinkish tinge), with posthumous, marginal Ag (white, bright); asp (pure grey) in idiomorphic crystals (prismatic and rhombic) disseminated in gg, whose white colour is perceived as grey next to Ag. The colour and R contrasts are much better appreciated in immersion (mau2). CR352, Cobalt, Ontario, Canada. **c, d** Same scene as mau1 (**d+p**) and (**o+p**), respectively. Strong anisotropism of bth, with greenish tones, and more moderate of nc, with bluish-green tones. The weak anisotropism of mau is not perceived in these images, due to oversaturation by the luminous intensity of the adjacent bth grains and, above all, of the internal reflections of gg. However, it is clearly visible in Fig. 82.3

and sulphoarsenides of Ni-Co, linnaeite, cobaltite, pentlandite.

Also Cr-Ni magmatic deposits, linked to ultramafic formations, although possibly the formation of maucherite is hydrothermal and late.

82.2.4 Identification

Typical features of mau are its pinkish-white colour, high reflectance, weak anisotropism, lamellar twinning and association with nickeline. Paragenesis and fibrous texture are additional criteria. It is distinguished:

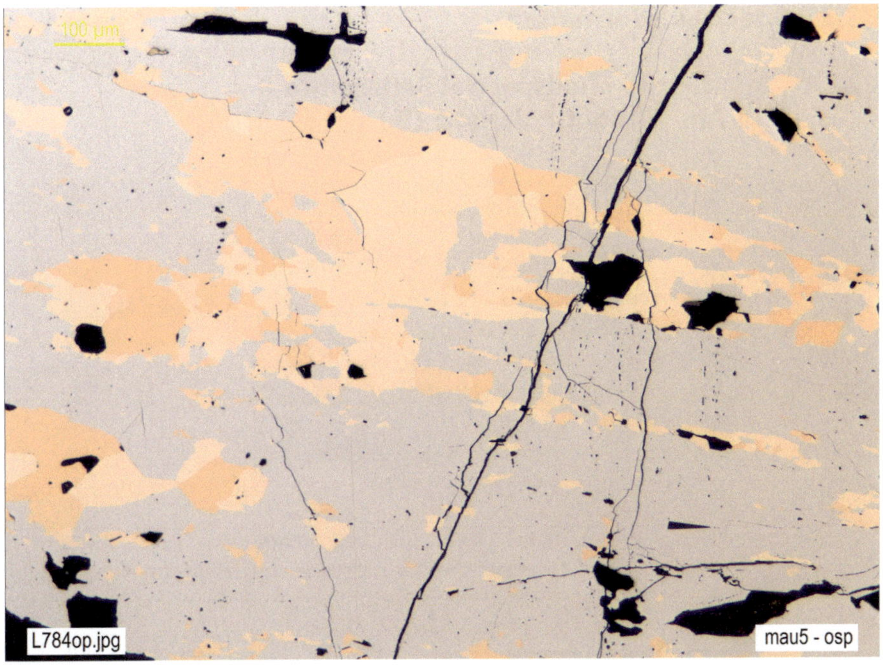

Fig. 82.2 Intergrowth of nc (orange-pink, pleochroic) and mau (grey, with a faint pinkish-brown hue). CR 326, Timiskaming, Ontario, Canada (**osp**)

Fig. 82.3 Same scene as mau6 (**o+p**): lamellar twins and weak anisotropism of mau, which contrasts with the bright colours typical of nc. The tabular habit of mau guides the deposition (probably replacement) of nc, in alignments parallel to the prismatic faces of mau

- from nickeline by the colour, pleochroism and strong anisotropism of nc;
- from skutterudite, gersdorffite, ullmannite, safflorite and rammelsbergite by their pure white colour and by the isotropic character of skutterudite, gersdorffite and ullmannite or the strong anisotropism of safflorite and rammelsbergite; and
- from cobaltite, by the strong automorphic, non-fibrous tendency and the complex twinning of cob.

82.3 Spectral Information

82.3.1 Maucherite Multispectral Reflectance R (in Air, Without Polariser) (%)

λ (nm)	370	400	425	450	475	500	525	550	575	600	625	650	675	700	750	800	850	900	950	1000
μ	42.8	44.5	44.8	46.3	46.6	47.7	48.2	49.4	50.8	52.4	53.7	55.1	56.4	57.5	59.3	60.4	62.0	62.7	63.7	64.4
P_5	41.4	43.1	43.0	45.1	45.0	46.5	46.7	48.3	49.5	51.0	52.2	53.5	54.4	55.6	57.2	58.4	60.3	61.1	61.8	62.5
P_{20}	42.0	43.7	43.8	45.7	46.0	47.2	47.7	48.9	50.4	52.0	53.3	54.8	56.0	57.1	58.9	59.9	61.6	62.1	63.2	63.8
P_{80}	43.4	45.1	45.6	46.9	47.3	48.2	48.8	49.9	51.3	52.9	54.3	55.7	57.0	58.1	60.0	61.0	62.6	63.4	64.3	65.3
P_{95}	43.8	45.3	45.9	47.2	47.7	48.5	49.1	50.2	51.6	53.2	54.6	56.0	57.3	58.3	60.3	61.3	62.9	63.7	64.5	65.6

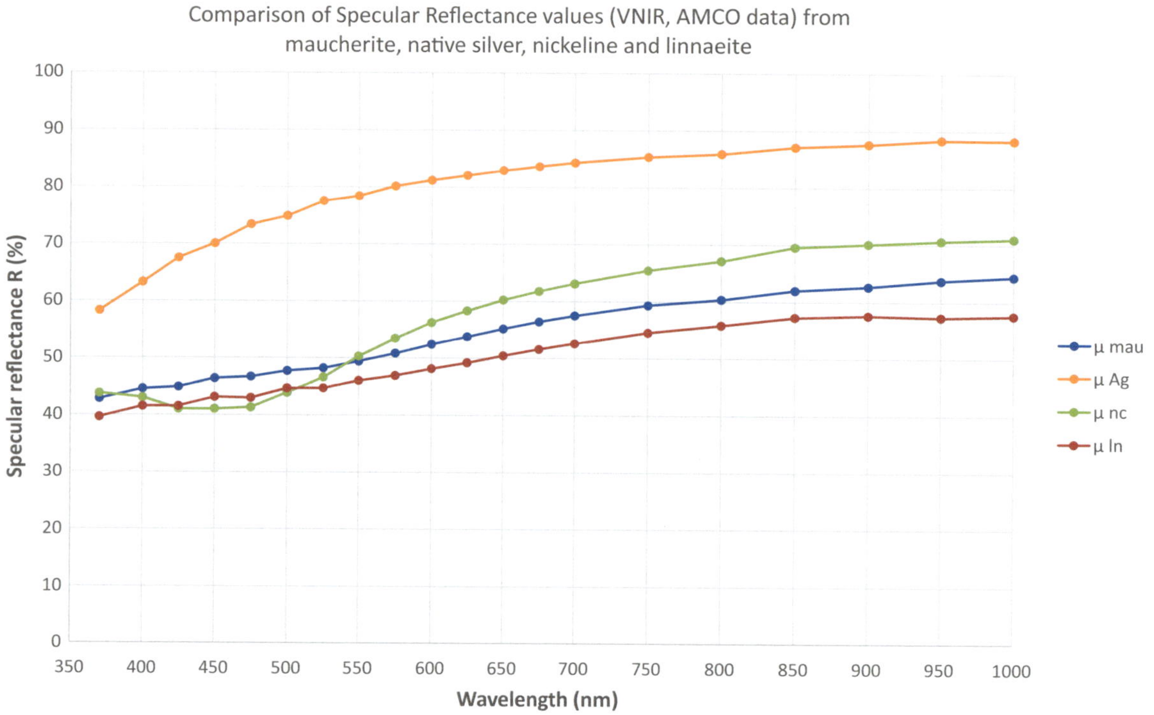

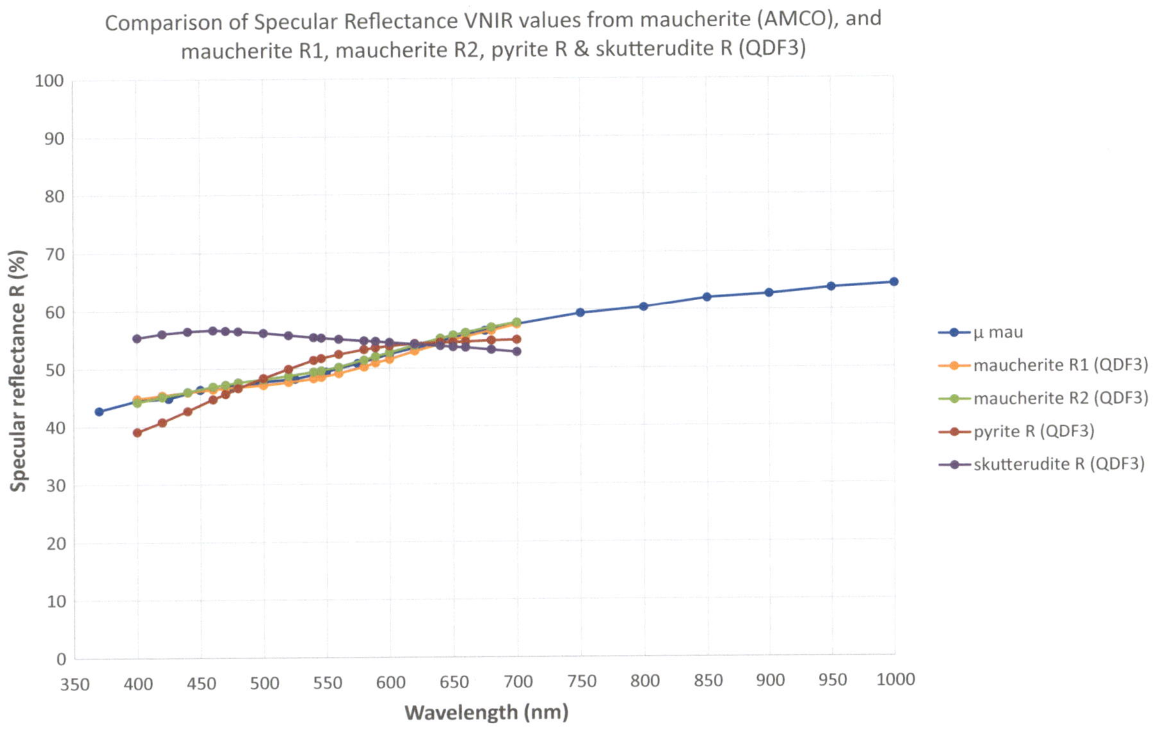

Reference

Picot P, Johan Z (1982) In: Atlas of ore minerals. B.R.G.M./Elsevier

83 Miargyrite (*my*)

$AgSbS_2$: Monoclinic

Miargyrite		R_{546} (%)	Y (%)	λ_d (nm)	P_e (%)	Colour sensation
Air	R_1	31.6	31.4	480	7.8	Greyish-white, bireflective
	R_2	34.5	34.2	480	7.5	
Oil	R_1	16.3	16.1	479	12.4	
	R_2	18.3	18.2	479	12.4	
Bir: Weak (3)				*Pleoc*: Very weak (2)		*Tarn*: Dim light attack (prg contacts)
Anis +P: Distinct, but masked by IR (4)				*IR*: Visible		*Twg*: Occasional polysynthetic
Mor: Masses, anhedral aggregates				*Clv*: Imperfect		*PolQ*: Good, easy
Hardness	*PolH*: Low, < gn (2)			*Mohs*: 2½	*Talm*: B	*VHN* $_{(100\ g)}$: 88–123

Obs: cf. Table and R graphs, multispectral (VNIR, AMCO) and visible ranges (QDF3; Picot and Johan 1982). VHN Uytenb & B 1971
Mineral abbreviations: ac—acanthite, Ag—native silver, asp—arsenopyrite, cc—chalcocite, cin—cinnabar, dg—digenite, dol—dolomite, dy—dyscrasite, fbg—freibergite, gg—gangue, gn—galena, my—miargyrite, plb—polybasite, pru—proustite, prg—pyrargyrite, py—pyrite, q—quartz, sbt—stibnite, sp—sphalerite, stb—sternbergite, stp—stephanite

83.1 General

Composition: $AgSbS_2$, with Se, As, Fe as occasional minor components (Se) or possible impurities.

83.2 Microscopic Description

83.2.1 SP Observation (*Single Polariser*)

Colour greyish-white, moderate bireflectance, distinct in oil (Fig. 83.1a, b), pleochroism practically imperceptible.

Relatively high *reflectance*: **R** *my* > stb, prg, stp; R *my* < gn and ≪ asp, py, Ag. Next to silver, arsenopyrite or pyrite, miargyrite looks dark (Fig. 83.2a, comp. Figs. 83.3a and 83.4a; cf. comparative R graphs, multispectral VNIR and visible ranges).

Good *polish*. Low *polishing hardness*: PolH my < cin, gn, cc, dg; > sbt, pru, prg, ac. Cleavage (010), (100) and (101) not very visible.

Light attack on edges intergrown with pyrargyrite (Fig. 83.6a, b, comp. Fig. 83.5).

It usually occurs as a relatively late hydrothermal filling, forming monomineral allotriomorphic aggregates or intergrown with other sulphosalts. Euhedral crystals in open spaces.

83.2.2 Observation +P (*Polariser + Analyser at 90°*)

Distinct anisotropism but without great colour effects and partly overprinted by IR. White and grey tones dominate, although bluish or purplish shades are also perceived (Figs. 83.1c, d, 83.2b, 83.3b, 83.4b and 83.6b).

Imperfect extinction.

Frequent red *internal reflections*, **IR**, although not as frequent as in pyrargyrite (Figs. 83.1c, d, 83.3b, 83.4b and 83.6b).

Occasional *twinning* (Fig. 83.6b), sometimes polysynthetic.

83.2.3 Paragenesis

A classic hydrothermal mineral of moderate temperature, my is typically found in epithermal and *five-element formation* (Ni–Co–Bi–Ag–U) *parageneses*.

R. Castroviejo, *A Practical Guide to Ore Microscopy—Volume 1*,
https://doi.org/10.1007/978-3-031-12654-3_83

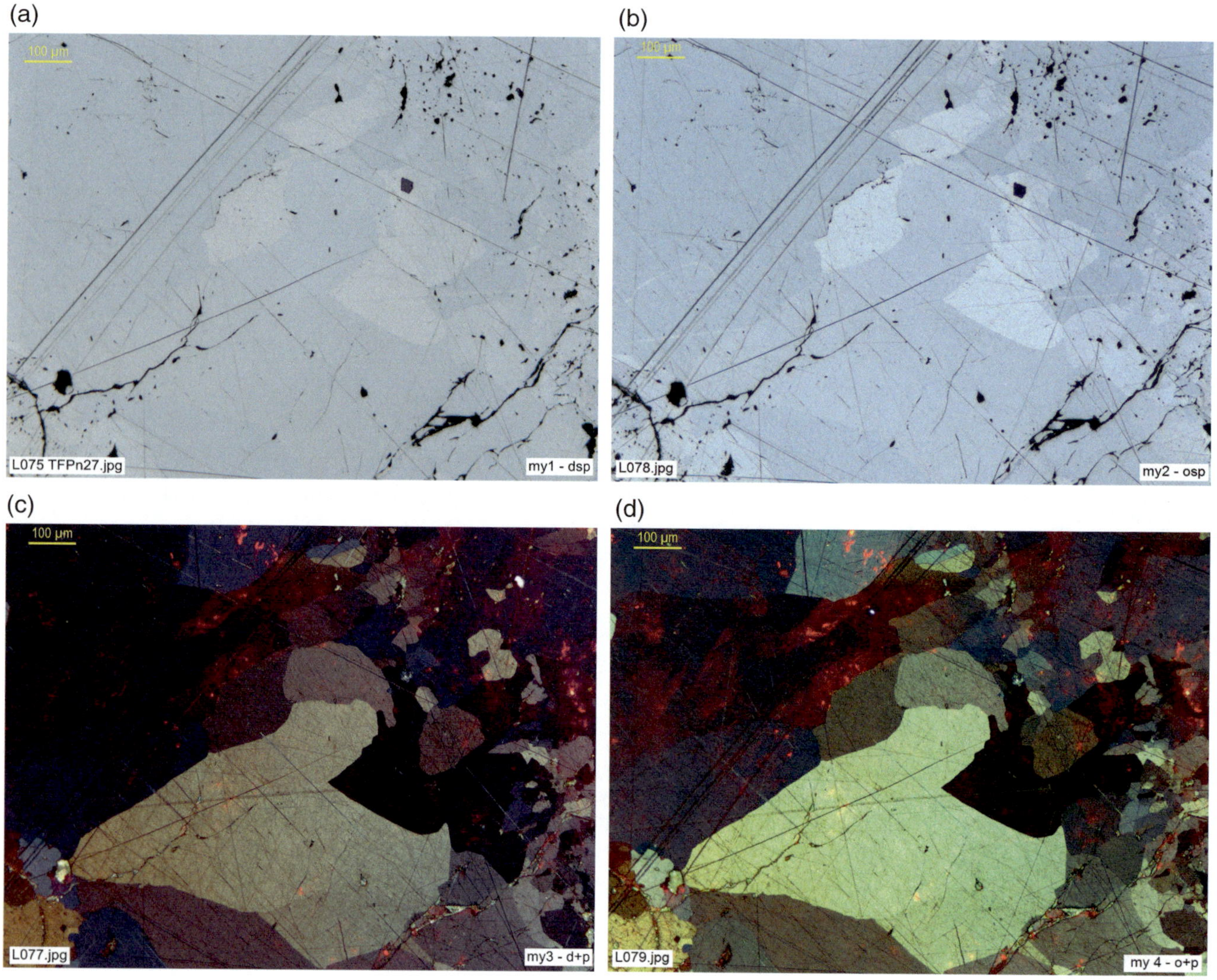

Fig. 83.1 **a, b** (**dsp** and **osp**, respectively)—Argentiferous ore, vein infill of massive my (bireflective, polishing scratches), with scarce gg of q (dark grey). TF Pn 27, Hiendelaencina, Guadalajara. **c, d** Same scene as my1 (**d+p** and **o+p**, respectively)

Most frequent accompaniments: native silver, allargentum, acanthite, dyscrasite, freibergite and silver sulphosalts, arsenopyrite, Ni–Co sulpho-arsenides and -antimonides, etc.

83.2.4 Identification

It can be recognized by its white colour, relatively high reflectance, distinct bireflectance, strong but faintly coloured anisotropism, red internal reflections, and light attack.

It is probably more frequent than thought, because it goes easily unnoticed due to possible confusions with other species, particularly with similar silver sulphosalts such as polybasite, pyrargyrite, stephanite and freibergite, or even with galena. The reflectance spectra are relatively close, if bireflectance is disregarded, which excludes isotropic freibergite and galena.

Careful observation of some characteristic features is required to avoid misidentification with these silver

(a)

(b)

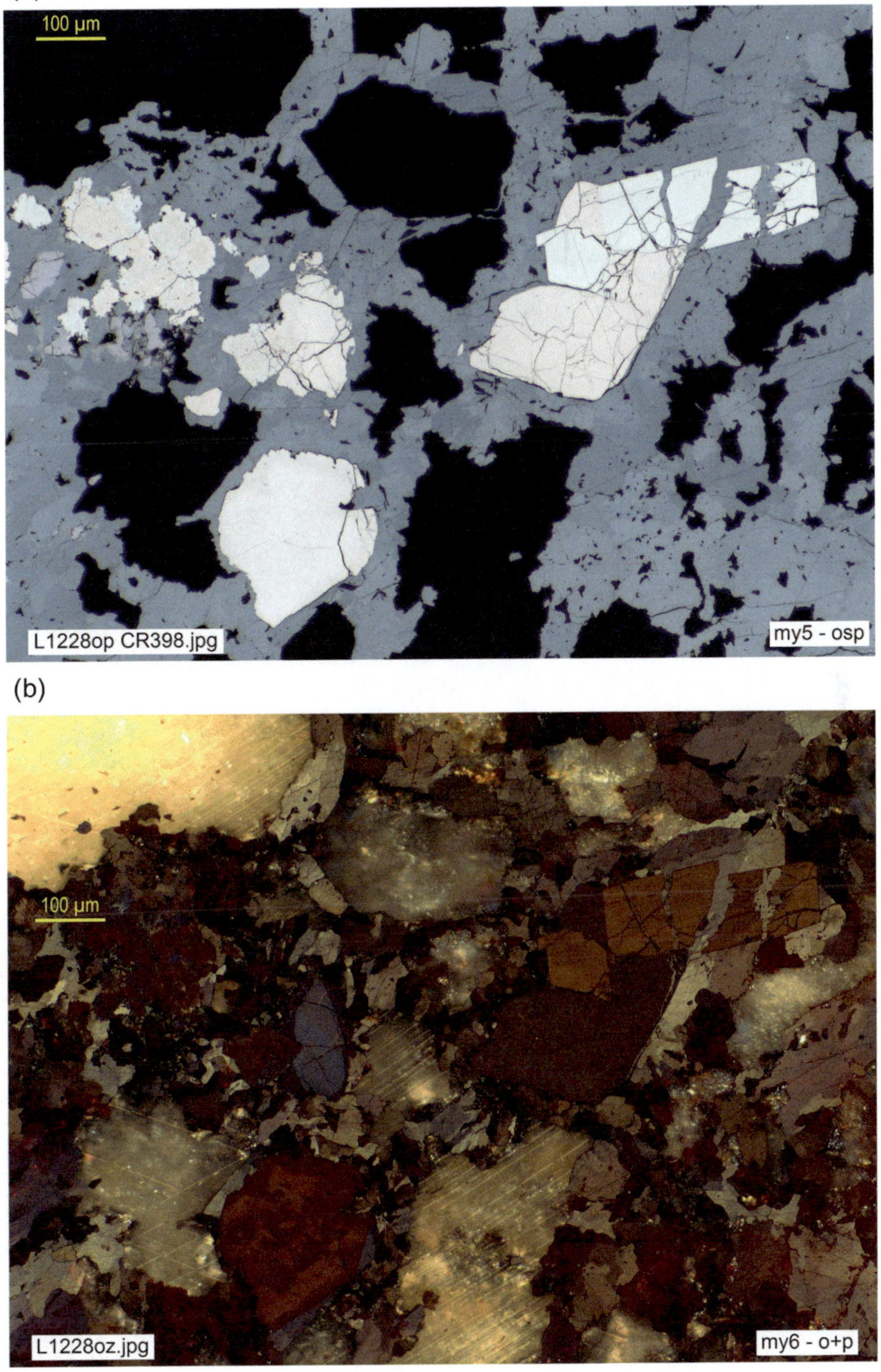

Fig. 83.2 **a, b** (**osp** and **o+p**, respectively)—Early, brecciated vein infill of dol (black, massive, colourless IR, polysynthetic twins), q (black, abundant IR) and asp (white, slightly pleochroic, anisotropic in variegated but poorly luminous shades, twinned), welded by later hydrothermal generation of my (grey, bireflective and anisotropic, with red IR) which corrodes and cements the pre-existing infill (dol, q, asp). CR398, Hiendelaencina, Guadalajara

Fig. 83.3 **a, b** (**osp** and **o+p**, respectively)—Hydrothermal filling of prg (bluish grey, very abundant red IR), my (greysh white with R > prg and scarcer IR), stb (pinkish brown, no IR), py (yellowish white) and gg. Replacement by prg of stb and py (only a few relicts left, included in prg) and of my (incipient). Note the differences in brightness, which indicate the progression of reflectance values: R gg < prg < stb < my < py. CR 336, Freiberg (Saxony, Germany)

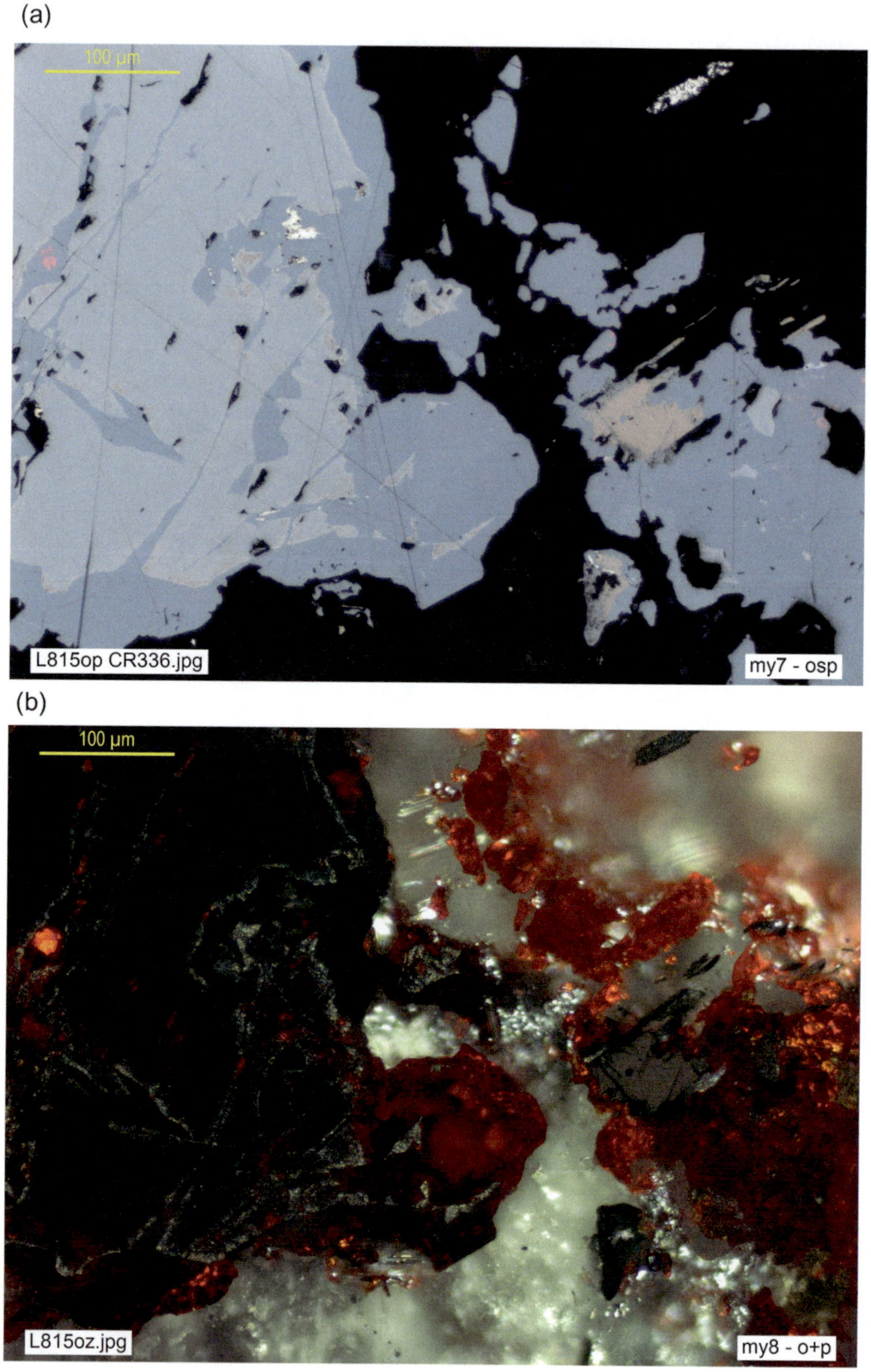

Fig. 83.4 **a, b (osp** and **o+p**, respectively—Aggregate of prg (bluish grey, with py microinclusions), stp (greyish brown, lower left third of image) and my (greyish white, centre). Besides faint hue and R nuances, the three Ag sulphosalts are distinguished by their bireflectance/pleochroism (my and prg, grey-white and bluish-grey shades, respectively; stp, slightly pleochroic, grey-brown to pinkish-brown), IR (very abundant in prg, frequent in my and absent in stp) and anisotropism (greenish in stp, poorly coloured or masked by IR in prg and my). CR 336, Freiberg (Saxony, Germany)

(a)

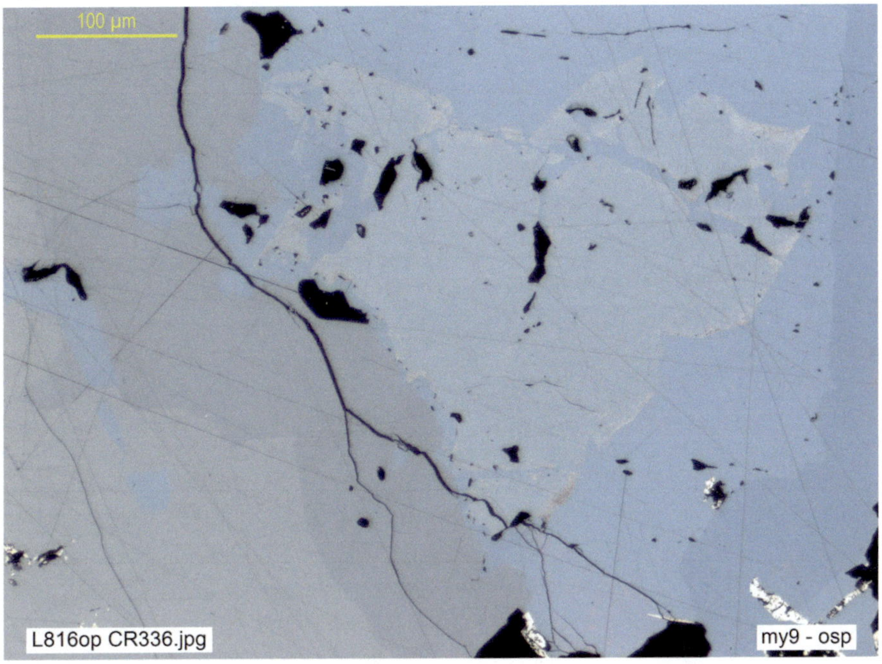

(b)

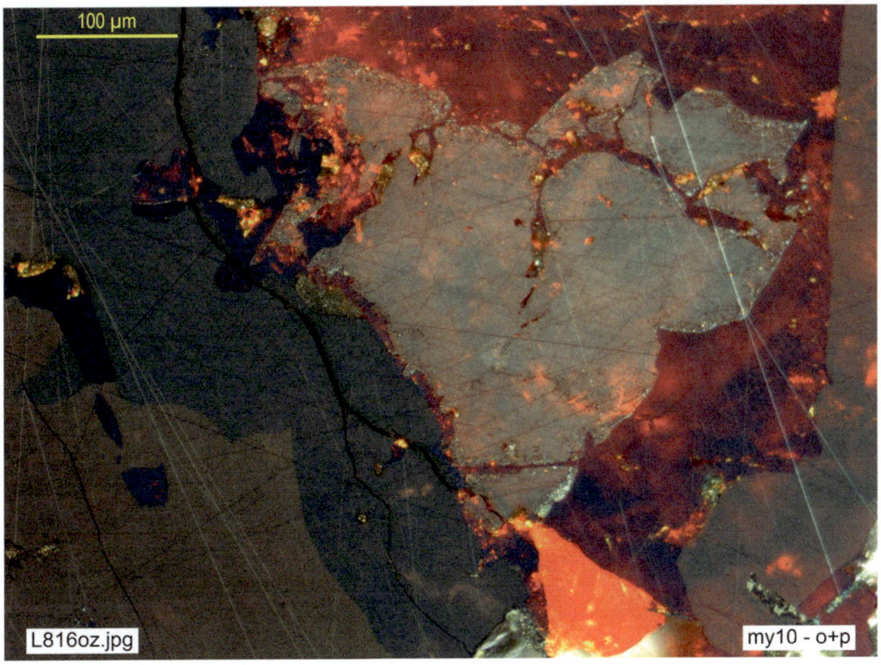

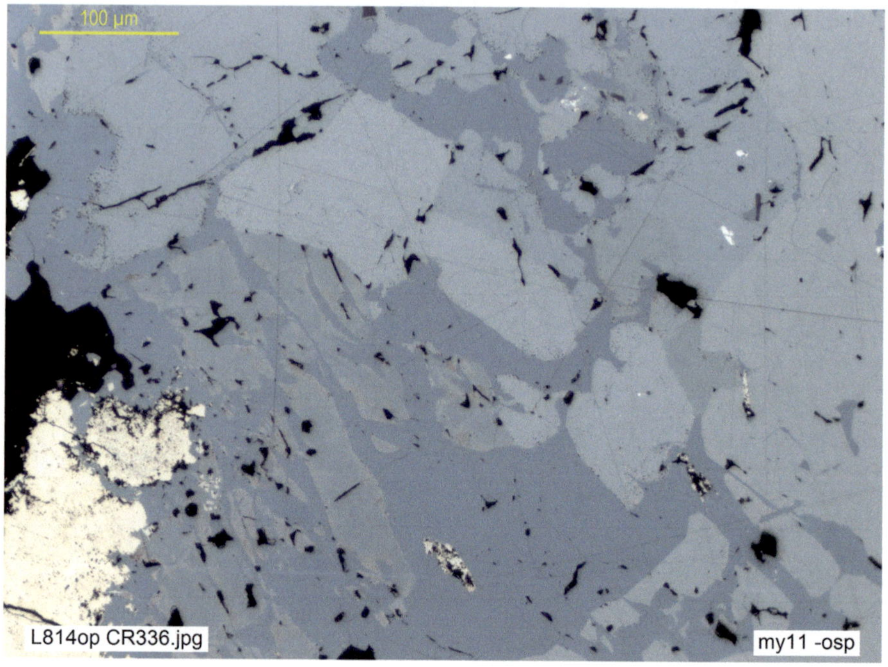

Fig. 83.5 Intergrowth of my (grey-white, distinctly bireflective), prg (bluish grey) and py (yellowish white, shiny), with partial replacement of my by prg (reaction edges). CR 336, Freiberg, Saxony, Germany (osp)

sulphosalts (cf. Figures 83.1a–d, 83.2a, b, 83.3a, b, 83.4a, b, 83.5 and 83.6a, b), *for example:*

- pyrargyrite is less reflective, has a bluish hue and many more IR.
- stephanite is less reflective but pleochroic, shows greenish anisotropism and lacks IR.
- polybasite is somewhat less reflective and more weakly anisotropic.
- freibergite is isotropic, as is galena.

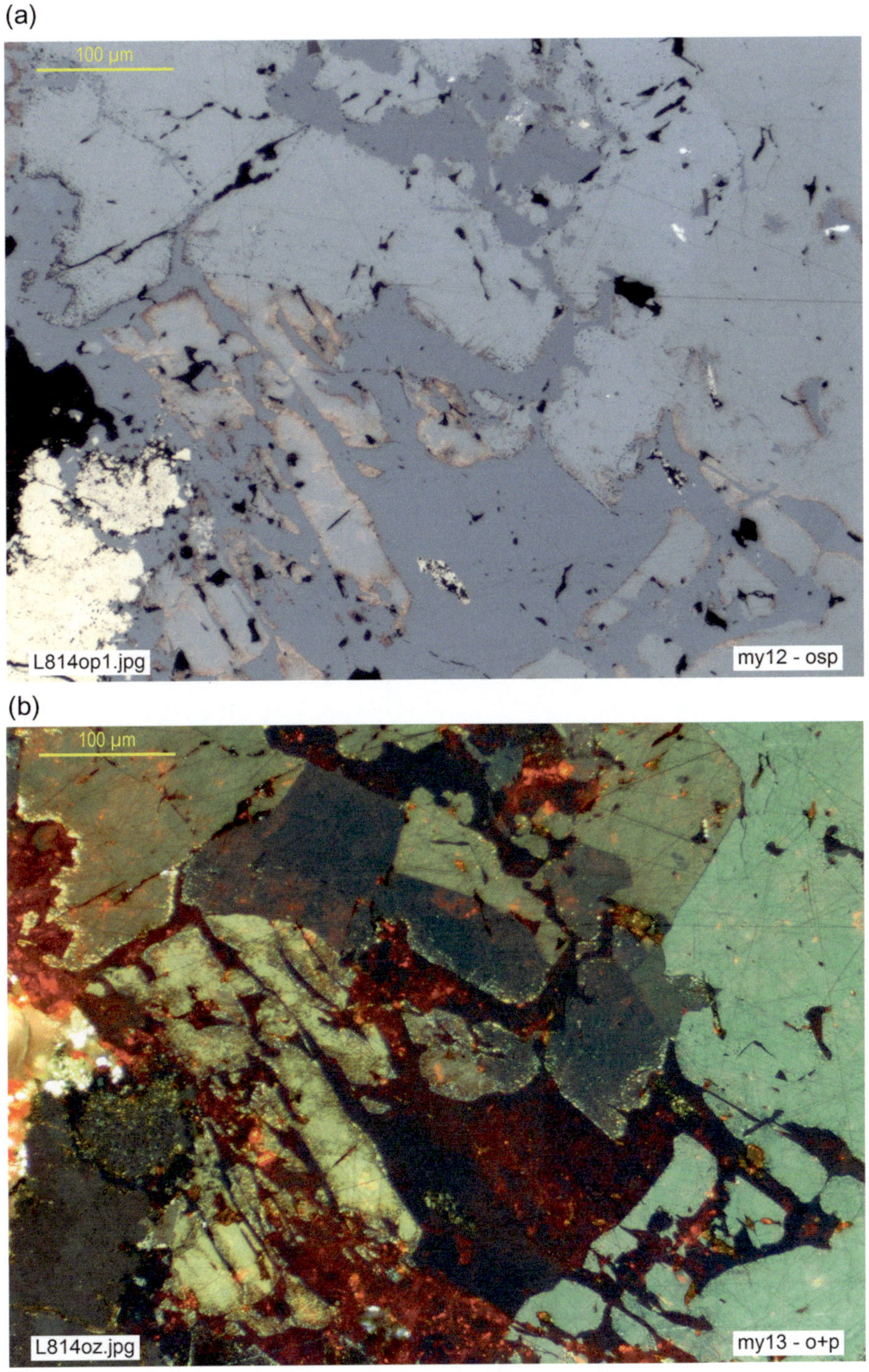

Fig. 83.6 abSame scene as my11, one minute later (**osp** and **o+p**, respectively). Light etching of my, bound to prg contacts, shows up as an irregular *corona* or etched rim, my darkens or changes colour along its margins (**osp**: my12). With crossed polars, it shows as a halo of bright spots (**o+p**, my13), which adds to the contrast of prg (massive IR) and my (more vivid anisotropism and coloured with bluish tones, less IR)

83.3 Spectral Information

83.3.1 Miargyrite Multispectral Reflectance R (in Air, Without Polariser) (%)

λ (nm)	370	400	425	450	475	500	525	550	575	600	625	650	675	700	750	800	850	900	950	1000
μ	35.7	36.0	35.5	35.5	34.9	34.3	33.6	32.7	32.1	31.3	30.5	29.7	29.2	28.8	28.0	27.5	27.3	26.9	26.8	26.6
P_5	34.0	34.3	33.8	33.7	33.0	32.5	31.8	31.0	30.3	29.6	28.8	28.2	27.8	27.5	26.8	26.3	26.1	25.8	25.6	25.4
P_{20}	34.9	35.2	34.7	34.7	34.1	33.5	32.7	31.8	31.1	30.4	29.5	28.8	28.4	28.1	27.3	26.9	26.7	26.3	26.2	25.9
P_{80}	36.5	36.9	36.3	36.3	35.7	35.1	34.4	33.5	32.9	32.2	31.4	30.6	30.0	29.5	28.6	28.0	27.7	27.4	27.3	27.1
P_{95}	37.4	37.8	37.5	37.8	37.3	36.5	35.5	34.5	33.9	33.1	32.1	31.2	30.4	29.9	29.0	28.3	28.1	27.7	27.6	27.4

Comparison of Specular Reflectance values (VNIR, AMCO data) from miargyrite, native silver, galena and goethite

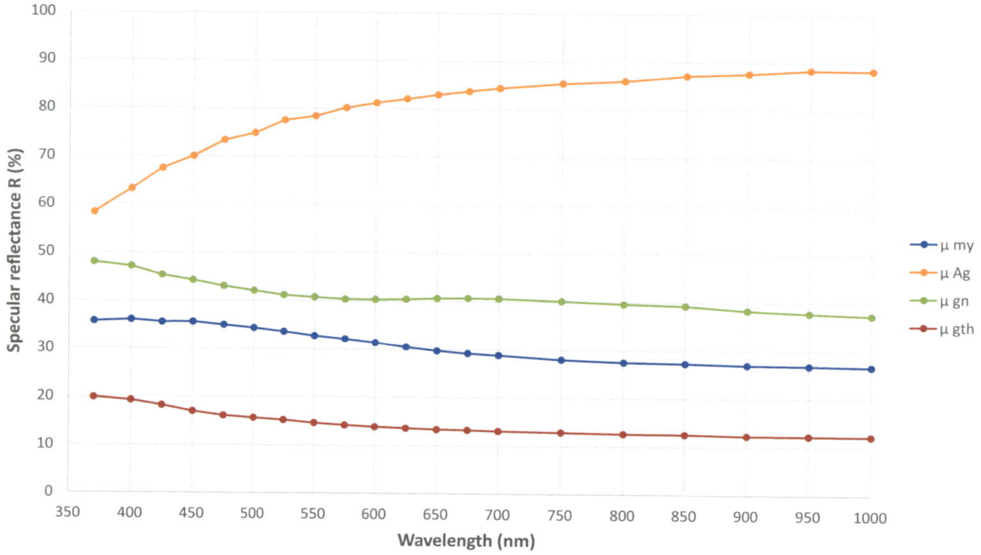

Specular Reflectance values in the visible spectrum (QDF3) from miargyrite R1, miargyrite R2, pyrargyrite Ro & pyrargyrite Re'

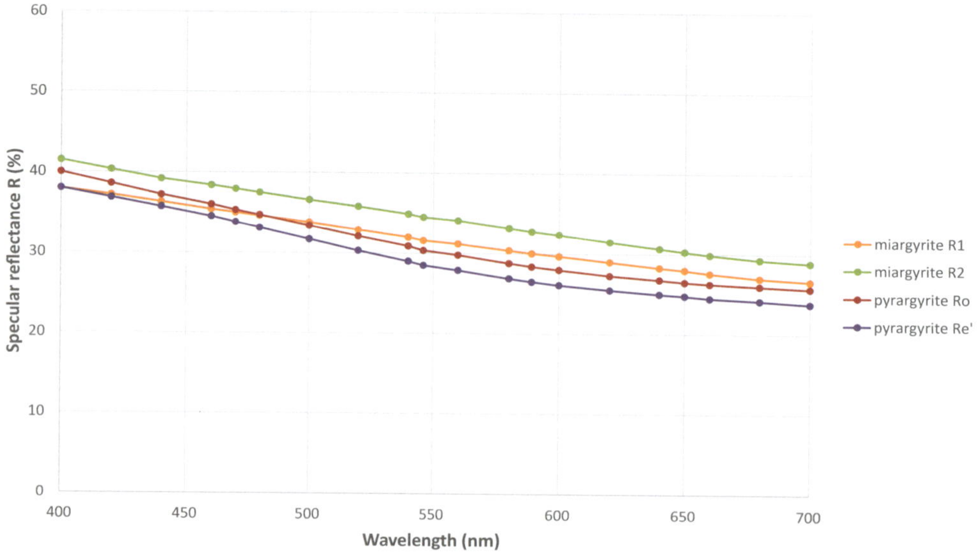

Specular Reflectance values in the visible spectrum (Picot and Johan, 1982) from miargyrite Rmax, miargyrite Rmin, sternbergite Rmax & sternbergite Rmin

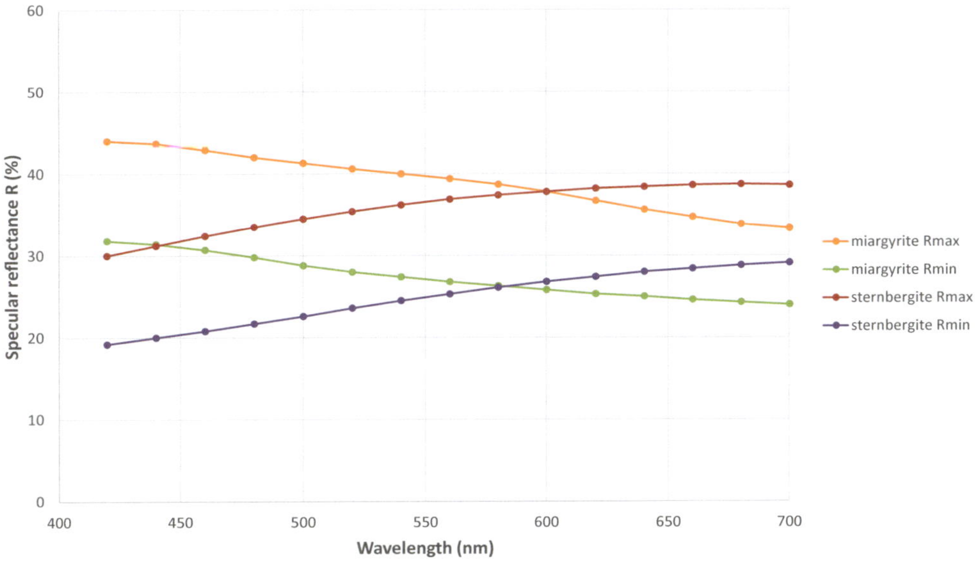

Comparison of Specular Reflectance VNIR values from miargyrite (AMCO), miargyrite R1, miargyrite R2, pyrargyrite Ro & pyrargyrite Re' (QDF3), and sternbergite Rmax & sternbergite Rmin (Picot and Johan, 1982)

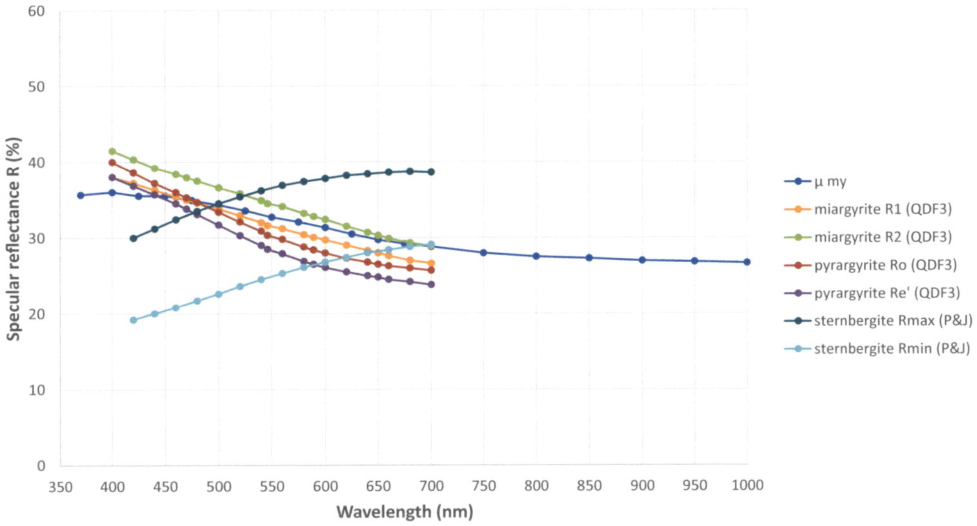

Comparison of Specular Reflectance VNIR values from miargyrite (AMCO), and miargyrite Rmax, miargyrite Rmin, polybasite R & freibergite R (Picot and Johan, 1982)

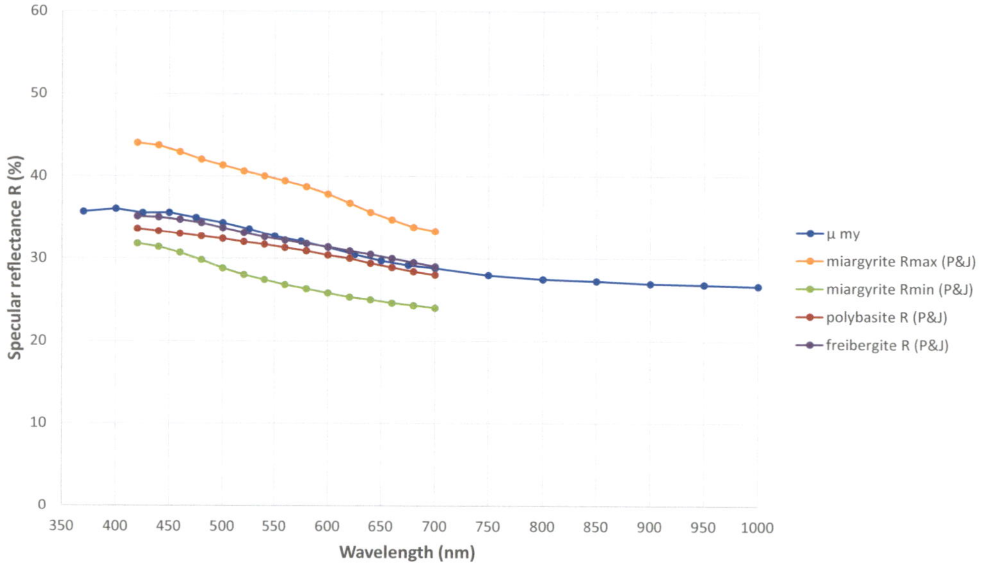

Specular Reflectance values in the visible spectrum (QDF3) from miargyrite R1, miargyrite R2, stephanite R1 & stephanite R2

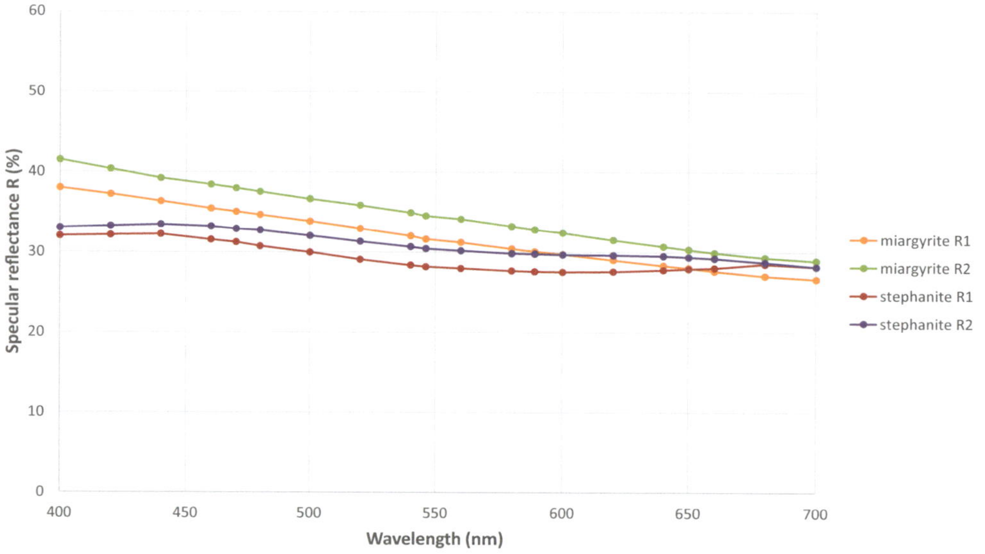

References

Picot P, Johan Z (1982) Atlas of ore minerals. B.R.G.M./Elsevier

Uytenbogaardt W, Burke EAJ (1971) Tables for microscopic identification of ore minerals, 2nd edn. Elsevier, Amsterdam

Millerite (*ml*)

84

NiS: Trigonal

Millerite		R_{546} (%)	Y (%)	λ_d (nm)	P_e (%)	Colour sensation
Air	R_o	50	50.3	575	12.5	Light yellow, slightly pleochroic
	R_e	54.2	54.5	575	19.4	
Oil	R_o	37.9	38.2	575	13.1	Yellow more intense and more pleochroic
	R_e	43.8	43.9	575	21.5	
Bir: Weak (3)				*Pleoc*: Weak but net(3)		*Tarn*: No
Anis + P: Strong (5)				*IR*: No		*Twg*: Lamellar
Mor: Acicular; sometimes massive				*Clv*: // (101–1)		*PolQ*: Good
Hardness	*PolH*: Medium >ccp, < sp (3)			*Mohs*: 3+	*Talm*: E	*VHN* $_{(20\ g)}$: 179

Obs: cf. Table and R charts, multispectral data (VNIR, AMCO) and visible range (QDF3)
Mineral abbreviations: brv—bravoite, cc—chalcocite, ccp—chalcopyrite, gf—gersdorffite, gn—galena, ln—linnaeite, ml—millerite, m—py, melnikovite py, pdm—polydymite, pn—pentlandite, po—pyrrhotite, py—pyrite, sp—sphalerite, ul—ullmannite, vio—violarite

84.1 General

Composition: NiS, may have some Co instead of Ni and traces of Fe and Cu.

84.2 Microscopic Description

84.2.1 SP Observation (*Single Polariser*)

Light yellow *colour*, similar to chalcopyrite but more clearly bireflective and *pleochroic* (from light yellow, close to chalcopyrite, to yellowish white, close to pyrite), especially in immersion (comp. Fig. 84.1a, b with Fig. 84.1c, d). Good *polish*, although acicular aggregates can create problems.

High *reflectance*, ***R***, comparable to that of pyrite but lower for $\lambda < 500$ nm (pyrite is whiter) and distinctly > R py in the NIR range (750–1000 nm). Curve parallel to that of chalcopyrite but higher (*cf* table and multispectral graphs).

Cleavage// $(10\bar{1}1)$ visible, even as relict in pseudomorphs (Figs. 84.2 and 84.3).

Polishing hardness medium, somewhat >ccp, <sp, ln, po, <<py (Fig. 84.3).

It occurs in acicular aggregates, of euhedral (trigonal) section, more rarely massive, in anhedral aggregates.

84.2.2 Observation +P (*Polariser + Analyser at 90°*)

Strongly *anisotropic*, with very bright tints (bright yellow to light grey with bluish-green tones, Fig. 84.1e, f).

Occasionally distinct twinning (Fig. 84.1f), very fine lamellar twins $(00\bar{1}2)$ visible in the massive aggregates; //c and pressure twins have also been reported.

84.2.3 Paragenesis

Millerite is commonly a late hypogene mineral, primary in hydrothermal veins (with chalcopyrite, linnaeite, galena, sphalerite, ullmannite) or secondary, associated with violarite and replacing pentlandite in metamorphic formations with Ni-pyrrhotite.

It may also form as a supergene mineral, due to weathering of sulphide masses (linnaeite, chalcopyrite, pyrite,

R. Castroviejo, *A Practical Guide to Ore Microscopy—Volume 1*,
https://doi.org/10.1007/978-3-031-12654-3_84

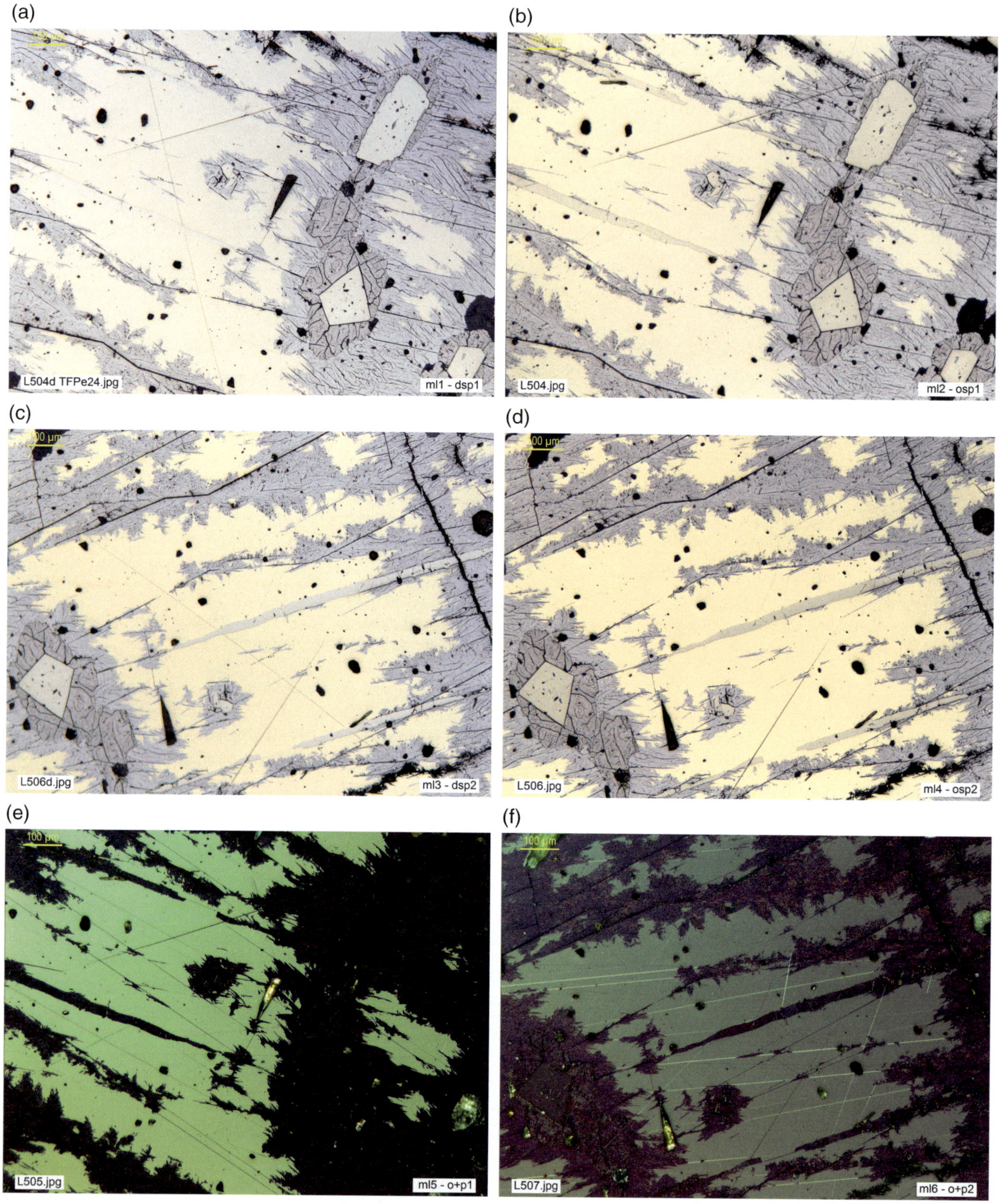

Fig. 84.1 **a, b** (Same scene: **dsp/osp**, *respectively*). Massive aggregate of ml (yellow), with euhedral or elongated inclusions of py (white, by comparison to ml). Partial alteration of ml to pdm (pinkish grey against ml) and total alteration of a missing mineral to Ni-vio (purplish grey against ml, darker than pdm; forms encrustations around intact py). TFPe24, Dry Nickel Mine, Bindura, Zimbabwe. **c, d** (Same scene as Fig. 84.1a, respectively, **dsp** and **osp**, with stage rotated 90°) Reinforcement of yellow colour evidences the pleochroism of ml which, in the first position, could be confused with py. The thin py microbands cutting through ml, clearly seen in Fig. 84.1c, d, are practically invisible in F ml1. **e** Same scene as ml1 (**o+p**): anisotropism of ml, in contrast with the isotropy of the other components. **f** Same scene as ml4, rotating the stage: the thin lamellar twins of ml are illuminated (**o+p**)

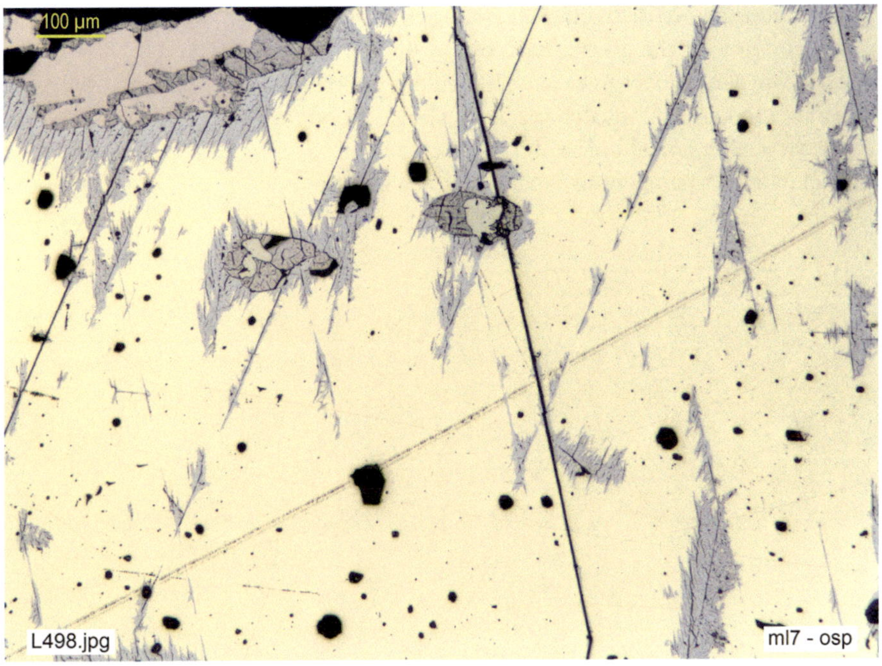

Fig. 84.2 Massive ml (yellow), partially altered to pdm (light purplish grey), with vio (pink, top left) partly replaced by Fe-pdm (or linnaeite-group mineral of intermediate composition between pdm and vio, after EMPA). Sparse relict s of py (white, relief) scattered in ml. Sample TFPe24, Dry Nickel Mine, Bindura, Zimbabwe (**osp**)

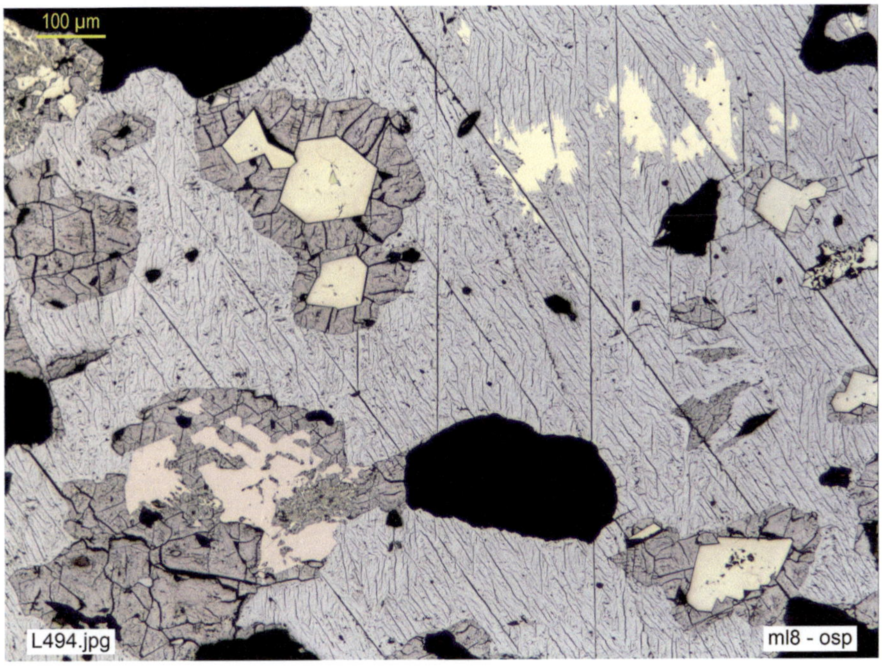

Fig. 84.3 Advanced pseudomorphic replacement of massive ml by pdm, preserving intact euhedral inclusions of py and relicts of vio (mostly replaced by Fe-pdm), relicts of py and secondary melnicovitic py (top left corner). Compared to Fig. 84.2, the replacement of ml by pdm is almost total, but the process is the same, progressing along the rhombohedral cleavage planes of ml, which are perfectly preserved in the secondary pdm. Note the relief of py, in contrast to the much softer ml. TFPe24, Dry Nickel Mine, Bindura, Zimbabwe (**osp**)

gersdorffite), and occurring in the transition to the secondary sulphide enrichment blanket or cementation zone, associated with bravoite, chalcocite.

84.2.4 Identification

Its high reflectance, colour, pleochroism and characteristic anisotropism (luminous and coloured from yellow to bluish-green / grey) ensure its recognition.

- It is immediately distinguished from pyrite by its anisotropism and by the greater hardness of pyrite.
- Chalcopyrite, which gives a similar colour impression, is softer, darker, less pleochroic and, above all, shows a much weaker anisotropism.
- Nickeliferous paragenesis is an additional criterion.

84.3 Spectral Information

84.3.1 Millerite Multispectral Reflectance R (in Air, Without Polariser) (%)

λ (nm)	370	400	425	450	475	500	525	550	575	600	625	650	675	700	750	800	850	900	950	1000
μ	20.6	24.2	29.8	36.3	41.5	46.2	49.4	51.7	53.6	54.8	55.8	56.7	57.4	58.0	59.0	59.9	61.5	62.2	62.6	63.5
P_5	19.2	22.7	28.0	34.4	39.5	44.4	47.8	50.4	52.3	53.7	54.9	55.8	56.6	57.2	58.4	59.3	60.9	61.6	61.9	63.0
P_{20}	20.1	23.7	29.2	35.8	40.9	45.7	48.9	51.3	53.2	54.5	55.5	56.4	57.1	57.7	58.8	59.6	61.2	61.9	62.3	63.3
P_{80}	21.2	24.9	30.5	37.1	42.2	46.9	50.0	52.3	54.1	55.3	56.3	57.1	57.8	58.3	59.3	60.2	61.8	62.6	63.0	63.8
P_{95}	21.5	25.2	30.9	37.5	42.6	47.3	50.4	52.7	54.4	55.5	56.6	57.4	58.1	58.6	59.7	60.5	62.0	62.8	63.2	64.0

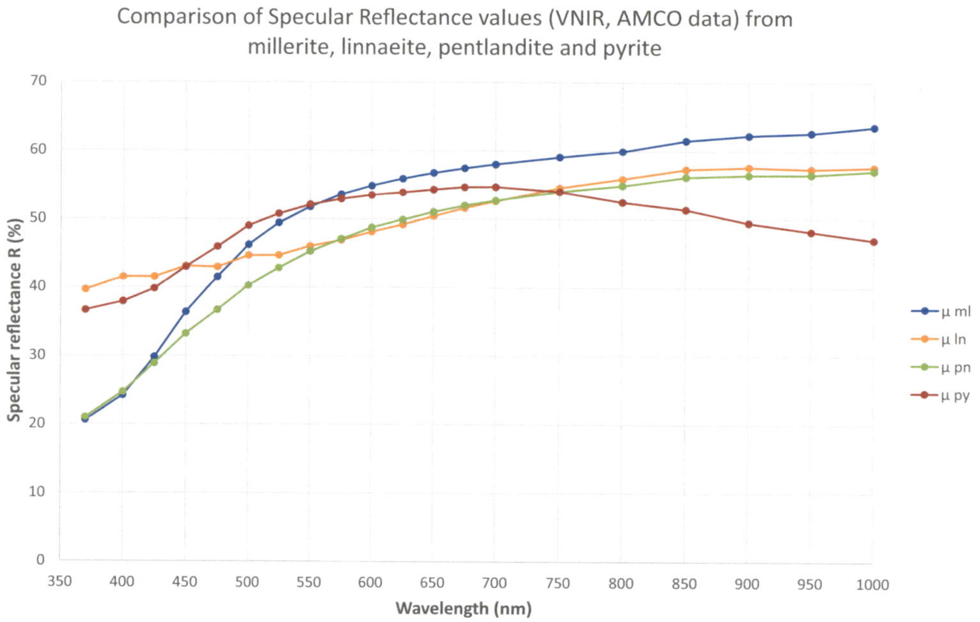

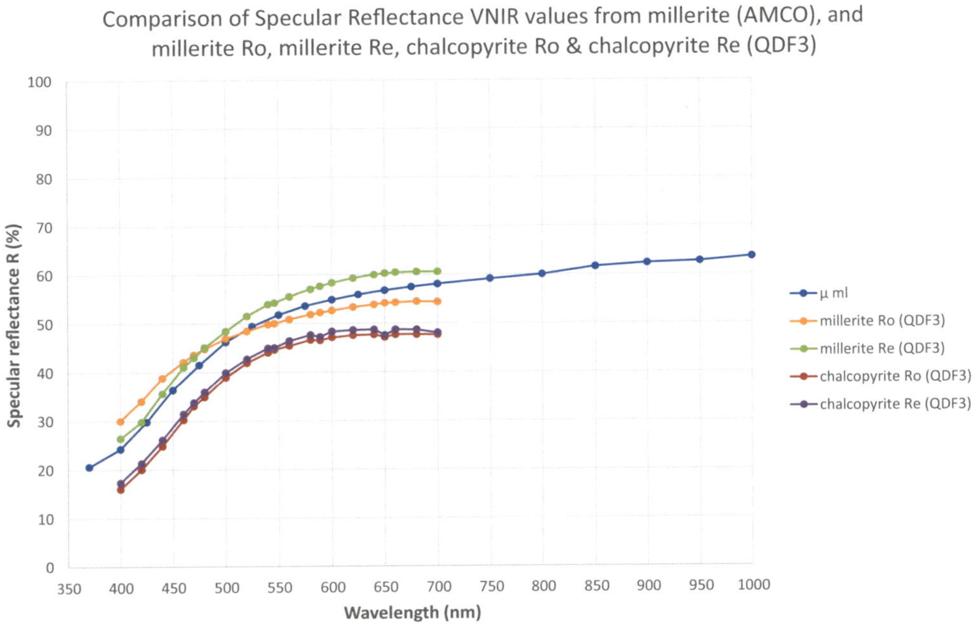
Comparison of Specular Reflectance VNIR values from millerite (AMCO), and millerite Ro, millerite Re, chalcopyrite Ro & chalcopyrite Re (QDF3)
Specular reflectance R (%)
Wavelength (nm)
100
90
80
70
60
50
40
30
20
10
0
350
400
450
500
550
600
650
700
750
800
850
900
950
1000
µ ml
millerite Ro (QDF3)
millerite Re (QDF3)
chalcopyrite Ro (QDF3)
chalcopyrite Re (QDF3)

Molybdenite (*mol*) 85

MoS_2: Hexagonal

Molybdenite		R_{546} (%)		Y (%)	λ_d (nm)		P_e (%)	Colour sensation
Air	R_o	38.6		39.3	468		6.6	White–grey, strongly bireflective
	$R_{e'}$	19.5		19.5	480		4.2	
Oil	R_o	24.1		24.9	470		11.6	
	$R_{e'}$	7.9		7.9	478		11.9	
Bir: Very strong (5)					*Pleoc*: Weak (3)			*Tarn*: Not observed
Anis +P: Very strong (5)					*IR*: No			*Twg*: nd
Mor: Leafy, tabular. Colom. (*jordisite*)					*Clv*: // (0001)			*PolQ*: Difficult (soft, unctuous)
Hardness		*PolH*: Medium, > gn, < gra (3)			*Mohs*: 1–1½	*Talm*: B+		*VHN* $_{(-)}$: 14.6–15.6 (basal sections)

Obs: cf. Table and R charts, multispectral data (VNIR, AMCO) and visible range (QDF3). VHN: Uytenbogaardt and Burke (1971)
Mineral abbreviations: Ag—native silver, asp—arsenopyrite, Au—native gold, Bi—native bismuth, bm—bismuthinite, bn—bornite, cbn—cubanite, ccp—chalcopyrite, cst—cassiterite, gg—gangue, gn—galena, gra—graphite, mol—molybdenite, mt—magnetite, po—pyrrhotite, py—pyrite, q—quartz, rt—rutile, sp—sphalerite, td—tetrahedrite, wf—wolframite

85.1 General

Molybdenite, ***mol***, is the main ore of Mo. Composition: MoS_2, occasionally with traces of Re. Polytypes, mol-2H (hexagonal) and mol-3R (trigonal). Dimorph, *jordisite* (amorphous, possibly a colloidal precursor of molybdenite; however, its supposed amorphous character is questioned, cf Kao et al. 2001).

85.2 Microscopic Description

85.2.1 SP Observation (*Single Polariser*)

Colour white to grey, strongly bireflective, but weakly or not at all pleochroic (R changes with no colour change, only from white to darker or lighter grey, Fig. 85.1a, b and multispectral comparative plots).

Moderately high *reflectance*, close to that of galena: R_{mol} > R wf, but < R bm, ccp.

It is the enormous *bireflectance that is* striking, as in the case of graphite. However, the minimum values (R_e) of molybdenite are always distinctly higher than graphite minimum R_e. They are also always > R of gangue and even of ores such as cassiterite (cf Fig. 85.1a, b). The strong bireflectance of molybdenite is clearly reflected in the variation of the multispectral values of R, even without polariser. This can be seen in the AMCO-QDF3 comparative graph, in which the P_5-P_{95} percentiles of AMCO have been plotted. The curves corresponding to these occupy a wide area between R_o and $R_{e'}$ of QDF3; however, all the curves always remain parallel, as corresponds to the constant colour of the mineral.

Difficult to *polish*, due to its low hardness and smoothness. Medium–low *polishing hardness* (near the limit between both ranges): PolH mol > PolH gn, bm; ≈ ccp; < gra, Au, td, sp.

Molybdenite occurs in sheets or leafy aggregates, with clear *cleavage*//(0001), rarely massive. The supposedly amorphous *jordisite*, described as a dimorph of molybdenite of colloidal origin, with globular or colomorphic morphology, is possibly a precursor of molybdenite precipitated in conditions of low temperature or strong disequilibrium. In one sample, Ramdohr (1982) observed a transition to

R. Castroviejo, *A Practical Guide to Ore Microscopy—Volume 1*,
https://doi.org/10.1007/978-3-031-12654-3_85

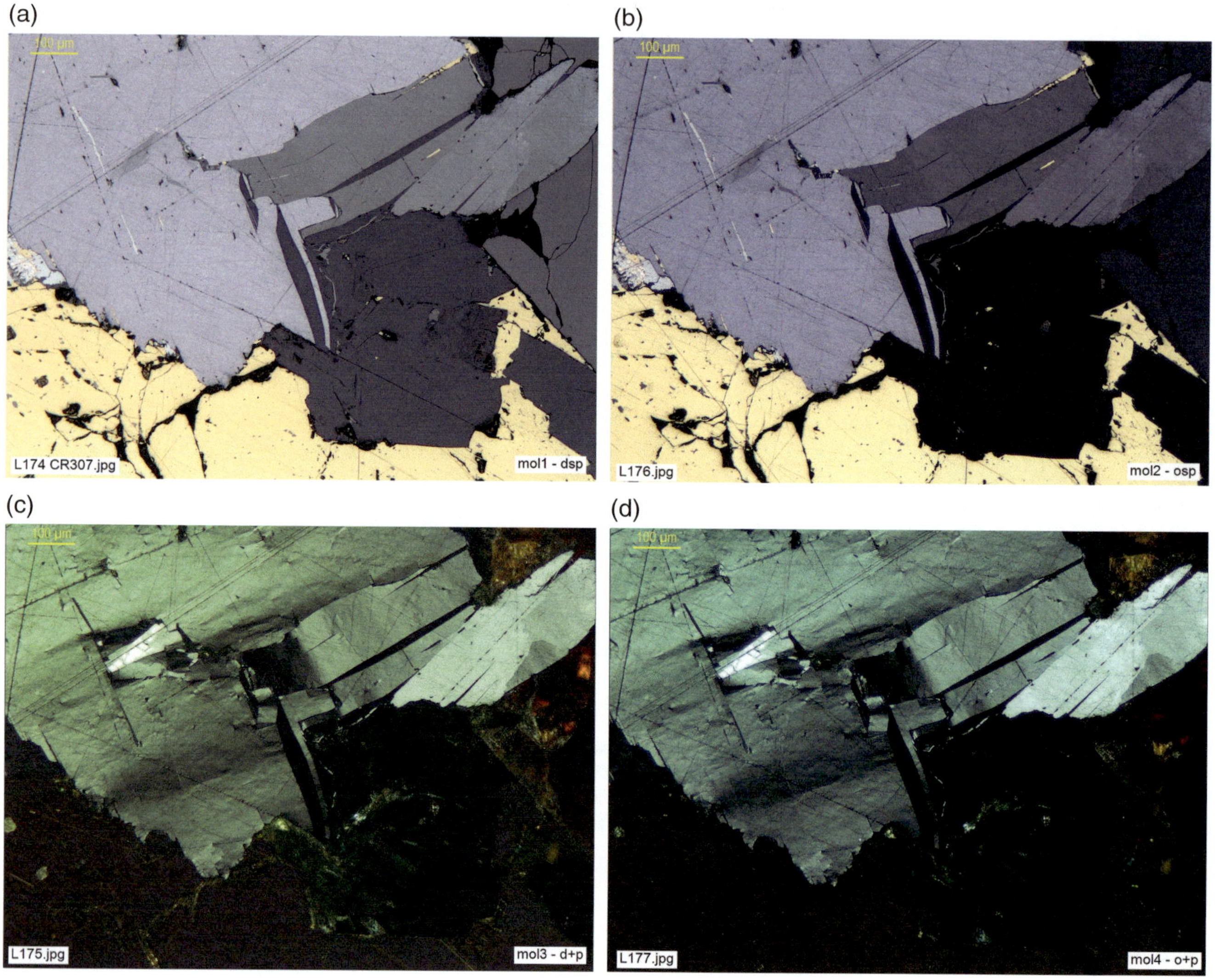

Fig. 85.1 **a**, **b** (**dsp** and **osp**, respectively). Molybdenite ore (white–grey, strongly bireflective), ccp (yellow), cst (dark grey, on the right in the image), bm (light bluish grey, center left and inclusions marking mol exfoliation), Bi (white, included in bm, left) and mica (dark grey/black in oil, lower right). CR 307, Las Sombras Mine, S[a] Jurés, Orense. **c**, **d** Same scene as mol1 (**d+p** and **o+p**, respectively). Strong anisotropism of mol. IR of cst and mica

anisotropic masses (?recrystallization) and, as mentioned, the amorphous character has been questioned in modern studies.

Due to its ductility, molybdenite easily undergoes *deformations*, which are recorded in the form of microfolds, kinks, undulose extinction, etc. These are strikingly visible due to the extreme bireflectance and anisotropism of the mineral.

85.2.2 Observation +P (*Polariser + Analyser at 90°*)

Extreme *anisotropism* but poorly coloured, without the bright yellowish tones of graphite (Fig. 85.1c, d).

85.2.3 Paragenesis

Paragenesis typically hydrothermal, pegmatitic-pneumatolitic (accompanied by cassiterite, bismutinite, wolframite, arsenopyrite, chalcopyrite, sphalerite, galena, etc.) and magmatic-hydrothermal (Cu-Mo porphyries, accompanied by pyrite, magnetite, chalcopyrite, bornite, cubanite, etc.).

85.2.4 Identification

The identification of molybdenite does not usually offer difficulties, given the peculiarity of its optical properties: low hardness, sheety or leafy appearance, with perfect exfoliation and frequent deformation, white–grey colour with extreme bireflectance and anisotropism, but poorly coloured. In particular, it is distinguished from graphite by its higher reflectance and by lacking the yellowish polarisation tints of graphite.

85.3 Spectral Information

85.3.1 Molybdenite Multispectral Reflectance R (in Air, Without Polariser) (%)

λ (nm)	370	400	425	450	475	500	525	550	575	600	625	650	675	700	750	800	850	900	950	1000
μ	26.8	28.4	29.2	30.7	30.6	29.4	28.2	27.5	27.0	27.4	28.0	27.6	28.5	28.2	27.1	26.7	26.5	26.2	26.1	26.1
P_5	18.4	20.6	21.9	23.5	23.7	23.0	22.2	21.7	21.4	21.8	22.3	22.1	23.0	22.7	21.8	21.6	21.5	21.4	21.4	21.5
P_{20}	21.6	23.7	25.0	26.8	26.9	26.0	25.1	24.5	24.1	24.4	25.0	24.7	25.4	25.2	24.2	23.8	23.7	23.4	23.3	23.4
P_{80}	32.6	33.2	33.4	34.4	34.0	32.5	31.1	30.2	29.6	30.0	30.7	30.4	31.3	31.1	29.8	29.4	29.3	28.8	28.7	28.6
P_{95}	33.8	34.5	35.0	36.2	36.3	34.4	33.1	31.8	31.3	31.6	32.3	31.8	32.8	32.6	31.1	30.6	30.3	29.9	29.7	29.9

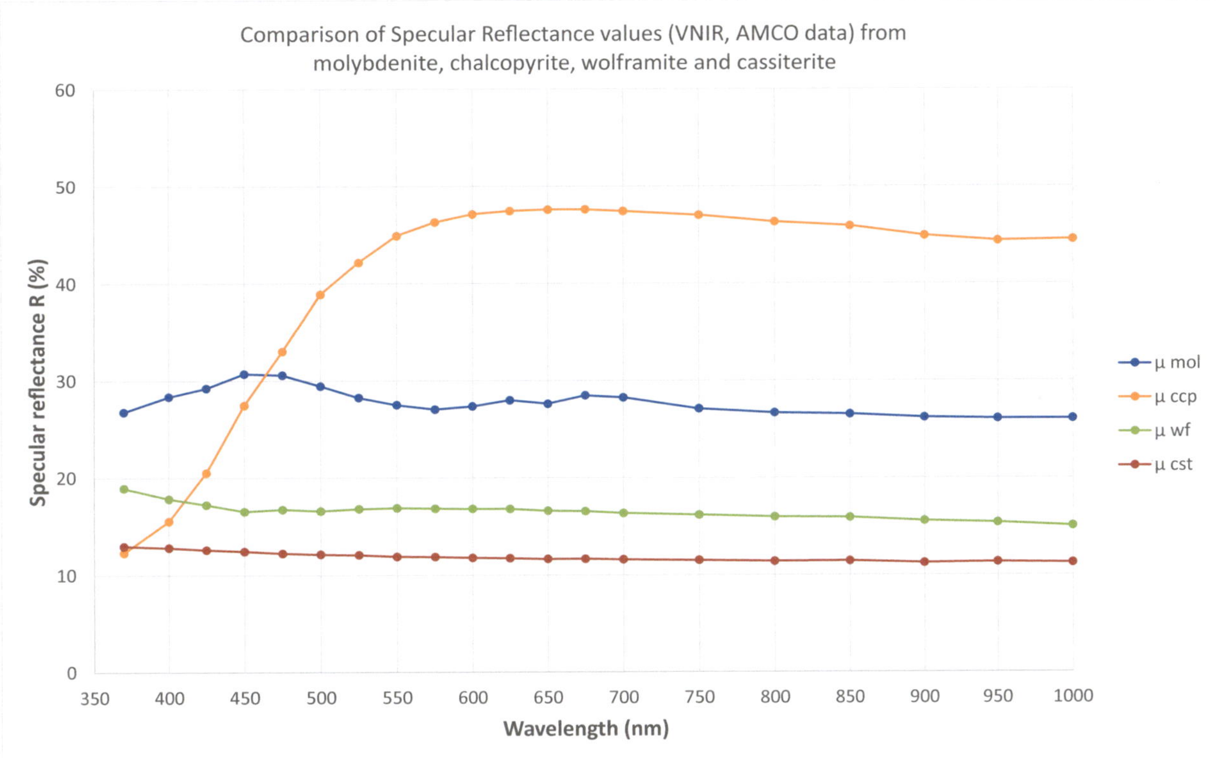

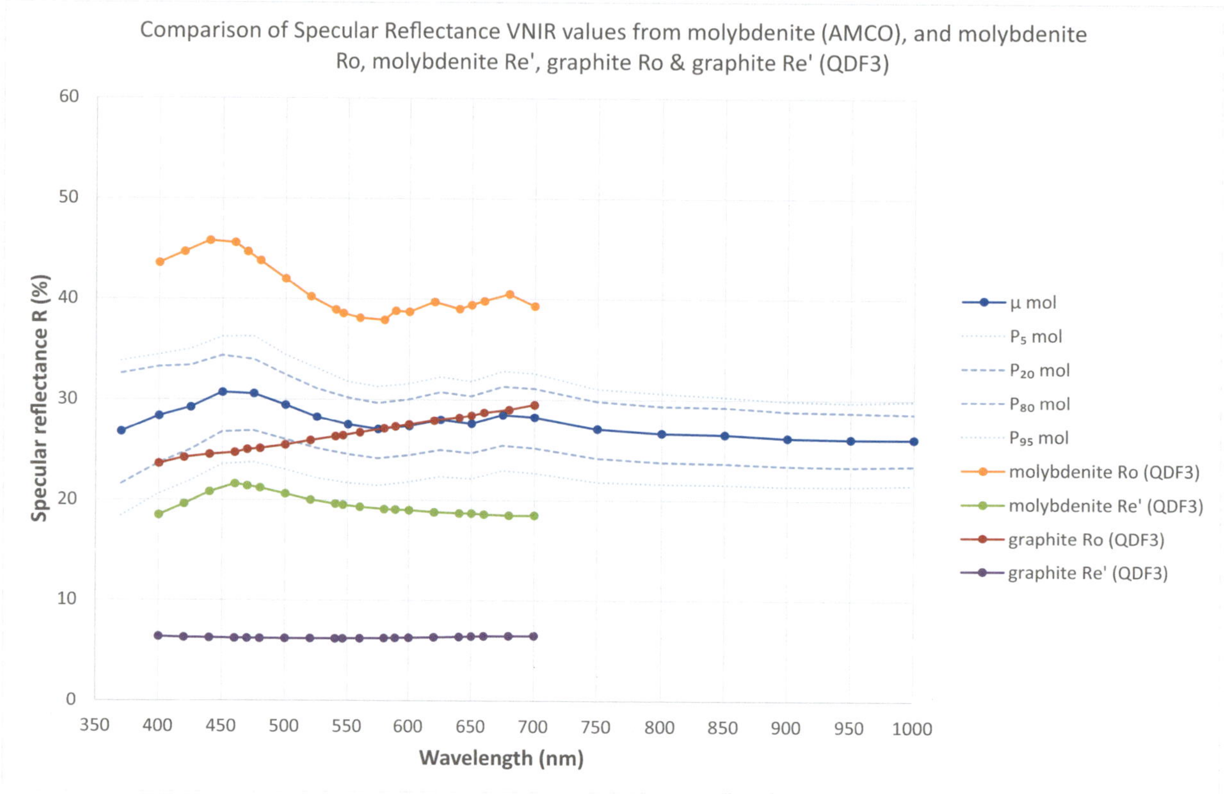

References

Kao L, Peacor DR, Coveney RM Jr, Zhao G, Dungey KE, Curtis MD, Penner-Hahn JE (2001) A C/MoS_2 mixed-layer phase (MoSC) occurring in metalliferous black shales from southern China, and new data on jordisite. Am Miner 86:852–861

Uytenbogaardt W, Burke EAJ (1971) Tables for microscopic identification of ore minerals, 2nd edn. Elsevier, Amsterdam

86 Monazite (*mz*)

(Ce,La,Nd,Th)PO_4: Monoclinic

Monazite			R_{546} (%)	Y (%)	λ_d (nm)	P_e (%)	Colour sensation
Air	R_1		8.7	nd	nd	nd	Grey
	R_2		nd	nd	nd	nd	
Oil	R_1		nd	nd	nd	nd	Dark grey
	R_2		nd	nd	nd	nd	
Bir: Weak / very weak (3 /2)					*Pleoc*: Imperceptible (2)		*Tarn*: No
Anis + *P*: Dubious, overprinted by IR (4)					*IR*: Yellowish, variegated		*Twg*: Varied
Mor: Idiomorphic, tabular, nodular…					*Clv*: (100)		*PolQ*: Acceptable
Hardness		*PolH*: Medium (3)			*Mohs*: 5	*Talm*: nd	*VHN* $_{(100\ p)}$: nd

Obs: cf. Table and R charts based on multispectral VNIR data (AMCO), no QDF3 data in the visible range
Mineral abbreviations: cst—cassiterite, gg—gangue, gth—goethite, ht—hematite, lim—limonite, mgh—maghemite, mt—magnetite, mz—monazite, q—quartz, REE—Rare Earth Element, rt—rutile, sch—scheelite, spn—spinel, tan—tantalite;,tit—titanite, tp—tapiolite, wf—wolframite, zir—zircon

86.1 General

Monazite, ***mz***, is a relatively scarce ore, but it iis of considerable *economic interest* these days (e.g., as a source of Nd).

Its composition is complex and varied. Different formulas have been proposed, varying the **REE**, Rare Earth Element, contained (Ce, La, Nd, Sm, Eu). There may also be a substitution of bi- or trivalent REE by Th (in combination with Ca, P, Si), replacing in part (PO_4) by (SiO_4, SO_4) or introducing U.

At present, several species of monazite are defined according to the dominant REE: Ce, Sm, Nd, Monazite-(Ce), Ce(PO_4) by far being the most common. All of them are part of the Monazite Group, comprising phosphates and arsenates, under the formula[1] **MTO_4**, where **M** = REE, Th, Ca, Bi and **T** = P, As.

With respect to the possible SiO_4 contents, experimental data from various researchers, summarized by Chang et al. (1998) suggest that monazite is isostructural with *cheralite*, (REE, Th, Ca, U) (P, SiO_4), and that *huttonite* ($ThSiO_4$) represents the final silicate term of the series, thus defining the ternary system $TR(PO_4)-Ca_3(PO_4)_2-ThSiO_4$.

In spite of these precisions and, as usually happens with REE minerals, the differentiation of monazite-(Ce), monazite-(Sm), etc. species from each other is not reliable by traditional optical methods, since it is based on chemical information that rarely translates into relevant differences under the microscope. Chang et al. (1998) distinguish variations from the typical light yellow colour for some specific types: dull grey in Eu-rich monazite or monazite-(Eu), light pink in monazite-(Nd), brown, yellow or grey in high metamorphic grade monazite, etc. However, from the practical point of view of ore microscopy, the classical concept of monazite as a mineral of variable composition is more realistic.

The infrared absorption spectrum shows a broad absorption band at 9–10 μm, as occurs in other phosphates. Strong absorption bands also appear at 5252 and ≈570 nm attributed to REE and particularly to Nd (Omori and Seki 1960, *in*: Chang et al. 1998).

[1] *cf* https://www.mindat.org/min-29313.html, *visit. 2020.12.12.*

R. Castroviejo, *A Practical Guide to Ore Microscopy—Volume 1*,
https://doi.org/10.1007/978-3-031-12654-3_86

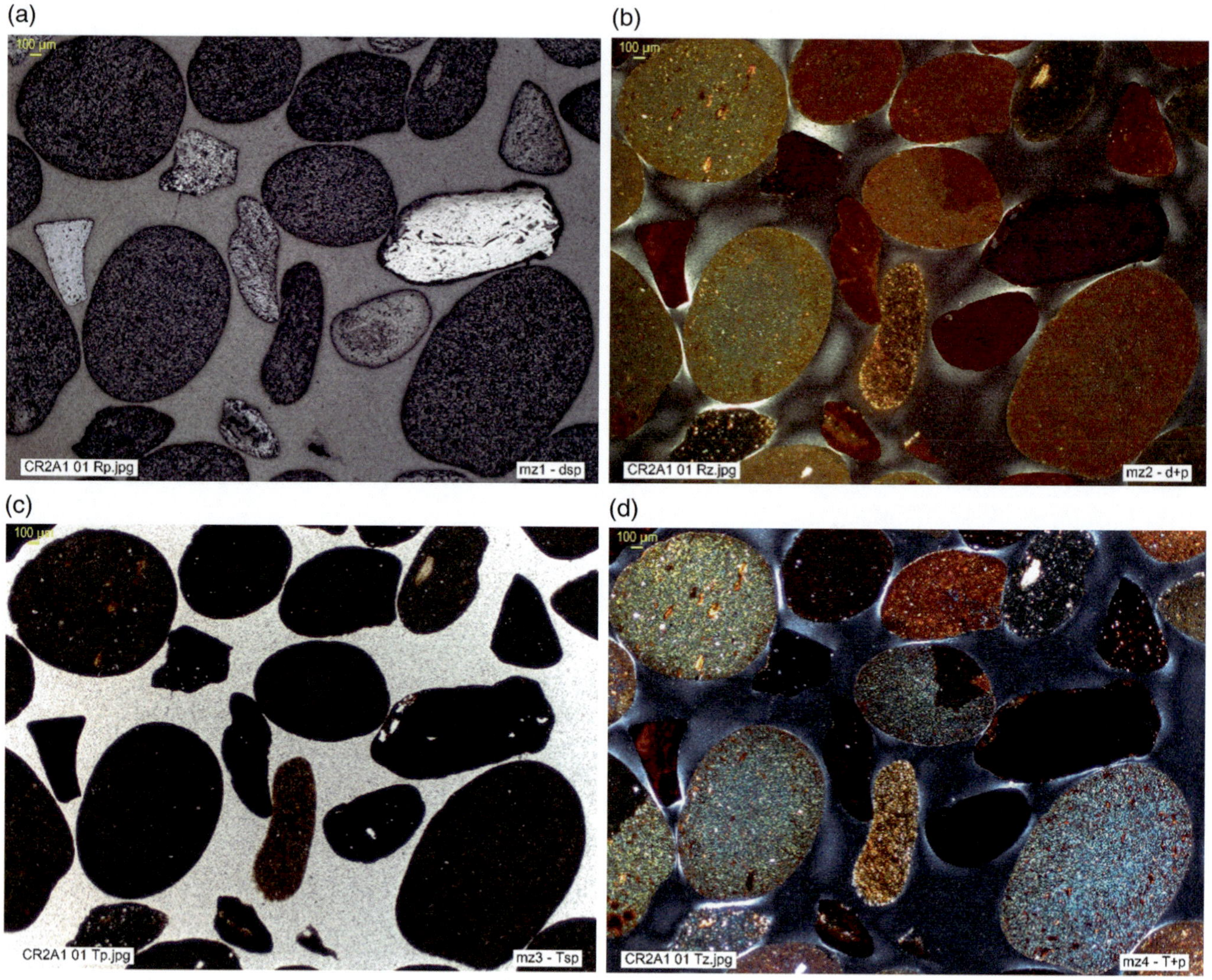

Fig. 86.1 **a, b** (Same scene: **dsp** and **d+p**, respectively)—Monazite concentrate (heavy fraction), mz nodules (grey, anisotropic, light-coloured IR) with silicate micro-inclusions (dark) and variable impregnation of limonite (light grey in Fig. 86.1a, reddish-brown IR in Fig. 86.1b) and maghemite (white, Fig. 86.1a, in extinction and without IR, Fig. 86.1b), monazite nodule twinned in the center (immediately left of mgh), central zone is enlarged in Fmz7-8. Sample CR2A1, Matamulas detrital deposit, Ciudad Real

Monazite is a **transparent mineral** and can be recognized with the petrographic microscope (Transmitted Light, **TL**, cf Figs. 86.2a, b, 86.6, 86.8a, b). For the purposes of this text, it is considered of interest to characterize it also, as the other ores, with reflected light, **RL**, even if summarily. As a compromise solution, *both types of observation* (TL and RL) have been documented with photomicrographs, substituting exceptionally the RL images in immersion (osp and o+p) for TL images (Tsp and T+p).

As far as the characterization by **reflected light** is concerned, no description of monazite has been found in any of the reference treatises commonly used for ore microscopy, nor its optical parameters (R%, Y%, λd, etc.), although Chang et al. (1998) offer very relevant information on the mineral (structure, chemistry, experimental data, etc.).

The following description is based on the observations by the author on monazite samples from a single deposit, Matamulas (Ciudad Real), recently investigated. These are

Fig. 86.2 **a, b** Same scene as Fmz1, seen with *transmitted light* (**Tsp** and **T+p**, respectively): limonite, sericite, biotite, chlorite and clay minerals included in monazite. Note, in mz4, bright interference colours (third order) and clearly visible twin (one element in extinction) in central nodule of monazite

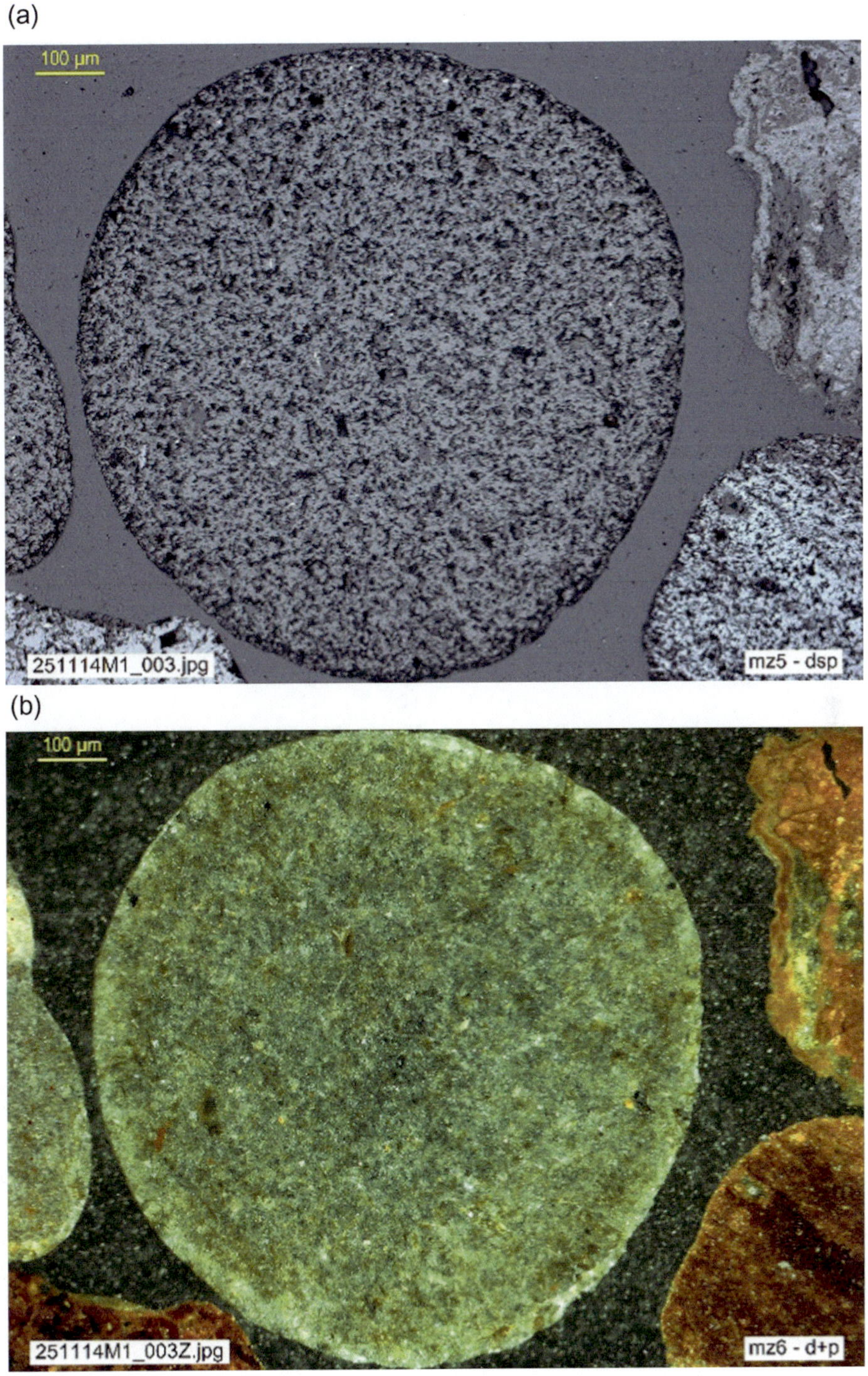

monazite nodules occurring in *detrital deposits* (Fig. 86.1). They are Nd-rich and non-radioactive monazites, free of U and Th, in spite of which their exploitation was not authorized, for environmental reasons.

In radioactive ores, some monazite occurs in a *metamict* state, which affects some properties such as birefringence or density and is reflected in a poorly defined XRD diffractogram. However, the structure can be recovered by thermal treatment. Metamict monazites seem to be an exception (*grey monazite*) since, in general, monazite retains its crystalline structure despite radiation damage. This has been explained by a mechanism, peculiar to the mineral, of defect restoration (*α-healing*, Seydoux-Guillaume et al. 2018). This highlights the relevance of its application as a *geothermometer*, based on REE fractionation between monazite and source rock.

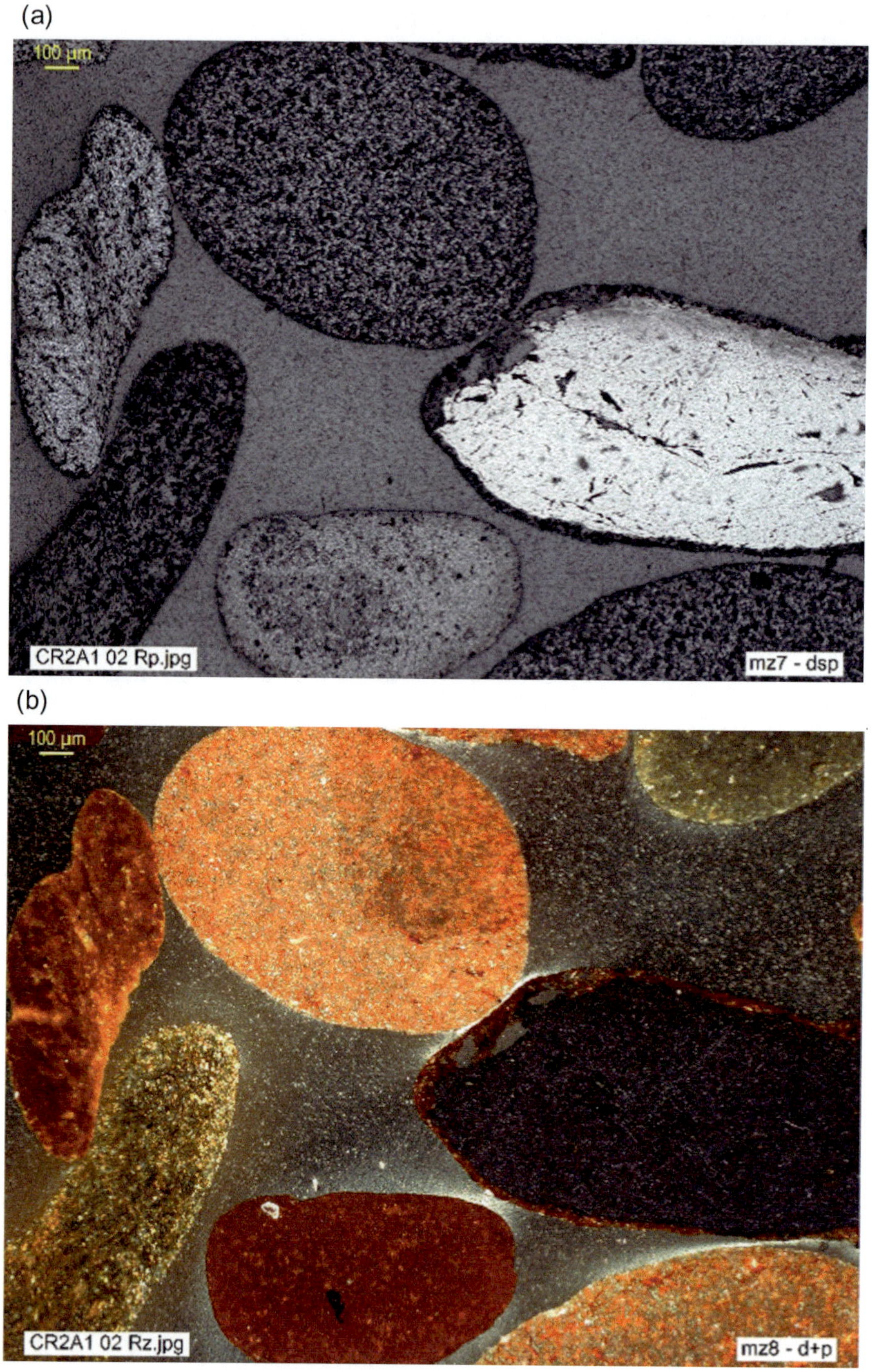

Fig. 86.3 **a, b** (Same scene: **dsp** and **d+p**, *respectively*)—Monazite nodule (grey, porous), with silicate micro-inclusions (dark grey) and traces of limonite (light grey or greyish white), which is abundant as impregnation in neighboring nodules. With analyzer, Fig. 86.3b shows the faint zoning in the nodule, the scarcely noticeable anisotropism and the light coloured internal reflections of monazite, becoming reddish-brown in the neighbouring limonite-impregnated nodules (e.g., botom right). Sample 251114_M1, Matamulas, Ciudad Real

86.2 Microscopic Description

86.2.1 SP Observation (*Single Polariser*)

Grey *colour*, *low reflectance* (Figs. 86.3a, 86.4b, 86.5 a and 86.7a. The excessively dark impression of monazite in Fig. 86.1a is due to imaging with wide-field objective and is not representative of the R value) Reflectance R ≈ 8–9% (AMCO measurements, coincident with those calculated from its refractive index n ≈ 1.8), slightly higher than gangue minerals and distinctly lower than limonite, goethite, maghemite, hematite (cf multispectral AMCO comparative plots).

It is not pleochroic and the bireflectance is virtually unnoticeable.

In transmitted light, **TL**, it shows the strong relief that corresponds to its high refractivity (n ≈ 1.72–1.86), and its

(a)

(b)

Fig. 86.4 **a, b** Detail of Fig. 86.1a (**dsp** and **d+p**, *respectively*). Low reflectance (grey shade) of mz, lighter than gg but darker than limonite and, above all, mgh (white). Perceptible but moderate anisotropism of monazite, whose bright internal reflections may acquire various reddish-brown shades due to limonite impregnation

colour is usually yellowish but may also be colourless, sometimes brownish by limonitic impregnation or opaque (Figs. 86.2a, 86.6 and 86.8a).

Euhedral crystals, generally small and with varied habits but larger in pegmatites, are common. However, the Ciudad Real monazites studied for this description always occur in nodules, sometimes slightly zoned and displaying radial or sectorial structures, apparently due to recrystallisation from precursor gels or cryptocrystalline aggregates.

In these nodules, monazite is frequently intergrown with silicates and impregnated with Fe ± Mn oxides. In RL, the former are darker and the latter are lighter than monazite (Figs. 86.1a and 86.3a). The silicates preserve the orientation of the host rock (of which occasional relicts may be preserved) and show reaction rims evidencing different degrees of replacement by monazite (comp. Figs. 86.5a, 86.7a and 86.3a, which show a progressively increasing enrichment in monazite), suggestive of some form of fluid-assisted mobilisation.

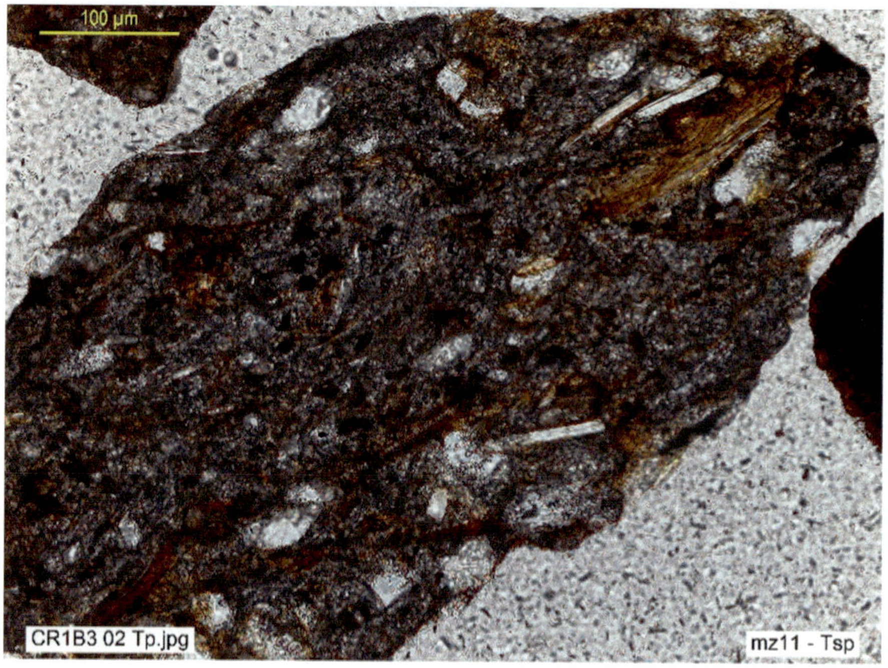

Fig. 86.5 **a, b** (**dsp** and **d+p**, respectively)—Incipient stage of formation of a monazite nodule in metalutite. The included silicates (muscovite/sericite, chlorite, quartz, biotite relicts, etc.) still comprise $\geq$ 50% (modal vol.) of the particle, but their corrosion (clearly visible in Fig. 86.5a) suggests that they are being replaced by mz. Sample CR1B3, Matamulas detrital deposit, Ciudad Real

86.2.2 Observation +P (*Polariser + Analyser at 90°*)

Monazite is *weakly anisotropic* and poorly coloured in RL, its anisotropism being partly overprinted by the abundant *internal reflections*, IR. These are colourless or slightly brownish to yellowish-brown, sometimes slightly greenish. They may appear darker and reddish to reddish-brown by contamination with Fe/Mn oxides or organic matter (intergrowths, impregnation, Figs. 86.1b, 86.3b, 86.4b, 86.5b and 86.7b).

In TL, its vivid and intense polarisation colours are characteristic, as corresponds to its birefringence ($\approx$ 0.05 −0.07, third-fourth order, Figs. 86.2b and 86.8b).

Frequent and varied *twins*, not always visible in reflected light but easily recognizable in transmitted light (comp. Figs. 86.1b and 86.4b with Fig. 86.2b).

86.2.3 Paragenesis

It appears, as an accessory mineral, in various igneous and metamorphic lithologies, such as acid granitoids, pegmatites, gneisses and, with its peculiar nodular morphology, in meta-lutites, palaeozoic slates or schists, as well as in the detrital deposits derived from them.

86.2.4 Identification

Its grey colour and low reflectance, slightly higher than gangue, its moderate anisotropism, its internal reflections and, in some cases, its nodular morphology are characteristic. In pegmatites, monazite tends to form coarser crystals, characterised by their reflectance, higher than common silicate gangue (e.g., R mz> quartz) *but lower than other possible companions* in this type of paragenesis: cassiterite, wolframite, tantalite, tapiolite, rutile, titanite, even typically transparent minerals such as zircon or scheelite (*cf* comparative graphs of multispectral reflectances).

However, for safe determination and given the importance of the mineral, it is generally advisable to use other complementary techniques such as transmitted light, XRD or EMPA. The spectroscopic eyepiece adaptable to the microscope devised by Smithson (1959) can be useful for the detection of the characteristic spectral absorption bands that identify monazite.

Fig. 86.6 Same scene as Fig. 86.5a, seen with *transmitted light* (**Tsp**). Because of its appearance, the mz nodule might be mistaken as a rock fragment (sand size), except for the high refractivity (relief) of the cement (mz)

(a)

(b)

Fig. 86.7 **a, b** (*reflected light*: **dsp** and **d+p**, respectively)—Advanced stage of mz concentration (light grey), in monazite nodule with silicate inclusions (darker grey). The corroded outlines of these and their scarcity suggest their replacement by mz, as in Fig. 86.5a, but now in an advanced stage, which reduces the silicate inclusions to scattered relicts in a matrix (or aggregate of crystallites) of very fine-grained mz, as evidenced by the **d+p** and, above all, **T+p** (transmitted light) images. Sample CR2A1, Matamulas detrital deposit, Ciudad Real

(a)

(b)

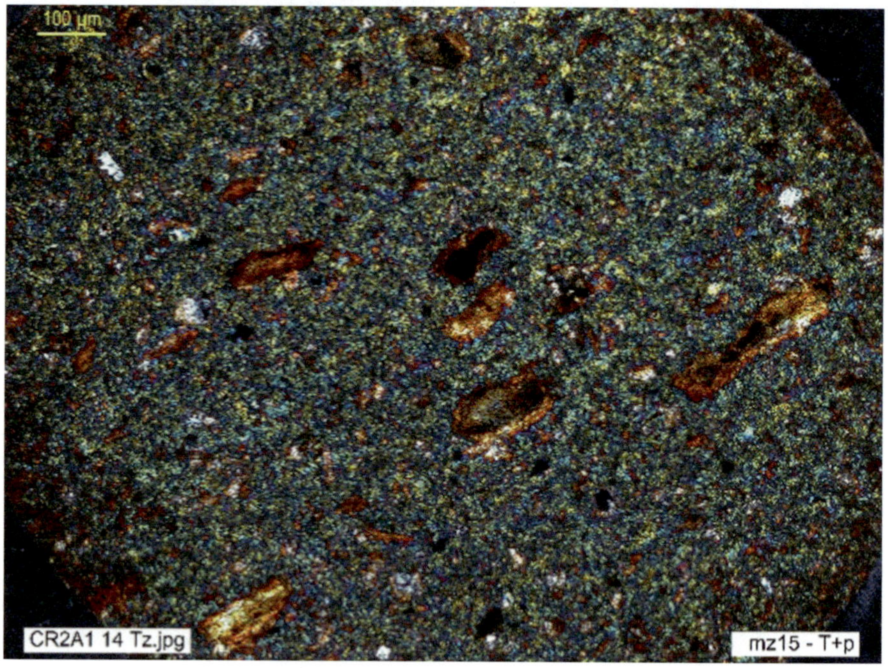

86.3 Spectral Information

86.3.1 Monazite Multispectral Reflectance R (in Air, Without Polariser) (%)

λ (nm)	370	400	425	450	475	500	525	550	575	600	625	650	675	700	750	800	850	900	950	1000
μ	9.0	8.6	8.6	8.7	8.5	8.7	8.6	8.7	8.6	8.8	8.6	8.7	8.6	8.7	8.9	9.0	9.2	9.4	9.7	10.0
P_5	8.3	7.9	7.9	7.9	7.8	8.0	7.9	8.0	7.8	8.0	7.9	8.0	7.8	7.9	8.2	8.3	8.6	8.7	9.0	9.3
P_{20}	8.6	8.3	8.3	8.4	8.2	8.4	8.3	8.4	8.3	8.4	8.3	8.4	8.3	8.3	8.6	8.7	8.9	9.1	9.4	9.6
P_{80}	9.3	8.9	8.9	9.0	8.8	9.0	9.0	9.1	8.9	9.1	9.0	9.0	9.0	9.0	9.2	9.3	9.6	9.8	10.0	10.3
P_{95}	9.7	9.2	9.2	9.3	9.2	9.4	9.3	9.4	9.3	9.5	9.4	9.4	9.3	9.4	9.6	9.6	9.9	10.1	10.4	10.7

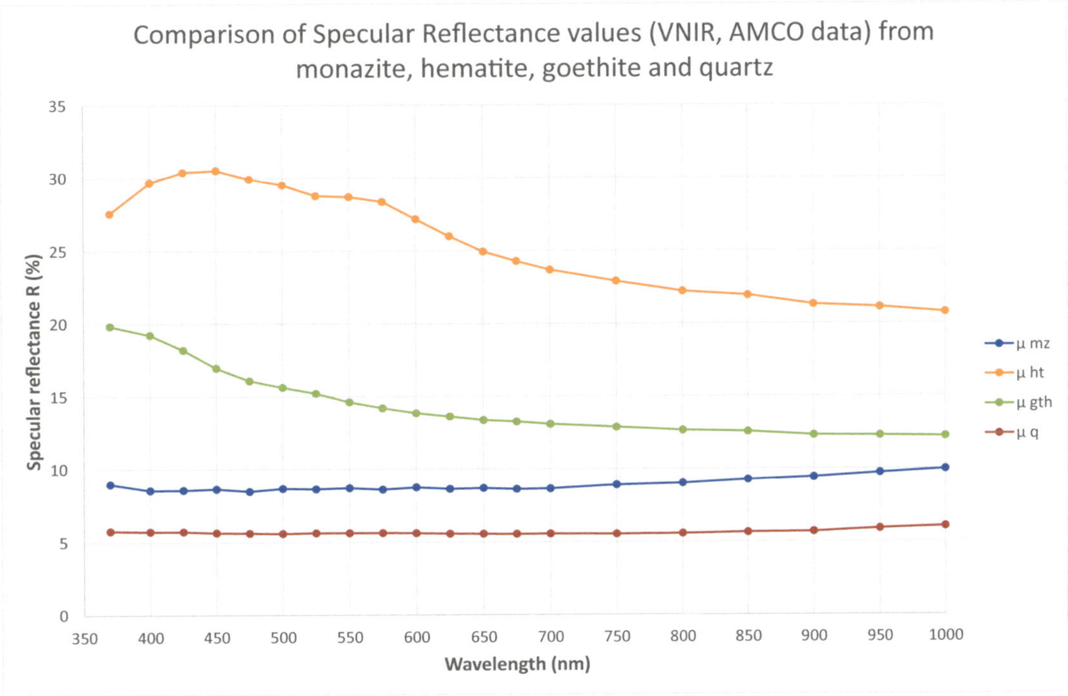

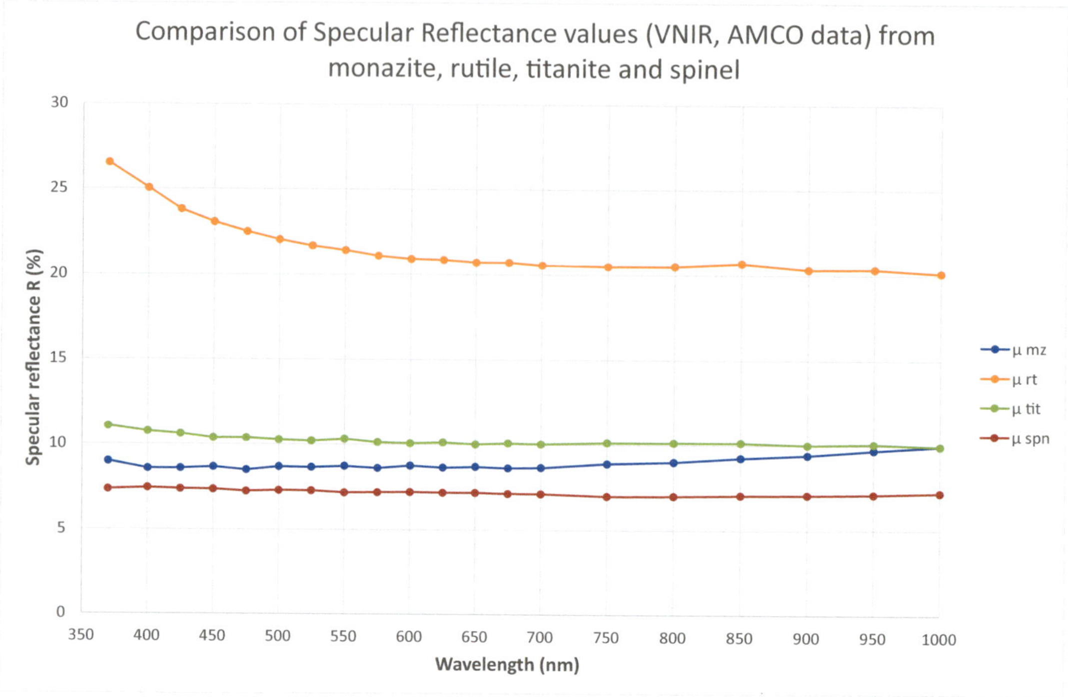
Comparison of Specular Reflectance values (VNIR, AMCO data) from monazite, rutile, titanite and spinel
Specular reflectance R (%)
Wavelength (nm)
0
5
10
15
20
25
30
350
400
450
500
550
600
650
700
750
800
850
900
950
1000
μ mz
μ rt
μ tit
μ spn

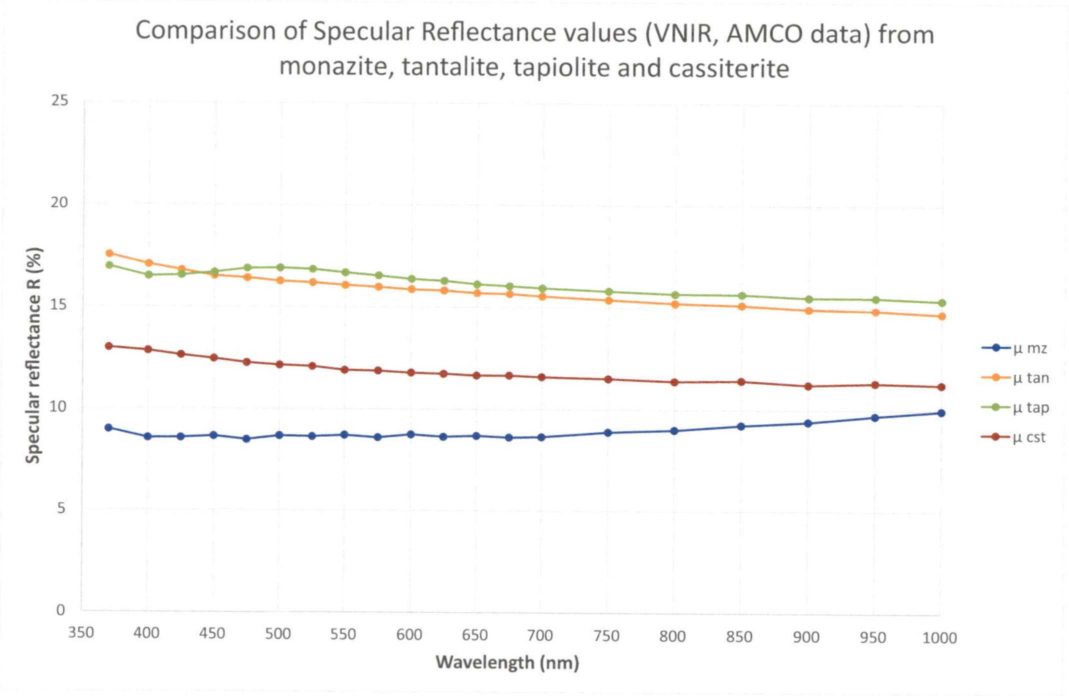
Comparison of Specular Reflectance values (VNIR, AMCO data) from monazite, tantalite, tapiolite and cassiterite
Specular reflectance R (%)
Wavelength (nm)
0
5
10
15
20
25
350
400
450
500
550
600
650
700
750
800
850
900
950
1000
μ mz
μ tan
μ tap
μ cst

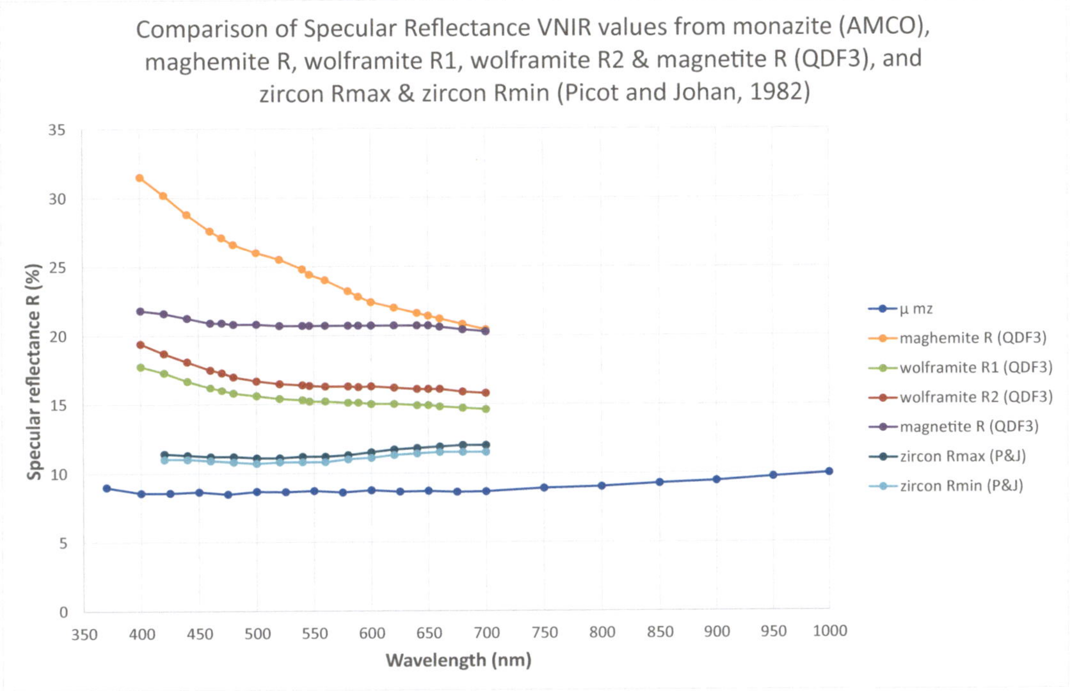

References

Chang LLY, Howie RA, Zussman J (1998) Rock forming minerals, v. 5B 2nd edn. Non-silicates: sulphates, carbonates, phosphates, and halides. The Geological Society, London, pp 383

Smithson F (1959) A simple spectroscopic eye-piece for testing monazite under the microscope. Min Mag 32:176

Seydoux-Guillaume AM, Deschanels X, Baumier C, Neumeier S, Weber WJ, Peuget S (2018) Why natural monazite never becomes amorphous: experimental evidence for alpha self-healing. Am Miner 103(5):824–827. https://doi.org/10.2138/am-2018-6447

87 Naumannite (*nau*)

Ag_2Se: Orthorhombic

Naumannite		R_{546} (%)	Y (%)	λ_d (nm)	P_e (%)	Colour sensation
Air	R_1	35.6	35.1	485	5.2	Grey
	R_2	36.5	36.4	478	5.5	
Oil	R_1	20.2	19.8	483	9.2	
	R_2	21.2	21.1	478	9.2	
Bir: Weak (3)				*Pleoc*: Weak (3)		*Tarn*: Slow darkening
Anis +P: Weak but coloured (xp) (3)				*IR*: No		*Twg*: Occasionally
Mor: Disseminated. Anhedral aggreg				*Clv*: No		*PolQ*: Demands care (soft, scratches)
Hardness	*PolH*: Low: > clt; < gn (2)			*Mohs*: 2½	*Talm*: B	*VHN* $_{(25\ g)}$: 33–84

Note**:** cf. Table and R charts, visible range data (QDF3)
Mineral abbreviations: ac—acanthite, Ag—native silver, agl—aguilarite, Au—native gold, bl—boulangerite, ccp—chalcopyrite, clt—clausthalite, el—electrum, fbg—freibergite, gg—gangue, gn—galena, nau—naumannite, py—pyrite, td—tetrahedrite

87.1 General

Composition: Ag_2Se. Some analyses with possible Pb or Cu impurities could actually correspond to intergrowths with other selenides.

Variable S contents, attributed to a solid solution series with Ag_4SeS (cf. *aguilarite* description).

87.2 Microscopic Description

87.2.1 SP Observation (*Single Polariser*)

Grey *colour*, similar to aguilarite but a little brighter (Fig. 87.1a, b).

Careful *polishing* needed because it is very soft, which makes it difficult to avoid scratches, but it can be polished better and be less scratched than aguilarite or acanthite.

Low or very low *polishing hardness*: $PolH_{nau}$ < gn, but different authors disagree when it comes to specifying more precisely. Schneiderhöhn (1952) places it between dyscrasite and boulangerite, i.e., near the low/medium limit. According to Uytenbogaardt and Burke (1971), naumannite would be among the softest ores, being the softest of all selenides after aquilarite. The difficulty lies in the scarcity of the mineral, which makes it difficult to find contacts of naumannite with the different ores, suitable for a safe determination.

Reflectance moderate, markedly lower in immersion (Fig. 87.2a). R_{nau} > ac; slightly > td or fbg and agl; < gn, py or ccp; ≪*el* (Figs. 87.1a, b and 87.2a, cf. comparative plots).

Pleochroism barely perceptible in immersion.

It occurs disseminated or as interstitial impregnation and in fine anhedral aggregates, associated with electrum and other sulphides or selenides.

87.2.2 Observation +P (*Polariser + Analyser at 90°*)

Weakly but clearly *anisotropic*, with inconspicuous polarisation tints, which become more vivid, varied and luminous by slightly uncrossing the analyser (Figs. 87.1c, d and

R. Castroviejo, *A Practical Guide to Ore Microscopy—Volume 1*,
https://doi.org/10.1007/978-3-031-12654-3_87

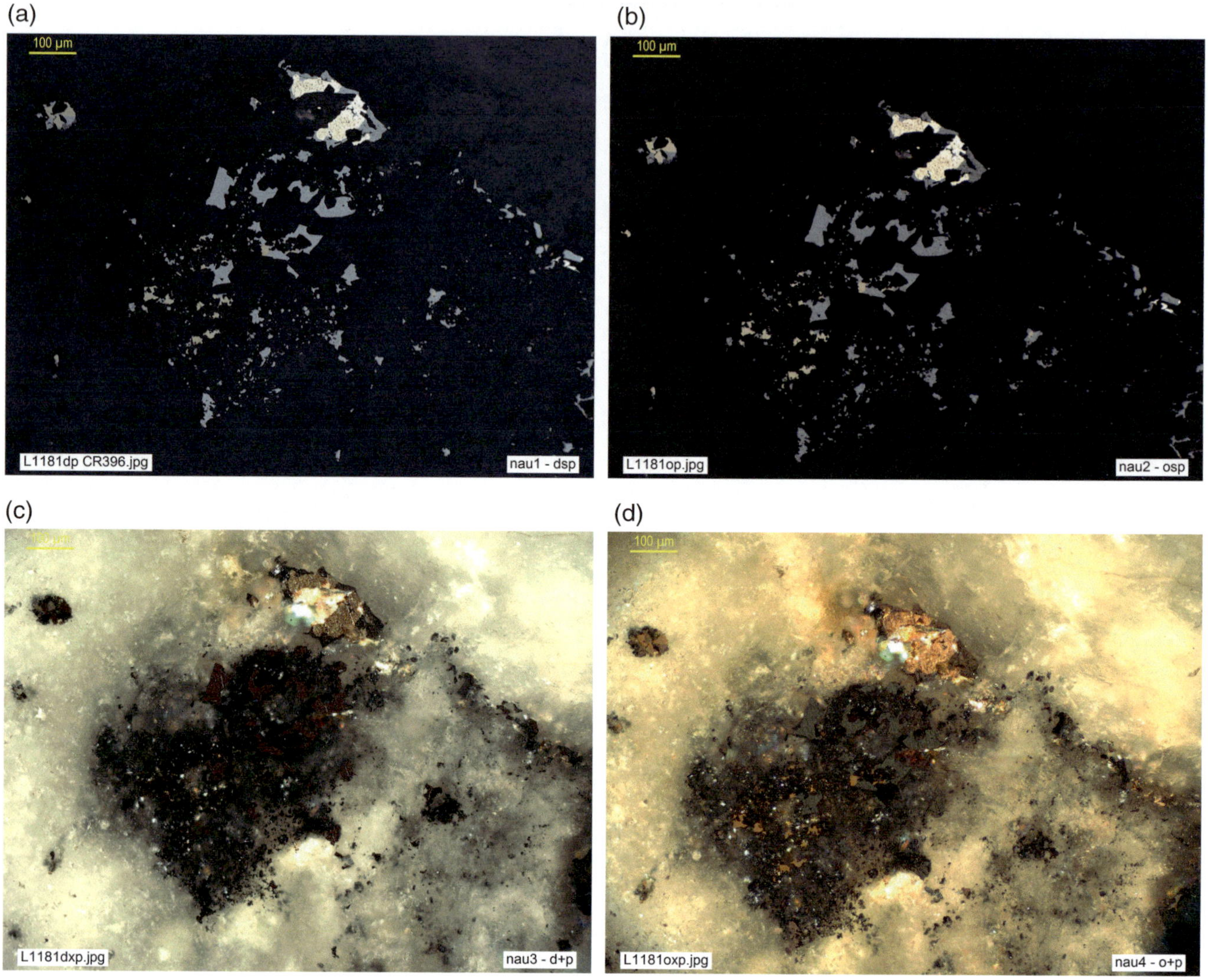

Fig. 87.1 **a, b** (Same scene: **dsp** and **osp**, respectively). Interstitial impregnation of nau, el, agl, ccp in vein of q. Here, nau (grey) is the dominant Ag ore and occupies the central part of the image, moulding idiomorphic crystals of q; agl (associated with el, top center) is less reflective (darker) than nau, and its polishing quality is poorer; ccp shows an atypical brownish hue, due to Se contamination. CR396, Midas Mine, NV, USA. **c, d** Same scene as nau1 (**d+p** and **o+p**, respectively). The anisotropism, more vivid and coloured in nau and duller in agl, helps to distinguish the two Se ores

87.2b, c). Greenish or bluish-green, reddish-brown tints appear, especially in immersion (Fig. 87.2).

87.2.3 Paragenesis

Low temperature hydrothermal *parageneses*, particularly *argentiferous* epithermal parageneses poor in S and rich in Se.

Companions: silver, gold, electrum, aguilarite, clausthalite, chalcopyrite and other Ag–Se selenides or sulphosalts.

87.2.4 Identification

It is characterized by its grey colour, moderate reflectance, low hardness and coloured anisotropism (especially with **oxp**), as well as by argentiferous paragenesis with selenides.

It is distinguished from aguilarite by being slightly more reflective, showing better polish (less scratching) and stronger, coloured anisotropism (Figs. 87.1a–d and 87.2a, b, c).

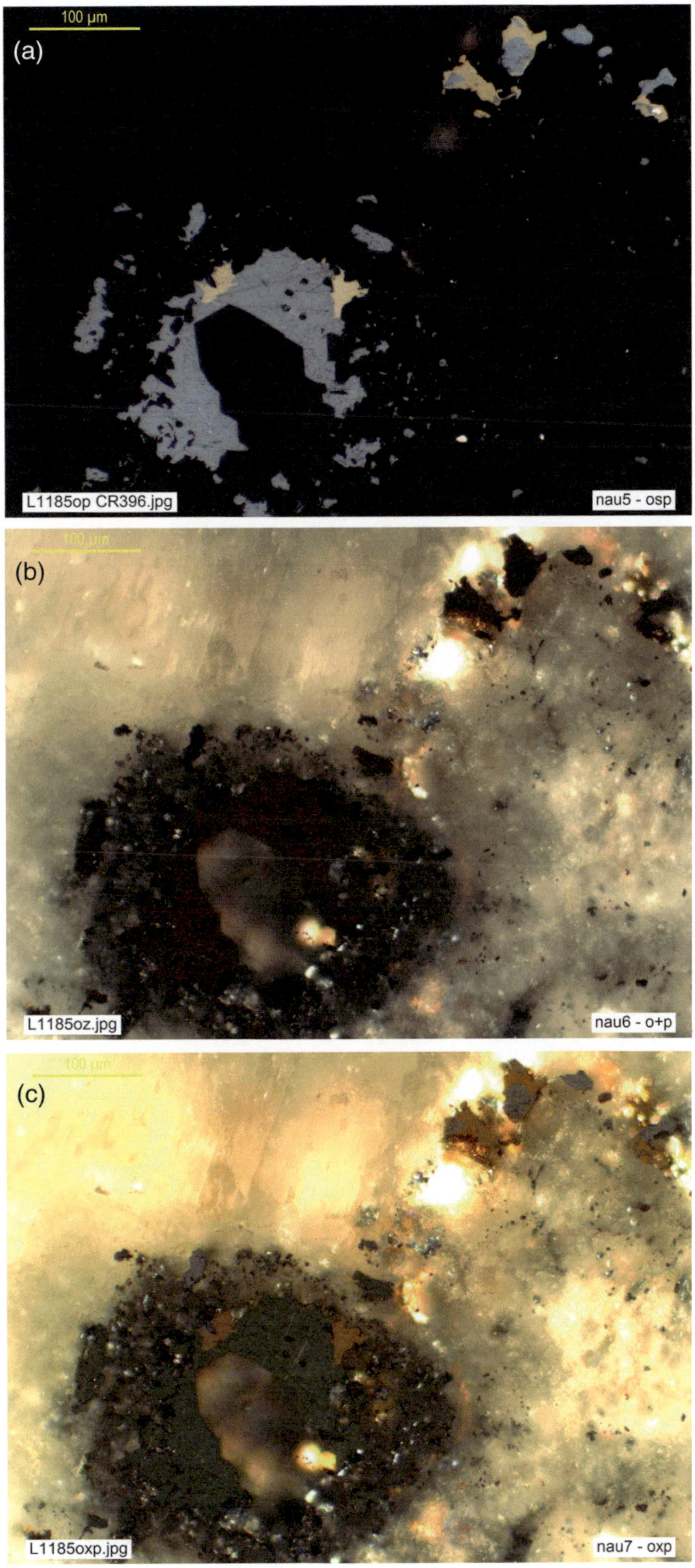

Fig. 87.2 **a, b, c** (**osp**, **o+p** and **oxp**, respectively)—Quartz-carbonate vein with impregnation of nau (grey, bottom left), ccp (brownish yellow), agl (grey, with ccp, top right) and scarce *el* particles (≈ 5–10 µm, very bright; bottom right). The anisotropism of nau is enhanced by slightly rotating the analyser (*xp*). More varied, luminous tints appear, in contrast to the dull tones of agl (comp. **b** and **c**). CR396, Midas Mine, NV, USA

87.3 Spectral Information

87.3.1 Naumannite Reflectance R (in Air, Visible Spectrum) (%)

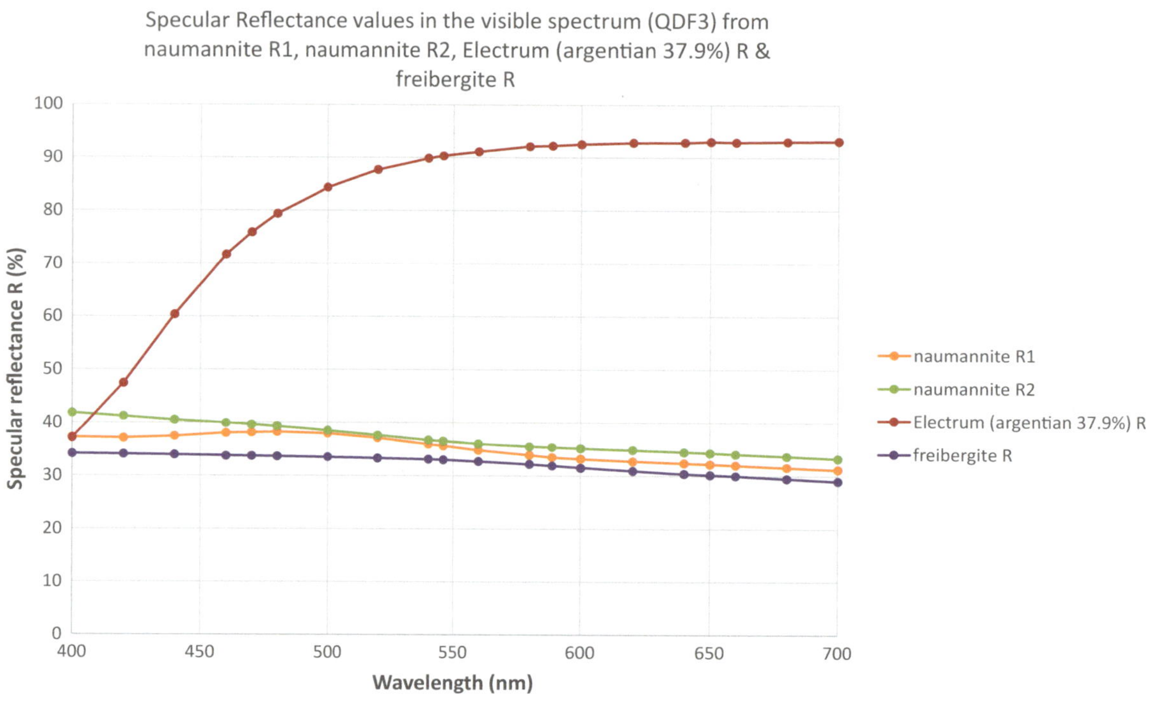

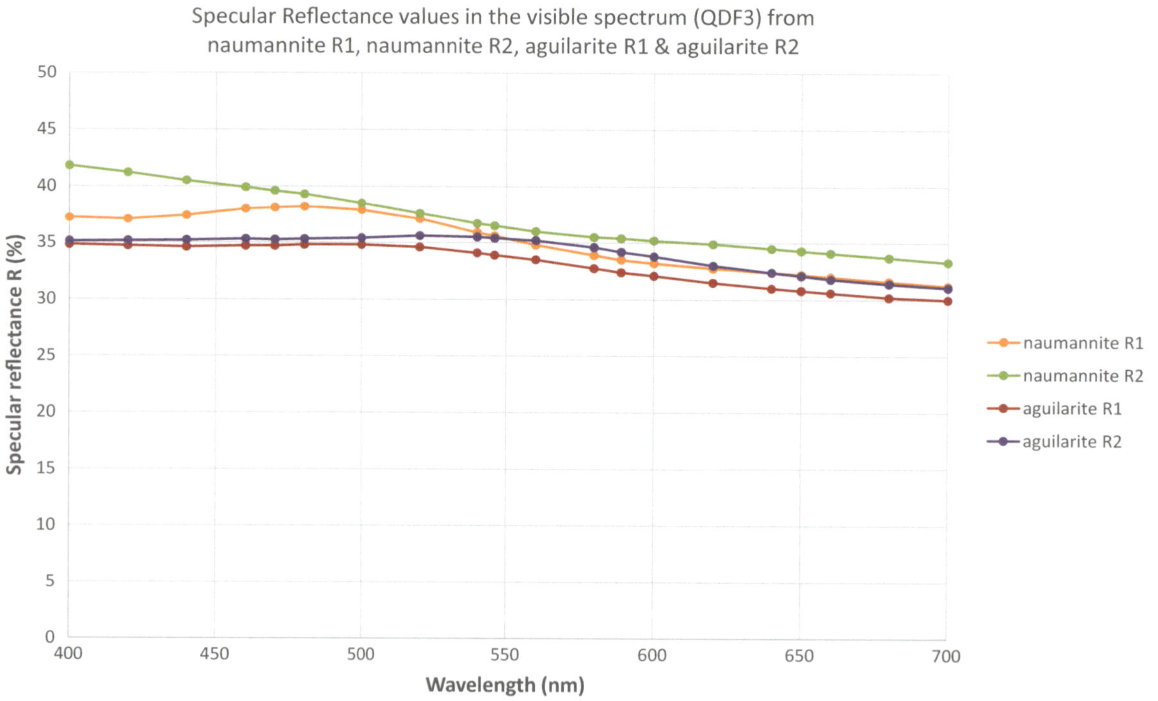

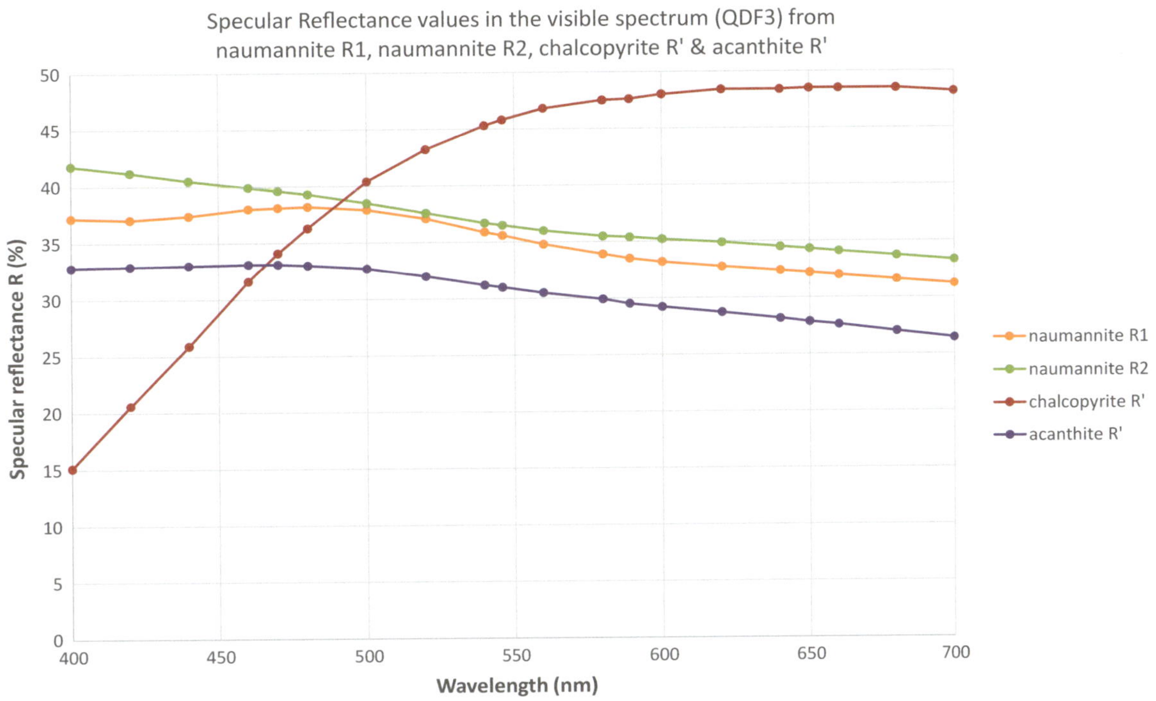

References

Schneiderhöhn H (1952) Erzmikroskopisches Praktikum. E. Schweizerbart'sche Verlagsbuchhandlung, Stuttgart, p 274

Uytenbogaardt W, Burke EAJ (1971) Tables for microscopic identification of ore minerals, 2nd edn. Elsevier, Amsterdam

Nickeline (*nc/nickelite*)

88

NiAs: Hexagonal

Nickeline		R_{546} (%)	Y (%)	λ_d (nm)	P_e (%)	Colour sensation
Air	R_o	51.4	52.4	582	11.5	Pinkish-yellowish white, bright
	R_e	46.1	47.7	583	16.2	Light brownish pink
Oil	R_o	38.3	39.6	582	17.7	Pinkish-yellowish white, bright
	R_e	33.2	35.1	583	25.5	Brownish pink, darker
Bir: Medium (4)				*Pleoc*: Distinct-strong (4)		*Tarn*: Weak (slowly darkening)
Anis +P: Strong, bright colours (5)				*IR*: -		*Twg*: Occasional
Mor: Anhedral; intergrown mau, bth				*Clv*: Occasional		*PolQ*: Good polishing
Hardness	*PolH*: Medium, ≈ bth, < sf (3)			*Mohs*: 5+	*Talm*: E	*VHN* $_{(100\ g?)}$: 308–533

Obs **:** cf. Table and R charts, multispectral data (VNIR, AMCO) and visible range (QDF3)
Mineral abbreviations: native Ag—silver, asp—arsenopyrite, native Bi—bismuth, bth—breithauptite, dy—dyscrasite, mau—maucherite, sf—safflorite, prm—pararammelsbergite, po—pyrrhotite, py—pyrite, rm—rammelsbergite, sf—safflorite, sk—skutterudite, urn—uraninite

88.1 General

Composition: NiAs, with occasional minor contents of Sb, Fe, Co. NiAs and NiSb are completely miscible in the laboratory at high temperatures. However, these solid solutions, although they exist, are very rare in nature (*arite*), even if pure NiAs and NiSb appear in contact. This suggests a generally moderate formation temperature.

In the Ni-Co-As system, nickeline can be associated with safflorite, skutterudite, rammelsbergite, maucherite and pararammelsbergite. Transitions and substitution relations between the species of the group are frequent, depending on the variable As contents of the system.

Nickeline is susceptible to substitution by other totally different species. This can significantly enhance the precipitation of other minerals of great economic interest, such as the silver ores in the *five-element* deposits (Ni–Co–Bi–Ag–U).

88.2 Microscopic Description

88.2.1 SP Observation (*Single Polariser*)

Characteristic pinkish white *colour*, bright (Fig. 88.1a, b) and with noticeable pleochroism in yellowish to light brownish tones. Excellent *polish*, free of pull-outs. T*arnishing* not very pronounced (the surface darkens slowly).

High *reflectance*, ≈ pyrite (cf. comparative R charts). Anisotropic, slightly bireflective and distinctly *pleochroic*, especially comparing adjacent grains and in immersion. Over the pinkish colour, alternating yellowish or darker, brownish tones appear (Fig. 88.2a).

Medium *polishing hardness*, like bth: $PolH_{nc}$>Ag, dy; mau; <sk, sf, po, asp; ≪py.

It occurs in varied aggregates and intergrowths, rarely in euhedral grains. Occasionally it develops singular textures, dendritic, skeletal. It shows frequent zoning.

R. Castroviejo, *A Practical Guide to Ore Microscopy—Volume 1*,
https://doi.org/10.1007/978-3-031-12654-3_88

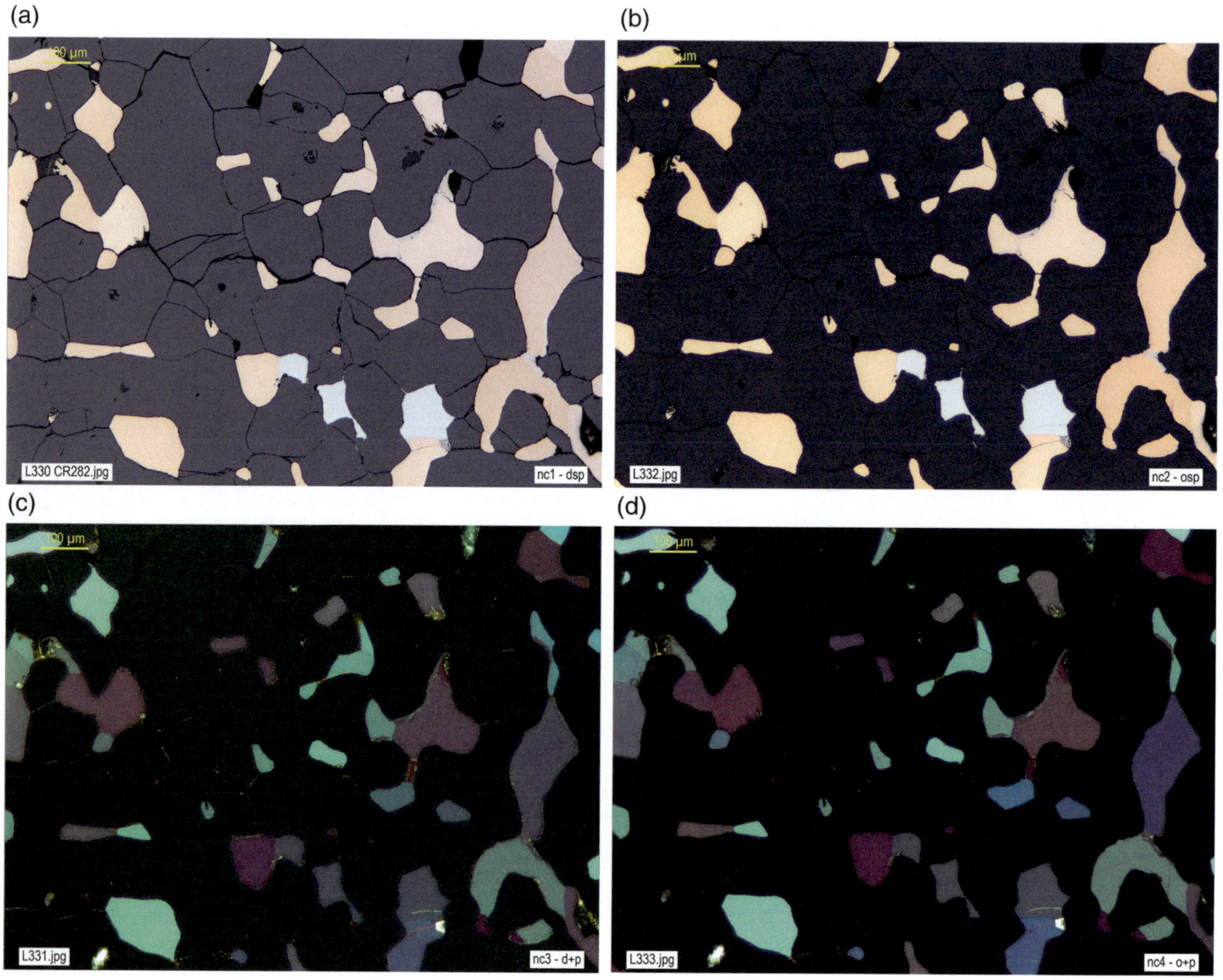

Fig. 88.1 **a** Cumulitic aggregate of chr (grey) with interstitial nc (pink) and gf (white). Traces of po (greyish brown), largely altered to *intermediate product* (*Zwischenprodukt*, Ramdohr, 1980, brownish grey) along nc–chr and nc–gf–chr contacts (*e.g.,* bottom right). CR282, Mina La Gallega (Sª Ronda), Málaga (**dsp**). **b** Same scene as nc1 (**osp**). The increased contrast allows us to distinguish po more clearly. **c, d** Same scene as nc1 (**d+p**) and (**o+p**). Vivid anisotropism of nc contrasts with isotropy of chr, which shows only scarce IR (reddish), strong anisotropism of *Zwischenprodukt* (*bottom* right)

88.2.2 Observation +P (*Polariser + Analyser at 90°*)

Strongly anisotropic, with bright and luminous polarisation colours (Figs. 88.1c, d and 88.2b), particularly greenish and bluish tones. There are noticeable changes when the analyser is slightly uncrossed. Distinct and parallel extinction (+p).

Twinning uncommon, although {1011} may be frequent at times.

88.2.3 Paragenesis

Paragenesis typical of the deposits known as the *five-element formations* (Ni–Co–Bi–Ag–U), with native Ag

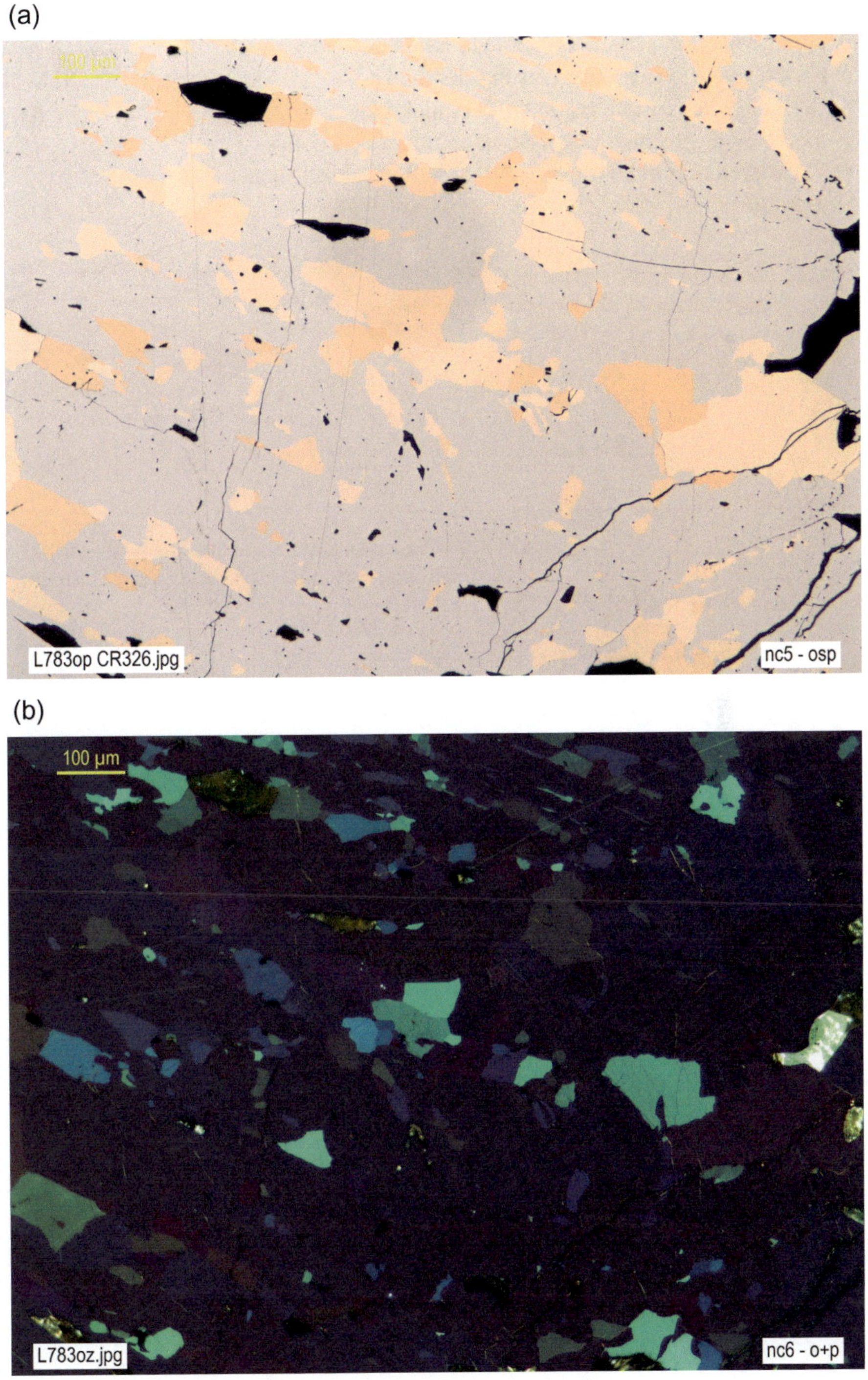

Fig. 88.2 **a, b** Disseminated nc inclusions in massive mau. CR 326, Timiskaming, Ontario, Canada, (**osp**) and (**o+p**), respectively

and Bi, Ag sulphosalts, other Ni–Co species, uraninite, native gold.

It often appears intergrown with safflorite, skutterudite, rammelsbergite, maucherite, pararammelsbergite, native silver.

Also found in Cr–Ni deposits associated with mafic and ultramafic rocks, *vbgr*. with chromite, as in the *La Gallega* mine (Fig. 88.1a–d).

88.2.4 Identification

It is easily recognized by its typical colour and pleochroism, its striking anisotropism with green and bluish luminous tones, its good polish, its hardness and its paragenesis.

However, there is a risk of confusion with maucherite and breithauptite, which appear in the same paragenesis:

- Maucherite has a less saturated, greyish colour, lacks pleochroism and has polysynthetic twins and very weak anisotropism (cf. **F mau** and Fig. 88.2a, b).
- Breithauptite is more coloured than nickeline and shows more vivid, greenish polarisation colours (cf. **F bth**).
- The distinction of pyrrhotite, darker and brownish, is not a problem.

88.3 Spectral Information

88.3.1 Nickeline Multispectral Reflectance R (in Air, Without Polariser) (%)

λ (nm)	370	400	425	450	475	500	525	550	575	600	625	650	675	700	750	800	850	900	950	1000
μ	43.7	43.0	41.0	41.0	41.3	43.9	46.6	50.3	53.4	56.2	58.3	60.2	61.8	63.1	65.4	67.1	69.5	70.1	70.6	71.0
P_5	41.7	41.0	38.9	38.8	39.0	41.8	44.8	48.9	52.1	54.5	56.2	57.9	59.3	60.7	63.0	64.6	66.4	67.2	67.2	67.6
P_{20}	42.8	42.1	39.9	39.9	40.0	42.8	45.6	49.6	52.8	55.7	57.8	59.8	61.3	62.6	64.9	66.5	68.5	68.9	69.2	69.2
P_{80}	44.8	43.9	42.4	42.4	43.0	45.3	47.7	50.8	54.0	56.9	59.0	60.9	62.5	63.8	66.1	67.8	70.4	71.0	71.8	72.2
P_{95}	45.7	45.2	43.6	43.8	44.2	46.6	49.1	52.2	55.0	57.5	59.4	61.4	63.0	64.4	67.1	69.2	72.1	73.3	74.0	74.5

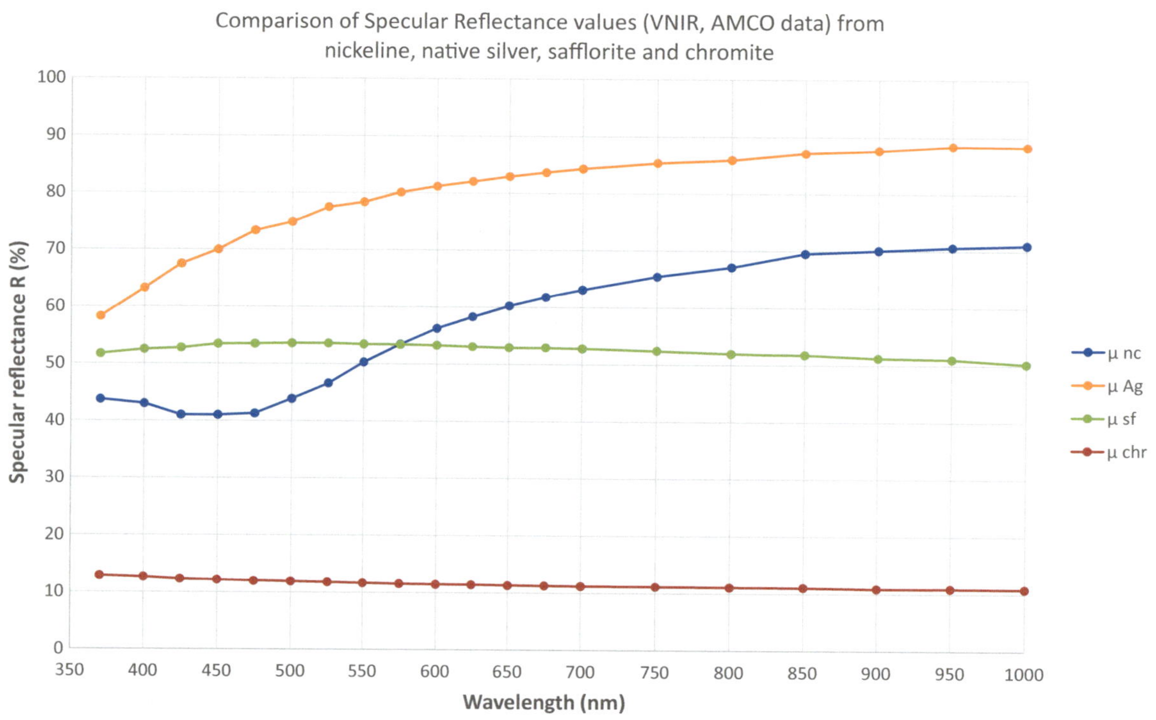

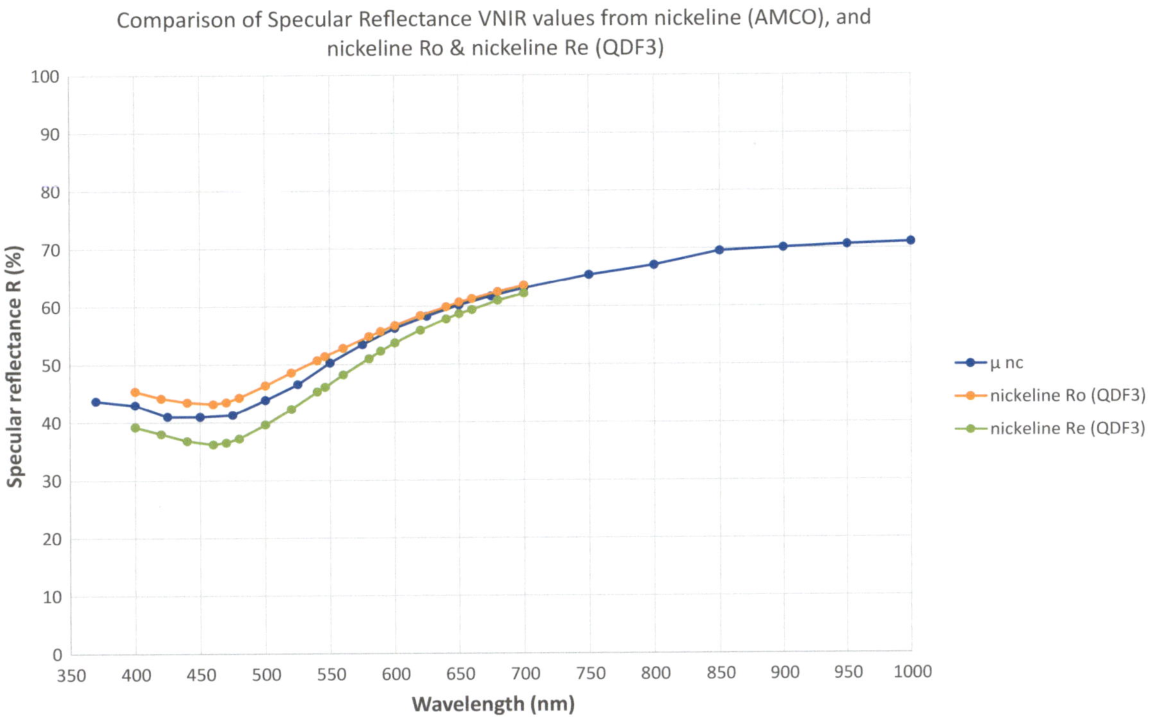
Comparison of Specular Reflectance VNIR values from nickeline (AMCO), and nickeline Ro & nickeline Re (QDF3)
Specular reflectance R (%)
Wavelength (nm)
µ nc
nickeline Ro (QDF3)
nickeline Re (QDF3)
0
10
20
30
40
50
60
70
80
90
100
350
400
450
500
550
600
650
700
750
800
850
900
950
1000

89 Orpiment (*orp, op*)

As_2S_3: Monoclinic

Orpiment		R_{546} (%)	Y (%)	λ_d (nm)	P_e (%)	Colour sensation
Air	R_1	23.1	23.1	474	10.1	
	R_2	27.5	27.6	473	8.4	
Oil	R_1	9.4	9.5	473	17.7	
	R_2	13.1	13.2	474	13.4	
Bir: Medium (4)				*Pleoc*: Very weak (2)		*Tarn*: nd
Anis +P: Overprinted by RI (4)				*IR*: Massive yellow		*Twg*: Occasionally
Mor: Acicular, spherulitic, encrust on rl				*Clv*: (010) fecuente		*PolQ*: Easy, but scratchy (very soft)
Hardness	*PolH*: Very low, but > rl (1)			*Mohs*: 1½–2	*Talm*: B+	*VHN* $_{(15\ g)}$: 29–58

Obs **:** cf. Table and R charts, visible range data: QDF3. VHN: Uytenbogaardt and Burke (1971)
Mineral abbreviations: As—native arsenic, asp—arsenopyrite, bar—baryte, cob—cobaltite, ery—erythrite, gg—gangue, mc—marcasite, py—pyrite, orp—orpiment, rl—realgar, sbt—stibnite, sk—skutterudite, sp—sphalerite, td—tetrahedrite

89.1 General

Composition: As_2S_3, sometimes with traces of Sb and Hg.

Orpiment, ***orp***, is a transparent mineral, yellow in hand specimen and in thin section. It can also be studied with TL (petrographic microscope).

89.2 Microscopic Description

89.2.1 SP Observation (*Single Polariser*)

Greyish-white *colour*, slightly pinkish (Figs. 89.1a and 89.2a) which, in immersion, turns to dominant yellow, due to the transparency of the mineral (Fig. 89.1b). Clearly bireflective.

Relatively high *reflectance*, **R**, for a transparent mineral, higher than realgar, *rl*, with which it is often associated (cf. comparative R graphs). However, R orp is much lower than skutterudite, arsenopyrite, or native arsenic. It can be formed by alteration of the former (skutterudite, arsenopyrite), from which some relicts may be preserved as inclusions in orp (Fig. 89.2a).

Polishing hardness very low, bordering on low: $PolH_{op}$ > rl, but < sbt. Exfoliation (010) frequent.

It occurs in encrustations on realgar or in fibrous-radial, botryoidal or acicular aggregates, in cryptocrystalline masses and, rarely, in well-developed, tabular or prismatic crystals (Figs. 89.1 a–d and 89.2a, b).

89.2.2 Observation +P (*Polariser + Analyser at 90°*)

Strongly *anisotropic*, but the effects of anisotropism are masked by IR, *internal reflections*, massive and luminous, in golden yellow to orange and ocher hues (Figs. 89.1c, d and 89.2b).

Occasional pressure *twins*, due to the low resistance of orp to deformation.

R. Castroviejo, *A Practical Guide to Ore Microscopy—Volume 1*,
https://doi.org/10.1007/978-3-031-12654-3_89

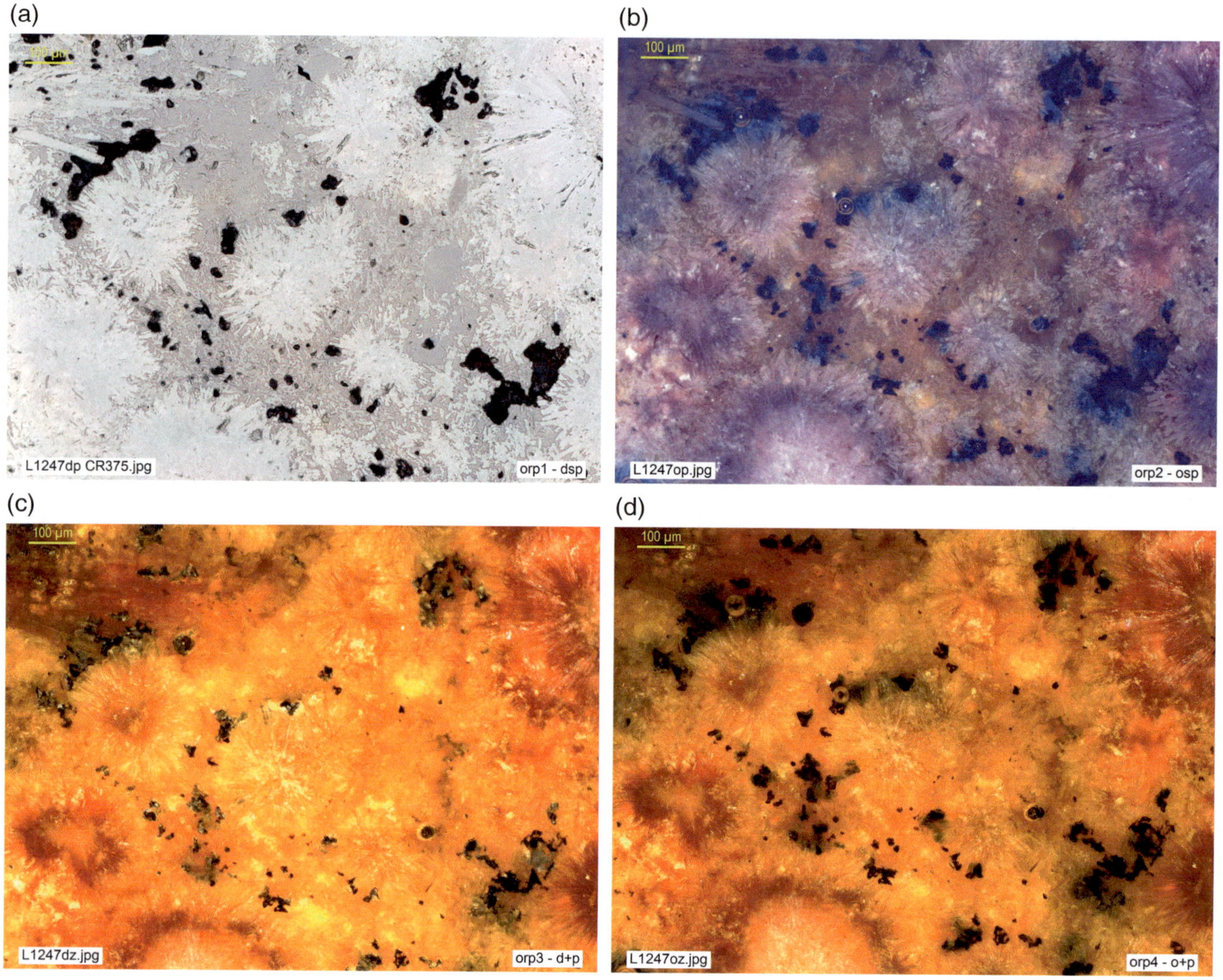

Fig. 89.1 **a, b** (**dsp** and **osp**, respectively. A magnification of the lower central zone of Fig. 89.2a). Aggregates of acicular to fibrous-spherulitic orp (in air, greyish white, bireflective; in immersion, the yellow to pink-orange colour of orp is directly visible by transparency), in anhedral mass of the same mineral. CR 375, Cobalt, Ontario, Canada. **c, d** Same scene as orp1 (**d+p** and **o+p**, respectively). The luminous and massive IR hide the anisotropism of orp

89.2.3 Paragenesis

Hydrothermal, low temperature *parageneses*, in which it is usually a late mineral. Also found in volcanic thermal emanations and fumaroles. Frequently a product of supergene alteration of realgar, on which it forms encrustations, as well as on arsenopyrite, skutterudite and other arsenides.

Companions: realgar, native arsenic, erythrite (Fig. 89.2a, b), skutterudite, cobaltite, antimonite, pyrite, marcasite, barite, scorodite, etc.

89.2.4 Identification

The identification of orpiment does not present difficulties, starting with its easy recognition in a hand sample, taking into account its paragenesis and confirming it by the observation of its properties. These are of particular diagnostic value:

- its abundant yellow IR,
- its very low hardness,

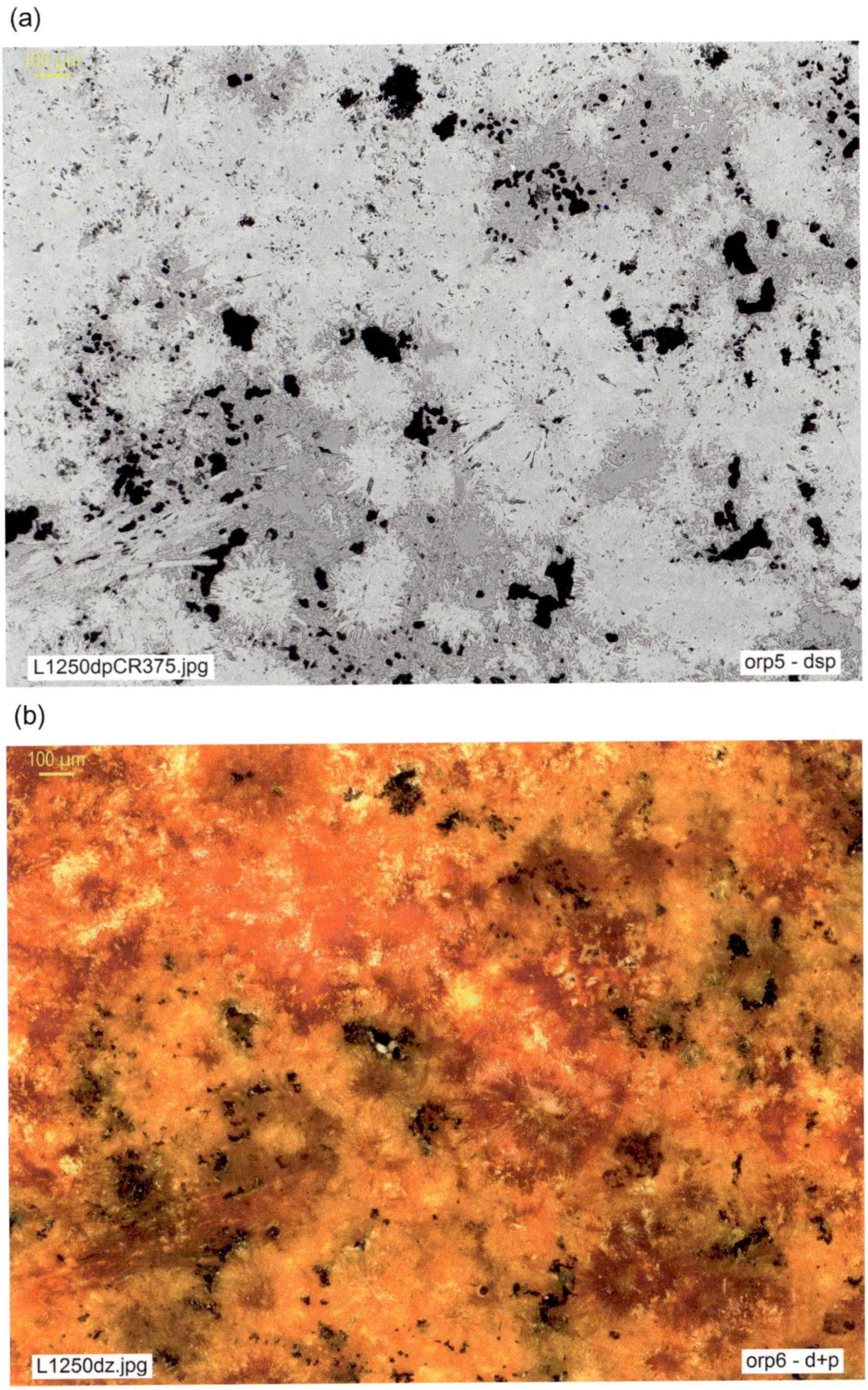

Fig. 89.2 **a, b** (Same scene: wider field view of Fig. 89.1a, **dsp** and **d+p**, respectively). Intergrowth of orp and erytrite, $Co_3(AsO_4)_2{\cdot}8H_2O$; ery is distinguished by its purple-red IR (**+p**). Relicts of probable sk (scattered, tiny white particles, possible source of Co for ery). CR 375, Cobalt, Ontario, Canada

- its transparency together with its relatively high reflectance, and
- its remarkable bireflectance.

89.3 Spectral Information

89.3.1 Orpiment Reflectance R (in Air, Visible Spectrum) (%)

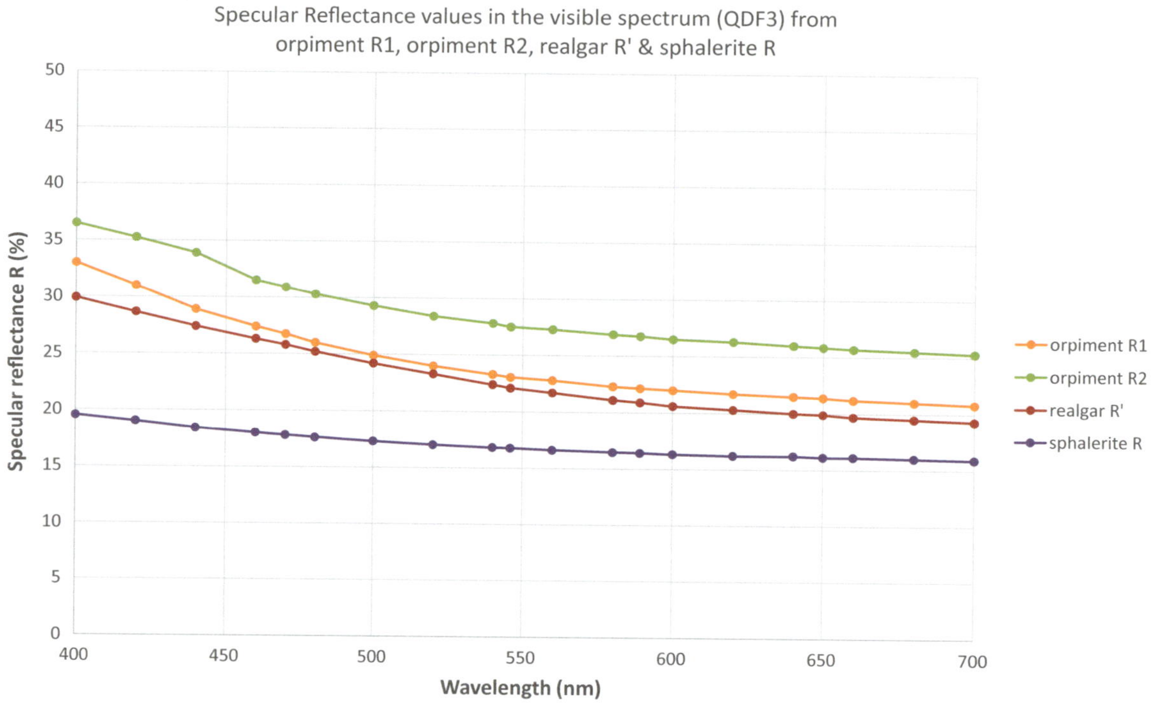

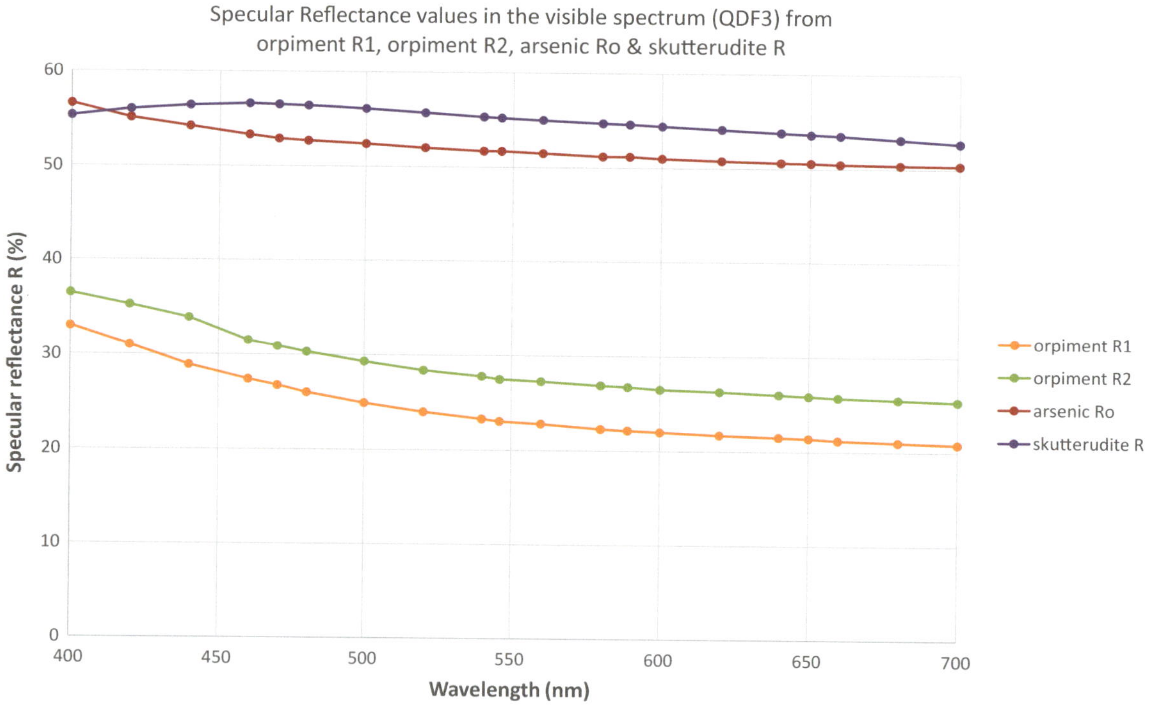

Reference

Uytenbogaardt W, Burke EAJ (1971) Tables for microscopic identification of ore minerals, 2nd edn. Elsevier, Amsterdam

Osmium (*Os*)

90

Os: Hexagonal

Osmium		R_{546} (%)	Y (%)	λ_d (nm)	P_e (%)	Colour sensation
Air	R_1	61.9	61.6	483	2.2	White, slightly bluish
	R_2	62.3	62.1	483	2.5	
Oil	R_1	nd	nd	nd	nd	
	R_2	nd	nd	nd	nd	
Bir: Very weak (2)				*Pleoc*: Very weak (2)		*Tarn*: No
Anis +P: Distinct (4)				*IR*: No		*Twg*: No
Mor: Lamellar (acicular section)				*Clv*: No		*PolQ*: Difficult (relief)
Hardness	*PolH*: Very high, > py (5)			*Mohs*: 7	*Talm*: nd	*VHN* $_{(100\ g)}$: 974–1074

Obs: cf. Table and R charts, multispectral data (VNIR, AMCO) and visible range (QDF3; Picot and Johan 1982)
Mineral abbreviations: Au—native gold, chr—chromite, el—electrum, erl—erlichmanite, Fe—native iron, gg—gangue, Ir—native iridium, Os—native osmium, Pt—native platinum, Pd—native palladium, pn—pentlandite, py—pyrite, spn—spinel, spy—sperrylite

90.1 General

Along with Pt, Pd, Rh, Ir and Ru, Os is one of the six metals known as **PGE** (*Platinum Group Elements*). It is also, as a native metal, one of 109 **PGM** (Platinum *Group Minerals)* recognized by the IMA (cf. comments on PGE and PGM in *Pt description* and more data in Cabri *ed.* 2002).

Composition: It admits, in its hexagonal lattice, Ru and Ir, forming broad solid-solution series extending into the Ru and Ir fields, the latter limited by the compositional field of *ruteniridosmin* (≈ 41 to 47% Ir, 33 to 43% Os, 19 to 6% Ru, Cabri 2002). However, the large variations possible by expanding the compositional field of Os as solid solution, as well as the new experimental data and IMA criteria, have led to discredit traditionally used terms such as *iridosmin* (now split between the Os and *ruteniridosmin fields*) and *osmiridium* (now with Ir). Os may also contain minor amounts of Pt, Rh, Pd, Fe and Cu.

90.2 Microscopic Description

90.2.1 SP Observation (*Single Polariser*)

White *colour*, with a slightly bluish grey tone in direct comparison with platinum (Figs. 90.1a, b and 90.2). Its high hardness makes *polishing* difficult and the mineral appears in relief. Very high *polishing hardness*: > pyrite; it is seen in relief on platinum (Fig. 90.3).

Reflectance is high but lower than that of Ir, Pt and Pd, variable with composition (cf. comparative graphs of R). Anisotropic, but its bireflectance and pleochroism are hardly noticeable.

The primary mineral is found in euhedral inclusions in platinum (or Pt-Fe alloy), generally tiny and of prismatic or acicular morphology (Figs. 90.1a, b and 90.3). This form actually corresponds to cross sections of hexagonal crystals of laminar habit.

R. Castroviejo, *A Practical Guide to Ore Microscopy—Volume 1*,
https://doi.org/10.1007/978-3-031-12654-3_90

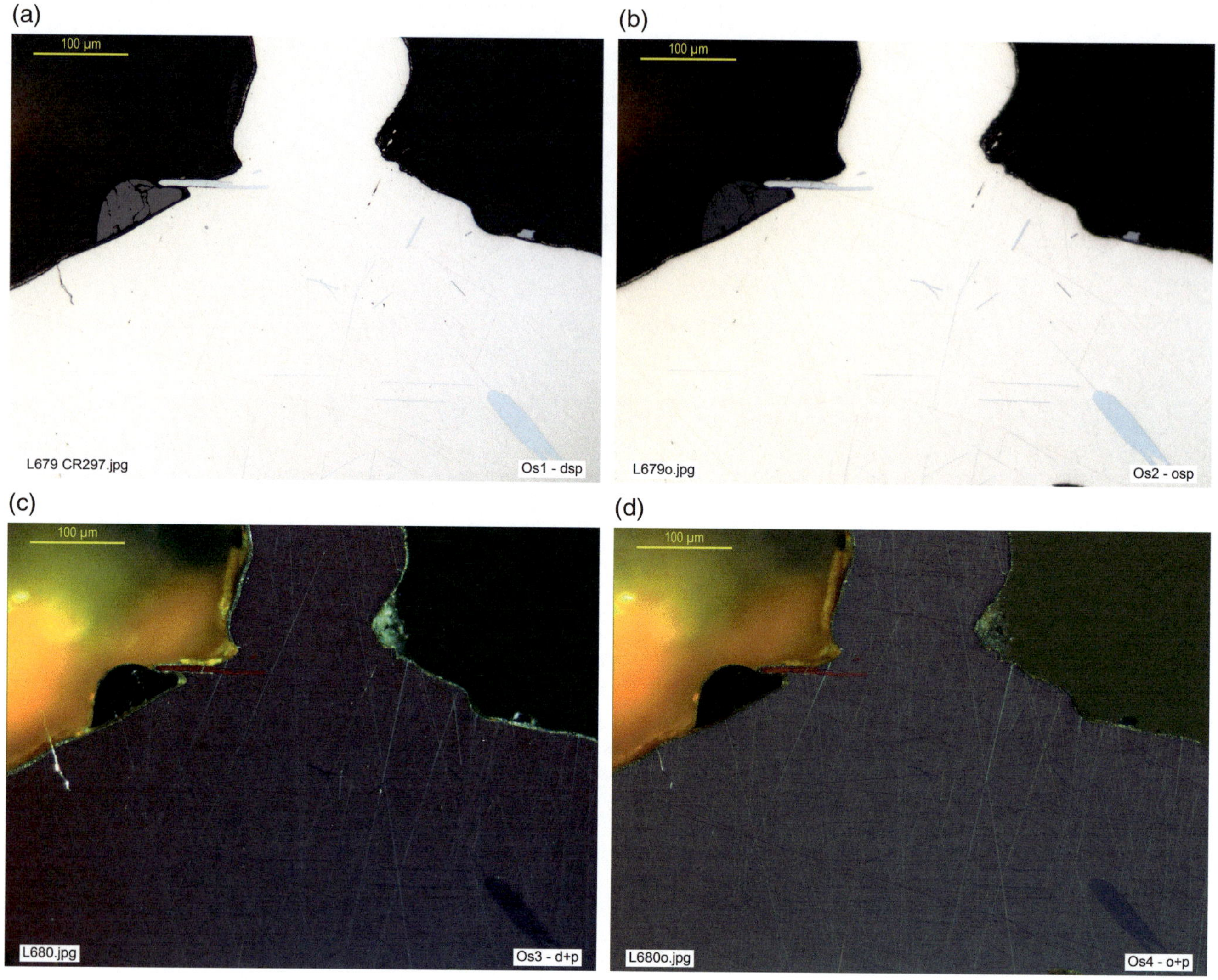

Fig. 90.1 **a, b** (**dsp** and **osp**, respectively)—Os inclusions (bluish white) in Pt (white, dominant component), with accessory chr (grey) and gangue (dark grey). Two types of Os inclusions: (i) elongated or lamellar crystals with no clear orientation pattern, well visible (e.g., bottom right, and to the left, thin E-W lamella, next to chr); (ii) very fine exsolutions with a more regular distribution throughout the Pt mass and following octahedral Pt directions, hardly visible except by interference contrast (cf. Fig. 90.3 and F Pt6). CR 297, Folx Gulch, Goodnews Bay, Alaska. **c, d** Same scene as Os1 (**d+p** and **o+p**, respectively): in contrast to the extinction of Pt and chr (cubic), Os anisotropism, with typical orange polarisation tints, visible in the lamellar inclusion next to chr (the other inclusion, bottom right, is in extinction position but shows its anisotropism in *F Pt3-4*)

90.2.2 Observation +P (*Polariser + Analyser at 90°*)

Clearly anisotropic, with bright and lively polarisation tints, in reddish and orange tones (Fig. 90.1c, d), in contrast to its usual immediate surroundings, dark due to the continuous extinction of platinum or chromite (isotropes).

90.2.3 Paragenesis

A scarce mineral, usually included in platinum. It occurs in *parageneses* typically related to ultramafic intrusions and ophiolites, as well as in placers derived from them. Given the scarcity of the primary mineral, Os is recovered from placers, usually as a co-product of platinum, as the main source.

Accompanying ores: platinum or Pt-Fe alloys, pentlandite, sperrylite and other PGM, chromite.

90.2.4 Identification

Its striking anisotropism in red–orange tones is typical, and, together with its bluish-white colour, laminar habit, high hardness and paragenesis, allows reliable identification.

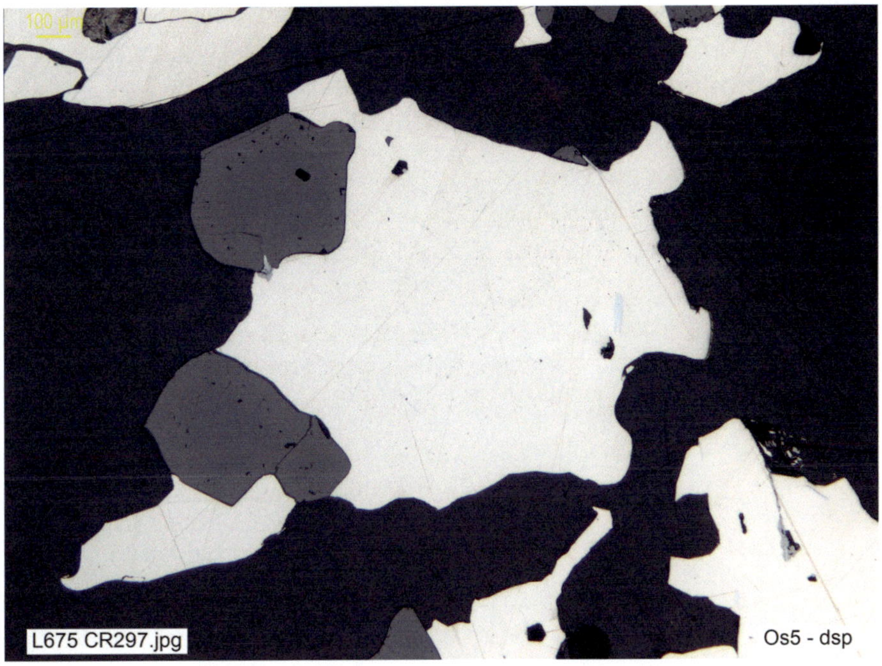

Fig. 90.2 PGE and chromite ore. Pt (white) is the main component. It is intergrown with chr (grey) and gg (dark grey) and shows inclusions of Os (bluish white) and erl (light grey). Wide field image of Fig. 90.1 a and also of Fig. 90.3, which are detailed views of the right-hand side of the coarser Pt grain. CR297, Folx Gulch, Goodnews Bay, Alaska (**dsp**)

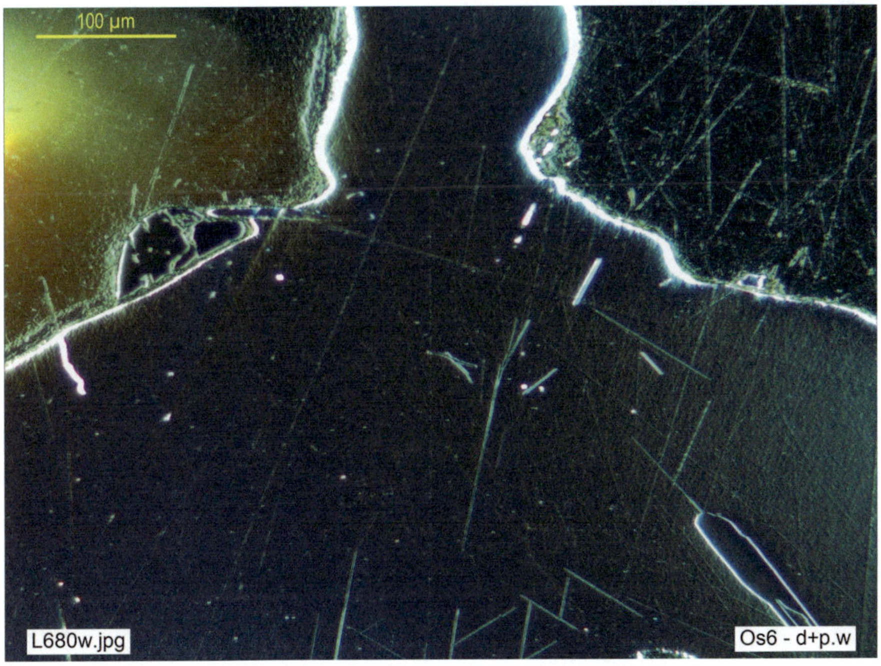

Fig. 90.3 Same scene as Fig. 90.1a (**d+p.w**). The interference contrast, with Wollaston Prism, makes visible the Os exsolutions, in thin lamellae oriented according to (111) of Pt (cf. F Pt6, for a wider field)

90.3 Spectral Information

90.3.1 Osmium Multispectral Reflectance R (in Air, Without Polariser) (%)

λ (nm)	370	400	425	450	475	500	525	550	575	600	625	650	675	700	750	800	850	900	950	1000
μ	60.6	61.2	61.5	62.1	62.4	62.7	62.6	62.2	61.7	61.3	61.1	60.8	60.8	60.5	60.9	61.3	62.4	61.9	62.9	63.2
P_5	56.3	57.5	57.8	58.3	58.8	59.1	59.0	58.6	58.1	57.7	57.5	57.4	57.6	57.5	58.3	59.2	60.8	60.6	61.8	62.0
P_{20}	58.9	59.8	60.2	60.6	61.0	61.0	60.7	60.0	59.3	58.8	58.5	58.3	58.5	58.3	59.0	59.8	61.4	61.3	62.5	62.6
P_{80}	62.4	62.7	63.0	63.9	64.3	64.8	65.1	64.8	64.4	64.0	63.7	63.4	63.2	62.7	62.6	62.7	63.4	62.5	63.4	63.9
P_{95}	63.3	63.5	63.9	64.7	65.0	65.5	65.6	65.3	64.9	64.6	64.2	63.9	63.7	63.2	63.1	63.2	64.0	63.1	64.0	64.2

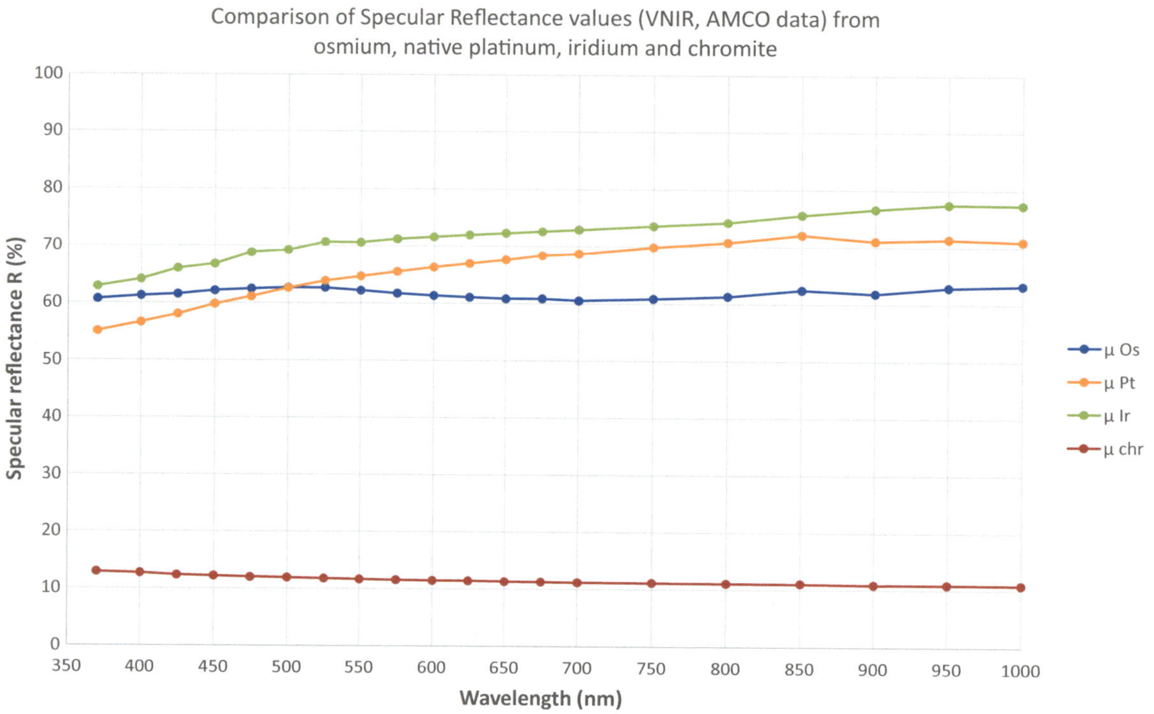

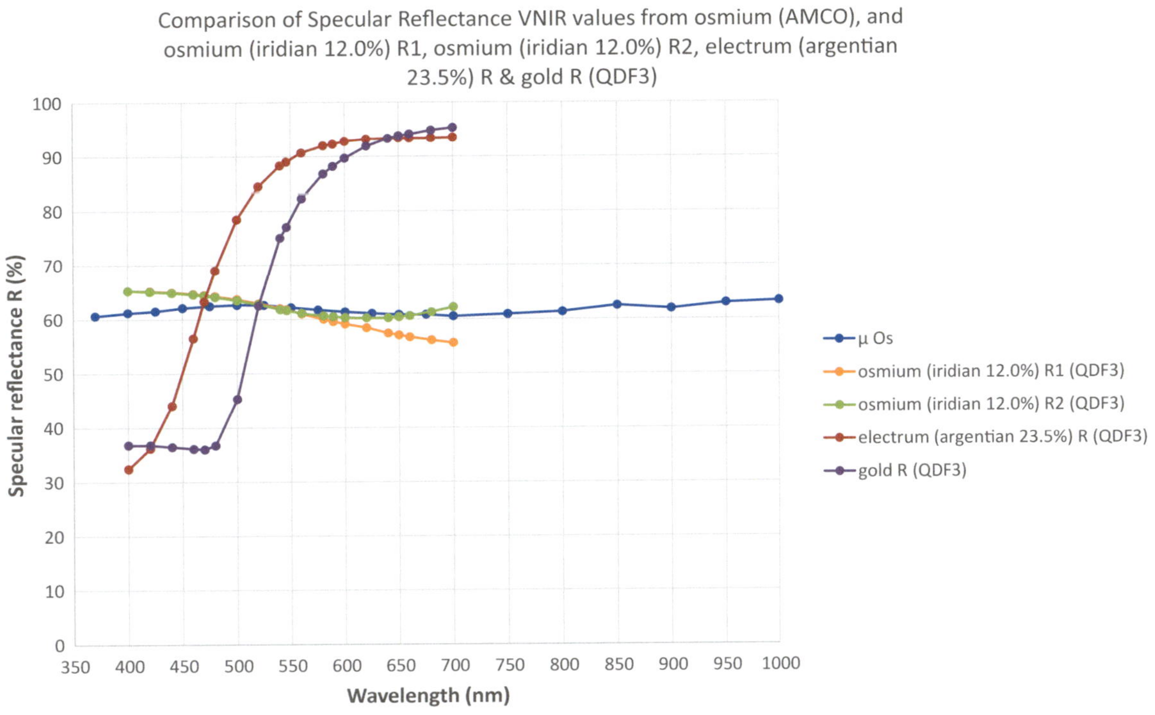
Comparison of Specular Reflectance VNIR values from osmium (AMCO), and osmium (iridian 12.0%) R1, osmium (iridian 12.0%) R2, electrum (argentian 23.5%) R & gold R (QDF3)
Specular reflectance R (%)
Wavelength (nm)
100
90
80
70
60
50
40
30
20
10
0
350
400
450
500
550
600
650
700
750
800
850
900
950
1000
µ Os
osmium (iridian 12.0%) R1 (QDF3)
osmium (iridian 12.0%) R2 (QDF3)
electrum (argentian 23.5%) R (QDF3)
gold R (QDF3)

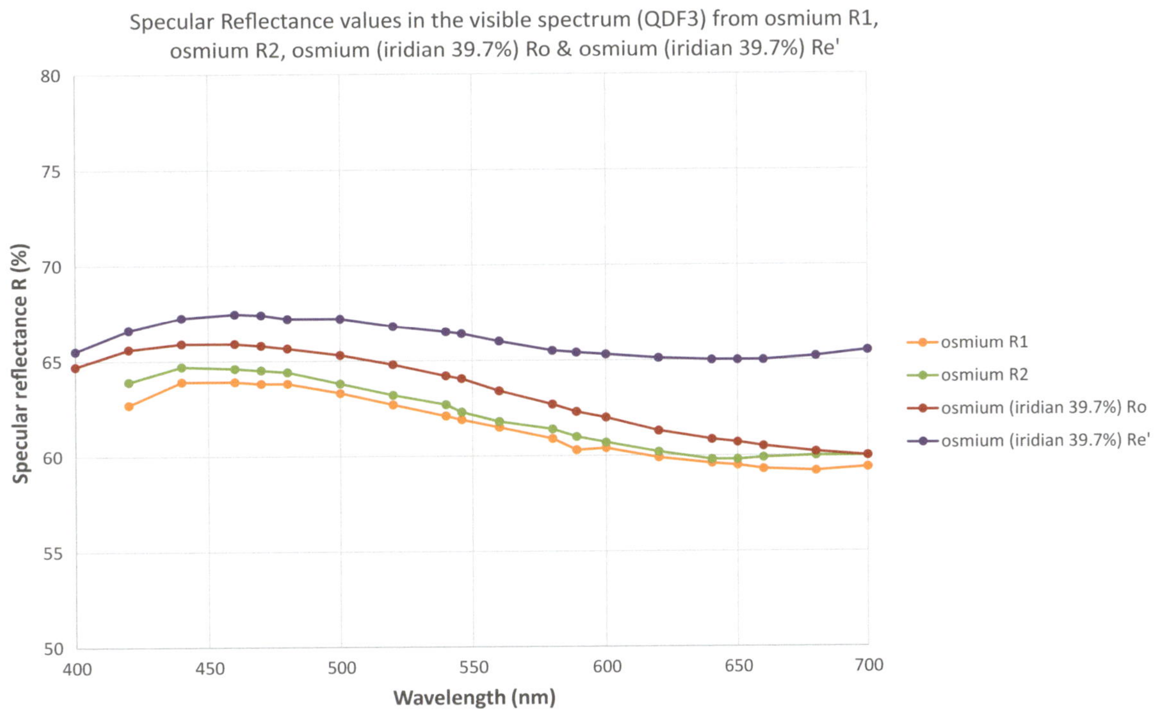
Specular Reflectance values in the visible spectrum (QDF3) from osmium R1, osmium R2, osmium (iridian 39.7%) Ro & osmium (iridian 39.7%) Re'
Specular reflectance R (%)
Wavelength (nm)
80
75
70
65
60
55
50
400
450
500
550
600
650
700
osmium R1
osmium R2
osmium (iridian 39.7%) Ro
osmium (iridian 39.7%) Re'

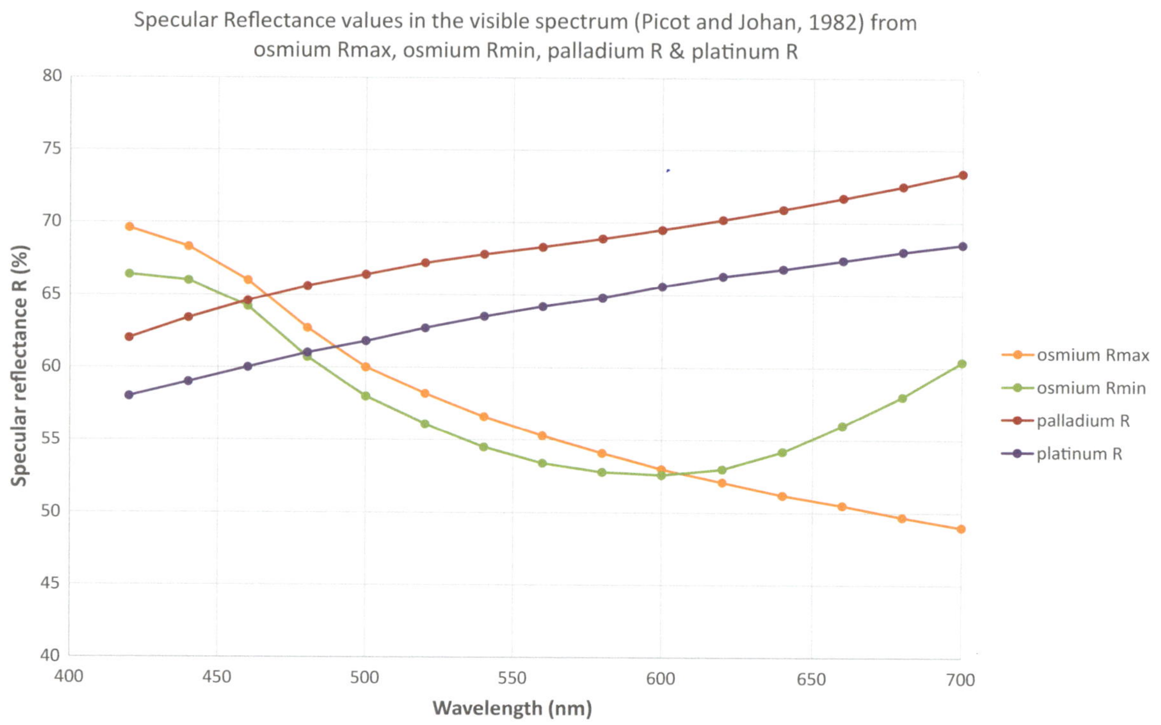

References

Cabri LJ (2002) The platinum-group minerals, pp. 13–129; en: Cabri LJ (2002) CIM special, vol 54. Canadian Institute Mining, Metallurgy and Petroleum, Calgary, Alberta, Canada, 852 p

Cabri LJ (2002) (ed) The geology, geochemistry, mineralogy and mineral beneficiation of platinum group minerals. Canadian Institute of Mining, Metallurgy and Petroleum, Calgary, Alberta, Canada, 852 p. ISBN 1-894495-27-5

Picot P, Johan Z (1982) Atlas of ore minerals. B.R.G.M./Elsevier

Pararammelsbergite (*prm*)

91

$NiAs_2$: Orthorhombic

Pararamelsbergite		R_{546} (%)	Y (%)	λ_d (nm)	P_e (%)	Colour sensation
Air	R_1	58.9	58.8	530	0.3	Bright white
	R_2	59.7	59.9	580	1.8	
Oil	R_1	45.5	45.3	495	0.6	
	R_2	46.8	47.1	582	2.1	
Bir: Weak (3)				*Pleoc*: Weak (3)		*Tarn*: Appears in humid environments
Anis +P: Strong, coloured (orange-brown)				*IR*: No		*Twg*: Seldom, no polysynthetic twins
Mor: Rect/romb euhedral shapes				*Clv*: Occasional		*PolQ*: Good
Hardness	*PolH*: Medium: > nc, < rm, po (3)			*Mohs*: 5.5	*Talm*: E	*VHN* $_{(100\ g)}$: 681–830

Obs: It is the low-temperature polymorph of $NiAs_2$, cf. Table and R plots, multispectral data (VNIR, AMCO) and visible range (QDF3; Picot and Johan 1982)

Mineral abbreviations: Ag—native silver, bth—breithauptite, asp—arsenopyrite, cob—cobaltite, gg—gangue, gud—gudmundite, gf—gersdorffite, gld—glaucodot, ln—linnaeite (group), mau—maucherite, lo—löllingite, nc—nickeline, prm—parammelsbergite, po—pyrrhotite, py—pyrite, rm—rammelsbergite, sf—safflorite, sk—skutterudite, ul—ullmannite

91.1 General

Composition: $NiAs_2$, may contain as impurities Co, Cu, S. Trimorph with rammelsbergite and krutovite.

91.2 Microscopic Description

91.2.1 SP Observation (*Single Polariser*)

Bright pure white *colour*, even more so than rammelsbergite, which it resembles; however, as usual in visual observation, it looks grey next to native silver (Figs. 91.1a, b).

Weak pleochroism.

High *reflectance* (highest of the Co–Ni arsenides), slightly > rammelsbergite.

Medium *polishing hardness* (> nc and ln, but < sk, rm, sf, po, lo, asp and ≪ py). Polishes very well.

Unlike rammelsbergite, it has a marked euhedral tendency, typically presenting tabular forms, with rectangular or rhomboidal sections (Fig. 91.1a, b), but it also forms fine-grained intergrowths with rammelsbergite.

91.2.2 Observation +P (*Polariser + Analyser at 90°*)

Strongly *anisotropic*, but somewhat less so than rammelsbergite, with different polarisation colours (typical is orange-brown), less vivid than those of rammelsbergite (Fig. 91.1c, d).

It is rarely twinned and it does not present the *polysynthetic twins* so typical of rammelsbergite.

No *internal reflections*.

R. Castroviejo, *A Practical Guide to Ore Microscopy—Volume 1*,
https://doi.org/10.1007/978-3-031-12654-3_91

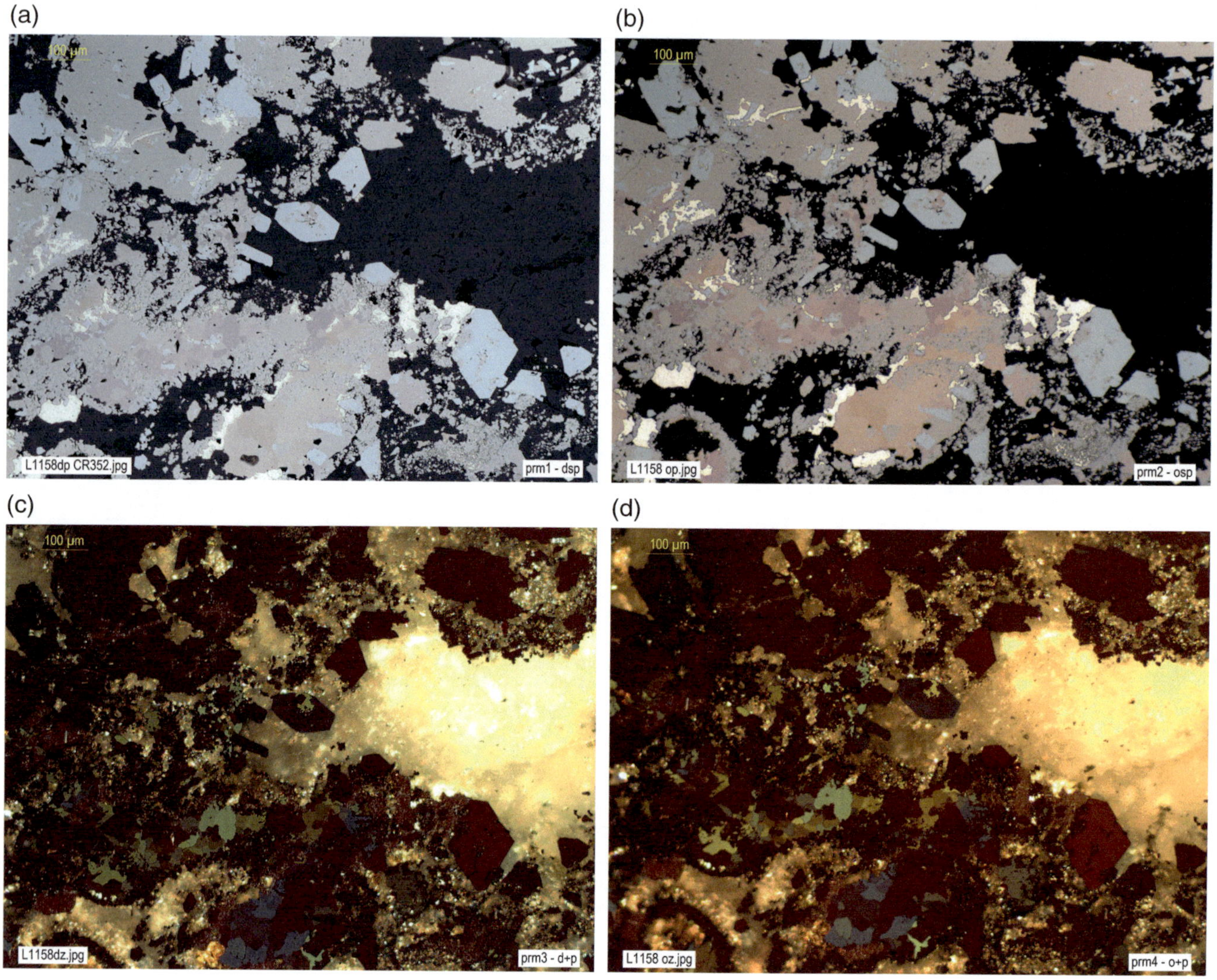

Fig. 91.1 **a** Ore composed of aggregate of prm, Ag, bth, nc and mau, with calcite gg. By comparison with Ag (bright white), prm looks grey, but is easily recognized by its euhedral shapes (rectangular and rhomboidal), as well as its anisotropism (Fig. 91.1c, d). The coloured shades allow the distinction of bth (purplish pink or *mauve*), nc (pink) and mau (pinkish grey). CR352, Cobalt, Ontario, Canada (**dsp**). **b** Same scene as prm1 (**osp**): note the higher contrast and saturation of the colour tones, which makes the differences in immersion much sharper. **c** Same scene as prm1 (**d+p**): clearly anisotropic character of prm (with typical orange colours), as well as nc and bth (greenish and bluish) but weak of mau. **d** Same scene as prm1 (**o+p**): brighter and more pronounced polarisation colours

91.2.3 Paragenesis

It typically occurs in hydrothermal veins of the *five-element formation* type (Ni–Co–Bi–Ag–U).

It is usually accompanied by other arsenides and Ni–Co minerals, such as pararammelsbergite, nickeline, safflorite, skutterudite, linnaeite, maucherite, Ag and Bi minerals, etc.

91.2.4 Identification

By its appearance it resembles other species of the so-called *white arsenides* (skutterudite, löllingite, safflorite, rammelsbergite) or even sulphoarsenides/sulphoantimonides (gersdorffite, arsenopyrite, glaucodot, cobaltite, gudmundite, ullmanite) and sulphides (linnaeite), which a hasty observer could confuse it with, particularly rammelsbergite, with which it is often associated.

Useful characteristics for its diagnosis are its pure white colour, high reflectance, medium hardness, very good polish and strong anisotropism, with orange-brown polarisation colours, different and less luminous than those of rammelsbergite. Also of note:

- The polysynthetic twins so frequent in rammelsbergite are missing in prm.
- Skutterudite, linnaeite and gersdorffite are isotropic.
- Arsenopyrite, löllingite and safflorite are harder.

Its marked tendency to idiomorphism and the paragenesis (association with other Ni and Co minerals) are additional criteria for identifying pararammelsbergite.

91.3 Spectral Information

91.3.1 Pararammelsbergite Multispectral Reflectance R (in Air, Without Polariser) (%)

λ (nm)	370	400	425	450	475	500	525	550	575	600	625	650	675	700	750	800	850	900	950	1000
μ	57.0	57.6	58.1	58.6	58.8	59.0	59.2	59.4	59.4	59.5	59.5	59.5	59.5	59.6	59.7	59.2	58.9	59.1	58.7	58.6
P_5	55.7	56.4	57.0	57.6	57.6	57.6	58.0	58.4	58.4	58.4	58.4	58.4	58.4	58.4	58.4	58.0	57.2	57.6	57.2	57.2
P_{20}	56.7	57.2	57.6	58.0	58.2	58.4	58.6	58.8	58.8	58.8	58.8	58.8	59.0	59.2	58.8	58.4	58.0	58.4	58.0	58.0
P_{80}	57.9	58.4	58.8	59.2	59.4	59.6	59.8	60.0	60.0	60.0	60.0	60.0	60.2	60.4	60.4	60.0	59.6	60.0	59.6	59.2
P_{95}	58.3	58.8	59.2	59.6	60.0	60.4	60.4	60.4	60.4	60.4	60.4	60.4	60.6	60.8	61.2	60.8	60.8	60.8	60.4	60.4

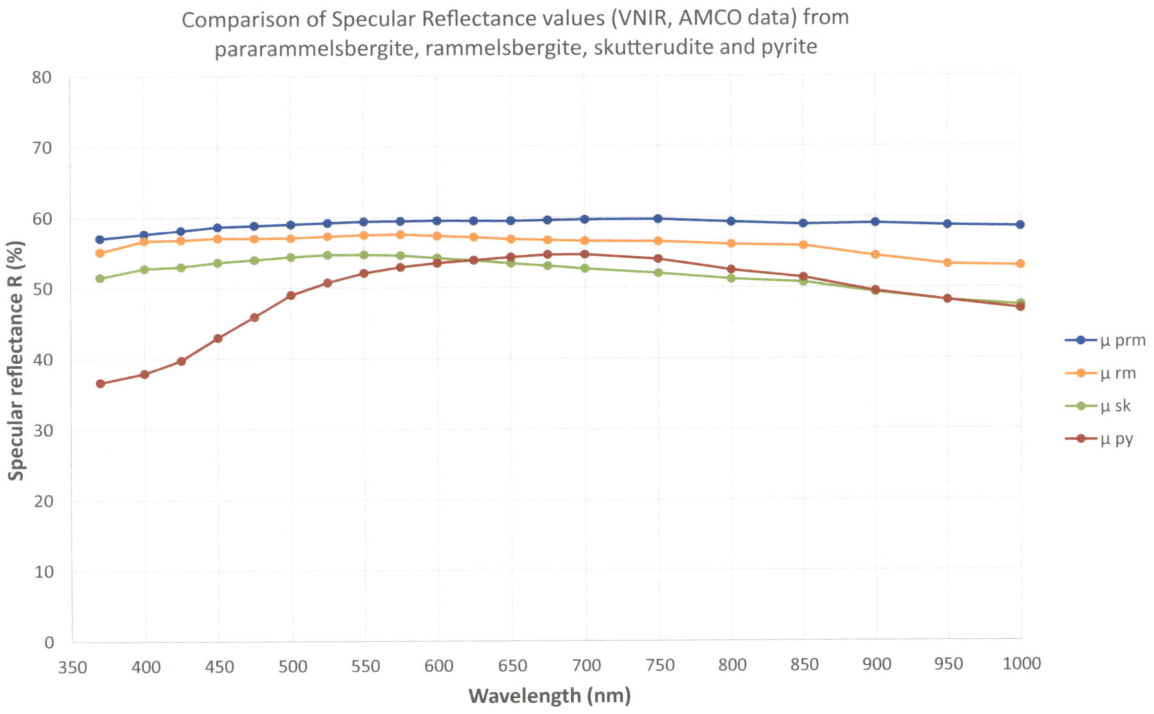

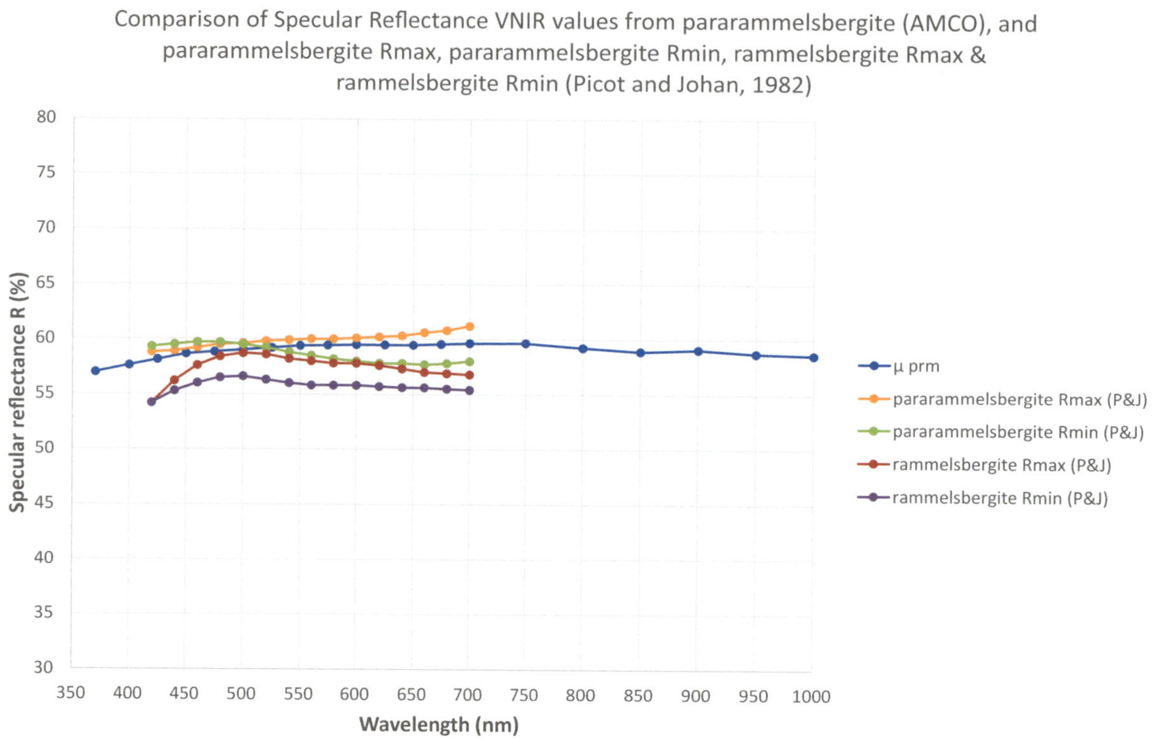

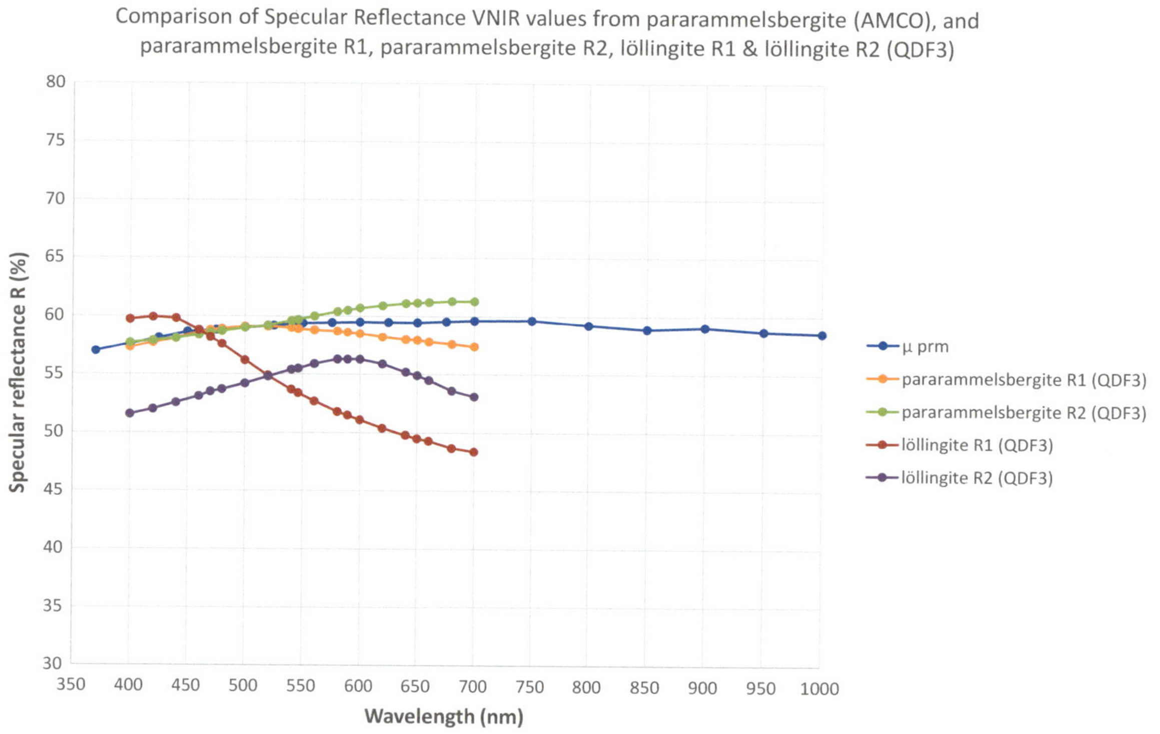

Reference

Picot P, Johan Z (1982) Atlas of ore minerals. B.R.G.M./Elsevier

Pearceite (*pc*) 92

$(Ag,Cu)_{16}As_2S_{11}$: Monoclinic

Pearceite		R_{546} (%)	Y (%)	λ_d (nm)	P_e (%)	Colour sensation
Air	R_1	29.1	29.4	469	5.2	Greyish-white, greenish
	R_2	32.2	32.0	481	3.4	
Oil	R_1	14.4	14.7	468	8.8	
	R_2	17.2	16.9	480	6.1	
Bir: Very weak (3)				*Pleoc*: Very weak (3)		*Tarn*: Light Attack
Anis +P: Distinct (4)				*IR*: Red, scattered		*Twg*: Not observed
Mor: Anhedral aggregates, lamellar				*Clv*: // (001) not very visible		*PolQ*: Defective (scratches)
Hardness	*PolH*: Very low (1)			*Mohs*: 3-	*Talm*: C-	*VHN* $_{(100\ g)}$: 130–142

Obs: cf. Table and R, multispectral (VNIR, AMCO) and visible range (QDF3 and Picot and Johan 1982). VHN Uyt-Burke (1971)
Mineral abbreviations: ac—acanthite, Ag—native silver, ccp—chalcopyrite, dy—dyscrasite, fbg—freibergite, gg—gangue, gn—galena, lo—löllingite, my—miargyrite, pc—pearceite, plb—polybasite, prg—pyrargyrite, pru—proustite, sbt—stibnite, sf—safflorite, sk—skutterudite, stm—stromeyerite, sp—sphalerite, stp—stephanite, tn—tennantite, td—tetrahedrite

92.1 General

Pearceite, ***pc***, has an ideal composition expressed as $[Ag_9CuS_4][Ag_6As_2S_7]$, but actually it is $[Ag_9CuS_4][(Ag,Cu)_6(As,Sb)_2S_7]$.

Isostructural with polybasite, ***plb***, pearceite is the arsenical term for the Pearceite-Polybasite Series: $[Ag_9CuS_4][(Ag,Cu)_6(As,Sb)_2S_7]$—$[(Ag,Cu)_6(Sb,As)_2S_7][Ag_9CuS_4]$. The reality is somewhat more complex, if the now known polytypes are taken into account, and some authors speak rather of the *Pearceite-Polybasite Group*.[1]

This series seems to admit an unlimited solid dissolution between the two terms, so that the microscopic distinction between them is questionable. An additional complication is that several terms of the series with different compositions may coexist in the same sample.

Nevertheless, an attempt will be made to characterize the main endmembers, pc and plb, separately, describing selected samples in which the mineralogy has been checked by electron microprobe and reflectance measurements. In order to not repeat too much, the description of shared properties is referred to that of polybasite.

92.2 Microscopic Description

92.2.1 SP Observation (*Single Polariser*)

Greyish-white *colour*, with a slightly greenish hue (whiter than polybasite) and very *weak pleochroism* and bireflectance (Fig. 92.1 a, b).

Moderate *reflectance*, slightly lower than polybasite and freibergite; much lower than white arsenides (cf. multispectral data and comparative R% plots), against which pearcite appears relatively dark grey (Figs. 92.2 a and 92.3 a).

Sensitive to *light attack*, although not as much as polybasite in the samples studied.

Polishing quality and *polishing hardness like* polybasite. Morphology as polybasite.

[1] For more information, cf. https://www.mindat.org/min-3138.html.

R. Castroviejo, *A Practical Guide to Ore Microscopy—Volume 1*,
https://doi.org/10.1007/978-3-031-12654-3_92

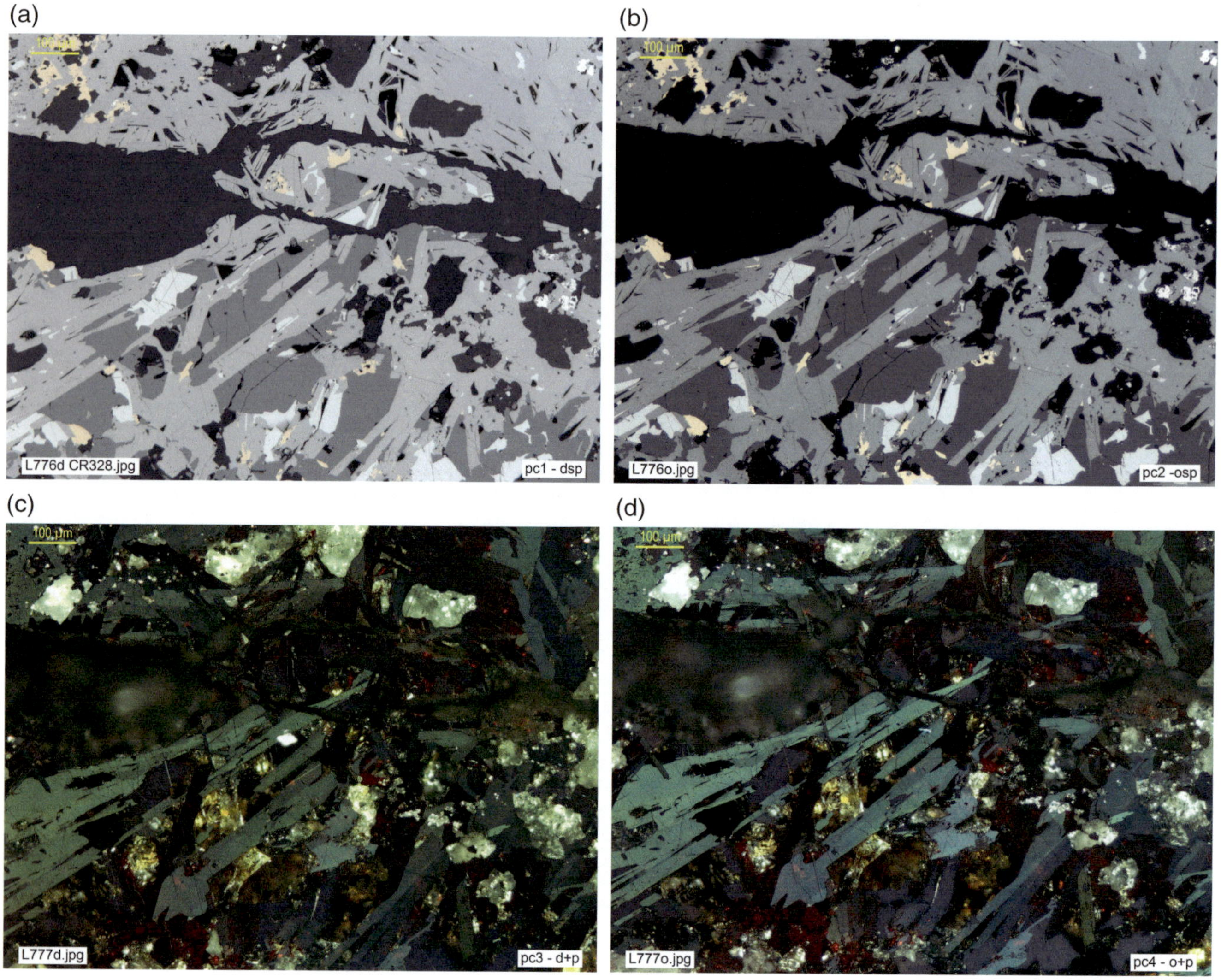

Fig. 92.1 **a**, **b** (**dsp** and **osp**, respectively). Ore of pc (prismatic-acicular, light greenish grey), gn (white), ccp (yellow), sp (dark grey), minor prg (light grey, slightly bluish against pc), with traces of Ag (top right) and with minor q—Carb gg (dark grey). CR 328, Freiberg, Saxen (Germany). **c**, **d** (Same scene as pc1: **d+p/o+p**, respectively). Polarisation clearly distinguishes pc (greenish or bluish-green anisotropism, sporadic red IR) and prg (abundant red IR, overprinting anisotropism), as well as sp (showing the luminous, light yellowish IR of pure ZnS composition, tested EMPA) and gg (very bright, colourless IR)

92.2.2 Observation +P (*Polariser + Analyser at 90°*)

Strongly *anisotropic*, more than polybasite, but less coloured, although greenish or bluish-green tones are perceptible (Figs. 92.1c, d, 92.2b and 92.3b). *Internal reflections like* polybasite.

92.2.3 Paragenesis

The paragenesis is similar to that of polybasite, but pc is much less frequent than plb, and it is accompanied by arsenical species (tennantite, proustite, Ni-Co arsenides, Fig. 92.3a, b) instead of antimonial species.

92.2.4 Identification

The very low hardness, mediocre polish, moderate reflectance, greyish-white colour with very faint greenish tones, *light attack*, anisotropism with bluish-green tones and arsenical paragenesis allow the identification of pearceite, although they do not assure the determination of the composition in the pearceite-polybasite series (cf. *polybasite*).

(a)

(b)

Fig. 92.2 **a**, **b** (same scene: **osp** and **o+p**)—Intergrowth of pc (light grey, tabular, greenish anisotropism), sp (dark grey, isotropic, caramel IR), pru (bottom left, grey, slightly bluer and darker than pc, duller anisotropism, more abundant red IR), fbg (grey slightly more greenish than pc, isotropic, top centre-right), gn, ccp and q (black, massive colourless IR). CR 328, Freiberg, Saxen (Germany)

Against stromeyerite and freibergite, these are minerals of similar aspect with which pearceite could be confused:

- Stromeyerite is distinguished by its more vivid anisotropism, absence of IR and weak light attack.
- Although colour and reflectance of freibergite are similar to those of pearceite, it is easily distinguished because it is isotropic.

Fig. 92.3 **a**, **b** (same scene: **osp** and **o+p**)—Arsenical argentiferous ore. Langisite (CoAs, very light pinkish-brown, anisotropism in greyish-brown and orange tones), coated by *corona* of intergrown white arsenides (sf, lo, anisotropic; sk, isotropic), which also form aggregates containing Ag inclusions (right) or are finely intergrown with pc (greenish anisotropism), pru (more abundant red IR), sp, ccp, gn. CR 328, Freiberg, Saxen (Germany)

(a)

92.3 Spectral Information

92.3.1 Pearceite Multispectral Reflectance R (in Air, Without Polariser) (%)

λ (nm)	370	400	425	450	475	500	525	550	575	600	625	650	675	700	750	800	850	900	950	1000
μ	32.6	32.2	31.9	31.6	31.2	30.8	30.6	30.4	30.1	29.8	29.3	28.8	28.4	28.1	28.1	28.1	27.8	27.7	27.5	27.5
P_5	31.0	30.8	30.6	30.4	29.8	29.2	29.2	29.2	28.8	28.4	28.0	27.6	27.0	26.4	25.2	24.8	24.0	24.0	23.6	23.6
P_{20}	32.1	31.6	31.2	30.8	30.4	30.0	29.8	29.6	29.4	29.2	28.6	28.0	27.4	26.8	26.0	25.2	24.8	24.4	24.4	24.4
P_{80}	33.0	32.8	32.6	32.4	32.0	31.6	31.4	31.2	30.8	30.4	30.0	29.6	29.2	28.8	30.0	33.2	33.6	34.0	34.0	33.6
P_{95}	34.1	33.6	33.2	32.8	32.6	32.4	32.0	31.6	31.2	30.8	30.4	30.0	30.4	30.8	34.0	36.0	36.8	36.8	37.2	36.8

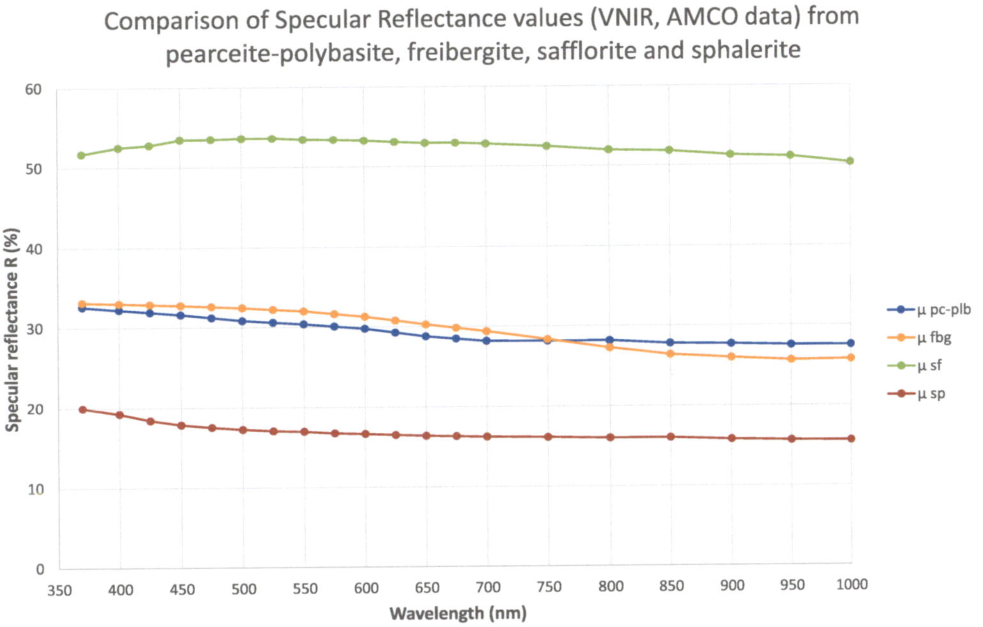

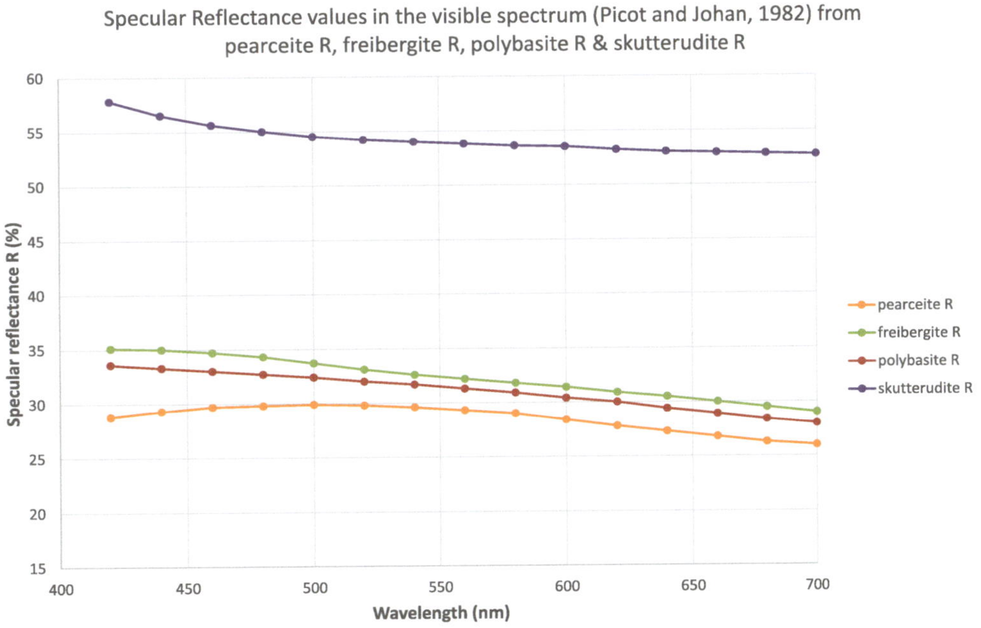

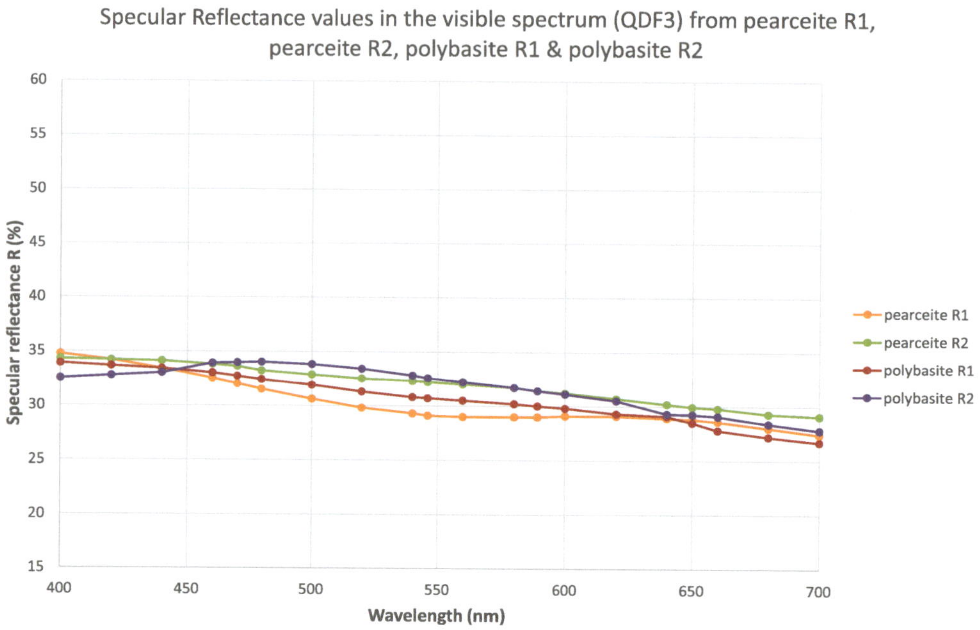

References

Picot P, Johan Z (1982) Atlas of Ore Minerals. B.R.G.M./Elsevier.
Uytenbogaardt W, Burke EAJ (1971) Tables for microscopic identification of ore minerals, 2nd edn. Elsevier, Amsterdam

Pentlandite (*pn*)

93

$(Fe,Ni)_9S_8$: Cubic

Pentlandite		R_{546} (%)	Y (%)	λ_d (nm)	P_e (%)	Colour sensation
Air		48.9	49.1	576	12.8	White
Oil		37.2	37.5	576	15.2	
Bir: Null (1)				*Pleoc*: No (1)		*Tarn*: No
Anis +P: Isotropic (1)				*IR*: No		*Twg*: // (111) not visible under microscope
Mor: Anhedral, included in po				*Clv*: // (111) distinct		*PolQ*: Excellent if not altered
Hardness	*PolH*: Medium, > ccp, < po (3)			*Mohs*: 4	*Talm*: D	*VHN* $_{(100\ g)}$: 174–238

Obs **:** cf. Table and R charts, multispectral data (VNIR, AMCO) and visible range (QDF3)
Mineral abbreviations: brv-bravoite, cbn-cubanite, ccp-chalcopyrite, chr-chromite, PGE-platinum-group elements, il-ilmenite, mc-marcasite, mck-mackinawite, mt-magnetite, pn-pentlandite, po-pyrrhotite, py-pyrite, vio-violarite, vll-valleriite

93.1 General

Important Ni ore, composition: $(Fe, Ni)_9S_8$. May contain Cu and Ag as impurities; by incorporation of Co, it forms a series with cobaltopentlandite $(Co, Fe, Ni)_9S_8$.

93.2 Microscopic Description

93.2.1 SP Observation (*Single Polariser*)

Colour bright white, but subjective perception varies according to the surroundings (Figs. 93.1a,b and 93.3); a faint brownish or yellowish tinge is usually discernible.

Isotropic.

Very good *polish*, except when altered (to violarite or bravoite, marcasite). It is typical then to see the appearance of secondary aggregates of very fine grain size, sometimes cryptocrystalline and darker, which are friable in the polishing process (Fig. 93.3). Medium *polishing hardness*, $PolH_{pn}$ > ccp and slightly < po, ≪ py.

High *reflectance*, between pyrrhotite and pyrite (cf. Table and R charts, multispectral VNIR data and visible range). In the visible range, QDF3 shows some reflectance variations with composition.

It is typically associated with pyrrhotite, as a late mineral that forms fine elongated inclusions (*pentlandite flames*—flaming bodies arranged along microcracks in pyrrhotite) or anhedral inclusions *of* various shapes, irregular, cordate, banded, interstitial or small masses occasionally (Fig. 93.2).

Prominent *octahedral* // (111) *cleavage*, very characteristic and attributed to cooling-related shrinkage. It is *diagnostic* of pentlandite and allows its recognition even in secondary aggregates in which pentlandite has been totally replaced or reduced to minimal relicts (Figs. 93.3 and 93.4a).

93.2.2 Observation +P (*Polariser + Analyser at 90°*)

Isotropic (Fig. 93.1c, d and 93.4). The octahedral *twin*, // (111), visible in hand samples, is imperceptible under the microscope by virtue of the isotropy.

93.2.3 Paragenesis

Almost exclusively liquid-magmatic *paragenesis*, associated with pyrrhotite and chalcopyrite ± magnetite, ilmenite, in mafic rocks. Occasionally also associated with chromite, cubanite, violarite, bravoite, mackinawite, valleriite, PGE, etc.

R. Castroviejo, *A Practical Guide to Ore Microscopy—Volume 1*,
https://doi.org/10.1007/978-3-031-12654-3_93

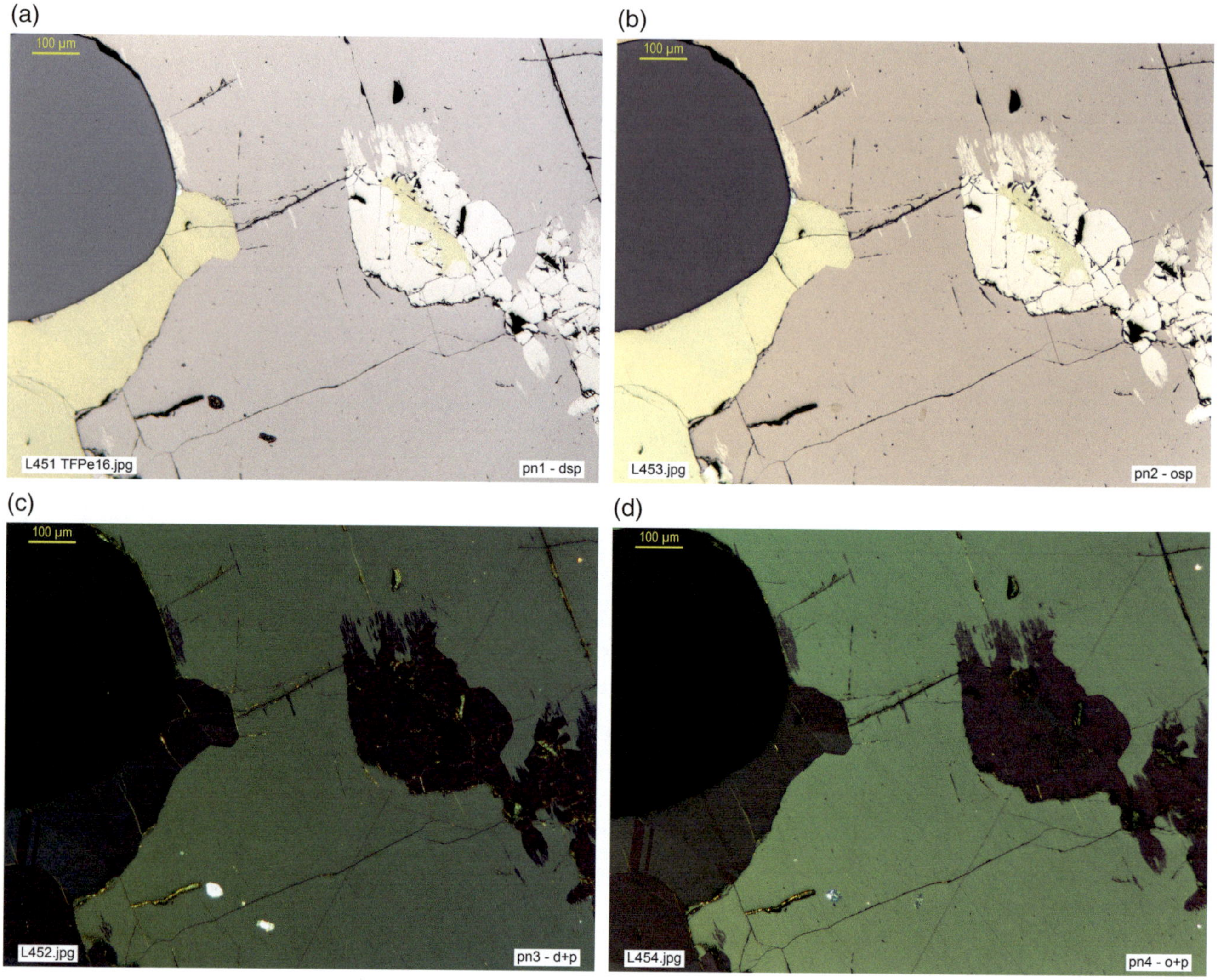

Fig. 93.1 **a, b** Cu-Ni ore with pn (white), ccp (yellow) and mt (grey), included in po (brownish-pink). TFPe16, Falconbridge Mine, Sudbury, Canada (**dsp**). **b** Same scene as pn1 (**osp**). **c, d** Same scene as pn1, (**d+p**) and (**o+p**). The net anisotropism of po and the weaker anisotropism and twins of ccp (well visible in immersion) contrast with the total extinction of mt and pn, which are isotropic

93.2.4 Identification

It is easily recognized by its colour, association with pyrrhotite and chalcopyrite, typical flaming exsolutions, octahedral exfoliation, isotropic character and alteration to bravoite, violarite. In isolated and scarce grains or in fine particles, it can look like pyrite or pyrrhotite, but there are criteria of distinction: pyrite is much harder and pyrrhotite, clearly anisotropic.

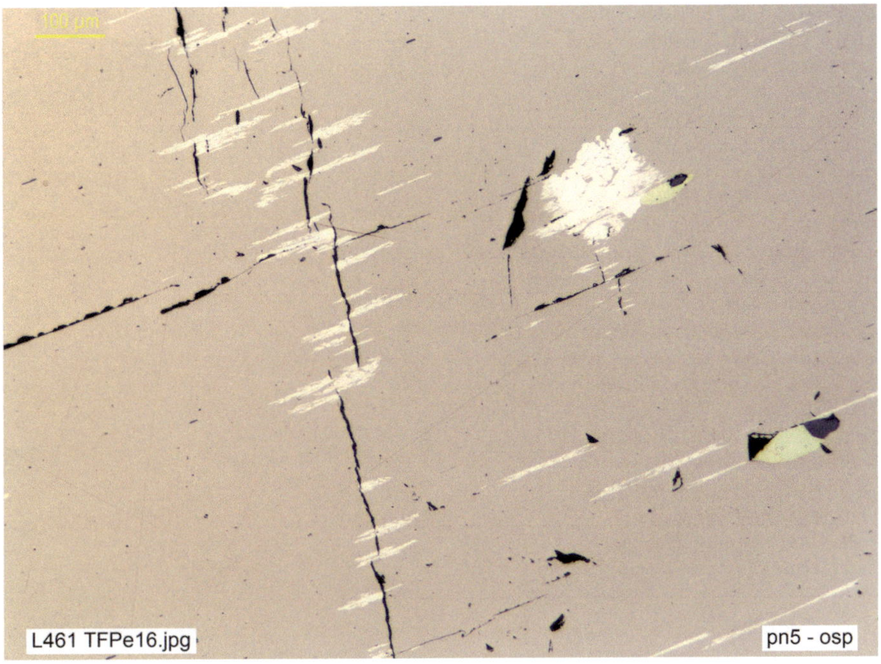

Fig. 93.2 Inclusions of pn (ccp, mt) in po, with *flames* along microcracks. Sample TFPe16, Falconbridge Mine, Sudbury, Canada (**osp**)

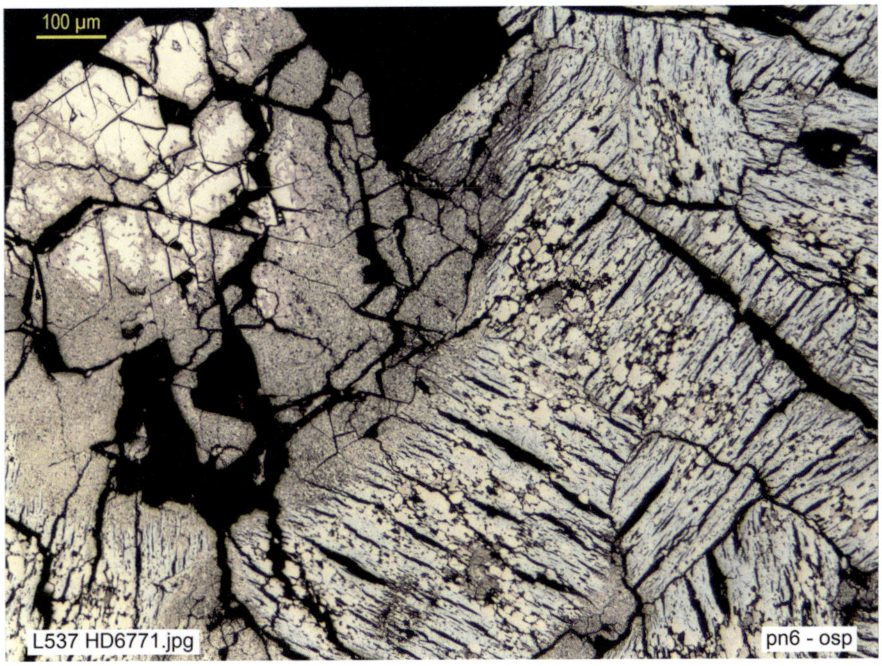

Fig. 93.3 Alteration of pn (white, upper left corner) and po (pinkish-brown relicts in the rest of the image): po is replaced by a thin oriented intergrowth (// [0001] of po) of py (white, slightly yellowish) and mc (bluish-white), inheriting the contours of po's original granoblastic texture; pn is replaced by a cryptocrystalline mass of vio (brownish, darker than pn) that inherits the prominent exfoliation of pn. The aggregate mc+py, secondary to po, is also partly replaced by vio, which establishes the chronology of the alteration: 1—py+mc on po; 2—vio on the whole pn+py +mc+po. HD 6771, Horbach, Schwarzwald, Germany (**osp**)

Fig. 93.4 **a, b** Centre of the image: pseudomorph of pn altered to vio (cryptocrystalline aggregate), with scarce pn relicts (white, lighter than vio) and octahedral cleavage inherited from pn. Periphery: mc+py aggregate with subordinate brv (purplish-brown, somewhat darker than po), secondary after po (pinkish-brown, anisotropic relicts and traces of // [0001] cleavage). The alteration to vio, which also affects mc, py, brv and po, is the posthumous episode. HD 6771, Horbach, Schwarzwald, Germany (**osp**). **b** Same scene as pn7 (**o+p**): isotropy of pn, vio, py, brv (dark), anisotropism of mc (vivid polarization colours, blue-green) and po (duller)

(a)

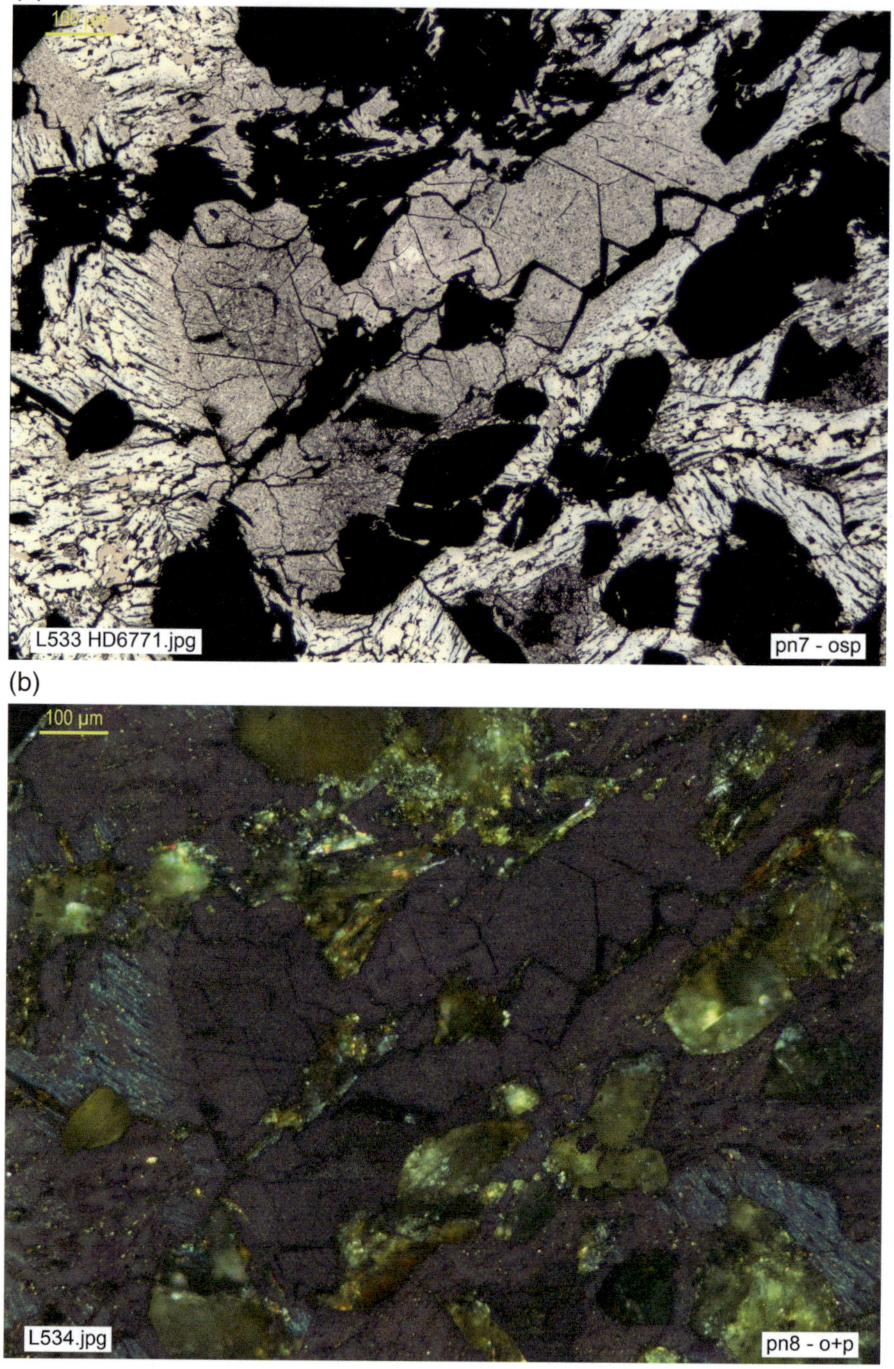

93.3 Spectral Information

93.3.1 Pentlandite Multispectral Reflectance R (in Air, Without Polariser) (%)

λ (nm)	370	400	425	450	475	500	525	550	575	600	625	650	675	700	750	800	850	900	950	1000
μ	21.0	24.7	28.9	33.2	36.7	40.2	42.8	45.2	47.1	48.7	49.9	51.0	52.0	52.7	54.0	54.9	56.1	56.4	56.5	57.0
P_5	18.3	21.7	26.0	30.2	34.1	37.8	40.7	43.3	45.3	47.1	48.4	49.6	50.6	51.3	52.6	53.5	55.0	55.3	55.4	56.0
P_{20}	19.0	22.4	26.6	31.0	34.9	38.5	41.4	43.9	45.9	47.7	49.0	50.2	51.1	51.8	53.2	54.1	55.4	55.8	55.7	56.3
P_{80}	25.4	30.0	33.7	38.0	40.2	43.5	45.1	47.5	48.9	50.2	51.2	52.1	53.1	53.9	55.0	56.0	57.2	57.3	57.6	58.1
P_{95}	26.4	30.8	34.5	38.8	41.2	44.6	46.3	48.5	50.0	51.3	52.4	53.4	54.2	54.9	56.0	56.8	57.9	57.9	58.1	58.5

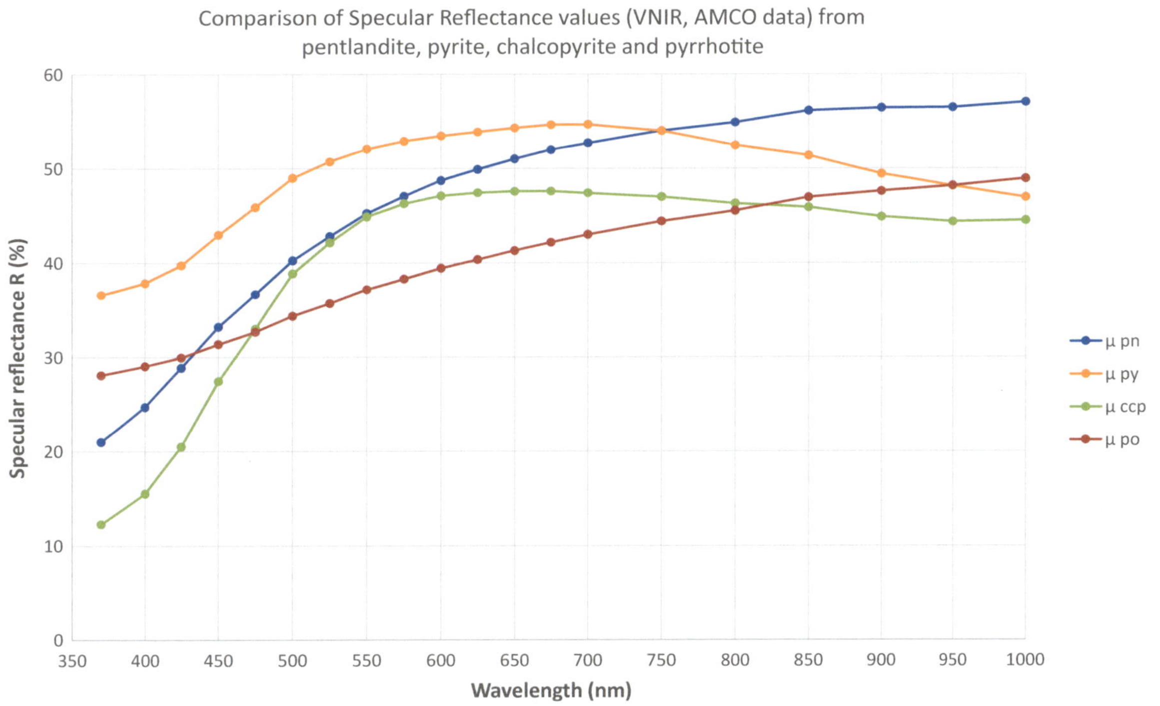

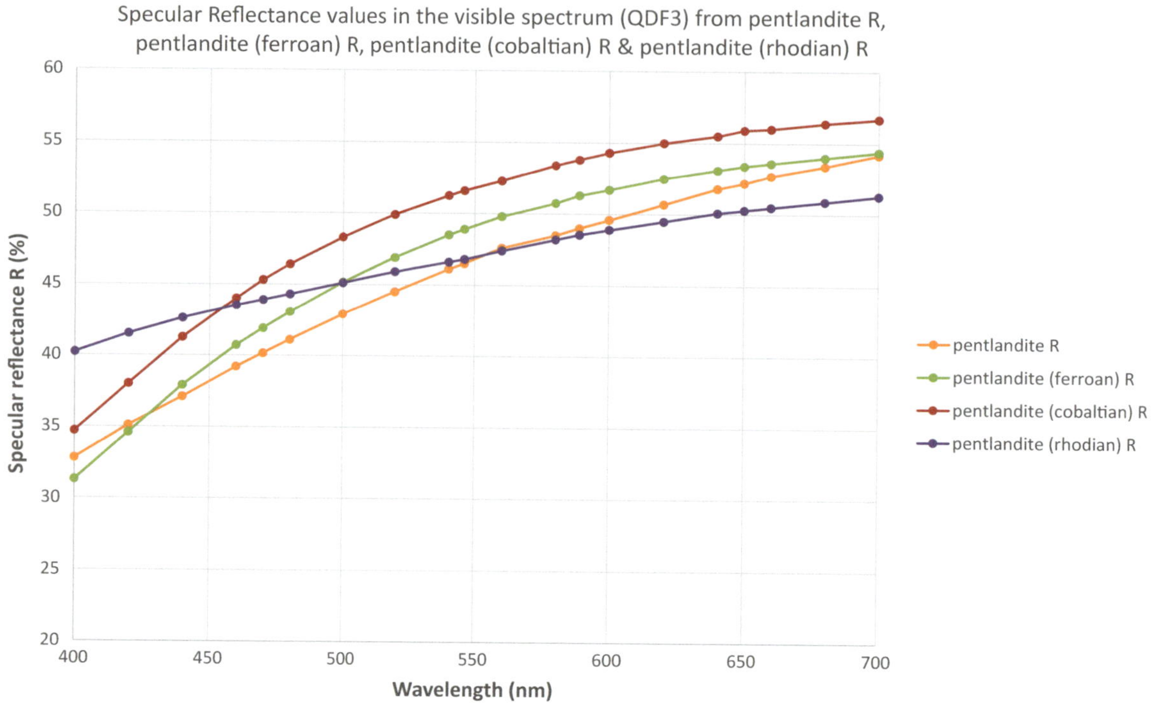
Specular Reflectance values in the visible spectrum (QDF3) from pentlandite R, pentlandite (ferroan) R, pentlandite (cobaltian) R & pentlandite (rhodian) R
Specular reflectance R (%)
60
55
50
45
40
35
30
25
20
400
450
500
550
600
650
700
Wavelength (nm)
pentlandite R
pentlandite (ferroan) R
pentlandite (cobaltian) R
pentlandite (rhodian) R

Platinum (*Pt*)

94

Pt: Cubic

Platinum		R_{546} (%)	Y (%)	λ_d (nm)	P_e (%)	Colour sensation
Air		69.7	69.8	576	4.7	White
Oil		59.2	59.3	576	6.3	
Bir: Null (1)				*Pleoc*: Null (1)		*Tarn*: No
Anis +P: Isotropic (1)				*IR*: No		*Twg*: (111) only visible by etching
Mor: Xenomorphic; idiomorphic (cubes)				*Clv*: Not observed		*PolQ*: Good
Hardness	*PolH*: Medium–high (3)			*Mohs*: 4–4½	*Talm*: nd	*VHN* $_{(100\ g)}$: 297–339

Obs **:** cf. Table and R charts, multispectral data (VNIR, AMCO) and visible range (Picot and Johan 1982; QDF3)
Mineral abbreviations: Ag—native silver, Au—native gold, chr—chromite, coh—cohenite, erl—erlichmanite, Fe—native iron, gg—gangue, Ir—native iridium, mt—magnetite, Os—native osmium, Pd—native palladium, pn—pentlandite, po—pyrrhotite, py—pyrite, Sb—native antimony, sf—safflorite, sk—skutterudite, sp—sphalerite, spn—spinel, spy—sperrylite

94.1 General

The best known of the **PGE** or *Platinum Group Elements* (Pt, Pd, Rh, Ir, Os, Ru), it was also the first to be "discovered", i.e., recognized by Europeans (Antonio de Ulloa 1735), although its use and metallurgy is documented in several pre-Columbian South American civilizations. It is also the most common of the 109 **PGM** or *Platinum Group Minerals* now credited by the IMA. Cabri (2002) presents a synthetic but very complete and updated treatment of the PGM, whose consultation is recommended for a better knowledge of any of the PGM, as well as the set of works in Cabri (2002) for the different aspects related to the PGM, their deposits and their metallurgy.

Composition: Pt, although it usually has notable contents of Fe (up to 20% according to Ramdohr 1982) and it is not uncommon that it also contains minor to trace amounts of Pd, Rh, Ir, Os, Au, Cu, Ni, Sb. At high temperatures, Pt and Fe form a disordered solid solution which, on cooling, becomes ordered towards the compositions Pt_3Fe (*isoferroplatinum*) and PtFe (*tetraferroplatinum*), according to Cabri (2002). Consequently, the term traditionally used to refer without further precision to platinum and iron alloys, *ferroplatinum,* is obsolete due to its ambiguity and is no longer accepted by IMA, although it may possibly continue to be used.

94.2 Microscopic Description

94.2.1 SP Observation (*Single Polariser*)

White, slightly yellowish *colour* (with variable shades, even slightly bluish tones, depending on the accompanying minerals and the composition), very bright and normally

well *polished*, although some persistent scratches are not uncommon (Figs. 94.1 a, b and 94.2).

Polishing hardness medium, near the medium–high limit: >sf, sk, sp; <po, Os, Ir, mt.

Very high *reflectance* but lower than silver or iridium; higher than osmium and sperrylite and much higher than chromite and spinel (cf multispectral comparison charts and Figs. 94.2 and 94.6). Isotropic.

It occurs in anhedral grains (Fig. 94.2), in euhedral crystals (cubic, Fig. 94.6a) and in more-or-less rounded

R. Castroviejo, *A Practical Guide to Ore Microscopy—Volume 1*,
https://doi.org/10.1007/978-3-031-12654-3_94

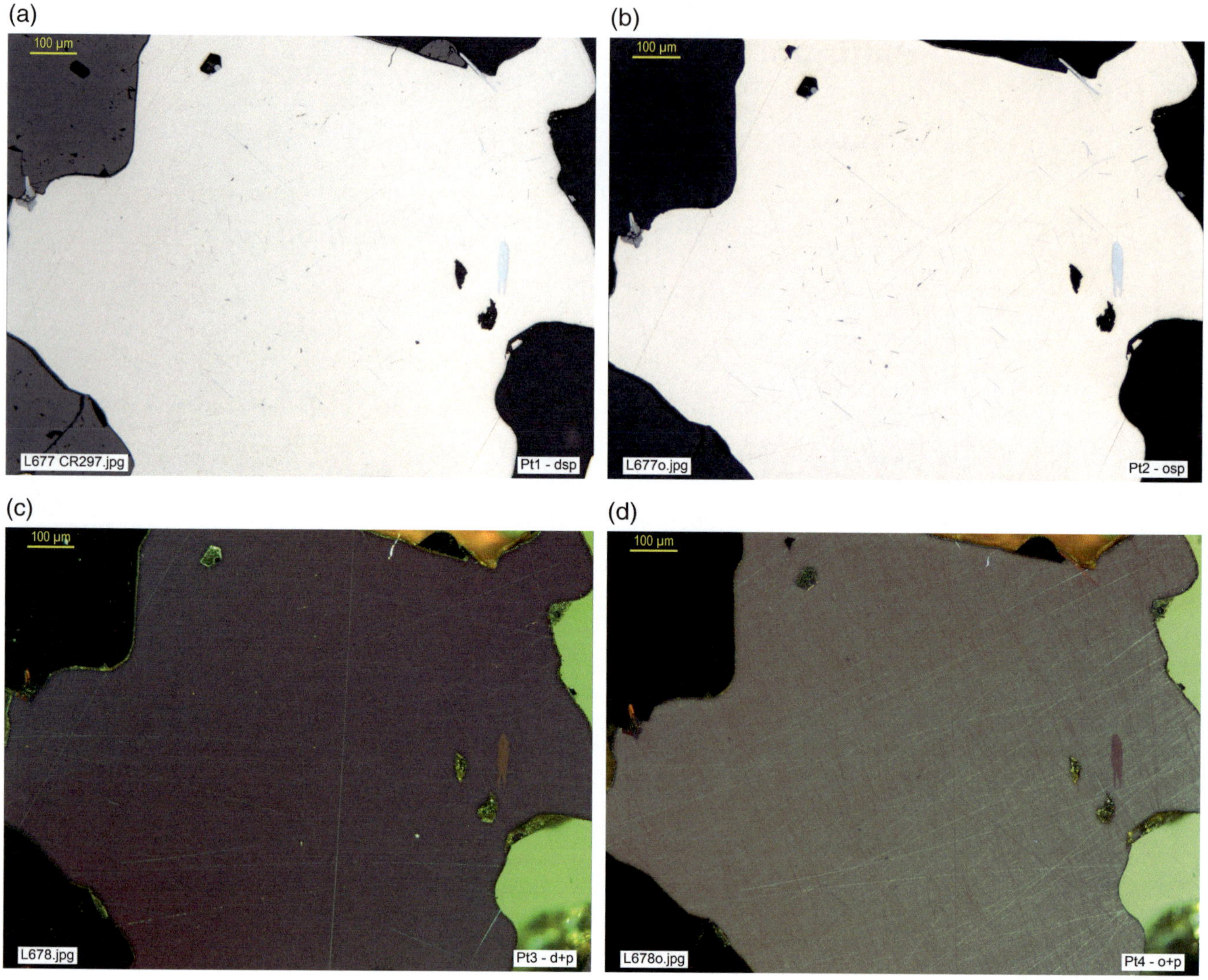

Fig. 94.1 **a**, **b** (**dsp** and **osp**, respectively)—PGE-Cr ore: Pt (white, almost filling the whole image) intergrown with chr (grey, left) and gg (dark grey, right); Os (bluish-white) and erl (light grey) inclusions in Pt. CR297, Folx Gulch, Goodnews Bay, Alaska. **c**, **d** (Same scene as Pt1: **d+p** and **o+p**, respectively) - Extinction of Pt (cubic) contrasts with Os anisotropy, well visible in elongated inclusion (right) and in thin lamella (top right border of Pt)

nuggets (Figs. 94.5 and 94.6a), depending on the transport. In grains of good size, various mineral inclusions can be observed (Figs. 94.1a, b, 94.2, 94.5 and 94.6a, b) and osmium or iridium exsolutions are not rare in lamellae marking octahedral orientations (111), Fig. 94.3, or showing varied and capricious forms like those illustrated by Ramdohr (1980) (Fig. 94.4).

94.2.2 Observation +P (*Polariser + Analyser at 90°*)

Isotropic (Figs. 94.1c, d and 94.6b). Observation with *Wollaston prism* is useful to reveal textural features, in particular exsolution, by interference contrast (Fig. 94.3). No visible *twinning*, but pressure twinning has been revealed by etching.

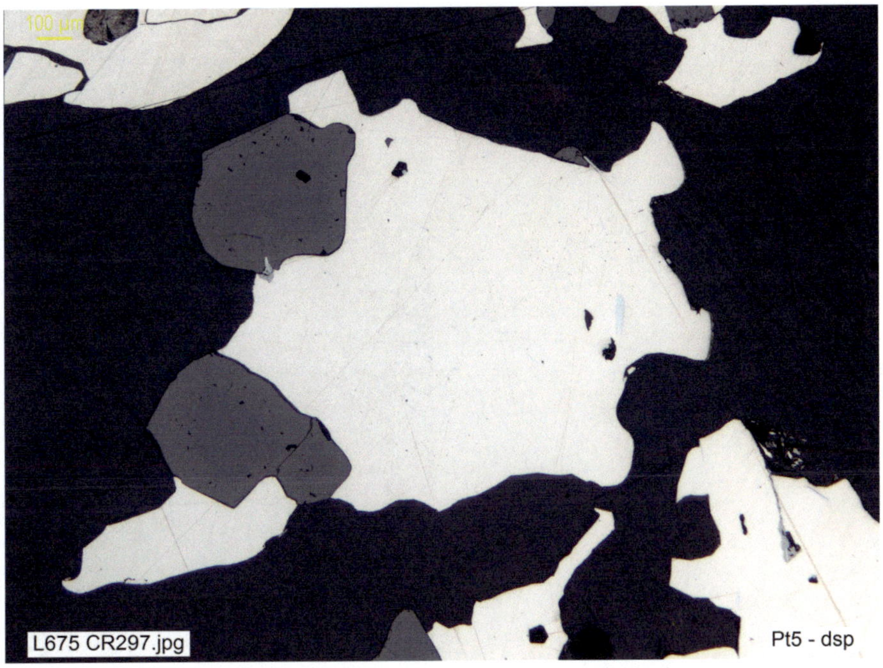

Fig. 94.2 PGE-Cr ore: Pt (white, most of the field) intergrown with chr (grey) and gg (dark grey) with Os (bluish white) and erl (light grey) inclusions. Wider image (**dsp**) of the Pt1 field (CR297, Fox Gulch, Goodnews Bay, Alaska)

Fig. 94.3 Same scene as Fig. 94.1a, seen with interference contrast (**d+p.w**): The enhanced relief image of the Pt surface makes visible the very thin lamellar exsolutions of Os, oriented after the octahedral directions of the Pt lattice which, with the usual microscope setting, would easily go unnoticed (comp Fig. 94.1a)

94.2.3 Paragenesis

Typical high-temperature *paragenesis*, related to ultramafic rocks (dunites, chromitites, peridotites, both in *stratified complexes* and in *ophiolitic series*), although the magmatic or hydrothermal genesis is a matter of debate. Also found in derived *detrital* deposits.

Platinum is often accompanied by one or more of the other five PGE (especially osmium, iridium) or the many known PGM (sperrylite, laurite, erlichmanite, braggite, etc.), as well as chromite, pentlandite, spinel, etc.

Minor but exploitable contents have also been found in other contexts, e.g., in the *Kupferschiefer* of Poland.

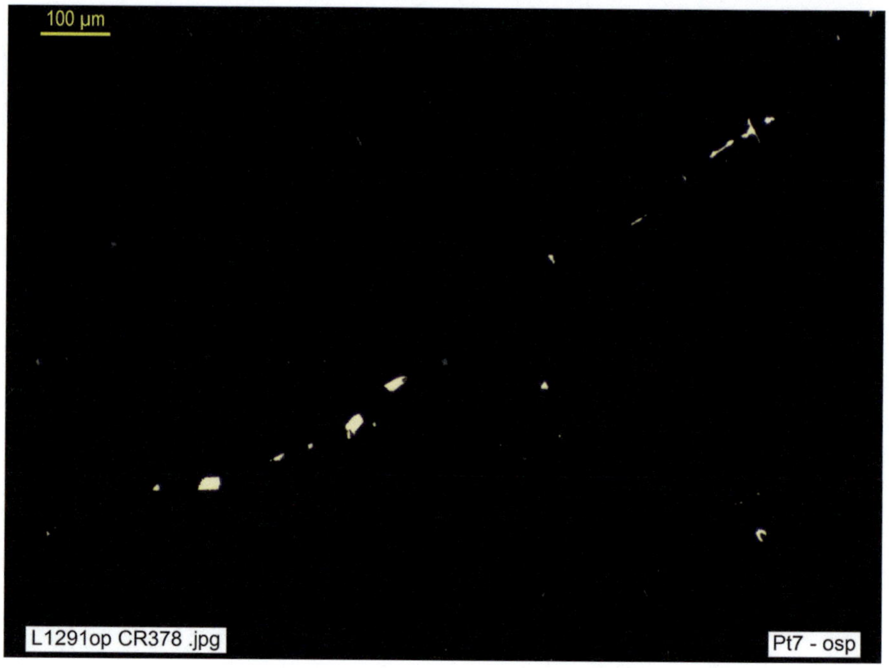

Fig. 94.4 Pt deposition along a microcrack in gg. CR 378, Merensky Reef, Bushveld, RSA (**osp**)

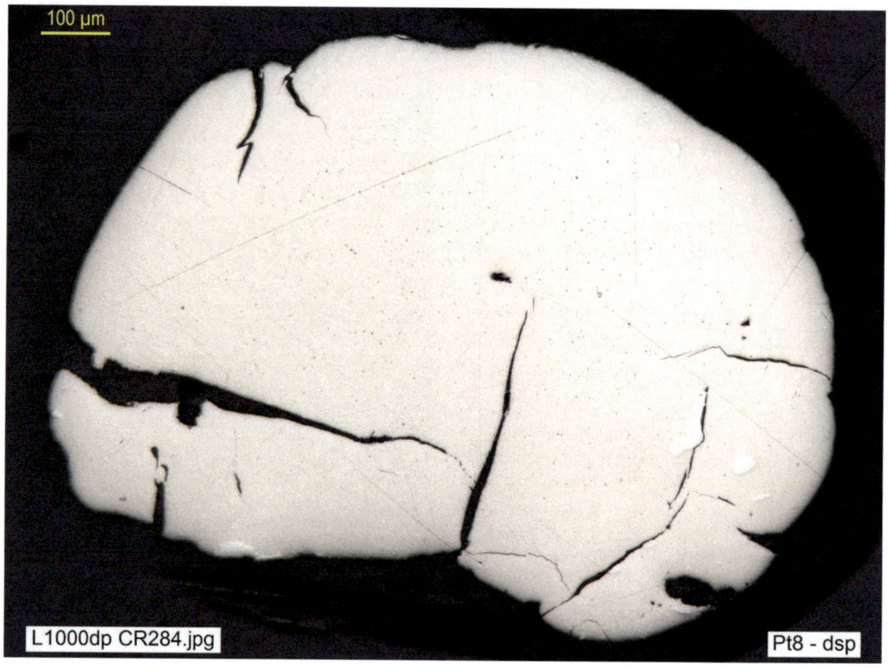

Fig. 94.5 Pt nugget, rounded by transport, with Ir inclusions (white, relief, lower right zone). CR 284, Urals, Russia (**dsp**)

94.2.4 Identification

Its bright white colour, high reflectance, isotropy, good polish, stability and remarkable hardness are typical properties which allow identification, taking into account the ultramafic paragenesis pointing to PGE.

Among the PGM, it is easy to distinguish Pt from osmium and iridium. Osmium is easily characterized by its strong anisotropism and its greater hardness, which makes it appear in relief on platinum. Iridium is more reflective, whiter and harder (also in relief).It is more difficult to differentiate it from palladium, slightly more yellowish,

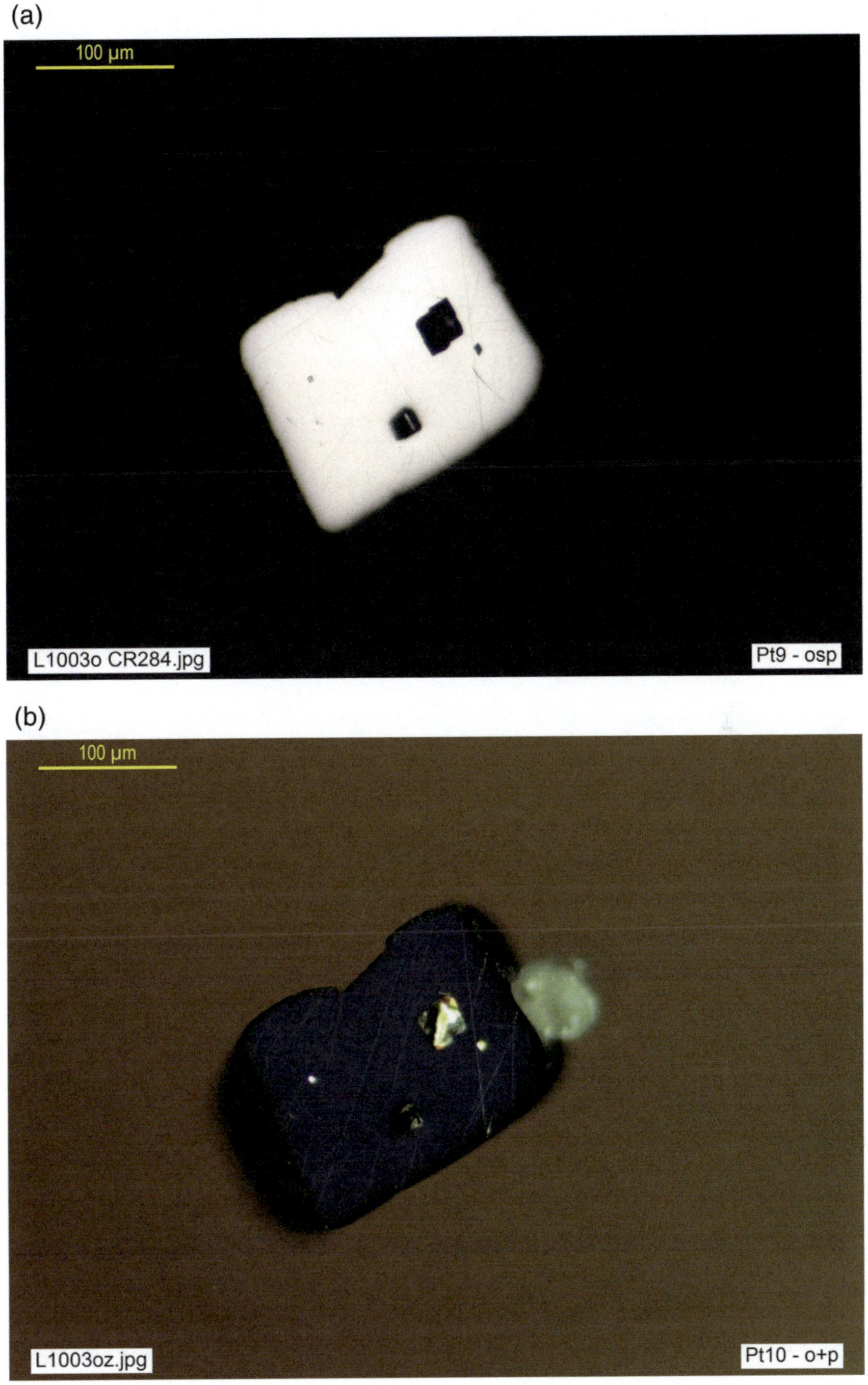

Fig. 94.6 **a**, **b** (same scene: **osp** and **o+p**, respectively) - Pt nugget with scarce transport (Pt grain still automorphic) and with cubic spn inclusions (dark, IR). CR 284, Urals, Russia

which may require chemical attack (palladium is much more easily etched). Apart from PGM, native iron and cohenite could also be mistaken for platinum, but these are less reflective and much more sensitive to acid attack, if in doubt.

It is easy to distinguish from other metals, such as silver or antimony, which are not to be expected in this paragenesis. Silver has higher reflectance, is much softer, shows abundant polishing scratches and is rapidly and conspicuously tarnished. Antimony is also much softer and anisotropic.

94.3 Spectral Information

94.3.1 Platinum Multispectral Reflectance R (in Air, Without Polariser) (%)

λ (nm)	370	400	425	450	475	500	525	550	575	600	625	650	675	700	750	800	850	900	950	1000
μ	55.0	56.6	58.0	59.8	61.1	62.6	63.8	64.7	65.5	66.3	67.0	67.6	68.4	68.7	69.9	70.8	72.1	71.0	71.4	71.0
P_5	51.0	52.9	54.2	56.2	57.6	59.4	60.7	61.8	62.8	63.7	64.5	65.2	66.0	66.4	67.7	68.6	69.8	68.7	69.2	68.6
P_{20}	53.5	55.2	56.6	58.5	59.9	61.5	62.8	63.7	64.5	65.3	66.0	66.7	67.5	67.8	69.1	70.0	71.3	70.4	70.5	70.5
P_{80}	56.7	58.1	59.5	61.2	62.4	63.8	65.0	65.8	66.5	67.3	67.9	68.6	69.3	69.6	70.8	71.7	73.0	71.8	72.3	71.8
P_{95}	57.6	58.8	60.2	61.9	63.0	64.4	65.6	66.3	67.0	67.8	68.4	69.1	69.8	70.1	71.2	72.1	73.4	72.2	72.7	72.1

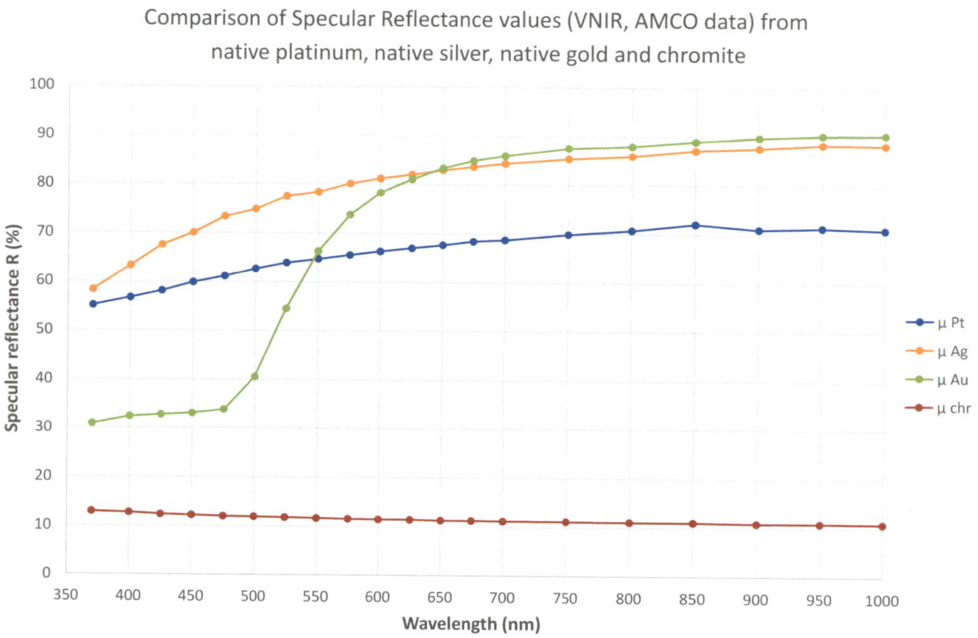

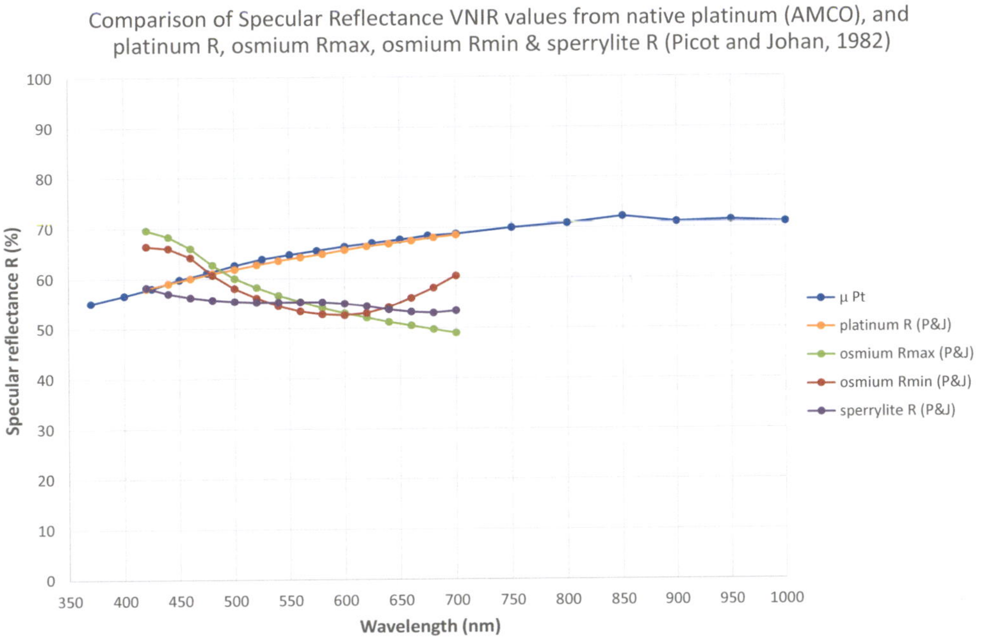

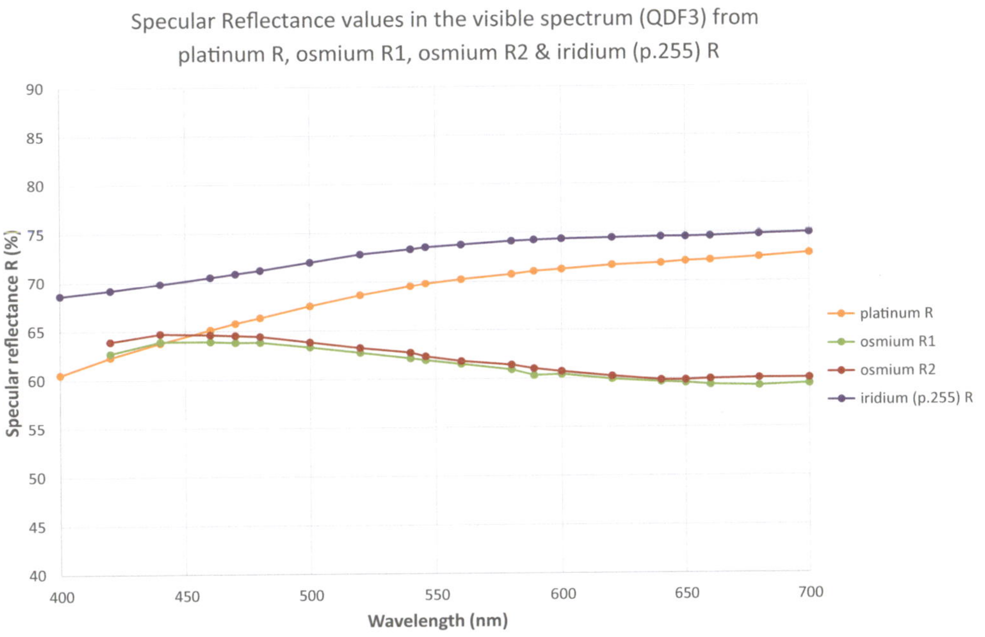

References

Cabri LJ (2002) The platinum-group minerals, pp 13–129. In: Cabri LJ (ed) CIM Special. Canad Inst Mining, Metall and Petroleum, vol 54. Calgary, Alberta, Canada, 852p

Cabri LJ (ed) (2002) The geology, geochemistry, mineralogy and mineral beneficiation of platinum group minerals. Canadian Institute of Mining, Metallurgy and Petroleum, Calgary, Alberta, Canada, 852p. ISBN 1-894495-27-5

Picot P, Johan Z (1982) Atlas of ore minerals. B.R.G.M./Elsevier

Ramdohr P (1980) The ore minerals and their intergrowths, 2nd edn, 2 volumes. Pergamon Press, Oxford, 1205p

Polybasite (*plb*)

95

$(Ag,Cu)_{16}Sb_2S_{11}$: Monoclinic

Polybasite		R_{546} (%)	Y (%)	λ_d (nm)	P_e (%)	Colour sensation
Air	R_1	30.7	30.6	479	4.5	Greyish-white, greenish
	R_2	32.6	32.2	487	3.4	
Oil	R_1	15.1	15.1	478	7.9	
	R_2	17.1	16.9	486	5.9	
Bir: Weak (3)				*Pleoc*: Weak (3)		*Tarn*: Fast light attack
Anis +P: Distinct, coloured (4)				*IR*: Red, scattered		*Twg*: Not observed
Mor: Anhedral/lamellar aggregates,…				*Clv*: //(001) not very visible		*PolQ*: Defective (scratches)
Hardness		*PolH*: Very low (1)		*Mohs*: 3	*Talm*: C $^-$	*VHN* $_{(50\ g)}$: 116–141

Obs **:** cf. R charts, according to AMCO spectral data, QDF3 and Picot and Johan (1971). VHN in Uytenbogaardt and Burke (1971)
Mineral abbreviations: ac—acanthite, Ag—native silver, ccp—chalcopyrite, dy—dyscrasite, fbg—freibergite, gg—gangue, gn—galena, my—miargyrite, pc—pearceite, plb—polybasite, prg—pyrargyrite, pru—proustite, sbt—stibnite, stm—stromeyerite, sp—sphalerite, stp—stephanite, tn—tennantite, td—tetrahedrite

95.1 General

The composition of polybasite, ***plb***, is $[Ag_9CuS_4][(Ag,Cu)_6(Sb,As)_2S_7]$, according to IMA.

Isostructural with *pc* (pearceite), plb is the antimonial term in the Pearceite-Polybasite Series: $[Ag_9CuS_4]$ $[(Ag,Cu)_6(As,Sb)_2S_7]$–$[(Ag,Cu)_6(Sb,As)_2S_7]$ $[Ag_9CuS_4]$. Although debated, this series seems to admit an unlimited solid solution between the two terms, pc-plb, so that the microscopic distinction between them is doubtful, as Ramdohr (1980) points out (p. 731), «it seems that many statements in the literature, which tell whether polybasite or pearceite is present, are made on the basis of good luck».

An additional complication is that several terms of the series with different compositions can coexist in the same sample. The complexity increases if the now known polytypes are taken into account; some authors speak rather of the *Pearceite-Polybasite Group.*[1]

While recognizing the limitations of optical microscopy in this case, an attempt will be made to characterize pearceite and polybasite separately, describing selected samples in which the mineralogy has been checked by electron microprobe and reflectance measurements.

95.2 Microscopic Description

95.2.1 SP Observation (*Single Polariser*)

Greyish-white *colour*, with a *weak pleochroism* that manifests itself in the passage from a slightly greenish tone to a purplish-violet or bluish tone, darker (Fig. 95.1a, b).

Moderate *reflectance*, **R**, practically like freibergite, although plb is weakly bireflective. R_{plb} is close to pyrargyrite and proustite, although differing in dispersion (cf. comparative R% graphs); somewhat lower than miargyrite, notably lower than galena.

Very sensitive to *light attack*, sometimes instantaneous (Figs. 95.2a, b, 95.3a, b and 95.4a, b), so it should be

[1] For further information, cf. https://www.mindat.org/min-3138.html.

R. Castroviejo, *A Practical Guide to Ore Microscopy—Volume 1*,
https://doi.org/10.1007/978-3-031-12654-3_95

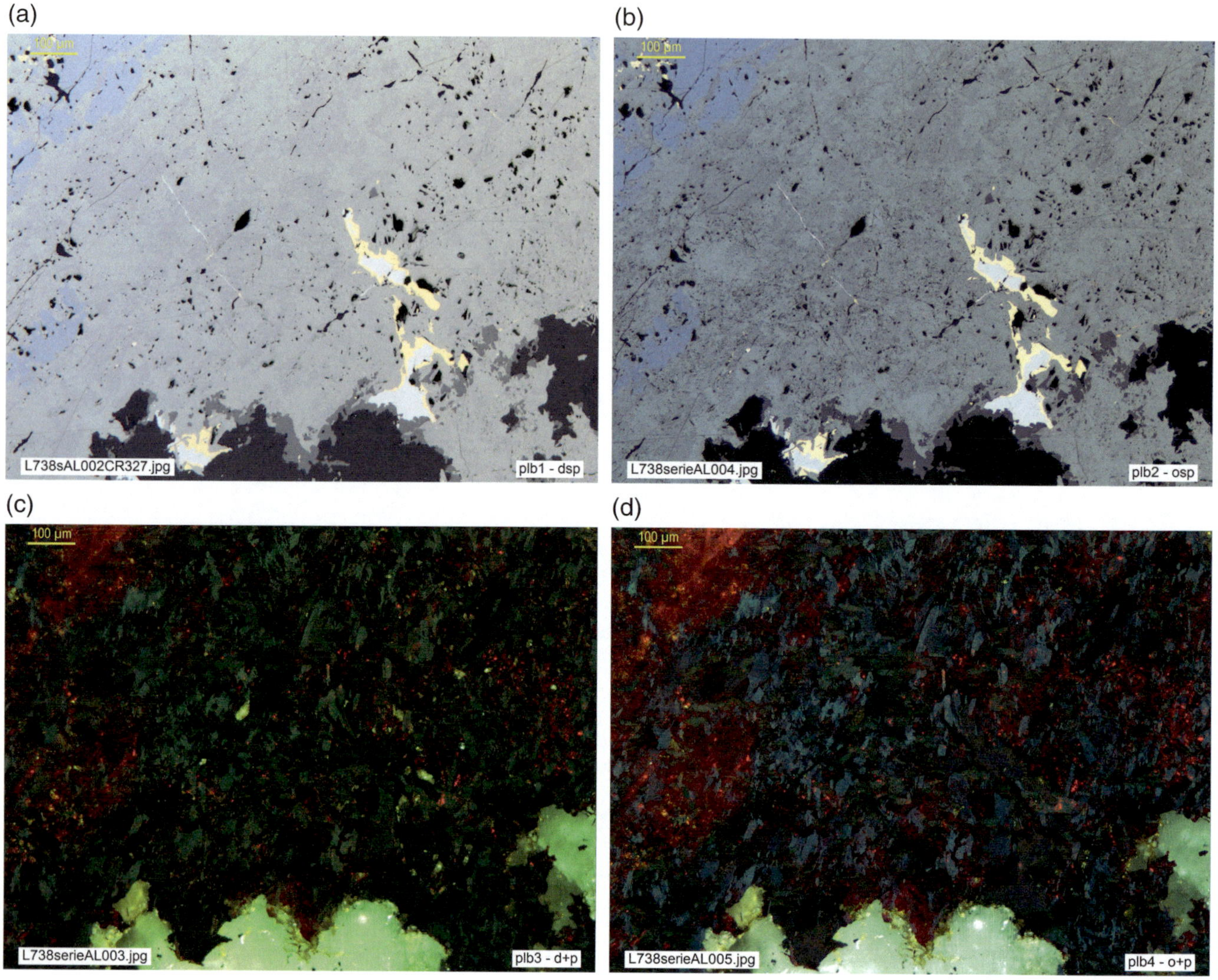

Fig. 95.1 **a, b** (**dsp** and **osp**, respectively). Ore of massive plb (slightly greenish grey to purplish-violet grey) intergrown with pru (bluish grey, top left), ccp (yellow), sp (dark grey), gn (white, inclusions in ccp) and calcite (black); incipient light etching of plb in (**b**). CR 327, Freiberg (Saxony, Germany). **c, d** Same scene as Fig. 95.1a (**d+p** and **o+p**, respectively): anisotropism of plb, more coloured than that of pru; IR in pru much more abundant than in plb

observed and described immediately after polishing and as soon as the microscope is turned on.

Because it is very soft, *polishing is* usually defective, with abundant scratches, although not as many as in acanthite. *Polishing hardness:* among the lowest known in metallic ores, PolH plb > ac, but < pru, prg, stp, sbt, my, Ag.

Euhedral tendency (lamellar crystallites) if space is available, but it usually occurs in anhedral inclusions in galena, tetrahedrite/freibergite, sphalerite and in apparent exsolutions or symplectitic intergrowths (Figs. 95.3a, b and 95.4a, b) with proustite, chalcopyrite, etc.

95.2.2 Observation +P (*Polariser + Analyser at 90°*)

Clearly *anisotropic*, with greenish and bluish polarisation colours well visible but not very luminous (Fig. 95.1c, d).

Light etching usually results (+p) in the appearance of a multitude of bright spots (Figs. 95.3b and 95.4b, probably native silver released by decay), which disappear if the surface is cleaned.

It usually shows red *internal reflections*, punctual, not abundant (much less than proustite or pyrargyrite) but visible

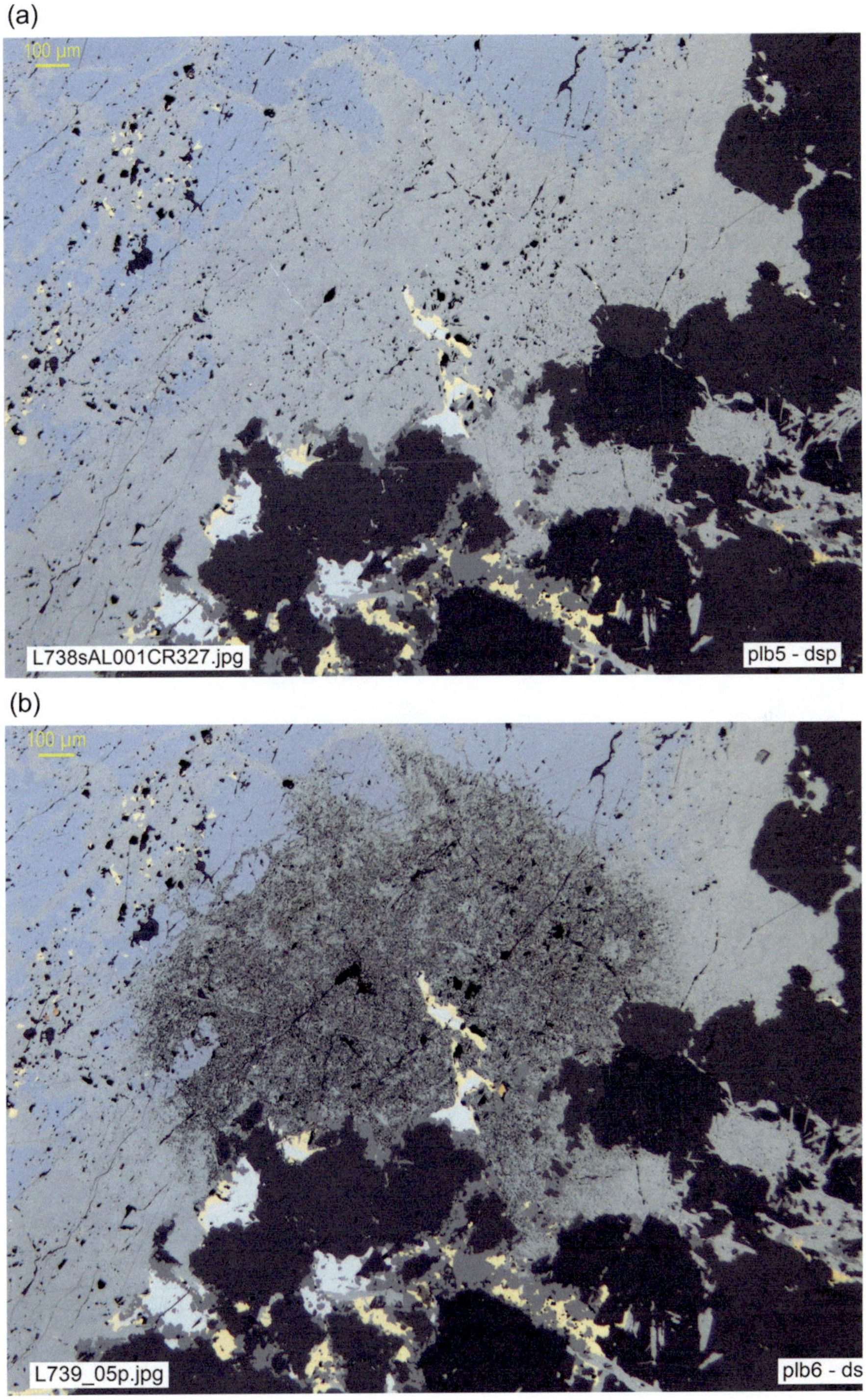

Fig. 95.2 **a, b** (Same scene, both **dsp**, taken with some minutes of interval, *respectively,* before and after attack by light with 20 × objective). Wider-field panoramic (obj. 5x) of the Fig. 95.1a scene: The etched (darkened) zone **b** marks the field illuminated by 20 × obj. (Fig. 95.1a). Same mineralogy as plb1: Only plb has been etched

especially in immersion (Fig. 95.1d). Their abundance (or absence) depends, according to Ramdohr (1980), on copper content and orientation but does not provide a criterion for distinguishing between polybasite and pearceite.

95.2.3 Paragenesis

Mineral of possible hypogene and supergene origins. The first case, the most frequent, is typical of hydrothermal

Fig. 95.3 **a, b** (same scene: **dsp** and **d+p**, respectively)—Symplectitic intergrowth of pru and plb, with scattered ccp inclusions. Pru appears intact (blue-grey colour, perfect polish, clean), while plb appears stippled (black dots, **a**) by almost instantaneous light etching. With analyser (**b**), the light attack shows up as a mottled pattern of bright dots (resulting from the surface decomposition of plb), which stand out against the green background or polarisation tint of plb. CR 327, Freiberg (Saxony, Germany)

(a)

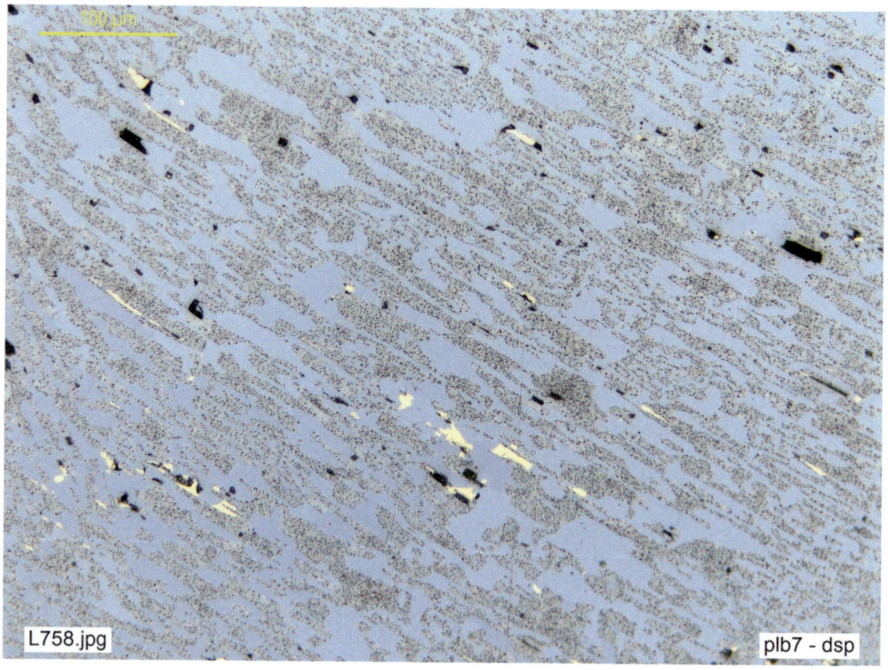

(b)

argentiferous *parageneses of* moderate and low temperatures, as the *five element* (Ni–Co–Bi–Ag–U) and the epithermal low sulphidation types. It may be accompanied by stephanite, pyrargyrite, dyscrasite, freibergite, acanthite, chalcopyrite, galena.

In supergene environments, plb usually occurs with the same accompaniment, but in this case as corroded relicts.

95.2.4 Identification

The very low hardness, mediocre polish, moderate reflectance, greyish colour with greenish tones, *light attack*, anisotropism with bluish-green tones and paragenesis provide sufficient criteria for the identification of polybasiteor at least of a member of the pearceite-polybasite series. Given the

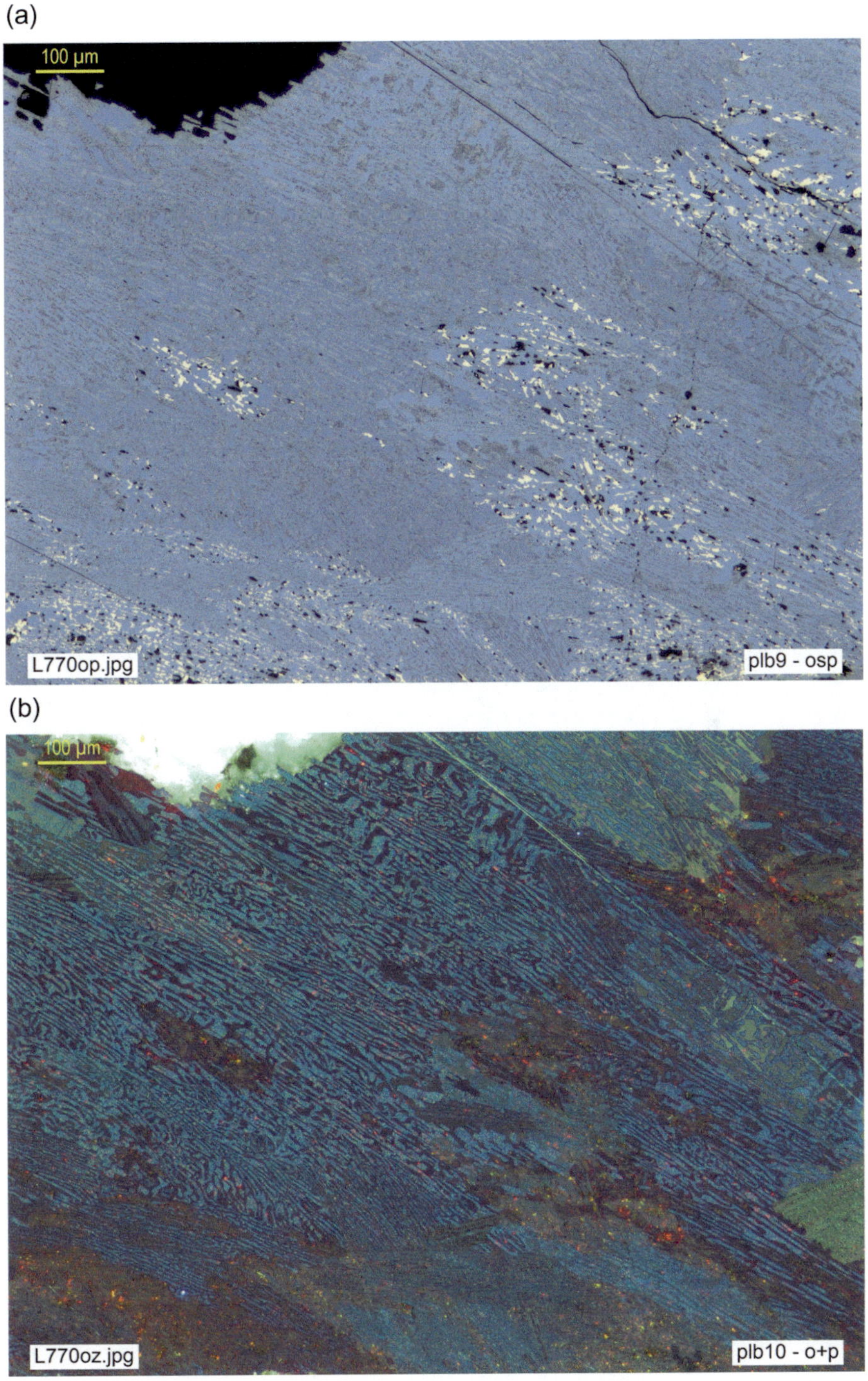

Fig. 95.4 **a, b** (Same scene: **osp** and **o+p**, respectively) Intergrown plb-pru-ccp-gn. Greenish anisotropism of plb, despite the inevitable light attack. Orange-red internal reflections of pru. CR 327, Freiberg (Saxony, Germany)

gradual compositional transition in this isomorphic series, the optical microscope is not sufficient to pinpoint the composition within it.

However, a useful criterion to distinguish polybasite and pearceite is the mineralogy of the associated paragenesis—antimonial in the case of polybasite, arsenical (with abundant tennantite, proustite and arsenides) in the case of pearceite.

Other similar minerals are stromeyerite and freibergite:

- Stromeyerite is distinguished by its more vivid anisotropism, absence of IR and weak etching by light.
- Freibergite is easily distinguished because it is isotropic, although its colour and reflectance are similar.

95.3 Spectral Information

95.3.1 Polybasite Multispectral Reflectance R (in Air, Without Polariser) (%)

λ (nm)	370	400	425	450	475	500	525	550	575	600	625	650	675	700	750	800	850	900	950	1000
μ	29.7	30.5	30.4	30.5	30.6	30.4	30.0	29.6	29.1	28.4	27.6	26.9	26.3	25.8	25.0	24.6	24.5	24.4	24.3	24.2
P_5	28.5	29.5	29.4	29.4	29.4	29.3	29.0	28.7	28.3	27.7	26.9	26.2	25.7	25.1	24.4	24.0	23.8	23.6	23.5	23.4
P_{20}	29.2	30.1	29.9	30.0	29.9	29.8	29.5	29.1	28.7	28.1	27.3	26.6	26.0	25.5	24.7	24.3	24.1	23.9	23.8	23.7
P_{80}	30.3	31.1	31.0	31.2	31.3	31.1	30.6	30.1	29.6	28.8	28.0	27.2	26.6	26.1	25.3	24.9	24.8	24.7	24.7	24.5
P_{95}	30.8	31.5	31.4	31.6	31.6	31.4	31.0	30.4	29.9	29.1	28.3	27.5	27.0	26.5	25.9	25.7	25.7	25.6	25.4	25.3

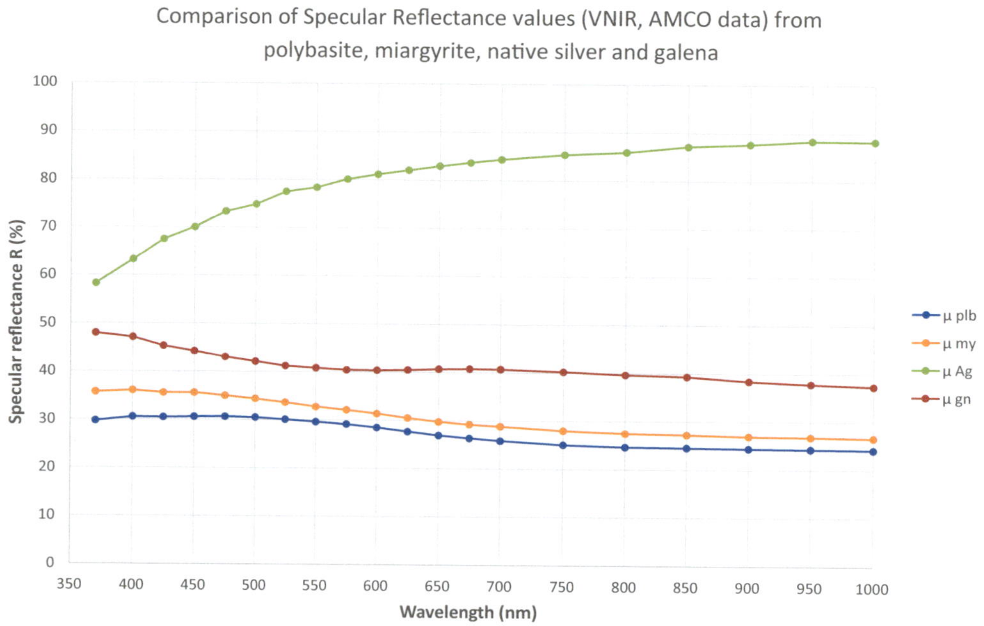

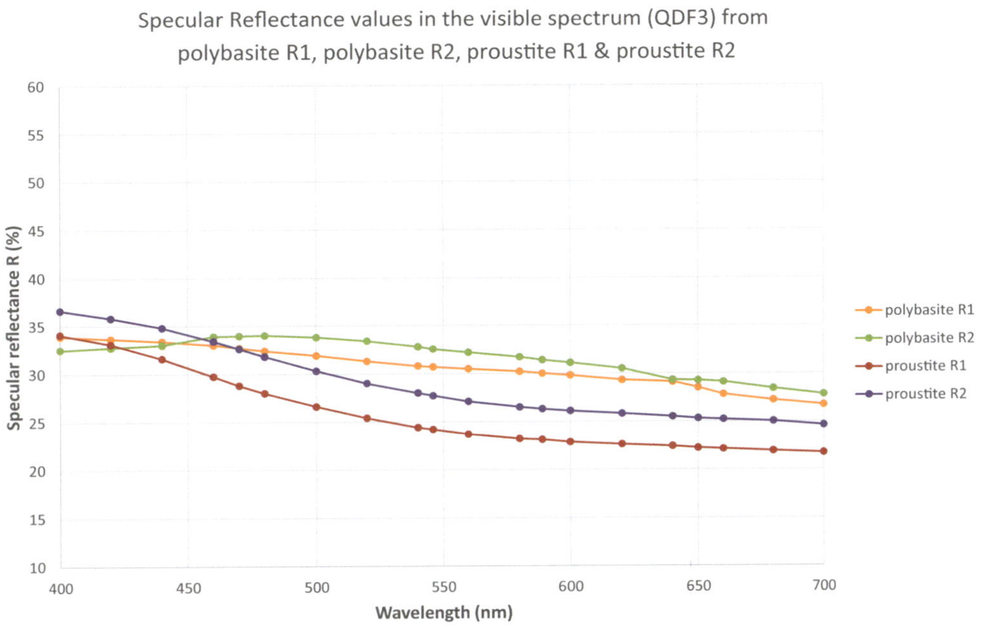
Specular Reflectance values in the visible spectrum (QDF3) from polybasite R1, polybasite R2, proustite R1 & proustite R2
Specular reflectance R (%)
60
55
50
45
40
35
30
25
20
15
10
400
450
500
550
600
650
700
Wavelength (nm)
polybasite R1
polybasite R2
proustite R1
proustite R2

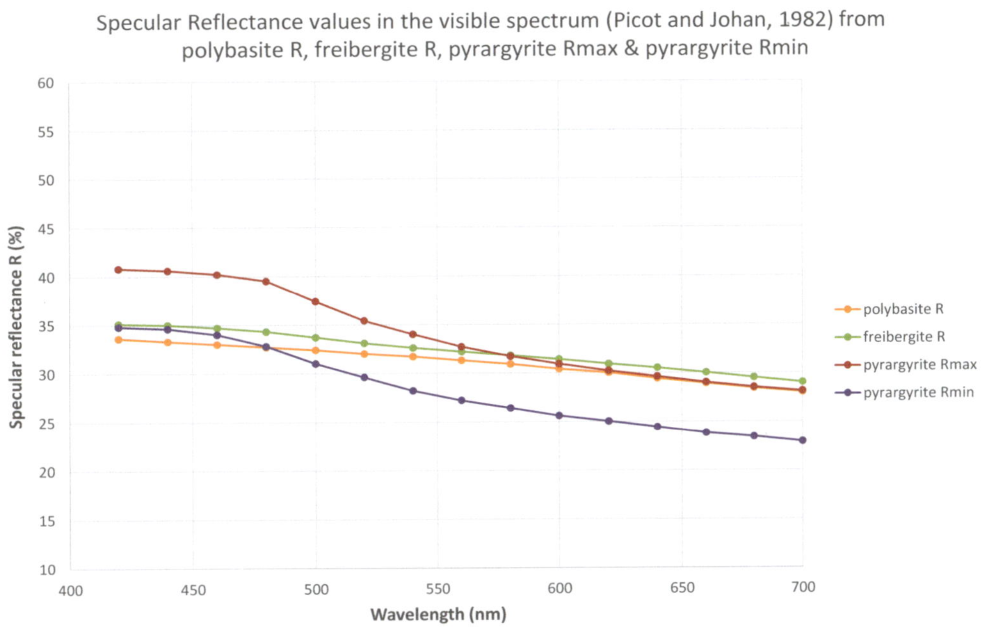
Specular Reflectance values in the visible spectrum (Picot and Johan, 1982) from polybasite R, freibergite R, pyrargyrite Rmax & pyrargyrite Rmin
Specular reflectance R (%)
60
55
50
45
40
35
30
25
20
15
10
400
450
500
550
600
650
700
Wavelength (nm)
polybasite R
freibergite R
pyrargyrite Rmax
pyrargyrite Rmin

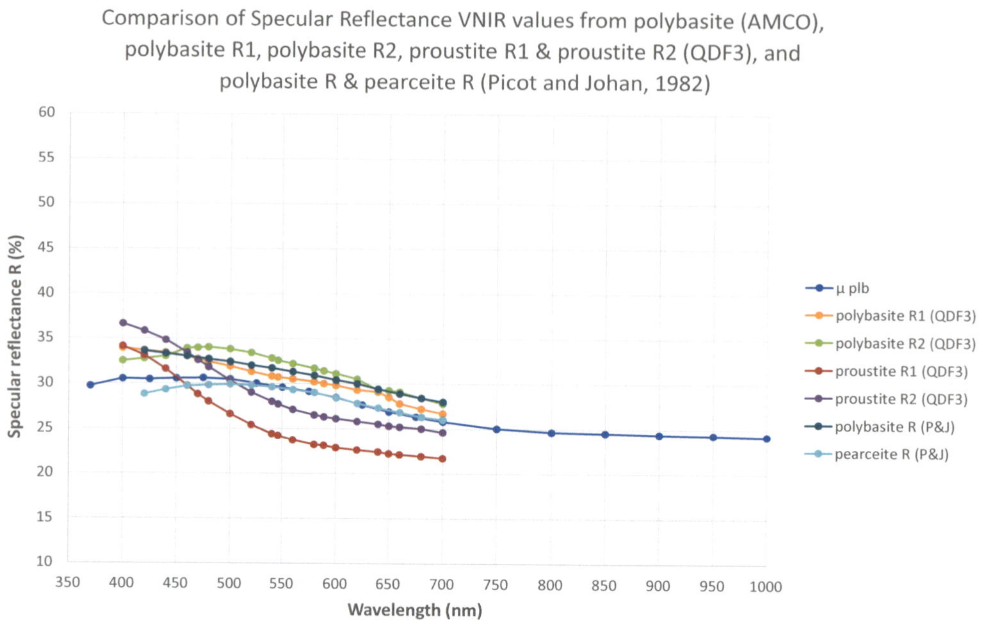

References

Ramdohr P (1980) The ore minerals and their intergrowths, 2nd edn, vol 2. Pergamon Press, Oxford, 1205 p

Uytenbogaardt W, Burke EAJ (1971) Tables for microscopic identification of ore minerals, 2nd edn. Elsevier, Amsterdam

Proustite (*pru/light red silver*)

96

Ag_3AsS_3: Trigonal

Proustite		R_{546} (%)	Y (%)	λ_d (nm)	P_e (%)	Colour sensation
Air	R_1	24.2	24.2	473	11.8	Bluish grey
	R_2	27.7	27.7	475	10.6	
Oil	R_1	10.2	10.4	472	19.9	
	R_2	12.9	13.1	474	17.8	
Bir: Weak (3)				*Pleoc*: Very weak (2)		*Tarn*: No light etching - very slow if any
Anis +P: Anisotropism masked RI (3)				*IR*: Very abundant, red		*Twg*: Yes
Mor: Massive filling, encrustations				*Clv*: {1011} poor		*PolQ*: Good
Hardness	*PolH*: Very low (1)			*Mohs*: 2½⁻	*Talm*: C	*VHN* $_{(25\ g)}$: 70–105

Obs: cf. Table and R charts, visible range data (QDF3 and Picot and Johan 1982)
Mineral abbreviations: ac—acanthite, native Ag—silver, native As—arsenic, native Bi—bismuth, cag—cerargyrite, ccp—chalcopyrite, cin—cinnabar, cup—cuprite, dy—dyscrasite, gg—gangue, gn—galena, gth—goethite, mc—marcasite, my—miargyrite, pc—pearceite, plb—polybasite, pru—proustite, prg—pyrargyrite, pru—proustite, py—pyrite, q—quartz, Sb—native antimony, sbt—stibnite, Ss—sulphosalts, stp—stephanite

96.1 General

Composition Ag_3AsS_3, with Sb impurities: proustite, ***pru***, is the As term of the Proustite-Pyrargyrite Series (Ag_3AsS_3–Ag_3SbS_3). However, both observational data in natural ores and laboratory experimental data suggest a very low miscibility between the extreme terms, proustite and pyrargyrite, ***prg***, at least at the low temperatures of the usual paragenesis. The optical properties of both two terms are similar, so the proustite description refers sometimes to pyrargyrite.

Monoclinic dimorph: *xanthoconite*.

96.2 Microscopic Description

96.2.1 SP Observation (*Single Polariser*)

Bluish-grey *colour* (compared to polybasite, bluer and darker, especially in oil: Figs. 96.1a, b and 96.2a). Comparison with galena, chalcopyrite, pyrite, marcasite and langisite: cf. Fig. 96.2a.

Slightly lower *reflectance* than pyrargyrite. *Bireflective*, described by some authors as clearly *pleochroic because of* the contrast between the R_1 shade (perceived as bluish, more intense than pyrargyrite) and R_2 (perceived as greyish white, slightly yellowish), although the two curves R_1 and R_2 are practically parallel and, therefore, do not differ in dispersion but only in magnitude of R (cf. comparative spectral graphs QDF3). In any case, the contrast is sufficient to be able to sometimes appreciate the twins with no analyser (cf. Fig. 96.1b).

Polish and *polishing hardness* as pyrargyrite. *Etching by light* not observed.

It usually occurs in encrustations and as fracture fillings, in aggregates intergrown with other silver sulphosalts or massive. The intergrowths can be varied and complex, with native silver, pyrargyrite, galena, symplectitic, irregular, replacement, etc. In favourable conditions (empty spaces), it forms well developed prismatic or scalenohedral crystals.

R. Castroviejo, *A Practical Guide to Ore Microscopy—Volume 1*,
https://doi.org/10.1007/978-3-031-12654-3_96

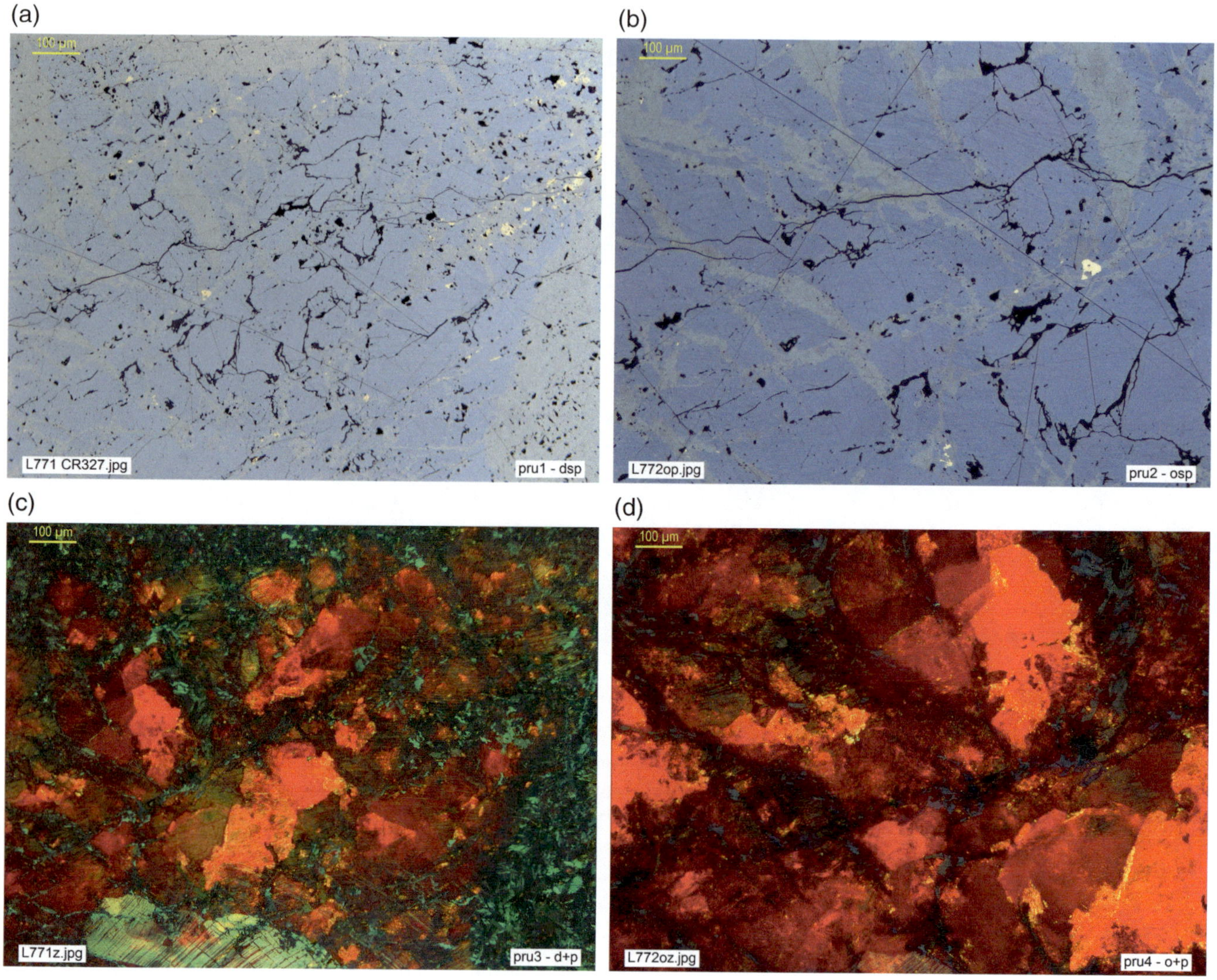

Fig. 96.1 **a, b** (**dsp** and **osp**, respectively)—Intergrowth of pru (bluish grey), plb (greenish grey) and ccp. Polysynthetic deformation twins of pru already visible in osp (Fig. 96.1b, obj. 10x, this image as well as F pr4 give an enlarged view of the bottom left quarter of Fig. 96.1a. Due to typographical error, the scale bar of Fig. 96.1a, obj. 5x, represents not 100 but 200 µm). CR 327, Freiberg (Saxony, Germany). **c, d** Same scene as pru1 and pru2 (**d+p** and **o+p**, respectively). Polysynthetic twins and anisotropism overprinted by massive IR in pru, more coloured anisotropism and less abundant IR in plb

96.2.2 Observation +P (*Polariser + Analyser at 90°*)

Anisotropic, with poorly coloured tints and usually masked by massive red *internal reflections*, lighter (orange) and more luminous than those of pyrargyrite, pearceite-polybasite (Figs. 98.1c, d and 98.2b).

Occasionally, polysynthetic or pressure *twins* (Figs. 96.1c and 96.2b).

96.2.3 Paragenesis

Like pyrargyrite but less frequent than pyrargyrite. Its typical *paragenesis* is low-temperature hydrothermal, particularly epithermal and *five-elements* (Ni–Co–Bi–Ag–U) veins. It can also form as a supergene mineral in the sulphide enrichment (or cementation) zone.

Accompanying: native silver, native arsenic, native antimony, native bismuth, acanthite, dyscrasite, miargyrite, pyrargyrite, pearceite, polybasite, other silver sulphosalts, Co–Ni arsenides, galena.

96.2.4 Identification

Characteristic features are bluish colour, moderate reflectance, bireflectance, very low hardness, and anisotropism often masked by massive and luminous red or orange IR.

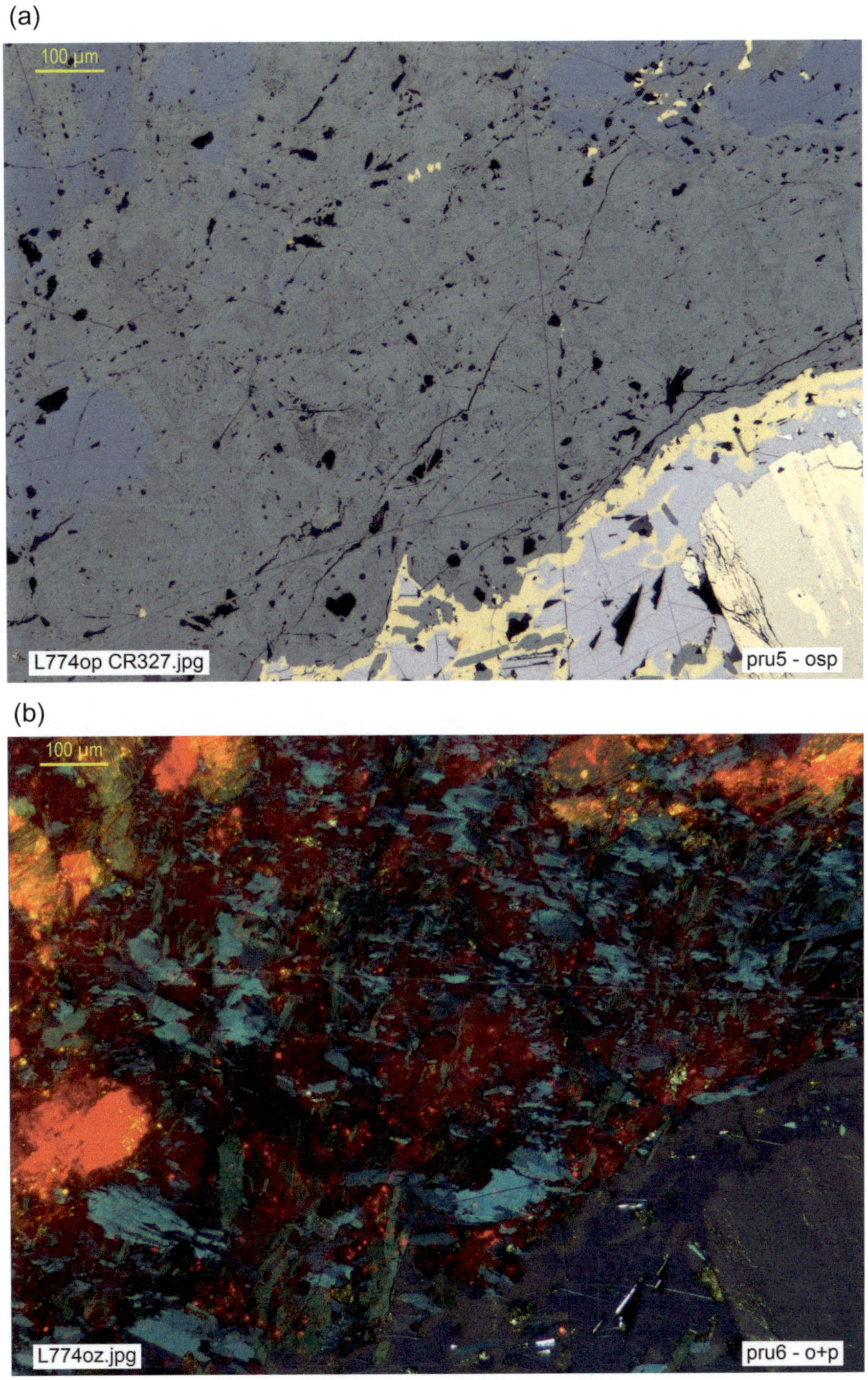

Fig. 96.2 **a, b** (**dsp** and **d+p**, respectively). Agentiferous ore, pru (bluish-grey, very abundant IR) largely replaced by abundant pc-plb (intergrown, centre of image, plb greenish and pc greyish hues, both dirty-looking due to incipient light etching). Comparison with the sulphides (ccp, gn, py, traces of mc) and Ni-langisite (light pinkish-brown, in Fig. 96.2a, EMPA composition: ([$Co_{\sim 0.6}$, $Ni_{\sim 0.4}$] As) intergrown, filling the bottom right corner of image, provides a reference for colour and reflectance estimation. CR 327, Freiberg (Saxony, Germany)

Paragenesis is an important data (e.g., to avoid confusion with cinnabar or cuprite), but it also includes similar accompanying minerals that must be distinguished:

- Pyrargyrite, which is more frequent, is distinguished by its slightly higher R and bireflectance, paler colours (less bluish), less abundant IR and association with Sb ores, instead of the As typical of proustite.
- Miargyrite is lighter, more anisotropic (with bluish tints) and has less abundant IR.
- Polybasite and pearceite are more greenish, are easily etched by light, have brighter blue-green or greenish polarization tints and less IR.
- Stephanite does not display IR.

96.3 Spectral Information

96.3.1 Proustite Reflectance R (%) (in Air, Visible Spectrum)

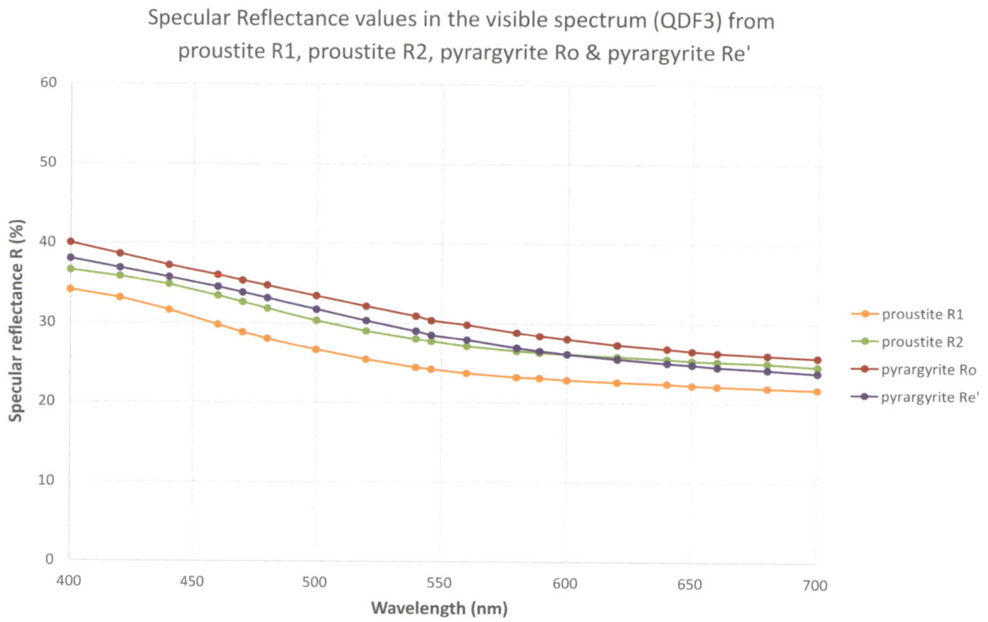

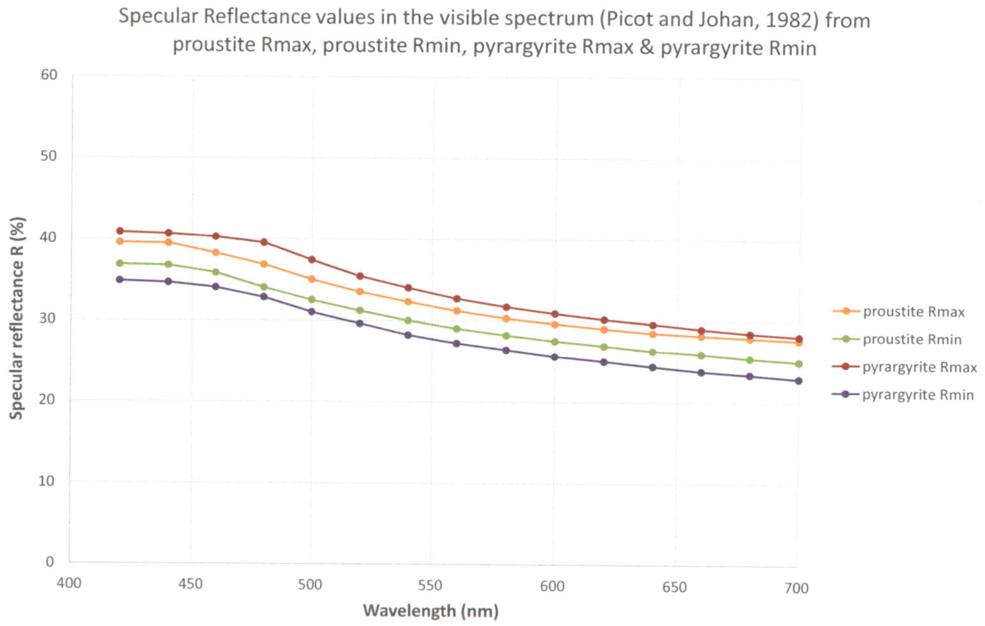

Reference

Picot P, Johan Z (1982) Atlas of ore minerals. B.R.G.M./Elsevier

Psilomelane (*psi*)

Mn-oxide (*s.l.*)

97

Psilomelane		R_{540} (%)	Y (%)	λ_d (nm)	P_e (%)	Colour sensation
Air	R_{max}	32.2	nd	nd	nd	Greyish white
	R_{min}	26.2	nd	nd	nd	Grey
Oil	R_{max}	nd	nd	nd	nd	Light grey
	R_{min}	nd	nd	nd	nd	Dark grey
Bir: Variable (2–4)				*Pleoc*: Very weak (2)		*Tarn*: No
Anis +P: Net, not coloured (4)				*IR*: No		*Twg*: No
Mor: Colomorph, banded, botryoidal				*Clv*: No		*PolQ*: Variable
Hardness	*PolH*: Variable (4)			*Mohs*: 5–6	*Talm*: G^-	*VHN* $_{(nd\ g)}$: 203–813

Obs: cf. Table and R charts, multispectral data (VNIR, AMCO) and visible range (QDF3, Picot and Johan 1982). VHN: Uytenbogaardt and Burke 1971

Mineral abbreviations: bra—braunite, cry—cryptomelane, gg—gangue, gth—goethite, lim—limonite, mng—manganite, mt—magnetite, prl—pyrolusite, psi—psilomelane, q—quartz, rom—romanechite

97.1 General

Psilomelane is an important ore of the metal Mn. However, the term seems destined to disappear, as it has been *discredited* by IMA (Black and Mandarino 2008) and discarded as *«obsolete group name for hard black Manganese Oxides, Formula : Mn, O»* ([1]). Even so, it will probably not disappear in practice, at least in the short term, in the field of mining or mineral resource research, for a very simple reason, the same as for the term limonite—these are important mineral substances for which no alternative scientific names exist, simply because it is not always possible to define them with due precision from an academic point of view. This is due to their natural occurrence: very fine grain size, often submicroscopic, heterogeneity, compositional variations, low crystallinity, etc.

It is therefore useful to recall the *preliminary remarks* by Ramdohr (1980), to «PSILOMELANE AND RELATED MINERALS *(Psilomelane in* sensu stricto*, cryptomelane , coronadite , hollandite)*»: «The name ‶psilomelane' is used in mineral collections and in the mining industry as a collective term for high grade, hard manganese ores of botryoidal layered structure and obviously of colloform origin. The properties of these masses vary within wide limits, especially their chemical composition. Attempts are made to subdivide ‶psilomelane' into a number of species, on the basis of chemical and other differences. Most of these ‶minerals' are not real or occur only as great rarities; however, several undoubtedly independent new members have been established with the aid of the microscope and by X-Ray analysis […] Microscopic examination has shown that many ‶psilomelanes' actually are mixtures of other minerals, e.g., pyrolusite, limonite, quartz, and braunite; on the other hand, it has also indicated that a ‶psilomelane s. str.' exists, or at least aggregates in which the latter is definitely a predominant constituent. Structural investigations have shown that the term ‶*psilomelane s.str*' has to be

[1] cf. https://www.mindat.org/min-3304.html, *visited 2020/07/01.*

R. Castroviejo, *A Practical Guide to Ore Microscopy—Volume 1*,
https://doi.org/10.1007/978-3-031-12654-3_97

subdivided also [*comprising*] a seemingly tetragonal group of minerals with body-centered lattice [*which includes*] one member requiring naming ``cryptomelane', the most common, and two isomorphous with it, *hollandite* and *coronadite*. The three latter and the ``original psilomelane' are very difficult to distinguish under the ore microscope; they represent what the present writer used formerly to call "Psilomelane Component 2". While hollandite occasionally occurs in coarse-spathic masses, the other members are almost exclusively extremelly fine-crystalline, usually at the limit of microscopic resolution, and seem to be dehydrated gels which may contain by adsorption Cu, Co, Ni, Mg, Ag, Ca and W» (*p.1030*).

Thus, for psilomelane, Ramdohr proposes the components *psilomelane s.str* , *cryptomelane*, *hollandite* and *coronadite*, which he characterises as $(Ba, H_2O)_2Mn_5O_{10}$ (monoclinic, pseudo-rhombic), the first; and as $XR^{II}Mn^{IV}{}_8O_{16}$, tetragonal, the following, with R = Mn, Fe, Cu, Zn (without counting H_20), X = K_2 (cryptomelane), Pb (coronadite), Ba (hollandite). He assimilates romanechite to cryptomelane, but he warns about the problems of recognition due to the extremely fine grain size. Similar is the position of Picot and Johan (1982), for whom the essential components of psilomelane are psilomelane *s.str*, cryptomelane and coronadite. In any case, the definition of the actual mineralogy of these mixtures requires additional techniques, such as XRD.

Essentially sharing the views of these authors and while recognising that it is not a *species* according to IMA criteria, the description of psilomelane is maintained in this text, corresponding to its importance as a mineral resource and to its specific properties, dictated by its deposit conditions. These, as discussed above, often make the precise microscopic definition of mineral species in natural deposits of Mn oxides impossible.

However, since all the components mentioned, except psilomelane *s.str*, have been recognised as independent species, the appropriate names (cryptomelane, coronadite, hollandite) should be used when these have been identified, reserving psilomelane for those cases in which this identification is not possible or is ambiguous, due to fine grain size, poor crystallinity and/or heterogeneity. Accordingly, **psilomelane**, *psi*, is here understood as a *group of Mn oxides of very fine grain size, with colomorphic , botryoidal or reniform banded textures, amorphous or aphanitic appearance in hand samples, notable hardness, conchoidal fracture and black colour*. It is frequently *heterogeneous from the mineralogical point of view and contains, apart from the main component, minor to trace amounts of pyrolusite , braunite , ramsdellite, limonite, quartz,* etc. Comparing the respective descriptions, the species with which the today discredited *psilomelane s.str* could be identified could be **romanechite**, *rom*, in most cases.

97.2 Microscopic Description

97.2.1 Observation SP (*Polariser Only*)

Greyish-white to grey or light bluish-grey but sometimes porosity or the presence of cryptocrystalline silicates can give a dark shade to some areas (Fig. 97.1a, b, 97.2a, 97.3a, 97.4 and 97.5). Noticeable *bireflectance* but only visible if grain size is sufficiently coarse; not pleochroic (only greyish-white tones).

Moderate *reflectance*, varying according to dominant components, texture, porosity, etc. In general R psi $\approx$ rom, ht. In any case, clearly lower than pyrolusite and higher than goethite or braunite (cf. multispectral comparative plots and Figs. 97.1a, b and 97.4).

Polishing quality is variable, depending on texture and components, but generally good. *Polishing hardness* is also variable.

It occurs in fine aggregates, even cryptocrystalline or submicroscopic, showing typical colomorphic, banded, concentric, spherulitic, botryoidal textures. It is frequently acicular and of various aspects: mossy, mouldy, felted, corona coatings with fibrous-radial structure (as romanechite) on cryptocrystalline nuclei of psilomelane or gangue.

97.2.2 Observation +P (*Polariser + Analyser at 90°*)

Distinctly *anisotropic*, but in grey-white or sometimes brownish tones, without the pyrolusite colours (Figs. 97.1c, d and 97.2b).

No *internal reflections* or *twinning* (however, Uytenbogaardt and Burke 1971, speak of occasional brownish IR. In any case, it is not to be confused with the possible effect of cryptocrystalline clay or limonite inclusions, Fig. 97.3b).

97.2.3 Paragenesis

A frequent mineral in virtually all Mn deposits, psilomelane is a typical weathering product of Mn-bearing minerals (pyrolusite, braunite, rhodochrosite, rhodonite, siderite, etc.). It is abundant in the oxidation zone of manganiferous deposits, where it can be an important ore.

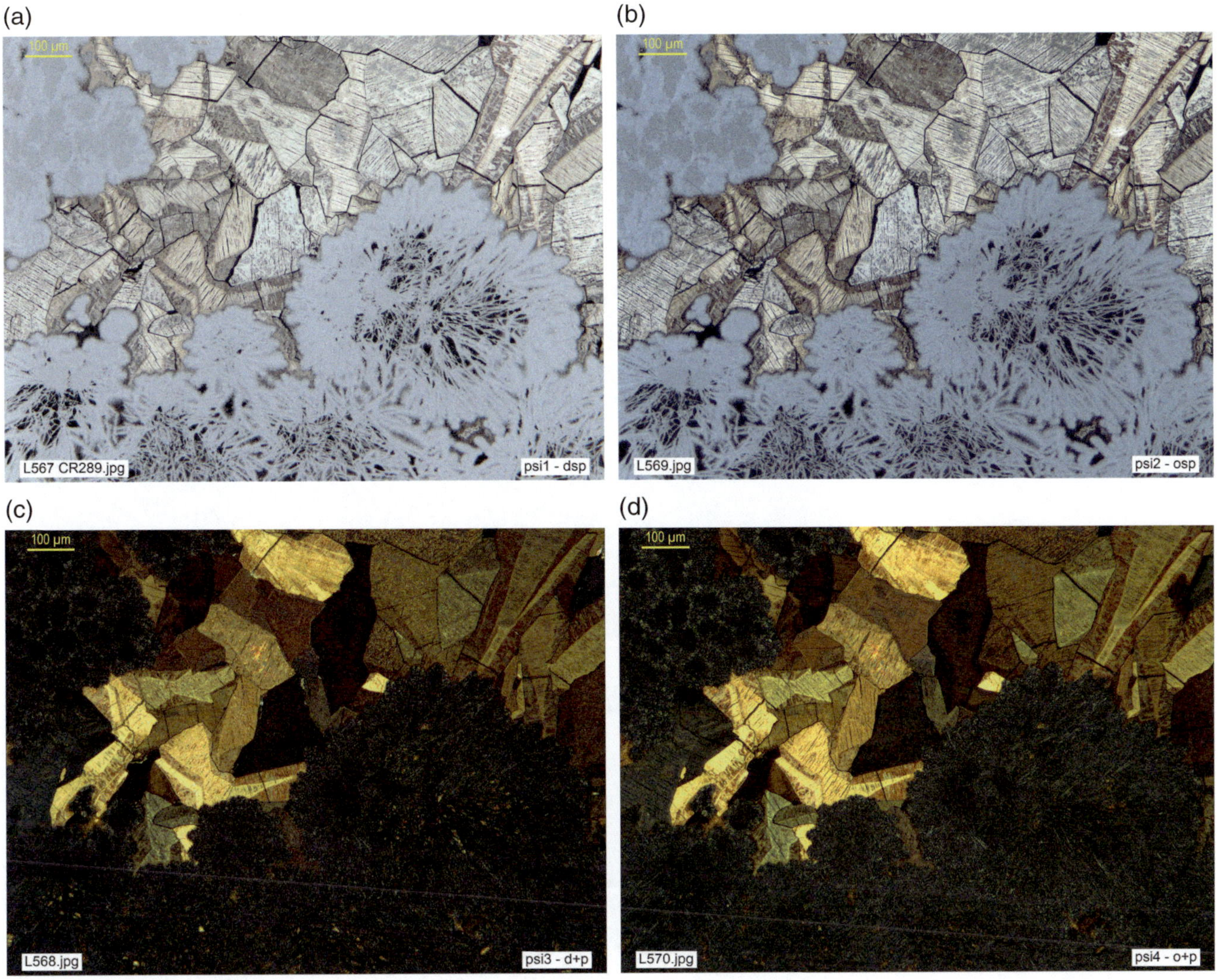

Fig. 97.1 **a, b** (**dsp** and **osp**, respectively)—Crystalline aggregate of prl (yellowish-white, pleochroic) showing alteration to psi (grey) in progress, with mouldy appearance, and fibrous-radial romanechite textures growing on cryptocrystalline cores of psi, of gg (dark grey) or leached (black pores). CR289, Amalienhohe Mine, Waldalgesheim, Bingerbrueck, Germany. **c, d** Same scene as psi1 (**d+p** and **o+p**, respectively). Coloured and luminous anisotropism of prl as opposed to the grey or dark tones of the psi mass; however, the contrast in the latter is sufficient to see the clearly anisotropic rom acicular forms, radiating from dark, cryptocrystalline cores

Companions: all Mn oxides, particularly pyrolusite, braunite, romanechite, plus quartz and frequently limonite, goethite and/or hematite.

97.2.4 Identification

Its fine grain size, colomorphic textures, grey colour, moderate reflectance, clear anisotropy and association with manganiferous paragenesis facilitate psilomelane recognition. In case of doubt, psilomelane is easily distinguished:

- from *goethite*, as psi is more reflective and lacks the IR so common in gth; and
- from *hematite*, also by the red IR frequent in ht.

Distinguishing the components in cryptocrystalline masses is more problematic when the grain size is submicroscopic, which prevents recognition of anisotropism. Even so, compared to psilomelane, *pyrolusite* is more reflective, unlike *goethite*, *limonite* and *braunite*, which appear noticeably darker. In these cases, attention should be paid to the porosity, which can give the surface a darker appearance than the one corresponding to the mineral present; the same applies to possible mixed submicroscopic silicates.

Some ancillary techniques such as X-Rays or EMPA are often essential means to define these ores.

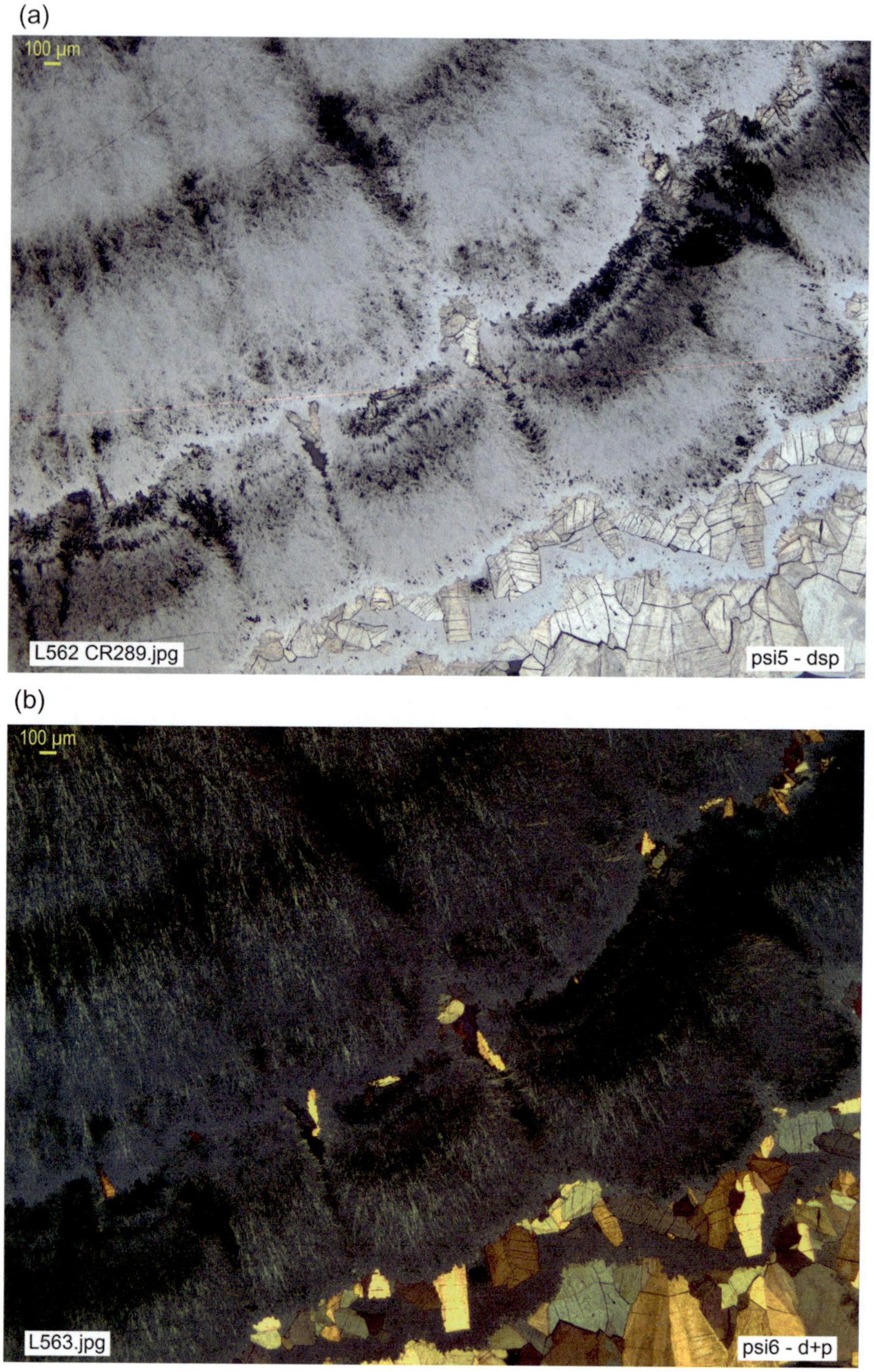

Fig. 97.2 **a, b** (**dsp** and **d+p**, resp.)—Advanced replacement of prl (strongly anisotropic) by banded psi, alternating light, fibrous bands (dominated by rom) with darker bands (silicate contaminated or leached, porous). Sample CR 289, Amalienhohe, Waldalgesheim, Bingerbrueck, Germany

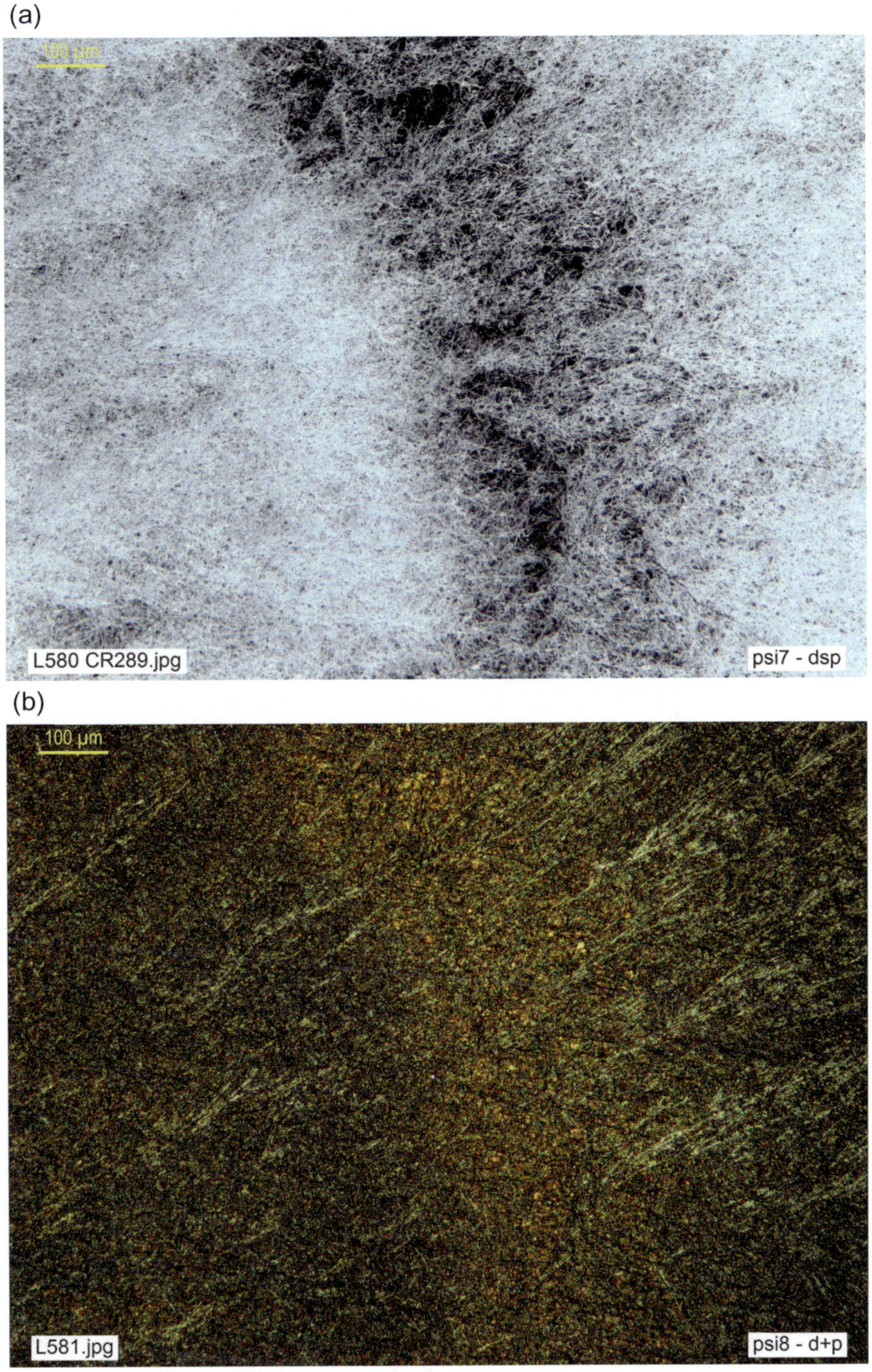

Fig. 97.3 **a, b** Same scene (**dsp** and **d+p**, *respectively*): banded psi with mossy texture and cryptocrystalline silicate intercalations in central band (very dark, but with massive internal reflections giving the false impression of vividly coloured anisotropism, +*p*). CR 289, Amalienhohe, Waldalgesheim, Bingerbrueck, Germany

Fig. 97.4 Substitution of prl by mouldy psi, radiating from spherulitic botryoidal aggregates of rom (light blue-grey) on cores of cryptocrystalline psi (darker), also secondary to prl. CR 289, Amalienhohe, Waldalgesheim, Bingerbrueck, Germany (osp)

Fig. 97.5 Spheroidal aggregates of network-fibrous and fibrous-radial psi (bluish-grey), with relict prl and dark interstitial spaces (silicates and leaching pores) between the fibres. CR 289, Amalienhohe, Waldalgesheim, Bingerbrueck, Germany (osp)

97.3 Spectral Information

97.3.1 Psilomelane Multi-spectral Reflectance R (in Air, Without Polariser) (%)

λ (nm)	370	400	425	450	475	500	525	550	575	600	625	650	675	700	750	800	850	900	950	1000
μ	22.8	23.4	23.5	23.3	23.1	22.6	22.3	21.7	21.3	20.8	20.5	20.2	20.0	19.8	19.5	19.3	19.2	19.0	18.9	18.8
P_5	19.1	19.4	19.2	18.8	18.4	18.0	17.6	17.3	17.1	16.8	16.7	16.5	16.4	16.2	16.1	16.0	15.9	15.6	15.5	15.3
P_{20}	21.0	21.4	21.3	20.8	20.5	19.8	19.5	19.0	18.7	18.4	18.2	18.0	17.8	17.6	17.4	17.2	17.1	16.6	16.5	16.3
P_{80}	24.6	25.3	25.6	25.6	25.5	25.0	24.5	23.9	23.3	22.6	22.2	21.9	21.6	21.4	20.9	20.7	20.6	20.3	20.3	20.3
P_{95}	25.9	26.5	26.7	26.7	26.5	26.0	25.5	24.8	24.1	23.4	22.8	22.5	22.2	22.0	21.3	21.1	21.1	20.9	20.9	20.9

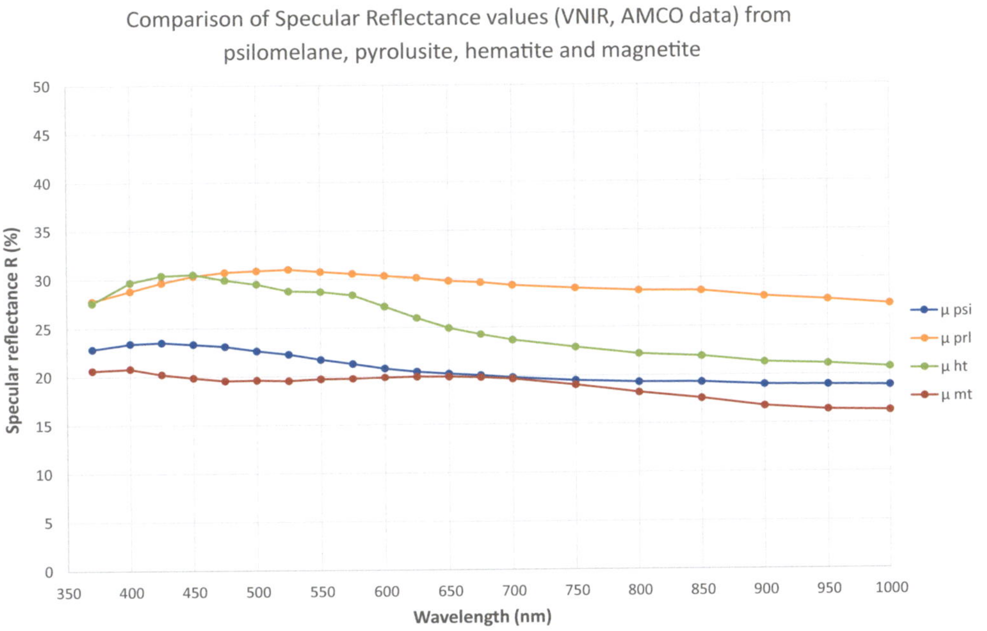

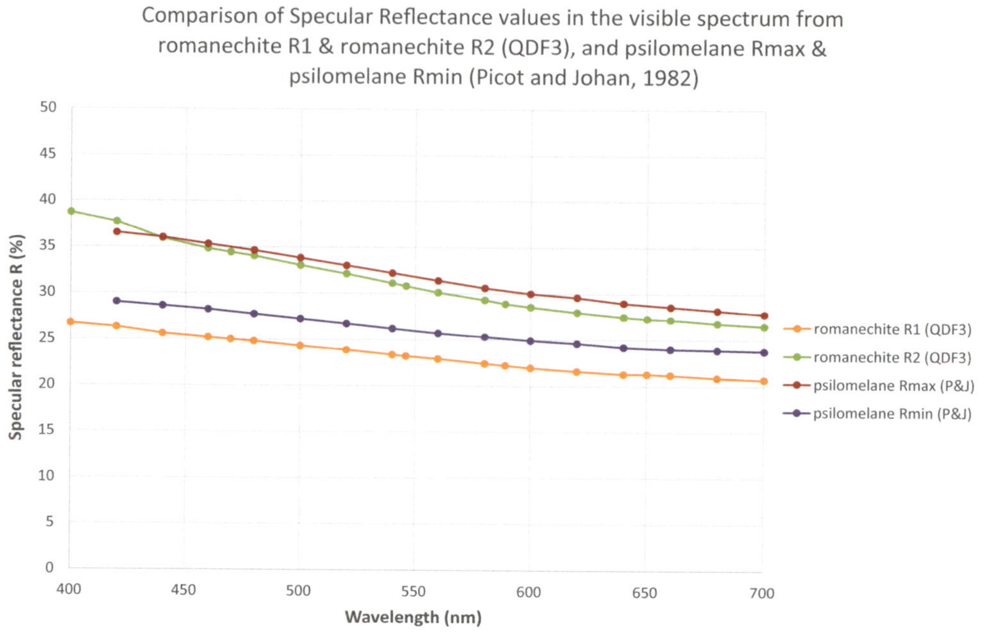
Comparison of Specular Reflectance values in the visible spectrum from romanechite R1 & romanechite R2 (QDF3), and psilomelane Rmax & psilomelane Rmin (Picot and Johan, 1982)
Specular reflectance R (%)
0
5
10
15
20
25
30
35
40
45
50
400
450
500
550
600
650
700
Wavelength (nm)
romanechite R1 (QDF3)
romanechite R2 (QDF3)
psilomelane Rmax (P&J)
psilomelane Rmin (P&J)

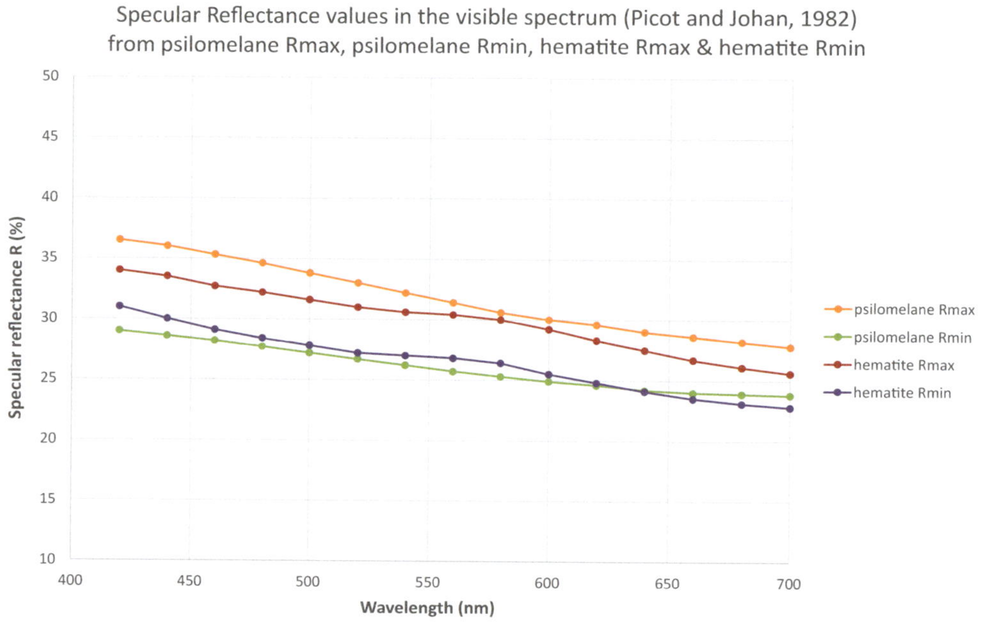
Specular Reflectance values in the visible spectrum (Picot and Johan, 1982) from psilomelane Rmax, psilomelane Rmin, hematite Rmax & hematite Rmin
Specular reflectance R (%)
10
15
20
25
30
35
40
45
50
400
450
500
550
600
650
700
Wavelength (nm)
psilomelane Rmax
psilomelane Rmin
hematite Rmax
hematite Rmin

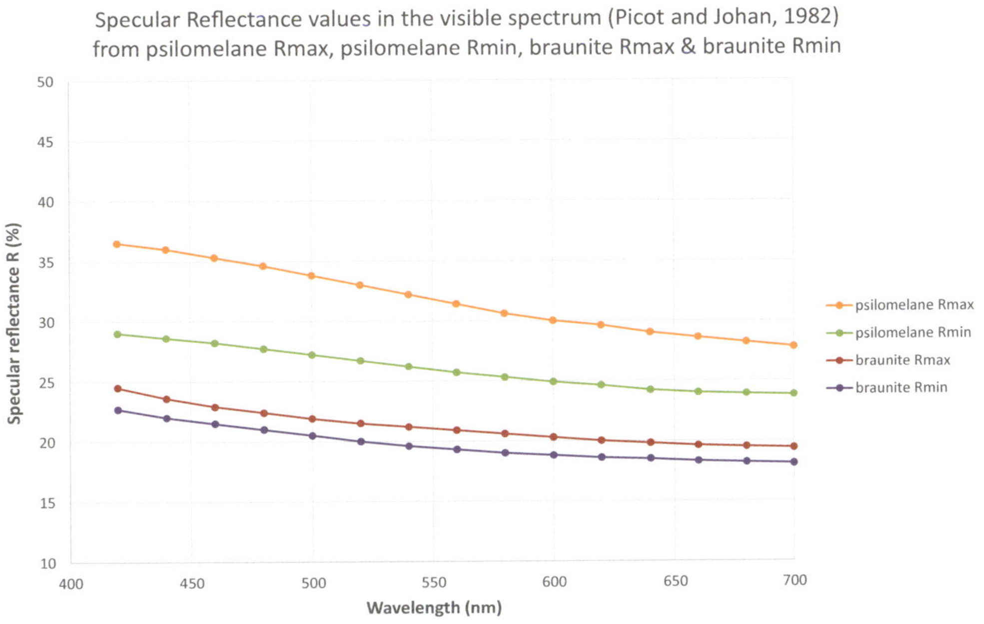

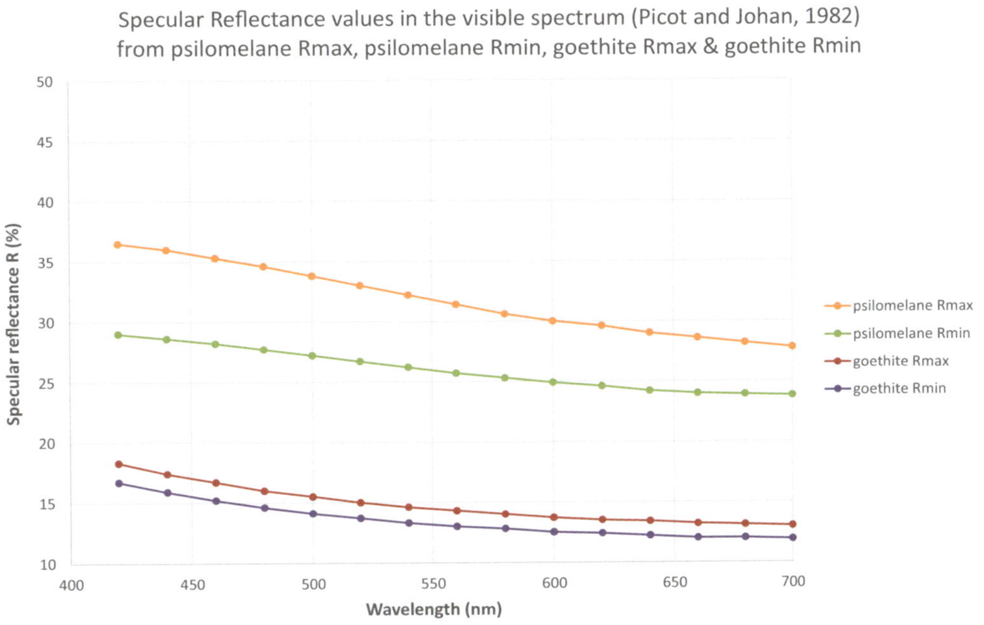

References

Back ME, Mandarino JA (2008) Fleischer's glossary of mineral species 2008. The Mineralogical Record Inc. Tucson, AZ, 345p

Picot P, Johan Z (1982) Atlas of ore minerals. B.R.G.M./Elsevier

Ramdohr P (1980) The ore minerals and their intergrowths, 2nd edn. Pergamon Press, Oxford. 2 volumes, 1205p

Uytenbogaardt W, Burke EAJ (1971) Tables for microscopic identification of ore minerals, 2nd edn. Elsevier, Amsterdam

Pyrargyrite (*prg/dark red silver*)

98

Ag_3SbS_3: Trigonal

Pyrargyrite		R_{546} (%)	Y (%)	λ_d (nm)	P_e (%)	Colour sensation
Air	R_o	30.4	30.2	478	10.7	Light bluish grey
	$R_{e'}$	28.5	27.7	472	11.0	
Oil	R_o	15.0	14.9	478	17.9	
	$R_{e'}$	13.4	13.3	477	19.5	
Bir: Weak (3)				*Pleoc*: Very weak (2)		*Tarn*: Very slow light attack
Anis +P: Distinct, little coloured (4)				*IR*: Abundant, red		*Twg*: Yes
Mor: Massive filling, encrustations				*Clv*: {1011} poor		*PolQ*: Good
Hardness	*PolH*: Very low (1)			*Mohs*: 2½	*Talm*: C	*VHN* $_{(100\ g)}$: nd

Obs **:** cf. Table and R charts, multispectral data (VNIR, AMCO) and visible range (QDF3)
Mineral abbreviations: ac—acanthite, Ag—native silver, As—native arsenic, Bi—native bismuth, cag—cerargyrite, cin—cinnabar, cup—cuprite, dy—dyscrasite, gg—gangue, gn—galena, gth—goethite, my—miargyrite, pc—pearceite, plb—polybasite, prg—pyrargyrite, pru—proustite, py—pyrite, q—quartz, Sb—native antimony, sbt—stibnite, sp—sphalerite, stp—stephanite

98.1 General

Composition Ag_3SbS_3, with As impurities: pyrargyrite is the Sb term of the Proustite–Pyrargyrite Series (Ag_3AsS_3–Ag_3SbS_3).

However, both observational data in natural ores and laboratory experimental data suggest a very low miscibility between the extreme terms, proustite and pyrargyrite, at least at the low temperatures of the common paragenesis. Nevertheless, the optical properties of both terms are similar.

Pyrargyrite has a monoclinic dimorph: pyrostilpnite.

Moderate *reflectance*: R > gth; ≈ pru; slightly < my; clearly ≥ gn and ≪ Sb, dy, Ag (cf. multispectral comparative plots and Figs. 98.1a, b and 98.2a).

Remarkable *bireflectance* (> pru) but hardly any pleochroism (no colour change).

It usually occurs in encrustations and as fracture fillings in aggregates intergrown with other silver sulphosalts or massive. The intergrowths can be varied and complex, with native silver, proustite, galena, symplectitic, irregular, replacement, etc. In favourable conditions (empty spaces), it forms well developed prismatic crystals.

98.2 Microscopic Description

98.2.1 SP Observation (*Single Polariser*)

Colour greyish-white, slightly bluish (especially in oil, Fig. 98.1a, b). Good *polish*, no (or few) scratches.

Polishing hardness very low, PolH < sbt, my, stp; slightly > plb, ac. Light attack irregular, generally slow or very slow, if not absent. Cleavage {1011} not very visible under the microscope.

98.2.2 Observation +P (*Polariser + Analyser at 90°*)

Clearly *anisotropic*, although weakly coloured and partially masked by the abundant red *internal reflections* (Figs. 98.1c, d and 98.2b).

Varied growth *twins* are sometimes observed; also, pressure twins (polysynthetic), although as a late mineral it is not usually very deformed.

R. Castroviejo, *A Practical Guide to Ore Microscopy—Volume 1*,
https://doi.org/10.1007/978-3-031-12654-3_98

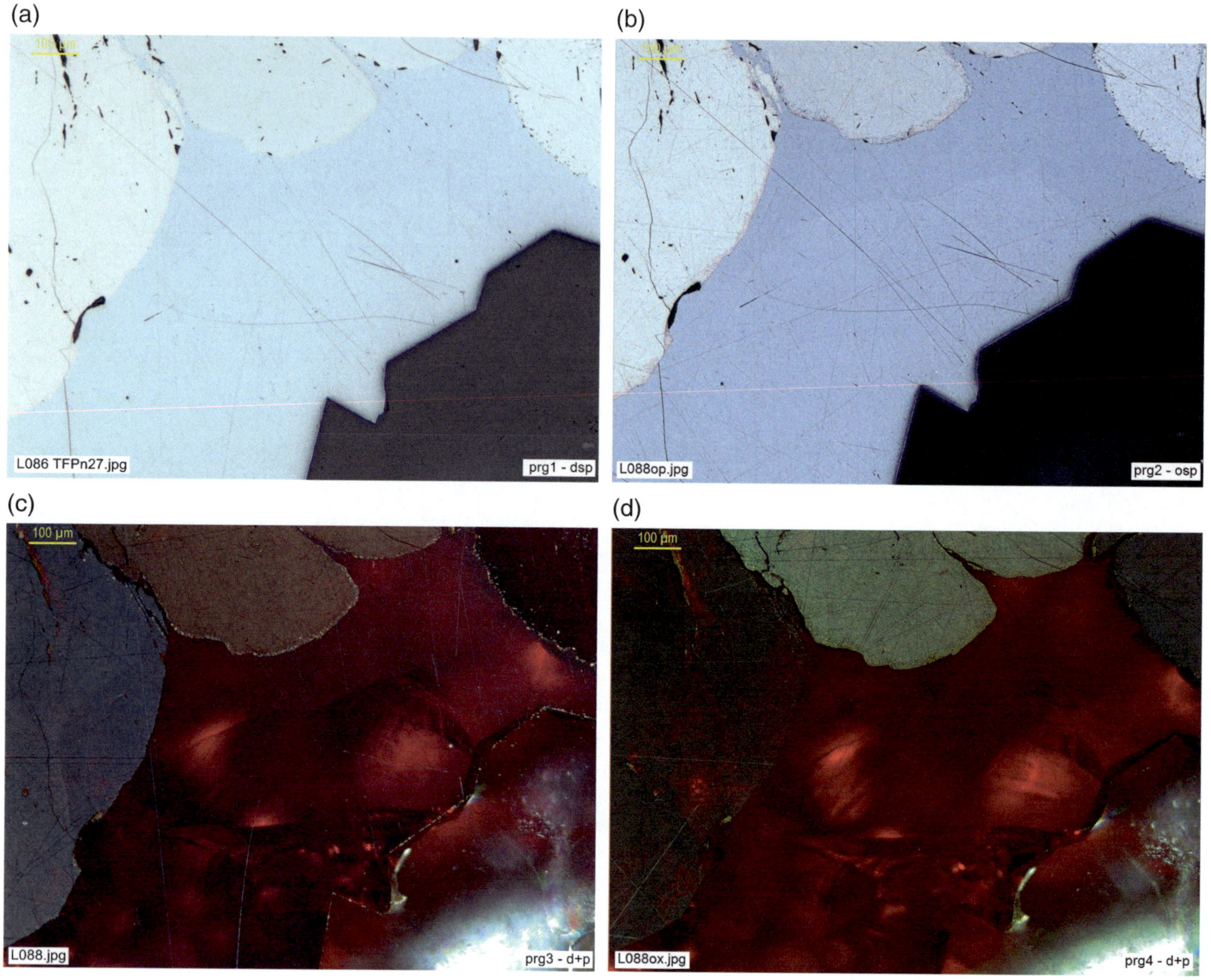

Fig. 98.1 **a, b** (**dsp** and **osp**, respectively)—Argentiferous ore, with prg (bluish-grey), my (greyish-white, bireflective) and q (greyish/black). TFPn27, Hiendelaencina, Guadalajara. **c, d** Same scene as prg1 (**d+p** and **o+p**, respectively). Massive red internal reflections of prg, much more abundant than in my, which is also distinguished by its higher bireflectance and more vivid anisotropism (with purplish-blue to greenish- blue tints). Light etching marks the prg and my contacts with a bright, thin halo in my. Contacts prg-q, shaded because of the q relief on prg (PolH of q $\gg$ PolH of prg)

98.2.3 Paragenesis

Its typical *paragenesis* is low-temperature hydrothermal, particularly epithermal and *five-element* (Ni–Co–Bi–Ag–U) veins. It can also form as a supergene mineral in the cementation zone or secondary sulphide enrichment blanket.

Accompanying ores: native silver, native arsenic, native antimony, native bismuth, acanthite, dyscrasite, miargyrite and other silver sulphosalts, Co–Ni minerals, galena.

98.2.4 Identification

The greyish-white, slightly bluish colour, moderate reflectance and bireflectance, very low hardness, net anisotropism and abundant IR are characteristic features of prg. Paragenesis is an important datum (e.g., to avoid confusion with cinnabar or cuprite), but it also includes similar minerals to be distinguished:

- Proustite, which is much scarcer, is distinguished by its bluer colour, slightly lower R (darker), brighter and more abundant IR and association with As ores, instead of the Sb ores typical of pyrargyrite.
- Miargyrite is lighter, more anisotropic (with bluish tones) and its IR are much scarcer.
- Polybasite and pearceite are more greenish, are easily attacked by light, have more vivid, greenish polarization colours and less IR.
- Stephanite does not display IR.

Fig. 98.2 **a, b (osp** and **o+p**, respectively)—Aggregate of Sb (white, bright), dy (in the centre, yellowish white, less reflective than Sb), Ag (tiny particles, marginal on dy, yellowish white, brighter than Sb and dy), prg (bluish grey), senarmontite/cerargyrite (secondaries to Sb/dy, dark grey) and gg (black), sealed by a crown of gn (grey, dull next to Sb and dy, but lighter than prg and my), with my (greyish white, darker than gn) on the outside (top right). Crossed analyser (prg6) shows the IR (abundant red in prg, sporadic in my, colourless, luminous, and massive in gg) and the anisotropism of Sb, dy, prg, and my. CR 398, Hiendelaencina, Guadalajara

(a)

(b)

98.3 Spectral Information

98.3.1 Pyrargyrite Multispectral Reflectance R (in Air, Without Polariser) (%)

λ (nm)	370	400	425	450	475	500	525	550	575	600	625	650	675	700	750	800	850	900	950	1000
μ	34.4	35.1	35.1	35.1	33.9	32.4	30.8	29.4	28.2	27.4	26.7	26.3	26.1	25.7	25.1	24.7	24.6	24.3	24.2	24.0
P_5	32.9	33.7	33.8	33.9	32.8	31.5	30.0	28.6	27.6	26.8	26.1	25.4	25.2	24.8	24.2	23.8	23.6	23.4	23.2	23.0
P_{20}	33.7	34.4	34.5	34.6	33.5	32.0	30.4	29.0	27.9	27.1	26.4	25.8	25.5	25.1	24.5	24.1	24.0	23.7	23.5	23.3
P_{80}	35.2	35.8	35.7	35.6	34.3	32.8	31.1	29.7	28.5	27.6	27.0	26.6	26.5	26.2	25.6	25.2	25.1	24.8	24.7	24.5
P_{95}	36.0	36.4	36.3	36.1	34.9	33.2	31.5	30.0	28.9	28.0	27.4	27.3	27.4	27.3	26.7	26.3	26.2	26.0	26.0	25.8

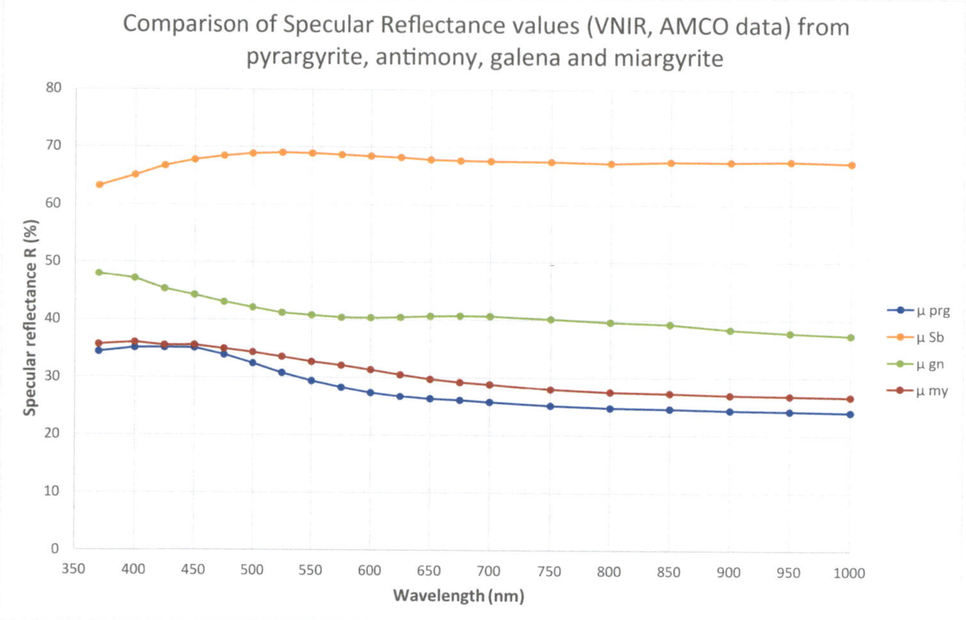

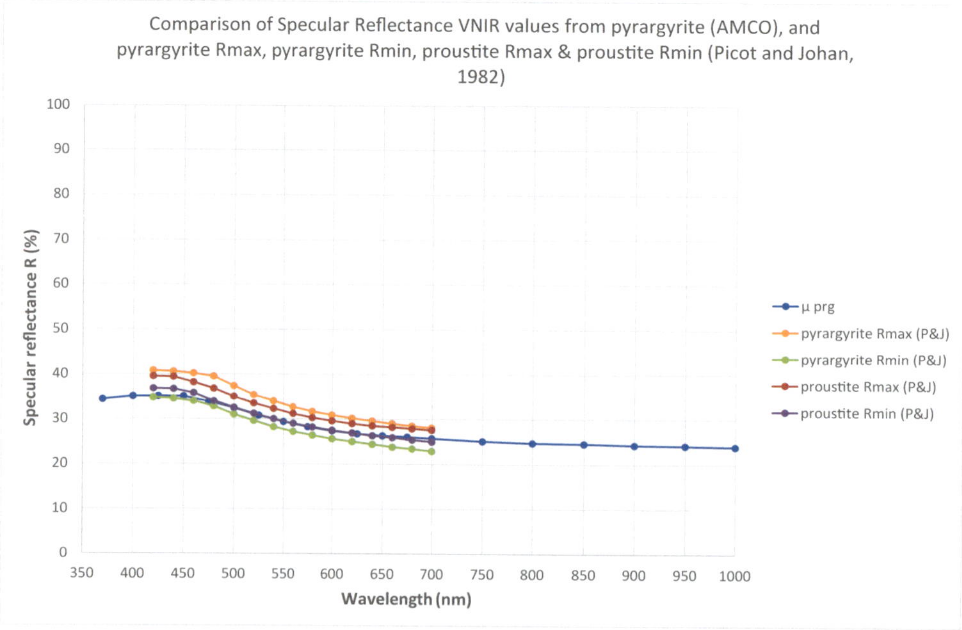

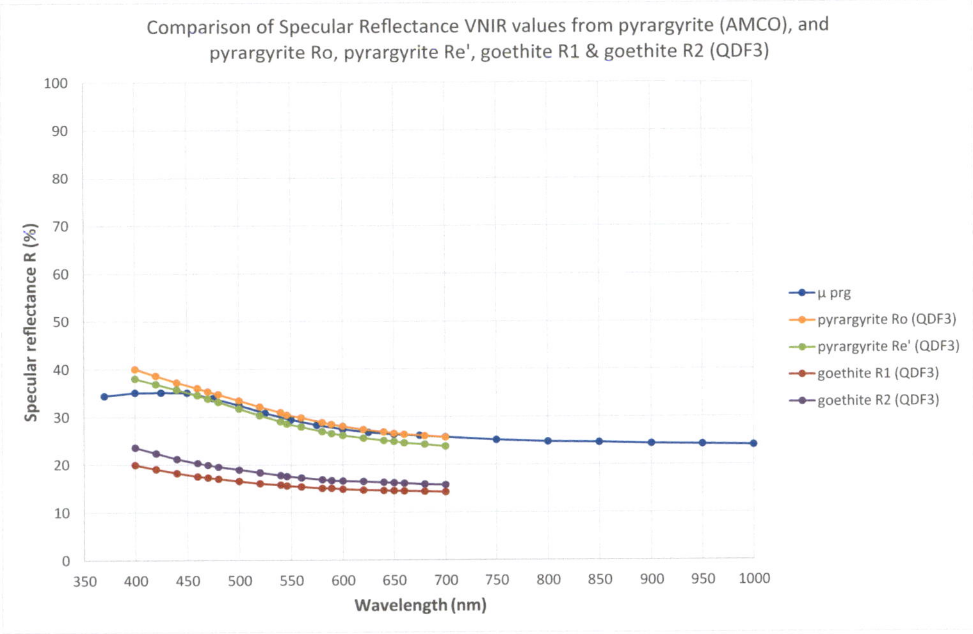
Comparison of Specular Reflectance VNIR values from pyrargyrite (AMCO), and pyrargyrite Ro, pyrargyrite Re', goethite R1 & goethite R2 (QDF3)
Specular reflectance R (%)
Wavelength (nm)
µ prg
pyrargyrite Ro (QDF3)
pyrargyrite Re' (QDF3)
goethite R1 (QDF3)
goethite R2 (QDF3)

Pyrite (*py*)

FeS_2: Cubic

99

Pyrite		R_{546} (%)	Y (%)	λ_d (nm)	P_e (%)	Colour sensation
Air		51.7	51.7	576	9.6	Light yellow-yellowish white, bright
Oil		39.3	39.0	573	17.5	Light yellow, bright
Bir: Null (1)				*Pleoc*: Null (1)		*Tarn*: Frequent, dark/assorted colours
Anis +P: Isotropic, ext imperfect (2)				*IR*: No		*Twg*: Not visible generally
Mor: Euhedral aggregates, massive, gel				*Clv*: // (110) and (001)		*PolQ*: Good, but frequent relief
Hardness	*PolH*: High/Very high (4–5)			*Mohs*: 6.5	*Talm*: F	*VHN* $_{(100\ g)}$: 1505–1620

Obs **:** cf. Table and R charts, multispectral data (VNIR, AMCO) and visible range (QDF3)
Mineral abbreviations: Ag—native silver, asp—arsenopyrite, Bi—native bismuth, bn—bornite, cb—columbite, cc—chalcocite, ccp—chalcopyrite, cob—cobaltite, cst—cassiterite, ctt—cattierite, cv—covellite, el—electrum, en—enargite, fm—famatinite, gf—gersdorffite, gg—gangue, gn—galena, gth—goethite, ht—hematite, lo—löllingite, lpc—lepidocrocite, prm—pararammelsbergite, po—pyrrhotite, py—pyrite, rm—rammelsbergite, sf—safflorite, sk—skutterudite, sp—sphalerite, spn—spinel, tit—titanite, tn—tennantite, td—tetrahedrite, ul—ullmannite, vae—vaesite

99.1 General

One of the most frequent minerals in nature, common in very numerous and varied types of deposits, although it is altered in oxidizing conditions and tends to disappear, replaced by Fe oxides (hematite, goethite, lepidocrocite) or limonite by the action of weathering.

Dimorph of marcasite (orthorhombic). Although both species may be co-genetic and paramorphic pyrite-marcasite transformations are known in both directions, pyrite is more stable and much more abundant than marcasite (Figs. 99.2a, b, 99.3, and 99.5a, b).

The disulphide pyrite is also more stable than the mono-sulphide (pyrrhotite, FeS), which it tends to replace in shallow crustal settings. However, it is replaced by pyrrhotite under conditions of high temperature and low S pressure. Thus, typically, in metamorphosed deposits (Fig. 99.1a, b) the proportion of pyrrhotite increases to the detriment of pyrite.

Hard and rigid mineral, with brittle response to deformation under the common conditions in the upper crust (Figs. 99.10, 99.12, 99.13, 99.14 and 99.15). Hence its frequent brecciation or, alternatively, its fluid-assisted mobilization, through combined mechanisms of *pressure solution*, *solution transfer*, fluid-assisted recrystallisation, oriented precipitation (*pressure shadows*), etc.

Pyrite may also occur as an extremely fine-grained, low-temperature, porous, often colomorphic variety known as *melnikovite*-pyrite. This may include other *invisible* components, as it is commonly of sub-microscopic grain size.

Composition: FeS_2, usually pure, although minor contents of Co, Ni, As, Se, Pb are possible. Other metals (Cu, Zn, Ag) detected chemically may correspond to discrete inclusions of the corresponding mineral phases (chalcopyrite, sphalerite, electrum, cf. Fig. 99.3) or to metallic nanoparticles, although it is possible to find Au disseminated in the crystal lattice. The name *bravoite* has been given to the nickeliferous pyrite, typically zoned, which can be considered a variety of pyrite (not recognized as a species by IMA/COM). With increasing Ni and Co contents, it forms the *vaesite-cattierite* series (cf. description *bravoite*).

R. Castroviejo, *A Practical Guide to Ore Microscopy—Volume 1*,
https://doi.org/10.1007/978-3-031-12654-3_99

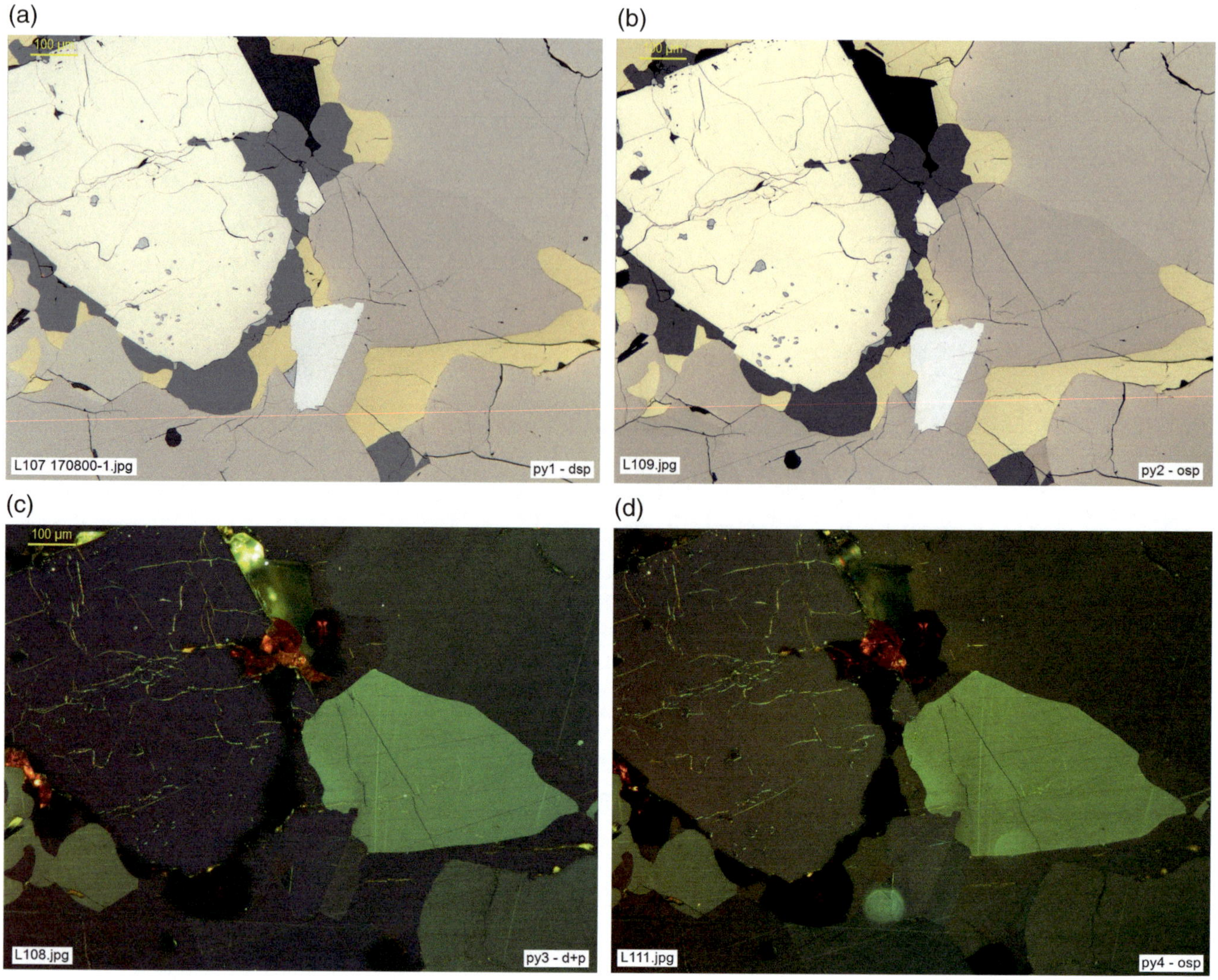

Fig. 99.1 **a, b** (Same scene, **dsp** and **osp**, respectively). Aggregate of py (yellowish-white porphyroblast), po (pinkish-brown), ccp (yellow), sp (grey), asp (white), gn (light grey, py inclusions) and gg (quartz, very dark grey). **b** Shows the effect of oil immersion, mainly higher colour saturation, e.g., in py, and strong decrease of reflectance of sp and gg. Sample 170800–1, Sulitjelma Mine, Norway. **c, d** Same scene as Fig. 99.1 (**d+p**) and (**o+p**), respectively: extinction of py, anisotropism of po and reddish internal reflections of sp (ferriferous)

99.2 Microscopic Description

99.2.1 SP Observation (*Single Polariser*)

Light yellow *colour*, very bright in well polished and fresh sections and more distinct in immersion (Fig. 99.1a, b). However, the perception of the colour is very conditioned by the adjacent ores. For example:

- it is much paler or greyish against native metals like Ag or Bi (Figs. 99.3 and 99.11) and also in contact with millerite (cf. Fig. 1.28a, vol. 2;, the identity of pyrite in this sample was checked with electron microprobe),
- it looks whiter against chalcopyrite (Fig. 99.10), enargite or famatinite (Fig. 99.4), and
- it is more intensely yellow against marcasite (Figs. 99.2a, b, 99.3, 99.4 and 99.5a, the contrast is accentuated in immersion).

Colour and reflectance also depend on composition and texture or crystallinity of py:

- lower R and brownish tones in nickeliferous pyrites and in colloform or cryptocrystalline pyrites (cf. bravoite and melnikovite, resp), and
- pinkish, reddish or violet tones (according to Uytenbogaardt and Burke 1971) if they contain Co, Ni, Cu.

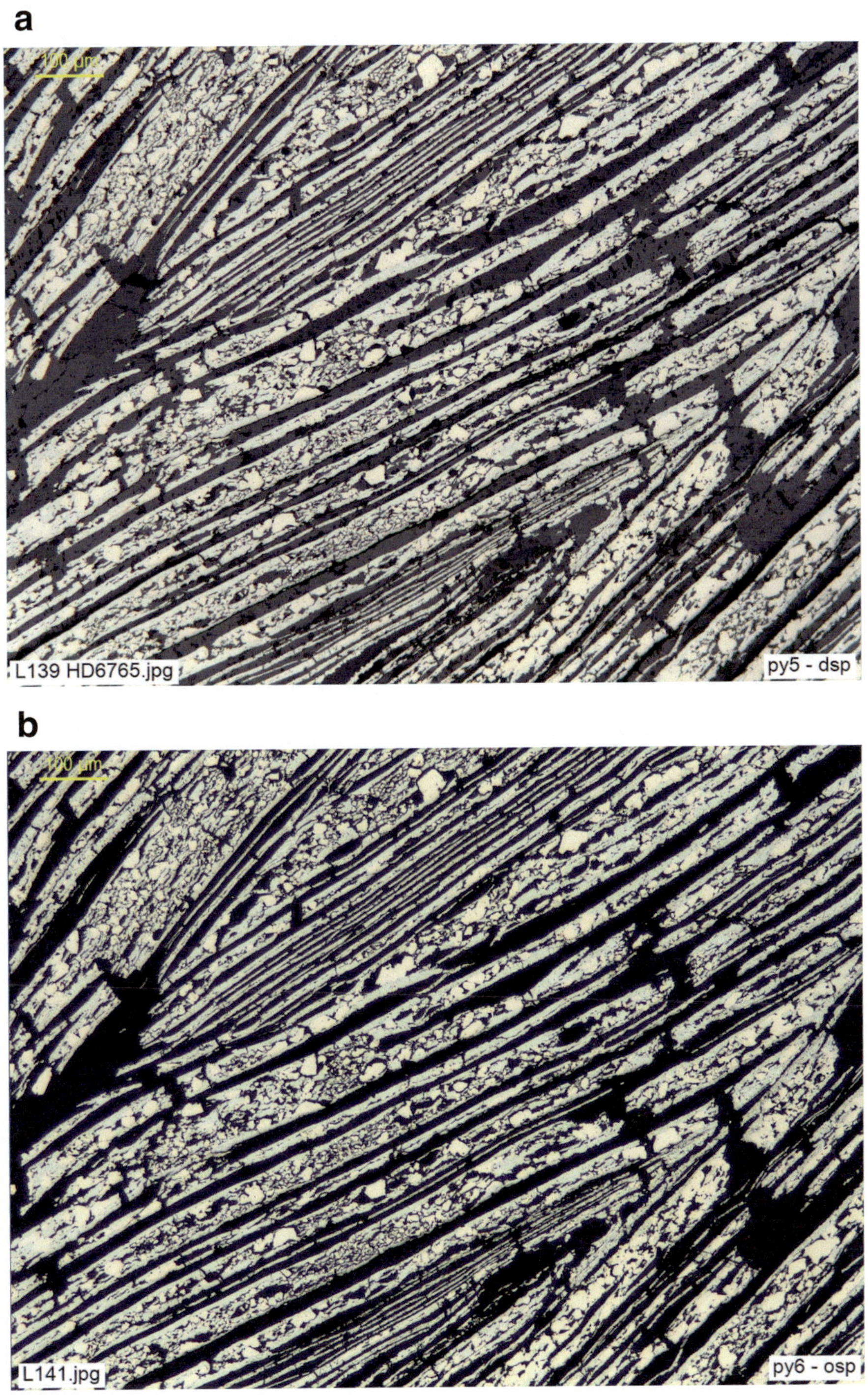

Fig. 99.2 **a, b** Intergrowth of lamellar py and mc (white, with slight greenish tint). HD_6765, Teufelsgrund, Schwarzwald (**dsp**). **b** Same scene as py5 (**osp**): contrast increase in immersion

Polishing is difficult because of its hardness, but the result is usually excellent and the slight relief it acquires, usually unavoidable, provides a useful identification criterion (Fig. 99.10). Its *polishing hardness* marks the high/very high limit: PolH py > marcasite and almost all ores but < columbite, cassiterite, spinel, corundum.

Pyrite frequently tarnishes, acquiring with time a dark brownish or sometimes multicoloured surface alteration *patina*. This can occasionally serve, instead of the classic acid attack, to reveal internal structures (Figs. 99.12, 99.13, 99.14 and 99.15).

Reflectance, ***R***, is high and constant in pure pyrite, but if it is not stoichiometric pyrite, R depends on the composition, decreasing notably with high Ni contents (cf. multispectral comparative plots and compare with Ni-pyrite of QDF3: 25.7% Ni).

Fig. 99.3 Aggregate of py (yellowish white) and mc (bluish-greenish white), with interstitial inclusions of el (very bright yellow), in banded ore of ccp (yellow), sp (dark grey), gn (light grey) and q (black). HD6153, Rammelsberg (Banderz), Niedersachsen, Germany (**d+p**)

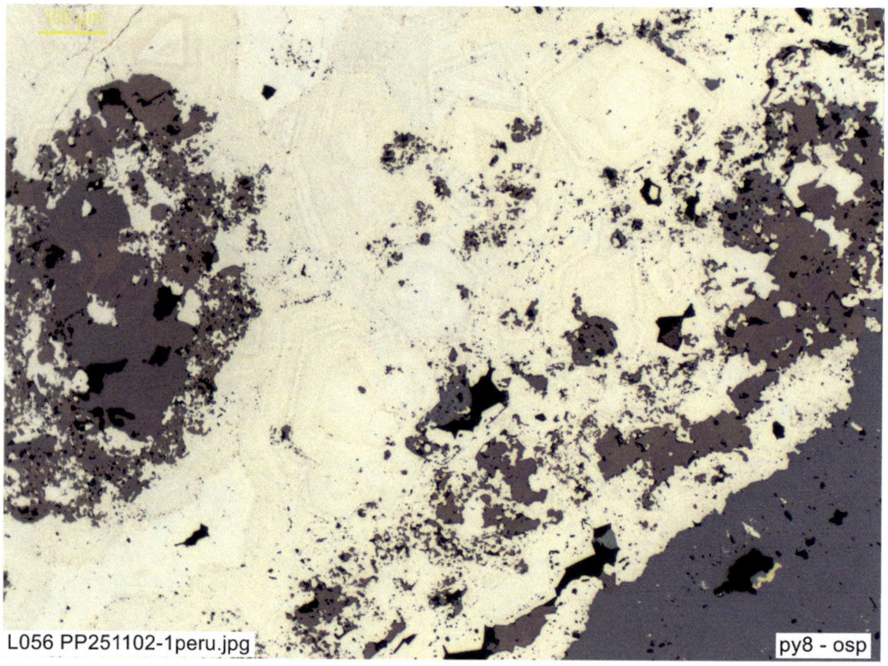

Fig. 99.4 Zoned py aggregate, with famatinite, fm, inclusions (brownish-violet-grey). Enargite, en (grey, bottom right). CR283, 251102–1 Peru (**osp**)

Isotropic (neither bireflectance nor pleochroism). *Cleavage:* frequent//(100),//(110), occasional (111). Enhanced by deformation or alteration. Occasional zoning (Fig. 99.4).

It occurs in very varied forms, depending on the genesis or type of deposit: massive, granular, holocrystalline aggregates, porphyroblasts, idioblasts, porphyroclasts, cryptocrystalline, colomorphic, framboidal (the supposed *pyritised bacteria*: Figs. 99.6 and 99.7), pseudomorphic (Figs. 99.7, 99.8 and 99.9), etc. Its idiomorphic tendency, with cubic forms (cube, pyritohedron, octahedron) is very pronounced, but as an early and brittle mineral it often shows cataclasis (cf. § **General** and Figs. 99.10, 99.12 and 99.13). In deposits of low temperature or very rapid precipitation (marked disequilibrium), it forms colomorphic to cryptocrystalline masses.

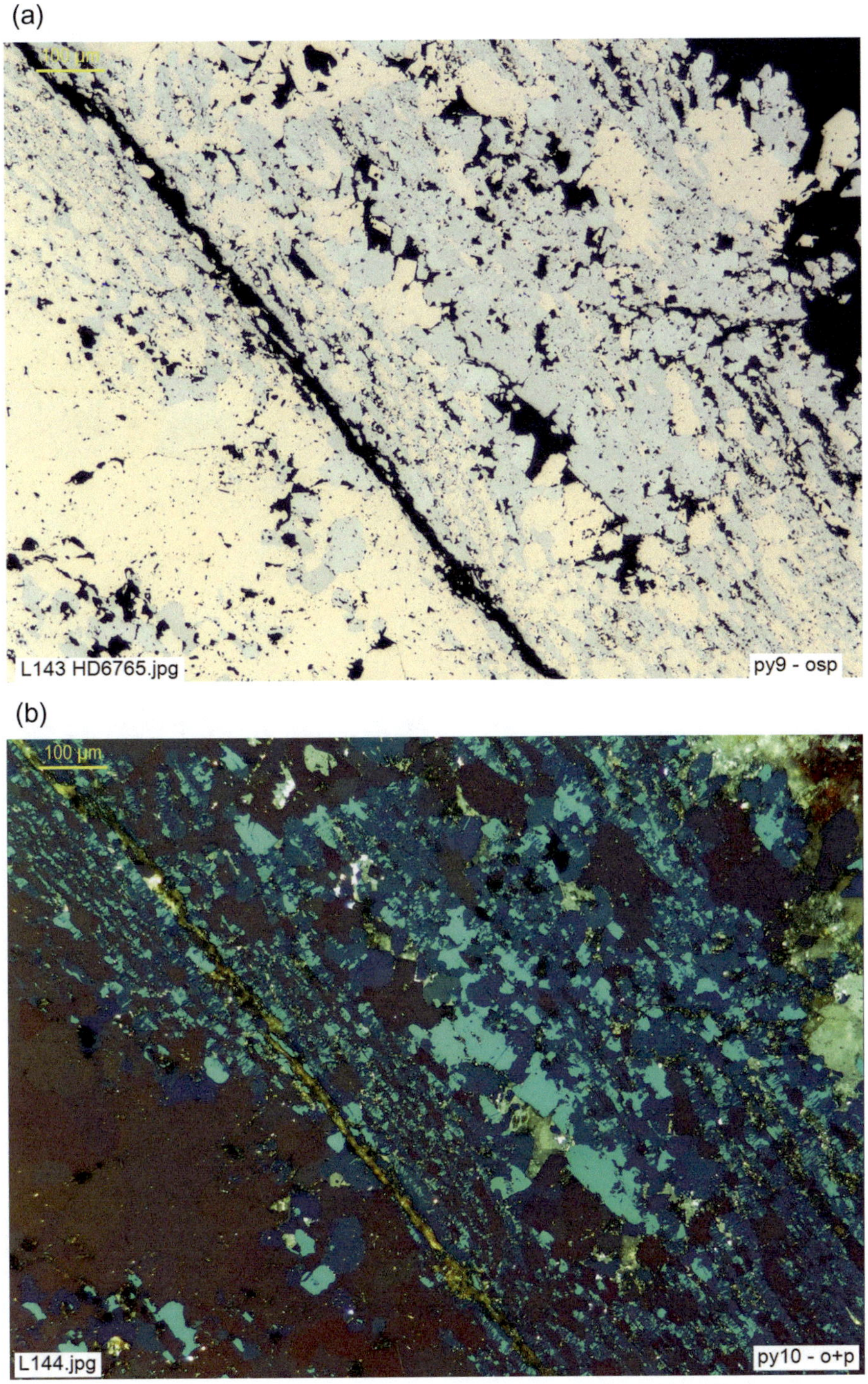

Fig. 99.5 **a, b** Intergrowth of py, with weak anomalous anisotropism, and mc, strongly anisotropic. HD_6765, Teufelsgrund, Schwarzwald, Germany (**osp**) and (**o+p**), respectively

Melnikovite or *colloform pyrite* is, according to Ramdohr (1980), a crystallized FeS_2 gel that incorporates FeS, As and even H_2O, forming cryptocrystalline, porous masses, usually darker than stoichiometric pyrite and of variable hardness. In the submicroscopic mass, other components can be adsorbed or can be hidden (e.g., chalcopyrite, sphalerite, galena), as *invisible* inclusions of potential economic interest.

99.2.2 Observation +P (*Polariser* + *Analyser at 90°*)

Although supposedly *isotropic* (as Fig. 99.1c, d), py often shows incomplete extinction and anomalous anisotropism, of variable but usually weak intensity (cf. comp mc, Fig. 99.5b). *Twinning* not visible under the microscope.

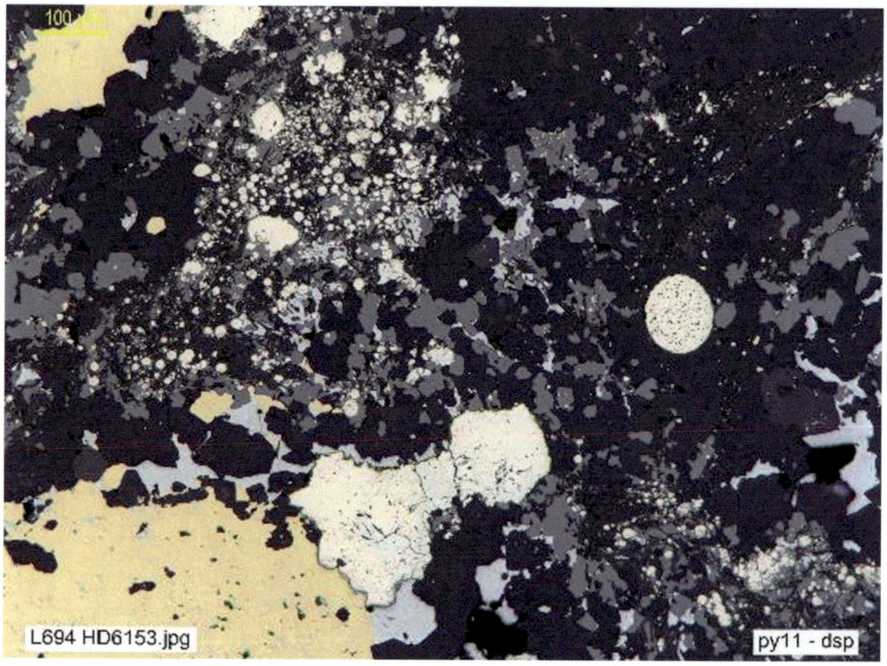

Fig. 99.6 Py framboids (generally ≈ 10–20 μm in diameter, but one of them exceptionally large, center right) in banded ccp, sp, gn, py, exhalative sedimentary ore. HD6153, Rammelsberg (Banderz), Niedersachsen, Germany (**dsp**)

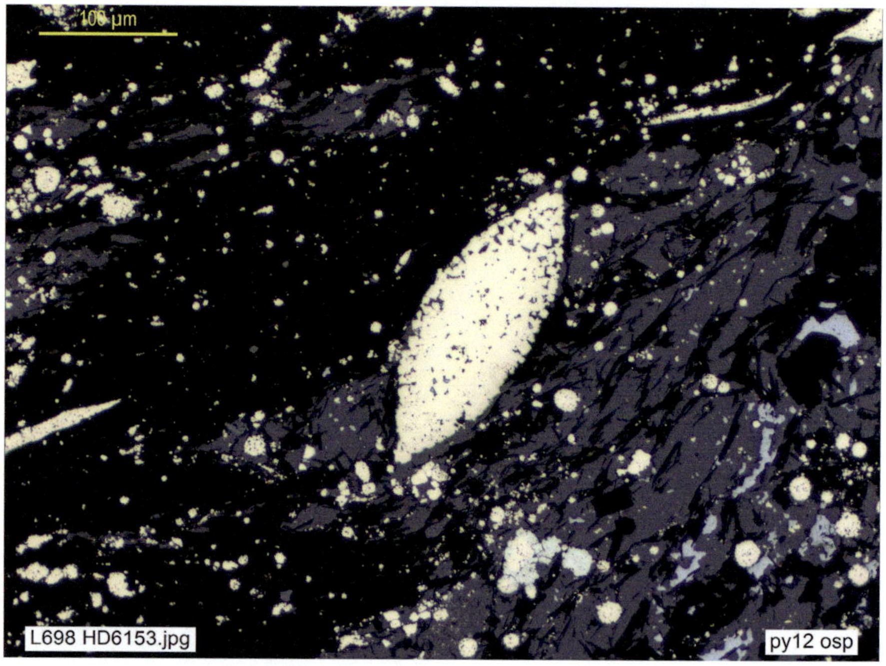

Fig. 99.7 Pyritised microfossil (possible fusuline) and disseminated py (and mc) framboids in banded sp and gn ore. HD6153, Rammelsberg (Banderz), Niedersachsen, Germany (**osp**)

99.2.3 Paragenesis

Paragenesis is varied, since it can be found in practically all types of deposits (cf. § *General*). Although it is not to be expected in oxidizing conditions, py relicts as well as limonitized py ghosts are still found in weathered bodies of deposits such as massive sulphides, copper porphyries, epithermal veins, etc.

The most frequent companions are chalcopyrite, sphalerite, pyrrhotite, galena, arsenopyrite, marcasite, chalcocite, digenite, and covellite. The supergene alteration products in the oxidation zone (goethite, lepidocrocite, hematite, limonite) often preserve traces of pyrite.

99.2.4 Identification

It is easily recognized by its colour, high reflectance, high polishing hardness, alteration and morphology. It can be distinguished:

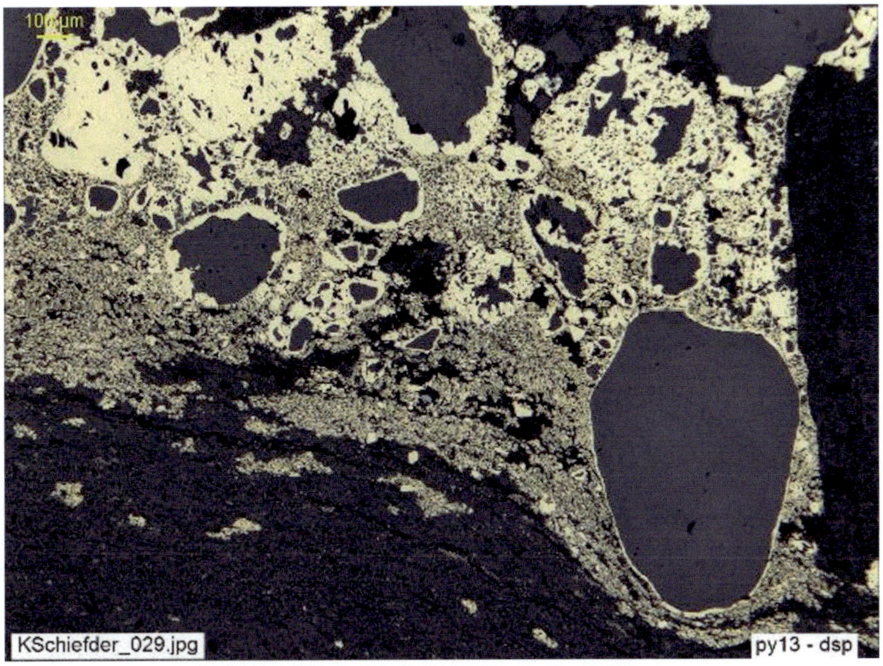

Fig. 99.8 Incipient pyritisation of shaly-sandy sediments. Kupferschiefer 010211-rub10/07, Brandenburg, Germany (**dsp**)

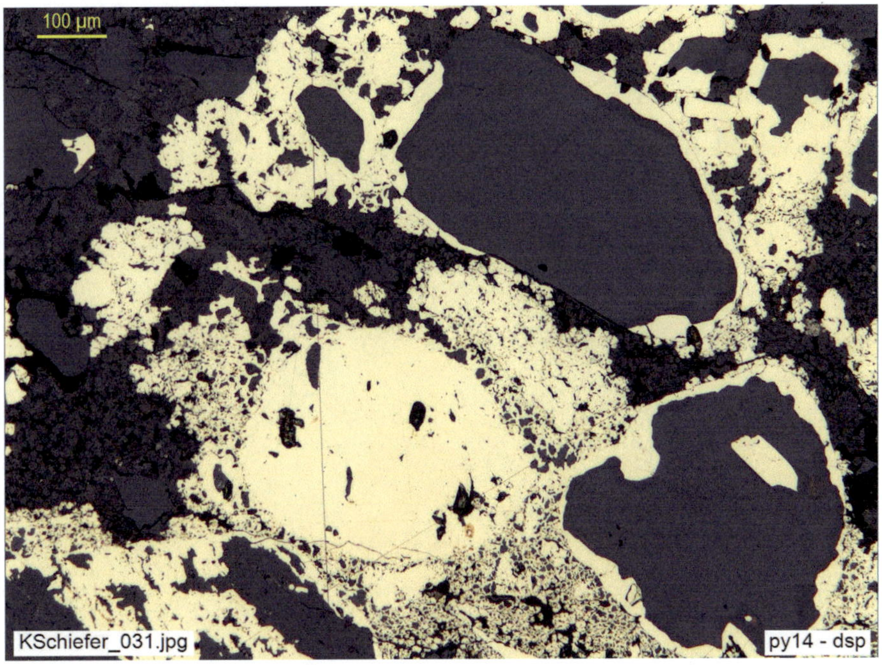

Fig. 99.9 More advanced pyritisation of detrital sediments (Rotliegendes), with almost complete replacement of the more reactive or porous sands. Kupferschiefer 010211-rub10/07, Brandenburg, Germany (**dsp**)

- from skutterudite, gersdorffite, ullmannite because they are whiter.
- from cobaltite, because of cob pink colour, anisotropism and twinning.
- from marcasite, as marcasite is strongly anisotropic and, compared to pyrite, shows a faintly greenish or bluish-greenish white shade (Figs. 99.2, 99.3 and 99.5a, b).
- from millerite, as millerite is pleochroic and clearly anisotropic.

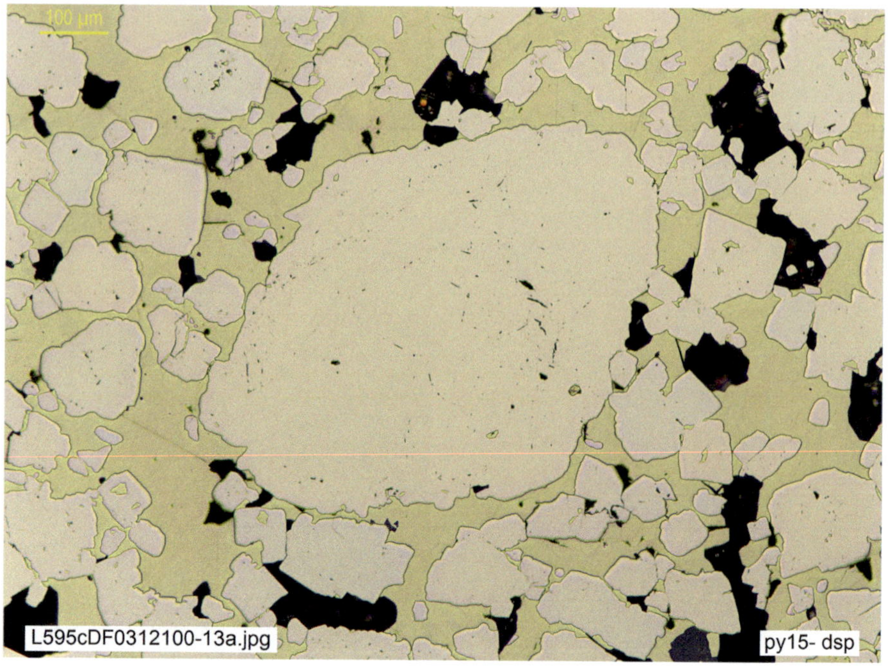

Fig. 99.10 Cemented pyrite microbreccia partially corroded by ccp, in VMS metamorphic ore. Relief of py enhanced with partial diaphragm closure. Piquito II-031200_13 a, Moeche, Coruña (**dsp**)

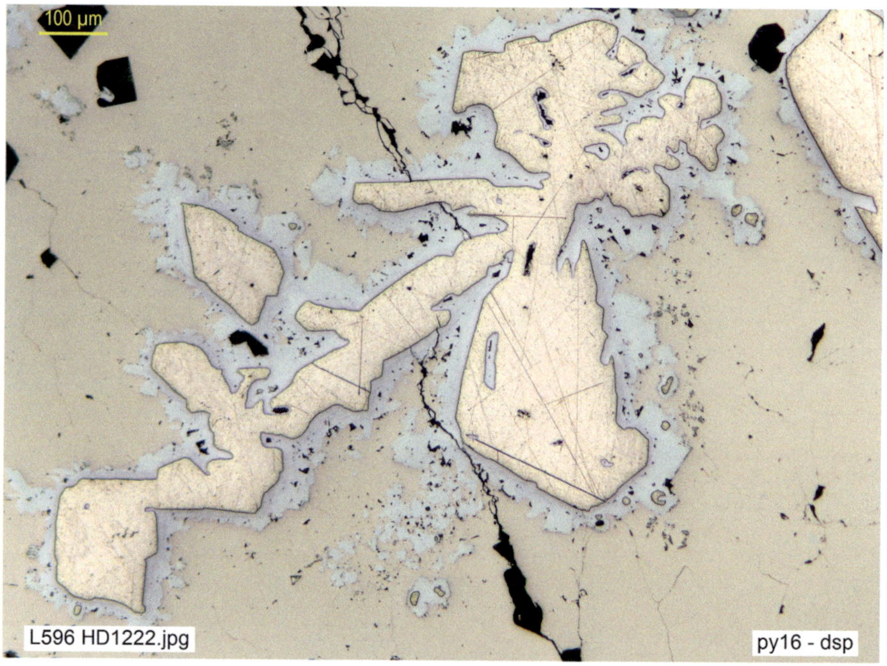

Fig. 99.11 Mass of py with skeletal Bi (brightest yellowish-white) inclusions coated by a greyish-white corona of finely intergrown rm/prm, sf, lo, sk (very similar aspect, white arsenides). HD 1222, Eisleben, Germany (**dsp**)

Fig. 99.12 Pyritic breccia, with minor ccp: py porphyroclast with zoning enhanced by tarnishing, and truncated by tectonic fragmentation. Piquito II-031200-2 m, Moeche, Coruña (**dsp**)

Fig. 99.13 Same scene as Fig. 99.12, with interference contrast to enhance relief (**d+p.w**)

Fig. 99.14 Py dissemination (with minor ccp, sp and tit) in mylonitic schist. Three py generations: (i) VMS primary ore microclasts (rounded core in cubic crystal) and relicts corroded by pressure-solution phenomena (internal indented outlines); (ii) mobilized py (solution transfer) cementing py (i); (iii) posthumous, post-tectonic py (intact cubes). Piquito II-031200-2 m, Moeche, Coruña (**dsp**)

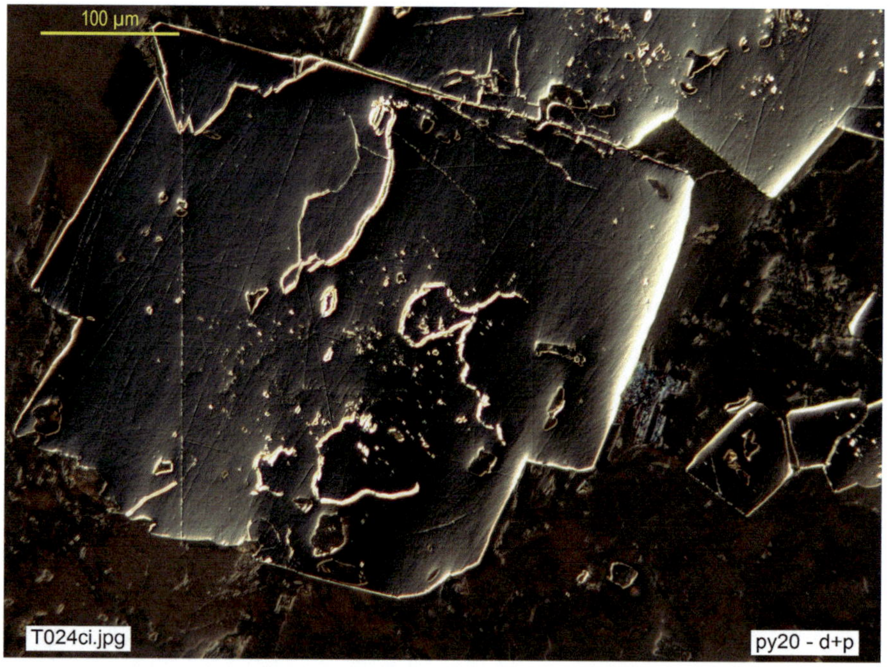

Fig. 99.15 Detail (bottom left cube of Fig. 99.14), in interference contrast setting to enhance the internal structure and show the pressure solution process in py (**d+p.w**)

99.3 Spectral Information

99.3.1 Pyrite Multispectral Reflectance R (in Air, Without Polariser) (%)

λ (nm)	370	400	425	450	475	500	525	550	575	600	625	650	675	700	750	800	850	900	950	1000
μ	36.6	37.9	39.8	43.0	45.9	49.0	50.7	52.1	52.9	53.5	53.9	54.3	54.6	54.7	54.0	52.4	51.4	49.4	48.2	46.9
P_5	33.6	35.4	37.2	40.6	43.5	46.8	48.7	50.1	51.2	51.8	52.3	52.7	53.0	53.0	52.3	50.9	49.8	48.1	46.9	45.6
P_{20}	35.8	37.1	38.9	42.1	44.9	48.0	49.7	51.1	52.0	52.6	53.1	53.5	53.9	53.9	53.3	51.8	50.8	48.8	47.4	46.1
P_{80}	37.5	38.8	40.9	44.0	47.0	50.0	51.8	53.1	53.8	54.3	54.7	55.1	55.5	55.5	54.7	53.1	52.0	50.1	48.9	47.8
P_{95}	39.2	39.6	41.6	44.5	47.7	50.6	52.4	53.6	54.3	54.8	55.2	55.6	56.1	56.1	55.5	53.8	52.8	50.7	49.8	48.3

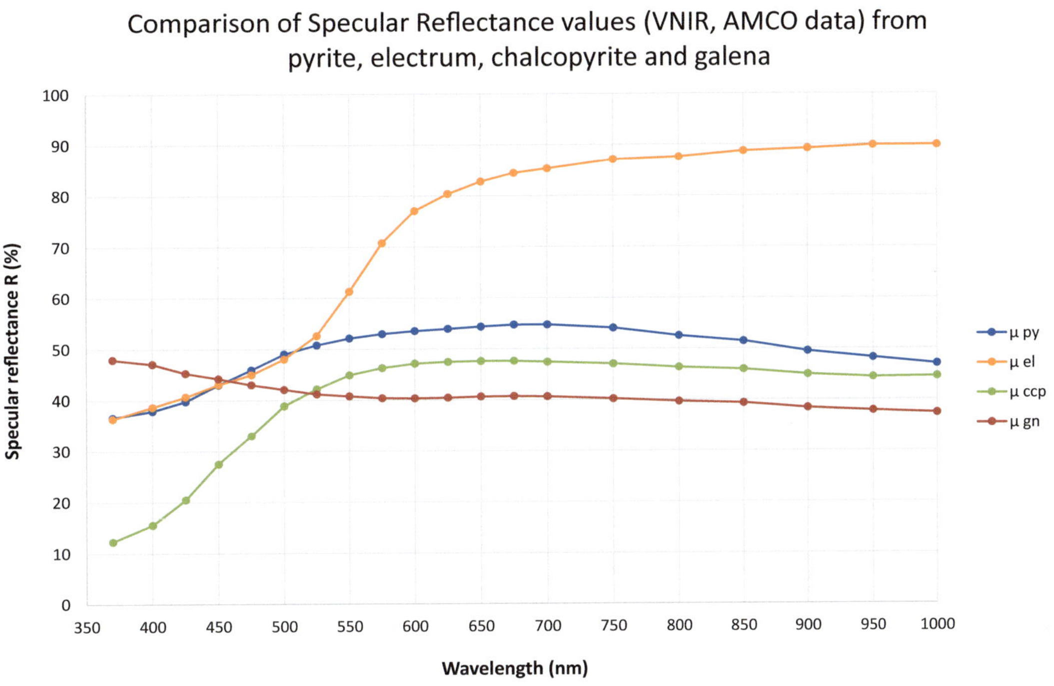

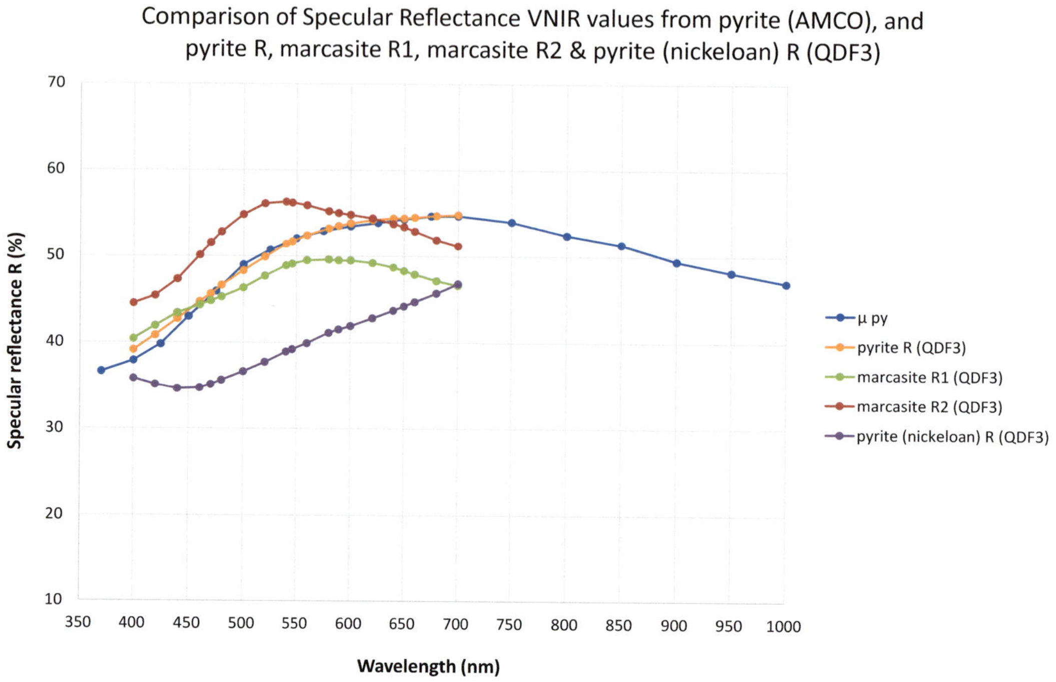

References

Ramdohr P (1980) The ore minerals and their intergrowths, 2nd edn, vol 2,. Pergamon Press, Oxford, 1205 p

Uytenbogaardt W, Burke EAJ (1971) Tables for microscopic identification of ore minerals, 2nd edn. Elsevier, Amsterdam

Pyrochlore (*pyc/koppite, columbomicrolite*) 100

$(Na,Ca)_2Nb_2O_6F$: Cubic

Pyrochlore		R_{546} (%)	Y (%)	λ_d (nm)	P_e (%)	Colour sensation
Air		12.8	12.8	472	4.7	Grey
Oil		2.5	2.5	475	4.5	Dark grey
Bir: Null (1)				*Pleoc*: No (1)		*Tarn*: No
Anis +P: Null (1)				*IR*: Abundant yellow–red		*Twg*: No
Mor: Massive. Idiomorphic, octahedral				*Clv*: No		*PolQ*: Good
Hardness	*PolH*: Medium (3)			*Mohs*: 5.5	*Talm*: –	*VHN* $_{(100\ g)}$: 542–665

Obs: cf. Table and R charts, multispectral data (VNIR, AMCO) and visible range (Picot and Johan 1982)
Mineral abbreviations: cb—columbite, cbt—columbotantalite, chr—chromite, cst—cassiterite, gg—gangue, lim—limonite, mt—magnetite, mlt—microlite, py—pyrite, pyc—pyrochlore, q—quartz, rt—rutile, sp—sphalerite, zir—zircon

100.1 General

Pyrochlore, ***pyc***, is an important ore of strategic value for the supply of critical substances (Nb, REE, Ta). It is an oxide of Nb, Na, REE and Ca as most typical components: $(Na,Ca)_2Nb_2O_6F$, although its composition is variable and is expressed by more complex formulas. For instance, the name *koppite*, applied in older works to pyc found in the alkali-carbonatite complexes of Kaiserstuhl (Germany), now refers to a calcium-pyc, with Ce, Nb, Ti.[1]

Traditionally, pyc is defined as the Nb term of the pyrochlore-microlite (*pyc-mlt*) solid solution series, in which *microlite*, ***mlt***, is the Ta-rich term, as expressed by its formula (IMA): $(Ca,Na)_2Ta_2$ $(O,OH,F)_7$. Ramdohr (1980) proposed for the series the formula $\mathbf{X_2Y_2O_7}$, in which **X** = Ca, Ce, Y, U; **Y** = Ti, Nb, Ta, recognizing that the optical distinction between the different members of the series is difficult because they have similar properties.

The compositions are actually more varied and, currently, not a single mineral species but a group of Nb oxides is designated as pyrochlore, the *pyrochlore group*, which is one of the seven groups that make up the *pyrochlore supergroup* (Atencio et al. 2017).

Today, the Nb-dominant members of the *pyrochlore supergroup* are called pyrochlore (Atencio et al. 2010). This supergroup, of formula: $\mathbf{A_2Nb_2(O,OH)_6Z}$, where **A** = Na, Ca, Sn^{2+}, Sr, Pb^{2+}, Sb^{3+}, Y, U^{4+}, H_2O and **Z** = OH, F, O, H_2O, comprises also the *microlite group*, of formula $\mathbf{A_{2\text{-}m}Ta_2X_{6\text{-}w}Z_{\text{-}n}}$, where **A** = Na, Ca, Sr, Pb^{2+}, Sn^{2+}, Sb^{3+}, Y, U, H_2O (Ag, Mn, Ba, Fe^{2+}, Bi^{3+}, Ce and other REE, Sc or Th), **X** = O(OH, F), and **Z** = anion, vacancy, H_2O or large monovalent cation (vbgr. OH^-, F, O, H_2O, K, Cs, Rb), while **m**, **w**, and **n** indicate incomplete occupancy of A, X and Z, respectively.

Obviously, the correct identification of the members of these groups cannot be done only by optical methods and requires other complementary techniques. Therefore, the following description contemplates only the optical recognition of the pyrochlore-microlite series in the classical sense, without going into the distinction of the individual species defined by IMA for the pyrochlore supergroup and its groups.

[1] Further information, cf. https://www.mindat.org/min-3316.html.

R. Castroviejo, *A Practical Guide to Ore Microscopy—Volume 1*,
https://doi.org/10.1007/978-3-031-12654-3_100

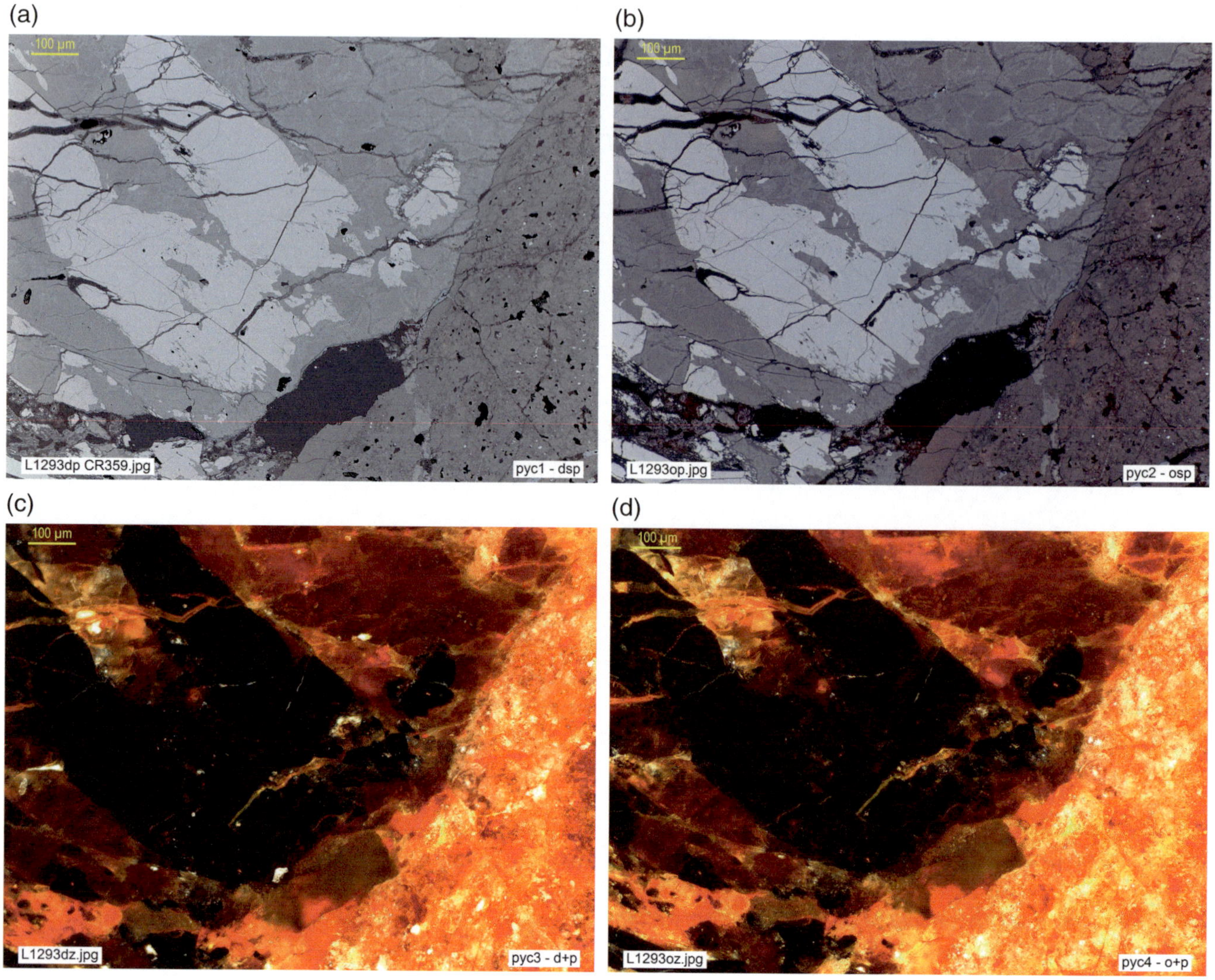

Fig. 100.1 **a** Mass of pyc (grey), with relicts of ferrocolumbite (light grey, partially replaced by pyc), in contact with q (dark grey) and with mass of zircon (somewhat lighter grey than q, lower right third of the image) intensely brecciated and with faint limonite impregnation. Tiny py inclusions in q provide a high reflectance reference. CR-359, Picahy, Paraiba do Norte, Brazil (**dsp**). **b** Same scene as pyc1 (**osp**): the higher contrast in immersion makes q disappear ($R \approx 0$, black) and enhances the reddish hue of lim (IR) in zir. **c**, **d** Same scene as pyc1, (d +p) and (o+p), respectively. Scarce IR in cb-(Fe) and abundant, reddish-brown, in pyc, which appears illuminated in spite of being isotropic and in extinction; zir characterized, as a transparent mineral, by intense, very luminous IR, which here show yellowish-brown tones due to the limonitic impregnation

100.2 Microscopic Description

100.2.1 SP Observation (*Single Polariser*)

Grey *colour*. As befits a translucent mineral, dark, especially in immersion (Figs. 100.1a, b). Good *polish*.

Reflectance low (cf comparative **R** graphs), similar to cassiterite, chromite: $R_{pyc} < R$ sp, mg. Higher than gg: $R_{pyc} > q$. The value of R varies with composition between the members of the series, depending on the value of n (refractive index). Isotropic.

It usually occurs in isolated or disseminated crystals with a marked *euhedral* tendency and *octahedral forms* but also in aggregates or irregular masses. Occasional zoning. Sometimes, the decay due to contained *radioactive elements*, such as U or Th, destroys the crystalline structure, modifying the aspect of the grain (loss of brightness, formation of perlitic microcracks, internal heterogeneities).

100.2.2 Observation +P (*Polariser + Analyser at 90°*)

Isotropic. *Internal reflections* abundant and coloured in reddish to yellowish-brown tones, sometimes colourless (Fig. 100.1c, d).

100.2.3 Paragenesis

Paragenesis: zircon, cassiterite, Nb–Ta minerals, ilmenite, rutile, magnetite, quartz. Typically found in *alkaline rocks*, as well as in associated metasomatites and pegmatites. Also found in greisen and contact metamorphic formations.

100.2.4 Identification

Its identification by optical properties alone is uncertain, but once the paragenesis is known or its presence has been identified by other analytical techniques, its recognition is evident, due to its colour, low reflectance (similar to or lower than cassiterite, but higher than gangue), morphology, hardness, isotropy and internal reflections (Figs. 100.1a–d).

100.3 Spectral Information

100.3.1 Pyrochlore Multispectral Reflectance R (in Air, Without Polariser) (%)

λ (nm)	370	400	425	450	475	500	525	550	575	600	625	650	675	700	750	800	850	900	950	1000
μ	13.1	12.7	12.6	12.4	12.3	12.2	12.1	12.1	12.1	12.1	12.2	12.2	12.3	12.2	12.2	12.2	12.1	12.0	12.1	12.1
P_5	12.4	12.1	11.9	11.7	11.6	11.4	11.4	11.3	11.2	11.2	11.2	11.1	11.1	11.0	11.0	11.0	11.0	11.0	11.0	11.0
P_{20}	12.7	12.4	12.3	12.0	12.0	11.8	11.7	11.7	11.6	11.6	11.6	11.6	11.6	11.5	11.5	11.5	11.5	11.4	11.4	11.4
P_{80}	13.3	12.9	12.8	12.7	12.6	12.5	12.5	12.4	12.5	12.6	12.8	12.9	13.0	13.1	13.1	13.0	12.9	12.8	12.8	12.8
P_{95}	13.6	13.2	13.1	12.9	12.8	12.7	12.7	12.7	12.8	12.9	13.1	13.2	13.3	13.4	13.5	13.4	13.2	13.2	13.3	13.4

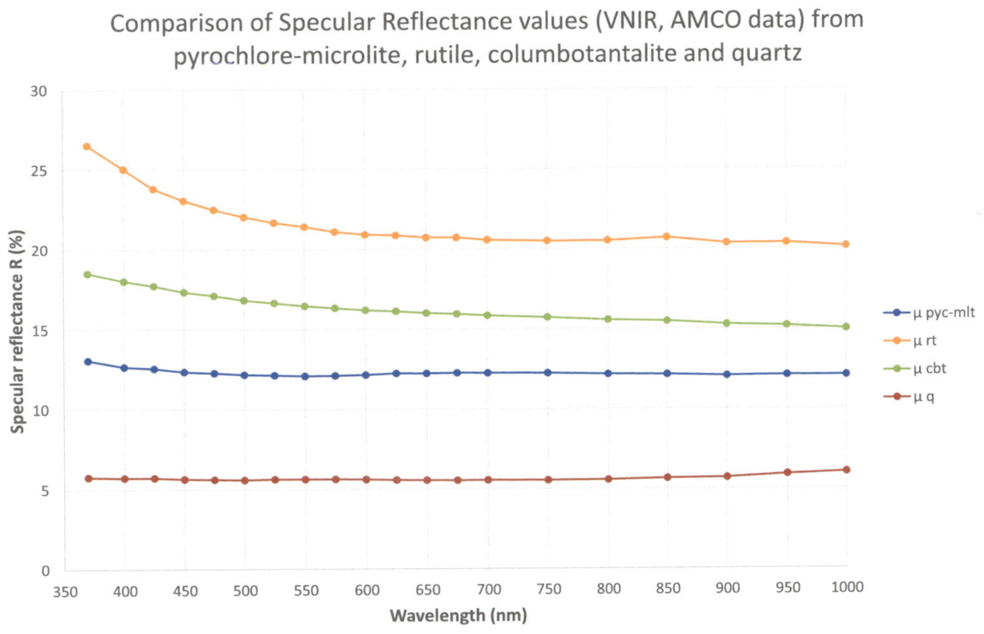

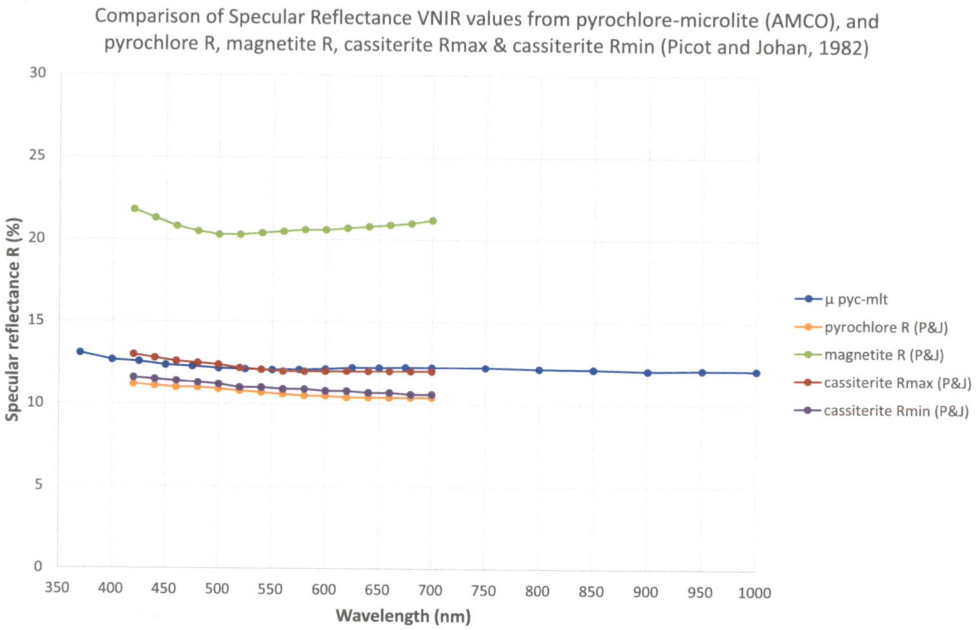

References

Atencio D, Andrade MB, Christy AG, Giere R, Kartashov PM (2010) The pyrochlore supergroup of minerals: nomenclature. Can Mineral 48:673–698

Atencio D, Andrade MB, Bastos Neto AC, Pereira VP (2017) Ralstonite renamed hydrokenoralstonite, coulsellite renamed fluornatrocoulsellite, and their incorporation into the pyrochlore supergroup. Can Mineral 55:115–120

Picot P, Johan Z (1982) Atlas of Ore minerals. B.R.G.M./Elsevier

Ramdohr P (1980) The ore minerals and their intergrowths, 2nd edn, vol 2. Pergamon Press, Oxford, p 1205

Pyrolusite (*prl/polianite*)

101

$Mn^{4+}O_2$: Tetragonal

Pyrolusite		R_{540} (%)	Y (%)	λ_d (nm)	P_e (%)	Colour sensation
Air	R_{max}	31.4	nd	nd	nd	Yellowish white
	R_{min}	18.8	nd	nd	nd	Greyish white
Oil	R_{max}	nd	nd	nd	nd	White
	R_{min}	nd	nd	nd	nd	White–grey
Bir: Medium (4)				*Pleoc*: Visible YlWh-Gr (4)		*Tarn*: No
Anis +P: Distinct, bright, yellowish (4)				*IR*: No		*Twg*: Occasionally visible
Mor: Varied, pseudom mng, etc				*Clv*: (110)		*PolQ*: Variable with texture/crystallinity
Hardness	*PolH*: Very high: >py (5)			*Mohs*: 2–6½	*Talm*: nd	*VHN* $_{(varied)}$: 76–1500

Obs: cf. Table and R charts: multispectral (VNIR, AMCO) and visible ranges (Picot and Johan 1982). VHN: Uytenbogaardt and Burke (1971)
Mineral abbreviations: chr—chromite, gn—galena, gth—goethite, mng—manganite, prl—pyrolusite, psi—psilomelane, py—pyrite, q—quartz, rom—romanechite, rt—rutile

101.1 General

Important Mn ore. Composition: MnO_2, frequently contaminated (easily incorporates impurities of other oxides and adsorbed water). It usually forms encrustations, colomorphic deposits and porous, fine-grained or friable masses; more rarely, crystalline aggregates. It is perhaps the best-known Mn mineral to the general public, but this popularity is based on an error—*pyrolusite dendrites* are actually composed of other Mn oxides.

In the literature, pyrolusite is also mentioned as *polianite*, a more recent term used to distinguish crystalline varieties, but both terms were later used in a contradictory, confusing or indistinct way by different authors.

Ramdohr (1980) argues that it is the same mineral and proposes to designate it as *polianite-pyrolusite*, reserving the name *pyrolusite s.str.* for the most common varieties, either cryptocrystalline/powdery and soft/friable, or pseudorhombic, pseudomorphs of manganite, as shown by the typical "*pyrolusite pits*" or shrinkage cracks parallel to the manganite cleavage. He suggests using *polianite s.str.*, for the crystalline varieties, lacking *pyrolusite pits and* very hard, which facilitate a better polishing quality.

This distinction explains the peculiar hardness data: while polianite shows a very high polishing hardness, the *PolH* of pyrolusite shows a huge range of variation, quantified on the Mohs scale between 1 and 7 according to the cited author, but it does not eliminate the ambiguity, as he himself acknowledges ("one variety may grade into the other one in the same sample, there may be some doubt in some cases as to which name is preferable", Ramdohr 1980, p. 1022).

Consequently, in this text and in accordance with the currently accepted nomenclature, the term polianite is dispensed with. Since it is the same mineral, the term used will be the common one—pyrolusite. The great hardness of polianite is that of crystalline pyrolusite, not friable.

101.2 Microscopic Description

101.2.1 SP Observation (*Single Polariser*)

Yellowish white, bireflective and pleochroic in yellowish and greyish tones (Figs. 101.1a, b, and 101.2a).

Reflectance, R, relatively high, the highest among the Mn oxides. However, the data in the literature vary considerably, most likely because of the usual heterogeneity or porosity of

R. Castroviejo, *A Practical Guide to Ore Microscopy—Volume 1*,
https://doi.org/10.1007/978-3-031-12654-3_101

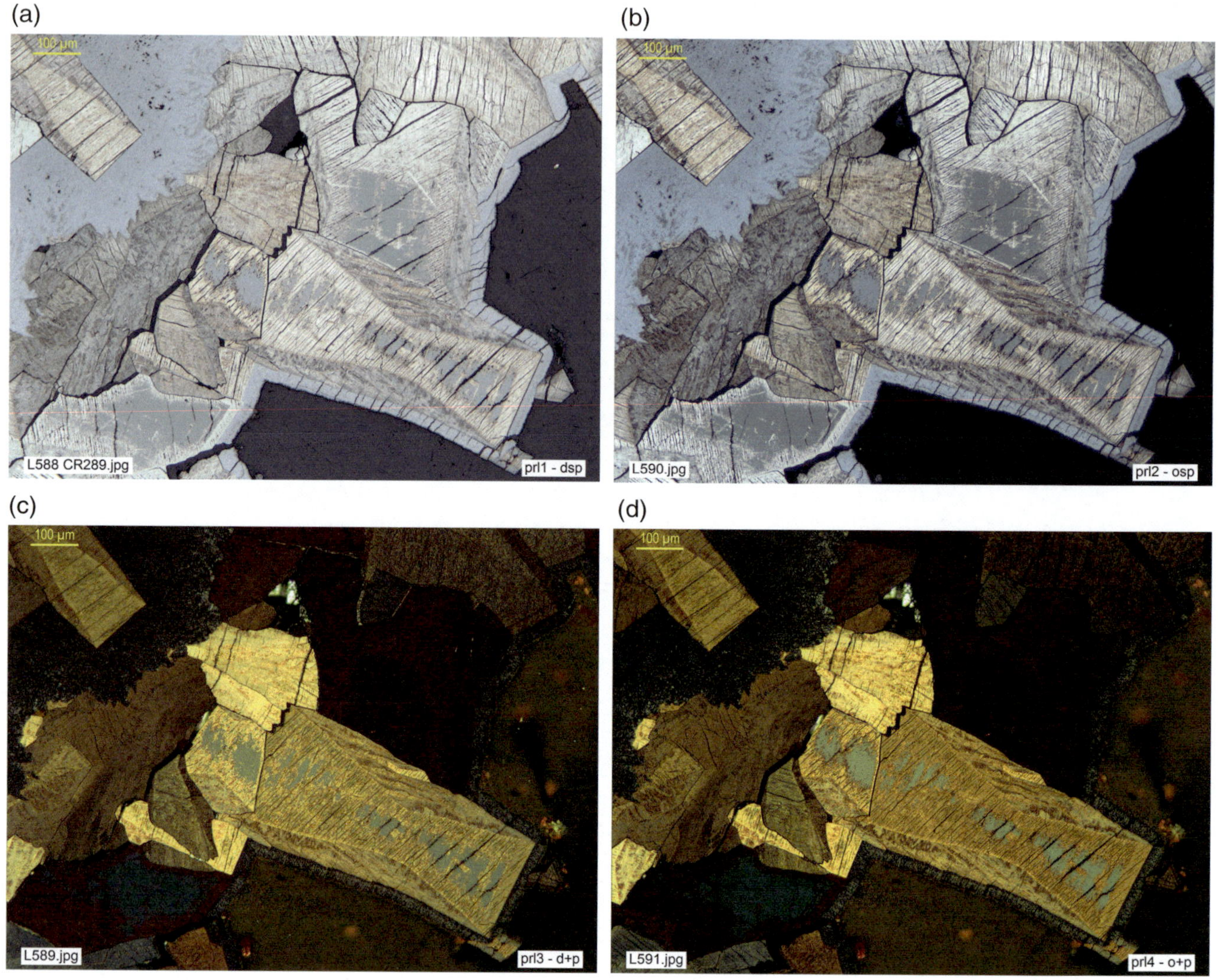

Fig. 101.1 a, b (**dsp** and **osp**, respectively)—Aggregate of prl (greyish-white, pleochroic), secondary to mng (grey relicts) and in the process of being replaced by psi (botryoidal aggregates and microcrystalline acicular rim coatings on prl, lighter grey than mng). CR 289, Amalienhohe Mine, Waldalgesheim, Bingerbrueck, Germany. **c, d** Same scene as prl1 (**d+p** and **o+p**, respectively). Bright, yellowish anisotropism, contrasting with the dull, greyish tones of the relict mng inclusions in prl, as well as the secondary psi aggregates in prl

the mineral, conditioned by its texture and genesis. Perhaps for this reason, pyrolusite is not among the minerals whose R is documented in QDF3. According to the sources, R would be in a range of 32.5–42.8% (in: Ramdohr 1980), but this author places it above galena, with values of 50–52%. In this text, the multispectral VNIR values measured with the AMCO system will be used. These coincide, in the visible spectrum, with the maximum values[1] measured by Picot and Johan (1982), ≈ 30%, and with direct observation (R prl > psi > mng, cf. Figs. 101.1a, b, 101.2a and 101.3).

Polishing hardness of crystalline varieties (former *polianite*) among the highest of all metallic ores: $PolH_{prl}$ > PolH q, py, rt, chr. In cryptocrystalline or powdery (friable) varieties, the hardness is not a reliable datum, as it usually cannot be determined with precision. Because of the hardness and/or friability, *polishing* is difficult, but in crystalline materials it can be good.

In general, prl polishing is made more difficult by the frequent transposition of the manganite cleavage in the so-called *pyrolusite pits* of pyrolusite, which are like channels or *shrinkage cracks* parallel to the manganite cleavage. This is related to the dehydration of manganite when it is altered to pyrolusite and inherited as ghost structures by prl (Fig. 101.1a, b).

[1] They also show the wide R value span found in prl, as the other AMCO set of measures is nearly like P&J's minimum R values (cf. AMCO *prl-D* values in the visible range and compare with P&J R plots).

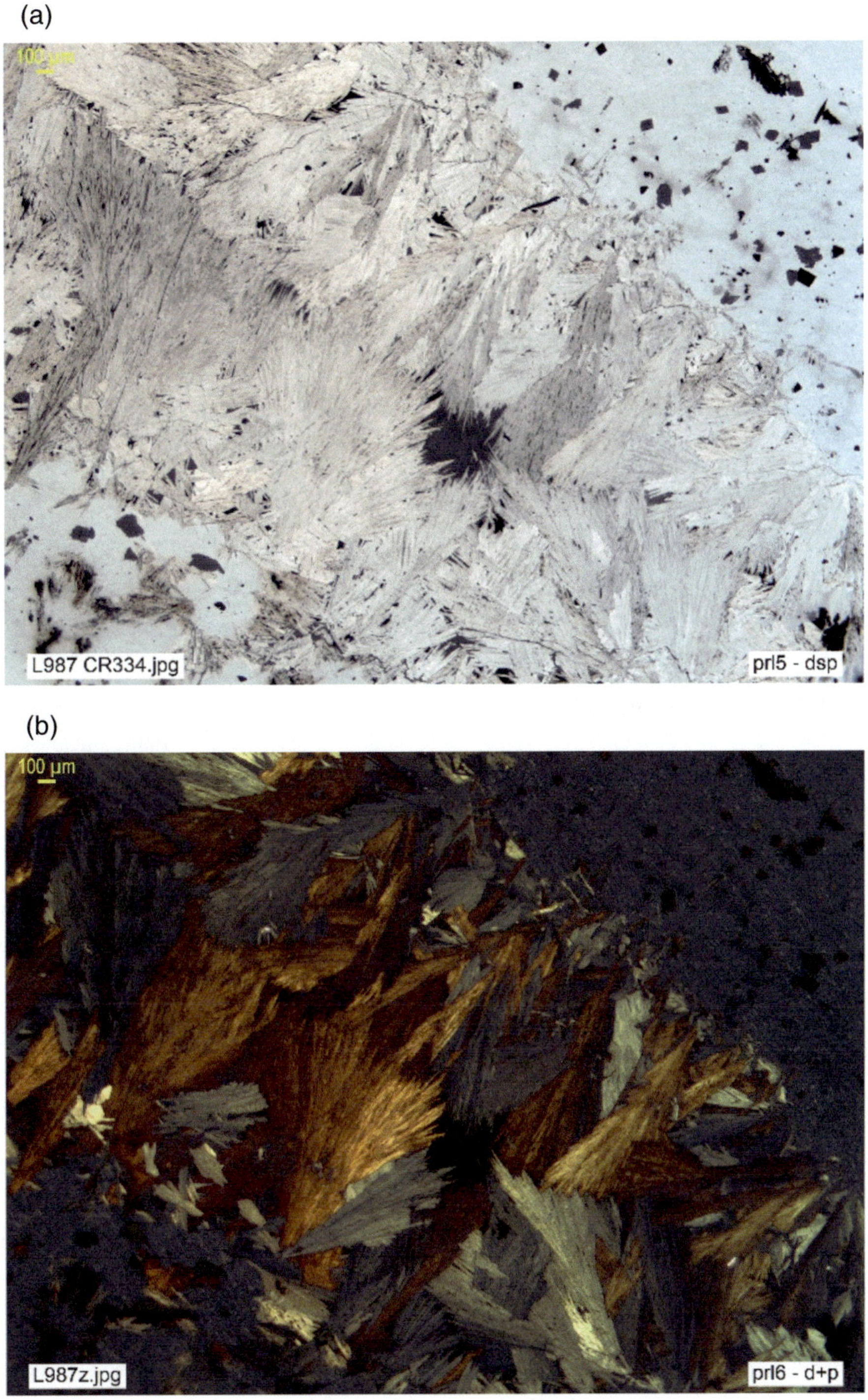

Fig. 101.2 a, b (dsp and d+p, respectively)—Fibrous-radial and prismatic-acicular texture of mng totally pseudomorphosed by prl. Then prl has been in part replaced by psi. Although the mng has disappeared, evidence of its presence remains in the inherited texture and, particularly, the prl shrinkage cracks, parallel to the mng cleavage. CR 334, Zelda (Morocco)

It usually occurs in encrustations, colomorphic aggregates and porous masses, fine-grained or powdery and friable; more rarely, in crystalline aggregates. The association with manganite, which it pseudomorphizes completely or partially and of which relicts or ghosts are often visible, is typical and frequent. Pyrolusite is also occasionally replaced by other oxides, such as psilomelane (Figs. 101.1a, b, 101.2a and 101.3).

101.2.2 Observation +P (*Polariser + Analyser at 90°*)

Clearly *anisotropic*, in typical yellowish luminous tones (Figs. 101.1c, d and 101.2b).

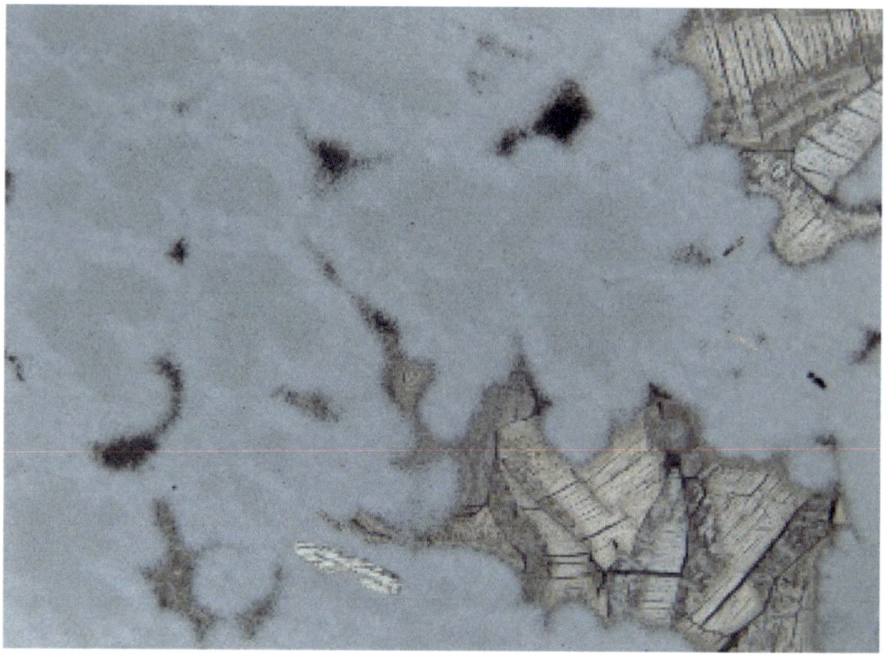

Fig. 101.3 Advanced replacement of prl, of which only relicts remain, by botryoidal psi. CR 289, Amalienhohe Mine, Waldalgesheim, Bingerbrueck, Germany (**dsp**)

101.2.3 Paragenesis

A frequent mineral in oxidizing environments common on the earth's surface, it occurs in sedimentary, continental, lacustrine and shallow marine deposits, often with colloidal or bacterial precipitation. It is also formed by weathering, concentrating either in residual, infiltrative or transported deposits.

Usual companions: manganite, other Mn oxides, goethite. It is typically associated with manganite, which it replaces, as it can also do with other primary Mn minerals (alabandite, braunite, rhodochrosite, rhodonite, etc.), sometimes producing economic concentrations, by secondary enrichment, from deposits of low grade.

101.2.4 Identification

The paragenesis is an important data. Pyrolusite is easily recognized by:

- its high reflectance, higher than that of the accompanying manganese oxides,
- its colour and pleochroism,
- its pseudomorphism after manganite, attested by relicts of manganite or by the unmistakable shrinkage cracks (*pyrolusite pits*),
- its luminous anisotropism in dominant yellow tones, and
- if the grain size allows the determination, by its great hardness.

However, its identification is problematic when it appears in micro- to cryptocrystalline masses, often associated with goethite or limonite, where it can be confused with psilomelane.

101.3 Spectral Information

101.3.1 Pyrolusite Multispectral Reflectance R (in Air, Without Polariser) (%)

λ (nm)	370	400	425	450	475	500	525	550	575	600	625	650	675	700	750	800	850	900	950	1000
μ	27.8	28.8	29.6	30.3	30.7	30.9	31.0	30.7	30.5	30.3	30.1	29.8	29.6	29.3	29.0	28.7	28.6	28.0	27.7	27.2
P_5	23.0	23.8	24.7	25.2	25.7	26.0	26.5	26.4	26.4	26.4	26.3	26.2	26.1	25.9	25.9	25.8	25.8	25.2	24.8	24.4
P_{20}	25.6	26.5	27.5	28.1	28.7	28.8	29.2	28.9	28.8	28.6	28.5	28.2	28.1	27.9	27.6	27.5	27.5	26.8	26.4	25.9
P_{80}	30.4	31.6	32.0	32.9	32.9	33.2	33.0	32.8	32.5	32.1	31.9	31.4	31.2	30.9	30.4	30.0	29.9	29.4	29.2	28.7
P_{95}	31.9	33.0	33.5	34.4	34.3	34.5	34.2	34.0	33.7	33.3	33.0	32.5	32.3	32.0	31.4	31.0	30.8	30.3	30.0	29.6

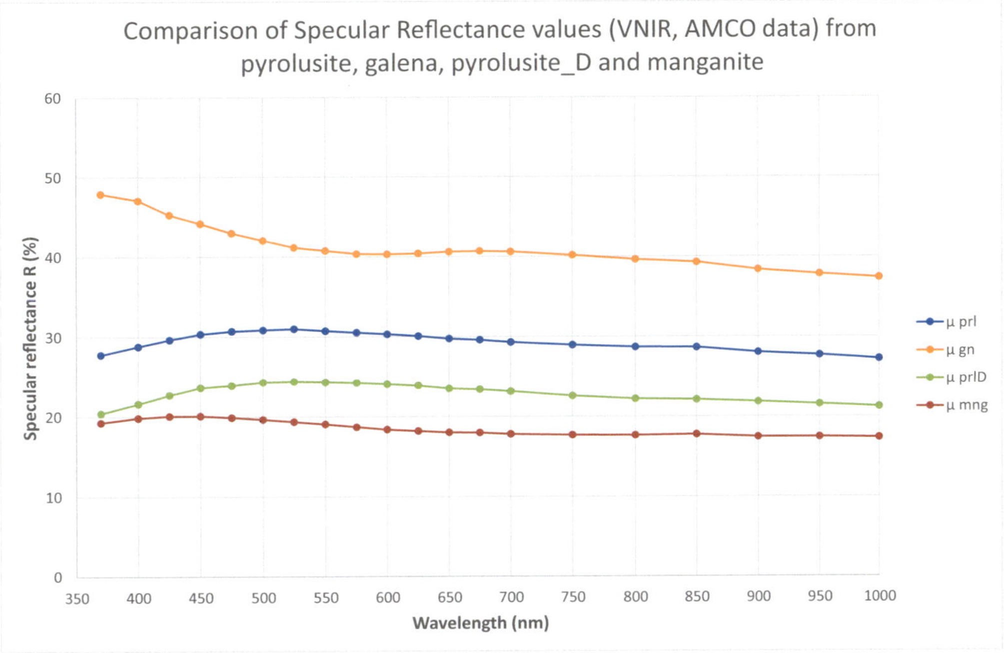

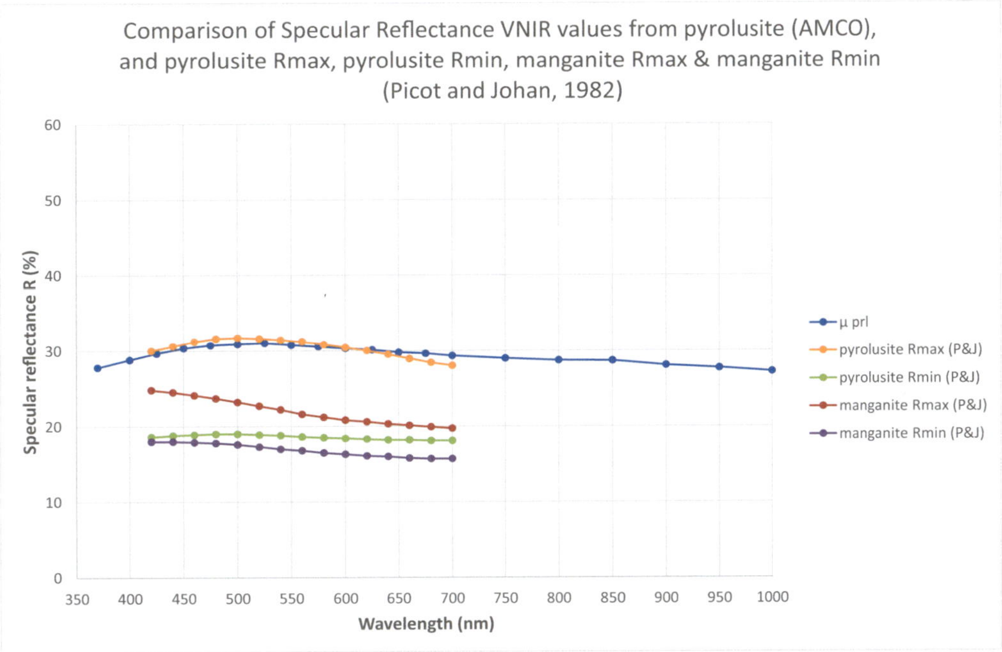

References

Picot P, Johan Z (1982) Atlas of ore minerals. B.R.G.M./Elsevier

Ramdohr P (1980) The ore minerals and their intergrowths, 2nd edn. Pergamon Press, Oxford. 2 volumes, 1205p

Uytenbogaardt W, Burke EAJ (1971) Tables for microscopic identification of ore minerals, 2nd edn. Elsevier, Amsterdam

Pyrrhotite (*po/pyrrhotite*) 102

$Fe_{1-x}S$ (x = 0 to 0.17): Hexagonal/Monoclinic

Pyrotite		R_{546} (%)	Y (%)	λ_d (nm)	P_e (%)	Colour sensation
Air	R_1	36.3	36.9	580	10.2	Pinkish-brown to yellowish-brown, slightly pleochroic
	R_2	41.4	41.8	578	9.4	
Oil	R_1	23.5	24.1	580	14.4	Pinkish brown to yellowish brown, distinctly pleochroic
	R_2	28.3	28.6	578	12.5	
Bir: Medium (4)				*Pleoc*: Medium (4)		*Tarn*: Frequent (brown)
Anis +P: Distinct (4)				*IR*: No		*Twg*: Lamellar, deformation, etc
Mor: Anhedral and granoblastic agrs				*Clv*: // (0001), (10$\bar{1}$0)		*PolQ*: Good
Hardness	*PolH*: Medium–high (3–4)			*Mohs*: 3.5–4	*Talm*: D	*VHN* $_{(100\ g)}$: 258–378 (average 332)

Obs: cf. Table and R charts, multispectral data (VNIR, AMCO) and visible range (QDF3)
Mineral abbreviations: Ag—native silver, agp—argentopyrite, Au—native gold, asp—arsenopyrite, bn—bornite, brv—bravoite, cbn—cubanite, ccp—chalcopyrite, chr—chromite, PGE—Pt-group elements, gg—gangue, ht—hematite, il—ilmenite, isc—isocubanite, mc—marcasite, mt—magnetite, pn—pentlandite, po—pyrrhotite, py—pyrite, sp—sphalerite, stb—sternbergite, vio—violarite

102.1 General

Crystallochemical investigation and the application of XRD have revealed significant compositional variations and crystalline symmetry in pyrrhotite that make it a mineral of remarkable complexity.

The traditionally accepted formula, the stoichiometric *FeS* (without Fe vacancies, non-magnetic), is rare in nature. It is found mainly in meteorites (*troilite*, cf. Figs. 64.1 to 64.3, with brief characterization of troilite, below).

In nature there are several *polytypes*, characterized by Fe deficits expressed by the nowadays accepted formulation, $Fe_{1-x}S$ (x ≈ 0.1–02), to which correspond different symmetries, among them, 11C, 11H, 4C, etc., *respectively,* orthorhombic, hexagonal, monoclinic.

While troilite is paramagnetic, vacancies in the structure of the Fe-deficient types give them ferromagnetic properties. However, from the optical point of view, these differences do not translate into diagnostic properties that would allow a reliable distinction of the different polymorphs under the reflection microscope; therefore, only one description of pyrrhotite is given. The type quantified in the introductory table is the monoclinic-4C, according to QDF3 data.

Compared to the di-sulphides pyrite and marcasite, FeS_2, the S content of pyrrhotite is ≈ half, so it tends to occur in deep deposits or linked to metamorphic devolatilization. Another consequence of pyrrhotite's composition and structure is its instability in shallow conditions, particularly under weathering. This makes it an undesirable companion for the miner, as it accelerates the production of acidic water in mines and waste dumps, forcing the intensification and urgency of *environmental protection* measures. In this respect, the *microscopist* can provide the plant engineer with *valuable early information* to anticipate the problem, take timely action and avoid irreparable damage.

R. Castroviejo, *A Practical Guide to Ore Microscopy—Volume 1*,
https://doi.org/10.1007/978-3-031-12654-3_102

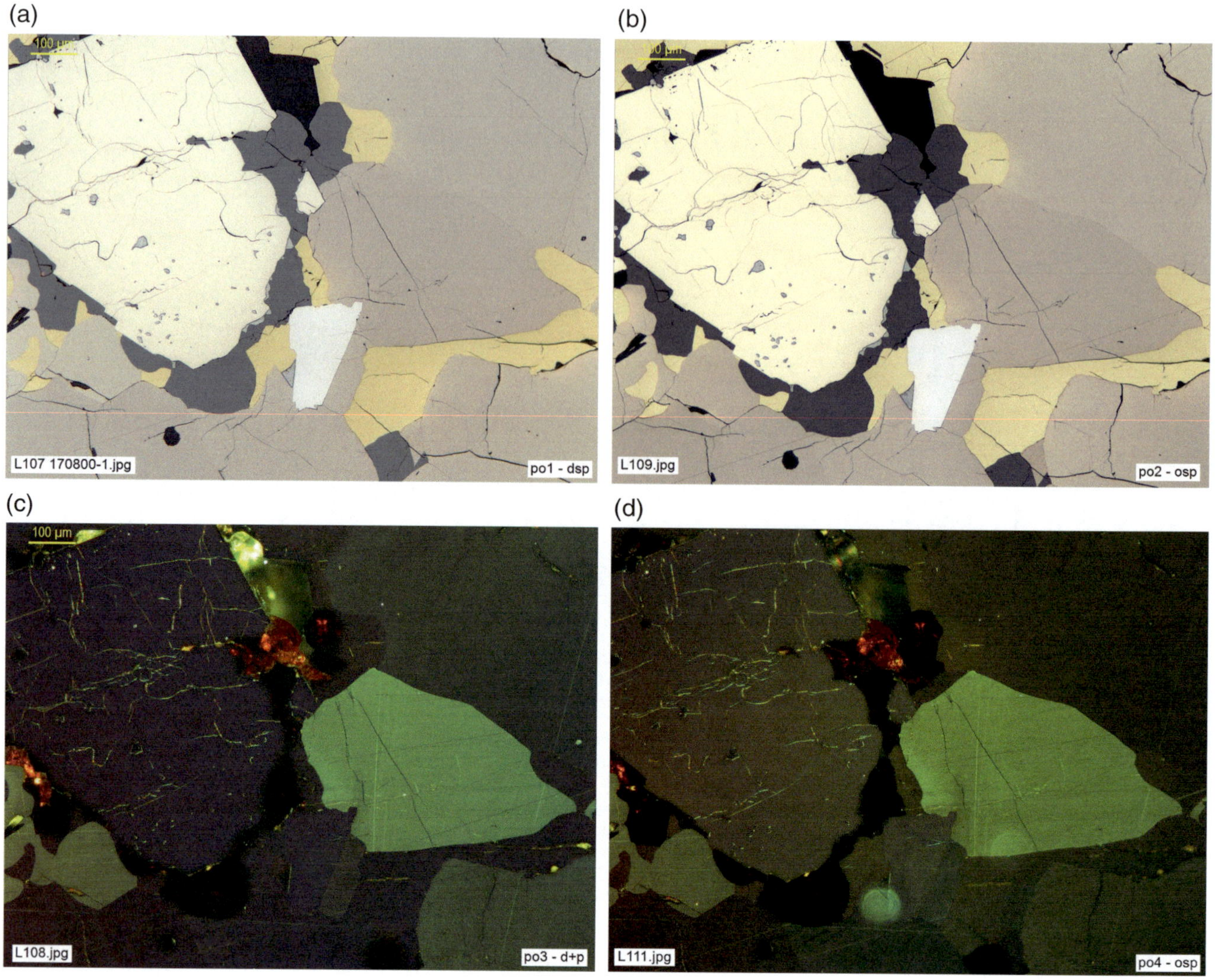

Fig. 102.1 **a** Aggregate of po (pinkish brown), py (yellowish-white), ccp (yellow), sp (grey), asp (white), gn (light grey, py inclusions) and gg (quartz, very dark grey). Sample 170800_1, Sulitjelma Mine, Norway (**dsp**). **b** Same scene as pol (**osp**): strong decrease of sp reflectance and higher sharpness of po pleochroism. **c**, **d** Same scene as pol (**d+p**) and (**o+p**), respectively: anisotropism of po and reddish internal reflections of sp (ferriferous); the anisotropism of asp and ccp is not seen because they are in extinction positions

102.2 Microscopic Description

102.2.1 SP Observation (*Single Polariser*)

Pinkish-brown *colour*, bright, but duller than pyrite or arsenopyrite. Characteristic pleochroism between darker reddish tones (R_1) and lighter yellowish tones (R_2), visible in air and very clear in oil (Figs. 102.1a, b and 102.2). With time it darkens, due to tarnishing.

Good *polish*. *Cleavage* // (0001) frequent (enhanced by alteration), // (10$\bar{1}$0) occasional.

Moderately high *reflectance*: R po clearly > R sp or mt but < R ccp, gn, py, pn; << R Au, Ag (cf multispectral comparative plots and Figs. 102.4 and 102.5).

Polishing hardness in the medium–high transition: PolH po > pn, sp, ccp; < mt, asp, py.

It usually occurs in allotriomorphic or granoblastic granular aggregates, with simple contacts between grains; more rarely in tabular idiomorphic crystals // (0001). *Inclusions* (pentlandite, magnetite, chalcopyrite) are frequent, especially in nickeliferous pyrrhotite. This one usually presents unmixed pentlandite, forming the typical *flames*, as well as varied inclusions, spindles, intergranular films, etc., whose distribution is conditioned by crystalline directions of pyrrhotite, contacts between grains, interstitial spaces or microcracks (cf Figs. 93.1 and 93.3). It is not rare, on the other hand, to find inclusions of pyrrhotite in other minerals, such as sphalerite or chalcopyrite.

Fig. 102.2 Granoblastic po aggregate (pinkish-brown) with interstitial sp and ccp and euhedral asp inclusions. Sample 170800_1, Sulitjelma Mine, Norway (**dsp**)

Fig. 102.3 Same scene as po5 (**d +p**): the wide-field image (obj. 2.5X) produces imperfect extinction, but the distinct anisotropism of po enhances the granoblastic texture

Pyrrhotite is prone to *alterations,* frequently supergene by oxidation in natural shallow environments but also hypogene, and it is commonly replaced by marcasite, pyrite, bravoite, violarite, magnetite. These alterations sometimes mirror crystalline directions of pyrrhotite, such as cleavage (0001), inherited by oriented aggregates of marcasite ± pyrite (Figs. 102.6 and 102.7). In other cases, the secondary product is composed of finely intergrown pyrite, marcasite and magnetite.

Typical aggregates, extremely fine-grained and with concentric structures, are also generated by alteration, the so-called *birds' eye* aggregates, arranged // (0001) and composed of a fine intergrowth of pyrite and marcasite, very characteristic of pyrrhotite (Fig. 102.5). The material that composes these *birds' eye* structures cannot always be defined under the optical microscope. At times, only a greyish cryptocrystalline aggregate is observed, powdery in appearance but with strong anisotropism (luminous yellowish to bluish-green tones), defined by Ramdohr (1955) as *Zwischenprodukt* (*intermediate product*). XRD investigations cited by Ramdohr (1980) detect the presence of marcasite (± pyrrhotite) and suggest an intermediate phase between pyrite and marcasite.

Fig. 102.4 Impregnation of po (anhedral), asp (euhedral) and Au (*fineness* 950 or 95% Au, 5% Ag, EMPA) in gg of quartz. Sample CR-295, La Rinconada Mine, Puno, Peru (**osp**)

Fig. 102.5 Quartz vein, with asp, po totally altered (*birds' eye*) and Au interstitial and in small inclusions in q. Note the distinct yellow colour of the Au (EMP analysis: 95% au and 5% Ag). Same sample as po7 (**dsp**)

102.2.2 Observation +P (*Polariser + Analyser at 90°*)

Strongly *anisotropic*, especially in immersion, varied polarization colours, with yellowish-grey, greenish, brownish, bluish-grey, reddish tones, Figs. 102.1c, d, and 102.3. Distinct extinction, except with wide field objectives (susceptible to introduce parasitic reflections).

Lamellar *twins*, in most cases a possible effect of deformation, are usually observed by translation or shearing // (0001). Lamellar inversion twins (or hexagonal-monoclinic transformation) are rare. Another type of lamellar forms, with peculiar discoidal, fusiform or flamboyant outlines, is attributed by Ramdohr (1980) to decomposition of high-temperature pyrrhotite into two distinct phases, which are differentiated mainly by chemical attack, but also by

Fig. 102.6 Pseudomorphs of granoblastic po replaced by mc (bluish-white) and py (yellowish-white), finely intergrown with the few preserved traces of primary po (light pinkish brown relicts). Further partial alteration of the assemblage to vio (cryptocrystalline, greyish-brown powdery mass), diffusing from grain boundaries and microcracks. Sample HD6771, Horbach, Schwarzwald, Germany (**osp**)

Fig. 102.7 Same scene as po9 (**o +p**). The striking blue-green anisotropism of mc highlights the orientation (// [0001] of po) of the secondary aggregates; distinct anisotropism of po and weak (anomalous) anisotropism of py

careful (+P) observation. They should not be confused with the exsolution of pentlandite *flames* in high-temperature pyrrhotite with excess Ni.

102.2.3 Paragenesis

This mineral is frequently found in various high-temperature, magmatic or metamorphic deposits. Among the former, the liquid-magmatic deposits of Cr-Ni and PGE related to mafic and ultramafic rocks are to be mentioned; also meteorites (troilite). Among the latter, orogenic gold deposits or contact metasomatic deposits. Moreover, po occurs in pegmatitic-pneumatolitic and hydrothermal formations as well.

Its possible formation at low temperatures is questioned, although its presence in sedimentary formations has been mentioned. However, it is typically found in volcano-sedimentary exhalative deposits that have undergone metamorphism (*vbgr* the copper-bearing pyrrhotite belt of Arinteiro, Fornás, Bama, Santiago de Compostela, or Röros, Lökken, Sulitjelma in Norway), but in these cases it is a high-temperature, metamorphogenic mineral.

Paragenesis: pyrite, chalcopyrite, cubanite, pentlandite, magnetite, ilmenite, arsenopyrite, marcasite, hematite, etc.

102.2.4 Identification

The recognition of pyrrhotite is immediate for an attentive observer by its diagnostic properties: brownish-pinkish-yellowish colour and pleochroism, moderately high reflectance and hardness, distinct and coloured anisotropism, alterations. It is sometimes confused with minerals such as cubanite, isocubanite, pentlandite, bornite, argentopyrite-sternbergite, but mistakes can be avoided by careful observation:

- Cubanite is softer and its polarization colours, bluish, are different.
- Isocubanite is softer and isotropic.
- Pentlandite is isotropic, brighter and shows octahedral cleavage.
- Bornite is much softer, quickly develops a dark patina and its anisotropism is much weaker; there seems to be an antipathy (*incompatibility*) between pyrrhotite and bornite (Gilbert 1925), so they would not appear together.
- Argentopyrites, argentopyrite-sternbergite are softer, darker, show a strong anisotropism even more vivid than pyrrhotite and are associated with argentiferous paragenesis.

The distinction between the different types of pyrrhotite requires the use of complementary XRD techniques.

102.3 Troilite: Brief Microscopic Characterization[1]

Troilite, *tr*, is stoichiometric FeS, of hexagonal symmetry, without the Fe deficit (vacancies) of pyrrhotite ($Fe_{1-x}S$). Troilite is paramagnetic and pyrrhotite, ferromagnetic, varying according to the different polymorphs (monoclinic or hexagonal symmetry). Although troilite is considered by some to be a pyrrhotite polytype, it maintains its status as an IMA *grandfathered* species.

The optical properties and parameters of troilite are (according to QDF3 data) similar to those of pyrrhotite, but its *reflectance*, ***R***, is slightly lower and it looks a little darker, as shown by the QDF3 data (Criddle and Stanley 1993, pp. 467 and 483):

- *Y % pyrrhotite* oriented after $R_1$and R_2: 36.9 and 40.5 'air', *respectively*; 23.6 and 27.1 'oil', *respectively*;
- *Y % troilite* oriented after R_1 and R_2: 33.9 and 38.7 'air', *respectively*; 22.1 and 26.3 'oil', *respectively.* Darkening is accentuated with time by its tendency to tarnish quickly.

Its colour and pleochroism (Figs. 64.1 to 64.3), in pinkish-brown to yellowish-brown tones as pyrrhotite, are characterized by identical λ_d (λ_d for R_1: 580 nm and λ_d for R_2: 578 nm, both in air and oil for both minerals), but higher P_e:

- P_e *% pyrrhotite (p. 467)*, oriented after R_1 and R_2: 9.7 and 8.9 'air', *respectively*; 14.1 and 12.3 'oil', *respectively*;
- P_e *% troilite*, oriented after R_1 and R_2: 14.6 and 11.2 'air', *respectively*; 18.1 and 13.2 'oil', *respectively*).

Its distinct anisotropism is, as in pyrrhotite, characteristic (Figs. 64.2 and 64.3, although here it is hardly visible because of the unavoidable light overexposure of the immediate surroundings due to the abundant internal reflections of the gangue).

Given the similarity of both minerals, the reader is referred to the description of pyrrhotite to illustrate the remaining properties of troilite and for its microscopic identification.

Troilite is typically found in meteorites, hence its synonym *cosmic pyrrhotite* , and its presence in terrestrial rocks —serpentinites, ultramafic complexes—is exceptional. The opposite occurs with pyrrhotite. The presence of troilite has also been observed in metallurgical slags and coal ashes. According to modern data from space probes, troilite could be a frequent mineral on some planets or satellites of the solar system, such as Mars or Ganymede.[2]

[1] The characterization of troilite is synthesized here, as a brief appendix that extends the information about pyrrhotite, as corresponds to the close relationship of both species. However, for practical reasons (so as not to repeat photomicrographs), the images cited in this description (Figs. 64.1–64.4a) are found with the description of iron (cf Fe), since the main subject of these images corresponds to Fe and other components that accompany it in meteorites (among them, troilite).

[2] More information: https://www.mindat.org/show.php?id=4029&ld=2&pho=

102.4 Spectral Information

102.4.1 Pyrrhotite Multispectral Reflectance R (in Air, Without Polariser) (%)

λ (nm)	370	400	425	450	475	500	525	550	575	600	625	650	675	700	750	800	850	900	950	1000
μ	28.1	29.0	30.0	31.4	32.7	34.4	35.7	37.2	38.3	39.4	40.4	41.3	42.2	43.0	44.4	45.5	46.9	47.6	48.2	48.9
P_5	24.3	25.1	26.0	27.2	28.8	30.7	32.3	33.9	35.2	36.5	37.4	38.4	39.1	39.7	40.4	40.9	41.9	42.4	43.1	44.0
P_{20}	26.3	27.3	28.4	29.4	30.9	32.3	34.0	35.4	36.5	37.7	38.5	39.4	40.1	40.8	42.0	42.9	44.1	44.7	45.4	46.2
P_{80}	29.4	30.9	31.7	33.6	34.7	36.7	37.8	39.4	40.5	41.6	42.5	43.3	44.2	45.1	46.5	47.8	49.2	49.8	50.4	51.0
P_{95}	31.4	32.4	33.5	35.2	36.5	38.2	39.6	41.0	42.2	43.4	44.4	45.4	46.3	47.1	48.6	49.7	51.0	51.6	52.0	52.7

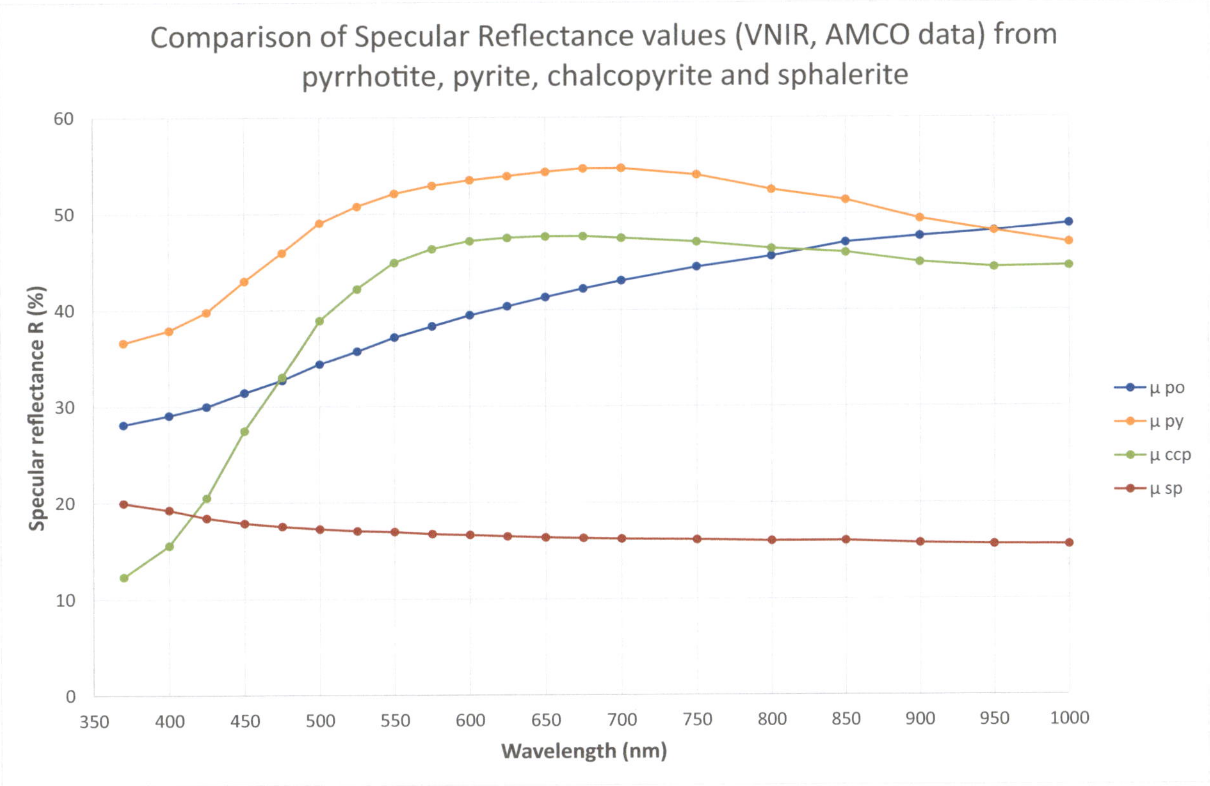

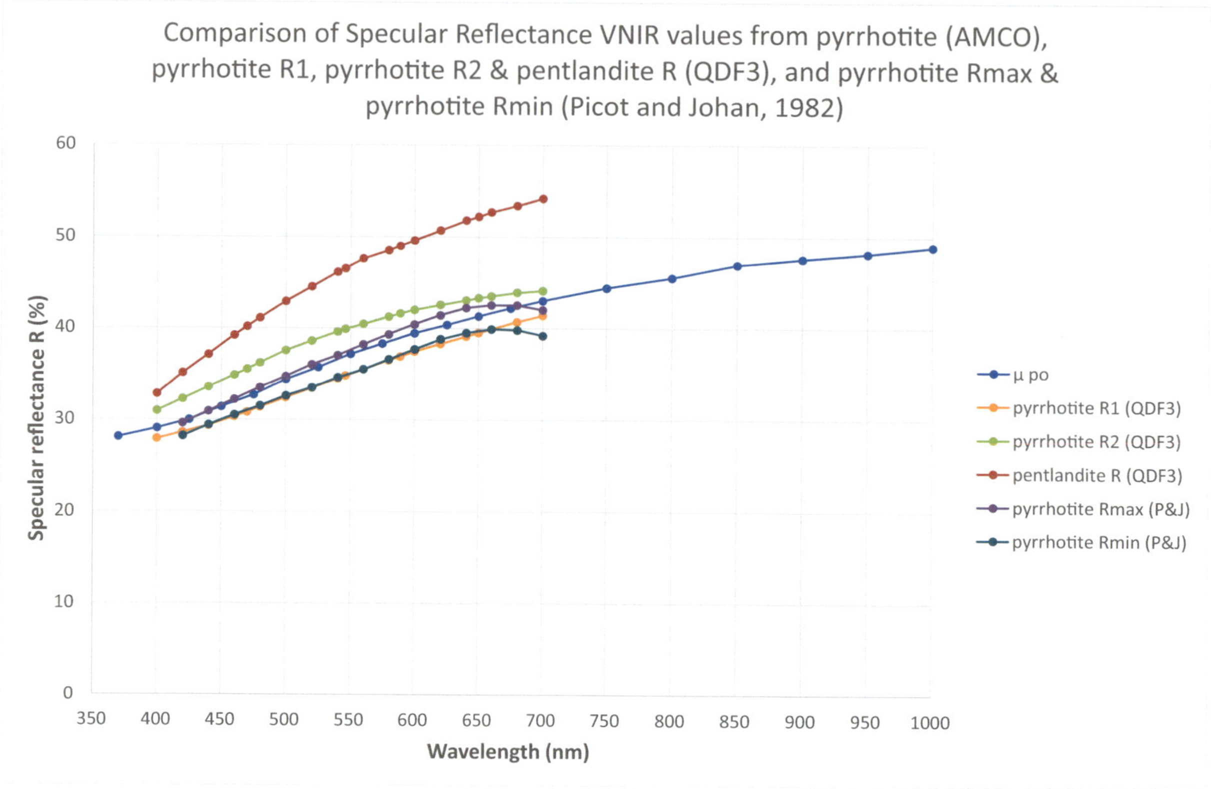

References

Criddle AJ, Stanley CJ (eds) (1993) Quantitative data file for Ore minerals, 3rd edn. Chapman & Hall, London, p 635

Gilbert G (1925) The antipathy of bornite and pyrrhotite. Econ Geol 20:364–370

Ramdohr P (1955) Die Erzmineralien und ihre Verwachsungen. Akademie Verlag, Berlin, p 875

Ramdohr P (1980) The ore minerals and their intergrowths, 2nd edn, vol 2. Pergamon Press, Oxford, p 1205

103 Rammelsbergite (*rm*)

$NiAs_2$: Orthorhombic

Rammelsbergite		R_{546} (%)	Y (%)	λ_d (nm)	P_e (%)	Colour sensation
Air	R_1	56.8	56.8	583	0.2	Bright white
	R_2	60.9	60.7	492	0.4	
Oil	R_1	43.1	43.0	553	0.2	
	R_2	48.1	47.9	482	1.4	
Bir: Weak (3)				*Pleoc*: Weak (3)		*Tarn*: Appears in humid environments
Anis +P: Strong, bright colours (5)				*IR*: –		*Twg*: Very frequent, polysynthetic
Mor: Intergrown aggregates				*Clv*: Rarely visible		*PolQ*: Very good
Hardness	*PolH*: Medium: > sk, < sf, po (3)			*Mohs*: 6^-	*Talm*: E	*VHN* $_{(100\ g)}$: 630–758
Obs: cf. Table and R charts, multispectral data (VNIR, AMCO) and visible range (QDF3) Mineral abbreviations: asp—arsenopyrite, cob—cobaltite, gd—gudmundite, gf—gersdorffite, gld—glaucodot, ln—linnaeite (group), lo—löllingite, nc—nickeline, prm—pararammelsbergite, po—pyrrhotite, py—pyrite, rm—rammelsbergite, sf—safflorite, sk—skutterudite, ul—ullmannite						

103.1 General

Composition: $NiAs_2$, may contain as impurities Co, Se, S, Fe, Sb. Trimorph with pararammelsbergite and krutovite.

103.2 Microscopic Description

103.2.1 SP Observation (*Single Polariser*)

Pure bright white, slightly yellowish against safflorite (Fig. 103.2). High *reflectance*, slightly > sf, sk (Fig. 103.2). *Weak pleochroism*, weaker than safflorite, hardly visible in air (white to bluish-white), clearer in immersion (Fig. 103.1a, b).

Polishes very well. Medium *polishing hardness* ($PolH_{rm}$ > nc, sk, ln; but $PolH_{rm}$ < sf, po, lo, asp, and << py).

It usually occurs in finely intergrown aggregates; rarely idiomorphic (isolated prismatic crystals).

103.2.2 Observation +P (*Polariser + Analyser at 90°*)

Strongly *anisotropic*, with bright and luminous polarisation, in bluish (or bluish-green, in air) and brownish to violet-blue tints, sharper in immersion (compare Figs. 103.1c, d and 103.3).

Very frequent *polysynthetic twins* (mineral characteristic: Figs. 103.1c, d and 103.3). No *internal reflections*.

103.2.3 Paragenesis

It typically occurs in hydrothermal veins of the *five-element formation* type (Ni–Co–Bi–Ag–U).

R. Castroviejo, *A Practical Guide to Ore Microscopy—Volume 1*,
https://doi.org/10.1007/978-3-031-12654-3_103

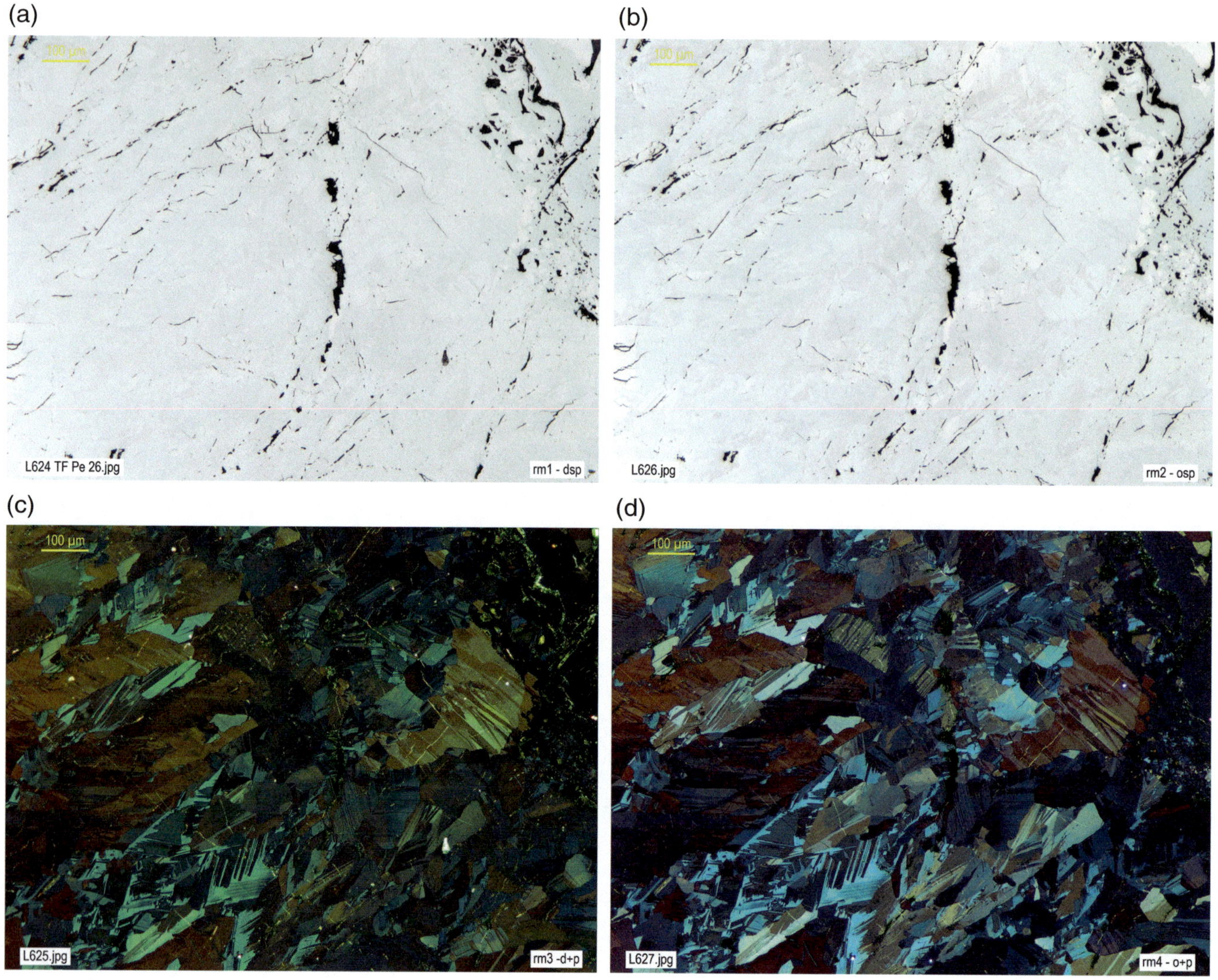

Fig. 103.1 **a** Aggregate of rm, with minor sk (bluish white, slightly darker, in top right corner). TF Pe26, ETSIME, Madrid (**dsp**). **b** Same scene as rm1 (**osp**): bireflectance/pleochroism of rm clearly perceptible. **c** Same scene as sf1 (**d+p**): isotropic character of sk and strong anisotropism of rm, with typical bluish, bluish-green and brownish colours, and polysynthetic twins. **d** Same scene as sf1 (**o+p**): brighter bluish and violet or purplish-blue to brownish polarisation colours of rm

It is usually accompanied by other arsenides such as pararammelsbergite, nickeline, safflorite, skutterudite, Ag and Bi minerals, cobaltite, etc.

103.2.4 Identification

By its appearance, rm resembles other species of the so-called *white arsenides* (mainly skutterudite, löllingite, safflorite, rammelsbergite, pararammelsbergite) or even of the sulphoarsenides/sulphoantimonides (gersdorffite, arsenopyrite, glaucodot, cobaltite, gudmundite, ullmannite) and sulphides (linnaeite), with which a hasty observer could confuse it.

- For rm diagnosis, some useful characteristics should be emphasized: its white colour, weak pleochroism (but visible in immersion), high reflectance, medium hardness, very good polishing and strong anisotropism, with bright and luminous polarisation colours and showy bluish tints.
- The very frequent polysynthetic twins distinguish it from pararammelsbergite, as well as the polarisation colours.
- Skutterudite, linnaeite and gersdorffite are isotropic; arsenopyrite, löllingite and safflorite are harder.
- The weak tendency to idiomorphism (preference for masses or aggregates of anhedral, finely intergrown grains) and the paragenesis (association with other Ni and Co minerals) are additional criteria for identifying rammelsbergite.

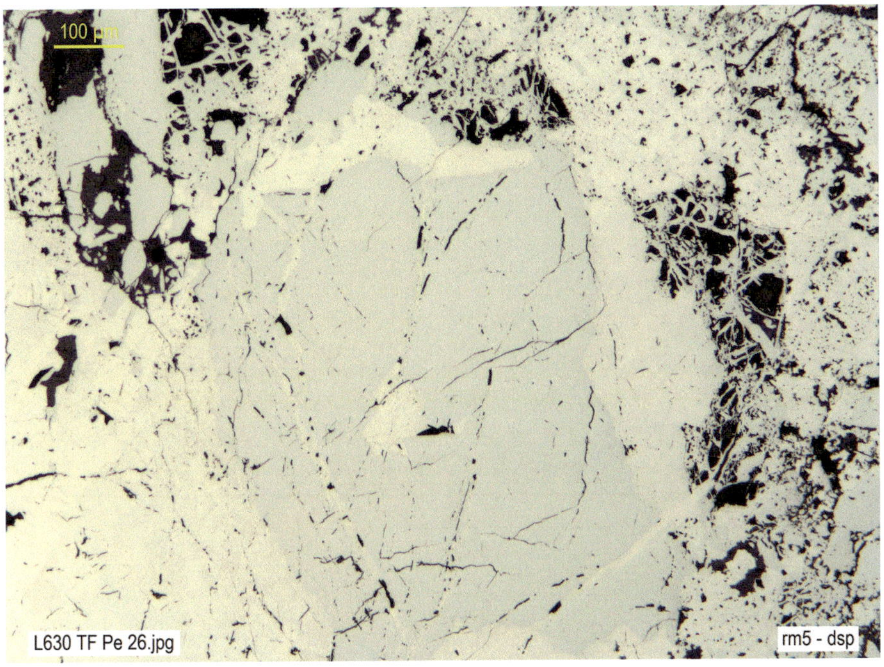

Fig. 103.2 Massive rm (yellowish white) which includes and cements microclasts of sf (bluish white or greyish white, by contrast as it is less reflective than rm). TF Pe26, ETSIME, Madrid (**dsp**)

Fig. 103.3 Microcrystalline aggregate of rm, comprising intergrown anhedral grains, twinned and strongly anisotropic (dominant blue tints), with relicts of sk (isotropic, dark; top right). HD 6784, Schneeberg, Saxony, Germany (**osp**)

103.3 Spectral Information

103.3.1 Rammelsbergite Multispectral Reflectance R (in Air, Without Polariser) (%)

λ (nm)	370	400	425	450	475	500	525	550	575	600	625	650	675	700	750	800	850	900	950	1000
μ	55.1	56.7	56.8	57.0	57.0	57.0	57.3	57.5	57.6	57.3	57.2	56.9	56.7	56.6	56.5	56.1	55.9	54.5	53.2	53.0
P_5	52.4	54.9	54.9	55.2	55.2	55.3	55.7	56.0	56.1	56.0	55.9	55.7	55.7	55.6	55.6	55.2	55.0	53.7	52.4	52.2
P_{20}	53.9	55.8	55.8	56.1	56.1	56.2	56.5	56.7	56.9	56.7	56.5	56.3	56.1	56.0	56.0	55.6	55.5	54.1	52.8	52.6
P_{80}	56.3	57.5	57.7	58.0	57.9	57.9	58.1	58.2	58.3	58.0	57.8	57.5	57.4	57.2	57.1	56.6	56.3	54.9	53.7	53.5
P_{95}	57.0	58.0	58.2	58.4	58.3	58.3	58.5	58.6	58.6	58.4	58.2	57.8	57.7	57.6	57.5	57.1	56.7	55.3	54.1	53.9

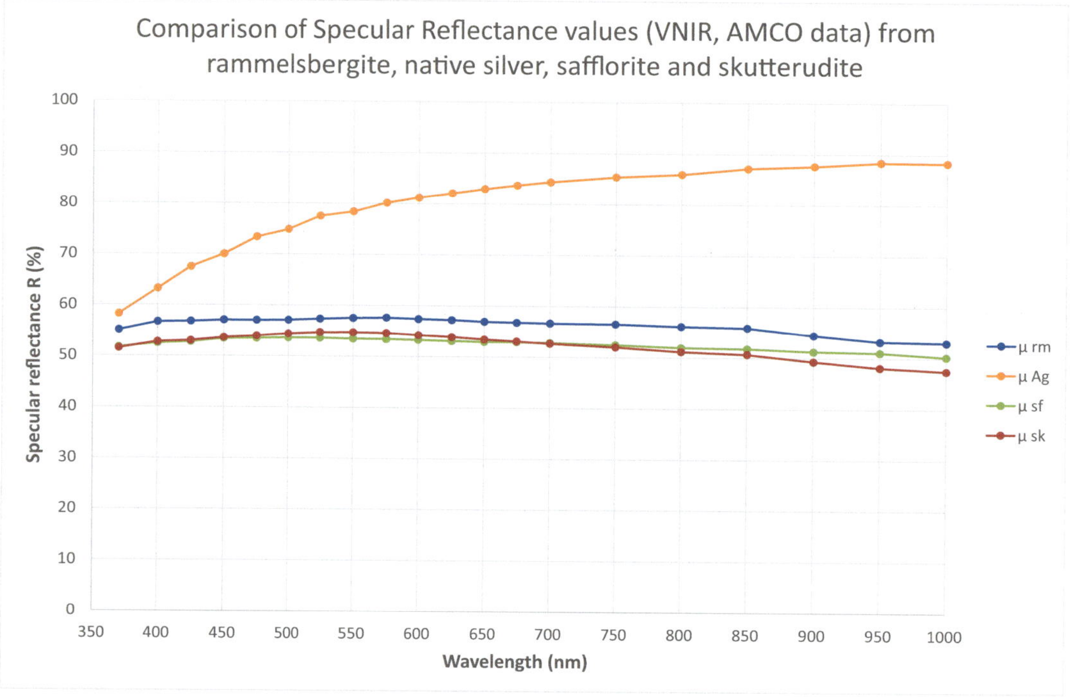

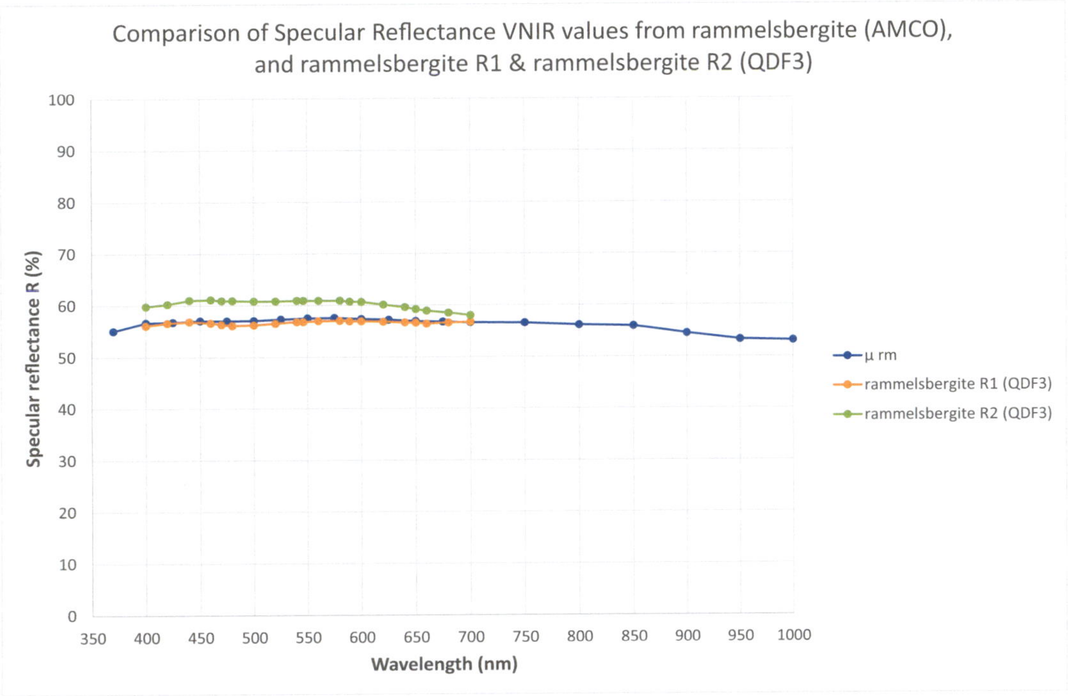
Comparison of Specular Reflectance VNIR values from rammelsbergite (AMCO), and rammelsbergite R1 & rammelsbergite R2 (QDF3)
Specular reflectance R (%)
100
90
80
70
60
50
40
30
20
10
0
350
400
450
500
550
600
650
700
750
800
850
900
950
1000
Wavelength (nm)
µ rm
rammelsbergite R1 (QDF3)
rammelsbergite R2 (QDF3)

Ramsdellite (*rdl*)

104

MnO_2: Orthorhombic

Ramsdellite		R_{548} (%)	Y (%)	λ_d (nm)	P_e (%)	Colour sensation
Air	R_{mx}	33.0	nd	nd	nd	Greyish white, slightly yellowish
	R_{mn}	11.7	nd	nd	nd	Light grey
Oil	R_{mx}	nd	nd	nd	nd	Grey
	R_{mn}	nd	nd	nd	nd	Dark grey
Bir: Medium (4)				*Pleoc*: YlWh-Gy (4)		*Tarn*: No
Anis +P: Net, very bright Yl tints (4)				*IR*: Occasional, red		*Twg*: nd
Mor: Varied, tabular, pseudom prl, etc				*Clv*: (110) and (010)		*PolQ*: Good
Hardness	*PolH*: Medium (3)			*Mohs*: 3	*Talm*: nd	*VHN* (miscellaneous): 93–1200

Note: cf. Reflectance, R, and VHN data: Uytenbogaardt and Burke (1971). No QDF3 data
Mineral abbreviations: bra—braunite, mng—manganite, prl—pyrolusite, psi—psilomelane, rdl—ramsdellite
Colour abbreviations: YlWh—Yellowish white, Gy—Grey, Yl—Yellow

104.1 General

Composition: MnO_2, like its polymorph pyrolusite, from which it is formed by inversion and into which it is transformed by heating above 300 °C. It closely resembles pyrolusite, so it is likely that its presence has often gone unnoticed and that ramsdellite is much more abundant than the published data suggest.

104.2 Microscopic Description

104.2.1 SP Observation (*Single Polariser*)

White *colour*, with a barely perceptible yellowish tone, clearly bireflective and slightly pleochroic in white or yellowish-white and greyish tones. In short, very similar to pyrolusite but less reflective and more greyish (Fig. 104.1a, b).

Reflectance similar to pyrolusite, but somewhat lower, especially R_{mn} (dark grey in ramsdellite). No full spectrum data found in the usual sources. Net exfoliation, like pyrolusite.

It occurs in crystalline or cryptocrystalline aggregates, sometimes with evidence of substitution by pyrolusite or vice versa (Figs. 104.2 and 104.3a).

104.2.2 Observation +P (*Polariser + Analyser at 90°*)

Clearly *anisotropic*, like pyrolusite, but with more vivid, luminous and varied colours, although yellows are still dominant (Figs. 104.1c, d and 104.3b). Red to purple IR, *internal reflections*, visible in immersion but not as abundant as in manganite (Fig. 104.1c, d and Fig. 104.3b).

R. Castroviejo, *A Practical Guide to Ore Microscopy—Volume 1*,
https://doi.org/10.1007/978-3-031-12654-3_104

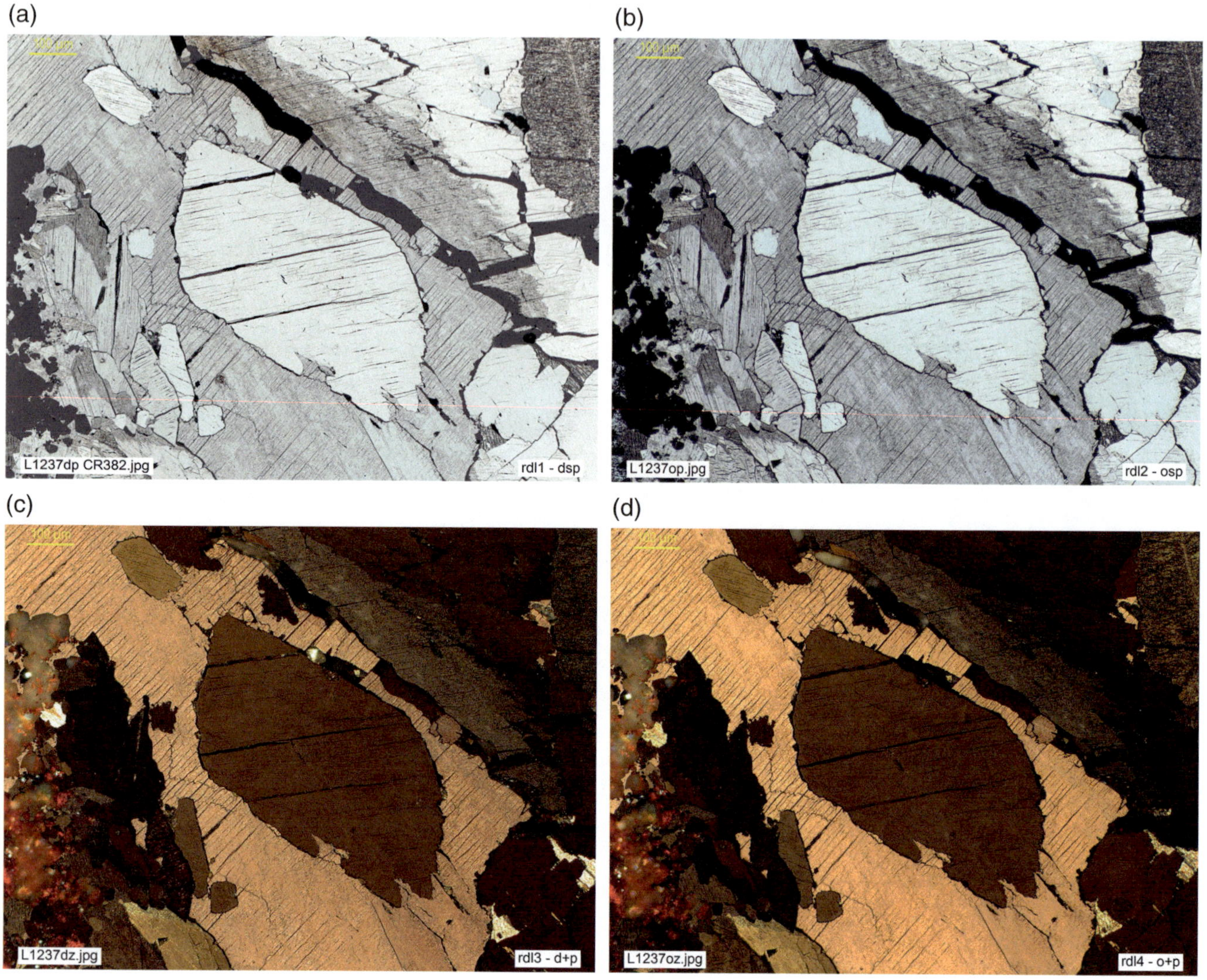

Fig. 104.1 **a**, **b** (**dsp** and **osp**, *respectively*)—Aggregate of rdl and prl, both grey and bireflective with marginal rhodonite (left, dark grey/black); rdl is darker than prl (especially in R_{mn} orientation, top right corner), and prl occurs in relicts suggesting its replacement by rdl. CR-382, Iron Dyke Mine, Plumas Co, California, USA. **c**, **d** Same scene as rdl1, (**d+p** and **o+p**, respectively). The brighter anisotropism and sparse—but visible in oil—red IR of rdl distinguish it from prl. Rhodonite shows abundant yellowish-brown to reddish IR (left)

104.2.3 Paragenesis

Like its polymorph pyrolusite, ramsdellite can be found in a variety of deposits, sedimentary, residual, infiltrative, formed under oxidizing conditions. In these, the relationships between pyrolusite and ramsdellite can be ambiguous, as both can coexist in equilibrium or replace each other.

Companions: pyrolusite, manganite, braunite, psilomelane and other manganese minerals.

Fig. 104.2 Ore of rdl (dark grey) and prl (lighter, relict), with reaction rims suggesting instability of prl, replaced by secondary rdl. More detailed view of central part in Fig. 104.3a, b. CR-382, Iron Dyke Mine, Plumas Co, California, USA (**dsp**)

104.2.4 Identification

The association with oxidized manganiferous paragenesis is the first criterion and, within these, ramsdellite can only be confused with pyrolusite, which it closely resembles. However, it can be easily distinguished, at least in direct comparison: pyrolusite is notably more reflective and shows more evident yellowish tones; its anisotropism is less vivid and luminous than that of ramsdellite.

Fig. 104.3 a, b Detail view of the central zone of rdl5 (**osp** and **o+p**, respectively): evidence for the textural relationships of rdl and prl and the vivid anisotropism characteristic of rdl

Reference

Uytenbogaardt W, Burke EAJ (1971) Tables for microscopic identification of ore minerals, 2nd edn. Elsevier, Amsterdam

105 Realgar (*rl*)

AsS: Monoclinic

Realgar	R'_{546} (%)		Y (%)	λ_d (nm)	P_e (%)	Colour sensation
Air	22.1		22.1	477	10.5	Greyish white
Oil	8.6		8.7	475	17.9	
Bir: Very weak to absent (2)				*Pleoc*: Very weak to absent (2)		*Tarn*: No
Anis +P: Scarcely visible (IR) (3)				*IR*: Massive		*Twg*: nd under the microscope
Mor: Prismatic. Impregnation, anhedral agrs				*Clv*: nd under the microscope		*PolQ*: Difficult, but sometimes good
Hardness	*PolH*: Very low, $\leq$ orp, <sbt (1)			*Mohs*: 1½–2	*Talm*: B	*VHN* $_{(nd)}$: 47–60

Obs: cf. Table and R charts, visible range data (QDF3). VHN: Uytenbogaadt and Burke (1971)
Mineral abbreviations: ac—acanthite, asp—arsenopyrite, cin—cinnabar, gg—gangue, gth—goethite, ht—hematite, mt—magnetite, orp—orpiment, pru—proustite, py—pyrite, rl—realgar, sbt—stibnite, sp—sphalerite, td—tetrahedrite

105.1 General

Composition: As_4S_4. By exposure to light, it alters to its *pararealgar* polymorph (also monoclinic, yellowish colour, gradually replacing the orange red of realgar).

105.2 Microscopic Description

105.2.1 SP Observation (*Single Polariser*)

Greyish white *colour.* Difficult *polishing (*scratched) but it can be achieved (Fig. 105.1a, b), although rl is a very soft mineral: *Polishing hardness* $PolH_{rl}$ > *red silvers* (pru-prg) or acanthite, but $\leq$ orp, < sbt.

Moderate *reflectance*, although if polished well (Fig. 105.1a, b), it may appear high when isolated in gangue. R_{rl} < orp, cin, ht, pru, td; > gth, mt (cf comparative charts).

It usually occurs as impregnations and anhedral aggregates (Fig. 105.1a, b), but in favourable conditions (pores, geodes) it can develop beautiful crystals (prismatic habit).

105.2.2 Observation +P (*Polariser + Analyser at 90°*)

Although clearly anisotropic, the effects of anisotropism are difficult to see because of the very abundant reddish (red–orange to red–yellowish, Fig. 105.1c, d) *internal reflections*, IR. In air, the anisotropism can be seen by slightly uncrossing the analyser, but in immersion the IR are so massive and ubiquitous that it is not visible.

R. Castroviejo, *A Practical Guide to Ore Microscopy—Volume 1*,
https://doi.org/10.1007/978-3-031-12654-3_105

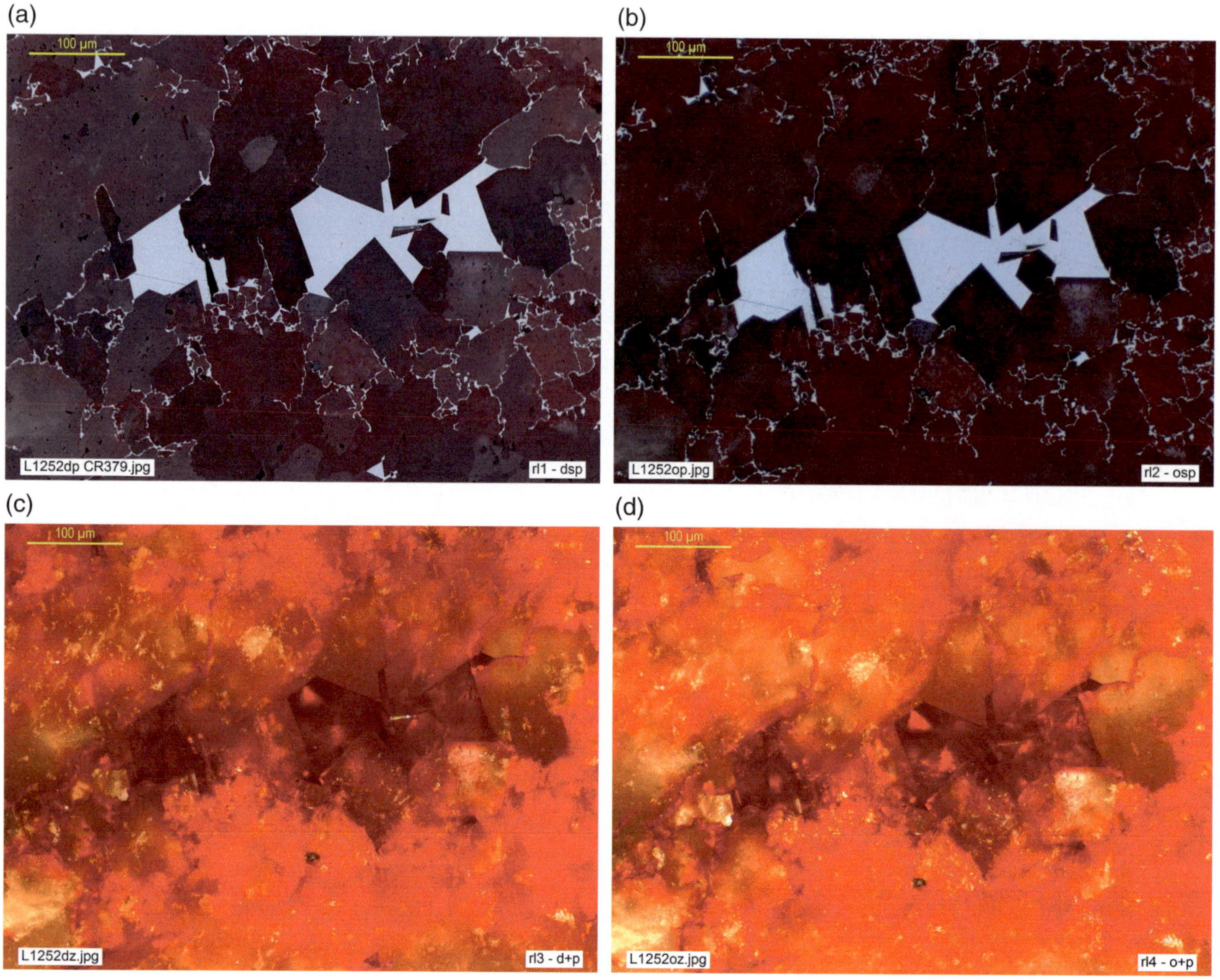

Fig. 105.1 **a**, **b** (**dsp** and **osp**, respectively)—Vein with posthumous rl (light grey), interstitial and in thin intergranular films, over carbonate hydrothermal infill (dark grey, bireflective). CR379, Cobalt, Ontario, Canada. **c**, **d** Same scene as rl1 (d+p and o+p, respectively): massive red–orange internal reflections hide rl anisotropism

105.2.3 Paragenesis

Posthumous or late mineral in hydrothermal *parageneses.* It can also be formed in shallow environments by weathering and fumarolic activity.

Accompanying ores: orpiment, pyrite, sphalerite, stibnite, arsenopyrite and other arsenic minerals.

105.2.4 Identification

Its properties (extremely low hardness, moderate reflectance, greyish–white colour, massive red–orange to yellowish-red IR hiding anisotropism) are so characteristic that identification should not be a problem, even in hand specimen (its striking red colour gives a first clue).

The lower brightness (i.e., lower reflectance, R) of realgar avoids possible confusion with cinnabar, proustite or hematite. Cinnabar also has many reddish IR, but not so many as to hide anisotropism; proustite IR are a deeper red than those of realgar; and hematite IR are by far scarcer. The R of orpiment is also higher than that of realgar.

105.3 Spectral Information

105.3.1 Realgar Reflectance R (in Air, Visible Spectrum) (%)

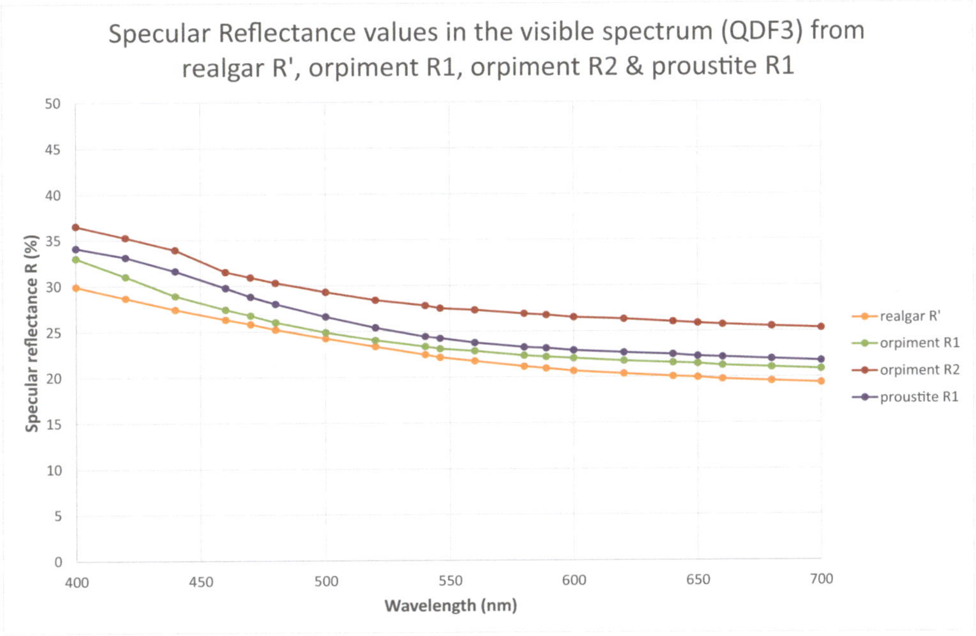

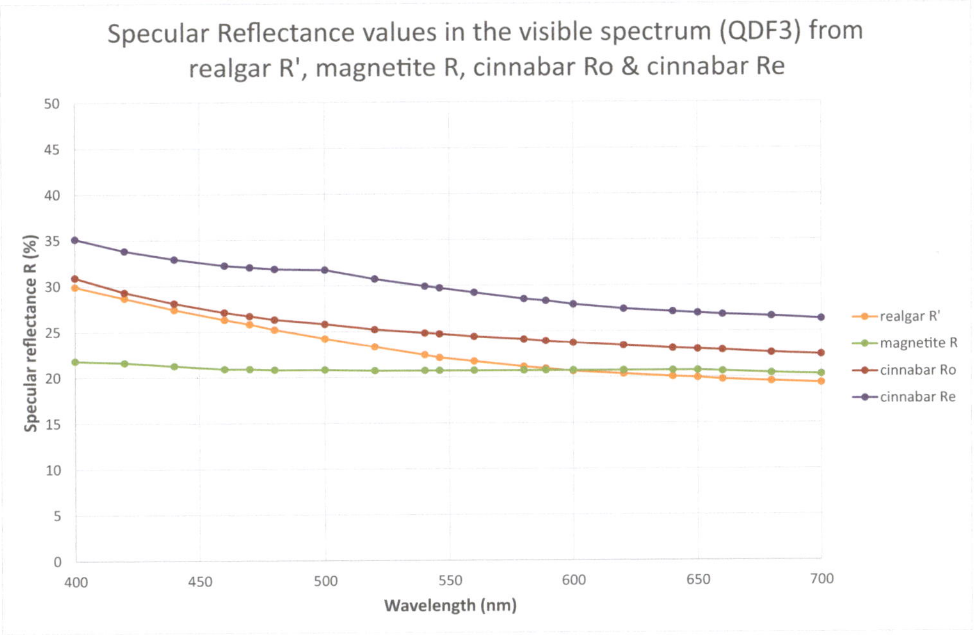

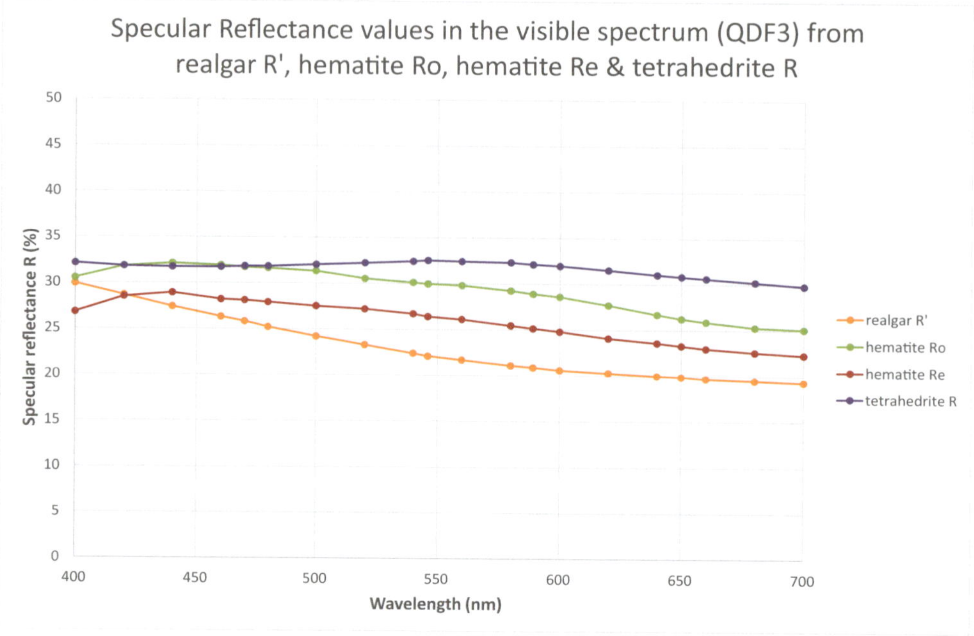

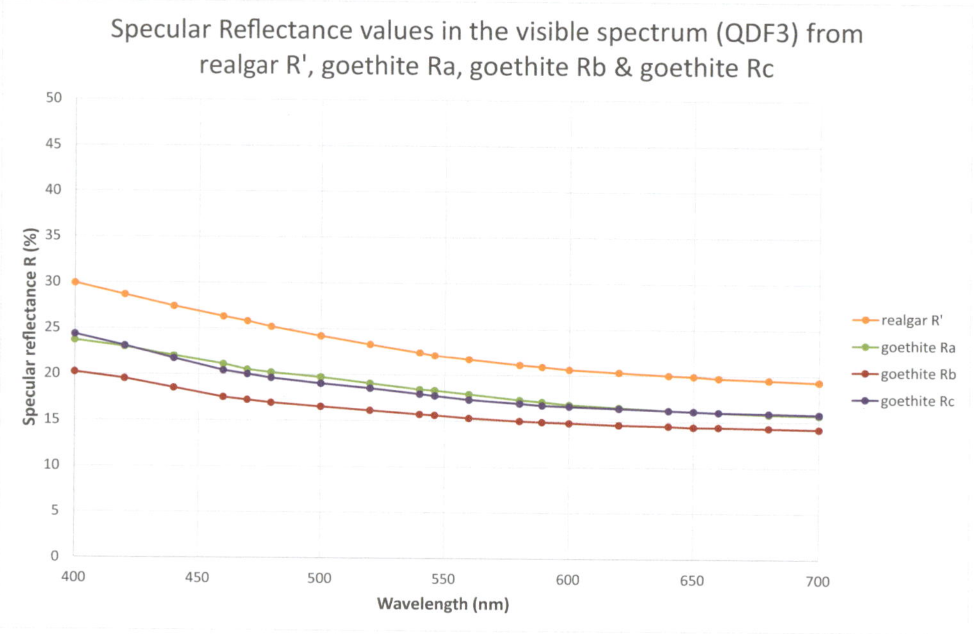

Reference

Uytenbogaardt W, Burke EAJ (1971) Tables for microscopic identification of ore minerals, 2nd edn. Elsevier, Amsterdam

Renierite (*ren*) 106

$(Cu, Zn)_{11}(Ge, As)_2Fe_4S_{16}$: Tetragonal (pseudocubic)
Briartite(bri): $Cu_2(Fe, Zn)GeS_4$: Tetragonal
Germanite(ger): $Cu_{13}Fe_2Ge_2S_{16}$: Cubic

Renierite		R_{546} (%)	Y (%)	λ_d (nm)	P_e (%)	Colour sensation
Air	R_o	27.7	28.4	580	24.0	Orange tone (approx. between ccp and fresh bn)
	$R_{e'}$	25.9	26.8	583	20.5	
Oil	R_o	15.8	16.2	579	37.9	Same orange tone, but noticeably sharper
	$R_{e'}$	14.1	14.7	581	34.7	
Bir: Very weak (2)				*Pleoc*: Very weak (2)		*Tarn*: No
Anis +P: Weak (3)				*IR*: No		*Twg*: Occasional
Mor: Grains / xenomorphic aggregates				*Clv*: -		*PolQ*: Good
Hardness	*PolH*: Medium, ≈ ccp; < sp (3)			*Mohs*: 4–5	*Talm*: nd	*VHN* $_{(100\ g)}$: 302–354

Obs: cf. Table and R charts, multispectral visible range data (QDF3; Picot and Johan 1982)
Mineral abbreviations: bn—bornite, bri—briartite, cc—chalcocite, ccp—chalcopyrite, cup—cuprite, cv—covellite, dg—digenite, en—enargite, ger—germanite, gg—gangue, gn—galena, IR—internal reflections, lz—luzonite, po—pyrrhotite, py—pyrite, sp—sphalerite, tn—tennantite, td—tetrahedrite

106.1 General

Renierite (***ren***), briartite (***bri***) and germanite (***ger***) are rare minerals, which are all Ge ores. They are known in very few deposits, essentially Kipushi (Katanga) and Tsumeb (Namibia), but they can be locally abundant. For this reason, and for being carriers of the uncommon metalloid Ge, their recognition is considered of interest, although they are not the exclusive carriers of this element, since Ge is also found associated with other ores, such as argyrodite-canfieldite or some Zn ores, and in coals.

The compositions attributed to these ores are complex and vary according to the authors.

- To the recognized stoichiometric components, such as As, Cu, Fe, Ge, S, Zn, are sometimes added other possible traces or impurities, such as Ga, Mo, V, Pb, Sn.
- Renierite is a mineral of the *Germanite Group*, for which Back and Mandarino (2008) propose the formula $(Cu, Zn)_{11}(Ge, As)_2Fe_4S_{16}$ or $(Cu^{1+},Zn)_{11}Fe_4(Ge^4,As^{5+})_2S_{16}$, with coupled substitution of Zn + Ge by Cu + As ([1]).
- Briartite, **bri**, is a mineral of the *Stannite Group*, although it is frequently associated with other species of the Germanite Group. Its composition is according to IMA Cu_3FeGeS_4, and it is also formulated as $Cu_2(Fe,Zn)GeS_4$ ([2]).
- The composition of **ger** (*Germanite Group*) is $Cu_3(Ge, Fe)S_4$ according to Picot and Johan (1982). However, $Cu_{13}Fe_2Ge_2S_{16}$ is proposed, similar to a renierite with no As or Zn ([3]).

Because of their frequent association in Tsumeb or Kipushi, these three ores will be described together, taking as reference the main one (ren) and pointing out the particularities of each of them, illustrated with the appropriate photomicrographs. Other complementary data can be found in the web addresses (*mindat*) cited above.

[1] Cf. *Mindat,* https://www.mindat.org/min-3389.html.
[2] Cf. *Mindat,* https://www.mindat.org/min-773.html.
[3] Cf. *Mindat,* https://www.mindat.org/min-1681.html.

R. Castroviejo, *A Practical Guide to Ore Microscopy—Volume 1*,
https://doi.org/10.1007/978-3-031-12654-3_106

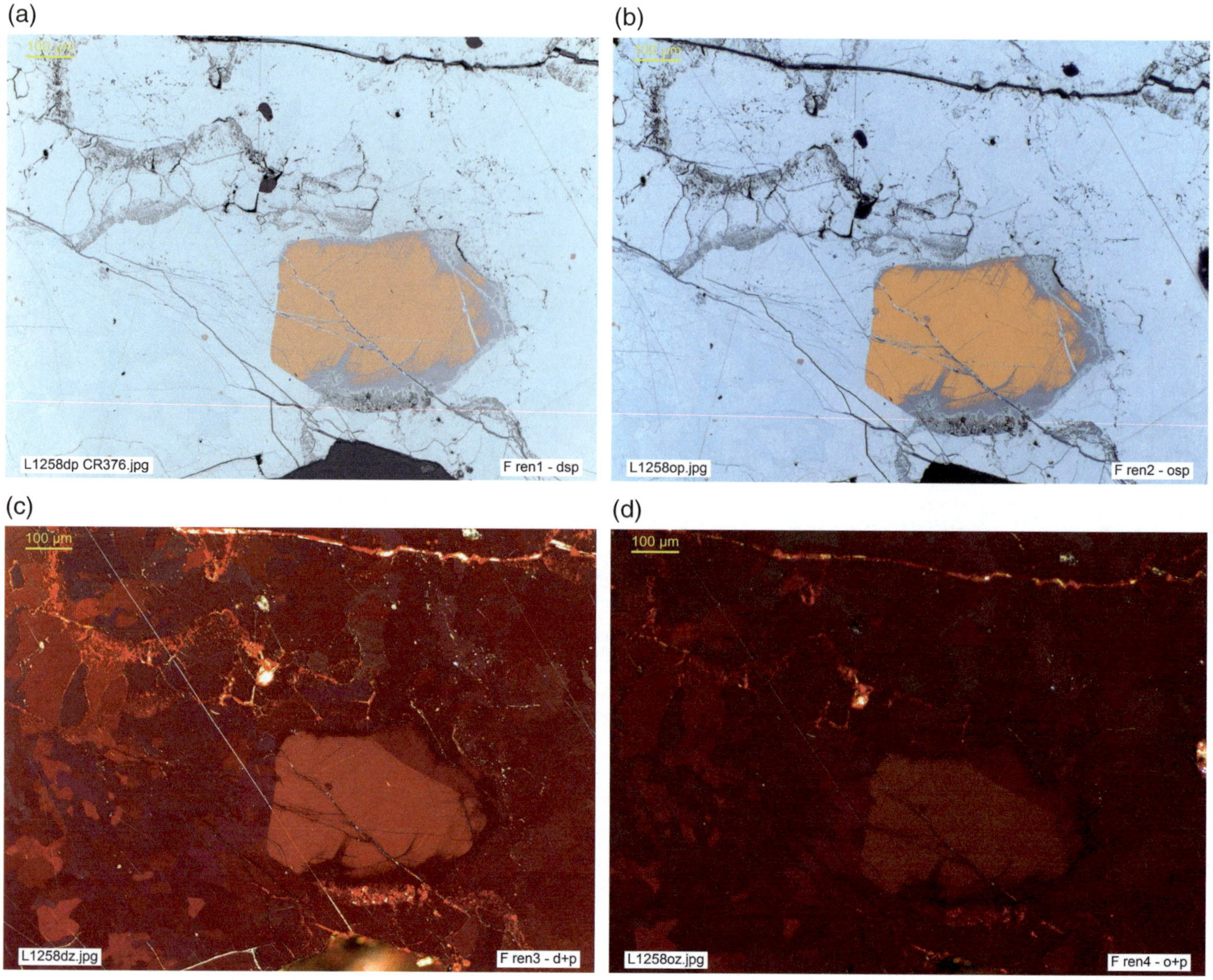

Fig. 106.1 **a**, **b** (**dsp** and **osp**, respectively). Inclusion of **ren** (orange tone-**osp**-to orange-brown-**dsp**) in massive *cc* (light bluish grey), both fragmented and slightly weathered (grey cryptocrystalline aggregates, porous, associated with microcracks; probably *cup*, cuprite, with some *lim*, cf. Fig. 106.1c, d). On ren, there is a thin rim/intergrowth of *bri* (briartite: grey, darker than cc and ren), also partly altered to cup (grey, porous and lighter than bri), mainly on the lower margin. Tiny inclusions of *ger* (germanite: purplish pink), scattered in cc. CR-376, Tsumeb, Namibia.**c**, **d** (same scene as Fig. 106.1a: **d+p** and **o+p**, respectively). Moderate anisotropism of ren, in brownish tones, and weak but coloured (typical blue-green and orange tones) of cc. The position of cup is marked by its typical red IR, while bri (in extinction position) and ger (isotropic) are both black

106.2 Microscopic Description

106.2.1 SP Observation (*Single Polariser*)

The colour of **renierite** is typically orange (YlOr to BrOr), between chalcopyrite and bornite. It can look very much like bornite, but bn is quickly darkened by tarnishing and then the more intense luster and orange tones of renierite clearly differentiate it. The (subjective) impression of colour varies notably according to the environment, as can be seen comparing the renierite of Fig. 106.1a, b (included in chalcocite) or Fig. 106.2a, b (in gangue) with that of Fig. 106.3, which looks much darker and brownish because it is next to pyrite. The colours are much more clearly perceived in immersion (comp. Figure 106.1a with Fig. 106.1b or Fig. 106.2a with Fig. 106.2b).

Very weak pleochroism, practically imperceptible (or only perceptible in oil on grain edges, if the orientations are suitable).

Moderate *reflectance*, between chalcocite and bornite (cf. Figs. 106.1a, b, 106.2a, b, 106.3 and 106.4a and corresponding comparative graphs of specular reflectance in the visible range: $R_{ren} \leq R_{cc}$; < R_{ccp}; < < R_{py}, but $R_{ren} > R_{bri}$ or R_{ger} and, above all, clearly > R_{bn}).

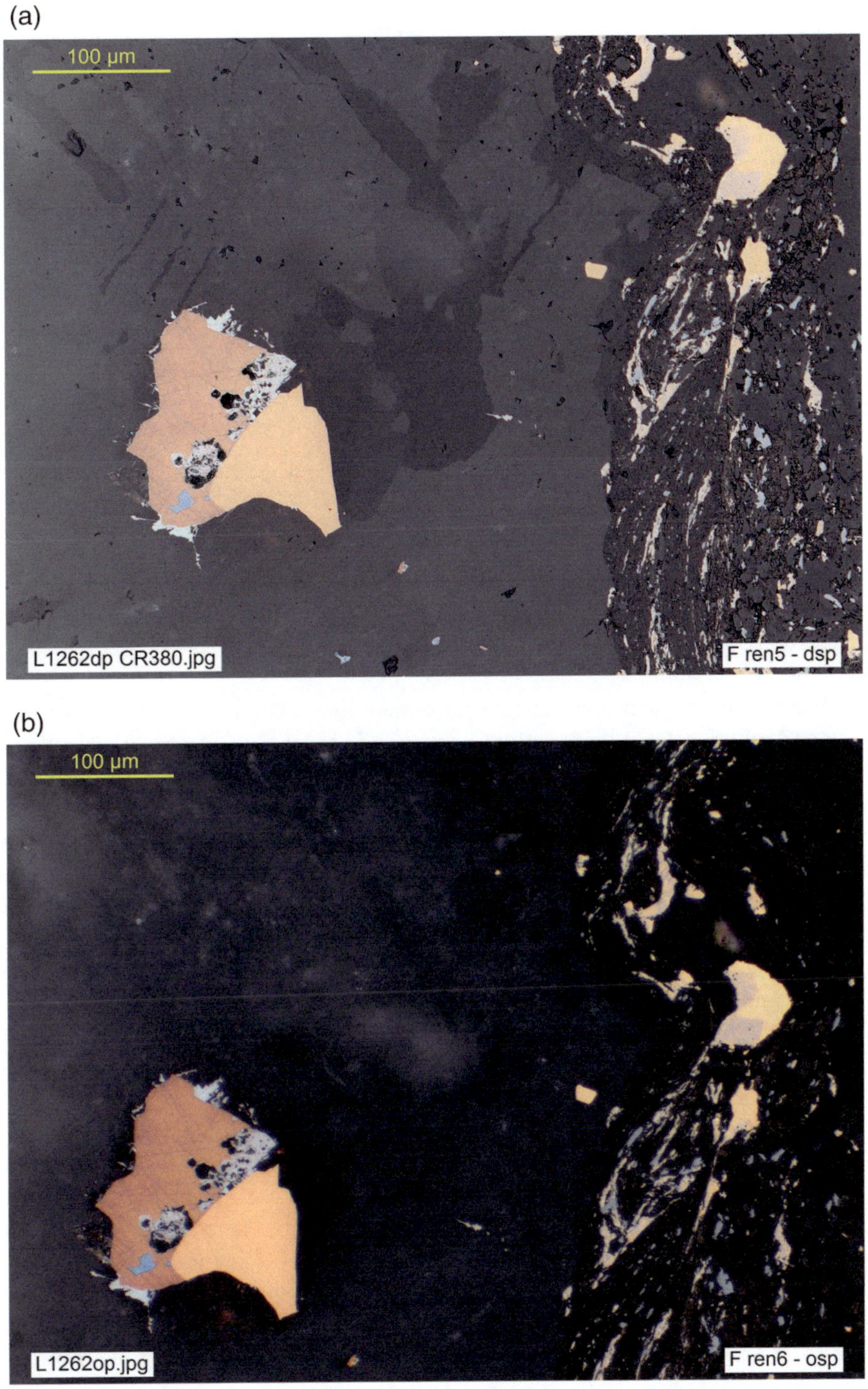

Fig. 106.2 **a**, **b** (Same scene, **dsp** and **osp**, respectively). View of carbonate-Ge-sulphide ore vein infill (left-hand side) and its hostrock contact (right-hand side). Veinfill is mostly carbonate gg (dark grey to black, bireflective), which includes ren (orange) and bn (brownish pink, somewhat darker), intergrown with minor dg, tn and cc. The hostrock shows a dissemination of ren (in the larger grains associated with ger, displaying its typical light pinkish-violet shade) and of tiny scattered particles of the other components seen in the veinfill. Figure 106.2b shows how colour definition and contrast of ren-bn-ger are enhanced if viewed in immersion. CR-380, Tsumeb, Namibia

Easy and good quality polishing. No tarnishing, which is an effective differentiation criterion against bornite.

Renierite usually occurs in small crystals or anhedral aggregates, frequently associated with briartite or germanite, with which it often shows replacements or exsolutions.

As for its companions briartite and germanite:

- **Briartite** is distinguished by its grey colour, slightly bluish and duller than renierite (R_{bri} lower than that of renierite, especially towards $\lambda \geq 600$ nm, close to red, cf. Fig. 106.1a, b and corresponding comparative specular reflectance plots). PolH bri ≈ chalcopyrite.
- The colour of **germanite** is typically pinkish grey with a peculiar purplish hue, somewhere between bornite and

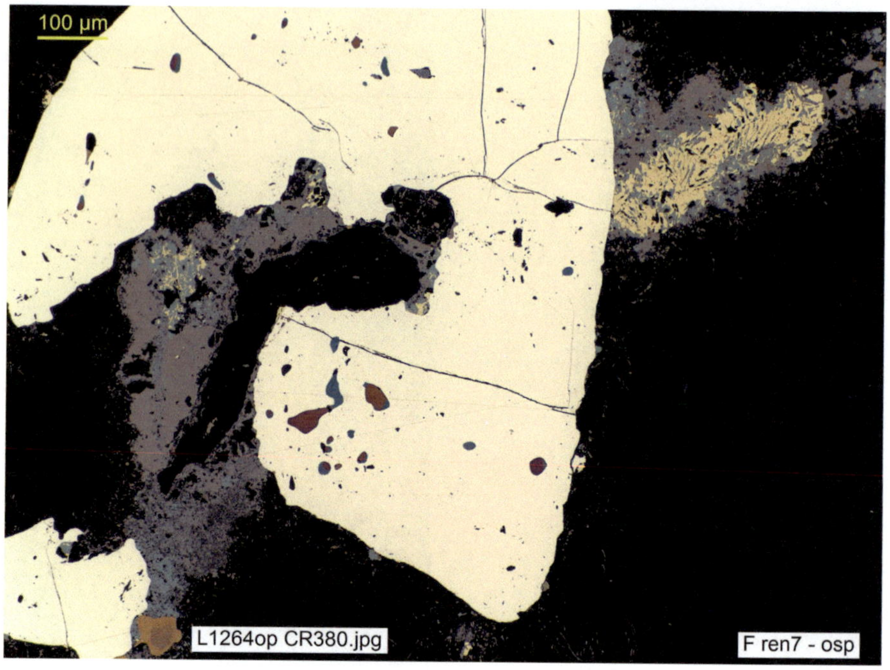

Fig. 106.3 Corroded pyrite phenocryst, with small inclusions of bn (purplish brown, darkened by tarnishing), dg (cv, ccp), on a sulphide microband, mainly composed of ger (purplish grey) intergrown with ccp and with subordinate tn (grey). At the lower left end of this band (next to py), mixed grain comprising a ren *corona* on a ger core, both clearly recognizable by their orange (ren) or violet (ger) colours. CR-380, Tsumeb, Namibia (**osp**)

enargite.Its frequent association with bornite may offer a good criterion for differentiation: germanite is more purplish than fresh bornite but less so than tarnished bornite, which looks darker (cf. Fig. 106.3 and corresponding comparative graphs of specular reflectance in the visible range: $R_{ger} < R_{cc}; < R_{ccp}; < < R_{py}$; but $R_{ger} > R_{tarnished\ bn}$ and $> R_{sp}$). However, with time, *germanite also tarnishes,* so it is convenient to study the freshly polished ore.

The impression of colour and brightness varies according to the surrounding minerals: germanite appears darker next to pyrite (Fig. 106.3) than next to chalcocite (Fig. 106.4a) and, in both cases, it appears much darker than when dispersed in gangue (Fig. 106.2a, b). PolH ger < tennantite and > galena.

106.2.2 Observation +P (*Polariser + Analyser at 90°*)

Weakly *anisotropic*, with brownish or greenish tones (Figs. 106.4b and 106.1c, d; in Fig. 106.1c, d there is a diffuse reddish tone that actually is an artifact due to the abundant internal reflections of the accompanying cuprite, resulting from the oxidation of the primary ore). It has no internal *reflections* and is occasionally *twinned.*

As for briartite and germanite:

- To perceive the weak anisotropism of **briartite,** it is sometimes necessary to slightly uncross the analyser (***xp***). Frequently twinned mineral.
- **Germanite** is always seen in extinction: it is isotropic, as corresponds to its cubic symmetry.

106.2.3 Paragenesis

Renierite occurrence restricted to a few deposits. Well documented in the *Central African Copperbelt*, at the type locality: Kipushi Mine, Katanga, DRC (S of the *Shaba Copperbelt*). It is related to the complex and debated genetic processes of Fe–Zn–Pb–Cu sulphide mineralization, with Ge, V, Mo, Ga, W accessories in veins and bodies associated with dolomitic formations of the *Katanga Supergroup* (Neoproterozoic). It has also been documented from the Tsumeb Mine (Namibia).

Paragenesis (accompanying ores): germanite, briartite, bornite, chalcocite, digenite, chalcopyrite, pyrite, sphalerite, tennantite, galena.

(a)

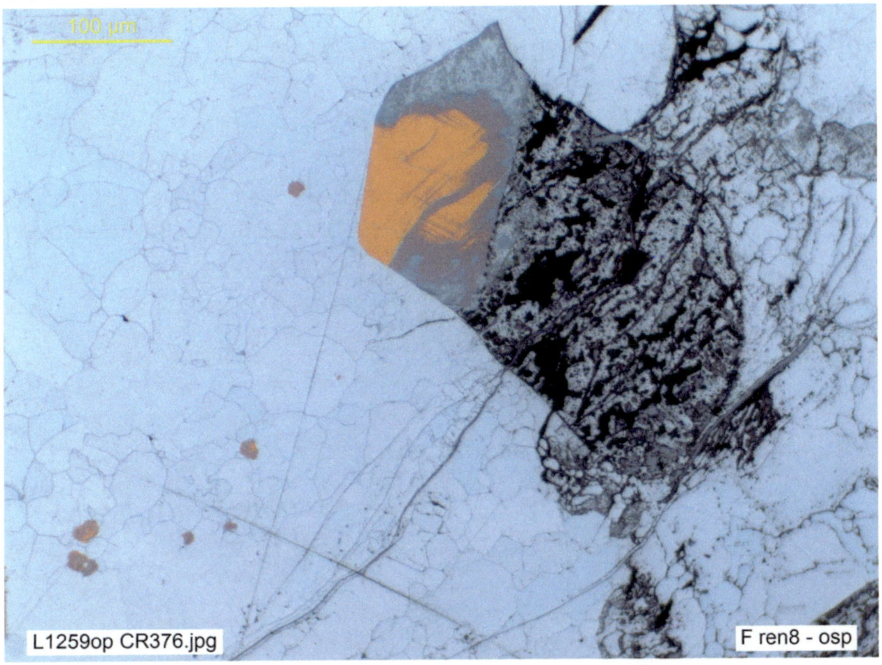

(b)

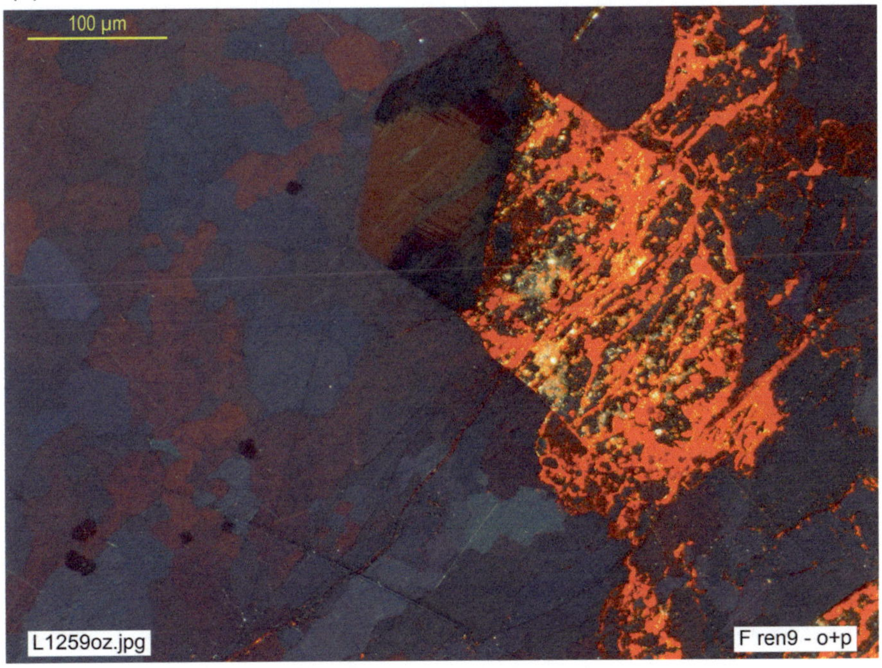

Fig. 106.4 **a**, **b** (Same scene: **dsp** and **osp**, respectively). Massive cc (grey, anisotropic), slightly crackled and weathered, including also a microbrecciated and altered aggregate, comprising: (i) a primary, relict sulphide part (upper left) and (ii) a secondary oxidized zone which includes most of the aggregate (lower right, grey-black tones). The former (i) is composed of primary ren (yellowish orange, anisotropic), rimmed by a secondary ger *corona* (pink-violet, isotropic), with marginal dg and cc. An oriented intergrowth of thin ger lamellae can be seen on ren, suggesting crystal lattice-controlled replacement of the latter. The oxidized zone (ii) contains cryptocrystalline cup and lim, which are distinguished by their red (cup) or yellowish-brown (lim) IR. CR-376, Tsumeb, Namibia

106.2.4 Identification

Characteristic of renierite are the peculiar orange colour (between chalcopyrite and bornite), not affected by tarnishing, its moderate reflectance and anisotropism, together with the important datum of its paragenesis or evidence of Ge contents.

- Paragenesis and presence of other Ge minerals are also important data to ensure the recognition of briartite, taking into account that its colour, reflectance curve and anisotropism distinguish it from renierite and germanite.
- The same general criteria (association with other Ge minerals) and its peculiar colour facilitate the recognition of germanite.

106.3 Spectral Information

106.3.1 Renierite Multispectral Reflectance R (in Air, Visible Spectrum) (%)

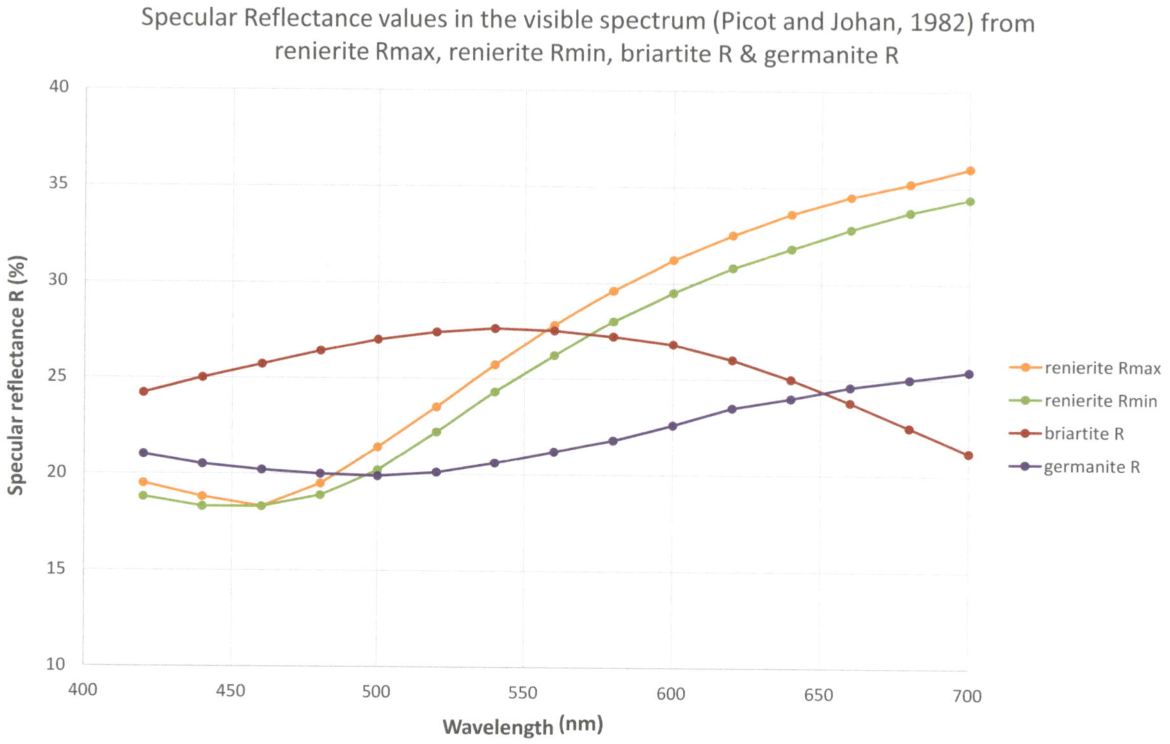

References

Back ME y Mandarino JA (2008) Fleischer's glossary of mineral species 2008. The Mineralogical Record Inc. Tucson, AZ, 345p

Picot P, Johan Z (1982) Atlas of ore minerals. B.R.G.M./Elsevier

Romanechite (*rom*)

107

$(Ba,H_2O)_2(Mn^{4+},Mn^{3+})_5O_{10}$: Monoclinic

Romanechite		R_{546} (%)	Y (%)	λ_d (nm)	P_e (%)	Colour sensation
Air	R_1	23.2	23.0	480	5.8	Greyish white
	R_2	30.8	30.4	479	8.7	White
Oil	R_1	9.9	9.7	478	12.2	Grey
	R_2	15.6	15.3	479	14.3	Light grey
Bir: Medium (4)				*Pleoc*: Weak (3)		*Tarn*: No
Anis +P: Distinct, weakly coloured (4)				*IR*: No		*Twg*: Uncommon
Mor: Colloform, acicular, fibrous-radial				*Clv*: No		*PolQ*: Good
Hardness	*PolH*: Variable (≈3–4)			*Mohs*: 5–6	*Talm*: nd	*VHN* $_{(100\ g)}$: 514–715

Obs **:** cf. Table and R charts, visible range data (QDF3 and Picot and Johan 1982)
Mineral abbreviations: cry—cryptomelane, gth—goethite, lim—limonite, mng—manganite, prl—pyrolusite, psi—psilomelane, rom—romanechite

107.1 General

Composition, $(Ba,H_2O)_2(Mn^{4+},Mn^{3+})_5O_{10}$, by its Ba content close to that of holandite and similar to that attributed to psilomelane (*s.str.*). Similar properties to the latter mineral, psilomelane, recently discredited as a species because of its heterogeneity, which led to ambiguous definitions. This suggests that many of the ores so far identified as psilomelane may, in fact, be of romanechite as the main component (cf. psilomelane description).

107.2 Microscopic Description

107.2.1 SP Observation (*Single Polariser*)

Colour greyish white, clearly bireflective, weak or imperceptible pleochroism (Figs. 107.1a, b and 107.2a). The only observable change on rotation is between shades of white and grey, although judging by its reflectance spectrum there would be a slight variation in colour.

Polishes well. Does not show exfoliation.

Relatively low *reflectance*, distinctly lower than pyrolusite, analogous to psilomelane and slightly higher than goethite (cf. comparative graphs of specular reflectance spectra). The impression of brightness depends largely on texture, grain size and, above all, porosity.

It occurs in fine colloform, botryoidal and radiate fibrous aggregates, in acicular aggregates or microbands structured in fine divergent, arborescent, palmate fibres.

107.2.2 Observation +P (*Polariser + Analyser at 90°*)

Clearly *anisotropic*, but faintly coloured (Figs. 107.1c, d and 107.2b): greyish-white polarisation colours, with sometimes slightly yellowish or bluish hues, the latter more visible when the analyser is slightly uncrossed, e.g., at ≈87°.

It shows no *internal reflections*.

107.2.3 Paragenesis

Typically, exogenous mineral, formed at low temperature in superficial, oxidising conditions, by weathering of primary concentrations and rocks containing manganese minerals (alabandite, braunite, rhodochrosite, rhodonite, etc.), or by

R. Castroviejo, *A Practical Guide to Ore Microscopy—Volume 1*,
https://doi.org/10.1007/978-3-031-12654-3_107

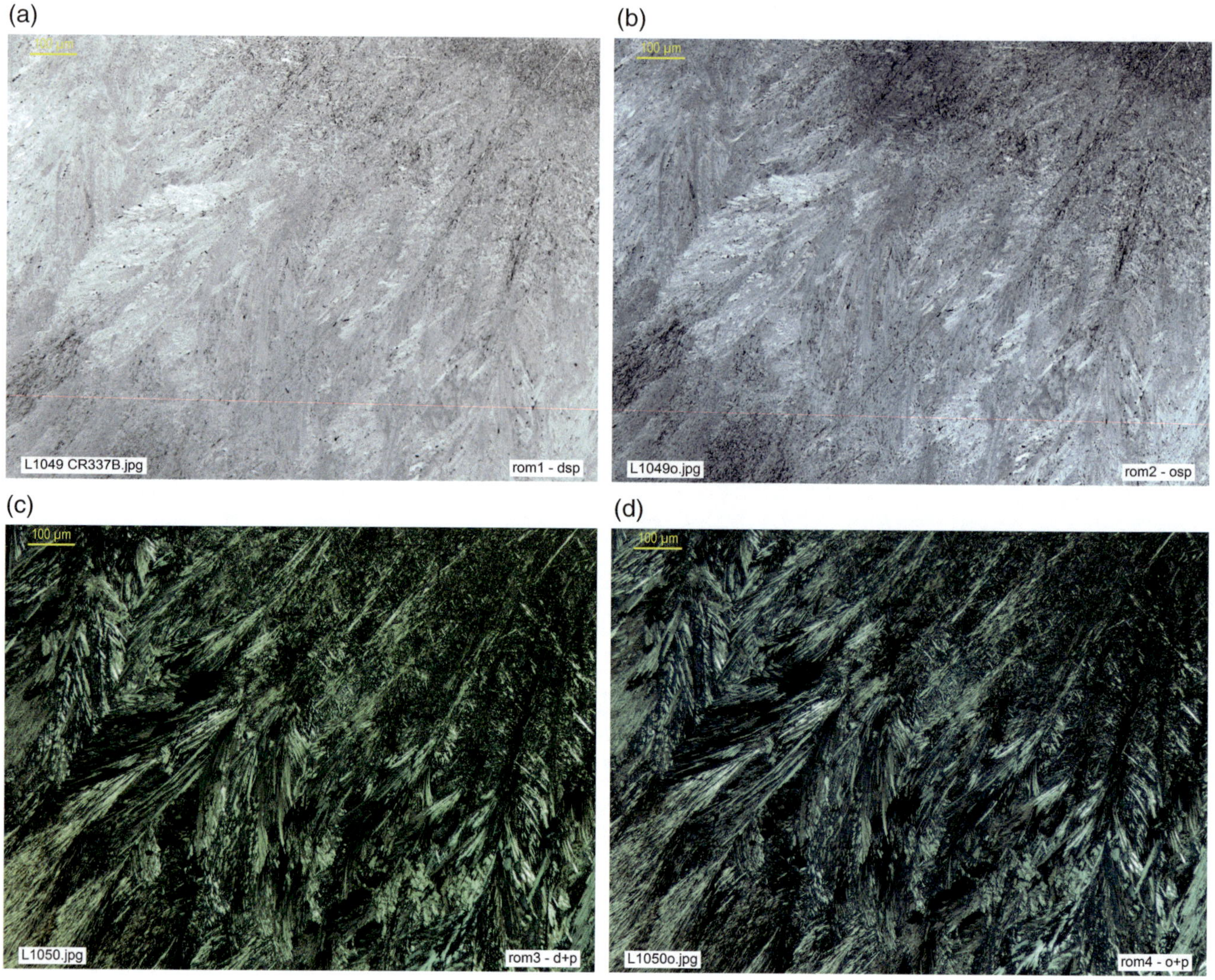

Fig. 107.1 **a, b** (**dsp** and **osp**, respectively)—Prismatic-acicular rom ore (detail of the central band in Fig. 107.2a): the dark zone in the upper right corner corresponds to the interstitial intergrowth of cryptocrystalline cry. The identity of both minerals has been confirmed by XRD. CR 337b, Clemira vein, El Remanso, Santiago del Estero, Argentina. **c, d** Same scene as Fig. 107.1a (**d+p** and **o+p**, respectively): the net anisotropism of rom, though poorly coloured, evidences its palmated fibrous texture. The very fine grain size of cry makes its anisotropism inconspicuous

direct precipitation in sedimentary deposits. Also possible hypogene (hydrothermal) origin, by cooling or mixing of ascending solutions, when accessing oxidising conditions at shallow levels; for example, in epithermal systems.

Usual companions: pyrolusite, cryptomelane, psilomelane, other manganese oxides, holandite, goethite, limonite, etc.

107.2.4 Identification

Under the microscope, low crystallinity or arborescent and fibrous-radiated acicular textures, together with reniform to botryoidal morphologies in hand specimen and associated paragenesis, are important criteria for identifying romanechite. It is further characterized by its greyish-white colour and relatively low reflectance, but with remarkable bireflectance, and by its distinct but poorly coloured anisotropism.

It differs from goethite in that gth is less reflective and has red IR. Compared to other common manganese minerals, rom is distinguished from pyrolusite because the latter is clearly more reflective and pleochroic, showing vividly coloured anisotropism (yellows) and, frequently, shrinkage cracks. However, when both minerals are cryptocrystalline or very fine-grained the distinction is problematic. The same is true of psilomelane and other minerals formerly treated as components of psilomelane (cryptomelane, coronadite, hollandite). Any or all of them may occur intergrown with romanechite, as well as goethite and limonite, in cryptocrystalline aggregates. In these cases, or when the grain size is submicroscopic, it is necessary to resort to additional techniques, such as XRD, to achieve a secure identification.

Fig. 107.2 **a, b** (**dsp** and **d+p**, respectively)—Colomorphic manganiferous banded ore, deposited on brecciated and partially replaced gangue (upper right corner: barite, ***bar***, dark grey aggregate, prismatic to tabular habit). Starting and away from this bar core, microbands of cry (darker and massive, microcrystalline to submicroscopic), arborescent fibrous rom (center) and colomorphic rom (or psi) (lower left) follow. Textures are most clearly seen in (+p), b; Fig. 107.2b; details in Figs. 107.1a, b, c, d. CR 337b, Veta Clemira, El Remanso, Santiago del Estero, Argentina

107.3 Spectral Information

107.3.1 Romanechite Reflectance R (in Air, Visible Spectrum) (%)

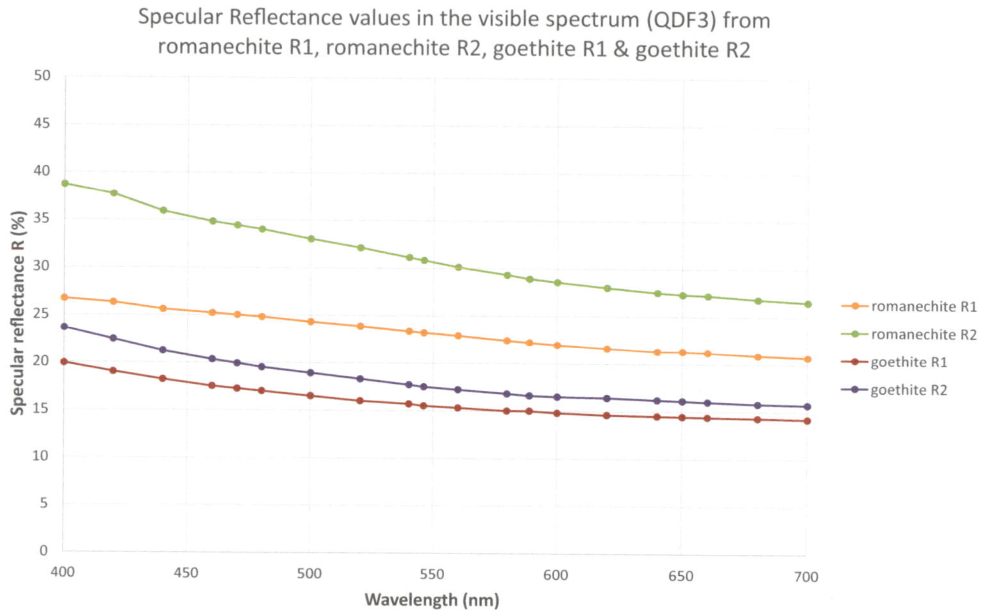

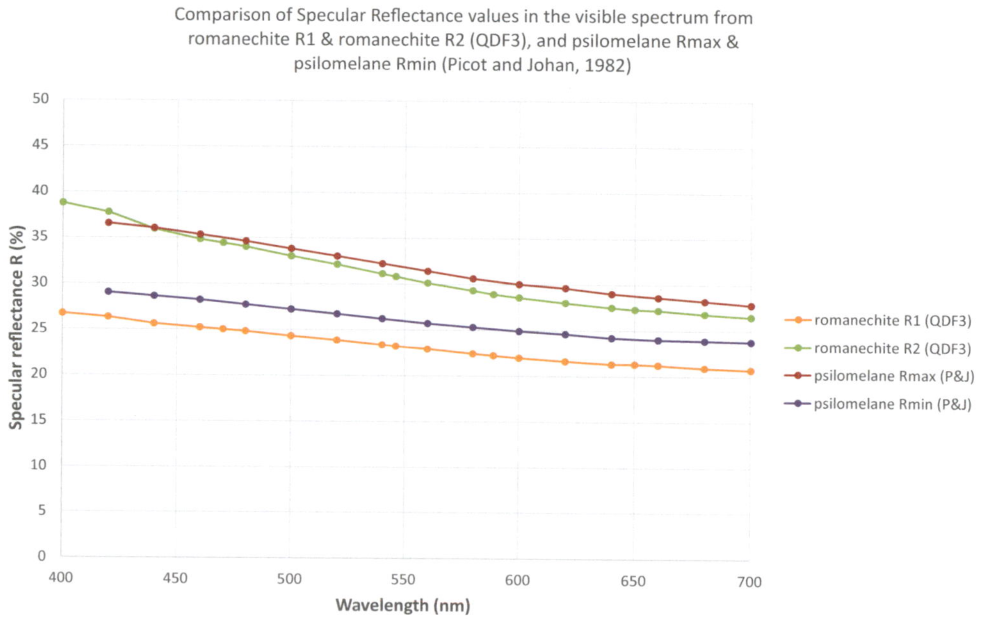

Reference

Picot P, Johan Z (1982) Atlas of ore minerals. B.R.G.M./Elsevier

Rutile (*rt*)

TiO_2: Tetragonal

108

Rutile		R_{546} (%)	Y (%)	λ_d (nm)	P_e (%)	Colour sensation
Air	R_o	19.8	19.7	475	5.8	Grey
	$R_{e'}$	23.1	23.0	476	4.5	
Oil	R_o	7.0	6.9	474	10.2	
	$R_{e'}$	9.3	9.3	474	7.4	
Bir: Medium (4)				*Pleoc*: Very weak (2)		*Tarn*: No
Anis +P: Net, little coloured (4)				*IR*: Abundant		*Twg*: Very frequent, multiple
Mor: Euhedral, prismatic, acicular				*Clv*: Occasional (110)		*PolQ*: Good
Hardness	*PolH*: High, ≥ ht, < py (4)			*Mohs*: 6.5–	*Talm*: G^+	*VHN* $_{(100\ g)}$: 894–974

Obs: cf. Table and R charts, multispectral data (VNIR, AMCO) and visible range (QDF3)
Mineral abbreviations: ana—anatase, cbn—cubanite, ccp—chalcopyrite, cst—cassiterite, gg—gangue, gra—graphite, gt—garnet, ht—hematite, il—ilmenite, lcx—leucoxene, mck—mackinawite, py—pyrite, q—quartz, rt—rutile, rtT—Ta-rich rutile, sp—sphalerite, tit—titanite, zir—zircon

108.1 General

Composition: TiO_2, occasionally with traces of Nb, Ta, Fe, Cr, V, Sn, W. Nb and Ta may become abundant, suggesting transitions to *ilmenorutile* and *strüverite*, respectively (cf. discussion and pertinent explanations in descriptions of ilmenorutile and tapiolite/strüverite), both now treated as varieties of rutile (IMA).

Trimorph with *brookite* (orthorhombic) and *anatase* (tetragonal). Transparent, so it can also be identified with transmitted light, characterized by both its high relief and birefringence ($n_\omega \approx 1.6$ and $n_\varepsilon \approx 1.9$).

108.2 Microscopic Description

108.2.1 SP Observation (*Single Polariser*)

Grey *colour*, relatively light for an oxide (Figs. 108.1a, b and 108.2a, b). It presents a dark tone next to sulphides like chalcopyrite, unlike the very light to white aspect against quartz or silicates (comp Figs. 108.2a and 108.4a).

Moderate *reflectance*, relatively high for an oxide (R rt > R sp or il); however, R rt is < R ht, and R rt is much lower than that of pyrite or chalcopyrite. The reflectance value may vary with composition: lower R values have been found in Ta-rich rutile (Fig. 108.5a, cf. AMCO multispectral plot, *rtT* curve).

Clearly bireflective (to the point that the twins may become visible without analyser: Figs. 108.1b and 108.3a) but not pleochroic (the two R_o and $R_{e'}$ curves vary in parallel, cf. comparative R-graphs in the visible spectrum and Fig. 108.3a).

Good *polishing quality* and visible cleavage, generally. Because it is a transparent mineral, IR are so abundant that they can sometimes be perceived without the analyser (Fig. 108.1b). *Polishing hardness* high, ≥ ht, but slightly < py.

It is usually well crystallised, with a marked euhedral tendency, typically in prisms or disseminated in acicular crystallites (Figs. 108.1a and 108.2a).

R. Castroviejo, *A Practical Guide to Ore Microscopy—Volume 1*,
https://doi.org/10.1007/978-3-031-12654-3_108

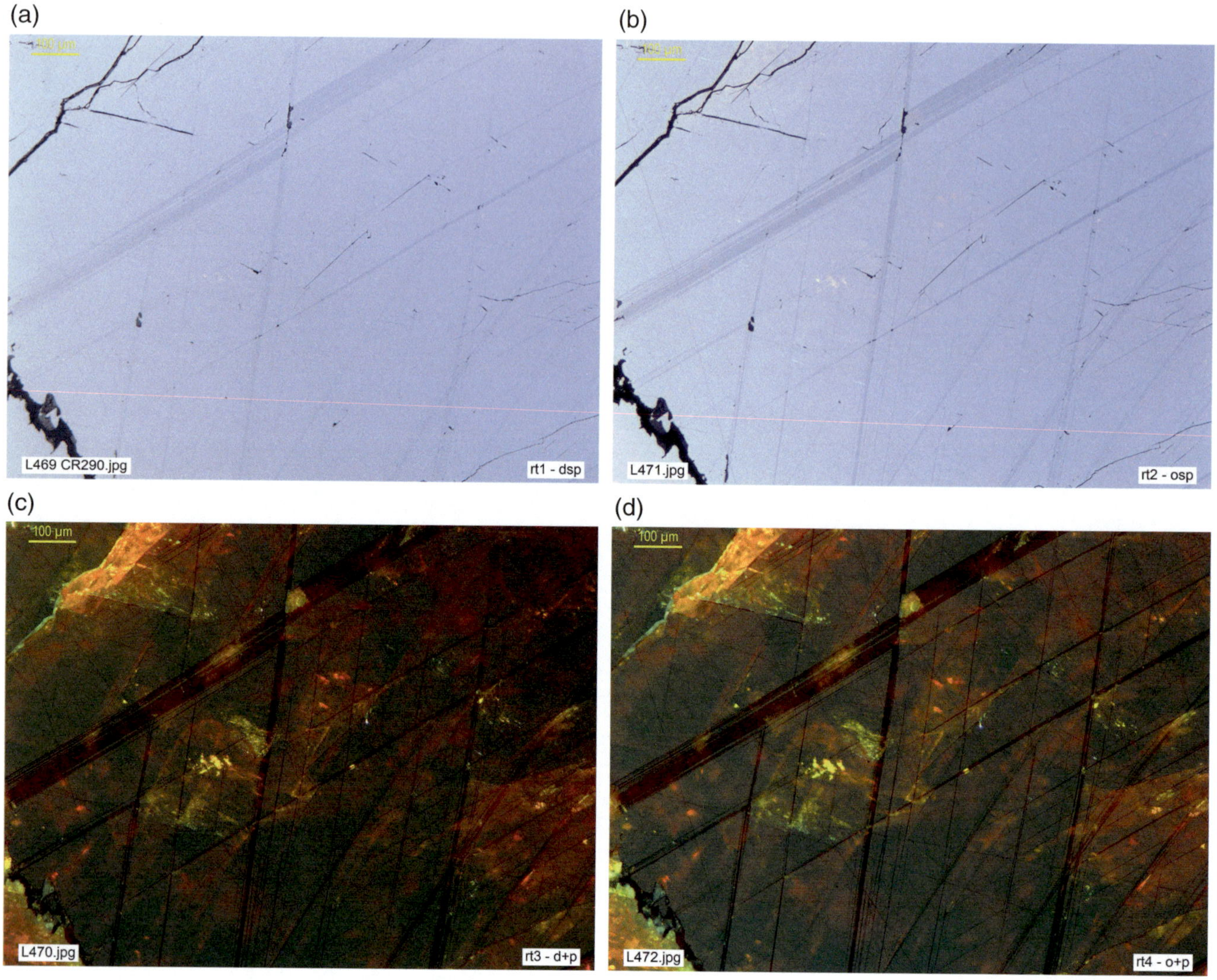

Fig. 108.1 **a** Single crystal of rt, microcracked and cemented by silicates (dark grey) and titanite (intermediate grey); rt polysynthetic twins in several directions, visible by bireflectance. CR290, Vegasheia, Norway (**dsp**). **b** Same scene as rt1 (**osp**). **c**, **d** Same scene as rt1 (**d+p**) and (**o+p**), respectively. Twinning, anisotropism and internal reflections of rt

108.2.2 Observation +P (*Polariser + Analyser at 90°*)

Clearly *anisotropic* but in poorly saturated or greyish tones (Figs. 108.1c, d, 108.2c, d, and 108.3b), usually masked by abundant *internal reflections*, IR. These are commonly massive and very luminous, typically reddish or reddish-brown, but can vary in colour and abundance depending on the composition (colourless, yellowish, brownish, greenish).

In the Ta-rich variety of Fig. 108.5b, the scarcity and low colour of the IR allows appreciation of the anisotropism in yellowish-brown to light olive-brown tones, poorly saturated. The Nb-rich varieties show weak anisotropism (according to Uytenbogaardt and Burke 1971). Ramdohr (1980) cites rutile with 15% Nb from metamorphic rocks and characterized by lower transparency, sparser and colourless IR and lower anisotropism.

It almost always presents multiple *polysynthetic twins* in several directions (Figs. 108.1c, d and 108.3b), partly due to deformation, although it can also show simple or contact twins (Figs. 108.4b and 108.5b). According to Ramdohr (1980), twins are only missing when rutile has formed by recrystallisation of ilmenite.

108.2.3 Paragenesis

As a minor, accessory mineral it is practically ubiquitous. In higher concentrations and as a primary mineral, it can be found in pegmatites and in various igneous and

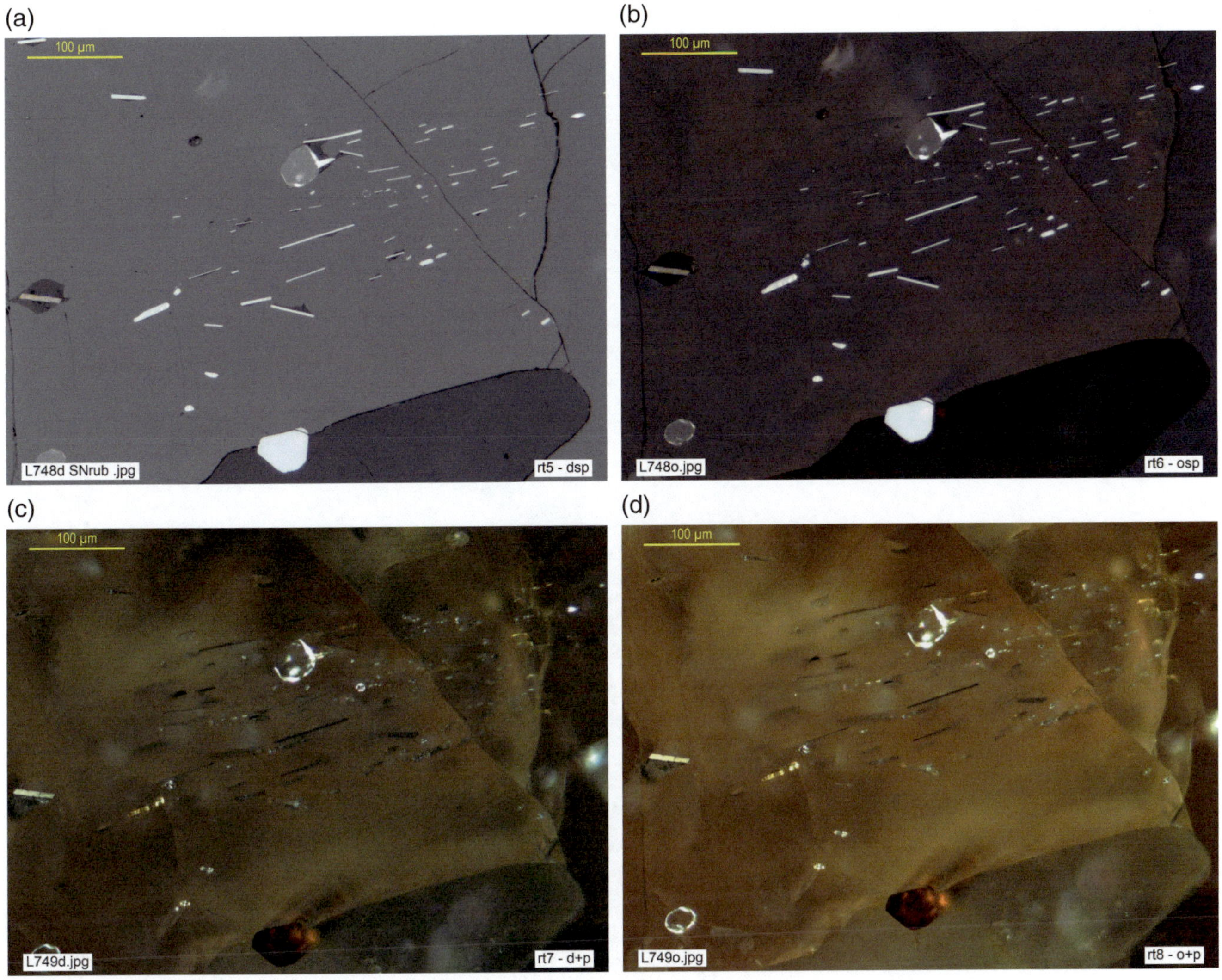

Fig. 108.2 **a** Acicular microinclusions of rt (white or very light grey) in phenoblast of garnet, grt (grey), also including q (dark grey), with a tiny lamellar crystal of gra (brownish grey, left-center) and two inclusions of zir (grey as grt, but lighter, with visible IR, top center and bottom left). SNrub, Sikkim, Himalaya (**dsp**). **b** Same scene as rt5 (**osp**). The enhanced contrast makes the bireflectance of rt more visible and makes q disappear. **c**, **d** Same scene as rt5 (**d+p**) and (**o+p**), respectively. Very strong anisotropism of gra. Massive IR of grt, q and zir ovderprint those of rt and prevent the extinction of grt from being perceived

metamorphic rocks, and, naturally, in its detrital derivatives.

Disseminated, it is usually associated—and often goes unnoticed—with hydrothermal alteration processes linked to the formation of various metal deposits. In cryptocrystalline form it can accompany titanite ± anatase as an integral part of *leucoxene*, a frequent secondary product derived from the liberation of Ti contained in the network of minerals such as biotite or inosilicates.

108.2.4 Identification

It is easily recognized by its relatively high reflectance (compared to gangue silicates, Fig. 108.2a, b), as well as by its hardness, good polishing, anisotropism, polysynthetic twinning and IR. Similar minerals include cassiterite, anatase, ilmenite, titanite, but these are generally easy to distinguish:

- Cassiterite is harder, polishes worse and is much less reflective, especially in immersion.
- Anatase is somewhat darker, lacks twins and its IR are colourless.
- Titanite, pseudobrookite and zircon are less reflective and less anisotropic.
- Ilmenite is more coloured and pleochroic, but darker (less reflective in the visible range); it usually lacks IR.

Fig. 108.3 **a**, **b** (Same scene: **dsp** and **osp**, respectively). Microbrecciated single crystal of rt and q. Deformation and twinning of rt, whose anisotropism is overprinted by its abundant to massive red IR. Sample CR-408

(a)

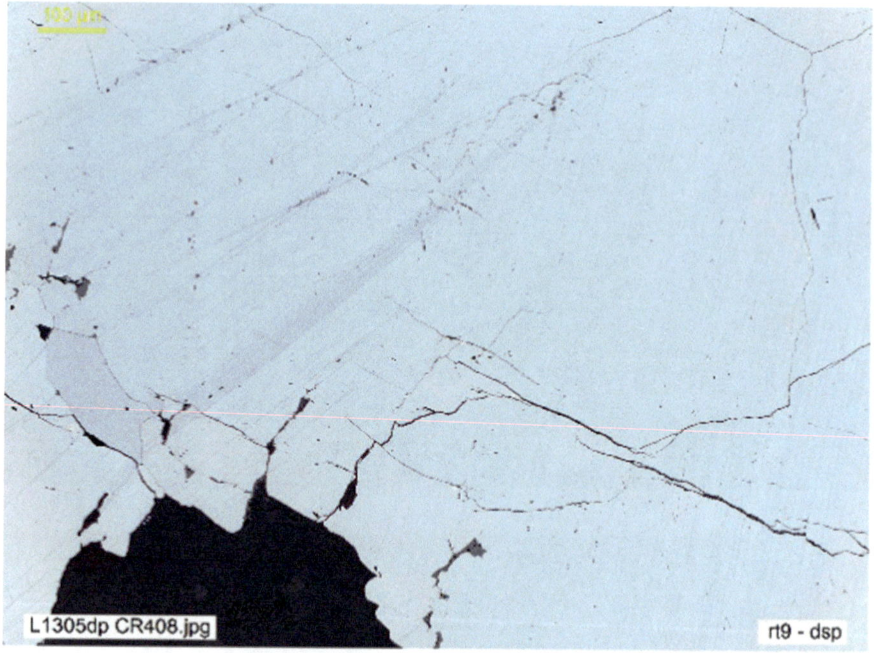

(b)

Fig. 108.4 **a**, **b** (Same scene: **d +p** and **o+p**, respectively). Massive ccp (yellow) with inclusions of sp (dark grey), rt (lighter grey, lower right), cbn (pinkish-brown lamellae) and mck traces. With *+p* setting, anisotropism and simple twinning of rt, easily visible due to the lack of IR, as rt is surrounded by the opaque ccp. Extreme anisotropism of mck allows it to be easily distinguished from cbn, of similar colour without analyser. Sample CR286, Gihen (Sulitjelma), Norway

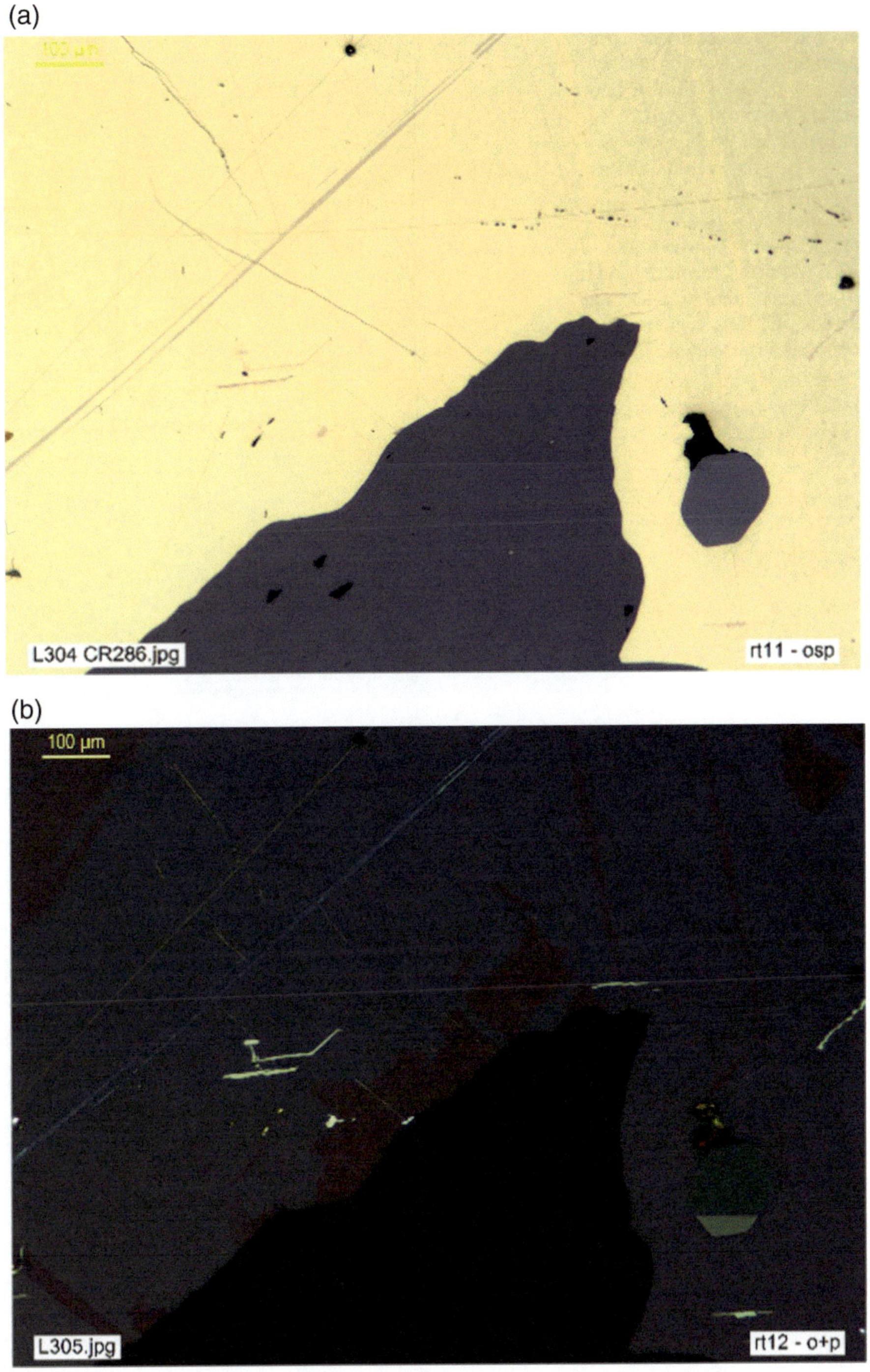

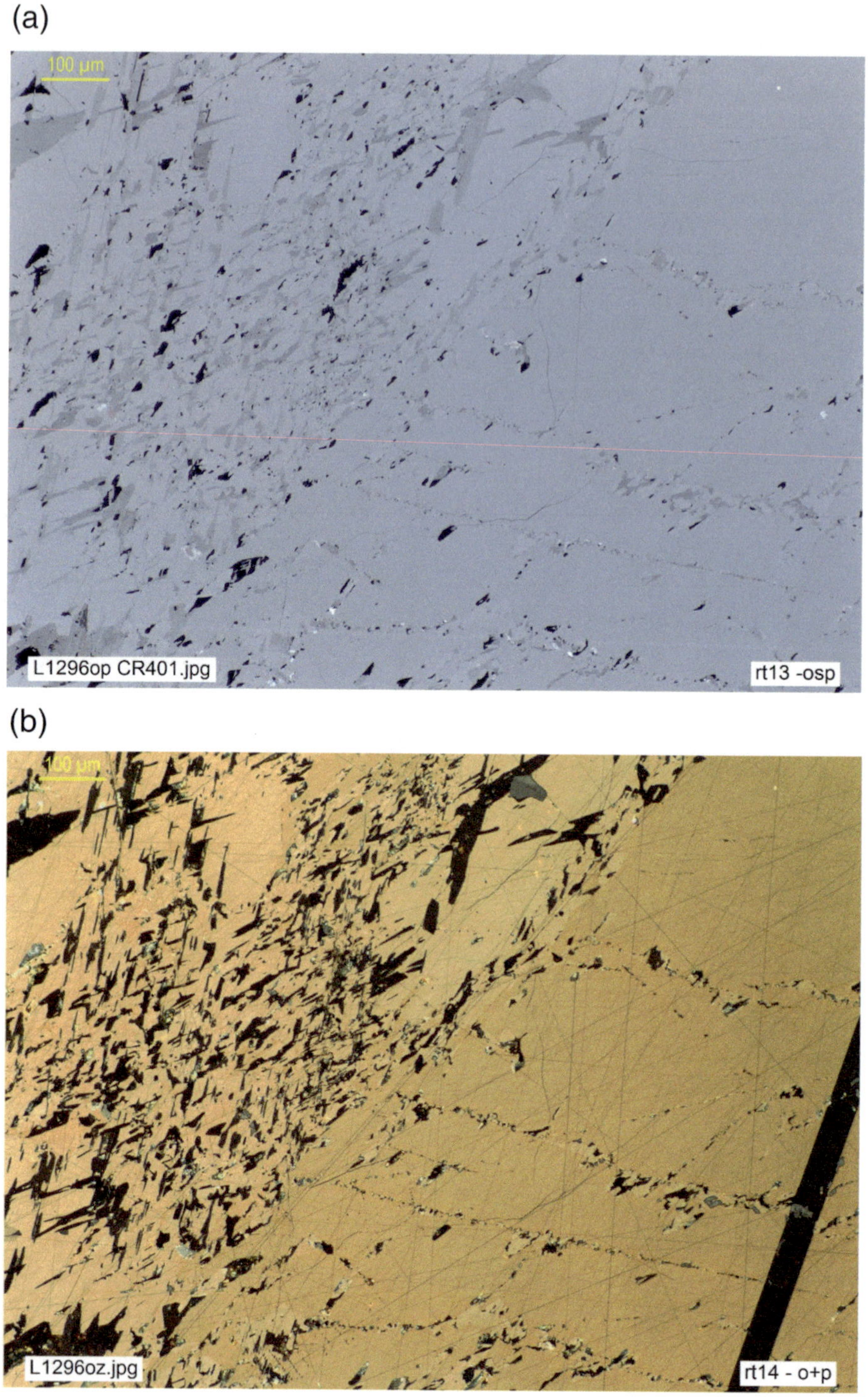

Fig. 108.5 **a**, **b** (Same scene: **osp** and **o+p**, respectively). Single crystal of Ta-rich rt (*strüverite*: grey), with exsolutions of ixiolite (darker grey) oriented after rt lattice planes. The scarcity of IR in this peculiar rt grain allows clear appreciation of the anisotropism of rt (*o+p*), from yellowish-orange-brown to olive-greenish-brown, depending on the orientation of each twin element. CR401, Canoa Pegmatite, Borborema, Brazil

108.3 Spectral Information

108.3.1 Rutile Multispectral Reflectance R (In Air, Without Polariser) (%)

λ (nm)	370	400	425	450	475	500	525	550	575	600	625	650	675	700	750	800	850	900	950	1000
μ	26.5	25.0	23.8	23.1	22.5	22.0	21.7	21.4	21.1	20.9	20.9	20.7	20.7	20.6	20.5	20.5	20.7	20.3	20.4	20.1
P_5	25.4	24.4	23.2	22.6	22.0	21.6	21.3	20.9	20.7	20.5	20.4	20.2	20.1	20.0	19.8	19.6	19.9	19.6	19.5	19.4
P_{20}	25.8	24.7	23.5	22.8	22.2	21.8	21.5	21.2	20.9	20.7	20.6	20.4	20.4	20.2	20.1	20.0	20.2	19.9	20.0	19.7
P_{80}	27.3	25.4	24.1	23.3	22.8	22.3	21.9	21.7	21.3	21.1	21.1	21.0	21.1	20.9	20.9	21.0	21.2	20.7	20.8	20.5
P_{95}	27.6	25.7	24.4	23.5	23.0	22.5	22.1	21.9	21.5	21.4	21.4	21.3	21.4	21.3	21.3	21.5	21.7	21.2	21.2	20.8

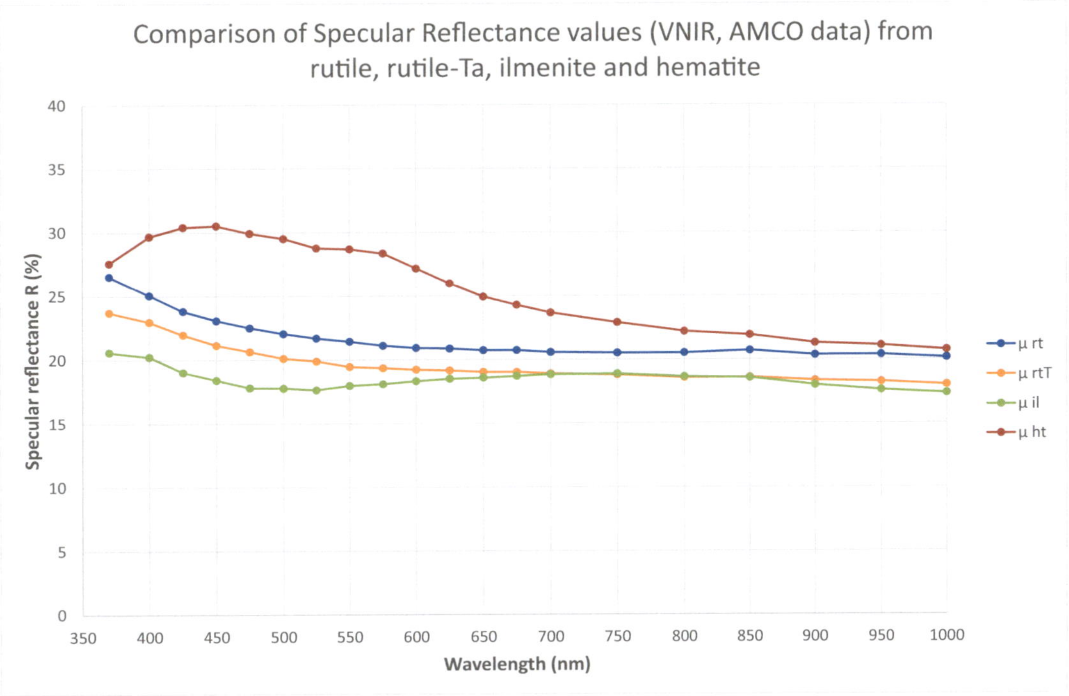

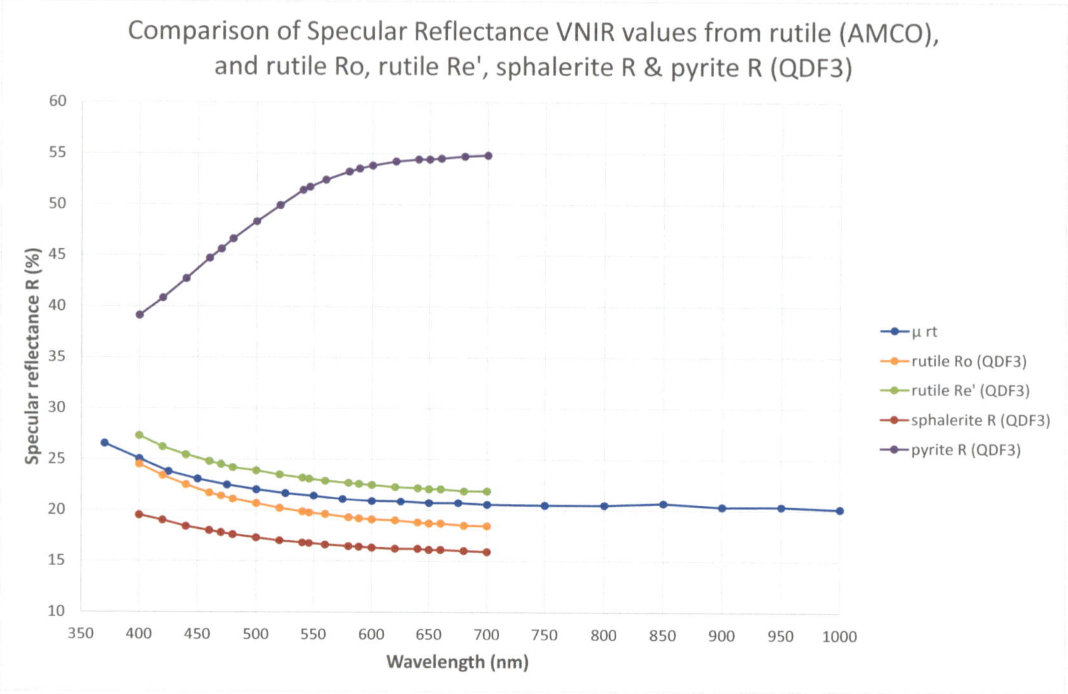

References

Ramdohr P (1980) The ore minerals and their intergrowths, 2nd edn, 2 vols. Pergamon Press, Oxford, 1205 p

Uytenbogaardt W, Burke EAJ (1971) Tables for microscopic identification of ore minerals, 2nd edn. Elsevier, Amsterdam

109 Safflorite (*sf*)

$CoAs_2$: Orthorhombic

Safflorite		R_{546} (%)	Y (%)	λ_d (nm)	P_e (%)	Colour sensation
Air	R_1	54.1	53.9	514	0.2	Bright white
	R_2	54.6	54.3	480	2.7	
Oil	R_1	39.6	39.3	495	0.6	
	R_2	40.15	39.8	479	4.5	
Bir: Weak (3)				*Pleoc*: Weak (3)		*Tarn*: -
Anis +P: Strong and coloured (5)				*IR*: –		*Twg*: Frequent, star-shaped
Mor: Idiomorphic, massive, mix-xx lo				*Clv*: –		*PolQ*: Good
Hardness	*PolH*: High, > sk, < asp, lo (4)			*Mohs*: ~5	*Talm*: E	*VHN* $_{(100\ g)}$: 792–882

Obs: cf. Table and R charts, multispectral data (VNIR, AMCO) and visible range (QDF3)
Mineral abbreviations: Ag—native silver asp—arsenopyrite, Bi—native bismuth, cob—cobaltite, fbg—freibergite, lo—löllingite, nc—nickeline, po—pyrrhotite, py—pyrite, rm—rammelsbergite, sf—safflorite, sk—skutterudite, td—tetrahedrite (s.l.)

109.1 General

Composition: $CoAs_2$, and in many cases (Co, Ni, Fe)As_2, as it often has important content of Fe, minor content of Ni, Bi and traces of Cu, S. Intergrowths and possible miscibility with löllingite, although the existence of a continuous series has not been demonstrated.

109.2 Microscopic Description

109.2.1 SP Observation (*Single Polariser*)

Pure bright white *colour*, without the yellowish hue of rammelsbergite. *Weak pleochroism*, hardly visible in air (white to bluish-white), more distinct in immersion (Figs. 109.1, 109.2 and 109.4).

Polishes very well, better than arsenopyrite. High *polishing hardness*, $PolH_{sf}$ > po, sk, rm, but < lo, asp, cob, py.

Reflectance high but < rm, distinctly < native Ag or Bi, against which safflorite appears grey, relatively dark, especially in immersion (Figs. 109.1a, b, and 109.2, cf. multispectral comparative R plots).

Tendency to euhedral development, stellate forms are typical (Figs. 109.1a, b, and 109.2), although not exclusive; occasionally it forms massive aggregates, also zoned—mixed crystals with löllingite.

109.2.2 Observation +P (*Polariser + Analyser at 90°*)

Strongly *anisotropic*, with vivid colours in bluish and brownish tones (when the analyser is partially uncrossed, slightly yellowish), often revealing the growth pattern. The polarisation tints can vary strongly with the composition, e.g., in nickeliferous varieties, reddish, purplish and dark blue tones appear.

The intensity and polarisation tints can vary widely, sometimes appearing almost isotropic zones or crystals.

As noted by Ramdohr (1980), among all the arsenides, safflorite shows the greatest variation in anisotropy effects, even within the same crystal (Figs. 109.1c, d and 109.5).

R. Castroviejo, *A Practical Guide to Ore Microscopy—Volume 1*,
https://doi.org/10.1007/978-3-031-12654-3_109

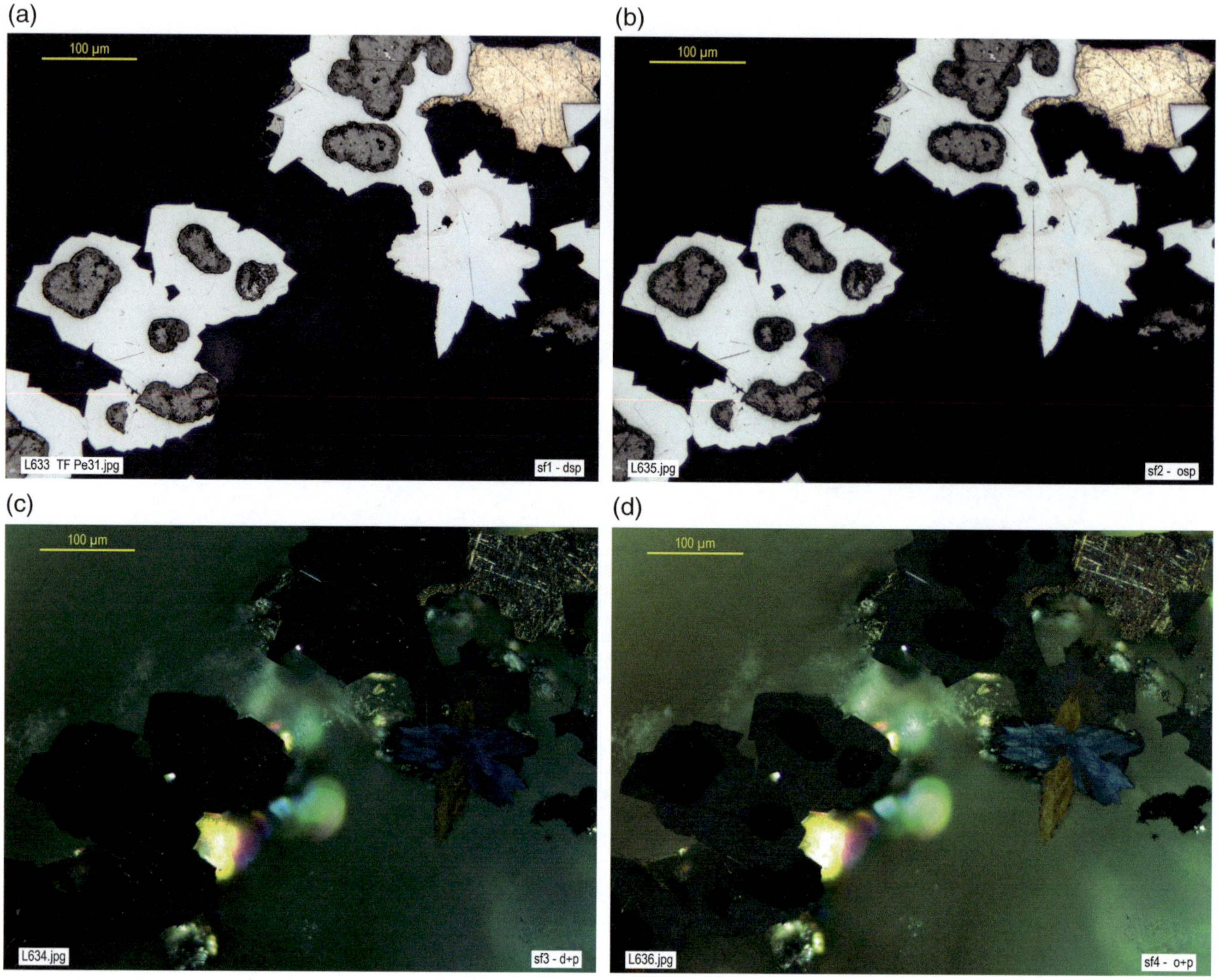

Fig. 109.1 **a** Vein infill of quartz (black), with impregnation of sk, sf and Bi. Bi (yellowish white, bright, very scratchy, top right corner) is associated with sk; colour and lustre of sk and sf are very similar, but sf shows a stellate morphology (and, unlike sk, is anisotropic, comp. Fig. 109.1c, d); colomorphic, secondary Bi-ochers form zoned grey inclusions in sk. Growth zonation discernible by faint colour variations in sk and sf (±lo), (**dsp**), **b** Same scene as Fig. 109.1a (**osp**): slight pleochroism of sf. TF Pe31, Schneeberg, Saxen, Germany. **c, d** (Same scene as Fig. 109.1: **d+p** and **o+p**, respectively). Isotropic character of sk, unlike distinctly anisotropic sf, which shows typical bluish and brownish colours and variations in the same crystal

Unlike löllingite, these effects are more pronounced in immersion than in air. It does not show *internal reflections*.

Its peculiar t*winning* is remarkable: triple *penetration twins* on (011), giving rise to the sf-typical six-pointed star forms (Fig. 109.2), whose development is more regular than in löllingite.

109.2.3 Paragenesis

More frequent mineral than commonly thought (in many classic collections it was confused, after Ramdohr, 1980, with cobaltiferous skutterudite). It typically occurs in silver deposits, particularly in hydrothermal veins of the *five-element formation* (Ni–Co–Bi–Ag–U). Also occurs in metamorphic types.

It may be accompanied by Ni/Co/Fe arsenides (particularly skutterudite, löllingite, nickeline), Ag and Bi minerals, cobaltite, etc.

109.2.4 Identification

Typical characteristics for its diagnosis: its tendency to idiomorphism and particularly to the appearance of star-shaped twins, white colour, weak pleochroism (but

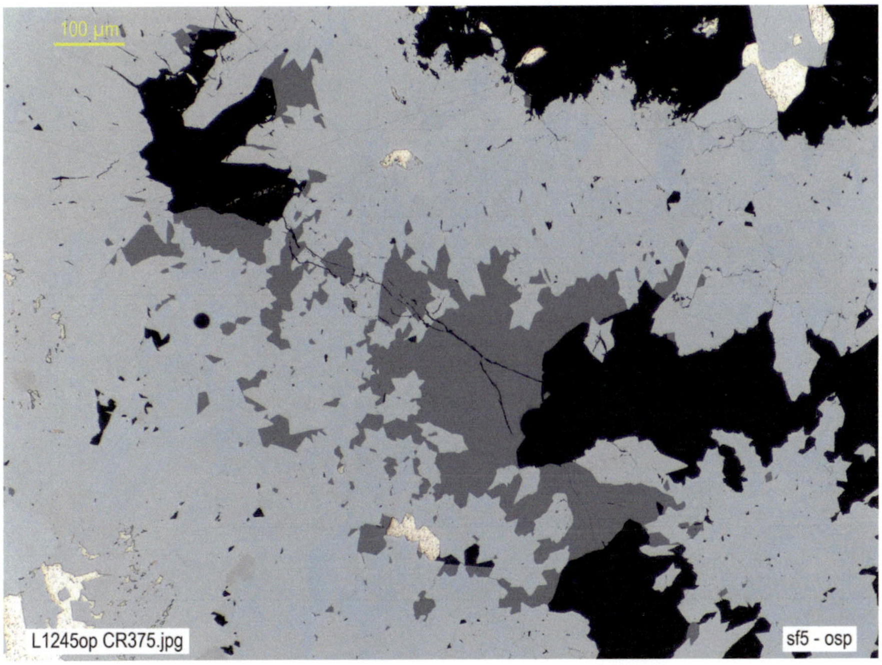

Fig. 109.2 Aggregate of sf, lo, fbg, cob, Ag. The white colour of sf and lo appears here grey by contrast with Ag (bright white); sf and lo form microcrystalline aggregates, slightly pleochroic, with typical star-twinned morphology (sf) or with thin twinned prismatic or radiating aggregates (both). Disseminated Ag inclusions in sg-lo mass, whose interstitial spaces are coated in part by td (dark grey, probably fbg); cob forms minute pink inclusions in the sf-lo aggregate. CR 375, Cobalt, Ontario, Canada (**osp**)

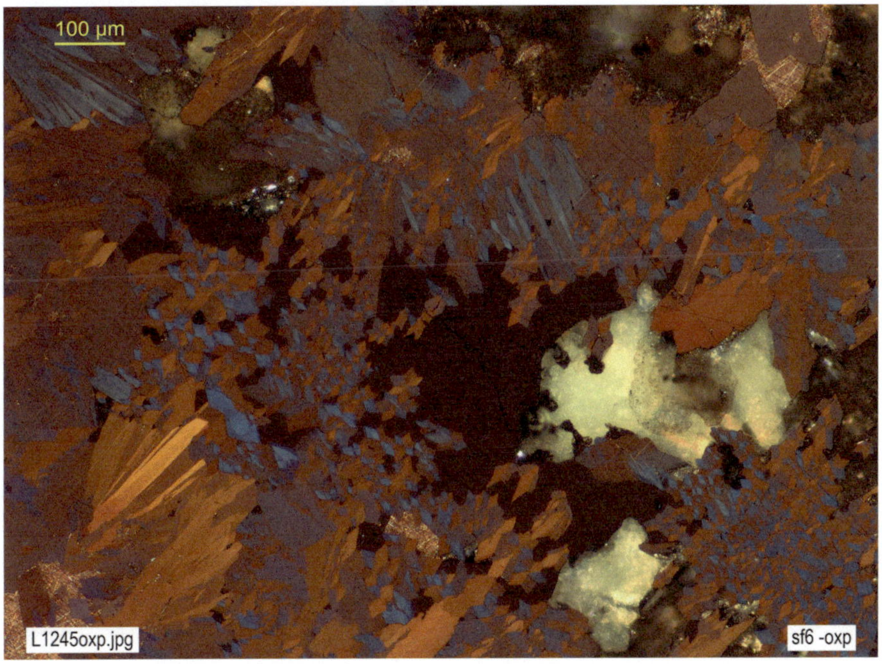

Fig. 109.3 Same scene as Fig. 109.2 (**o+p**)

visible in immersion), high reflectance, very good polishing and strong anisotropism, with pronounced polarisation tints.

However, sf can be confused with other white arsenides, particularly löllingite and rammelsbergite. The latter is more reflective, shows a yellowish tone and abundant polysynthetic twins, absent in safflorite, while the former has more vivid polarisation tints and its stellate twins are more imperfect than those of safflorite. Nevertheless, the distinction between safflorite and löllingite is not always possible in mixed or zoned crystals whose compositions may vary between the two species. In general, the determination of polishing hardnesses can provide a useful criterion.

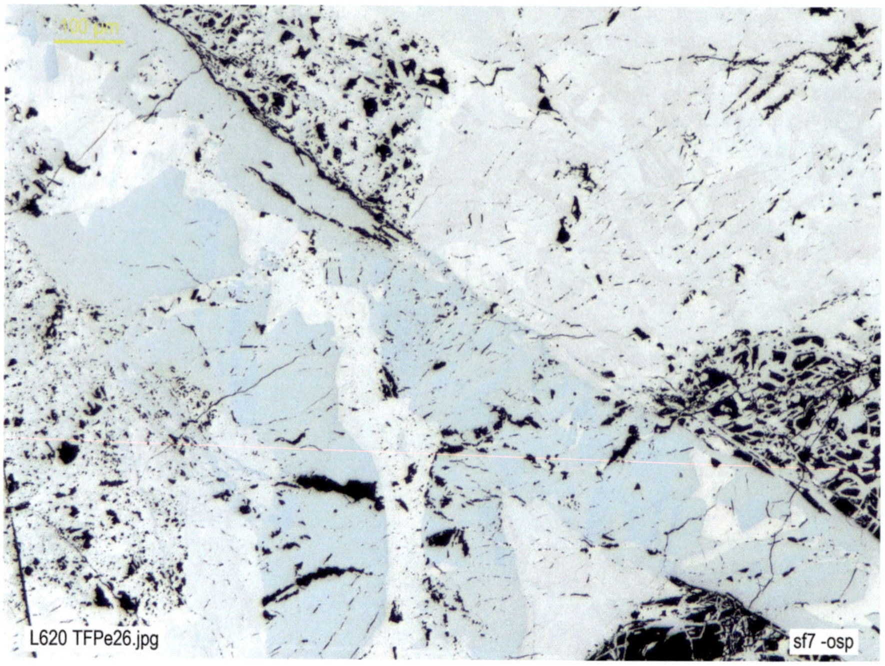

Fig. 109.4 Massive rm (white, bright, pleochroic), cut by sk + sf veinlets (bluish-white tones, less reflective than rm). Later rm generation in microveinlets (*lower left half of image*), crosscutting the former ores. Weak but visible pleochroism of rm (very faintly pinkish-bluish-white) and sf (slightly bluish-white). TFPe.26 ETSIME, Madrid (**osp**)

Fig. 109.5 Same scene as Fig. 109.4 (**o+p**): the main mass of rm has zones of very different grain size and presents abundant polysynthetic twins in grains of sufficient size; striking and vivid polarisation tints of rm and sf, unlike sk (isotropic)

109.3 Spectral Information

109.3.1 Safflorite Multispectral Reflectance R (in Air, Without Polariser) (%)

λ (nm)	370	400	425	450	475	500	525	550	575	600	625	650	675	700	750	800	850	900	950	1000
μ	51.7	52.5	52.8	53.4	53.5	53.6	53.6	53.4	53.4	53.3	53.1	52.9	52.9	52.8	52.4	51.9	51.8	51.3	51.1	50.3
P_5	36.5	38.0	39.2	40.3	41.5	42.7	44.0	45.3	46.6	47.9	49.3	50.7	51.0	50.8	50.3	49.7	49.4	48.8	48.6	47.5
P_{20}	51.7	52.6	52.5	53.3	53.1	53.3	53.1	53.0	52.8	52.6	52.3	52.0	52.1	52.0	51.4	50.7	50.3	49.6	49.3	48.2
P_{80}	55.1	55.9	56.2	56.2	55.9	55.7	55.5	55.2	54.9	54.7	54.4	54.0	54.0	53.8	53.4	52.7	52.4	51.9	51.9	51.1
P_{95}	56.6	56.5	57.0	56.7	57.4	56.2	56.4	55.6	55.6	55.4	55.1	54.9	54.6	54.4	55.8	57.1	58.7	59.0	59.2	59.5

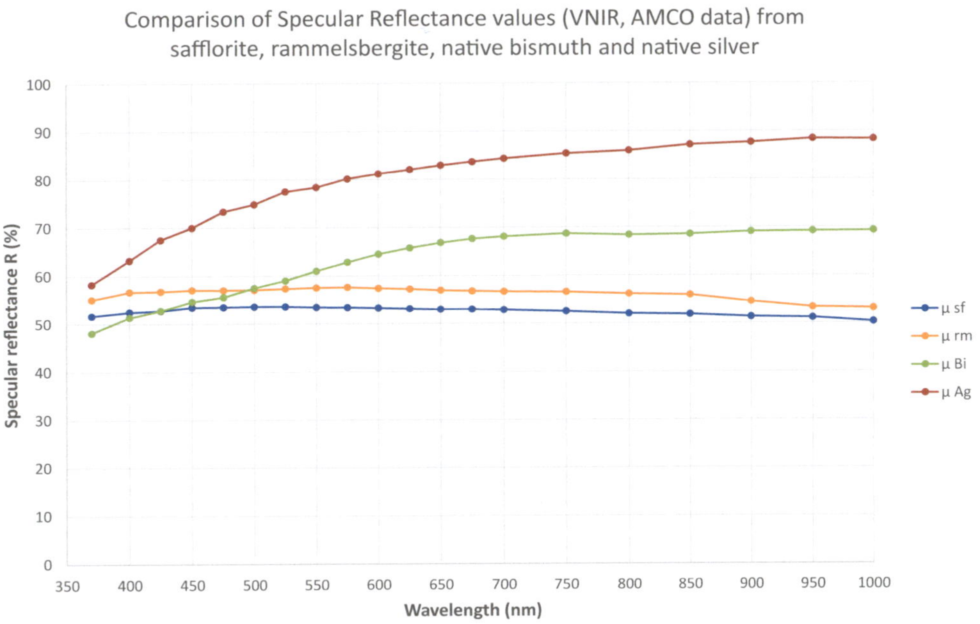

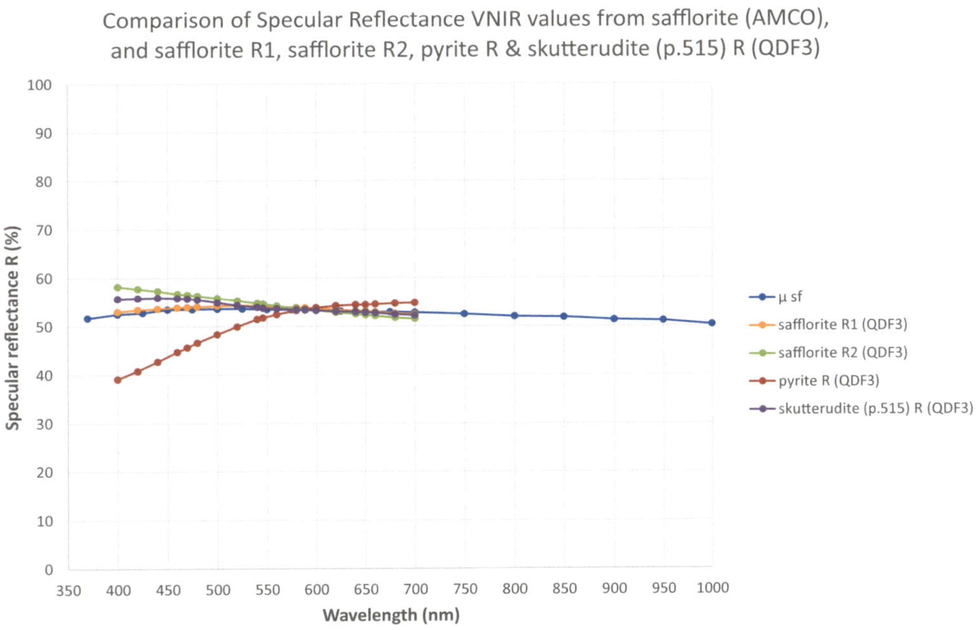

Reference

Ramdohr P (1980) The ore minerals and their intergrowths, 2nd edn. Pergamon Press, Oxford. 2 volumes, 1205p

Scheelite (*sch*)

110

$Ca(WO_4)$: Tetragonal

Scheelite		R_{546} (%)	Y (%)	λ_d (nm)	P_e (%)	Colour sensation
Air	R_o	9.8	9.8	472	2.8	Grey, somewhat lighter than gg
	$R_{e'}$	10.1	10.2	472	2.5	
Oil	R_o	1.4	1.4	472	4.1	Dark grey, but lighter than gg
	$R_{e'}$	1.5	1.5	471	4.1	
Bir: Imperceptible (2)				*Pleoc*: Imperceptible (2)		*Tarn*: No
Anis +P: Imperceptible by IR (3)				*IR*: Massive, colourless		*Twg*: (110), (001)
Mor: Massive, sometimes tabular				*Clv*: (101), not very visible		*PolQ*: Good
Hardness	*PolH*: High, but < wf (4)			*Mohs*: 5–	*Talm*: D–	*VHN* $_{(100\ g)}$: 383–464

Obs: cf. Table and R charts, multispectral data (VNIR, AMCO) and visible range (QDF3)
Mineral abbreviations: bm—bismuthinite, cb—columbite, ccp—chalcopyrite, cst—cassiterite, gg—gangue, gn—galena, mol—molybdenite, po—pyrrhotite, py—pyrite, wf—wolframite

110.1 General

Important ore of W. Composition $Ca(WO_4)$, may contain impurities of Ta, Nb, Cu, Mo. Isostructural with powellite, $Ca(MoO_4)$, powellite-scheelite series. Epitaxial with wolframite, with {001} [110] of scheelite parallel to {010} [001] of wolframite.

Thermoluminescent. Bluish-white fluorescence (under short-wave UV), yellowish if containing Mo, inhibited by Fe.

Transparent mineral, it can also be characterized by transmitted light in petrographic microscope.

110.2 Microscopic Description

110.2.1 SP Observation (*Single Polariser*)

Grey *colour*, dark but a little lighter than gangue, with which it could be confused if reflectances, **R**, are not compared. R makes a noticeable difference in perceptible brightness: R gangue ≈ 5%; R scheelite ≈ 10% (Figs. 110.1a, b and 110.4).

Good *polish*, although with occasional scratches and pull-outs. High *polishing hardness*, but lower than wolframite, cassiterite, quartz and columbite.

Low *reflectance*, R sch ≈ R cst (R_o cst ≈ $R_{e'}$ sch > R_o sch) or slightly lower ($R_{e'}$ cst > $R_{e'}$ sch > R_o sch, cf. QDF3) and R sch < wf; R sch between: R_1 and R_2 of titanite (cf. comparative plots R). R sch is much lower than R of sulphides (Fig. 110.2).

It occurs in massive or granoblastic aggregates, occasionally developing crystalline forms (pseudo-octahedral, tabular). Occasional zoning.

110.2.2 Observation +P (*Polariser + Analyser at 90°*)

Anisotropic, but the effects of anisotropism are imperceptible under the RL microscope, owing to the ubiquity and intensity of the *internal reflections*, usually colourless, massive and very luminous (Figs. 110.1c, d, 110.3, and 110.5).

Occasional *twinning*, {110} contact or penetration.

R. Castroviejo, *A Practical Guide to Ore Microscopy—Volume 1*,
https://doi.org/10.1007/978-3-031-12654-3_110

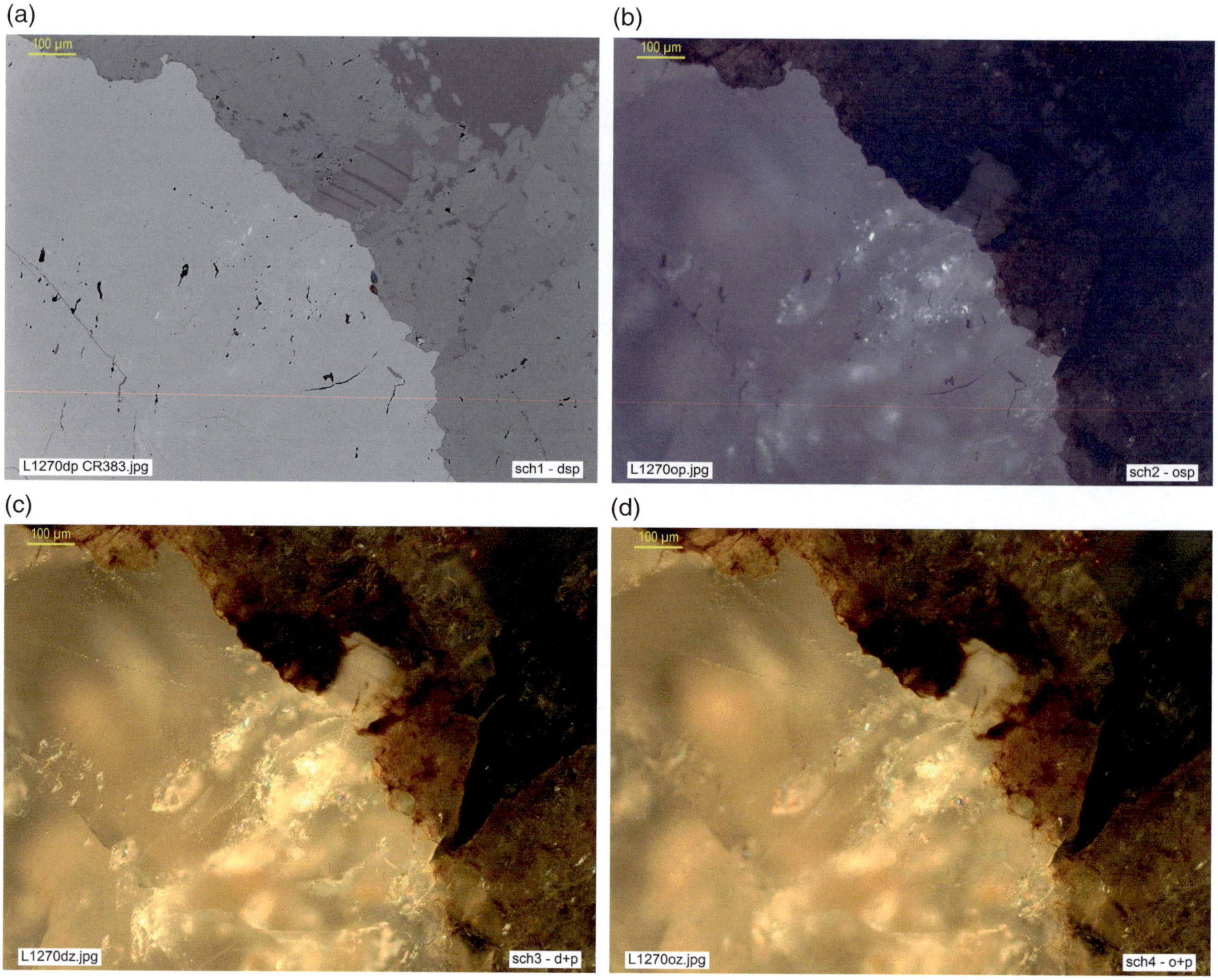

Fig. 110.1 **a** Aggregate of sch (light grey, large crystal occupying lower left half of image), with intergrown calcite (grey, bireflective, polysynthetic twins) and quartz (grey). CR-383, Dinkey Creek, Fresno County, CAL, USA (**dsp**). **b** Same scene as sch1 (**osp**). Faint sch fluorescence. **c**, **d** Same scene as sch1 (**d+p**) and (**o+p**), respectively. The generalized internal reflections of sch bring to light several planes of secondary fluid inclusions

110.2.3 Paragenesis

Metamorphogenic deposits, typically contact metamorphic, skarn, and metasomatic (greisen); pegmatites and hydrothermal seams. Frequently associated with wolframite, which it tends to replace, although the reverse process also occurs, as well as the replacement of scheelite by stolzite, Pb(WO_4), by reaction with galena.

Other companions: molybdenite, arsenopyrite, pyrrhotite, bismuthinite, as well as calc-silicate skarn associations.

As a detrital mineral, it is concentrated in alluvial placers.

110.2.4 Identification

Its grey colour, low reflectance, internal reflections and fluorescence, together with the paragenesis, allow it to be identified.

Fig. 110.2 Replacement of wf (light grey) by sch (darker grey) and ccp (yellow). In black, pull-outs due to difficult polishing of wf. CR306, As Sombras Mine, Lobios, Orense (**dsp**)

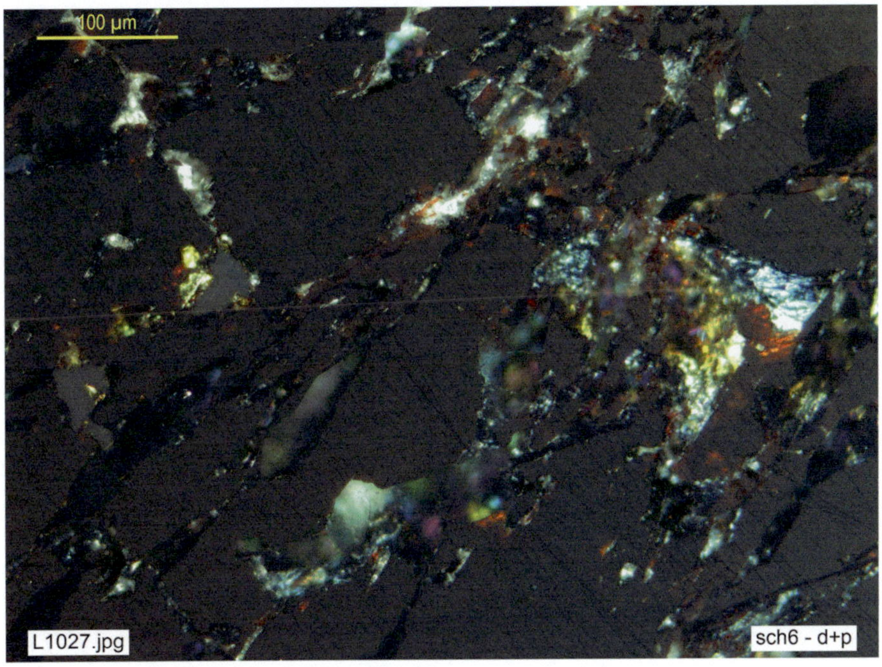

Fig. 110.3 Same scene as sch5 (**d+p**). Red IR from wf, colourless and abundant from sch

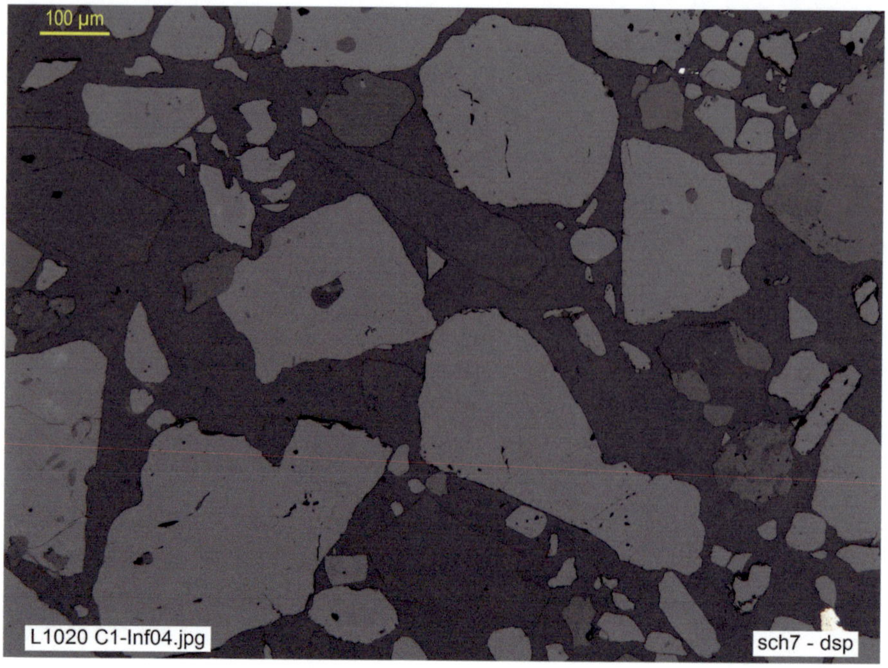

Fig. 110.4 Concentrate of sch (light grey), with darker grey gg (q, the darkest, like resin; calcite, bireflectant; diopside, the lightest) and with a small grain of py (bottomright). Sample C1-LMA04, Los Santos Mine, Cáceres (**dsp**)

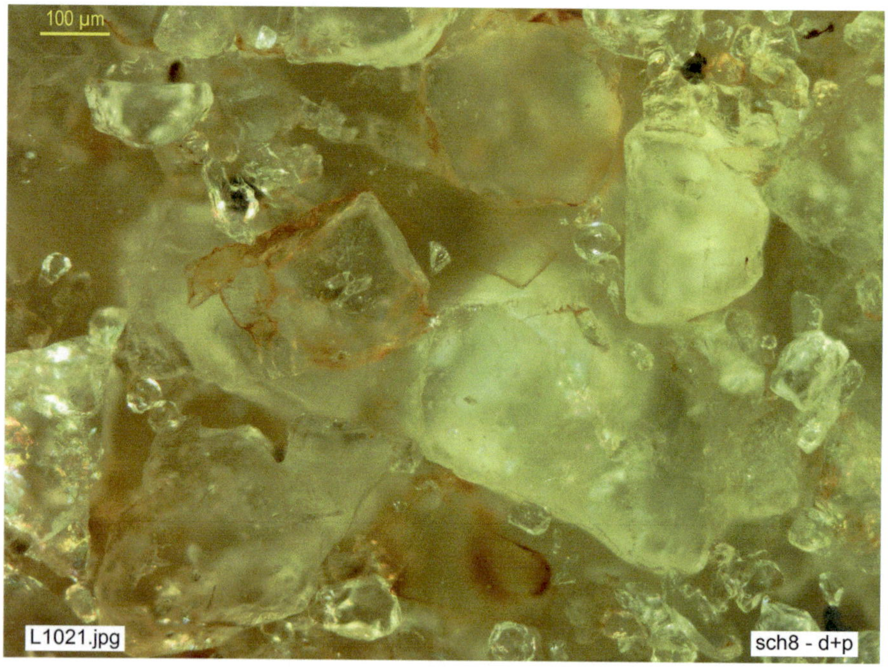

Fig. 110.5 Same scene as sch7 (**d+p**)

110.3 Spectral Information

110.3.1 Scheelite Multispectral Reflectance R (in Air, Without Polariser) (%)

λ (nm)	370	400	425	450	475	500	525	550	575	600	625	650	675	700	750	800	850	900	950	1000
μ	12.0	11.9	11.8	11.7	11.7	11.6	11.6	11.5	11.4	11.4	11.4	11.3	11.3	11.3	11.2	11.2	11.3	11.3	11.3	11.3
P_5	11.2	11.0	10.9	10.7	10.6	10.5	10.4	10.3	10.3	10.2	10.2	10.1	10.1	10.0	10.1	10.2	10.5	10.5	10.5	10.5
P_{20}	11.5	11.3	11.2	11.1	11.0	10.9	10.8	10.7	10.7	10.6	10.6	10.5	10.5	10.5	10.5	10.6	10.7	10.7	10.7	10.7
P_{80}	12.5	12.3	12.2	12.1	12.1	12.0	12.0	12.0	12.0	12.0	11.9	11.9	11.9	11.8	11.7	11.8	12.0	12.0	12.0	11.9
P_{95}	13.4	13.2	13.1	13.1	13.1	13.0	13.0	12.9	12.9	12.8	12.8	12.7	12.7	12.6	12.6	12.6	12.5	12.7	12.8	12.7

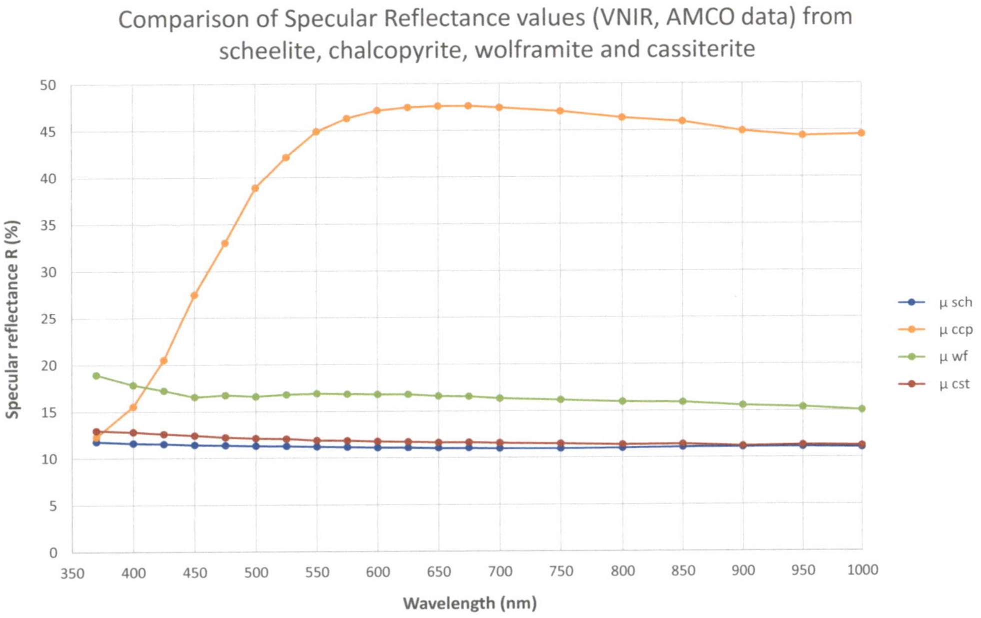

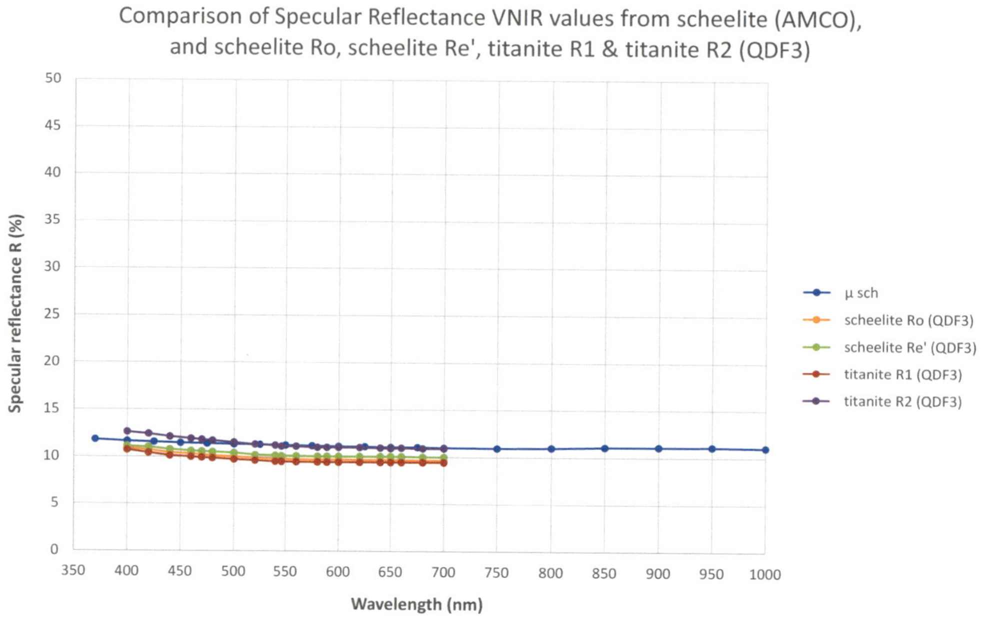
Comparison of Specular Reflectance VNIR values from scheelite (AMCO), and scheelite Ro, scheelite Re', titanite R1 & titanite R2 (QDF3)
Specular reflectance R (%)
Wavelength (nm)
µ sch
scheelite Ro (QDF3)
scheelite Re' (QDF3)
titanite R1 (QDF3)
titanite R2 (QDF3)

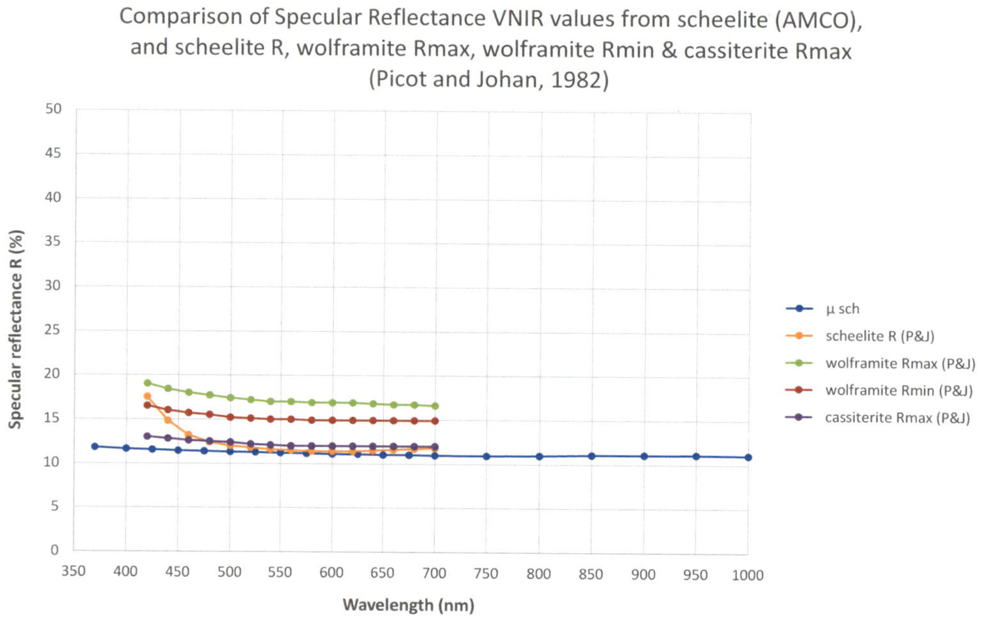
Comparison of Specular Reflectance VNIR values from scheelite (AMCO), and scheelite R, wolframite Rmax, wolframite Rmin & cassiterite Rmax (Picot and Johan, 1982)
Specular reflectance R (%)
Wavelength (nm)
µ sch
scheelite R (P&J)
wolframite Rmax (P&J)
wolframite Rmin (P&J)
cassiterite Rmax (P&J)

Comparison of Specular Reflectance values in the visible spectrum from scheelite Ro & scheelite Re' (QDF3), and scheelite R & wolframite Rmax (Picot and Johan, 1982)

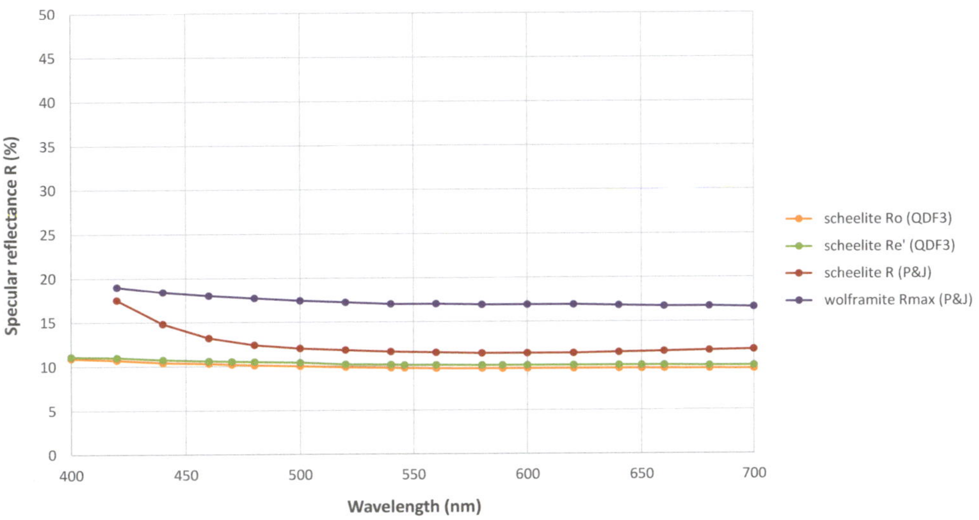

Silver (*Ag, native silver*) 111

Ag (±Au, Sb, Hg, Cu, Bi): Cubic

Silver		R_{546} (%)	Y (%)	λ_d (nm)	P_e (%)	Colour sensation
Air		93.4	92.9	574	3.9	White to yellowish white, very bright
Oil		91.5	90.9	574	4.2	
Bir: No (1)				*Pleoc*: No (1)		*Tarn*: Fast, dark and coloured
Anis +P: Isotropic, no total extinction (2)				*IR*: No		*Twg*: Occasionally
Mor: Fine aggregates, allotriomorphic				*Clv*: No		*PolQ*: Easy, but many scratches
Hardness	*PolH*: Medium (3)			*Mohs*: 3^-	*Talm*: B	*VHN* $_{(100\ g)}$: 60–65

Obs: cf. Table and R charts, multispectral data (VNIR, AMCO) and visible range (QDF3; Picot and Johan 1982). Hydrothermal deposits, particularly epithermal, also *Kupferschiefer*. Secondary (supergene) enrichment

Mineral abbreviations: Ag—native silver, ac—acanthite, asp—arsenopyrite, Au—native gold, ccp—chalcopyrite, gn—galena, plb—polybasite, pru—proustite, sp—sphalerite, td—tetrahedrite

111.1 General

Composition: Ag, often with traces or variable contents of Au. Both metals can form unlimited solid solution, the compositions: <1%, ≈20% and >60% Au (Ramdohr, 1980) being more frequent in nature.

The compositions 30–70% Ag are known as *electrum*, a name now abandoned by IMA-COM in favour of *aurian silver* Ag-Au, but still in frequent use (cf. discussion on electrum).

Ag may also contain Sb, Hg (up to 12%), Cu, Bi, with possible solid solution up to 1% (Cu) and 8% (Bi); Sb contents ≤5% define the *antimonian silver* variety. With higher Sb contents different species appear, well defined in the Ag-Sb system, as dyscrasite or allargentum (cf. respective descriptions).

In atmospheric conditions, it tends to darken relatively fast due to surface alteration.

111.2 Microscopic Description

111.2.1 SP Observation (*Single Polariser*)

Bright white *colour* (Fig. 111.1a), with a slightly yellowish tinge that becomes more accentuated with higher Au contents (electrum).

It is easily polished but, due to its low hardness (VHN 60–65), polishing scratches are unavoidable and very numerous (Fig. 111.1b).

In polished section, Ag tarnishes quickly, showing a multi-coloured *patina* which masks its appearance and can make it unrecognizable. However, this is easily and quickly removed by cleaning or repolishing with fine abrasive (diamond 0.25 μm), comp. Figs. 111.2 and 111.5.

Ag *reflectance* ***R*** (≈93%, 546 nm; 85%, white light) is the highest known (visible range), so that next to it almost all other minerals appear dark (Figs. 111.2, 111.3 and 111.4), particularly in oil because of the higher contrast (Fig. 111.1b). Nevertheless, R (or brightness) is lower with growing Sb contents (Ag Y%, air, according to QDF3 data, is 92.9 for 0.87% Sb but 86.1, for 4.8% Sb, i.e., the *antimonial silver* variety). In Ag-Au compounds (*gold-silver alloys*), careful measurement of the reflectance allows determination of the Ag-Au ratio by its correlation with the spectral composition.[1] In addition to composition-related variations, the difficulty of achieving a perfect polish and the rapid tarnishing of Ag can lower the measured R-values (cf. Table and R-graphs). Isotropic, although reflections in the different polishing scratches may give a false idea of anisotropism when the stage is rotated.

[1] cf. Ramdohr 1980: v.1, p. 322, Fig. 384.

R. Castroviejo, *A Practical Guide to Ore Microscopy—Volume 1*,
https://doi.org/10.1007/978-3-031-12654-3_111

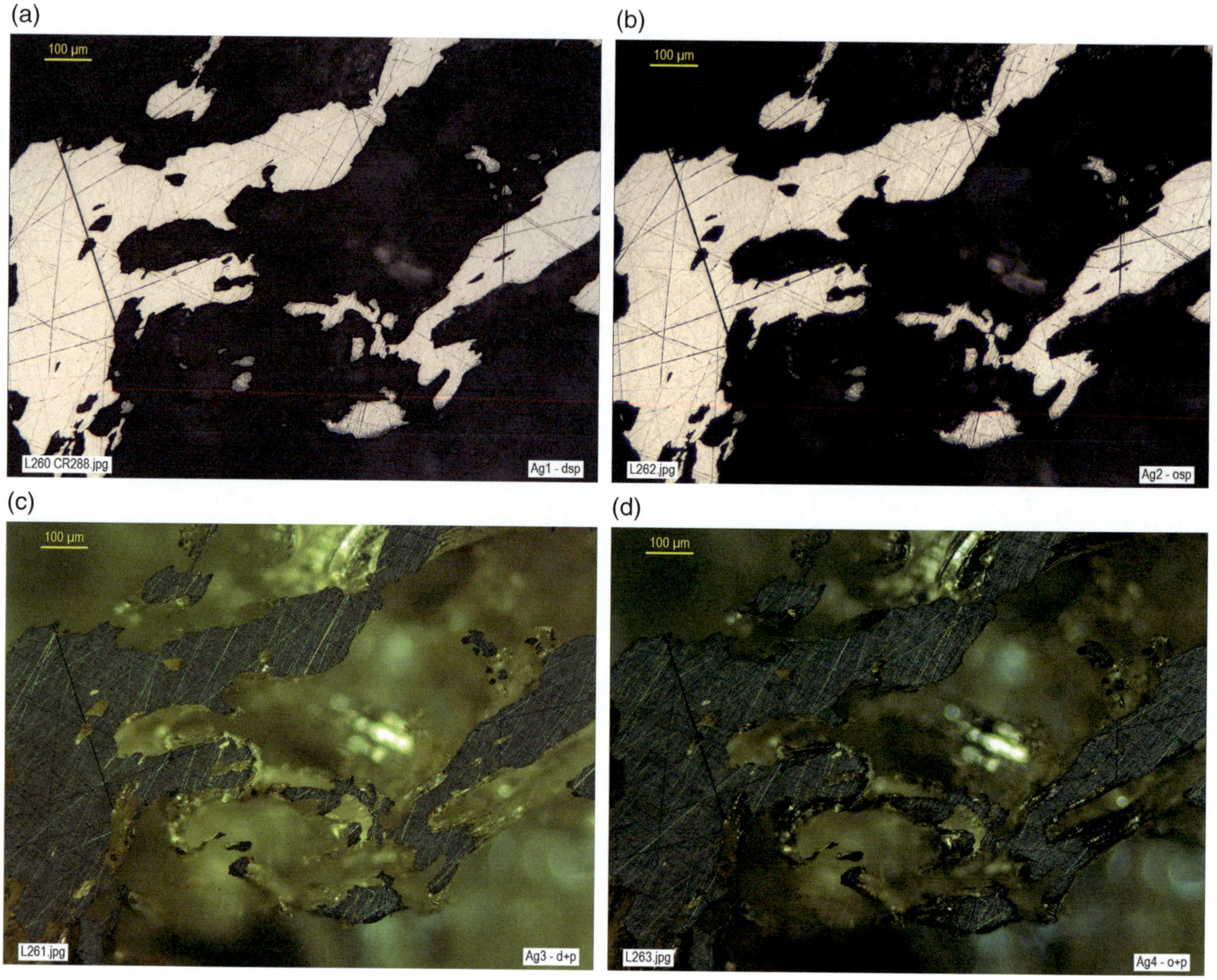

Fig. 111.1 **a** Native silver (white) on quartz gangue (dark grey). CR 288: Schneeberg Mine, Erzgebirge, Germany (**dsp**). **b** Same scene (**osp**). **c** Same scene (**d+p**). **d** Same scene (**o+p**)

Medium *polishing hardness*, similar to chalcopyrite, between that of galena and tetrahedrite, much lower than sphalerite ($PolH_{Ag} \approx$ ccp; $PolH_{Ag} >$ gn. $PolH_{Ag} <$ td, and << sp).

It usually occurs disseminated or as hydrothermal infill, in both cases in allotriomorphic grains or aggregates, occasionally developing peculiar dendritic or filiform textures, but its morphology can be varied.

111.2.2 Observation +P (*Polariser + Analyser at 90°*)

Isotropic, as cubic minerals but, with *imperfect extinction* (*+p*), showing a striking, irregular luminosity reinforced by the polishing scratches and the false polarization effects associated with them (Fig. 111.1c, d).

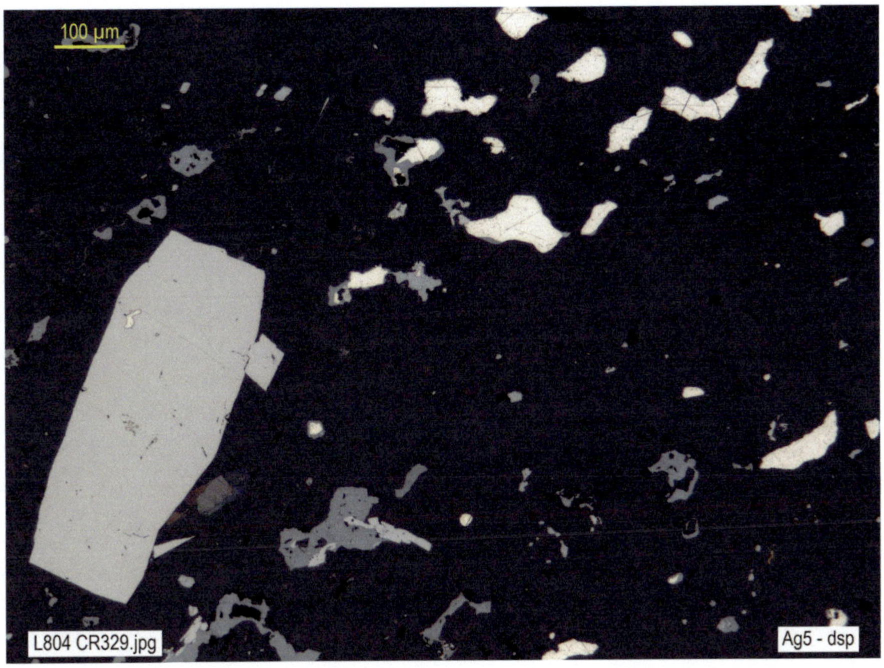

Fig. 111.2 Native silver (white) disseminated in gangue, with arsenopyrite (light grey, left), acanthite (grey) and traces of polybasite and proustite (grey). CR-329: Imiter, Morocco (**dsp**)

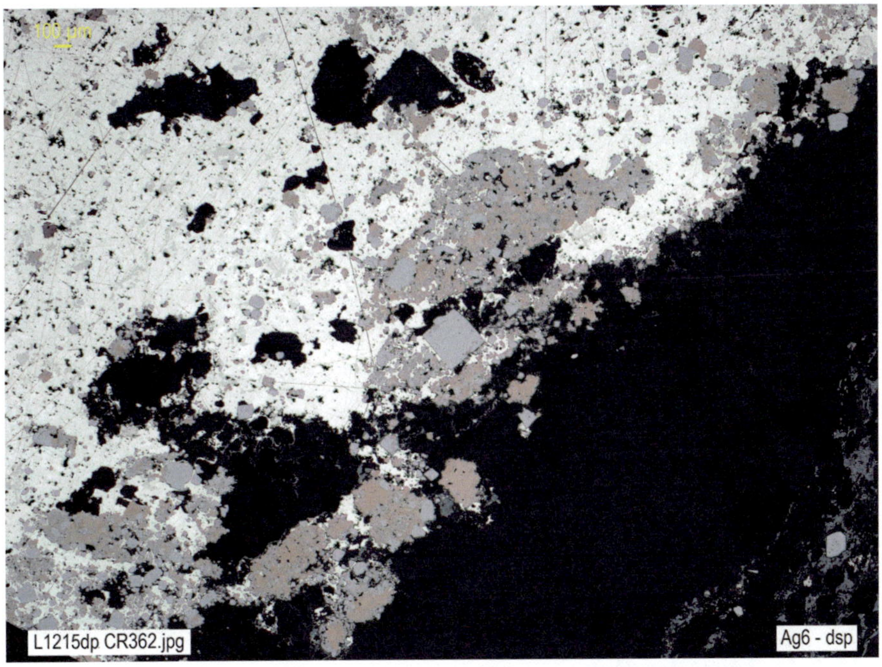

Fig. 111.3 Native silver (white) with skutterudite (grey, cubic euhedral tendency) and nickeline (pinkish brown) on quartz gangue (dark grey). CR-362: Nipissing Mine, Cobalt, Ontario, Canada (**dsp**)

No internal reflections. Occasional twinning and zoning visible, especially in coarse-grained aggregates.

111.2.3 Paragenesis

Rare mineral but of great economic interest. Paragenesis generally complex, typical of hydrothermal and especially epithermal deposits.

Although native silver is usually found, at least in traces, in many argentiferous deposits, most of the argentiferous ores normally occur in accompanying sulphide or sulphosalt phases (acanthite, red silvers, freibergite, etc., Figs. 111.2, 111.3 and 111.4) and also in antimonial compounds such as allargentum and dyscrasite. The oxidation and surface mobilization of silver from these primary ores gives rise to remarkable secondary or supergene enrichment concentrations, in which native silver can become the dominant ore.

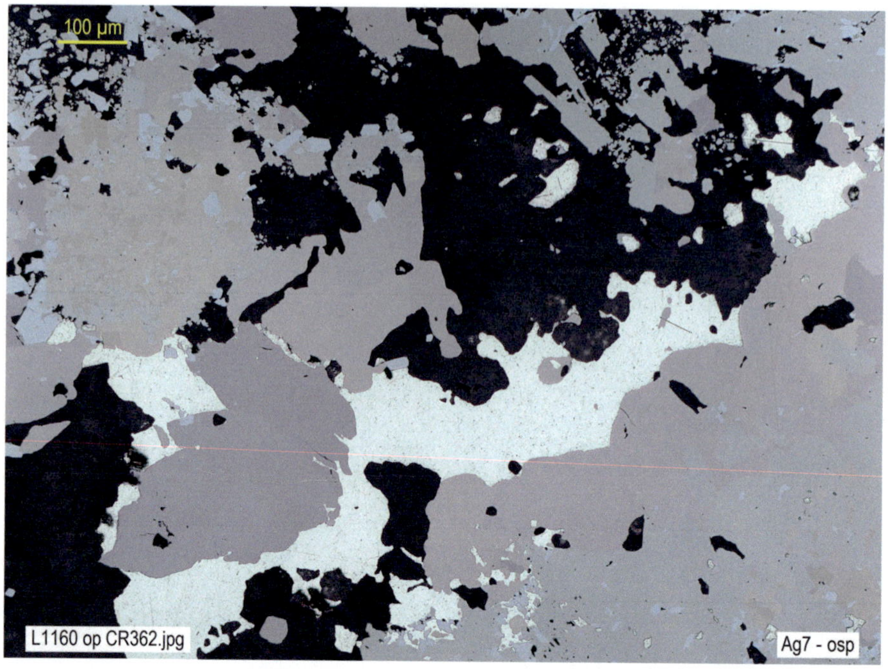

Fig. 111.4 Native silver (white) with breithauptite, nickeline, maucherite, violarite (different brownish tones, pinkish, purplish or orange), pararammelsbergite (pure grey, no pinkish shades, euhedral tendency) and quartz gangue (dark grey). CR-352: Cobalt, Ontario, Canada (**osp**)

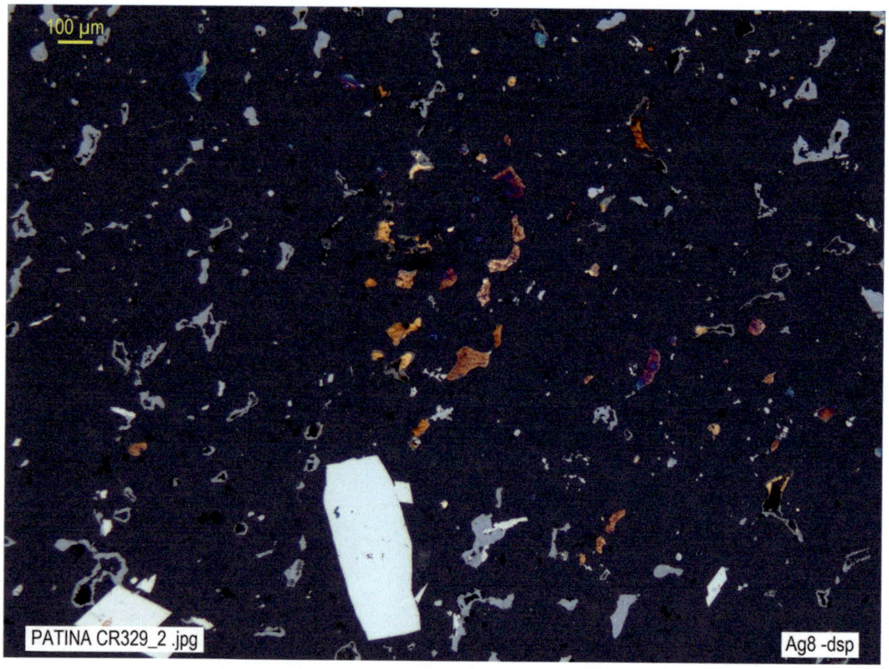

Fig. 111.5 Native silver, darkened and coloured by tarnishing, disseminated in gangue with asp, ac, plb and pru (described in Fig. 111.2, which is an easily recognizable detail view of the central zone of Fig. 111.5, photographed after cleaning the tarnishing patina). CR-329: Imiter, Morocco (**dsp**)

111.2.4 Identification

Native silver is, in general, easily recognized by its extreme reflectance (the highest of all known ores), white colour, frequent polishing scratches, fast tarnishing to dark often multicoloured patina, medium polishing hardness, isotropic character (but with incomplete extinction) and paragenesis. The colour becomes more yellow with increasing gold contents (electrum), which also prevent or diminish tarnishing.

To avoid possible confusion with other high reflectance compounds or native metals of similar appearance, the following observations can be of assistance:

- Platinum is whiter but less reflective, shows no tarnishing and hardly any polishing scratch, its polishing hardness being clearly higher than PolH Ag.
- Native Bi, Sb and As are softer, less reflective (darker) and anisotropic.
- The antimonial compounds allargentum and dyscrasite are less reflective and anisotropic (Fig. 111.1c, d).

111.3 Spectral Information

111.3.1 Silver Multispectral Reflectance R (in Air, Without Polariser) (%)

λ (nm)	370	400	425	450	475	500	525	550	575	600	625	650	675	700	750	800	850	900	950	1000
μ	58.2	63.2	67.5	70.0	73.3	74.8	77.5	78.3	80.1	81.2	82.0	82.9	83.6	84.3	85.3	85.9	87.1	87.6	88.4	88.3
P_5	45.2	50.5	54.3	57.0	59.2	62.2	64.6	66.5	68.4	69.9	71.0	72.1	73.3	74.3	76.0	77.1	78.9	79.5	80.8	80.8
P_{20}	48.5	54.0	57.7	61.4	64.3	67.1	69.4	71.2	72.9	74.1	75.1	76.1	77.2	78.1	79.6	80.7	82.4	82.7	84.0	83.9
P_{80}	64.3	69.3	73.9	76.0	79.6	80.6	83.4	83.9	85.7	86.6	87.4	88.3	88.8	89.4	90.2	90.6	91.6	92.2	92.7	92.6
P_{95}	66.7	71.6	76.3	78.3	81.8	82.6	85.4	85.7	87.4	88.3	89.1	89.9	90.4	90.9	91.6	91.9	92.9	93.4	93.8	93.7

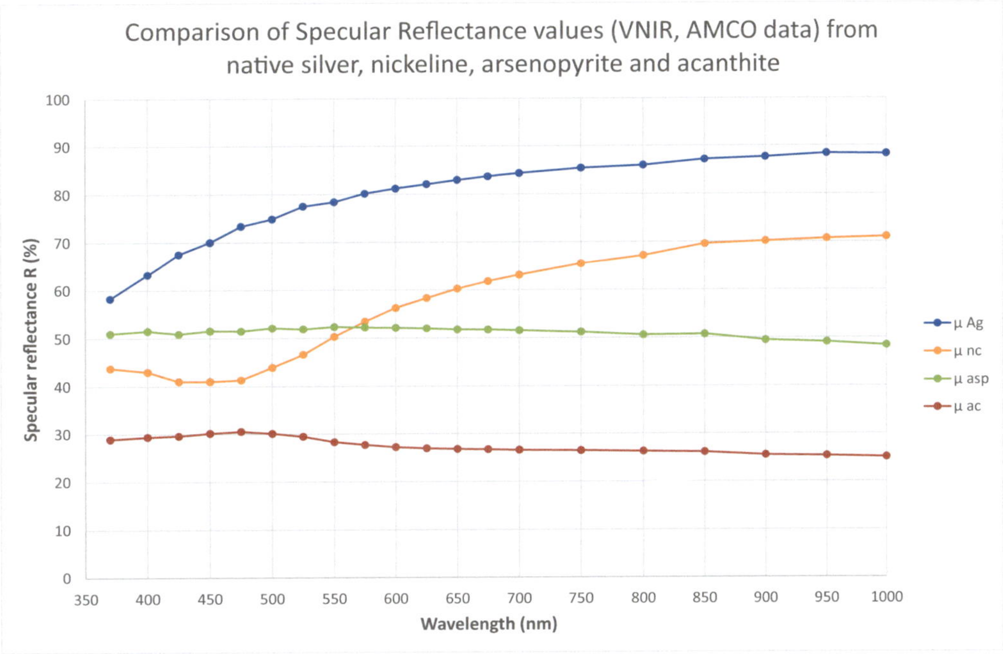

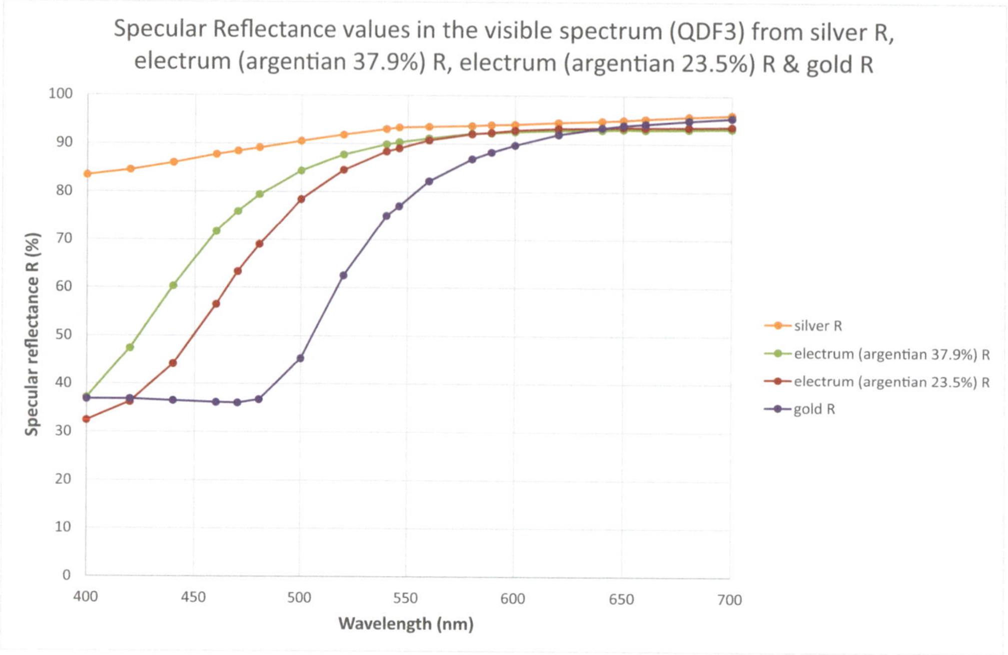

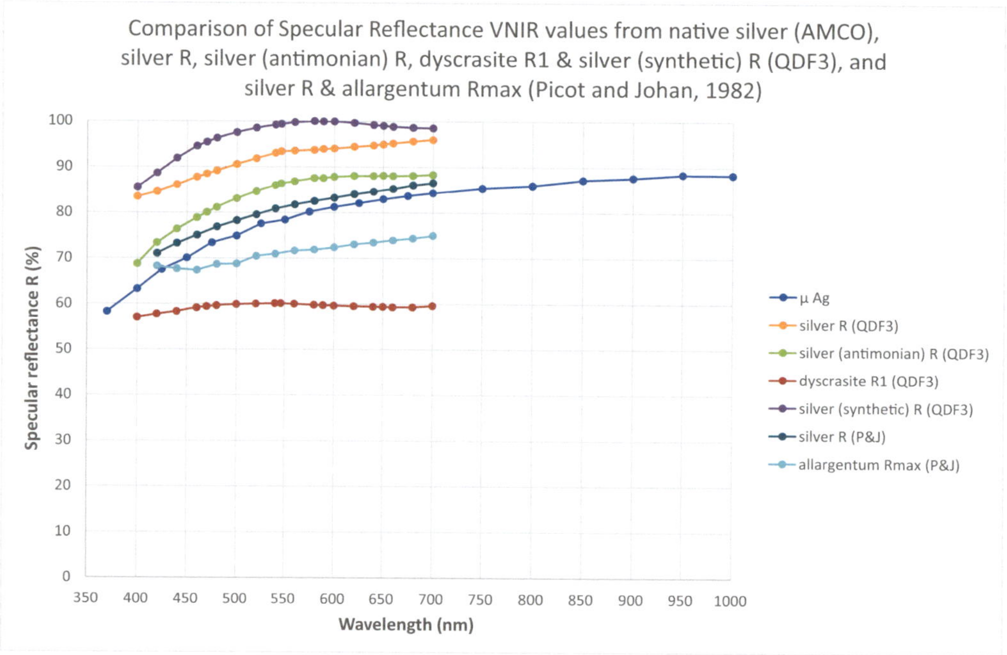

References

Picot P, Johan Z (1982) Atlas of ore minerals. B.R.G.M./Elsevier

Ramdohr P (1980) The ore minerals and their intergrowths, 2nd edn. Pergamon Press, Oxford. 2 volumes, 1205p

Skutterudite (*sk; smaltite-chloantite*) 112

$CoAs_3$: Cubic

Skutterudite	R_{546} (%)	Y (%)	λ_d (nm)	P_e (%)	Colour sensation
Air	55.2	55.1	483	1.5	Bright white
Oil	40.7	40.5	480	2.9	
Bir: Null (1)			*Pleoc*: Null (1)		*Tarn*: Variable (depending on zoning, etc.)
Anis +P: Isotr, imperfect extinction (2)			*IR*: -		*Twg*: -
Mor: Cubic Idiomorphic, massive			*Clv*: // (100)		*PolQ*: Good
Hardness	*PolH*: Medium, > nc, < po (3)		*Mohs*: 6^-	*Talm*: E–F^-	*VHN* $_{(100\ g)}$: 606–824

Obs: cf. Table and R charts, multispectral data (VNIR, AMCO) and visible range (QDF3). Previously, sk was assimilated to smaltite-chloanthite, now discredited as mineral species

Mineral abbreviations: Ag—native silver, asp—arsenopyrite, Bi—native bismuth, cob—cobaltite, lo—löllingite, nc—nickeline, po—pyrrhotite, py—pyrite, rm—rammelsbergite, sf—safflorite, sk—skutterudite

112.1 General

Composition: $CoAs_3$, often deficient in As down to $As_{2.5}$, sk is actually $(Co,Fe,Ni)As_{2\text{-}3}$, as it may contain considerable amounts of Fe, Ni or both and is now accepted to form a series with the nickeliferous term $(Ni,Co,Fe)As_{2\text{-}3}$ or *nickelskutterudite*.

Traditionally, it was described as the smaltite-chloanthite series (respectively, $CoAs_2$-$NiAs_2$, with excess As up to $As_{2.5}$), but these terms are no longer credited.

112.2 Microscopic Description

112.2.1 SP Observation (*Single Polariser*)

Pure white *colour* like safflorite, bright (more than arsenopyrite), slightly bluish against rammelsbergite (or without the yellowish tone of the latter, Figs. 112.1a, b, 112.2 and 112.4).

Polishes very well, better than arsenopyrite. Medium *polishing hardness*,

- PolH sk > nc, but
- PolH sk < rm, sf, po, asp, << py.

High *reflectance*, but < rammelsbergite, and distinctly < native Ag or Bi, against which skutterudite looks grey. More information: cf. comparative R table and plots (multispectral VNIR data and visible range).

Marked tendency to idiomorphic development, with typical cubic shapes and re-entrant angles enhanced by the characteristic zoning (*fortress structure*), with differentiated alteration (Figs. 112.2 and 112.3). Alernatively, it may also occasionally develop skeletal textures (with Ag or Bi) or form massive or brecciated aggregates (Fig. 112.4).

112.2.2 Observation +P (*Polariser + Analyser at 90°*)

Isotropic, although with imperfect extinction (Figs. 112.1c, d and 112.3).

No *internal reflections* or twinning.

R. Castroviejo, *A Practical Guide to Ore Microscopy—Volume 1*,
https://doi.org/10.1007/978-3-031-12654-3_112

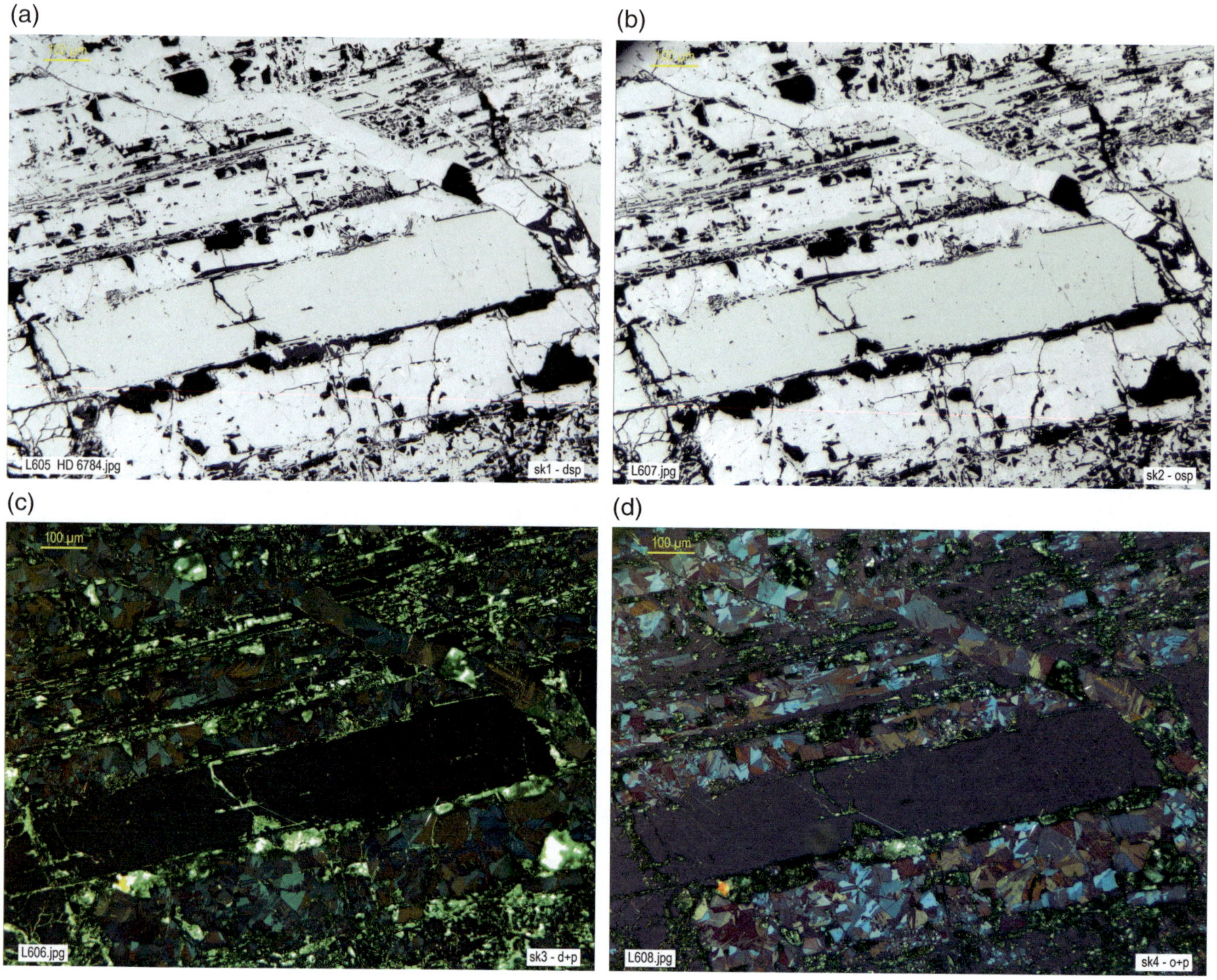

Fig. 112.1 **a**, **b** (Same scene: **dsp** and **osp**). Detail view of large, zoned sk crystal (Fig. 112.2 gives a wider view), partially replaced by rm and with rm vein cutting relict zoning. Slightly bluish hue of sk versus rm (white, lighter). Weak rm pleochroism, with yellowish-white to pinkish tones (very faint but noticeable when comparing adjacent grains), enhanced in oil (osp settting). Abundant black spots are gg inclusions, or pores due to leaching of sk in the replacement process. HD 6784, Schneeberg, Saxen, Germany. **c**, **d** (Same scene as Fig. 112.1**a**: **d + p** and **o+p,** *respectively*). The isotropic character of sk and the clearly anisotropic character of rm distinguish them perfectly. Polarization tints and polysynthetic twins of rm, reinforced in immersion

112.2.3 Paragenesis

It occurs in high-temperature hydrothermal and metasomatic deposits but also typically in silver deposits, particularly in hydrothermal veins of the *five-element formation* type (Ni-Co-Bi-Ag-U).

It may be accompanied by Ni-Co-Fe arsenides, Ag and Bi minerals, cobaltite, etc.

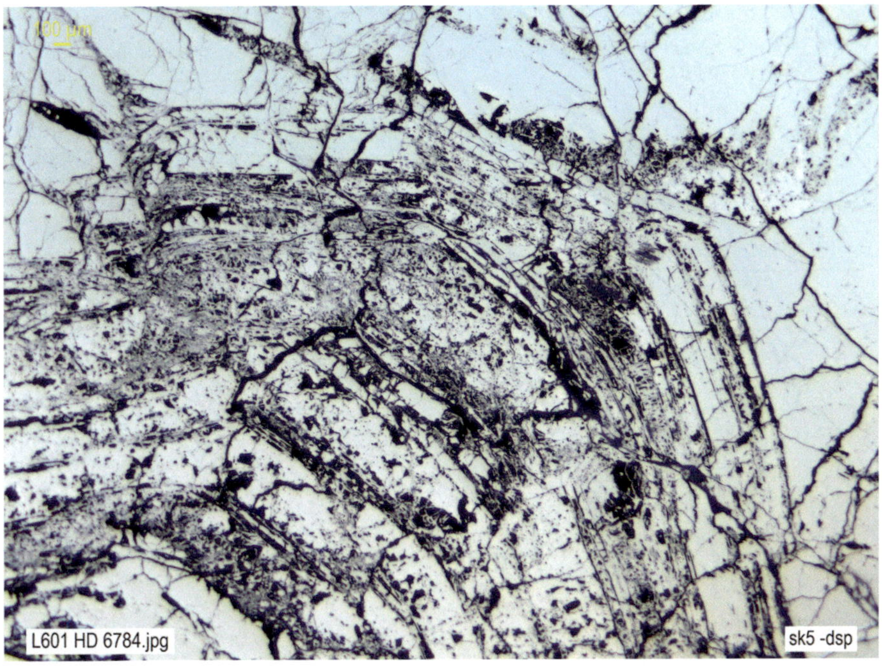

Fig. 112.2 Zoned structure of sk, enhanced by dissolution (porous zones) and substitution by rm (white). Same sample as FIg. 112.1 (**dsp**)

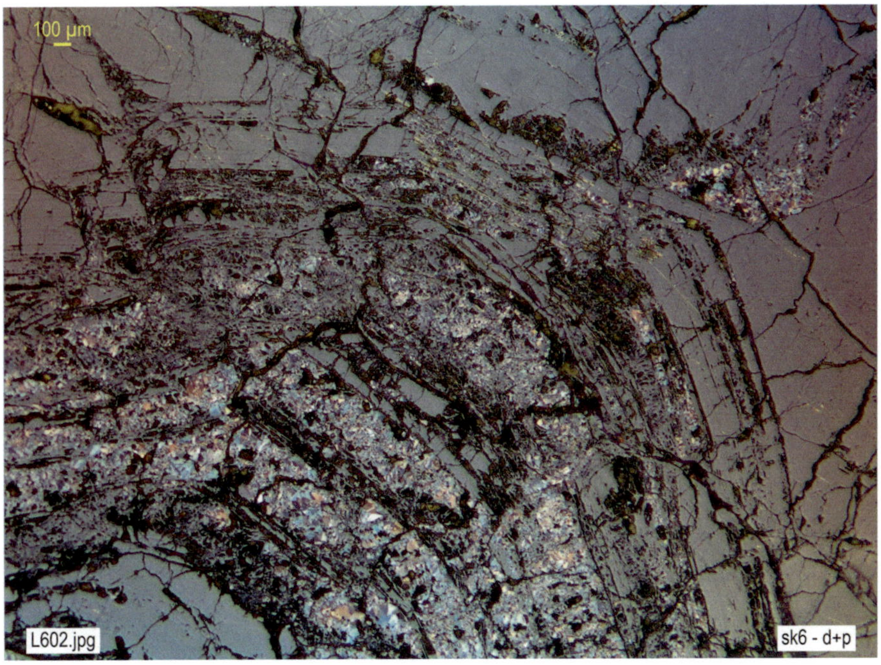

Fig. 112.3 Same scene as Fig. 112.2 (**d+p**). Imperfect extinction of sk, anomalous polarization tints, enhanced by the chromatic aberrations of widefield image. Still, they are easily distinguished from the vivid anisotropism colours of rm

112.2.4 Identification

Paragenesis, white colour, medium hardness, morphology and isotropy offer good recognition criteria.

- Among the *white arsenides*, its isotropy easily distinguishes it from *safflorite, rammelsbergite, löllingite*, etc.
- In colour, it resembles *gersdorffite* and *ullmanite*, but these are less reflective and often show exfoliation.
- Its pure white colour, its lower hardness and its easier polishing distinguish it from *pyrite* (Fig. 112.1c, d).

Fig. 112.4 Sk microbreccia (white). HD 6784, Schneeberg, Saxen, Germany (**osp**)

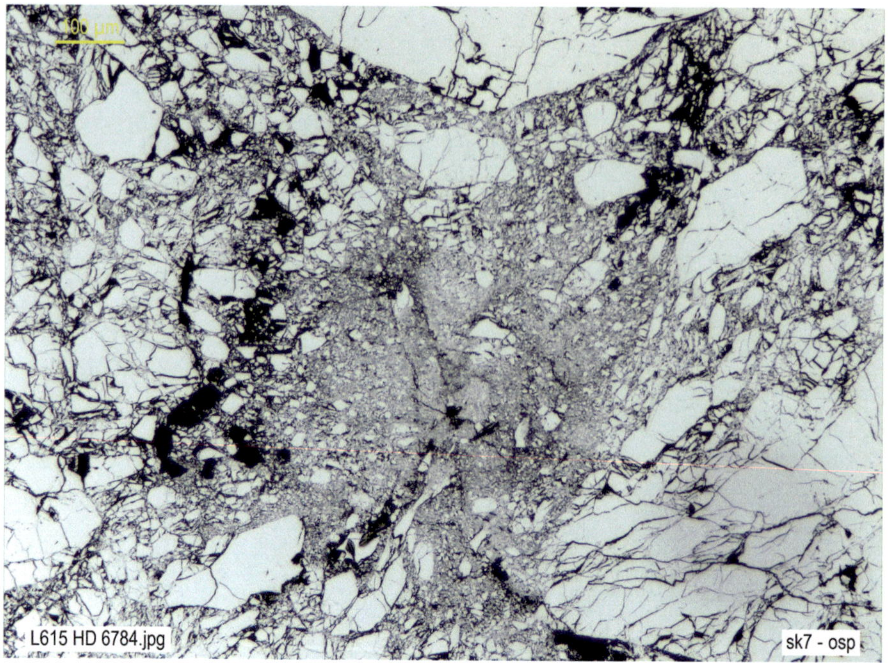

112.3 Spectral Information

112.3.1 Skutterudite Multispectral Reflectance R (in Air, Without Polariser) (%)

λ (nm)	370	400	425	450	475	500	525	550	575	600	625	650	675	700	750	800	850	900	950	1000
μ	51.5	52.7	53.0	53.6	54.0	54.4	54.7	54.7	54.6	54.2	53.9	53.4	53.1	52.7	52.0	51.1	50.7	49.3	48.1	47.5
P_5	49.0	51.0	51.2	52.0	52.5	53.1	53.6	53.8	53.7	53.4	53.0	52.6	52.2	51.8	51.2	50.3	49.8	48.1	46.6	45.9
P_{20}	49.9	51.6	51.8	52.5	53.0	53.6	54.1	54.2	54.1	53.7	53.4	52.9	52.5	52.1	51.5	50.6	50.1	48.5	47.0	46.4
P_{80}	52.6	53.4	53.5	54.2	54.5	54.9	55.1	55.2	55.0	54.6	54.3	53.8	53.5	53.0	52.4	51.6	51.2	50.0	49.2	48.5
P_{95}	55.2	55.9	56.7	56.7	57.0	56.3	56.4	55.8	55.7	55.3	55.0	54.8	54.4	54.1	53.3	52.5	52.0	51.3	51.0	50.3

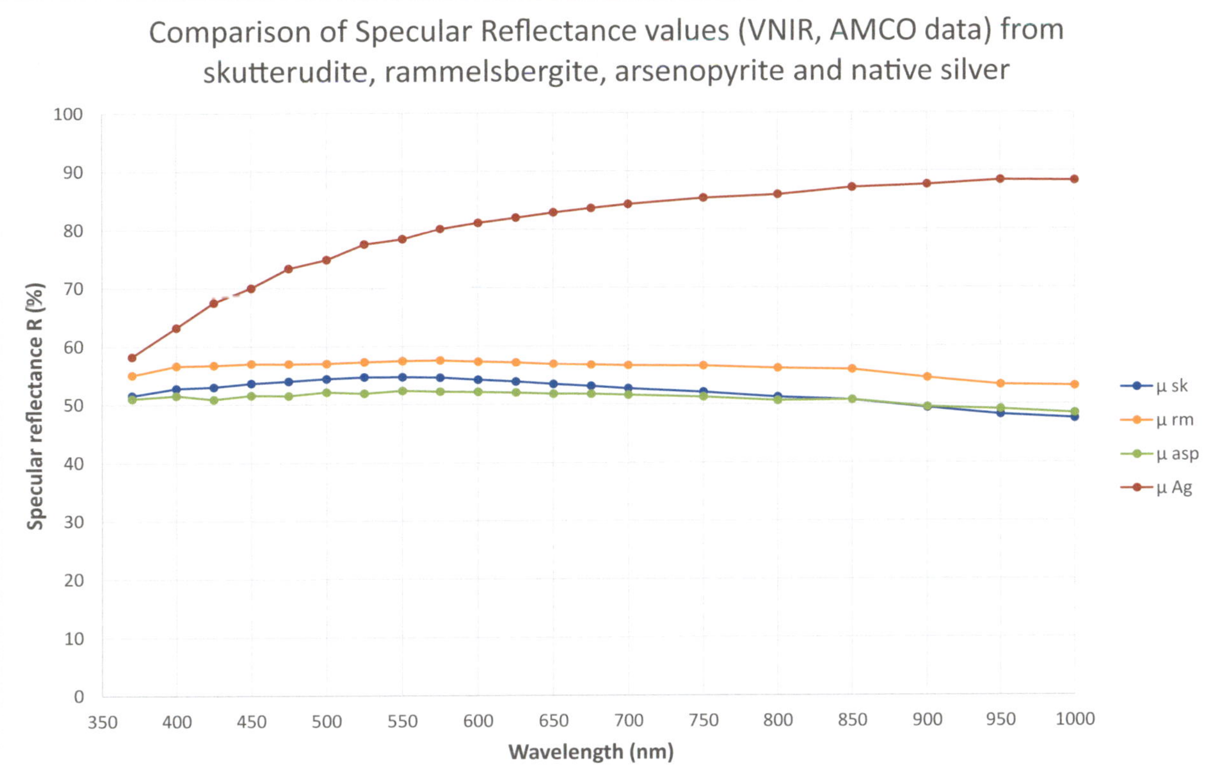
Comparison of Specular Reflectance values (VNIR, AMCO data) from skutterudite, rammelsbergite, arsenopyrite and native silver
Specular reflectance R (%)
0
10
20
30
40
50
60
70
80
90
100
350
400
450
500
550
600
650
700
750
800
850
900
950
1000
Wavelength (nm)
μ sk
μ rm
μ asp
μ Ag

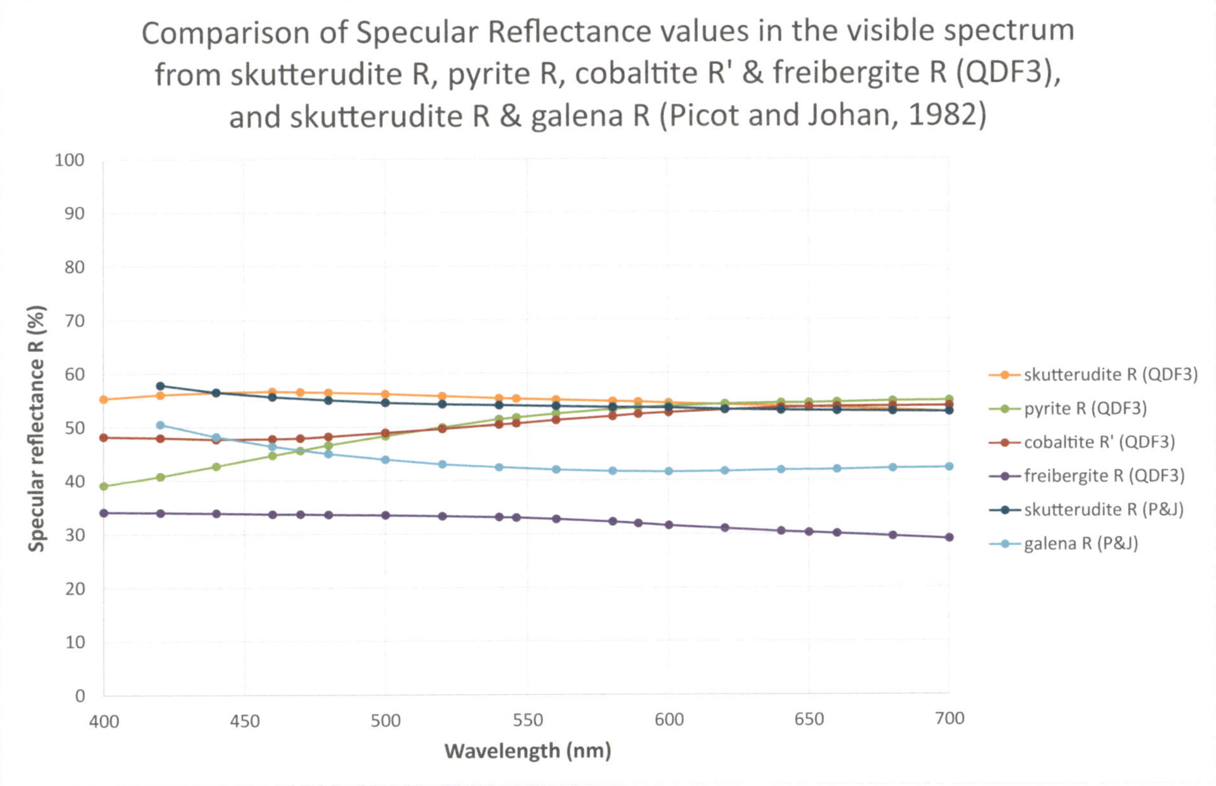
Comparison of Specular Reflectance values in the visible spectrum from skutterudite R, pyrite R, cobaltite R' & freibergite R (QDF3), and skutterudite R & galena R (Picot and Johan, 1982)
Specular reflectance R (%)
0
10
20
30
40
50
60
70
80
90
100
400
450
500
550
600
650
700
Wavelength (nm)
skutterudite R (QDF3)
pyrite R (QDF3)
cobaltite R' (QDF3)
freibergite R (QDF3)
skutterudite R (P&J)
galena R (P&J)

Sperrylite (*spy*)

113

$PtAs_2$: Cubic

Sperrylite	R_{546} (%)	Y (%)	λ_d (nm)	P_e (%)	Colour sensation
Air	53.9	53.7	493	0.4	White
Oil	38.8	38.7	488	0.8	Slightly bluish white
Bir: Null (1)			*Pleoc*: No (1)		*Tarn*: No
Anis +P: Isotropic (1)			*IR*: No		*Twg*: No
Mor: Euhedral trend (cub, oct)			*Clv*: // (100) occasional		*PolQ*: Well, as py
Hardness	*PolH*: Very high, ≥ py, > Pt (4)		*Mohs*: 6–7	*Talm*: nd	*VHN* $_{(100\ g)}$: 960–1277 (on 100); 1048–1225 (on 111)

Note: cf. R charts, visible range data (QDF3). VHN according to Cabri (2002)
Mineral abbreviations: Au—native gold, ccp—chalcopyrite, chr—chromite, Cu—native copper, gg—gangue, gth—goethite, Ir—native iridium, lim—limonite, Os—native osmium, pn—pentlandite, po—pyrrhotite, Pt—native platinum, py—pyrite, spy—sperrylite

113.1 General

One of thc most ubiquitous, among the 109 **PGMs** (*Platinum Group Minerals)* credited by IMA (cf. comments on PGE and PGMs in *Pt description* and more data in Cabri 2002).

Composition: $PtAs_2$; may contain Sb (up to 21.39%, exceptionally) and minor to trace contents of Rh, Ru, Ir, Fe, Ni, Co, Cu, S (Cabri 2002).

113.2 Microscopic Description

113.2.1 SP Observation (*Single Polariser*)

Pure white, bright (Fig. 113.1a, b and 113.2). Good *polishing*. Due to its hardness, it usually appears in relief even more than pyrite (Fig. 113.2).

Very high *polishing hardness*, slightly higher than pyrite; *PolH* spy > chr, Os, Ir, po, Pt; >> ccp.

Reflectance high but < Pt. R spy ≈ py; > ccp (Figs. 113.1a, b and 113.2, cf comparative graphs of reflectance spectra). Isotropic.

It occurs in crystallites with a pronounced idiomorphic tendency (cubes, octahedrons), disseminated or frequently included in chalcopyrite, pyrrhotite, pentlandite and also in limonite (Figs. 113.3 and 113.4), in gossan deposits derived from primary sulphides, in which sperrylite is preserved by its remarkable resistance to weathering.

113.2.2 Observation +P (*Polariser + Analyser at 90°*)

Isotropic, remains in extinction (Fig. 113.1c, d).

113.2.3 Paragenesis

Given the variety of *parageneses* in which spy can be found, it is one of the most ubiquitous PGMs, as it is found in any of the known types of platinum deposits.

Besides the typical deposits linked to ultramafic rocks with chromite (intrusions, stratified complexes, ophiolites), it is also found associated with chalcopyrite, pentlandite and pyrrhotite in the large Cu-Ni sulphide deposits in gabbros and related rocks (such as Sudbury, its *locus typicus*, or

R. Castroviejo, *A Practical Guide to Ore Microscopy—Volume 1*,
https://doi.org/10.1007/978-3-031-12654-3_113

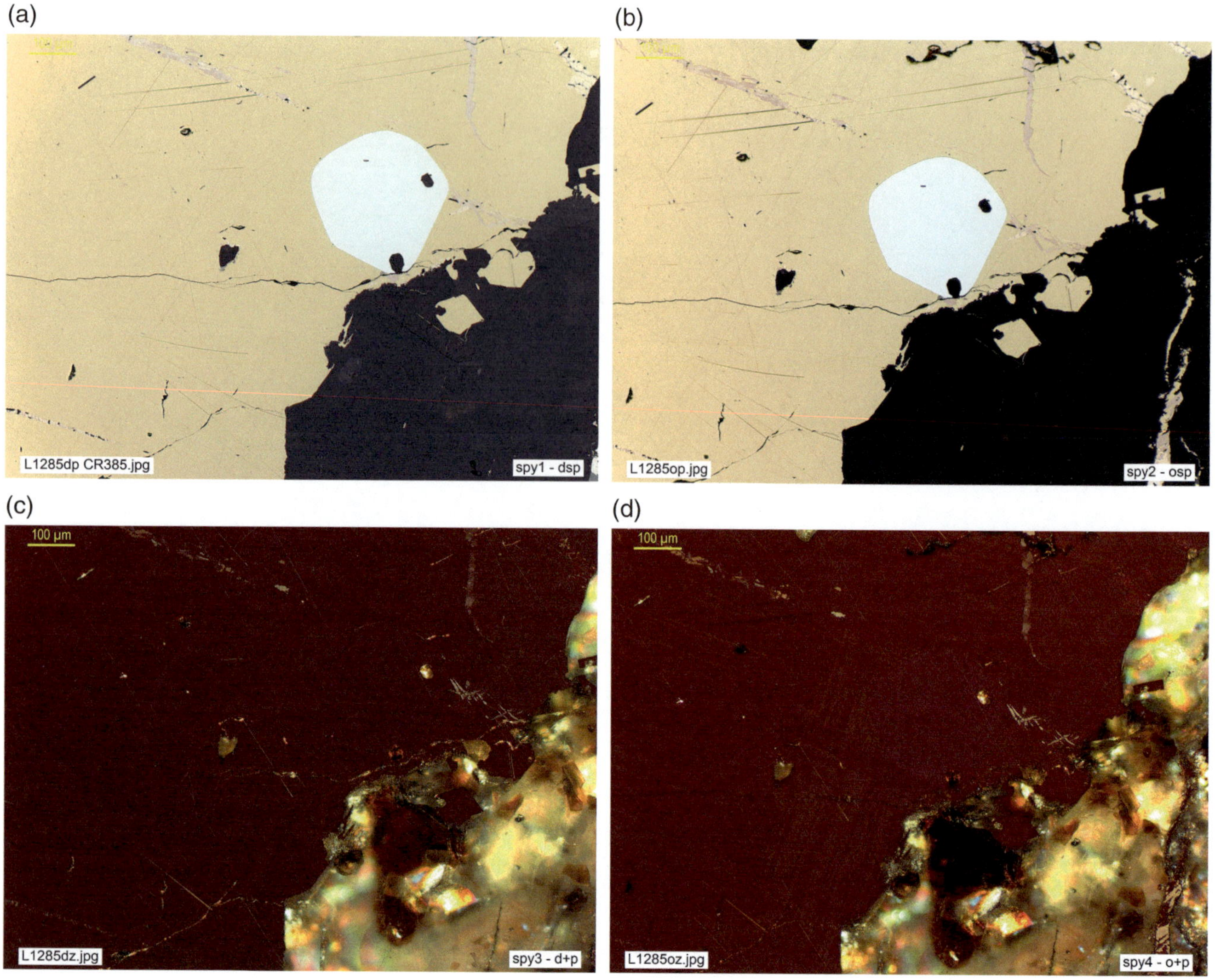

Fig. 113.1 **a**, **b** (**dsp** and **osp**, resp)—Inclusion of spy (euhedral, white, relief) in ccp, which also includes po with pn "flames' and sparse py (top right). CR 385, Frood Mine, Sudbury, Ontario, Canada. **c**, **d** (same scene as Fig. 113.1a: **d+p** and **o+p**, respectively)—Extinction of spy, pn and py (isotropic); ccp, weakly anisotropic and twinned; po, more distinctly anisotropic

Noril'sk), in the oxide concentrations derived from these and in the placers derived from all of them.

113.2.4 Identification

In grains of sufficient size, sperrylite is well identified by its pure white colour, its strong automorphic tendency with isometric forms, its very high hardness and its isotropy, always taking into account the reference to paragenesis.

In cases of doubt with, also white, Ni–Co arsenides, particularly gersdorfite, sperrylite differs by its resistance to acid attack. Confusion with pyrite is possible in small, isolated grains whose colour is difficult to pinpoint (Figs. 113.3 and 113.4), but acid attack and specular reflectance spectra (especially in the 400–550 nm range) differentiate them.

Common accompanying ores: native platinum, native iridium, pentlandite, chromite, chalcopyrite, pyrrhotite, laurite and others, such as native gold and occasional native copper.

Fig. 113.2 Chalcopyrite ore with inclusions of spy (white, idiomorphic, cube section, top left) and py (bottom right). The purely white colour, unlike the weakly yellowish py, and the relief clearly identify spy. CR 385, Frood Mine, Sudbury, Ontario, Canada (**dsp**)

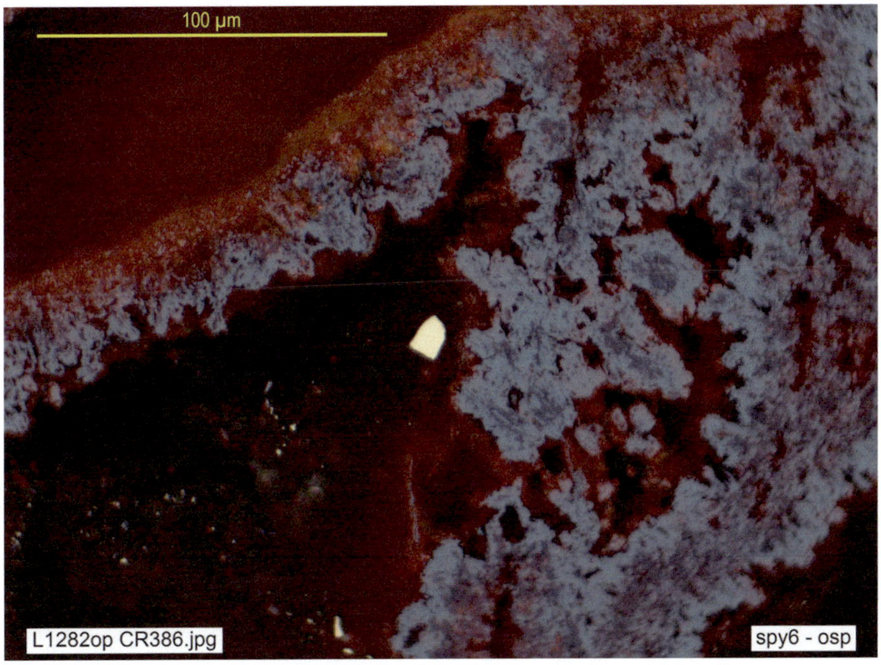

Fig. 113.3 Isolated, hypautomorphic spy crystal, in oxidized ore with gth and lim (bluish grey); in contrast and lacking the ccp reference, here spy looks somewhat yellowish. CR-386, Tweefontein Pt Mine, Bushveld, RSA (**osp**)

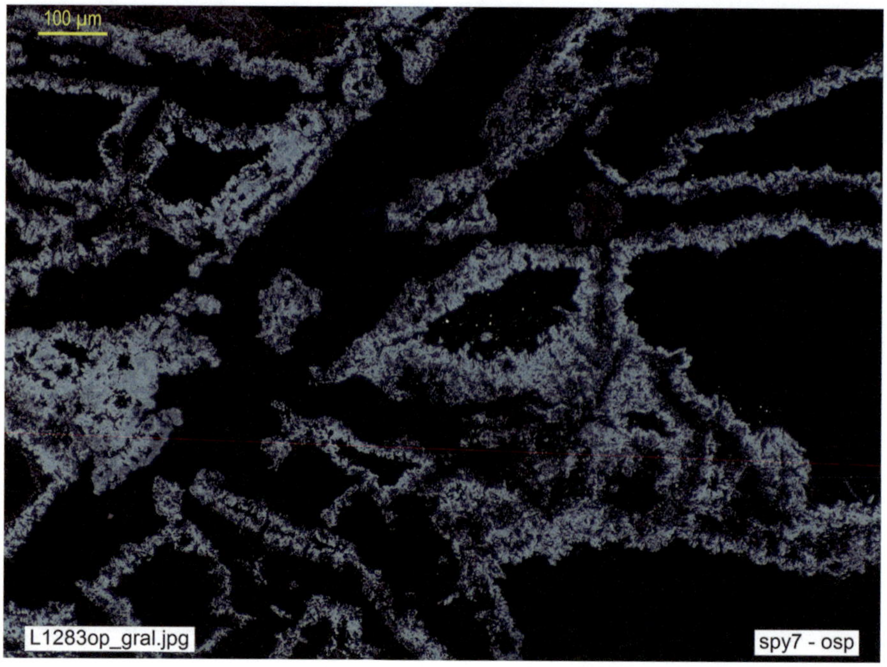

Fig. 113.4 Limonitic *boxwork* with unaltered spy dissemination (tiny bright crystals, in the central and right-hand side boxwork cells), inherited from the primary paragenesis. CR-386, Tweefontein Pt Mine, Bushveld, RSA (**osp**)

113.3 Spectral Information

113.3.1 Sperrylite Reflectance R (in Air, Visible Spectrum) (%)

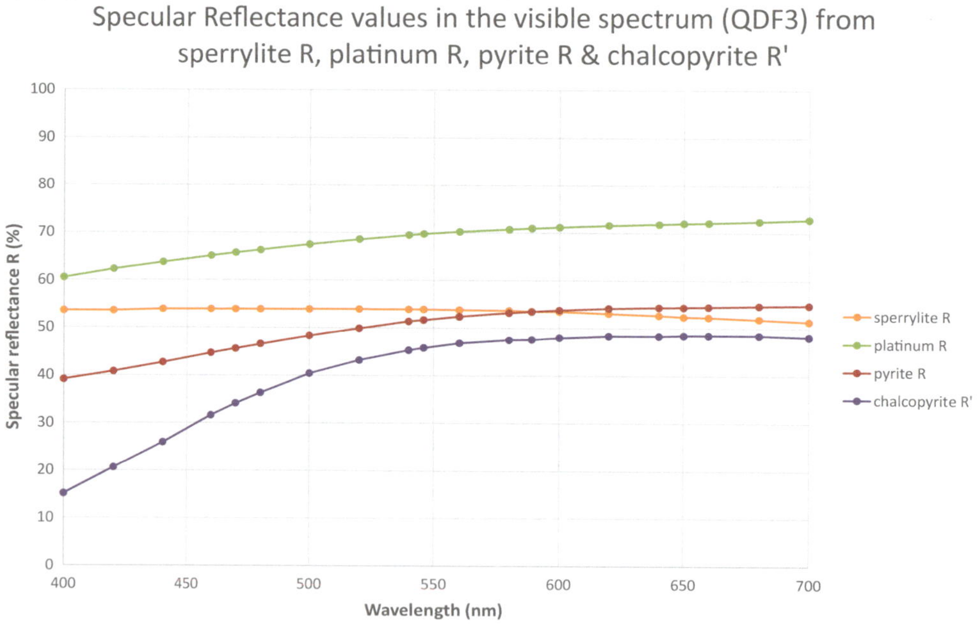

References

Cabri L (2002) The platinum-group minerals. In: Cabri L (ed) CIM special, vol 54. Canadian Institute of Mining, Metallurgy and Petroleum, Calgary, Alberta, Canada, p 852, pp 13–129

Cabri LJ (ed) (2002) The geology, geochemistry, mineralogy and mineral beneficiation of platinum group minerals. Canadian Institute of Mining, Metallurgy and Petroleum, Calgary, Alberta, Canada, p 852, (ISBN 1-894495-27-5)

Sphalerite (*sp/blende*)

114

ZnS: Cubic

Sphalerite (*sp/blende*)			ZnS		Cubic
	R_{546} (%)	Y (%)	λ_d (nm)	P_e (%)	Colour sensation
Air	16.8	16.7	475	4.4	Grey
Oil	5.1	5.0	475	7.5	Dark grey
Bir: Null (1)			*Pleoc*: Null (1)		*Tarn*: No
Anis +P: Null (1)			*IR*: Generally frequent		*Twg*: Not visible without chemical attack
Mor: Cubic automorphic, massive, botryoidal			*Clv*: Depends on grain size		*PolQ*: Medium top poor
Hardness	*PolH*: Medium, > ccp; < po (3)		*Mohs*: 3½–4	*Talm*: –	*VHN* $_{(100\ g)}$: 208–224

Obs: cf. Table and R charts, multispectral data (VNIR, AMCO) and visible range (QDF3; Picot and Johan 1982)
Mineral abbreviations: Ag—native silver, alb—alabandite, asp—arsenopyrite, Au—native gold, bnn—bournonite, cbn—cubanite, cc—chalcocite, ccp—chalcopyrite, cst—cassiterite, cv—covellite, dg—digenite, en—enargite, gg—gangue, gn—galena, gth—goethite, gtt—gratonite, hub—hübnerite, jd—jordanite, ln—linnaeite, mc—marcasite, mt—magnetite, po—pyrrhotite, py—pyrite, q—quartz, Sb—native antimony, sf—safflorite, sk—skutterudite, sp—sphalerite, stn—stannite, tn—tennantite, td—tetrahedrite, wf—wolframite, wz—wurtzite

114.1 General

Important mineral for its economic interest (main ore of Zn and other metals), for its frequency in various environments, and for its scientific significance, in particular for experimental mineralogy and as a geobarometer.

Composition: ZnS, although rarely stoichiometric, mainly because of the easy incorporation of Fe in substitution of Zn. Pure sphalerite (100% ZnS) is seldom found, e.g., the colourless to light yellow (honey or *caramel*-like) *blenda acaramelada* ("*caramel blende*") from the Picos de Europa or Reocín deposits (Santander, Spain) or the *cleiophane* from Franklin Furnace (N. Jersey, USA), nearly pure ZnS.

Fe-rich varieties (up to ≈25% Fe in natural sp) are named *marmatite*, although in the studied samples from Marmato, Colombia (*l.t.* marmatite), part of the Fe is contained in micro-inclusions of pyrrhotite or chalcopyrite. According to data compiled by Bowles et al. (2011), the following solid solutions have also been found: complete, between ZnS and ZnSe, CdS or HgS; partial, with MnS (up to 52%) and CoS (34%).

Many other metals can also be found in sphalerite as minor or trace elements, *vbgr*: Cu (5%), Cd (4.4%), Sb (1%), Hg (1%), Sn (1%), In (1%), Ag (1%), As (1%), Ti (5000 ppm), Ga (3000 ppm), Co (3000 ppm), Ge (1000 ppm), Se (900 ppm), Ni (3000 ppm), Ba (150 ppm), Mo (100 ppm), V (100 ppm), Cr (10 ppm). The numbers in brackets indicate the maximum contents quoted in sphalerite from all over the world, although, as Bowles et al. (2011) point out, a part of these contents may be found as inclusions and not only as structural substitutions. Higher contents are possible in compositions transitional to other species, *e.g.*, *Mn (transition to alabandite*, MnS) or Hg (15%, *polhemusite*, [Zn, Hg]S).

R. Castroviejo, *A Practical Guide to Ore Microscopy—Volume 1*,
https://doi.org/10.1007/978-3-031-12654-3_114

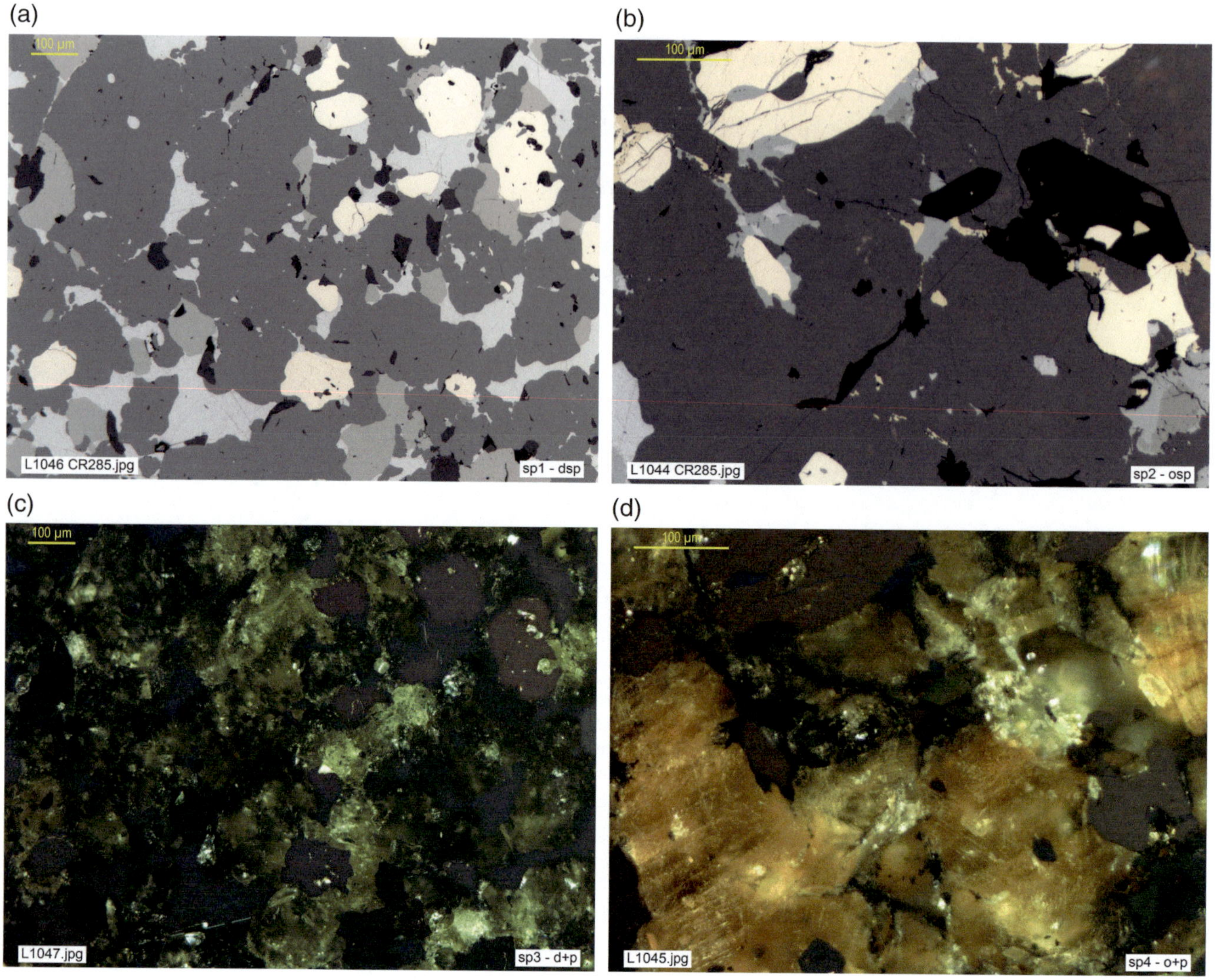

Fig. 114.1 **a** Ore of sp (main component, grey) intergrown with gn (greyish white) and td (olive grey), inclusions of py (yellowish white) and q (very dark grey). CR 285, Linares, Jaén (**dsp**), **b** Ore of sp (dark grey), with interstitial gn (greyish white), bnn (greenish grey) and td (olive grey, included in bnn and somewhat darker than it) and with inclusions of py (yellowish white) and q (black, hexagonal crystals). Same sample of Fig. 114.1 (detail, **osp**), **c** Same scene as Fig. 114.1a (**d+p**). Extinction of all sulphides (isotropic), but sp illuminated by internal reflections, **d** Same scene as Fig. 114.1b (**o+p**). Abundance and luminosity of sp IR in immersion. Weak bnn anisotropism

In addition to its importance as a Zn ore, sphalerite is the *main source of Cd, In, Ge, Ga, despite* the usual limited contents of these metals.

Dimorph of *wurtzite*, wz, of which several polymorphs are known, the most common being 2H (hexagonal). The relations between sphalerite and wurtzite have been the subject of numerous investigations since Friedel's description of the latter (1861) but have not yet been fully elucidated. However, the non-stoichiometry of ZnS, with Zn vacancies in sphalerite and S vacancies in wurtzite, is generally accepted. Wurtzite is considered to be the stable form at high temperature (T > 1020 °C) and low pressure (1 atm). In fact, the sp-wz transition also depends on f_{S2} and there are experimental data suggesting that both species can stably coexist below this temperature, if fs_2 is low enough to maintain the S deficit in wurtzite. Wurtzite is found in nature as well in low fs_2 environments with formation T well below 1020 °C, e.g., associated with sphalerite in the so-called *schalenblende*. Other interpretations suggest that wurtzite is metastable at low T.

114.2 Microscopic Description

114.2.1 SP Observation (*Single Polariser*)

Grey *colour* (Figs. 114.1a, b and 114.10a). Difficult to *polish*. Isotropic.

Fig. 114.2 Recrystallized massive sulphide ore, with py, sp, gn, td, ccp and q, intergrown, py corroded by sp. Sample 170,315, Mina Magdalena, Huelva (**dsp**)

Low *reflectance*, among the lowest of the sulphides, $R_{sp} \approx$ R *wf* (almost identical, cf. Fig. 114.4a), similar but lower than *mt*, slightly > *gth* and distinctly > *cst* (cf. multispectral comparative plots). In addition:

- the subjective perception depends very much on the environment. It looks very dark next to Ag, Au, Sb, dark next to *py*, *ccp*, *gn* (Figs. 114.2 and 114.7), even *stn*, *td*, *tn* (Figs. 114.6 and 114.5) but light next to *cst* (Fig. 114.5) or in comparison with *gg*. Increasing contrast by observation in oil is very useful, e. g., to distinguish sphalerite from magnetite (comp. Figs 114.3a and 114.3b).
- the R value depends on the composition of the grain measured. In general R increases slightly with Fe content, as indicated by the curves of the third R graph, which plots R values in the visible spectrum of four sphalerite samples, with Fe contents increasing in the same order as R in the curves. This agrees with the luminance data, **Y** %, according to QDF3[1]:

Y % variation according to Fe content (% weight)				
Sample (%)	(p. 525) R	(p. 522) R	(p. 526) (Mn-Fe) R	(p. 524) R
Fe	0	2.6	8.6	13.3
Y	16.3	16.6	17.1	17.4

[1] Undoubtedly, other cations not considered in this simple example can also play a role.

Medium *polishing hardness*: PolH sp > en, td, stn, cbn, Au, ccp, bn; PolH sp < po, mt, ln, sk, sf. Upon *deformation*, sphalerite is relatively rigid, without the ductility of galena or chalcopyrite, so its response is usually brittle, appearing fractured or crushed (cf. *§4*, Fig. 1.52f, vol. 2), although at high temperatures and pressures its behaviour is more ductile. Exfoliation rarely visible under the microscope, in any case only in good-sized grains.

Zoning visible in well-developed crystals, due variations of composition (e.g., % Fe) of growth rate, of type or frequency of inclusions or pores (Fig. 114.6). Best observed in doubly polished thin sections, as well as some particular textures, like rhythmic precipitation or botryoidal and spherulitic textures, known as *schalenblende*, showing successive concentric layers, in which very fine sphalerite and wurtzite usually alternate, sometimes with marcasite or pyrite (Figs. 114.8, 114.9 and 114.10a, b).

Sphalerite may occur in a variety of forms and environments. In open spaces (veins, breccias, karstic cavities) sphalerite shows its euhedral tendency (cubic forms with tetrahedral dominant habit), but in general, in common ores, it usually occurs in masses or anhedral aggregates; also disseminated, more rarely in banded layers and botryoidal or spherulitic encrustations, with or without wurtzite (*schalenblende*).

Fine inclusions (*emulsion*) of chalcopyrite (Figs. 114.3b and 114.7) are frequent. Conversely, sphalerite can form fine, typically stellate inclusions (*sphalerite stars*) in chalcopyrite (cf Fig. 1.30b, vol. 2). Emulsion textures are usually interpreted as exsolutions, but they should be carefully

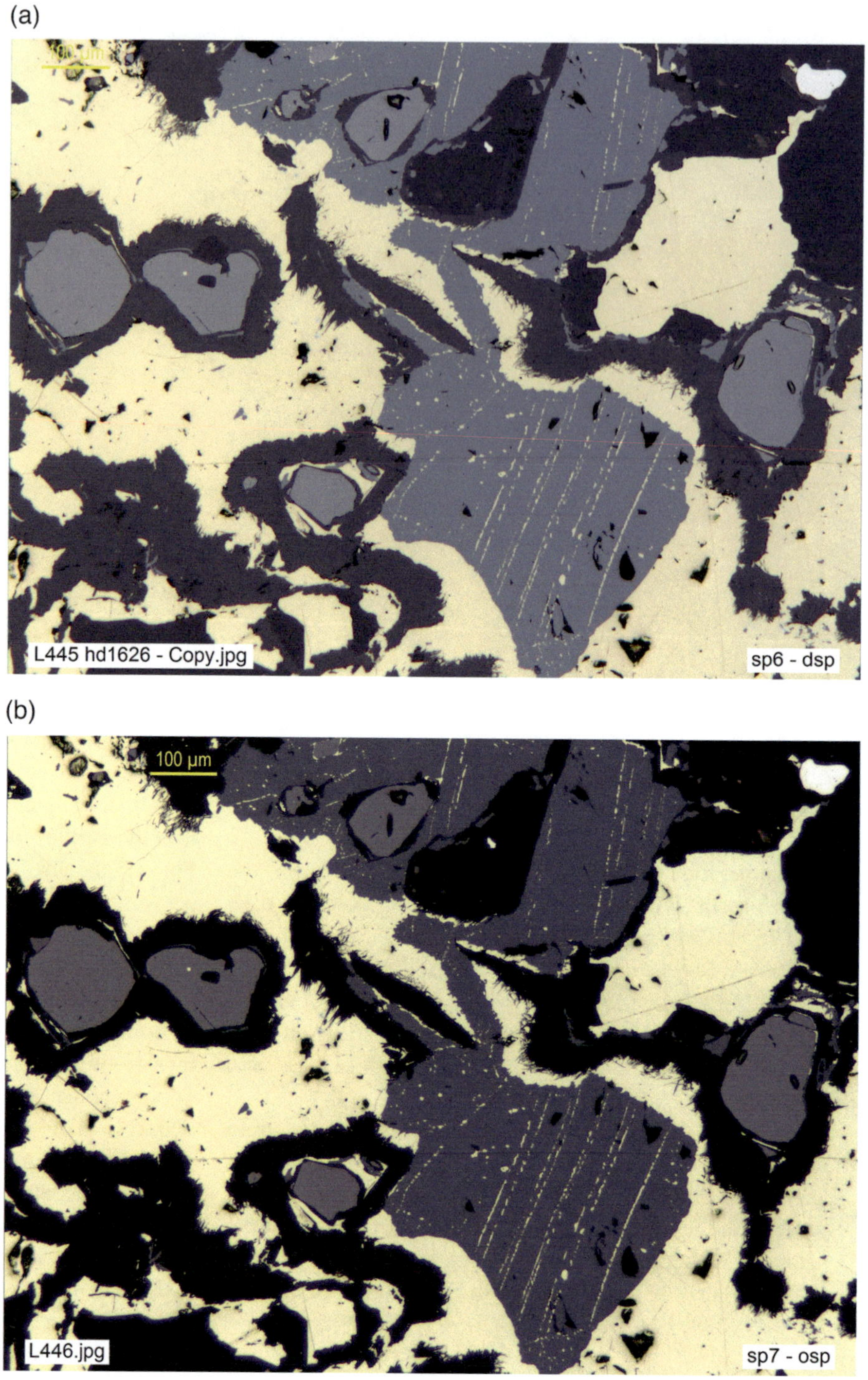

Fig. 114.3 **a, b** (same scene: **dsp** and **osp**, respectively) Ore of ccp (yellow), with sp (grey, with ccp emulsion) and mt (grey, slightly brownish, somewhat lighter than sp) inclusions, both corroded by ccp. Comparison of Figs. 114.3a and Fig. 114.3b shows how contrast between sp and mt increases in oil. HD 1626, Bayernland Mine, Germany

observed to avoid confusing them with replacement textures (*chalcopyrite disease*, cf. Figs. 1.44c and d, vol. 2). Their occurrence is usually not good news for the plant engineer, because the separation of chalcopyrite and sphalerite in these fine intergrowths would require too fine a milling, which is undesirable.

114.2.2 Observation +P (*Polariser + Analyser at 90°*)

Isotropic, but the extinction is often masked by abundant IR, *internal reflections* (Figs. 114.1c, d, 114.8, 114.9 and 114.10b). Tectonic deformation can lead to local anisotropism.

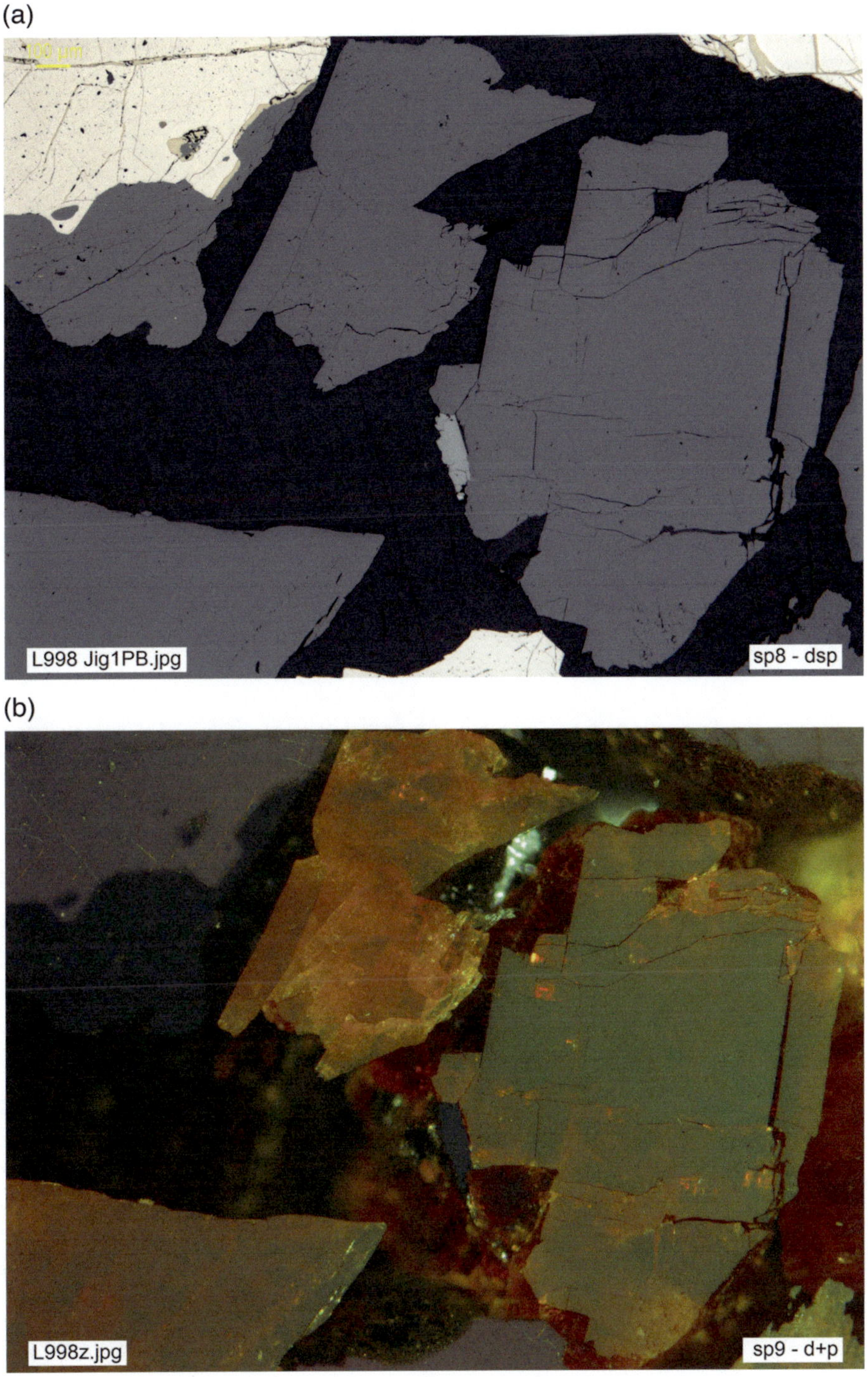

Fig. 114.4 a, b (same scene: **osp** and **o+p**, respectively)—Tungsten concentrate: hübnerite (*hub*, grey, main ore), py, sp (grey, very similar to hb, rimming py border, top left), ccp, td (light grey, centre, on hub border) and traces of q. The delicate distinction of hub and sp is resolved with analyzer (+*p*, Fig. 114.4b). Hub is anisotropic and shows abundant yellowish-brown to orange internal reflections, while sp is isotropic and remains in extinction, despite its IR (scarcer and less coloured than those of hb). Jig1.PB, Pasto Bueno Mine, Ancash, Peru

IR can have varied colours: colourless to light yellowish-brown (luminous and massive typically in pure ZnS varieties); reddish-brown to blood red in more ferriferous varieties (becoming scarcer and more intense red as Fe content increases, until almost disappearing in *marmatite*, which may look opaque); occasionally green in Ga-bearing varieties (Picot and Johan 1982) or blue-green in varieties with <1% Fe and a very slight Zn substitution by Co (Rager et al. 1966, *in*: Bowles et al. 2011).

Twinning (111) and (211) very frequent but rarely visible under the microscope without HNO_3 attack.

114.2.3 Paragenesis

Sphalerite is a very frequent mineral, present in varied *paragenesis*: sedimentary or *Sedex*, volcano-sedimentary (massive sulphides), magmatic and metamorphic, and it has also

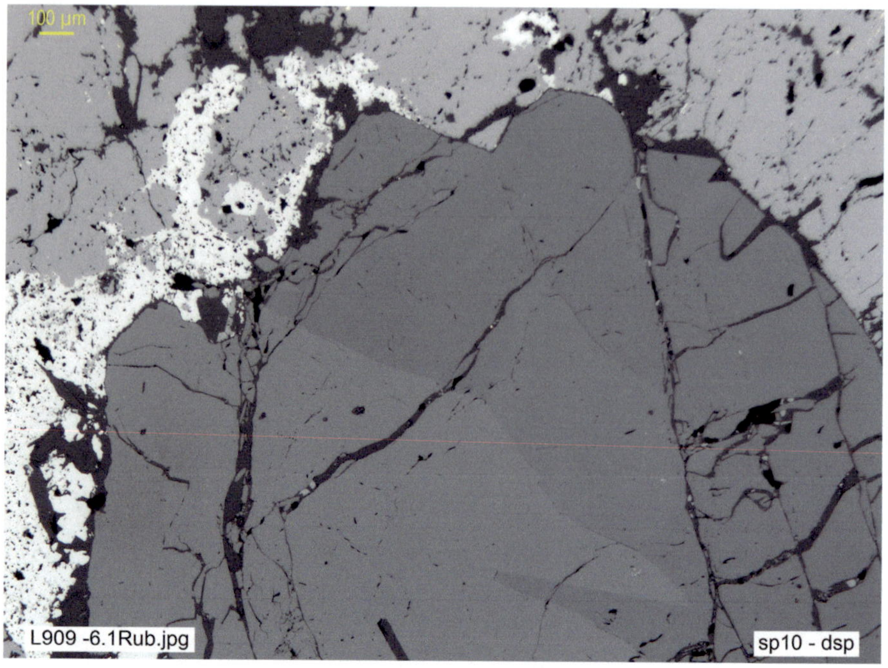

Fig. 114.5 Ore of cst (grey, bireflective and twinned, largest crystal, lower right), sp (light grey, upper third of the image) and tn (greyish white, thin band along cst-sp contact). Rub.6.1, Böhm. Zinnwald, Erzgebirge (**dsp**)

Fig. 114.6 Intergrowth of sp (dark grey) and stn (brownish-olive grey, lighter), with sparse gn, po and ccp. Growth zoning of sp (large crystal, center) marked by alignment of minute inclusions of ccp and stn and interrupted by subsequent corrosion of sp by stn. Rub483/71, Davischacht, Saxen, Germany (**osp**)

been found in meteorites. Therefore, it can occur *accompanied by a* great number of minerals, the following being particularly frequent: galena, pyrite, marcasite, chalcopyrite, pyrrhotite, magnetite, stannite, arsenopyrite, tetrahedrite.

It is also frequently associated with wurtzite, particularly in colomorphic or concentric banded deposits (*schalenblende*, cf. Figs. 114.8, 114.9 and 114.10a, b), with common paramorphic transformation of wurtzite into sphalerite, preserving the original texture (*icefern* or "ice fern').

114.2.4 Identification

Microscopic recognition of sphalerite is easy and, in general, should present no problems, despite its possible confusion with magnetite. The grey colour, the medium polished hardness, the low reflectance (and its transparency that makes its characterization possible in transmitted light), the isotropic character and the IR allow identification of sphalerite without difficulty.

Fig. 114.7 *Complex sulphide* ore concentrate, intergrown py, sp, gn and (ccp). Emulsion of ccp in sp and reaction rims sp-py and gn-sp. ISLA1.01bis RT2014, Riotinto, Huelva (**dsp**)

In typical *volcanogenic massive sulphide* ores (Figs. 114.2 and 114.7), such as those of the Iberian Pyritic Belt (Neves-Corvo, Riotinto, La Zarza, etc.) or their metamorphic equivalents (Norway, Canada, Australia, etc.), the distinction between sphalerite and its usual companions (pyrite, chalcopyrite, galena, tetrahedrite, marcasite, arsenopyrite, chalcocite, digenite, covellite and also cassiterite in Neves-Corvo, Portugal) is obvious and poses no problem. In other cases, it is convenient to check the recognition by diagnostic comparison with distinctive properties of some similar ores:

- **magnetite**: despite the close colour and reflectance and the isotropic character of both, sphalerite is easily distinguished from magnetite by the absence of IR in the latter, by the lower *PolH* of sphalerite and by the notable decrease in sphalerite of the reflectance in immersion, which clearly increases the contrast with magnetite (comp. Figs. 114.3a and 114.3b). Unlike sphalerite, magnetite does not show the typical association with chalcopyrite in *emulsion*, unmixing or substitution textures (*ccp disease*).
- **alabandite** is brighter (higher reflectance) and softer (shows lower PolH than sphalerite), its IR are scarcer, darker and greenish and it is associated with manganiferous paragenesis (*vbgr.* with rhodochrosite gangue), cf. *description and* Figs. 5.1a, b and 5.2.
- **wurtzite** has optical properties which, for the most part (colour, R, IR), practically coincide with those of its polymorph sphalerite, with which it sometimes forms cryptocrystalline intergrowths not resolvable under the optical microscope. In this case, the two polymorphs cannot be distinguished without resorting to auxiliary techniques, such as XRD. However, the anisotropic character of wurtzite (hexagonal) establishes a differentiation criterion for oriented monomineral fibers (fibrous-radial aggregates in *schalenblende*, Figs. 114.8, 114.9 and 114.10a, b), and also, in general, for microscopically resolvable sp-wz intergrowths. However, there are cases, with wurtzite of weak or imperceptible anisotropism or with sphalerite weakly anisotropic by deformation, in which the difficulty persists and the observer has to resort to observation by transmitted light, XRD or textural attack, which reveals either the peculiar fibrous texture (*icefern*) of wurtzite or the polysynthetic twins of sphalerite (non-existent in wurtzite).
- **cassiterite and wolframite** are both clearly anisotropic. Cassiterite is also much darker than sp (Fig. 114.5), and wolframite IRs are generally darker and sparser, with the notable exception of the manganiferous wolframite term (**hübnerite**), whose abundant IRs are reminiscent of those of *blenda acaramelada (*although they are more orange) and whose resemblance to sphalerite is striking at first sight but, even so, they are certainly distinguished by the prismatic habit and distinct anisotropism of hübnerite (Fig. 114.4a, b).

Fig. 114.8 Deposition of botryoidal sphalerite (*schalenblende*) on dendriform aggregates of jordanite/gratonite, jd/gtt, (in extinction, dark, diagonal band SW-NE). HD 8, Wiesloch, Baden-Wuerttenberg, Germany (**d+p**)

Fig. 114.9 Botryoidal *schalenblende* deposited on tabular aggregates of *jd*, pseudomorphic of gratonite, *gtt*, relict. The spherulitic, fibrous-radial, anisotropic structure of the concentric zones showing cross-extinction suggests the presence of wz. HD 8, Wiesloch, Baden-Wuerttenberg, Germany (**d+p**)

Fig. 114.10 **a, b** Same scene (**osp** and **o+p**, respectively): inclusions of gtt (white, almost isotropic, euhedral with hexagonal habit, top and bottom, left) and jd (white, anisotropic, pseudomorphic of gtt) in *schalenblende* (grey). As seen with crossed polars (Fig. 114.10b), the *Schalenblende* comprises zones and layers in extinction (isotropic, *sp*) and spherulitic zones with cross-extinction (anisotropic, *wz*). HD 8, Wiesloch, Baden-Wuerttenberg, Germany

(a)

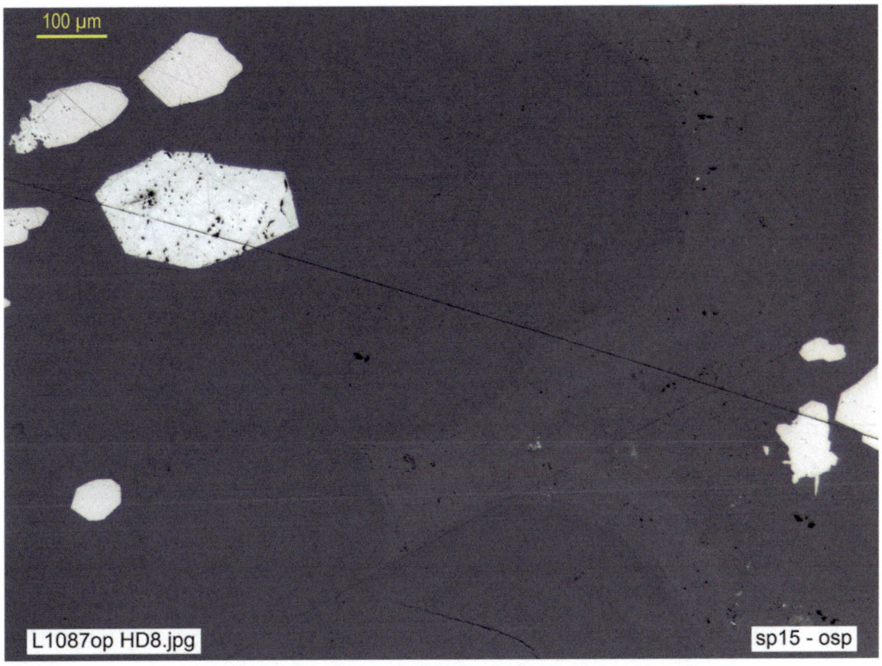

(b)

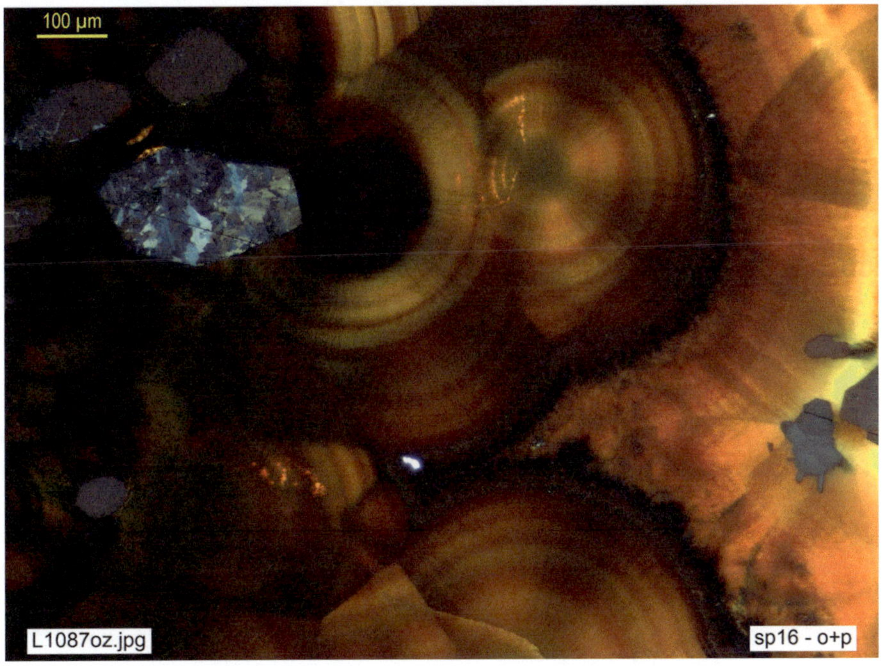

114.3 Spectral Information

114.3.1 Sphalerite Multispectral Reflectance R (in Air, Without Polariser) (%)

λ (nm)	370	400	425	450	475	500	525	550	575	600	625	650	675	700	750	800	850	900	950	1000
μ	19.9	19.2	18.4	17.8	17.5	17.2	17.0	16.9	16.7	16.6	16.5	16.3	16.3	16.2	16.1	16.0	16.0	15.7	15.6	15.6
P_5	18.4	17.9	17.1	16.7	16.4	16.4	16.2	16.3	16.1	16.0	15.9	15.8	15.7	15.7	15.6	15.4	15.5	15.2	15.1	15.0
P_{20}	19.3	18.6	17.8	17.3	16.9	16.7	16.5	16.5	16.3	16.2	16.1	16.0	15.9	15.9	15.8	15.6	15.7	15.4	15.3	15.2
P_{80}	20.6	20.0	19.0	18.5	18.1	17.8	17.6	17.4	17.1	17.0	16.8	16.7	16.6	16.5	16.4	16.2	16.2	15.9	15.8	15.8
P_{95}	21.0	20.2	19.4	18.7	18.4	18.1	17.8	17.6	17.3	17.2	17.0	16.9	16.8	16.7	16.6	16.6	16.7	16.5	16.4	16.4

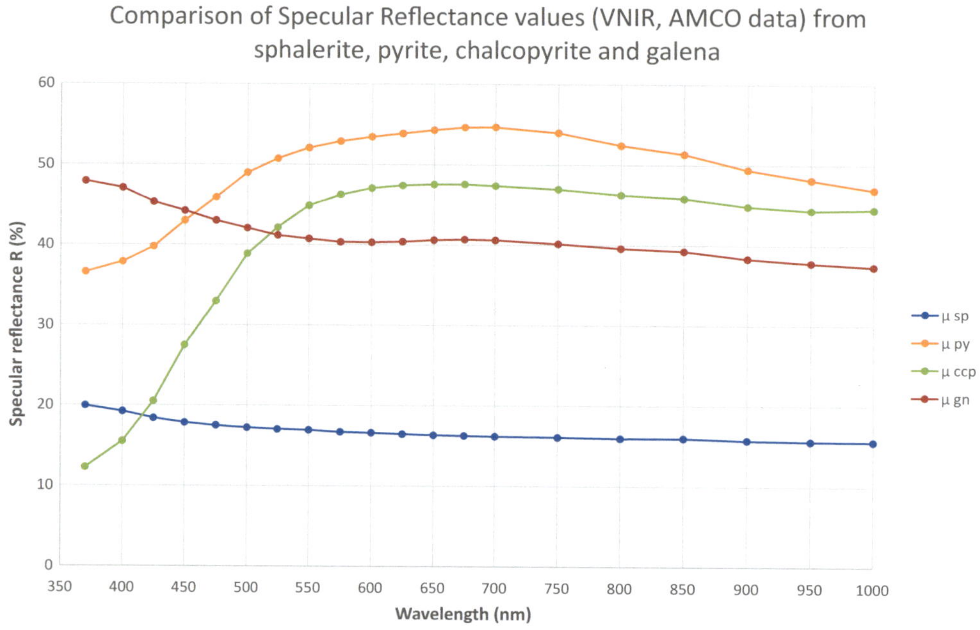

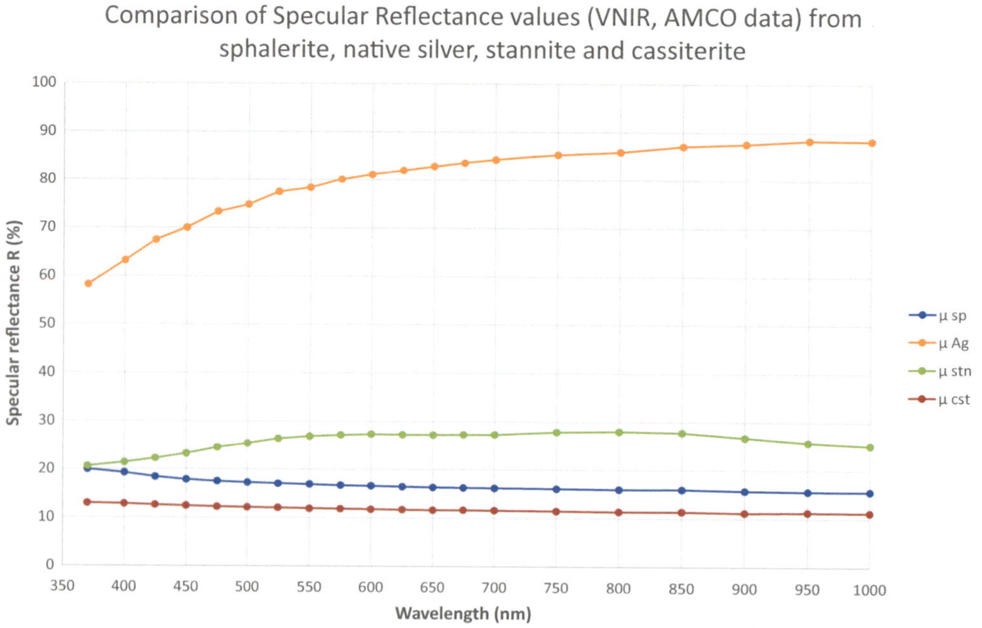

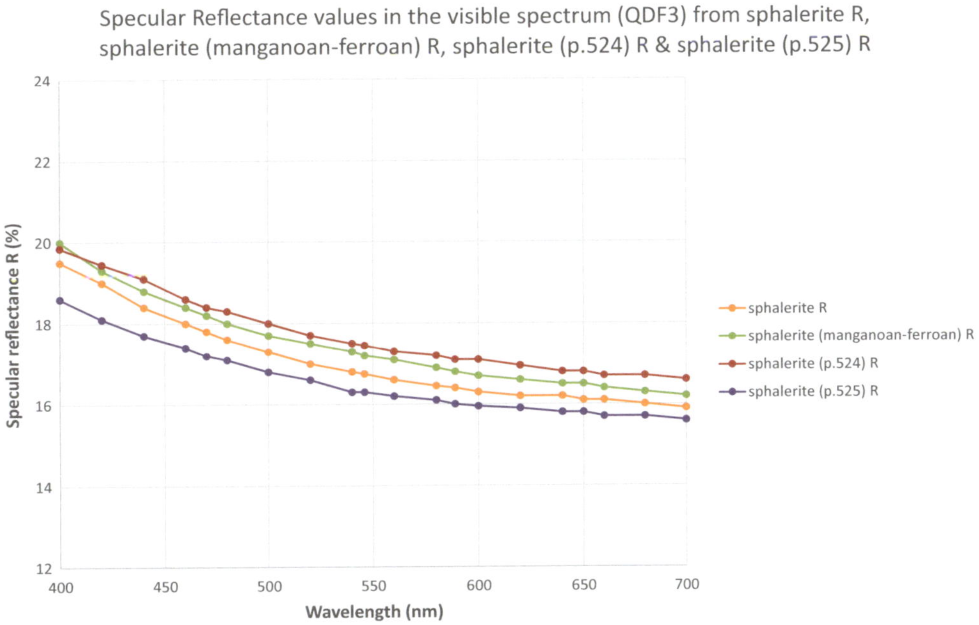

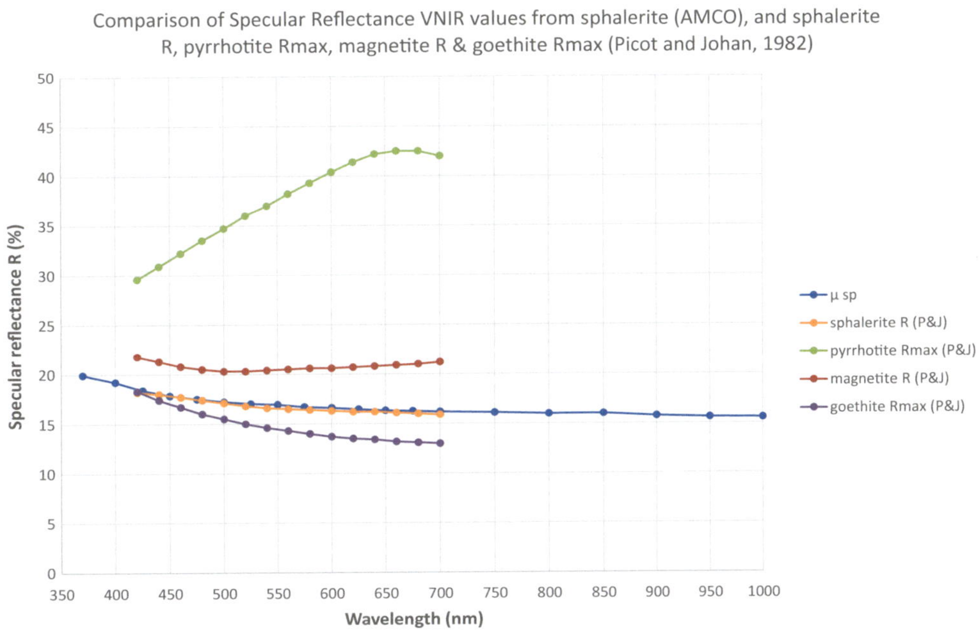

References

Bowles JFW, Howie RA, Vaughan DJ, Zussman J (2011) Rock-forming minerals, vol. 5A (2nd edn.) Non-silicates: oxides, hydroxides and sulphides. The Geological Society, London, 920p

Picot P, Johan Z (1982) Atlas of ore minerals. B.R.G.M./Elsevier

115 Spinel (*spn*)

$MgAl_2O_4$: Cubic

Spinel	R_{546} (%)	Y (%)	λ_d (nm)	P_e (%)	Colour sensation
Air	6.9	6.9	475	1.4	Grey
Oil	0.4	0.4	463	0.7	Dark grey
Bir: Null (1)			*Pleoc*: No (1)		*Tarn*: No
Anis +P: Null (1)			*IR*: Frequent		*Twg*: (111)
Mor: Idiomorphic (octahedral), massive			*Clv*: No		*PolQ*: Good
Hardness	*PolH*: Very high (5)		*Mohs*: 8	*Talm*: nd	*VHN* $_{(100\ g)}$: 1479–1548

Obs **:** cf. Table and R charts, multispectral data (VNIR, AMCO) and visible range (QDF3)
Mineral abbreviations: chr—chromite, dol—dolomite, gg—gangue, ht—hematite, il—ilmenite, mt—magnetite, py—pyrite, q—quartz, rt—rutile, spn—spinel, ulv—ulvöspinel

115.1 General

Traditionally, spinel, ***spn***, was considered as the type species ($MgAl_2O_4$) of the Spinel Group (AE_2O_4), composed of three isomorphic series of cubic oxides whose composition can vary widely: spinel series ($E = Al^{3+}$), magnetite series ($E = Fe^{3+}$) and chromite series ($E = Cr^{3+}$), in which A represents the divalent cation (Fe, Mg, Mn, Zn). However, according to a proposal approved by IMA in 2018,[1] a new classification has been established that does not recognize the Spinel Group, integrating spinel into the Spinel Subgroup. The latter is part of a wider and compositionally more varied supergroup, the Spinel Supergroup which, in addition to oxides, also comprises sulphides and selenides with spinel-type structure.

The composition of the **Spinel Supergroup** is expressed by the formula $\mathbf{AD_2X_4}$, where:

- **A** = Fe, Mn, Mg, Si, Ge, Co, Cu, Sb, Zn, Ti, Ni
- **D** = Fe, Cr, V, Mn, Al, Co, In, Ir, Rh, Pt, Ni
- **X** = O^{2-}, S^{2-} or Se^{2-}

Depending on the dominant anion X, this supergroup is divided into the three groups of *Oxyspinel* (O^{2-}), *ThioSpinel* (S^{2-}), *SeleniumSpinel* (Se^{2-}), which in turn are divided into two pairs of subgroups, according to the valence and identity of the cations. The spinel mineral is now integrated into the *Spinel Subgroup* which, together with the *Ulvöespinel Subgroup,* constitutes the *Oxyspinel Group*. *Ulvöespinel*, formerly *ulvite* ($TiFe_2O_4$) is the type species of the Ulvöespinel Subgroup. *Carrollite* and *Linnetite* give their names to the thio-spinel subgroups, as do *Bornhardtite* and *Tyrrellite* in the selenospinels.

The formula $\mathbf{A^{2+}D^{3+}{}_2O_4}$ is attributed to the **Spinel Subgroup**, where:

- A = Fe^{2+}, Mn^{2+}, Mg^{2+}, Co^{2+}, Zn^{2+}, Ni^{2+}
- D = Fe^{3+}, Cr^{3+}, Al^{3+}, V^{3+}
- X = O^{2-}

Therefore, in addition to spinel itself ($MgAl_2O_4$), this subgroup includes other species such as chromite ($Fe^{2+}Cr^{3+}{}_2O_4$), magnesiochromite ($MgCr_2O_4$), magnetite ($Fe^{2+}Fe^{3+}{}_2O_4$), franklinite ($Zn^{2+}Fe^{3+}{}_2O_4$), gahnite ($ZnAl_2O_4$), galaxite ($Mn^{2+}Al_2O_4$), hausmannite ($Mn^{2+}Mn^{3+}{}_2O_4$), hercynite ($Fe^{2+}Al_2O_4$), jakobsite ($Mn^{2+}Fe^{3+}{}_2O_4$), maghemite ($(Fe^{3+}{}_{0.67}\square_{0.33})Fe^{3+}{}_2O_4$) or trevorite ($Ni^{2+}Fe^{3+}{}_2O_4$), which formed the former spinel group. For spinel, the *three-series*

[1] *Proposal IMA 17-H*: CNMNC Newsletter 42, April 2018, p. 451; Mineralog. Mag. 82, 445–451; cf: https://www.mindat.org/min-52865.html.

R. Castroviejo, *A Practical Guide to Ore Microscopy—Volume 1*,
https://doi.org/10.1007/978-3-031-12654-3_115

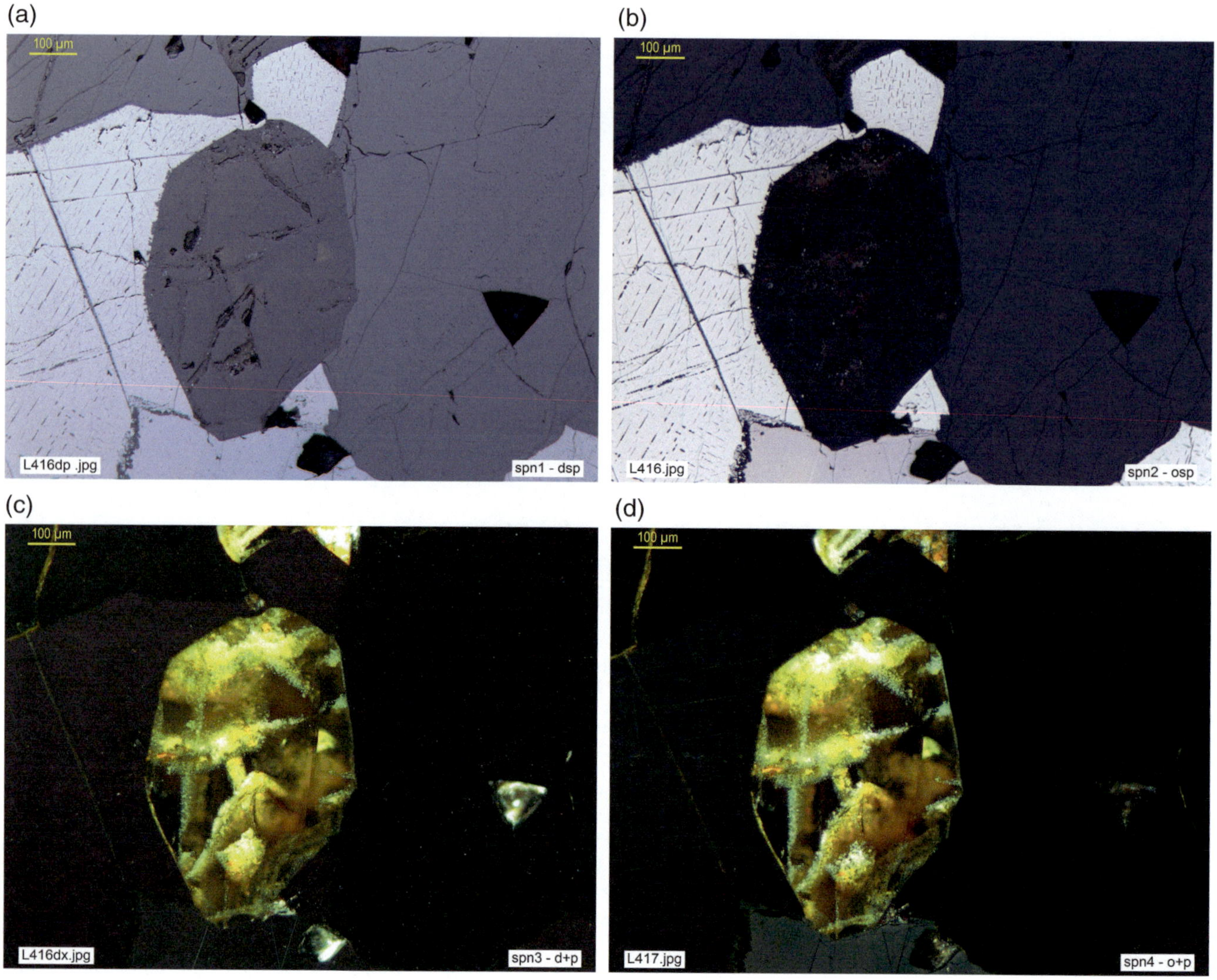

Fig. 115.1 **a**, **b** (Same scene, **dsp** and **osp**, respectively). Titanomagnetite-ilmenite ore (light grey), with large spn inclusions (darker grey), in two generations distinguished by their reflectance R and by their IR: type 1 (less dark, probably FeCr-spn, right half and top left corner) and type 2 (darker, MgAl-spn, centre). Brownish tone of Ti-mt (lower left), against which the typical pinkish tone of ilm (lower center) looks pale; reaction edge (indented contact, with spn segregation) between both minerals. In mt, tiny exsolutions of spn (octahedral, aligned), of ulv (lamellar, lighter) oriented according to (100) of mt and longer, acicular and anisotropic, of il. Oil immersion (Fig. 115.1b) enhances R contrast of the two spn types and of Ti-mt and il colours. CR253, Lac Berta, Quebec, Canada). **c**, **d** Same scene as Fig. 115.1a, (**d+p**) and (**o+p**) respectively. The greenish anisotropism of il clearly distinguishes it from mt and spn (isotropic). Abundance of IR, light-coloured and luminous, in MgAl-spn (whose transparency makes visible numerous families of fluid inclusions in microcracks), unlike type 1 spn (scarce, greenish IR)

spinel-hercynite, gahnite-spinel and magnesiochromite-spinel, already recognized in this group, are maintained (Back and Mandarino 2008).

The spinel species composition can vary by incorporating cations such as Fe, Cr, Ti, Mn, Zn, Ca in appreciable quantities, which results in variations in certain properties, such as increased reflectance for high contents of Cr and Fe, concomitant with a lower transparency. Certain impurities give the mineral beautiful colours that can give it value as a gem. Spinel *s.str.* ($MgAl_2O_4$) is a transparent, refractive (n = 1.719), optically isotropic mineral, which is easily identified with petrographic microscope. Classical mineralogical works or texts may refer to spinels (*s.l.*), meaning the whole ancient group (present spinel subgroup), i.e., including magnetite, chromite, franklinite, etc., but these species are clearly differentiated. To avoid confusion in what follows, when speaking of spinel, the species proper ($MgAl_2O_4$) or spinel (*s.str.*) should be understood.

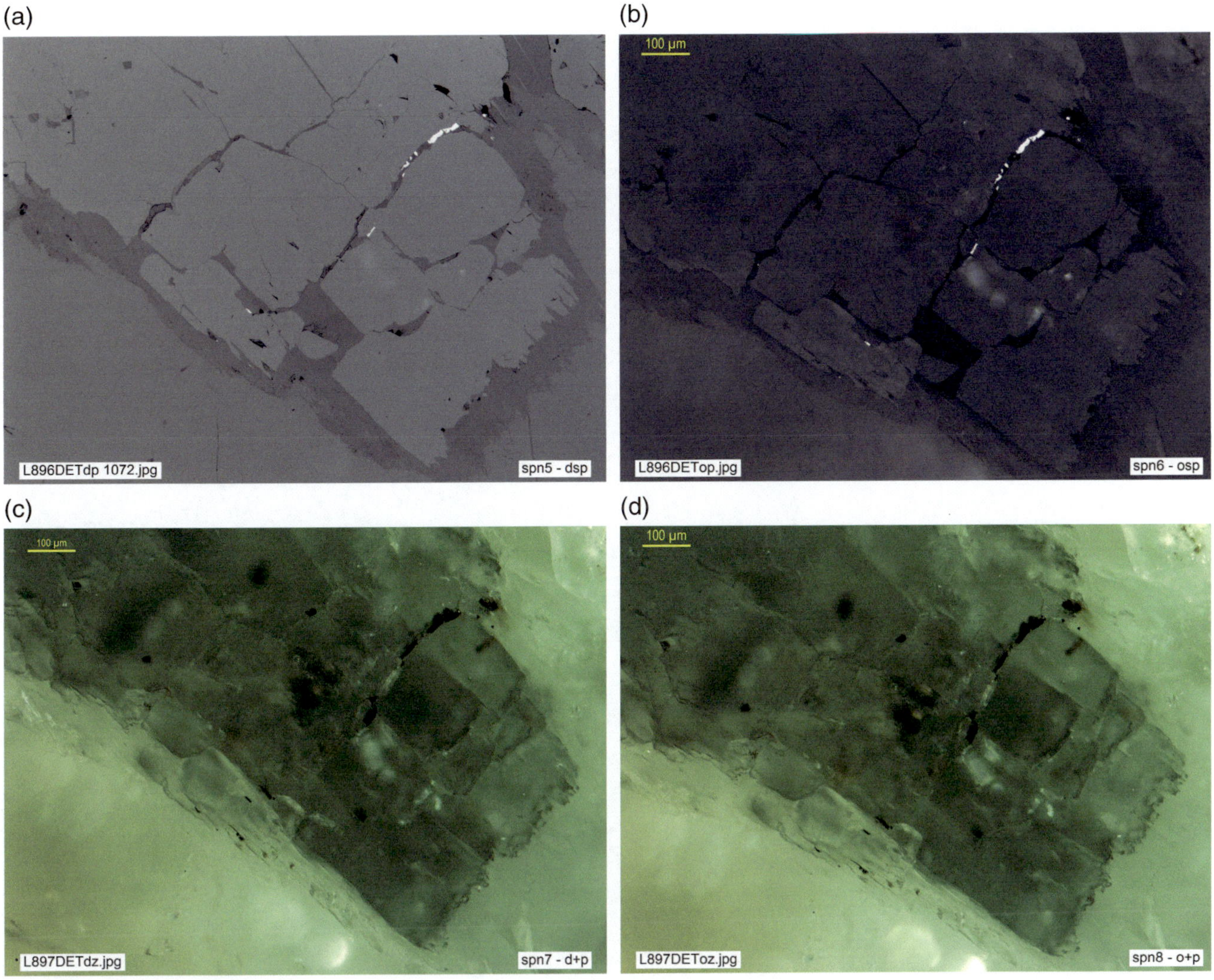

Fig. 115.2 **a**, **b** Idioblast of spn (lighter grey) crackled and partly cemented by mt (white), in dolomitic marble (dark grey background). R spn slightly >dol and <<mt. Sample MinII-rub, Val Malenco, Italy (**dsp** and **osp**, respectively). **c**, **d** Same scene as Fig. 115.2a (**d+p** and **o+p**, respectively). The transparency of spn and dol allow the idioblast to be seen in 3D

115.2 Microscopic Description

115.2.1 SP Observation (*Single Polariser*)

Grey colour, only slightly lighter than common gangue (comp. Figs. 115.1a, b, 115.2a, b and 115.4a). *Good polish*, despite its *very high polishing hardness* (> pyrite and quartz).

Reflectance almost as low as gangue and distinctly lower than that of other minerals of the subgroup, such as chromite or magnetite, the value of R depends on the composition and increases with Cr and/or Fe contents (cf multispectral plots: the comparison of the two visible spectra QDF3 evidences the difference between pure spinel or MgAl-spinel and FeCr-spinel, as can also be seen in Fig. 115.1a, b; the higher R of the latter implies a lower transparency and, therefore, also a higher scarcity of IR). Isotropic.

It occurs crystallized (cubic or octahedral forms Figs. 115.1a and 115.4a) and in massive or granoblastic aggregates (Fig. 115.3a) or in exsolutions in magnetite (Fig. 115.5a and 115.6). It does not show exfoliation.

115.2.2 Observation +P (*Polariser + Analyser at 90°*)

Isotropic but illuminated by abundant *internal reflections*, green, colourless or, more rarely, brownish (colours and abundance may vary according to composition, comp. Figs. 115.1c, d, 115.2c, d, 115.3b, 115.4b and 115.5b). IR are typically clear

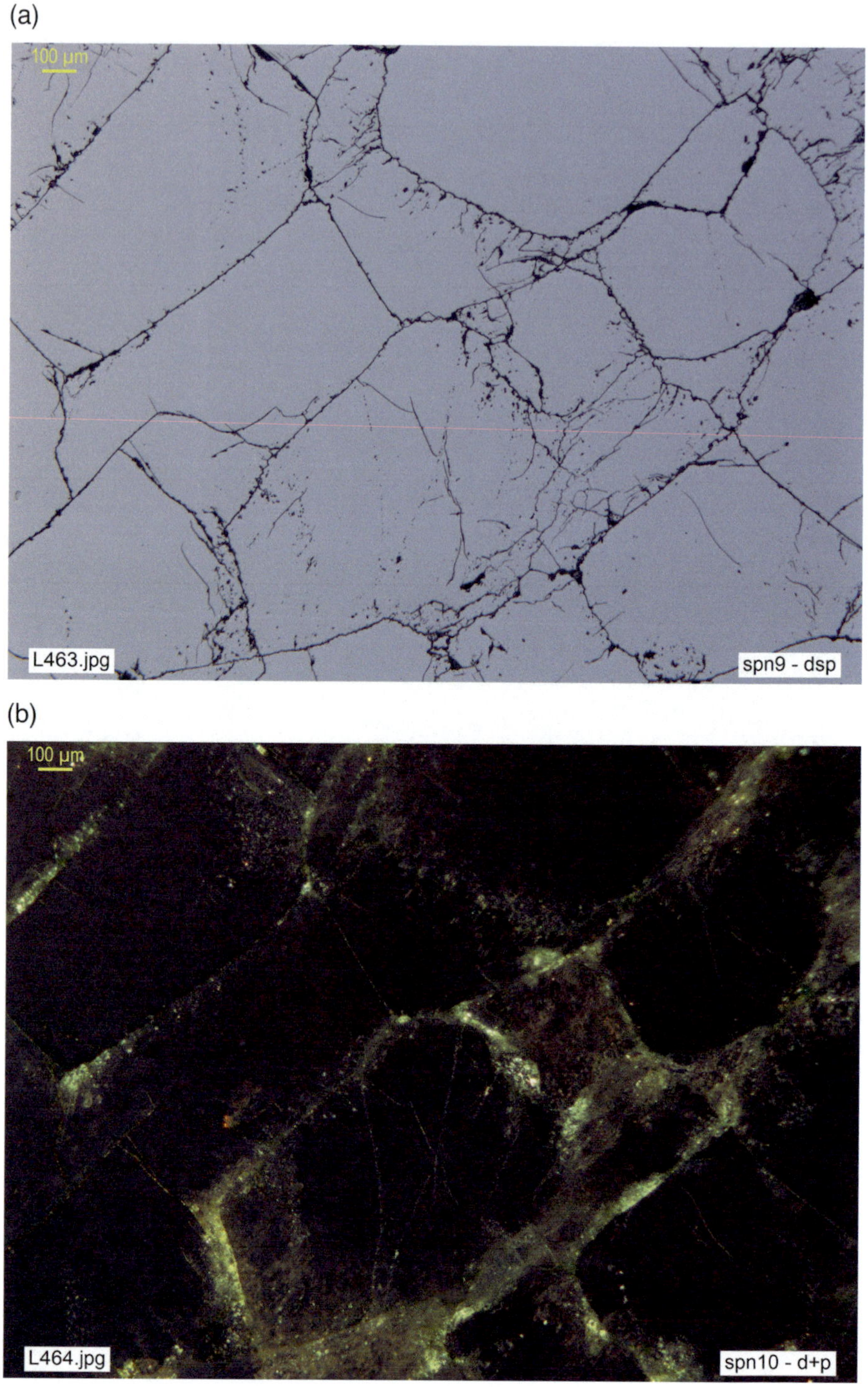

Fig. 115.3 **a**, **b** Granoblastic aggregate of spn. Sample R8, S[a]. Ronda, Málaga (**dsp** and **osp**, respectively)

and luminous (Fig. 115.1c, d) in the spinel proper (or Mg-rich terms). Greenish tones dominate in compositions rich in Zn (gahnite) or Fe^{2+} (hercynite) and brownish tones in those rich in Mn^{2+} (galaxite). Varied *twins*, but rarely visible due to their optical isotropy.

115.2.3 Paragenesis

High-temperature mineral, common as a minor component in certain igneous and metamorphic rocks, as well as in derived detrital deposits. Exsolutions of magnetite, ilmenite or

Fig. 115.4 **a**, **b** Relict spn phenoblast in dolomitic marble. Strong bireflectance, twinning and anisotropism of dol, in contrast to isotropy of spn. R dol always < R spn. Sample 1072-rub, Val Malenco, Italy (**dsp** and **d+p**, respectively)

magnetite ± ilmenite ± ulvöspinel are frequent in magmatic spinels. Fe-spinel (hercynite) ± ilmenite exsolutions are also observed in high-temperature magnetite (Fig. 115.6). Common spinel can be found in pegmatites, but also in peridotites, meteorites, kimberlites, basalts, marbles. In mineral deposits, only hercynite (titaniferous ores, titanomagnetites) and gahnite (sulphide ores of Pb–Cu–Zn) are common.

115.2.4 Identification

Spinel does not have many diagnostic properties, but its characterization is not difficult. Its idiomorphic tendency (Figs. 115.4a and 115.6), grey colour, low reflectance, very high hardness, abundant IR (depending on composition) and optical isotropy are important. Additionally, two useful

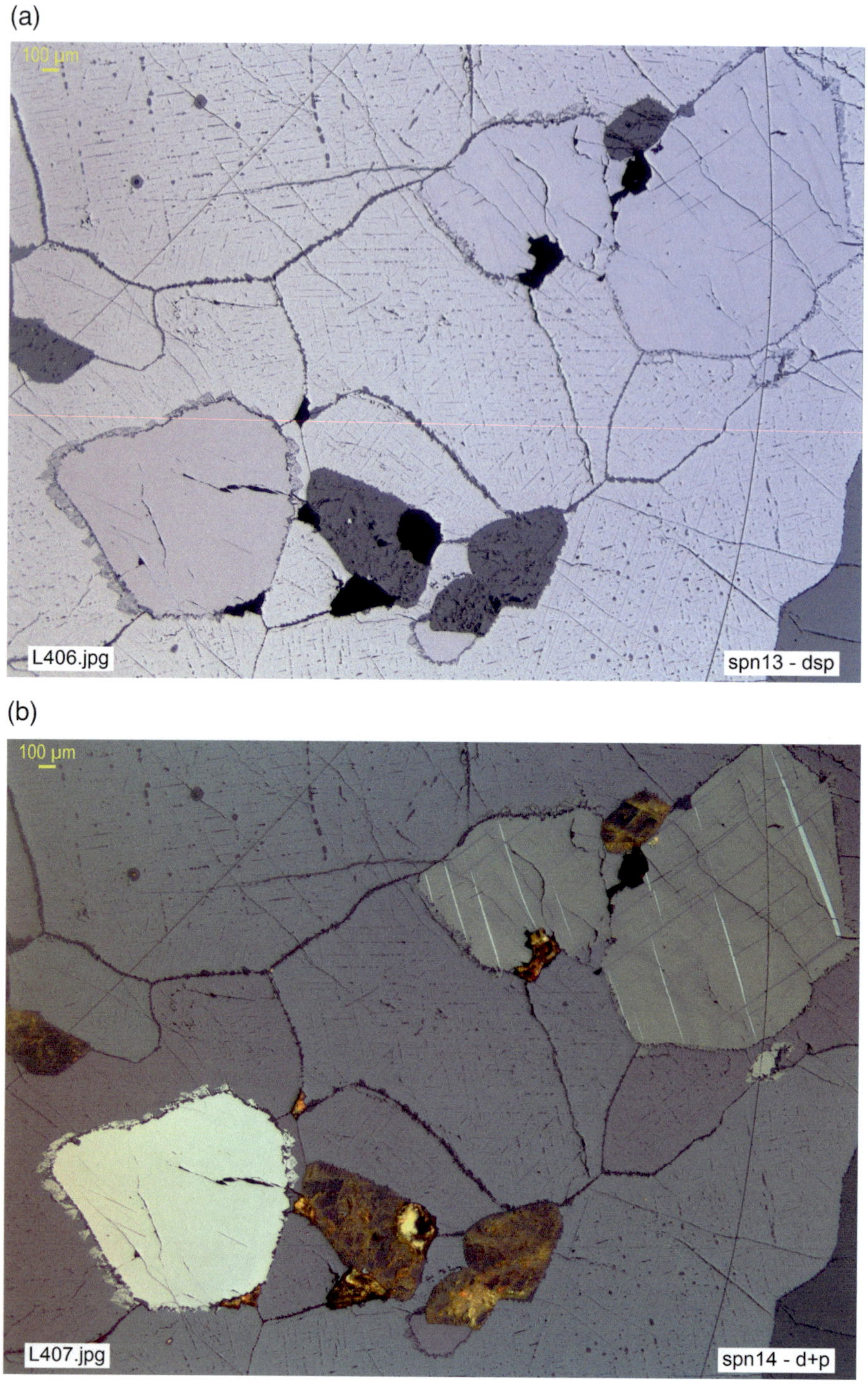

Fig. 115.5 **a**, **b** (**dsp** and **d+p**, respectively)—Inclusions of ilm (pinkish grey, anisotropic, greenish polarisation-tints and twinned) and spn (dark grey, isotropic) in granoblastic mt (light grey), with thin spn intergranular films. Reaction edges between il and mt. Both types of spn from Figs. 115.1a–c appear (type 1, only bottom right). Exsolutions in mt are described in Fig. 115.6. CR253, Lac Berta, Quebec, Canada

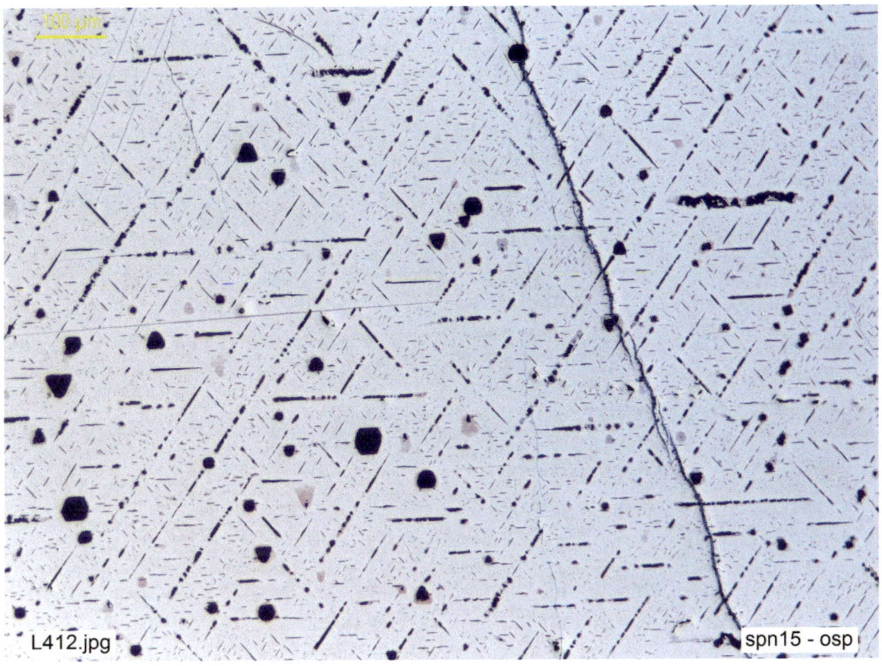

Fig. 115.6 Microinclusions in mt, fine exsolution of spn (dark grey) as cubic crystallites, scattered or aligned with octahedral directions of mt (sometimes associated with il unmixing) and as thin lamellae oriented //111 of mt; long, acicular segregations of il (pinkish grey), also oriented //111 mt but scarce. Detail of Fig. 115.5a (**osp**)

criteria can be applied: paragenesis and the presence of exsolutions (cf. § Sect. 115.2.3: *Paragenesis*). In any case, examination with transmitted light may resolve doubts.

115.3 Spectral Information

115.3.1 Spinel Multispectral Reflectance R (in Air, Without Polariser) (%)

λ (nm)	370	400	425	450	475	500	525	550	575	600	625	650	675	700	750	800	850	900	950	1000
μ	7.3	7.4	7.4	7.3	7.2	7.3	7.3	7.2	7.2	7.2	7.2	7.2	7.1	7.1	7.0	7.0	7.1	7.1	7.1	7.2
P_5	7.2	7.3	7.3	7.2	7.1	7.2	7.2	7.1	7.1	7.1	7.1	7.1	7.0	7.0	6.9	6.9	6.9	7.0	7.0	7.1
P_{20}	7.3	7.4	7.3	7.3	7.2	7.2	7.2	7.1	7.1	7.2	7.1	7.1	7.1	7.1	6.9	7.0	7.0	7.0	7.1	7.2
P_{80}	7.4	7.5	7.4	7.4	7.3	7.3	7.3	7.2	7.2	7.2	7.2	7.2	7.2	7.2	7.0	7.0	7.1	7.1	7.2	7.3
P_{95}	7.5	7.5	7.5	7.4	7.3	7.4	7.4	7.2	7.3	7.3	7.3	7.3	7.2	7.2	7.1	7.1	7.2	7.2	7.2	7.3

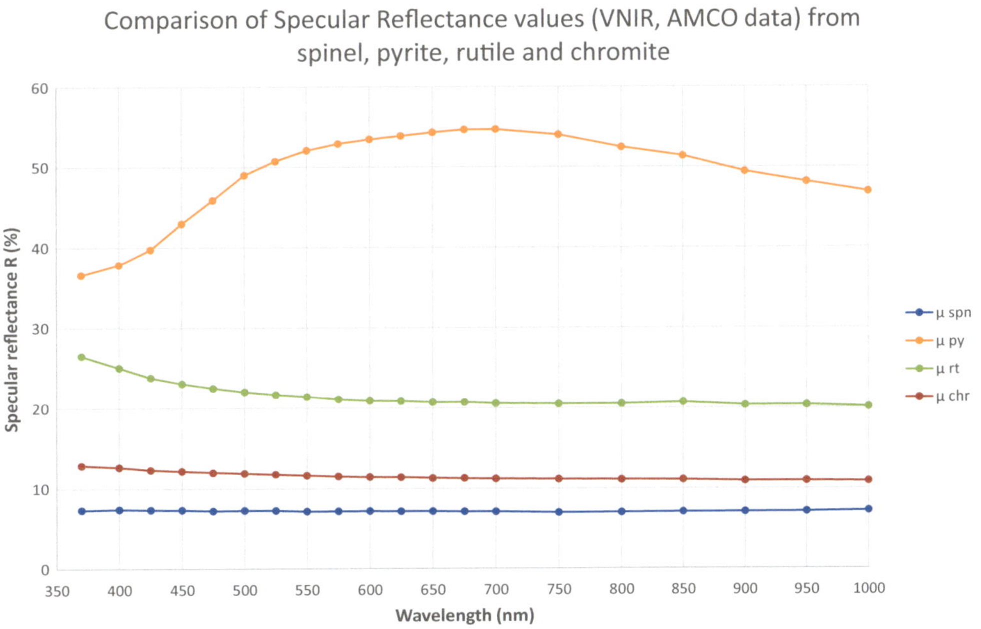

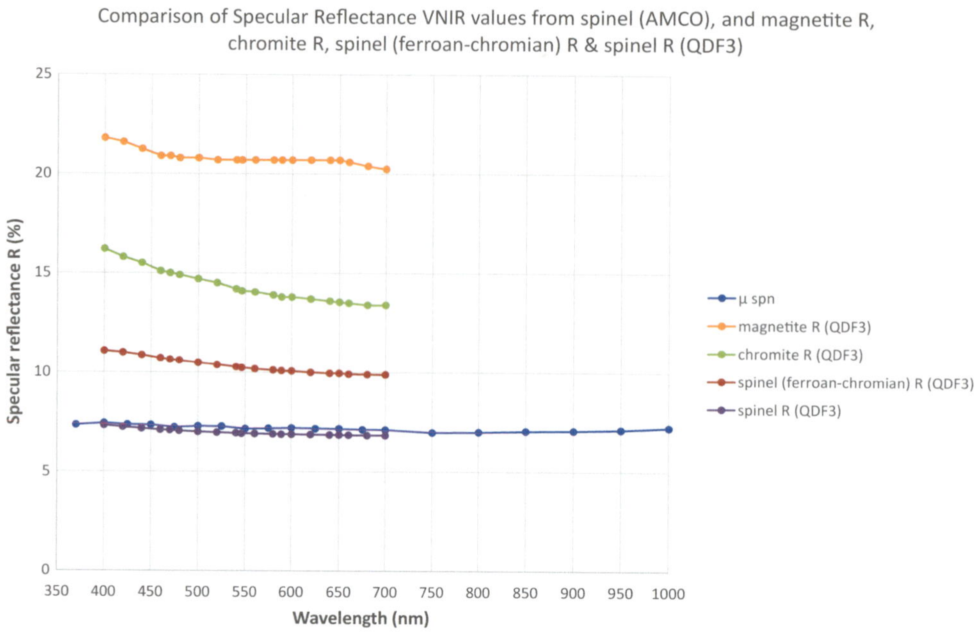

Reference

Back ME, Mandarino JA (2008) Fleischer's glossary of mineral species 2008. The Mineralogical Record Inc, Tucson, AZ, 345p

Stannite (*stn/stannine*)

116

Cu_2FeSnS_4: Tetragonal

Stannite		R_{546} (%)	Y (%)	λ_d (nm)	P_e (%)	Colour sensation
Air	R_o	27.3	27.1	571	4.2	Brownish-olive grey
	$R_{e'}$	26.0	25.8	572	7.4	Greenish grey
Oil	R_o	14.1	13.9	566	3.4	Olive-brown, more saturated
	$R_{e'}$	13.0	12.9	569	8.4	Greenish-brownish grey
Bir: Weak (3)				*Pleoc*: Weak (3)		*Tarn*: Bluish (in hand sample)
Anis +P: Distinct, bluish, violet (4)				*IR*: No		*Twg*: Occasional lamellar, parquetry
Mor: Anhedral / granoblastic aggregates				*Clv*: (110), *triangular pits*		*PolQ*: Easy, good
Hardness	*PolH*: Medium, <sp, >ccp (3)			*Mohs* 4	*Talm*: D+	*VHN* $_{(25\ g)}$: 216–265

Obs: cf. Table and R charts, multispectral data (VNIR, AMCO) and visible range (QDF3)
Mineral abbreviations: Ag—native silver, asp—arsenopyrite, Bi—native bismuth, ccp—chalcopyrite, cst—cassiterite, gg—gangue, gn—galena, gth—goethite, ks—kësterite, po—pyrrhotite, py—pyrite, q—quartz, sp—sphalerite, stn—stannite, td—tetrahedrite, wf—wolframite
Colour abbreviations

116.1 General

The composition of stannite, ***stn***, can be derived from chalcopyrite by substituting half of the Fe atoms for Sn, resulting in Cu_2FeSnS_4. Frequently, there is a partial substitution of Fe by Zn in stannite. There is a complete solid solution between the two terms stannite and *kësterite* (***ks***, Cu_2ZnSnS_4), defining the isomorphous series stn-ks. Other possible impurities are In, Ge, Cd, Ag.

116.2 Microscopic Description

116.2.1 SP Observation (*Single Polariser*)

Colour olive grey. Its *reflectance*, **R**, is low (R stn < td, gn), though higher than sp (*cf* comparative charts and Figs. 116.1 and 116.4).

Slightly bireflective and pleochroic, especially in immersion, with slight variation from dominant grey to olive shades (Fig. 116.3).

Good *polishing*. Medium *polishing hardness*: PolH stn > Ag, ccp, gn; PolH stn < td, sp, po, asp. Cleavage (110) and more rarely (001), can lead to triangular *pits* in the polishing process.

Usually xenomorphic, it occurs as fissure filling, interstitial or in intergranular films and, when abundant, forms granular aggregates. Frequently associated with chalcopyrite, sphalerite, kësterite, with which it can present exsolution textures.

It usually replaces early minerals such as arsenopyrite, pyrrhotite, cassiterite, quartz, sphalerite.

116.2.2 Observation +P (*Polariser + Analyser at 90°*)

Clearly *anisotropic*, with typical violet to bluish (or bluish-greenish) colours, luminous in diagonal position, especially vivid with immersion (Figs. 116.1c, d and 116.3).

The anisotropism makes the internal structures, oriented intergrowths, twins and exsolutions very visible. It presents oriented *intergrowths* with chalcopyrite and sphalerite.

R. Castroviejo, *A Practical Guide to Ore Microscopy—Volume 1*,
https://doi.org/10.1007/978-3-031-12654-3_116

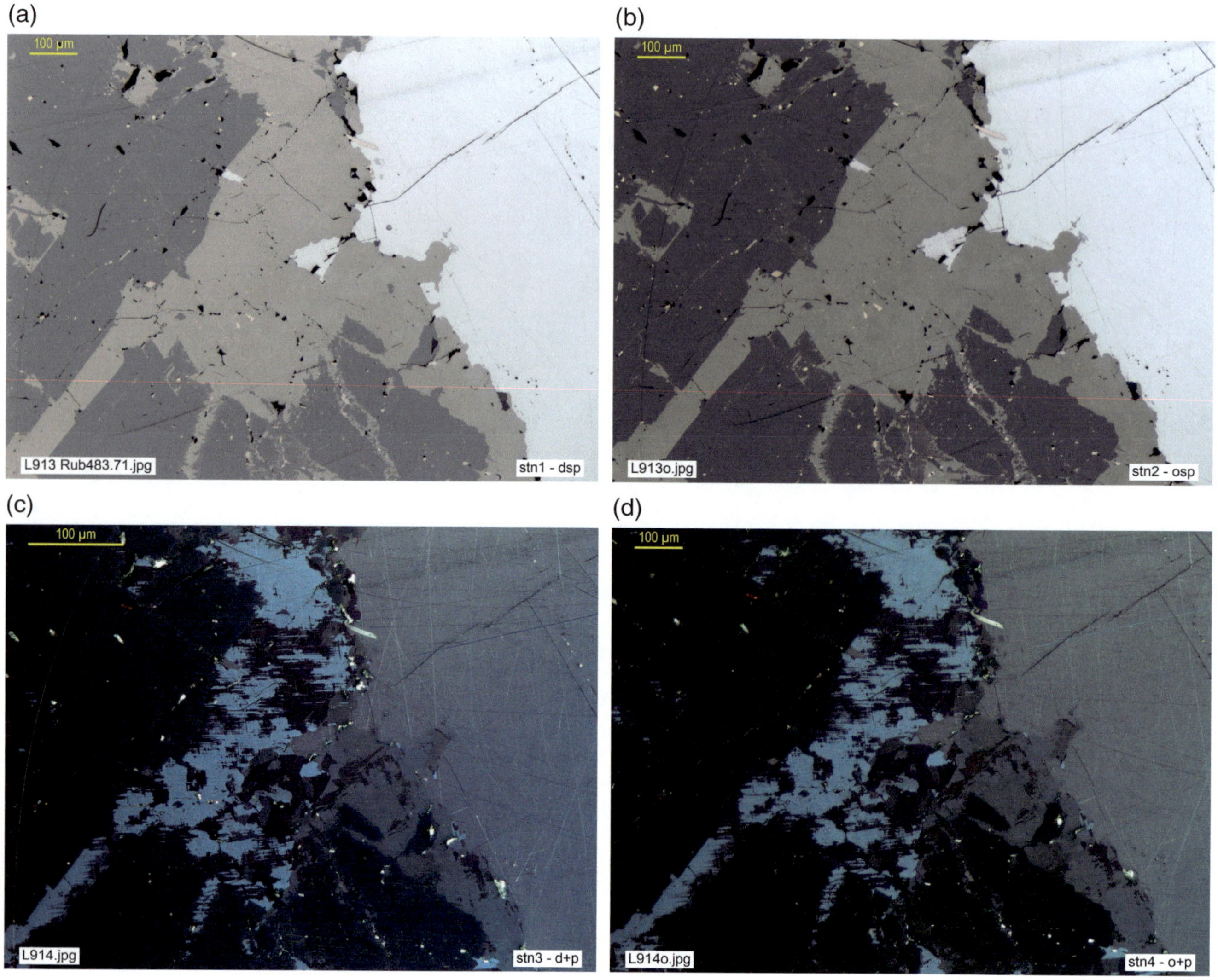

Fig. 116.1 **a, b** (**dsp** and **osp**, respectively)—Intergrowth of stn (olive-brownish grey), sp (grey) and gn (white), with traces of ccp (yellow, minute inclusions in sp and stn) and po (pinkish-brown); sp is an early mineral, fragmented and welded—corroded by stn. Sample Rub483.71, Davidschacht, Saxen, Germany. **c, d** Same scene as Fig. 116.1a, (**d+p** and **o+p**, respectively): Anisotropism and twinning of stn; IR sp

Varied *twinning*, not as frequent as in chalcopyrite or sphalerite but, unlike sphalerite, well visible without attack. Growth (the most common, thick lamellar {111}), pressure and inversion twins can be observed. Characteristic inversion twins show a *parquetry* morphology, with pseudocubic orientations, // (100) and (001), or an orthogonal polysynthetic, *microcline-like*, morphology (Fig. 116.1c, d).

Exsolutions of chalcopyrite, sphalerite or kësterite, frequently enhanced by polarization (Fig. 116.5). Stannite exsolutions are also found in chalcopyrite, in sphalerite and, rarely, in pyrrhotite.

116.2.3 Paragenesis

Pegmatitic-pneumatolitic and high-temperature hydrothermal *paragenesis* related to cassiterite vein formations, in which stannite is almost always present in the sulphide phases, also in subvolcanic deposits (*porphyry tin*).

Usual companions: chalcopyrite, sphalerite, pyrite, cassiterite, wolframite, arsenopyrite, pyrrhotite, bismuth, tetrahedrite, kësterite, silver sulphosalts.

Fig. 116.2 Granoblastic aggregate of stn, intergrown with asp, sp, ccp and q. With an immersion objective, stn appears darker and slightly pleochroic in olive-grey tones (comp. Fig. 116.1a); slight pleochroism (white to pinkish-white) of asp is also visible. Sample CR 319 (**osp**)

Fig. 116.3 Same scene as Fig. 116.2, (**o+p**): the bright polarization colours make the granoblastic texture of stn very visible, whose recrystallization is probably the cause of the appearance of thick lamellar or polysynthetic twins

116.2.4 Identification

It is easily recognized by its greenish- grey to olive-brown colour, with weak pleochroism in immersion, its vivid anisotropism with violet and bluish tones and, occasionally, typical twinning.

Its colour might be confused with some species of the *fahlore* series (the less reflective, Fe-rich tetrahedrites, *s.l.*), but misidentification is easily avoided since, unlike stn, all of them are isotropic and their R is higher than that of stannite.

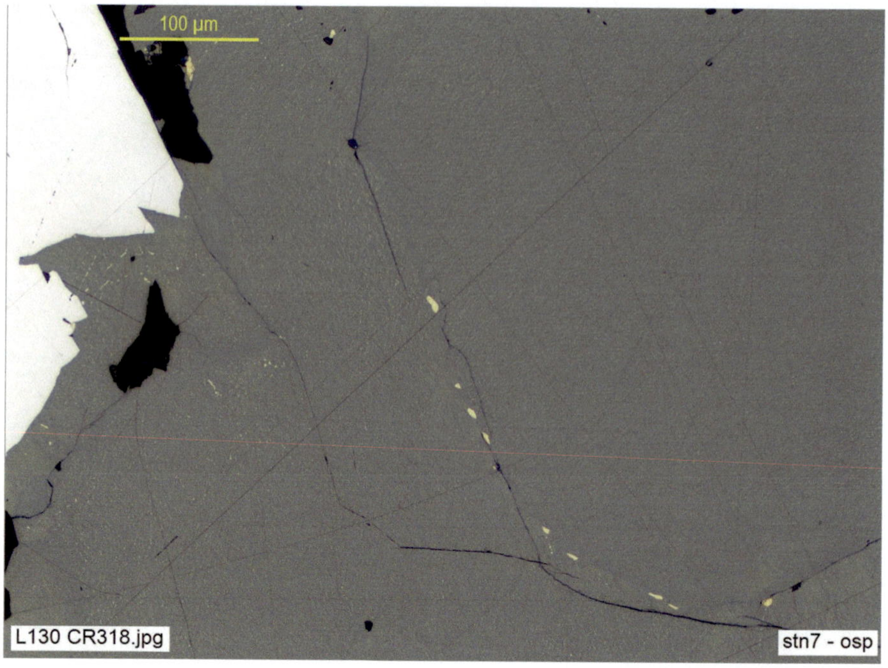

Fig. 116.4 Intergrowth of stn and asp. Emulsion of ccp and probable kësterite (**ks**) in stn, ccp dominant in bottom-left grains and ks in large top-right crystal, which looks slightly darker, as corresponds to the lower reflectance of ks. CR 318, Hingston Downs, Cornwall, UK (**osp**)

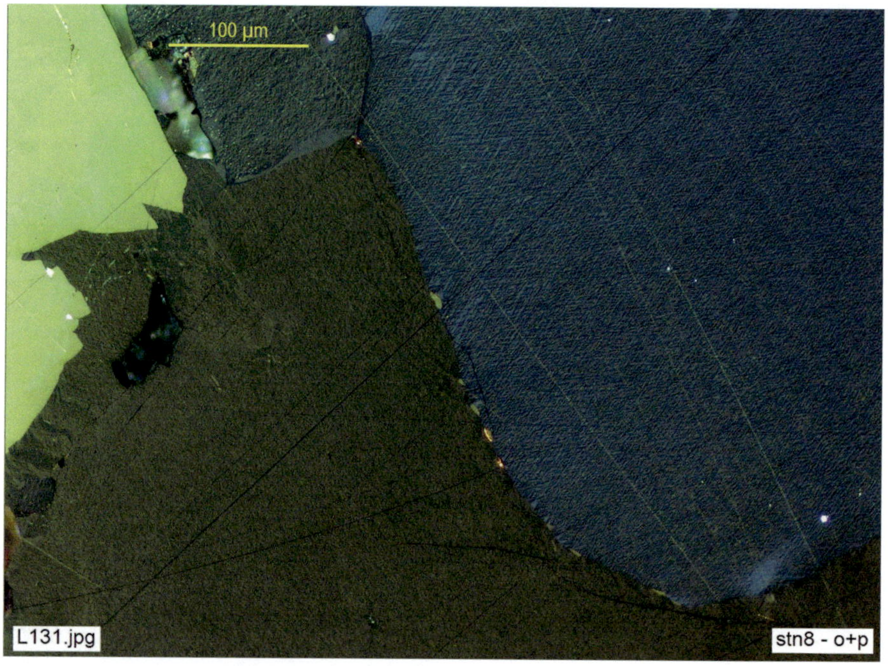

Fig. 116.5 Same scene as Fig. 116.4, (**o+p**): anisotropism highlights the felted textures of the emulsion in stn, suggesting unmixing oriented by the crystal lattice of stn

116.3 Spectral Information

116.3.1 Stannite Multispectral Reflectance R (in Air, Without Polariser) (%)

λ (nm)	370	400	425	450	475	500	525	550	575	600	625	650	675	700	750	800	850	900	950	1000
μ	20.6	21.4	22.3	23.2	24.5	25.4	26.4	26.9	27.2	27.4	27.3	27.2	27.3	27.3	27.9	28.0	27.8	26.8	25.8	25.2
P_5	20.0	20.8	21.3	22.3	23.4	24.3	25.2	25.7	26.0	26.2	26.1	26.1	26.2	26.3	26.8	27.0	26.8	25.8	24.9	24.4
P_{20}	20.3	21.1	21.9	22.9	24.2	25.1	26.1	26.6	26.9	27.1	27.0	27.0	27.0	27.1	27.6	27.7	27.5	26.5	25.5	24.9
P_{80}	20.9	21.7	22.6	23.7	25.0	25.8	26.8	27.3	27.6	27.8	27.7	27.6	27.7	27.7	28.3	28.4	28.2	27.2	26.1	25.6
P_{95}	21.1	21.9	22.9	23.9	25.2	26.1	27.1	27.6	27.9	28.0	27.9	27.8	27.9	27.9	28.5	28.6	28.4	27.4	26.4	25.8

Comparison of Specular Reflectance values (VNIR, AMCO data) from stannite, arsenopyrite, sphalerite and cassiterite

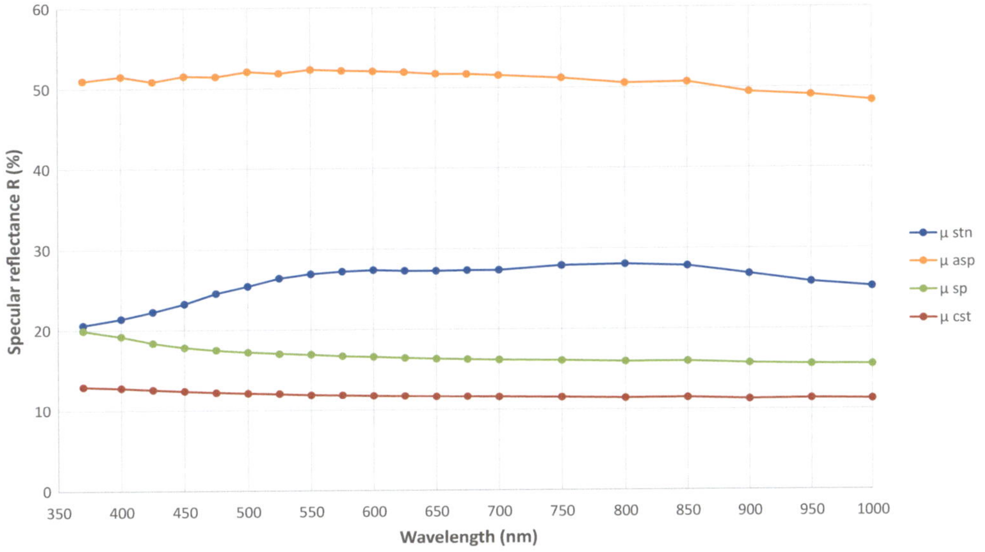

Comparison of Specular Reflectance VNIR values from stannite (AMCO), and stannite Ro, stannite Re', galena R & chalcopyrite R' (QDF3)

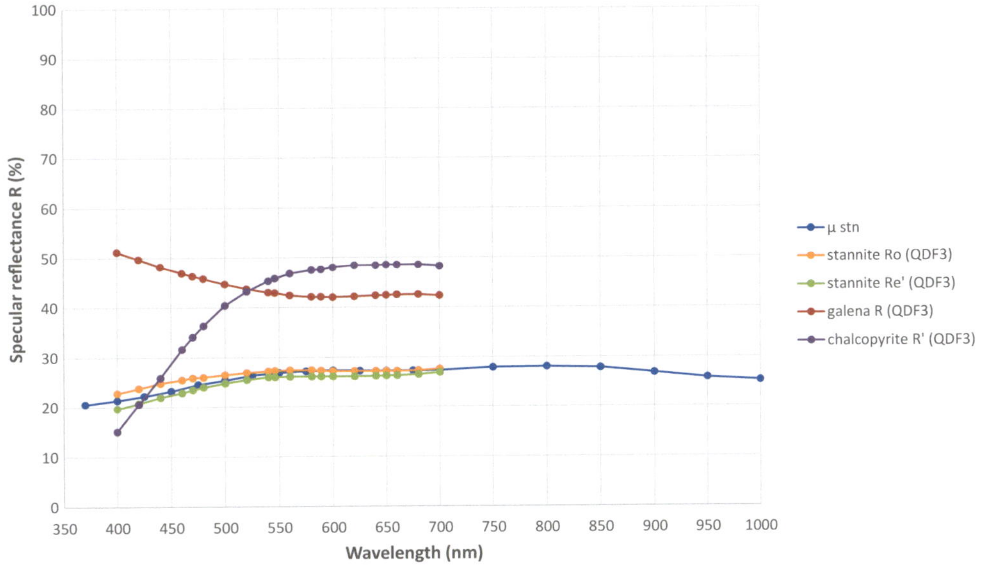

117 Stephanite (*stp/black silver, brittle silver ore*)

Ag_5SbS_4: Orthorhombic

Stephanite		R_{546} (%)	Y (%)	λ_d (nm)	P_e (%)	Colour sensation
Air	R_1	28.2	28.3	474	5.7	Light grey, slight brownish tone
	R_2	30.4	30.5	478	4.6	
Oil	R_1	13.7	13.9	473	10.6	
	R_2	15.5	15.6	478	7.9	
Bir: Weak (3)				*Pleoc*: Weak (3)		*Tarn*: Light etching
Anis +P: Net (4)				*IR*: No		*Twg*: Frequent
Mor: Anhedral aggr. Prismatic, tabular				*Clv*: Imperfect		*PolQ*: Good
Hardness	*PolH*: Very low, but > ac (1)			*Mohs*: 2–2½	*Talm*: B	*VHN* $_{(nd)}$: 45–107

Obs: cf. R charts, visible range (QDF3; Picot and Johan 1982). VHN: Uytenbogaardt and Burke (1971)
Mineral abbreviations: ac—acanthite, Ag—native silver, ccp—chalcopyrite, cv—covellite, dg—digenite, dy—dyscrasite, fbg—freibergite, gg—gangue, gn—galena, ht—hematite, my—miargyrite, pc—pearceite, plb—polybasite, prg—pyrargyrite, pru—proustite, py—pyrite, sbt—stibnite, sp—sphalerite, stp—stephanite, td—tetrahedrite

117.1 General

Stephanite, ***stp***, is a mineral very rich in Ag, commonly scarce although widespread; from an economic point of view it is not usually relevant.

Composition: Ag_5SbS_4, usually stoichiometric, but with possible traces of As, Fe, Cu. A relatively low-temperature mineral, unstable above ≈ 200 °C (197 ± 5 °C), which gives it value as a geological thermometer.

117.2 Microscopic Description

117.2.1 SP Observation (*Single Polariser*)

Light grey *colour*, similar to acanthite but, unlike ac, stp is pleochroic, moderately but distinctly visible, especially in oil, light grey to brownish-grey or slightly pinkish tones (Figs. 117.1a, b, 117.2a and 117.3a).

Light etching can be very rapid (Figs. 117.2a and 117.3a), although this does not always seem to be the case: Ramdohr (1980) describes it as weak to absent, although he cites sources that present it as conspicuous. For Picot and Johan (1982), it is slower than in polybasite.

Very low *polishing hardness* (PolH stp < sbt; << ccp, td, sp), though > pru, prg, plb, ac.

Moderate *reflectance*, R stp ≈ pc; ≤ ac, fbg, plb; < gn, py (cf multispectral comparative plots). *Weak bireflectance*. *Weak pleochroism* (light grey to brownish grey or slightly pinkish) but discernible in aggregates, by comparison of neighbouring grains.

It forms, under favourable conditions, columnar or tabular crystals. Alternatively, it occurs in anhedral aggregates but may develop euhedral, prismatic or acicular forms against acanthite or polybasite (Figs. 117.1a, b and 117.2a).

R. Castroviejo, *A Practical Guide to Ore Microscopy—Volume 1*,
https://doi.org/10.1007/978-3-031-12654-3_117

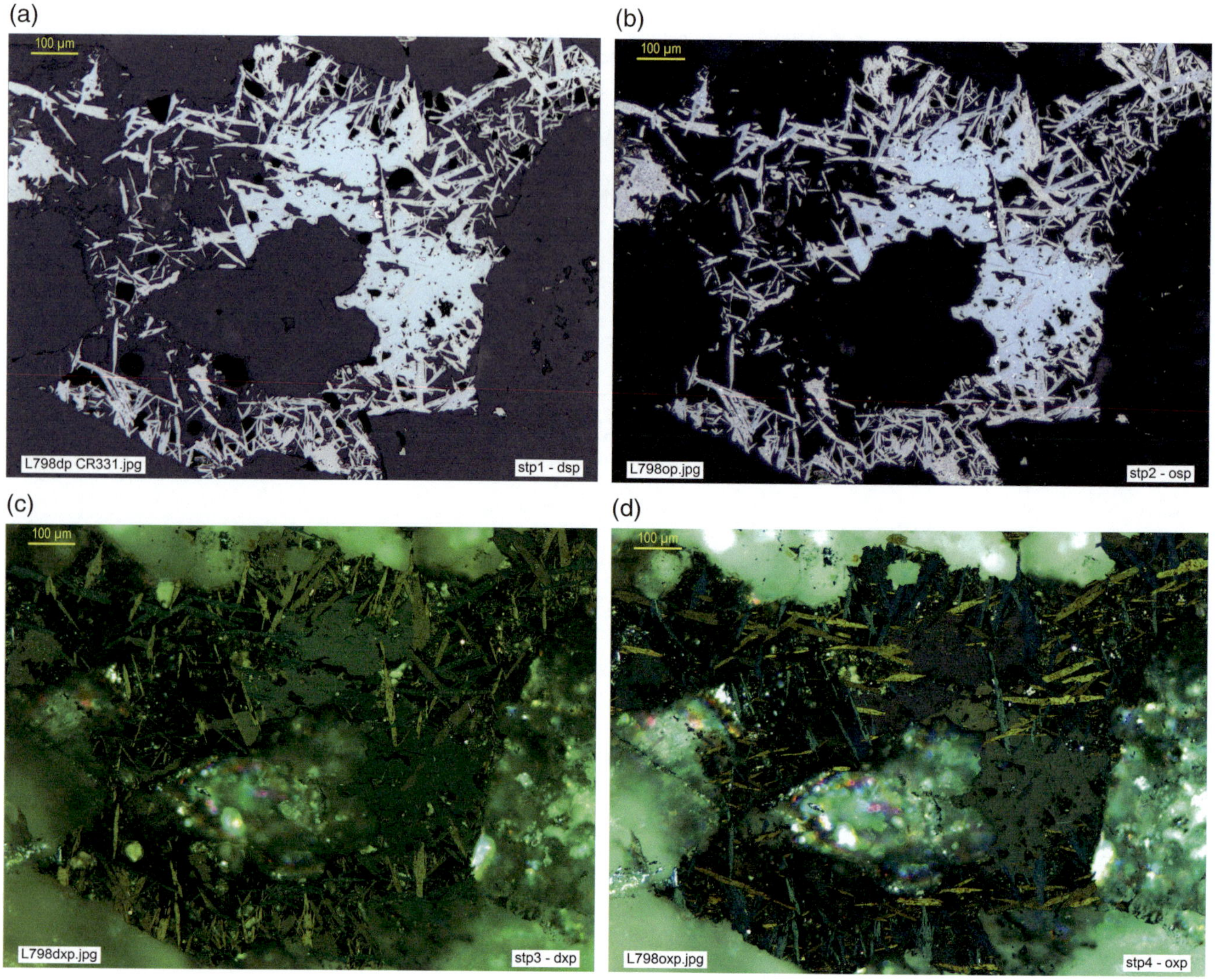

Fig. 117.1 **a**, **b** (**dsp** and **osp**, respectively)—Acicular aggregate of stp (slightly brownish grey) intergrown with ac (grey, slightly bluish-green), with small inclusions of py, gg of q (dark grey/black). CR 331, Fresnillo, Zacatecas, Mexico. **c**, **d** (same scene as Figa: **dxp** and **oxp**, respectively)—Anisotropism of stp and ac, with analyser at 87°

117.2.2 Observation +P (*Polariser + Analyser at 90°*)

Clearly anisotropic, with characteristic greenish to yellowish-brown tints (Figs. 117.1c, d, 117.2b and 117.3b), visible especially in immersion and slightly uncrossing the analyser (*xp*).

Distinct extinction.

Frequent *twinning*.

It does not show *internal reflections*, which facilitates its identification compared to other silver sulphosalts.

117.2.3 Paragenesis

Parageneses: epithermal Ag-Pb-Zn and hydrothermal *five-element* (Ni-Co-Bi-Ag-U) type.

Typical mineral of silver deposits, hypogene low temperature, late hydrothermal. Possibly also supergene.

It can occur as a minor component, accompanying native silver, acanthite, polybasite, tetrahedrite, pyrargyrite and other silver sulphosalts, as well as skutterudite or Co–Ni minerals.

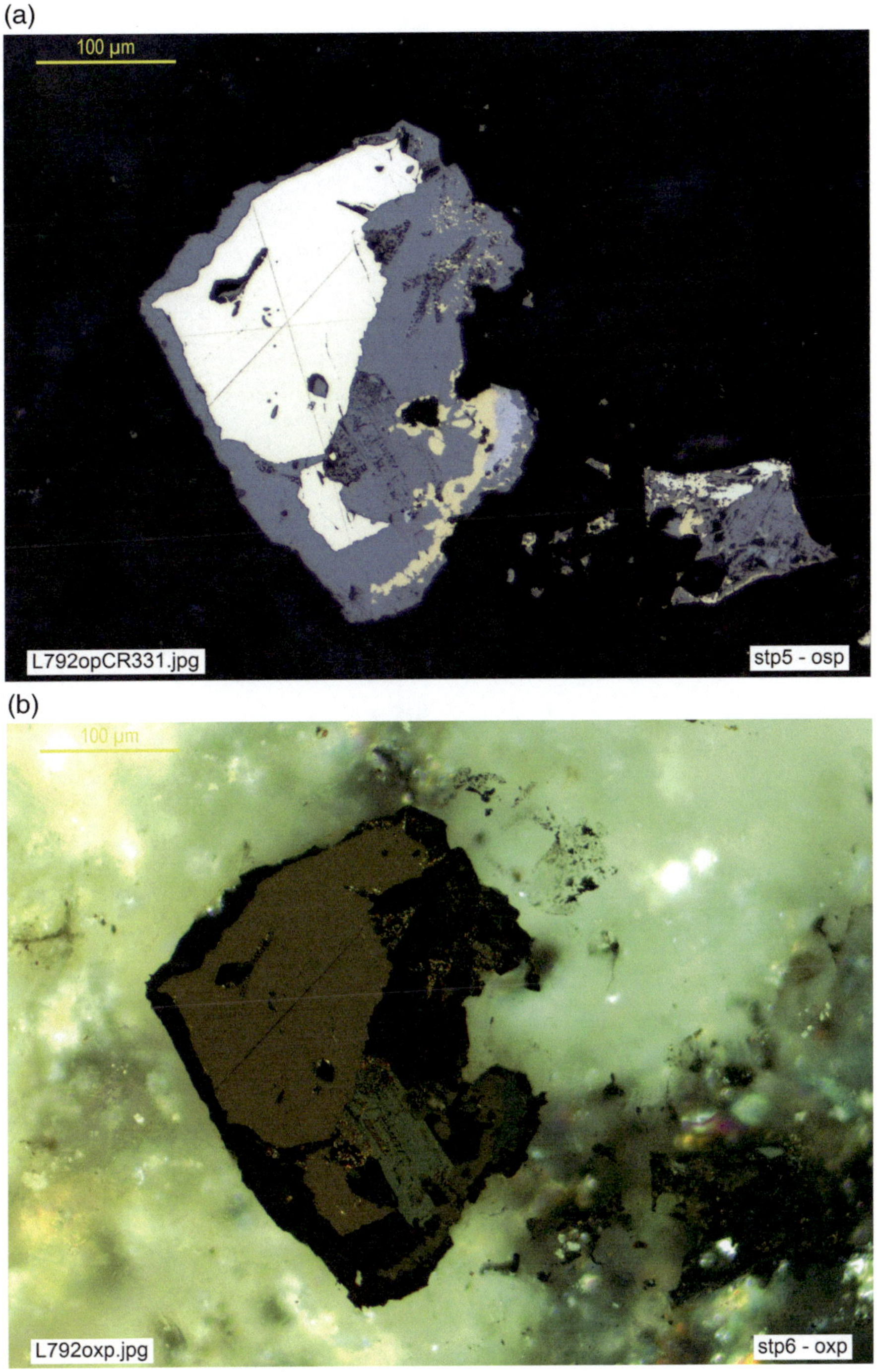

Fig. 117.2 **a**, **b** (Same scene: **osp** and **oxp**, respectively)—Aggregates of py, ac, stp, ccp and gn (in the centre) and py, ccp, stp (bottom right). Stp instantly light-etched acquires a black stippling, resulting in a dirty appearance, unlike ac, still unetched, which remains limpid. CR 331, Fresnillo, Zacatecas, Mexico

117.2.4 Identification

Stephanite is recognized by its colour and pleochroism, light corrosion, and particularly by its anisotropism, with uncommon greenish to yellowish-brownish tones (*xp*) and absence of internal reflections, in addition to paragenesis.

The lack of IR helps to avoid possible confusions with proustite, pyrargyrite, miargyrite, polybasite, pearceite.

It is harder than acanthite and polybasite.

Fig. 117.3 **a**, **b** (Same scene: **osp** and **oxp**, respectively)—Aggregate of stp, ccp and gn, with sparse sp and py, in gg of q. Stp shows unavoidable, immediate light etching, but its light grey to brownish-slightly pinkish grey pleochroism is still evident, as well as its anisotropism with green and yellow shades. CR 331, Fresnillo, Zacatecas, Mexico

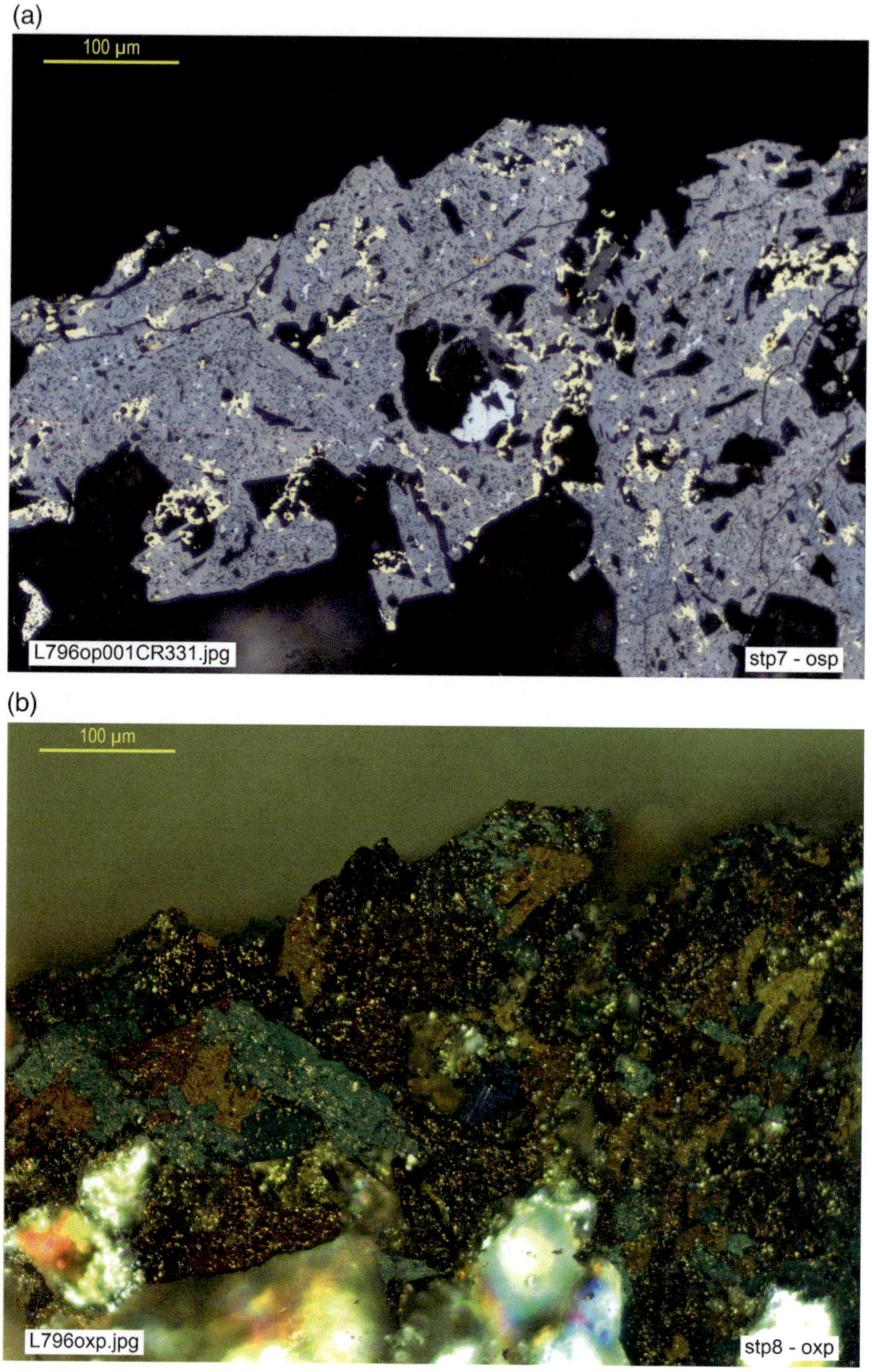

117.3 Spectral Information

117.3.1 Stephanite Reflectance R (in Air, Visible Spectrum) (%)

Specular Reflectance values in the visible spectrum (QDF3) from stephanite R1, stephanite R2, freibergite R & acanthite R'

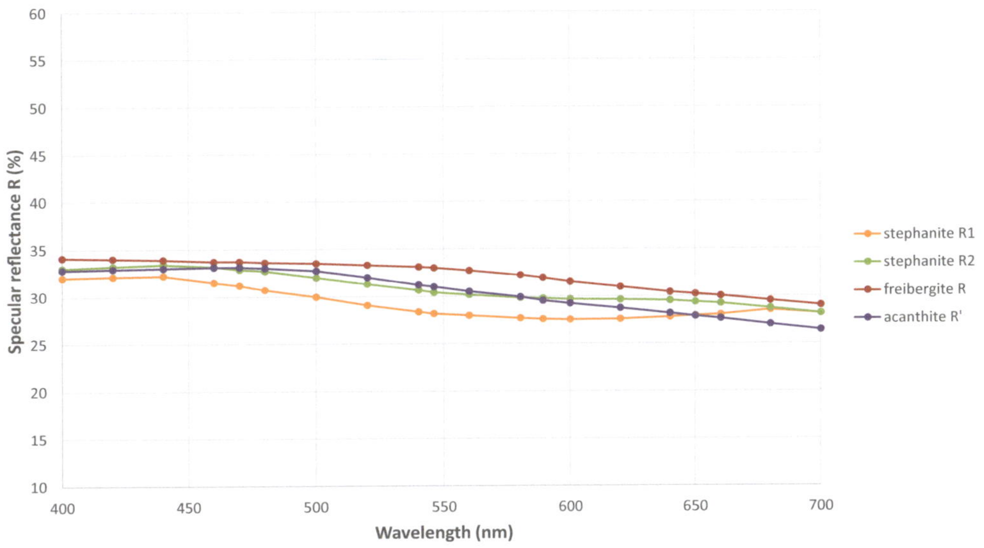

Comparison of Specular Reflectance values in the visible spectrum from stephanite R1 & stephanite R2 (QDF3), and pearceite R & polybasite R (Picot and Johan, 1982)

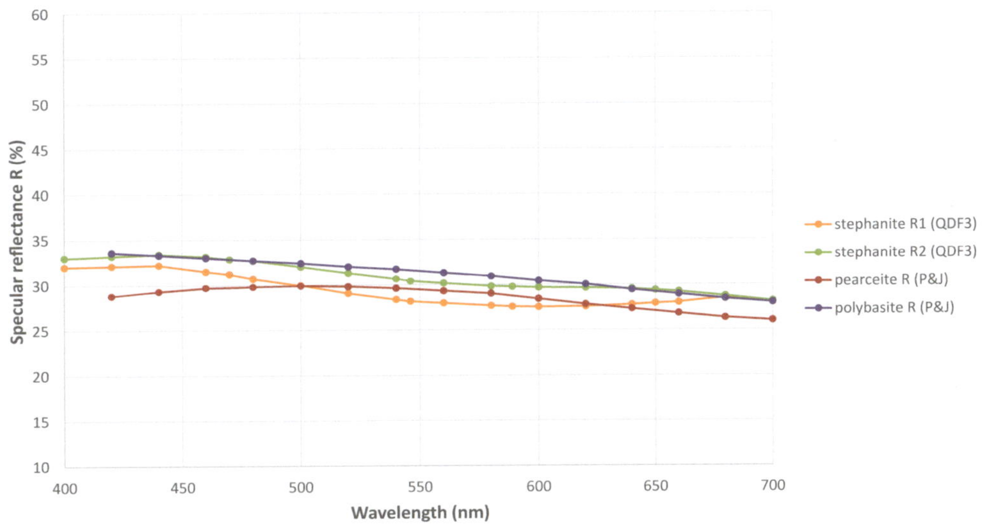

References

Picot P, Johan Z (1982) Atlas of ore minerals. B.R.G.M./Elsevier

Ramdohr P (1980) The ore minerals and their intergrowths, 2nd edn, vol 2. Pergamon Press, Oxford, p 1205

Uytenbogaardt W, Burke EAJ (1971) Tables for microscopic identification of ore minerals, 2nd edn. Elsevier, Amsterdam

Sternbergite (*stb*) 118

$AgFe_2S_3$: Orthorhombic

Sternbergite		R_{546} (%)	Y (%)	λ_d (nm)	P_e (%)	Colour sensation
Air	R_1	24.0	24.4	579	11.2	Pinkish-brown, bireflective
	R_2	36.6	36.7	576	6.9	
Oil	R_1	11.4	11.7	579	16.8	
	R_2	22.7	22.8	576	7.6	
Bir: Medium (4)				*Pleoc*: Apparent (4)		*Tarn*: Occasional light etching
Anis +P: Strong and coloured (5)				*IR*: No		*Twg*: Lamellar, complex
Mor: Tabular, prismatic, inclusions				*Clv*: Perfect		*PolQ*: Difficult (soft) but good
Hardness	*PolH*: Very low (1)			*Mohs*: 1–1½	*Talm*: -	*VHN* $_{(50\ g)}$: 31–44

Note: cf. R charts, visible range data (QDF3, and Picot and Johan 1982)
Mineral abbreviations: aag—allargentum, ac—acanthite, Ag—native silver, agp—argentopyrite, cc—chalcocite, ccp—chalcopyrite, fbg—freibergite, gg—gangue, gn—galena, lo—löllingite, mc—marcasite, nc—nickeline, my—miargyrite, pc—pearceite, plb—polybasite, po—pyrrhotite, prg—pyrargyrite, pru—proustite, py—pyrite, q—quartz, sbt—stibnite, sf—safflorite, sk—skutterudite, sp—sphalerite, stb—sternbergite, stp—stephanite

118.1 General

Mineral with a complicated history. This is still perceived in the ambiguities or contradictions of the descriptions available today, whose information has been updated for this text.[1]

Sternbergite, ***stb***, is a species of composition $AgFe_2S_3$, included in the old *Silberkiese* of the German miners, together with its *dimorph argentopyrite*, **agp**, with which it shares many properties.

Given the similarity of both dimorphs, reference to argentopyrite is made to complete the description of sternbergite that follows.

118.2 Microscopic Description

118.2.1 SP Observation (*Single Polariser*) (cf *agp Description*)

Colour like argentopyrite, pinkish brown, similar to pyrrhotite, but somewhat darker and notably more *bireflectant*, cf comparative R charts.

The bireflectance of sternbergite is still somewhat higher than that of argentopyrite, especially in immersion. Therefore, its apparent pleochroism is even more pronounced (the noticeable difference in brightness on rotation is perceived as a distinct colour change from deeper brown to lighter

[1] With data from https://www.mindat.org/min-3767.html (*visit.* 2020/05/30).

R. Castroviejo, *A Practical Guide to Ore Microscopy—Volume 1*,
https://doi.org/10.1007/978-3-031-12654-3_118

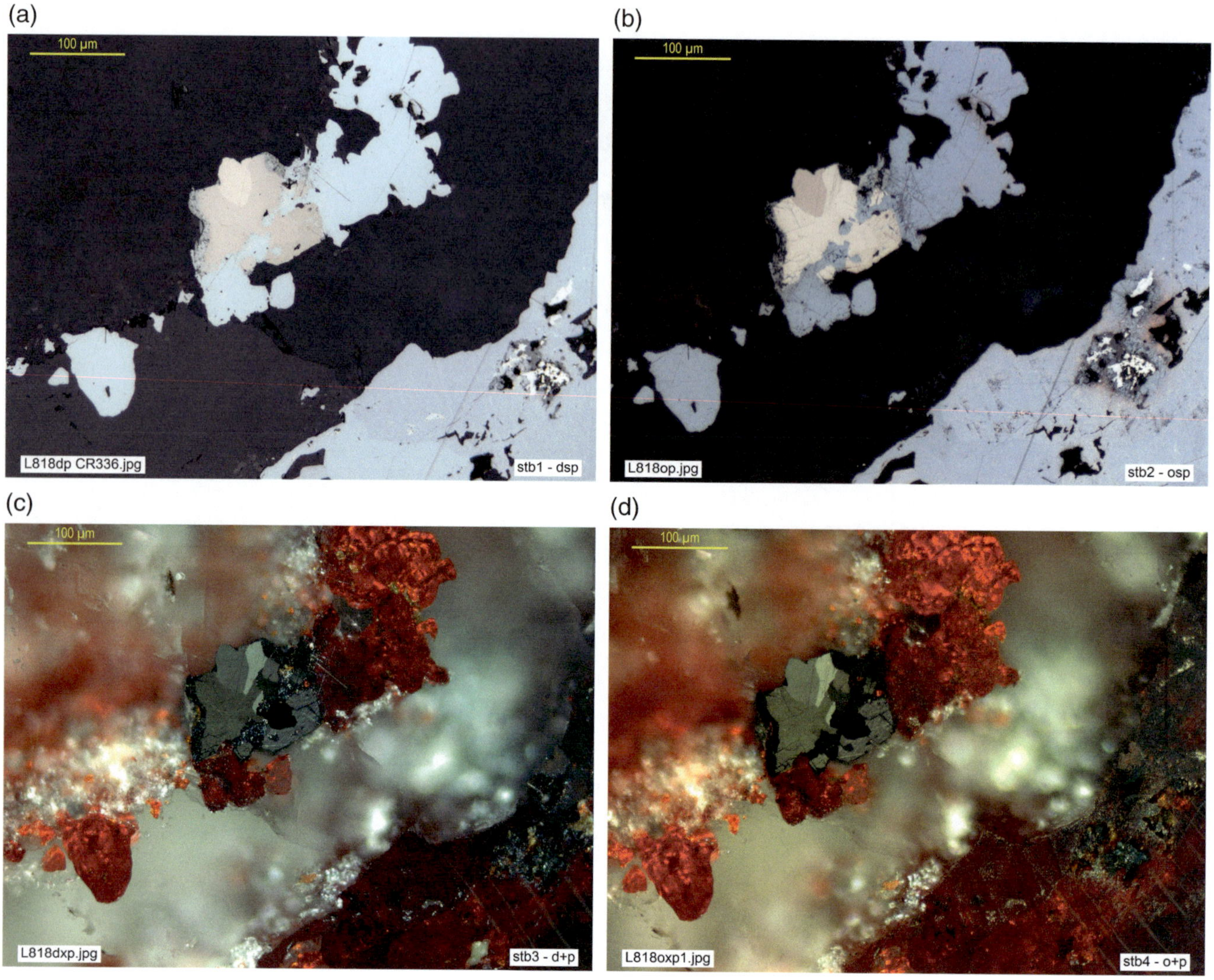

Fig. 118.1 **a**, **b** (**dsp** and **osp**, respectively)—Argentiferous ore of prg (grey), stb (pinkish brown, noticeable pleochroism, especially in immersion) and ac (thin grey ribbon between stb, prg and gg, slightly darker than prg); py inclusions in prg. The short time since the Fig. 118.1a image was taken has been sufficient for ac to appear completely black in Fig. 118.1b, because of light etching, while light attack on prg is very limited. CR 336, Freiberg, Saxony, Germany. **c**, **d** Same scene as Fig. 118.1a (**d+p** and **o+p**, respectively). Anisotropism of stb and IR of prg (comp. Figure 118.4)

yellowish-brown. Apparent brightness depends also on the adjacent ores, *comp*. Figs. 118.1a, b, 118.3 and 118.4).

Moderate *reflectance*, slightly lower than pyrrhotite (cf comparative R plots). According to Picot and Johan (1982), in white light:

- R stb (34.5%) ≈ agp (35%)
- R stb > fbg (32%), ac (32%), cc (31.5%), plb (31%), prg (31%), pc (30.5%)
- R stb < po (39%), << nc (50.5%), py (51%), sk (54%); <<< aag (68%) and Ag (85%)

These quantitative data should not be forgotten, because the subjective visual perception can be very different according to the accompanying minerals.

Good *polishing*, although more difficult than argentopyrite.

Very low *polishing hardness* but PolH stb > prg, agp.[2] Cleavage (001) perfect.

It occurs in thin tabular crystals, pseudohexagonal, in aggregates, in which cleavage is usually perceived. All of them, in general, in small inclusions in argentiferous ores (pyrargyrite, miargyrite, acanthite). Sometimes it forms oriented intergrowths with argentopyrite.

[2] After Ramdohr (1980), stb PolH data from other sources may be confusing or inconsistent with the very low Mohs hardnes of this ore.

Fig. 118.2 Same scene as Fig. 118.1c, with another orientation to compare with Fig. 118.1c and observe the variety of polarization colours of stb (**o+p**)

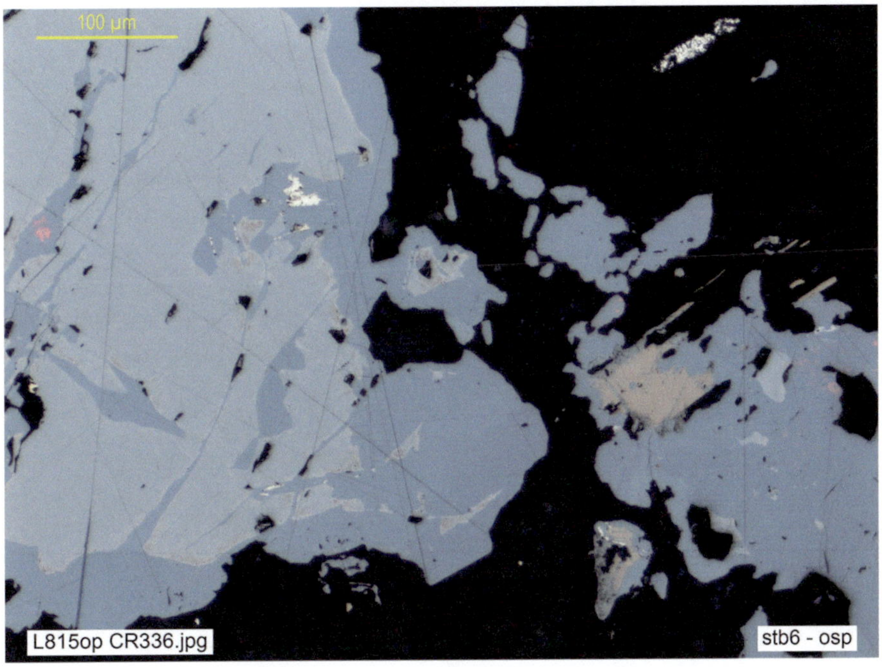

Fig. 118.3 Argentiferous ore vein, with my (very light grey) and prg (bluish grey) infill; py (yellowish white, bright) and stb (pinkish brown) inclusions in prg. Decreasing reflectance by the order R py >> R stb ≈ R my > R prg >> R gg. CR 336, Freiberg, Saxony, Germany (**osp**)

118.2.2 Observation +P (*Polariser + Analyser at 90°*)

Strongly anisotropic (Figs. 118.1c, d and 118.2), with more vivid and varied tones than argentopyrite, light greenish or yellowish-green, brownish or reddish-brown, and purplish to bluish (especially when the analyser is slightly uncrossed).

No *internal reflections*. Lamellar or complex *twinning*.

118.2.3 Paragenesis (cf *Description Argentopyrite*)

118.2.4 Identification

In general, what was said for argentopyrite (cf *description argentopyrite) applies*. The argentopyrite-sternbergite distinction is more questionable, given their great similarity, but

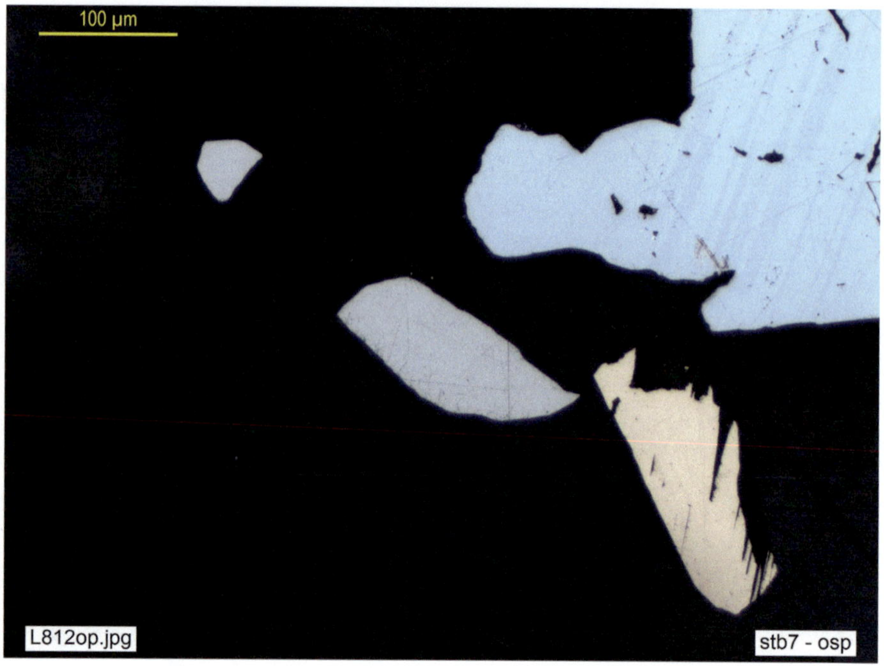

Fig. 118.4 Ore of prg (bluish grey, twinned, top right), stb (pinkish, showing cleavage) and stp (slightly brownish grey, two isolated grains) in carbonate gangue (black). CR 336, Freiberg, Saxony, Germany (**osp**)

the anisotropism of sternbergite is more vivid and varied, its cleavage more perfect and its polishing poorer.

118.3 Spectral Information

118.3.1 Sternbergite Reflectance R (in Air, Without Polariser) (%)

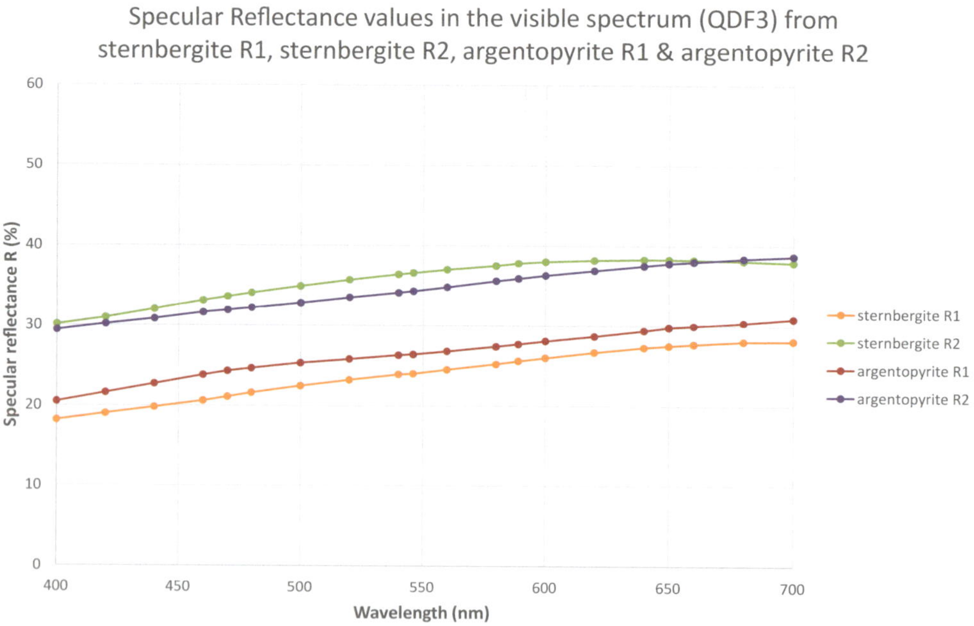

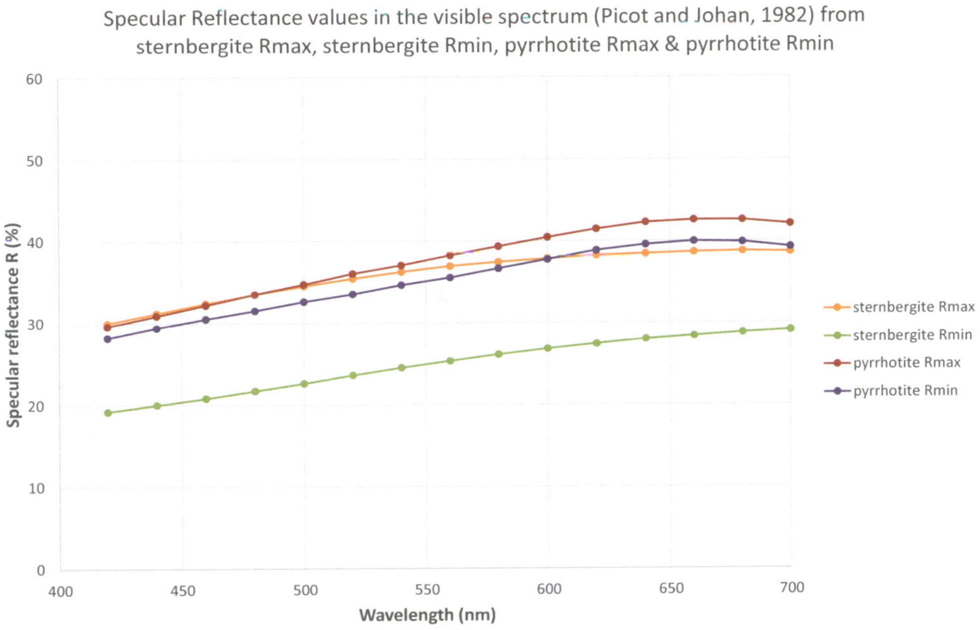

References

Picot P, Johan Z (1982) Atlas of ore minerals. B.R.G.M./Elsevier
Ramdohr P (1980) The ore minerals and their intergrowths, 2nd edn, vol 2. Pergamon Press, Oxford, p 1205

119 Stibarsen (*sba/allemontite*)

$Sb_{0.5}As_{0.5}$: Trigonal

Stibarsen		R_{546} (%)	Y (%)	λ_d (nm)	P_e (%)	Colour sensation
Air	R_1	61.6	*nd*	*nd*	*nd*	White, glossy
	R_2	66.3	*nd*	*nd*	*nd*	
Oil	R_1	*nd*	*nd*	*nd*	*nd*	
	R_2	*nd*	*nd*	*nd*	*nd*	
Bir: Imperceptible (2)				*Pleoc*: Very weak (3)		*Tarn*: No next to As, yes next to Sb
Anis +P: Distinct brown-orange (4)				*IR*: No		*Twg*: Polysynthetic
Mor: Fine intergrowths (As, Sb…)				*Clv*: -		*PolQ*: Good
Hardness	*PolH*: Low (>Sb, < As) (2)			*Mohs*: 3–4	*Talm*: *nd*	*VHN* $_{(100\ g)}$: *nd*

Note: *Id. Allemontite II* of Ramdohr (1980). R data according to Picot and Johan (1982; *nd* = *no* data in QDF3)
Mineral abbreviations: As—native arsenic, lo—löllingite, Sb—native antimony, sba—stibarsen

119.1 General

Originally the name *allemontite* was given to what was considered a "mineral" consisting of mixed crystals of As and Sb, apparently miscible in different proportions. However, clear unmixing textures were revealed by microscopic observation. IMA-COM only accepts as a mineral species, under the name of *stibarsen,* the alloy with 50 *at%* of each component.

Ramdohr describes three types of *allemontite: I, II and III*, with compositions rich in Sb (I), in As (III) or As–Sb (II, which he identifies as the stibarsen mineral, *sba*). The latter, stibarsen, may appear unmixed with native arsenic or with native antimony, depending on the predominance of one or the other in the original composition of the As–Sb gel.

119.2 Microscopic Description

119.2.1 SP Observation (*Single Polariser*)

White *colour*, very similar to that of native antimony (Figs. 119.1a, b and 119.2a) but somewhat less bright.

Reflectance, R, between those of As and Sb (cf. R charts, in the visible range, according to Picot and Johan, 1982; no stibarsen information has been found in QDF3). Stibarsen may show a peculiar "tarnish" or "*air etching*", as Ramdohr (1980) notes. Stibarsen darkens quickly when it is intergrown with antimony (i.e., a dark *patina* quickly appears on stibarsen) whereas, when it is intergrown with arsenic, it is the latter that darkens, while stibarsen remains unaltered (comp. Figs. 119.1a and 119.3). He attributes this effect to "air *etching*" (oxidation). This means that, on first impression, stibarsen can be mistaken for arsenic or antimony, depending on the predominance of one or the other.

Very good *polishing*. *Polishing hardness* slightly higher than antimony but lower than arsenic, which makes the unmixed textures sharper.

It typically forms fine vermiform or symplectic *intergrowths* with As or Sb, depending on the dominant element. These are explained by *unmixing* from a primary As–Sb gel (Figs. 119.1a, b, and 119.2a), in which it is not uncommon to find relicts (colomorphic or botryoidal ghost textures, observed in samples from Engineer Mine).

R. Castroviejo, *A Practical Guide to Ore Microscopy—Volume 1*,
https://doi.org/10.1007/978-3-031-12654-3_119

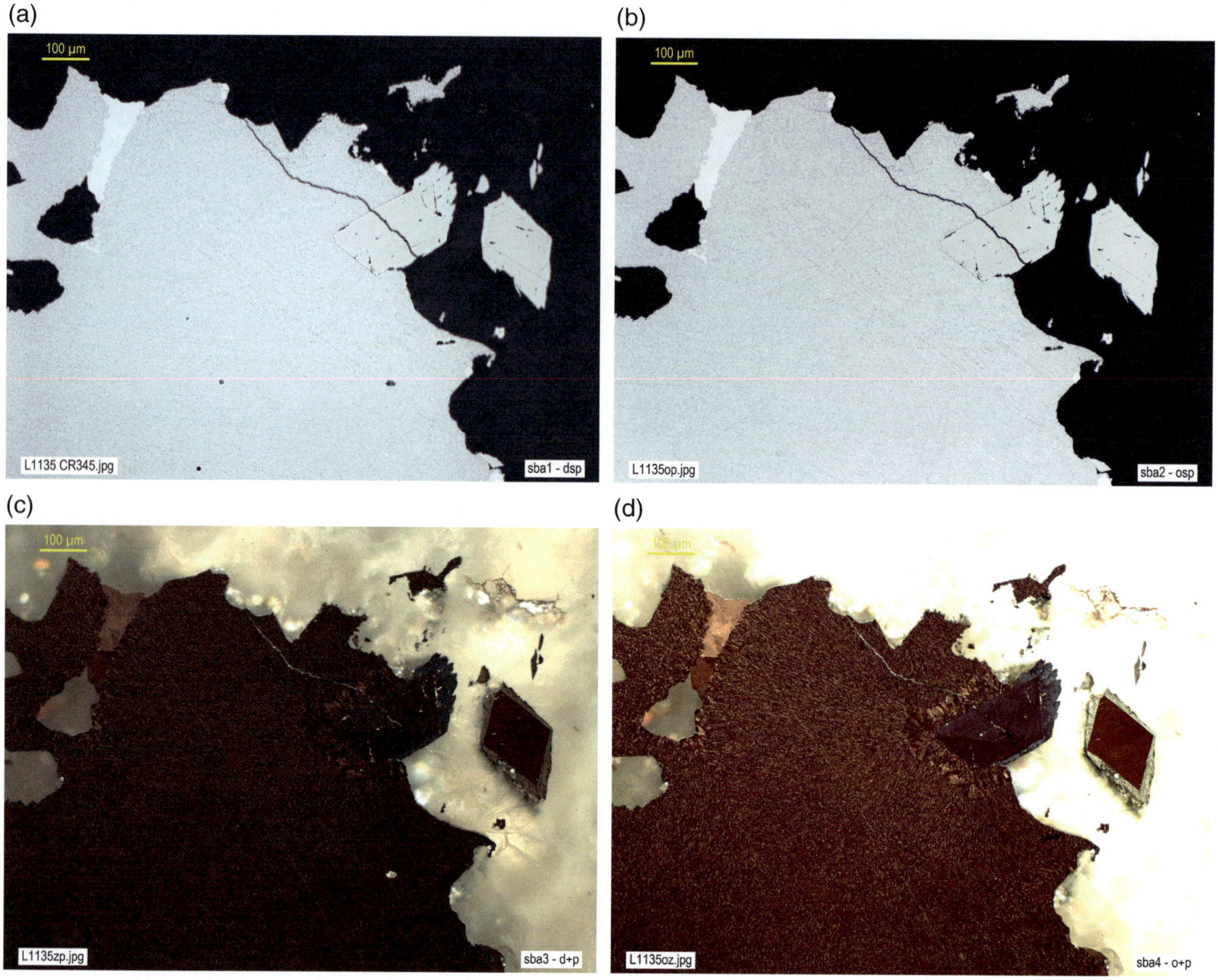

Fig. 119.1 **a, b** (Same scene: **dsp** and **osp**). Mass of stibarsen (white) and finely intergrown native As (grey), with microcracks cemented by sba (top left); to the right, adjacent löllingite crystals (light grey, euhedral rhombic forms). Similar reflectance of As and lo, relatively high but clearly lower than that of sba, so that both As and lo are perceived as grey in contact with the brighter white-appearing sba. CR 345, Engineer Mine, Atlin District, BC, Canada. **c, d** (Same scene as Fig. 119.1a: **d+p**-and **o+p**, respectively). Polarisation evidences symplectitic intergrowth of sba and As (probable unmixing texture from a primary Sb–As gel. See details, Fig. 119.2a, b)

119.2.2 Observation +P (*Polariser* + *Analyser at 90°*)

Clearly anisotropic, with tints similar to those of antimony, although somewhat duller (Figs. 119.1c, d and 119.2b).

Frequent polysynthetic twins, like antimony. No internal reflections.

119.2.3 Paragenesis

It occurs in hydrothermal *parageneses*, typically in argentiferous ores, associated with antimony or arsenic, löllingite, Ag ores, Co–Ni ores, etc.

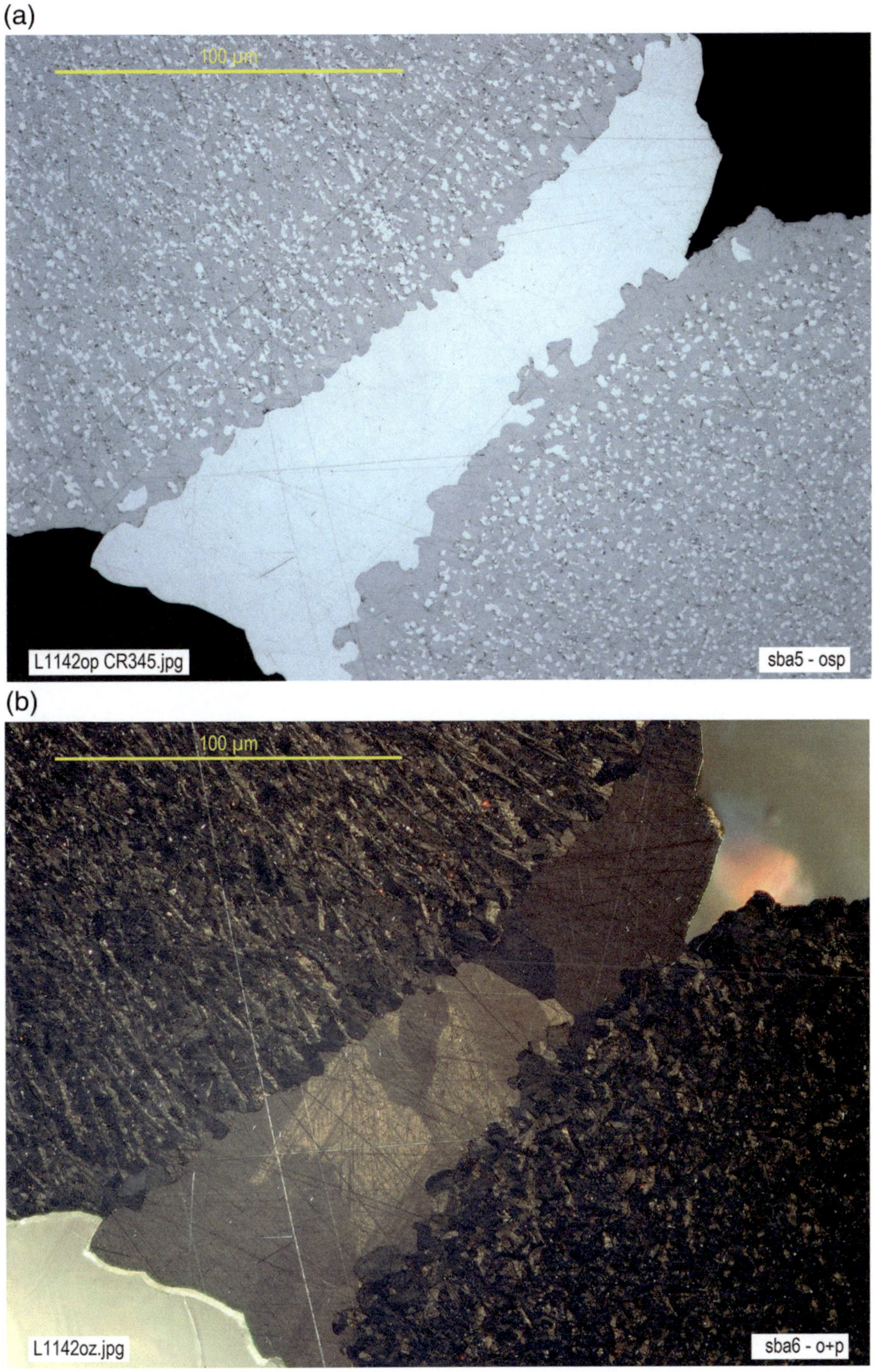

Fig. 119.2 **a, b** (Same scene: detail of Fig. 119.1a-top left corner, **osp** and **o+p**, respectively). Microcrack cemented by sba (white) in intergrown mass of sba (white) +As (grey) with symplectitic texture, likely result of Sb–As gel unmixing

119.2.4 Identification

It is very similar to native arsenic and antimony, with which it forms the typical symplectic intergrowths. The paragenesis, its intermediate reflectance between those of As and Sb, the appearance or not of dark *patina*, depending on the association with one or the other, and the polishing hardness, slightly higher than that of antimony but lower than that of arsenic, allow safe identification.

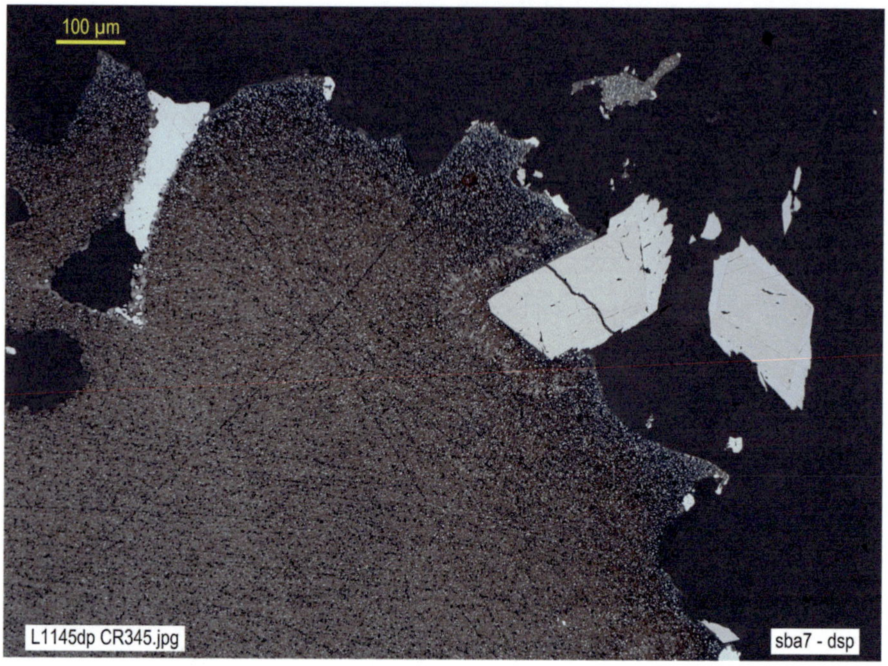

Fig. 119.3 Same scene as Fig. 119.1a, showing As darkening two days after polishing (**dsp**). Sba and lo show no tarnish, while As darkens noticeably. Unmixed sba stands out as white dots in the dark arsenical mass

119.3 Spectral Information

119.3.1 Stibarsen Reflectance Spectrum R (in Air, Visible Range) (%)

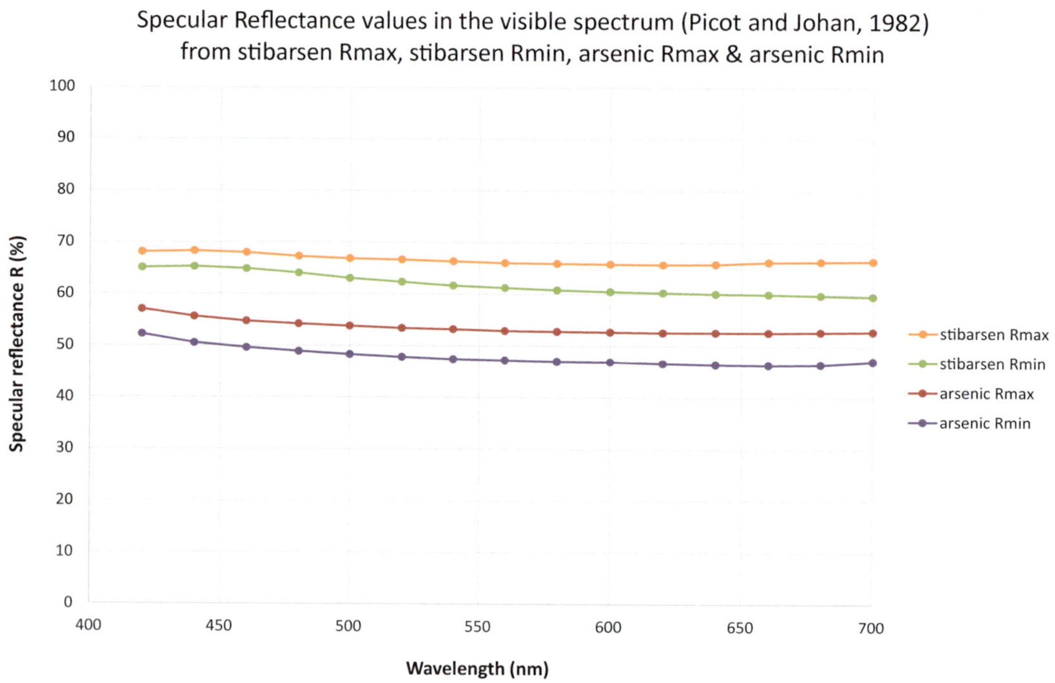

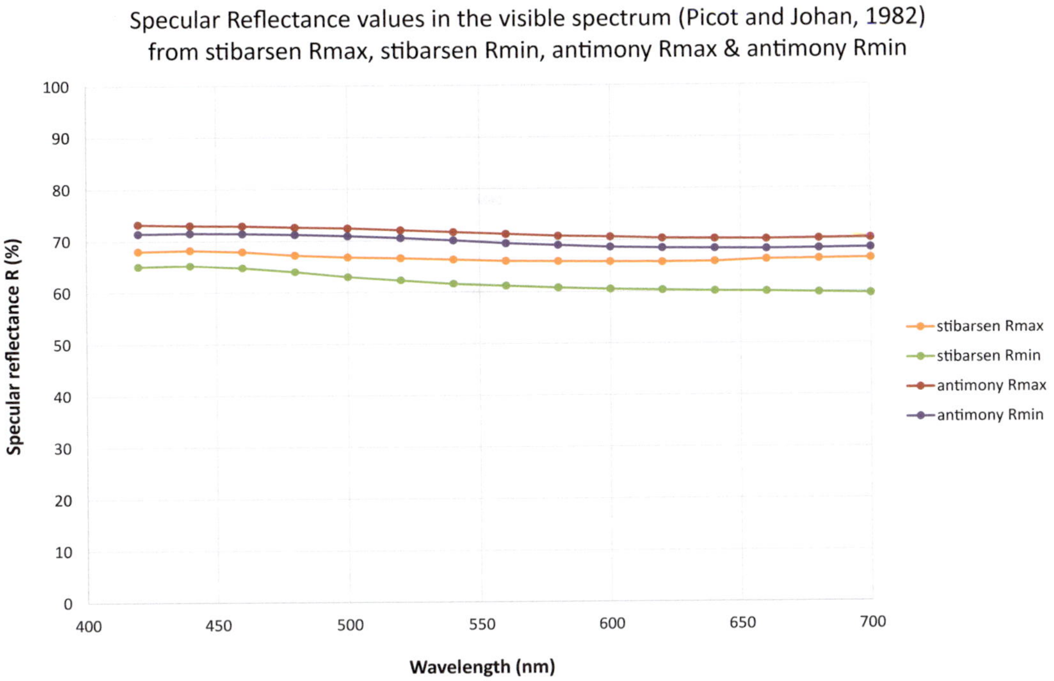

References

Picot P, Johan Z (1982) Atlas of ore minerals. B.R.G.M./Elsevier

Ramdohr P (1980) The ore minerals and their intergrowths, 2nd edn, vol 2. Pergamon Press, Oxford 1205 p

Stibnite (*sbt*)

120

Sb_2S_3: Orthorhombic

Stibnite		R_{546} (%)	Y (%)	λ_d (nm)	P_e (%)	Colour sensation
Air	R_a	42.2	41.8	481	4.0	White to brownish grey (pleochroic)
	R_b	31.1	30.6	494	1.4	
	R_c	48.1	47.3	482	6.9	
Oil	R_a	26.8	26.6	479	7.1	
	R_b	16.1	15.7	490	3.1	
	R_c	33.4	32.7	481	11.3	
Bir: Strong (5)				*Pleoc*: Strong (5)		*Tarn*: Occasional, slow darkening
Anis +P: Strong (5)				*IR*: No		*Twg*: Polysynthetic, very frequent
Mor: Varied, frequent deformation				*Clv*: (010)		*PolQ*: Easy, good polishing
Hardness	*PolH*: Low, < gn (2)			*Mohs*: 2	*Talm*: -	*VHN* $_{(100\ g)}$: 71–86

Obs: cf. Table and R charts, multispectral data (VNIR, AMCO) and visible range (QDF3)
Mineral abbreviations: As—native arsenic, Au—native gold, Bi—native bismuth, bm—bismuthinite, brt—berthierite, ccp—chalcopyrite, cv—covellite, gg—gangue, gn—galena, orp—orpiment, po-pyrrhotite, prg-pyrargyrite, pru-proustite, py-pyrite, rl-realgar, Sb—native antimony, sem—semseyite, stp—stephanite, td—tetrahedrite

120.1 General

Stibnite is the principal sulphide of Sb. It shares the composition Sb_2S_3 with its dimorph *metastibnite*, which is rare and amorphous but accepted (*grandfathered*) by IMA (*locus typicus*, Steamboat Hot Springs, Washoe Co., Nevada, USA).

120.2 Microscopic Description

120.2.1 SP Observation (*Single Polariser*)

Colour white to brownish-olive grey, distinctly *pleochroic*, soft but *well polished*, though with unavoidable scratches (Fig. 120.1a, b).

Moderate *reflectance* (R sbt < gn and > td), with a wide bireflectance range, between galena and tetrahedrite. However, in its direction of maximum R value (R_c), Sb reflectance is > R of galena, which gives an idea of the magnitude of its bireflectance (cf. multispectral comparison graphs). The bireflectance is sufficient to make the twins (or the contours of adjacent, differently oriented grains) clearly visible without analyzer (Fig. 120.1a, b).

Polishing hardness low, nearly very low:

- PolH sbt > orp, rl, stp, red silvers (pru-prg)
- but PolH sbt < brt, Bi, cv, Sb, As gn, ccp.

It forms prismatic crystals, occasionally acicular, also massive, fibrous or granoblastic aggregates, often deformed. Exfoliation (010) occasionally visible under the microscope.

R. Castroviejo, *A Practical Guide to Ore Microscopy—Volume 1*,
https://doi.org/10.1007/978-3-031-12654-3_120

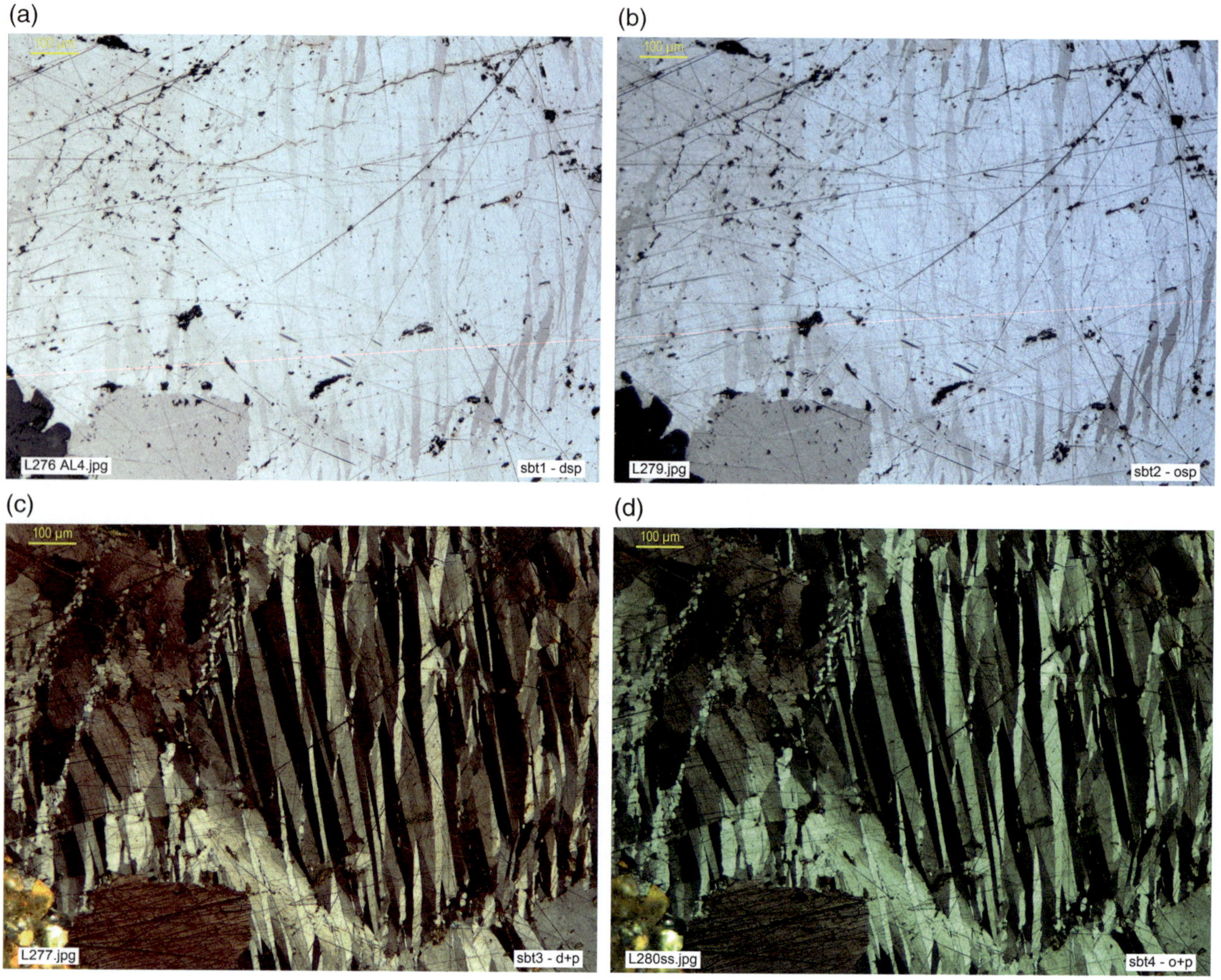

Fig. 120.1 **a**, **b** Vein infill of stibnite (grey twinned, pleochroic) and quartz (dark grey), with relict berthierite microclast (olive-brownish-grey, bottom left) and tiny pyrite inclusions in sbt. The strong pleochroism of sbt makes its twins visible even without analyzer. Sample AL4, Almuradiel, Ciudad Real (**dsp** and **osp**, respectively). **c**, **d** Same scene as Fig. 120.1a (**d+p** and **o+p**, respectively). Strong but poorly coloured anisotropism of sbt, showing lamellar deformation twinning

120.2.2 Observation +P (*Polariser + Analyser at 90°*)

Strongly *anisotropic* but with inconspicuous polarization colours, grey, with slightly bluish to brownish-pinkish tints (Fig. 120.1c, d). Extinction straight.

Almost always *twinned*, polysynthetic or wedge-shaped twins, which easily show deformation (Figs. 120.1a–d).

120.2.3 Paragenesis

Paragenesis typically hydrothermal and generally related to low-temperature, posthumous episodes of vein filling.

Accompanying ores: berthierite, pyrite, marcasite, cinnabar, native antimony, native gold, proustite-pyrargyrite and other argentiferous ores, arsenopyrite, gudmundite, etc.

In addition, in zones of supergene oxidation, Sb ochre (pseudomorphic of stibnite, described with oxide gangue minerals), realgar, orpiment.

120.2.4 Identification

In general, its recognition is easy, taking into account the paragenesis, its low hardness (and polishing scratches), its medium reflectance, its strong and poorly coloured pleochroism and anisotropism, the abundance of twins and their sensitivity to deformation.

Frequently associated with *berthierite*, with which it might be confused. Misidentification may be avoided, since brt can be distinguished by:

- its greater hardness. Consequently, brt appears in relief on stibnite and brt shows a greater tendency to brittle response to deformation as well, Fig. 120.1a.
- its pleochroism with a more intensely brownish tone (almost like pyrrhotite).
- its more varied anisotropism (with greenish tones, cf. F brt).

Further, stibnite is distinguished from.

- *semseyite* by the much stronger pleochroism and anisotropism of stibnite, as well as by the presence in stibnite of deformed twins.
- *bismuthinite* as this is brighter, less pleochroic and less anisotropic than stibnite.

Finally, its amorphous dimorph *metastibnite* is bluish grey, darker than sbt, shows frequent red IR, and occurs rarely (e.g., Oruro, Bolivia). As other "sulphide glasses", it seems to be an aging, slowly crystallizing gel and may have been mistaken with *kermesite*, according to Ramdohr (1980).

120.3 Spectral Information

120.3.1 Stibine Multispectral Reflectance R (in Air, Without Polariser) (%)

λ (nm)	370	400	425	450	475	500	525	550	575	600	625	650	675	700	750	800	850	900	950	1000
μ	38.6	38.6	38.2	38.1	37.9	37.7	37.3	36.7	36.3	35.8	35.3	35.3	35.2	34.8	34.0	33.4	33.7	33.5	33.4	32.9
P_5	33.9	34.3	34.2	34.2	34.1	34.0	33.7	33.1	32.6	32.0	31.6	31.6	31.7	31.6	30.8	30.4	31.0	30.7	30.7	30.3
P_{20}	35.7	35.8	35.5	35.5	35.3	35.1	34.9	34.5	34.2	33.7	33.3	33.3	33.4	33.2	32.3	31.7	32.3	32.0	31.9	31.4
P_{80}	41.2	41.0	40.5	40.3	40.1	39.7	39.3	38.7	38.2	37.8	37.3	37.2	37.0	36.5	35.8	35.0	35.0	34.7	34.8	34.4
P_{95}	43.0	42.8	42.5	42.6	42.6	42.1	41.6	40.6	39.8	39.1	38.6	38.4	38.2	37.6	36.5	35.8	36.2	36.2	36.2	35.8

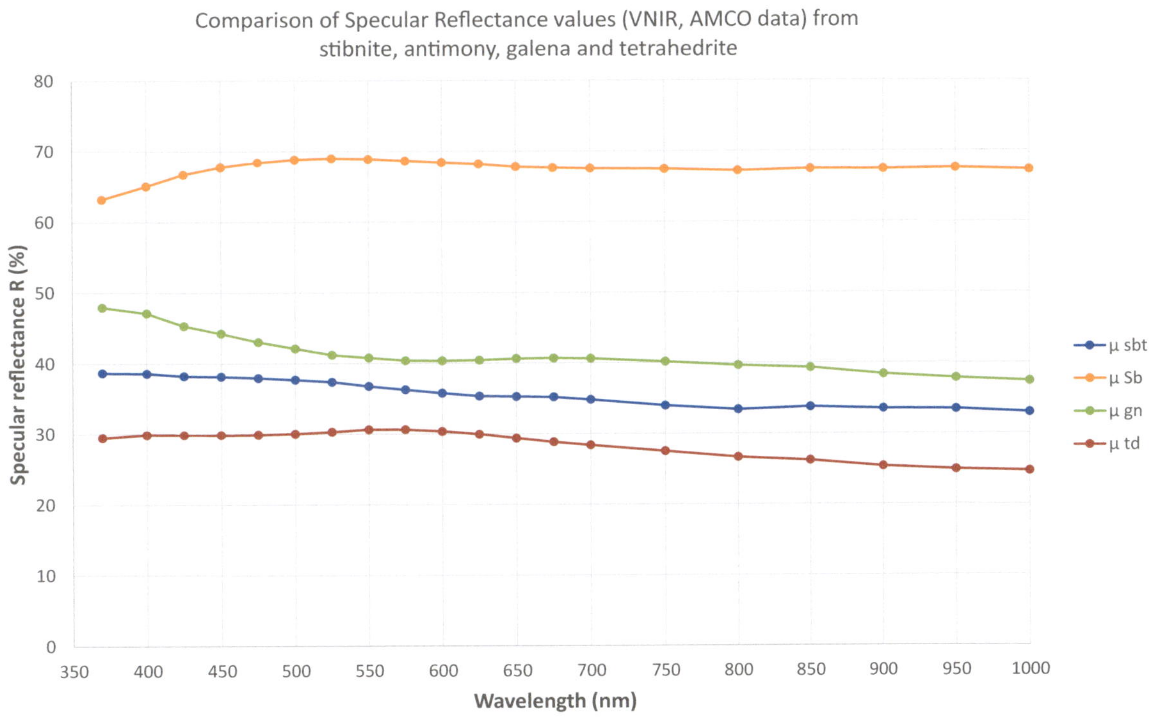

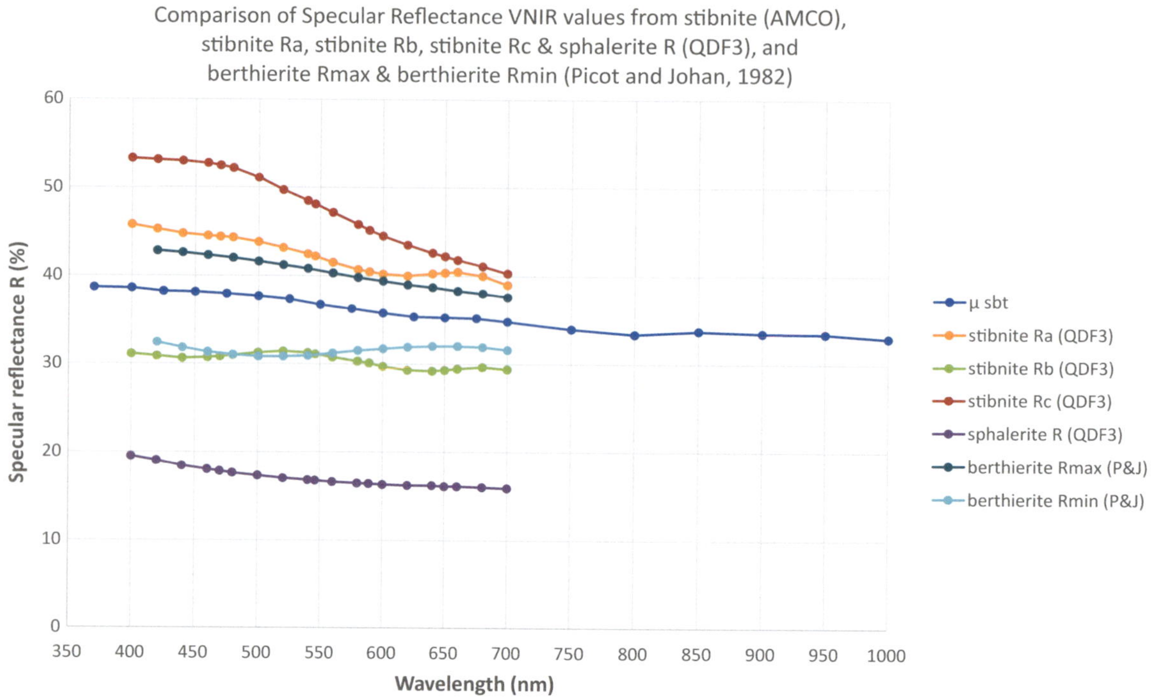

Reference

Ramdohr P (1980) The ore minerals and their intergrowths, vol 2, 2nd edn. Pergamon Press, Oxford, 1205p

Stromeyerite (*stm*)

121

CuAgS: Orthorhombic

Stromeyerite		R_{546} (%)	Y (%)	λ_d (nm)	P_e (%)	Colour sensation
Air	R_1	26.6	26.7	470	4.3	Light grey (peculiar pleochroism, with pinkish-brown to bluish-green tones)
	R_2	30.9	31.0	473	12.5	
Oil	R_1	12.4	12.5	468	7.1	
	R_2	15.7	15.9	471	21.2	
Bir: Medium (4)				*Pleoc*: Visible (4)		*Tarn*: Yes
Anis + P: Strong and coloured (5)				*IR*: No		*Twg*: Only from transformation (93°)
Mor: Generally allotriom. (late cement)				*Clv*: No		*PolQ*: Good (with slow disc rotation)
Hardness	*PolH*: Low, ≈ cv; > stp, < cc (2)			*Mohs*: 2½–3	*Talm*: B	*VHN* $_{(25g)}$: 70–72

Obs: cf. R charts, visible spectrum data (QDF3 and Picot and Johan 1982)
Mineral abbreviations: ac—acanthite, Ag—native silver, bn—bornite, cc—chalcocite, ccp—chalcopyrite, cv—covellite, dg—digenite, dy—dyscrasite, fbg—freibergite, gg—gangue, gn—galena, ht—hematite, my—miargyrite, pc—pearceite, plb—polybasite, prg—pyrargyrite, pru—proustite, sbt—stibnite, snr—schachnerite (Ag amalgam), stm—stromeyerite, stp—stephanite, td—tetrahedrite

121.1 General

Composition: CuAgS or $Cu_2S{\cdot}Ag_2S$, with possible Fe impurities, although chemical analysis may give different compositions due to inclusions or mixtures of other species. Stromeyerite is a low temperature mineral, usually supergene.

According to experimental determinations of Ramdohr (1982), the orthorhombic stromeyerite undergoes inversion by heating to a cubic form. When it cools down again, it undergoes inversion back to the orthorhombic form, but now some particular, lanceolate structures appear (*oleander leaf*). This inversion, which occurs at 93 °C, is an interesting *geological thermometer*. Nevertheless, as evidence of high-temperature genesis, it is only reliable if it can be ensured that the sample has not been heated in the process of preparing the polished section, otherwise it would be an artifact generated in the laboratory.

121.2 Microscopic Description

121.2.1 SP Observation (*Single Polariser*)

Light grey *colour*, with a peculiar and delicate pleochroism (Figs. 121.1a, b and 121.2a), from darker pinkish-brown or greyish-brown tones (R_1) to much lighter, violet and slightly bluish or bluish-green tones (R_2), although to display all their chromatic variety, grains must be observed with the appropriate orientation (Fig. 121.2a).

Moderate *reflectance*, R ≈ ht; ≤ cc, > bn (cf. Fig. 121.3 and comparative R graphs, which show the peculiar dispersion of R_2 stromeyerite, with a pronounced maximum at 440 nm). However, in the presence of highly reflective phases (such as native silver, dyscrasite or schachnerite), stromeyerite is perceived as a dark mineral, of apparently deceptive low R (cf Fig. 121.4).

R. Castroviejo, *A Practical Guide to Ore Microscopy—Volume 1*,
https://doi.org/10.1007/978-3-031-12654-3_121

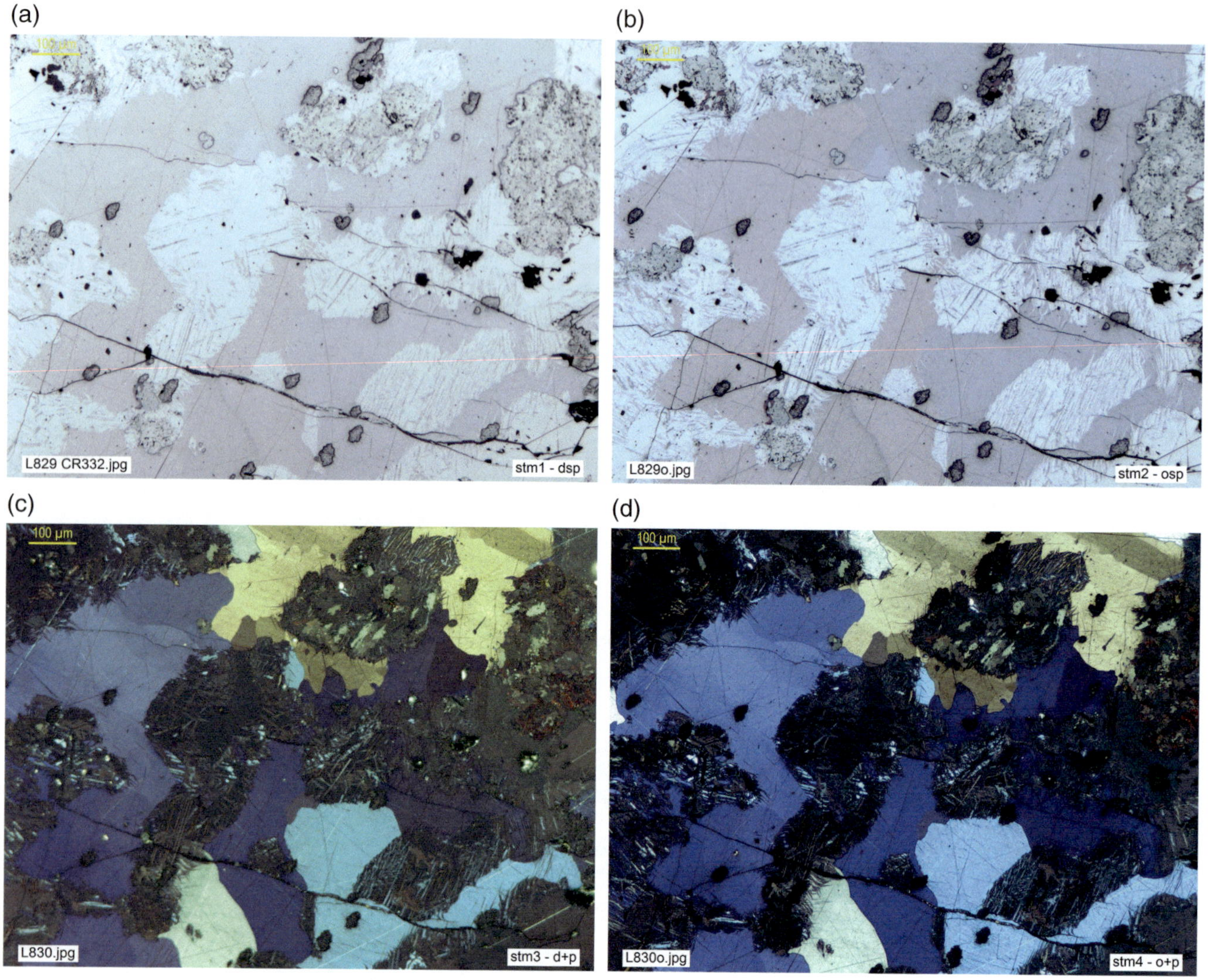

Fig. 121.1 a, b (Same scene: **dsp** and **osp**, respectively)—Intergrowth of stm (pinkish) and cc (greyish white), with relicts of ht (grey) and probable mt (dark grey, porous), both in relief on the sulphide phases (very soft) and with reaction rims; stm also forms multiple lamellar inclusions in cc, oriented according to cc lattice planes. CR 332, Vrancice, Pribram (Czech Rep). **c**, **d** (same scene as Fig. 121.1a: **d+p** and **o+p**, respectively)—The strong anisotropism of stm with vivid colours, in particular the characteristic bluish-green and purplish-violet blue bright hues, allows immediate recognition; ht shows colourless anisotropism and red IR

Good polish, if care is taken to avoid scratches. Occasional tarnish, dark or multicoloured *patina*. Low *polishing hardness*, between those of covellite and miargyrite (> sbt, stp, pru, plb; < my, dy, cc, dg, gn).

It usually occurs as a late filling in microcracks, or infiltrates along the cleavage planes of other ores (particularly argentiferous tetrahedrites, as a secondary mineral in the cementation zone of silver deposits). Also in myrmekitic intergrowths.

121.2.2 Observation +P (*Polariser + Analyser at 90°*)

Strongly anisotropic, with bright polarisation tints, among which bluish-green and violet are characteristic but also yellowish-brown or purplish (Figs. 121.1c, d and 121.2b).

No twins have been observed, except for the aforementioned lamellar lanceolate (*oleander leaf*) inversion twins, a possible geological thermometer (cf. § Sect. 121.1: *General*).

(a)

(b)

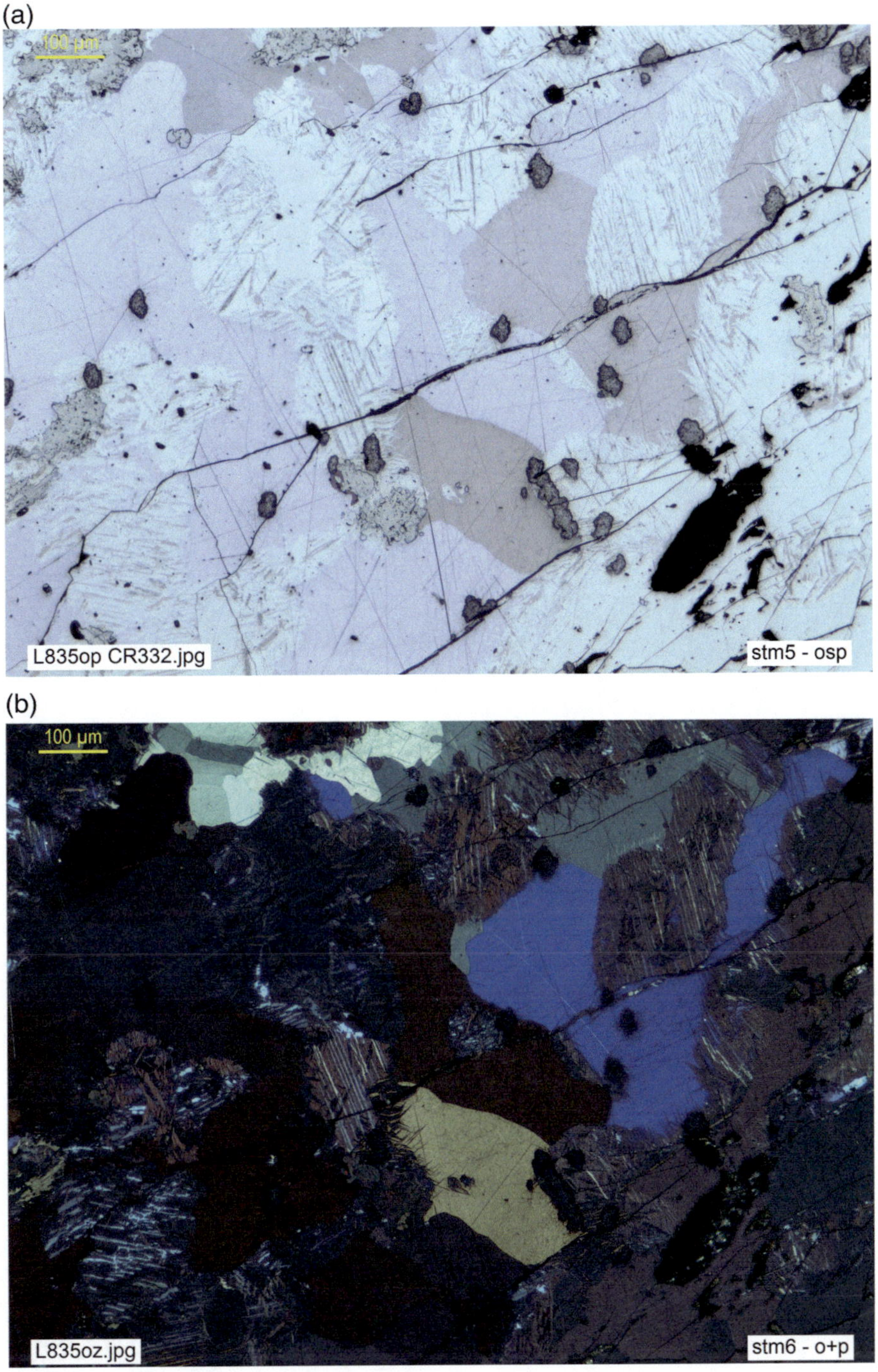

Fig. 121.2 **a**, **b** (same scene: **osp** and **o+p**, respectively)—Intergrowth of stm and cc, with relict ht and mt (as Fig. 121.1a). Note the striking pleochroism of stm from pinkish-brown or greyish-brown (R_1) to very light and slightly bluish violet-grey (R_2), which distinguishes the mineral perfectly from its companions. Note also the strong and coloured anisotropism of stm in the center of the image. CR 332, Vrancice, Pribram (Czech Rep.)

121.2.3 Paragenesis

As an accessory mineral it is relatively frequent, although generally scarce. It is a characteristic mineral of the supergene enrichment zone of argentiferous deposits, so it can occur with various relicts, even of oxides (cf. Figs. 121.1a, 121.2a, 121.3 and 121.4) but especially of fahlores (tetrahedrite *s.l.*). However, its hypogene formation has also been demonstrated in silver-rich deposits.

The most frequent companions are native silver, tetrahedrite (*s.l.*), acanthite, bornite, chalcocite, chalcopyrite.

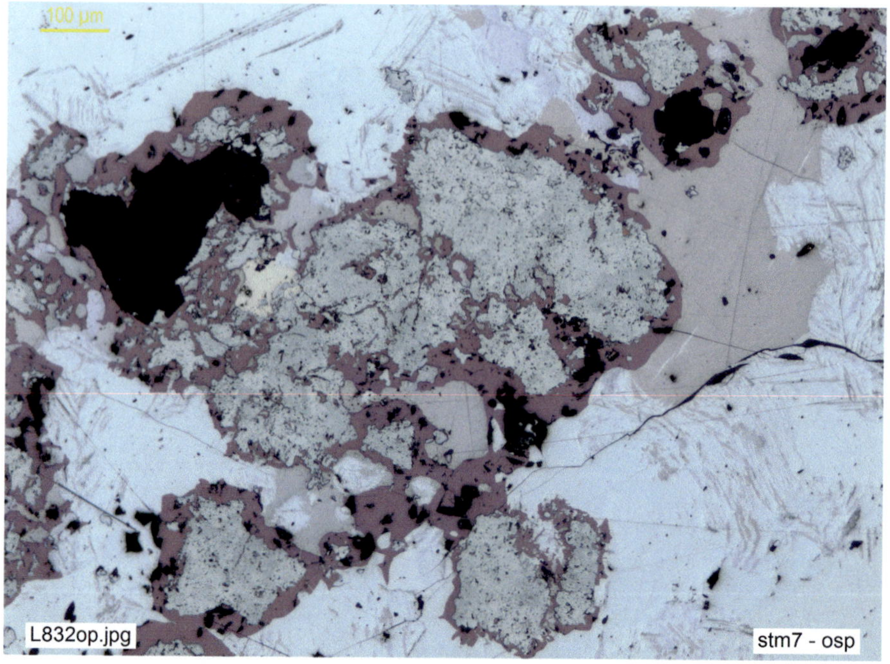

Fig. 121.3 Reaction rim of bn (dark pinkish brown) on ht, as an interface between the oxidized Fe (ht) and the Cu-Ag sulfide phases (stm, cc). Note the low reflectance of bn compared to stm, and the relief of ht on the soft sulphides. CR 332, Vrancice, Pribram, Czech Rep. (**osp**)

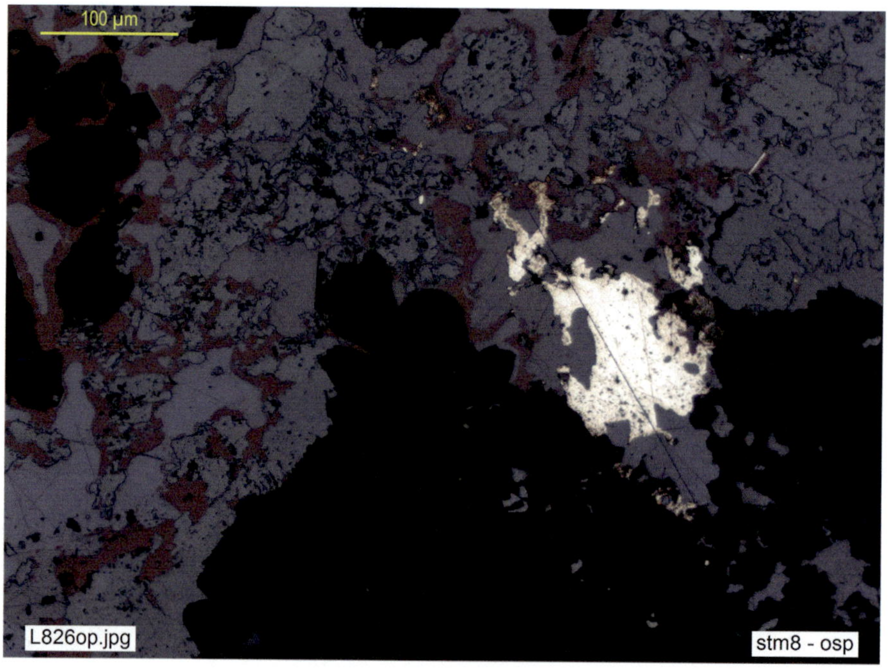

Fig. 121.4 Hydrothermal infill of q (black), corroded and cemented by bn, stm and amalgam (*schachnerite*, ***snr***, AgHg, white, very bright), with relict ht. Unlike Fig. 121.3, contrast with the highly reflective snr makes stm and the other ores appear dark and dull to the eye. CR 332, Vrancice, Pribram, Czech Rep. (**osp**)

121.2.4 Identification

Its optical properties (such as its pleochroism and particularly its strong anisotropism with vivid bluish, purplish-violet and yellowish-brown polarisation tints) are practically unmistakable. However, its paragenesis can make it go unnoticed when it is accompanied by native silver, which, because of its high luster, makes stromeyerite look much darker and dull.

121.3 Spectral Information

121.3.1 Stromeyerite Reflectance R (in Air, Visible Spectrum) (%)

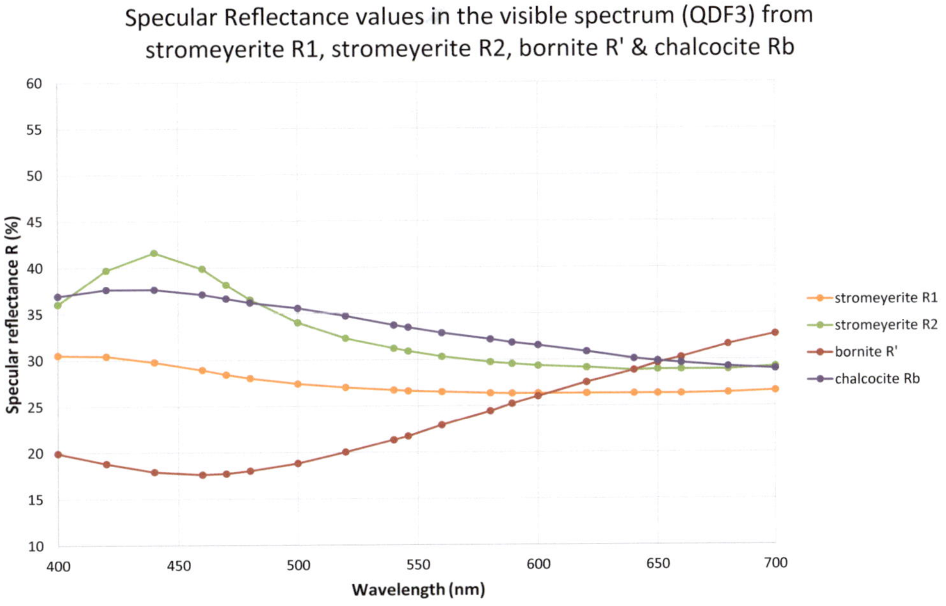

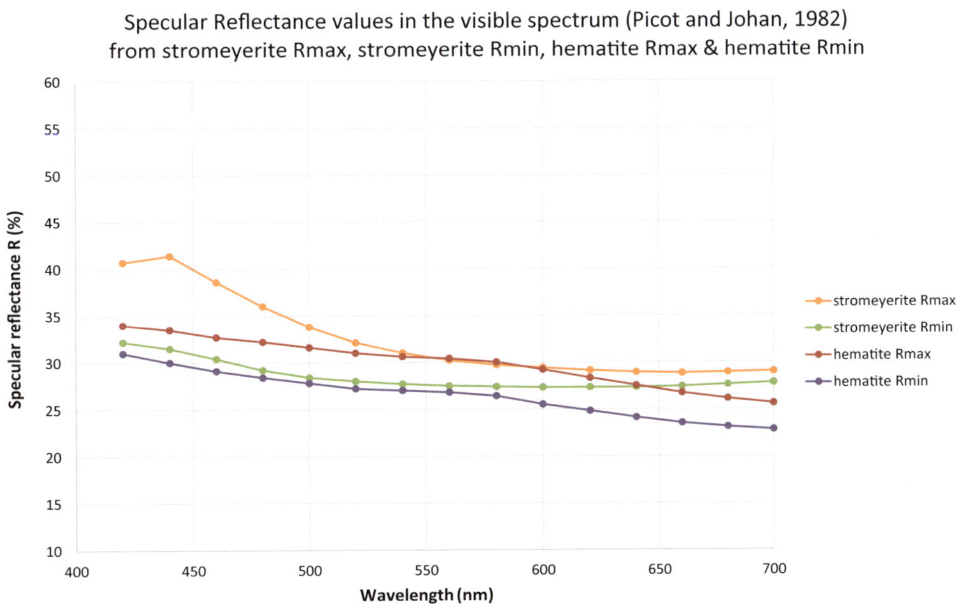

Reference

Picot P, Johan Z (1982) Atlas of ore minerals. B.R.G.M./Elsevier

122 Tapiolite (*tp*)

$(Fe,Mn)(Ta,Nb)_2O_6$: Tetragonal

Tapiolite		R_{546} (%)		Y (%)	λ_d (nm)	P_e (%)	Colour sensation
Air	R_1	16.0		nd	nd	nd	Grey (light brownish-grey to darker bluish-grey pleochroism)
	R_2	18.8		nd	nd	nd	
Oil	R_1	nd		nd	nd	nd	
	R_2	nd		nd	nd	nd	
Bir: Medium (4)					*Pleoc*: Notable (4)		*Tarn*: Not observed
Anis +P: Distinct and coloured (4)					*IR*: Reddish		*Twg*: Frequent, single or polysynthetic
Mor: Euhedral, short prismatic habit					*Clv*: Not observed		*PolQ*: Slow, but good
Hardness		*PolH*: Very high, ≈ cb (5)			*Mohs*: 6	*Talm*: nd	*VHN* $_{(nd\ g)}$: 796–1132

Obs: cf. Table and R charts, multispectral data (VNIR, AMCO) and visible range (Picot and Johan 1982; no QDF3 data)
VHN according to Uytenbogaardt and Burke (1971)
Mineral abbreviations: cb—columbite, cbt—columbotantalite, cst—cassiterite, gg—gangue, il—ilmenite, q—quartz, rt—rutile, sch—scheelite, sp—sphalerite, tan—tantalite, tp—tapiolite, wf—wolframite

122.1 General

Tapiolite, ***tp***, is traditionally $(Fe,Mn)(Ta,Nb)_2O_6$, that is, a mineral of variable composition between a ferriferous term, *ferrotapiolite*, and a manganiferous term, *manganotapiolite*. In nature, the first is more common and the second, very rare; likewise, Ta dominates over Nb.

Tapiolite contains up to 84% Ta_2O_5, while Nb-rich varieties (*mossite*) contain up to 35% Nb_2O_5; there is no pure Nb-mossite ($FeNb_2O_6$). At high temperature (1100 °C), however, the solid solution range is expanded and $FeTa_2O_6$ can accept up to 65% $MnTa_2O_6$ and up to 56% $FeNb_2O_6$ and is fully miscible with TiO_2. The titaniferous varieties are known as *strüverite*.

The traditional terminology, as in the case of columbo-tantalite, has been updated by IMA/COM, referring to the Fe and Mn terms of the Tapiolite-(Fe)—Tapiolite-(Mn) series:

- Tapiolite-(Fe): $Fe^{2+}Ta_2O_6$, formerly ferrotapiolite, dimorphous with tantalite-(Fe)
- Tapiolite-(Mn): $Mn^{2+}Ta_2O_6$, formerly manganotapiolite, dimorphous with tantalite-(Mn).

However, the application of this nomenclature requires chemical composition data (electron microprobe) not always available so, for the optical description, the reference to the general term tapiolite (***tp***) is maintained here. The photomicrographs shown (Figs. 122.1a–d, 122.2a, b and 122.3a, b) correspond to Tapiolite-(Fe), analyzed by Beurlen et al. (2008).

122.2 Microscopic Description

122.2.1 SP Observation (*Single Polariser*)

Grey *colour*, with medium–low *reflectance*, **R**, like sp, close to or slightly higher than wf and distinctly higher than gangue: R tp ≈ R sp; R tp ≥ tan, cbt; R tp > cs (cf. multispectral VNIR and visible range comparison graphs).

Visible *bireflectance*, simply by comparison of twins in the same grain. Distinct *pleochroism*, especially in

R. Castroviejo, *A Practical Guide to Ore Microscopy—Volume 1*,
https://doi.org/10.1007/978-3-031-12654-3_122

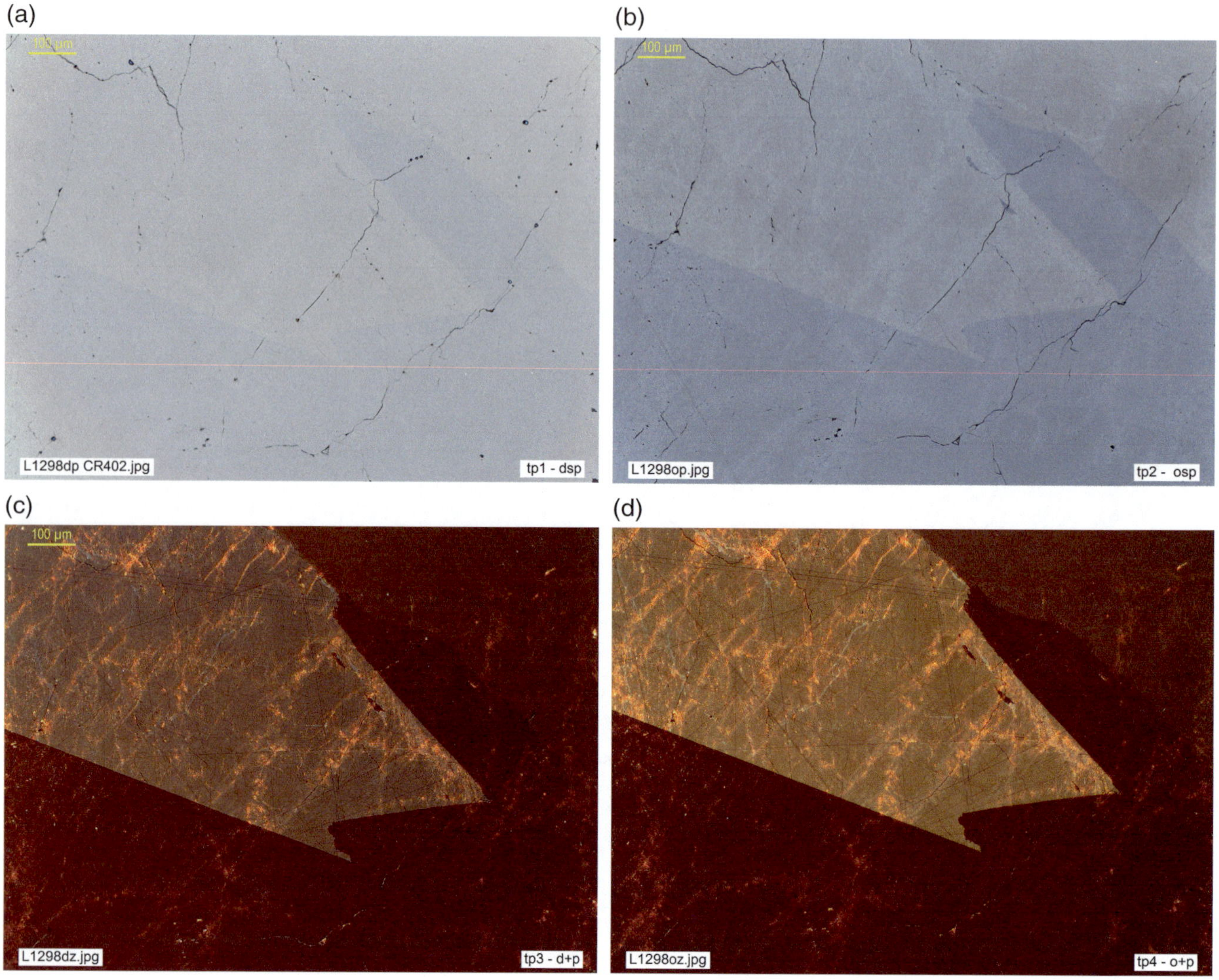

Fig. 122.1 **a**, **b** Tapiolite, twinned. Same scene (**dsp**) and (**osp**), respectively. CR 402, Borborema, Brazil. **c**, **d** Tapiolite, twinned. Same scene as Fig. 122.1a (**d+p**) and (**o+p**), respectively

immersion (Fig. 122.1a, b), from light grey, with slight brownish tinge (R_2) to bluish grey, darker (R_1).

Good polishing, very high *polishing hardness*: $PolH_{tp} \approx$ cb, but < cst.

It presents prismatic euhedral forms (Fig. 122.2a), but it also appears in small inclusions in cassiterite or intergrown with tantalite, which may replace it (Fig. 122.3a, b).

122.2.2 Observation +P (*Polariser + Analyser at 90°*)

Strongly *anisotropic* (Figs. 122.2c, d, 122.2b and 122.3b), with dominant greenish and bluish-green to yellowish colours, although sometimes (especially in immersion) the abundant reddish *internal reflections*, **IR**, can hinder its perception.

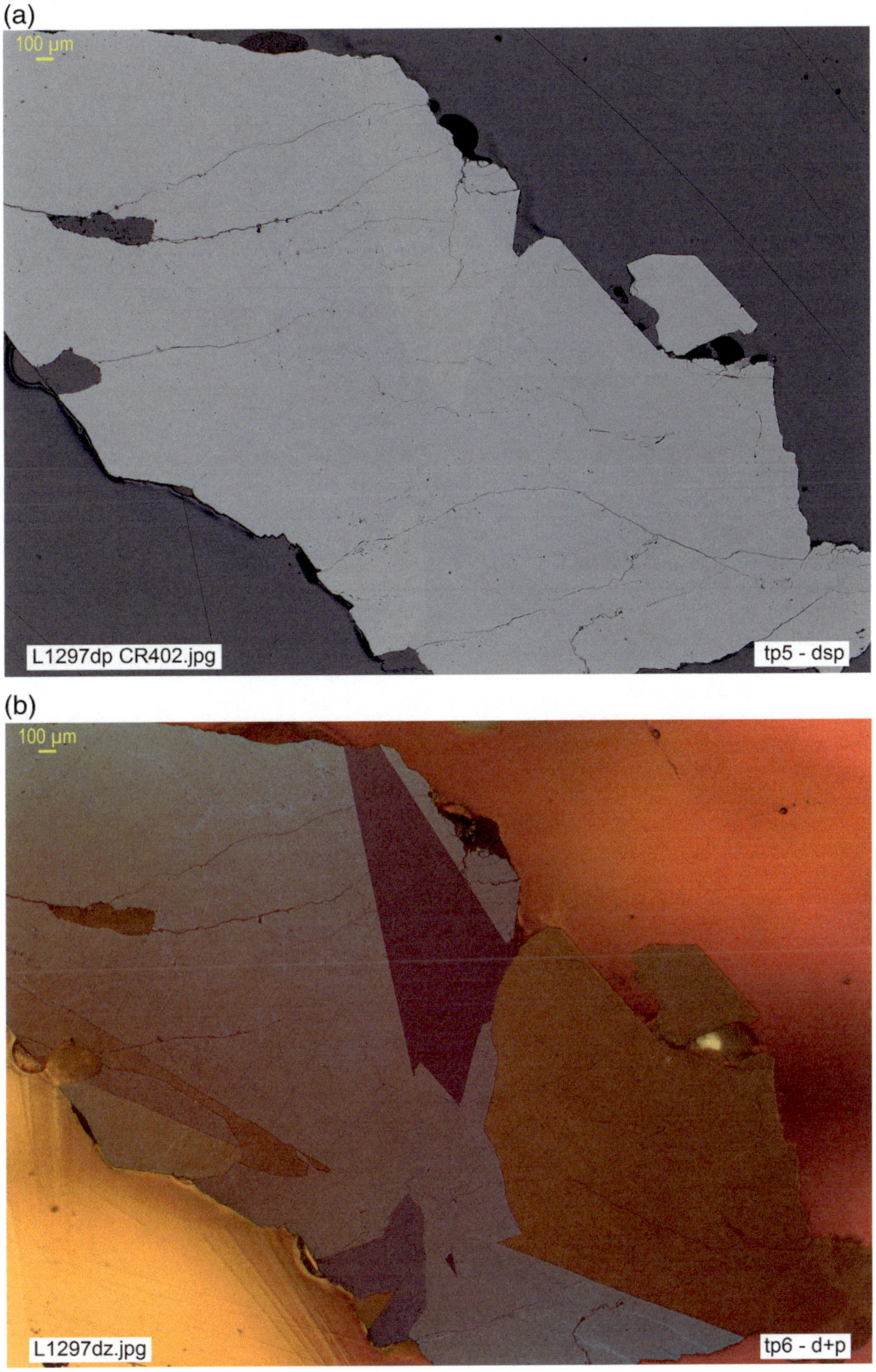

Fig. 122.2 **a**, **b** Concentrate of tp: twinned prismatic crystal (same wide-field image, obj. 2.5x, **dsp** and **d+p**, respectively). Figure 122.2a shows tp bireflectance and slight pleochroism, visible comparing twin elements, grey with slightly brownish (R_2) to bluish (R_1) tones; outer resin and q inclusions ($R \approx 5\%$) serve as reference for estimating tp reflectance ($R \approx 18\%$). Figure 122.2b—abundant IR and strong anisotropism, coloured in bluish-green and yellowish tones; outer, coloured, luminous zones are resin. CR 402, Borborema, Brazil

Frequently *twinned* (Figs. 122.1a–d and 122.2a, b), with single // (011) twins as cassiterite or, sometimes, polysynthetic as ilmenite.

122.2.3 Paragenesis

In pegmatites and, by weathering of these, in detrital deposits, the same as cassiterite.

Paragenesis: cassiterite, columbotantalite, tantalite.

122.2.4 Identification

Its hardness, pleochroism, coloured anisotropism, internal reflections and twinning, taking into account the paragenesis, allow reliable identification.

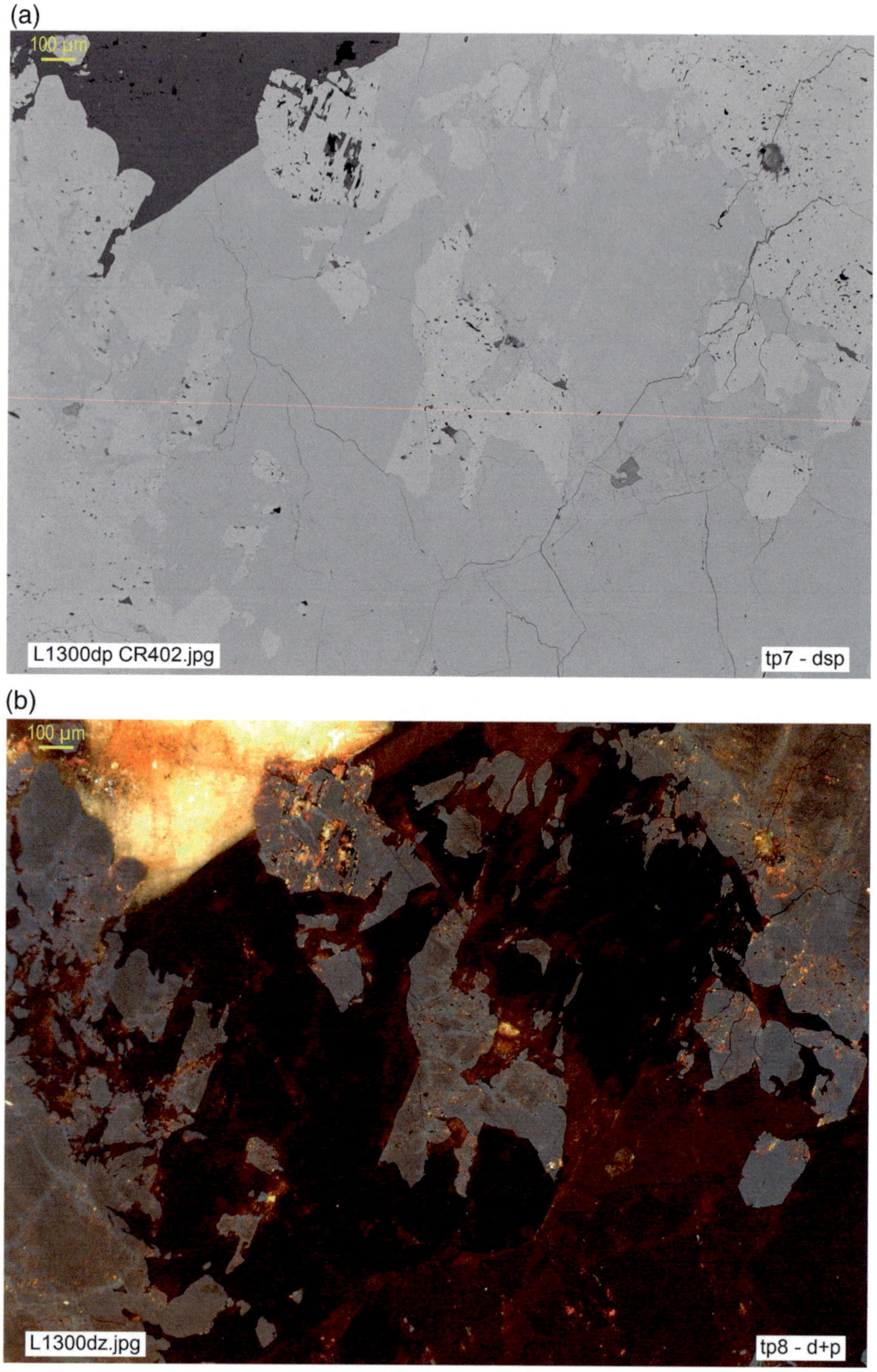

Fig. 122.3 **a, b** (Same image: **dsp** and **d+p**, *respectively*). Complex intergrowth of tp (light grey, strongly anisotropic, abundant light-coloured IR) and tan (darker grey, weak anisotropism, scarcer reddish IR), suggesting partial replacement of tp by tan. The image, partial view of the core from a tp phenocryst, mirrors the textural relationships observed inside it, whose complexity might sustain contested interpretations. However, it also offers a few useful keys: (i) the tp-tan contacts, extremely irregular, are typical reaction borders pointing to dis-equilibrium relationships between both phases; (ii) most re-entrants and concave lobes advance into tp and are filled by tan, suggesting leaching of tp prior to tan precipitation; (iii) while there are no microcracks in tan, reentrants in tp mirror microfractures or cleavage planes as flow-paths for secondary Ta-rich fluids; (iv) all tp patches included in tan share the same orientation (as show their identical polarisation tints, common also to the external tp); therefore, they all belong together, as relicts of a single phenocryst. In all, this evidence points to a two-stage sequence: 1st-primary tp phenocryst (now relict); 2nd-tp un-stability and tan crystallization, secondary to tp. CR 402, Borborema, Brazil.

It is distinguished from cassiterite by its pleochroism, its more coloured anisotropism and its IR, more reddish than those of cassiterite. Wolframite and columbotantalite show a weaker anisotropism and do not usually present twins. Ilmenite is browner and less anisotropic, and it lacks IR, except in varieties rich in Mg or Mn.

122.3 Spectral Information

122.3.1 Tapiolite Multispectral Reflectance R (in Air, Without Polariser) (%)

λ (nm)	370	400	425	450	475	500	525	550	575	600	625	650	675	700	750	800	850	900	950	1000
μ	17.0	16.5	16.5	16.7	16.9	16.9	16.8	16.7	16.5	16.4	16.3	16.1	16.0	15.9	15.8	15.7	15.7	15.5	15.5	15.4
P_5	16.6	16.1	16.1	16.1	16.3	16.3	16.3	16.1	16.1	15.9	15.8	15.7	15.6	15.6	15.5	15.4	15.4	15.3	15.3	15.1
P_{20}	16.8	16.3	16.3	16.4	16.6	16.6	16.5	16.4	16.3	16.1	16.0	15.9	15.8	15.7	15.6	15.5	15.5	15.4	15.4	15.2
P_{80}	17.1	16.7	16.8	17.0	17.2	17.3	17.2	17.0	16.8	16.6	16.5	16.4	16.3	16.1	16.0	15.8	15.8	15.6	15.6	15.5
P_{95}	17.3	16.8	16.9	17.2	17.4	17.4	17.4	17.1	17.0	16.8	16.7	16.5	16.4	16.3	16.1	16.0	15.9	15.7	15.7	15.6

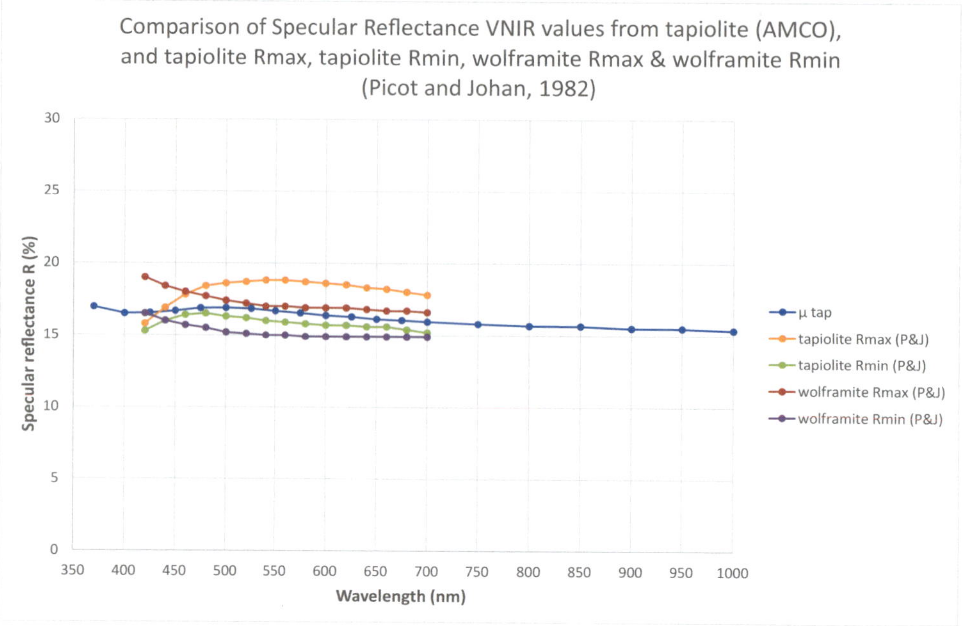

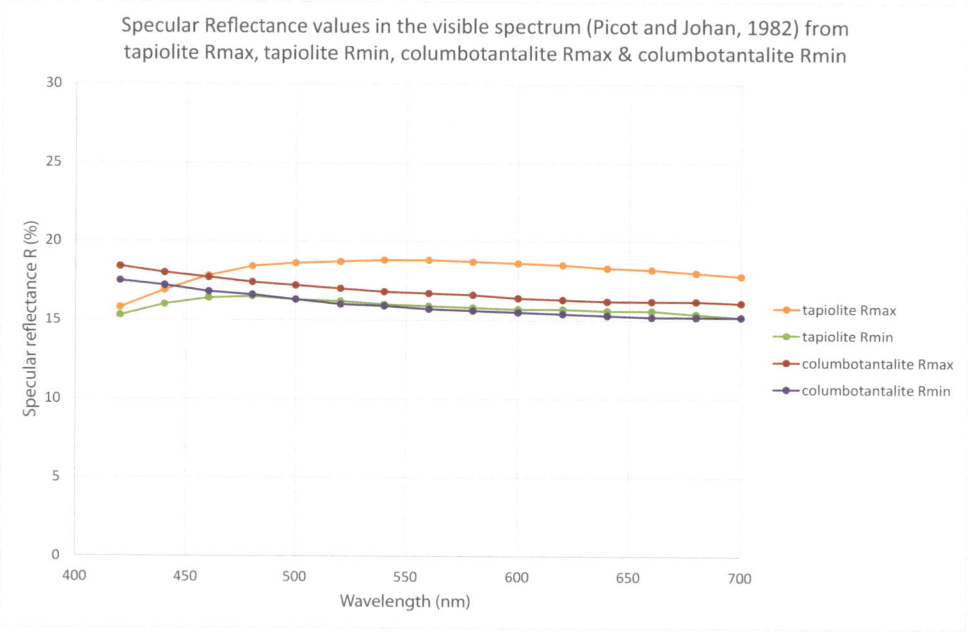

References

Beurlen H, Da Silva M, Thomas R, Soares D, Olivier P (2008) Na-Ta-(Ti-Sn) oxide mineral chemistry as tracer of rare-element granitic pegmatite fractionation in the Borborema Province, NE Brazil. Miner Deposita 43:207–228

Picot P, Johan Z (1982) Atlas of ore minerals. B.R.G.M./Elsevier

Uytenbogaardt W, Burke EAJ (1971) Tables for microscopic identification of ore minerals, 2nd edn. Elsevier, Amsterdam

123 Tellurium (*Te*)

Te: Trigonal

Tellurium		R_{546} (%)	Y (%)	λ_d (nm)	P_e (%)	Colour sensation
Air	R_o	60.2	59.8	546	0.6	White, glossy
	R_e'	67.0	66.4	538	0.6	
Oil	R_o	48. 0	47.6	491	1.3	
	R_e'	54.8	54.2	496	0.9	
Bir: Weak (3)				*Pleoc*: Weak (2)		*Tarn*: No
Anis +P: Net, not coloured (4)				*IR*: No		*Twg*: Not observed
Mor: Anhedral aggregates, impregnation				*Clv*: Occasional		*PolQ*: Difficult to avoid scratches
Hardness	*PolH*: Low (2)			*Mohs*: 2–2½	*Talm*: B	*VHN* $_{(10\ g)}$: 48–66

Note: cf. Table and R charts, visible range data (QDF3)
Mineral abbreviations: Ag—native silver, Au—native gold, clv—calaverite, gg—gangue, Pt—native platinum, q—quartz, syl—silvanite, Te—native tellurium

123.1 General

Composition: Te, occasionally with Se (isomorph), more rarely with Fe, Bi.

123.2 Microscopic Description

123.2.1 SP Observation (*Single Polariser*)

Colour bright white (Fig. 123.1a, b), bright, despite frequent *polishing* scratches.

Very high *reflectance* but less than native gold or native platinum and much less than native silver:

- R Te $\geq$ R clv, syl (cf comparative R graphs).

Weak *bireflectance* and *pleochroism*, visible comparing grains with different orientations (Fig. 123.3a). Low *polishing hardness.*

It usually occurs in xenomorphic aggregates, as impregnation in veins (Fig. 123.2a), although it is also found in prismatic or acicular crystals.

123.2.2 Observation +P (*Polariser + Analyser at 90°*)

Clearly *anisotropic*, although without striking colours, it has silvery to slightly brownish tones (Figs. 123.1c, d, 123.2b and 123.3b).

R. Castroviejo, *A Practical Guide to Ore Microscopy—Volume 1*,
https://doi.org/10.1007/978-3-031-12654-3_123

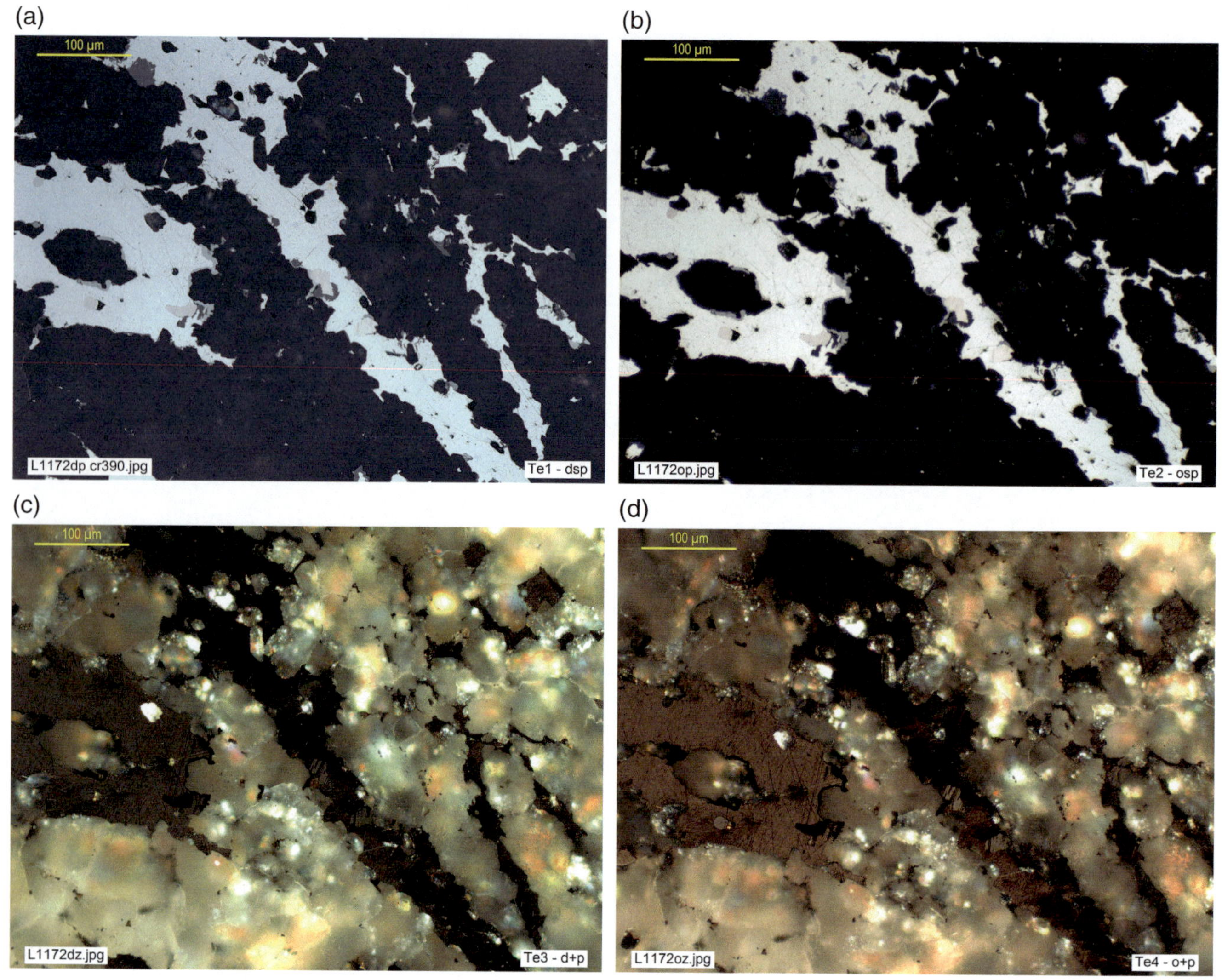

Fig. 123.1 a, b (**dsp** and **osp**, respectively)—Impregnation of Te (white), with small inclusions of syl (very light pinkish brown) in gg of q (grey). CR390, Teine Mine, Milsuyama-Sanko Vein, Hokkaido, Japan. **c**, **d**—Same scene as Fig. 123.1a (**d+p** and **o+p**, respectively). Anisotropism of Te and syl, this one twinned

123.2.3 Paragenesis

Restricted *paragenesis*: hydrothermal veins with tellurides, sulphides, arsenides and/or sellenides and precious metals.

Companions: other tellurides, such as sylvanite or calaverite, native gold.

123.2.4 Identification

Paragenesis is a fundamental datum. Native tellurium is typical of some hydrothermal paragenesis with native gold, tellurides, etc., particularly epithermal (Romania).

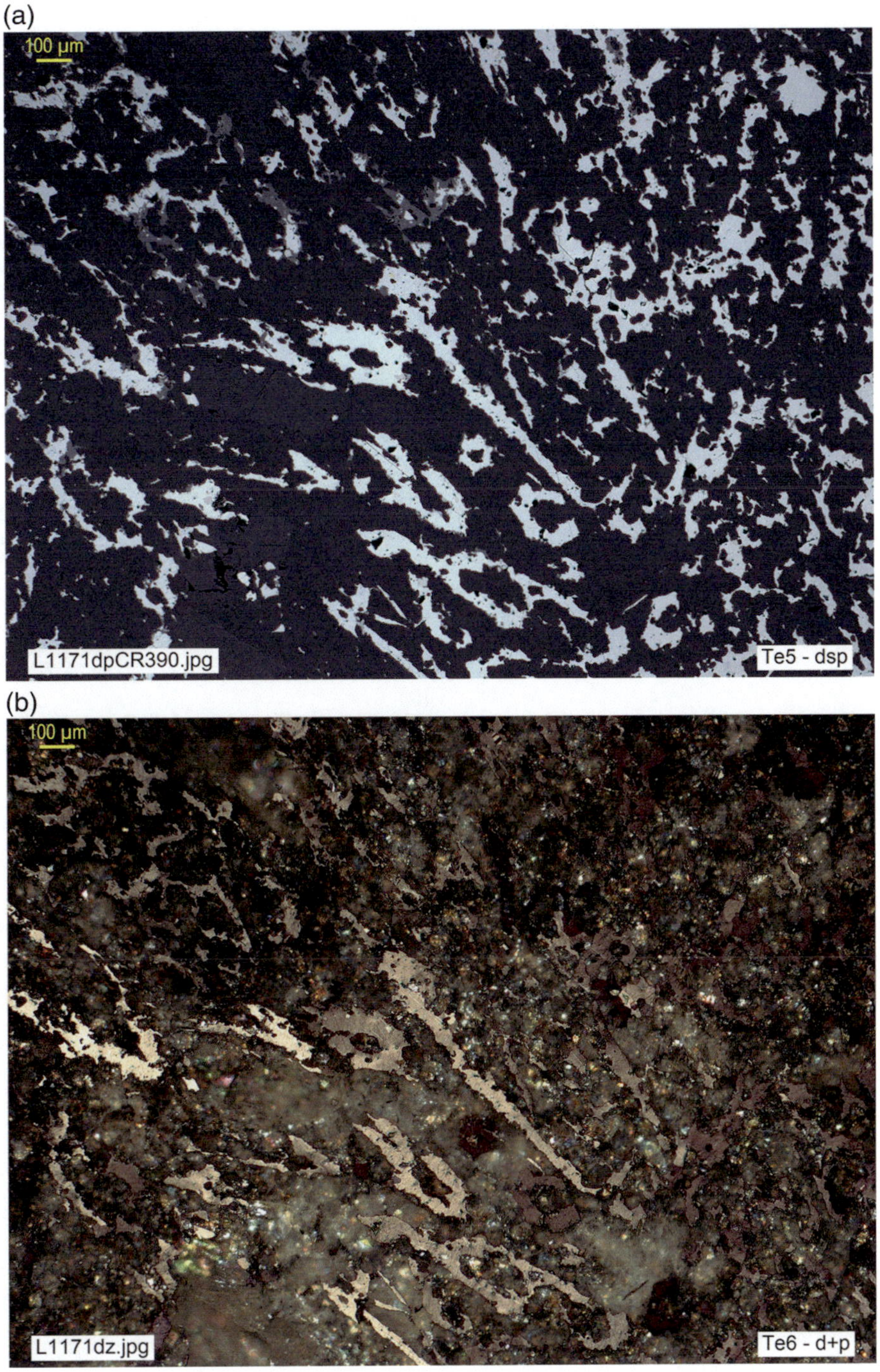

Fig. 123.2 **a**, **b** (same scene: **dsp** and **d+p**, respectively)—Hydrothermal filling of q with Te impregnation (wider view of Fig. 123.1a, whose field corresponds to the central zone)

It is easily distinguished from native gold by its colour and by its lower reflectance, R.

In general, Te is characterized by its high R, which makes it appear very bright against most companions. In addition, its bireflectance and pleochroism, its remarkable but poorly coloured anisotropism and frequent polishing scratches characterize it.

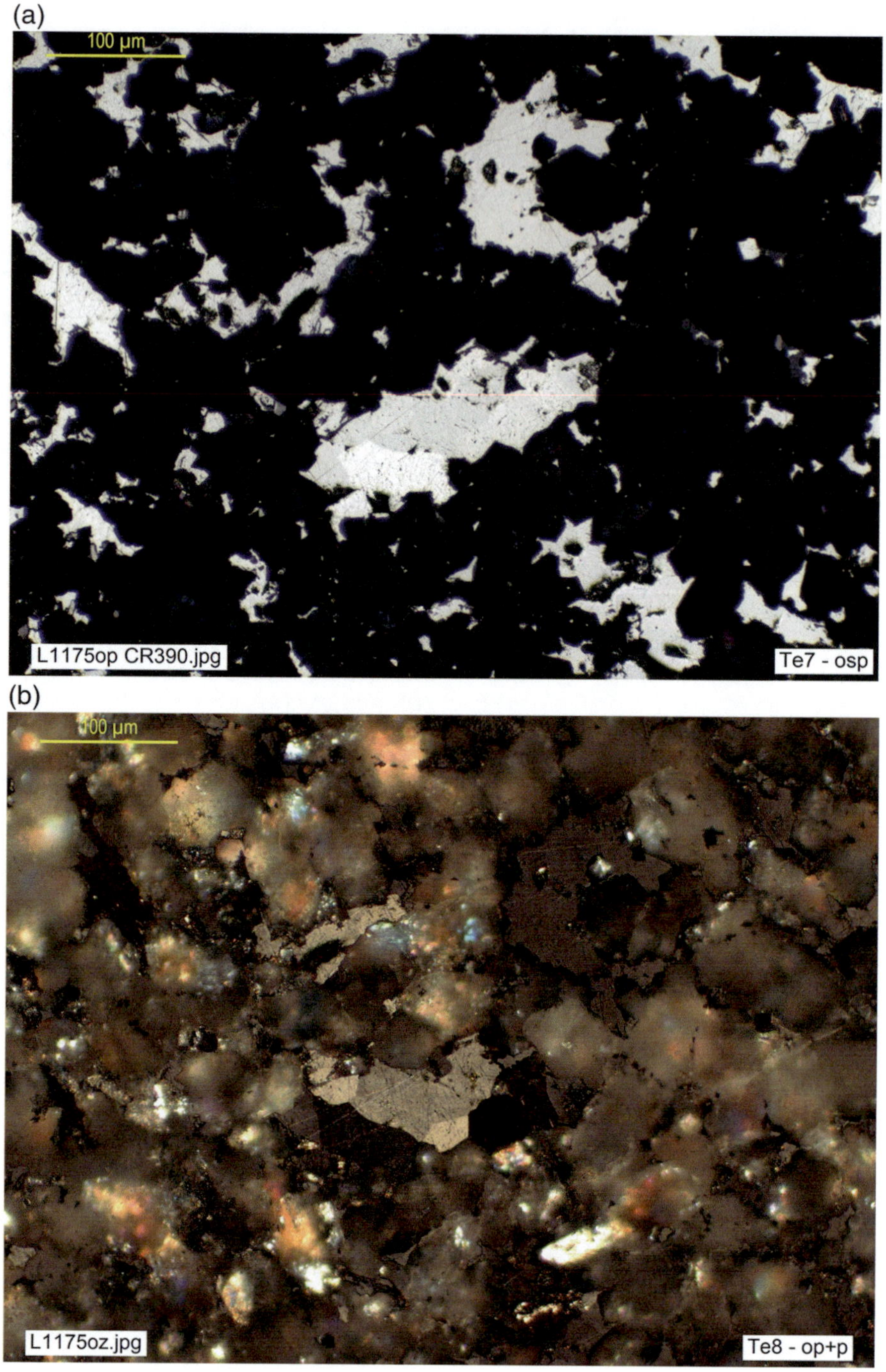

Fig. 123.3 **a**, **b** (same scene: **osp** and **o+p**, respectively)—Birreflectance and anisotropism of Te (immersion) in q vein. CR390, Teine Mine, Milsuyama-Sanko Vein, Hokkaido, Japan

123.3 Spectral Information

123.3.1 Tellurium Reflectance R (in Air, Visible Spectrum) (%)

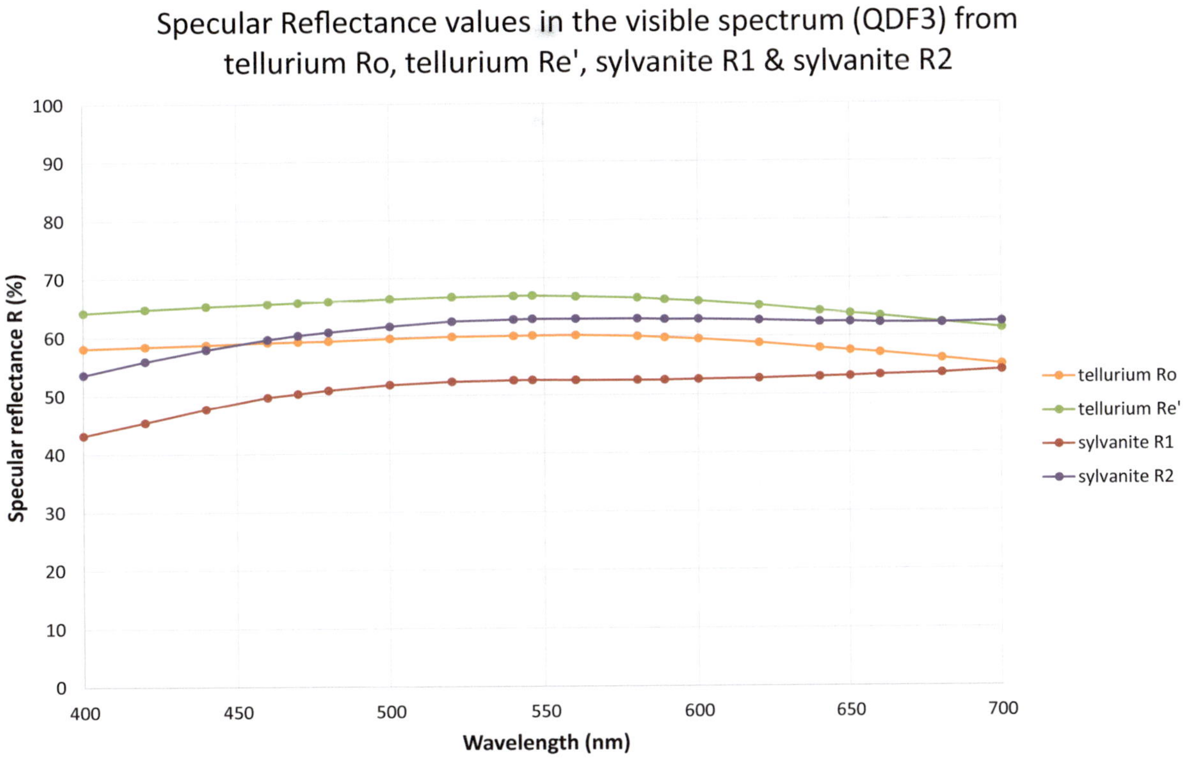

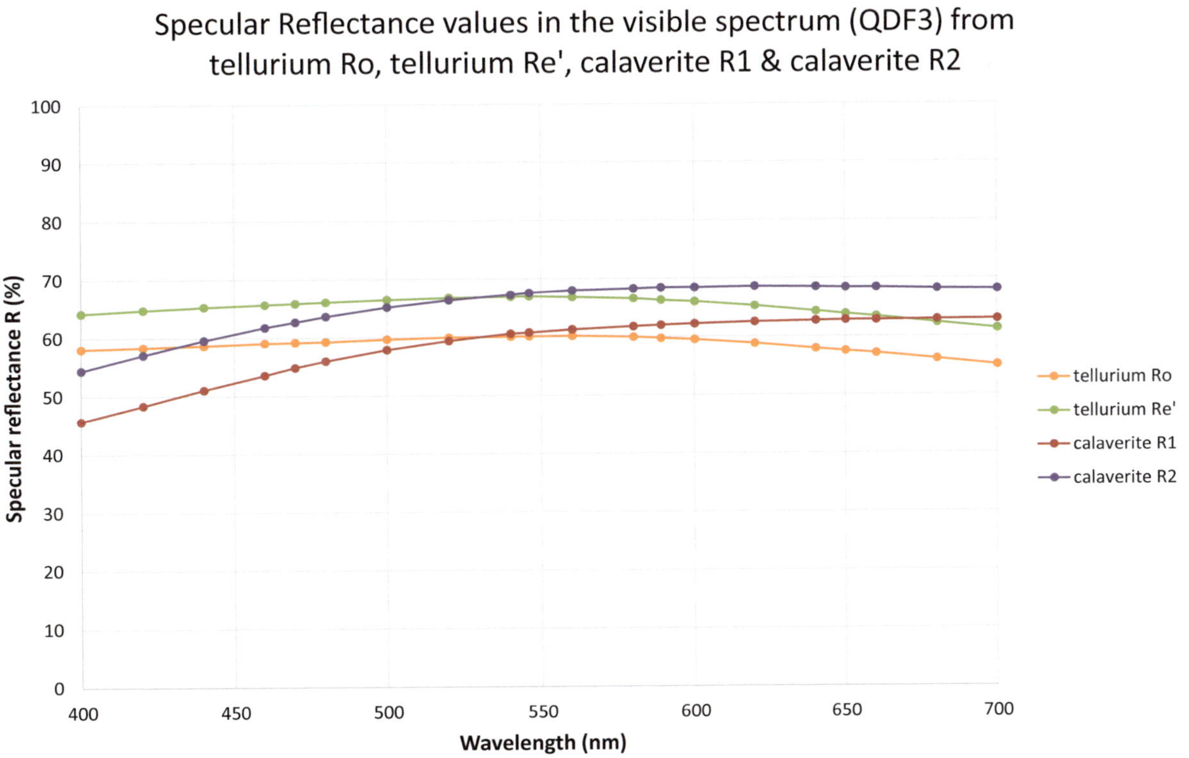

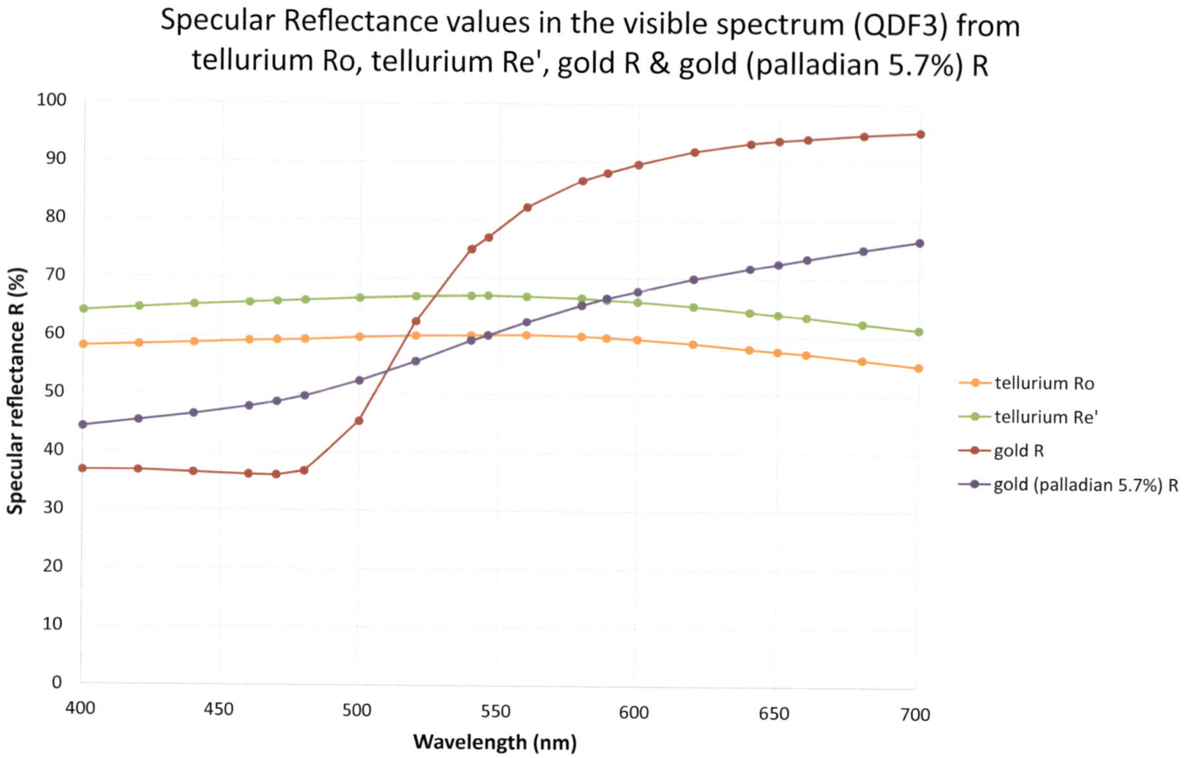
Specular Reflectance values in the visible spectrum (QDF3) from tellurium Ro, tellurium Re', gold R & gold (palladian 5.7%) R
Specular reflectance R (%)
Wavelength (nm)
0
10
20
30
40
50
60
70
80
90
100
400
450
500
550
600
650
700
tellurium Ro
tellurium Re'
gold R
gold (palladian 5.7%) R

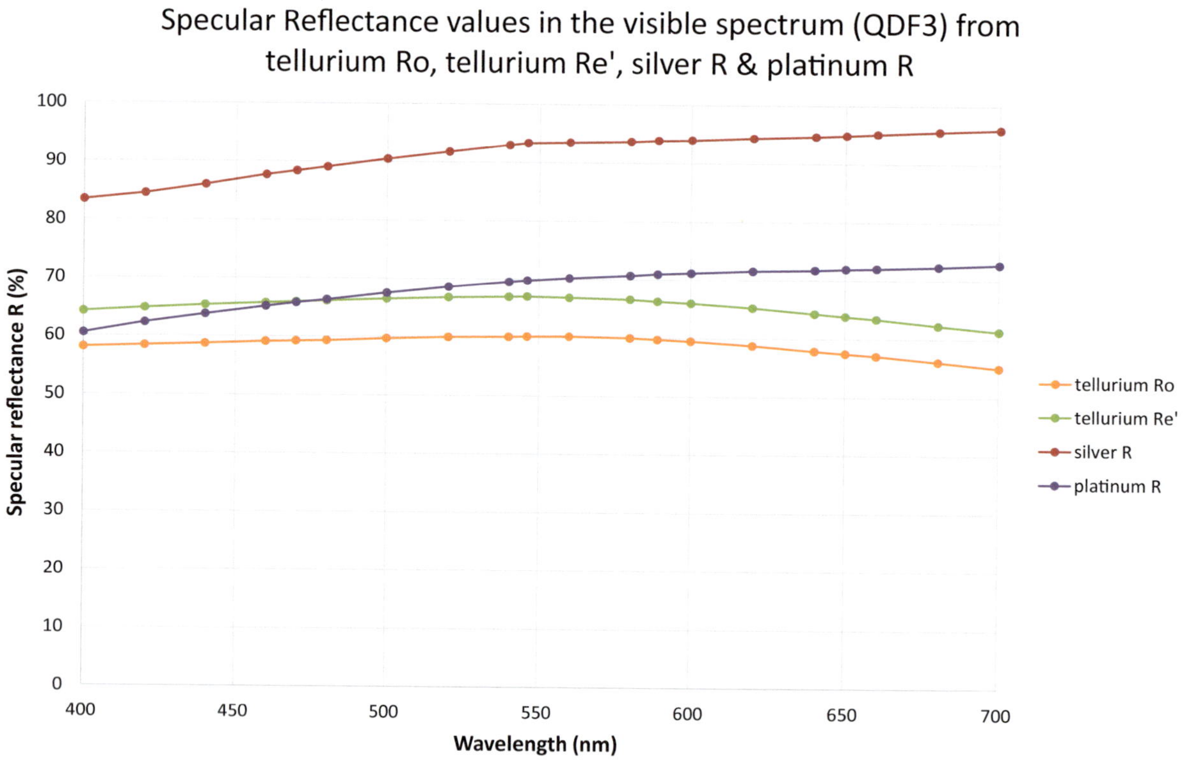
Specular Reflectance values in the visible spectrum (QDF3) from tellurium Ro, tellurium Re', silver R & platinum R
Specular reflectance R (%)
Wavelength (nm)
0
10
20
30
40
50
60
70
80
90
100
400
450
500
550
600
650
700
tellurium Ro
tellurium Re'
silver R
platinum R

124 Tetrahedrite (*td/fahlore*)

$Cu_6[Cu_4(Fe,Zn)_2]Sb_4S_{13}$: Cubic

Tetrahedrite		R_{546} (%)	Y (%)	λ_d (nm)	P_e (%)	Colour sensation
Air		32.5	32.2	541	0.6	Light grey, olive shade
Oil		17	16.8	498	0.8	Darker grey (BlGn, OlBr)
Bir: Null (1)				*Pleoc*: Null (1)		*Tarn*: Seldom
Anis +P: Isotropic, with exceptions (1)				*IR*: Scarce		*Twg*: Rarely observable
Mor: Aggregates, zoned xx, inclusions				*Clv*: Occasional (100)		*PolQ*: Good, sometimes *triangular pits*
Hardness	*PolH*: Medium, > gn, < sp (3)			*Mohs*: 4 $^-$	*Talm*: B	*VHN* $_{(100\ g)}$: 312–351

Obs: cf. Table and R charts, multispectral data (VNIR, AMCO) and visible range (QDF3)

Mineral abbreviations: ac—acanthite, Ag—native silver, bl—boulangerite, bn—bornite, bnn—bournonite, cc—chalcocite, ccp—chalcopyrite, cs—cassiterite, en—enargite, fbg—freibergite, gg—gangue, gn—galena, ht—hematite, plb—polybasite, po—pyrrhotite, prg—pyrargyrite, pru—proustite, py—pyrite, sbt—stibnite, sp—sphalerite, stp—stephanite, tn—tennantite, td—tetrahedrite

124.1 General

Tetrahedrite is the most frequent species in the group of the so-called *Fahlores*, or isomorphic series tetrahedrite-tennantite, but the term is also commonly used to refer to the whole group. In this sense, the group will be referred to as tetrahedrite (*s.l.*), specifying tetrahedrite (*s. str.*) to refer to the species. In the absence of indication, tetrahedrite is understood as an undifferentiated species of the series (*s.l.*).

The components of the tetrahedrite group (*s.l.*) have a remarkable compositional variation, implying transitional terms. Terminology is still under revision, but at least the arsenical (*tennantite, tn*) and argentiferous (*freibergite, fbg*) terms should be cited; *schwazite* is Hg-rich; *annivite*, Bi; and *goldfieldite*, Te.

The composition of tetrahedrite can also be formulated by expressing the Fe (ferrous)-Zn substitution, $Cu_6(Cu_4x_2)Sb_4S_{13}$, where x = Fe^{2+} or Zn. In tetrahedrite Zn dominates over Fe, while in tennantite, $Cu_6[Cu_4(Fe,Zn)_2]As_4S_{13}$, Fe dominates over Zn, as it does in freibergite, $Ag_6[Cu_4Fe_2]Sb_4S_{12}$. The actual compositions are even more complex if the contents of minor elements or impurities are taken into account, e.g., Ag, Pb, Fe, As in tetrahedrite; Sb, Bi, Pb, Zn, Sb in tennantite; Zn, Hg, Bi in freibergite.

This enormous compositional variety influences the optical properties in a way that is not easy to pin-point in qualitative terms, as will be seen when describing colour.

From the quantitative point of view, Hall et al. (1974, *in*: Cervelle and Moello 1990) and Charlat and Levy (1976, *in ibid.*) compare the reflectance curves of fahlores, in samples of different compositions, synthesized in laboratory and natural, respectively. Although they show that the substitution of Sb by As or Cu by Ag reduces the reflectance in the whole visible spectrum, while Zn or Fe increases it in Cu-rich tetrahedrites, the very varied and complex substitution possibilities in the fahlore group question the practical usefulness of the method for determining the composition in real ores.

124.2 Microscopic Description

124.2.1 SP Observation (*Single Polariser*)

Light grey, with a slight olive, greenish to brownish tinge compared to galena or bluish, if compared with chalcopyrite. Darker than galena, very similar to chalcocite, but slightly darker (cf. Fig. 19.7), distinctly lighter than sphalerite (Figs. 124.1a, b). The colour impression is difficult to define

R. Castroviejo, *A Practical Guide to Ore Microscopy—Volume 1*,
https://doi.org/10.1007/978-3-031-12654-3_124

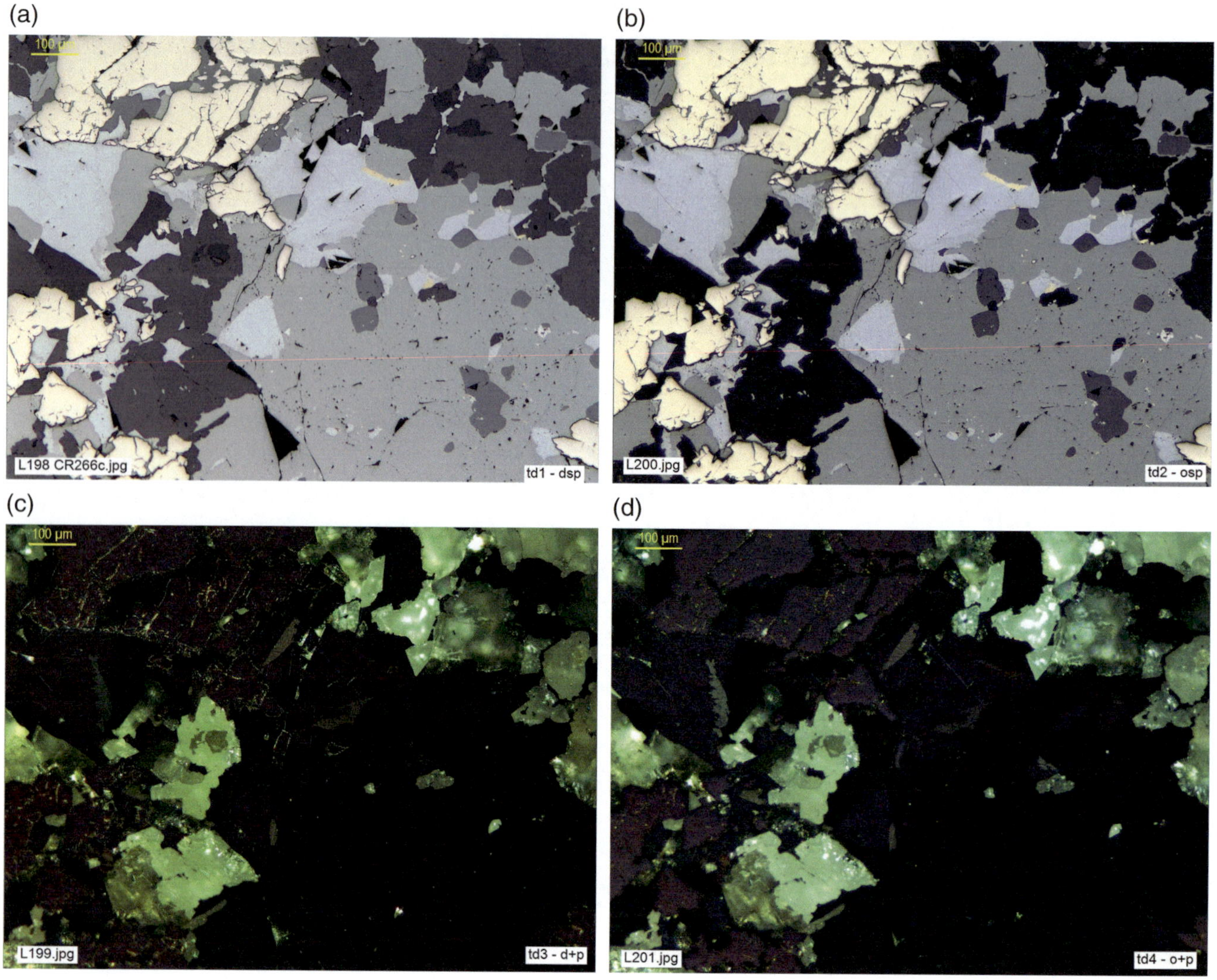

Fig. 124.1 **a** Aggregate of td (olive grey), gn + bl (very light grey), sp (grey), ccp (trace, yellow) and gg (dark grey), cementing py microclasts (pale yellow, relief). Polishing pull-outs (*triangular pits*) in gn, td, sp. CR266c, Lucky Friday Mine, Coeur d'Alene, Idaho, USA (**dsp**). **b** Same scene as td1 (**osp**): in immersion, gg (black) disappears and the higher contrast makes easier to distinguish the inclusions of bl (slightly darker) in gn. **c**, **d** Same scene as td1 (**d+p**) and (**o+p**), respectively. The polarization shows the anisotropism of bl (faintly coloured) and of gg (with internal reflections), while the other cubic components remain in extinction

and varies with composition and environment. In immersion and following Ramdohr (1980), it could be described as dirty olive-brown, for common tetrahedrite (*s.str.*, Cu–Sb fahlore); dull grey, colourless, for very Fe-rich tetrahedrites-CuSb; greenish-grey, slightly bluish, for tennantite (CuAs fahlore, Fig. 124.3, cf. also Fig. 46.1a, b); light olive-grey, slightly yellowish-brown, for freibergite (silver-rich Cu–Sb-fahlore).

Medium–low *reflectance*: R td $\approx$ or slightly > R ht; distinctly > sp; < gn, < py (cf. comparative R plots). The R values depend on the composition (cf. § General, above) but the spectral differences are small and it does not seem possible to establish a precise correspondence, due to the multiple simultaneous substitutions (As, Sb, Fe, Zn, Cu, Ag, Hg, Bi) that occur in fact. According to the spectral data available, freibergite is the most reflective species and tennantite, the least reflective, with tetrahedrite in an intermediate position but overlapping R values of freibergite or tennantite, depending on whether $\lambda > 500$ nm or < 500 nm, respectively (cf. comparative graphs QDF3). In any case, it should not be forgotten that the subjective impression of brightness and colour, conditioned by the adjacent ores, can be misleading (compare Figs. 124.1a, b, 124.2 and 124.3).

As cubic minerals, the fahlores are isotropic and neither bireflective nor pleochroic, but in some cases anisotropism is observed (cf. +**p**).

Polishing is generally good, although small pull-outs may occur, even triangular pits (Figs. 124.1a, b),

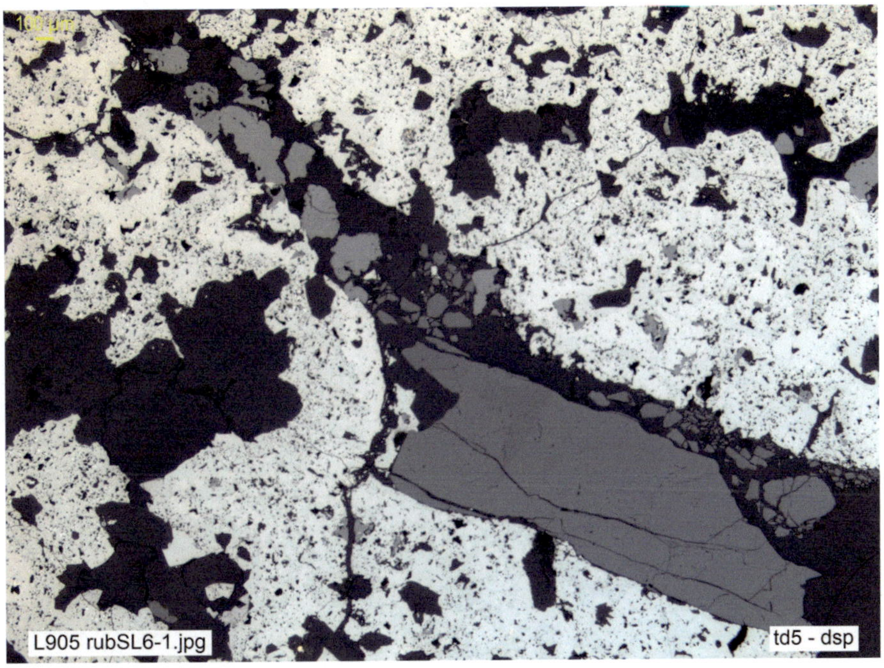

Fig. 124.2 Microbreccia with clasts of cst (dark grey) and sp (light grey), cemented by tn (greyish white) and quartz (very dark grey). A wide-field lens and contrast with low-reflectance minerals makes td appear very clear. Sample SL6.1-rub, Böhm. Zinnwald, Erzgebirge, Germany (**dsp**)

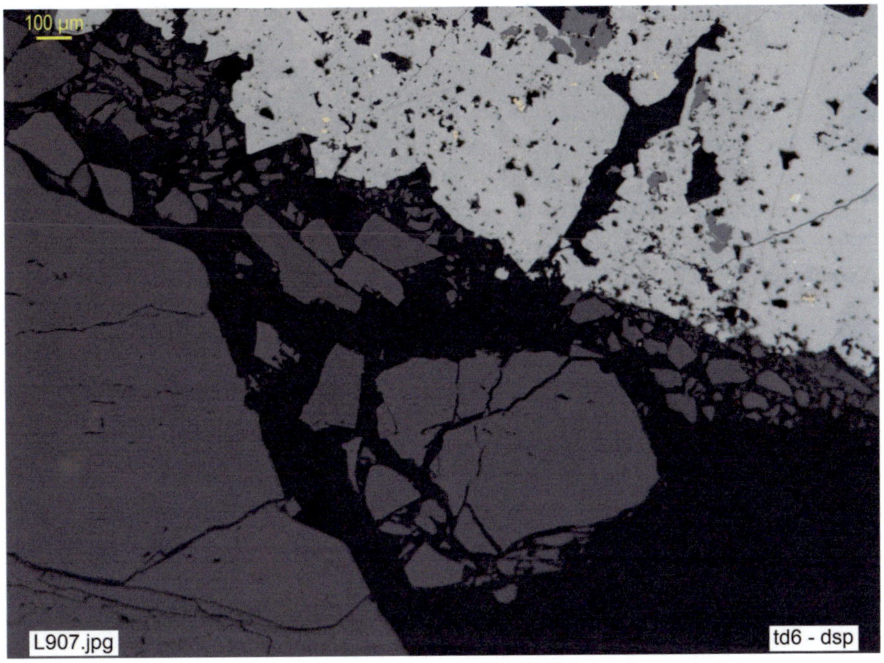

Fig. 124.3 Detail of Fig. 124.2 (bottom right corner): sp and ccp inclusions in tn (olive grey, slightly greenish), contrasting with the very low reflectance of the dark grey cst and almost black quartz clasts (**dsp**)

occasionally favored by the occurrence of (100) cleavage. Medium *polishing hardness*, between those of ccp and sp:

- PolH td > gn, cc, sbt, Ag sulphosalts, ccp, Ag; but
- PolH td < sp, po, py, varying with composition (lower in freibergite).

Fahlores occur frequently included in galena (freibergite) or in chalcopyrite, sphalerite, also as massive veinfill, sometimes brecciated, in aggregates and disseminations. Frequent compositional zoning, sometimes resulting in colour zoning. Also, symplectitic (myrmektitic) intergrowths or decomposition textures, with various minerals, including Ag sulphosalts.

Fig. 124.4 Intergrowth of tn (in extinction, scarce and faintly visible red internal reflections) and en (strong anisotropism, coloured). CR 389, Tsumeb Mine, Namibia (**o+p**)

124.2.2 Observation +P (*Polariser + Analyser at 90°*)

Isotropic in general due to its cubic symmetry (Fig. 124.1c, d), but it can be anisotropic due to deformation or compositional peculiarities (Te: goldfieldite).

Red *internal reflections*, scarce but usually visible in immersion, especially in tennantite (Fig. 124.4) and in As-rich species; much scarcer in Ag, Bi, Hg-rich species.

124.2.3 Paragenesis

Varied *parageneses*, particularly frequent in epithermal argentiferous deposits. It can be replaced by a multitude of minerals, either by hypogene alteration (most common, proustite, pyrargyrite, miargyrite, polybasite) or supergene alteration (acanthite, native silver). It can also replace numerous species (cf. Fig. 46.1a, b). It is also found in massive sulphides, orogenic gold deposits, *Kupferschiefer*, etc.

Characteristic companions: galena, chalcopyrite, bornite, chalcocite, sphalerite, bournonite, Ag–Co–Ni ores.

124.2.4 Identification

Its good polish, colour, moderate reflectance, medium hardness, isotropic character, internal reflections (commonly scarce or missing) and paragenesis allow safe identification. Silver sulphosalts such as miargyrite, polybasite or stephanite, similar in appearance, are much softer and anisotropic; bournonite is also anisotropic, as is chalcocite. Moreover, in this case an additional test can be used, the acid attack (1 drop of HNO_3 for 5 s, cf. explanation in *Description cc*). If it is chalcocite, it is covered with a blue patina (of covellite), which is not the case if it is a *fahlore*.

More difficult is to determine the fahlore species, although paragenesis may help; freibergite if argentiferous, tetrahedrite or tennantite if dominated by Sb or As minerals, respectively (tennantite also characterized by a higher abundance of IR).

124.3 Spectral Information

124.3.1 Tetrahedrite Multispectral Reflectance R (in Air, Without Polariser) (%)

λ (nm)	370	400	425	450	475	500	525	550	575	600	625	650	675	700	750	800	850	900	950	1000
μ	29.5	29.9	29.9	29.9	29.9	30.0	30.3	30.6	30.6	30.4	30.0	29.4	28.9	28.4	27.5	26.7	26.2	25.4	24.9	24.7
P_5	28.3	28.8	28.6	28.6	28.9	28.9	29.0	28.9	28.6	28.4	28.2	28.0	27.7	27.4	26.7	26.0	25.5	24.7	24.4	23.9
P_{20}	29.1	29.6	29.6	29.6	29.6	29.8	30.0	30.5	30.5	30.2	29.8	29.3	28.7	28.2	27.3	26.5	25.9	25.2	24.7	24.5
P_{80}	30.0	30.4	30.4	30.3	30.3	30.4	30.7	31.0	31.0	30.8	30.3	29.7	29.2	28.7	27.8	26.9	26.5	25.6	25.2	24.9
P_{95}	30.4	30.6	30.6	30.5	30.6	30.7	31.2	31.4	31.4	31.2	30.7	30.2	29.7	29.3	28.4	27.5	27.0	26.1	25.7	25.3

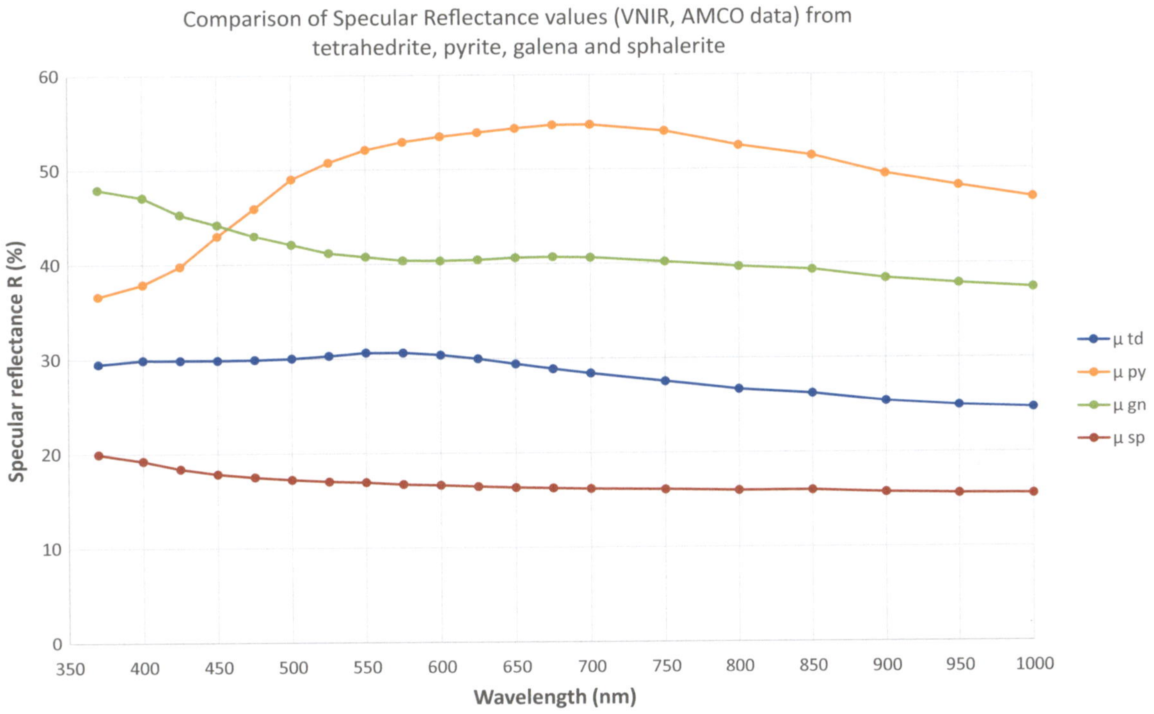

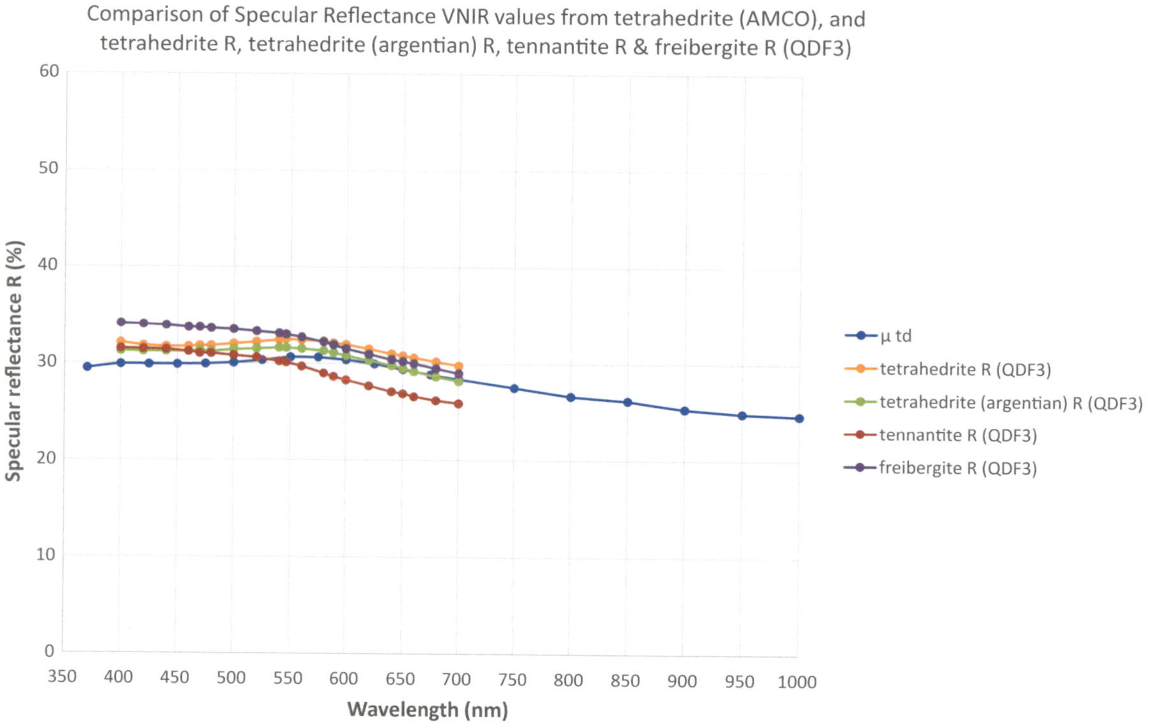

References

Cervelle B, Moello Y (1990) Reflected-light optics, pp 87–108. In: Jambor JL, Vaughan DJ (eds) Advanced microscopic studies of ore minerals. Short course handbook, vol 17. Mineralogical Association of Canada, Ottawa, p 426

Ramdohr P (1980) The ore minerals and their intergrowths, vol 2, 2nd ed. Pergamon Press, Oxford, p 1205

Titanite (*tit/sphene*)

$CaTiO(SiO_4)$: Monoclinic

125

Titanite		R546 (%)	Y (%)	λ_d (nm)	P_e (%)	Colour sensation
Air	R_1	9.5	9.5	470	2.6	Grey
	R_2	11.1	11.2	473	3.6	
Oil	R_1	1.3	1.3	469	3.8	Dark grey
	R_2	2.0	2.0	473	6.1	
Bir: Weak (3)				*Pleoc*: Weak (3)		*Tarn*: No
Anis +P: Weak (3)				*IR*: Massive		*Twg*: Occasional
Mor: Euhedral tendency, rhomboids				*Clv*: Occasional		*PolQ*: Good
Hardness	*PolH*: High (4)			*Mohs*: 5.5	*Talm*: nd	*VHN* $_{(100\ g)}$: 707–865

Obs: cf. Table and R charts, multispectral data (VNIR, AMCO) and visible range (QDF3)
Mineral abbreviations: ana—anatase, gg—gangue, ht—hematite, lcx—leucoxene, mt—magnetite, py—pyrite, rt—rutile, tit—titanite

125.1 General

Composition: $CaTiO(SiO_4)$, with possible exchange of Ca for REE, Mn, Th, Y, Sr and Ti for Fe. It can also contain Nb, Ta, Sn, Al, Zr, V. As a gangue mineral, it is transparent and easily recognized with transmitted light. However, as it is a common minor mineral in many metallic deposits, it should also be recognized with reflected light.

125.2 Microscopic Description

125.2.1 SP Observation (*Single Polariser*)

Grey *colour*. Low *reflectance*, particularly in immersion (Fig. 125.1a, b) but remarkably high for a silicate. Close to cassiterite, it appears distinctly brighter than common gangue (cf. multispectral comparative graphs and Fig. 125.2 and 125.3a). Weakly bireflective/pleochroic.

It occurs typically disseminated, in small euhedral crystals with rhomboidal or wedge-shaped habit (hence its synonym *sphene*) but also massive. As a pseudomorph or in cryptocrystalline aggregates resulting from hydrothermal alteration processes that release Ti, it may pass to *leucoxene*, lcx, very common in the gangue of metallic deposits, where its inconspicuous appearance, much darker than sulphides, usually goes unnoticed.

125.2.2 Observation +P (*Polariser + Analyser at 90°*)

Although it is *anisotropic*, polarization effects are hardly noticeable, due to overprinting by the abundance and luminosity of its *internal reflections* (colourless to yellowish-brown or yellow, Figs. 125.2a, b and 125.3b).

125.2.3 Paragenesis

Common accessory mineral in igneous and metamorphic rocks, as well as in detrital deposits. It is frequently a product of alteration of ilmenite or Ti-magnetite, which it can pseudomorphize. It is also found disseminated in the host rock of various types of deposits, associated with numerous hydrothermal alteration processes. Typical component of leucoxene, together with anatase.

R. Castroviejo, *A Practical Guide to Ore Microscopy—Volume 1*,
https://doi.org/10.1007/978-3-031-12654-3_125

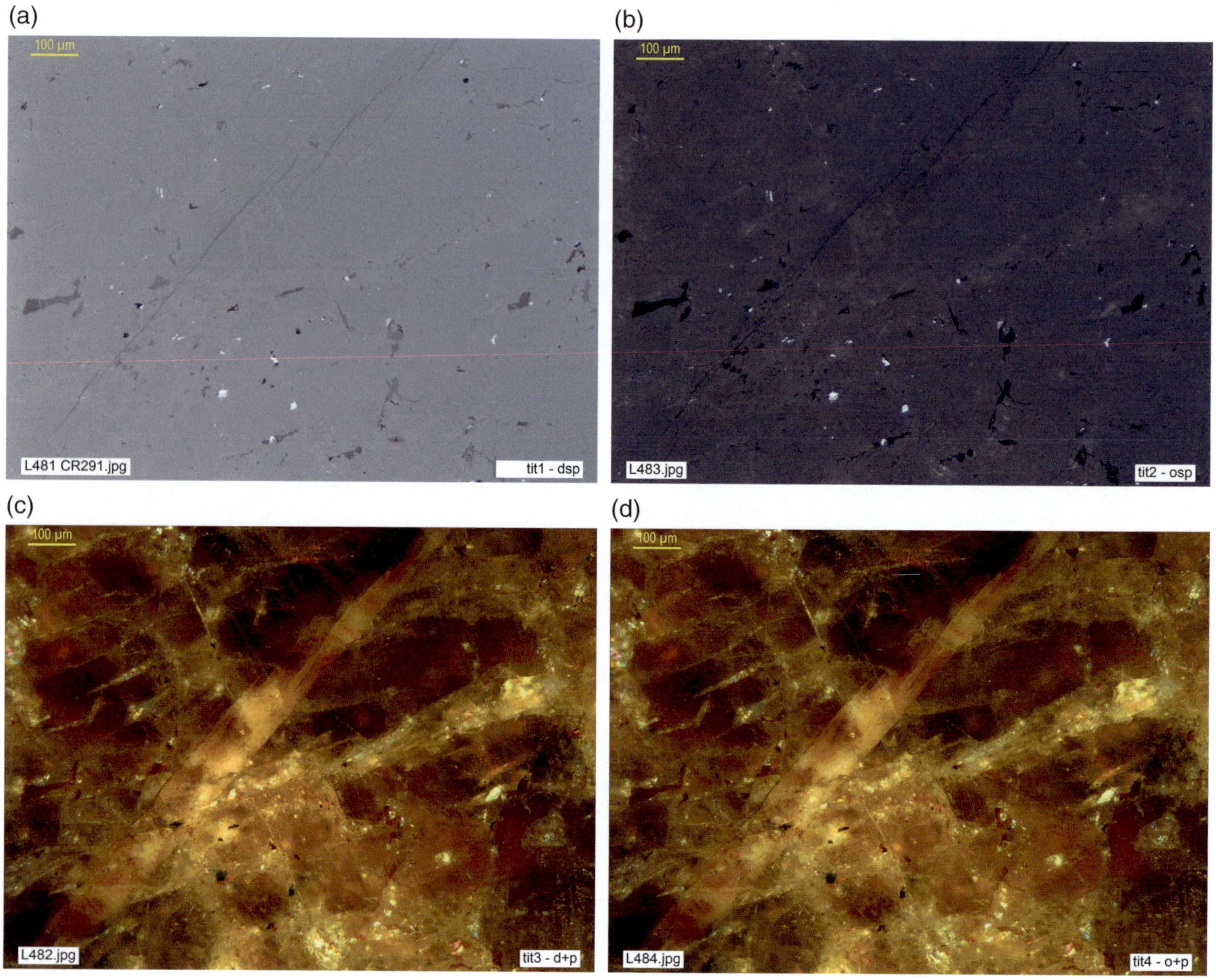

Fig. 125.1 **a** Single crystal of tit, crackled and partially altered, with small inclusions of q (dark grey), ht, mt, rt (light grey to white). CR291, Fogue Mine, Gjerstad, Norway (**dsp**) **b** Same scene as tit1 (**osp**). **c**, **d** Same scene as tit1 (**d+p**) and (**o+p**)

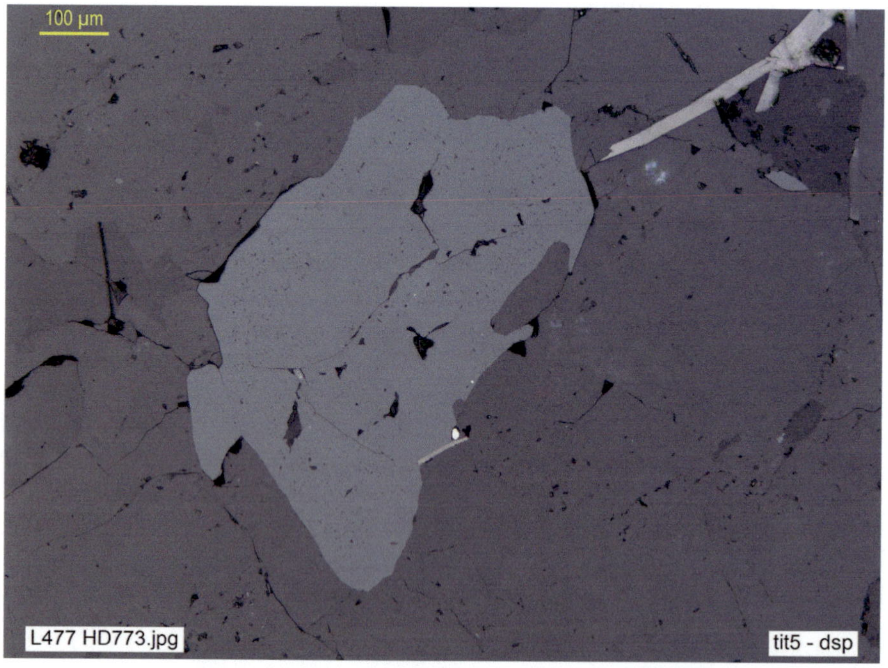

Fig. 125.2 Metamorphic carbonate and quartz gangue (darker grey) with tit (light grey), graphite (lamellar, light brown) and py particle (white, bright), providing a reference for tit reflectance estimation. HD 773, Kropfmuehl, Passau, Germany (**dsp**)

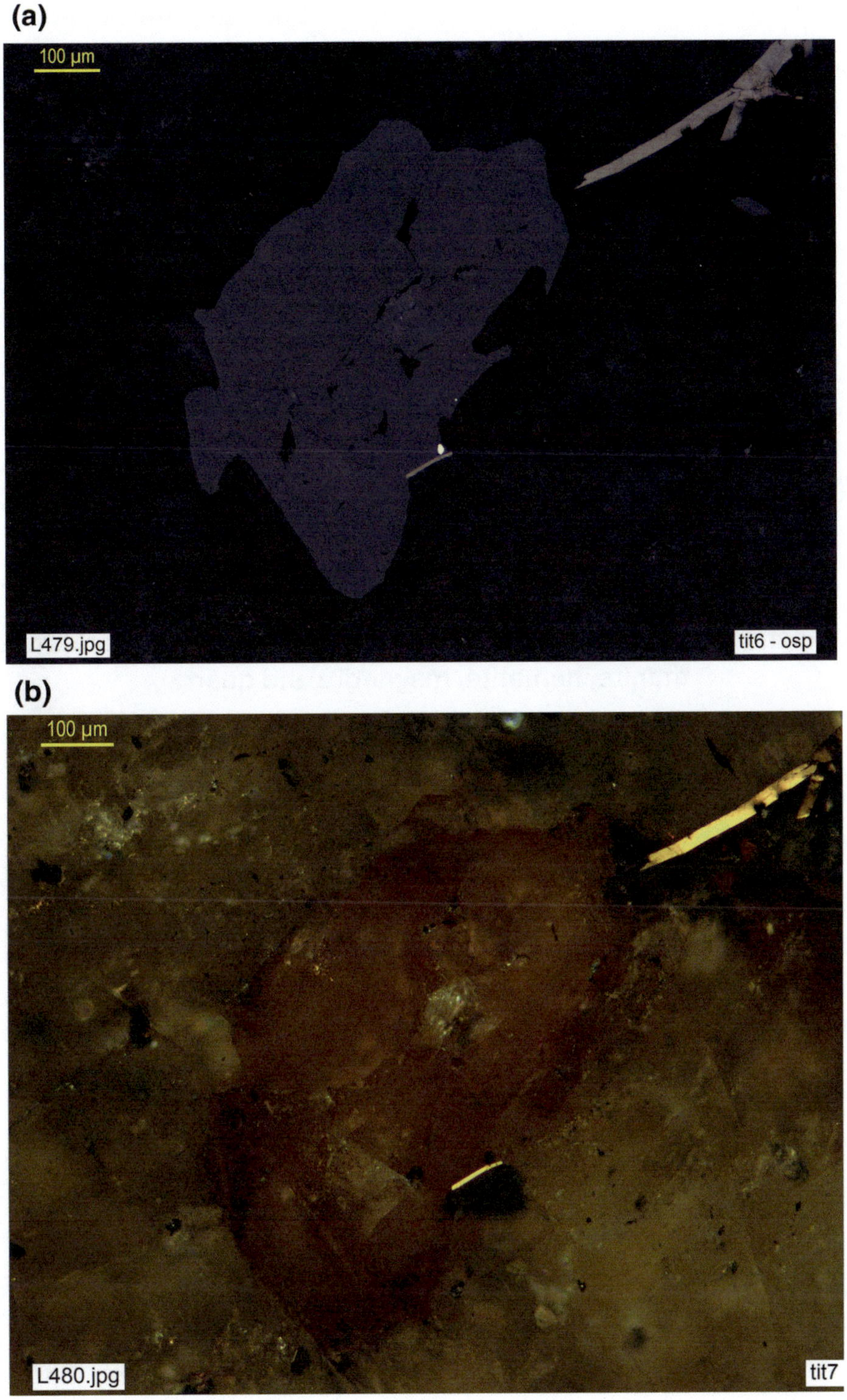

Fig. 125.3 **a**, **b** Same scene as Fig. 125.2: (**osp**) and (**o+p**)

125.2.4 Identification

It is easily recognized by its habit, its low reflectance, which results in an inconspicuous grey colour but noticeably brighter than gangue, its poorly coloured anisotropism and its abundant and luminous internal reflections. Association with other titaniferous species is an additional criterion.

125.3 Spectral Information

125.3.1 Titanite Multispectral Reflectance R (in Air, Without Polariser) (%)

λ (nm)	370	400	425	450	475	500	525	550	575	600	625	650	675	700	750	800	850	900	950	1000
μ	11.1	10.7	10.6	10.4	10.4	10.3	10.2	10.3	10.1	10.1	10.1	10.0	10.1	10.0	10.1	10.1	10.1	10.0	10.1	10.0
P_5	10.4	10.1	10.0	9.7	9.8	9.6	9.6	9.7	9.5	9.4	9.5	9.4	9.4	9.4	9.5	9.6	9.6	9.5	9.5	9.4
P_{20}	10.9	10.6	10.4	10.2	10.2	10.1	10.0	10.1	10.0	9.9	10.0	9.9	10.0	9.9	10.0	10.0	10.0	9.9	9.9	9.8
P_{80}	11.3	11.0	10.8	10.6	10.6	10.5	10.4	10.5	10.3	10.3	10.3	10.2	10.3	10.2	10.3	10.3	10.3	10.2	10.3	10.2
P_{95}	11.3	11.0	10.9	10.6	10.7	10.5	10.5	10.6	10.4	10.3	10.4	10.3	10.4	10.4	10.5	10.5	10.6	10.4	10.5	10.4

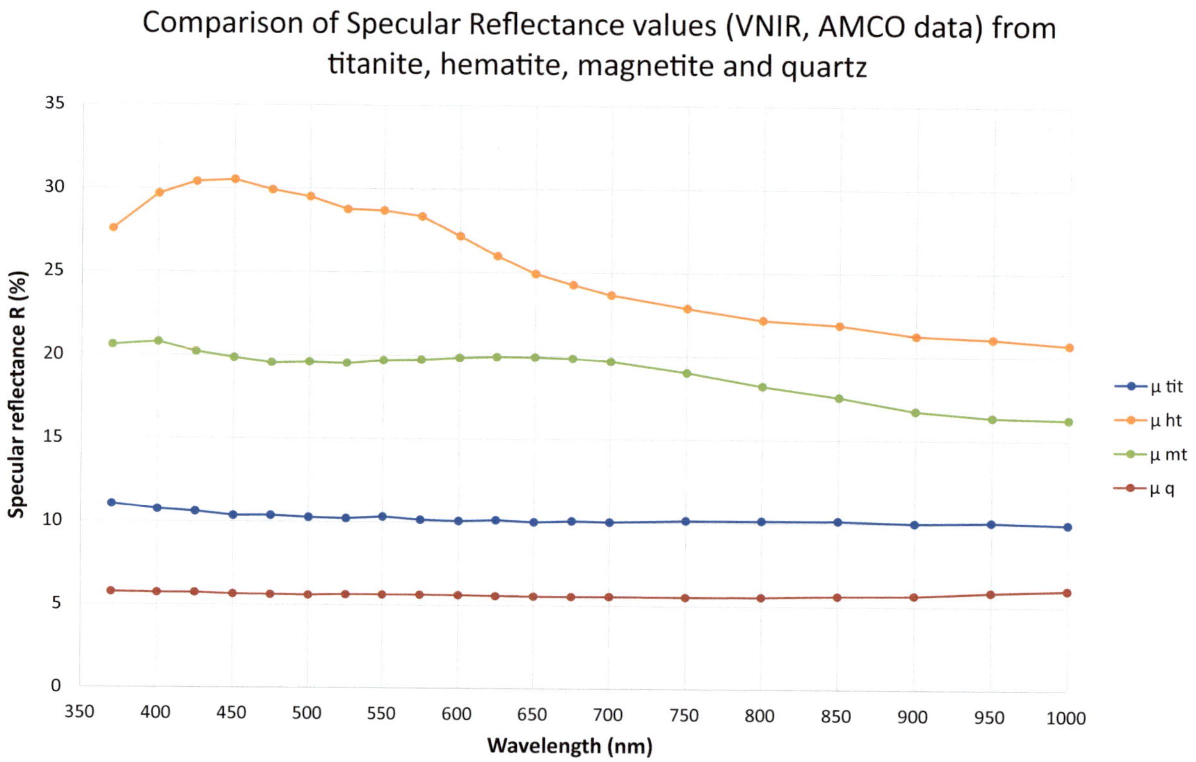

Comparison of Specular Reflectance VNIR values from titanite (AMCO), and titanite R1, titanite R2, rutile Ro & rutile Re' (QDF3)

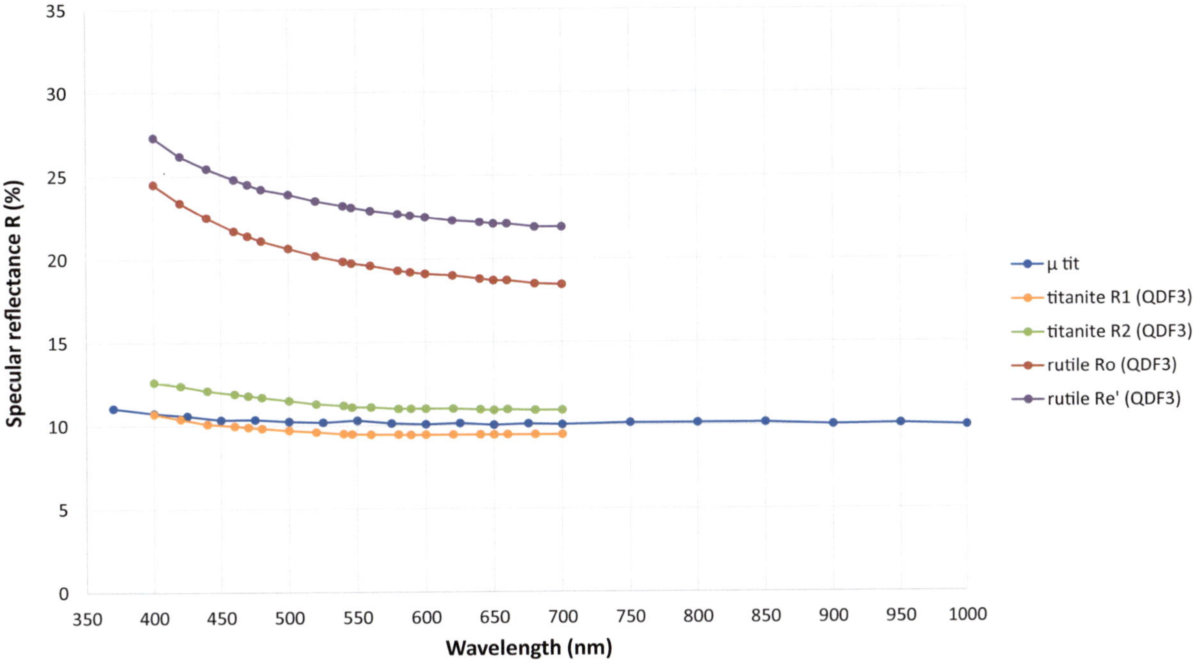

Comparison of Specular Reflectance VNIR values from titanite (AMCO), titanite R1, titanite R2, hematite Ro & hematite Re (QDF3), and cassiterite Rmax & cassiterite Rmin (Picot and Johan, 1982)

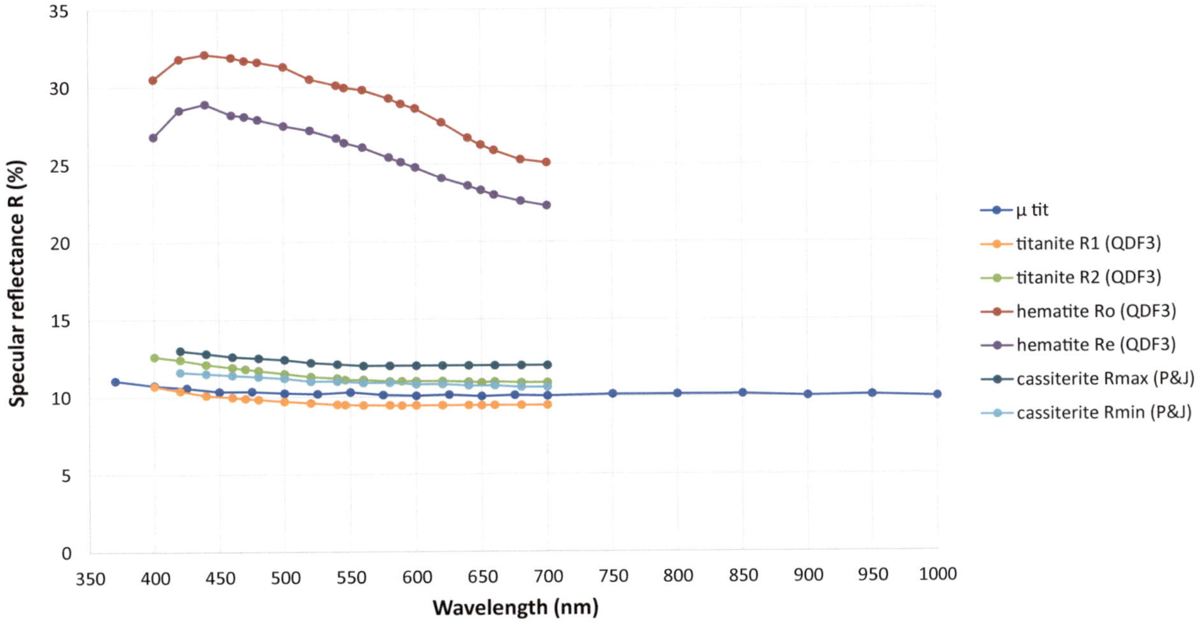

Ullmannite (*ul*)

126

NiSbS: Cubic

Ullmannite	R_{546} (%)	Y (%)	λ_d (nm)	P_e (%)	Colour sensation
Air	47.3	47.3	475	1.2	Bright white
Oil	32.8	32.8	472	2.3	Bright white, slightly bluish
Bir: Null (1)			*Pleoc*: Null (1)		*Tarn*: –
Anis +P: Null (1)			*IR*: No		*Twg*: –
Mor: Cubic idiomorphic, sometimes zon			*Clv*: // (100) visible		*PolQ*: Good
Hardness	*PolH*: High, > po, ≪ py (4)		*Mohs*: 5.5	*Talm*: E	*VHN* $_{(100\ g)}$: 592–627

Note: cf. comparative R spectra of ul and some ores of similar reflectance (visible range: QDF3; Picot and Johan 1982)
Mineral abbreviations: asp—arsenopyrite, bth—breithauptite, ccp—chalcopyrite, cob—cobaltite, gf—gersdorffite, gg—gangue, gn—galena, ln—linnaeite, lo—löllingite, nc—nickeline, po—pyrrhotite, py—pyrite, sk—skutterudite, sp—sphalerite, ul—ullmannite

126.1 General

Rare mineral. Composition: NiSbS, may contain, as impurities, Fe, Co, As, Bi. Higher As contents (*korynite*) mark a gradual transition to gersdorffite.

126.2 Microscopic Description

126.2.1 SP Observation (*Single Polariser*)

Bright white *colour*. Good *polish*, despite occasional pull-outs (*triangular pits*), due to its cubic cleavage (Fig. 126.1a, b), which also controls fracturing (Fig. 126.2a).

Relatively high *reflectance*, close to that of gersdorffite in the visible range (cf. R spectra compared: R ul > R gn, but < R cob, sk, py). Isotropic.

Relatively high *polishing hardness*, PolH ul > PolH cp and sp; PolH ul ≥ PolH gf, slightly higher than po, ln; PolH ul < PolH lo, asp ≪ PolH py.

Usually euhedral, typically in cubes, with distinct cleavage parallel to the cube faces // (100). Sometimes also // (110), rarely visible. Occasional zoning.

126.2.2 Observation +P (*Polariser + Analyser at 90°*)

Isotropic, net extinction (+P, Fig. 126.1c, d, imperfect extinction with wide field objective, Fig. 126.2b).

126.2.3 Paragenesis

It is found in some uncommon hydrothermal formations of magmatic derivation, with nickeline, breithauptite,

R. Castroviejo, *A Practical Guide to Ore Microscopy—Volume 1*,
https://doi.org/10.1007/978-3-031-12654-3_126

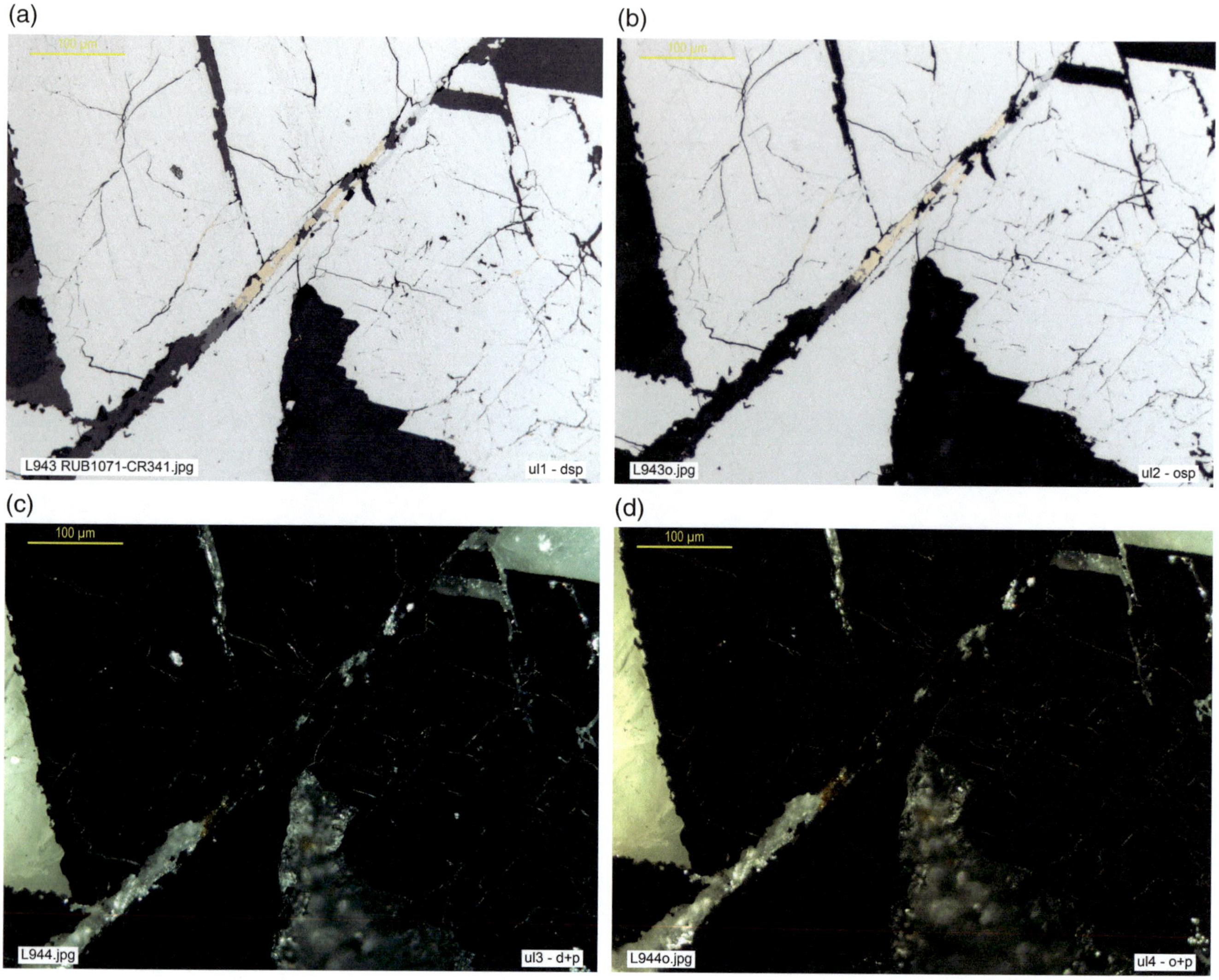

Fig. 126.1 **a**, **b** (Same scene: **dsp** and **osp**, respectively). Main ore ul (white), showing cleavage and microcracks cemented by ccp (yellow), gn (greyish white), sp (grey) and carbonate gg (dark grey); enhanced contrast and strong darkening of sp and gg in oil (Fig. 126.1b). 1071rub/CR341, Alte Petersbach Mine, Altenkirchen, Germany. **c**, **d** Same scene as ul1 (**d+p**) and (**o+p**), respectively. Isotropism of ul, gn and sp; IR of sp and gg; weak anisotropism of ccp

gersdorffite, skutterudite, linnaeite, chalcopyrite, galena, precious metals, Ag–Bi sulphosalts.

126.2.4 Identification

Its appearance is similar to that of gersdorffite, skutterudite, linnaeite, but careful observation can provide adequate identification criteria:

- Ul can be distinguished from skutterudite by the yellowish hue of ullmannite in direct comparison and by the higher reflectance of skutterudite, which also has a very visible zoning and does not show the cubic cleavage of ullmannite.
- The minerals of the linnaeite group are softer, usually somewhat coloured (e.g., with pinkish shades, as seen in spectral curves. R of linnaeite decreases for $\lambda < 600$ nm) and show octahedral forms.
- Ul distinction from gersdorffite can be more difficult, especially considering the existence of gradual transitions between the two, with intermediate compositions and properties, *korynite*, *Ni(Sb,As)S*.
- However, compared to ullmannite, gersdorffite is less reflective, shows a sharper cleavage and far more frequent *triangular pits*, i.e., pull-outs due to easier failures along cubic cleavages during polishing, and it often presents a weak anisotropism, not very striking but enough to make its complex twins visible. The association with Sb or As minerals is also a possible aid.

Fig. 126.2 **a**, **b** (Same image: **dsp** and **d+p**, respectively). Crackled ul ore, with microfractures favoured by ul cubic cleavage, cemented by carbonate and td traces. Figure 126.2b shows striking anisotropism of Carb; also, incomplete ul extinction, due to wide field objective (2.5x) aberrations. CR341, Alte Petersbach Mine, Altenkirchen, Germany

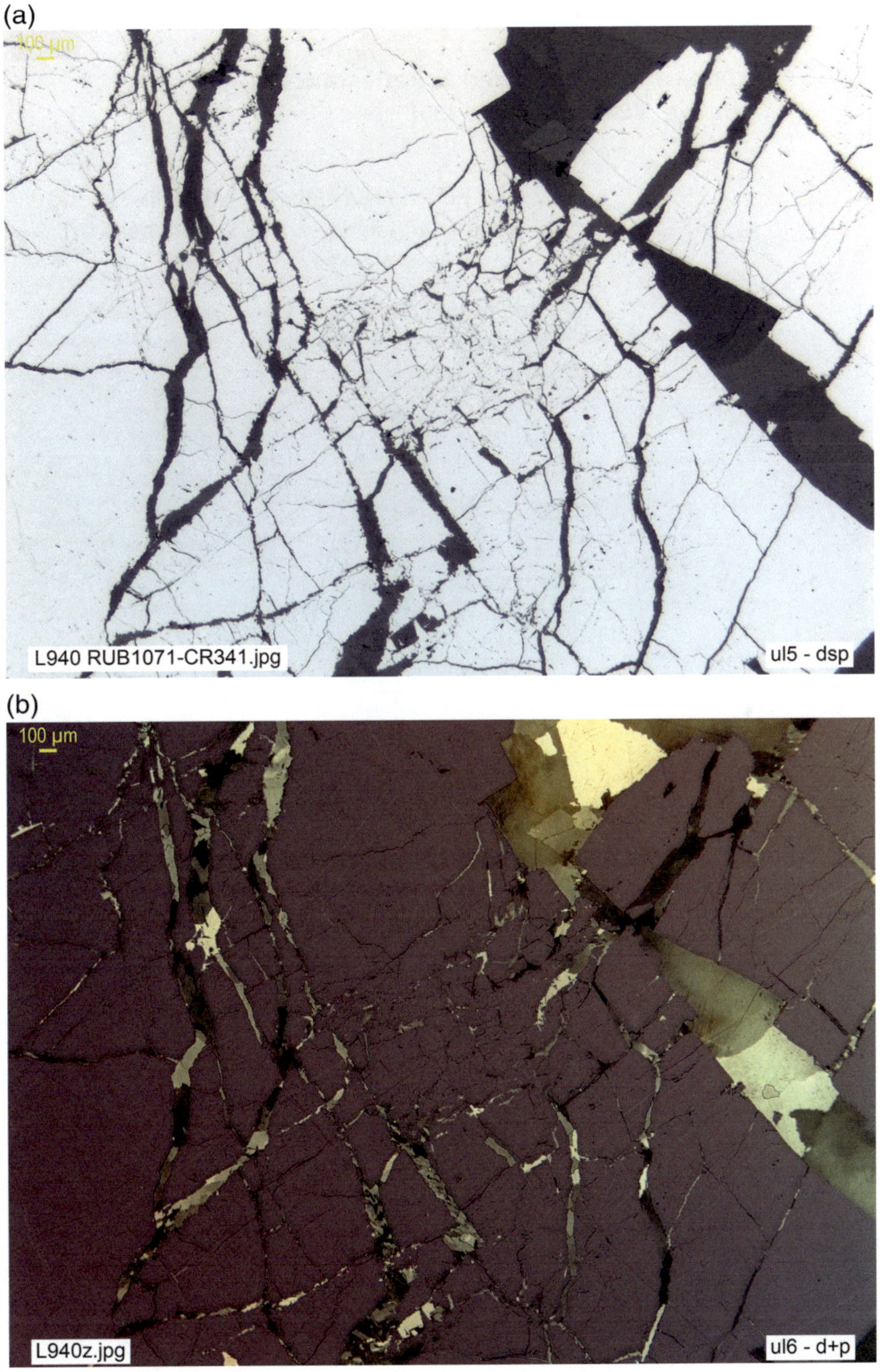

126.3 Spectral Information

126.3.1 Ullmannite Compared R Reflectance (in Air, Visible Range) (%)

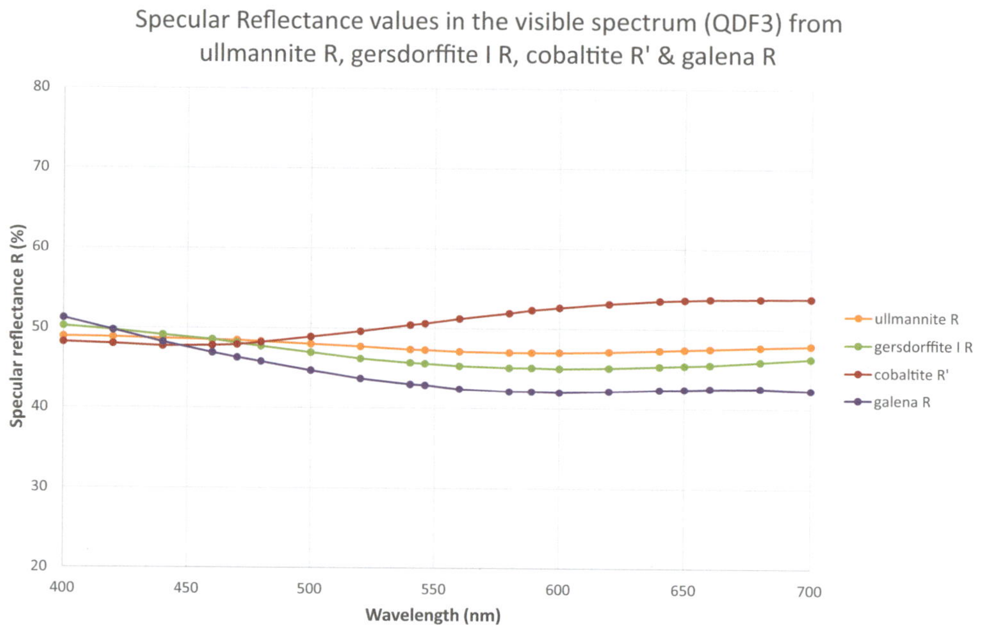

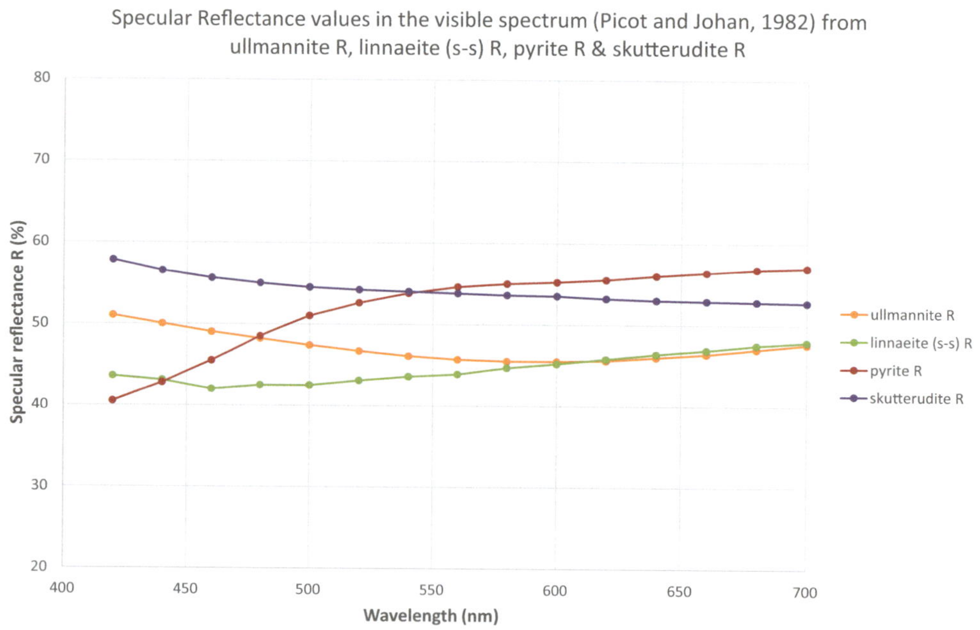

Reference

Picot P, Johan Z (1982) Atlas of ore minerals. B.R.G.M./Elsevier

Ulvöspinel (*ulv/ulvite*)

127

$TiFe_2O_4$: Cubic

Ulvöspinel		R_{546} (%)	Y (%)	λ_d (nm)	P_e (%)	Colour sensation
Air		15.4	15.7	c504	2.9	Greyish brown
Oil		4.5	4.7	c493	4.3	
Bir: Null (1)				*Pleoc*: No (1)		*Tarn*: No
Anis +P: Isotropic (1)				*IR*: No		*Twg*: nd
Mor: Tiny exsolutions in mt				*Clv*: nd		*PolQ*: Good
Hardness	*PolH*: High (4)			*Mohs*: ≈ 6	*Talm*: nd	*VHN* $_{(100\ g)}$: ≈ 650

Obs: *cf.* R charts (visible range data, QDF3). Hardness data according to Bowles et al. (2011)
Mineral abbreviations: chr—chromite, gg—gangue, il—ilmenite, isc—isocubanite, mt—magnetite, pn—pentlandite, po—pyrrhotite, rt—rutile, spn—spinel, ulv—ulvöspinel

127.1 General

Although it is an uncommon and poorly documented mineral, ulvöspinel, ***ulv***, will be briefly discussed, as its recognition may be of interest due to its possible interference in the beneficiation of titaniferous magnetite ores, given its variable magnetic properties. It is treated succinctly by Ramdohr (1980) and by Picot and Johan (1982). Bowles et al. (2011) provide more up-to-date information, although optical microscopy data are scarce.

Its composition is $TiFe_2O_4$, but it may contain appreciable amounts of Cr (up to 24% Cr_2O_3 in lunar rocks), Al and V, as well as minor amounts of Mg, Mn, Ca, Si and trace amounts of Os and Re (Bowles et al. 2011). Within the spinel supergroup and the oxyspinel group, it gives its name to the *ulvöspinel subgroup*.[1]

In terrestrial rocks, it is practically never found as a primary mineral (of direct magmatic precipitation) but in fine exsolutions in titaniferous magnetite. As a rare exception, Ramdohr (1980) cites certain basalts with native Fe and graphite. It is typical of reducing conditions, hence the sporadic occurrence of spherulitic graphite (*Retortengraphit*) in titanomagnetite ores with unmixed ulvöspinel and the rapid oxidation of the latter to ilmenite under oxidizing conditions.

127.2 Microscopic Description

127.2.1 SP Observation (*Single Polariser*)

Grey or brownish *colour*, similar to that of ilmenite oriented $//R_e$, but unlike ilmenite it is not pleochroic (Figs. 127.1a, b, 127.2a, 127.3, 127.4 and 127.6). Low *reflectance*, clearly lower than magnetite, but higher than spinel. Its spectral curve is practically coincident with R_e of ilmenite (cf. comparative graphs in the visible spectrum, QDF3). However, given its natural compositional variation, these data must be seen with caution, especially considering that the QDF3 spectrum has been measured on synthetic material only, not supposed to show such compositional variations. These perhaps cause the slight differences between the values measured by Picot and Johan (1982) and those of QDF3 (cf. comparative ulvöspinel plots). They may also influence the colour, described as brownish by Picot and Johan (1982) but "brownish or grey" by Bowles et al. (2011). The sparsity

[1] *cf* introduction to description of spinel, *spn*; also, https://www.mindat.org/min-52868.html.

R. Castroviejo, *A Practical Guide to Ore Microscopy—Volume 1*,
https://doi.org/10.1007/978-3-031-12654-3_127

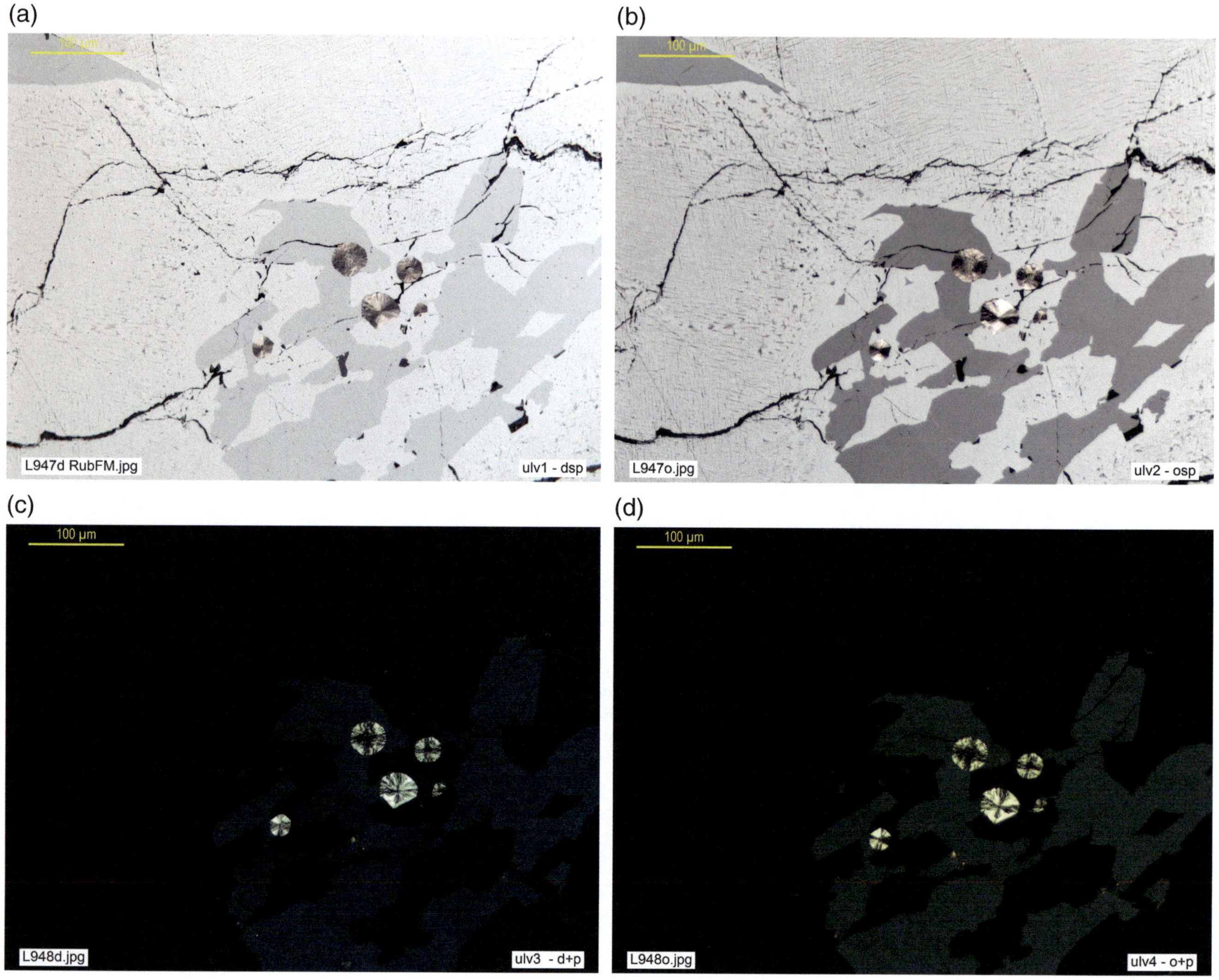

Fig. 127.1 **a, b** (**dsp** and **osp**, respectively)—Aggregate of titaniferous magnetite (light brownish grey) and ilmenite (Mg/Mn-rich, grey), with exsolution of ulvöspinel in titanomagnetite and spherulitic graphite inclusions. RubFM, Foskor Mine, Palabora, South Africa. **c, d** Same image as ulv1 (**d+p** and **o+p**, respectively). Anisotropism of ilmenite and graphite, which shows spherulitic structure

and the fine grain size of natural terrestrial samples do not facilitate accurate measurements but, according to Halfen (1978), R increases continuously at all wavelengths as the Ti content decreases. As far as Cr-rich lunar ulvöspinels are concerned, their colour is described by Bowles et al. (2011) as "cinnamon" (*tan*), pink or brownish.

It occurs associated with titaniferous magnetite (±ilmenite, spinel), in minute exsolutions (Figs. 127.2a and 127.4) usually oriented // 100 of magnetite, appearing as a network of orthogonal fibers that resembles a mesh or a cloth —the *cloth texture* of Ramdohr.

127.2.2 Observation +P (*Polariser + Analyser 90°*)

Isotropic: remains in extinction and has no *internal reflections* (Figs. 127.1c, d, 127.2b, 127.5a, b, 127.7).

127.2.3 Paragenesis

Other than moon rocks and meteorites, the known occurrences of ulvöspinel in terrestrial rocks are in mafic rocks and related magmatic Ti-Fe or gabbro-anorthosite deposits,

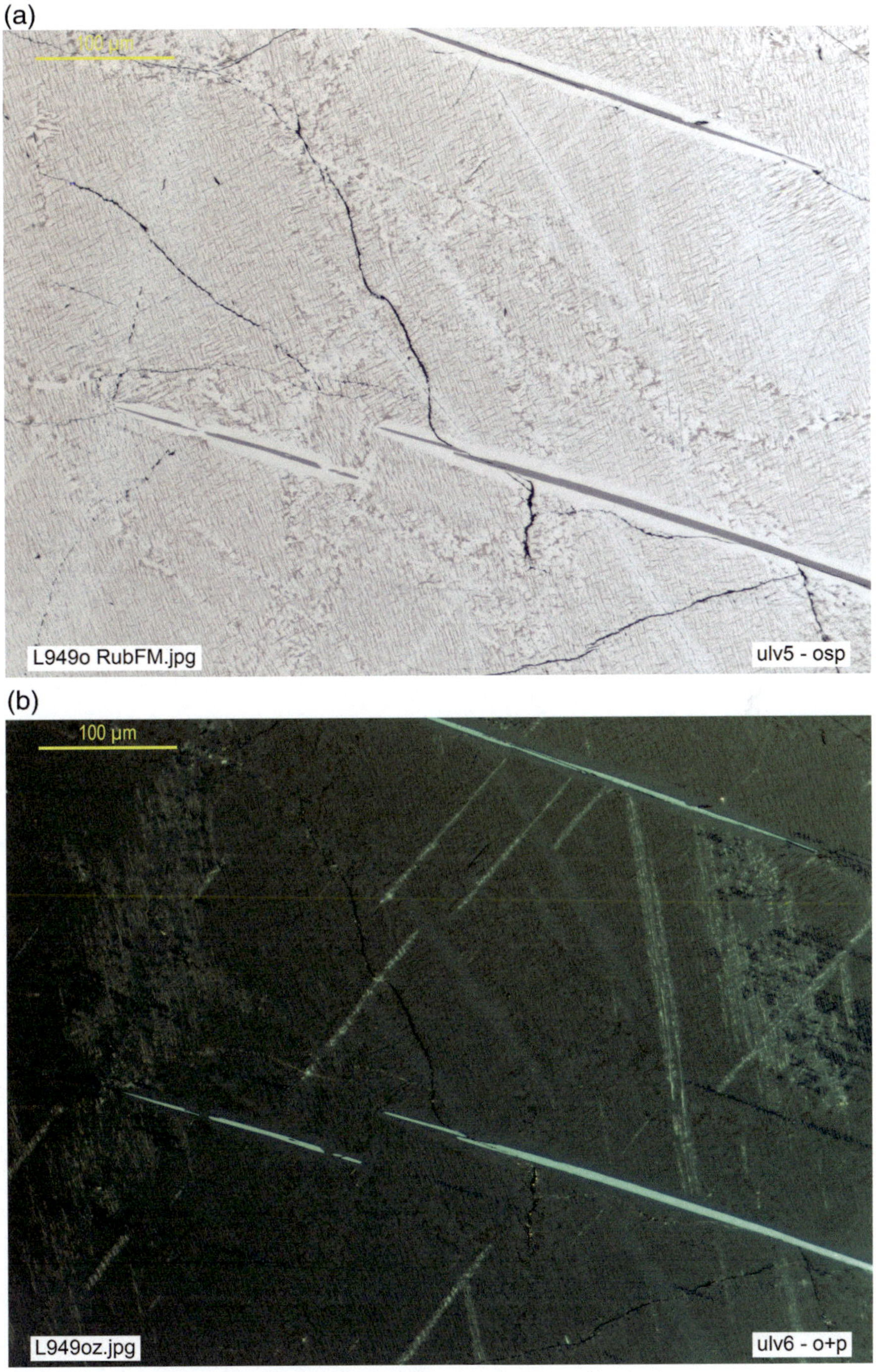

Fig. 127.2 **a, b** Titanomagnetite with fine, brownish, isotropic exsolutions of ulvöspinel //100 (displaying the typical *cloth-like* texture) and long, coarser, spindle-like //111 exsolutions of ilmenite (purplish grey, anisotropic). RubFM, Foskor Mine, Palabora, South Africa (**osp** and **o+p**, respectively)

where ulv is found as finely intergrown exsolutions within titaniferous magnetite (±ilmenite). Ramdohr (1980) suggests that ulvöspinel is probably the original exsolution in most magmatic titanomagnetites and then commonly transforms into ilmenite. This would explain the common // (100) cloth-like pattern of ilmenite exsolution textures in Ti-magnetite.

127.2.4 Identification

It is not frequent in common ores, but it can be found in titanomagnetite ores. Therefore, paragenesis is an important criterion for its recognition. Its scarcity and usual very fine grain size means that its optical properties are poorly documented, but its isotropy, its low reflectance and, above all, its exsolution texture associated with Ti-magnetite allow it to be identified.

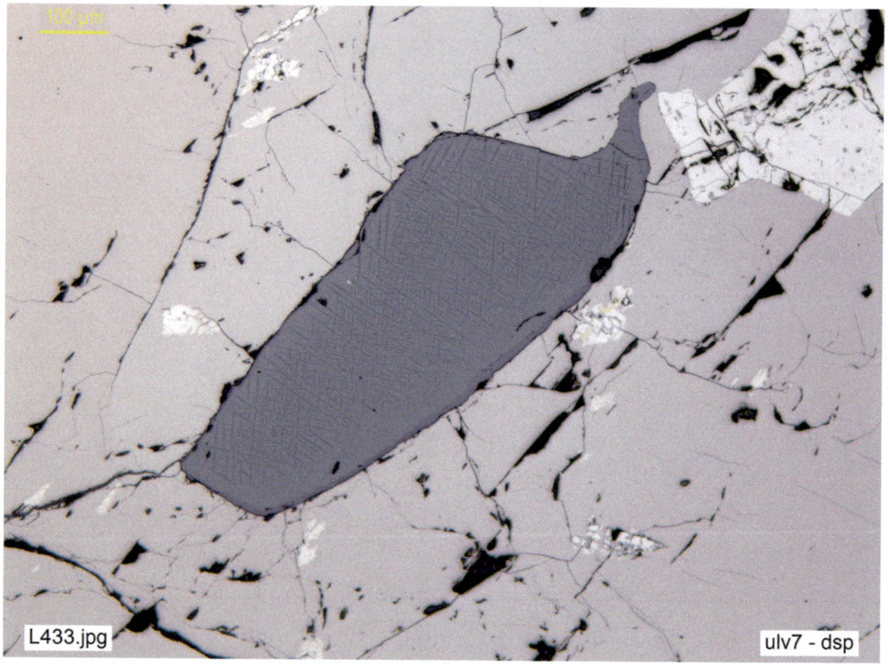

Fig. 127.3 Inclusion of titaniferous magnetite (grey) in Ni-sulphide ore, comprising po +isc (pinkish brown) and pn (white). Fine *corona* of il rimming mt and oriented exsolutions of ulv (//100) and il (acicular, very fine) on mt. HD 6896, Sudbury, Ont., Canada (**dsp**)

Fig. 127.4 Detail of central zone of Fig. 127.3 (**dsp**): next to po/isc, the brownish tones of mt, il and ulv are practically not visible. They look grey, with a very slight violet (il) or brownish shade (mt and ulv). Of the three oxides, ulv is the darkest.

The occasional presence of graphite suggests a reducing environment. Ulvöspinel is altered to ilmenite even in slightly oxidizing conditions. Unmixed ulvöespinel can be located in a *box* of lamellar exsolutions of ilmenite or spinel (Bowles et al. 2011). The ilmenite secondary to ulvöspinel can be distinguished by its extinction, oblique to the length of the lamella.

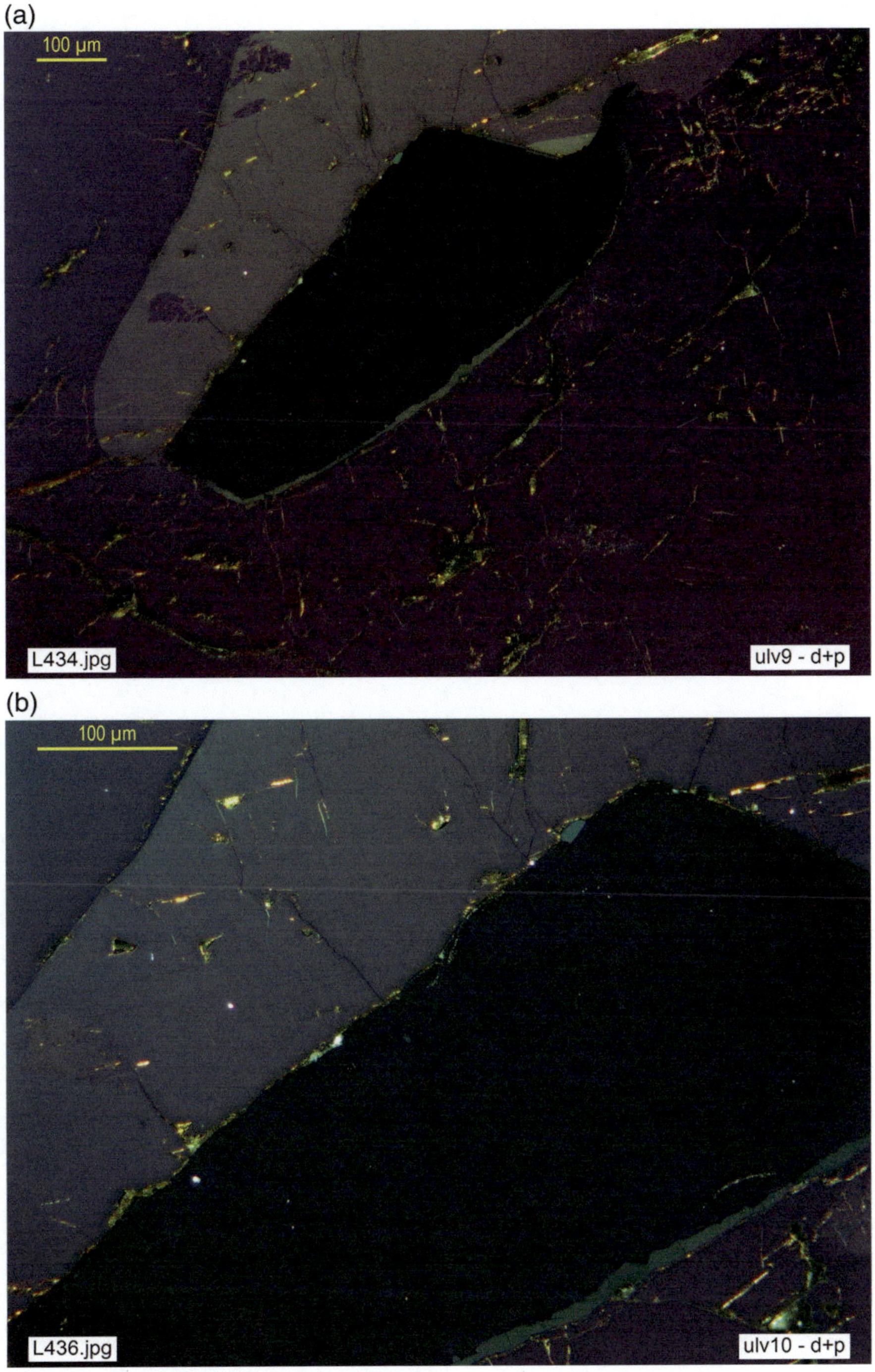

Fig. 127.5 **a, b** Same scene as Figs. 127.3 and 127.4, respectively (both d+p). Anisotropism of po and il distinguishes them from isc and mt, respectively; pn and ulv remain extinct

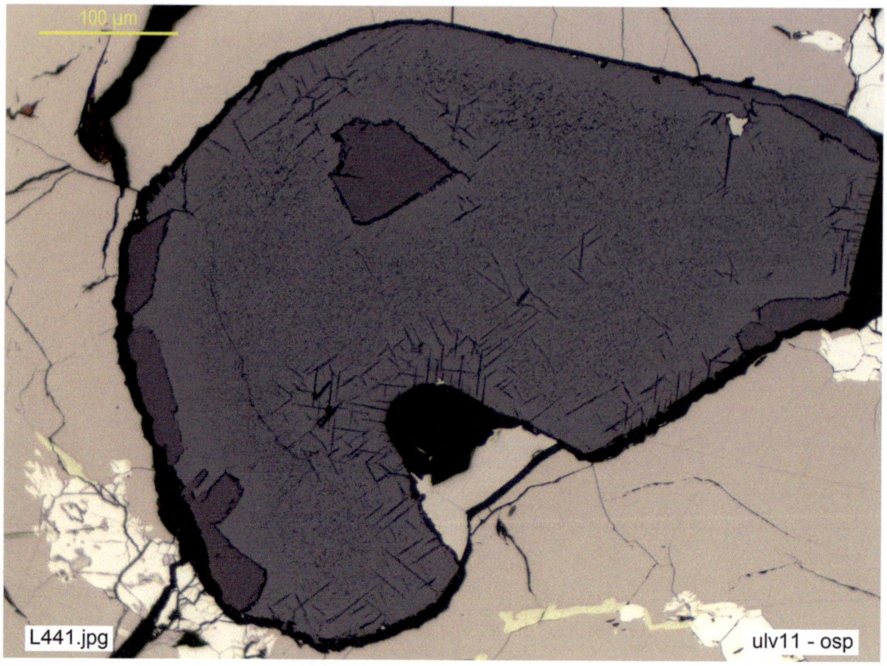

Fig. 127.6 Inclusion of titaniferous mt and il in Cu–Ni sulphide ore (po, isc, ccp, pn), with ulv (*cloth texture*) and acicular il exsolutions in mt. HD 6896, Sudbury, Ont., Canada (**osp**)

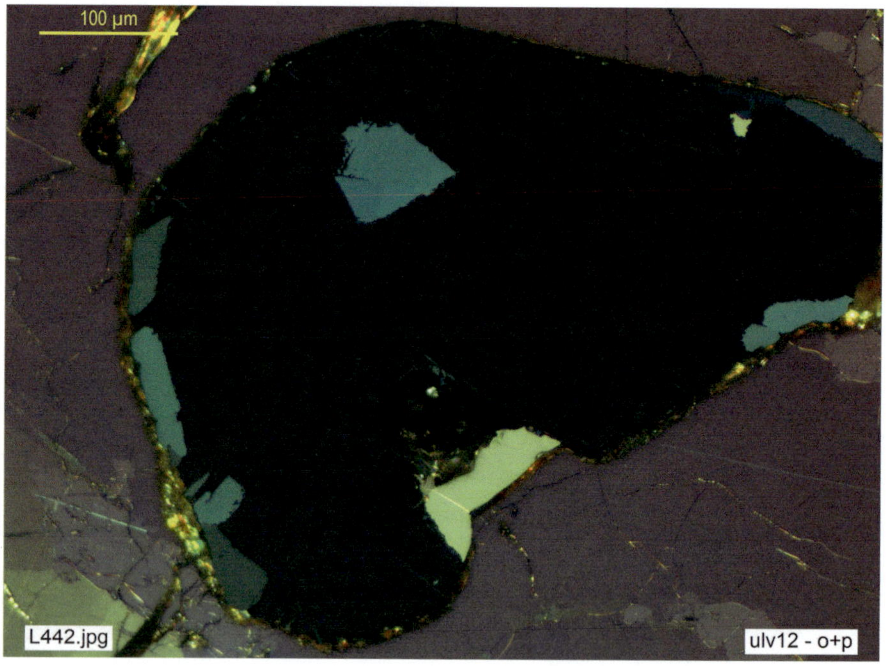

Fig. 127.7 Same scene as Fig. 127.6 (**o+p**). The anisotropism of po (bottom left corner) and il (greenish tones) clearly distinguishes them from isc (dominant) and from mt and ulv, resp

127.3 Spectral Information

127.3.1 Ulvöspinel Reflectance R (in Air, Visible Spectrum) (%)

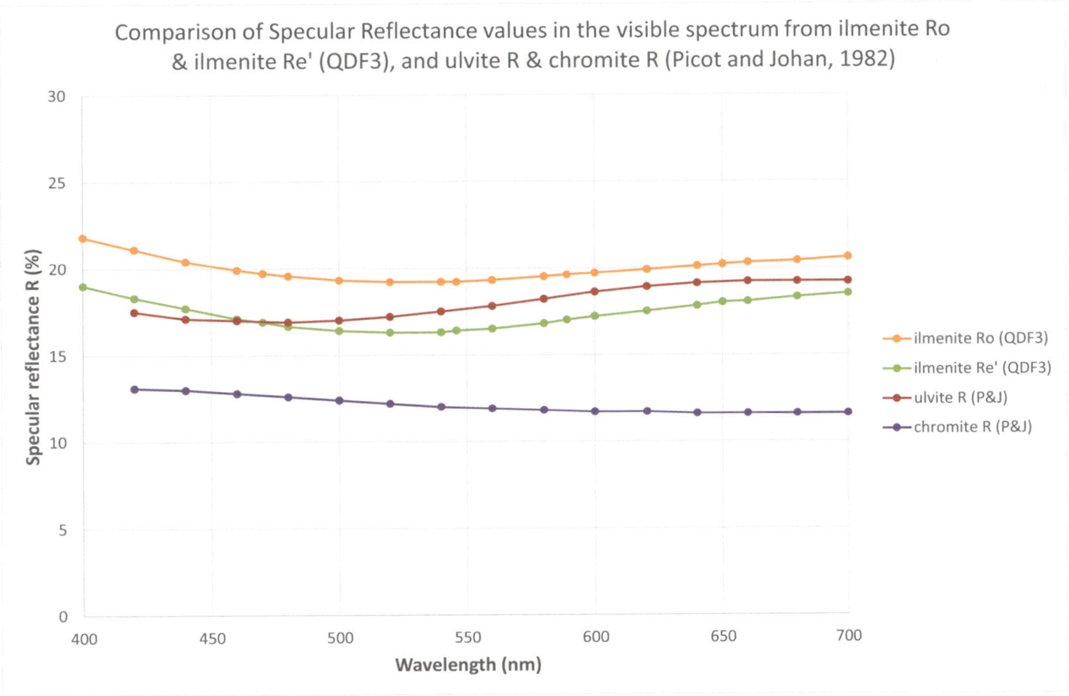

References

Bowles JFW, Howie RA, Vaughan DJ, Zussman J (2011) Rock-forming Minerals, vol. 5A (2nd ed) Non-silicates: Oxides, Hydroxides and Sulphides. The Geological Society, London, 920p

Halfen B (1978) Relations between the spectral reflectance and the chemical composition in the magnetite-ulvöspinel series. Min Mag 42:265–269

Picot P, Johan Z (1982) Atlas of ore minerals. B.R.G.M./Elsevier

Ramdohr P (1980) The ore minerals and their intergrowths, vol 2 (2nd edn) Pergamon Press, Oxford, 1205p

Umangite (*um*)

128

Cu_3Se_2: Tetragonal

Umangite			R_{546} (%)	Y (%)	λ_d (nm)	P_e (%)	Colour sensation
Air	R_o		12.7	14.0	c563	16.1	Purplish- or violet-brown
	R_e'		15.9	16.2	466	10.1	Greyish brown
Oil	R_o		4.0	5.4	c513	27.1	Reddish- to orange-brown
	R_e'		5.3	5.7	c564	13.7	Olive grey
Bir: Medium (4)					*Pleoc*: Strong (5)		*Tarn*: On polished surface, very slow
Anis +P: Strong (5)					*IR*: No		*Twg*: Thin polysynthetic
Mor: Allotriomorphic aggregates					*Clv*: Occasional		*PolQ*: Good
Hardness		*PolH*: Low, $\leq$ clt (2)			*Mohs*: 3	*Talm*: B	*VHN* $_{(100\ g)}$: 88–100

Note: cf. Table and R charts, visible range data (QDF3)

Mineral abbreviations: agl—aguilarite, az—azurite, bm—bismuthinite, bn—bornite, cal—calcite, cc—chalcocite, ccp—chalcopyrite, clt—clausthalite, cv—covellite, el—electrum, fse—ferroselite, gg—gangue, gn—galena, mch—malachite, my—miargyrite, nau—naumannite, py—pyrite, q—quartz, um—umangite, urn—uraninite

128.1 General

Composition: Cu_3Se_2, with possible substitution of Cu by Ag in small proportion.

128.2 Microscopic Description

128.2.1 SP Observation (*Single Polariser*)

Violet-brown or purplish-brown to greyish-brown *colours*. *Strongly pleochroic* (Figs. 128.1a and 128.2a), with more striking tones (copper-like or orange and olive hues) in immersion (Figs. 128.1b and 128.3a). The colour in the R_o orientation is reminiscent of bornite. The bireflectance is moderate, contrasting with the strong pleochroism. This corresponds to the particular shape of the R_o and R_e' reflectance spectral curves, which intersect (as do those of covellite, another mineral of peculiar pleochroism). Moderate to low *reflectance*, clearly lower than clausthalite, ferroselite (Fig. 128.3a) or chalcopyrite, similar to bornite (cf. comparative R graphs).

Good *polishing*. As noted by Ramdohr (1982), in contrast to the rapid darkening of fracture surfaces in hand specimen, in polished section tarnishing is much slower. Low *polishing hardness*, PolH um $\approx$ klockmannite and > my; $\leq$ clt, < gn, cc or bm.

It occurs in masses and fine aggregates, almost always xenomorphic.

128.2.2 Observation +P (*Polariser + Analyser at 90°*)

Strongly *anisotropic*, with bright and striking colours (luminous, orange-red to yellowish-brown tones), particularly vivid in immersion (Figs. 128.1c, d, 128.2b and 128.3b).

Frequent polysynthetic *twins*, in several directions (Figs. 128.1c, d, 128.2b and 128.3b).

R. Castroviejo, *A Practical Guide to Ore Microscopy—Volume 1*,
https://doi.org/10.1007/978-3-031-12654-3_128

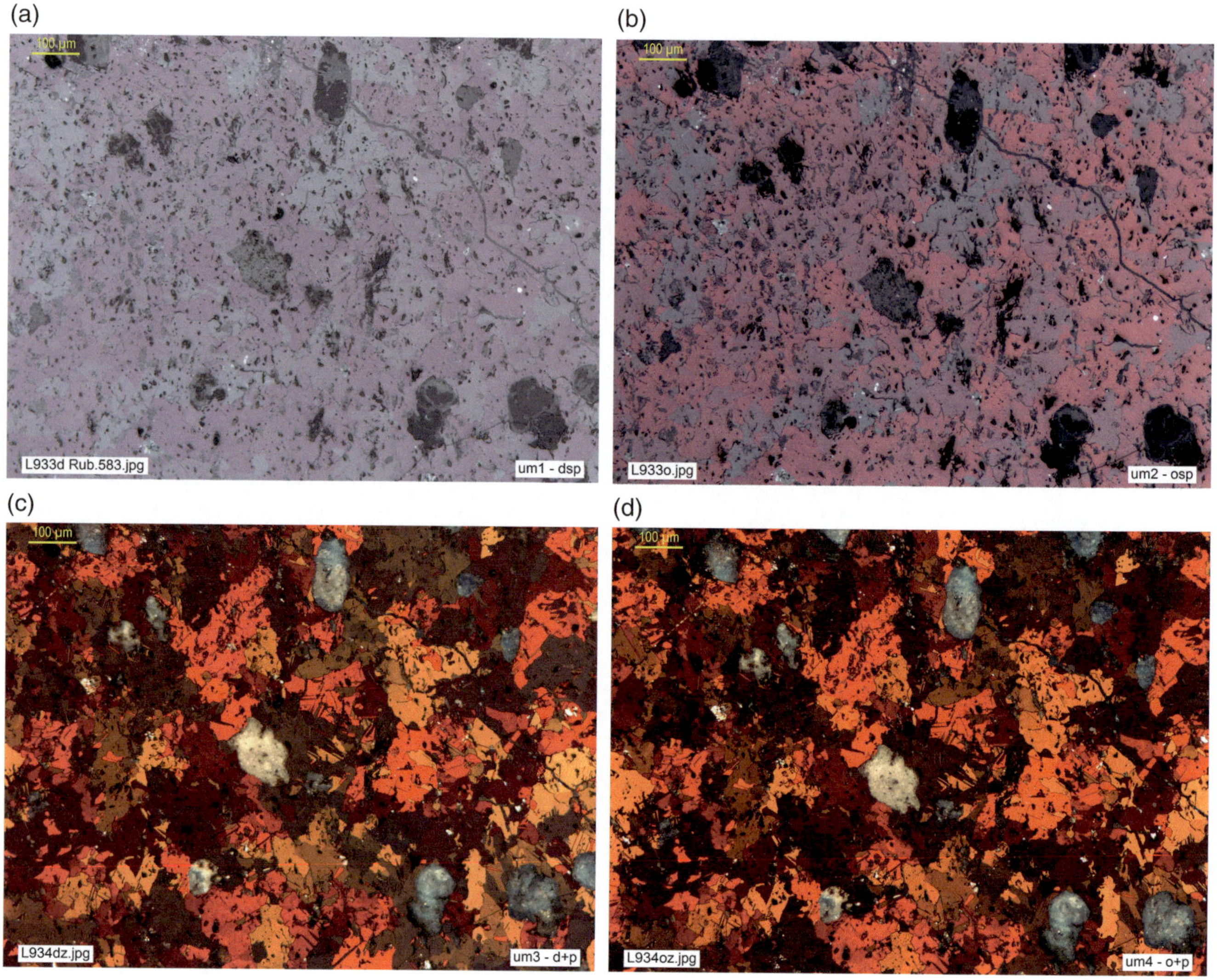

Fig. 128.1 **a, b** (**dsp** and **osp**, respectively)—Um aggregate, bireflective and strongly pleochroic (especially in immersion, Fig. 128.1b) in greyish and violet-brown to purplish shades, with traces of disseminated clt (white, very bright, tiny crystallites $\leq$ 10 µm); pores filled with gg (q, likely az). Rub.583, Mina San Francisco, La Rioja, Argentina. **c, d** Same scene as um1 (**d+p** and **o+p**, respectively), displaying typical um anisotropism (strong with orange-red polarisation tints, particularly vivid in immersion) and polysynthetic twins

128.2.3 Paragenesis

Rare mineral, um occurs in certain hydrothermal veins with selenide *parageneses*, in some of which it is abundant (Sª de Umango, Argentina). It has also been found in abundance associated with certain uraniferous deposits (Canada). Companions: clausthalite, ferroselite and other selenides, uraninite or pitchblende, calcite, azurite, malachite, etc.

128.2.4 Identification

It is a mineral whose identification offers no doubt to the ore microscopist, as its optical properties are unmistakable: low reflectance but very striking colour and pleochroism with orange or purplish, copper-like tones, vivid anisotropism (orange-red to brownish-yellowish). The twins and paragenesis are additional data.

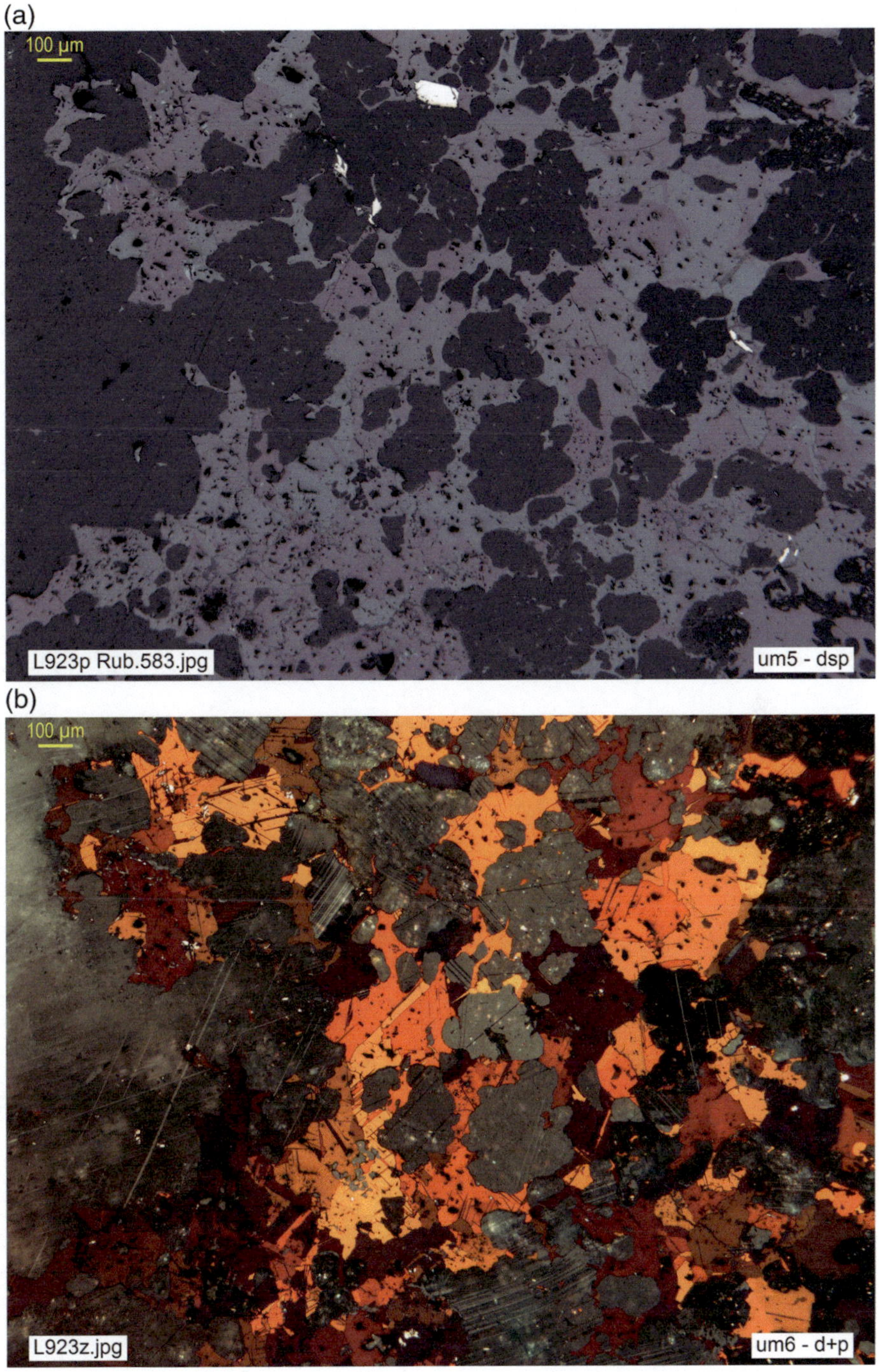

Fig. 128.2 **a, b** (same scene: **dsp** and **d+p**, respectively)—Aggregate of anhedral um and carbonate (grey), both with polysynthetic twins; fse (yellowish-white) disseminated rhombic or prismatic tiny crystals. Rub.583, Mina San Francisco, La Rioja, Argentina

Fig. 128.3 **a, b** (same scene: osp and o+p, respectively)—Massive allotriomorphic aggregate of um, crossed by healed microfracture cemented by *clt* (white, bright, isotropic) and *fse* (bright but less than clt, automorphic, yellowish-white, pleochroic in light yellowish-brown to greyish tones, anisotropic with orange to bluish-grey tones and twinned), with quartz and carbonate gangue. Rub.583, Mina San Francisco, La Rioja, Argentina

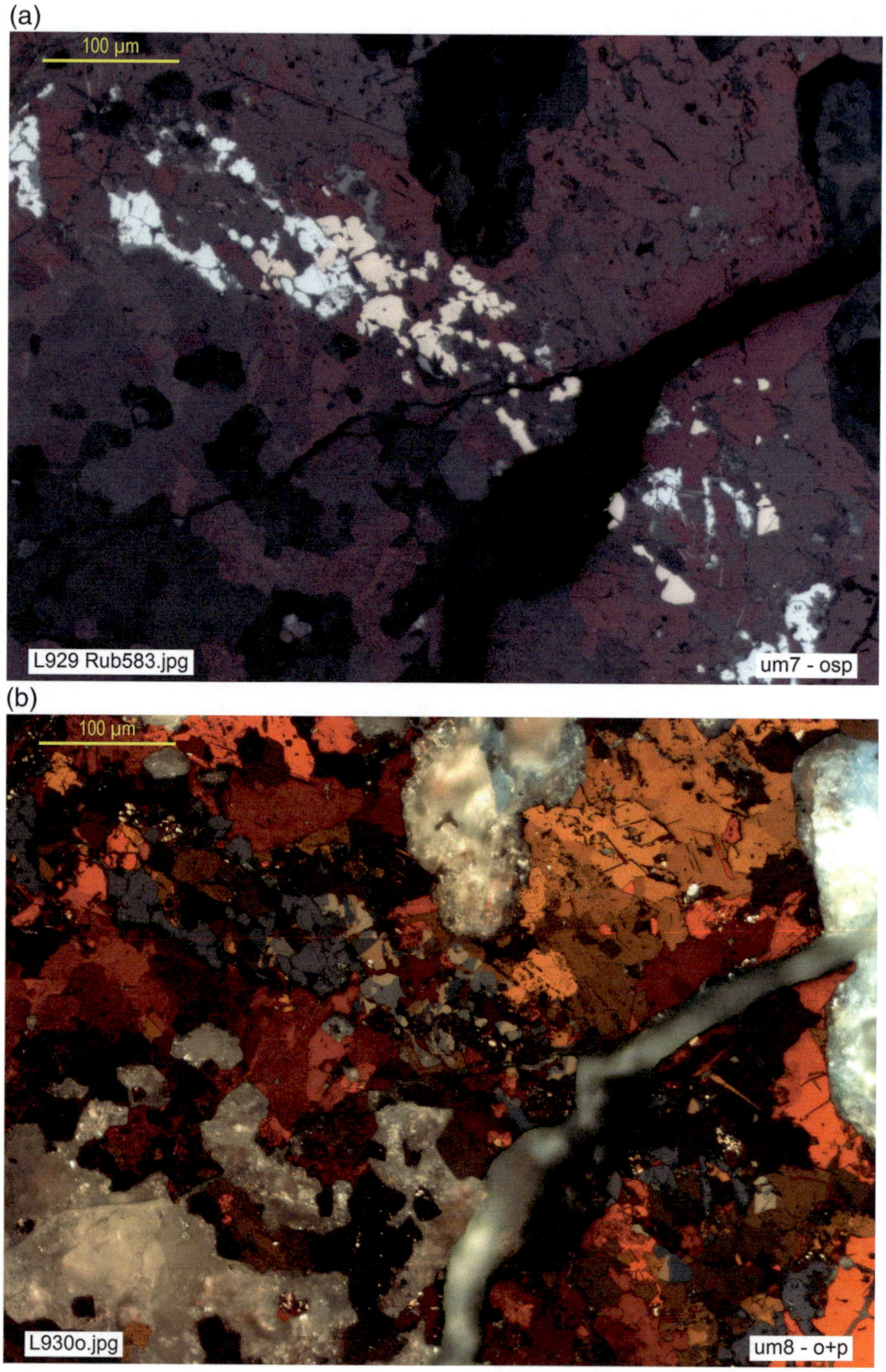

128.3 Spectral Information

128.3.1 Umangite Reflectance R (in Air, Visible Spectrum) (%)

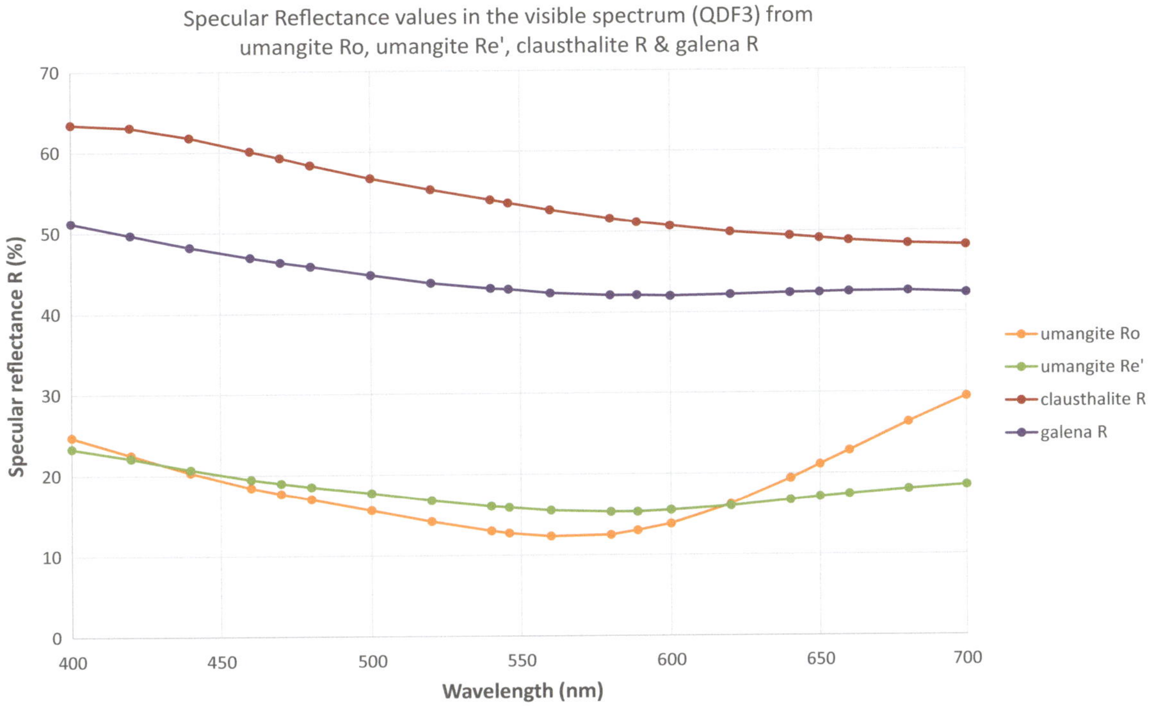

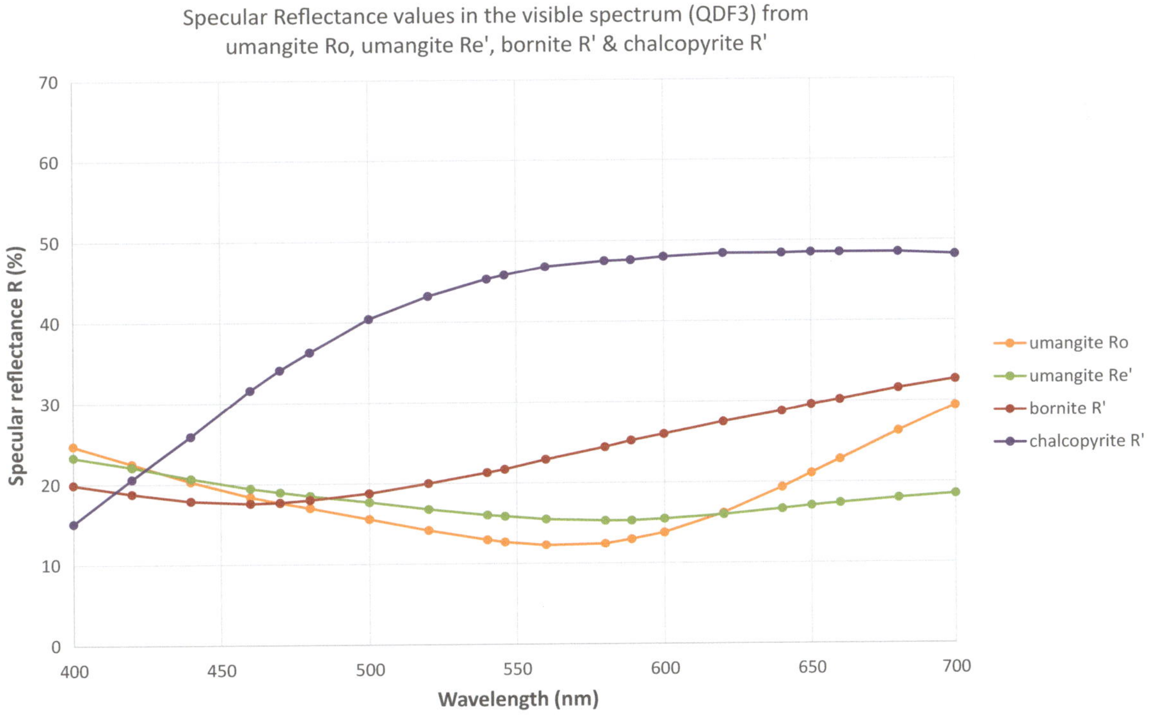

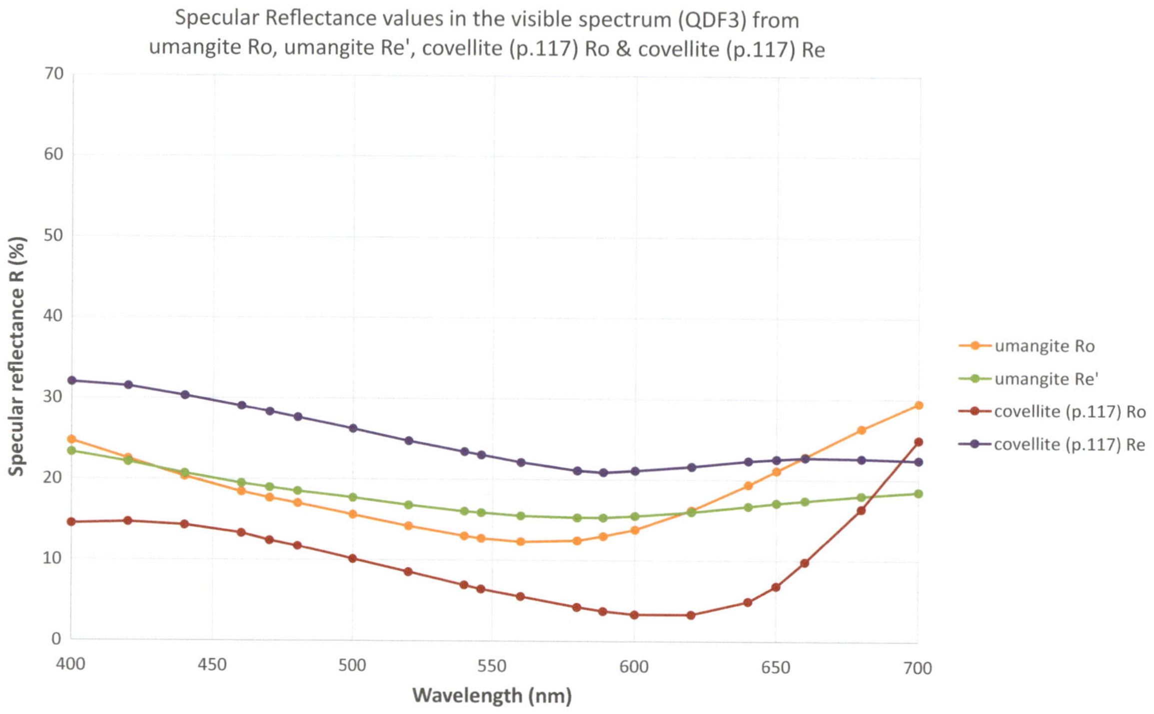
Specular Reflectance values in the visible spectrum (QDF3) from
umangite Ro, umangite Re', covellite (p.117) Ro & covellite (p.117) Re
Specular reflectance R (%)
70
60
50
40
30
20
10
0
400
450
500
550
600
650
700
Wavelength (nm)
umangite Ro
umangite Re'
covellite (p.117) Ro
covellite (p.117) Re

Uraninite (*urn/pitchblende*)

129

UO_2: Cubic

Uraninite	R_{546} (%)		Y (%)	λ_d (nm)	P_e (%)	Colour sensation
Air	13.6		13.7	469	2.9	Grey
Oil	nd		nd	nd	nd	Dark grey
Bir: Null (1)				*Pleoc*: No (1)		*Tarn*: Oxidation
Anis +P: No (1)				*IR*: Very dark brown		*Twg*: Rare (111)
Mor: Massive, x-aggregates, spheroids				*Clv*: Cubic		*PolQ*: Variable
Hardness	*PolH*: High: >chr, ht; < py, q (4)			*Mohs*: 5–6	*Talm*: G	*VHN* $_{(50\ g)}$: 499–548

Obs: cf. Table and R charts, multispectral data (VNIR, AMCO) and visible range (QDF3)
Mineral abbreviations: Ag—native silver, Au—native gold, Bi—native bismuth, chr—chromite, gf—gersdorffite, gg—gangue, gn—galena, lgs—langisite, ln—linnaeite, lo—löllingite, mc—marcasite, mt—magnetite, nc—nickeline, py—pyrite, sf—safflorite, sk—skutterudite, sp—sphalerite, urn—uraninite

129.1 General

Composition: UO_2. It may contain radiogenic Pb (or galena derived from it), Th, Ce and, in the colomorphic varieties, numerous adsorbed elements. By incorporation of ThO_2 it forms the solid solution continuous series thorianite—uraninite (ThO_2–UO_2).

It may occur crystallised but is also frequently colomorph. The first variety, referred to as *uraninite, urn*, is typically idiomorphic and presents hard cubic shapes. The colomorphic varieties (*pitchblende*), characterized by their botryoidal morphology, are often oxidized (up to almost UO_3), softer, even friable and, recovering Kobell's terminology, have been redefined as *nasturan* (Soboleva and Pudovkina, *in:* Ramdohr 1980, p. 1050) and classified (*nasturan I to* IV) after their degree of oxidation and their reflectances, which decrease as oxidation increases (from $UO_{2.33}$ to $UO_{2.92}$ for the *nasturan II to IV* stages of alteration).

129.2 Microscopic Description

129.2.1 SP Observation (*Single Polariser*)

Grey *colour* (Figs. 129.1a, b, 129.2a, 129.3); it is very dark against white arsenides, pyrite and especially native gold and native silver. *Polishing quality* varies according to the degree of oxidation but is normally good.

High *polishing hardness*, if fresh uraninite: PolH urn > chr, ht or accompanying ores, only lower than very hard minerals: PolH urn <py, q. However, decomposed or oxidized samples can be considerably softer and polish worse,

R. Castroviejo, *A Practical Guide to Ore Microscopy—Volume 1*,
https://doi.org/10.1007/978-3-031-12654-3_129

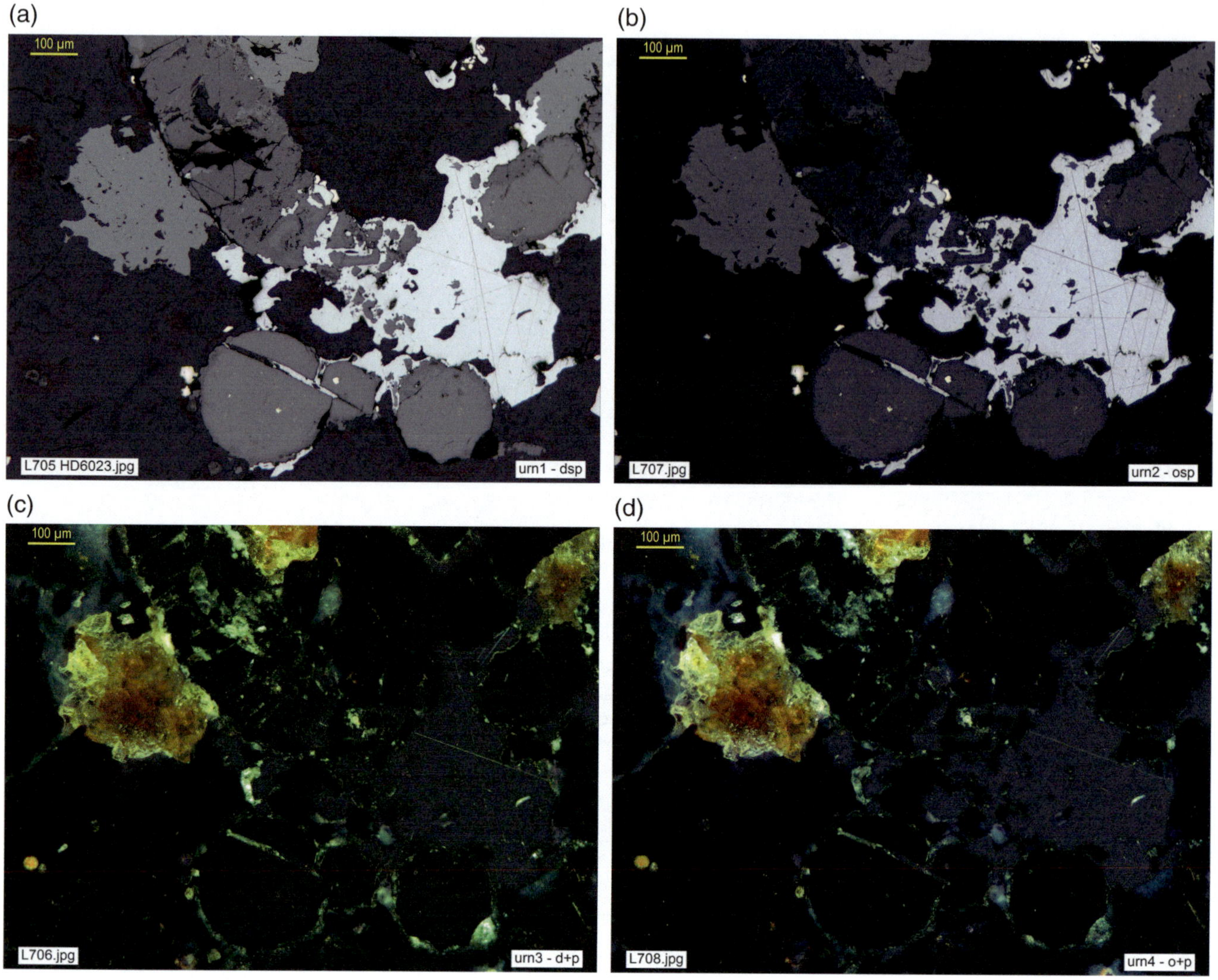

Fig. 129.1 **a, b** (**dsp** and **osp**, respectively)—Ore of urn (grey) brecciated and cemented-corroded by gn (white) and sp (light grey), small py inclusions in fluorite gg (dark grey, black in immersion) and in urn. Spheroidal pitchblende (bottom) and partially oxidized urn band (darker, top centre-left). HD 6023, Wolsendorf, Bavaria, Germany. **c, d** Same scene as urn1 (**d+p** and **o+p**, respectively): urn, gn, sp, py isotropic (extinct), sp contrasts with urn by its abundant and luminous internal reflections, IR, which are scarce and dark in urn

according to the cited authors: PolH urn > mt and sp, for *nasturan I* and *II* but PolH urn <mt for *nasturan III* and *IV*. For the same reason, the *Talmage Hardness* varies from G to C.

Reflectance low, only slightly higher than chromite. R urn < sp, mt, gn, py (*cf* multispectral comparison plots); R decreases by oxidation (Figs. 129.1a, b, 129.2a). Isotropic.

It occurs in crystalline masses, as a filling of fractures or impregnation, with a tendency to develop euhedral forms—octahedrons, cubes, generally with net exfoliation and frequent *zoning*. In low temperature deposits, it usually forms reniform or colloform aggregates, with characteristic botryoidal and spheroidal textures.

129.2.2 Observation +P (*Polariser + Analyser at 90°*)

Isotropic, complete extinction (Figs. 129.1c, d and 129.2b).

Dark brown to yellowish-brown *internal reflections*, not abundant, but generally visible in immersion. Crystalline forms may be *twinned*.

129.2.3 Paragenesis

As a high-temperature hypogene mineral, uraninite is found in granites, pegmatites and hydrothermal *veins*, e.g., of the

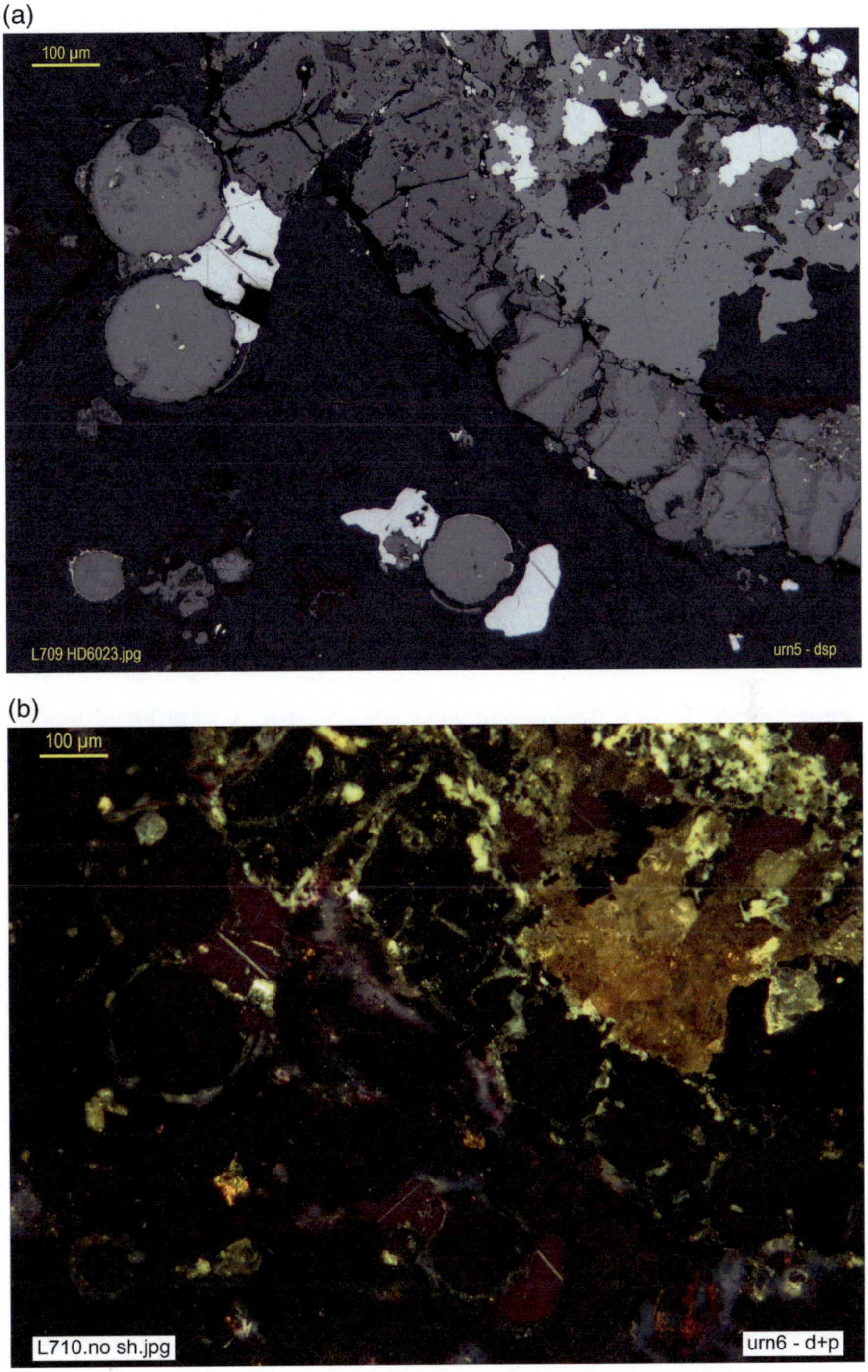

Fig. 129.2 **a, b** (Same image: **dsp** and **d+p**, respectively) - Partial oxidation of urn, well visible by the decrease of reflectance in diagonal band; mineralogy is the same as in urn1-4. HD 6023, Wolsendorf, Bavaria, Germany

five-element or *Ni-Co-Bi-Ag-U* type (Fig. 129.3). Pitchblende occurs in low temperature deposits.

Also in detrital concentrations, such as Witwatersrand-type auriferous conglomerates, involving non-oxidizing environmental conditions that prevent easy dissolution of uraninite. In strongly reducing conditions, it can even neo-form from the retrograde reduction of oxidized ores (Colorado Plateau type deposits).

Accompanying ores: galena, pyrite, marcasite, Ni-Co arsenides, linnaeite, native bismuth, etc. Secondary uranium minerals (oxides).

129.2.4 Identification

The microscopic identification of urn relies on its isotropy, its grey colour with scarce brightness (which give urn a dull

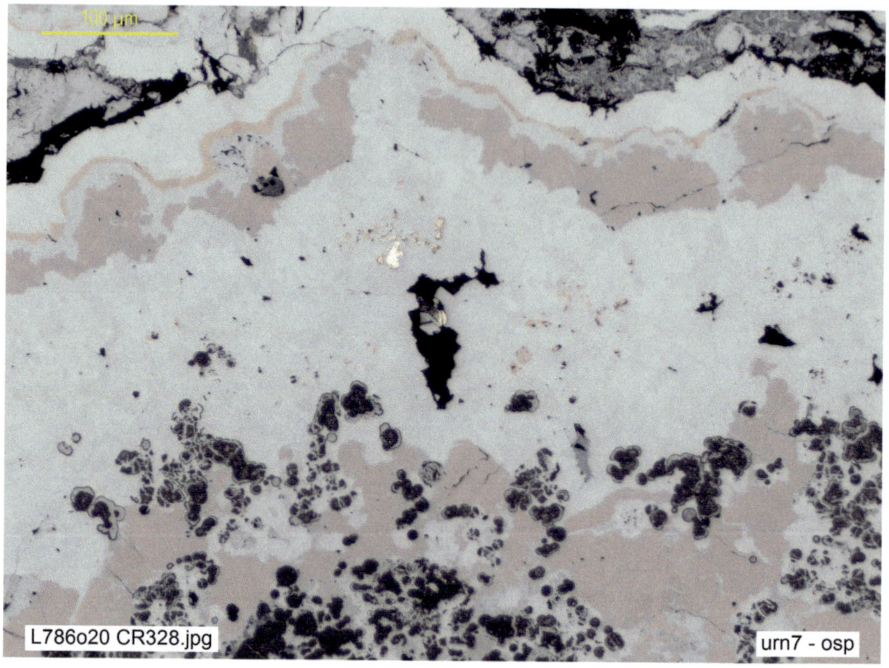

Fig. 129.3 Nickel–cobalt-iron arsenide intergrowth (main mass, light grey and pinkish to pinkish-brownish shades, sk, sf-lo, lgs, gf, nc), with minute interstitial Ag inclusions (very bright white) and pitchblende (numerous disseminated grey spheroids, ≈ 5–10 μm sized). CR 328, Ni-Co-Ag-U ore, Freiberg, Saxony, Germany (**osp**)

appearance due to the low reflectance and scarcity of IR), its high hardness (when fresh), the paragenesis, the formation of radioactive halos on the minerals of its environment and the impregnation of multicoloured oxidized products due to its weathering.

However, because of possible variations in its properties due to alteration, recognition of uraninite can be problematic, except in the *pitchblende-nasturan* varieties, whose textures are characteristic. If the polishing quality is not good, which easily happens in oxidized and friable samples, uraninite can be confused with several minerals, such as magnetite, wolframite and manganese oxides-hydroxides. For this reason and, given the enhanced uranium solubility and mobilisation in oxidizing media and also because U often occurs as a non-visible impregnation of the rock, the ancillary use of a *Geiger counter* or *scintillometer* is recommended.

129.3 Spectral Information

129.3.1 Uraninite Multispectral Reflectance R (in Air, Without Polariser) (%)

λ (nm)	370	400	425	450	475	500	525	550	575	600	625	650	675	700	750	800	850	900	950	1000
μ	16.4	15.8	15.0	14.6	14.3	14.2	14.0	13.9	13.9	13.9	14.0	14.0	14.1	14.2	14.3	14.2	14.4	14.3	14.4	14.4
P_5	16.0	15.4	14.6	14.2	13.9	13.8	13.6	13.5	13.5	13.5	13.6	13.6	13.7	13.7	13.8	13.8	13.9	13.8	13.9	13.9
P_{20}	16.2	15.6	14.8	14.5	14.1	14.0	13.9	13.7	13.7	13.7	13.8	13.8	13.9	14.0	14.0	14.0	14.1	14.1	14.2	14.1
P_{80}	16.6	16.0	15.2	14.8	14.5	14.3	14.2	14.1	14.1	14.1	14.2	14.2	14.3	14.4	14.4	14.4	14.6	14.6	14.6	14.6
P_{95}	16.8	16.1	15.4	15.0	14.6	14.5	14.4	14.3	14.3	14.3	14.4	14.4	14.5	14.6	14.7	14.7	14.9	14.9	15.0	15.0

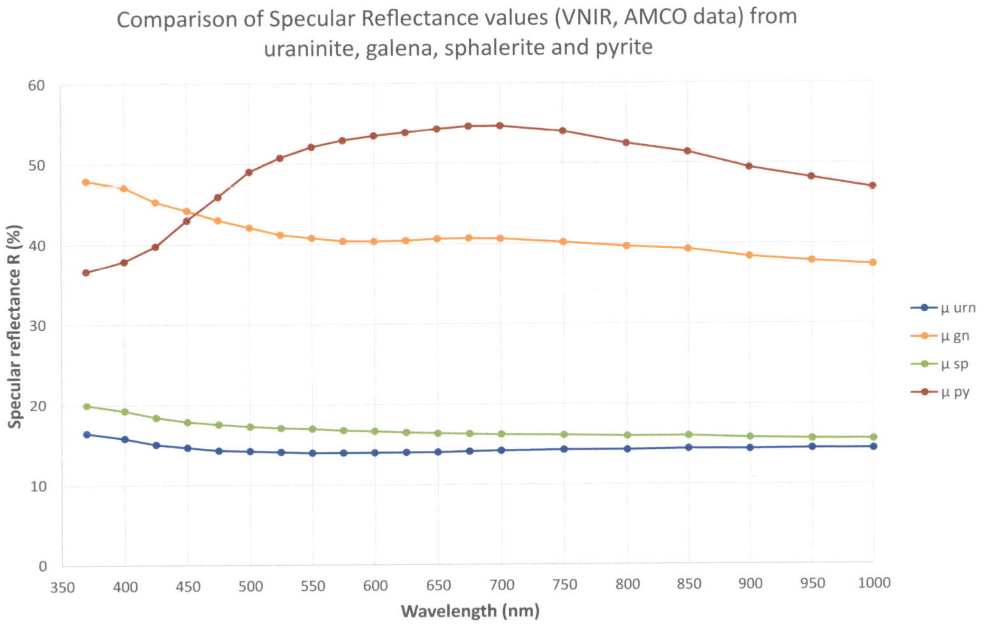

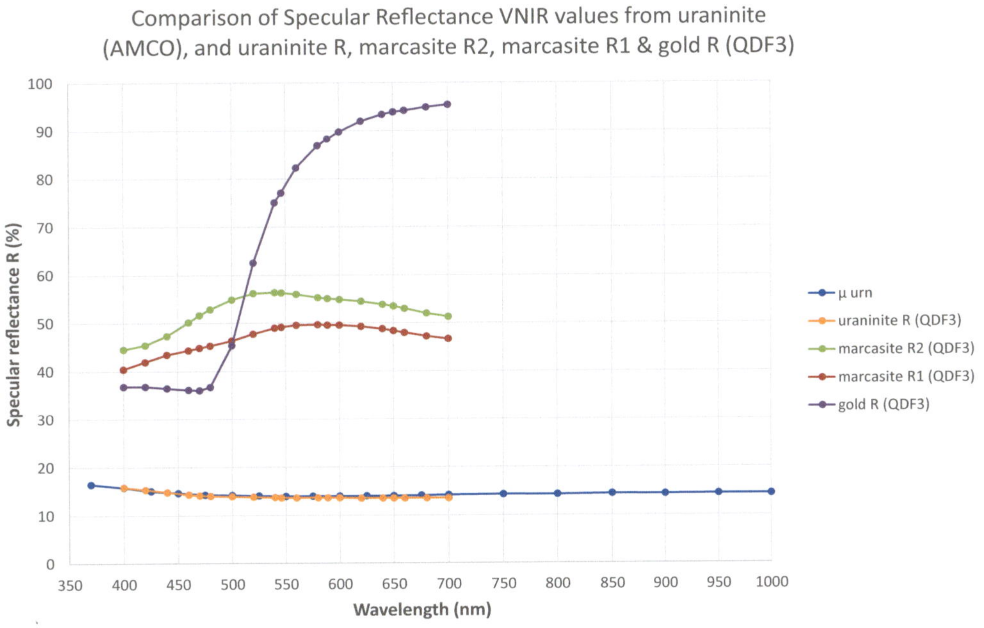

Reference

Ramdohr P (1980) In: The ore minerals and their intergrowths. vol 2. 2nd edn, Pergamon Press, Oxford, pp 1205

Valleriite (*vll*)

130

$4[(Fe,Cu)S]\cdot3[(Mg,Al)(OH)_2]$: Hexagonal

Valleriite		R_{546} (%)	Y (%)	λ_d (nm)	P_e (%)	Colour sensation
Air	R_o	20.5	20.9	578	24.5	Light yellowish brown
	R_e'	10.3	10.3	466	1.5	Yellowish grey
Oil	R_o	10.7	11.0	578	30.3	Yellowish brown
	R_e'	1.6	1.7	457	0.9	Dark grey
Bir: Very strong (5)				*Pleoc*: Very strong (5)		*Tarn*: nd
Anis+P: Very strong and bright (5)				*IR*: No		*Twg*: No
Mor: Leafy or lamellar aggregates				*Clv*: Distinct		*PolQ*: Difficult (very soft)
Hardness	*PolH*: Very soft (1)			*Mohs*: 1½	*Talm*: -	*VHN* $_{(100\ g)}$: nd (not measurable)

Obs: cf. Table and R charts: multispectral data, visible range (QDF3 and Picot and Johan 1982)
Mineral abbreviations: calc—calcite, Carb—carbonate, ccp—chalcopyrite, chr—chromite, gg—gangue, gra—graphite, il—ilmenite, mck—mackinawite, ml—millerite, mt—magnetite, pn—pentlandite, py—pyrite, srp—serpentine

130.1 General

The composition of valleriite is, according to Ramdohr (1980), $CuFeS_2$, as chalcopyrite, although almost always with intercalations (0001) of $Mg(OH)_2 \pm Ni(OH)_2$, Fe $(OH)_2$, but it is not very desirable as a Cu ore because of the concentration problems caused by its very low hardness.

Strictly speaking, the formula varies according to the sources.[1] The formula $(Fe^{2+},Cu)_4(Mg,Al)_3S_4(OH,O)_6$ or, according to IMA, $2[(Fe,Cu)S]\cdot1.53[(Mg,Al)(OH)_2]$ is currently accepted.

130.2 Microscopic Description

130.2.1 SP Observation (*Single Polariser*)

Light yellowish brown to yellowish grey, strongly pleochroic (Fig. 130.1a, b), very similar to graphite on first impression. Nevertheless, the bireflectance of gra is more pronounced (cf comparative R graphs vll-gra) and the minimum R_e' value of valleriite (Fig. 130.2a, b) is not as low as that of graphite. Compared with common gg, vll is always brighter (even if oriented after R_e'), while gra oriented according to R_e' is very dark and can be mistaken for gangue (cf Fig. 57.3a).

Reflectance highly variable with orientation (Figs. 130.1 a, b and 130.3a), with extreme values between those of magnetite (maximum, R_o, of valleriite) and chromite (minimum, R_e', of valleriite). Compared with graphite, it lies between a maximum < R_o gra and a minimum R_e' which is > R_e' gra, but < R chr. Compared to mackinawite, the bireflectance of valleriite is lower, as are both R_o and R_e' values (cf multispectral comparison plots).

Its *polishing* is difficult because vll is a very soft mineral (to the point of precluding measurement of the Vickers hardness value, VHN).

It usually occurs in aggregates of fine flakes, with very pronounced {0001} *cleavage*, often bent or deformed. It may be frequent in tiny inclusions in ultrabasites.

[1] cf. https://www.mindat.org/min-4136.html, *visit*. 20,210,318.

R. Castroviejo, *A Practical Guide to Ore Microscopy—Volume 1*,
https://doi.org/10.1007/978-3-031-12654-3_130

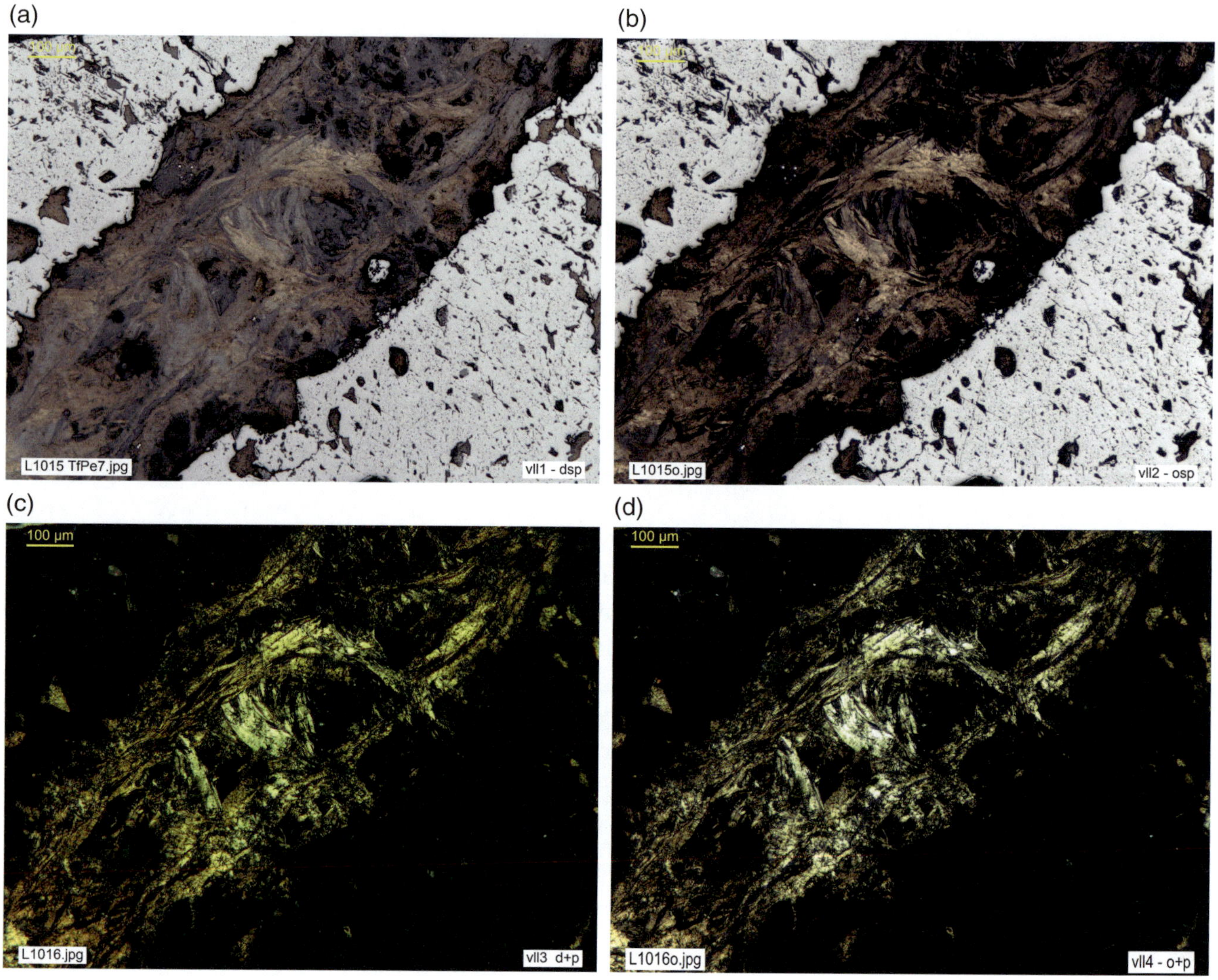

Fig. 130.1 **a, b** (Same scene: **dsp** and **osp**, respectively). Mass of mt (light grey) fractured and cemented by vll (diagonal veinlet, SW-NE, consisting of leafy aggregates, strongly pleochroic in yellowish-brown to grey tones), including sparse microclasts of carbonate gg (dark grey) and tiny py particles (very bright against vll). Extreme bireflection and pleochroism of vll, reminiscent of gra. TF-PE7, Palabora (Rep. South Africa). **c, d** (Same scene as Fig. 130.1a: **d+p** and **o+p**, respectively). Complete extinction of mt, in contrast to the extreme anisotropism of vll

130.2.2 Observation +P (*Polariser + Analyser at 90°*)

Strongly *anisotropic* (Fig. 130.1c, d and 130.3b), very similar to graphite and more coloured than mackinawite. Compared to the white–grey shades of mackinawite, valleriite shows dominant yellowish tints, very luminous and striking, which facilitate its identification even in tiny inclusions (common form of occurrence in ultramafites). No internal reflections.

130.2.3 Paragenesis

Valleriite is a rare mineral in general, restricted to certain parageneses (copper deposits or prospects in serpentinites; ophiolites, chromitites; carbonatites), where it can be frequent or even abundant (e.g., with magnetite in Palabora).

Commonly accompanied by: magnetite, pentlantite, chromite, chalcopyrite, ilmenite, millerite, pyrite.

(a)

(b)

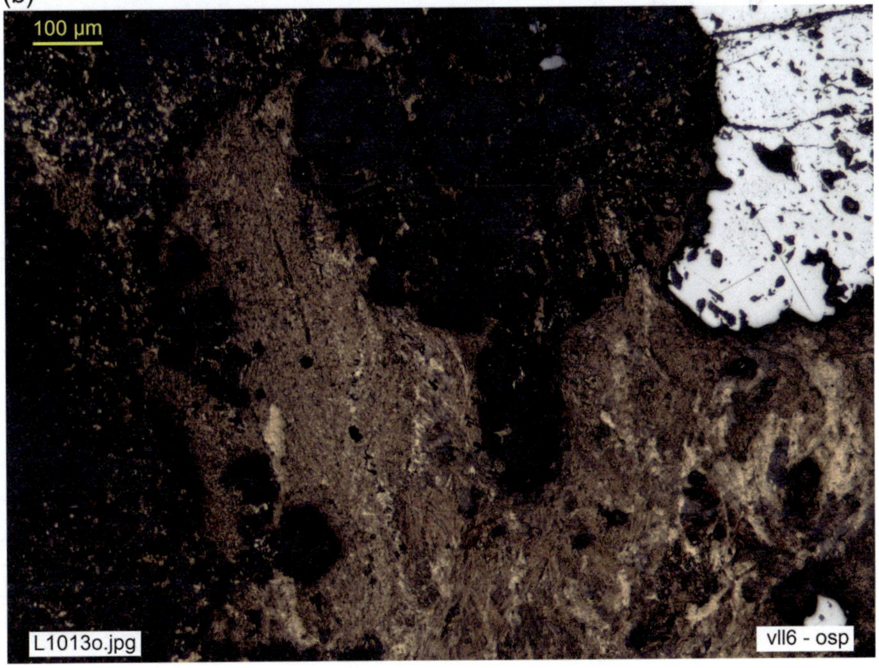

Fig. 130.2 a, b (Same scene: **dsp** and **osp**, respectively). Microbreccia with mt (light grey) and Carb (dark grey, bireflective), cemented by massive vll, whose strong bireflectance and pleochroism are evident when comparing grains or lamellae of different orientation, especially in immersion (Fig. 130.2b). However, vll never looks as dark as the Carb gg (possibly calcite), because vll reflectance (also R_e') is always higher. Tf-Pe7, Palabora (South African Rep.)

130.2.4 Identification

Valleriite is easily identifiable by its resemblance to graphite but without reaching such extreme values of bireflectance, which provides an easily applied criterion to distinguish them. R_{min} of gra is ≈ common gg, while R_{min} of vll is never so low, it remains >10%. The paragenesis, anisotropism and very low hardness of vll generally allow recognition without difficulty.

At first, vll was not distinguished from mackinawite but, once mackinawite was described and recognized as a new species, the optical properties of vll distinguish them clearly—valleriite is darker, but its bireflectance is stronger and its anisotropism, more coloured.

Fig. 130.3 **a, b** (Same scene: osp and o + p, respectively). Microbreccia with chromite (grey) as main ore, cemented by silicates and altered. Tiny inclusions of vll in chr (the two largest ≈ 10–15 μm, top centre-left; several < 5 μm, bottom centre), which are barely visible *-osp*—because of their low reflectance (Fig. 130.3a) but very *striking -o+ p*—by their luminous anisotropism, in contrast to chr extinction (Figs. 130.3b, comp. 130.1c and d). The silicates are essentially srp (with chlorite), very dark to black (osp, Fig. 130.3a) but luminous, with abundant to massive greenish internal reflections (o+p, Fig. 130.3b). Alteration produces redistribution of Fe, Cr and Al from chr, *e.g.*, iron-enriched *coronas* on chr microclasts, with neo-formed mt and ferritchromite, both distinctly brighter (more reflective) than chr. TP110607_21, Tapo Mine, Tarma (Dept. Junín, Perú)

(a)

(b)

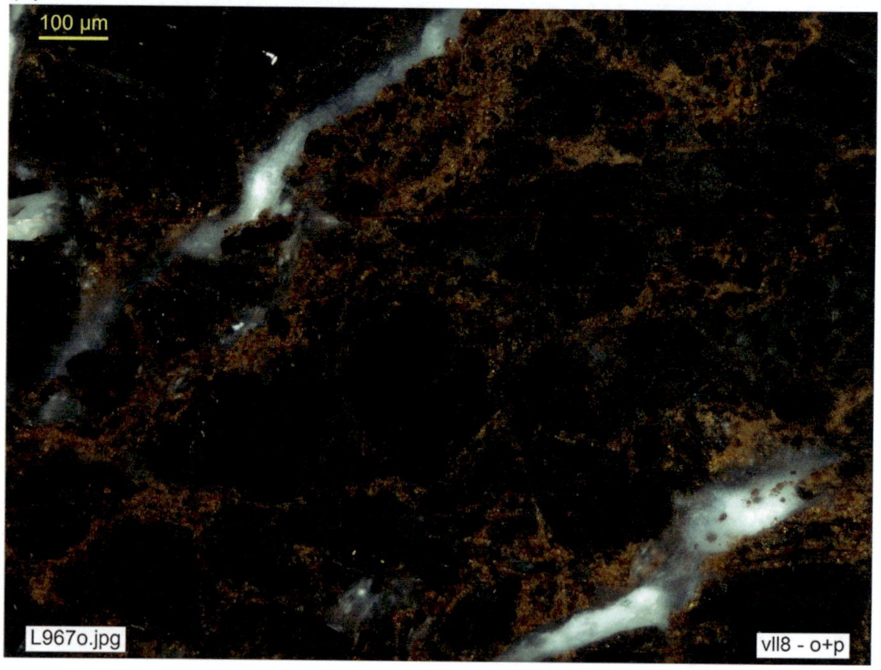

130.3 Spectral Information

130.3.1 Valleriite Spectral Reflectance R (in Air, Visible Range) (%)

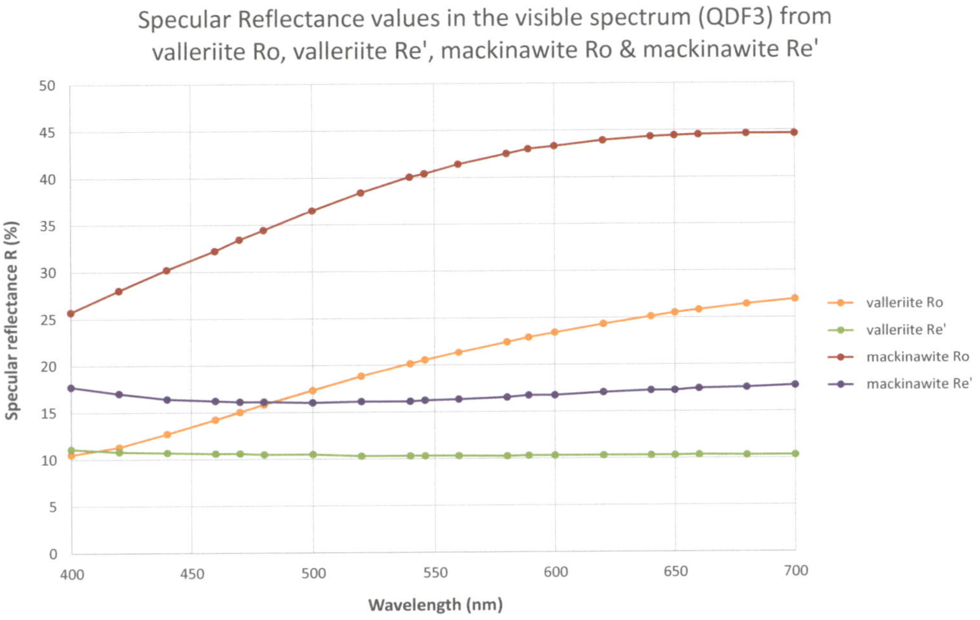

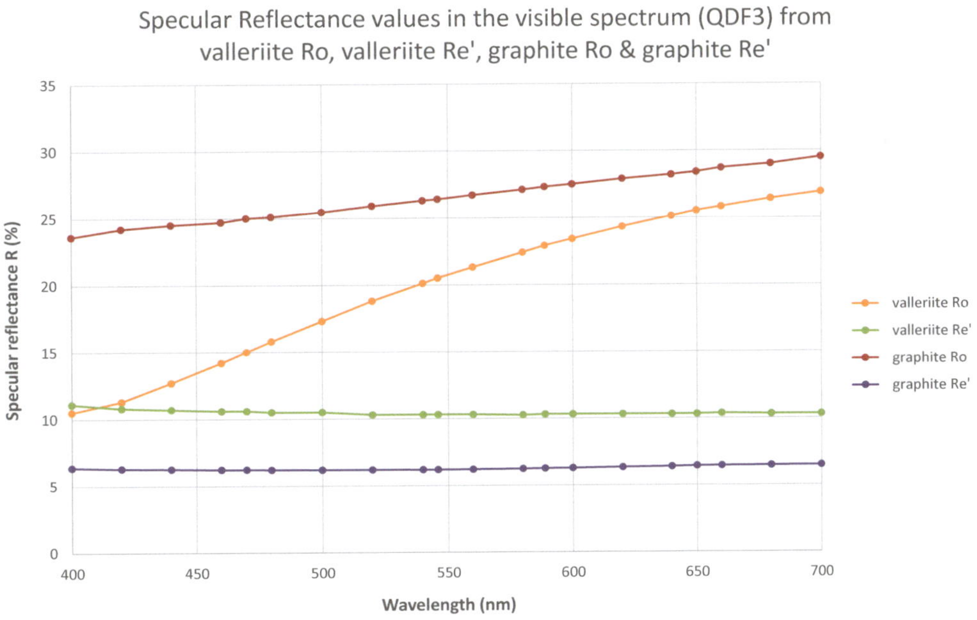

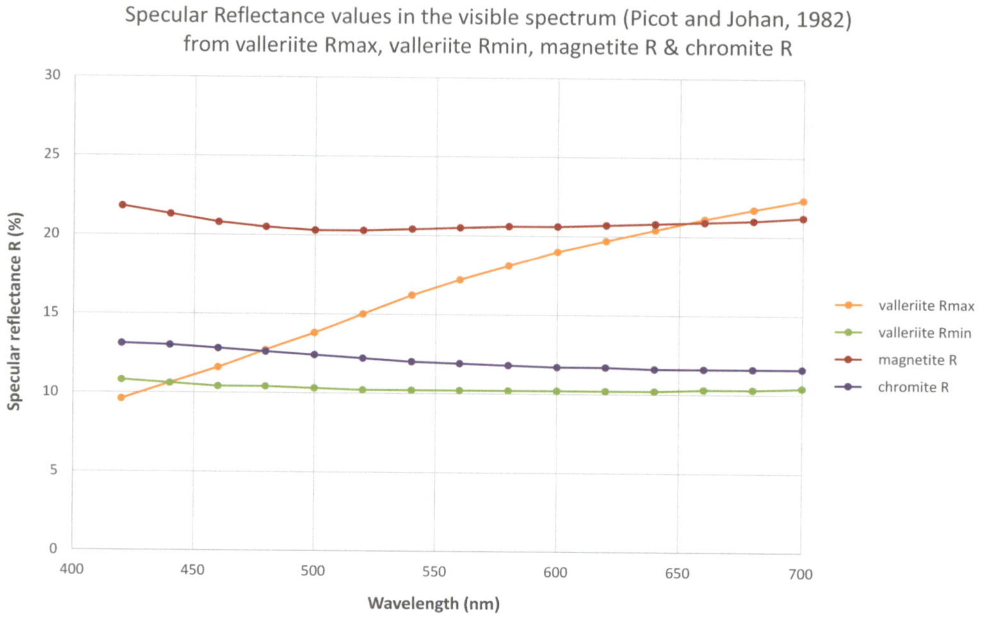

References

Picot P, Johan Z (1982) In: Atlas of ore minerals. B.R.G.M./Elsevier

Ramdohr P (1980) In: The ore minerals and their intergrowths. vol 2. 2nd edn, Pergamon Press, Oxford, 2 pp 1205

131 Wolframite (*wf*)

(Fe, Mn)WO_4: Monoclinic

Wolframite		R_{546} (%)	Y (%)	λ_d (nm)	P_e (%)	Colour sensation
Air	R_1	15.2	15.3	472	3.7	Grey to brownish-grey
	R_2	16.4	16.4	468	3.8	
Oil	R_1	4.1	4.1	470	6.5	Darker grey to brownish-grey
	R_2	4.8	4.8	465	6.6	
Bir: Weak (3)				*Pleoc*: Weak (3)		*Tarn*: No
Anis +P: Weak but visible (3)				*IR*: Red		*Twg*: Occasional (100)
Mor: Prismatic, tabular, acicular				*Clv*: Visible (010), (100)		*PolQ*: Difficult, but acceptable
Hardness	*PolH*: High, but < cs (4)			*Mohs*: 4.5	*Talm*: E	*VHN* $_{(100\ g)}$: 319–390

Obs **:** cf. Table and R charts, multispectral VNIR data (AMCO) and visible range (QDF3; Picot and Johan 1982)
Mineral abbreviations: asp—arsenopyrite, bm—bismuthinite, bn—bornite, cb—columbite, ccp—chalcopyrite, cst—cassiterite, ferb—ferberite, gg—gangue, ht—hematite, hub—hübnerite, il—ilmenite, mt—magnetite, mol—molybdenite, po—pyrrhotite, py—pyrite, q—quartz, rt—rutile, sch—scheelite, sp—sphalerite, td—tetrahedrite, tur—tourmaline, wf—wolframite, zir—zircon

131.1 General

Variable composition between the ferriferous and manganiferous terms of (Fe, Mn) WO_4, i.e., the species *ferberite*, (Fe^{2+})WO_4, and *hübnerite*, (Mn^{2+})WO_4, respectively. At very high temperature the miscibility between both terms is total and they form a continuous series but, at the temperatures common in hydrothermal deposits, the solubility is limited.

Therefore, commonly a manganiferous ferberite (i.e., (**Fe**, Mn) WO_4) and a ferriferous hübnerite (i.e., (**Mn**, Fe) WO_4) can coexist. In some cases, *e.g.*, Pasto Bueno, Peru, hübnerite may contain considerable amounts of Zn.

131.2 Microscopic Description

131.2.1 SP Observation (*Single Polariser*)

Colour grey, sometimes with a slight brownish tinge, depending on the accompanying minerals; slight pleochroism, more noticeable in immersion (Figs. 131.1, and 131.2

Low *reflectance,* R wf ≈ sp (Fig. 131.3) but always somewhat higher than its frequent companions cassiterite or scheelite (Figs. 131.1; cf. comparative plots R) and distinctly higher than zir. However, R wf < rt, ht or td (Fig. 131.5) and ≪py.

Polishing difficult due to high hardness but results are usually acceptable.

Polishing hardness high: > mt, po; < rt, py, q, cs. Relatively low *Vickers microhardness*: VHN < 400, which is a notable difference with cassiterite (VHN > 1200) and columbite (VHN ≈ 700–800).

It occurs in euhedral crystalline aggregates, frequently coarse-grained and usually with tabular, prismatic or acicular habit; rarely colomorphic. Common cleavage (010) and (100). Frequent zoning.

131.2.2 Observation +P (*Polariser + Analyser at 90°*)

Weakly *anisotropic*, with poorly defined, brownish to greenish colours (masked in hübnerite by the bright, reddish IR, *internal reflections*, omnipresent).

R. Castroviejo, *A Practical Guide to Ore Microscopy—Volume 1*,
https://doi.org/10.1007/978-3-031-12654-3_131

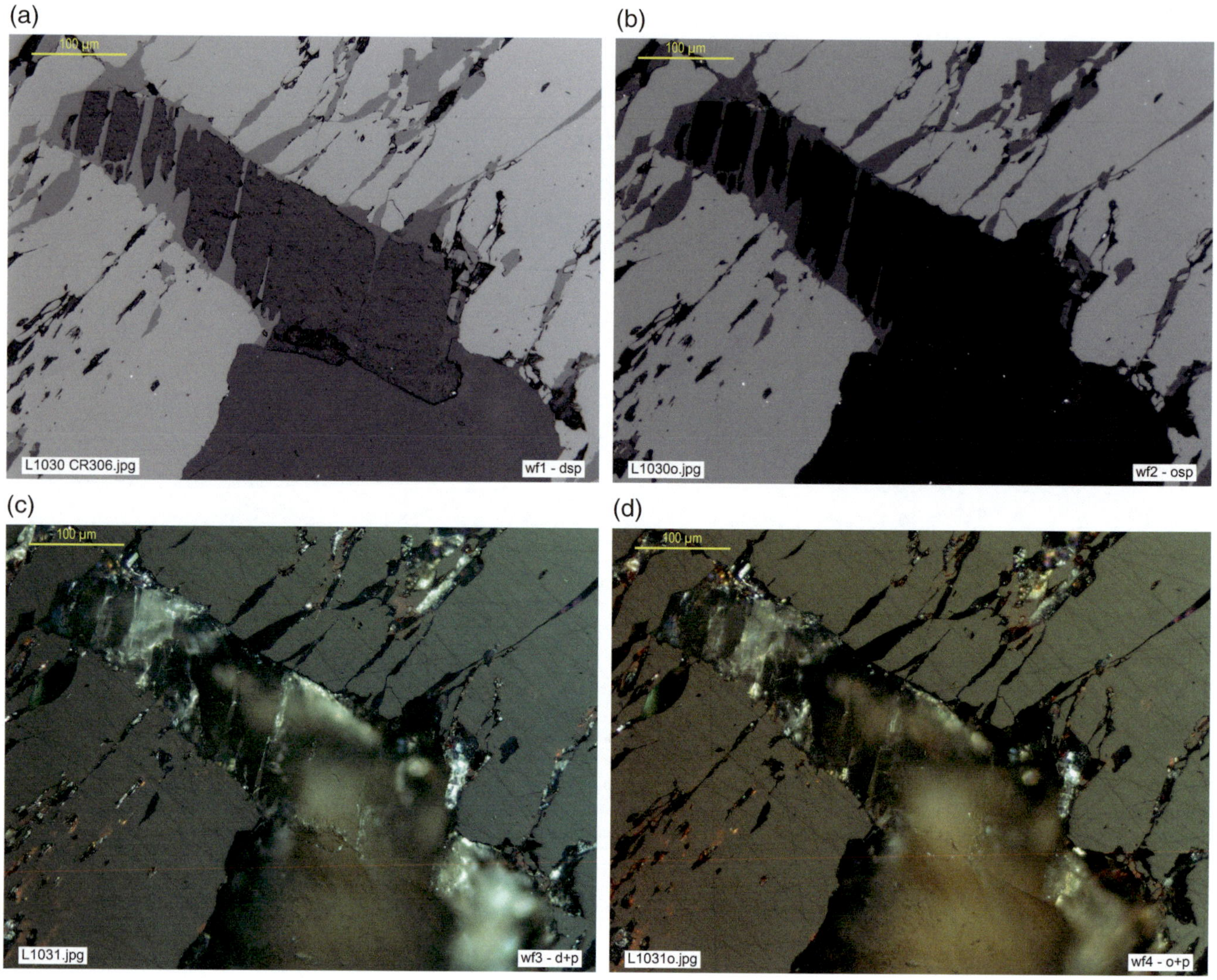

Fig. 131.1 **a** Vein infill of wf (light grey), partially replaced by late sch (grey), with q (dark grey, good polish) and relict mica (dark grey, rough polishing). CR 306, As Sombras Mine, Lobios, Orense, (**dsp**). **b** Same scene as wf1 (**osp**): sch much darker; gg black. **c, d** Same scene as wf1 (**d+p**) and (**o+p**), respectively. Anisotropism and sparse IR of wf

Red to orange-red IR, not very abundant in ferberite, massive and luminous in hübnerite (Figs. 131.1c, d, 131.2b and 131.4).

Oblique extinction. *Twinning* (100) visible.

131.2.3 Paragenesis

Typically, pegmatitic, pneumatolytic and hydrothermal *parageneses*, where wf usually occurs as holocrystalline vein filling or as impregnations (greisen), related to evolved granites, and also in skarn and metamorphogenic deposits.

Accompanying minerals: arsenopyrite, cassiterite, molybdenite, bismuthinite, scheelite, quartz, columbite, tourmaline, etc. Associated with scheelite, which sometimes replaces it.

131.2.4 Identification

Wolframite is easily identified by its *paragenesis*, *low reflectance*, *grey colour*, difficult polishing, high polishing hardness (but moderate VHN), prismatic or acicular habit, *weak anisotropism*, *internal reflections* (massive in hübnerite) and *oblique extinction*.

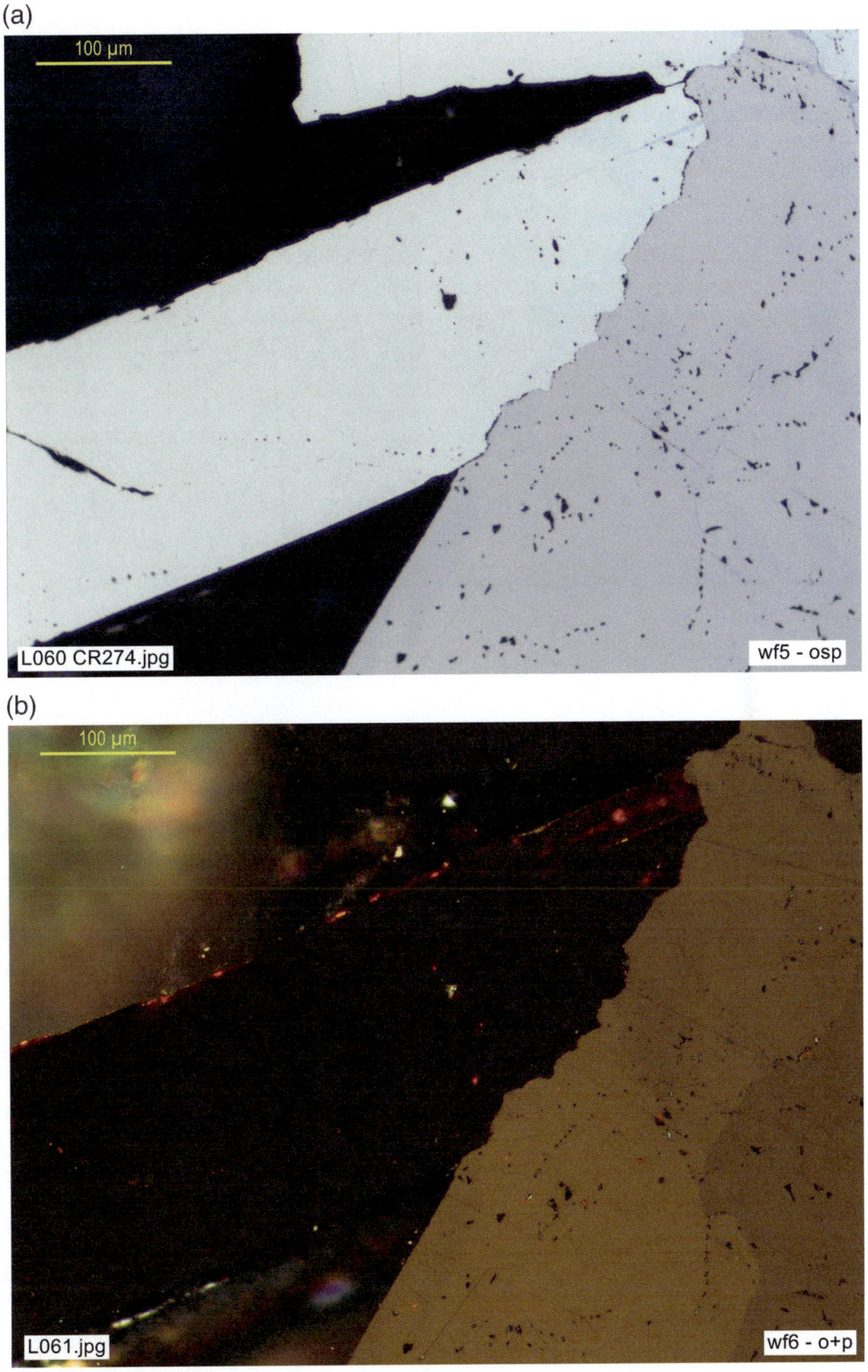

Fig. 131.2 **a, b** (Same image: **osp** and **o+p**, respectively). Vein infill of prismatic wf (grey, pleochroic, anisotropic, red IR) and q (black). CR 274, Monte Neme (Coruña)

The frequent association of wf with scheelite should not pose recognition problems:

- scheelite is softer and darker, has abundant colourless IR and often shows textural evidence of replacement with wolframite.

It is easily distinguished from magnetite, hematite, rutile and ilmenite:

- *Magnetite* is isotropic and has no IR.
- *Hematite* is much brighter and more distinctly anisotropic.
- *Rutile* is brighter and harder.
- *Ilmenite* is browner, more pleochroic and, except in the rare Mg- or Mn-rich varieties, lacks IR.
- Compared to *cassiterite*, wolframite is lighter, less anisotropic, has more coloured IR and is softer.

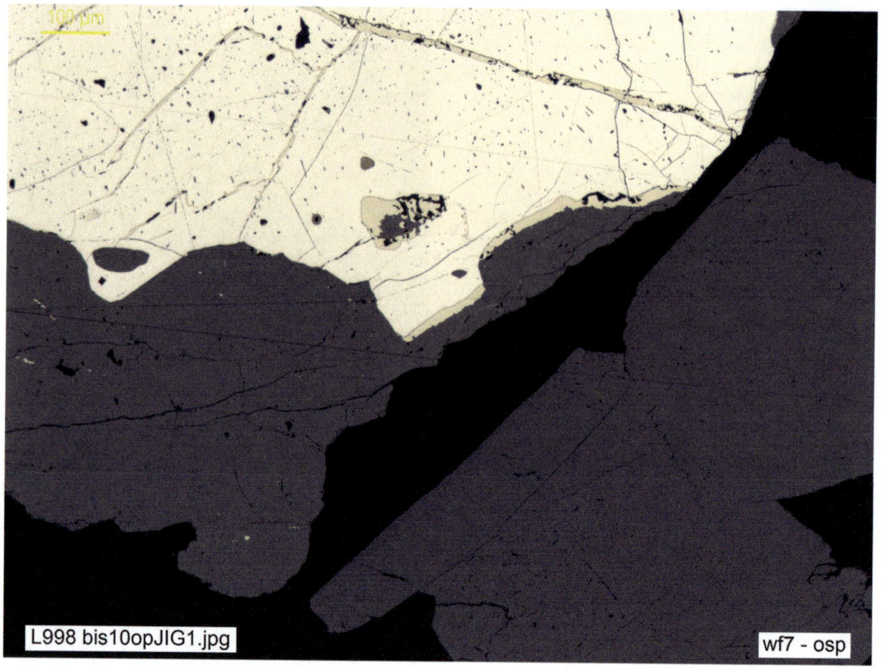

Fig. 131.3 Aggregate of py, wf (hub, bottom right corner), sp (center, in contact with py), ccp, (bn), q (black). Close resemblance of hub and sp, both dark grey. LMA01-JIG-1, Pasto Bueno Mine, Ancash, Peru (**osp**)

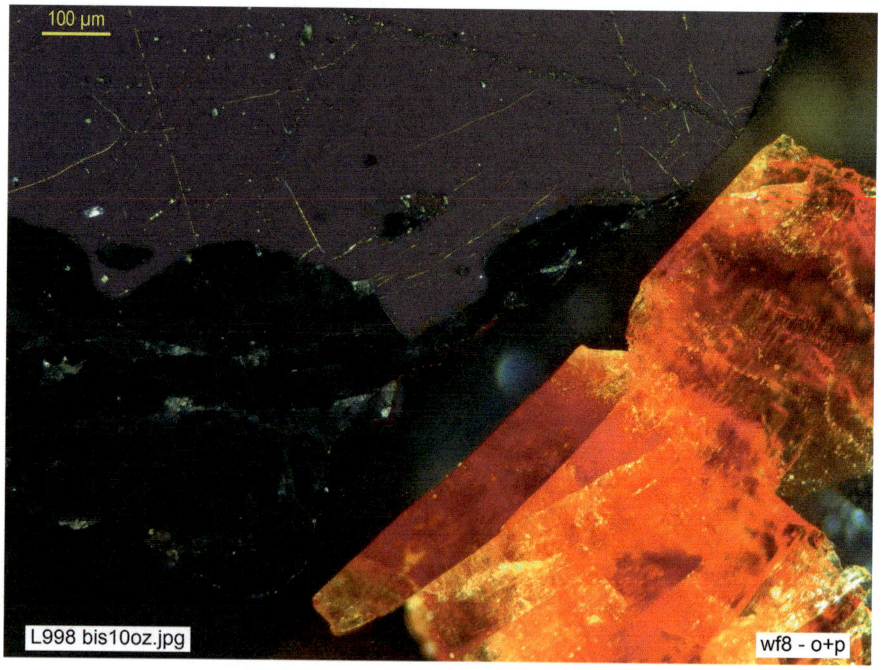

Fig. 131.4 Same scene as Fig. 131.3 (**o+p**). Reddish-orange IR, abundant and luminous, in hub, unlike the more scarce and colourless IR in sp

- The distinction of wolframite and *columbite* can be more difficult, but columbite is harder, more weakly anisotropic and has parallel extinction. In both cases, determination of *Vickers microhardness*, when possible, is diagnostic: VHN < 400 for wolframite, while VHN ≈ 700–800 for columbite and VHN > 1200 for cassiterite.

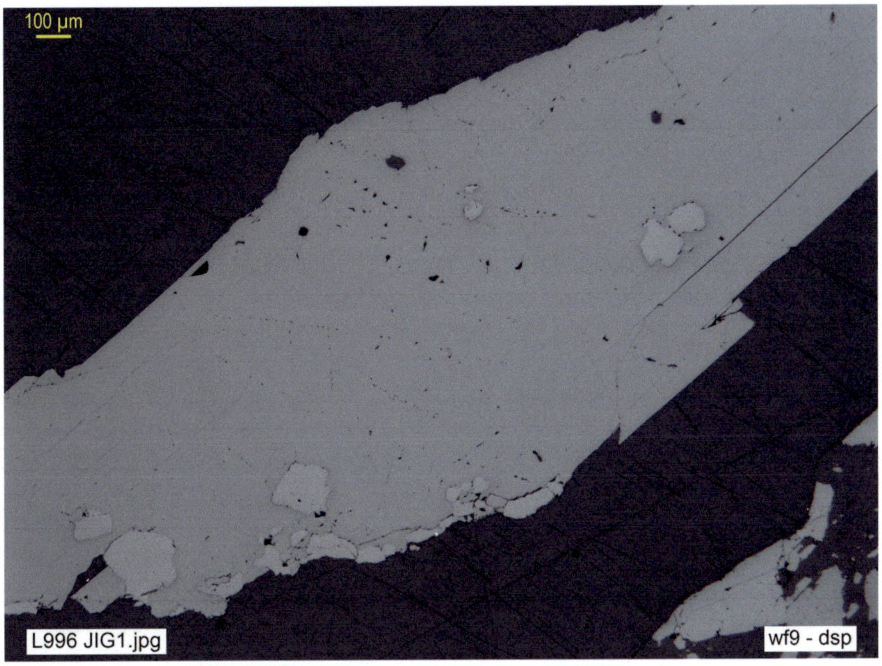

Fig. 131.5 Relict inclusions of rt (light grey) in prismatic hub (grey), and gg (dark grey, soft, likely fluorite). The relief of rt evidences its higher polishing hardness. LMA01-JIG-1, Pasto Bueno Mine, Ancash, Peru (**dsp**)

131.3 Spectral Information

131.3.1 Wolframite Multispectral Reflectance R (in Air, Without Polariser) (%)

λ (nm)	370	400	425	450	475	500	525	550	575	600	625	650	675	700	750	800	850	900	950	1000
μ	18.9	17.8	17.2	16.5	16.7	16.6	16.8	16.9	16.8	16.8	16.8	16.6	16.5	16.3	16.2	16.0	15.9	15.5	15.4	15.0
P_5	17.7	17.3	16.7	15.9	16.1	16.0	16.2	16.3	16.3	16.2	16.2	16.0	15.9	15.7	15.6	15.4	15.3	15.0	14.8	14.5
P_{20}	18.4	17.5	16.9	16.2	16.4	16.2	16.4	16.5	16.4	16.4	16.4	16.2	16.1	15.9	15.8	15.5	15.5	15.2	15.0	14.7
P_{80}	19.5	18.2	17.7	16.9	17.2	17.0	17.2	17.3	17.2	17.2	17.2	17.0	16.9	16.7	16.6	16.4	16.3	15.9	15.7	15.3
P_{95}	19.8	18.5	17.9	17.1	17.4	17.2	17.5	17.6	17.5	17.5	17.5	17.3	17.2	17.0	16.8	16.6	16.5	16.1	15.9	15.5

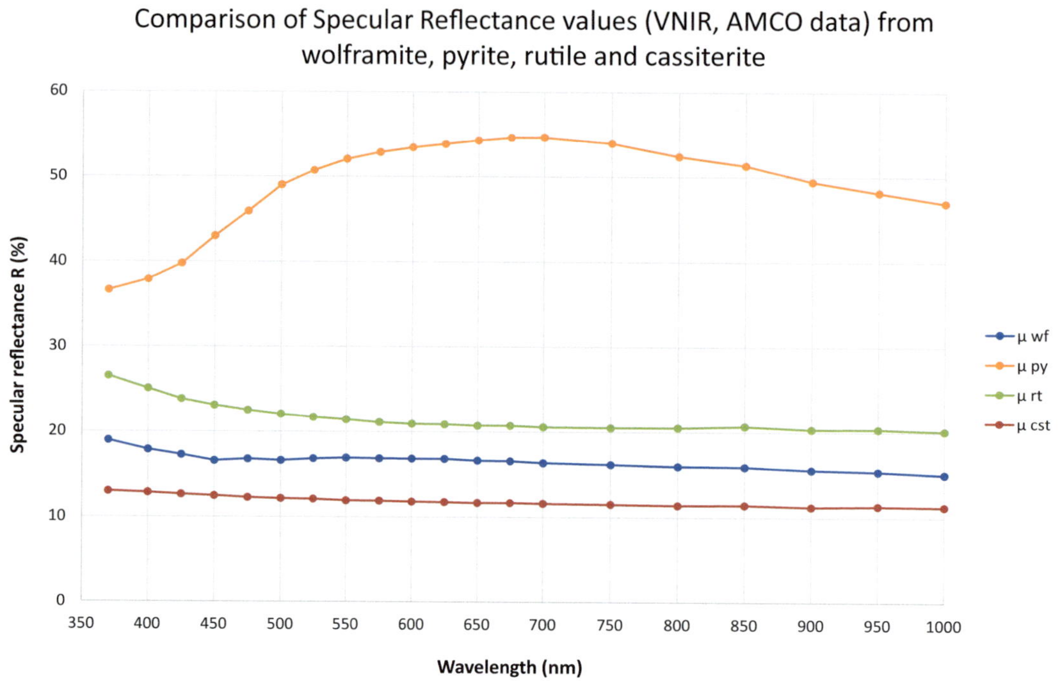

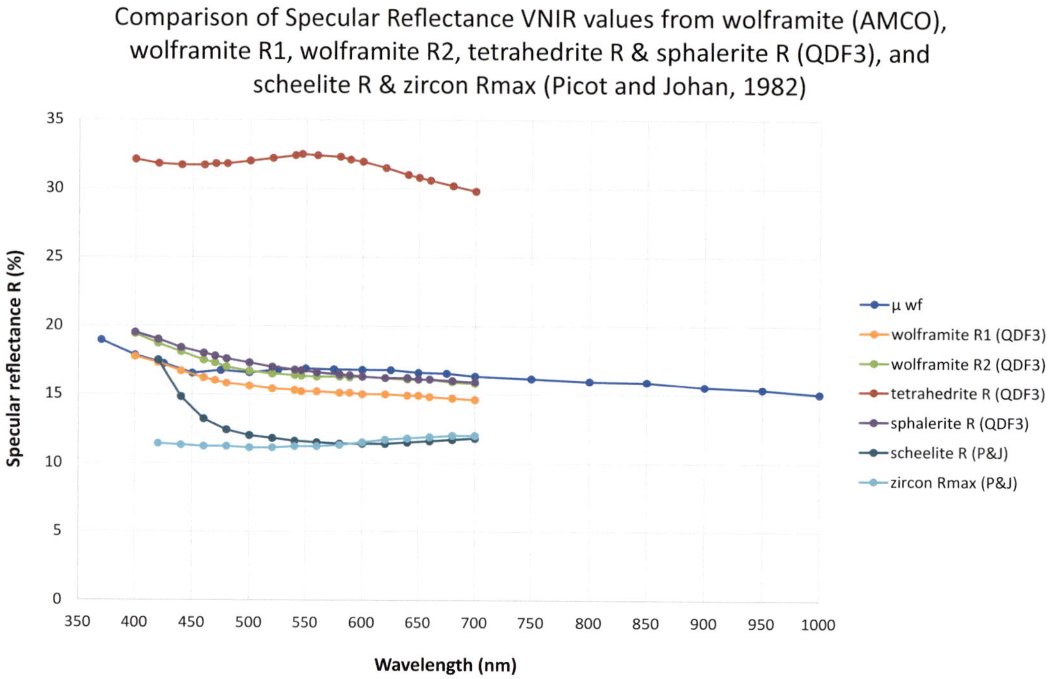

Reference

Picot P, Johan Z (1982) Atlas of ore minerals. B.R.G.M./Elsevier

Wurtzite (*wz*)

132

ZnS: Hexagonal

Wurtzite	R'_{546} (%)	Y (%)	λ_d (nm)	P_e (%)	Colour sensation
Air	≈sp: 16.8	≈sp: 16.7	≈sp: 475	≈sp: 4.4	Grey
Oil	≈sp: 5.1	≈sp: 5.0	≈sp: 475	≈sp: 7.5	Dark grey
Bir: Very weak (2)			*Pleoc*: No (1)		*Tarn*: No, but paramorphic alteration to sp
Anis +P: Weak (3)			*IR*: ≈sp		*Twg*: Not visible
Mor: Acicular, icefern, colloform, spherulitic			*Clv*: Occasional//(0001)		*PolQ*: Medium to poor
Hardness	*PolH*: Mean ≈sp; >ccp; <po (3)		*Mohs*: 3½–4	*Talm*: –	*VHN* $_{(100\ g)}$: 146–264

Note: cf. comparative R charts, visible spectrum data (Picot and Johan 1982). VHN after Uytenbogaardt and Burke 1971
Mineral abbreviations: ccp—chalcopyrite, gg—gangue, gn—galena, gth—goethite, jd—jordanite, mc—marcasite, mt—magnetite, po—pyrrhotite, py—pyrite, q—quartz, sp—sphalerite, stn—stannite, tn—tennantite, td—tetrahedrite, wf—wolframite, wz—wurtzite

132.1 General

Wurtzite, *wz*, is a dimorph of sphalerite, hexagonal and with compositional variations more restricted than in sphalerite, according to Ramdohr (1980), ZnS, with possible isomorphic incorporation of Fe or Mn (scarce) or of Cd (more significant) instead of Zn.

Several wurtzite polymorphs are known, the most common being 2H (hexagonal). With *greenockite* (CdS), wurtzite is the best known member of the Wurtzite Group, which comprises five densely packed hexagonal hemimorphic single sulphide species of general formula **MX** (**M** = Cd, Mn, Zn; **X** = S, Se).[1]

Since the description of wurtzite by Friedel (1861),[2] its relations to sphalerite have been the subject of numerous investigations but are not yet fully understood. Nevertheless, the non-stoichiometry of ZnS, with Zn vacancies in sphalerite and S vacancies in wurtzite, is generally accepted. Further, wurtzite was considered to be the stable form at high temperature (T > 1020 °C) and low pressure (1 atm). However, the sphalerite-wurtzite transition also depends on fs_2 and experimental data suggest that both species can stably coexist below this T, if fs_2 is low enough to maintain the S deficit of wurtzite. Wurtzite is found in nature in low fs_2 environments with formation T well below 1020 °C, for example, associated with sphalerite in the concentric rhythmic or banded variety known as *schalenblende* (German, *layer* blende). Other interpretations suggest that wurtzite is metastable at low T, which explains the frequent paramorphic inversion of wurtzite to sphalerite, irreversible under these conditions, making it possible for sphalerite to inherit the typical *icefern* texture. Wurtzite is considered a common component of the *schalenblende*, the main one for some authors or minor for others. This is a question that cannot always be clarified, given the difficult distinction of both species under the microscope and the scarcity of samples in which the mineral wurtzite can be isolated for reliable quantitative determinations.

In general, the properties of wurtzite that can be routinely observed with **RL** (reflected light) microscopy are practically coincident with those of sphalerite and, surprising as it may seem, the anisotropy of wurtzite (undoubtable because it is hexagonal) is rarely visible with RL as anisotropism and never as pleochroism, which deprives the observer of an indispensable differential criterion in many cases. For all these reasons, the description of wurtzite often relies on sphalerite.

[1] cf: https://www.mindat.org/min-10922.html, *visit.* 2020/06/09.

[2] cf: https://www.mindat.org/min-4318.html, *visit.* 2021/02/12.

R. Castroviejo, *A Practical Guide to Ore Microscopy—Volume 1*,
https://doi.org/10.1007/978-3-031-12654-3_132

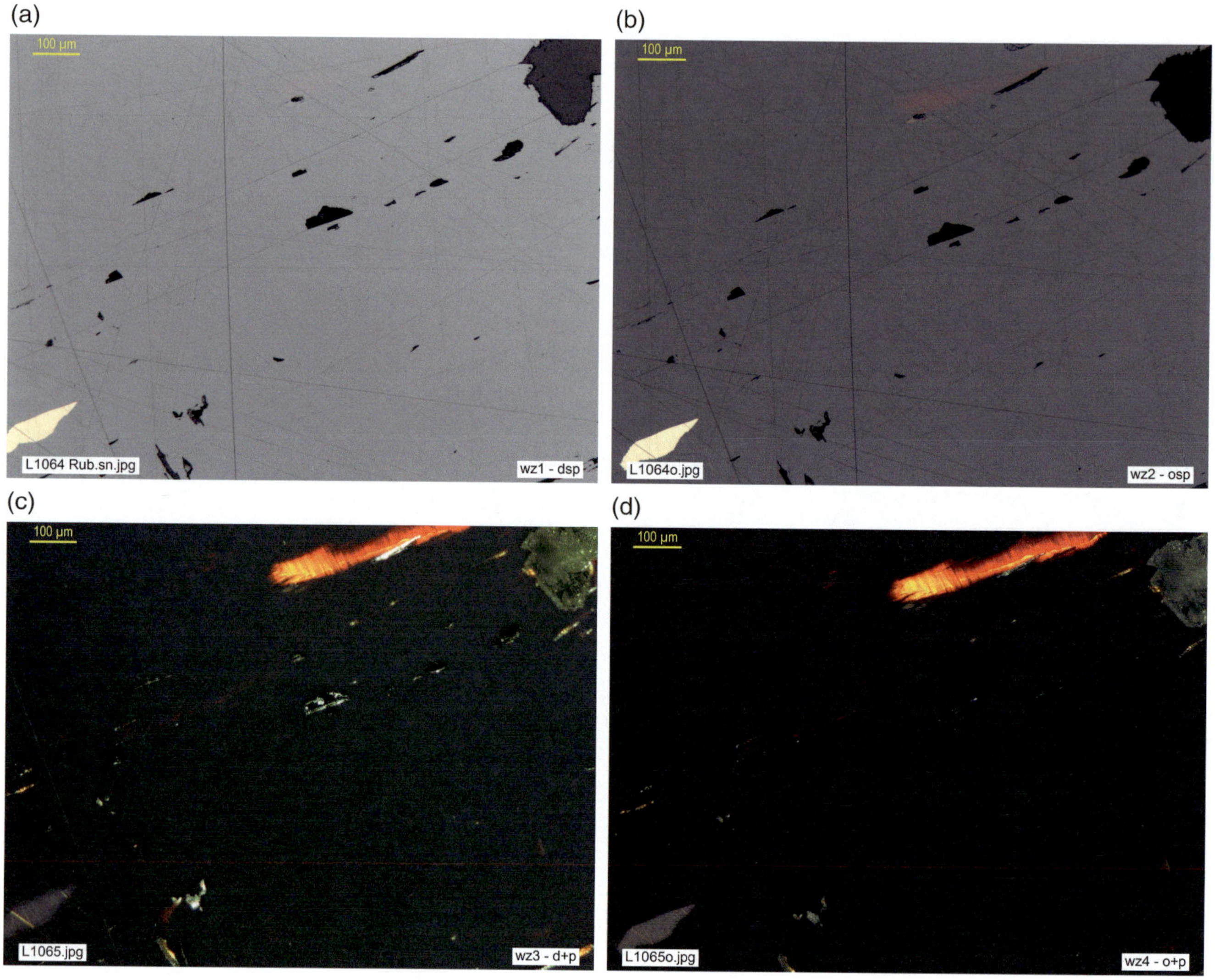

Fig. 132.1 **a, b** (Same scene: **dsp** and **osp**, respectively). Fibrous wz mass with small inclusions of ccp and q. Although the fibrous texture itself is not visible without chemical attack, it can be inferred from the alignment of inclusions parallel to the elongation of the fibers and triangular pores (by intersection of fibers). Rub.sn sample, Nagymuzsaly, Ukraine **c, d** (Same scene as wz1, **d+p** and **o+p**, respectively). Imperceptible anisotropism of wz and IR similar to those of sp but allowing glimpses of the striations (0001), perpendicular to the elongation of the wurtzite fibers, ascertained by the alignment of inclusions

132.2 Microscopic Description

132.2.1 SP Observation (*Single Polariser*)

Grey *colour*, like sphalerite (Figs. 132.1 a, b, 132.2a, b and 132.3a), difficult to *polish*. Anisotropic, but shows neither bireflectance nor pleochroism visible with reflected light microscopy. Low *reflectance*, somewhat lower than magnetite, practically identical to sphalerite and slightly higher than goethite (cf. comparative graphs R). Medium *polishing hardness*: PolH wz $\approx$ sp. Therefore, the basic properties coincide with those of sphalerite (cf. sphalerite description).

Although it can form very beautiful crystals,[3] wz usually occurs intergrown with sphalerite, in typical acicular aggregates but also in cryptocrystalline intergrowths or masses, whose grain size escapes the resolution power of the optical microscope (Bowles et al. 2011), making it necessary to resort to XRD to distinguish the two species.

[3] As examples, cf. selected photographs: https://www.mindat.org/gm/4318, *visit.* 2022/01/30.

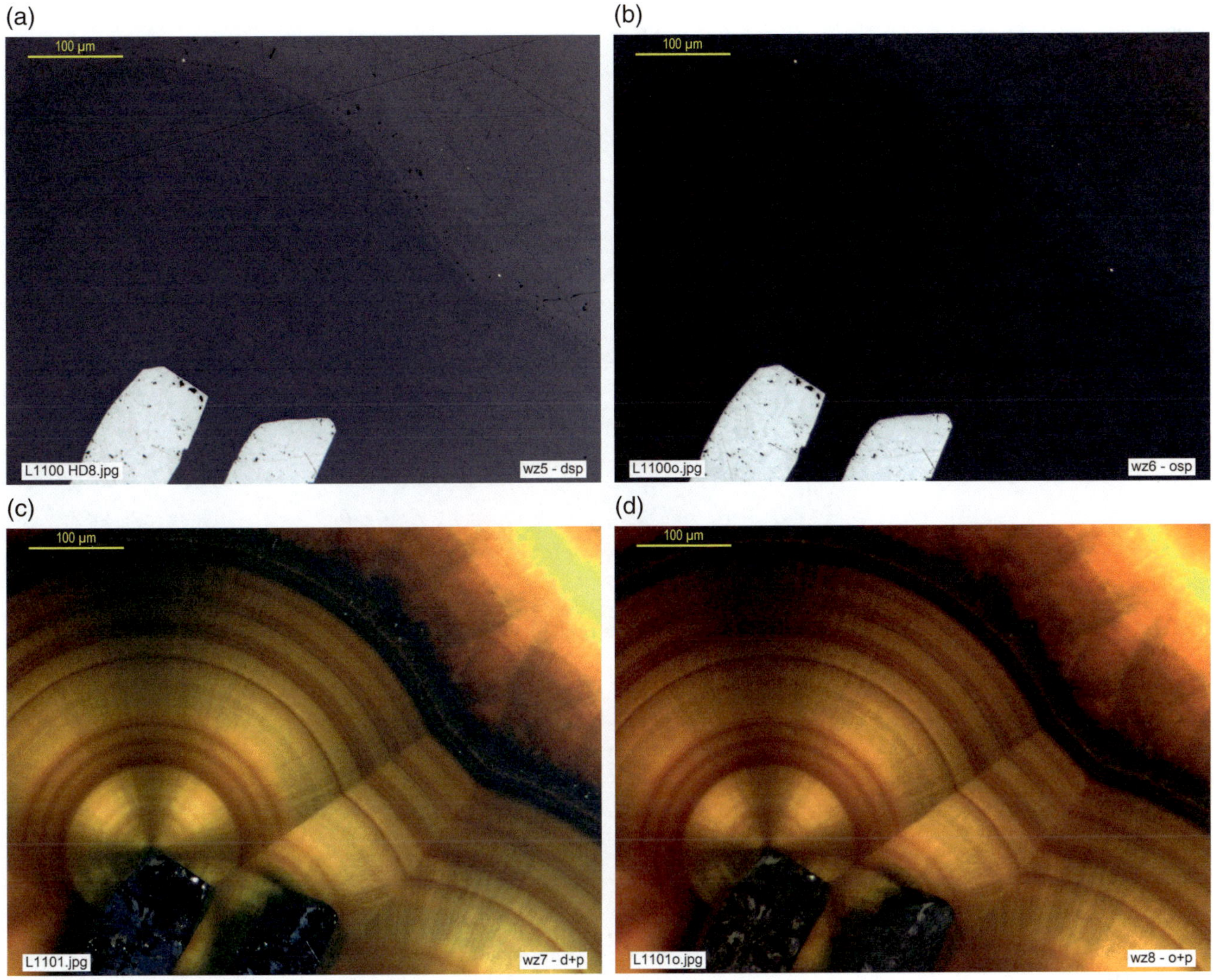

Fig. 132.2 **a, b** (**dsp** and **osp**, respectively)—Concentric rhythmic deposition of *schalenblende* (grey) over tabular aggregates of jd (white). HD 8, Wiesloch, Baden-Wuerttenberg, Germany (**d+p**) **c, d** (same scene as wz5, **d+p** and **o+p**, respectively)—Dual *schalenblende* structure, discernible despite generalized IR; mostly radial fibrous, with wz whose anisotropy is inferred from cross-extinction of radial fibers (straight extinction of wz); minor sphalerite areas, isotropic (total extinction)

The most characteristic texture of the wurtzite species is the acicular texture, with a dendritic-palmated appearance, known as *icefern texture (*so called because it recalls the shape of the acicular aggregates of ice crystals that form on windows by cold nights). Unfortunately, the distinct observation of this texture under the microscope requires a previous acid attack (HNO_3), so the microscopist who does not resort to etching is forced to infer it from collateral data such as the alignment of inclusions or pores parallel to the longitudinal axis of the fibres or the appearance of triangular hollow spaces that indicate the interstices between intersecting fibres (Figs. 132.1a and 132.3a). Given the metastable character of wurtzite at low temperatures, it is not rare that paramorphic inversion to sphalerite occurs, preserving the icefern texture in the neoformed mineral (sphalerite).

Very noticeable are the textures of rhythmic, botryoidal and spherulitic precipitation, known as *schalenblende*, with successive concentric micro-bands, in which sphalerite and wurtzite of very fine grainsize usually alternate, sometimes including also some marcasite or pyrite (Figs. 132.3a, b, c, d and 132.3a.

132.2.2 Observation +P (*Polariser + Analyser at 90°*)

Anisotropic, but the anisotropism is rarely perceptible (Fig. 132.3b), so it can be easily confused with sphalerite, whose extinction is often masked by **IR**, *internal reflections*, analogous to those of wurtzite (Figs. 132.1a, b, Fig. 132.3b and Fig. 132.2a, b).

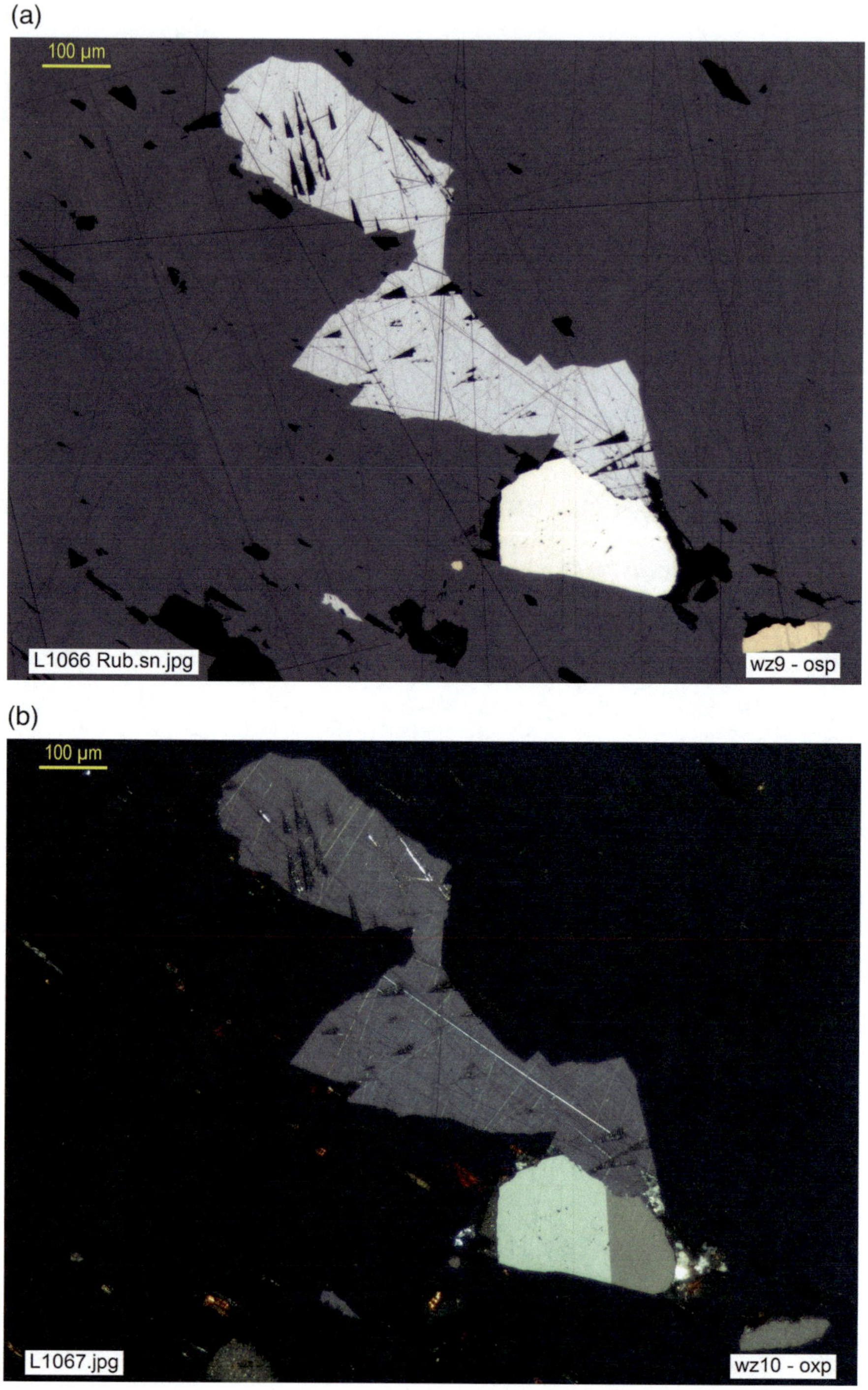

Fig. 132.3 **a, b** (dsp and oxp, respectively)—Mass of supposed wz whose prismatic habit is recognizable by the orientation (// c) of inclusions (gn, mc, ccp, q) and polishing pull-outs. With analyser b, although the anisotropism of mc and ccp is clearly recognizable, wz maintains extinction, despite the slightly uncrossed analyser (*xp*: partial illumination of gn). Its IR are like those of sphalerite. Sample Rub.sn, Nagymuzsaly, Ukraine

Examination with transmitted light, **TL**, can be an aid in identifying wurtzite, as well as other features that might be also visible in RL, as cross extinction in spherulitic aggregates (Fig. 132.2 a, b), not to be expected in a purely isotropic material. Also the presence of striations perpendicular to the elongation of the fibre, which are occasionally visible, thanks to the IR (Fig. 132.1a, b).

The *absence of the polysynthetic twins,* so frequent in sphalerite, is another distinctive criterion, although this observation requires structural etching.

132.2.3 Paragenesis

Wurtzite is found mainly associated, almost always as a late mineral, with shallow, *low pressure*, sulphide parageneses. These include Pb–Zn–Ba–F deposits in carbonate rocks and other stratabound types in sedimentary rocks, as well as submarine hydrothermal concentrations currently forming on ocean floors (precursors of volcanogenic massive sulphide deposits). It also forms as a posthumous mineral (in geodes or as a secondary sulphide) in various sulphide concentrations. In some cases, it can give rise to interesting economic concentrations (*e.g.,* Silesian *schalenblende* deposits).

Typical *accompanying ores*: sphalerite, galena, chalcopyrite, marcasite, pyrite, gratonite, jordanite, other sulphosalts of Pb, Sb, Ag (Figs. 132.1a, b, 132.2a, b and 132.5a, b).

132.2.4 Identification

The microscopic RL characterization of wurtzite is practically like that of sphalerite, with which it shares most of the visible properties (grey colour, medium polishing hardness, low reflectance, transparency allowing TL examination, IR). Therefore, the main difficulty for its effective recognition is the easy *confusion with sphalerite*. Distinguishing features are:

- wurtzite anisotropism (when it can be observed),
- the absence of the polysynthetic twins so common in sphalerite (but this observation requires structural etching, as does that of the icefern texture),
- the *icefern* texture of wz (although it can be pseudomorphized by sphalerite, if the frequent paramorphic inversion of wurtzite to sphalerite occurs),
- the fibrous-radial, spherulitic textures in *schalenblende*, when they show the anisotropic character of wurtzite by the cross-like extinction of radially oriented monomineral fibers—not to be confused with the very similar botryoidal, isotropic textures of sphalerite,
- in general, the fibrous textures inferred from some features, such as alignment of small sulphide inclusions or the presence of triangular interstitial pores (by intersection of oblique fibers), and
- the restriction of wurtzite to shallow or *low T/low P* paragenesis, as opposed to the ubiquity of sphalerite.

Nevertheless, the sphalerite-wurtzite distinction with optical microscopy is often impossible, which is not surprising considering that the intergrowths between the two polymorphs can occur even at the *nm scale* (Bowles et al. 2011). In doubtful cases, additional characterization with TL may be useful, but XRD determination or *nanoscale* studies by SEM-TEM may be indispensable for reliable determination.

132.3 Spectral Information

132.3.1 Wurtzite Reflectance R (in Air, Visible Spectrum) (%)

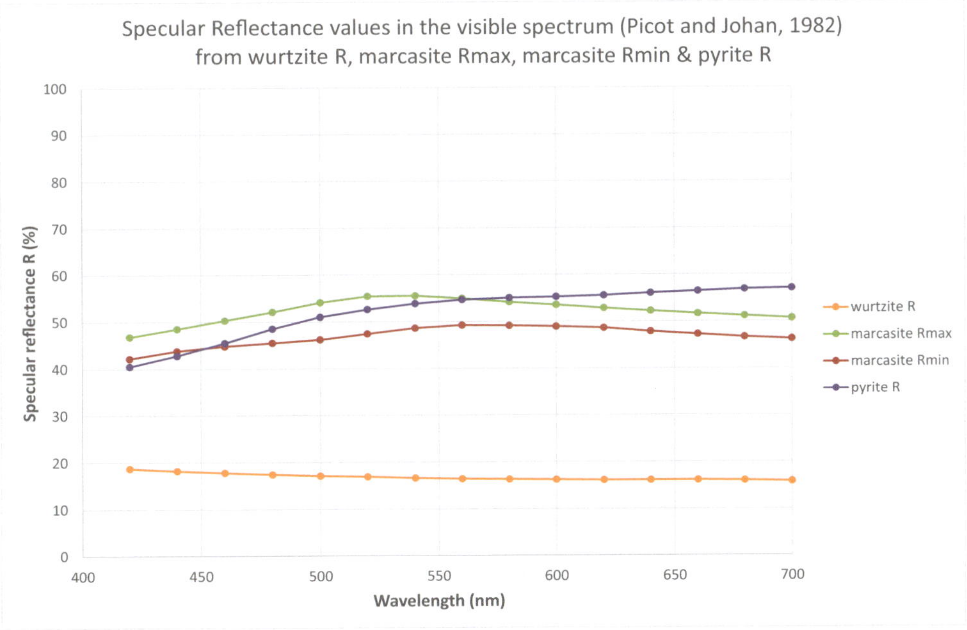

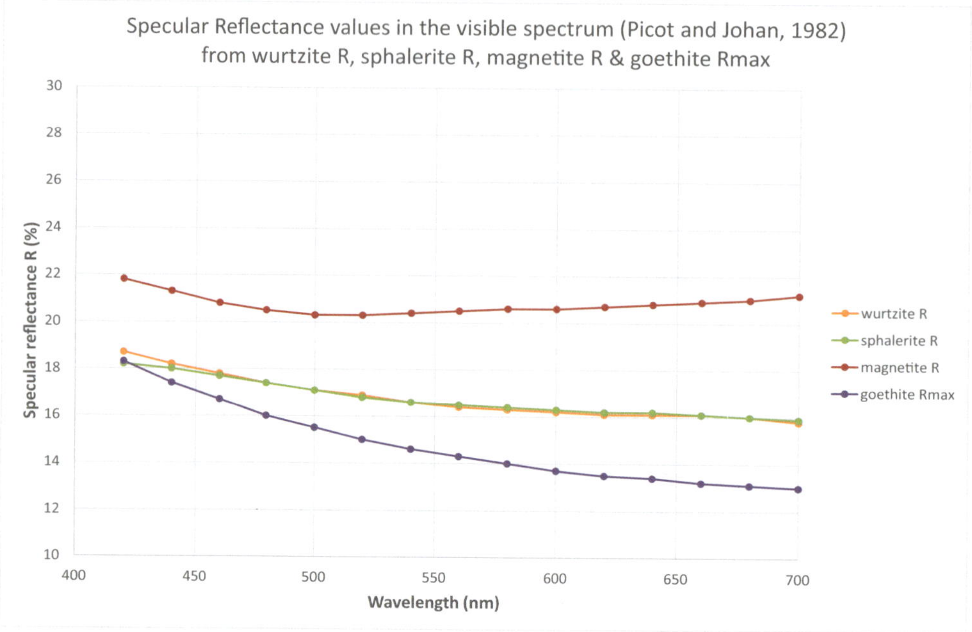

References

Bowles JFW, Howie RA, Vaughan DJ, Zussman J (2011) Rock-forming minerals, vol. 5A (2nd ed.) Non-silicates: oxides, hydroxides and sulphides. The Geological Society, London, 920p
Picot P, Johan Z (1982) Atlas of ore minerals. B.R.G.M./Elsevier
Uytenbogaardt W, Burke EAJ (1971) Tables for microscopic identification of ore minerals, 2nd edn. Elsevier, Amsterdam

Part III
Description of Gangue

Common Gangue Minerals

133

The positive identification of non-opaque minerals in reflected light is very difficult if not impossible, especially for the great number of minerals of relatively low refractive indices. The methods which are developed in this chapter give little promise of truly practical results [...] In spite of this, one will naturally wish to try to identify the non-opaque minerals in a polished section, at least approximately, using all available criteria. (Ramdohr 1980, v.2, p. 1101)

In Mineral Resources, gangue—abbreviated **gg**—is understood as the set of accompanying minerals of the ore that have no market value and that the miner or plant engineer must separate in order to make the ore economic. In fact, the raw ore extracted from the mine (known as ***ROM***, i.e., *run-of-the-mine ore*) very rarely has a metallic content (*grade*) sufficient for metallurgical processing. It is necessary to process it in the plant (*concentration* or *mineral processing plant*), usually at the foot of the mine, in order to achieve a grade high enough for market values. In this process, ideally, the aim is to eliminate all the gangue, without losing any of the ore, i.e. a 100% ore recovery, producing a concentrate with 0% gangue content, as well as pure gangue tailings with 0% ore. However, actual plants can't be so efficient. How far their results deviate from the recovery goal qualifies their performance.

The minerals discarded in the plant are known by names that vary according to the process or the region: *tailings, sludge, effluents*. These, due to their potential environmental impact, can represent a serious problem for the viability of the mine in sustainable conditions. However, there is currently a tendency to develop alternatives, to remove them and make them useful, recycling them as mine filler, aggregates or industrial minerals, or at least neutralize them. For this, it is necessary to mineralogically characterize the gangue, in addition to the ore, which may also contain deleterious or toxic components such as As, Hg, Se, Bi, Th, U, whose dumping must be avoided. The environmental impact has become an essential conditioning factor, as much or even more than the market value, for the viability of a mining operation. All of these factors make mineralogical control (predictive and follow-up) of the process indispensable, which must be integrated with all the information from drilling, geology, rock-mechanics, chemistry, processing plant, etc., following the approaches of *Geometallurgy* (cf. § 1.1).

133.1 Methodological Approach

Considering the variety of processes generating mineral deposits, as well as their hypo- and supergene alterations, gangue minerals can also be very varied and should be studied with transmitted light (cf. § 1.2.2), with possible support of micro-analytical techniques (*vbgr. Scanning electron microscope*, ***SEM***,[1] and *Electron probe micro-analyzer*, ***EPMA***[2]).

However, it should not be forgotten that:

- The gangue minerals of the ROM are the most closely related to the ore. They usually comprise a limited number of species (cf. Table 133.1). Most of these are very common and can be easily identified, with only a minimum of petrographic support, by any experienced observer. They comprise a few silicates (mainly silica or quartz and phyllosilicates/clays) and a few carbonates (calcite, dolomite, ankerite, seldom including minor species as malachite or azurite which, when abundant, may be worked as oxidized Cu ores), occasional sulphates

[1] https://serc.carleton.edu/research_education/geochemsheets/techniques/SEM.html.

[2] https://serc.carleton.edu/research_education/geochemsheets/techniques/EPMA.

R. Castroviejo, *A Practical Guide to Ore Microscopy—Volume 1*,
https://doi.org/10.1007/978-3-031-12654-3_133

Table 133.1 Common Gangue minerals and symbols (*abbreviations*)

Common GANGUE minerals & Symbols (*abbreviations*)			
Symbol	Name	Nombre (Spanish)	Classification
ab	Albite	Albita	Fs (Group)
adu	Adularia	Adularia	Fs (Group)
an	Anorthite	Anortita	Fs (Group)
ana	Anatase	Anatasa	Ox/Hydroxides
Anf	Amphibole (Group)	Anfíbol (Group)	Amphibole (Group)
ang	Anglesite	Anglesita	Sulfates (Class)
anh	Anhydrite	Anhidrita	Sulfates (Class)
ank	Ankerite	Ankerita	Carb (Class)
Arc	Clay	Arcilla	FSlc (Subclass)
bar	Barite	Barita	Sulfates (Class)
bio	Biotite	Biotita	FSlc (Subclass)
calc	Calcite	Calcita	Carb (Class)
Carb	**Carbonates** (Class)	**Carbonatos (Class)**	**Carb Class**
ch	Chert	Chert	Silica (Group)
Chl	Chlorite (Group)	Clorita	FSlc (Subclass)
dol	Dolomite	Dolomita	Carb (Class)
Ep	Epidote (Group)	Epidota (Group)	Slc (Class)
fl	Fluorite	Fluorita	Haluros (Class)
Fs, Felds	Feldspar	Feldespato (Group)	Fs (Group)
PSlc, FSlc	Phyllosilicates	Filosilicatos (Subcl)	FSlc (Subclass)
Grt	Garnet (Group)	Granate-Grupo	Slc (Class)
hb	Hornblende	Hornblenda	Amphibole (Group)
ili	Illite	Illita	FSlc (Subclass)
jar	Jarosite	Jarosita	Sulfates (Class)
jas	Jasper	Jaspe	Silica (Group)
Kfs	K Feldspar	Feldesp Potásico	Fs (Group)
lcx	Leucoxene	Leucoxeno	Ox/Hydroxides Class
lim	Limonite	Limonita	Ox/Hydroxides Class
Mica	Mica	Mica	FSlc (Subclass)
or	Orthoclase	Ortoclasa, ortosa	Fs (Group)
ov	Olivine	Olivino	Slc (Class)
pie	Piemontite	Piemontita	Slc (Class)
Plag	Plagioclase (Group)	Plagioclasa (Group)	Fs (Group)
Px, Pyx	Pyroxene (Group)	Piroxeno (Group)	Px (Group)
q, (qz, qtz)	Quartz	Cuarzo	q (Group)
rhc	Rhodochrosite	Rodocrosita	Carb (Class)
ser	Sericite	Sericita	FSlc (Subclass)
sid	Siderite	Siderita	Carb (Class)
Slc	**Silicates**	**Silicatos**	**Slc Class**
sms	Smithsonite	Smithsonita	Carb (Class)
srp	Serpentine (Group)	Serpentina (Group)	FSlc (Subclass)
tit (sph)	Titanite	Titanita	Slc (Class)
tur	Tourmaline	Turmalina	Slc (Class)
zir	Zircon	Zircón	Slc (Class)

COLOUR EXPLANATION
SILICATES
CARBONATES
OXIDES & HYDROX.
HALIDES
SULFATES

(barite, anhydrite, gypsum, jarosite), halides (fluorite, cerargyrite), mineraloids (mainly oxidized phases such as limonite or leucoxene) and organic or carbonaceous matter, in different degrees of evolution. Of course, other gangue minerals can be found, but they are usually associated with particular types of deposits (e.g., olivine, pyroxenes, serpentine, spinel, as companions to chromite or MGP), so their tentative recognition is usually no problem, taking the association into account (cf. **Annex 2**).

- The low value of Reflectance, **R**, of common (transparent) gangue minerals, corresponding to their typical non-absorbing character (cf. § 1.2.2), in general allows them to be easily distinguished from ores (much higher **R**). As established by the Fresnel equation and clearly shown in Fig. 1.2, the range for **R** of gangue minerals is between ≈ 3% (fluorspar, ***fl***) and ≈ 12% (cerussite maximum, ***crs***), the most frequent values being ≈ 4–5% (quartz, feldspar, sericite). The **R** values of ores, on the other hand, lie in the range ≈ 15–95%, with few exceptions (vbgr. monazite, **R** ≈ 8%, cassiterite, **R** ≈ 12%). In Fig. 1.2, a typical **low-R** ore is depicted -sphalerite, ***sp***, with **R** ≈ 17% (for more data, cf. § 2, ore descriptions). If the observation is complemented with an oil immersion objective, the contrast between ore and gangue is magnified since, in immersion and according to the Fresnel equation (Fig. 1.2), the common gangue (quartz **q**, clays **Arc**, phyllosilicates **Fslc**, calcite **calc**) disappears. As its reflectance **R** goes to ≈ 0%, it looks black, like a hole, which is not the case for any ore.

These considerations show that, in general, an approximate characterization by reflected light, **RL**, of common gangue minerals (or, at least, their categorization into broad types with similar properties) can be simplified and is possible. The series Figs. 66.1 and 66.3 illustrates the potential of simple observation with **RL** for characterization, in favorable cases, not only of the ore (jacobsite ***jac***, braunite ***bra***) but also of the gangue (quartz ***q***, piemontite ***pie***). This sample is composed of two well-crystallized minerals with remarkable contrast in color, brightness (i.e., reflectance, **R**) and internal reflections (**IR**). The quartz, colorless, dark (**R** ≈ 4–5%, like resin), appears furrowed by innumerable microcracks lined with fluid inclusions and exhibits its anisotropic character (*d+p*) and its typical undulatory extinction in response to deformation. In contrast, the mineral piemontite, of the epidote group, stands out in apparent relief and appears lighter, as corresponds to its higher reflectance (**n** ≈ 1.75–1.8, thence calculated **R** ≈ 7–8%). It is distinctly idiomorphic, with pronounced prismatic habit and exhibits twinning (*d+p*) and its typical reddish-brown/purplish colors transmitted, above all, by its **IR**.

Basic training and, if possible, a few years of experience in metallogeny and petrology are invaluable aids in this task, as is the quick preliminary examination with a stereomicroscope (binocular microscope). However, even if reflected light examination usually provides a valuable general approach to gangue characterization, its conclusions should be confirmed by rapid examination of a representative sample with transmitted light, **TL**, as an ad hoc technique for transparent minerals.

The most practical solution, however, is very easy. The sample preparation techniques current today (cf. **Annex 3**) allow us to make *polished thin sections* (**PTS**) and also *doubly polished thin sections* (**DPTS**, used for the study of fluid inclusions) which can be studied both with TL and with RL. The series of photomicrographs Figs. 86.1, 86.2, 86.5 and 86.6, which directly compare in TL and RL the same field, are practical examples of application of these complementary techniques, combining both types of observation on PTS, to characterize a very low reflectance ore (***mz*** monazite, with **R** ≈ 8%).

133.2 Selection of Gangue Minerals

The most common gangue minerals have been selected in Table 133.1, identifying them with the same symbols (abbreviations) as in **Annex 1.3** (Gangue). As Table 133.1 shows, most of the gangue species are classified in only two of the Berzelius classes, namely (in approximate order of decreasing abundance and noting the **main families** within the classes):

- *Class VIII: **Silicates***
- *Class V: **Carbonates**, Nitrates, Borates*

As for *Class VI* (***sulphates***, *chromates, molybdates, tungstates*), the presence of sulphates such as jarosite ***jar*** (associated with limonite ***lim*** in oxidizing environments) is sporadic, while others such as barite ***bar*** or anhydrite ***anh***, are abundant only in some types of deposits, *vbgr. Mississippi Valley* and, apart from their presence in evaporite deposits, *copper porphyries*, respectively.

There are also a few representatives of *Class III* (***Oxides and Hydroxides***) and one—fluorite—of *Class IV* (***Halides***). The scarce presence of oxides in Table 133.1 is due not so much to their actual scarcity as to the fact that almost all of them, when abundant, are beneficiated as metallic ores (vbgr. cassiterite ***cst***, hematite ***ht***, manganite ***mng***, magnetite ***mt,*** pyrolusite ***prl***, chromite ***chr***, cuprite ***cup***, rutile ***rt***, etc.) and, therefore, are not treated as gangue, but are described as ores. This can also occur with other species, such as, among the **carbonates**, bastnäsite ***bst***, a valuable REE (rare earth) ore, malachite ***mch*** and azurite ***az***, valuable Cu ores in oxide-type deposits. Likewise, for scheelite ***sch*** among the **tungstates** and cerargyrite ***crg*** among the **halides**. The group of *oxy-spinels* represents a peculiar case, since it includes about thirty minerals, from the typically transparent minerals like spinel (*s.str.*) to the clearly opaque minerals like magnetite ***mt***. They cover a wide compositional range which (together with its abundance in the ROM ore), determines its value as ore—vbgr. chromite ***chr***—or its disposal as gangue.

Most of the **silicates** are described by grouping those with similar properties into subclasses (e.g., *Phyllosilicates*) or groups (e.g., Silica or Quartz, Feldspars), to facilitate the exposition by systematizing it (subdivisions according to Berry et al. 1982).

References

Berry LG, Mason L, Dietrich RV (1982) Mineralogy, 2nd edn. Freeman & Co., New York, p 561

Ramdohr P (1980) The ore minerals and their intergrowths, 2nd edn, 2 vols. Pergamon Press, Oxford, 1205 p

134 Gangue Characterisation

134.1 Criteria for Reflected Light Characterisation

The brightness responds to the visual perception (subjective) of the Specular Reflectance (which itself is an objective measure[1]: **R** = I/I_0). The brightness is a very important data for the recognition of the gangue. In fact, **R** is the property whose systematic analysis offers the most useful and immediate results to appreciate the possibilities of characterisation with reflected light.

Table 134.1 presents complementary information to that shown in Fig. 1.2. It gives the values of **R** (according to Criddle 1998: Table 5, p. 46, modified and simplified in Table 134.1[2]), measured or estimated according to the Fresnel equation (§ 1.2.2):

$$R = \frac{(n-N)^2}{(n+N)^2}$$

as an equation of the refractive indices **N** (of the medium) and **n** (of the mineral), for transparent minerals (gangue) or weakly absorbing minerals (ores with low reflectance, such as pyrochlore, cassiterite or chromite). In the latter (more or less absorbing ores), when the absorption coefficient **k** acquires significant values, it must be included in the Fresnel equation[3] and **R** no longer depends only on **n**, both with dry objectives (air, **N** ≈ 1) and immersion in oil (with **N** = **1**.515).

The first thing that stands out in Table 134.1 is the immediate natural separation of the gangue minerals (marked with the same background colours as in Table. 3.1) and the ores[4] (highlighted in green) into two groups. These are characterised by their lower reflectances (gangue minerals), and by their higher reflectances (the ores), the threshold being at **R** ≈ 10% (air). Another fact to take into account is the disappearance of most of the gangue minerals in immersion (**R** ≈ 0%), which allows differentiation of them immediately from the ore. This shows how useless the immersion may be to distinguish from each other the common gangue minerals. They are all equally black, as empty holes.

There is also a strong predominance of silicates and, to a lesser extent, carbonates among the gangue components, in contrast to the scarcity of sulphates, oxides and halides, although these may be abundant in certain deposits. Some examples are found in the supergene oxidation zone of sulphide deposits, where sulphates and oxides are typical, or in stratabound deposits of the *Mississippi Valley type*, where fluorite can become dominant and can even be extracted as the main ore (Illinois-Kentucky Metallogenic Province, USA).

Finally, Table 134.1, ordered according to increasing **R** values (1% increments) and corresponding **n** values, shows once again that the usual gangue occupies a very

[1] The *specular reflectance*, **R**, is defined by the quotient $R = I/I_0$, where I_0 is the luminous intensity incident on a reflective surface; **I** is the intensity of the light reflected by it; in both cases the incidence being orthogonal (i.e., the angle between each ray—incident and reflected—and the surface is 90°).

[2] The values presented in this Table 134.1 are only approximate, indicative, to facilitate discussion. Strictly speaking, not only the values of the absorption coefficient ***k*** but also the *birefringence must be taken into account*. The latter, which is remarkable in many minerals, is so pronounced in some (*as* ***Carb***, cf **n** values in Table 134.1, for example, ***calc***: 1.49–1.66) that it translates into a manifest *birefringence*, directly perceptible by the human eye, without the need of a reflectometer.

[3] When dealing with absorbent or opaque (non-transparent) media, the absorption coefficient is significant and its value must be entered in the Fresnel equation to calculate R:

$$R = \frac{(n-N)^2 + k^2}{(n+N)^2 + k^2}$$

[4] Only some of the less reflective ores are shown, which are almost all oxides.

R. Castroviejo, *A Practical Guide to Ore Microscopy—Volume 1*,
https://doi.org/10.1007/978-3-031-12654-3_134

Table 134.1 Gangue minerals: reflectances (R %: in air & in oil) as a function of n

n	R%: air	R%: oil imm.	Minerals
1.00	0	0.4	Air
1.33	2	0.04	Water (1.33), Ice (1.31)
1,42	3	0,01	Fluorite, opal (chalcedony, jasper: 1.43)
1.50	**4**	**0.0**	**Quartz**, or, **calcite** (1.49-1.66), **Mica** (1.54-1.61)
1.58	5	0.04	Barite, biotite, rhc (1.6-1.8), sms (1.62-1.85)
1.65	6	0.2	Tourmaline, olivine, malachite (1.66-1.91)
1.72	7	0.4	Hornblende, Grt (pyrope), azurite (1.73-1.84)
1.79	8	0,7	Grt (almandine, uvarovite), cerussite (1.8-2.08)
1.86	9	1.0	Garnet-Grt (andradite)
1.93	10	1.4	Titanite (1.9-2.03), zir (1.93-1.97), lepidocrocite
1.99	11	1.9	Pyrochlore, sulphur (1.96-2.25)
2.06	12	2.3	Cassiterite (≈2.0), chromite (≈2.07 a 2.16)
2.13	13	2.8	Baddeleyite (2.1-2.2)
2.20	14	3.4	Tantalite (2.19-2.43)
2.26	15	3.9	Goethite (2.26-2.39)

COLOUR EXPLANATION	*SILICATES*
	CARBONATES
	ORES (OXIDES, etc.)
	HALIDES
	SULFATES

restricted range of reflectances: R 3% to 10% (even more restricted for Ramdohr 1980, p. 1101[5]: 4%–9 1/2%). This restriction is even more pronounced if one considers the gangue minerals really frequent in most orebodies—quartz (or silica group), phyllosilicates and carbonates of the calcite-dolomite type, whose reflectances are between 3 and 5%. In the particular case of carbonates, their high birefringence typically shows up as bireflectance (5–6%), visible if contiguous grains are observed. The remaining gangue minerals can reach up to ***R*** ≈ 10%, with rare exceptions (*vbgr.* anatase or brookite: 18–19%, not represented in Table 134.1, since their presence is commonly in trace amounts or is comprised in leucoxene ***lcx***).

In general, the differences in brightness that the observer must perceive to visually distinguish the gangue species defy the threshold of sensitivity of the human eye. Most of them appear as dull, dark minerals, whose apparent brightness can be masked by internal reflections, IR. Abundant IR may increase the measured values of **R**, exceeding by ≈ 1–2% of the theoretical values. All this makes it necessary to resort to other properties as complementary information, in order to achieve reliable visual characterisation, as far as it is possible with reflected light.

In the following, this characterisation is approached, by means of a simplified description of transparent gangue minerals as they are observed under the RL microscope. It is done in a free format, illustrated with photomicrographs but without following a prefixed model, as was done to describe ores (cf. Chapter 1: *Boxes 1.1 and 1.2*). A systematic description of transparent minerals that takes into account all the relevant optical properties can only be made with transmitted light, as is done for the petrographic study.

Because of their greater abundance, the silicates and carbonates will be described first. It is important to remember that the reflectance of the ***epoxy* resins** typically used for the preparation of polished sections is precisely at the level **R** ≈ 4% to 5%, the same as the most frequent gangue species, such as quartz, orthoclase, albite, sericite, calcite, etc. This is not a problem for the human observer since, although the colour (grey) and lustre are identical, a minimally trained eye easily recognizes the grain contours and thus can distinguish gangue from resin without error. For automated systems (*Part II*, vol. 2) a specifically designed preparation method must be used (cf. *Annex 3*). Pores, **voids** or pull-outs in the preparation are black, due to their zero reflectance (void) and are therefore usually easily distinguishable.

[5] According to this author, with a little practice it is possible to estimate refractive indices from the relative reflectance observed first in air and then in oil, using a tabulation of the type shown in Table 134.1. In his opinion, with very careful observation one can recognize, *as a slight veil*, a reflectance as low as ¼ %, the **R** equivalent of the n-index of calcite in immersion, while an **R** of ½ % is already detectable. These comparative observations help to distinguish carbonates.

134.2 Recognition of Most Frequent Types of Gangue

To introduce the discussion of the different gangue species, the application of the above criteria is illustrated by the analysis of a practical example, to characterize a sample with varied mineralogy (cf. photomicrographs Fig. 134.1a, b). It is a mineralurgical concentrate of scheelite (***sch***), a low Reflectance ore (**R** ≈ 11%) and is, therefore, susceptible to confusion with gangue for a hasty observer. However, its higher brightness and its **IR**, abundant and colourless, clear any doubt.

The problem now posed is the **RL**, Reflected Light, distinction of gangue species from each other. An outstanding difficulty for their RL characterisation is the inadequacy of RL to provide rigorous information about important diagnostic optical properties of transparent minerals. These properties are typically documented by routine **TL**, Transmitted Light, observation:

- *Color and pleochroism*: the gangue species in **RL** usually look uniformly gray, although there are cases in which some color is perceived. *vbgr.* malachite, due to the effect of extremely intense and colored internal **IR** reflections.
- *Cleavage*: seldom clearly visible on the gangue polished surface.
- *Refractive index*, **n**: cannot be determined directly as in **TL** (Becke's line) but can be estimated from the brightness, which depends on the value of **R**. This depends on **n** according to the Fresnel equation (cf. *§ 1.2.2*). In this way, species with strong *birefringence*, such as carbonates, can also be distinguished, since this is seen in **RL** as bireflectance, i.e., variation of the grey level with orientation.
- *Extinction and interference colors* (*d+p*): often completely masked by the intense diffuse light reflected from within the transparent minerals (by the **IR**).
- *Twinning*: the same happens as with extinction. In spite of this, there are minerals, such as carbonates, in which it is frequently seen (*d+p*) and can be diagnostic.

As explained (cf. *§1.4.1* and *§1.5.4*), a relative strength of **RL** is the possibility of easily determining the objective and precise (multispectral) values of **R**, which allow the automated ore characterisation by artificial vision and its industrial applications in problems of Mineral Resources and Geometallurgy (*AMCO System,* cf. *Part II*, vol. 2).

It is precisely the visual estimation of the Reflectance **R** value in gangue species that is the usual starting point for their characterization. The comparison of the brightness (or *grey level*) provides an immediate selection criterion to assign each component to one of the established ranges or categories for the **R** values appreciated in the sample.

In the case analyzed, the range of variation of **R** is relatively large: from 4% (quartz, resin) to about 60% (steel, resulting from the contamination of the concentrate by shrapnel detached in the milling process, three tiny, white and very bright particles in Fig. 134.1a), as can be deduced from the minerals present (cf. Fig. 134.1a) and their **R** values (cf. *Specular Reflectance plots*). However, only **R-values** ≤ 11% are frequent. These correspond to the ore, scheelite ***sch*** (R ≈ 11%), and the other grey minerals, darker than sch (**R** < 11%).

The exceptions are very few. Apart from steel an artifact of external origin, they comprise only the three particles described as *greyish-white* (Fig. 134.1a), two towards the centre (one of them, tiny, marked with an arrow in Fig. 134.2a, b) and one towards the lower right corner of the image. For these, visual examination estimates an **R-value** between 20 and 25%, certainly >11% but <<60%. The spectral curve of two of these particles—the larger, euhedral particle in the center, and the particle in the lower right corner of the image—suggests probable rutile, **rt** (or possible brookite, **bk**[6]), while the third (tiny, Ø ≈ 0.1 mm, somewhat darker) is probable wolframite, **wf**.

The remaining **R** charts (with reference also to hematite ***ht***, titanite ***tit***, cassiterite ***cst***, absent in the image), delimit all the mineral species present, **R** < ht. Disregarding the scarce rutile, ***rt***, and wolframite, ***wf***, the common gangue species (discussed below) lie between the thresholds:

- **R** < R titanite ***tit***, cassiterite ***cst*** or scheelite ***sch***, and
- **R** ≥ R quartz **q**, and resin, i.e., between ≈ 5 and 10%.

To recognize the gangue minerals, successive steps are described. In order to facilitate the follow-up, the species now discussed are marked in duplicate images (Fig. 134.2a, b) for comparison.

As a first step, the darkest phases (*minimum* **R** around 4%) are distinguished. They comprise the *epoxy* resin and minerals such as quartz and orthoclase ***or*** or alkali feldspars ***Fs***, as shown in Table 134.1. Three grains of ***q*** corresponding to this R value, identical to that of the resin, have been joined with a line across (white, vertical line, Fig. 134.2a, left). Both materials, q and resin, would be confused by their brightness. They are only distinguished by the human eye, because the eye can easily outline the contour of each grain.

---

[6] The two minerals, rutile ***rt*** and brookite ***bk***, are polymorphs (TiO_2), similar in appearance under RL. The measured spectral values—cf. *R graphs*—could be valid for both, but the remaining properties suggest rather ***rt***. Their **IR** are not seen, either because of opacity (W-rich varieties, Bowles et al. 2011), or because of possible luminous supersaturation of the environment (minerals with very abundant **IR**, cf. Fig. 134.1b).

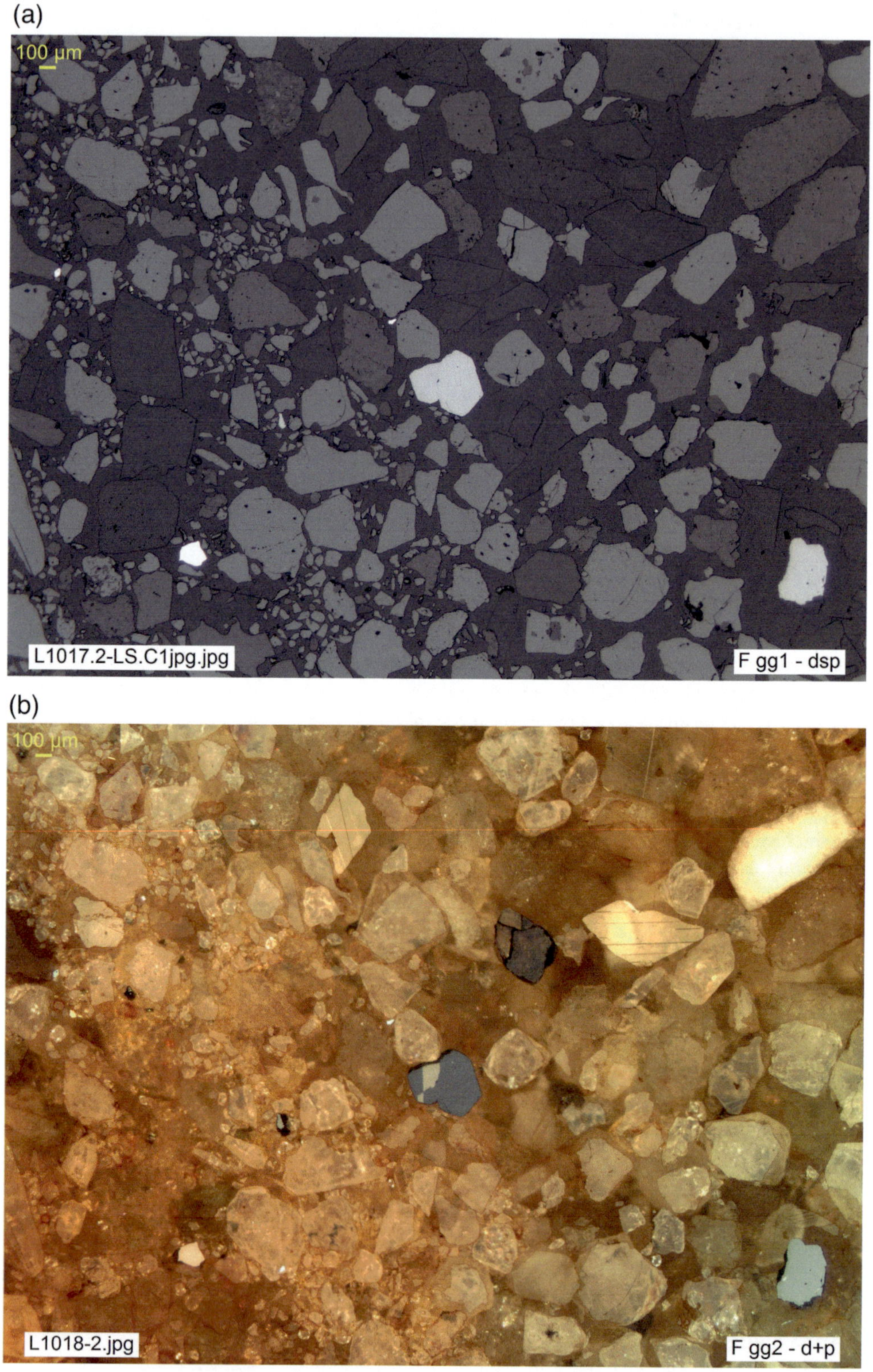

Fig. 134.1 **a**, **b** (Same scene, **dsp** and **d+p**). W-skarn milling concentrate, composed essentially of transparent minerals, including the major one—scheelite ore (***sch***). In order of decreasing **R** (grey tone, getting darker and darker): traces of mill steel (three white, shiny micro-shards), rutile and wolframite (***rt***, ***wf***: three grains, greyish-white), ***sch*** (light grey), ***Px***, ***Ep, Carb***, ***Plag*** and finally ***q*** and ***epoxy*** resin. With analyzer (Fig. 134.1b), the anisotropism is clearly seen in ***rt*** and ***wf*** but not in the other components (it is masked by the **IR**). Twinning is observed in ***rt*** and ***Carb*** (cf. Fig. 134.2b) and the steel shows typical pseudo-anisotropism. *Sample C1, Los Santos Mine, Cáceres*

Then, the top central line marks the two grains located at each end, which have different brightness. The one on the right, darker, looks like ***q*** (R ≈ 4%). The one on the left is clearly lighter, although still relatively dark (in comparison with the minerals marked by the third line and with respect to scheelite ***sch***), its **R** can be estimated at ≈ 5%. It might seem that they are different species, but if they are observed with crossed analyzer (Fig. 134.2b, *d+p*), it is found that the

Fig. 134.2 a, b Same images as above, on which the minerals discussed in the text have been marked with white lines and arrow. The arrow indicates a tiny ***wf*** particle. The ends of the WNW-ESE line (upp. centr.) rest on individual ***Carb*** grains (calc or dol), showing their characteristic bireflectance (different shades of gray, Fig. 134.2a) and fine polysynthetic twins (Fig. 134.2b). The NNW-SSE line (upper right) joins three gangue grains with higher **R** (intermediate gray tone, Fig. 134.2a) and colorless **IR** (Fig. 134.2b): ***CPx*** and/or ***Ep*** (clinozoisite) and the N-S line (left), three grains of **q** (very dark gray, like resin)

two have identical twins, typical of calcite ***calc***, which is bireflective, as are carbonates in general. The variation of these **R** values is not due to a mineralogical difference but to the different orientation of the two grains and is diagnostic of ***calc***, showing bireflectance (due to its birefringence, with **n** values between $\approx 1.5^{-}$ and $\approx 1.6^{+}$).

The third line (upper rightmost corner) joins three grains of gangue which are relatively light, although always darker than scheelite ***sch***. They correspond to minerals with an **n-value** in the higher range of the gangue and without the pronounced birefringence of the carbonates. Their **R** can be estimated to be around 7%, which would correspond to a range of **n** around 1.7 (≈ 1.65–1.75). They represent relatively frequent species in the gangue of this sample: clinopyroxene ***CPx***, epidote (or clinozoisite) ***Ep,*** garnet, clinoamphibole and their hydrothermal alterations, whose identification would require petrographic study.

Finally, a careful examination reveals some grains of intermediate brightness between these last two groups (**R** ≈ 5–6%), which can sometimes be identified (apatite, plagioclase) by their morphology or other properties. Some aggregates or microbreccias with clayey or cryptocrystalline matrix (sericite), related to deformation and secondary alteration processes, are also evident by their texture.

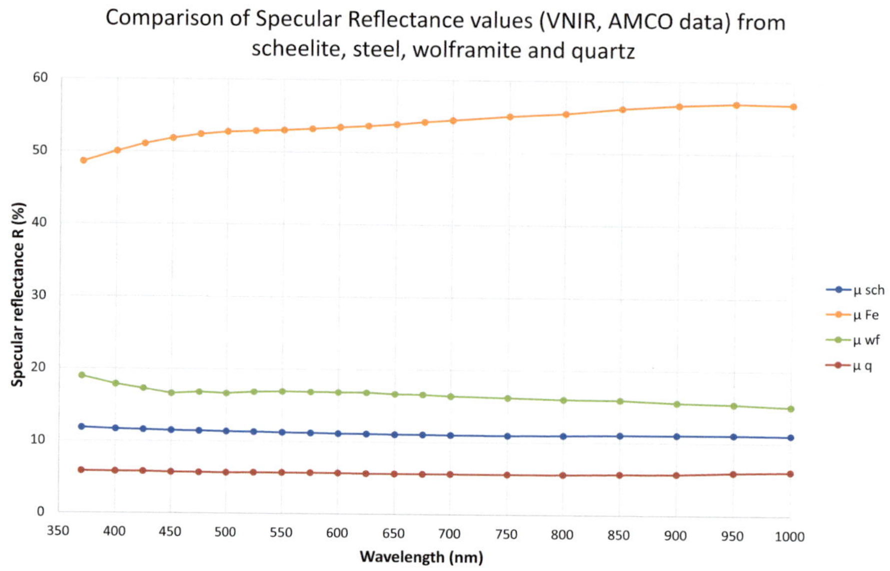

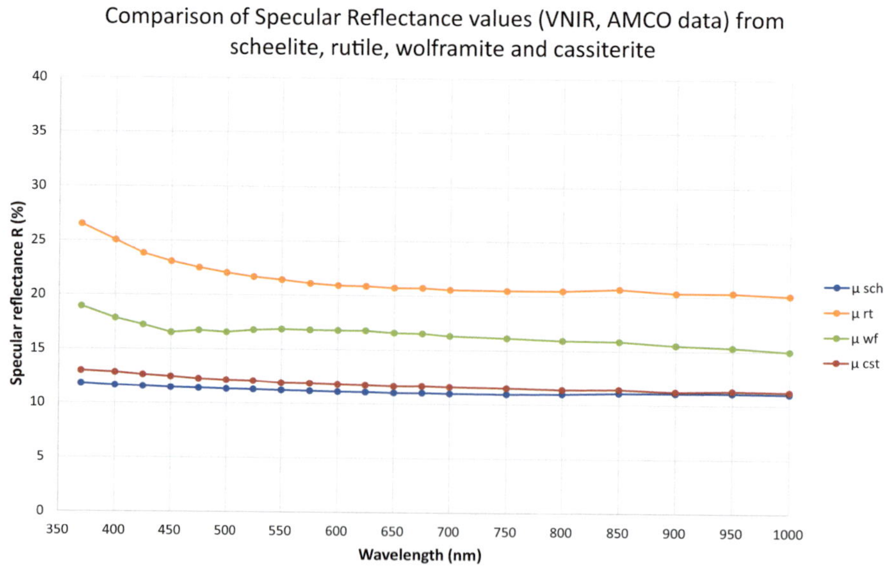

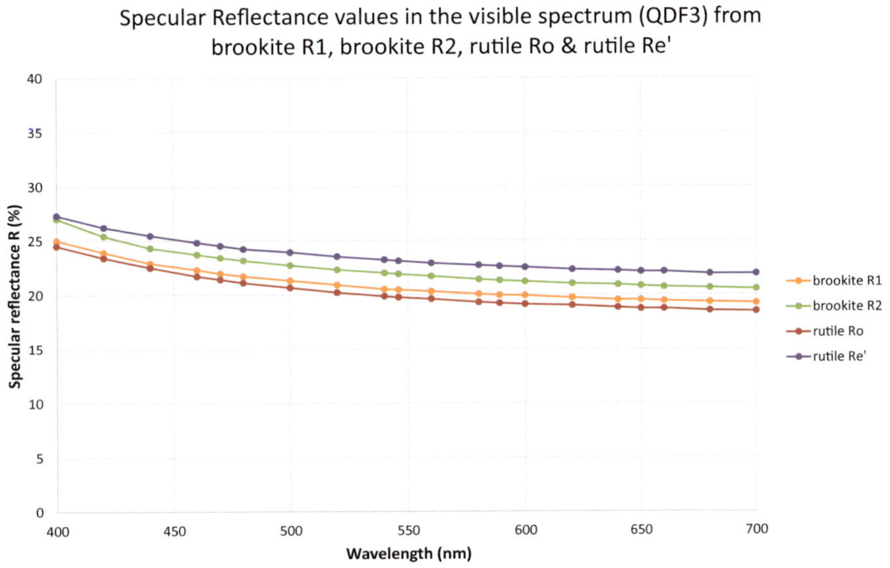
Specular Reflectance values in the visible spectrum (QDF3) from brookite R1, brookite R2, rutile Ro & rutile Re'
Specular reflectance R (%)
Wavelength (nm)
brookite R1
brookite R2
rutile Ro
rutile Re'

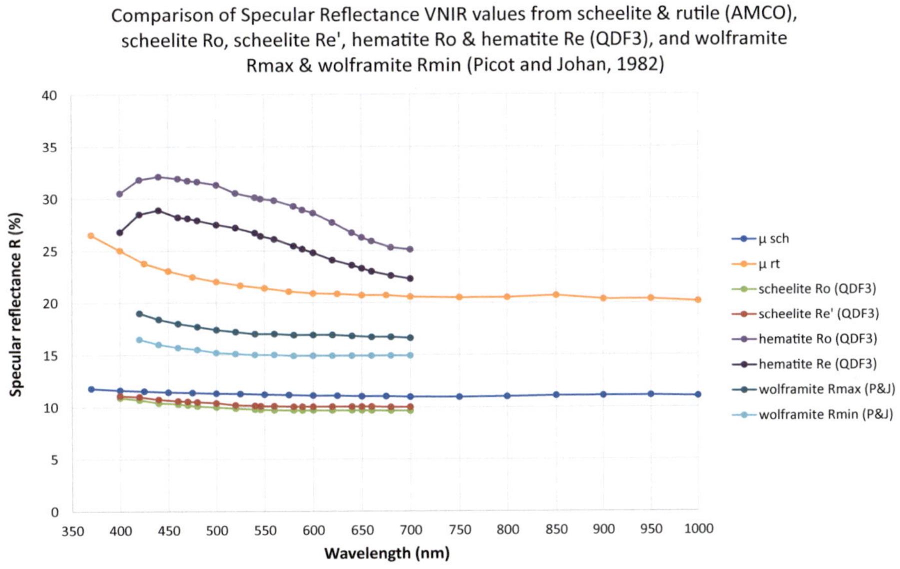
Comparison of Specular Reflectance VNIR values from scheelite & rutile (AMCO), scheelite Ro, scheelite Re', hematite Ro & hematite Re (QDF3), and wolframite Rmax & wolframite Rmin (Picot and Johan, 1982)
Specular reflectance R (%)
Wavelength (nm)
µ sch
µ rt
scheelite Ro (QDF3)
scheelite Re' (QDF3)
hematite Ro (QDF3)
hematite Re (QDF3)
wolframite Rmax (P&J)
wolframite Rmin (P&J)

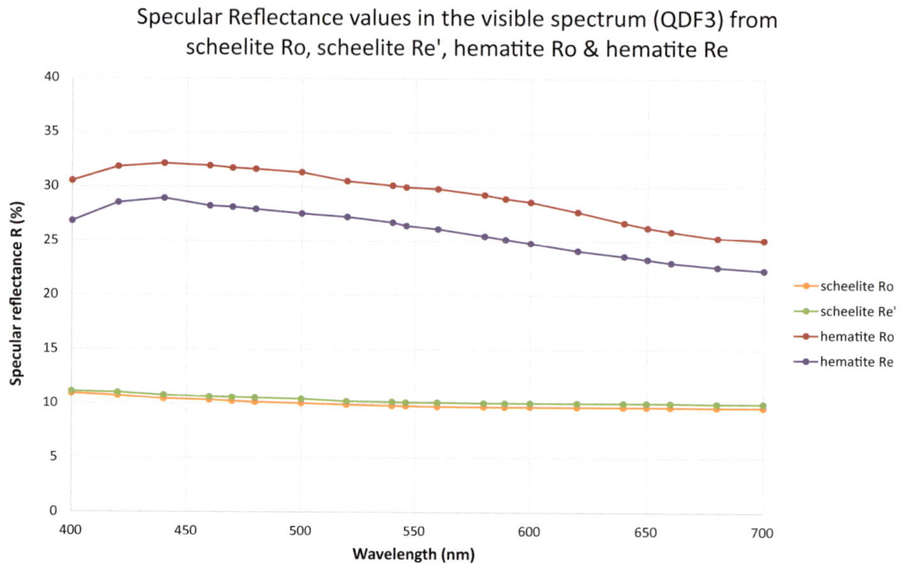

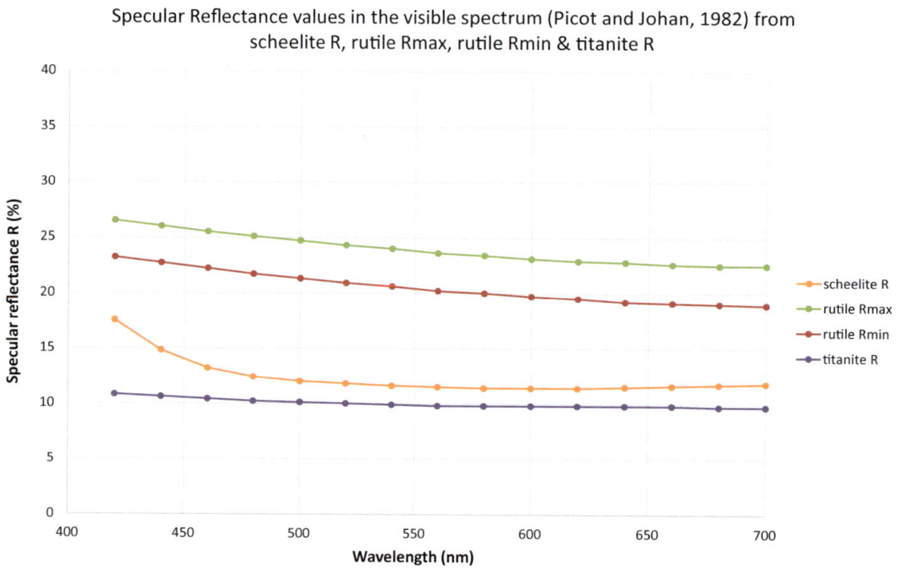

References

Bowles JFW, Howie RA, Vaughan DJ, Zussman J (2011) Rock-forming Minerals, vol 5A, 2nd ed. Non-silicates: oxides, hydroxides and sulphides. The Geological Society, London, p 920

Criddle AJ, Stanley CJ (eds) (1993) Quantitative data file for ore minerals, 3rd edn. Chapman & Hall, London, p 635

Criddle AJ (1998) Ore microscopy and photometry (1890–1998). In Cabri LJ, Vaughan DJ (eds) Modern approaches to ore and environmental mineralogy, vol 27. COM/IMA Short Course Series, Ottawa, Ontario, p 421

Ramdohr P (1980) The ore minerals and their intergrowths, vol 2, 2nd edn. Pergamon Press, Oxford, p 1205

Silicates 135

135.1 Tectosilicates. Silica. Feldspar Group

Among the tectosilicates, only the silica and feldspar groups comprise common gangue minerals. In the first group, quartz, ***q*** (including its cryptocrystalline varieties, such as chalcedony, jasper and chert, and opal, an amorphous and hydrated variety of silica) is very frequent, but not its polymorphs such as tridymite, cristobalite or the very rare coesite, stishovite and lechatelierite. Among the feldspars, ***Fs***, much less frequent in gangue than quartz, the alkaline ones usually predominate in epithermal environments and in deep deposits linked to acid magmatism—greisen formations, Sn–W, Li, Nb, Ta, TR, etc. veins—and calcic plagioclases, in skarn deposits or also in those associated with mafic rocks (Ti-mt, Cu–Ni Sulfides), as well as with ultramafic formations (Cr, EGP, Ni–Cu, etc.).

135.2 Silica Group

Quartz (*q*, SiO_2, Trigonal) is the most frequent silicate among the gangue components. It occurs in practically all hydrothermal deposits and in most sedimentary and metamorphic deposits. If we take into account its micro to cryptocrystalline varieties, abundant in some types of deposits such as precious metal epithermal deposits (*chalcedony*, *opal*, *sinter*), BIF (*Banded Iron Formations*) or ferriferous jaspilites (*jasper*) and massive sulphides (*chert*), among others, it can be concluded that the Silica Group, either as quartz or as any of the varieties mentioned, is practically ubiquitous among the gangue minerals, with few exceptions (*vbgr.* chromitites).

The recognition of crystalline quartz is not difficult. Its *morphology* shows up in typical prismatic and hexagonal sections. These are visible when it has grown in open space; for example, in low sulphidation epithermal environments. Being a stable mineral, it may keep a clear, *limpid* appearance in most environments, even in altered or weathered environments. Its high *polishing hardness* can give rise to occasional relief but does not prevent a *good polishing*. Some optical properties stand out:

- *Low reflectance*, ***R***, which makes it look dark grey, like resin, in air (*dsp*), and disappear (look black) in immersion (*osp*); compare Fig. 135.1a, b with Fig. 135.2. Thus, *neighbouring grains* of low reflectance ores, such as cassiterite ***cst***, scheelite ***sch***, wolframite ***wf***, magnetite ***mt*** or spinel ***sp***, appear much brighter (Figs. 135.3a, 135.5a and, above all, Fig. 135.4a, cf. Fig. 135.15a).
- *Abundance of* ***IR*** (internal reflections, Figs. 135.3b and 135.4b, cf. Fig. 134.1b, visible with crossed analyser, *d +p/o+p*), since it is transparent but its *anisotropic* structure gives rise to notable differences in luminosity depending on the orientation of the reticular planes intercepted by the light beam, or else mirrored by internal deformations. These may also be evidenced by *undulatory extinction* (cf. Fig. 66.3b; compare Fig. 135.22b with TL, Fig. 135.23b); however, fine-grained aggregates and cryptocrystalline varieties usually show a more uniform appearance (cf. Fig. 138.5a, b). Deformed grains often show cemented microcracks, with fluid relics captured in **FI** (*fluid inclusions*, cf. Figs. 135.3b and 135.4b).

It is found in the most varied paragenesis. Being very abundant, it is not unusual for it to occur in successive generations in the same deposit, showing euhedral or anhedral forms, depending on the conditions (growth in geodes and free spaces or competition with other phases to occupy space, respectively), and also depending on the companions (e.g., euhedral with galena ***gn*** or silver sulfosalts, Figs. 135.1a, b and 135.2; anhedral with cassiterite ***cst*** or wolframite ***wf***, which impose their own crystalline forms, Figs. 135.3a and 135.4a). cit can replace early ore generations, but it can also be replaced by different processes (*vbgr.* Fig. 1.43f or Fig. 1.47e, both in vol. 2). Its response to deformation under upper crustal conditions is brittle and, as an early mineral, brecciation is not uncommon (Fig. 135.5a, b).

R. Castroviejo, *A Practical Guide to Ore Microscopy—Volume 1*,
https://doi.org/10.1007/978-3-031-12654-3_135

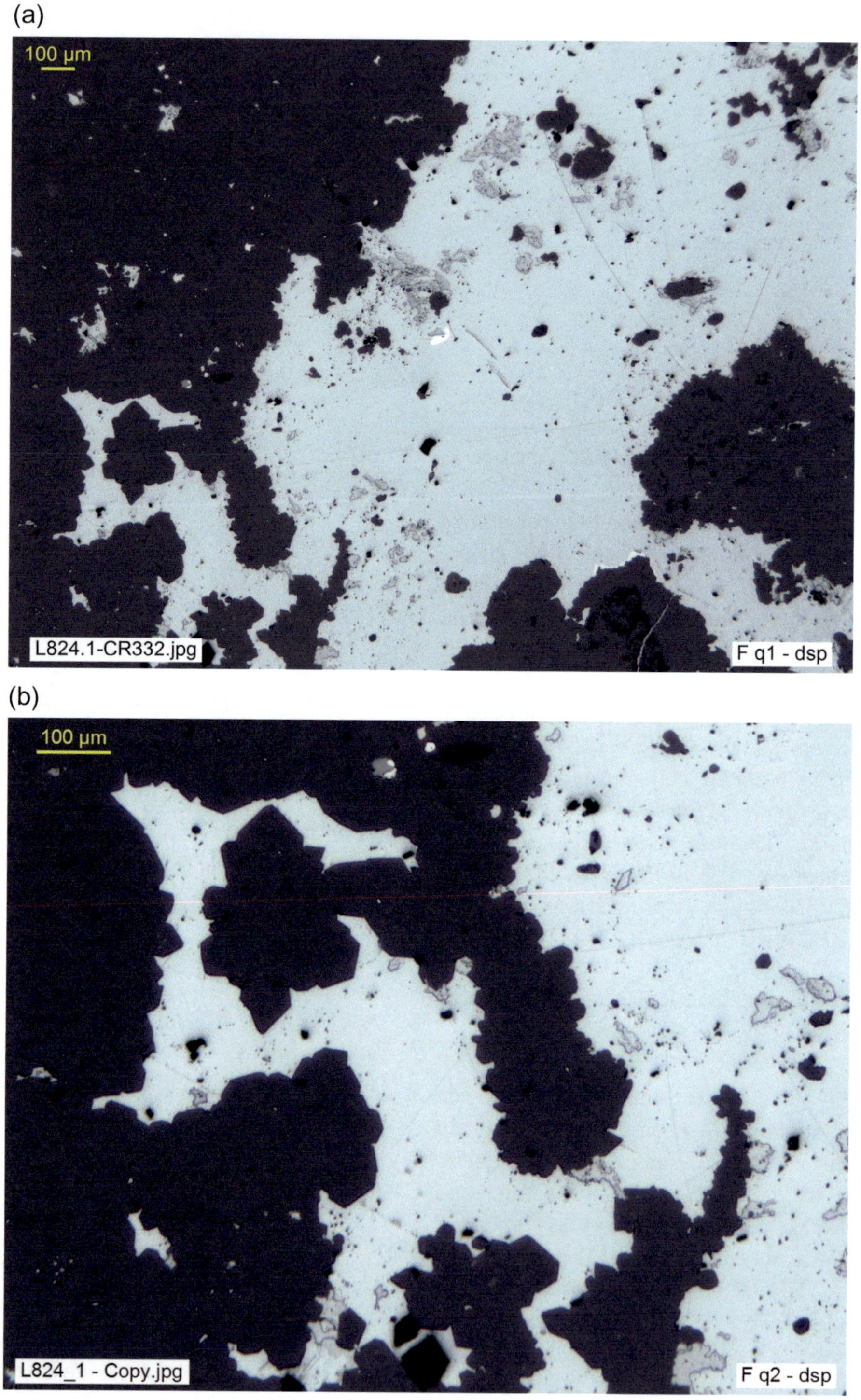

Fig. 135.1 **a**, **b** Ore of galena ***gn*** (white, with grey relics of hematite ***ht***) intergrown with quartz ***q*** (dark). While ***gn*** is anhedral, the ***q constantly*** shows an automorphic, although fine-grained, development. This is apparent in the prismatic and hexagonal forms, clearly perceptible increasing magnification (Fig. 135.1b). CR-332, Vrancice, Pribram, Czech Rep. (**dsp**)

To start, prior examination of the hand sample with the naked eye or, if possible, with a stereomicroscope or binocular magnifying lens is always recommended. This is usually enough to clear up any doubt for a mineral as common as quartz, even in cases where not all of these properties are clearly seen under the microscope, as may happen with micro to cryptocrystalline silica varieties. In these cases, its outstanding hardness, low **R** (dark aspect) and transparency (**IR**) are still useful criteria.

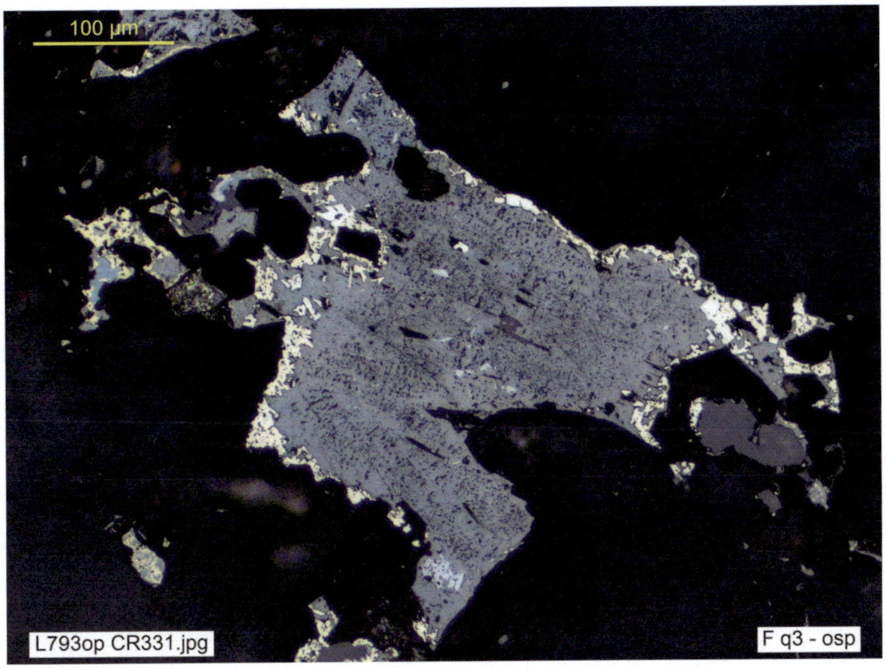

Fig. 135.2 Automorphic *q* intergrown with argentiferous sulfide ore (stephanite ***stp***, pyrite ***py***, chalcopyrite ***ccp***, sphalerite ***sp***, proustite ***pru***). Relatively early **q** shows incipient corrosion by the ore (e.g., in the indented hexagonal section, right). Stephanite ***stp*** shows a widespread black stippling, unavoidable result of light etching. CR331, Fresnillo, Zacatecas, Mexico (osp)

135.3 Feldspar Group

The minerals of the broad feldspar group, **Fs**, are more abundant than quartz, ***q***, in the earth's crust (common rocks) but much less in the gangue of metallic ores.

Fs share with q some properties such as a relatively high hardness (although Fs are somewhat softer than q) and transparency, low brightness or dark grey color in **RL** (reflected light), due to its low reflectance.[1] They differ in their morphology. Instead of the hexagonal shapes of hydrothermal q fill, Fs show occasionally other shapes, such as the pseudo-rhombohedral form of the tiny adularia crystals dispersed in epithermal ores. They differ, above all, in their appearance—instead of the clean transparency of q, Fs show often a cloudy aspect, due to their instability in the environment of numerous ore deposits; that is, in typical hypogene alteration processes, particularly in pneumatolitic-hydrothermal environments. Fs also alter under surface conditions (weathering).

This cloudy or "pitted" appearance contrasts with the usually limpid aspect of q (cf. Fig. 135.6a, b) and is due to the minute inclusions of secondary minerals (***q***, clay ***Arc***, sericite ***ser*** and other phyllosilicates ***FSlc***, carbonates ***Carb***) typical of the various alteration processes (propylitic, argillic, phyllic, etc.). However, there are also processes characterized by posthumous formation of feldspar, such as the potassic alteration of porphyry copper deposits, with late-magmatic deposition of potassium feldspar.[2]

Since Fs *twinning*, one of the most distinctive features of the various ***Fs*** species, is seldom distinguishable in RL, the presence of *Fs* can hardly be recognized with reflected light. However, sometimes typical features appear, such as rock textures implying plagioclases ***Plag*** (cf. Fig. 135.7a, b) which, with appropriate typological data (e.g., Fe–Ti concentrations in mafic rocks, Cr deposits in ultramafites, meteorites), make their identity obvious.

135.4 Phyllosilicates. Mica Group. Clay. Serpentine. Chlorite Group

The clay minerals (***Arc***), serpentine (***srp***), micas (***Mica***) and chlorites (***Chl***) share a leafy structure to which the Subclass Phyllosilicates (**FSlc**, **PSlc**) owes its name. This structure is manifested by a distinct cleavage, in parallel planes, whose intersection with the polished section is occasionally visible in **RL** (***Mica***, ***Chl***), and may be enhanced by IR (internal reflections). As a whole, they are soft minerals (comparable to galena ***gn***) whose polishing is often problematic, resulting in a rough or pitted appearance, especially in secondary

[1] R_{Fs} is somewhat more varied than R_q, considering the slightly lower values of refractive index, *n*, in the alkali feldspars ($n \approx 1.52$, ***or***) and somewhat higher in the calcic plagioclases (from $n \approx 1.53$, ***ab***, to $n \approx 1.59$, ***an***), as opposed to the intermediate value ($n \approx 1.55$) of ***q***, but converted to **R** values—cf. § 1.2.2, Fresnel function—these small differences are negligible and generally imperceptible to the human eye.

[2] However, it is not uncommon for this relatively early potassic alteration to be obliterated to varying degrees by subsequent q-sericitic or phyllic alterations, in which case the neoformed ***Fs*** will in turn be replaced, partially or totally, by secondary ***FSlc***.

Fig. 135.3 **a**, **b** (Same scene: **dsp** and **d+p**, *respectively*). Vein infill of quartz ***q*** (dark grey, anhedral) and cassiterite ***cst*** (light grey, bireflective, euhedral). With analyser, Fig. 135.3b, both minerals show abundant **IR** and some microfractures stand out (sharper in ***q***), lined with secondary fluid inclusions, **FI**. Twins visible in the distinctly anisotropic **cst**, while ***q*** anisotropism is not apparent due to abundant IR. TF Pn9, Montánchez, Caceres

micro- or cryptocrystalline aggregates (sericite, illite, smectites, etc.).

Their theoretical reflectances are broadly similar to those of quartz and feldspar. Although the maximum values, n_γ of the refractive indices of biotite ***bio*** and chlorite ***Chl*** (n_γ up to ≈ 1.69 and 1.60, resp) exceed those of ***q*** and slightly those of Plagioclase, only under conditions of exceptional contrast, good polishing and proper orientation does the difference become apparent as a lighter shade or higher brightness, perceived by the eye (Fig. 135.8a). Usually, on the contrary, they look even darker than ***q*** (Figs. 135.9a, 135.10 and 135.13a).

The very fine grain size of ***Arc*** and **srp** does not make it easy to distinguish their crystalline properties under the optical microscope (Fig. 135.14a, b). Neither does the reflectance (calculated from their n-values) make an appreciable difference to that of ***q*** and ***Fs***, but they can often be differentiated by the poorer *polishing quality*. Polishing is often deficient in sericite **ser** and *Arc* minerals because of

Fig. 135.4 **a**, **b** (Same scene: **dsp** and **d+p**, *respectively*). W ore in ***q*** vein (dark grey): prismatic crystal composed of relic wolframite ***wf*** (very light brownish grey tone) and secondary scheelite ***sch*** (light grey), pseudomorphic of ***wf***. Anisotropism (Fig. 135.4b) barely distinguishes ***q*** and ***sch***. Microcracks with secondary **IF** in both minerals. CR320, m. ETSIMM (42-67-2)

their low hardness and the consequent pull-outs produced in the polishing process. The result is an irregular and porous surface, which necessarily looks darker under the microscope (Fig. 135.13a), because a perfect specular reflection of the incident beam is not possible, as it would be on a smooth, well-polished surface.

The *color* transmitted by IR—*d+p* or *o+p*—sometimes helps to identify serpentine ***srp*** (greenish IR: Fig. 135.11a, b), chlorite ***Chl*** (blue-green to colorless IR, Fig. 135.12a, b), and clays ***Arc*** (colorless—Fig. 135.13b—to yellowish-brown, intensely colored sometimes by impregnation of iron oxides, Figs. 135.11b, 135.12b and 135.14b). Likewise, the *cleavage* may help to identify mica if the right orientations are found to make it visible (cf. Figs. 135.8a, b and 135.9a, b). Nevertheless, it is often imperceptible or can only be inferred from other observations (e.g., IR, alteration

Fig. 135.5 **a**, **b** (Same scene: **dsp** and **d+p**, *respectively*). Tennantite ore (***tnn***, light grey, upper right, intergrown with sphalerite ***sp***, intermediate grey, chalcopyrite ***ccp***, pyrite ***py***) and cassiterite ***cst*** (dark grey, lower left), ***q*** gangue (darker). Microbrecciation of the **cst-q-tnn** contact. RubSL6.1, Böhm. Zinnwald, Erzgebirge

(a)

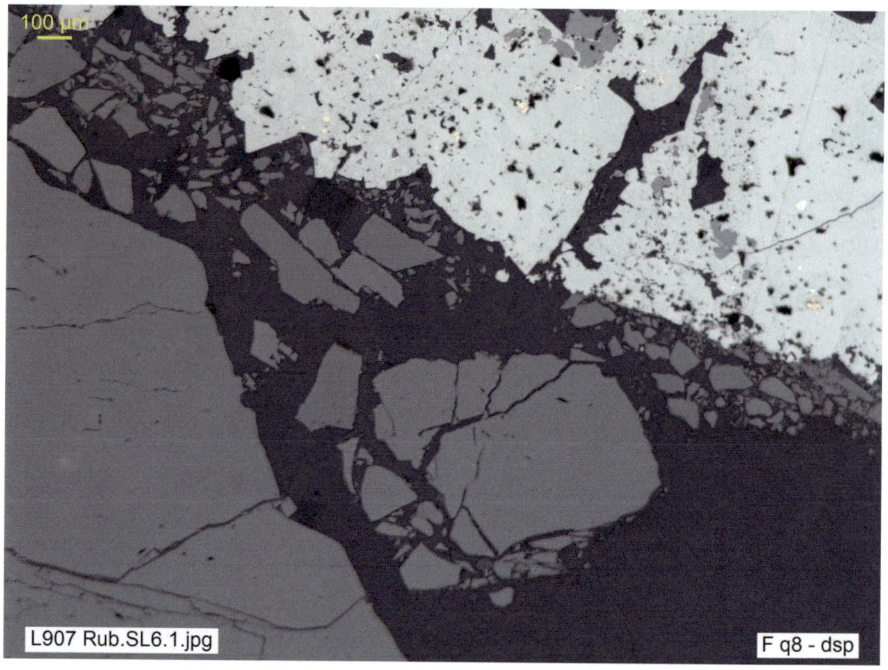

(b)

features, Fig. 135.10a, b). In addition, tabular or leafy *morphology* can be a useful aid (e.g., ***Chl***: Fig. 135.12a; ***bio***: Figs. 135.8a and 135.9b).

In comparing biotite ***bio*** and *q* (Figs. 135.8a, b and 135.9a, b), it is interesting to note that when there are small differences in *reflectance* between ***PSlc*** and the other components of gg, these are much better seen in air than in oil. In fact, with dry lens (Fig. 135.8a) the higher brightness of bio can be clearly perceived (it looks brighter than *q*, even if the difference is small), while in immersion (Fig. 135.9a) the reduction of **R** is so marked that both minerals *disappear* (they look black), and the difference between them becomes imperceptible to the eye. They can no longer be distinguished by brightness, not even by morphology. On the contrary, morphology and cleavage are better perceived with crossed analyser (comp. Figs. 135.8a, b and 135.9a, b) and, because the **IR** are reinforced, more clearly in immersion (Fig. 135.9b) than in air (Fig. 135.8b).

(a)

(b)

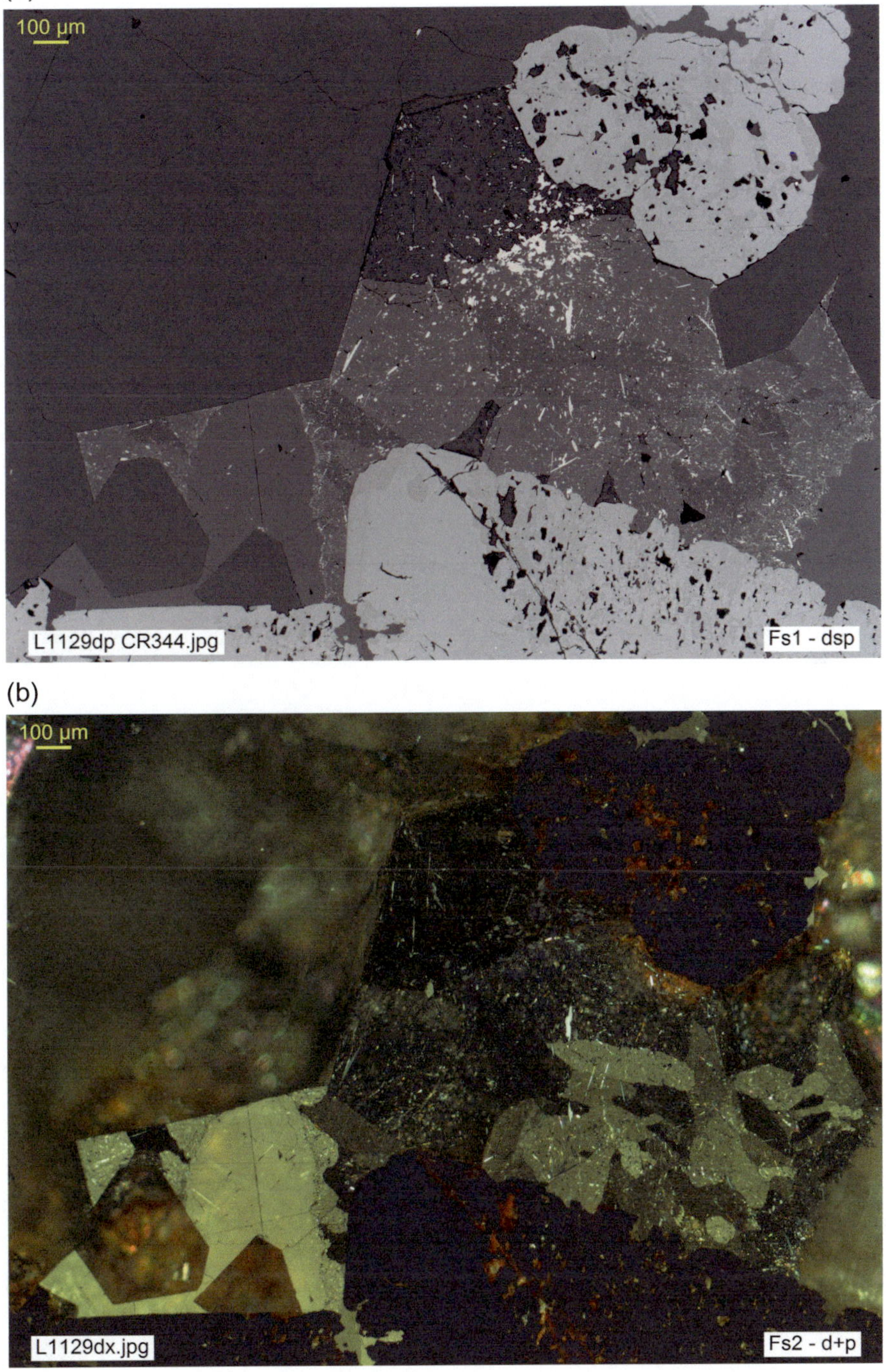

Fig. 135.6 a, b (Same image: **dsp** and **d+p**, *respectively*). Sequential vein fill (1) alabandite ***alb*** (light grey) and quartz ***q*** (dark grey), both idiomorphic, leaving a microgeode (center of image) sealed by (2) a posthumous generation of ***Carb*** (bireflective, occupies almost the entire space of the geode) and adularia ***adu*** (dark grey as ***q*** but cloudy or "pitted" by alteration, upper corner), both allotriomorphic and with numerous acicular inclusions of ***sbt*** stibnite (white, anisotropic). The morphology and the absence of ***sbt*** in ***q*** and ***alb*** are evidence of the chronology of deposition. CR344, Inakuraishyi Mine, Hokkaido, Japan

To sum up:

- In general, ***FSlc*** can be characterised by their tabular or leafy morphology (though it is sometimes inappreciable, due to their fine grain size), their low hardness, their dark grey tone (enhanced by poor polishing), their cleavage (not always visible) and, in some cases, by their frequent secondary character (e.g., *q-ser* alteration of *Fs* in epithermal and porphyry environments).
- However, distinguishing them from each other for precise identification of the ***FSlc*** species is beyond the possibilities of **RL** observation.

(a)

(b)

Fig. 135.7 **a**, **b** (Same scene: **dsp** and **d+p**, *respectively*). Impregnation of iron ***Fe*** and cohenite ***coh*** (white), with traces of pyrrhotite ***po*** and ilmenite ***il*** disseminated in basalt. The rock shows a pilotaxitic (flow) texture, marked by the orientation of probable ***Px*** (light grey) and plagioclase ***Plag*** (dark grey), which with crossed analyzer (Fig. 135.7b) reveal their morphologies—thin oriented prisms of inosilicates and Plag microlites, respectively. Sample DI.Rub, Basalt Grunland, Greenland

135.5 Inosilicates. Pyroxene Group. Amphibole Group

Pyroxenes, *Px*, **and amphiboles**, *Amp/Anf*, have in common their characteristic chain structures, which are manifested in a prismatic habit or tabular morphology, frequently euhedral or subhedral, quite different from the flaky or bladed habit of phyllosilicates. They comprise a great diversity of species, rhombic or monoclinic, and a great compositional variety which, if it already defies precise definition with **TL** requiring the support of other analytical techniques (XRD, EMPA, SEM, etc.), makes identification with RL in general questionable and often impossible (cf. Fig. 135.15a, b).

In fact, the diagnostic properties with TL (such as color, refractive index or relief, habit, pleochroism, cleavage, extinction angle, **n**, interference colors, twinning, alterations, and determinations of properties such as interference figure

(a)
(b)

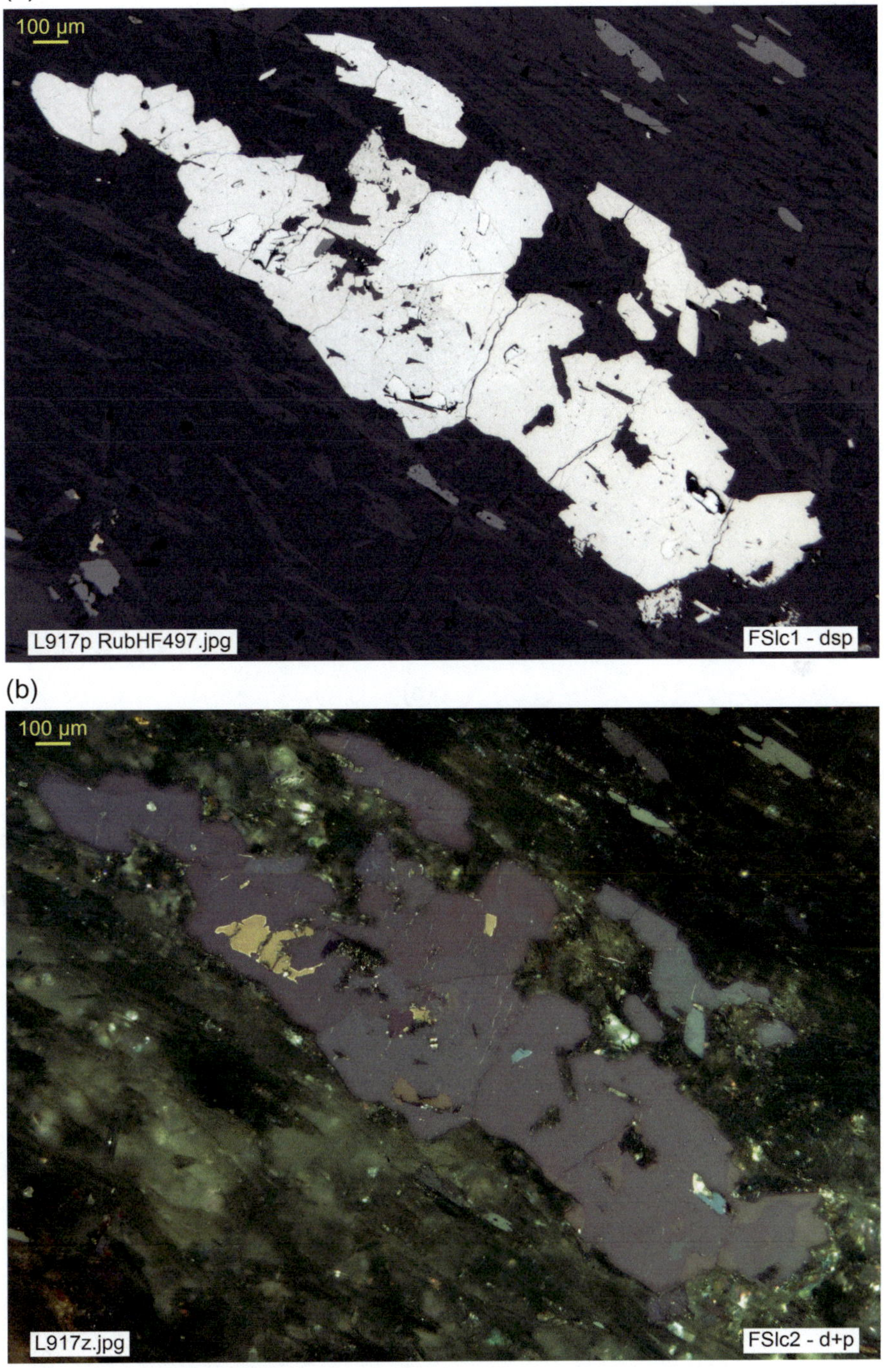

Fig. 135.8 **a**, **b** (Same scene: **dsp** and **d+p**, *respectively*). Mica schist (quartz, dark grey; biotite, slightly lighter), with aggregate of glaucodote ***gld*** and safflorite ***sf*** (bright white aggregate; anisotropic). Some small, prismatic ilmenite ***il*** (greyish brown; anisotropic) crystals are scattered, following the schistosity (mica orientation). RubHF497, Przecznica, Lower Silesia (Poland)

and parameters such as 2V angle and optical sign) cannot be determined with RL. Only a few can be estimated:

- the *morphology*, when its euhedral tendency is manifested,
- the *brightness* (qualitatively estimated, as a *proxy* of the reflectance and this, in turn of the refractive index, n: § 1.2.2, Fresnel equation), and
- the *cleavage* and, in particular, the typical *angle of intersection* of the two cleavage planes: 57° in amphiboles and 93° in pyroxenes (in these, rarely visible).

Since no further details are possible, they are described together, as they can be observed under the reflection microscope (Figs. 135.15a, b, 135.16 and 135.17):

(a)

(b)

Fig. 135.9 **a**, **b** (Same scene: **osp** and **o+p**, *respectively*). Detail of Fig. 135.8a (right upper quadrant, enlarged), in oil immersion. The brightness of gg disappears and turns to black, so ***q*** and ***bio*** can no longer be distinguished (in *osp*), but with crossed analyser (*o+p*) the mica cleavage shows up due to IR (comp. Fig. 135.8b)

- Px and Anf generally polish well, as befits their high hardness and cohesion.
- Its most typical morphology is prismatic, tabular or acicular and, if basal sections appear, the cleavage angles can sometimes be seen (intersections at ≈ 57°: Anf; intersections at ≈ 93°: Px, more difficult to observe).
- Dull appearance: dark, grey colour with no remarkable pleochroism, darker than most ores (even chromite ***chr***) but somewhat lighter than resin or quartz, and also usually lighter than plagioclase or apatite.
- The reflectances, **R**, depending on the values of **n** (refractive indices, cf. § 1.2.2), which vary widely according to species and composition, might be estimated with a roughly reasonable approximation following the descending sequence:

(a)

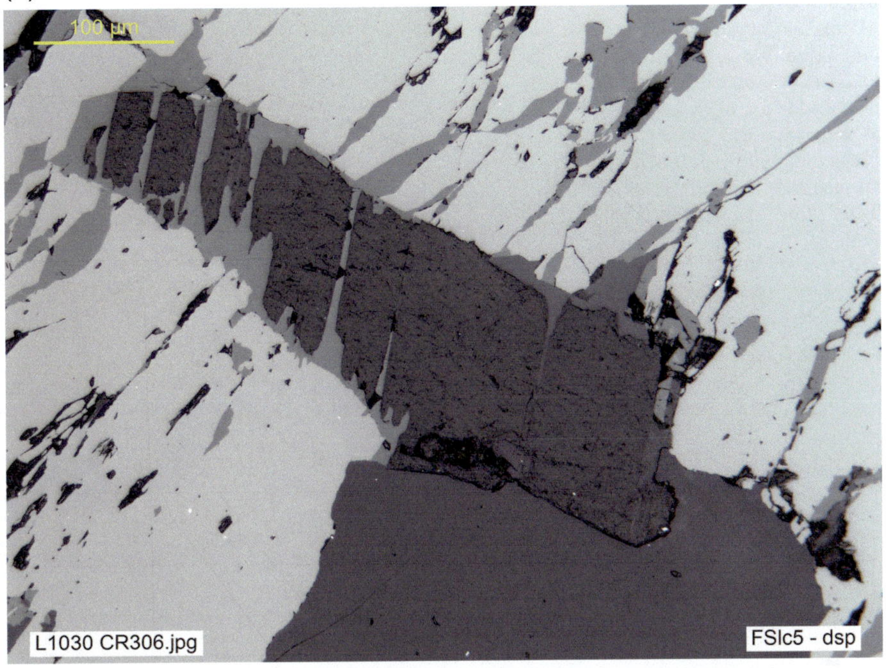

(b)

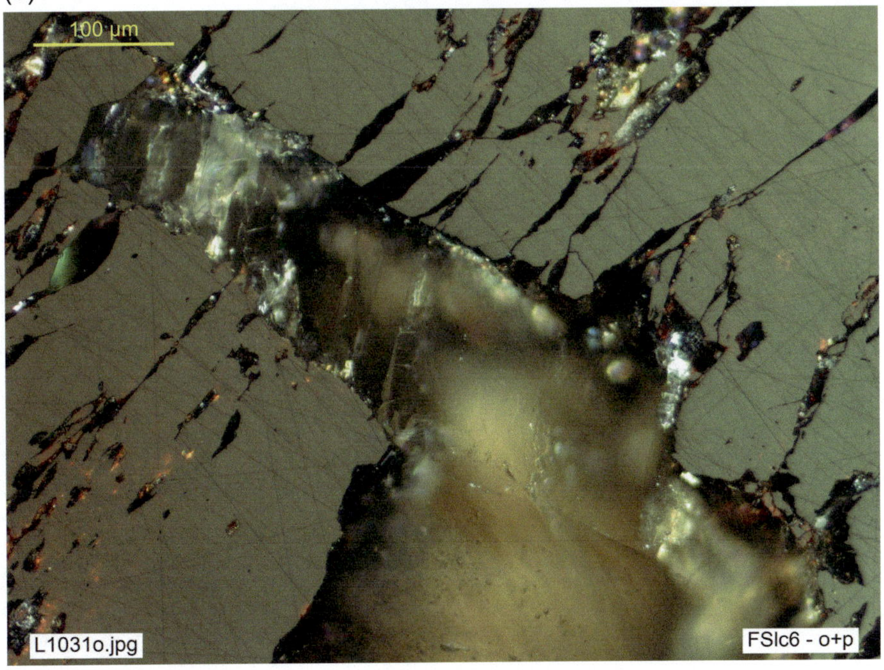

Fig. 135.10 **a**, **b** (Same scene: **dsp** and **o+p**, *respectively*). Ore of W (wolframite ***wf***, very light grey; scheelite ***sch***, light grey) in silicate gangue ***q*** (dark grey, good polish) and ***mica*** (dull, darker grey, bad polish or irregular aspect). As a late, secondary mineral, *sch* partially replaces *wf* and mica, enhancing the cleavage planes of the latter. CR306, Mina As Sombras, Lovios (Orense)

$$R_{Px}(\approx 7.4\%) > R_{Anf}(\approx 6\text{–}6.5\%) > R_{Plag,ap}(\approx 5\%) > R_{q,resin}(\approx 4.5\%),$$

i.e., they look progressively darker grey in the order Px-q. With crossed polars (+p), internal reflections mask anisotropism and twinning.

An important hint for their recognition is paragenesis. These minerals, although not rare in the earth's crust, have a presence in mineral deposits restricted to a few types: mainly contact metamorphic and metasomatic deposits (*skarn*, amphiboles, often accompanied by garnet, epidote, pyroxenes, etc.) and magmatic deposits (Fe, Ni–Cu, Cr, EGP) linked to mafic and ultramafic rocks (pyroxenes, with olivine, serpentine, plagioclase, etc.).

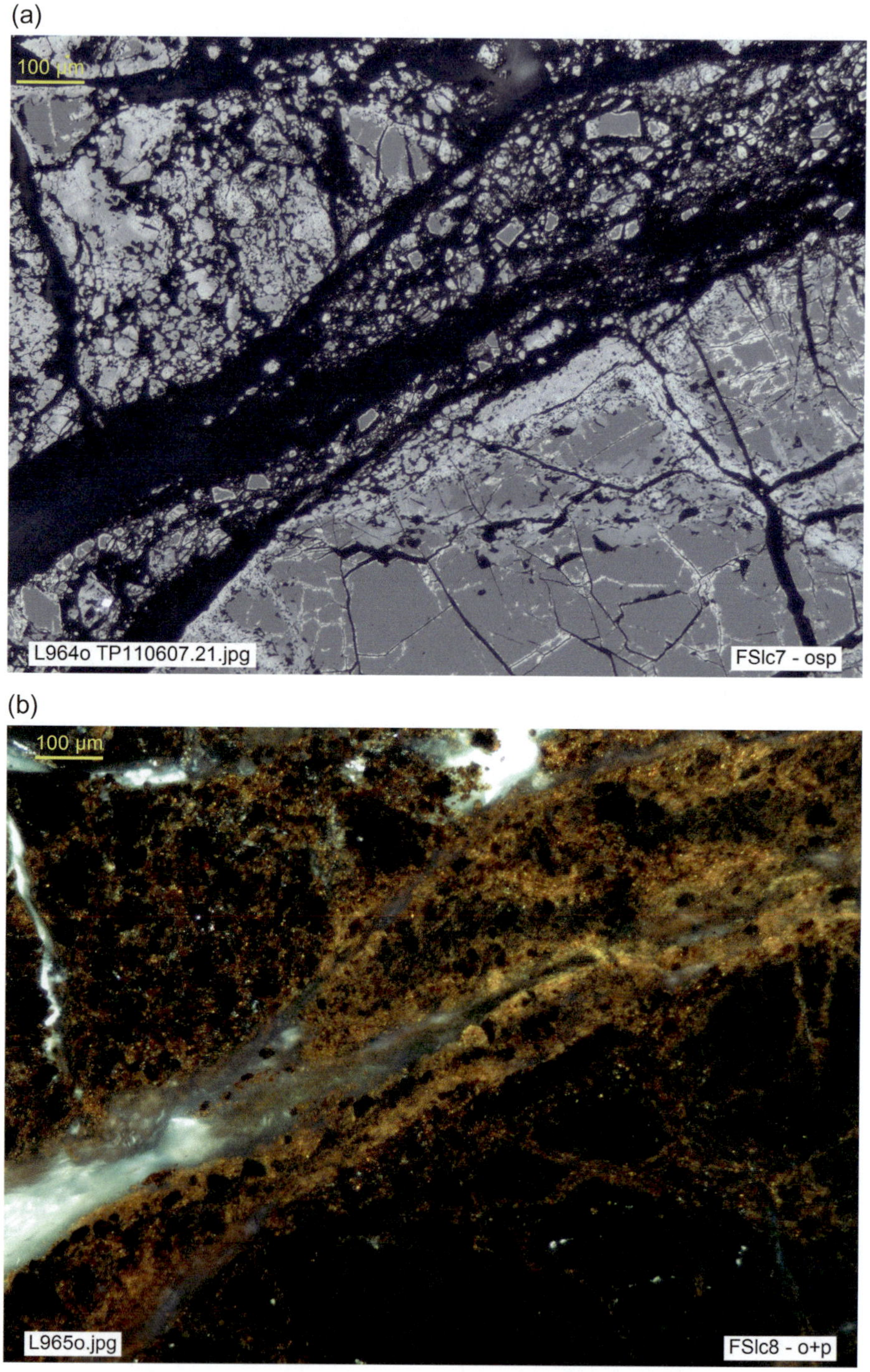

Fig. 135.11 **a**, **b** (Same scene: **osp** and **o+p**, *respectively*). Microshear in podiform chromitite (grey), altered and cemented by *invisible* PSlc (black as a void in oil, *osp*), recognizable as serpentine and clay by their green and yellowish-brown IR (*respectively*, *o+p*). TP110607.21, Tapo Mine (Tarma, Dept. Junin), Peru

135.6 Other Silicates. Olivine. Garnet Group. Epidote Group. Willemite. Chrysocolla

Olivine (*ov*, $(Fe^{2+}, Mg)_2SiO_4$, orthorhombic) shows in **RL**, like the inosilicates, few diagnostic features:

- High polishing hardness ($\geq$ mt, < py), good polishing quality
- Dull grey colour, without appreciable bireflectance or pleochroism
- Reflectance, **R**, low but remarkable for being gg (Fig. 135.18a) and very composition dependent, as expected from the difference between n values (refractive

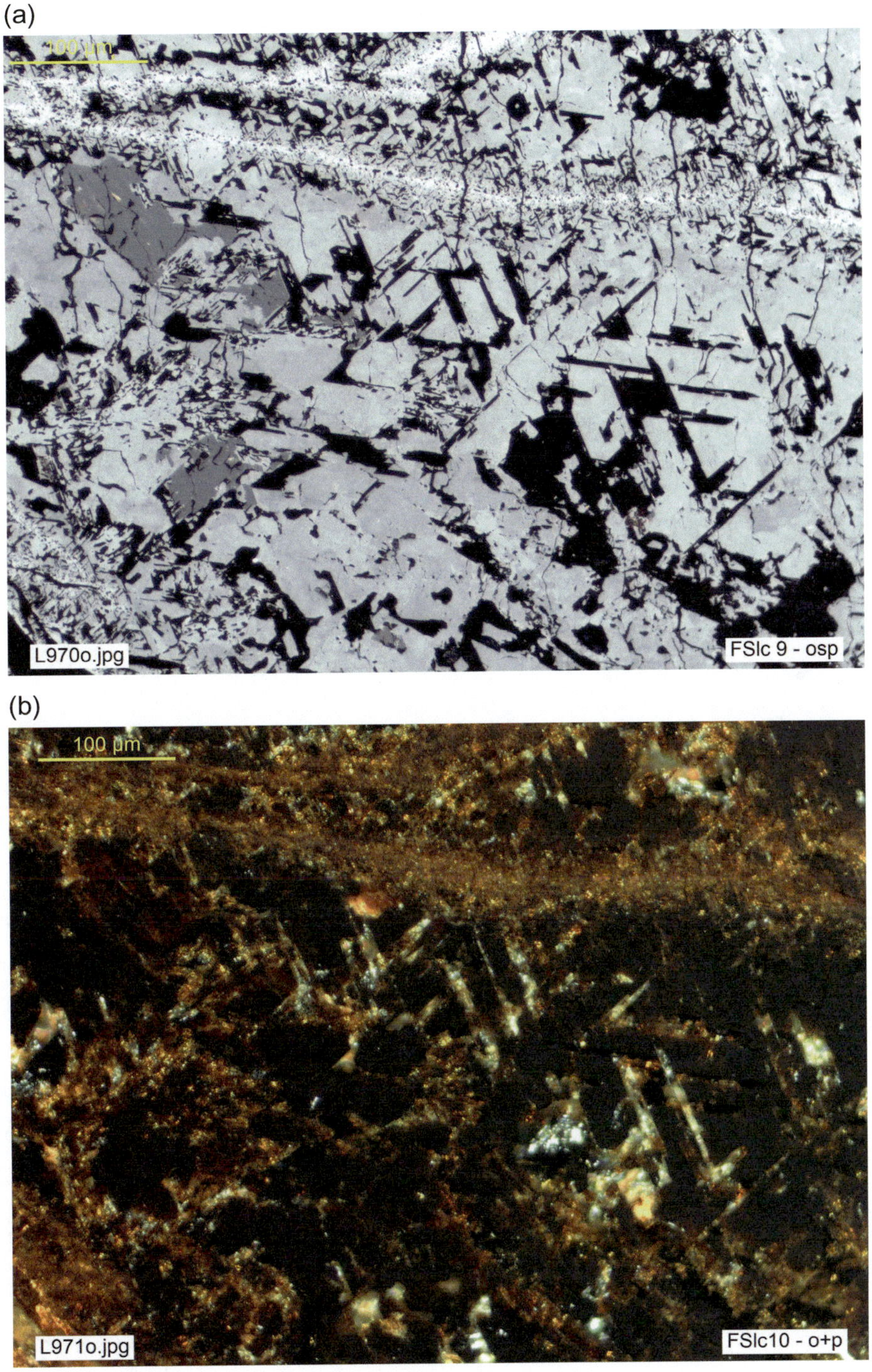

Fig. 135.12 **a**, **b** (Same scene: **osp** and **o+p**, *respectively*). Metasomatic alteration of chromite (***chr***, grey relics) to ferritchromite (light grey) and magnetite ***mt*** (white), with formation of chlorite ***Chl*** (black, euhedral, *osp*; colorless to yellowish-brown RI, *o+p*. In small oriented blades, following octahedral directions of *mt*), and of interstitial *srp* (black, greenish RI, *o+p*). TP110607.21, Tapo Mine (Tarma, Dept. Junin), Peru

indices) of the magnesian term, forsterite ($n_\beta \approx 1.65$, and $R \approx 6\%$, comparable to barite and > serpentine ***srp***), and the ferriferous term, fayalite ($n_\beta \approx 1.87$, and $R \approx 9\%$, comparable to Grt)

- Abundant IR (Fig. 135.18b)
- No perceptible effects of anisotropism
- Frequent alteration to *srp*

However, its restricted paragenesis is an important help. It is typically found associated with chr and srp in orthomagmatic deposits linked to ultramafic rocks (intrusive complexes and ophiolites) or in contact metamorphic ferriferous deposits (fayalite). The Mn term—tephroite—is found in some manganiferous paragenesis (Sterling Hill Mine, New Jersey, USA).

Fig. 135.13 **a**, **b** (Same scene: **osp** and **oxp**, *respectively*). Columbotantalite ***cbt*** ore (greyish-white) with sericitised quartz-feldspathic gg (dark grey, with poor polish and pull-outs in altered zones, ser-Arc, osp; bright, due to luminous IR, *oxp*). CR400, Borborema, Brazil

The frequent cumulate texture of ultramafic formations with chr and ov makes the development of euhedral forms of ov difficult, but it usually presents granoblastic textures, in which chromite ***chr***, olivine ***ov*** and serpentine ***srp*** (Fig. 135.18a, b) are associated, offering two useful clues for recognition:

- The brightness or grey level of these three components can be directly compared, increasingly darker from *chr* to *srp*: R_{chr} ($\approx$ 12%) $\gg$ R_{ov} ($\approx$ 7%) > R_{srp} ($\leq$ 5%), with a strong contrast in R between *chr* and *ov* and, more moderately, between *ov* and *srp*.
- The noticeably higher hardness of *chr* and *ov*, which appear in relief on *srp*, is visible.

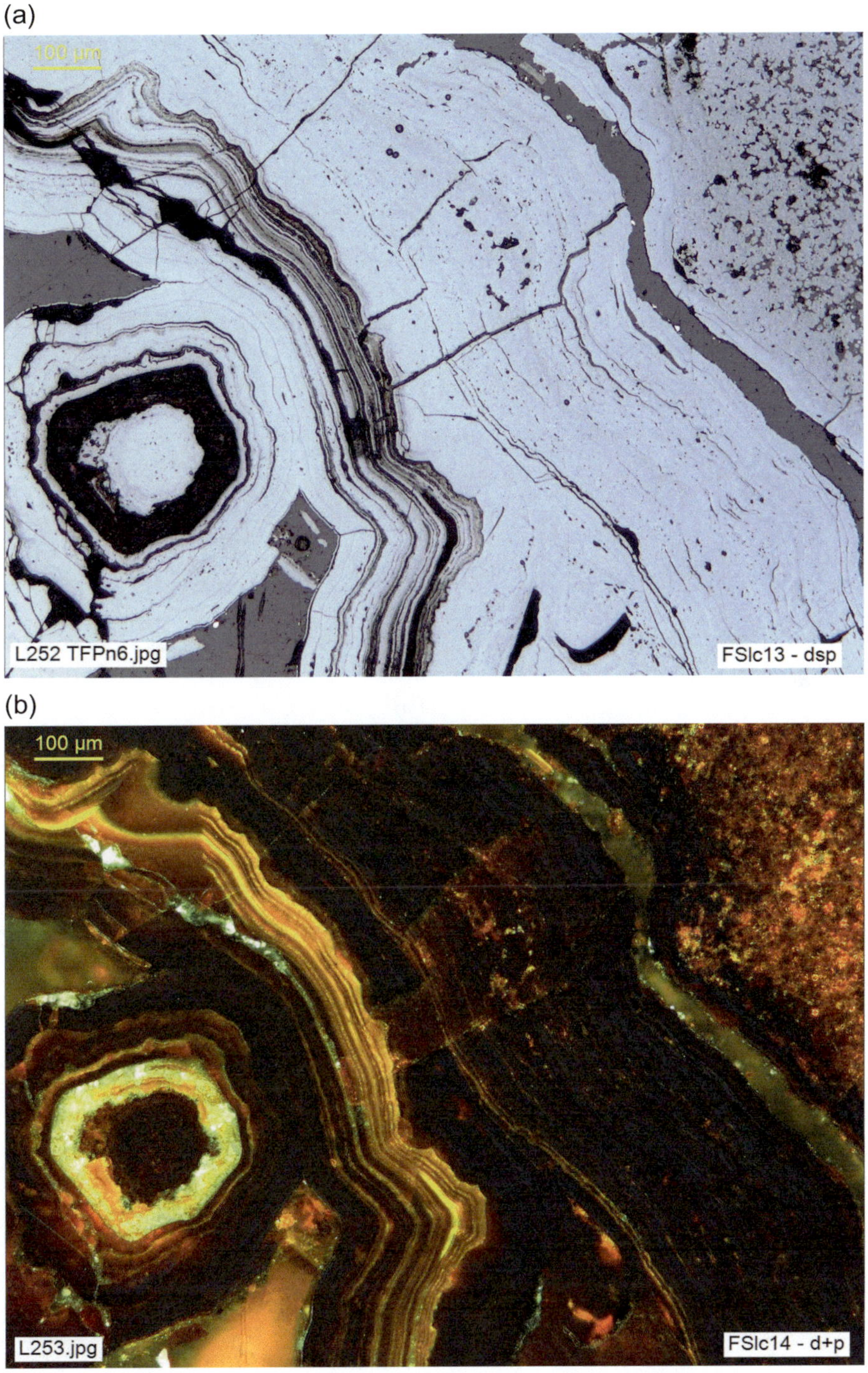

Fig. 135.14 a, b (Same scene: **osp** and **o+p**, *respectively*). Clay intercalations (dark grey to black, *dsp*; very luminous IR, colourless to yellowish-brown, *dxp*) in limonite ***lim*** (light grey). TF Pn6, Mina Sar, Santiago de Compostela

These textures are frequently modified by deformation, especially in podiform or ophiolitic deposits, with cataclastic *ov* and *chr* (Fig. 135.18a). Although they are often relict, they can be easily recognized as pseudomorphs, because the secondary paragenesis usually replicates the primary *ov* forms:

- either by hypogene alteration to serpentine, at depth, or
- by supergene alteration, under oxidizing conditions, generally to *iddingsite*, ferruginous aggregate, with sub-microscopic *gth* goethite and *ht* hematite.

The **garnets**, ***Grt*** ($X_3Z_2[SiO_4]_3$, cubic), with

$$X = Mg, Ca, Fe^{2+}, Mn^{2+}, \text{etc.}$$
$$Z = Al, Fe^{3+}, Cr^{3+}, V^{3+}, \text{etc.,}$$

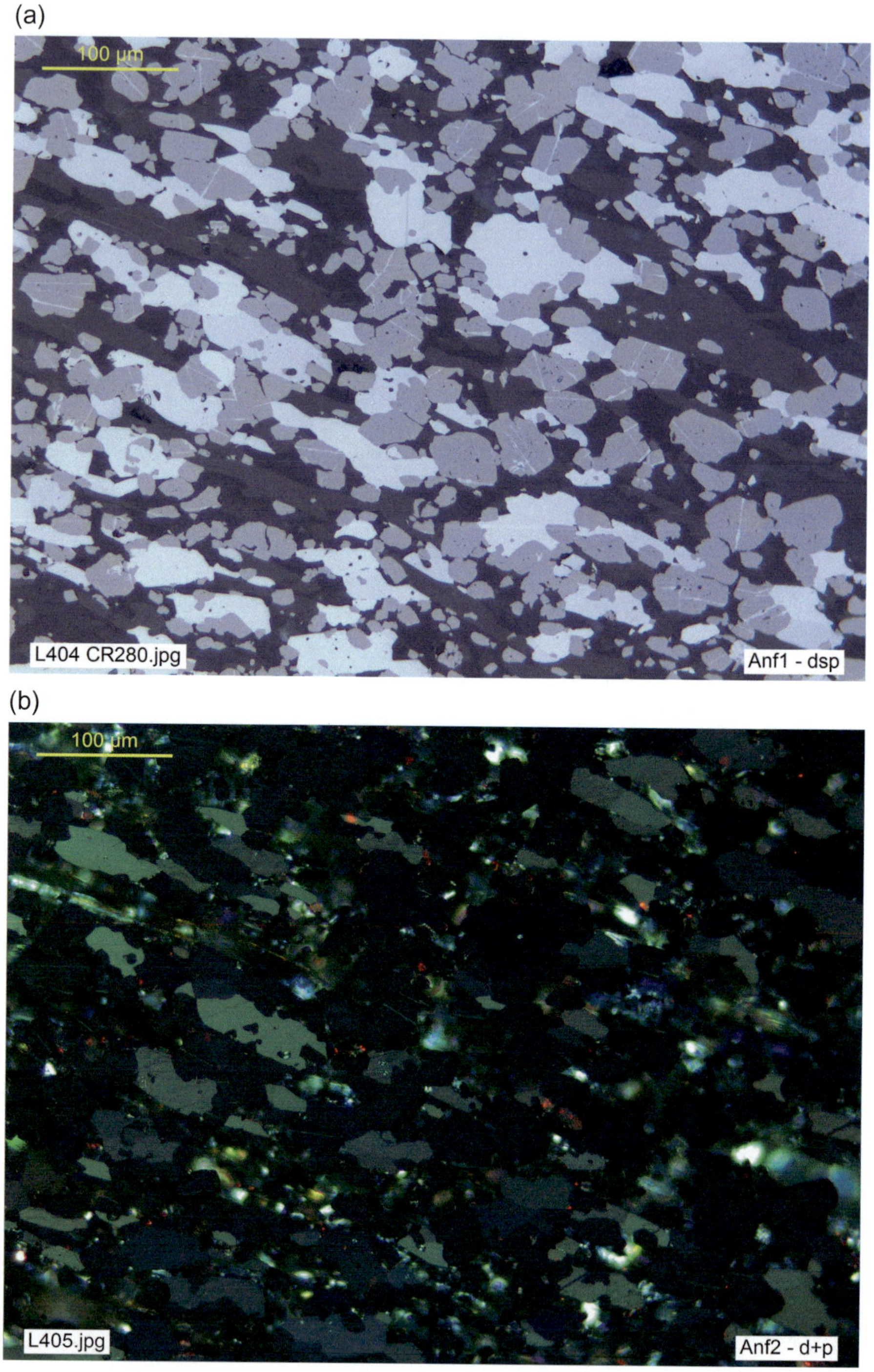

Fig. 135.15 **a**, **b** (Same image: **dsp** and **d+p**, *respectively*). Quartz **q**-actinolite-magnetite **mt**-hematite **ht** schist (Iron Banded Formation, *BIF*): amphiboles display a slightly brighter luster (R>) than q and the prismatic habit of the inosilicates, which contributes together with ht to the oriented, grano-nematoblastic texture of the rock. However, showing no diagnostic properties under **RL**, the precise determination of the species (actinolite) had to be made with XRD. CR280, BIF Harare, Zimbabwe

form a large group with great compositional variations, whose species cannot be properly identified with RL but can be characterized as a whole by some common properties (Figs. 135.19a, b and 135.20a, b), such as:

- high scratch (Mohs 7–71/2) and polishing hardnesses,
- euhedral tendency but with globular, isometric morphology (cubic symmetry),
- monotonous grey color (not pleochroic) without luster, but lighter than usual gg minerals, as corresponds to its refractive indexes $n > 1.7$,
- relatively wide range of reflectances from $R \approx 7\%$ (pyrope, $n \approx 1.71$) to $R \approx 9.5\%$ (andradite, $n \approx 1.89$; cf. Table 3.2, and
- with crossed analyzer (Fig. 135.21a, b) (**d+p** and **o+p**), frequent IR and absence of anisotropism.

Fig. 135.16 Alkali gabbro pegmatoid, comprising as main components: *Px* (Ti-augite, intermediate grey, upper left), Ti-mt (lighter brownish grey, center) and *Plag* (darker grey, ubiquitous), characterized by their Reflectances, **R**, decreasing in the order R_{mt} (≈ 21%) > R_{Px} (≈ 7.4%) > R_{Plag} (≈ 4.6%). Their morphology is also characteristic: mt occurs in skeletal aggregates; Px, in large phenocrysts with prismatic habit; and Plag, intergrown with both Px and mt (occasionally with symplectitic textures) or interstitial, anhedral and apparently late. Minor accessories: *ap*, generally euhedral (hexagonal or acicular sections) and included in Px or Plag, as well as likely amphibole (hornblende, with typical hexagonal and tabular sections, less regular than ap), both grey and lighter than Plag but darker than Px, as expected from its R (≈ 6–6.5%), intermediate between R_{Px} and R_{Plag}. Sample 070513, Lanzarote (**dsp**)

Fig. 135.17 (Same image as Fig. 135.16: **d+p**). *Widmanstätten* structures of *mt*. The abundance and luminosity of **IR**, Internal Reflections hide other properties of Px, Plag, ap and hb, while evidencing incipient weathering or oxidation of mt, with intense limonitic impregnation recognizable by typical red and yellowish-brown IR (*respectively*: hematite ***ht*** and goethite ***gth***) in their surroundings

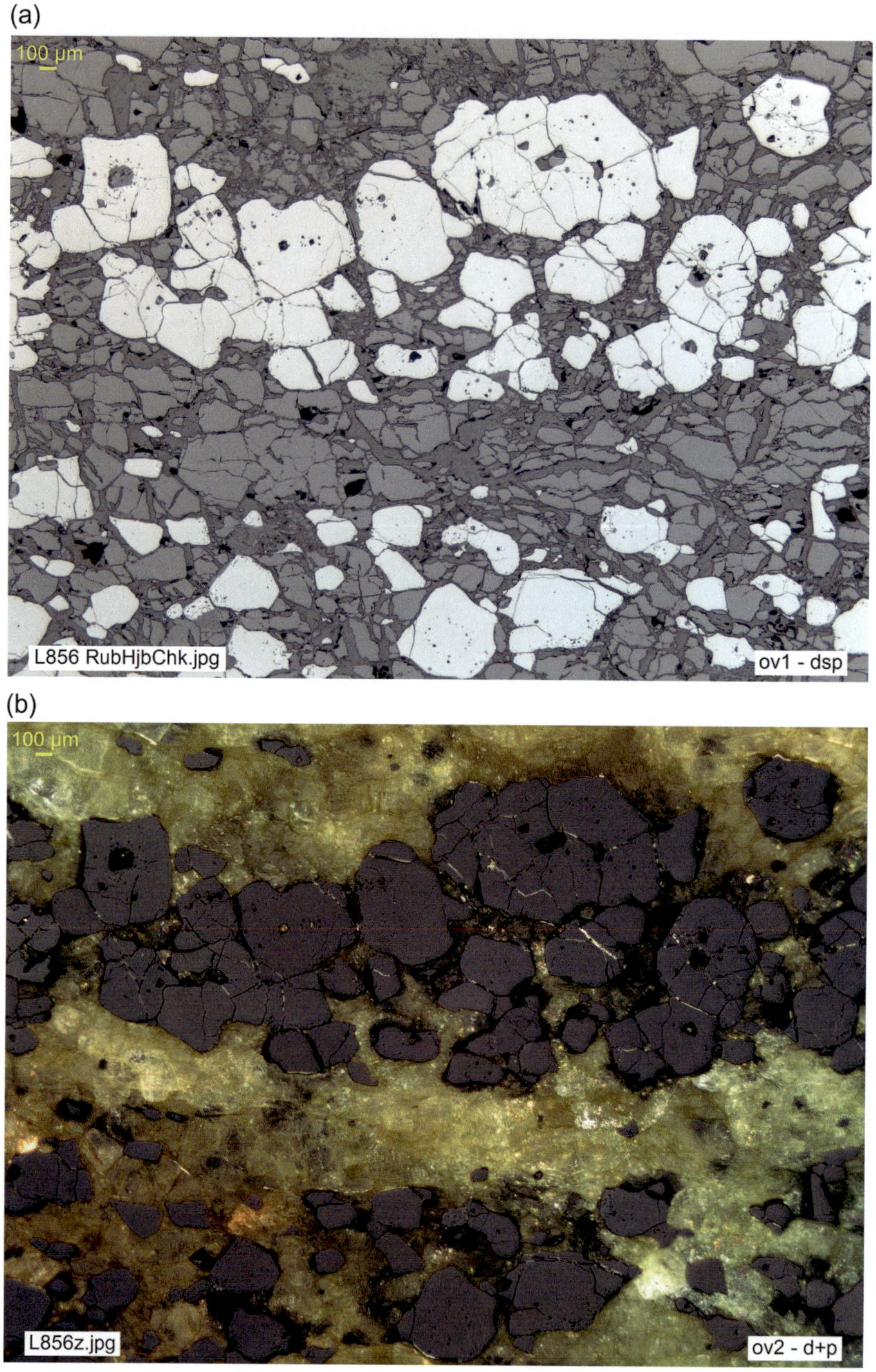

Fig. 135.18 **a**, **b** (Same scene: **dsp** and **d+p**, *respectively*). Chromite microbands (very light grey, *dsp*; in extinction, *d+p*) and olivine (grey) cataclastic and serpentinized (serpentine matrix *srp*: dark grey, *dsp*). Ov is recognised by its intermediate grey tone, much darker than *chr* and less than *srp* (*dsp*), by its high polishing hardness, which makes it stand out in relief over *srp*, and by the abundant colorless to greenish yellow IR, which mask extinction (*d+p*). Sample RubHjb-20, Chalkidiki, Greece

They can be altered, mainly to ***Chl*** chlorite, but also to Fe oxy-hydroxides or limonite, usually pseudomorphs.

Epidote, *Ep*, gives its name to a group of several rhombic and monoclinic silicates with Ca, Al, Fe, Mg, (Ti) and, in some species, also with rare earths (Ce, Y: *allanite*) or Mn (*piemontite*). Any of these species may be usually meant by the term "epidote", used in a general, non-specific sense (Ep *s.l.*). Nevertheless, epidote is also a species of the group - with this meaning, it is referred to as "ep *s.str.*". Although with TL at least the most common of these species can be identified, such a distinction is not possible with reflected light (except in some cases, such as piemontite, ***pie***). As they share the properties that can be observed with RL, the RL description also refers to the group, like for plagioclase, inosilicate and garnet.

In RL it is not easy to distinguish ***Ep*** from the inosilicates, particularly from px, with which it shares several properties, such as:

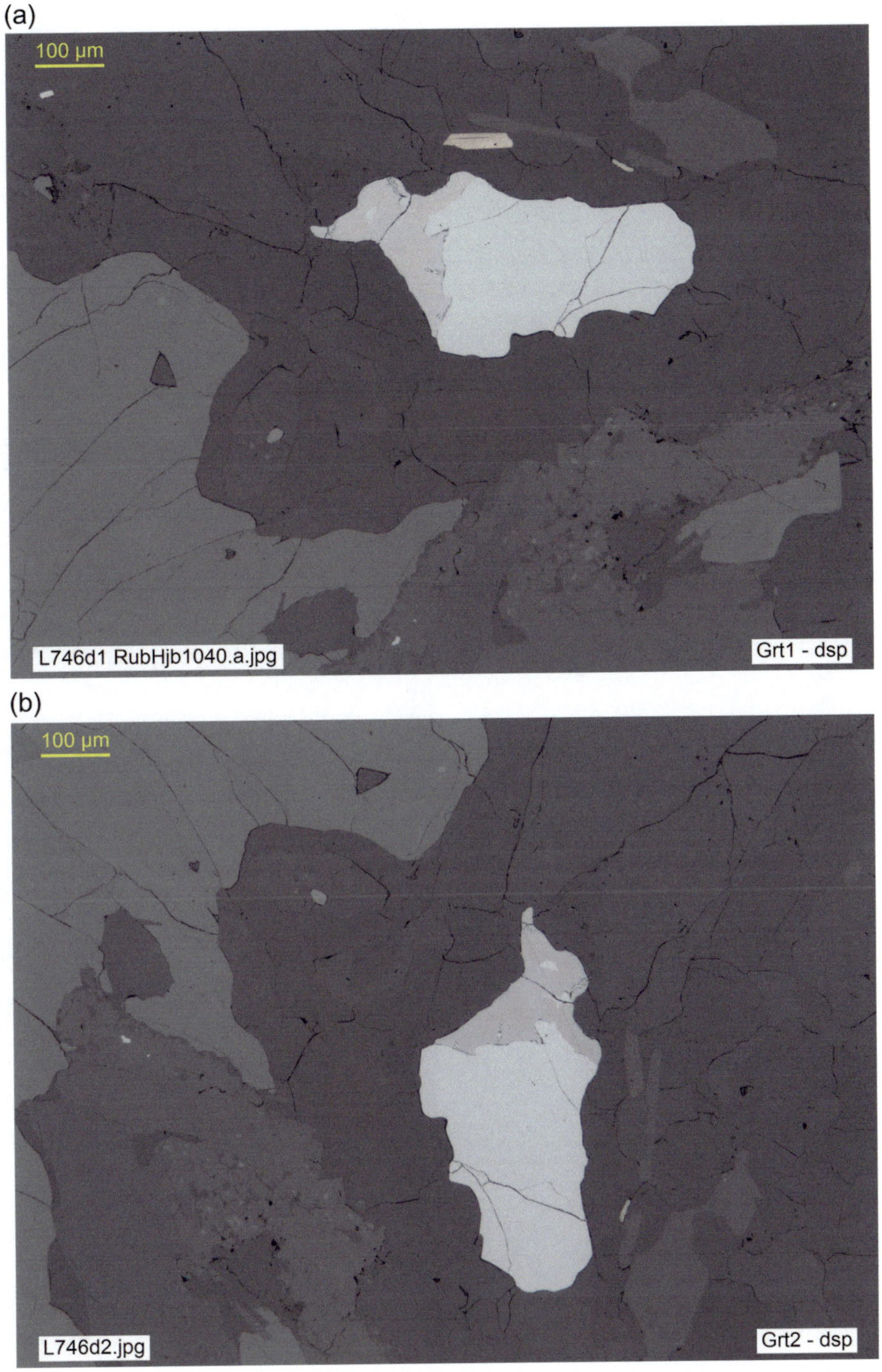

Fig. 135.19 **a**, **b** (Same field, rotating the microscope stage 90° when going from 1 to 2, **dsp**). Granetiferous migmatite: quartz (***q***, dark grey, main component), garnet (***Grt***, light grey, lower left corner), biotite (***bio***, leafy, intermediate grey between *q* and *Grt*), rutile (***rt***, white), ilmenite (***il***, pinkish brown and anisotropic), graphite (***gra***, lamellar, greyish brown to grey, extreme pleocroism and anisotropism). Note R (brightness) decreasing by the order *rt* > *gra*-R_o > *il* > *Grt* > *bio* > *gra*-$R_{e'}$ > *q*. Rotation confirms the isotropy of *Grt*, as opposed to the strong bireflectance of *gra*, which goes from R_o > *il* (Fig. 135.19a) to $R_{e'}$ ≤ *bio* (Fig. 135.19b). Sample *RubHjb1040.a*, Sikkim, Himalaya

- tabular, prismatic or acicular habit,
- dark grey color, though not as dark as the resin and somewhat lighter than px,
- reflectance, R, low (cf. Fig. 134.2a, b), variable according to orientation and species. It is estimated—on an intermediate value of the refractive index of *ep*, *s.str.*: n_β 1.75—at a theoretical value R ≈ 7.4%, comparable to, but slightly higher than the R ≈ 7% obtained for px (augite, for $n_\beta \approx 1.72$), and
- IRs often make the effects of anisotropism unnoticeable, unlike the bright interference colors common in TL.

In some fortunate cases, the colors of Ep (either the natural color of the mineral or the interference colors, which can be very vivid and striking) become visible with RL because of the IR and facilitate identification. Figure 135.22a, b illustrate the properties of a mineral of the Ep group that can be recognized with RL. It is **piemontite**,

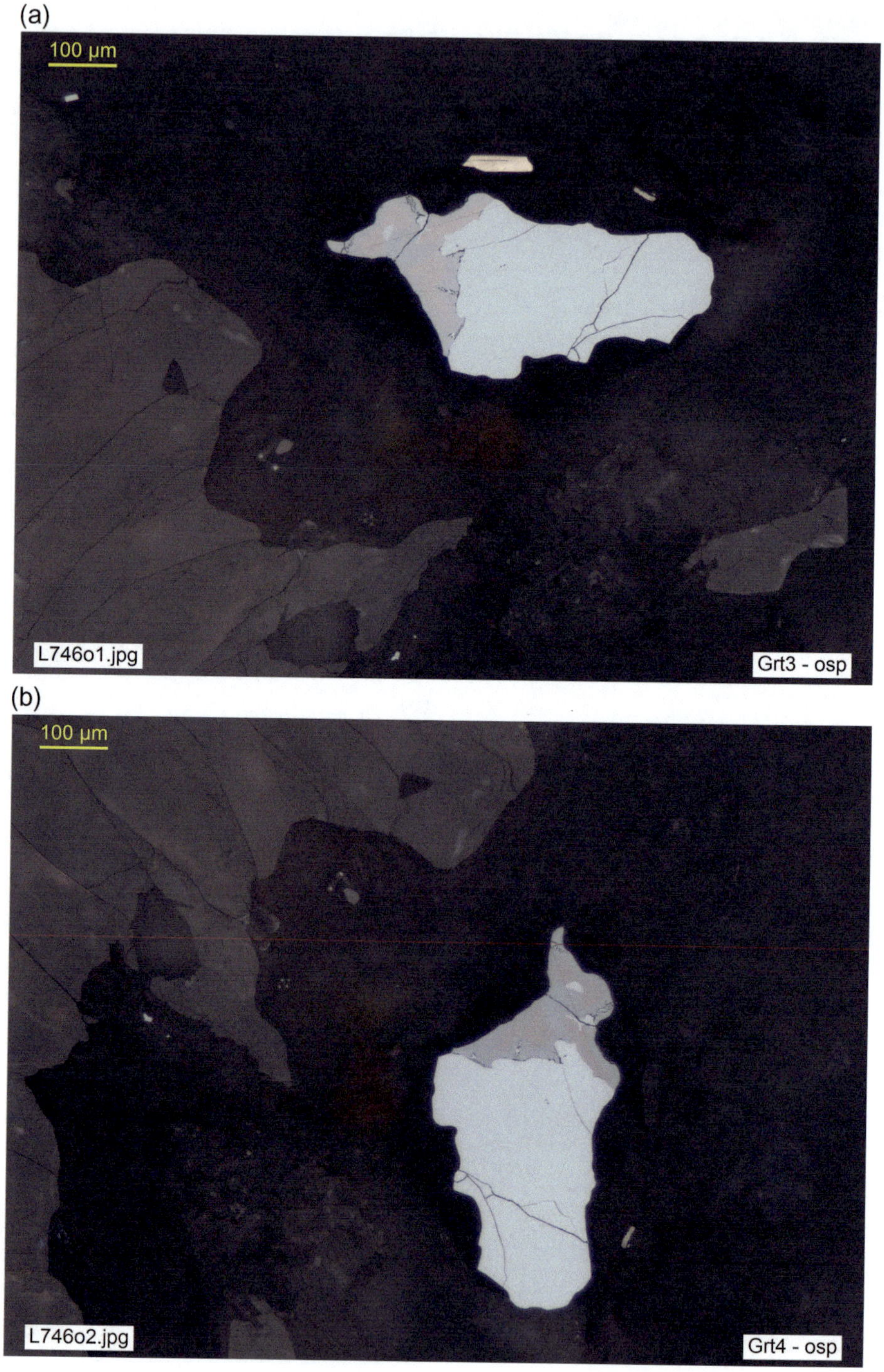

Fig. 135.20 **a**, **b** (Same scenes as Fig. 135.19a and b, *respectively*: **osp**). The reduction of R by immersion makes q and bio practically *invisible* but increases the contrast of Grt, rt, il and gra

Ep-pie, the manganiferous species of Ep. As a reference, the same images are shown with TL (Figs. 135.22b and 135.23b), confirming the identification.

In practical terms, the recognition of Ep generally suggests calc-silicate paragenesis, which can be related to important and varied metal deposits, as is typically the case with skarn concentrations.

The silicate **willemite** (***wil***, Zn_2SiO_4, trigonal) is not common as a gangue mineral, except in weathered Zn deposits, mainly as a secondary oxidation product of the sulphide sphalerite. However, it reaches economic concentrations, along with other oxidized Zn minerals, in deposits of the type known as *oxidized Zn deposits*. Exceptionally (Sterling Hill, N. Jersey), as a presumably primary, metamorphic, and massive ore in significant concentrations.

It usually occurs in oxidation zones of sulphide deposits and is formed, in general but not exclusively by downward weathering, in iron caps or gossan blankets. As companions,

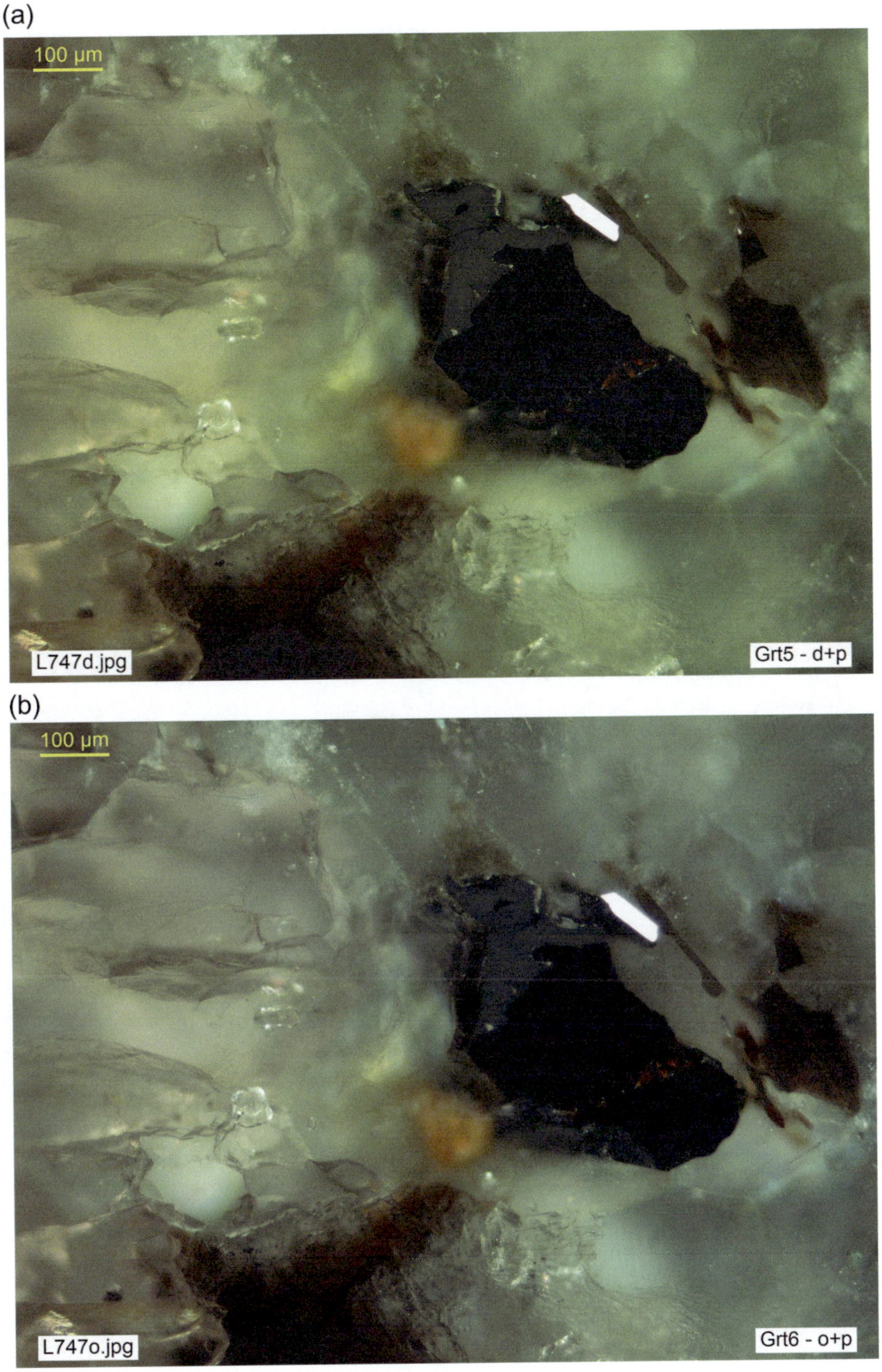

Fig. 135.21 **a**, **b** (Same field as Fig. 135.19a and Fig. 135.20a, **d +p** and **o+p**, *respectively*, rotating the stage 45°). There is no visible extinction of Grt, hidden by the brightness of its IR. In this position the extreme anisotropism of ***gra*** is observed

apart from relict primary Zn ores that may have been preserved or the typical gossan components (limonite *lim*, goethite *gth*, jarosite *jar*, etc.): franklinite, $ZnFe_2O_4$ (Sterling Hill, Figs. 135.24a, b and 135.25a, b); zincite, ZnO; hemimorphite, $Zn_4Si_2O_7(OH)_2 \cdot H_2O$; hydrozincite, $Zn_5(CO_3)_2(OH)_6$; smithsonite ($ZnCO_3$), among others.

In RL, reflected light, ***wil*** appears grey, noticeably darker than its frequent companion franklinite (Fig. 135.24a) but not as dark as *q*. Its reflectance, R, is that of a transparent but refractive mineral, theoretically, $R \approx 6.7\%$ (according to Fresnel's equation, for a value $n \approx 1.7$), i.e., between *ov* olivine and pyrope garnet. With analyzer (*d+p*), as expected, the IR mask anisotropism (Fig. 135.24b).

Its recognition with RL is therefore generic and imprecise, but possible in broad outline taking into account its companions: the specific paragenesis of oxidized Zn ores is, together with the geological environment, a decisive datum. Nevertheless, being a silicate and, as such, a transparent

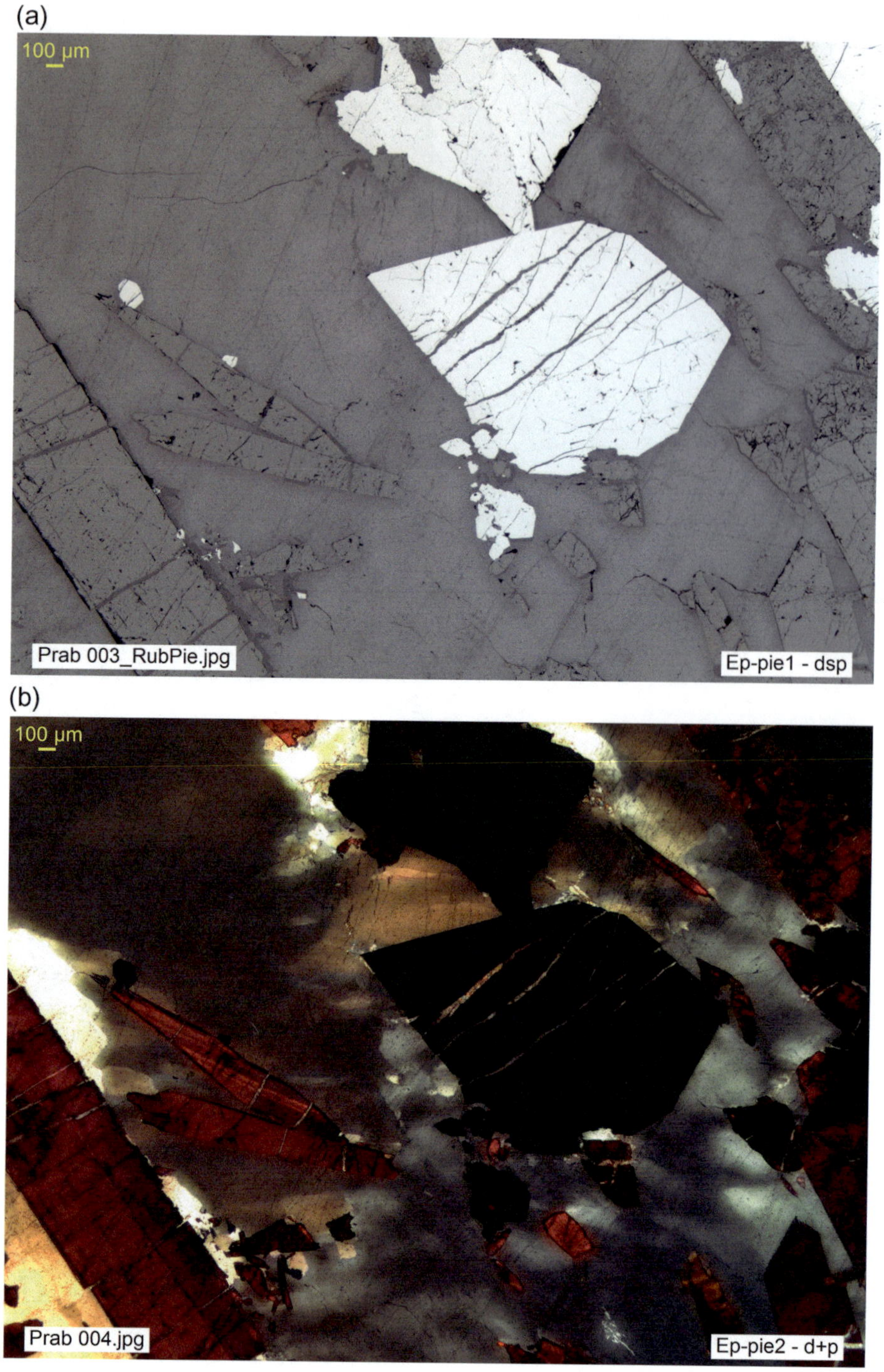

Fig. 135.22 **a**, **b** (Same scene: **dsp** and **d+p**, *respectively*). Quartz vein (***q***, grey, main mass), with piemontite (***pie***, lighter grey, prismatic or acicular) and jacobsite ***jac*** (idiomorph, greyish white, center). Crossed polars (Fig. 135.22b) show the extinction of *jac* (isotropic) and the partial illumination by IR of *q* (colourless to yellowish-grey) and *pie* (typical reddish-brown colors). *RubPie*, Praborna, Val d'Aosta, Italy

mineral, ***wil*** should be characterized with TL (petrographic microscope): trigonal; colourless; high relief ($n_\omega \approx 1.69$, $n_\varepsilon \approx 1.72$); birefringence $\delta = 0.028$ (second and third order colours), anisotropic and with vivid interference colours, uniaxial +. For reference, Fig. 135.25a, b show with **TL** the same scene that in Fig. 135.24a, b is observed with RL.

Chrysocolla (***crc***, ‹Cu, Al›$_2$H$_2$Si$_2$O$_5$‹OH›$_4$. nH$_2$O,[3] orthorhombic), although treated here as silicate gangue, may

[3] $(Cu_{2-x}Al_x)H_{2-x}Si_2O_5(OH)_4 \cdot nH_2O$, según IMA, cf. https://www.mindat.org/min-1040.html (*visit*. 20,210,501).

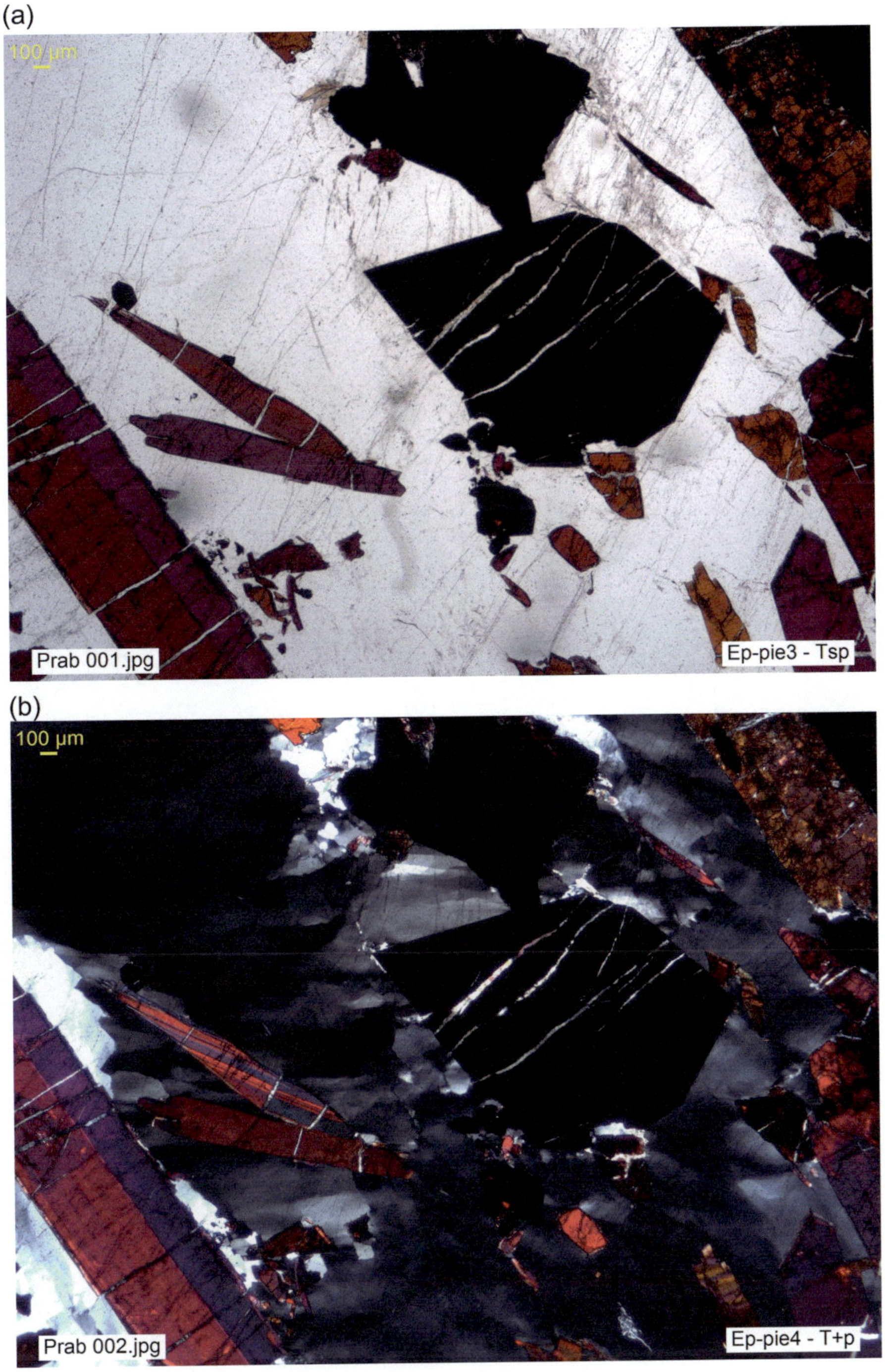

Fig. 135.23 **a**, **b** (Same scene as Fig. 135.22a, seen with TL: **Tsp** and **T+p**, *respectively*). With *Tsp*, easy recognition of ***q*** (white, microcracks with secondary fluid inclusions) and *pie* (prismatic habit, relief, cleavage, pleochroism in reddish and violet-brown to yellowish tones). With *T+p* (Fig. 135.23b), *pie* shows polysynthetic twins and vivid interference colors (partly masked by the color of the mineral). Remarkable tectonic deformation, undulatory extinction of q, microcracks welded by muscovite in *pie* and *jac*

form part of the ore, in the important type known as oxidized copper deposits, together with minerals such as cuprite, **cup**, malachite, **mch**, or azurite, **az**, among others, which have been described as ores (Part II).

Despite its unmistakable green or bluish-green (turquoise-like) colour in hand specimen, under the microscope (*dsp*) it appears dark grey (Fig. 135.26a), as expected from its estimated reflectance (**R** ≈ 5–5.5%, for **n** ≈ 1.6).

Fig. 135.24 **a**, **b** (Same scene, **dsp** and **d+p**). Massive ore of willemite ***wil*** (dark grey, center) and franklinite ***fk*** (greyish white). CR394, Sterling Hill Mine, Sussex County, New Jersey, USA

(a)

(b)

Fig. 135.25 **a**, **b** (Same scene as Fig. 135.24a, seen with transmitted light: **Tsp** and **T+p**, *respectively*). Note the relief (high refringence, n) of ***wil*** (*Tsp*) and its birefringence (*T+p*, interference colors up to third order), which contrast with the opacity of ***fk***

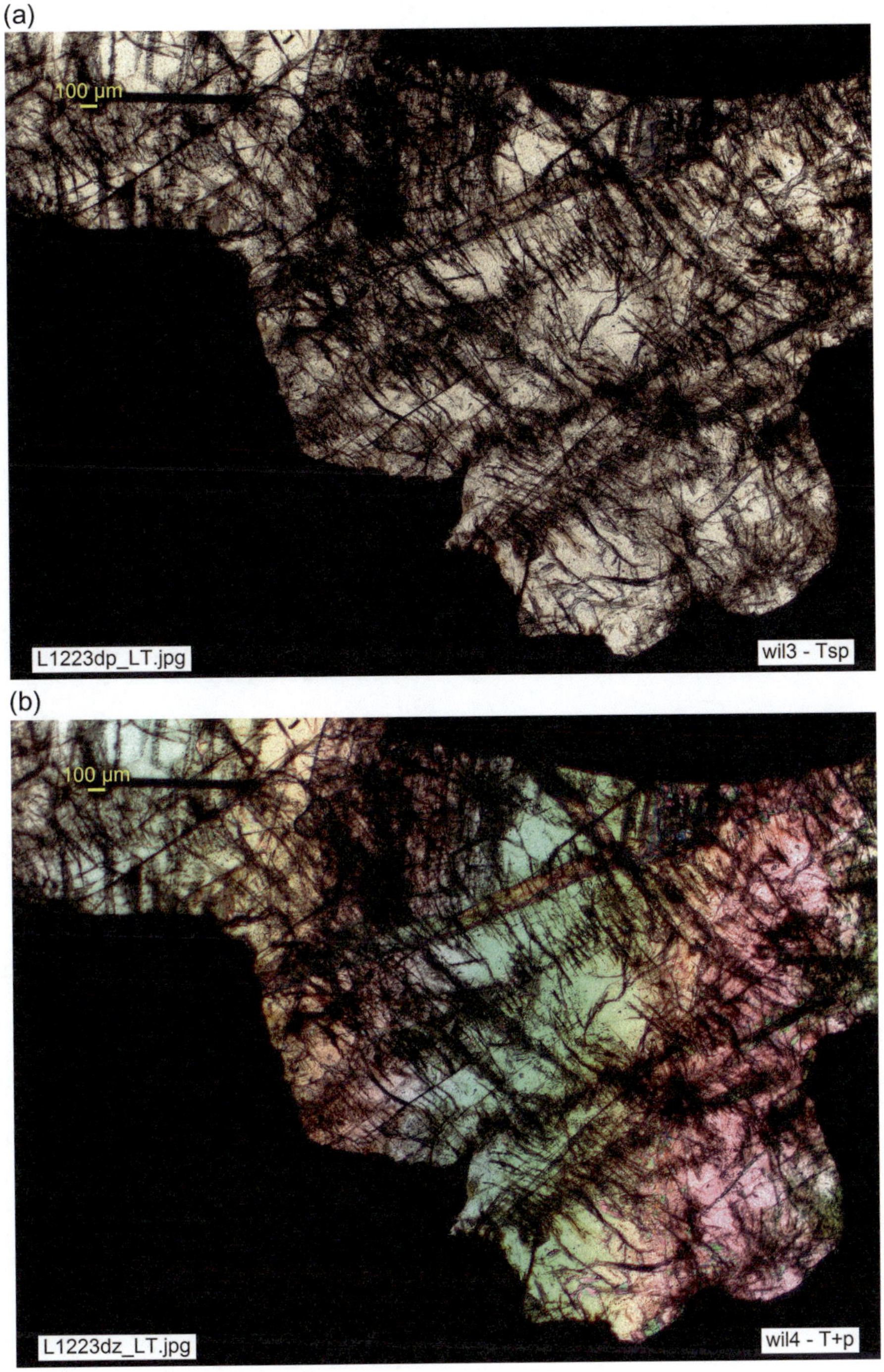

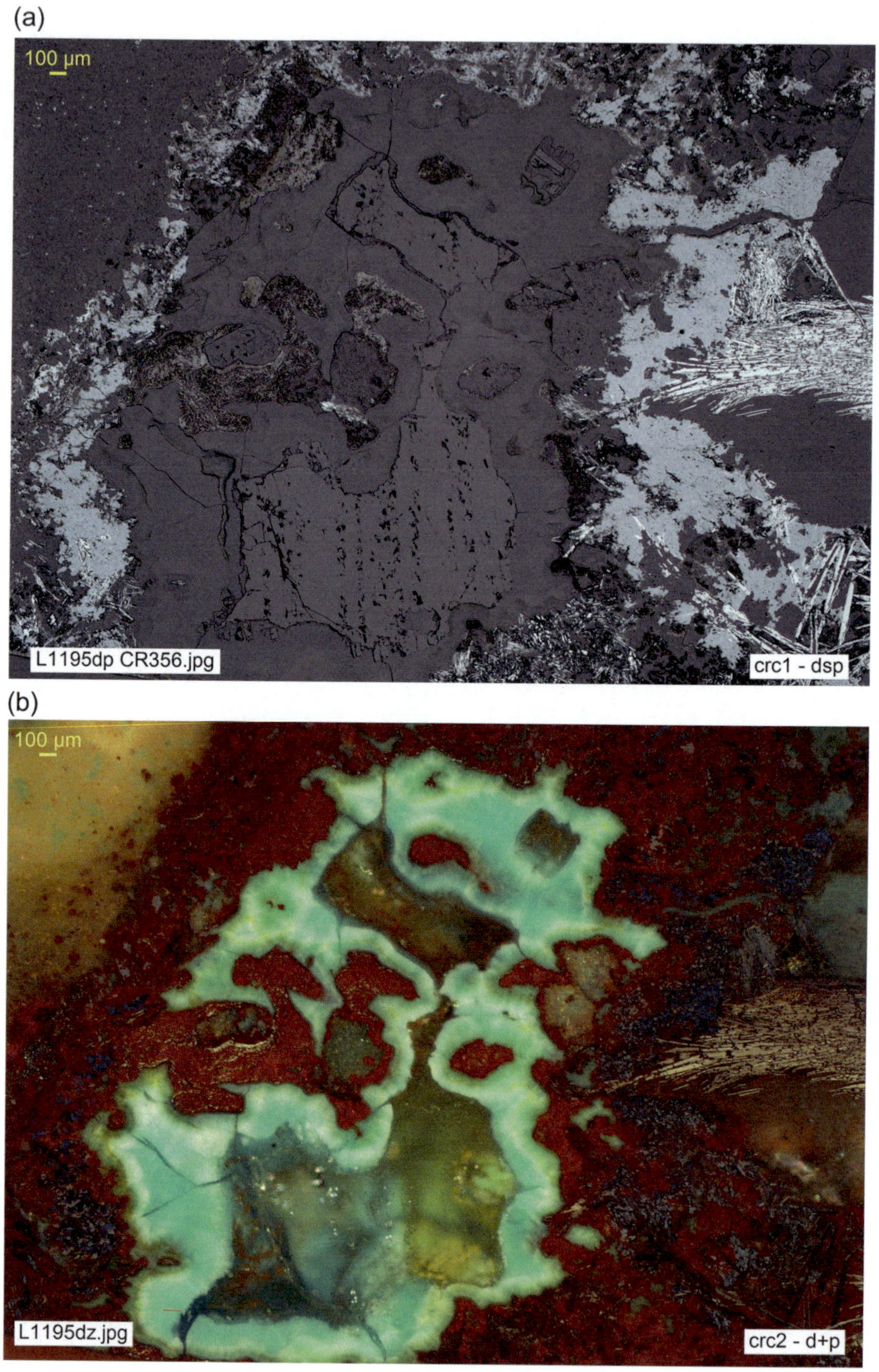

Fig. 135.26 **a**, **b** (Same scene: **dsp** and **d+p**, *respectively*). Mineralized breccia with copper and iron oxides. Geode with sequential filling by a fine aggregate of cuprite (light bluish-grey) and hematite (white, acicular and powdery), sealed by encrustation of cryptocrystalline chrysocolla (crc, dark grey, *dsp*) with pores finally filled by limonite (grey-black, *dsp*). Crossed polars (*d+p*) evidence the widespread turquoise-green internal reflections of *crc*, which contrast with the deep red IR hue of limonite, cuprite and powdery hematite, **ht**. The anisotropism of *cup* (with bluish-green tones) and *ht* (intense but weakly colored) confirm their identity. CR356, San Antonio Mine, Sonora, Mexico

With analyzer (*d+p*, Fig. 135.26b), the intense diffuse luminosity with vivid green or turquoise color of its **IR** (internal reflections) distinguish it immediately. IR also hide its anisotropism, which would, however, almost always be imperceptible in spite of being an orthorhombic mineral, due to the low or null crystallinity of its aggregates and incrustations, usually botryoidal, cryptocrystalline or amorphous.

136 Carbonates. Borates

136.1 Carbonate Ore and Gangue Minerals

The realm of **carbonates**, ***Carb***, extends mainly in the sedimentary and metamorphic environments, where calcite, ***calc***, and dolomite, ***dol***, are the main components of limestones, dolomites and marbles. They are much more scarce in igneous rocks, although they become main components in carbonatites.

They can also be abundant in certain metasomatic rocks (calc-silicate hornfelses, etc.) and hydrothermal veins, as well as in related mineral deposits (skarn, epithermal systems, etc.); hence, the presence of some carbonates (mainly calcite and dolomite but also ankerite, malachite, rhodochrosite, etc.) as relatively frequent gangue minerals and even, if the tonnage and grade are sufficient, as ores of some metals (Cu, Zn, Fe, Mn, REE).

The classification of the mineral in one or another of the three structural groups defined for *Carb* (those of calcite and dolomite, trigonal, that of aragonite, orthorhombic) is irrelevant from the point of view of its recognition by reflected light, so the discussions below refer only to the individual species found in metallic deposits.

The description of ***Carb*** has been broken down between **ores** (azurite, **az**; bastnäsite, ***bst***; malachite, ***mch***: described in the corresponding chapter, cf. Part II) and **gangue** components (*gg*, the common case). It is well known, however, that the distinction between ore and gangue is not strictly mineralogical but depends on economic criteria, and these depend on the evolution of the markets. Therefore, the ascription of a mineral to one or another category may change over time:

- Thus, rhodochrosite ***rhc***, now treated as a gangue because it occurs as such in epithermal precious metal deposits, has been an important Mn ore (e.g., at Butte, Montana, USA), is considered the main ore at Molango, Mexico and is currently beneficiated as an ornamental rock (e.g., Capillitas Mine, Argentina).
- The same could be said of siderite ***sid***, typical of the *Carbonate Iron Formation* in the Banded Iron Formations (BIF), and frequent gangue in many hydrothermal veins (e.g., Coeur D'Alene, Idaho, USA), exploited until recently as Fe ore in other formations (e.g., Bilbao, Spain).
- Also, cerussite ***crs***, common in the oxidation zone of Pb deposits, was mined in the past as *white lead ore* at Leadhills, Scotland, UK.

136.2 General RL Characterisation of Carbonates

Only carbonates considered as gangue are the subject of this section, but before dealing with their individual description, it is useful to consider some **characteristics common to all carbonates**, regardless of whether they are considered as ore or gg. These are easily observable properties,[1] derived from their high birefringence, which allow characterization, in general, of any carbonate, whatever its economic meaning (ore or gangue) or its structural group and distinguish it from the common silicate gg:

- Remarkable **bireflectance** (Figs. 136.1a,b and 136.4a; Figs. 136.9a and 136.14a), perceptible even with dry objective (in air, *dsp*), provided that the grain size is sufficient to be able to individualize and compare adjacent grains with different orientation;
- Distinct **anisotropism** (despite the abundance of ***IR***, internal reflections, common in transparent minerals, Figs. 136.2, 136.4b, 136.5b, 136.14b and 136.15b), the

[1] These distinctive properties can only be observed if the resolution of the microscope is sufficient to distinguish the individual grains and compare them with each other. In masses or aggregates of carbonates that are submicroscopic or too fine to clearly individualize the grains, only the gray level (reflectance) and, hopefully, the **IR** or their colors are perceived.

R. Castroviejo, *A Practical Guide to Ore Microscopy—Volume 1*,
https://doi.org/10.1007/978-3-031-12654-3_136

(a)

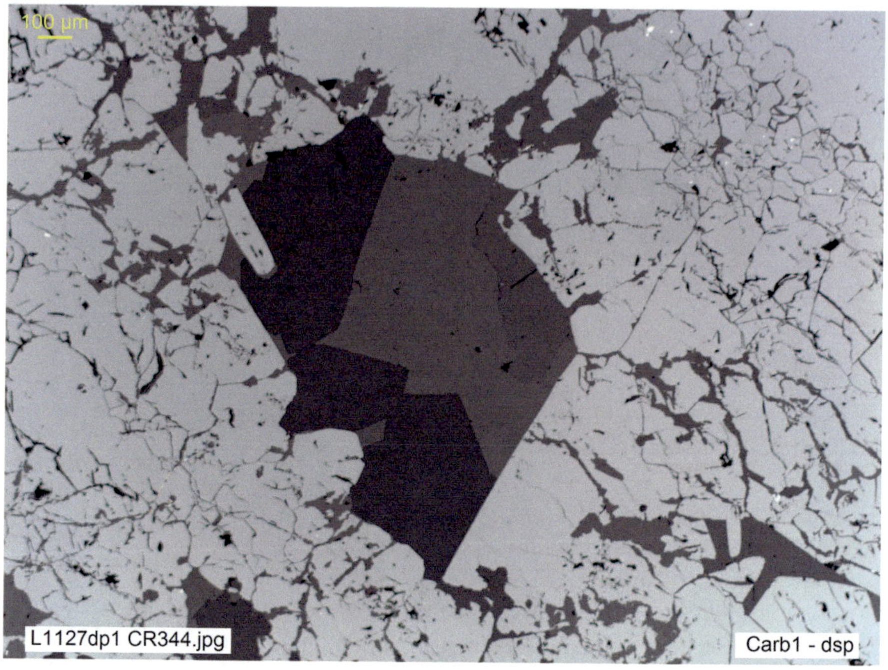

(b)

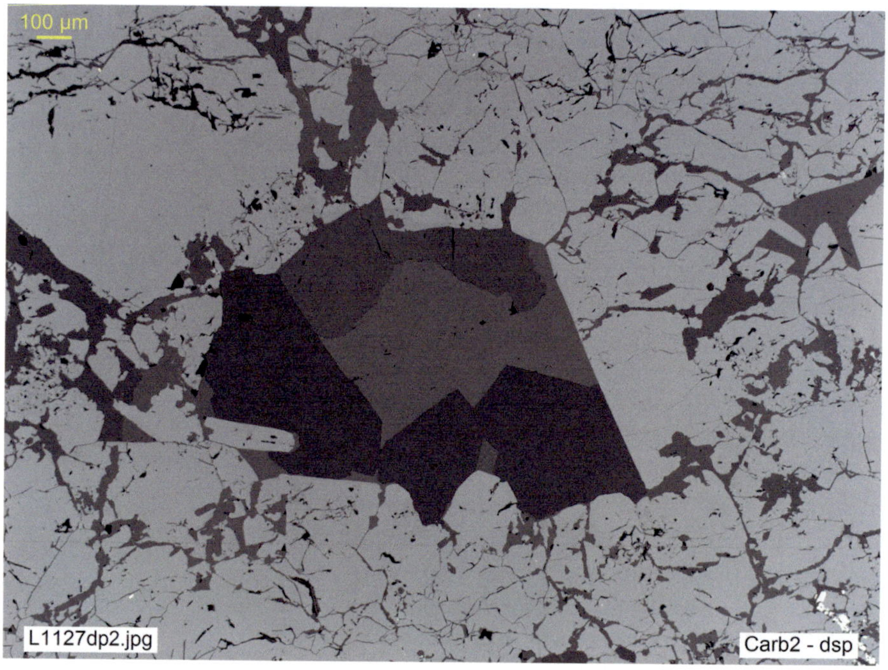

Fig. 136.1 **a**, **b** (Same scene, rotating the slide 90° from one to the other). Microgeode filled with *q* (dark gray, euhedral) and *Carb* (*calc*, somewhat lighter gray, anhedral), in ***alb*** (very light gray, subhedral) alabandite crystal aggregate with interstitial spaces filled by the same minerals. The changes of gray level in *Carb* on rotation (*comp.* Fig. 136.1a, b) denote its bireflectance. CR 344, Inakuraishyi Mine, Hokkaido, Japan (***dsp***)

effects of anisotropism are striking but little coloured and are generally best seen with dry, low magnification objective (*d*+*p*); as in immersion (*o*+*p*), the IR are reinforced.

- Visible **twins** (Figs. 136.5b and 136.6b), sometimes even without analyzer (Fig. 136.15a), due to the effect of the notorious bireflectance. In some species (e.g., calcite), very conspicuous, polysynthetic *pressure twins* (Figs. 136.6b and 136.7b).

In addition, *Carb* share with silicates their character of transparent minerals and, therefore,

- their appearance is that of an anodyne gangue; because of their grey colour and low lustre they are dark in general (Fig. 136.7a, *dsp*), more so against common ores (Figs. 136.1a, b and 136.7a), much more so against native metals (Ag, Au, Fe, Pt, Bi, etc. Fig. 136.4a) and

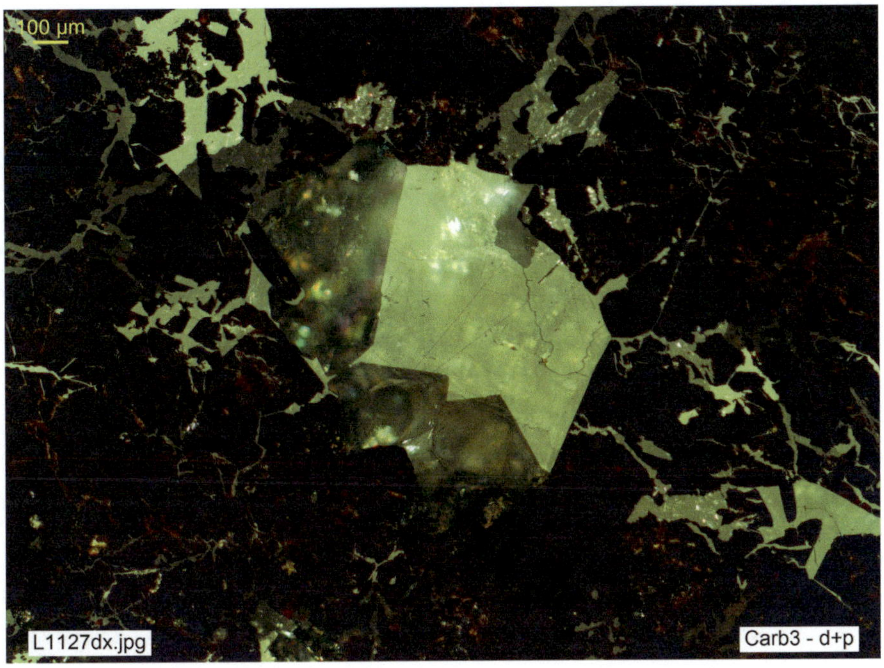

Fig. 136.2 (Same scene as Fig. 136.1, ***d+p***). Anisotropism and IR of *Carb*, as opposed to the barely perceptible anisotropism of *q* and extinction of *alb*

Table 136.1 Optical data of Carbonate gangue[(*)]

Refringence (n), Birefringence (δ) and Reflectance estim. (R %, in air)				
Mineral	n_{ω}	n_{ε}	$\delta = n_{\omega} - n_{\varepsilon}$	**R** (Estimated reflectance, air: d)
Calcite, ***calc***	1.658	1.486	0.172	≈ 4.7 %
Magnesite, ***mag***	1.700	1.509	0.191	≈ 5.3 %
Rhodochrosite, ***rhc***	1.816	1.597	0.219	≈ 6.7 %
Smithsonite, ***sms***	1.850	1.625	0.225	≈ 7.2 %
Siderite, ***sid***	1.875	1.635	0.240	≈ 7.5 %
Dolomite, ***dol***	1.679	1.500	0.179	≈ 5.2 %
Ankerite, ***ank***	1.690-1.750	1.510-1.548	0.182-0.202	≈ 5.7 % (5.3-6.0 %)
Cerussite, ***crs***	$n_{\alpha} = 1.803$; $n_{\beta} = 2.074$; $n_{\gamma} = 2.076$		0.273	≈ 9.6 % (8.2-12.2 %)

[(*)] *Data sources*: refractive indices (n) and birefringence (δ) according to Chang et al. (1998). Reflectance in air (R) calculated by Fresnel equation on estimated n values. Additionally, for ank and crs, also calculated minimum and maximum (in brackets). Measurements published in QDF3 for crs (546 nm, *dsp*, 8.3% and 13.4%, *respectively*) are similar to those estimated here

Carbonate groups: Calcite—without raster. Dolomite—shaded. Aragonite (crs)—dotted raster.

Symbols for n: n_{ω} and n_{ε}, refractive indices of ordinary and extraordinary rays for rhombohedral carbonates (equivalent to n_o and n_e of Ramdohr 1980, and other authors); n_{α}, n_{β}, n_{γ} for cerussite, orthorhombic.

extremely dark or black—they practically disappear—in immersion (osp: Figs. 136.5a and 136.6a).

- this dark appearance is typical for less refringent carbonates such as calcite (refractive index, **n**≈ 1.49 to 1.66) or dolomite (cf. Table 136.1), but it is not so pronounced in the more refractive carbonates, such as rhodochrosite ***rhc***, ankerite ***ank***, siderite ***sid***, bastnäsite ***bst*** (Figs. 136.8a and 136.15a) and especially cerussite ***crs*** (**n**≈ 1.80–2.08, Fig. 136.16a), whose Reflectance, **R**, is higher. Applying the Fresnel equation (§ 1.2.2, cf Table 134.1) results in an estimated value of R ≈ 4.7% for ***calc*** (assumed n ≈ 1.55) and R ≈ 9.6% (more than double) for ***crs*** (assumed n ≈ 1.90).
- they show *abundant IR*, internal reflections,
  - generally **colourless** (Figs. 136.2, 136.4b, 136.5b and 136.6b), calcite, ***calc***; dolomite, ***dol***; ankerite, ***ank***; magnesite, ***mag***; smithsonite, ***sms***; cerussite, ***crs***; bastnäsite, ***bst;***
  - but sometimes slightly **colored** (Figs. 136.8b, 136.9b and 136.15b), rhodochrosite, ***rhc***; ankerite, ***ank***; siderite, ***sid***;

Fig. 136.3 Oxidized Cu ore (azurite ***az***, malachite ***mch***). The IR of *mch* (green) and *az* (blue) are so intense and luminous that they blur the grain outlines. CR 349, Grand Gulch, Coconino Co, Arizona, USA (***o+p***)

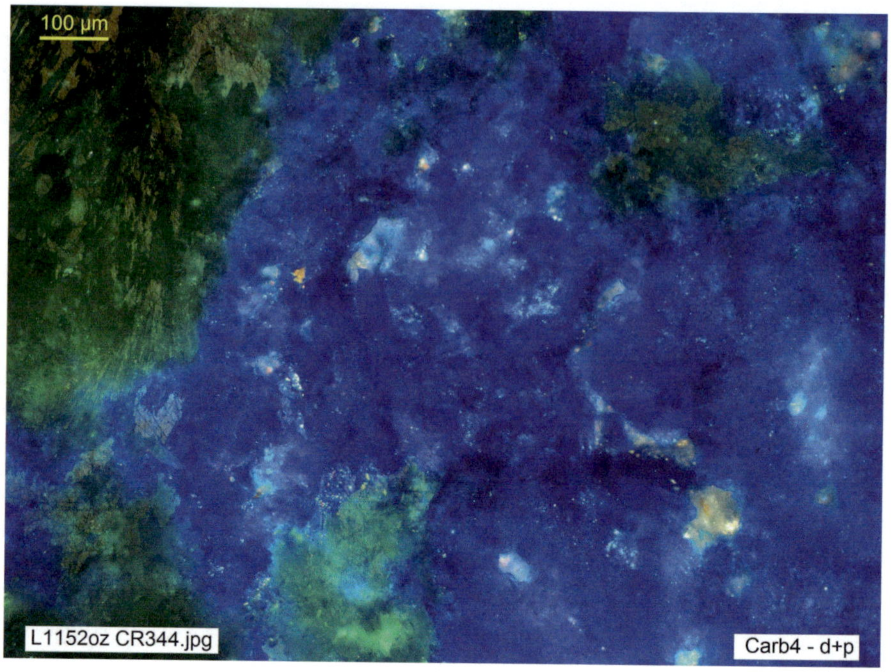

– or intensely **coloured** (Fig. 136.3, cf also F az and F mch), azurite, ***az***; malachite, ***mch***.

The anisotropism effects themselves are not coloured but may appear so because of the high birefringence of ***Carb***. When it is not the **IRs** themselves that are coloured (as in ***mch***, ***az*** or ***rhc***), what is perceived in colour are actually the interference colours produced by the polarisation of the light beam derived from the **IRs** on its way through the interior of the mineral to the surface (Fig. 136.9b, cf. Figs. 15.3b and 15.4b).

The **polishing quality** is usually good. Carbonates are generally polished without difficulty due to their moderate hardness.

Taking into account these general data, we characterize the most common species of Carb by their unique features.[2]

136.3 Calcite. Rhodocrosite. Magnesite. Smithsonite. Siderite.

Calcite (*calc*, $CaCO_3$, trigonal) is the most common ***Carb*** and also the least refractive, i.e., the one with the darkest appearance (Fig. 136.7a), R ≈ 4.7%, like ***q*** or resin. This value is an estimated average for **R** applying the Fresnel formula, assuming an average value of the refractive index **n** between $\mathbf{n_o}$ and $\mathbf{n_e}$ (*ordinary and extraordinary indices*, resp) but, in reality, in ***calc*** there is not a unique value of **n** but a maximum value (n_o = 1.658) and a minimum value (n_e = 1.486). Between them, are all the intermediate values corresponding to the different possible orientations of the crystal.

Despite its low refringence, the difference between the n_o and n_e values, i.e., the **birefringence** of calcite ($\delta_{calc} = n_o - n_e = 0.172$) is high (Table 136.1). It is very striking if compared with that of common gg-silicates, such as quartz, ***q***—almost 20 times lower ($\delta_q = 0.009$). A consequence of this strong birefringence of ***calc*** is its notorious **bireflectance**, variation of **R** (or its subjective equivalent—the *brightness* perceived by the eye of the observer) with orientation, due to the effect of the strong birefringence of the mineral. Hence, although its reflectance is very low (approximately like that of quartz), its bireflectance is clearly perceived. This property, practically diagnostic of calcite, as in general of the other ***Carb***, is preferably evidenced in air (dry objective, *dsp*, better with low magnification) by the variation of the gray level between contiguous grains, randomly oriented (Fig. 136.1a, b).

In immersion (*osp*) it is easy for the bireflectance to go *unnoticed* since, due to its low refractive index, the mineral practically disappears, i.e., it is almost always black, and it is then difficult to distinguish shades (Figs. 136.5a and 136.6a). However, although the extraordinary ray disappears completely ($R_{e\text{-oil}} \approx 0$), the ordinary ray is still visible as a faint veil, as noted by Ramdohr (1980). Indeed, applying the Fresnel function, the result is an immersion reflectance (*osp*), $R_{o\text{-oil}} \approx 0.25\%$, but to perceive this faint difference at such a low gray level, at the limit of the sensitivity of the human eye, one must find sections with the appropriate orientation.

[2] For more information, the synthesis of Ramdohr (1980, p. 1103–1109), comprising a detailed exposition of the main Carb species (calc, dol, sid, rhc, sms, crs, mch and az), is highly recommended.

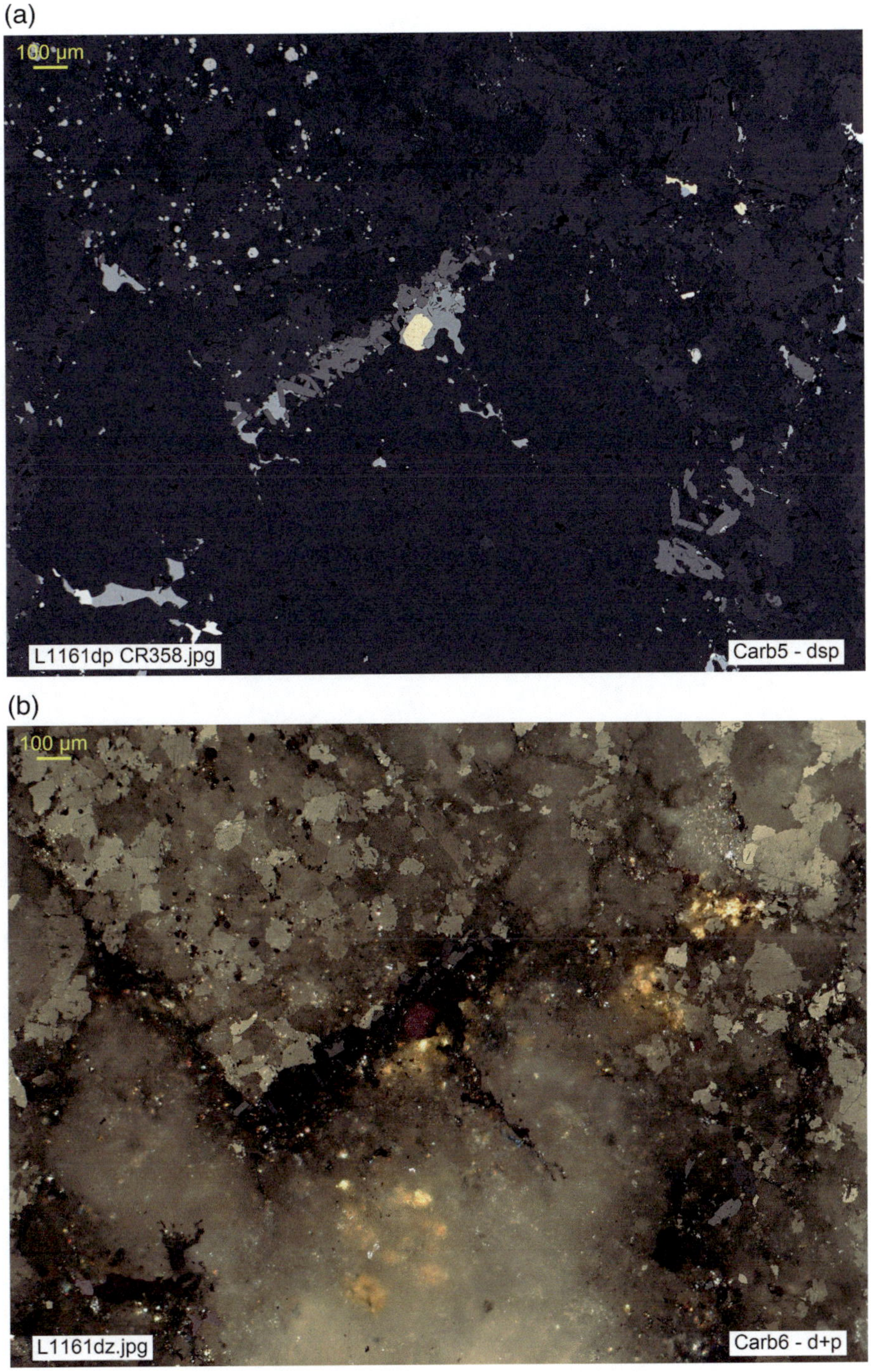

Fig. 136.4 **a**, **b** (Same scene: ***dsp*** and ***d+p***, *respectively*). Auriferous ore (Au, center), with tetrahedrite, ilmenite, pyrite and tellurides (coloradoite, calaverite), in brecciated quartz-carbonate breccia. Bireflectance, anisotropism and IR are well visible in the large carbonate clast occupying the upper left corner of the image. CR358, Oroya-Brownhill Mine, Kalgoorlie, Australia (***d+p***)

With analyzer (*d+p*, *o+p*), its properties are typical of ***Carb.*** Anisotropism is usually visible (Fig. 136.2), despite its abundant **RI**, colorless and luminous (Figs. 136.2 and 136.5b). Polysynthetic pressure twins are very frequent in ***calc*** (Figs. 136.5b, 136.6b and 136.7b).

In the same group of rhombohedral carbonates are **magnesite**, ***mag***, $Mg(CO_3)$; **rhodochrosite**, ***rhc***, $Mn(CO_3)$; **smithsonite**, ***sms***, $Zn(CO_3)$ and **siderite**, ***sid***, $Fe(CO_3)$, isostructural with ***calc***. These minerals have been listed in increasing order of their refractive indices, **n** (cf. Table 136.1). In practical terms, the increasing value of **n** implies that, although when observed with **RL** they all appear grey, they appear increasingly lighter or brighter (increasing reflectances) from left to right. All of them are more reflective than ***calc***.

The brightness is therefore an important data for recognition, which must be complemented with geological and paragenesis information (e.g., ***sms***: paragenesis of

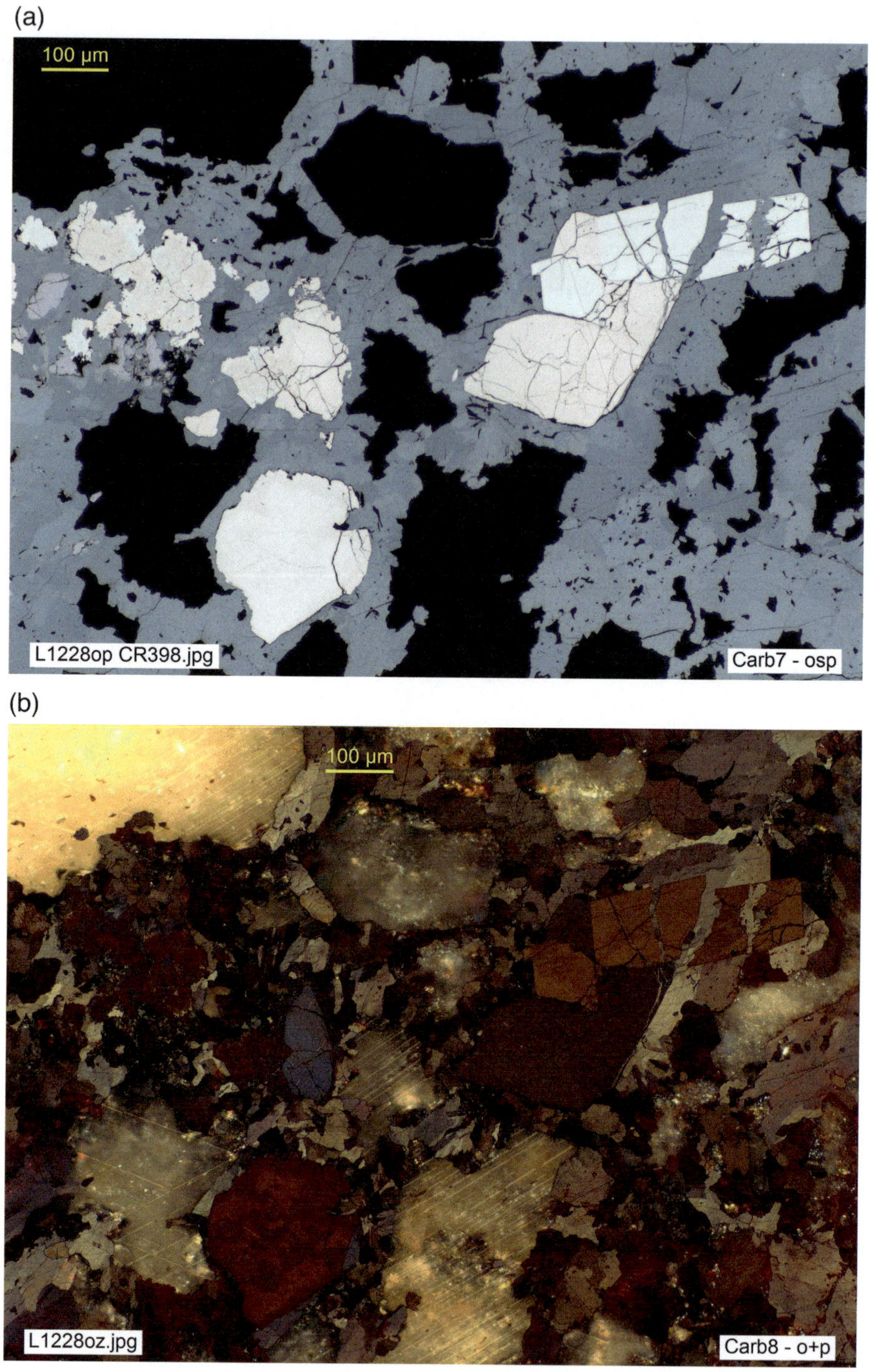

Fig. 136.5 **a**, **b** (Same scene: ***osp*** and ***o+p***, *respectively*). Silver ore vein, with miargyrite (grey) cementing and corroding early filling of arsenopyrite (white) and gangue carbonate and quartz, brecciated. The gangue becomes invisible (black, as a hole) in immersion (*osp*). However, with crossed polars (***o+p***), shows abundant colourless IR and the typical anisotropism and polysynthetic twins of the carbonate calcite, whose low reflectance is evident (*osp*) by its black color, like quartz. CR398, Hiendelaencina, Guadalajara

oxidized Zn minerals; ***sid***, ubiquitous; ***mag***, in particular geological or metasomatic environments: associated with alteration or metamorphism of Mg-rich igneous rocks, such as peridotites and ophiolitic complexes, metasomatism, evaporite deposits). However, the quantification of the brightness by eye is not easy in isolated grains and, to be more reliable in cases like these, in which the differences are small and gradual, it is convenient to estimate it by comparison of contiguous grains of the considered species. This contiguity between similar carbonate species is not exactly frequent, with the exception of ***calc*** and ***dol***, which appear associated in many rocks. Thus, in the absence of other diagnostic properties, brightness is insufficient for a sure identification with **RL**, although it is the only data available in some cases. When the mineral is colourless and appears only grey, with **RI** also colourless, as occurs

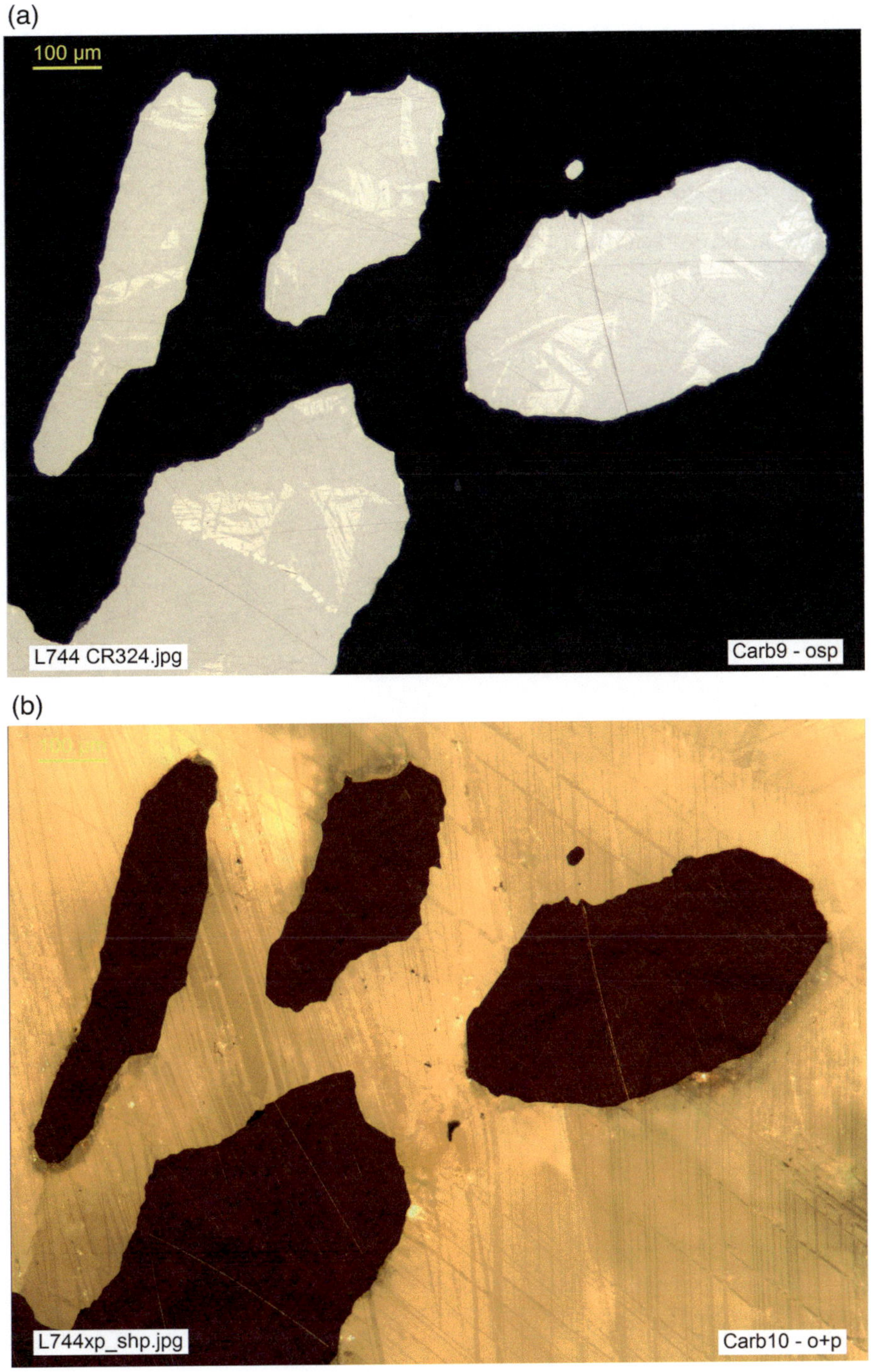

Fig. 136.6 a, b (Same scene: *osp* and *o+p*, *respectively*). Argentiferous ore (allargentum and eugenite, bright) with carbonate gangue, probable calcite (black, *osp*; IR and twins, *o+p*). CR 324, Bouismas Mine, Bou Azzer (Morocco)

especially in ***mag*** or ***sms***, it is convenient to resort to the support of other techniques (petrographic, staining, acid attack, microanalysis, XRD). However, ***rhc*** and ***sid*** can generally be recognized with an acceptable degree of confidence.

Rhodochrosite, ***rhc***, in addition to the typical properties of *Carb* that it shares with *calc*, has the following as distinguishing characteristics:

- its relatively high reflectance, much more than that of ***calc***; hence, unlike ***calc***, it does not disappear in immersion (*osp*), although it is barely visible in the n_ε direction, where it looks very dark ($R_{\varepsilon/\mathrm{rhc\text{-}oil}} \approx 0.1\%$; but: $R_{\varepsilon/\mathrm{calc\text{-}oil}} \approx 0.002\%$!).
- its strong bireflectance, evident under the microscope by the notorious difference in shades of grey of the different grains (Fig. 136.8a).

Fig. 136.7 **a**, **b** (Same scene: ***dsp*** and ***dxp***, *respectively*). Brecciated chalcopyrite (yellow) ore, with acicular hematite (gray) and calcite cement with scarce quartz (both, dark gray, *dsp*). With analyzer, calcite is immediately identified by its clear, apparent anisotropism enhanced by IR, and well visible pressure twins. RepPL (031200 g, Maruxa Mine, Moeche, Coruña

- Typically, pinkish color to the naked eye, although it can be brown to black by alteration, mainly to Mn oxides.
- in **RL**, the colour is grey as in the other carbonates, but in fresh samples, due to the effect of the **IR**, it shows a slight pinkish hue superimposed (*dsp*).
- with analyser (*d+p*) the RI are deep pink, massive and luminous (Fig. 136.8b); however, dark, brown to black alteration products are often interposed and may obscure the **RI**.
- the alteration, dominated by Mn oxides such as pyrolusite or psilomelane which can be recognized with **RL**, distinguishes ***rhc*** from other similar carbonates such as ***sid*** or ***sms***.
- paragenesis is often epithermal and related to precious metal deposits but also pegmatitic, varied hydrothermal or sedimentary.

(a)

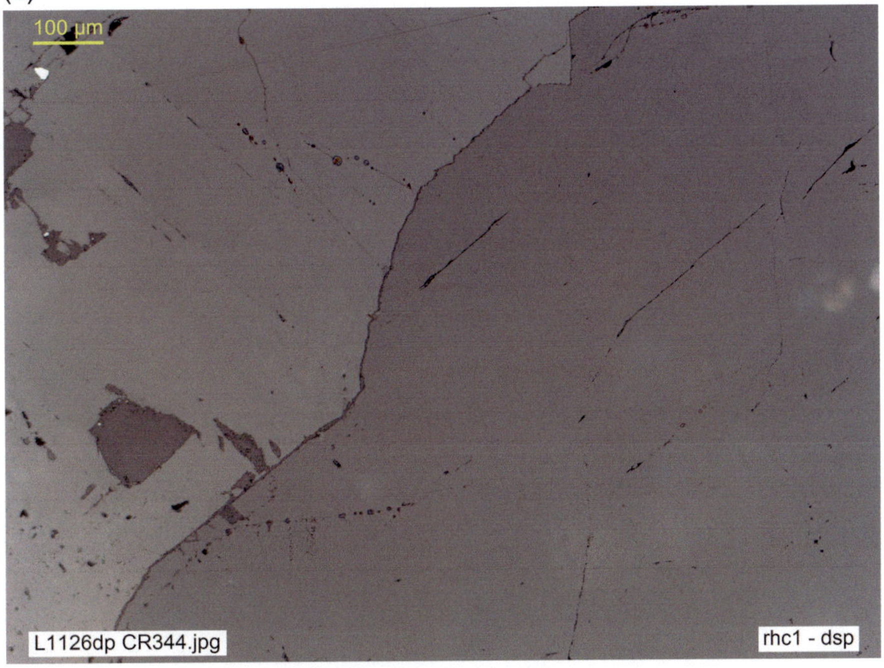

(b)

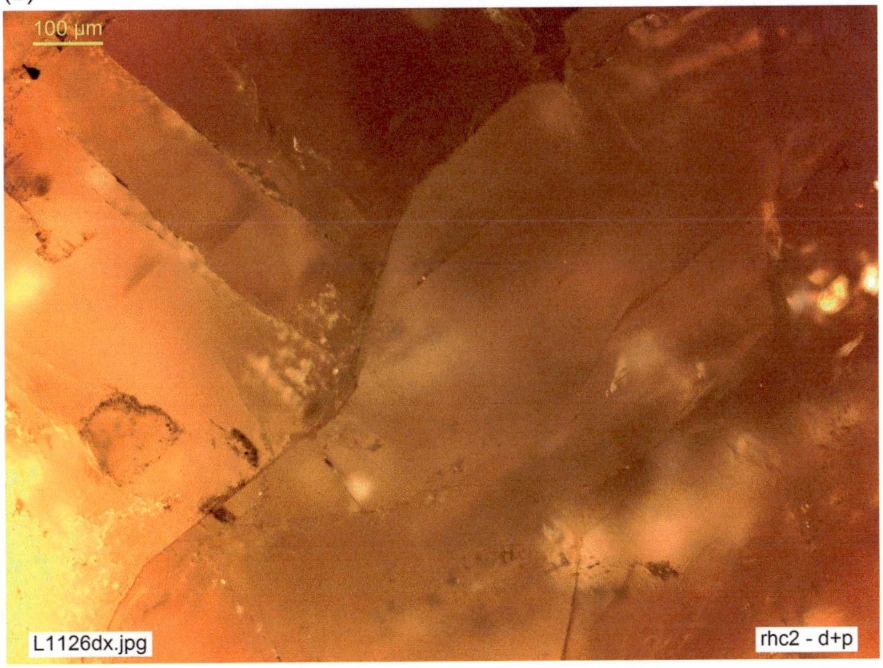

Fig. 136.8 **a**, **b** (Same image: **dsp** and **d+p**, *respectively*). Holocrystalline rhodochrosite ***rhc***, grey with pinkish hue (*dsp*): the tiny pyrite inclusion (upper left corner) provides a reference for estimating its reflectance; the different shades of grey of *rhc* display its bireflectance. With analyzer (*d+p*), the pinkish hue of *rhc* is enhanced by IR and its transparency allows us to distinguish the secondary fluid inclusions trails trapped in the cement of microcracks. CR 344, Inakuraishyi Mine, Hokkaido, Japan

Siderite, ***sid***, very clearly shows the carbonate properties, enhanced by peculiar characteristics that facilitate recognition:

- Its relatively high reflectance (Figs. 136.10a and 136.11a) only surpassed by *crs*, among the *Carb* considered here—siderite stands out for its brightness against *calc*, *dol* or *mag* and remains always visible (even in immersion, *osp*, siderite is very dark, hardly visible, in the n_ε direction, but not totally black like calcite), compare Fig. 136.12a, b, as well as the values of Reflectance in immersion, $\boldsymbol{R_{oil}}$, minimum (ε) and maximum (ω), of siderite: $R_{\varepsilon/sid} \approx 0.2\%$ and $R_{\omega/sid\text{-}oil} \approx 1.2\%$, with those of calcite, whose near-zero R_ε makes it practically invisible in the n_ε-*direction*: $R_{\varepsilon/calc\text{-}oil} \approx 0.002\%$, and $R_{\omega/calc\text{-}oil} \approx 0.25\%$.
- Its high birefringence, the highest among the *Carb* of the *calc* group (cf. Table 136.1), notorious in RL for the resulting *strong birefringence* (Figs. 136.10a, 136.11a and 136.12a).

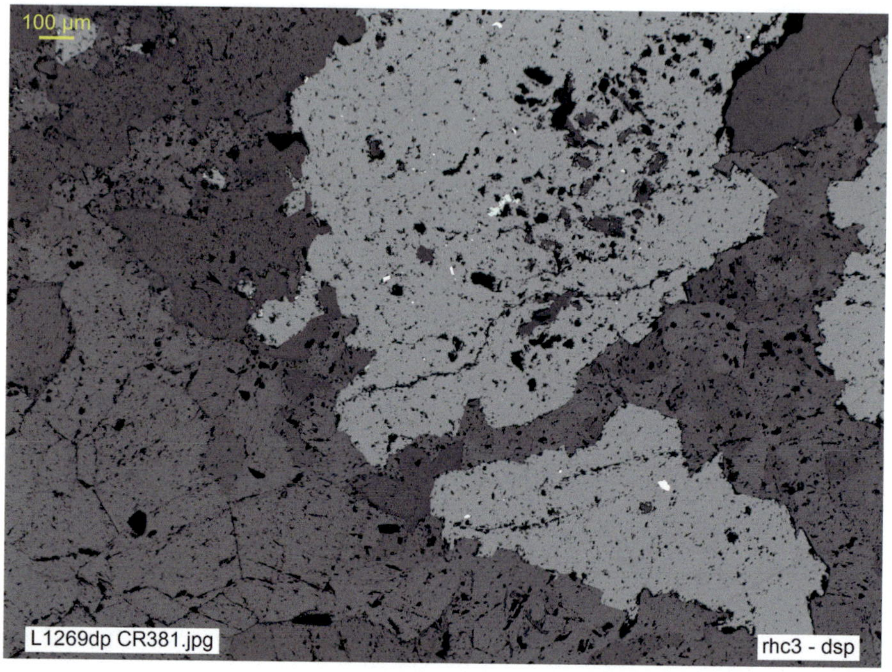

Fig. 136.9 **a**, **b** (Same image: **dsp** and **d+p**, *respectively*). Sphalerite *sp* ore with massive *rhc*, minor *q* and traces of galena *gn*; *sp* looks light grey by comparison with gangue, but the tiny inclusions of *gn* (white, bright, R ≈ 43%) evidence by contrast its low reflectance ($R_{sp} \approx$ 17%). Comparison with the inclusion of q in rhc (dark grey, relief, upper right corner) evidences the remarkable bireflectance of *rhc*, from a value close to *q* ($R_q \approx 4.5\%$) to a distinctly higher value, though always lower than *sp*. With analyzer (*d+p*), the abundant and luminous IR of *rhc* partly obliterate the IR of *sp* but not the anisotropism of *rhc*. CR-381, True Fissure Mine, Philipsburg district, Montana, USA

- Its frequent lamellar or polysynthetic twins, unlike *mag* or *sms*, which lack them.
- With analyzer (*d+p*), in addition to its striking anisotropism, which with a dry objective and low magnification can be spectacular for its luminosity and contrasts, its colorless or yellowish to reddish-brown RI (Figs. 136.10b and 136.11b) stand out.
- The frequent alteration, meteoric or hypogene, of ***sid*** to limonite, which intensifies the reddish color of the **IR** by meteoric oxidation or other processes. Among these, the metasomatic-hydrothermal substitution by several minerals, such as magnetite, as observed in Fig. 136.13 (Minas La Unión, Cartagena).

Fig. 136.10 **a**, **b** (Same scene, 2.5 × objective, **dsp** and **d+p**, *respectively*). Veinlet fill of siderite (crystalline, euhedral, very bireflective, *dsp*: different shades of gray, strong anisotropism, *d+p*), which cements and replaces magnetite (brownish white, isotropic). SV-1, San Valentín Quarry, La Unión, Cartagena

Limonitic alteration is also found in ***ank*** (cf. Fig. 136.15a, b), a ***dol*** group mineral with which siderite could be confused, when the former is present in its more Fe-rich (more reflective) varieties, although in direct comparison its brightness is clearly lower (Table 136.1).

The paragenesis of ***sid*** is varied. Its presence is typical in sedimentary series, e.g., in the *carbonate iron formations* of the BIF's and also in various types of pegmatitic, hydrothermal (e.g., Coeur d'Alene, Idaho, USA) or metasomatic deposits, but it is not frequent in oxidizing environments (Fig. 136.13).

136.4 Dolomite. Ankerite. Cerussite.

Dolomite (***dol***, $CaMg\langle CO_3\rangle_2$, trigonal), is also a frequent mineral, although not as frequent as *calc*, which it resembles (though it is harder and denser than *calc*) and with which it is

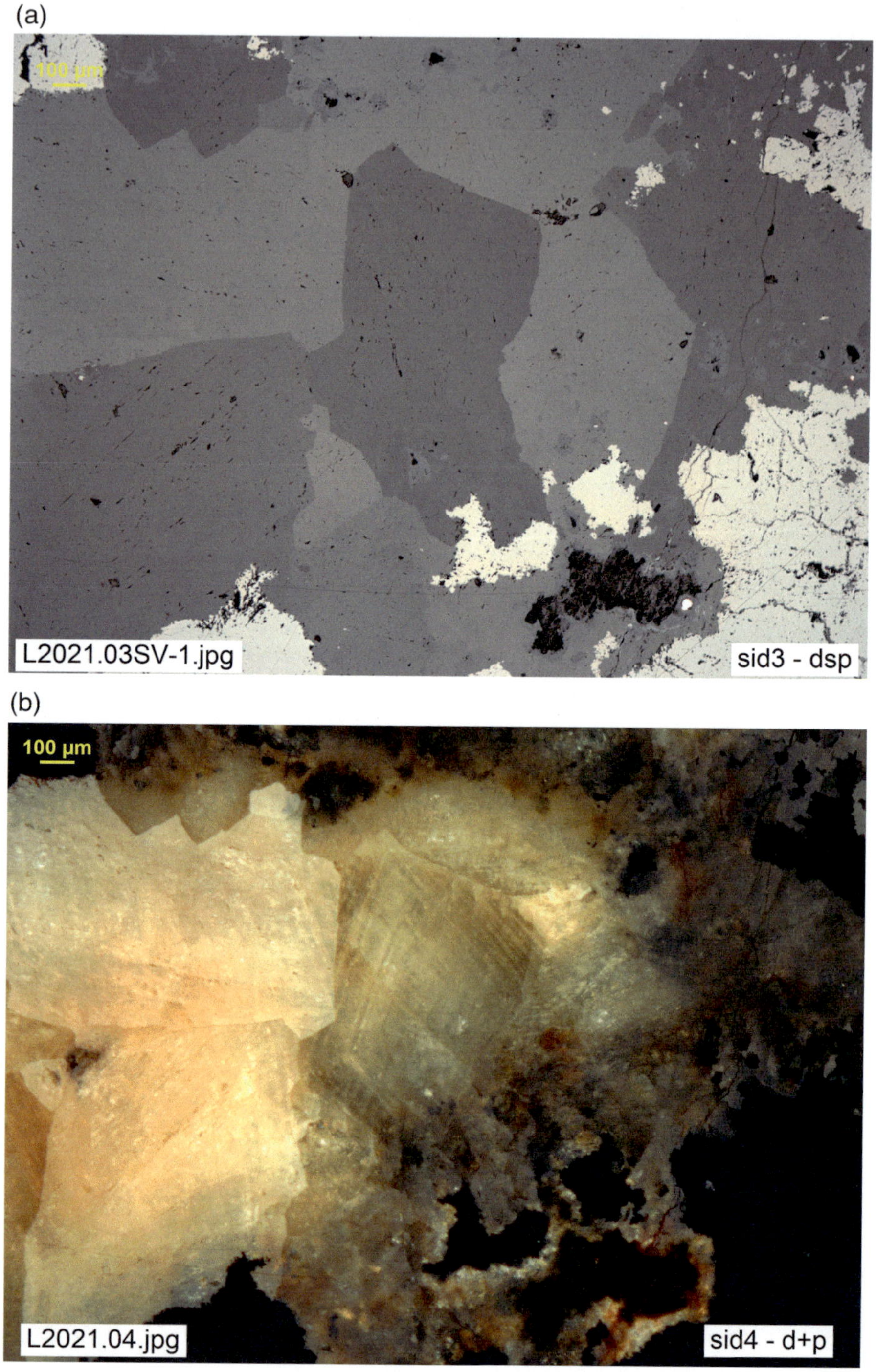

Fig. 136.11 **a**, **b** (Detail of Fig. 136.10a, central zone, **dsp** and **d+p**, *respectively*). At higher magnification (obj. 5x), the siderite bireflectance is still clearly visible, but its anisotropism is partly obliterated by the IR, which instead show the zoning more clearly

closely related, to the point of being both, with its rhombohedral lattice, almost isostructural. However, in *calc* all the cations are equivalent (Ca ions). This does not occur in *dol*, which incorporates Mg ions alternating with Ca. Therefore, in *dol* the incorporation of Fe in the lattice is frequent, by isomorphous substitution of Mg. This substitution of Mg by Fe can occur in different degrees and may reach compositions transitional to ankerite, which define the ankerite-dolomite series, $CaFe(CO_3)_2$—$CaMg(CO_3)_2$.

In terms of optical properties, it shares with calc the remarkable bireflectance, anisotropism, and abundance of generally colorless IR (Fig. 136.14a, b). The refringence ($n_\omega = 1.679$, $n_\varepsilon = 1.500$) and birefringence ($\delta = 0.179$) of *dol* are slightly higher than those of *calc*, but the differences in reflectance (gray level) or bireflectance, resulting from that slight variation, are too small for the human eye to distinguish with certainty by gray level. Therefore, the microscopic distinction of *calc* and *dolc*, which are often

Fig. 136.12 **a**, **b** (Same scene: **dsp** and **osp**, *respectively*). Sulfide infill (tetrahedrite, galena, sphalerite ores -various shades of gray, pyrite, yellowish white), with siderite gangue. Sid looks very dark by comparison with the ore, distinctly darker, even darker than sphalerite (inclusions in tetrahedrite), especially in oil (*osp*), but even so it is still visible (*comp.* galena *triangular pits*, *osp*). CR266.c, Lucky Friday Mine, Coeur d'Alene, Idaho, USA

associated in some rocks such as limestones or dolomites, is uncertain.

However, apart from the diagnostic staining thin sections or the well-known and useful use of acid etching in hand specimen, to distinguish from *calc* the *dol-ank* series, which is much more resistant to etching, the microscope allows, also with **RL,** the observance of subtle differences:

- From the point of view of morphology, *dol* has a much stronger tendency than *calc* to develop crystalline forms, often euhedral (Fig. 136.14a, b).
- Unlike *calc*, *dol* shows rhombohedral exfoliation $(10\bar{1}1)$ perfect.
- The refractive index, **n**, of *dol* grows with increasing Fe contents.

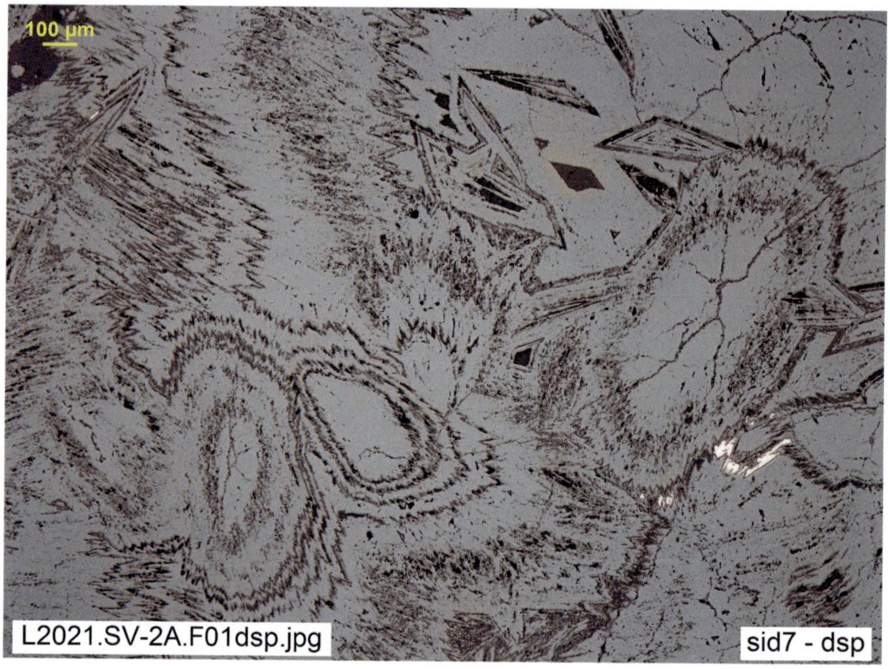

Fig. 136.13 Pseudomorphic replacement of siderite (represented by rhombohedral *ghosts* and indented aggregates, with scarce carbonate relics, dark gray) and greenalite (fibrous or acicular) by magnetite (light gray). SV-2.A, San Valentín Quarry, La Unión, Cartagena (**dsp**)

- *dol* IR, initially colourless as in *calc*, may gradually turn yellowish-ochre or reddish-brown (ankerite, cf Fig. 136.15a, b) as Fe content increases.
- Pressure twins, very frequent in *calc*, are scarce in dolomite.

Ankerite ‹***ank***, ($CaFe^{2+}$, Mg, Mn) $(CO_3)_2$, trigonal›, belongs to the ***dol*** structural group, with which it forms a series (*cf.* dol). In fact, ***ank*** has been defined as an Fe-enriched ***dol***, with a ratio $Fe^{2+}/(Mg + Fe^{2+}) \geq 0.2$ (Chang et al. 1998), identifying as ferroan *dolomite* the compositions between ***dol*** and ***ank*** ($0 < Fe^{2+}/(Mg + Fe^{2+}) < 0.2$).

As a rhombohedral *Carb*, like *dol*, ank shows a perfect rhombohedral exfoliation ($10\bar{1}1$), but its birefringence is much higher than that of *dol*. *Ank* is much scarcer in nature than *dol*. Especially in the most Fe-rich terms of the series, it looks rather like *sid*, with which it could be confused, by its reflectance and bireflectance, higher than those of *dol,* though varying with ank composition (reflectance increases with Fe content). In addition, there is the yellowish-brown coloration of its IR, locally reinforced with reddish tones due to the frequent limonitic alteration (Fig. 136.15a, b) or with pinkish tones, in varieties enriched in Mn.

Cerussite

Among the *orthorhombic* carbonates (*Aragonite Group*), only cerussite, ***crs***, $PbCO_3$, is relatively frequent as a supergene manifestation of Pb ores, typically weathered galena. It is precisely the association with oxidized Pb ores, corroborated by the spatial relationship with relicts of galena ***gn*** or other Pb sulphide ores visible in the sample that is a first indication of the probable presence of ***crs***. The identity of ***crs*** can then be recognized (Fig. 136.16a, b) right away by:

- its very strong refringence and birefringence (Table 136.1)
- its remarkable reflectance, distinctly higher than usual for gg minerals, including the other carbonates (Table 136.1)
- its strong bireflectance
- its manifest anisotropism in air, partly obliterated by IR in immersion
- the IR generally colourless, but often showing diffuse and coloured edges due to the effect of strong dispersion, according to Ramdohr (1980)
- the tendency to form crystalline aggregates, although these may mimic the shapes, the cleavage or the zoning of ***gn*** by pseudomorphic replacement (Fig. 136.16a)
- the preserved small relics of *gn* or other Pb sulphide ores, often surrounded by an intermediate corona (somewhat darker than *crs* and less crystalline) of anglesite, a sulphate that represents a first stage in the oxidation of the sulphide (Fig. 136.16b).

To conclude, other important carbonates such as **malachite**, **azurite** or **bastnäsite** have been described in the corresponding chapters on ores (§ Part II: *14-az, 15-bst, 79-mch*), as already indicated.

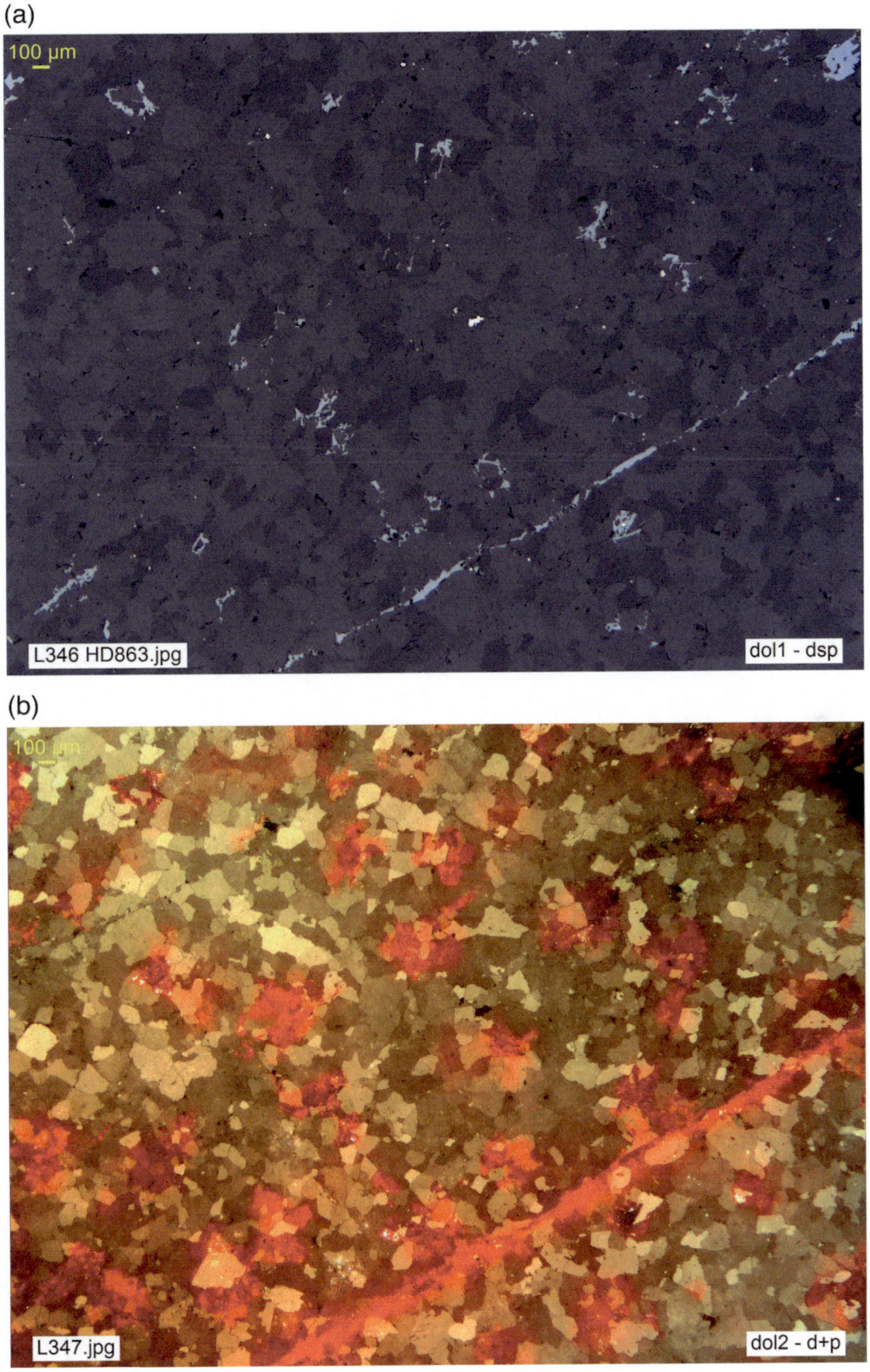

Fig. 136.14 **a**, **b** (Same image: **dsp** and **d+p**, *respectively*). Interstitial impregnation of cinnabar ***cin*** (light bluish grey) and scarce pyrite (yellowish white), in crystalline dolomite (grey, in different dark tones, showing the distinct bireflectance of the mineral, *dsp*). With analyzer (*d+p*) the anisotropism and abundant colorless RI of *dol* is immediately perceived, unlike the *cin*, whose anisotropism is masked by its striking RI, of intense red tone. HD 863, Idria, Yugoslavia

136.5 Borates. Vonsenite

Vonsenite ‹***von***, $Fe^{2+}{}_2Fe^{3+}(BO_3)O_2$, orthorhombic› is an uncommon mineral but is typically present in certain parageneses (contact metamorphic or metasomatic ferriferous deposits). It belongs to the ludwigite group, with which it forms the **ludwigite-vonsenite series**, $Mg_2Fe^{3+}(BO_3)O_2$–$FeFe_2{}^{2+}_{2}Fe^{3+}(BO_3)O_2$. From its appearance and colour, shiny black in hand specimen, almost opaque, grey under the microscope, with reflectance (R ≈ 12–14%, for 546 nm) of the order of chromite (slightly higher than cassiterite and

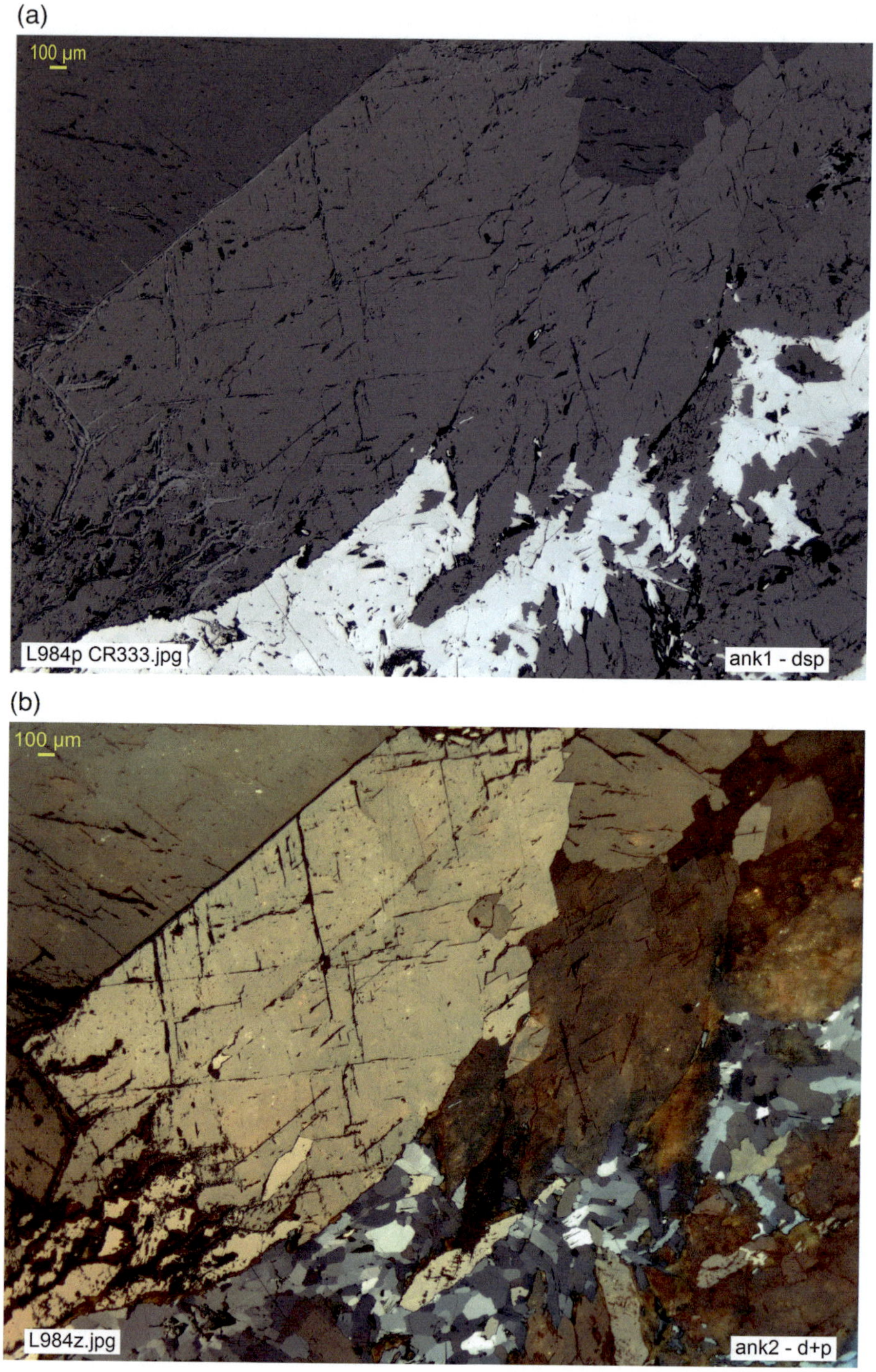

Fig. 136.15 **a**, **b** (Same scene: ***dsp*** and ***d+p***, *respectively*). Holocrystalline ankerite (grey), slightly deformed, scarcely altered along microcracks to limonite *lim* (lower left) and cemented by stibnite (white, strongly anisotropic). Ankerite shows its relatively high reflectance, remarkable bireflectance (gray to dark gray) and perfect cleavage (*dsp*), as well as its distinct anisotropism (*d+p*) and yellowish-white IR, turning reddish by oxidation (*lim*). CR 333, Carrok Mine, Devon (UK)

distinctly higher than the usual gangue), it might appear to be an ore. However, it has no known industrial applications and is, therefore, conceptually a gangue mineral.

Apart from the typical paragenesis of Fe-rich, contact metamorphic or *skarn* deposits with magnetite, which may be a first hint to its presence, it is recognized by:

- their often prismatic or acicular habit (Fig. 136.17a, b)
- its pleochroism in shades of grey to brownish-grey (lighter), perceptible in air and more pronounced in immersion (Fig. 136.17a, b)
- its low reflectance, compared to the magnetite that usually accompanies it (Fig. 136.17a, b)

(a)

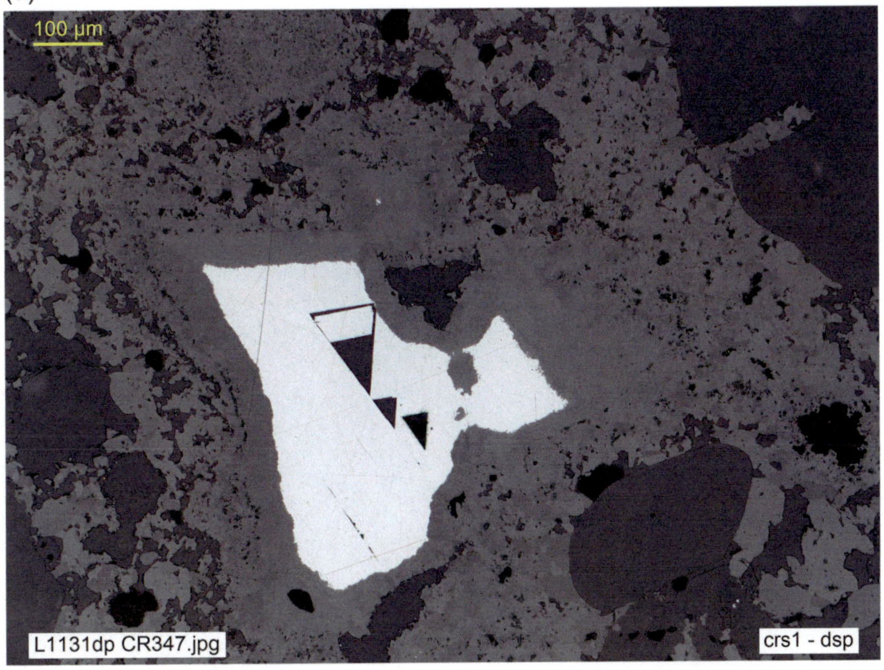

(b)

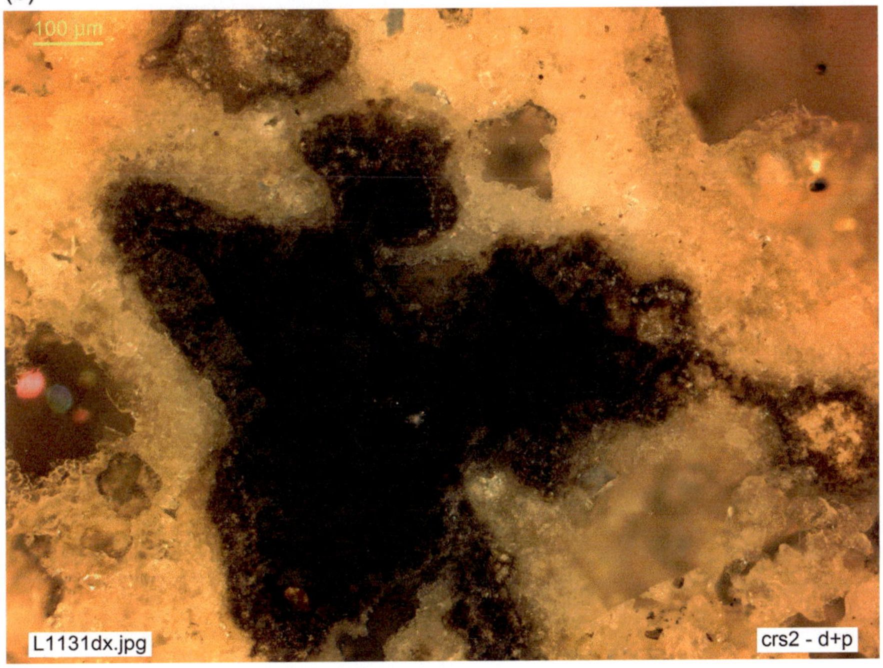

Fig. 136.16 **a**, **b** (Same image: ***dsp*** and ***d+p***, *respectively*). Zoned alteration of galena *gn* (white, *triangular pits*) to anglesite *ang* and cerussite *crs*. The innermost zone is a *corona* or rim of microcrystalline *ang* (grey, *dsp*) on *gn*. Its poor transparency (dark shade, *d+p*) suggests that it still includes numerous submicroscopic, opaque relics of the primary mineral (*gn*). The outer zone contains essentially *crs* (gray, more crystalline than *ang* and more distinctly bireflective, *dsp*) and minor quartz (dark gray). The abundant and luminous RI of *crs* indicate its transparency and the almost complete oxidation of *gn*, hinting to the disappearance of primary sulfide relics in the *crs* zone. CR 347, Webb's Consols Mine, New England, Australia

- its striking anisotropism, even in air, with vivid reddish-brown and grey, slightly purplish or greenish tones (Fig. 136.18a, b)
- the scarcity of reddish-brown IR, which are instead frequent in the Mg-rich terms of the series (ludwigite)

By its appearance, it could be confused with dark varieties of tourmaline (*chorlo*), but tourmaline has much more abundant and luminous IR, besides being more transparent (lower reflectance).

Fig. 136.17 **a**, **b** (Same scene: ***dsp*** and ***osp***, *respectively*). Nematoblastic aggregate of vonsenite *von* (gray to brownish-gray, pleochroic) and magnetite *mt* (very light brownish gray), with minor undif gangue inclusions (dark gray, black in oil, *osp*) and pyrite *py* (yellowish white). Rub.81.15, Monchi Mine, Burguillos del Cerro, Badajoz

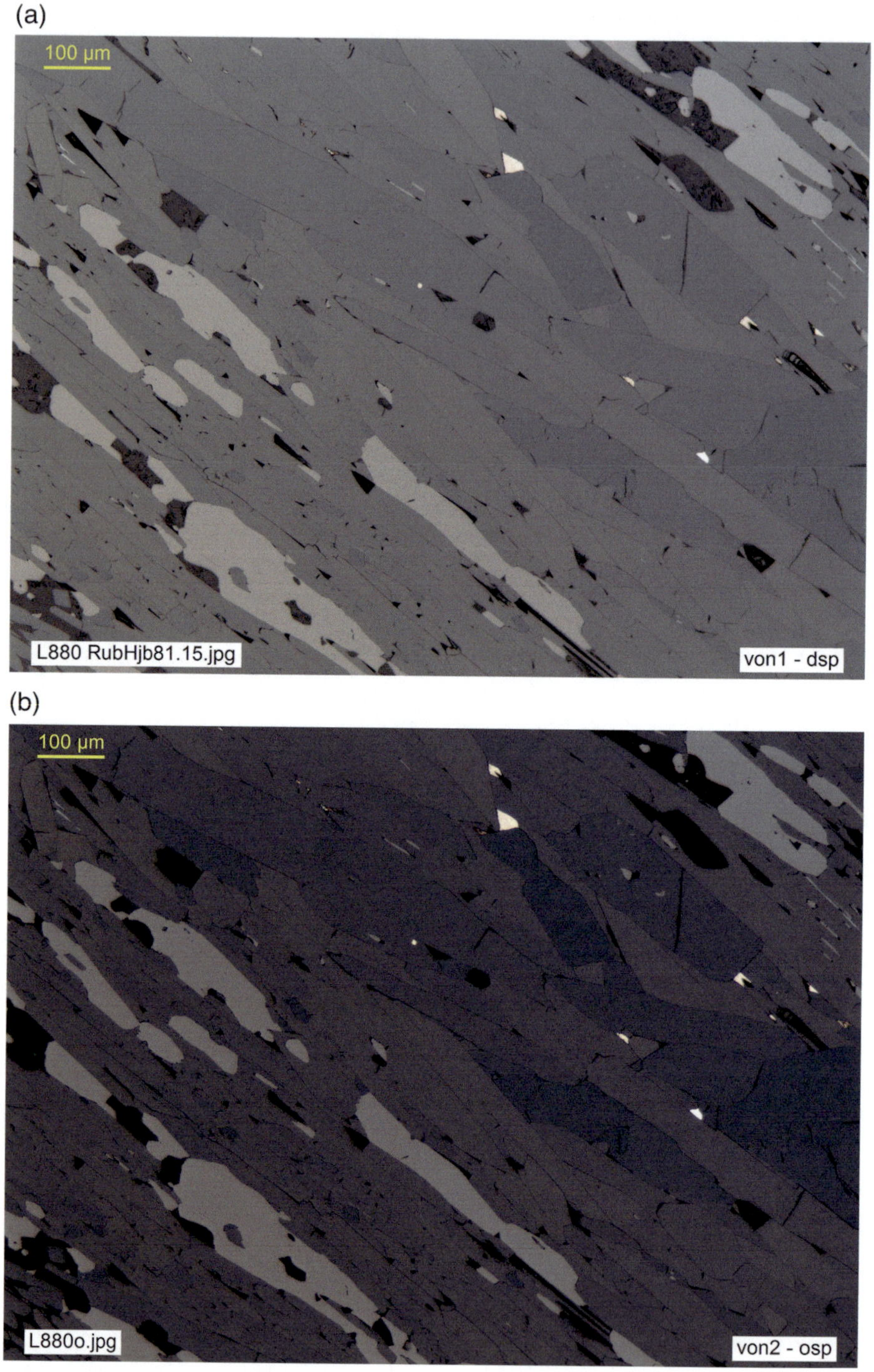

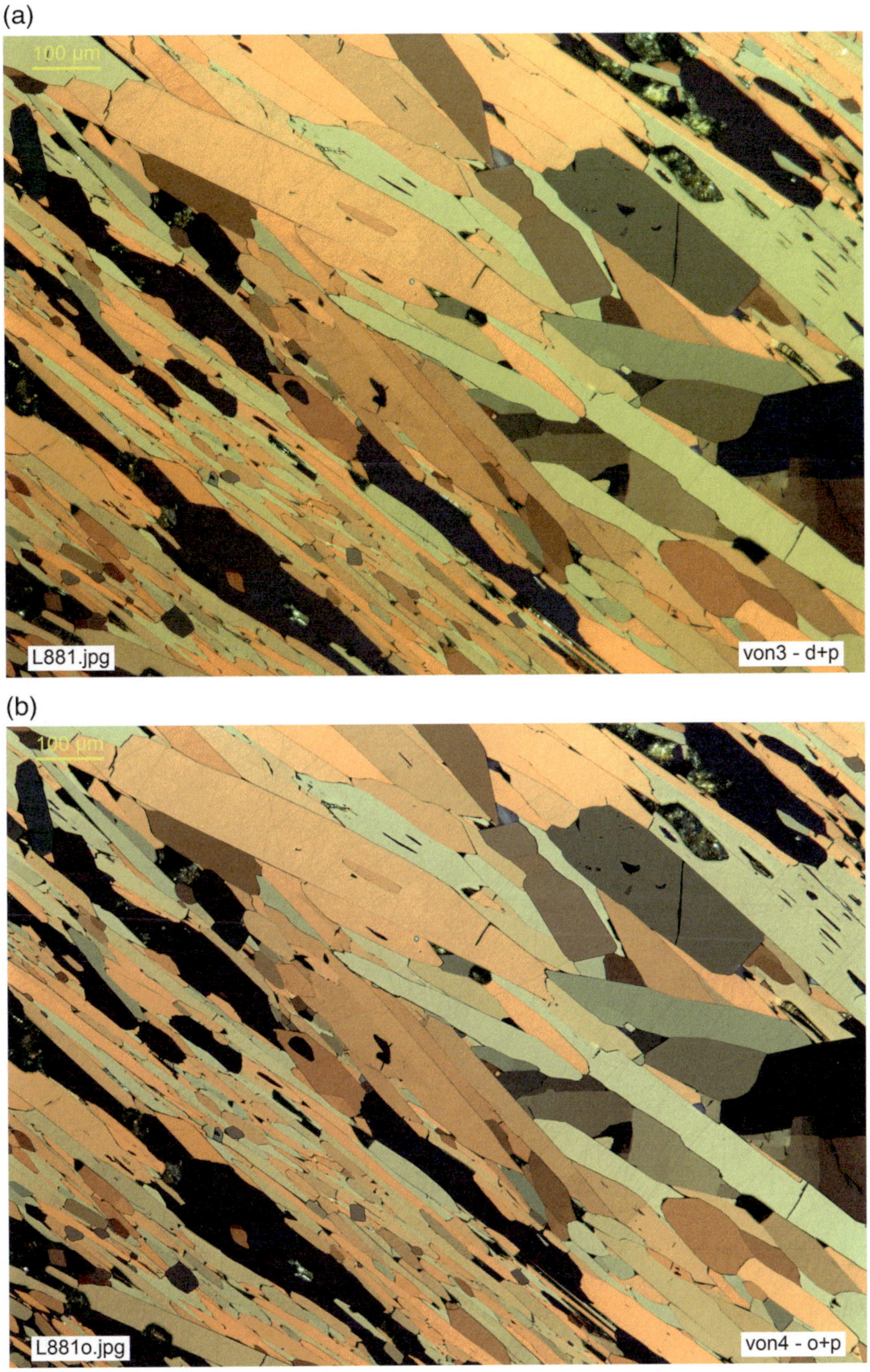

Fig. 136.18 **a**, **b** (same scene as Fig. 136.17a, **d+p** and **o+p**, *respectively*). Vivid, coloured *von* anisotropism, in contrast to *mt* and *py* (isotropic) extinction

References

Chang LLY, Howie RA, Zussman J (1998) Rock forming minerals, vol 5B, 2nd ed. Non-silicates: sulphates, carbonates, phosphates, and halides. The Geological Society, London, p 383

Ramdohr P (1980) The ore minerals and their intergrowths, vol 2, 2nd ed. Pergamon Press, Oxford, p 1205

Sulfates. Oxides and Hydroxides 137

137.1 Sulphates. Barite. Anglesite. Jarosite

Sulfates are to be expected, in principle, as supergene companions of sulfide ores, formed by weathering in the oxidation zone. These are sulphates of metals such as Cu, Fe, Zn, Pb, Ag. They are very numerous and frequently hydrated, like *calcantite*, $CuSO_4.5H_2O$, *brochantite*, $Cu_4(SO_4)(OH)_6$, *langite*, $Cu_4(SO_4)(OH)_6.2H_2O$, *jarosite*, $KFe_3(SO_4)_2(OH)_6$, *goslarite*, $ZnSO_4.7H_2O$, or anhydrous, like *anglesite*, $PbSO_4$, among many others, of typically secondary origin. Although some may have industrial use if occurring in high-grade oxidized deposits—e.g., of Cu or of Ag, in the case of *argentojarosite*, $AgFe_3(SO_4)_2(OH)_6$, its presence is in most cases anecdotal, with the possible exception of jarosite.

However, primary sulphates are also found in nature, such as *barite*, $BaSO_4$, *celestite*, $SrSO_4$, or *anhydrite*, $CaSO_4$. Barite, mostly of hypogene origin, can be found as a typical companion among the gangue components of some hydrothermal concentrations, while celestite and anhydrite are typical of sedimentary environments, although anhydrite is also occasionally found—sometimes altered to *gypsum*—in hydrothermal or magmatic-hydrothermal deposits (*vbgr* some zones of copper porphyries).

These minerals are transparent or weakly absorbent and must be studied with **TL**, *transmitted light* microscopy, or by other techniques more suitable than **RL**, *reflected light* microscopy—XRD, SEM, etc. However, given the relative frequency with which some of them occur associated to ores (such as barite, anglesite and jarosite), it seems useful to briefly outline their broad characterisation, also with RL.

Barite (*bar*, $BaSO_4$, orthorhombic) is a relatively common sulfate in the gangue of hydrothermal deposits. In **RL** (Reflected Light) it appears dark grey and is characterised by:

- its low reflectance, only slightly higher than that of quartz which, in air, looks a little darker still (Fig. 137.1), $R_{bar} \approx 5\%$ in air; in oil, $R_{bar} \approx 0.04\%$, it is nearly zero, so that in immersion it looks black, like quartz (Table 134.1)
- and by the absence of striking features—neither pleochroism nor anisotropism effects, nor striking features in its **IR**, *internal reflections* (colorless)

Further features helping to recognize bar are:

- low polishing hardness, lower than that of siderite and most of the silicates, similar to that of the phyllosilicates and somewhat higher than that of calcite, which it accompanies in many cases
- imperceptible bireflectance, unlike carbonates
- morphology—often tabular or prismatic to acicular, although it is not uncommon, when it has formed as an early mineral (Fig. 137.2) for bar to be fragmented and corroded or replaced by later sulphides, calcite, quartz, etc.
- paragenesis—evidence of mineral assemblages typical of any one of the ore deposit classes in which the mineral is known to occur frequently (e.g., Pb–Zn–F–Ba strata-bound concentrations in carbonate formations of the Mississippi Valley type, Pb–Zn–Ag, Cu, Mn veins and replacements or mantos, low sulfidation and intermediate sulfidation epithermal deposits, sub-sea floor *black smoker* deposits).

However, these features cannot always be observed clearly and, in addition, barite is very similar in RL to other possible gangue components, such as **anhydrite**, **celestite** or the phosphate **apatite**, so it is convenient to confirm its identification with **TL** (transmitted light) or other techniques. Its appearance in terms of the subjective perception of its brightness (estimation of reflectance) depends very much on its companions, as can be seen by comparing the shades of gray in Figs. 137.1 and 137.2. The barite in Fig. 137.2 appears darker, simply because it is next to minerals much more reflective than those in Fig. 137.1.

R. Castroviejo, *A Practical Guide to Ore Microscopy—Volume 1*,
https://doi.org/10.1007/978-3-031-12654-3_137

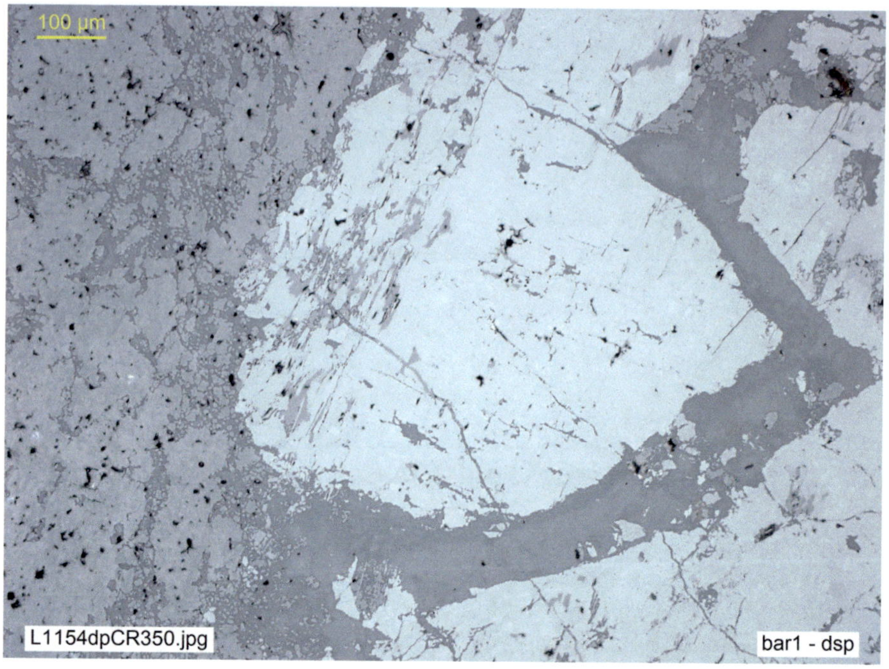

Fig. 137.1 Brecciated bastnäsite ore (light grey), with barite (intermediate grey) and quartz (darker grey). Cataclasis prevents the recognition of the original barite morphology. It is evident from the luster comparison that these minerals follow the order of reflectance (decreasing): $R_{bst} > R_{bar} > R_q$. CR350, Mountain Pass, Calif., USA (**dsp**)

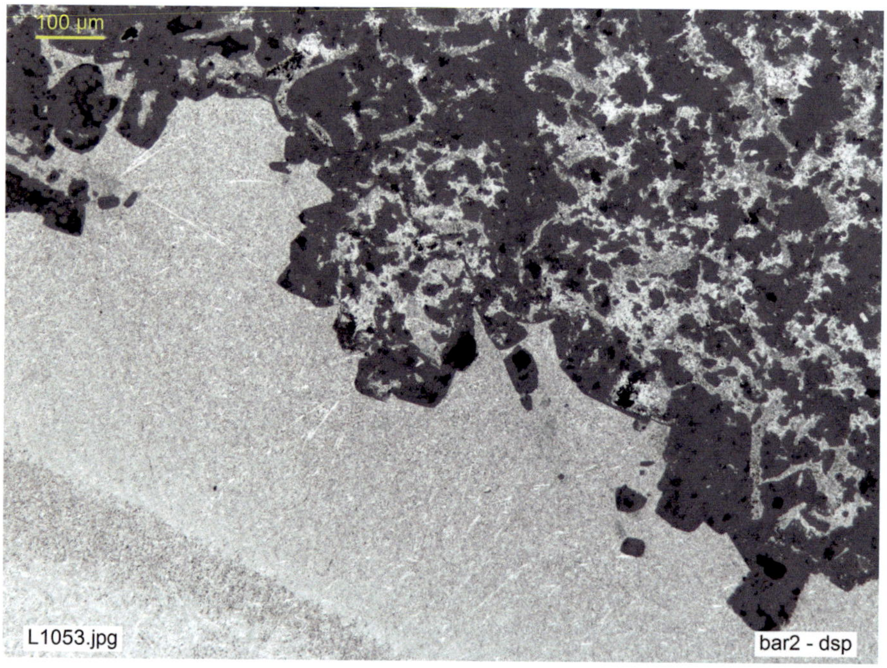

Fig. 137.2 Manganiferous ore (mass of cryptocrystalline cryptomelane and acicular romanechite, with light gray and grayish-white tones, *resp*), with dark gray barite gangue, partially replaced by the Mn ore, deleting the prismatic habit of many of the bar crystals. CR 337.b, Veta Clemira, El Remanso, Santiago del Estero, Argentina (**dsp**)

Anglesite (*ang*, $PbSO_4$, orthorhombic) is also a dull-looking, gray mineral, although brighter than barite and the common quartz-silicate gangue. Its estimated reflectance is ≈ 9.4% (calculated according to the Fresnel equation, with n assumed ≈1.88–1.89), similar to that of cerussite but with more moderate bireflectance (cf. Figs. 136.16a, b). Anglesite can easily go unnoticed by an inattentive observer, but its origin provides a helpful hint. As it is typically formed by alteration of Pb sulphide ores in the supergene oxidation zone, its association with galena (or Pb minerals) in oxidized ores, as well as with the carbonate cerussite or other secondary minerals, is common. It usually preserves relics of the primary sulphide (galena), around which a secondary zoning develops, with a first rim of anglesite in immediate contact with the sulphide and an external zone of cerussite, as shown in images Figs. 136.16a, b .

Fig. 137.3 Aggregate of pyrite ***py*** and marcasite ***mc*** (yellowish white and slightly greenish white), with inclusions of quartz ***q*** (dark gray) and pores and interstices covered by totally limonitized *jar* (bluish gray with reflected light), recognizable by the tabular or rhombohedral morphology inherited by *lim*. FPI.0, Huelva (**dsp**)

137.1.1 Jarosite

Jarosite ‹***jar*** , $KFe^{3+}{}_3(SO_4)_2(OH)_6$, hexagonal› is an alteration product frequently associated in superficial environments with Fe-containing sulphides. It is not, however, a stable mineral. In temperate and humid climates, it is easily altered to limonite, in one of its many varieties, or to goethite ± hematite. For this reason, although it is a sulphate of the *alunite* group, which should properly be identified with **TL** (transmitted light), it often appears as an opaque mineral. It then occurs frequently as pseudomorph in assemblages of secondary Fe oxyhydroxides. It is comprised, either as ***jar*** proper or as pseudomorphized relics of ***jar***, among the possible components of the "mineral", imprecisely defined but ubiquitous on the earth's surface, known as limonite ***lim***.[1]

Figures 137.3, 137.4 and 137.5 show a porous aggregate of pyrite ***py*** and marcasite ***mc,*** whose voids are largely occupied by a mineral, mostly opaque in TL but poorly absorbing in RL (dark grey, only slightly lighter than *q*). Optical properties (brightness, color, multispectral reflectance measurements VNIR, IR) are those of *lim*, in its hydrated, subcrystalline varieties, a poorly structured mass of mostly amorphous Fe oxy-hydroxides. The morphology of this product (aggregates of small tabular and rhombohedral or fibrous crystallites, lining pores and cavities), as well as the (0001) cleavage (visible with condenser in TL, Fig. 137.5), the straight extinction and the pleochroism preserved in some fragments are evidence that the secondary product replicates the morphology and crystalline orientation of a primary crystalline material, whose relict properties coincide with those of *jar*. They are, therefore, *lim* pseudomorphs after jar, as suggested by these jar *ghosts*.

Apart from the fact that some minerals of the group may be of economic interest in themselves (*vbgr*, argentojarosite, because of its Ag content), the significance of the ***jar*** as a guide in mineral exploration, i.e., as a visible indicator of potential orebodies of altered metallic sulphides, is also interesting.

[1] ***Lim*** is a mixture composed of cryptocrystalline aggregates of minerals such as ***gth***, ***ht***, ***lpc***, ***jar***, silica or opal, clays and Fe oxides or hydroxides, amorphous and in different degrees of hydration, which makes its reflectance, **R**, vary considerably, generally remaining below that of its main component, ***gth***. In the sample of Fig. 137.3, its multispectral **R** has been measured in the VNIR range, resulting in a spectrum similar to that of ***gth*** but with lower values, going from **R**≈ 8% (visible) to **R** ≈ 12% (**IR**).

137.2 Oxide/Hydroxide Ore and Gangue Minerals. Ochers

Oxides and hydroxides are, in general, relatively abundant and of great economic interest, to the point that Fe and ferrous metals represent the priority sector for mining investment. For this reason, almost all of them (cassiterite ***cst***, hematite ***ht***, rutile ***rt***, anatase ***ana***, magnetite ***mt***, pyrolusite ***prl***, goethite ***gth***, etc.) have been described as ores (§ Part II) and those described exclusively as gangue are rare. This does not mean that many of them do not appear as gangue, when they occur in small quantities or when they are not beneficiated in a deposit. Thus, for example, the

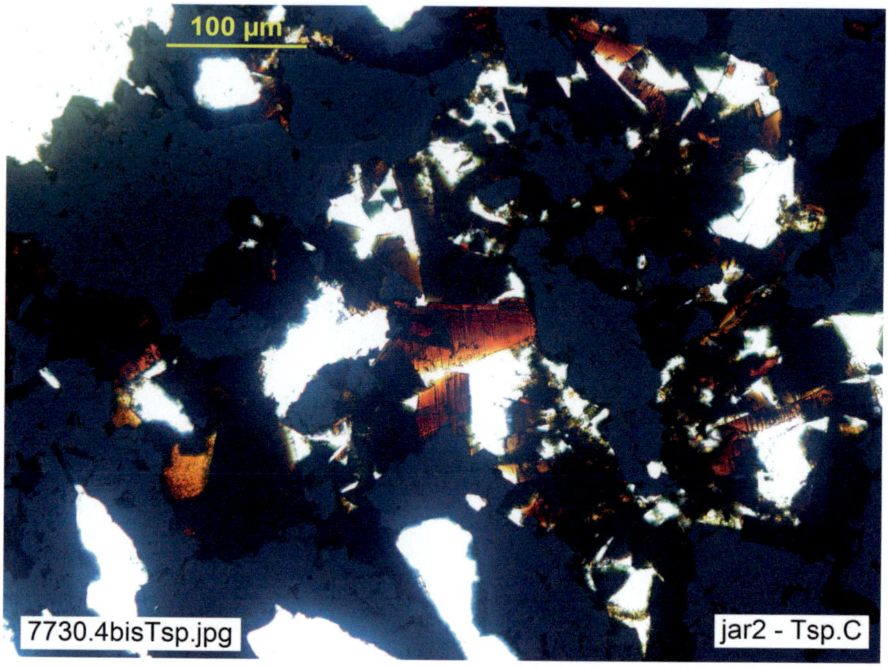

Fig. 137.4 (Same scene as Fig. 137.3, seen with transmitted light and condenser, **Tsp.C**). Alteration to *lim* has reduced the transparency of *jar*, perceptible (only with condenser) in some areas as yellowish-brown relics, contrasting with the conspicuous luminosity of *q* (white). The pseudomorphic *lim* is mostly opaque (black with TL, like *py*)

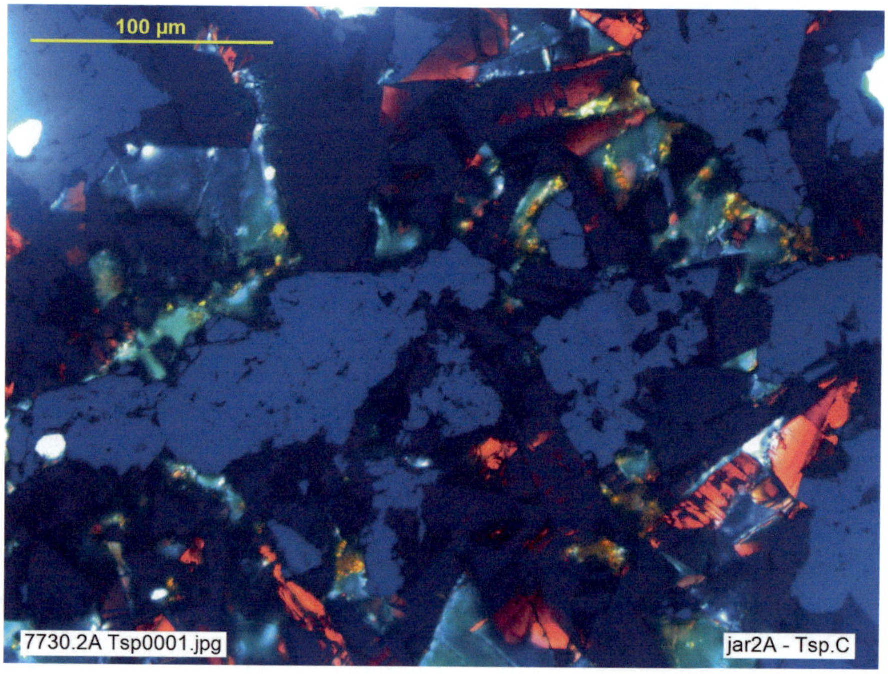

Fig. 137.5 Detail of the central part of Fig. 137.3 (**Tsp.C**), after rotating the stage 90°. The anisotropism of *lim* and the cleavage still inherited from *jar* can be seen

common oxides of Fe and Mn, such as *goethite, hematite, pyrolusite, romanechite*, etc. are usually found in any sulphide deposit that has undergone weathering, even if only as thin infiltrative coatings or incrustations and of an imprecise nature (limonite, psilomelane, ochres); but, in these conditions they are far from being a viable resource for the extraction of Fe or Mn metals.

The distinction is economic, rather than mineralogical, although in general a mineralogical criterion is also applied. Ores are essentially opaque or absorbent minerals and gangue, transparent or not very absorbent. Even so, *spinel* has also been treated with ores, because of its affinity with chromite, magnetite, etc., typical ores, belonging to its group and closely related from the mineralogical point of view.

Obviously, in these cases, the identification of the mineral refers to the corresponding description of the ore, presented in Part II. An exception will be made only in the case of certain iron oxides/hydroxides. These are mentioned as gangue in the ochre group, although they could be considered as a peculiar form of limonite, already characterized (Part II). Therefore, they do not require a new description, unlike other oxides (Bi or Sb ochres) which, as such, have not been described among the ores.

In summary, most of the minerals of interest in the *Class Oxides and Hydroxides* have already been treated as ores (Part II), although in many deposits where they are found as accessories they are not of the slightest economic interest. In these cases, since they are neither beneficial nor desired because they interfere with the plant process, they are to be considered in practice as gangue but are not described in this section. Consequently, there are few minerals of this class selected as *common* gangue and described below: Fe, Bi, Sb ochres (limonite ochres or pigments, bismite, bismuthite,[2] cervantite, senarmontite, valentinite). Last, but not least, leucoxene, which is very common, although minor and not credited as mineral species.

137.2.1 Ochres (Fe, Sb, Bi)

Many metallic ores (sulphides, sulphosalts, native Bi or Sb, U minerals, etc.), subject to weathering in superficial conditions, are gradually transformed into earthy masses that are generally known as **ochre**, due to their frequent *ochre* colour (yellowish to reddish-brown, although the colour of the mineral is actually variable, depending on the composition and degree of hydration). The process begins as a superficial patina or coating and progresses, forming incrustations, filling pores, etc. until the ore is transformed into a crumbly mass or a fine aggregate of cryptocrystalline oxides/hydroxides. Among these are very abundant *iron ochres* and frequent, in their respective deposits, those of *antimony or bismuth*. Although in many cases they are subcrystalline or amorphous mixtures, sometimes defined mineral species can be identified among their components.

Iron ochres can be considered as a variety of limonite, already described (ch. ***lim***) among the ores, which find industrial use as a pigment due to their intense and stable colour.

[2] Bismutite is a carbonate, but it is described in this section because of its affinity with Bi ochres, whose mineralogical composition can be complex and varied.

137.3 Bismite. Bismutite

Bismuth ochres include the mineral species bismutite (***bis***) and bismite (***bim***), together with other possible by-products, which occur in yellow incrustations or masses.

Bismutite ‹***bis***, $Bi_2O_2(CO_3)$, orthorhombic[3]› can be described as a carbonate, a group whose properties have already been described in general, but it is worth noting for ***bis:***

1. its high refractive index (n ≈ 2.1 − 2.3), which allows rough estimation for an intermediate value of n ≈ 2.2, the theoretical reflectance R ≈ 10%
2. its strong birefringence, typical of the ***Carb*** group and appreciable in **RL** as bireflectance, although it is a transparent mineral

Bismite, (***bim***, Bi_2O_3, monoclinic), a mineral not described in the usual reference sources of ore microscopy and with contradictory data in others, has the following appearance under the reflection microscope (Figs. 137.6a b and 137.7a, b):

- Gray color, noticeably lighter than the usual gangue, *vbgr. q*, but dark next to native Bismuth **Bi**, which, as alteration progresses, is usually reduced to relicts in the secondary mass of bismite
- Reflectance, **R**, relatively high. It can be estimated, for a value n ≈ 2.42[4], as R ≈ 17.2%, but is perhaps actually somewhat higher, if the absorption coefficient **k** is taken into account, as befits a semi-transparent, i.e., moderately absorbent mineral (cited as transparent or opaque depending on the sources). The subjective perception (brightness) is comparable to those of sphalerite ***sp*** or goethite ***gth***.
- **IR** abundant and luminous, colourless or slightly coloured in yellowish or greenish tones, which hide the possible anisotropism.

These properties, as well as paragenesis and relics of Bi, bismuthinite ***bm*** or other bismuth-bearing species, facilitate the recognition of bismite.To be sure, it is not superfluous to check by electron microprobe.

[3] More information at: https://www.mindat.org/min-687.html (*visit.* 20210426).

[4] More information at: https://www.mindat.org/min-682.html (*visit.* 20210427).

(a)

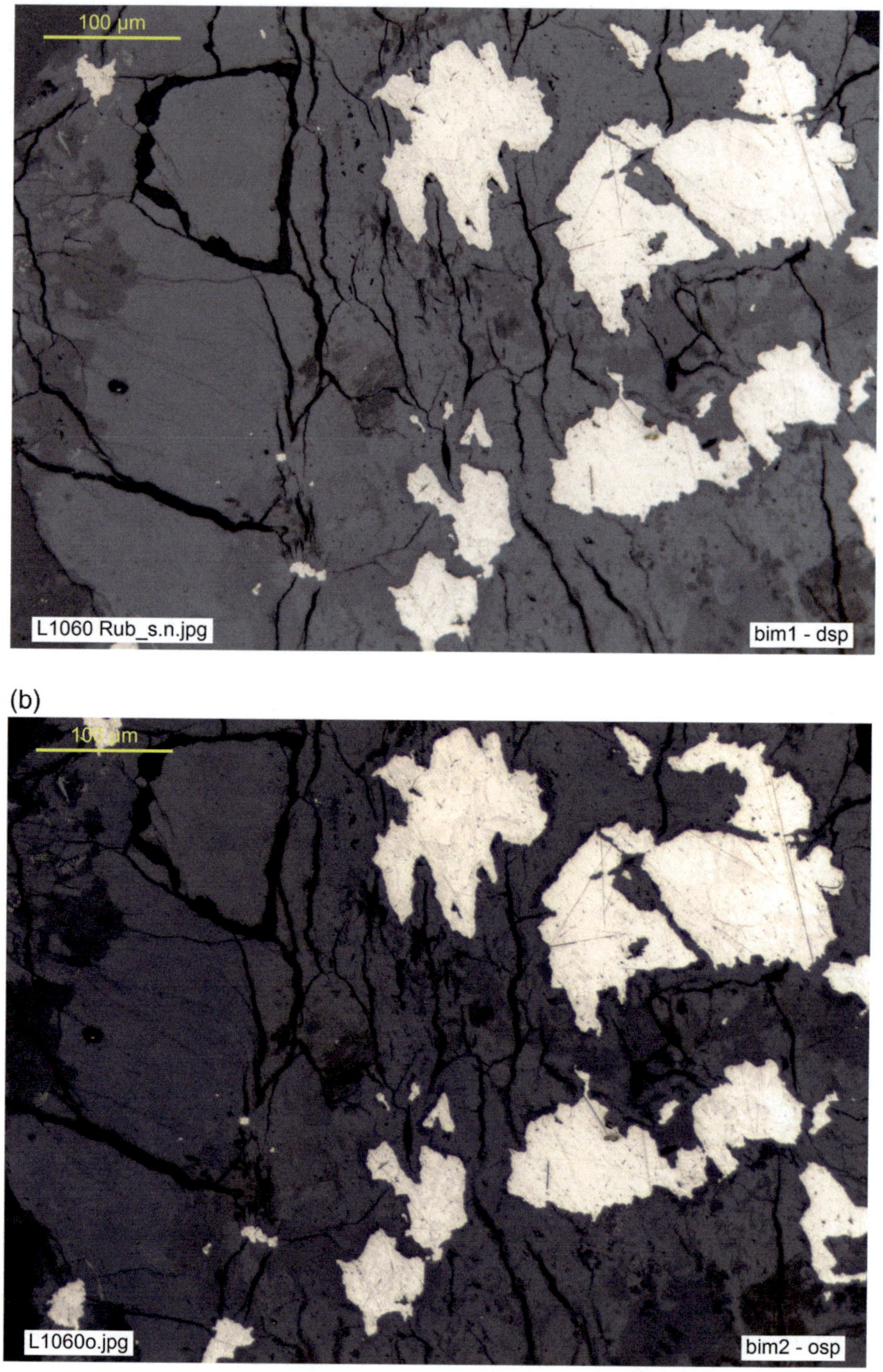

Fig. 137.6 a, b (Same scene: **dsp** and **osp**, *respectively*). Mass of ***bim*** bismite (grey) with relicts of Bi (white, bright) and minor inclusions of oxycalciomicrolite, ***ocm*** (darker grey than *bim*, e.g., left margin of *bim* mass). The compositions of *bim* and *ocm* have been confirmed by electron microprobe. Rub-s.n, Sowiesky Pty Ltd, Springbock Area, Namaqualand (South Africa)

137.4 Cervantite. Sénarmontite

Antimony ochres,[5] a common result of weathering of Sb or stibnite, ***sbt***, include, among others, the mineral species:

- **Cervantite**, ***cerv***, $Sb^{3+}Sb^{5+}O_4$ (orthorhombic, with remarkable refringence **n** ≈ 2–2.1, strong birefringence ≈ 0.1 and **R** ≈ 11–12%), occurs in masses and in fibrous or acicular aggregates, which may preserve relicts of the primary ore, ***sbt*** or **Sb.**
- **Sénarmontite**, ***sen***, Sb_2O_3 (cubic, n ≈ 2.01, anomalous birefringence), and
- its *dimorph* **Valentinite**, ***val*** (orthorhombic, n ≈ 2.1–2.3)

[5] More information at: https://www.mindat.org/min-209.html (visit. 20210427).

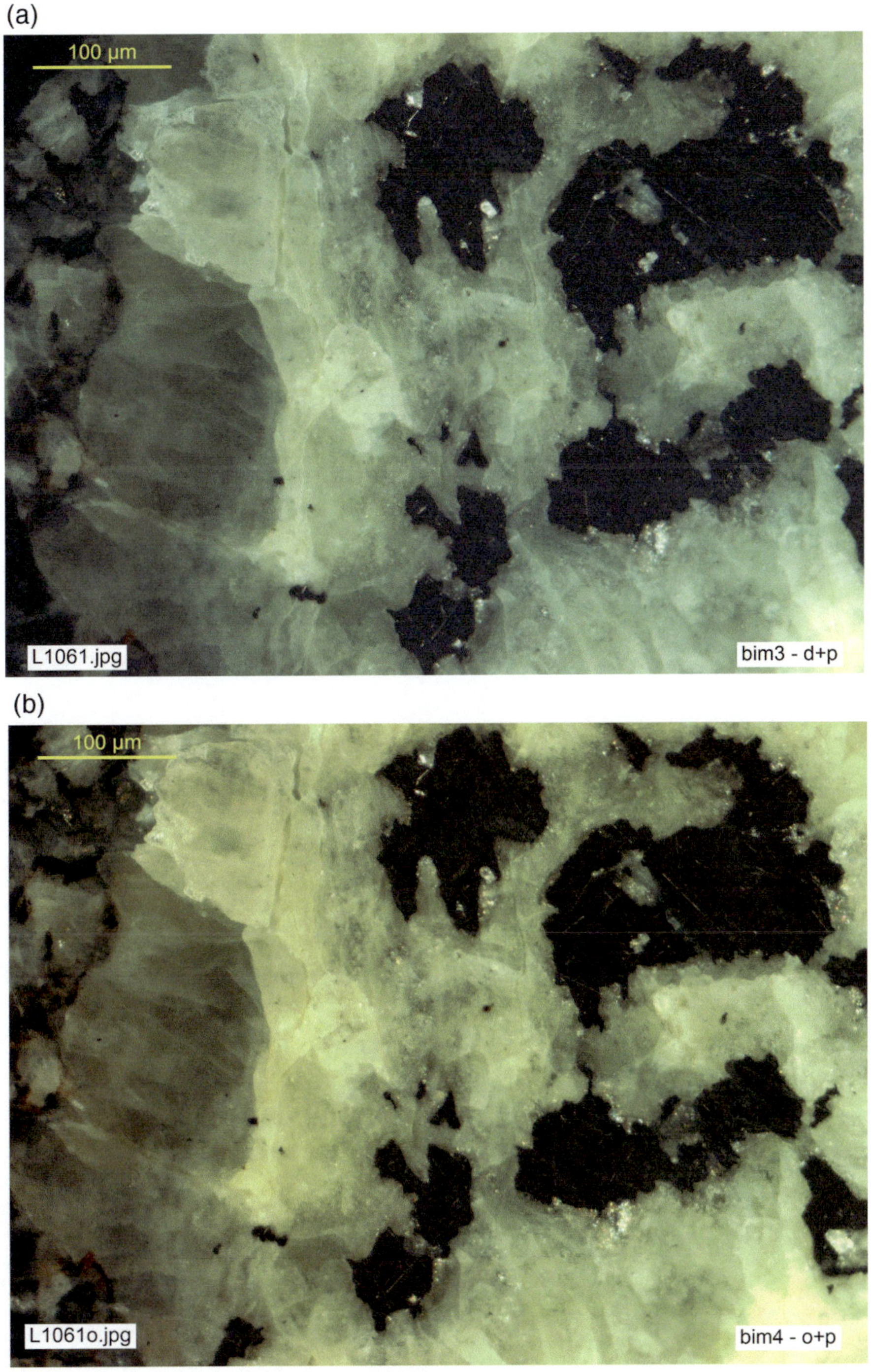

Fig. 137.7 **a, b** (Same scene: **d+p** and **o+p**, *respectively*). Mass of ***bim*** marked by abundant and luminous internal reflections. These are so intense that, to avoid overexposure, the exposure time has to be kept to a minimum, and as a result the anisotropism of **Bi** is hardly visible in the images (*+p*)

Although colorless or yellowish in hand sample, with **RL** these oxides appear gray, relatively light against the usual gangue[6] but very dark next to **Sb** or ***dy*** dyscrasite (Figs. 137.8a, b), as corresponds to their reflectances, estimated between a minimum R ≈ 11% for sénarmontite ***sen*** and a maximum R ≈ 15% for valentinite ***val***. Their **IR**, though colourless or poorly coloured (especially when compared to the vivid colours of kermesite ***ker***, ch. Figs. 69.1a–69.2b), are abundant and luminous, so they hide the effects of anisotropism.

Bearing these data in mind and always with the endorsement of paragenesis, it is not difficult to identify the

[6] In the image Fig. 137.8a, the quartz-carbonate gg looks very dark. It actually appears practically black, as in immersion, although the photos have been taken in air. This is due to the very low exposure time, forced by the very high reflectance of Sb, which would produce overexposure in normal conditions. However, the abundant colorless IR of both minerals and the *Carb* twins (Fig. 137.8b, *d+p*) put them in evidence.

(a)

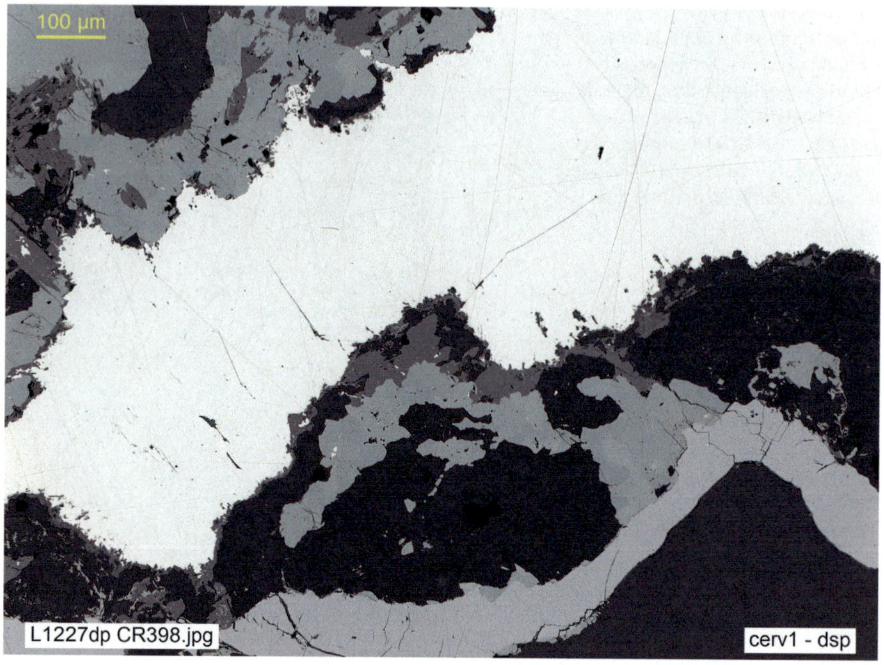

(b)

Fig. 137.8 **a, b** (Same scene: **dsp** and **d+p**, *respectively*). Antimonial silver ore: mass of Sb (center, white, very bright, anisotropic), partially altered at the borders to cervantite ***cerv*** (dark gray, fibrous, abundant colorless **IR**), intergrown with a marginal rim of pyrargyrite ***prg*** (bluish gray, red IR, anisotropic). External concentric band or *corona* of galena ***gn*** (lower right, grayish white, isotropic). The gg of *q* and *Carb*, very dark gray, is characterized (*d+p*) by abundant colorless IR of both minerals, as well as by the *Carb* twins. CR398, Hiendelaencina, Guadalajara

antimony ochres as a group, but it can be very difficult to distinguish with certainty the species of the group from each other by optical methods only.

137.5 Leucoxene

Leucoxene, ***lcx***, is a very peculiar oxide or, rather, a mixture of oxides (anatase, rutile), frequently cryptocrystalline or sub-microscopic, so that its appearance under the microscope is usually that of an amorphous mass, and its mineralogical composition has to be determined with XRD. It is considered a secondary product, generally the result of the alteration of titaniferous minerals such as ilmenite, pseudo-ilmenite, titanite, Ti-mt, and perovskite. It also results many times from the segregation of Ti from silicates such as biotite ***bio*** or amphibole ***Anf***. If these contain Ti in their crystal lattices and undergo hydrothermal, deuteric, retrograde metamorphic alteration processes to lower

Fig. 137.9 **a, b** (Same scene: **dsp** and **d+p**, *respectively*). Impregnation of chalcopyrite ***ccp*** (yellow, bright), leucoxene ***lcx*** (light grey) and traces of galena (white) and sphalerite (grey), in a carbonate and *OM*-rich marly lutite (***OM***: organic matter). Note the enhanced *contrast between lcx (disseminated) and sphalerite (included in ccp) in immersion* (*osp*). Although in air (*dsp*) the brightness of sphalerite and *lcx* could be confused, sphalerite is clearly darker in immersion (**b**). Carbonate (calcite or dolomite) shows up with its typical bireflectance in air but disappears in immersion. Disseminated OM shows very low reflectance (*dsp* , *osp*) and, unlike gg, no IR (*d+p*, *o +p*, Fig. 137.10a y -4). HD 6615-Kupferschiefer, Richelsdorf, Germany

temperature minerals, *vbgr*. chlorite ***chl***, whose lattice does not admit Ti, this is expelled and must accommodate as neo-formed (secondary) oxides.

lcx is not considered a mineral species—and as such has been discredited by IMA—as it does not have a defined crystal structure and composition. However, its recognition is of interest to the prospector, as it is present, even in trace amounts, in most of the processes associated with the generation of metal deposits (also in various igneous and metamorphic rocks, as well as in derived detrital deposits). As a typical product of hydrothermal alteration, it may occur in the ore or be disseminated in the host rock.

Its appearance under the microscope is generally unremarkable, but a careful observation easily discovers it. It is characterized (*dsp*) by its light gray color (Fig. 137.9a, b), with a relatively high reflectance, which contrasts with the dark tone of the surrounding gangue and which could be confused with that of sphalerite (but in immersion sphalerite

(a)

(b)

Fig. 137.10 **a, b** (Same scene as Fig. 137.9a: **osp** and **o+p**, *respectively*). Chalcopyrite is easily recognized by its anisotropism, ***lcx*** by its strong, abundant, colorless IR (which hide its anisotropism), ***Carb*** by its IR, twinning and anisotropism, while OM shows no IR

is darker, ch. Fig. 137.9a, b). With analyzer (*d+p*), lcx shows abundant and luminous, colorless or faintly colored (sometimes greenish to yellowish-brown) internal reflections, which hide the anisotropism, giving the appearance of an isotropic or cryptocrystalline mineral (Fig. 137.10a, b). This surprising duplicity of properties, simultaneous aspects of gg (color impression, gray, IR that hide the anisotropism) and of ore (relatively high R, brightness $\gg$gg), together with the usual relation with hydrothermal or alteration processes, allow its quick recognition.

Other Minerals 138

138.1 Halides. Fluorite

The halide **chlorargyrite** (or **cerargyrite)** has already been described with the ores (**Part II**, *cag*), as this mineral, which is not uncommon in the gangue of weathered argentiferous ores, was an important source of silver in the past.

Likewise, **fluorite** (*fl*, cubic CaF_2), besides being a common component in the gangue of a varied typology of deposits, is currently a demanded mineral when it is present in sufficient abundance for its economic extraction (vbgr. some deposits of the *Pb–Zn-F-Ba or Mississippi Valley type*). In reflected light, it is among the species with the lowest reflectance (cf. Table 134.1), with hardly any other distinguishing features, apart from its isotropy and the **IR** typical of a transparent mineral, colourless in principle but often showing the varied colours typical of the mineral (Figs. 138.1a and 138.1b).

138.2 Phosphates, Arsenates and Vanadates

138.2.1 Phosphates. Monazite

Among the **phosphates**, attention should be paid to **monazite**, ***mz*** ‹(Ce, L, Nd, Th)PO_4, monoclinic›, because of its economic significance and because its characterization with reflected light is possible. As it is a carrier of rare earth elements demanded by the industry, it is actually a REE ore and, as such, it has already been described in the corresponding chapter (cf. Part II). To identify other phosphates that may be found in the gangue, such as **pyromorphite** ‹***pymp***, $Pb_5(PO_4)_3Cl$, hexagonal› or **apatite**, ***ap***, ‹$Ca_5(PO_4)_3$(F, Cl, OH), hexagonal›, recognition with transmitted light or other techniques is recommended, as well as for **xenotime** ‹**xtm**, Y(PO_4), tetragonal› or **amblygonite** ‹*amb,* (Li, Na)Al(PO_4)(F, OH), triclinic›, potential sources of REE or Li, respectively.

138.2.2 Arsenates and As Ores

There are several **As-bearing** minerals (sulphides, sulpharsenides, arsenides and sulphosalts, in addition to native arsenic and its alloys, stibarsen or allemontite) that have been described as ores, however surprising this qualification may seem if we take into account their market value, since their presence is often undesirable and is penalised in the market due to their environmental impact. In general, these ``ores' have in common their instability in the surface oxidation zone, which makes them dangerous because they pollute the environment with As. Some of them, such as **realgar**, ***rl*** (As_4S_4), and **orpiment**, ***orp*** (As_2S_4), are usually formed in the late, waning phases of hydrothermal deposition but possibly also by incipient weathering. Nevertheless, they have also been described with the ores (cf. Part II), as this fits better with their reflectance values (R > 20%, 546 nm), notably higher than those commonly found in gangue minerals. Therefore, the practical criterion applied here is to refer to arsenical gangue when dealing with arsenates or oxidized As minerals, commonly of secondary origin, related to weathering or oxidation of As-bearing ores.

138.3 Scorodite. Eryhtrite. Annabergite

Among the minerals that can be formed by oxidation of arsenical ores are the well-known arsenates **scorodite** (***sco***, $FeAsO_4.2H_2O$, orthorhombic), **erythrite** ‹***ery***, $Co_3(AsO_4)_2.8H_2O$, monoclinic› and **annabergite** ‹***annbg***, $Ni_3(AsO_4)_2.8H_2O$, monoclinic›. All of them are transparent, so the suitable means to characterize them, apart from the examination in hand specimen or with stereomicroscope, is the petrographic microscope (**TL**). However, under favorable conditions, it is possible in some cases to recognize them with **RL**, Reflected Light, or at least to deduce their hypothetical presence by observing their internal reflections, **IR**,

R. Castroviejo, *A Practical Guide to Ore Microscopy—Volume 1*,
https://doi.org/10.1007/978-3-031-12654-3_138

Fig. 138.1 **a**, **b** (Same scene: **dsp** and **o+p**, *respectively*). Ore of fluorite (dark grey) and uraninite (grey, spheroids surrounded by alteration rim, including tiny pyrite corpuscles, yellow), with inclusions of galena (white). In Fig. 138.1b *(o+p)*, all of them (urn, fl, gn, py) are in extinction (cubic minerals). Only a diffuse luminosity is seen, due to the internal reflections of fluorite (purple) and of the alteration halo of uraninite (bluish-green, likely torbernite). HD6023, Wolsendorf, Bavaria, Germany

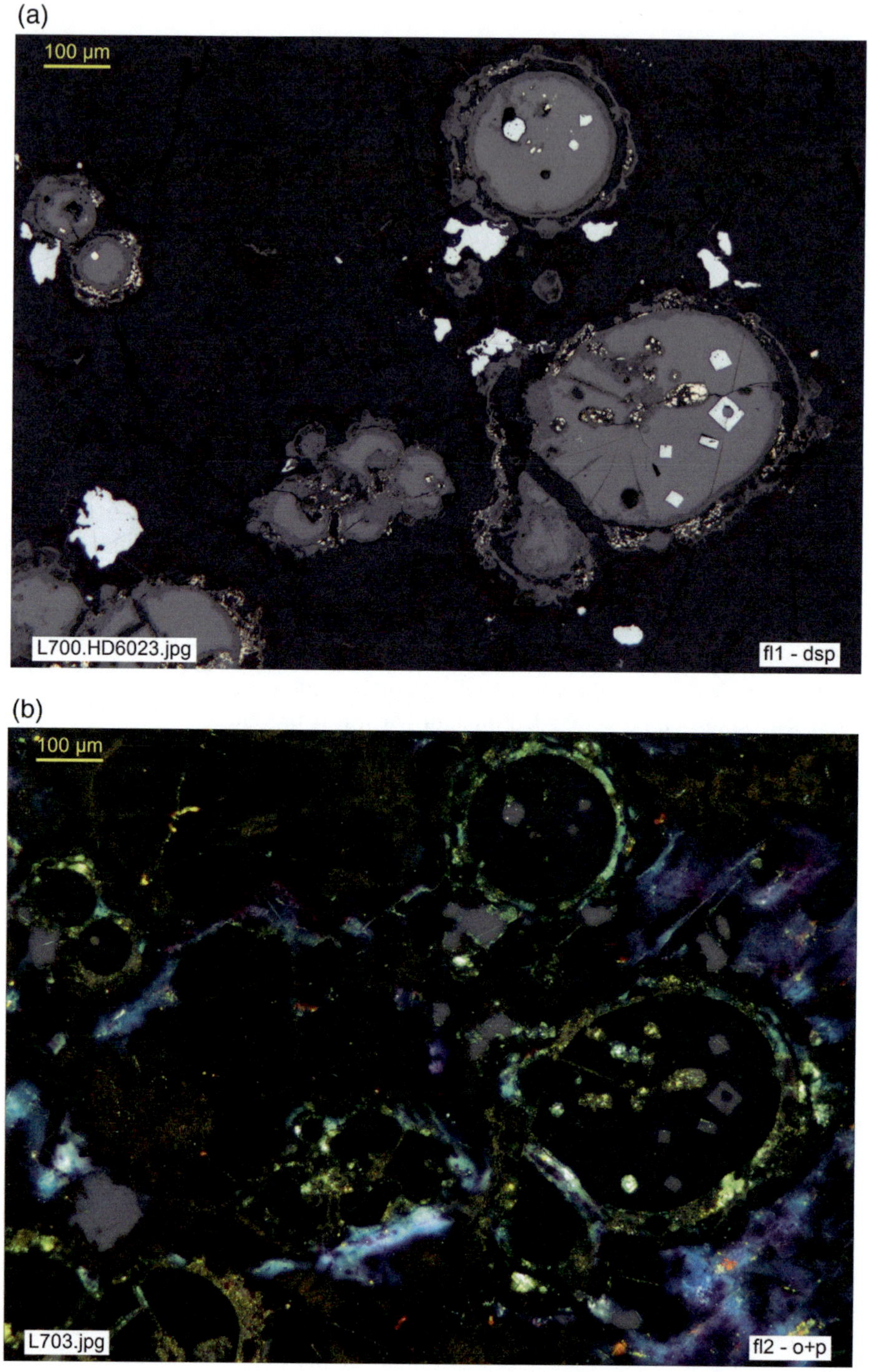

which appear colored, *respectively*, with *bluish-green to yellowish-brown* (or reddish, due to limonitic contamination), *pinkish* or *greenish tones*. It is also useful to take into account their relationship with arsenopyrite or arsenical Co or Ni ores, respectively.

As an example Figs. 138.2a, b, 138.3a, b, 138.4a and 138.4b illustrate a favourable case where **RL** recognition of **erythrite** is possible. This is a sample of arsenical ore of the *five-element formation* or Ni–Co–Bi–Ag–U type, coated by an alteration crust, in which pink incrustations of *cobalt*

Fig. 138.2 (Same scene: **dsp** and **d+p**, *respectively*). Crystal aggregate of löllingite (white, prismatic, anisotropic), cobaltite (pinkish white), skutterudite (white, isotropic) and native silver (white, very bright, lower right corner) partially replaced by erythrite (grey, luminous pink RI). CR 375, Cobalt, Ontario, Canada

(a)

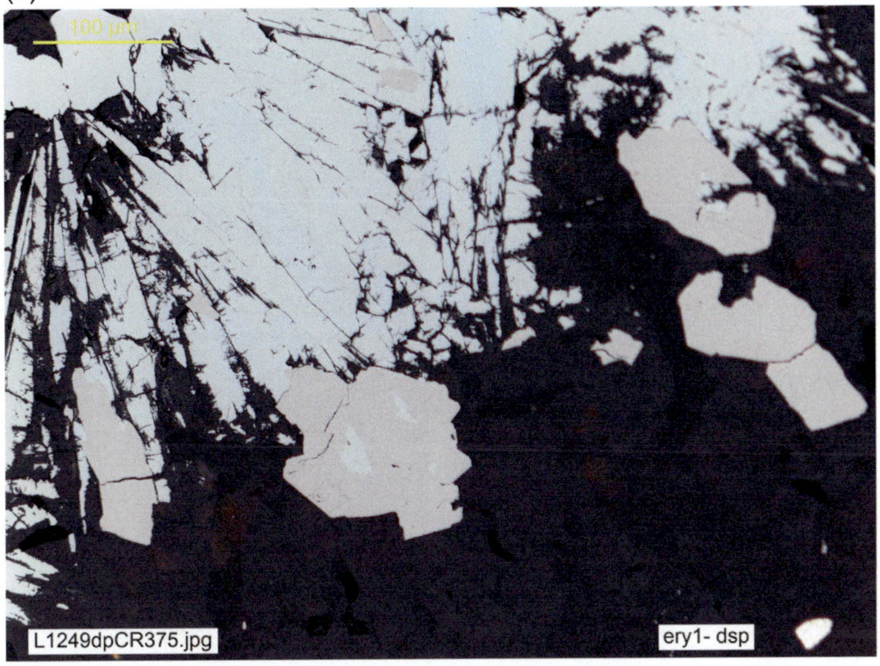

(b)

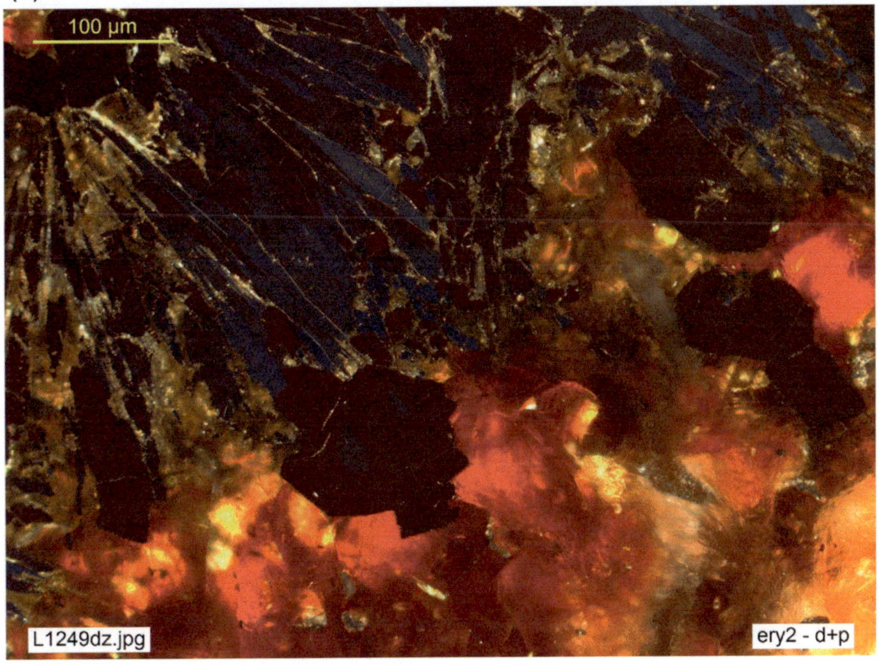

*flowers (*a popular name for *erythrite,* hinting to its appearance in hand sample*)* can be seen.

As deduced from its indented borders in Fig. 138.2a, the ore is partially replaced by a greyish gangue, as is typical for surface alteration processes. The reflectance, **R**, of the gangue is indeed low if compared to the ore (composed essentially of *white arsenides* and silver: löllingite, safflorite, native silver, cobaltite, skutterudite, etc., not all in the image) but not so low as to prevent perceiving some bireflectance (or variation in the gray level between contiguous grains) and a pinkish shade. The dark aspect is enhanced in the image by contrast, because the **gg**, gangue,

(a)

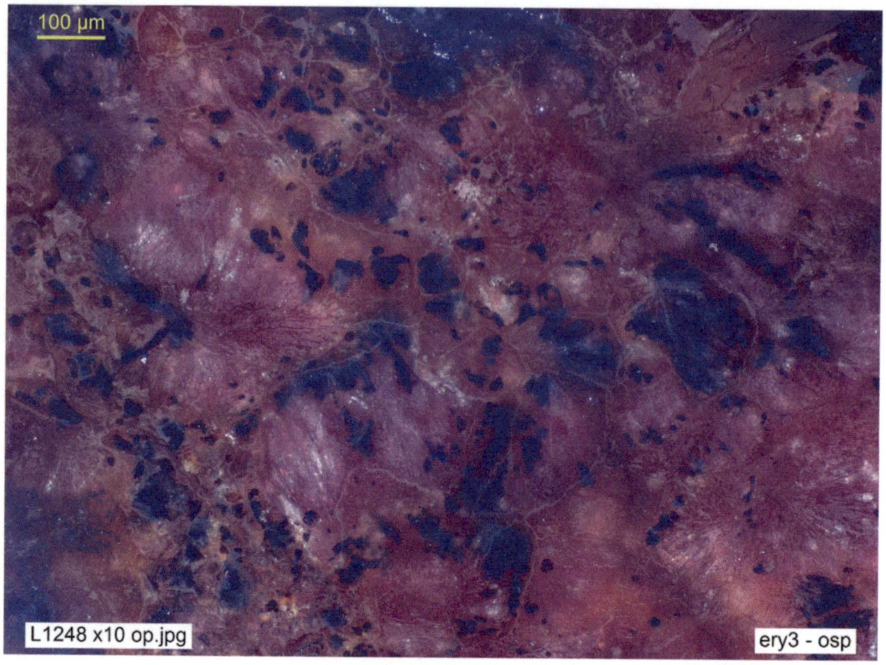

(b)

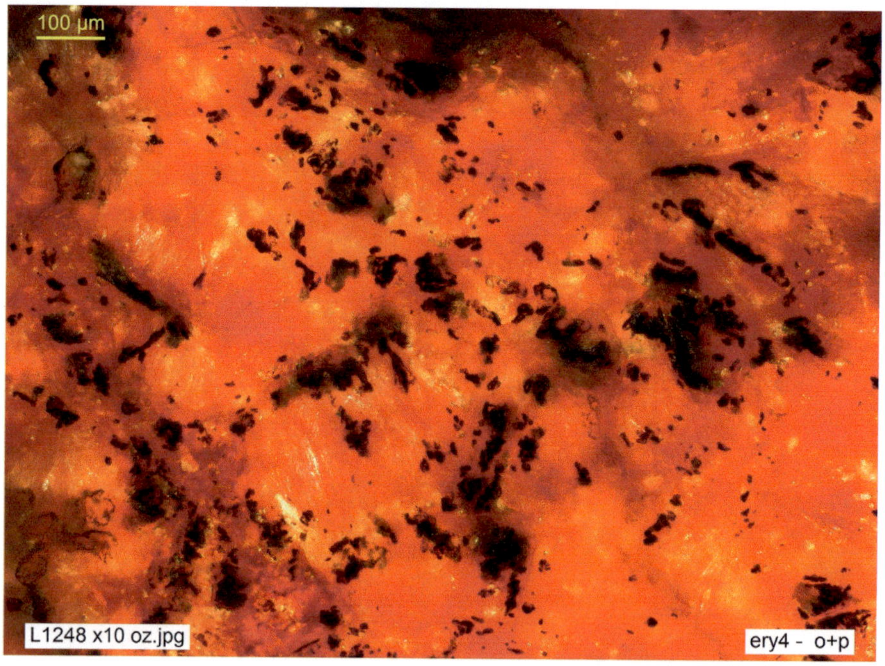

Fig. 138.3 **a**, **b** (Same scene: **osp** and **o+p**, *respectively*). Massive erythrite gangue, secondary to Ni–Co–Bi–Ag–U ore. Tiny relict particles (disseminated in ery, white, bright), the only preserved vestige of the primary arsenical mineralization. In oil immersion, the original gray tone shown in Fig. 138.2a appears overprinted by the pink, diffuse IR luminosity (*osp*, Fig. 138.3a). IR are still more intensive with crossed polars (*o+p*, Fig. 138.3b). CR375, Cobalt, Ontario, Canada

is accompanied by very reflective ores, among them the ore with the highest possible R—a native silver inclusion (lower right corner).

The visual estimation of gg **R** is compatible with the approximate value **R** $\approx$ 6% which, according to the Fresnel function, corresponds (in air) to a refractive index **n** $\approx$ 1.65, consistent with the known values of erythrite ($n_{ery} \approx 1.63$–1.70). The pink colour, perceptible even without analyser (*dsp*) and enhanced by immersion (cf. Fig. 138.3a, *osp*), highlights intense and massive IR, which with analyzer (*d+p* and *o+p*) show a strong luminosity (Figs. 138.2b and 138.3b) and obliterate the effects of anisotropism.

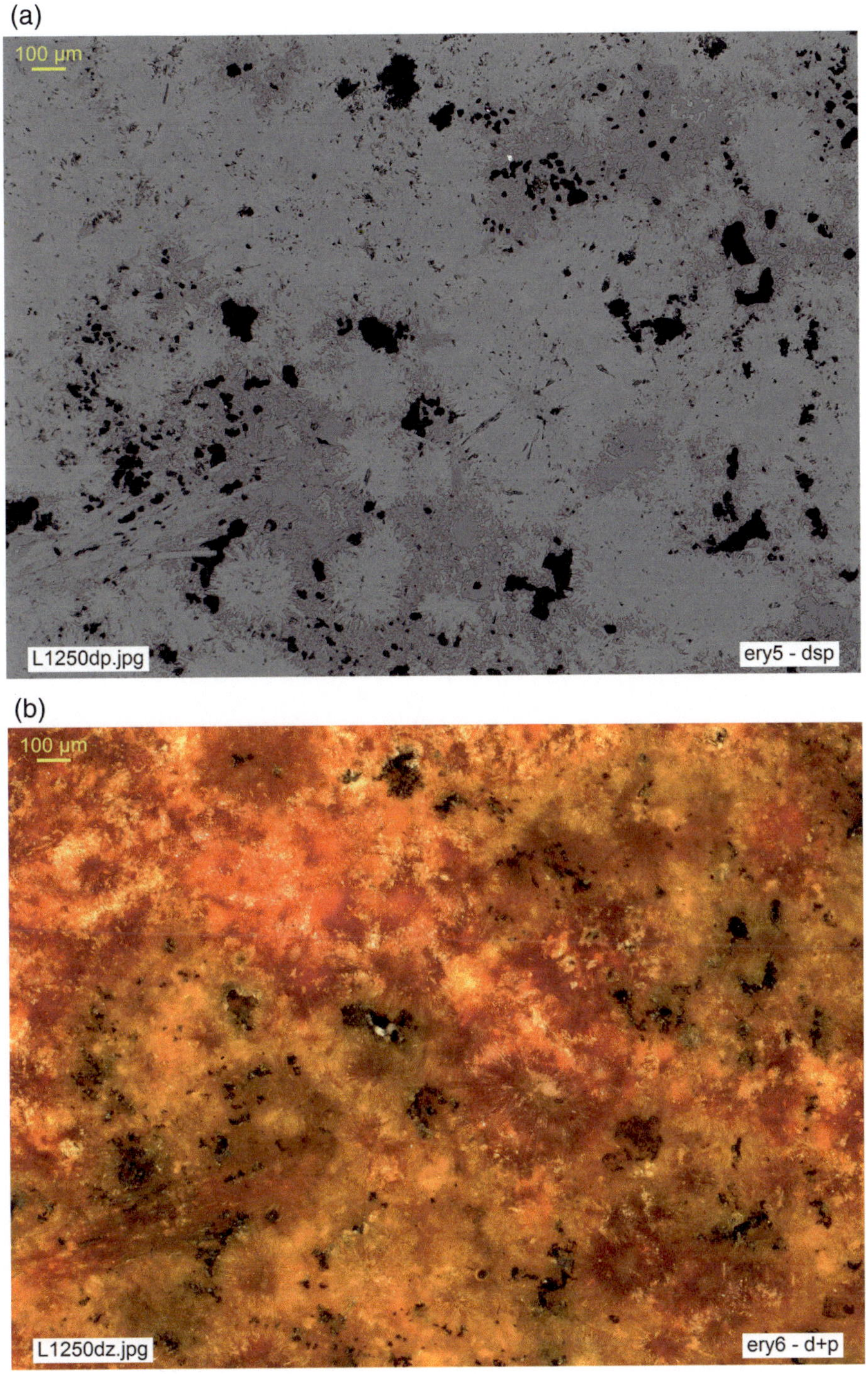

Fig. 138.4 **a**, **b** (Same scene: **osp** and **o+p**, *respectively*). Intergrown, fibrous-radial aggregates of ***orpiment*** (light gray, yellow IR) and erythrite (darker gray, pink IR), with sparse relics of arsenide (white, bright dots, possible safflorite). CR375, Cobalt, Ontario, Canada

The gangue also comprises orpiment intergrown with erythrite and forming fibrous-radial aggregates; orp is recognizable by its higher reflectance (looks lighter grey) and yellow IR (Fig. 138.4a, b). The scarcity of arsenides, reduced to tiny scattered relics in these two photomicrographs reduces the contrast effect which, in the previous photos, made the gangue appear very dark (underexposed). Thus, here appears a more realistic picture of ery (or As-gg) true reflectance, higher than that of quartz or resin, but lower than that of orpiment ($R_{orp} \approx 23$–27%).

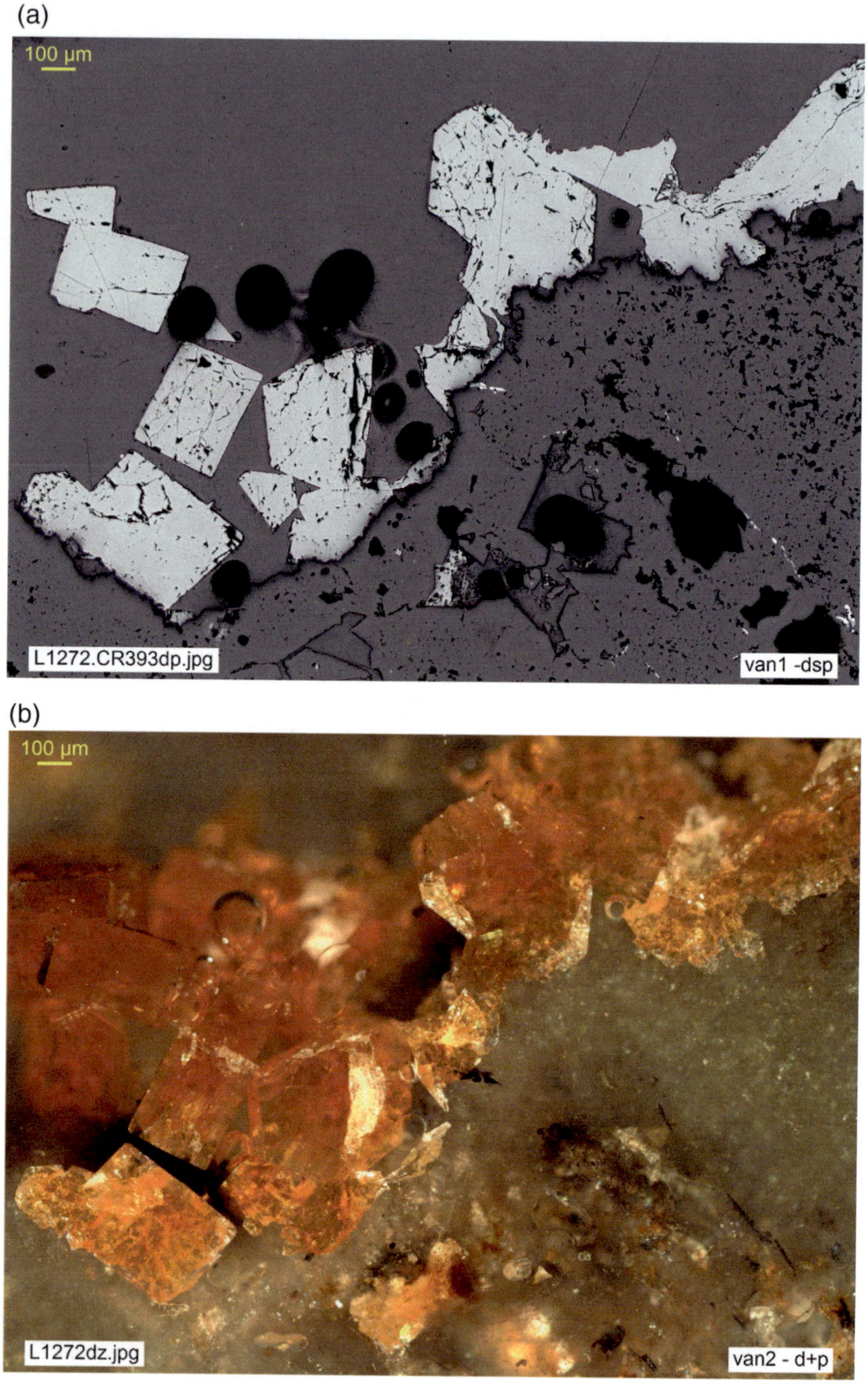

Fig. 138.5 **a**, **b** (same scene: **dsp** and **d+p**, *respectively*). Vanadinite crust (gray aggregate of euhedral crystals, *dsp*, massive, orange IR, *d+p*) on chalcedony clast (dark gray, lower right half, colorless IR, faintly lustrous) with faint hematite impregnation (white). CR 393, Globe, Arizona, USA

138.4 Vanadinite

The vanadate **Vanadinite** ‹*van*, Pb_5 (VO_4) $_3Cl$, hexagonal› must be considered in some cases as an ore, since it is or has been an important source of V in some mines, such as Tsumeb (Namibia), Sulejman Sai (Kazakstan), Broken Hill (Zambia) or Chihuahua (Mexico). It is described, however, among the gangue minerals because it is more frequently found in small quantities, as a secondary, sporadic mineral in the surface oxidation zone of various sulphide deposits with galena, Pb sulphosalts and others that, upon weathering,

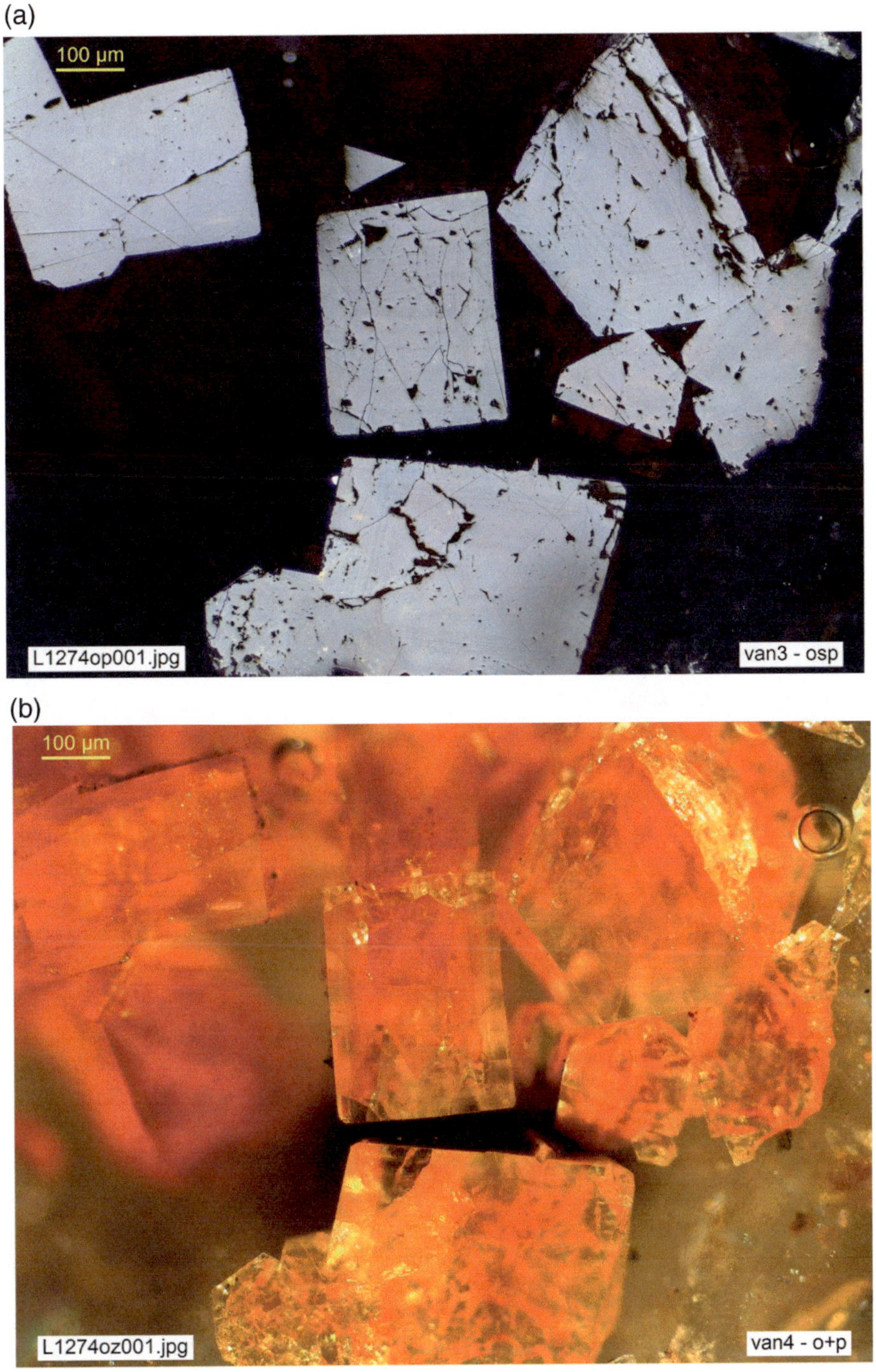

Fig. 138.6 **a**, **b** (Same scene: **osp** and **o+p**, *respectively*). Detail of Fig. 138.5a, in immersion: notable increase of the contrast between quartz (which disappears completely) and vanadinite, which maintains the aspect of an ore, gains in sharpness and shows compositional or growth zoning and orange to yellowish or reddish RI, perceptible even without analyzer (*osp*) due to the greater luminosity and orange tone in some points (lower central grain)

release the V that they may supposedly contain in traces. In these cases, vanadinite appears disseminated, as impregnation in cavities, microcracks or incrustations (Figs. 138.5a, b, 138.6a, b), accompanied by other neoformed lead carriers such as pyromorphite, wulfenite, cerussite, anglesite, as well as limonite, jarosite and other species of the oxidation zone.

Its relationship with the geological context of the oxidation zone is an important data for its microscopic recognition, which is supported by the following peculiarities:

- its morphology, showing in cavities or incrustations, aggregates and beautiful euhedral crystals, with prismatic, tabular and pyramidal forms (Figs. 138.5a, b, 138.6a, b).
- its transparent but relatively absorbent character, almost opaque at times
- its light grey colour, with occasional orange reflections (IR effect, locally perceptible even without analyser, Fig. 138.6a)
- its reflectance[1] ($R_{van} \approx 16.2\%$ to $\approx 17.2\%$), moderate or low in absolute terms but very high in relation to the usual values of gangue (commonly $R_{gg} \leq 10\%$), in correspondence with its very high refringence (refractive indices: $n_\omega = 2.416$ and $n_\varepsilon = 2.350$; for comparison, the refractive index of sphalerite, a typical ore that can also vary from transparent to almost opaque, is only n = 2.369, while its reflectance varies, depending on the composition, between $R_{sp} \approx 16.3\%$ and 17.4%);
- its low bireflectance, imperceptible despite its remarkable birefringence ($\delta = 0.066$; cf. Figs. 138.5a and 138.6a)
- its massive and luminous internal reflections (Figs. 138.5b and 138.6b), with yellow to orange and reddish-brown tones, which hide the effects of anisotropism but occasionally enhance internal structures, such as the growth zonation (Fig. 138.6a, b)
- its usually imperceptible anisotropism, masked by IR

[1] The reflectance, **R**, of vanadinite has been estimated according to the Fresnel function, given the known values of its refractive indices ($n_\omega = 2.416$, $n_\varepsilon = 2.350$), taking as reference an intermediate value n ≈ 2.383, it results R ≈ 16.7% (air). This is an approximate value for random orientations, although in reality one could expect, in correspondence with the actual values of n (n_ω and n_ε), values of R, varying according to grain orientation between $R_\varepsilon \approx 16.2\%$ and $R_\omega \approx 17.2\%$. However, in practice, these theoretical values should be adjusted slightly upwards, if one takes into account the contribution of the IR, perceptible even without analyzer (dsp) as lighter, orange zones on the surface of the mineral. They could be adjusted even more, if the Fresnel formula is corrected by introducing the value of the absorption coefficient k, not documented but which should not be disregarded, given the semi-absorbent or almost opaque character of vanadinite.

138.5 Organic Matter and Coals

> Volatile or liquid carbohydrates can be polymerized and converted into materials similar to coal through various known and unknown natural processes. [...] In many cases the conversion of carbohydrates to carbonaceous matter is a normal process of various stages of metamorphism. In others the process is chemically activated through some catalytic effect. Of great importance, however, is the conversion by means of radioactivity. In all cases the first effect is compaction to a splinty, conchoidally breaking mass, with a gradual increase in reflectivity. At a certain stage the material begins to assume a graphitic pre-orientation which can lead to distinct anisotropy almost like that of graphite. In extreme cases, coaly substance may be difficult to distinguish from graphite -perhaps there actually is a transition [...] "Coaly substance' as defined above is surprisingly widespread, occurring even in granite, pegmatite and hydrothermal veins where one would truly not expect it. The great mobility of carbohydrates at higher temperatures causes extraordinary migration. (Ramdohr 1980, p. 1121–1122)

The materials that, as a whole, we designate by the term **"Coal and Organic Matter"**, ***COM***, comprise the deposits and organic remains preserved in sediments and rocks, coming from living beings, usually lithified plants, including the products of their natural transformation (maturation, carbonification, pre-grafitisation) throughout the geological history of these deposits. It also applies, exceptionally, to coal-like products of debatable origin, possibly due to polymerization processes of various hydrocarbons or even of inorganic origin. It does not refer to the fossils that are the subject of paleontology.

Coal has been a fundamental resource for the development of modern industry and transport, although since the last quarter of the twentieth century there has been an increasing opposition to its use, due to its contribution to atmospheric CO_2. Its enormous importance and complexity have required the development of a specific discipline, **coal petrography**, for its study and exploitation.

Organic matter, *OM*, is present in many mineral concentrations and may even play an important role in their genesis. Even if its role is only passive, from the moment of its deposition it participates in the geological history of the environment, and its evolution can, therefore, gather information—geochronological or geo-thermometric—of interest in various fields. Among them are included the investigation of deposits, from sedimentary and fossil energy resources (hydrocarbons, coals) to metamorphic and graphite.

This means that the study of ***COM*** has developed and diversified enormously, drawing on disciplines as disparate as palynology, isotope geochemistry, mineral genesis, palaeontology, metamorphic petrology and, of course, the aforementioned coal petrography.

Obviously, to treat these topics with a minimum of rigour would require several specific monographs to be devoted to them, due to their breadth and variety. However, without departing from the objectives of the present text and, given that many rocks and ores contain ***COM***, the microscopist must be able to:

- avoid confusion of ***COM***, in any form, with metallic ores
- recognize and distinguish the presence and possible significance of ***COM*** in an ore
- characterize, broadly speaking, the evolution of the material comprising COM or the conditions to which it might have been subjected, from immature and recognizable organic or vegetal remains or their derivatives (*organic matter* of Picot and Johan 1982) through possible polymers or carbonaceous materials of not so clear origin (*coaly substance* or *carbonaceous matter* of Ramdohr 1980) to progressively more carbonified types, reaching anthracites (metamorphic) and graphite.

The **organic matter, *OM*, is** normally grey (or brownish), usually with polishing scratches (due to its low hardness) and dark.

Its reflectance, **R**, in the less evolved types, may appear as low as that of the common gangue (Figs. 138.7 and 138.8).

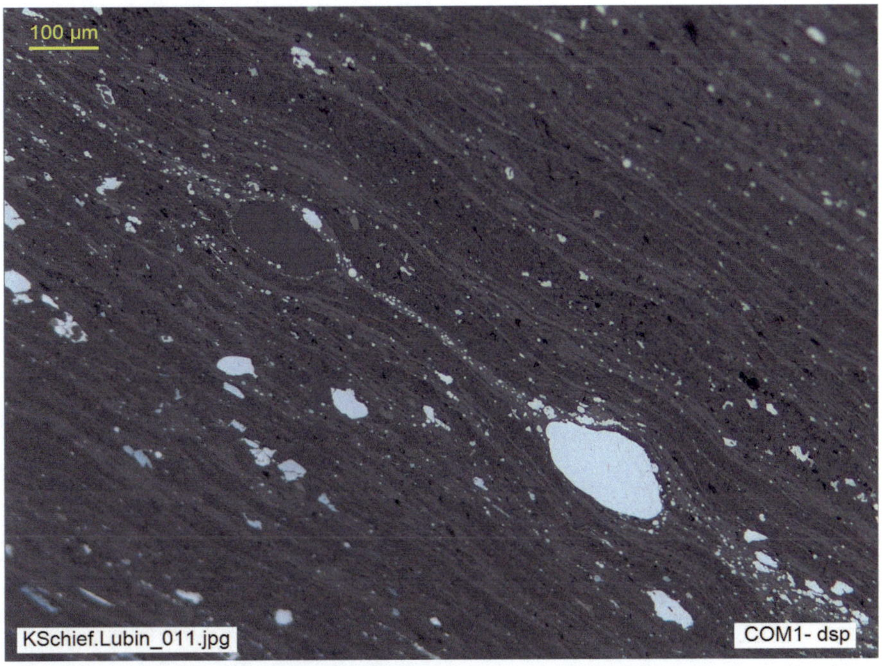

Fig. 138.7 *Kupferschiefer* (cupriferous schist) with calcocite (bluish-white) and *COM* (brownish-gray, slightly lighter than silicate matrix), disseminated and oriented in microbands parallel to the schistosity. M. Rub_ltp, Lubin, Poland (**dsp**)

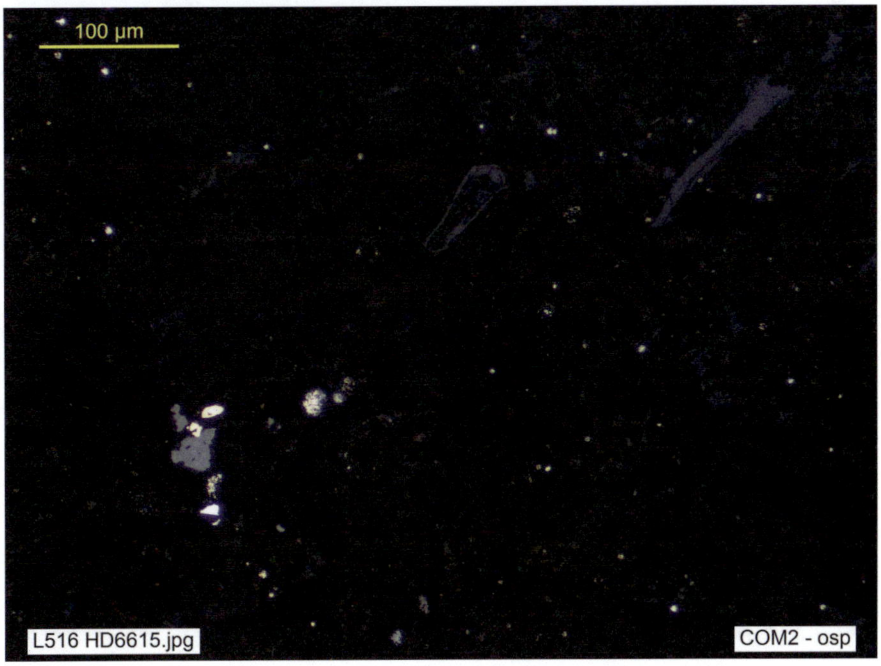

Fig. 138.8 Faint dissemination of sulfides (pyrite, chalcopyrite, digenite) in *Kupferschiefer*, with abundant scattered and very poorly evolved organic matter (dark to very dark gray), which has preserved the shape of some plant remains (upper right). HD 6615, Richelsdorf, Germany (**osp**)

Fig. 138.9 **a**, **b** (Same scene, **dsp** and **osp**, *respectively*). Bituminous marly lutite, with detrital quartz framework (dark grey, black in immersion oil), calcitic cement (dark like quartz, but with frequent spherulites enhanced in Fig. 138.10; calc shows bireflectance—*dsp*—or disappears—*osp*) and interstitial filling of *COM*. *COM is* gray, slightly lighter than the quartz-carbonate gangue and much darker than the few disseminated sulfides (chalcopyrite, galena, pyrite), but it does not disappear in immersion, unlike quartz and calcite. HD 6615, Richelsdorf, Germany (Kupferschiefer)

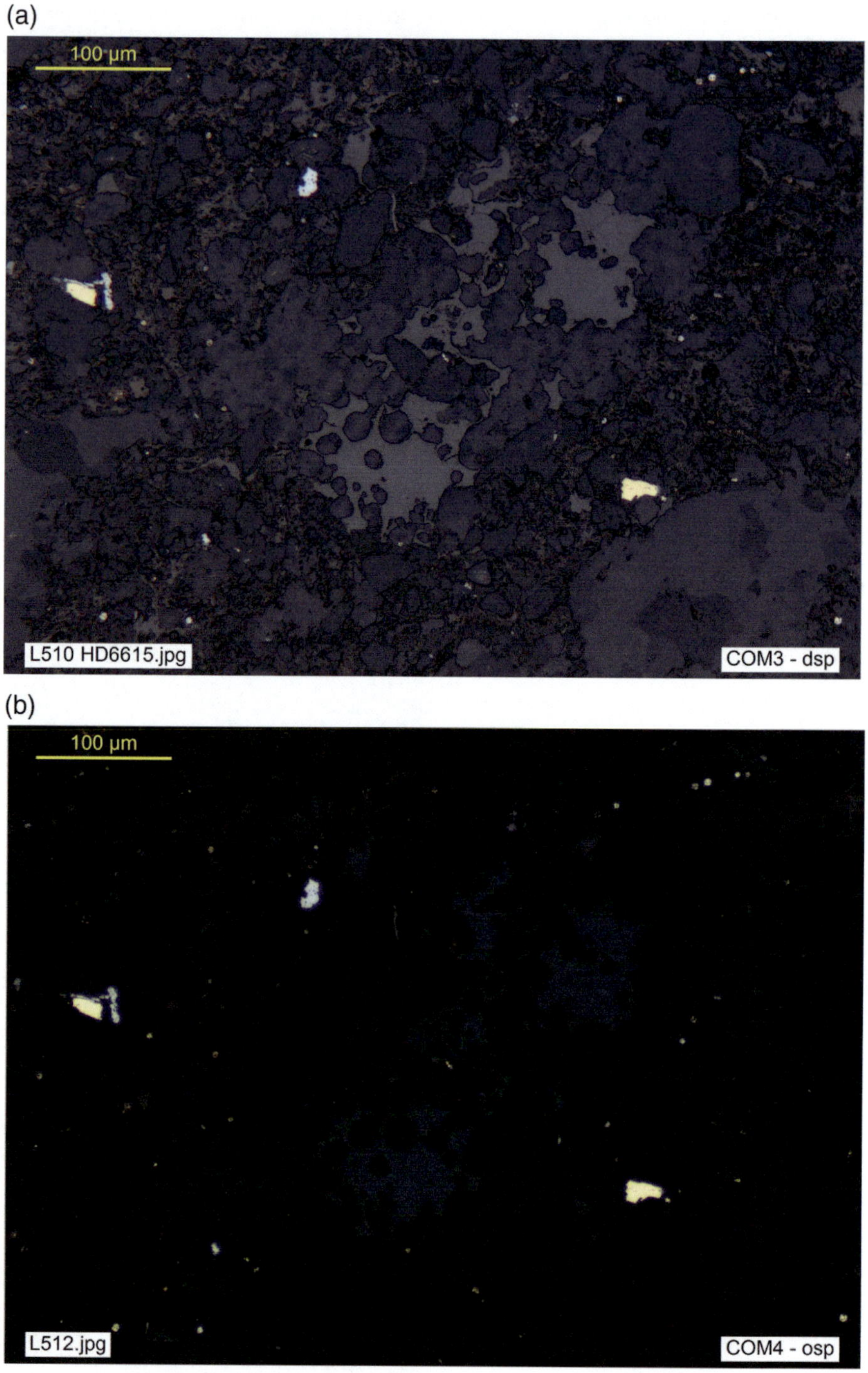

In case of doubt, observation in immersion allows a quick estimation. Thus, for example, although the *OM* of Fig. 138.9a looks very dark, in immersion it does not disappear (Fig. 138.9b), unlike the common gangue. This shows the higher R of *OM* (certainly, in this case, $R_{OM} >$ 5%). Gg and OM should never be confused since, unlike the common gangue, *OM* is opaque and therefore lacks**IR**, internal reflections (Fig. 138.10).

Fig. 138.10 Same scene as Fig. 138.9a (**o+p**, magnified). Absence of internal reflections in *COM*, unlike the quartz-carbonate gangue which shows abundant IR and spherulites. COM (isotropic, always in extinction) is scarcely evolved here

With analyzer, apart from proving the absence of **RI** in all cases, the less evolved COM terms are always seen in extinction (Fig. 138.10), i.e., they are isotropic. However, as the carbonification progresses or even enters the field of metamorphism, not only does the reflectance increase but the material becomes progressively anisotropic, even strongly bireflective and pleochroic, as well as extremely anisotropic, like *graphitoid* and *graphite* ***gra*** (cf description *gra* and photomicrographs Figs. 57.1–57.4, in Part II). The association with metamorphic materials may indicate the presence of graphite, although graphite may also occur as a detrital component (cf. Fig. 57.4). In any case, confusion with **valleriite *vll*** (Figs. 138.11a, b, cf. Part II, description ***vll***), very similar to graphite, should be avoided.

The analysis of **coals** shows that they have a complex composition, structure and evolution. Different components or *macerals* are distinguished (inertinite, liptinite, vitrinite) and different ranges, the study of which is the object of coal petrography. The measurement of the reflectance, in immersion, is an important data.

The following criteria should help the beginner to characterise ***COM***:

- *Isotropy versus Anisotropy.* Less mature organic matter is isotropic. As it evolves by maturation, it becomes increasingly anisotropic, following a process called by Ramdohr (1980) *pre-graphitization*. According to this author, radioactivity can also produce the same effect. At the end of this evolution, graphite (described as ore, *gra,* cf *Part II*) is already strongly anisotropic.
- *Reflectance.* Another important criterion is the reflectance, **R**, which is variable, but tends to grow in the carbonification process from values similar to those of the gangue ($\approx$5%) to those of graphite or even higher (Picot's *graphitoids*, 1982).
- *Morphology*. In the less evolved types, the forms and structures of plant remains (stems, spores, cellular structures) and fossils can be recognized, although sometimes the *OM*, because of its plasticity and its ductile response to deformation, appears in very thin strands following the orientation or schistosity of the rock, or flows and segregates in pores or interstices (Figs. 138.7 and 138.8). As the process of carbonification and metamorphism progresses, the morphological features of the original organisms progressively disappear.

Fig. 138.11 **a**, **b** (Same scene, **dsp** and **d+p**, *respectively*). Carbonatite (calcite, gray, anisotropic) with disseminated magnetite (white, isotropic) and valleriite (light gray to grayish-brown, bireflective, pleochroic and strongly anisotropic, like graphite but not as bireflective as graphite). TFPe7, Palabora (South African Rep.)

(a)

(b)

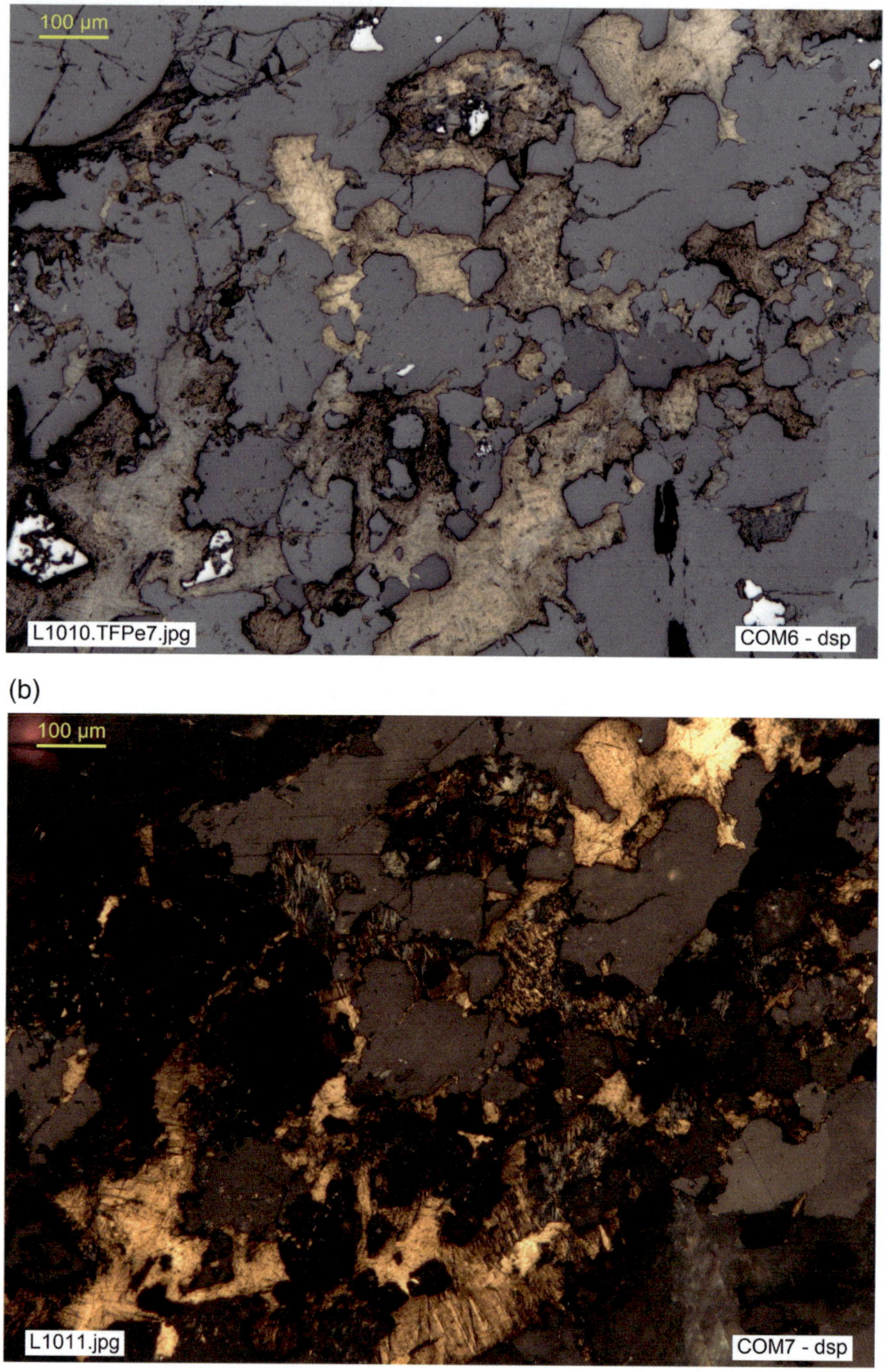

References

Berry LG, Mason L, Dietrich RV (1982) Mineralogy, 2nd edn. Freeman & Co, New York, p 561

Bowles JFW, Howie RA, Vaughan DJ, Zussman J (2011) Rock-forming Minerals, vol 5A, 2nd ed. Non-silicates: Oxides, Hydroxides and Sulphides. The Geological Society, London, 920p

Chang LLY, Howie RA, Zussman J (1998) Rock forming minerals, v 5B, 2nd ed. Non-silicates: Sulphates, Carbonates, Phosphates, Halides. The Geological Society, London, 383p

Criddle AJ (1998) Ore microscopy and photometry (1890–1998). In: Cabri LJ, Vaughan DJ (eds) Modern approaches to Ore and Environmental Mineralogy. COM/IMA Short Course Series, vol 27, Ottawa, Ontario, 421p

Criddle AJ, Stanley CJ (eds) (1993) Quantitative data file for ore minerals, 3rd edn. Chapman & Hall, London, p 635

Picot P, Johan Z (1982) Atlas of ore minerals. B.R.G.M./Elsevier
Ramdohr P (1980) The ore minerals and their intergrowths, vol 2 (2nd edn.) Pergamon Press, Oxford, 1205p

Websites

Mindat: https://www.mindat.org/min-1040.html (*visit.* 20,210,501).
https://www.mindat.org/min-687.html (*visit.* 20,210,426).
https://www.mindat.org/min-682.html (*visit.* 20,210,427).
https://www.mindat.org/min-209.html (visit. 20,210,427).

Mineral Indices, Annexes

Analytical Index of Minerals Described*

Name	Abbreviation	Explanation
Acanthite	ac	+
Adularia	adu	gg
Aguilarite	agl	+
Aikinite	aik	cf. bm
Alabandite	alb	+
Albite	ab	gg
Allargentum	aag	+
Alloclasite	alc	cf. cob
Amalgam	am (amal)	+
Amblygonite	amb	cf. Phos
Amphibole (Group)	Anf, Amp	gg
Anatase	ana	+
Anglesite	ang	gg
Anhydrite	anh	gg
Anilite	anil	cf. cc, dg
Ankerite	ank	gg
Annabergite	anbg	gg
Anorthite	an	gg
Antimony	Sb	+
Apatite	ap	gg
Argentite	arg	+
Argentopyrite	agp	+
Arsenic	As	+
Arsenopyrite	asp	+
Aurian silver Ag–Au	*ausi*	cf. *el*
Azurite	az	+
Baddeleyite	bad	+
Barite	bar	gg
Bastnäsite	bst, (bas)	+
Berthierite	brt	+
Berzelianite	bzl	cf. dg
Biotite	bio	gg

(continued)

R. Castroviejo, *A Practical Guide to Ore Microscopy—Volume 1*,
https://doi.org/10.1007/978-3-031-12654-3

(continued)

Name	Abbreviation	Explanation
Bismite	bim	gg
Bismuth	Bi	+
Bismuthinite	bm	+
Bismutite	bis	gg
Bornite	bn	+
Boulangerite	bl	+
Bournonite	bnn	+
Braggite	brg	cf. erl
Braunite	bra (br)	+
Bravoite	brv	+
Breithauptite	bth	+
Briartite	bri	cf. ren
Brookite	bk	gg
Calaverite	clv	+
Calciotantite	cta	cf. mlt, pyc
Calcite	calc	gg
Carrollite	car	cf. ln
Cassiterite	cst (cs)	+
Cattierite	ctt	cf. brv
Cerargyrite	crg	cf. cag
Cerussite	crs	cf. gn
Cervantite	cerv	cf. ochres Sb
Chalcedony	cld	cf. q
Chalcocite	cc	+
Chalcopyrite	ccp	+
Chert	ch	gg
Chloanthite (disc.)	cloa	cf. sk
Chlorargyrite	cag	+
Chlorite (Series)	Chl	gg
Chromite	chr	+
Chrysocolla	crc	+
Cinnabar	cin	+
Clausthalite	clt (cla)	+
Clay	Arc	gg
Clay minerals (Gr.)	CM, MArc	gg
Coal—organic matter	COM	cf. gg & gra
Cobaltite	cob	+
Cohenite	coh	cf. Fe
Coloradoite	colo (cld)	+
Columbite	cb	cf. cbt
Columbotantalite	cbt	+
Cooperite	cpr	cf. erl
Copper	Cu	+
Coronadite	crn	cf. rom, psi
Corundum	cor	gg
Cosalite	cos	+
Covellite	cv	+
Cryptomelane	cry	+
Cubanite	cbn	+
Cuprite	cup	+
Cylindrite	cyl	cf. F 41.04
Digenite	dg	+
Djurleite	dju (dj)	cf. cc, dg

(continued)

(continued)

Name	Abbreviation	Explanation
Dolomite	dol	gg
Dyscrasite	dy	+
Electrum	el	+
Enargite	en	+
Epidote (Group)	Ep	gg
Erlichmanite	erl	+
Erythrite	ery	gg
Eugenite	eug	cf. am, aag
Euxenite	eux	cf. REE
Falkmanite	flk	cf. bl
Famatinite	fm	+
Feldspar (Group)	Fs, Felds	gg
Ferberite	fb	cf. wf
Ferroselite	fse	cf. clt, um
Fletcherite	fle	cf. ln
Fluorite	fl	gg
Franklinite	fkl (fk)	+
Freibergite	fbg	cf. td
Freieslebenite	flb	+
Galena	gn	+
Gangue (undiff.)	gg	gg
Garnet (Group)	grt	gg
Geocronite	gc	+
Germanite	ger	cf. ren
Gersdorffite	gf	+
Glaucodot	gld	+
Glauconite	gln	gg
Goethite	gth	+
Gold	Au	+
Gossan	gos	cf. lim
Graphite	gra	+
Gratonite	gtt	+
Greenockite	gk	gg
Greigite	grg	cf. ln
Gudmundite	gud (gd)	+
Hematite	ht	+
Hessite	hss	
Hollandite	hol	cf. psi
Hornblende	hb	gg
Horobetsuite	hrb	cf. bm
Hübnerite	hub	cf. wf
Illite	ili	gg
Ilmenite	il	+
Ilmenorutile	ilr	+
Iridium	Ir	+
Iridosmine (discr.)	iros	cf. Os, Ir
Iron	Fe	+
Isocubanite	isc	+
Ixiolite	ix	cf. rt (F13-14)
Jacobsite	jac	+
Jalpaite	jl	cf. Sulf
Jamesonite	jm	+
Jarosite	jar	cf. lim
Jasper	jas	gg

(continued)

(continued)

Name	Abbreviation	Explanation
Jordanite	jd	+
K Feldspar	Kfs	gg
Kermesite	ker, (km)	+
Kësterite	ks	cf. stn
Krennerite	krn	+
Laurite	lrt	cf. erl
Lepidocrocite	lpc	+
Leucoxeno	lcx	cf. ana (gg)
Limonite	lim, lm	+
Linnaeite	ln	+
Löllingite	lo	+
Ludwigite	ldw	cf. von
Luzonite	lz	+
Mackinawite	mck	+
Maghemite	mgh	+
Magnesite	mag	cf. Carb, calc
Magnetite	mt, (mg)	+
Malachite	mch	+
Manganite	mng	+
Marcasite	mc	+
Maucherite	mau	+
Melnicovite	mln	cf. py
Mercury/Quicksilver	Hg	cf. cin
Metacinnabar	mcin	cf. cin
Metastibnite	msbt	cf. sbt
Miargyrite	my	+
Mica (Group)	Mica	gg
Microlite	mlt	cf. pyc
Millerite	ml	+
Molybdenite	mol	+
Monazite	mz	+
Moschellandsbergite	mobg	cf. am
Muscovite	mv, mus	gg
Naumannite	nau	+
Nickeline, Niccolite	nc, (nk)	+
Ochre, Ocher	Ochre	cf. Oxides
Olivin (Series)	ov	gg
Organic matter	OM	gg
Orpiment	orp, op	+
Orthoclase	or	gg
Orthopyroxene (Gr.)	Opx	gg
Osmium	Os	+
Oxycalciomicrolite	ocm	cf. mlt, bim
Palladium	Pd	cf. PGE
Pararammelsbergite	prm	+
Pearceite	pc	+
Pentlandite	pn	+
Perovskite	pvk	gg
Phosphate	Phos	gg
Phyllosil. (Subclass)	FSlc, PSlc	gg
Piemontite	pie	cf. bra
Pitchblende	pch	cf. urn
Plagioclase (Series)	Plag	gg
Platinum	Pt	+

(continued)

(continued)

Name	Abbreviation	Explanation
Platinum Group Elements	PGE	cf. Pt, Os, Ir
Platinum Group Minerals	PGM	cf. spy
Polianite	poli	cf. prl
Polybasite	plb	+
Polydymite	pdm	cf. ln
Proustite	pru	+
Psilomelane	psi	+
Pyrargyrite	prg	+
Pyrite	py	+
Pyrochlore	pyc (pyl)	+
Pyrolusite	prl	+
Pyromorphite	pymp	gg
Pyrostilpnite	pyst	cf. prg
Pyroxene (Group)	Px, Pyx	gg
Pyrrhotite	po	+
Quartz	q, qz…	gg
Rammelsbergite	rm	+
Ramsdellite	rdl	+
Realgar	rl	+
Renierite	ren	+
Rhodochrosite	rhc	gg
Rhodonite	rhn	gg
Romanechite	rom	+
Rutheniridosmine	ruiro	cf. Os, Ir
Rutile	rt	+
Rynersonite	ryn	cf. mlt, pyc
Safflorite	sf	+
Schachnerite	snr	cf. am, F stm8
Scheelite	sch	+
Scorodite	sco	gg (cf. asp)
Selenides	Sel	cf. nau, um
Senarmontite	sen	cf. ochres Sb
Sericite	ser	gg
Serpentine	srp	gg
Siderite	sid (sd)	cf. Carb
Siegenite	sg	cf. ln
Silica	Sic	cf. q
Silicate (Class)	Slc	gg
Silver	Ag	+
Skutterudite	sk	+
Smaltite (≈sk)	smal	cf. sk
Smithsonite	sms	gg
Souxite	sx	cf. cst
Sperrylite	spy	+
Sphalerite	sp	+
Spinel	spn	+
Spionkopite—Bbl.cv	spk	cf. cv, F dg
Stannite	stn	+
Stephanite	stp	+
Sternbergite	stb	+
Stibarsen	sba	+
Stibioluzonite	sbl	cf. fm

(continued)

(continued)

Name	Abbreviation	Explanation
Stibnite	sbt	+
Stromeyerite	stm	+
Sulfur	S	gg
Sylvanite	syl (sv)	cf. Tel
Talnakhita	tlk	cf. ccp
Tantalite	tan (tt)	cf. cbt
Tapiolite	tp, (tap)	+
Tedradymite	tdy	cf. Tel
Tellurium	Te	+
Tennantite	tn	cf. td
Tetrahedrite	td	+
Titanite	tit (sph)	+
Tourmaline	tur	gg
Troilite	tr (tro)	cf. po, Fe
Ullmannite	ul	+
Ulvöspinel (Ulvite)	ulv	+
Umangite	um	+
Uraninite	urn	+
Vaesite	vae	cf. ctt
Valentinite	val	cf. ochres Sb
Valleriite	vll	+
Vanadinite	van	+
Vanadinite	van	gg
Violarite	vio	cf. ln
Vonsenite	von	gg
Willemite	wil	cf. Slc
Wolframite	wf	+
Wurtzite	wz	+
Xanthoconite	xan	cf. pru
Xenotime	xtm	cf. Phos
Zeolite (Group)	Zeo	gg
Zincite	znc	gg (cf. fkl)
Zinkenite	zk	cf. Ss
Zircon	zir	gg
Other name—Synonym		Observations
Allemontite		cf. sba
Antimonite		cf. sbt
Blende		cf. sp
Chalcopyrrhotite		cf. isc
Fahlore		cf. td
Hollandite		cf. rom, psi
Martite		cf. ht, mt
Mispickel		cf. asp
Polianite		cf. prl
Schalenblende		cf. sp
Sphene		cf. tit
Steel		cf. Fe
Strüverite		cf. rt
Ulvite		cf. ulv

*__Key__ to locating the descriptions of ores and minerals in the text
+ Ore is fully described in the chapter under its own name, symbol or abbreviation
***cf m1, m2*, etc.** Ore or mineral described/commented on in the chapter under mineral *m1*, *m2*, etc.
gg Described in the corresponding chapter in Part III (Gangue)

Selected Ore Minerals—Learning Priority 1

Name (EN)	Abbreviation	Nombre (ES)	Observations	Priority
Acanthite	**ac**	**Acantita**	+	**1**
Antimony	**Sb**	**Antimonio**	+	**1**
Arsenopyrite	**asp**	**Arsenopirita**	+	**1**
Bastnäsite	**bst (bas)**	**Bastnäsita**	+	**1**
Bornite	**bn**	**Bornita**	+	**1**
Cassiterite	**cst (cs)**	**Casiterita**	+	**1**
Chalcocite	**cc**	**Calcocita = Calcosina**	+	**1**
Chalcopyrite	**ccp**	**Calcopirita**	+	**1**
Chromite	**chr**	**Cromita**	+	**1**
Cinnabar	**cin**	**Cinabrio**	+	**1**
Cobaltite	**cob**	**Cobaltita**	+	**1**
Columbotantalite	**cbt**	**Columbotantalita**	+	**1**
Copper	**Cu**	**Cobre**	+	**1**
Covellite	**cv**	**Covellita**	+	**1**
Cuprite	**cup**	**Cuprita**	+	**1**
Digenite	**dg**	**Digenita**	+	**1**
Galena	**gn**	**Galena**	+	**1**
Goethite	**gth**	**Goethita**	+	**1**
Gold	**Au**	**Oro**	+	**1**
Graphite	**gra**	**Grafito**	+	**1**
Hematite	**ht**	**Hematites**	+	**1**
Ilmenite	**il**	**Ilmenita**	+	**1**
Iron	**Fe**	**Hierro**	+	**1**
Limonite	**lim, lm**	**Limonita**	+	**1**
Linnaeite	**ln**	**Linneíta**	+	**1**
Magnetite	**mt, (mg)**	**Magnetita**	+	**1**
Manganite	**mng**	**Manganita**	+	**1**
Marcasite	**mc**	**Marcasita**	+	**1**
Miargyrite	**my**	**Miargirita**	+	**1**
Millerite	**ml**	**Millerita**	+	**1**
Molybdenite	**mol**	**Molibdenita**	+	**1**
Nickeline, Niccolite	**nc, (nk)**	**Niquelita, Niquelina**	+	**1**
Pentlandite	**pn**	**Pentlandita**	+	**1**
Platinum	**Pt**	**Platino**	+	**1**
Proustite	**pru**	**Proustita**	+	**1**
Psilomelane	**psi**	**Psilomelana**	+	**1**
Pyrargyrite	**prg**	**Pirargirita**	+	**1**
Pyrite	**py**	**Pirita**	+	**1**
Pyrolusite	**prl**	**Pirolusita**	+	**1**
Pyrrhotite	**po**	**Pirrotita (Pirrotina)**	+	**1**
Rammelsbergite	**rm**	**Rammelsbergita**	+	**1**
Romanechite	**rom**	**Romanechita**	+	**1**
Rutile	**rt**	**Rutilo**	+	**1**
Safflorite	**sf**	**Safflorita**	+	**1**
Silver	**Ag**	**Plata**	+	**1**
Skutterudite	**sk**	**Skutterudita**	+	**1**
Sphalerite	**sp**	**Esfalerita**	+	**1**
Stibnite	**sbt**	**Estibina, Antimonita**	+	**1**
Tetrahedrite	**td**	**Tetraedrita-Tennant**	+	**1**
Uraninite	**urn**	**Uraninita**	+	**1**
Wolframite	**wf**	**Wolframita**	+	**1**

Selected Ore Minerals—Learning Priority 2

Name (EN)	Abbreviation	Nombre (ES)	Observations	Priority
Alabandite	*alb*	*Alabandita*	+	*2*
Allargentum	*aag*	*Allargentum*	+	*2*
Amalgam	*am, (amal)*	*Amalgama*	+	*2*
Argentopyrite	*agp*	*Argentopirita*	+	*2*
Arsenic	*As*	*Arsénico*	+	*2*
Aurian silver Ag–Au	*ausi*	*Plata áurea Ag–Au*	cf. *el*	*2*
Azurite	*az*	*Azurita*	+	*2*
Berthierite	*brt*	*Berthierita*	+	*2*
Bismuth	*Bi*	*Bismuto*	+	*2*
Bismuthinite	*bm*	*Bismutinita*	+	*2*
Boulangerite	*bl*	*Boulangerita*	+	*2*
Braunite	*bra (br)*	*Braunita*	+	*2*
Bravoite	*brv*	*Bravoíta*	+	*2*
Breithauptite	*bth*	*Breithauptita*	+	*2*
Calaverite	*clv*	*Calaverita*	+	*2*
Chlorargyrite	*cag*	*Clorargirita*	+	*2*
Cohenite	*coh (chn)*	*Cohenita*	cf. *Fe*	*2*
Cosalite	cos	Cosalita	+	*2*
Cryptomelane	*cry*	*Criptomelana*	+	*2*
Cubanite	*cbn*	*Cubanita*	+	*2*
Dyscrasite	*dy*	*Discrasita*	+	*2*
Electrum	*el*	*Electrum*	+	*2*
Enargite	*en*	*Enargita*	+	*2*
Famatinite	*fm*	*Famatinita*	+	*2*
Franklinite	*fk (fkl)*	*Franklinita*	+	*2*
Gersdorffite	*gf*	*Gersdorffita*	+	*2*
Glaucodot	*gld*	*Glaucodoto*	+	*2*
Gudmundite	*gud (gd)*	*Gudmundita*	+	*2*
Ilmenorutile	*ilr*	*Ilmenorutilo*	+	*2*
Iridium	*Ir*	*Iridio*	+	*2*
Isocubanite	*isc*	*Isocubanita*	+	*2*
Jamesonite	jm	Jamesonita	+	*2*
Lepidocrocite	*lpc*	*Lepidocrocita*	+	*2*
Löllingite	*lo*	*Löllingita*	+	*2*
Luzonite	*lz*	*Luzonita*	+	*2*
Mackinawite	*mck*	*Mackinawita*	+	*2*
Maghemite	*mgh*	*Maghemita*	+	*2*
Malachite	*mch*	*Malaquita*	+	*2*
Maucherite	*mau*	*Maucherita*	+	*2*
Monazite	*mz*	*Monazita*	+	*2*
Orpiment	*orp, op*	*Oropimente*	+	*2*
Osmium	*Os*	*Osmio*	+	*2*
Pararammelsbergite	*prm*	*Pararamelsbergita*	+	*2*
Pearcite	*pc*	*Pearceíta*	+	*2*
Platinum Group Minerals	*PGM*	*Minls del Grupo del Pt*	cf. *spy*	*2*
Polybasite	*plb*	*Polibasita*	+	*2*
Pyrochlore	*pyc (pyl)*	*Pirocloro*	+	*2*
Renierite	*ren*	*Renierita*	+	*2*
Scheelite	*sch*	*Scheelita*	+	*2*
Sperrylite	*spy*	*Sperrylita*	+	*2*

(continued)

(continued)

Name (EN)	Abbreviation	Nombre (ES)	Observations	Priority
Spinel	*spn*	*Espinela*	+	*2*
Stannite	*stn*	*Estannita*	+	*2*
Stephanite	*stp*	*Stephanita*	+	*2*
Stibarsen	*sba*	*Stibarsen*	+	*2*
Stromeyerite	*stm*	*Stromeyerita*	+	*2*
Tapiolite	*tp, (tap)*	*Tapiolita*	+	*2*
Tellurium	*Te*	*Teluro*	+	*2*
Titanite	*tit (sph)*	*Titanita*	+	*2*
Umangite	*um*	*Umangita*	+	*2*
Valleriite	*vll*	*Valleriíta*	+	*2*
Wurtzite	*wz*	*Wurtzita*	+	*2*

Most Common Gangue Minerals

Name	Abbreviation	Observations
Adularia	adu	cf. Slc, Fs
Albite	ab	cf. Slc, Fs
Amblygonite	amb	cf. Phos
Amphibole (Group)	Anf, Amp	cf. Slc
Anglesite	ang	cf. Sulfates
Anhydrite	anh	cf. Sulfates
Ankerite	ank	cf. Carb
Annabergite	anbg	Arsenates
Anorthite	an	cf. Slc, Fs
Apatite	ap	cf. Phos
Barite	bar	cf. Sulfates
Biotite	bio	cf. Slc, FSlc
Bismite	bim	cf. Bi ochre
Bismutite	bis	cf. Bi ochre
Brookite	bk	Oxides, cf. lcx
Calcite	calc	cf. Carb
Carbonate (Class)	Carb	cf. Carb, gg
Cerussite	crs	cf. Carbonates
Cervantite	cerv	cf. Sb ochre
Chalcedony	cld	cf. q
Chert	ch	cf. Slc, Sic
Chlorite (Series)	Chl	cf. Slc, FSlc
Chrysocolla	crc	cf. Slc
Clay	Arc	cf. FSlc
Clay Minerals	CM, MArc	cf. Slc, FSlc
Coal—organic matter	COM	cf. Org.Matter
Corundum	cor	
Dolomite	dol	cf. Carb
Epidote (Group)	Ep	cf. Slc
Erythrite	ery	Arsenates
Feldspar (Group)	Fs, Felds	cf. Slc

(continued)

(continued)

Name	Abbreviation	Observations
Fluorite	fl	cf. Halides
Gangue (undiff.)	gg	
Garnet (Group)	grt	cf. Slc
Glauconite	gln	cf. Slc, FSlc
Hornblende	hb	cf. Slc, Inoslc
Illite	ili	cf. Slc, FSlc
Jarosite	jar	cf. lim
Jasper	jas	cf. q
K Feldspar	Kfs	cf. Slc, Fs
Leucoxene	lcx	cf. tit, ana… +
Ludwigite	ldw	cf. von
Magnesite	mag	cf. Carb
Mica (Group)	Mica	cf. Slc, FSlc
Muscovite	mv, mus	cf. Mica
Ochre, Ocher	Ocre	cf. Oxides
Olivine (Series)	ov	cf. Slc
Organic matter	OM	
Orthoclase	or	cf. Slc, Fs
Perovskite	pvk	Oxides
Phosphate	Phos	gg
Phyllosilicates (Subclass)	FSlc, PSlc	cf. Slc
Piemontite	pie	cf. bra, Ep
Plagioclase (Series)	Plag	cf. Slc, Fs
Pyromorphite	pymp	cf. Phos
Pyroxene (Group)	Px, Pyx	cf. Slc, Inoslc
Quartz	q (qz, qtz)	cf. Slc, Sic
Rhodochrosite	rhc	cf. Carb
Rhodonite	rhn	cf. rdl F 1-4
Scorodite	sco	cf. asp-Arsenates
Senarmontite	sen	cf. Sb ochre
Sericite	ser	cf. Slc, FSlc
Serpentine	srp	cf. Slc, FSlc
Siderite	sid (sd)	cf. Carb
Silica	Sic	cf. Slc, q
Silicate (Class)	Slc	gg
Smithsonite	sms	cf. Carb
Souxite	sx	cf. cst
Sulfur	S	gg
Tourmaline	tur	cf. Slc
Valentinite	val	cf. Sb ochre
Vanadinite	van	cf. Vanadates
Vonsenite	von	cf. Carb-Borates
Willemite	wil	cf. Slc
Xenotime	xtm	cf. Phos
Zeolite (Group)	Zeo	cf. Slc
Zincite	znc	Oxides
Zircon	zir	cf. Slc
Other name (synonym)		**Observations**
Sphene		cf. tit

Annex 1: Abbreviations and Criteria

The following list of mineral abbreviations is based on that used at the Laboratory of Applied Microscopy and Image Analysis (LMA) of the Madrid School of Mines, Universidad Politécnica de Madrid (UPM). It responds to the practical need to annotate quickly, accurately and easily the minerals identified in images, graphs, technical reports and texts, such as those used in the daily mineralogy work of the LMA. It refers mainly to the most common ores, without excluding certain rock-forming minerals (transparent) that stand out for their importance and frequency either as companions of metallic ores or for their interest to industry. The criteria for mineral selection in the text (§ 1.2) have been taken into account. Given the practical and necessarily restrictive character of these criteria, we cannot aim to include all possible species: the most common are selected, not the scarcer or highly specific ones from peculiar deposit types, for instance those many rare and as yet not well-known mineral species of the PGE (Platinum Group Elements) or REE (Rare Earth Elements).

As there is no unanimously accepted and complete international proposal containing at least all those considered as common ores, the list has been drawn up according to our own criteria.[1] In the first place, the classical or most widespread proposals (especially for **metallic ores**) in international circles have been consulted and, as far as reasonable, taken into account. *Vbgr.* Chace (1956), Kretz (1983), Fontboté (2006), Whitney and Evans (2010), Neumann (2019), or those applied unilaterally by journals, texts or institutions specialized in the field of mineral resources (such as that published just for names "sanctioned by the International Mineralogical Association" by *The Canadian Mineralogist*, cf. https://canminportal.files.wordpress.com/2018/03/symbols2.pdf, visit. 2020/12/31). These proposals, generally disparate from each other, do not entirely satisfy the criteria appropriate to our needs, which are summarized below. Nevertheless, Chace's approaches (1956) have proved a particularly useful reference, and many of the abbreviations that he proposes for minerals have been incorporated into the list.

These are the criteria followed to elaborate the list of abbreviations used in this text (cf. Tables of Selected Minerals and Abbreviations, **Annexes 1.1**, **1.2** and **1.3.a**, **m**):

1. As far as possible, advantage should be taken of pre-existing contributions and criteria whose rationality is supported by their use in the international literature.
2. Abbreviations should be short (ideally two characters, and a minimum of one and a maximum of four, the first coinciding with the first letter of the mineral's name) and, as far as possible, easy to remember by evoking the name of the mineral.
3. Any confusion should be avoided, so that the abbreviation-mineral correspondence is, as far as possible, bi-univocal.
4. In order to speed up its use and to maintain reference to the international literature, the direct correspondence between the mineral symbol and the names of the mineral in English and Spanish is established by means of a simple database. This becomes a table with three columns: Abbreviation, Name (EN) and Nombre (ES), thus offering instant information from any of the three.
5. Any coincidence of an abbreviation or symbol of a mineral with a symbol for a chemical element of the Periodic System should be avoided (unless the mineral species is a native element, in which case it is obligatory to use the same symbol).
6. This confusion is ideally avoided by introducing a differential criterion, such as the requirement to write the abbreviation of the mineral species in lower case (unless it is an element); moreover, it is convenient that there is no literal coincidence. For example, Ag, Au or S are valid for native silver, gold or sulfur, respectively; yet *fe* for ferberite, *mo* for molybdenite, *ba* for barite or *cr* for chromite are **not** valid.

[1] After finishing the book, a new proposal (Warr 2021) was published, which unfortunately could not be considered here for obvious scheduling reasons.

R. Castroviejo, *A Practical Guide to Ore Microscopy—Volume 1*,
https://doi.org/10.1007/978-3-031-12654-3

7. Exceptionally, the coincidence of mineral abbreviations sanctioned by a tradition of extended use in the mineralogical community with symbols of elements is admitted when nature excludes any real risk of confusion. Among others, this happens with minerals such as *po* (pyrrhotite), *ac* (acanthite), *mc* (marcasite) or *mt* (magnetite), which no minimally informed reader could confuse with the elements *Po* (Polonium), *Ac* (Actinium), *Mc* (Moscovium) or *Mt* (Meitnerium), respectively. These are radioactive and extremely rare (ppb) in nature, have no known mineralogical expression and are only found as traces in radioactive ores (uraninite), are the product of the decay of a radioactive series or are known only as a synthetic product in the laboratory—some of them for only a few decades.
8. To distinguish mineral species from groups, the former use only lower-case letters and the latter have a capital initial letter. For example: ab = albite, but Plag = plagioclase (group); hb = hornblende, but Anf = amphibole (group); likewise, Pt = Platinum, PGM = *Platinum Group Minerals*, and so on.
9. When a mineral has several names, the preference is for defining the abbreviation to the most commonly used term (usually in English) and accepted by IMA/COM (International Mineralogical Association, Commission on Ore Mineralogy) to facilitate as far as possible equivalence with international usage.
10. As long as they are compatible with the other criteria, the usual pre-existing usages must be respected, for instance py = pyrite, po = pyrrhotite. Exceptionally, the possibility of more than one symbol for the same species is allowed when their use is very widespread (for example: q or qz for quartz); then the recommended name is given as the first option and, as the second, and in brackets, the other.

In compiling this Table of Abbreviations we have sought the greatest possible consensus and maximum simplicity, avoiding any proliferation of superfluous symbols that would generate confusion. Discussions and exchanges with colleagues and students over the years have been a welcome contribution to the development of this table, and the input of Dr. César Cánepa (Lima, Peru) deserves special recognition. If the result meets our aims, these colleagues share any merit; if not, the errors are mine alone.

For ease of use, the table is presented in two versions, depending on the column chosen as an entry (in alphabetical order): **Index of Minerals** (labelled **m**); or **Index of Abbreviations** (labelled **a**). Each can be consulted through three search options: **all minerals** (**Annex 1.1**); only **ores** (**Annex 1.2**); or only **gangue** (**Annex 1.3**).

Finally, the **M**eanings of the **A**cronyms, **A**bbreviations and **S**ymbols used are summarized in **Annex 1.4** MAAS Glossary, to make the explanations in the text and tables easier.

Selected Minerals and Use of the Table of Abbreviations

The species chosen (*common minerals*, collected in the **Abbreviations** table) have been selected by applying the criteria explained in the text (§ 1.2), taking into account the interest in the ore, according to its abundance or its importance to industry (§ 1.2.1). As for the gangue (silicates, carbonates, oxides, etc.), the species chosen are those most frequently found in common ore bodies (§ 1.2.2). They result in about 300 entries, of which ≈ 230 correspond to ores and ≈ 80 to gangue.

Since their significance can vary widely, the entries have been further sorted and grouped by importance. The main ores are marked + in the Remarks column of the table, meaning that they are described fully. They are further classified in the *Priority* column into a key group (Priority 1: fundamental ores) and a collateral group (Priority 2: important, but subordinate). This is intended to guide the beginner in following the text, to ensure orderly progress and effective learning. This will always rely on the practitioner's personal observations under the microscope.

- The ≈ 50 ores whose learning is considered essential for the beginner or junior practitioner **to get started** are identified as **Priority 1**.
- For a more advanced level, another ≈ 60 ores follow (**Priority 2**), with which it is recommended to become **familiar**.
- Finally, the remaining species described are marked +, and are also worthy of attention. They might not be in the priority groups because they are less frequent or seldom cited in the literature (actually, some are sometimes confused or uncited simply because they are unknown to the observer: their supposed scarcity in nature seems an information gap).

Apart from the microscopic descriptions, most of the ores in the table, due to their importance or industrial interest, have been investigated for their automated identification by artificial vision (CAMEVA and AMCO systems: cf. **Part 5**). Therefore, their VNIR specular reflectance spectra (Visible and Near Infrared ranges) have been measured and incorporated into the AMCO database created at the LMA_UPM. These data are also available in this text as Tables and R-λ plots to support the description of each ore.

Annex 1.1.a All Described Minerals' Symbols or Abbreviations and Comments

Abbreviation	Name (EN)	Nombre (ES)	Observations
aag	Allargentum	Allargentum	+
ab	Albite	Albita	gg
ac	Acanthite	Acantita	+
adu	Adularia	Adularia	gg
Ag	Silver	Plata	+
agl	Aguilarite	Aguilarita	+
agp	Argentopyrite	Argentopirita	+
aik	Aikinite	Aikinita	cf. bm
alb	Alabandite	Alabandita	+
alc	Alloclasite	Aloclasita	cf. cob
alt	Altaite	Altaíta	
am (amal)	Amalgam	Amalgama	+
amb	Amblygonite	Ambligonita	cf. Phos
an	Anorthite	Anortita	gg
ana	Anatase	Anatasa	+
anbg	Annabergite	Annabergita	gg
Anf, Amp	Amphibole (Group)	Anfíbol (Grupo)	gg
ang	Anglesite	Anglesita	gg
anh	Anhydrite	Anhidrita	gg
anil	Anilite	Anilita	cf. cc, dg
ank	Ankerite	Ankerita	gg
ap	Apatite	Apatito	gg
Arc	Clay	Arcilla	gg
arg	Argentite	Argentita	+
As	Arsenic	Arsénico	+
asb	Asbolane	Asbolana	
asp	Arsenopyrite	Arsenopirita	+
Au	Gold	Oro	+
ausi	*Aurian silver Ag–Au*	*Plata áurea Ag–Au*	cf. *el*
az	Azurite	Azurita	+
bad	Baddeleyite	Baddeleyita	+
bar	Barite	Barita	gg
baux	Bauxite	Bauxita	
Bi	Bismuth	Bismuto	+
bim	Bismite	Bismita	gg
bio	Biotite	Biotita	gg
bis	Bismutite	Bismutita	gg
bk	Brookite	Brookita	gg
bl	Boulangerite	Boulangerita	+
bm	Bismuthinite	Bismutinita	+
bn	Bornite	Bornita	+
bnn	Bournonite	Bournonita	+
bra (br)	Braunite	Braunita	+
brg	Braggite	Braggita	cf. erl
bri	Briartite	Briartita	cf. ren
brt	Berthierite	Berthierita	+
brv	Bravoite	Bravoíta	+
bst (bas)	Bastnäsite	Bastnäsita	+
bth	Breithauptite	Breithauptita	+
buk	Bukovite	Bukovita	
bzl	Berzelianite	Berzelianita	cf. dg

(continued)

(continued)

Abbreviation	Name (EN)	Nombre (ES)	Observations
cag	Chlorargyrite	Clorargirita	+
calc	Calcite	Calcita	gg
car	Carrollite	Carrolita	cf. ln
Carb	Carbonate (Class)	Carbonato (Clase)	
cb	Columbite	Columbita	cf. cbt
cbn	Cubanite	Cubanita	+
cbt	Columbotantalite	Columbotantalita	+
cc	Chalcocite	Calcocita = Calcosina	+
ccp	Chalcopyrite	Calcopirita	+
cerv	Cervantite	Cervantita	cf. ocres Sb
ch	Chert	Chert	gg
Chl	Chlorite (Series)	Clorita (Serie)	gg
chr	Chromite	Cromita	+
cin	Cinnabar	Cinabrio	+
cld	Chalcedony	Calcedonia	cf. q
cloa	Chloanthite (disc. IMA)	Cloantita	cf. sk
clt (cla)	Clausthalite	Clausthalita	+
clv	Calaverite	Calaverita	+
CM, MArc	Clay minerals (Group)	Minerales Arcillosos (Grupo)	gg
cob	Cobaltite	Cobaltita	+
coh	Cohenite	Cohenita	cf. Fe
colo (cld)	Coloradoite	Coloradoíta	+
COM	Coal—organic matter	Carbón—materia orgánica	cf. gg and gra
cor	Corundum	Corindón	gg
cos	Cosalite	Cosalita	+
cpr	Cooperite	Cooperita	cf. erl
crc	Chrysocolla	Crisocola	+
crg	Cerargyrite	Cerargirita	cf. cag
crn	Coronadite	Coronadita	cf. rom, psi
crs	Cerussite	Cerusita	cf. gn
cry	Cryptomelane	Criptomelana	+
cst (cs)	Cassiterite	Casiterita	+
cta	Calciotantite	Calciotantita	cf. mlt, pyc
ctt	Cattierite	Cattierita	cf. brv
Cu	Copper	Cobre	+
cup	Cuprite	Cuprita	+
cv	Covellite	Covellita	+
cyl	Cylindrite	Cilindrita	cf. F 41.04
dg	Digenite	Digenita	+
dju (dj)	Djurleite	Djurleíta	cf. cc, dg
dol	Dolomite	Dolomita	gg
dom	Domeykite	Domeiquita	
dy	Dyscrasite	Discrasita	+
el	Electrum	Electrum	+
en	Enargite	Enargita	+
Ep	Epidote (Group)	Epidota (Grupo)	gg
erl	Erlichmanite	Erlichmanita	+
ery	Erythrite	Eritrina	gg
eug	Eugenite	Eugenita	cf. am, aag
eux	Euxenite	Euxenita	cf. REE

(continued)

(continued)

Abbreviation	Name (EN)	Nombre (ES)	Observations
fb	Ferberite	Ferberita	cf. wf
fbg	Freibergite	Freibergita	cf. td
Fe	Iron	Hierro	+
fkl (fk)	Franklinite	Franklinita	+
fl	Fluorite	Fluorita	gg
flb	Freieslebenite	Freieslebenita	+
fle	Fletcherite	Fletcherita	cf. ln
flk	Falkmanite	Falkmanita	cf. bl
fm	Famatinite	Famatinita	+
Fs, Felds	Feldspar (Group)	Feldespato (Grupo)	gg
fse	Ferroselite	Ferroselita	cf. clt, um
FSlc, PSlc	Phyllosilicates (Subclass)	Filosilicatos (Subclase)	gg
gal	Gallite	Galita	
gc	Geocronite	Geocronita	+
ger	Germanite	Germanita	cf. ren
gf	Gersdorffite	Gersdorffita	+
gg	Gangue (undifferentiated)	Ganga (indiferenciada)	gg
gk	Greenockite	Greenockita	gg
gld	Glaucodot	Glaucodoto	+
gln	Glauconite	Glauconita, Glauconia	gg
gn	Galena	Galena	+
gnb	Galenobismuthite	Galenobismutina	
gos	Gossan	Gossan	cf. lim
gra	Graphite	Grafito	+
grg	Greigite	Greigita	cf. ln
grt	Garnet (Group)	Granate (Grupo)	Gg
gth	Goethite	Goethita	+
gtt	Gratonite	Gratonita	+
gud (gd)	Gudmundite	Gudmundita	+
hb	Hornblende	Hornblenda	gg
Hg	Mercury (quicksilver)	Mercurio	cf. cin
hol	Hollandite	Hollandita	cf. psi
hrb	Horobetsuite	Horobetsuita	cf. bm
hs	Hausmannite	Hausmanita	
hss	Hessite	Hessita	
ht	Hematite	Hematites	+
htg	Heterogenite	Heterogenita	
hub	Hübnerite	Hübnerita	cf. wf
il	Ilmenite	Ilmenita	+
ili	Illite	Illita	gg
ilr	Ilmenorutile	Ilmenorutilo	+
Ir	Iridium	Iridio	+
iros	Iridosmine (discredited)	Osmiridio (desacreditado)	cf. Os, Ir
isc	Isocubanite	Isocubanita	+
ix	Ixiolite	Ixiolita	cf. rt (F 13-14)
jac	Jacobsite	Jacobsita	+
jar	Jarosite	Jarosita	cf. lim
jas	Jasper	Jaspe	gg
jd	Jordanite	Jordanita	+
jl	Jalpaite	Jalpaíta	cf. Sulf
jm	Jamesonite	Jamesonita	+

(continued)

(continued)

Abbreviation	Name (EN)	Nombre (ES)	Observations
ker (km)	Kermesite	Kermesita	+
Kfs	K Feldspar	Feldespato Potásico	gg
ko	Kobellite	Kobellita	
krn	Krennerite	Krennerita	+
ks	Kësterite	Kësterita	cf. stn
kut	Kutinaite	Kutinaita	
lcx	Leucoxeno	Leucoxeno	cf. tit, ana (gg)
ldw	Ludwigite	Ludwigita	cf. von
lgs	Langisite	Langisita	
lim, lm	Limonite	Limonita	+
ln	Linnaeite	Linneíta	+
lo	Löllingite	Löllingita	+
lor	Lorandite	Lorandita	
lpc	Lepidocrocite	Lepidocrocita	+
lph	Lithiophorite	Litioforita	
lrt	Laurite	Laurita	cf. erl
lz	Luzonite	Luzonita	+
M, min, mnl	Mineral (undifferentiated)	Mineral (indiferenciado)	
m1, m2, …	M not identified	M no identificado	
mag	Magnesite	Magnesita	cf. Carb, calc
mau	Maucherite	Maucherita	+
mc	Marcasite	Marcasita	+
mch	Malachite	Malaquita	+
mcin	Metacinnabar	Metacinabrio	cf. cin
mck	Mackinawite	Mackinawita	+
mgh	Maghemite	Maghemita	+
Mica	Mica (Group)	Mica (Grupo)	gg
ml	Millerite	Millerita	+
mln	Melnicovite	Melnicovita	cf. py
mlt	Microlite	Microlita	cf. pyc
mng	Manganite	Manganita	+
mngn	Manganosite	Manganosite	
mobg	Moschellandsbergite	Moschellandsbergita	cf. am
mol	Molybdenite	Molibdenita	+
msbt	Metastibnite	Metastibnita	cf. sbt
mt (mg)	Magnetite	Magnetita	+
mv, mus	Muscovite	Moscovita	gg
my	Miargyrite	Miargirita	+
mz	Monazite	Monazita	+
nau	Naumannite	Naumannita	+
nc (nk)	Nickeline, Niccolite	Niquelita, Niquelina	+
Oanf, Oamp	Orthoamphibole (Group)	Ortoanfíbol (Grupo)	gg
ocm	Oxycalciomicrolite	Oxicalciomicrolita	cf. mlt, bim
Ocre	Ochre, Ocher	Ocres	cf. Óxidos
OM	Organic matter	Materia Orgánica	gg
OpM	Opaque minerals	Minerales Opacos	
Opx	Orthopyroxene (Group)	Ortopiroxeno (Grupo)	gg
or	Orthoclase	Ortoclasa, ortosa	gg
orp, op	Orpiment	Oropimente	+
Os	Osmium	Osmio	+
ov	Olivin (Series)	Olivino (Serie)	gg
pc	Pearceite	Pearceíta	+

(continued)

(continued)

Abbreviation	Name (EN)	Nombre (ES)	Observations
pch	Pitchblende	Pechblenda	cf. urn
Pd	Palladium	Paladio	cf. PGE
pdm	Polydymite	Polidimita	cf. ln
PGE	Platinum Group Elements	Elementos del Grupo del Platino	cf. Pt, Os, Ir
PGM	Platinum Group Minerals	Minerales del Grupo del Platino	cf. spy
Phos	Phosphate	Fosfato	gg
pie	Piemontite	Piemontita	cf. bra
Plag	Plagioclase (Series)	Plagioclasa (Grupo)	gg
plb	Polybasite	Polibasita	+
pn	Pentlandite	Pentlandita	+
po	Pyrrhotite	Pirrotita (Pirrotina)	+
poli	Polianite	Polianita	cf. prl
prg	Pyrargyrite	Pirargirita	+
prl	Pyrolusite	Pirolusita	+
prm	Pararammelsbergite	Pararammelsbergita	+
pru	Proustite	Proustita	+
psi	Psilomelane	Psilomelana	+
Pt	Platinum	Platino	+
pvk	Perovskite	Perovskita	gg
Px, Pyx	Pyroxene (Group)	Piroxeno (Grupo)	gg
py	Pyrite	Pirita	+
pyc (pyl)	Pyrochlore	Pirocloro	+
pymp	Pyromorphite	Piromorfita	gg
pyst	Pyrostilpnite	Pirostilpnita	cf. prg
q (qz, qtz)	Quartz	Cuarzo	gg
rdl	Ramsdellite	Ramsdellita	+
REE (ETR)	Rare Earth Elements	Tierras Raras (Elementos de)	
ren	Renierite	Renierita	+
rhc	Rhodochrosite	Rodocrosita	gg
rhn	Rhodonite	Rodonita	gg
rl	Realgar	Rejalgar	+
rm	Rammelsbergite	Rammelsbergita	+
rob	Robinsonite	Robinsonita	
rom	Romanechite	Romanechita	+
rt	Rutile	Rutilo	+
ruiro	Rutheniridosmine	Ruteniridosmina	cf. Os, Ir
ryn	Rynersonite	Rynersonita	cf. mlt, pyc
S	Sulfur	Azufre	gg
Sb	Antimony	Antimonio	+
sba	Stibarsen	Stibarsen	+
sbl	Stibioluzonite	Estibioluzonita	cf. fm
sbt	Stibnite	Estibina, Antimonita	+
sch	Scheelite	Scheelita	+
sco	Scorodite	Escorodita	gg (cf. asp)
Sel	Selenides	Seleniuros	cf. nau, um
sen	Senarmontite	Senarmontita	cf. ocres Sb
ser	Sericite	Sericita	gg
sf	Safflorite	Safflorita	+
sg	Siegenite	Siegenita	cf. ln
Sic	Silica	Sílice	cf. q
sid (sd)	Siderite	Siderita	cf. Carb

(continued)

(continued)

Abbreviation	Name (EN)	Nombre (ES)	Observations
sk	Skutterudite	Skutterudita	+
sl	sensu lato	sensu lato	
Slc	Silicate (Class)	Silicato (Clase)	gg
smal	Smaltite (≈ sk)	Esmaltina	cf. sk
sms	Smithsonite	Smithsonita	gg
smtt	Smithite	Smithita	cf. Ss
snr	Schachnerite	Schachnerita	cf. am, F stm8
sp	Sphalerite	Esfalerita	+
spk	Spionkopite (Blaubl. cv)	Spionkopita	cf. cv and Fig. 43.1a, b
spn	Spinel	Espinela	+
spy	Sperrylite	Sperrylita	+
srp	Serpentine	Serpentina	gg
Ss	Sulfosalt	Sulfosal	
stb	Sternbergite	Sternbergita	+
stm	Stromeyerite	Stromeyerita	+
stn	Stannite	Estannita	+
stnd	Stannoidite	Estannoidita	
stp	Stephanite	Stephanita	+
Sulf	Sulfide	Sulfuro	
sx	Souxite	Souxita	cf. cst
syl (sv)	Sylvanite	Silvanita	cf. Tel
tan (tt)	Tantalite	Tantalita	cf. cbt
td	Tetrahedrite	Tetraedrita (Tetrahedrita)	+
tdy	Tedradymite	Tetradimita	cf. Tel
Te	Tellurium	Teluro	+
Tel (Tell)	Telluride (Group)	Telururo (Grupo)	
tit (sph)	Titanite	Titanita	+
tlk	Talnakhita	Talnakhita	cf. ccp
tn	Tennantite	Tennantita	cf. td
tno	Tenorite	Tenorita	
tod	Todorokite	Todorokita	
tp (tap)	Tapiolite	Tapiolita	+
tr (tro)	Troilite	Troilita	cf. po, Fe
tur	Tourmaline	Turmalina	gg
ul	Ullmannite	Ullmannita	+
ulv	Ulvöspinel (Ulvite)	Ulvöespinela (Ulvita)	+
um	Umangite	Umangita	+
urn	Uraninite	Uraninita	+
vae	Vaesite	Vaesita	cf. ctt
val	Valentinite	Valentinita	cf. ocres Sb
van	Vanadinite	Vanadinita	+
van	Vanadinite	Vanadinita	gg
vio	Violarite	Violarita	cf. ln
vll	Valleriite	Valleriíta	+
von	Vonsenite	Vonsenita	gg
wf	Wolframite	Wolframita	+
wil	Willemite	Willemita	cf. Slc
wz	Wurtzite	Wurtzita	+
xan	Xanthoconite	Xantoconita	cf. pru
xtm	Xenotime	Xenotimo	cf. Phos
Zeo	Zeolite (Group)	Zeolita (Grupo)	gg

(continued)

(continued)

Abbreviation	Name (EN)	Nombre (ES)	Observations
zir	Zircon	Circón (Zircón)	gg
zk	Zinkenite	Zinkenita	cf. Ss
znc	Zincite	Zincita	gg (cf. fkl)

Other names	Otros nombres	Observations
Allemontite	Allemontita	cf. sba
Antimonite	Antimonita	cf. sbt
Blende	Blenda	cf. sp
Chalcopyrrhotite	Calcopirrotita	cf. isc
Hollandite	Holandita	cf. rom, psi
Martite	Martita	cf. ht, mt
Mispickel	Mispíquel	cf. asp
Polianite	Polianita	cf. prl
Schalenblende	–	cf. sp
Sphene (see: Titanite)	Esfena (ver: Titanita)	cf. tit
Steel	Acero	cf. Fe
Strüverite	Struverita	cf. rt
Ulvite	Ulvita	cf. ulv

Annex 1.1.m All Described Minerals and their Symbols or Abbreviations and Comments

Name (EN)	Abbreviation	Nombre (ES)	Observations
Acanthite	ac	Acantita	+
Adularia	adu	Adularia	gg
Aguilarite	agl	Aguilarita	+
Aikinite	aik	Aikinita	cf. bm
Alabandite	alb	Alabandita	+
Albite	ab	Albita	gg
Allargentum	aag	Allargentum	+
Alloclasite	alc	Aloclasita	cf. cob
Altaite	alt	Altaíta	
Amalgam	am (amal)	Amalgama	+
Amblygonite	amb	Ambligonita	cf. Phos
Amphibole (Group)	Anf, Amp	Anfíbol (Grupo)	gg
Anatase	ana	Anatasa	+
Anglesite	ang	Anglesita	gg
Anhydrite	anh	Anhidrita	gg
Anilite	anil	Anilita	cf. cc, dg
Ankerite	ank	Ankerita	gg
Annabergite	anbg	Annabergita	gg
Anorthite	an	Anortita	gg
Antimony	Sb	Antimonio	+
Apatite	ap	Apatito	gg
Argentite	arg	Argentita	+
Argentopyrite	agp	Argentopirita	+

(continued)

(continued)

Name (EN)	Abbreviation	Nombre (ES)	Observations
Arsenic	As	Arsénico	+
Arsenopyrite	asp	Arsenopirita	+
Asbolane	asb	Asbolana	
Aurian silver Ag–Au	*ausi*	*Plata áurea Ag–Au*	cf. *el*
Azurite	az	Azurita	+
Baddeleyite	bad	Baddeleyita	+
Barite	bar	Barita	gg
Bastnäsite	bst (bas)	Bastnäsita	+
Bauxite	baux	Bauxita	
Berthierite	brt	Berthierita	+
Berzelianite	bzl	Berzelianita	cf. dg
Biotite	bio	Biotita	gg
Bismite	bim	Bismita	gg
Bismuth	Bi	Bismuto	+
Bismuthinite	bm	Bismutinita	+
Bismutite	bis	Bismutita	gg
Bornite	bn	Bornita	+
Boulangerite	bl	Boulangerita	+
Bournonite	bnn	Bournonita	+
Braggite	brg	Braggita	cf. erl
Braunite	bra (br)	Braunita	+
Bravoite	brv	Bravoíta	+
Breithauptite	bth	Breithauptita	+
Briartite	bri	Briartita	cf. ren
Brookite	bk	Brookita	gg
Bukovite	buk	Bukovita	
Calaverite	clv	Calaverita	+
Calciotantite	cta	Calciotantita	cf. mlt, pyc
Calcite	calc	Calcita	gg
Carbonate (Class)	Carb	Carbonato (Clase)	
Carrollite	car	Carrolita	cf. ln
Cassiterite	cst (cs)	Casiterita	+
Cattierite	ctt	Cattierita	cf. brv
Cerargyrite	crg	Cerargirita	cf. cag
Cerussite	crs	Cerusita	cf. gn
Cervantite	cerv	Cervantita	cf. ocres Sb
Chalcedony	cld	Calcedonia	cf. q
Chalcocite	cc	Calcocita = Calcosina	+
Chalcopyrite	ccp	Calcopirita	+
Chert	ch	Chert	gg
Chloanthite (disc. IMA)	cloa	Cloantita	cf. sk
Chlorargyrite	cag	Clorargirita	+
Chlorite (Series)	Chl	Clorita (Serie)	gg
Chromite	chr	Cromita	+
Chrysocolla	crc	Crisocola	+
Cinnabar	cin	Cinabrio	+
Clausthalite	clt (cla)	Clausthalita	+
Clay	Arc	Arcilla	gg
Clay minerals (Group)	CM, MArc	Minerales Arcillosos (Grupo)	gg
Coal—organic matter	COM	Carbón—materia orgánica	cf. gg and gra

(continued)

(continued)

Name (EN)	Abbreviation	Nombre (ES)	Observations
Cobaltite	cob	Cobaltita	+
Cohenite	coh	Cohenita	cf. Fe
Coloradoite	colo (cld)	Coloradoíta	+
Columbite	cb	Columbita	cf. cbt
Columbotantalite	cbt	Columbotantalita	+
Cooperite	cpr	Cooperita	cf. erl
Copper	Cu	Cobre	+
Coronadite	crn	Coronadita	cf. rom, psi
Corundum	cor	Corindón	gg
Cosalite	cos	Cosalita	+
Covellite	cv	Covellita	+
Cryptomelane	cry	Criptomelana	+
Cubanite	cbn	Cubanita	+
Cuprite	cup	Cuprita	+
Cylindrite	cyl	Cilindrita	cf. F 41.04
Digenite	dg	Digenita	+
Djurleite	dju (dj)	Djurleíta	cf. cc, dg
Dolomite	dol	Dolomita	gg
Domeykite	dom	Domeiquita	
Dyscrasite	dy	Discrasita	+
Electrum	el	Electrum	+
Enargite	en	Enargita	+
Epidote (Group)	Ep	Epidota (Grupo)	gg
Erlichmanite	erl	Erlichmanita	+
Erythrite	ery	Eritrina	gg
Eugenite	eug	Eugenita	cf. am, aag
Euxenite	eux	Euxenita	cf. REE
Falkmanite	flk	Falkmanita	cf. bl
Famatinite	fm	Famatinita	+
Feldspar (Group)	Fs, Felds	Feldespato (Grupo)	gg
Ferberite	fb	Ferberita	cf. wf
Ferroselite	fse	Ferroselita	cf. clt, um
Fletcherite	fle	Fletcherita	cf. ln
Fluorite	fl	Fluorita	gg
Franklinite	fkl (fk)	Franklinita	+
Freibergite	fbg	Freibergita	cf. td
Freieslebenite	flb	Freieslebenita	+
Galena	gn	Galena	+
Galenobismuthite	gnb	Galenobismutina	
Gallite	gal	Galita	
Gangue (undif.)	gg	Ganga (indiferenciada)	gg
Garnet (Group)	grt	Granate (Grupo)	gg
Geocronite	gc	Geocronita	+
Germanite	ger	Germanita	cf. ren
Gersdorffite	gf	Gersdorffita	+
Glaucodot	gld	Glaucodoto	+
Glauconite	gln	Glauconita, Glauconia	gg
Goethite	gth	Goethita	+
Gold	Au	Oro	+
Gossan	gos	Gossan	cf. lim

(continued)

(continued)

Name (EN)	Abbreviation	Nombre (ES)	Observations
Graphite	gra	Grafito	+
Gratonite	gtt	Gratonita	+
Greenockite	gk	Greenockita	gg
Greigite	grg	Greigita	cf. ln
Gudmundite	gud (gd)	Gudmundita	+
Hausmannite	hs	Hausmanita	
Hematite	ht	Hematites	+
Hessite	hss	Hessita	
Heterogenite	htg	Heterogenita	
Hollandite	hol	Hollandita	cf. psi
Hornblende	hb	Hornblenda	gg
Horobetsuite	hrb	Horobetsuita	cf. bm
Hübnerite	hub	Hübnerita	cf. wf
Illite	ili	Illita	gg
Ilmenite	il	Ilmenita	+
Ilmenorutile	ilr	Ilmenorutilo	+
Iridium	Ir	Iridio	+
Iridosmine (discredited)	iros	Osmiridio (desacreditado)	cf. Os, Ir
Iron	Fe	Hierro	+
Isocubanite	isc	Isocubanita	+
Ixiolite	ix	Ixiolita	cf. rt (F 13-14)
Jacobsite	jac	Jacobsita	+
Jalpaite	jl	Jalpaíta	cf. Sulf
Jamesonite	jm	Jamesonita	+
Jarosite	jar	Jarosita	cf. lim
Jasper	jas	Jaspe	gg
Jordanite	jd	Jordanita	+
K Feldspar	Kfs	Feldespato Potásico	gg
Kermesite	ker (km)	Kermesita	+
Këesterite	ks	Këesterita	cf. stn
Kobellite	ko	Kobellita	
Krennerite	krn	Krennerita	+
Kutinaite	kut	Kutinaita	
Langisite	lgs	Langisita	
Laurite	lrt	Laurita	cf. erl
Lepidocrocite	lpc	Lepidocrocita	+
Leucoxeno	lcx	Leucoxeno	cf. ana (gg)
Limonite	lim, lm	Limonita	+
Linnaeite	ln	Linneíta	+
Lithiophorite	lph	Litioforita	
Löllingite	lo	Löllingita	+
Lorandite	lor	Lorandita	
Ludwigite	ldw	Ludwigita	cf. von
Luzonite	lz	Luzonita	+
M not identified	m1, m2	M no identificado	
Mackinawite	mck	Mackinawita	+
Maghemite	mgh	Maghemita	+
Magnesite	mag	Magnesita	cf. Carb, calc
Magnetite	mt (mg)	Magnetita	+
Malachite	mch	Malaquita	+

(continued)

(continued)

Name (EN)	Abbreviation	Nombre (ES)	Observations
Manganite	mng	Manganita	+
Manganosite	mngn	Manganosite	
Marcasite	mc	Marcasita	+
Maucherite	mau	Maucherita	+
Melnicovite	mln	Melnicovita	cf. py
Mercury (quicksilver)	Hg	Mercurio	cf. cin
Metacinnabar	mcin	Metacinabrio	cf. cin
Metastibnite	msbt	Metastibnita	cf. sbt
Miargyrite	my	Miargirita	+
Mica (Group)	Mica	Mica (Grupo)	gg
Microlite	mlt	Microlita	cf. pyc
Millerite	ml	Millerita	+
Mineral (undif.)	M, min, mnl	Mineral (indiferenciado)	
Molybdenite	mol	Molibdenita	+
Monazite	mz	Monazita	+
Moschellandsbergite	mobg	Moschellandsbergita	cf. am
Muscovite	mv, mus	Moscovita	gg
Naumannite	nau	Naumannita	+
Nickeline, Niccolite	nc (nk)	Niquelita, Niquelina	+
Ochre, Ocher	Ocre	Ocres	cf. Óxidos
Olivin (Series)	ov	Olivino (Serie)	gg
Opaque minerals	OpM	Minerales Opacos	
Organic matter	OM	Materia Orgánica	gg
Orpiment	orp, op	Oropimente	+
Orthoamphibole (Group)	Oanf, Oamp	Ortoanfíbol (Grupo)	gg
Orthoclase	or	Ortoclasa, ortosa	gg
Orthopyroxene (Group)	Opx	Ortopiroxeno (Grupo)	gg
Osmium	Os	Osmio	+
Oxycalciomicrolite	ocm	Oxicalciomicrolita	cf. mlt, bim
Palladium	Pd	Paladio	cf. PGE
Pararammelsbergite	prm	Pararammelsbergita	+
Pearceite	pc	Pearceíta	+
Pentlandite	pn	Pentlandita	+
Perovskite	pvk	Perovskita	gg
Phosphate	Phos	Fosfato	gg
Phyllosilicates (Subclass)	FSlc, PSlc	Filosilicatos (Subclase)	gg
Piemontite	pie	Piemontita	cf. bra
Pitchblende	pch	Pechblenda	cf. urn
Plagioclase (Series)	Plag	Plagioclasa (Grupo)	gg
Platinum	Pt	Platino	+
Platinum Group Elements	PGE	Elem. Grupo del Platino	cf. Pt, Os, Ir
Platinum Group Minerals	PGM	Minerales Grupo del Platino	cf. spy
Polianite	poli	Polianita	cf. prl
Polybasite	plb	Polibasita	+
Polydymite	pdm	Polidimita	cf. ln
Proustite	pru	Proustita	+
Psilomelane	psi	Psilomelana	+
Pyrargyrite	prg	Pirargirita	+
Pyrite	py	Pirita	+
Pyrochlore	pyc (pyl)	Pirocloro	+

(continued)

(continued)

Name (EN)	Abbreviation	Nombre (ES)	Observations
Pyrolusite	prl	Pirolusita	+
Pyromorphite	pymp	Piromorfita	gg
Pyrostilpnite	pyst	Pirostilpnita	cf. prg
Pyroxene (Group)	Px, Pyx	Piroxeno (Grupo)	gg
Pyrrhotite	po	Pirrotita (Pirrotina)	+
Quartz	q (qz, qtz)	Cuarzo	gg
Rammelsbergite	rm	Rammelsbergita	+
Ramsdellite	rdl	Ramsdellita	+
Rare Earth Elements	REE (ETR)	Tierras Raras (Elementos de)	
Realgar	rl	Rejalgar	+
Renierite	ren	Renierita	+
Rhodochrosite	rhc	Rodocrosita	gg
Rhodonite	rhn	Rodonita	gg
Robinsonite	rob	Robinsonita	
Romanechite	rom	Romanechita	+
Rutheniridosmine	ruiro	Ruteniridosmina	cf. Os, Ir
Rutile	rt	Rutilo	+
Rynersonite	ryn	Rynersonita	cf. mlt, pyc
Safflorite	sf	Safflorita	+
Schachnerite	snr	Schachnerita	cf. am, F stm8
Scheelite	sch	Scheelita	+
Scorodite	sco	Escorodita	gg (cf. asp)
Selenides	Sel	Seleniuros	cf. nau, um
Senarmontite	sen	Senarmontita	cf. ocres Sb
sensu lato	*sl*	sensu lato	
Sericite	ser	Sericita	gg
Serpentine	srp	Serpentina	gg
Siderite	sid (sd)	Siderita	cf. Carb
Siegenite	sg	Siegenita	cf. ln
Silica	Sic	Sílice	cf. q
Silicate (Class)	Slc	Silicato (Clase)	gg
Silver	Ag	Plata	+
Skutterudite	sk	Skutterudita	+
Smaltite (≈ sk)	smal	Esmaltina	cf. sk
Smithite	smtt	Smithita	cf. Ss
Smithsonite	sms	Smithsonita	gg
Souxite	sx	Souxita	cf. cst
Sperrylite	spy	Sperrylita	+
Sphalerite	sp	Esfalerita	+
Spinel	spn	Espinela	+
Spionkopite (Blaubl. cv)	spk	Spionkopita	cf. cv and Fig. 43.1a, b
Stannite	stn	Estannita	+
Stannoidite	stnd	Estannoidita	
Stephanite	stp	Stephanita	+
Sternbergite	stb	Sternbergita	+
Stibarsen	sba	Stibarsen	+
Stibioluzonite	sbl	Estibioluzonita	cf. fm
Stibnite	sbt	Estibina, Antimonita	+
Stromeyerite	stm	Stromeyerita	+
Sulfide	Sulf	Sulfuro	

(continued)

(continued)

Name (EN)	Abbreviation	Nombre (ES)	Observations
Sulfur	S	Azufre	gg
Sulfosalt	Ss	Sulfosal	
Sylvanite	syl (sv)	Silvanita	cf. Tel
Talnakhita	tlk	Talnakhita	cf. ccp
Tantalite	tan (tt)	Tantalita	cf. cbt
Tapiolite	tp (tap)	Tapiolita	+
Tedradymite	tdy	Tetradimita	cf. Tel
Telluride (Group)	Tel (Tell)	Telururo (Grupo)	
Tellurium	Te	Teluro	+
Tennantite	tn	Tennantita	cf. td
Tenorite	tno	Tenorita	
Tetrahedrite	td	Tetraedrita (Tetrahedrita)	+
Titanite	tit (sph)	Titanita	+
Todorokite	tod	Todorokita	
Tourmaline	tur	Turmalina	gg
Troilite	tr (tro)	Troilita	cf. po, Fe
Ullmannite	ul	Ullmannita	+
Ulvöspinel (Ulvite)	ulv	Ulvöespinela (Ulvita)	+
Umangite	um	Umangita	+
Uraninite	urn	Uraninita	+
Vaesite	vae	Vaesita	cf. ctt
Valentinite	val	Valentinita	cf. ocres Sb
Valleriite	vll	Valleriíta	+
Vanadinite	van	Vanadinita	+
Vanadinite	van	Vanadinita	gg
Violarite	vio	Violarita	cf. ln
Vonsenite	von	Vonsenita	gg
Willemite	wil	Willemita	cf. Slc
Wolframite	wf	Wolframita	+
Wurtzite	wz	Wurtzita	+
Xanthoconite	xan	Xantoconita	cf. pru
Xenotime	xtm	Xenotimo	cf. Phos
Zeolite (Group)	Zeo	Zeolita (Grupo)	gg
Zincite	znc	Zincita	gg (cf. fkl)
Zinkenite	zk	Zinkenita	cf. Ss
Zircon	zir	Circón (Zircón)	gg

Other names	Otros nombres	Observations
Allemontite	Allemontita	cf. sba
Antimonite	Antimonita	cf. sbt
Blende	Blenda	cf. sp
Chalcopyrrhotite	Calcopirrotita	cf. isc
Hollandite	Holandita	cf. rom, psi
Martite	Martita	cf. ht, mt
Mispickel	Mispíquel	cf. asp
Polianite	Polianita	cf. prl
Schalenblende	–	cf. sp
Sphene (see: Titanite)	Esfena (ver: Titanita)	cf. tit
Steel	Acero	cf. Fe
Strüverite	Struverita	cf. rt
Ulvite	Ulvita	cf. ulv

Annex 1.2.a ORE Mineral Symbols or Abbreviations and Comments

Abbreviation	Name	Nombre	Observations	Priority
aag	*Allargentum*	*Allargentum*	+	2
ac	**Acanthite**	**Acantita**	+	**1**
Ag	**Silver**	**Plata**	+	**1**
agl	Aguilarite	Aguilarita	+	
agp	*Argentopyrite*	*Argentopirita*	+	2
aik	Aikinite	Aikinita	cf. bm	
alb	*Alabandite*	*Alabandita*	+	2
alc	Alloclasite	Aloclasita	cf. cob	
alt	Altaite	Altaíta		
am (amal)	*Amalgam*	*Amalgama*	+	2
ana	Anatase	Anatasa	+	
anil	Anilite	Anilita	cf. cc, dg	
arg	Argentite	Argentita	+	
As	*Arsenic*	*Arsénico*	+	2
asb	Asbolane	Asbolana		
asp	**Arsenopyrite**	**Arsenopirita**	+	**1**
Au	**Gold**	**Oro**	+	**1**
ausi	*Aurian silver Ag–Au*	*Plata áurea Ag–Au*	cf. *el*	2
az	*Azurite*	*Azurita*	+	2
bad	Baddeleyite	Baddeleyita	+	
baux	Bauxite	Bauxita	Mezcla	
Bi	*Bismuth*	*Bismuto*	+	2
bl	*Boulangerite*	*Boulangerita*	+	2
bm	*Bismuthinite*	*Bismutinita*	+	2
bn	**Bornite**	**Bornita**	+	**1**
bnn	Bournonite	Bournonita	+	
bra (br)	*Braunite*	*Braunita*	+	2
brg	Braggite	Braggita	cf. erl F 1, 3	
bri	Briartite	Briartita	cf. ren	
brt	*Berthierite*	*Berthierita*	+	2
brv	*Bravoite*	*Bravoíta*	+	2
bst (bas)	**Bastnäsite**	**Bastnäsita**	+	**1**
bth	*Breithauptite*	*Breithauptita*	+	2
buk	Bukovite	Bukovita		
bzl	Berzelianite	Berzelianita	cf. dg	
cag	*Chlorargyrite*	*Clorargirita*	+	2
car	Carrollite	Carrolita	cf. ln	
cb	Columbite	Columbita	cf. cbt	
cbn	*Cubanite*	*Cubanita*	+	2
cbt	**Columbotantalite**	**Columbotantalita**	+	**1**
cc	**Chalcocite**	**Calcocita = Calcosina**	+	**1**
ccp	**Chalcopyrite**	**Calcopirita**	+	**1**
chr	**Chromite**	**Cromita**	+	**1**
cin	**Cinnabar**	**Cinabrio**	+	**1**

(continued)

(continued)

Abbreviation	Name	Nombre	Observations	Priority
cloa	Chloanthite (disc. IMA)	Cloantita	cf. sk	
clt (cla)	Clausthalite	Clausthalita	+	
clv	*Calaverite*	*Calaverita*	+	*2*
cob	**Cobaltite**	**Cobaltita**	**+**	**1**
coh (chn)	*Cohenite*	*Cohenita*	cf. *Fe*	*2*
colo (cld)	Coloradoite	Coloradoíta	+	
COM	Coal—organic matter	Carbón—materia orgánica	cf. gra and gg	
cos	Cosalite	Cosalita	+	*2*
cpr (coo)	Cooperite	Cooperita	cf. erl F 1, 3	
crc	Chrysocolla	Crisocola	+	
crg	Cerargyrite	Cerargirita	cf. cag	
crn	Coronadite	Coronadita	cf. rom, psi	
cry	*Cryptomelane*	*Criptomelana*	+	*2*
cst (cs)	**Cassiterite**	**Casiterita**	**+**	**1**
cta	Calciotantite	Calciotantita	cf. mlt, pyc	
ctt	Cattierite	Cattierita	cf. brv	
Cu	**Copper**	**Cobre**	**+**	**1**
cup	**Cuprite**	**Cuprita**	**+**	**1**
cv	**Covellite**	**Covellita**	**+**	**1**
cyl	Cylindrite	Cilindrita	cf. F 41.04	
dg	**Digenite**	**Digenita**	**+**	**1**
dju (dj)	Djurleite	Djurleíta	cf. cc, dg	
dom	Domeykite	Domeiquita		
dy	*Dyscrasite*	*Discrasita*	+	*2*
el	*Electrum*	*Electrum*	+	*2*
en	*Enargite*	*Enargita*	+	*2*
erl	Erlichmanite	Erlichmanita	+	
eug	Eugenite	Eugenita	cf. am, aag	
eux	Euxenite	Euxenita	cf. REE	
fb	Ferberite	Ferberita	cf. wf	
fbg	Freibergite	Freibergita	cf. td	
Fe	**Iron**	**Hierro**	**+**	**1**
fk (fkl)	*Franklinite*	*Franklinita*	+	*2*
flb	Freieslebenite	Freieslebenita	+	
fle	Fletcherite	Fletcherita	cf. ln	
flk	Falkmanite	Falkmanita	cf. bl	
fm	*Famatinite*	*Famatinita*	+	*2*
fse	Ferroselite	Ferroselita	cf. clt, um	
gal	Gallite	Galita		
gc	Geocronite	Geocronita	+	
ger	Germanite	Germanita	cf. ren	
gf	*Gersdorffite*	*Gersdorffita*	+	*2*
gg	Gangue (undiff.)	Ganga (indiferenciada)		
gk	Greenockite	Greenockita	cf. Sulf	
gld	*Glaucodot*	*Glaucodoto*	+	*2*
gn	**Galena**	**Galena**	**+**	**1**
gnb	Galenobismuthite	Galenobismutina		
gos	Gossan	Gossan	cf. lim	
gra	**Graphite**	**Grafito**	**+**	**1**
grg	Greigite	Greigita	cf. ln	

(continued)

(continued)

Abbreviation	Name	Nombre	Observations	Priority
gth	**Goethite**	**Goethita**	+	**1**
gtt	Gratonite	Gratonita	+	
gud (gd)	*Gudmundite*	*Gudmundita*	+	*2*
Hg	Mercury (quicksilver)	Mercurio	cf. cin	
hol	Hollandite	Hollandita	cf. psi	
hrb	Horobetsuite	Horobetsuita	cf. bm	
hs	Hausmannite	Hausmanita		
hss	Hessite	Hessita		
ht	**Hematite**	**Hematites**	+	**1**
htg	Heterogenite	Heterogenita		
hub	Hübnerite	Hübnerita	cf. wf	
il	**Ilmenite**	**Ilmenita**	+	**1**
ilr	*Ilmenorutile*	*Ilmenorutilo*	+	*2*
Ir	*Iridium*	*Iridio*	+	*2*
iros	Iridosmine (discredited)	Osmiridio (desacreditado)	cf. Os, Ir	
isc	*Isocubanite*	*Isocubanita*	+	*2*
ix	Ixiolite	Ixiolita	cf. rt (F 13-14)	
jac	Jacobsite	Jacobsita	+	
jd	Jordanite	Jordanita	+	
jl	Jalpaite	Jalpaíta		
jm	Jamesonite	Jamesonita	+	2
ker (km)	Kermesite	Kermesita	+	
ko	Kobellite	Kobellita		
krn	Krennerite	Krennerita	+	
ks	Kësterite (kösterite)	Kësterita (kesterita)	cf. stn	
kut	Kutinaite	Kutinaita		
lgs	Langisite	Langisita	Figures 92.4a, b and 96.3a, b	
lim, lm	**Limonite**	**Limonita**	+	**1**
ln	**Linnaeite**	**Linneíta**	+	**1**
lo	*Löllingite*	*Löllingita*	+	*2*
lor	Lorandite	Lorandita		
lpc	*Lepidocrocite*	*Lepidocrocita*	+	*2*
lph	Lithiophorite	Litioforita		
lrt (lau)	Laurite	Laurita	cf. erl	
lz	*Luzonite*	*Luzonita*	+	*2*
M, min, mnl	Mineral (undiff.)	Mineral (indiferenc.)		
m1, m2	M not identified	M no identificado		
mau	*Maucherite*	*Maucherita*	+	*2*
mc	**Marcasite**	**Marcasita**	+	**1**
mch	*Malachite*	*Malaquita*	+	*2*
mcin	Metacinnabar	Metacinabrio	cf. cin	
mck	*Mackinawite*	*Mackinawita*	+	*2*
mgh	*Maghemite*	*Maghemita*	+	*2*
ml	**Millerite**	**Millerita**	+	**1**
mln	Melnicovite	Melnicovita	cf. py	
mlt	Microlite	Microlita	cf. pyc	
mng	**Manganite**	**Manganita**	+	**1**
mngn	Manganosite	Manganosite		
mobg	Moschellandsbergite	Moschellandsbergita	cf. am	
mol	**Molybdenite**	**Molibdenita**	+	**1**

(continued)

(continued)

Abbreviation	Name	Nombre	Observations	Priority
msbt	Metastibnite	Metaestibina	cf. sbt	
mt (mg)	**Magnetite**	**Magnetita**	+	**1**
my	**Miargyrite**	**Miargirita**	+	**1**
mz	*Monazite*	*Monazita*	+	*2*
nau	Naumannite	Naumannita	+	
nc (nk)	**Nickeline, Niccolite**	**Niquelita, Niquelina**	+	**1**
ocm	Oxycalciomicrolite	Oxicalciomicrolita	cf. mlt, bim	
OpM	Opaque Minerals	Minerales Opacos		
orp, op	*Orpiment*	*Oropimente*	+	*2*
Os	*Osmium*	*Osmio*	+	*2*
pc	*Pearcite*	*Pearceíta*	+	*2*
pch	Pitchblende	Pechblenda	cf. urn	
Pd	Palladium	Paladio	cf. PGE	
pdm	Polydymite	Polidimita	cf. ln	
PGE	Platinum Group Elements	Elementos Gr. Platino	cf. Pt, Os, Ir	
PGM	*Platinum Group Minerals*	*Minls del Grupo del Pt*	cf. *spy*	*2*
Phos	Phosphate	Fosfato		
plb	*Polybasite*	*Polibasita*	+	*2*
pn	**Pentlandite**	**Pentlandita**	+	**1**
po	**Pyrrhotite**	**Pirrotita (Pirrotina)**	+	**1**
poli	Polianite	Polianita	cf. prl	
prg	**Pyrargyrite**	**Pirargirita**	+	**1**
prl	**Pyrolusite**	**Pirolusita**	+	**1**
prm	*Pararammelsbergite*	*Pararamelsbergita*	+	*2*
pru	**Proustite**	**Proustita**	+	**1**
psi	**Psilomelane**	**Psilomelana**	+	**1**
Pt	**Platinum**	**Platino**	+	**1**
py	**Pyrite**	**Pirita**	+	**1**
pyc (pyl)	*Pyrochlore*	*Pirocloro*	+	*2*
pyst	Pyrostilpnite	Pirostilpnita	cf. prg	
rdl	Ramsdellite	Ramsdellita	+	
REE (ETR)	Rare Earth Elements	Tierras Raras (Elem)		
ren	*Renierite*	*Renierita*	+	*2*
rl	Realgar	Rejalgar	+	
rm	**Rammelsbergite**	**Rammelsbergita**	+	**1**
rob	Robinsonite	Robinsonita		
rom	**Romanechite**	**Romanechita**	+	**1**
rt	**Rutile**	**Rutilo**	+	**1**
ruiro	Rutheniridosmine	Ruteniridosmina	cf. Os, Ir	
ryn	Rynersonite	Rynersonita	cf. mlt, pyc	
Sb	**Antimony**	**Antimonio**	+	**1**
sba	*Stibarsen*	*Stibarsen*	+	*2*
sbl	Stibioluzonite	Estibioluzonita	cf. fm	
sbt	**Stibnite**	**Estibina, Antimonita**	+	**1**
sch	*Scheelite*	*Scheelita*	+	*2*
Sel	Selenides	Seleniuros	cf. nau, um	
sen	Senarmontite	Senarmontita	cf. Fig. 98.3a, b	cf. ocres Sb
sf	**Safflorite**	**Safflorita**	+	**1**
sg	Siegenite	Siegenita	cf. ln	
sk	**Skutterudite**	**Skutterudita**	+	**1**

(continued)

(continued)

Abbreviation	Name	Nombre	Observations	Priority
smal	Smaltite (≈ sk)	Esmaltina	cf. sk	
smtt	Smithite	Smithita	cf. Ss	
snr	Schachnerite	Schachnerita	cf. F atm8	cf. am
sp	**Sphalerite**	**Esfalerita**	+	**1**
spk	Spionkopite (Blaub. cv)	Spionkopita	cf. cv and Fig. 43.1a, b	
spn	*Spinel*	*Espinela*	+	*2*
spy	*Sperrylite*	*Sperrylita*	+	*2*
Ss	Sulfosalt	Sulfosal		
stb	Sternbergite	Sternbergita	+	
stm	*Stromeyerite*	*Stromeyerita*	+	*2*
stn	*Stannite*	*Estannita*	+	*2*
stnd	Stannoidite	Estannoidita		
stp	*Stephanite*	*Stephanita*	+	*2*
Sulf	Sulfide	Sulfuro		
syl (sv)	Sylvanite	Silvanita	cf. Tel	
tan (tt)	Tantalite	Tantalita	cf. cbt	
td	**Tetrahedrite**	**Tetraedrita/Tetrahed**	+	**1**
tdy	Tedradymite	Tetradimita	cf. Tel	
Te	*Tellurium*	*Teluro*	+	*2*
Tel (Tell)	Telluride (Group)	Telururo (Grupo)		
tit (sph)	*Titanite*	*Titanita*	+	*2*
tlk	Talnakhita	Talnakhita	cf. ccp	
tn	Tennantite	Tennantita	cf. td	
tno	Tenorite	Tenorita		
tod	Todorokite	Todorokita		
tp (tap)	*Tapiolite*	*Tapiolita*	+	*2*
tr (tro)	Troilite	Troilita	cf. po, Fe	
ul	Ullmannite	Ullmannita	+	
ulv	*Ulvöspinel (Ulvite)*	*Ulvöespinela (Ulvita)*	+	
um	*Umangite*	*Umangita*	+	*2*
urn	**Uraninite**	**Uraninita**	+	**1**
vae	Vaesite	Vaesita	cf. brv	
vio	Violarite	Violarita	cf. ln	
vll	*Valleriite*	*Valleriita*	+	*2*
wf	**Wolframite**	**Wolframita**	+	**1**
wz	*Wurtzite*	*Wurtzita*	+	*2*
xan	Xanthoconite	Xantoconita	cf. pru	
zk	Zinkenite	Zinkenita	cf. Ss	

Other names	Otros nombres	Observations	Priority
Allemontite	Allemontita	cf. sba	
Antimonite	Antimonita	cf. sbt	
Blende	Blenda	cf. sp	
Chalcopyrrhotite	Calcopirrotita	cf. isc	
Hollandite	Holandita	cf. rom, psi	
Martite	**Martita**	**cf. ht, mt**	**1**
Mispickel	**Mispíquel**	**cf. asp**	**1**
Polianite	**Polianita**	**cf. prl**	**1**
Schalenblende	**–**	**cf. sp**	**1**
Sphene (see: Titanite)	**Esfena (ver: Titanita)**	**cf. tit**	**1**
Steel	**Acero**	**cf. Fe**	**1**
Strüverite	Struverita	cf. rt	
Ulvite	Ulvita	cf. ulv	

Annex 1.2.m Ore Minerals and their Symbols or Abbreviations Comments

Name (EN)		Nombre (ES)	Observations	Priority
Acanthite	**ac**	**Acantita**	+	**1**
Aguilarite	agl	Aguilarita	+	
Aikinite	aik	Aikinita	cf. bm	
Alabandite	*alb*	*Alabandita*	+	*2*
Allargentum	*aag*	*Allargentum*	+	*2*
Alloclasite	alc	Aloclasita	cf. cob	
Altaite	alt	Altaíta		
Amalgam	*am (amal)*	*Amalgama*	+	*2*
Anatase	ana	Anatasa	+	
Anilite	anil	Anilita	cf. cc, dg	
Antimony	**Sb**	**Antimonio**	+	**1**
Argentite	arg	Argentita	+	
Argentopyrite	*agp*	*Argentopirita*	+	*2*
Arsenic	*As*	*Arsénico*	+	*2*
Arsenopyrite	**asp**	**Arsenopirita**	+	**1**
Asbolane	asb	Asbolana		
Aurian silver Ag–Au	*ausi*	*Plata áurea Ag–Au*	cf. *el*	*2*
Azurite	*az*	*Azurita*	+	*2*
Baddeleyite	bad	Baddeleyita	+	
Bastnäsite	**bst (bas)**	**Bastnäsita**	+	**1**
Bauxite	baux	Bauxita	Mixture	
Berthierite	*brt*	*Berthierita*	+	*2*
Berzelianite	bzl	Berzelianita	cf. dg	
Bismuth	*Bi*	*Bismuto*	+	*2*
Bismuthinite	*bm*	*Bismutinita*	+	*2*
Bornite	**bn**	**Bornita**	+	**1**
Boulangerite	*bl*	*Boulangerita*	+	*2*
Bournonite	bnn	Bournonita	+	
Braggite	brg	Braggita	cf. erl F 1, 3	
Braunite	*bra (br)*	*Braunita*	+	*2*
Bravoite	*brv*	*Bravoíta*	+	*2*
Breithauptite	*bth*	*Breithauptita*	+	*2*
Briartite	bri	Briartita	cf. ren	
Bukovite	buk	Bukovita		
Calaverite	*clv*	*Calaverita*	+	*2*
Calciotantite	cta	Calciotantita	cf. mlt, pyc	
Carrollite	car	Carrolita	cf. ln	
Cassiterite	**cst (cs)**	**Casiterita**	+	**1**
Cattierite	ctt	Cattierita	cf. brv	
Cerargyrite	crg	Cerargirita	cf. cag	
Chalcocite	**cc**	**Calcocita = Calcosina**	+	**1**
Chalcopyrite	**ccp**	**Calcopirita**	+	**1**
Chloanthite (discr.)	cloa	Cloantita	cf. sk	
Chlorargyrite	*cag*	*Clorargirita*	+	*2*
Chromite	**chr**	**Cromita**	+	**1**
Chrysocolla	crc	Crisocola	+	

(continued)

(continued)

Name (EN)		Nombre (ES)	Observations	Priority
Cinnabar	**cin**	**Cinabrio**	+	**1**
Clausthalite	*clt (cla)*	Clausthalita	+	
Coal—organic matter	COM	Carbón—mat. orgánica	cf. gra and gg	
Cobaltite	**cob**	**Cobaltita**	+	**1**
Cohenite	*coh (chn)*	*Cohenita*	*cf. Fe*	*2*
Coloradoite	colo (cld)	Coloradoíta	+	
Columbite	cb	Columbita	cf. cbt	
Columbotantalite	**cbt**	**Columbotantalita**	+	**1**
Cooperite	cpr (coo)	Cooperita	cf. erl F 1, 3	
Copper	**Cu**	**Cobre**	+	**1**
Coronadite	crn	Coronadita	cf. rom, psi	
Cosalite	cos	Cosalita	+	2
Covellite	**cv**	**Covellita**	+	**1**
Cryptomelane	*cry*	*Criptomelana*	+	*2*
Cubanite	*cbn*	*Cubanita*	+	*2*
Cuprite	**cup**	**Cuprita**	+	**1**
Cylindrite	cyl	Cilindrita	cf. F 41.04	
Digenite	**dg**	**Digenita**	+	**1**
Djurleite	dju (dj)	Djurleíta	cf. cc, dg	
Domeykite	dom	Domeiquita		
Dyscrasite	*dy*	*Discrasita*	+	*2*
Electrum	*el*	*Electrum*	+	*2*
Enargite	*en*	*Enargita*	+	*2*
Erlichmanite	erl	Erlichmanita	+	
Eugenite	eug	Eugenita	cf. am, aag	
Euxenite	eux	Euxenita	cf. REE	
Falkmanite	flk	Falkmanita	cf. bl	
Famatinite	*fm*	*Famatinita*	+	*2*
Ferberite	fb	Ferberita	cf. wf	
Ferroselite	fse	Ferroselita	cf. clt, um	
Fletcherite	fle	Fletcherita	cf. ln	
Franklinite	*fk (fkl)*	*Franklinita*	+	*2*
Freibergite	fbg	Freibergita	cf. td	
Freieslebenite	flb	Freieslebenita	+	
Galena	**gn**	**Galena**	+	**1**
Galenobismuthite	gnb	Galenobismutina		
Gallite	gal	Galita		
Gangue (undiff.)	gg	Ganga (indif.)		
Geocronite	gc	Geocronita	+	
Germanite	ger	Germanita	cf. ren	
Gersdorffite	*gf*	*Gersdorffita*	+	*2*
Glaucodot	*gld*	*Glaucodoto*	+	*2*
Goethite	**gth**	**Goethita**	+	**1**
Gold	**Au**	**Oro**	+	**1**
Gossan	gos	Gossan	cf. lim	
Graphite	**gra**	**Grafito**	+	**1**
Gratonite	gtt	Gratonita	+	
Greenockite	gk	Greenockita	cf. Sulf	
Greigite	grg	Greigita	cf. ln	
Gudmundite	*gud (gd)*	*Gudmundita*	+	*2*

(continued)

(continued)

Name (EN)		Nombre (ES)	Observations	Priority
Hausmannite	hs	Hausmanita		
Hematite	**ht**	**Hematites**	+	**1**
Hessite	hss	Hessita		
Heterogenite	htg	Heterogenita		
Hollandite	hol	Hollandita	cf. psi	
Horobetsuite	hrb	Horobetsuita	cf. bm	
Hübnerite	hub	Hübnerita	cf. wf	
Ilmenite	**il**	**Ilmenita**	+	**1**
Ilmenorutile	*ilr*	*Ilmenorutilo*	+	*2*
Iridium	*Ir*	*Iridio*	+	*2*
Iridosmine (discr.)	iros	Osmiridio (desacr.)	cf. Os, Ir	
Iron	**Fe**	**Hierro**	+	**1**
Isocubanite	*isc*	*Isocubanita*	+	*2*
Ixiolite	ix	Ixiolita	cf. rt (F 13-14)	
Jacobsite	jac	Jacobsita	+	
Jalpaite	jl	Jalpaíta		
Jamesonite	jm	Jamesonita	+	*2*
Jordanite	jd	Jordanita	+	
Kermesite	ker (km)	Kermesita	+	
Kësterite (Kösterite)	ks	Kësterita (kesterita)	cf. stn	
Kobellite	ko	Kobellita		
Krennerite	krn	Krennerita	+	
Kutinaite	kut	Kutinaita		
Langisite	lgs	Langisita	Figures 92.4a, b and 96.3a, b	
Laurite	lrt (lau)	Laurita	cf. erl	
Lepidocrocite	*lpc*	*Lepidocrocita*	+	*2*
Limonite	**lim, lm**	**Limonita**	+	**1**
Linnaeite	**ln**	**Linneíta**	+	**1**
Lithiophorite	lph	Litioforita		
Löllingite	*lo*	*Löllingita*	+	*2*
Lorandite	lor	Lorandita		
Luzonite	*lz*	*Luzonita*	+	*2*
M not identified	m1, m2, …	M no identificado		
Mackinawite	*mck*	*Mackinawita*	+	*2*
Maghemite	*mgh*	*Maghemita*	+	*2*
Magnetite	**mt (mg)**	**Magnetita**	+	**1**
Malachite	*mch*	*Malaquita*	+	*2*
Manganite	**mng**	**Manganita**	+	**1**
Manganosite	mngn	Manganosite		
Marcasite	**mc**	**Marcasita**	+	**1**
Maucherite	*mau*	*Maucherita*	+	*2*
Melnicovite	mln	Melnicovita	cf. py	
Mercury (quicksilver)	Hg	Mercurio	cf. cin	
Metacinnabar	mcin	Metacinabrio	cf. cin	
Metastibnite	msbt	Metaestibina	cf. sbt	
Miargyrite	**my**	**Miargirita**	+	**1**
Microlite	mlt	Microlita	cf. pyc	
Millerite	**ml**	**Millerita**	+	**1**
Mineral (undiff.)	M, min, mnl	Mineral (indiferenc.)		
Molybdenite	**mol**	**Molibdenita**	+	**1**

(continued)

(continued)

Name (EN)		Nombre (ES)	Observations	Priority
Monazite	*mz*	*Monazita*	+	*2*
Moschellandsbergite	mobg	Moschellandsbergita	cf. am	
Naumannite	nau	Naumannita	+	
Nickeline, Niccolite	**nc (nk)**	**Niquelita, Niquelina**	+	**1**
Opaque minerals	OpM	Minerales Opacos		
Orpiment	*orp, op*	*Oropimente*	+	*2*
Osmium	*Os*	*Osmio*	+	*2*
Oxycalciomicrolite	ocm	Oxicalciomicrolita	cf. mlt, bim	
Palladium	Pd	Paladio	cf. PGE	
Pararammelsbergite	*prm*	*Pararamelsbergita*	+	*2*
Pearcite	*pc*	*Pearceita*	+	*2*
Pentlandite	**pn**	**Pentlandita**	+	**1**
Phosphate	Phos	Fosfato		
Pitchblende	pch	Pechblenda	cf. urn	
Platinum	**Pt**	**Platino**	+	**1**
Platinum Group Elements	PGE	Elem. Grupo Platino	cf. Pt, Os, Ir	
Platinum Group Minerals	*PGM*	*Minls. Grupo del Pt*	cf. *spy*	*2*
Polianite	poli	Polianita	cf. prl	
Polybasite	*plb*	*Polibasita*	+	*2*
Polydymite	pdm	Polidimita	cf. ln	
Proustite	**pru**	**Proustita**	+	**1**
Psilomelane	**psi**	**Psilomelana**	+	**1**
Pyrargyrite	**prg**	**Pirargirita**	+	**1**
Pyrite	**py**	**Pirita**	+	**1**
Pyrochlore	*pyc (pyl)*	*Piroclоro*	+	*2*
Pyrolusite	**prl**	**Pirolusita**	+	**1**
Pyrostilpnite	pyst	Pirostilpnita	cf. prg	
Pyrrhotite	**po**	**Pirrotita (Pirrotina)**	+	**1**
Rammelsbergite	**rm**	**Rammelsbergita**	+	**1**
Ramsdellite	rdl	Ramsdellita	+	
Rare Earth Elements	REE (ETR)	Tierras Raras (Elem)		
Realgar	rl	Rejalgar	+	
Renierite	*ren*	*Renierita*	+	*2*
Robinsonite	rob	Robinsonita		
Romanechite	**rom**	**Romanechita**	+	**1**
Rutheniridosmine	ruiro	Ruteniridosmina	cf. Os, Ir	
Rutile	**rt**	**Rutilo**	+	**1**
Rynersonite	ryn	Rynersonita	cf. mlt, pyc	
Safflorite	**sf**	**Safflorita**	+	**1**
Schachnerite	snr	Schachnerita	cf. F atm8	cf. am
Scheelite	*sch*	*Scheelita*	+	*2*
Selenides	Sel	Seleniuros	cf. nau, um	
Senarmontite	sen	Senarmontita	cf. Fig. 98.3a, b	cf. Sb ocr
Siegenite	sg	Siegenita	cf. ln	
Silver	**Ag**	**Plata**	+	**1**
Skutterudite	**sk**	**Skutterudita**	+	**1**
Smaltite (≈ sk)	smal	Esmaltina	cf. sk	
Smithite	smtt	Smithita	cf. Ss	
Sperrylite	*spy*	*Sperrylita*	+	*2*
Sphalerite	**sp**	**Esfalerita**	+	**1**

(continued)

(continued)

Name (EN)		Nombre (ES)	Observations	Priority
Spinel	*spn*	*Espinela*	+	*2*
Spionkopite (Blub. cv)	spk	Spionkopita	cf. cv, Fig. 43.1a, b	
Stannite	*stn*	*Estannita*	+	*2*
Stannoidite	stnd	Estannoidita		
Stephanite	*stp*	*Stephanita*	+	*2*
Sternbergite	stb	Sternbergita	+	
Stibarsen	*sba*	*Stibarsen*	+	*2*
Stibioluzonite	sbl	Estibioluzonita	cf. fm	
Stibnite	**sbt**	**Estibina, Antimonita**	**+**	**1**
Stromeyerite	*stm*	*Stromeyerita*	+	*2*
Sulfide	Sulf	Sulfuro		
Sulfosalt	Ss	Sulfosal		
Sylvanite	syl (sv)	Silvanita	cf. Tel	
Talnakhita	tlk	Talnakhita	cf. ccp	
Tantalite	tan (tt)	Tantalita	cf. cbt	
Tapiolite	*tp (tap)*	*Tapiolita*	+	*2*
Tedradymite	tdy	Tetradimita	cf. Tel	
Telluride (Group)	Tel (Tell)	Telururo (Grupo)		
Tellurium	*Te*	*Teluro*	+	*2*
Tennantite	tn	Tennantita	cf. td	
Tenorite	tno	Tenorita		
Tetrahedrite	**td**	**Tetraedrita o Tetrah**	**+**	**1**
Titanite	*tit (sph)*	*Titanita*	+	*2*
Todorokite	tod	Todorokita		
Troilite	tr (tro)	Troilita	cf. po, Fe	
Ullmannite	ul	Ullmannita	+	
Ulvöspinel (Ulvite)	*ulv*	*Ulvöespinela (Ulvita)*	+	
Umangite	*um*	*Umangita*	+	*2*
Uraninite	**urn**	**Uraninita**	**+**	**1**
Vaesite	vae	Vaesita	cf. brv	
Valleriite	*vll*	*Valleriita*	+	*2*
Violarite	vio	Violarita	cf. ln	
Wolframite	**wf**	**Wolframita**	**+**	**1**
Wurtzite	*wz*	*Wurtzita*	+	*2*
Xanthoconite	xan	Xantoconita	cf. pru	
Zinkenite	zk	Zinkenita	cf. Ss	

Other names	Otros nombres	Observations	Priority
Allemontite	Allemontita	cf. sba	
Antimonite	Antimonita	cf. sbt	
Blende	Blenda	cf. sp	
Chalcopyrrhotite	Calcopirrotita	cf. isc	
Hollandite	Holandita	cf. rom, psi	
Martite	**Martita**	**cf. ht, mt**	**1**
Mispickel	**Mispíquel**	**cf. asp**	**1**
Polianite	**Polianita**	**cf. prl**	**1**
Schalenblende	**–**	**cf. sp**	**1**
Sphene (see: Titanite)	**Esfena (ver: Titanita)**	**cf. tit**	**1**
Steel	**Acero**	**cf. Fe**	**1**
Strüverite	Struverita	cf. rt	
Ulvite	Ulvita	cf. ulv	

Annex 1.3.a Gangue Mineral Abbreviations and Comments

Abbreviation	Name (EN)	Nombre (ES)	Observations
ab	Albite	Albita	cf. Slc, Fs
adu	Adularia	Adularia	cf. Slc, Fs
amb	Amblygonite	Ambligonita	cf. Phos
an	Anorthite	Anortita	cf. Slc, Fs
anbg	Annabergite	Annabergita	Arseniatos
Anf, Amp	Amphibole (Group)	Anfíbol (Grupo)	cf. Slc
ang	Anglesite	Anglesita	cf. Sulfatos
anh	Anhydrite	Anhidrita	cf. Sulfatos
ank	Ankerite	Ankerita	cf. Carb
ap	Apatite	Apatito	cf. Phos
Arc	Clay	Arcilla	cf. FSlc
bar	Barite	Barita	cf. Sulfatos
bim	Bismite	Bismita	cf. ocres Bi
bio	Biotite	Biotita	cf. Slc, FSlc
bis	Bismutite	Bismutita	cf. ocres Bi
bk	Brookite	Brookita	Óxidos, cf. lcx
calc	Calcite	Calcita	cf. Carb
Carb	Carbonate (Class)	Carbonato (Clase)	cf. Carb, gg
cerv	Cervantite	Cervantita	cf. ocres Sb
ch	Chert	Chert	cf. Slc, Sic
Chl	Chlorite (Series)	Clorita (Serie)	cf. Slc, FSlc
cld	Chalcedony	calcedonia	cf. q
CM, MArc	Clay minerals	Mins Arcillosos	cf. Slc, FSlc
COM	Coal—organic matter	Carbón—materia orgán	cf. Organ. Mat
cor	Corundum	Corindón	
crc	Chrysocolla	Crisocola	cf. Slc
crs	Cerussite	Cerusita	cf. Carbonatos
dol	Dolomite	Dolomita	cf. Carb
Ep	Epidote (Group)	Epidota (Grupo)	cf. Slc
ery	Erythrite	Eritrina	Arseniatos
fl	Fluorite	Fluorita	cf. Haluros
Fs, Felds	Feldspar (Group)	Feldespato (Grupo)	cf. Slc
FSlc, PSlc	Phyllosilicates (Subclass)	Filosilicatos (Subclase)	cf. Slc
gg	Gangue (undifferenciated)	Ganga (indiferenciada)	
gln	Glauconite	Glauconita, Glauconia	cf. Slc, FSlc
grt	Garnet (Group)	Granate (Grupo)	cf. Slc
hb	Hornblende	Hornblenda	cf. Slc, Inoslc
ili	Illite	Illita	cf. Slc, FSlc
jar	Jarosite	Jarosita	cf. lim
jas	Jasper	Jaspe	cf. q
Kfs	K Feldspar	Feldespato Potásico	cf. Slc, Fs
lcx	Leucoxene	Leucoxeno	cf. tit, ana
ldw	Ludwigite	Ludwigita	cf. von
mag	Magnesite	Magnesita	cf. Carb
Mica	Mica (Group)	Mica (Grupo)	cf. Slc, FSlc
mv, mus	Muscovite	Moscovita	cf. Mica

(continued)

(continued)

Abbreviation	Name (EN)	Nombre (ES)	Observations
Ocre	Ochre, Ocher	Ocres	cf. Óxidos
OM	Organic matter	Materia Orgánica	
or	Orthoclase	Ortoclasa, ortosa	cf. Slc, Fs
ov	Olivine (Series)	Olivino (Serie)	cf. Slc
Phos	Phosphate	Fosfato	gg
pie	Piemontite	Piemontita	cf. bra, Ep
Plag	Plagioclase (Series)	Plagioclasa (Serie)	cf. Slc, Fs
pvk	Perovskite	Perovskita	Óxidos
Px, Pyx	Pyroxene (Group)	Piroxeno (Grupo)	cf. Slc, Inoslc
pymp	Pyromorphite	Piromorfita	cf. Phos
q (qz, qtz)	Quartz	Cuarzo	cf. Slc, Sic
rhc	Rhodochrosite	Rodocrosita	cf. Carb
rhn	Rhodonite	Rodonita	cf. rdl F 1-4
S	Sulfur	Azufre	gg
sco	Scorodite	Escorodita	cf. asp; Arseniato
sen	Senarmontite	Senarmontita	cf. ocres Sb
ser	Sericite	Sericita	cf. Slc, FSlc
Sic	Silica	Sílice	cf. Slc, q
sid (sd)	Siderite	Siderita	cf. Carb
Slc	Silicate (Class)	Silicato (Clase)	gg
sms	Smithsonite	Smithsonita	cf. Carb
srp	Serpentine	Serpentina	cf. Slc, FSlc
sx	Souxite	Souxita	cf. cst
tur	Tourmaline	Turmalina	cf. Slc
val	Valentinite	Valentinita	cf. ocres Sb
van	Vanadinite	Vanadinita	cf. Vanadatos
von	Vonsenite	Vonsenita	cf. Carb-Boratos
wil	Willemite	Willemita	cf. Slc
xtm	Xenotime	Xenotimo	cf. Phos
Zeo	Zeolite (Group)	Zeolita (Grupo)	cf. Slc
zir	Zircon	Circón (Zircón)	cf. Slc
znc	Zincite	Zincita	Óxidos

Other names	Otros nombres	Observaciones
Sphene	Esfena	cf. tit

Annex 1.3.m Gangue Minerals and their Abbreviations and Comments

Name (EN)	Abbreviation	Nombre (ES)	Observations
Adularia	adu	Adularia	cf. Slc, Fs
Albite	ab	Albita	cf. Slc, Fs
Amblygonite	amb	Ambligonita	cf. Phos
Amphibole (Group)	Anf, Amp	Anfíbol (Grupo)	cf. Slc
Anglesite	ang	Anglesita	cf. Sulfates
Anhydrite	anh	Anhidrita	cf. Sulfates
Ankerite	ank	Ankerita	cf. Carb
Annabergite	anbg	Annabergita	Arsenates
Anorthite	an	Anortita	cf. Slc, Fs
Apatite	ap	Apatito	cf. Phos
Barite	bar	Barita	cf. Sulfates
Biotite	bio	Biotita	cf. Slc, FSlc
Bismite	bim	Bismita	cf. ocres Bi
Bismutite	bis	Bismutita	cf. ocres Bi
Brookite	bk	Brookita	Oxides, cf. lcx
Calcite	calc	Calcita	cf. Carb
Carbonate (Class)	Carb	Carbonate (Clase)	cf. Carb, gg
Cerussite	crs	Cerusita	cf. Carbonates
Cervantite	cerv	Cervantita	cf. ocres Sb
Chalcedony	cld	Calcedonia	cf. q
Chert	ch	Chert	cf. Slc, Sic
Chlorite (Series)	Chl	Clorita (Serie)	cf. Slc, FSlc
Chrysocolla	crc	Crisocola	cf. Slc
Clay	Arc	Arcilla	cf. FSlc
Clay minerals	CM, MArc	Mins Arcillosos	cf. Slc, FSlc
Coal—organic matter	COM	Carbón—mat. orgánica	cf. Mat Organ
Corundum	cor	Corindón	
Dolomite	dol	Dolomita	cf. Carb
Epidote (Group)	Ep	Epidota (Grupo)	cf. Slc
Erythrite	ery	Eritrina	Arsenates
Feldspar (Group)	Fs, Felds	Feldespato (Grupo)	cf. Slc
Fluorite	fl	Fluorita	cf. Halides
Gangue (undiff.)	gg	Ganga (indiferenciada)	
Garnet (Group)	grt	Granate (Grupo)	cf. Slc
Glauconite	gln	Glauconita, Glauconia	cf. Slc, FSlc
Hornblende	hb	Hornblenda	cf. Slc, Inoslc
Illite	ili	Illita	cf. Slc, FSlc
Jarosite	jar	Jarosita	cf. lim
Jasper	jas	Jaspe	cf. q
K Feldspar	Kfs	Feldespato Potásico	cf. Slc, Fs
Leucoxene	lcx	Leucoxeno	cf. tit, ana, …
Ludwigite	ldw	Ludwigita	cf. von
Magnesite	mag	Magnesita	cf. Carb
Mica (Group)	Mica	Mica (Grupo)	cf. Slc, FSlc
Muscovite	mv, mus	Moscovita	cf. Mica
Ochre, Ocher	Ocre	Ocres	cf. Oxides

(continued)

(continued)

Name (EN)	Abbreviation	Nombre (ES)	Observations
Olivine (Series)	ov	Olivino (Serie)	cf. Slc
Organic matter	OM	Materia Orgánica	
Orthoclase	or	Ortoclasa, ortosa	cf. Slc, Fs
Perovskite	pvk	Perovskita	Oxides
Phosphate	Phos	Fosfato	gg
Phyllosilicates (Subclass)	FSlc, PSlc	Filosilicatos (Subclase)	cf. Slc
Piemontite	pie	Piemontita	cf. bra, Ep
Plagioclase (Series)	Plag	Plagioclasa (Serie)	cf. Slc, Fs
Pyromorphite	pymp	Piromorfita	cf. Phos
Pyroxene (Group)	Px, Pyx	Piroxeno (Grupo)	cf. Slc, Inoslc
Quartz	q (qz, qtz)	Cuarzo	cf. Slc, Sic
Rhodochrosite	rhc	Rodocrosita	cf. Carb
Rhodonite	rhn	Rodonita	cf. rdl F 1-4
Scorodite	sco	Escorodita	cf. asp; Arsenate
Senarmontite	sen	Senarmontita	cf. Sb ochers
Sericite	ser	Sericita	cf. Slc, FSlc
Serpentine	srp	Serpentina	cf. Slc, FSlc
Siderite	sid (sd)	Siderita	cf. Carb
Silica	Sic	Sílice	cf. Slc, q
Silicate (Class)	Slc	Silicato (Clase)	gg
Smithsonite	sms	Smithsonita	cf. Carb
Souxite	sx	Souxita	cf. cst
Sulfur	S	Azufre	gg
Tourmaline	tur	Turmalina	cf. Slc
Valentinite	val	Valentinita	cf. Sb ochers
Vanadinite	van	Vanadinita	cf. Vanadates
Vonsenite	von	Vonsenita	cf. Borates
Willemite	wil	Willemita	cf. Slc
Xenotime	xtm	Xenotimo	cf. Phos
Zeolite (Group)	Zeo	Zeolita (Grupo)	cf. Slc
Zincite	znc	Zincita	Oxides
Zircon	zir	Circón (Zircón)	cf. Slc

Other names	Otros nombres	Observations
Sphene	Esfena	cf. tit

Annex 1.4 Glossary MAAS: Meanings of Acronyms, Abbreviations and Symbols

Acronyms and Abbreviations

+nic	Microscope setting +p
+p	Setting with double polarization, i.e. polarizer and analyzer crossed at 90° (previously called "crossed nicols")
+p.w.	Setting with polarizer, analyzer and interference contrast (i.c.) inserting Wollanston prism
Abrv	Abbreviations
AMCO	Automated microscopic characterization of ores. Automated ore mineral characterization system developed at the LMA (Applied Microscopy Laboratory, ETSI Minas y Energia, UPM)
Anis +P	Description of polarization colors or anisotropism (+p setting)
Bir	Bireflectance, or variation of R with orientation
Bk	Black color (no colour, actually)
Blu	Blue color
BP	Block (ore or rock) polished
Br	Brown color
BSE	Backscattered-electron imaging
C illuminant	Type of illumination similar to daylight
ca (circa)	Approximately
CAMEVA	VNIR database
cf. (confer)	View, compare or consult
cf. ibid.	Consult the same place
CIE	Commission Internationale de l'Éclairage, or International Commission on Illumination
COM-IMA	Commission on Ore Mineralogy of the International Mineralogical Association
CRIRSCO	Committee for Mineral Reserves International Reporting Standards, organization that controls the various bodies that certify quality standards
cx	Cryptocrystalline
d = dry	Observation with dry lens (in air)
d+p	Observation in air with polarizer and analyzer crossed at 90°
d+pw	Observation in air with polarizer and analyzer crossed and with i.c. or interference contrast
discr	*Discredited* (term, name or mineral rejected by IMA)
DPTS	Double polished thin section, the usual one for the study of fluid inclusions
dsp	Observation with dry lens (in air) with only the polarizer (*single polarizer*)
dxp	Observations with dry objective (in air) and with polarizer and analyzer crossed a little less than 90° (about 87°)
EIT raw materials	EU organization promoting technological innovation in raw materials and mineral resources
EMP (EMPA)	Electron microprobe (analyzer or analysis)
ETSI	Escuela Tecnica Superior de Ingenieros (Mines and Energy, UPM) de Minas y Energía, Universidad Politecnica de Madrid
EU	European Union
Gn	Green color
Gy	Grey colour
H	Hardness
i.c.	Interference contrast, obtained by inserting the Wollanston prism and rotating the analyzer until the optimum contrast is obtained
IMA	International Mineralogical Association
IMA/COM	Commission on Ore Mineralogy of the IMA: Commission on Ore Mineralogy of the International Mineralogical Association
IMEB	Organization that certifies quality standards in Spain and Portugal
IR	Internal reflection, also described by some authors as diffuse reflectance or diffuse light: light reflected by reticular planes, microcracks or defects from inside the crystal. Perceived by the eye as additional luminosity, showing the mineral's own color, which interferes with the polarization colors and can even mask them
JORC	Joint Ore Reserves Committee, an Australian organization that certifies quality standards

l.t. (*locus typicus*)	Place of first definition of a species, mineral or rock, considered typical
LAS	Leica Application System v3.8, image acquisition software
LED	Light emitting diode
LLE	Life-long-education
LLL	Life-long-learning
m, min, mnl	Mineral (undifferentiated)
m1, m2, etc.	Unidentified minerals
MAN	Mean atomic number or average atomic number
MicroH	Microhardness
MLA	Commercial system for automated mineralogy
Mohs	Quantitative Mohs hardness
MS-DOS	MicroSoft Disk Operating System
mx	Microcrystalline
N	Refractive index of the medium
n	Refractive index of the mineral
N// (or parallel nicols)	Obsolete term for sp setting (polarizer only)
NI 43 101	National Instrument: Canadian Organization that certifies quality standards
NX (or crossed nicols)	Obsolete designation for +p (polarizer and analyzer crossed at 90°)
o (= *oil*)	Observation with oil immersion lens
o+p	Observation with oil and polarizer and analyzer crossed at 90°
o+pw	Observation in oil with crossed polarizer and analyzer and with i.c. or interference contrast
Og	Orange colour
Ol	Olive color
OM	Optical microscopy
osp	Observation with oil and with *single polarizer*
oxp	Observations with objective immersed in oil and with polarizer and analyzer crossed a little less than 90° (about 87°)
P_e	Excitation purity
PERC	European organization that certifies quality standards
PGE	Platinum group elements
PGM	Platinum group minerals
Pi	Pink color
PolH	Qualitative polishing hardness
PPL	Plane polarized light, sp setting (polarizer only)
PS	Polished section
PTS	Polished thin section
Pu	Purple color
QDF3	*Quantitative data fil*e, for reflectance values in the visible range (Criddle and Stanley 1993)
QemScan	Commercial automated mineralogy system
R	Specular reflectance
Rd	Red color
REE	Rare earth elements
REM	Rare earth minerals
resp	Respectively
RL	Reflected light
rltv	Relative
ROM (run-of-mine)	Raw ore, as extracted from the mine and prior to any kind of selection or processing
RRUFF	Project to create a catalogue of spectral data of well-characterized minerals
s.l. (sensu lato)	In a general, approximate sense
s.str. (sensu stricto)	In the strict, precise sense
SAMREC	South African Organization that certifies quality standards
SEM	Scanning electron microscope
sp	Single polarizer setting (no analyzer, formerly referred to as *parallel nichols*, now obsolete), equivalent to the PPL (plain polariser light) setting
sRGB	Color space, based on sRGB color definition
T, or TL	Transmitted light
T+p	Transmitted light setting, with polarizer and analyzer crossed at 90°
Talm	Talmage scratch hardness
TESCAN TIMA	Commercial automated mineralogy system
TL	Transmitted light
Tsp	Transmitted light setting, with only polarizer
undiff	Undifferentiated
UPM	Universidad Politécnica de Madrid
UV	Ultra violet wavelength range
VHN	Vickers microhardness number
Vi	Violet color
VNIR	Wavelength range between 370 and 1000 nm (visible and near infrared)
Wh	White color
x/xx	Crystal/crystals
xp	Observations with polarizer and analyzer crossed a little less than 90° (about 87°)

XPL (*crossed polars*)	+p, microscope setting with polarizer and analyzer crossed at 90°
XRD	X-ray diffraction
Y (%)	Luminance
Yl	Yellow
α, β, γ, n_α, n_β, $n\gamma$	Values of n (or *refractive index*) in the crystallographic directions α, β, γ
ε, ω = n_ε, n_ω or n_e, n_o	Values of n (or *refractive index*) in the directions of the extraordinary and ordinary rays (uniaxial minerals)
λ	Wavelength

Symbols

§	Chapter, paragraph or subparagraph of text
ø	Grain size, diameter measurement
$\pm$	With or without (a mineral or object)
$\approx$	Approximate value, approximately
$>$	Greater than (used to compare measurements, e.g. ø, or properties of minerals)
$\geq$	Equal to or greater than
$\gg$	Much larger than
$<$	Less than (used to compare measurements, e.g. ø, or properties of minerals)
$\leq$	Equal to or less than
$\ll$	Much smaller than
$\sim$	Similar to, nearly

References

Chace FM (1956) Abbreviations in field and mine geological mapping. Econ Geol 51:712–723

Criddle AJ, Stanley CJ (eds) (1993) Quantitative data file for ore minerals, 3rd edn. Chapman & Hall, London, 635 p

Fontboté L (2006) Recommended abbreviations for studies on ore deposits. https://www.unige.ch/sciences/terre/research/Groups/mineral_resources/opaques/ore_abbreviations.php (visit. 2019/11/07)

Neumann U (2019) Guide for the microscopical identification of ore and gangue minerals: mineral profiles with photomicrographs. Tübingen University Press, 319 p

Kretz R (1983) Symbols for rock-forming minerals. Am Min 68:277–279

Whitney DL, Evans BW (2010) Abbreviations for names of rock-forming minerals. Am Min 95:185–187

Warr LN (2021) IMA-CNMNC approved mineral symbols. Mineral Mag 85:291–320. 10.1180/mgm.2021.43

Annex 2: Common Mineral Associations

> «A mineral most commonly occurs in nature as a part of an assemblage, and this association itself is of considerable value in identification, because there are numerous examples of characteristic assemblages». (Vaughan 1990, p. 85)

The many types of deposits comprise a variety of mineral associations, knowledge of which is an important aid to identifying an ore. To facilitate this help, from a simplified typology of metal deposits limited to the most common classes (**Annex 2.1**) a table of common mineral associations (**Annex 2.2**) has been compiled. Both are based on my own experience and on published data (Castroviejo 2004; Castroviejo and Berrezueta 2009; Edwards and Atkinson 1986; Laznicka 2006; Ridley 2013; Robb 2005; Schneiderhöhn 1962; Vaughan 1990).

Annex 2.1 Main Classes of Metal Deposits: Simplified Typology

Magmatic environment		
1	In ultramafic rocks and ophiolitic complexes (Cr, PGE, Ni, Cu, Fe-Ti)	
	1.1	Chromitite/Podiform chromitite (ophiolitic)
	1.2	PGE (platinum group elements)
	1.3	Ni–Cu sulfides (cf. Mafic R, *class 2.1*). Ophiolitic VMS (cf. *cl. 8.1*)
2	In mafic rocks (Ni, Cu, Fe, Ti, PGE)	
	2.1	Ni–Cu sulfides (± PGE)
	2.2	Fe–Ti oxides ± V
3	In granitic rocks and alkaline complexes	
	3.1	Pegmatitic–pneumatolytic (Sn, W, Nb–Ta, Au…)
	3.2	Alkali complexes and carbonatites (REE)
Hydrothermal and magmatic-hydrothermal environments		
4	Hydrothermal veins and *mantos*	
	4.1	Pb–Zn–Ag and other veins
5	Porphyries (Cu, Mo, Au)	
	5.1	Cu and Cu–Au deposits
	5.2	Mo and Cu–Mo deposits
6	Epithermal (Ag, Au, Pb, Zn, Cu, Hg…)	
	6.1	High sulfidation
	6.2	Low sulfidation
7	Disseminated Carlin type (Au)	
Sedimentary environment		
8	Volcanic-sedimentary deposits. Fe/Mn banded formations	
	8.1	Volcanogenic massive sulfides (VMS: Cu, Pb, Zn, Ag, Au)
	8.2	Banded iron formations (BIF) and Mn formations (BMF)
9	Cu–Pb–Zn deposits in sedimentary rocks	
	9.1	SEDEX and Mississippi valley types (Pb, Zn, Ag)
	9.2	Kupferschiefer/Copperbelt types (Cu and polymetallic)
10	Detrital and residual types	
	10.1	Placers, paleoplacers, meta-conglomerate (Au, Sn, W, U, etc.)
	10.2	Laterites (Ni, Al, Au, etc.)
Metamorphic and complex environments		
11	Skarn and metasomatic processes (Cu, Au, Zn, Ag, U, REE, etc.)	
	11.1	Skarn and contact metasomatic deposits (Cu, Zn, Au, Sn, W, Ag, etc.)
	11.2	IOCG: iron oxide–copper–gold deposits (Cu, Au, Ag, U, REE)
12	In greenstone belts and metamorphic terrains (Au, Cu, Zn, Fe, Mn, etc.)	
	12.1	Orogenic gold deposits (Au)
	12.2	Metamorphosed deposits: VMS, SEDEX, BIF/Itabirites (§ 8, 9)

R. Castroviejo, *A Practical Guide to Ore Microscopy—Volume 1*,
https://doi.org/10.1007/978-3-031-12654-3

Annex 2.2 Mineral Associations by Deposit Class (cf. Annex 2.1)

Type of deposit	Class (**Annex 2.1**)	Mineral associations: ores(*)		Examples	Remarks
		Principal ores	Minor ores		
Chromitite	1.1	chr (± mt/spn, il)	MGP, sulfides	Bushveld, RSA; Moa, Cuba	Very restricted PGM
Platinum or PGE	1.2	Pt, spy, Ir, Os, etc.	Sulf, erl, chr…	Merensky Reef (RSA)	Exceptional mineralogy
Ni–Cu sulfides associated with ultramafites	1.3.a	po, pn, py, ccp, cbn, brv	ml, chr, vll, mt, il, PGM	Kambalda, Australia	Sulf. subclass Ni–Cu
VMS (Cu–Zn type)	1.3.b	py, ccp, Au	sp-po-mc-cc-cv-mt	Cyprus	Subcl. *Ophiolitic VMS* ± Au
Ni–Cu sulfides associated with gabbros	2.1	po, pn, py, mt, ccp	cbn, PGM, nc, ml	Sudbury; Noril'sk (Russia)	Occasional PGM
Titano-magnetite	2.2	mt, il, ht, rt	po, ccp, py	Allard Lake (Quebec, Can.)	Gabro association
Pegmatitic and pneum-hydrothermal	3.1	cst, wf, py, asp	mol, po, ccp, stn, REM, etc.	Kalima, R.D. Congo; Cornwall UK; Finland; Erzgebirge, Germany	Very varied mineralogy (pegmatites)
Carbonatite complexes	3.2	RE minerals: very varied mineralogy		Tsumeb, Namibia	Strategic interest REE
Pb–Zn–Ag veins and *mantos*	4.1	py, sp, gn, ccp, td	Ag, bn, cc, cv, Pb-Ag Ss	Coeur d'Alène (USA); Linares (Spain); Peru	Mineralogy can be varied; Sb–As and Ag sulfosalts
Copper porphyries (Cu–Au and Cu–Mo)	5.1/5.2	py, ccp, mol, bn, cc, cv, dg, Ss	mt, ht, en, cbn, il, rt, Au–Ag, gth, mch	Chuquicamata (Chile); Toquepala (Peru)	Hypo- and superg. minerals; potential secondary enrichment
High sulfidation epithermal	6.1	Au ± Ag, py, ccp, en	cv, sp, gn, gth	Yanacocha (Peru); Rodalquilar (Almeria)	≈ Quartz-alunitic or acid-sulfate type
Low sulfidation epithermal	6.2	py, gn, sp, Au/Ag/el, td, ac	ccp, ht, Sel, Ss Ag	Creede, Colorado, USA; Cabo de Gata (Almeria)	Formerly known as the adularia-sericite type
Disseminated Au	7	Au, py, asp, Ss	sbt, rl, orp, gth	Carlin Trend (Nevada, USA)	*Invisible Au* (tiny particles)
Volcanogenic massive sulfides (VMS)	8.1	py, ccp, sp, po, gn, td	cc, cv, bn, dg, mc, cst, bm, el, asp, Ss, gth, ht	Faja Pirítica Iberica (Río Tinto, Neves Corvo, etc.)	Varied subclasses. Potential supergene enrichment (Au/Ag)
Banded Iron Formations (*BIF*)	8.2	mt, ht, chert	sid, py, martite, gth	Guyana; Liberia; Superior, USA; Hammersley B, Australia; India	Large tonnages in Precambrian shields. Possible Au metallotect
Pb, Zn, Cu stratabound deposits (various types: Sedex/Carbonate rocks/Kupferschiefer/Copperbelt)	9.1/9.2	ccp, bn, cc, cv, gn, sp, dg	td/Ss, ln, py	• Sedex: Rammelsberg, German • Carb R: Miss. Valley, USA • Kupferschiefer: Lubin, Pol.; Mansfeld, Germany	Dominant metals • Cu/Cu–Co (Copperbelt: Shaba, DR Congo) • Pb–Zn (Sedex/Miss. Vall.) • Cu–Ag (Pt) (Kupferschiefer)

(continued)

(continued)

Type of deposit	Class (**Annex 2.1**)	Mineral associations: ores(*)		Examples	Remarks
		Principal ores	Minor ores		
				• Copperbelt: Mufulira (Zambia), Shaba (DR Congo)	
Detrital, placer and paleoplacer deposits	10.1	Au, PGE, urn, rt, cst, il, rt, mz, wf, diamonds, etc.	ht, mt, cbt, zir, silicates, etc.	Malaysia (cst placers), Witwatersrand, RSA (Au, U)	Witwatersrand: the world's largest Au concentration (meta-conglomerates)
Lateritic	10.2	gth, lim, jar, residual Al-Ni minerals: garnierite, gibbs., boehmite, Ni-nontronite, etc.	Traces of PGE, Au	Laterites, bauxites and nickeliferous laterites • New Caledonia • Moa (Cuba), etc.	Mineralogy unsuitable for reflected light microscopy techniques
Skarn	11.1	mt, ccp, wf, sch, Au, asp, mol, sp, cst, etc.	po, ht, Ss, gth; Ca–Mg–Mn Silicates and Carbonates, etc.	• Tintaya (Cu–Au), Antamina (Zn), Peru • Cala, Badajoz (Fe, Au) • Carlés, Asturias (Au–Cu)	Very varied and complex deposits and mineralogy
IOCG	11.2	py, bn, ccp, Au, cc	mt, ht, fl, sid	Olympic Dam (Australia)	Large tonnage, complex gen
Orogenic gold deposits	12.1	Au, Tel, el, py, asp, po, etc.	gn, sbt, sp, ccp, gra, rt, mol, Ss	Val d'Or, Hemlo, etc. Quebec, Canada	Synonyms: *greenstone, archaean, shear zone, gold-only, vein type, metamorphic, mesothermal,* etc., *gold deposits,* etc.
Metamorphosed and polymetamorphic deposits	12.2	Generally analogous mineralogy to Sedex, SMV, BIF, etc., protoliths, although re-equilibrated under metamorphic conditions. Catazonal metamorphic changes—Broken Hill, Arinteiro—may mask protolith		Pb: Broken Hill (Australia) Fe: Itabira, Quadrilatero Ferrífero (Brazil) SMV: Arinteiro (Santiago de Compostela, Spain); Norway	Metamorphism recognizable by modified textures

(*)**ORES**: due to space constraints, the ores are represented by their corresponding symbols or abbreviations (cf. **Annex 1.1**)

References

Castroviejo R (2004) Gold in ophiolites. In: Ophiolithic complexes in iberoamerica. CYTED, Madrid, pp 25–70

Castroviejo R, Berrezueta E (2009) Automated recognition of metallic ores by digital image analysis: a support to the mineralogical process. II: Discriminating metallogenic criteria. Rev Metal 45 (6):439–456. CSIC, Madrid; ISSN: 0034-8570

Edwards R, Atkinson K (1986) Ore deposit geology. Chapman, London

Laznicka P (2006) Giant metallic deposits—future sources of industrial metals. Springer, Berlin

Ridley J (2013) Ore deposit geology. Cambridge University Press

Robb L (2005) Introduction to ore-forming processes. Blackwell, London

Schneiderhöhn H (1962) Erzlagersttaetten. Fischer-Verlag, Stuttgart, 4e. Auflage

Vaughan D (1990) Advanced microscopic studies of ore minerals. In: Jambor JL, Vaughan DJ (eds) Short course handbook, vol 17. Mineral Association of Canada, Ottawa, p 85

Annex 3 Preparation of Polished Sections

Ursula Grunwald-Romera, MSc Geological Eng.
LMA_UPM (Technical University of Madrid)

1. **Preliminary Considerations**

The preparation of samples, in this case polished sections, is an important yet often neglected preliminary step in mineralogical analysis. However, we must bear in mind that the quality of an analysis will inevitably depend on the quality of the polished section. Indeed, whether the object to be worked on makes visible or loses the information that determines the success of the final results depends on it.

Although the procedures to follow to prepare a polished section are simple, the reality is that the quality of polish ultimately depends on the skill and responsibility of the technician, whether the polishing is manual or automatic.

It is important that the polishing is adjusted to the hardness of the various minerals present in the sample. In particular, special attention must be paid to the intergrowths between hard and soft minerals: poor polishing can lead to the creation of pronounced reliefs around the harder minerals. Experience is undoubtedly what helps the expert to choose the best polishing cloths and techniques according to the equipment, which means that the preparation techniques for polished sections differ according to the laboratory preparing them and the degree of automation of the process.

For this reason, in what follows, rather than systematizing the method through a series of fixed steps (difficult to adapt to the specific needs of each user), essential guidelines are given on the preparation of samples that, in any case, must be adjusted to the conditions of each. Following the basic steps to be taken into account to prepare a polished section, the most common problems derived from polishing that affect mineralogical analysis are explained, trying to anticipate a practical solution. These indications are especially relevant when the objective is to perform digital image analysis, either by means of electron microscopy equipment or by using innovative automatic characterization systems based on optical microscopy, as is the case of the AMCO system.

The main difference between doing an analysis manually with an expert in mineralogy and doing it with an automated system is that the human eye is able to discriminate certain flaws—and even benefit from them—and reconstruct the ideal aspect of the grains or mineral particles that it is observed. A good example can be found in the acicular aggregates of wolframite, with totally irregular relief due to its difficult polishing; or in the shadow or bevelling produced by a simple edge around a mineral particle, irregular in appearance due to differential polishing. However, what is relatively easy for a person to understand (sometimes even unconsciously) can be an insurmountable barrier for an automated system, and the only way to overcome it is to take the utmost care in sample preparation.

The sample preparation process can be carried out either completely manually, by polishing and grinding with carborundum and glass, or fully automated, with purpose-designed machinery. A basic automation stage is considered here, for which a vacuum hood, a precision saw and a semi-automated polishing machine are essential. Whenever possible, the highest degree of automation available is recommended, to ensure, above all, that the polished surface is flat and remains parallel to the base, without the sometimes unavoidable surface undulations due to unevenly applied pressure at different points on the sample by manual polishing.

The basic process of preparing a polished section, either from a hand sample or from a mineral concentrate, can be summarized in four phases:

I. Sample preparation
II. Embedding
III. Cutting, grinding and pre-polishing
IV. Polishing.

As a first step, however, it is necessary to make some general considerations on the use of abrasives and polishing cloths. These apply both to roughening and polishing, and also in the previous phases of the embedding, when a *polished thin section* (**PTS**) is to be produced.

R. Castroviejo, *A Practical Guide to Ore Microscopy—Volume 1*,
https://doi.org/10.1007/978-3-031-12654-3

- Nature of the polished section: before starting, the predominant mineral composition and hardness must be taken into account. As it is normal to find mixed typologies in the same sample (hard minerals such as pyrite and soft minerals such as galena), it is recommended to make careful polishing at low revolutions and use hard polishing surfaces.
- Type of abrasive: it must be selected according to the result to be obtained and the nature of the mineral sample. During the grinding and polishing process, the type of abrasive can be changed, if necessary. The most commonly used are sandpapers impregnated with carborundum and those marketed in paste or grain form, to be mixed with a lubricating liquid.
- Grain size of the abrasive: in each phase of preparation it is necessary to use the correct grain size (whatever of the type used), always taking care to move from the coarser to the finer sizes. The sample and the supports used must be washed thoroughly before moving on to the next grain size, to avoid contamination, which can scratch the polished surface or even cause material to be removed. It must be ensured that in each polishing phase all the scratches produced in the previous phase are removed, otherwise they will remain in all subsequent phases and the polishing will be defective.

These considerations are documented with additional information (see Appendix at the end) about the materials currently available on the market.

2. **Sample Preparation Process**

In the following, the many cases that may arise when dealing with each phase of the process are developed in detail.

I. Sample preparation

If the polished section is made from a hand specimen, simply cut a section with the disc to the desired shape and size, clean it well to remove all traces of abrasive or oils that may have been left in the pores and dry it. The cleaning can be done with acetone (which will facilitate the drying process), immersing the sample in an ultrasonic bath. Finally, all moisture is removed in a 60 °C oven. It is important not to exceed this temperature so as not to compromise the stability of the minerals present in the sample. Above 75 °C, transformation processes can occur, which must be avoided at all costs so as not to alter it and thus be able to study the sample with its original ore assemblages.

This is the first step in preparing a polished block (**PB**) from a rock sample, and the process is somewhat more complex if a polished thin section (**PTS**) is to be prepared from a sample of milled ore, as required for the usual analyses needed to support processing plants. Therefore, the techniques described below will focus on this type of preparation although they generally cover both cases, as detailed below.

When the objective is to characterize the metallic ore fed to a concentration plant, or even to study the mineral content of the effluents, it is normal to start from a sample of mineral concentrate, previously crushed or milled, to make a PTS.

The first thing to consider when dealing with a milled sample is the particle size. Although it is true that the study of the particle size is one of the aspects to be studied through a mineralogical analysis, it is easy to answer some of the questions with a careful observation of the sample: to recognize if it is a milled concentrate (powder) or only a crushed or ground sample (visible fragments), or a mixture without previous sieving; that is, both options at the same time. Something as simple as anticipating the basic characteristics of the concentrate in advance can determine the success or failure of the final result. Whenever possible, it is recommended to do a grain-size screening to make the sample size as homogeneous as possible. However, we are often faced with heterogeneous samples, without prior screening. In these cases, the presence of fines and their quantity will be key in the preparation of the sample.

The fines tend to agglomerate, forming lumps that hinder the application of digital image analysis programs, as they produce a large number of "unclassified" pixels. It is also important that the particles do not touch each other (Fig. 3.1) otherwise the particles in contact might be processed as a single particle during digital segmentation. This is a fairly common problem in electron microscopy systems. It introduces errors in determining the real particle size; moreover, and perhaps even more importantly, it gives inaccurate or erroneous information about the degree of liberation of the sample, an essential aspect in the study of a concentrate for processing in a mineral concentration plant.

Then there is the problem of segregation. Almost inevitably, after mixing with the resin the particles will stratify according to size and density. This stratification means that the particle size results obtained by studying the surface of the polished specimen may be biased, and not true to reality. In fact, at the time of setting this surface corresponded to the bottom of the mold, where a certain segregation by gravity is to be expected, so there will be a greater presence of large (or denser) particles and a smaller presence of fines (or lighter particles) deposited on the larger particles (i.e. when the specimen is inverted to proceed to polishing, they will remain below the surface to be studied or, at best, in the interstices only). They can thus give the false impression of

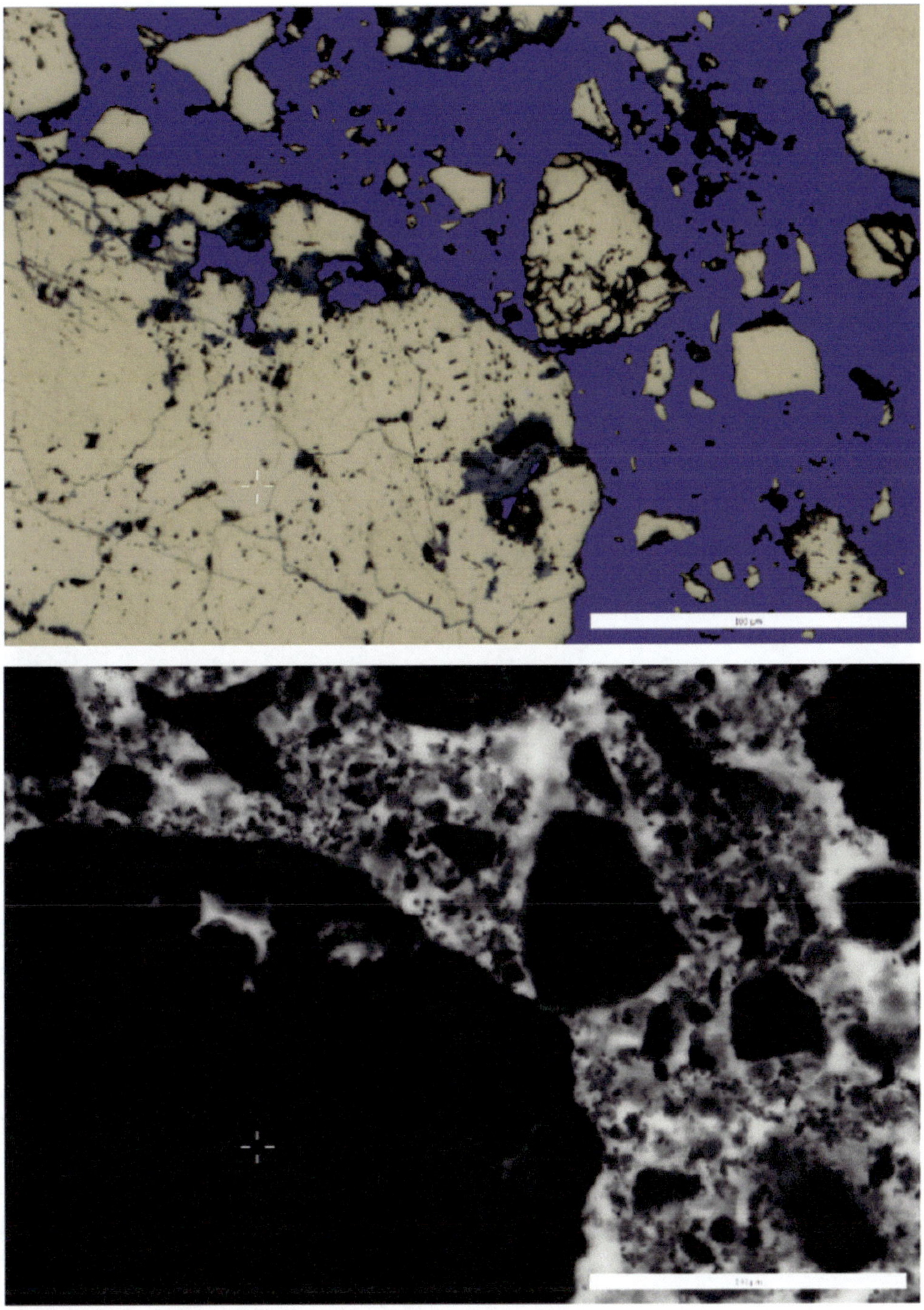

Fig. 3.1 Two independent particles touching each other. On the left, false color image generated with the AMCO system with a blue mask on the resin. On the right, a fluorescence image. It can be seen that these are two independent pyrite particles (a large one in lower left corner and a smaller one beside its upper-right border) but, when touching each other, if the image is not properly treated digitally, the automatic analysis systems could consider it as a single particle

having an average particle size that is coarser (or with a higher proportion of dense components) than the actual size.

Although there is no definitive and completely effective solution to this problem, some actions can be performed in this first phase to facilitate the process in subsequent phases. First, it is recommended to wash the sample, preferably with acetone, to remove possible traces of oils or abrasives and to facilitate drying. The objective is to disintegrate the agglomerations of material as much as possible (Fig. 3.2), aided by a spatula. Once the acetone has evaporated, the sample should be dried in an oven at 60 °C so that it loses all moisture. As a general rule, the sample should be left to dry for two hours, but this can be extended depending on the initial state of the concentrate. In general, the greater the amount of fines, the longer the drying time required.

For some years now, moreover, the addition of pure graphite particles to the mineral concentrate has been spreading in the preparation of samples for electron microscopy studies Although it has not been shown to have a great effect in reducing particle segregation, there is evidence,

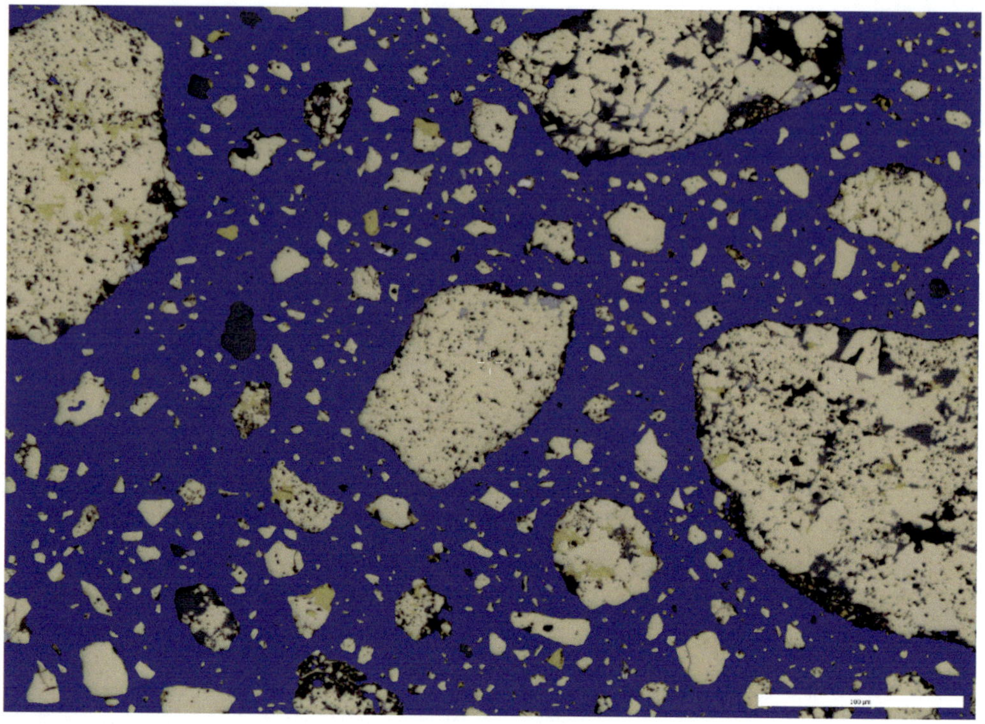

Fig. 3.2 Example of well-distributed particles of different sizes. Blue mask on the resin generated with the AMCO system

both empirical and theoretical, of its effectiveness in preventing the formation of particle agglomerates (Røisi and Aasly 2018), which is directly reflected in a lower contact ratio between them. Graphite, as it is not detected by EDS sensors, does not affect the final results of electron microscopy analysis.

Regardless of the system used, lumps are a widespread problem in the study of metal ore concentrates through automatic image analysis programs because, above all, they make it difficult or even impossible to carry out particle size and mineral liberation studies. This is because the contours of each individual particle cannot be accurately determined.

These problems disappear if the object is to make a preparation from a rock fragment (to make a PB), because the only thing to take care of during the preparation is that no bubbles are left on the surface to be polished and, later on, to avoid differential pull-outs and polishing.

II. Embedding

For embedding samples, currently the most widespread option is to use silicone molds that allow the sample to be easily extracted once the resin has set (Fig. 3.3). If it is a concentrate, it is essential to use a fluid epoxy resin, although it is advisable to use this type of resin for blocks as well.

Although there is a wide range of resins on the market, a widespread example is the Epofix kit by Struers, which consists of two parts: part A is the epoxy resin itself; and part B is the hardener.

To increase the fluidity of the resins, it is important to work them hot. Ideally (if working with a resin with hardener such as Epofix) it is best to put aside in a smaller pot the amount of resin required for the work. In this way, this pot can be kept warm in an oven at 40–50 °C during the sample preparation process.

Before mixing the resin with the hardener, 2–3 g of the ground sample to be analysed are put in each mold. The sample is spread at the bottom of the molds as evenly as possible, without piling up at one point. Next, a small amount of resin (parts A and B already mixed according to the manufacturer's instructions) is poured gradually to cover the concentrate (Fig. 3.5). The molds are placed in a vacuum hood to remove the bubbles by slowly bringing the resin to effervescence. The decrease in surface tension will wet all the particles, thus avoiding agglomeration of the material. Special care must be taken to ensure that it does not overflow, reducing the vacuum by allowing air to enter, if necessary. Then, the test molds are removed and filled with the resin, making the vacuum again, if necessary. In order to concentrate the particles at the bottom, the samples can be centrifuged (Castroviejo 1980).

In this step, to make a PB it is necessary only to cut the rock fragment to a suitable size to fit the mold and cover it with the resin, making sure that at the base of the cylinder the face of the fragment is as flat and as large as possible.

The samples should then be left in the oven at 60 °C for at least three hours to accelerate the setting of the resin. After

Fig. 3.3 Cylindrical silicon molds. This format allows the same sample to be studied both in the AMCO system and in an electron microscopy system or by traditional methods

this time, they are left at room temperature for half an hour before being removed from the mold, to allow them to finish setting. Once out of the mold, the test piece is sufficiently hardened to allow preliminary roughening to even out the surface.

Nowadays, resins are available that, with the help of an ultraviolet lamp, can harden in a few minutes. This type of technique makes it possible to have a polished section ready to be analysed on the same day that the sampl is taken.

Box Annex 3.1 Automated Gangue-Resin Discrimination (Sample Preparation for the AMCO System)

The AMCO system for automated ore characterisation, based on **RL** (reflected light) microscopy, requires polished sections, either PB (from rocks/run-of-mine-ores) or PTS (from milled ore, plant feed or ore concentrates of effluents). To take full advantage of the system and perform particle size and particle analysis, the latter requires the addition of a fluorescent dye to the resin to ensure its automated distinction from the gangue.

To discriminate between gangue and resin, a procedure was developed in LMAAI, because both resin and gangue have very similar reflectances and cannot be distinguished with RL. The goal is for the image analysis program (AMCO, in this case) to be able to create a mask over the resin so that it can be discounted from the image classification process. This involves altering some aspect of the resin so that AMCO can detect it and distinguish it. While this problem is negligible in whole-rock samples directly embedded in resin to form polished blocks, ground mineral samples have to be mixed with epoxy resin to which a fluorescent additive (e.g. EpoDye, also from Struers) is added. Subsequently, they need to be cut and polished as thin transparent slides (about 30 μm thick), so that no fluorescent resin remains under the

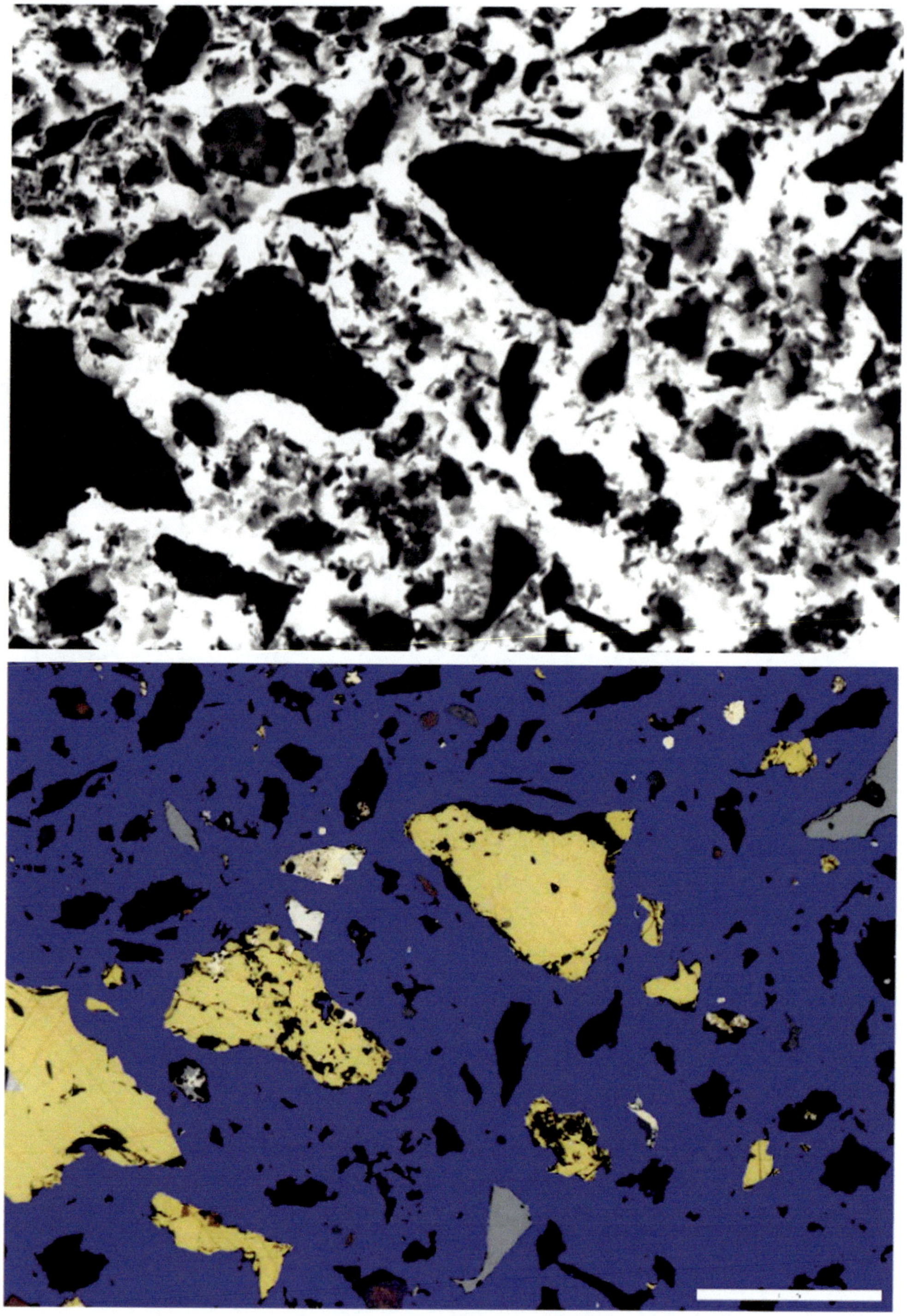

Fig. 3.4 Above, fluorescence image of a polished thin section. Below, blue mask on the resin in the final digital image. Images acquired with AmcoCapture and processed with AmcoAnalysis

transparent gangue minerals; otherwise, it will shine from below and interfere with distinguishing resin (Grunwald et al. 2019).

This sample preparation method allows effective discrimination between the fluorescent resin matrix and mineral particles, including gangue minerals. The solution is applied automatically by the program. First, an additional fluorescence band is acquired from the polished transparent thin section (Fig. 3.4, left), which highlights all fluorescein-stained resin regions. Processing with AmcoAnalysis generates a mask on the resin, removing those pixels from the final modal analysis (Figs. 3.4, right and 3.5).

Fig. 3.5 Embedding of ground samples in fluorescent resin

If, in the preparation of the PTS, it is suspected that the segregation of the material has been significant, it is advised to make two samples from the same concentrate: one as explained above; and another by cutting the specimen vertically. In this way it is possible to study the segregation of particles according to their size and density, and to make a complementary or more reliable particle-size study.

So far the process for making a polished block or a PTS is the same. Now, to make PTS, it is necessary to glue the bottom surface of the block that has just been prepared onto a glass slide, after an initial roughening of the block's surface. If you wish to continue making a PB, go directly to the explanation of the **pre-polishing** process (§ III).

One of the possible roughening sequences to prepare the surface to be bonded is as follows:

1. Manual grinding is performed on a polishing machine with medium-grit sandpaper (P600–P1200, depending on

Fig. 3.6 Preparation of the specimen surface to be glued to the object holder glass. A light roughening is done to even the surface, and to facilitate gluing the edges are rounded with the same polishing machine

the hardness of the predominant material in the sample), using water as a lubricant. The aim is to achieve a flat surface and to round the edges of the block to facilitate gluing to a slide (Fig. 3.6). Before proceeding to the next step, it is important to clean the samples to remove the abrasive residues.

2. The second grinding is done on an automatic polisher, using 6–7 μm carborundum for six minutes, or fine grit sandpaper (P2400).
3. Finally, the blocks are cleaned with the help of an ultrasonic bath and dried with a compressed air nozzle.

To glue the samples to the slide, the easiest way is to use the same epoxy resin used during embedding, in this case without the fluorescent additive. It is important to glue with an undyed epoxy resin in order not to modify the behaviour of the transparent gangue grains under fluorescent light.

For the glass to fit the polishers optimally, it is recommended to polish the bottom side of the glass (the side opposite to where the sample will be glued), as well as rounding the corners and edges. In addition, in order to permanently mark the name of each sample and not to confuse them, it is advisable to scratch the glass with the corresponding code with an awl or similar tool.

To prepare the glue, parts A and B of the resin mentioned above are mixed again. One gram of the final product is sufficient to glue 12 samples. To avoid the production of new bubbles, it is very important to pay attention to the process of gluing the block to the glass. Perfect technique requires practice but, as general advice, place a few drops of the resin both on the slide and on the surface of the sample and then centre the sample. Carefully put them together while making small circular movements, pressing the centre of the preparation to remove all the bubbles. Check that there are no air bubbles by carefully lifting and turning the sample over. If necessary, peel off the specimen and start again (Fig. 3.7). If the surface is extremely porous, use the vacuum hood.

To finish using the resin, place the glass on a hot plate at 45° (a higher temperature could create bubbles) with weights

Fig. 3.7 Example of ineffective bonding of the specimen to the glass. It can be seen that, while in half of the specimen the epoxy resin is adhering the surfaces correctly, in the other half it is missing. The name of the sample is marked on the glass with a chisel

Fig. 3.8 Drying test specimens glued to slides

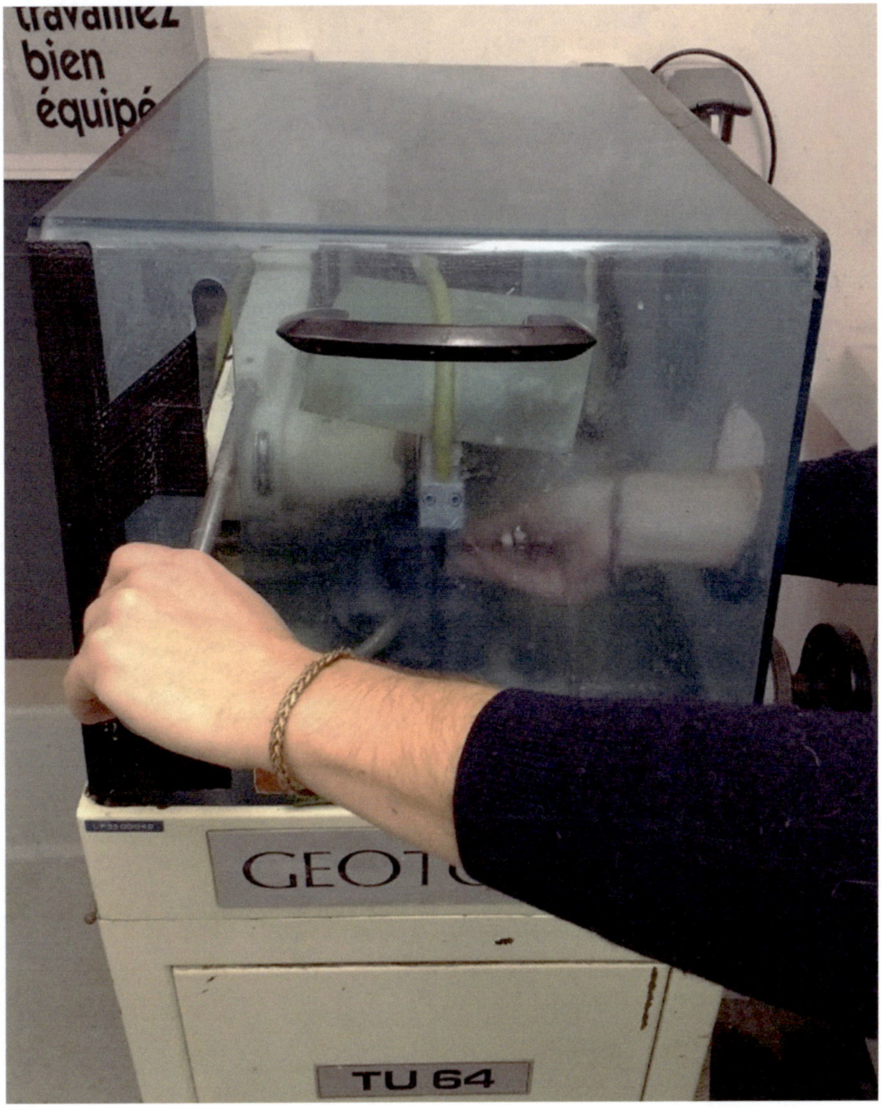

Fig. 3.9 Precision slicer to reduce the specimen thickness to 1 mm

in the centre of the preparation (Fig. 3.8). Leave to dry for one to two hours.

III. Slicing, grinding and pre-polishing

To speed up the process and keep the sample surface as flat as possible, it is essential to cut the specimen using a precision slicer (Fig. 3.9). Ideally, if the machine allows it, it should be cut to a thickness of approximately 1 mm (Fig. 3.10).

From this point on, there is a wide range of machinery to facilitate the process. The most complete is the multiplate grinder type (Fig. 3.11), a grinding machine that allows automatic thinning of the section adhered to a slide, down to a thickness determined by the technician (normally 100 μm, if it is a PTS, to prevent the preparation from disappearing during the polishing process). This type of technology ensures that the surface to be polished remains parallel to the base of the sample in order to guarantee its parallelism to the stage when, once polished, it is studied under the microscope. This is an indispensable condition for the validity of reflectance measurements, which require the orthogonality of the incident beam on the sample. For the reader's convenience, a commercial example of this type of machine is the multiplate grinder 1.03.12 G, by BROTLAB, although there is a wide variety on the market.

Fig. 3.10 Cutting result with the precision-slicing blade

Fig. 3.11 Multiplate grinder

If only a classic *lapping-machine* is available, the most common procedure is to use carborundum mixed with water on the specific metal plate on the polishing rotor; discs with abrasive; or sandpaper glued on magnetic surfaces that adhere to the polishing rotor. The advice given below is for a general case, using sandpaper on magnetic support.

The general sequence of **roughening** or **pre-polishing** is done with water or a commercial preparation as a lubricant, progressing from larger carborundum diameters to finer and finer grit sizes of abrasive: P120, P240, P320, P400, P600, P800, P1000, P1200 and P1500 (this sequence is only an example. Depending on the manufacturer, the variety of abrasive grit sizes may change. As a guideline, Buehler offers a larger number of grit sizes in its catalog than Struers, Table 3.1).

When working with a PTS, it is recommended to start directly with P800 or P1000, depending on the starting thickness. This sequence is not rigid and can be adapted as needed, taking care always to go from coarser to finer grain size and cleaning the sample well with water and an ultrasonic bath before changing abrasive size. Several short cycles will be carried out, especially with the coarser grits, some five to ten minutes each, taking care that there is no lack of lubrication or abrasive (in the case of using carborundum powder). In each cycle the result should be checked with an auxiliary microscope to decide whether to repeat or to move on to the next particle size.

The boundary between roughening or pre-polishing and polishing is very thin, and will depend more on the grain size of the abrasive, the revolutions and pressure exerted by the machine, and the duration of each cycle than by the materials used. In the first stages of grinding it is normal to use between 120 and 200 rpm (less if the minerals are soft) and to decrease the speed as the polishing and final finishing progresses.

Table 3.1 Equivalence between abrasive grit sizes in μm and FEPA P

Grain size (μm)	200	125	82	68	46	30	22	18	15	7	4
FEPA P (Europe)	P80	P120	P180	P220	P320	P500	P800	P1000	P1200	P2000	P4000

Source Struers

As an empirical orientation, the grinding ends when the thin section starts to become transparent, starting the more delicate phases of polishing and finishing. In the case of a polished specimen, polishing begins when, with the sample completely dry (as an observation for a quick intuitive check; it is always better to check with a microscope) the surface begins to shine when placed under the light at a certain inclination.

IV. Polishing

This is the most delicate step. If a PTS is being prepared, it is recommended to start polishing when the section is between 60 and 40 μm thick. Before starting, it is necessary to check the thickness and parallelism with a precision digital micrometer. The indications given here are valid for PB and PTS, with finishing using cloths and abrasive suspension with diamond powder of 6, 3 and 1 μm, but the general features are easily exportable to other materials.

Depending on the thickness and the scratches present, start the process with 6 or 3 μm abrasive. The pressure of the sample on the cloth should be minimal, and the abrasive fluid should form a constant film between the preparation and the cloth. It is recommended not to exceed 100 rpm. If, due to the number and depth of the scratches, it is necessary to start with 6 μm abrasive, it is recommended that the section be about 50 μm thick to avoid excessive wear and to prevent it going below 30 μm. Short cycles of five minutes (shorter if there are many contacts between soft and hard minerals) will be carried out until the deepest scratches have disappeared, taking care and observing that relief is not created.

The objective is to start with the abrasive of 1 μm when the sample is practically polished, so that this last step serves only as finishing, at 50 rpm, for the shortest time possible, making cycles of no more than ten minutes. Of course, the cloth must be changed with each grain size. The ideal is to have three cloths already mounted on a magnetized support, marked for which grain size that each is intended, to avoid cross-contamination.

If the pre-polishing has left deep marks or relief, it is recommended to continue polishing on a rigid support with alumina, carborundum or very fine grit paper, between P2000 and P4000 (equivalent to a grit size between 7 and 4 μm). However, if the damage is deep and long-lasting, it is necessary to return to pre-polishing.

The final polishing step, if necessary, can be done with diamond paste, applied with a lubricant. The diameter of the diamond monocrystals in the pastes offered by the commercial companies ranges from 15 to 0.25 μm. However, for polishing or finishing, it is recommended to start with 6 or 3 μm and gradually decrease the particle size of the diamond paste towards finer and finer sizes, ending with 0.25 μm. The finish should always be fine-grained ($\leq$ 1 μm).

3. **Polishing Defects and Artefacts**

When sample preparation or polishing is faulty or insufficient, various artefacts may appear. Below are some examples of the various problems caused by polishing defects that mineralogists may encounter during the study of polished sections.

I. Lumps or aggregates of material

As indicated, lumping is the main problem faced by automated image analysis systems on finely ground samples (Fig. 3.12).

II. Poor resin setting

In Fig. 3.13, two comparative images of a polished section are shown, the top image taken with polarized light and it lower image with interference contrast, using the Wollaston prism. The prism highlights the differences in relief. In this case, an apparently well-polished sample reveals a rough surface due to poor resin set, which contrasts with the smooth, well-polished surface of ore and gangue.

III. Loss of material, relief and pores

If care is not taken, the presence of minerals of different hardnesses leads to the appearance of reliefs on the edges of the grains. It is also very common for particle pits or pull-outs to occur. Generally, as shown in Figs. 3.14, 3.15 and 3.16, these problems are associated with insufficient polishing.

IV. Exudate and patina

Figures 3.17 and 3.18 show the same particle in air (Fig. 3.17) and with oil immersion objective (Fig. 3.18). Figure 3.17 shows a zonation that does not disappear with crossed polars (d+p setting), which could be mistakenly understood as a secondary marginal zonation. However, after immersion in oil, the zonation disappears completely, both with single polarizer and with crossed polars, because the oil exudates from the polishing have dissolved in the immersion oil. This is precisely why the exudations appear on edges, pores and microfissures.

A similar type of zoning may occur due to the presence of a patina that is removed after cleaning with hexane, so it is very important to clean the samples whenever in doubt (Fig. 3.19).

V. Final finishing

Finally, Fig. 3.20 presents, as an example, good-quality polishing of a composite ore (pyrite, pyrrhotite, chalcopyrite, sphalerite, galena and quartz) with intergrown species of different hardnesses and no artefacts.

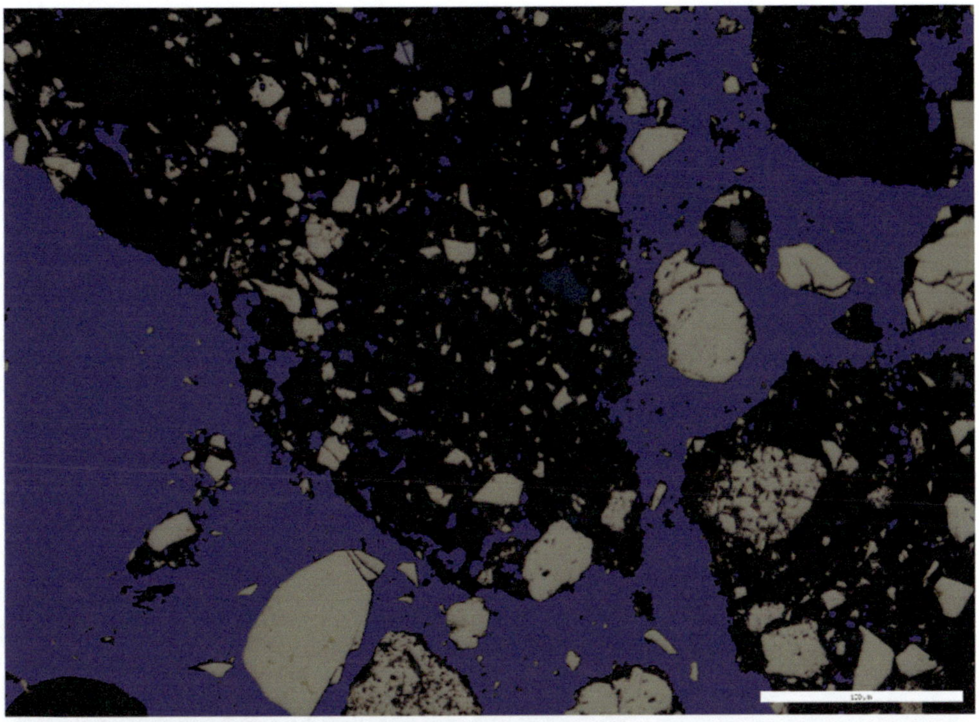

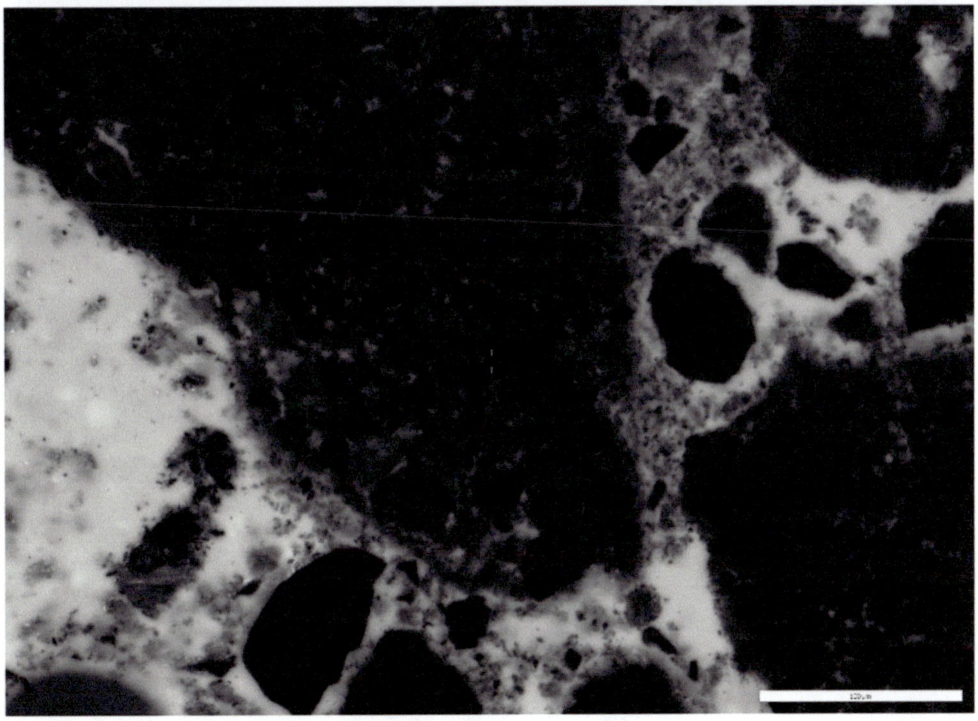

Fig. 3.12 A field of an automated sample scan processed with amcoAnalysis. Top, image with the blue mask covering the area assigned to the resin. Below, image acquired in fluorescence observation mode to identify the fluorescent resin areas. It can be seen that large lumps have formed, composed of many particles stuck together, because there are interior areas where the fluorescent resin has entered

4. **Final Considerations**

As we have seen in the previous pages, there is no single method or formula for making good-quality polished sections. In fact, only some of the methods and products available on the market have been discussed here. If well chosen, they can facilitate and improve the process.

Keep in mind that once the polished section is finished, tarnishing starts, and darkening or an oxidation patina may begin to develop. The speed depends on the minerals present (for example, bornite can create patina in a few hours; arsenic, in minutes). Dust or even dirt and grease are also deposited. Because of this, it is important to check the quality of the polish and, if necessary, clean the sample

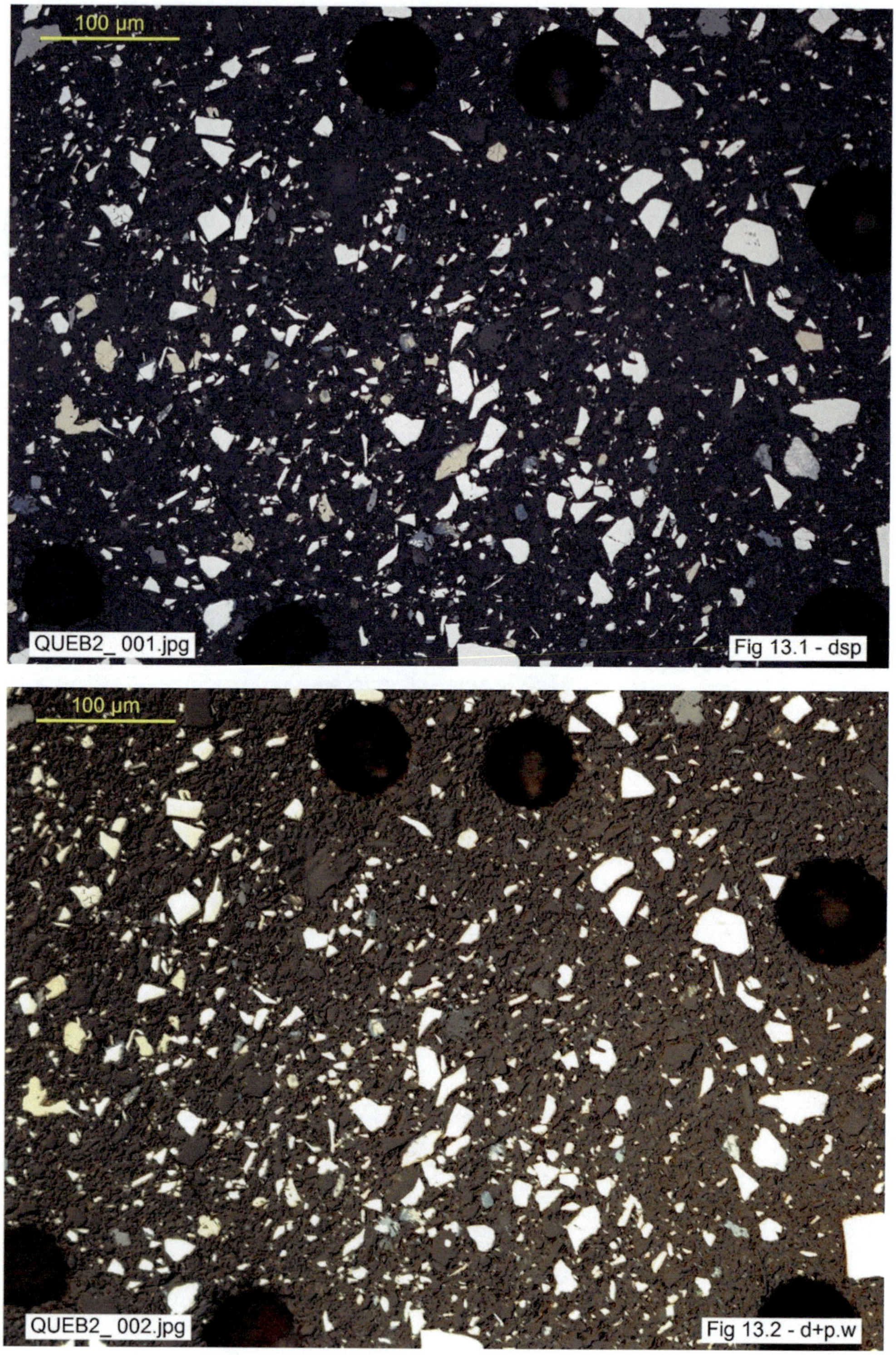

Fig. 3.13 Above, image of a crushed mineral sample taken with the 10x objective and polarizer. Below, same sample with the Wollaston prism, where the relief can be seen

properly before studying it. It is recommended to study it freshly polished.

As an example of a general cleaning sequence giving very good results in removing the patina from diverse mineralogies, the following is recommended:

1. After washing hands, place a small amount (such as a grain of rice or a lentil) of ¼ µ (0.25 µm) diamond paste on the polished surface of the sample to be cleaned. Gently spread it loosely, with the fingertip, over the entire preparation, paying particular attention to the

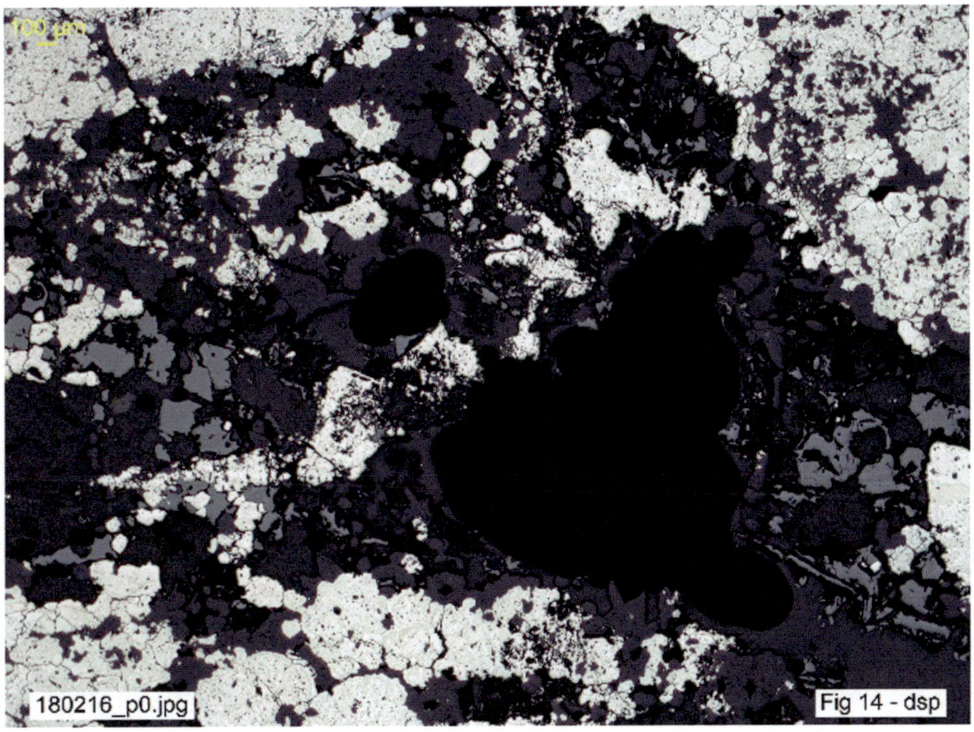

Fig. 3.14 Presence of pores and relief in the polishing

Fig. 3.15 Excessive relief on the edge of the particles

corners. Once it is well distributed, with a bit of good quality toilet paper or cloth that does not leave lint, gently rub the surface, pressing as evenly as possible with your thumb to avoid creating relief. If necessary, apply another drop of diamond paste and repeat the operation. It is important to make sure that you are getting to the contour of the sample, as this is usually the most difficult area to clean and the most neglected.

Fig. 3.16 Irregularities on the surface of the preparation due to different hardnesses in the material: loss of mineral grains, pores and relief

Exceptionally, if the patina is very marked and widespread and cannot be removed by hand, a manual polisher can be used with a clean billiard cloth and a small amount of diamond paste. At the lowest possible speed, hold the sample firmly by pressing gently and evenly in the centre over the diamond paste, and polish for a maximum of one to two minutes in small circles.

2. Using an eyedropper, pour a small amount of hexane onto the surface to remove the diamond paste. Hexane evaporates very quickly, so just leave it for a few seconds to work well before removing it, carefully and without rubbing, with a piece of soft toilet paper. Make small circles with uniform pressure to reach the entire surface.

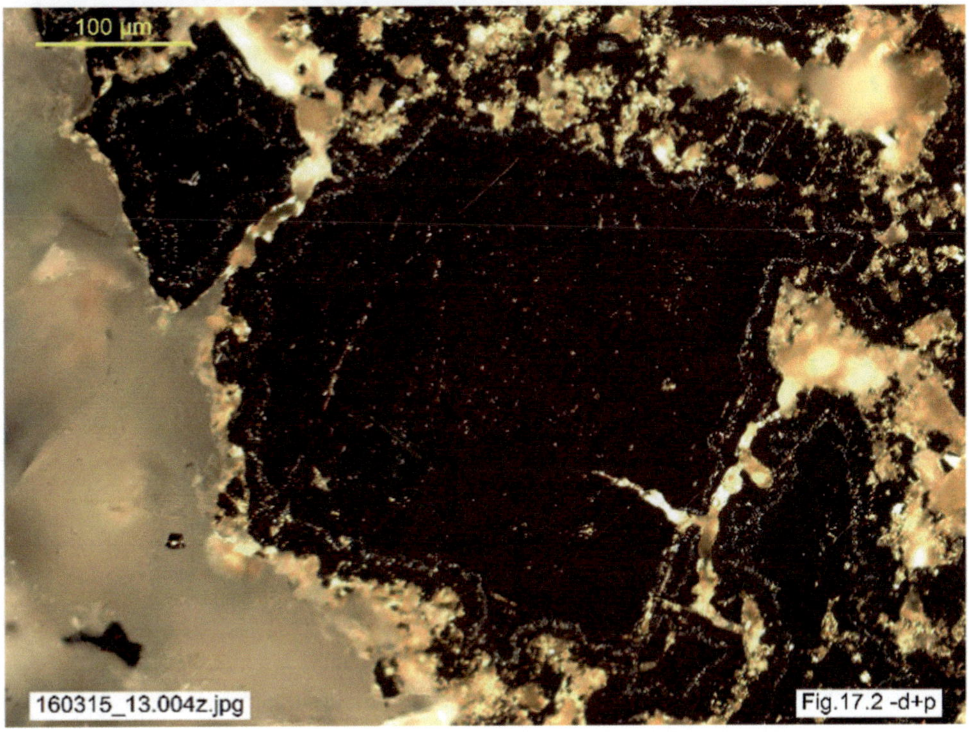

Fig. 3.17 Theoretical secondary marginal zoning at the edge of the grains. The effect of the patina is removed by cleaning the sample (Fig. 3.19). Upper image with single polariser (dsp); below, with crossed polars (dsp)

3. Finally, it is necessary to remove any residues that may remain on the surface. To do this, use pure alcohol, following the same procedure as with the hexane. Once they are removed, dry the surface with another piece of clean toilet paper to remove any marks that may remain.
4. Blow the surface of the preparation with an air pump or compressed air canister to remove any dust, lint or paper fibres that may have adhered to the preparation.

Once the process is completed, it is important not to touch the surface of the sample again and to handle it

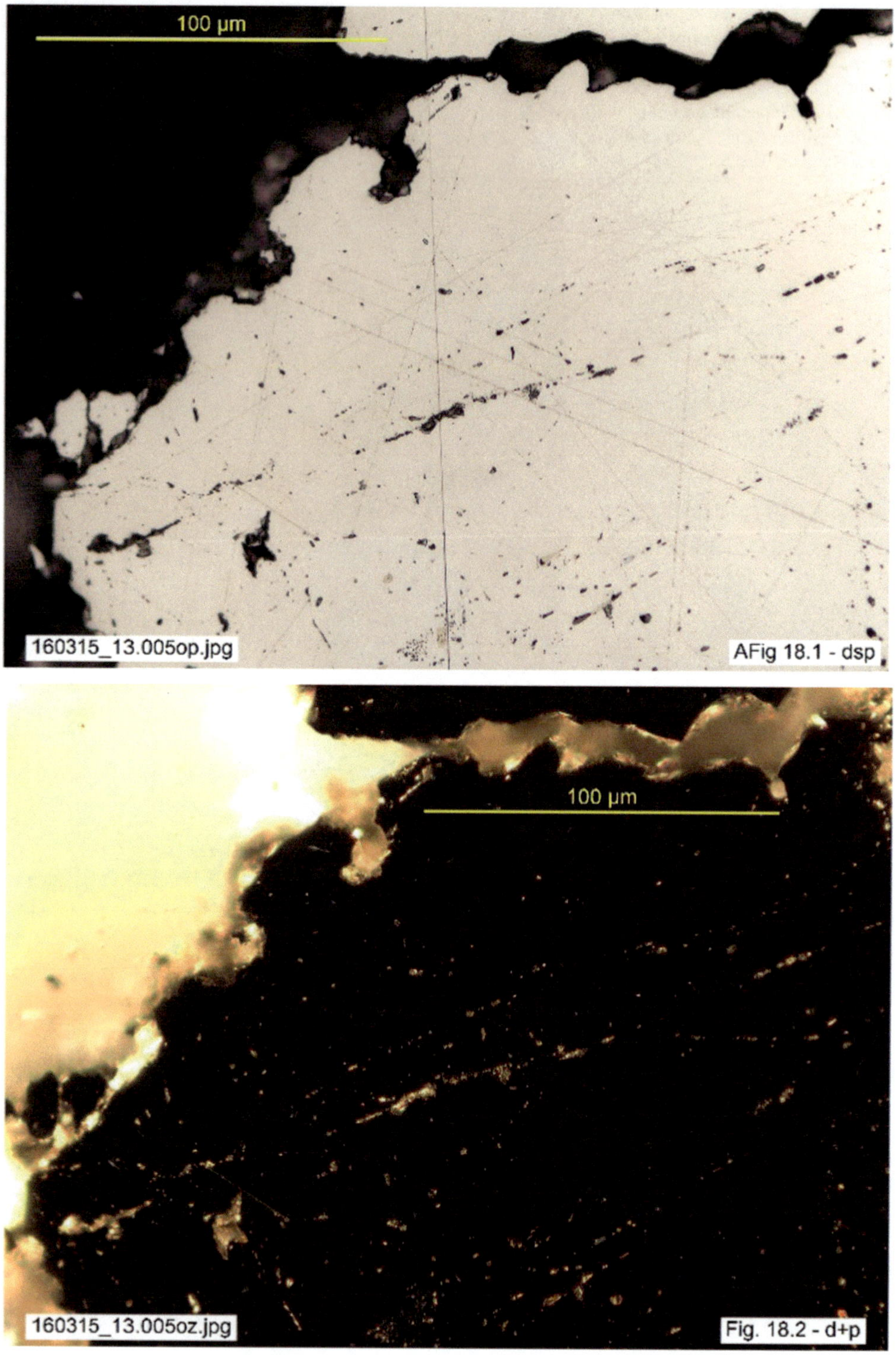

Fig. 3.18 Detail of the field in Fig. 3.17 after immersion in oil. With the oil, the zoning disappears both without the analyser (upper image, dsp) and with it crossed (lower image, d+p)

carefully. It is necessary to always check the cleaning result under the microscope. If drops, fingerprints or small drag marks are observed, go back to step 2. On the other hand, if a patina is still visible on the mineral, return to step 1.

Depending on the initial condition of the polished section, sometimes it is only necessary to clean from step 2 or 3 (if the marks are slight fingerprints).

Cleaning is an essential process that should be carried out whenever the previous examination indicates the need

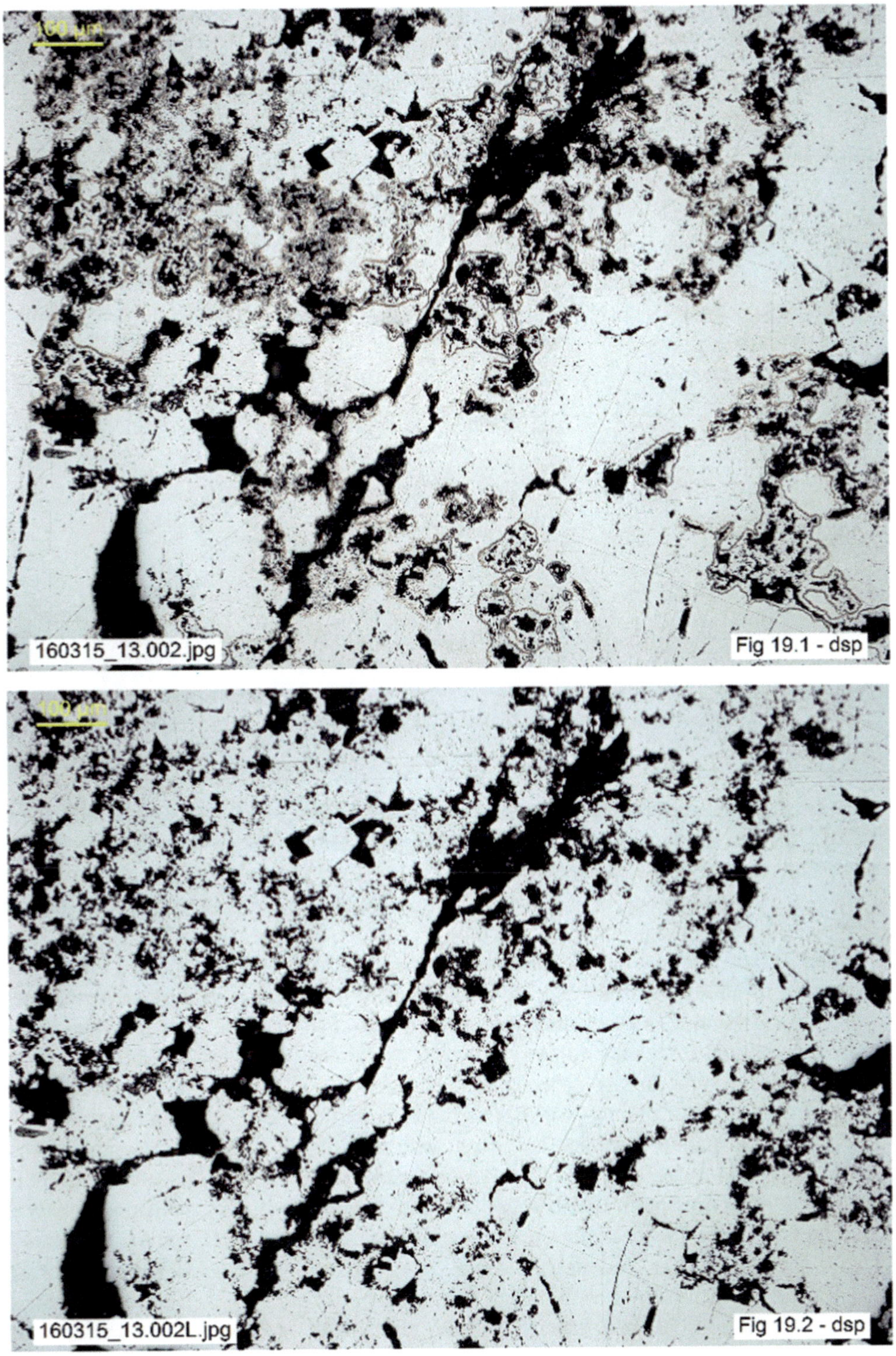

Fig. 3.19 Zoning artefact (upper image) that disappears after cleaning with hexane when the patina is removed (lower image)

(which is often the case when dealing with bornite, chalcopyrite or silver ores). Therefore, it should always be done with great care, because, although ¼ µ diamond paste will not thin the surface, it can create relief at the contact between hard and soft grains. If the samples are stored correctly, certain mineralogies that do not develop a great deal of patina may not need the first pass with diamond paste, which keeps their surface in good condition for longer.

Fig. 3.20 Example of quality finished polishing (without artefacts)

5. **Appendix: Introduction to Abrasive Types and Surfaces**

Three fundamental aspects must be taken into account: the type of abrasive; the size of the abrasive; and the nature of the minerals present in the polished section.

1. Type of abrasive: depending on the time and materials available and the nature of the sample, one type of abrasive or another is more suitable. According to their characteristics and finishes they are divided into five groups, ordered from most to least effective, explaining the advantages and disadvantages of each choice:

   1st. In the form of grains—whether carborundum, alumina or diamond—which are mixed with lubricating liquid (water or oil) to form a more or less fluid paste on the hard metal plate prepared for this purpose on the polishing machine rotor, or used directly on glass for manual grinding or polishing. This is the method with which the most perfect polishing is obtained, especially when minerals of different hardness are treated. As the abrasive is on a hard, flat surface, this does not deform during the work, so that the surface of the section remains horizontal and the contacts between grains are not rounded, as they are all ground equally. If the abrasive were spread on a soft surface, it would tend to wear when polishing hard minerals. Therefore, irregularities would be created on the surface, causing the abrasive to sink deeper into the soft minerals and leaving deep scratches, and accentuating the relief.
   Despite being the method giving the best results, it is a very slow process lasting about 20 h divided into four phases, the third being the longest (14–16 h) with a very fine grain size of abrasive (0.5 μm). In addition, it is important to ensure that the abrasive forms a thin, continuous film on the disc and on the surface of the sections during all phases. Although the method is very time-consuming for manual polishing, performance increases considerably with polishing machines that work on several samples simultaneously. For example, the classic Rehwald-Depiereux polishes six specimens at a time, with excellent results (cf. Castroviejo 1980), reaching a polishing quality that is almost without equal (vbgr. that of the Ramdohr-Rehwald collection, used in Heidelberg for the fundamental descriptions by Ramdohr 1980).

   2nd. Rather than a category in their own right, these are a variant of the first option: ready-made commercial abrasive liquids, such as Struers DiaDuo-2, a water-based solution with monocrystalline diamond and lubricant. Like the powder abrasives, they can be used on a rigid metal surface or on cloth. The implications of each choice are discussed later.
   In our experience, commercial abrasive suspensions are most useful in the final stages of cloth polishing.

3rd. Discs formed from a special alloy of metals (usually lead and antimony) that embed the grains of the abrasive material. These types are very common and give good results with samples containing minerals of varying hardness, but it should be borne in mind that their optimal use involves rigorous maintenance. This involves periodic roughening and levelling of the surface so that it remains horizontal and without protrusions that can create relief in the polished section.
Usually, sample preparation laboratories have several grinders and polishers and have a standardized preparation process, because the constant change of discs in each grain size is slower than working with sandpaper on a magnetic backing. Instead of changing the abrasive, they change the preparations to another machine previously prepared with a disc of a lower grain size. This method is useful and saves time in production lines, but it is not very flexible when it comes to adjusting to the particularities of each sample or batch of samples.
The roughening process is faster, and two or three phases of about 15 min per cycle is sufficient. The number of cycles required depends on the initial state of the sample.

4th. Sanding papers (usually impregnated with carborundum) on magnetic backing that are easily fixed on the polisher rotor. The results are somewhere between the first and the third option, provided they are used in good condition. In addition, they are quicker and easier to maintain. When the surface starts to wear (usually after five hours of use), simply peel the paper off the magnetic support and replace with a new one.
They are very useful for use in small laboratories with little infrastructure and to perform the initial coarser roughening (especially when the objective is to make a PB), due to their versatility and ease of changing from one to another. It is only necessary to take care that the sandpaper is well stuck to the support, without wrinkles, and that the machine is free of the previous abrasive, cleaning it with a damp cloth.
It is a fairly fast system, but it tends to leave numerous polishing streaks. It is normal to carry out three or four phases, in a similar way to the previous method, taking care not to make cycles of more than one or two minutes when using the coarser grits in the first stages of roughening (P120–600–800). From P1000 onwards, longer cycles are possible. For polished thin sections, it is recommended to dispense with grits than P800 in order to prevent the sample from disappearing during roughening. In preparations with soft minerals or minerals of different hardness, our own experience advises to start directly with P1000–1200 paper to avoid creating coarse grooves in the softer minerals and excessive relief.

5th. Exempt cloths and sanding papers. Cloth or billiard cloth is an alternative to the use of hard metal discs with abrasives in powder or suspension. In general, with certain exceptions that are discussed in the next point, it is recommended to avoid generalized use of this method. Personally, I reserve it for finishing polishing or even for cleaning samples with oxidation patinas, at low revolutions and in short cycles to avoid creating relief as much as possible. This is because it eliminates a large number of the scratches that may remain from the last phase of polishing with abrasive on a rigid support.
Finally, although not recommended, there is the option of using a carborundum-embedded but exempt sandpaper; that is, it is not placed on a rigid backing, rather the preparation is rubbed directly onto it. Sometimes it can be attached to the polisher rotor with a metal ring that fits into the rotor. However, whenever possible, use of this option is discouraged because it is easy for the paper to wrinkle or move around, resulting in grain-edge relief and even particle pull-outs. It is a technique that wears quickly. In return, it produces much streaking and embossing. Its use is only recommended in situations demanding quick action, where the minerals present are of high and homogeneous hardness.

2. Abrasive size: Using the correct abrasive grit size at each stage of both roughening and polishing is the basis of any well-polished preparation. Always start with the largest grit size required by the sample, always wash it thoroughly before moving on to the next abrasive and always check with an optical reflection microscope that each phase has been carried out correctly, paying particular attention to the horizontality of the sample (i.e. the parallelism of the surface with the base) and the appearance of scratches. In each phase, the streaks left by the previous one must be removed. If the diameter of the abrasive is reduced too quickly or the correct abrasive is not used, the process may take too long, may not progress or may even spoil the preparation by creating relief or causing particles to jump off the surface of the sample. Keep in mind that if, for example, we are grinding with carborundum with a P800–P1000–P1500 sequence, the thicker polishing scratches created with P800 that were not removed with by grinding with P1000 will not be

removed with P1500. This is why it is important to periodically check the grinding, and not to change phases until you are sure that the previous step is complete. It is not recommended to go back to coarser grits once the abrasive type has been changed.

3. Nature of the polished section: Each preparation has to be treated according to the predominant mineral composition. However, it is very common to find preparations with high contents of hard and soft minerals, such as a sample composed mainly of pyrite, bornite, galena and gangue. This heterogeneity usually makes it impossible to choose a polishing method for exclusively hard or soft minerals, so a compromise has to be reached. This usually involves softer polishing with finer grits, increasing the duration of cycles at low revolutions (maximum 50 revolutions per minute if soft and difficult to polish minerals are present).

   Although it is generally advisable to polish on hard surfaces, if we know in advance that the sample contains minerals that are difficult because of their tendency to generate pull-outs or their low hardness, it may be interesting to use a cloth with a suspension of abrasive in water and lubricant, as long as there are no significantly harder minerals present (for the reasons explained above). Also some hard minerals, such as pyrite, can benefit from this type of material during the finishing phase (Galopin and Henry 1972).

References

Castroviejo R (1980) Sobre el desarrollo de técnicas auxiliares para la investigación de menas metálicas: preparación de secciones pulidas para su estudio microscópico con luz reflejada. In: Jornadas Minerometalúrgicas (VI national and IV international), pp 529–559. Huelva

Galopin R, Henry N (1972) Microscopic study of opaque minerals. W. Heffer & Sons, Cambridge

Grunwald Romera U, Catalina J, Alarcón D, López Benito A, Castroviejo R (2019) A reliable method for the automated distinction of quartz gangue and epoxy resin with reflected light microscopy for geometallurgical characterization. In: 15 Biennial SGA meeting—life with ore deposits on earth, vol 4. Society for Geology Applied to Mineral Deposits, Glasgow, pp 1528–1531

Ramdorh P (1980) The ore minerals and their intergrowths. Pergamon Press

Røisi I, Aasly K (2018) The effect of graphite filler in sample preparation for automated mineralogy—a preliminary study. Mineralproduksjon 8:A1–A23

Annex 4: Ore Properties Compared

Annex 4.1 Specular Reflectance of Common Ores(*) (Estimated Values in Air, White Light)

Group	Abbrv.	Mineral	R (%)
Very high R (R $\geq$ 60%)	el	Electrum	≈ 90–88
	Ag	Silver	85
	Cu	Copper	81
	Sb	Antimony	74
	Au	Gold	72
	Bi	Bismuth	63.5
	aag	Allargentum	70.9–69.5
	Te	Tellurium	67
	Ir	Iridium	65
	sba	Stibarsen	64
	dy	Discrasita	62.7–59.9
	Pt	Platinum	63
	krn	Krennerite	62
	clv	Calaverita	62
	prm	Pararammelsbergite	60
High R (60 > R $\geq$ 40)	Os	Osmio	59
	Fe	Iron	59
	rm	Rammelsbergite	58
	sk (Ni)	Skutterudite (Ni)	57
	gud	Gudmundite	54
	sk (Co)	Skutterudite (Co)	54
	lo	Löllingite	53.5
	sf	Safflorite	53
	ml	Millerite	53 (48)
	asp	Arsenopyrite	52
	mc	Marcasite	52 (46)
	As	Arsenic	52
	spy	Sperrylite	52
	py	Pyrite	51
	nc	Nickeline	50.5 (45)
	gld	Glaucodot	50, 1

(continued)

R. Castroviejo, *A Practical Guide to Ore Microscopy—Volume 1*,
https://doi.org/10.1007/978-3-031-12654-3

(continued)

Group	Abbrv.	Mineral	R (%)
	cob	Cobaltite	50
	mau	Maucherite	49.5 (39)
	bth	Breithauptite	49
	clt	Clausthalite	49
	pn	Pentlandite	49
	bm	Bismuthinite	49 (37)
	ln	Linneite	47.5
	sbt	Stibnite	47 (30)
	ul	Ullmannite	46.5
	gf	Gersdorffite	46
	ccp	Chalcopyrite	45
	gn	Galena	43
	brv	Bravoite	43–35
	erl	Erlichmanite	≈ 42.7–41.8
	bl	Boulangerite	42
	mck	Mackinawita	42 (24)
	mol	Molybdenite	42 (19)
	brt	Berthierite	41.5 (30)
	hss	Hessite	40.5
Medium R (40 > R $\geq$ 25)	gc	Geochronite	39.5
	flb	Freieslebenite	39.5
	po	Pyrrhotite	39
	cbn	Cubanite	39
	colo	Coloradoite	36.5
	bnn	Bournonite	36
	my	Miargyrite	36
	nau	Naumannite	36
	prl	Pyrolusite	35
	agp	Argentopyrite	35
	hol	Hollandite	33
	agl	Aguilarite	33
	ac	Acanthite	32
	cc	Chalcocite	31.5
	stm	Stromeyerite	31
	plb	Polybasite	31
	prg	Pyrargyrite	31
	td	Tetrahedrite	31
	isc	Isocubanite	35.5
	pc	Pearceite	30.5
	pru	Proustite	30
	stp	Stephanite	30
	tn	Tennantite	29
	ht	Hematite	29
	stn	Stannite	28
	en	Enargite	28
	cin	Cinnabar	28
	lz	Luzonite	27.5
	cup	Cuprite	27
	fm	Famatinite	27
	psi	Psilomelane	27

(continued)

(continued)

Group	Abbrv.	Mineral	R (%)
	cry	Cryptomelane	≈ 26.7
	orp	Orpiment	26
	rom	Romanechite	25.5
Low R (25 > R ≥ 12)	mgh	Maghemite	24
	rt	Rutile	23.5
	orp	Orpiment	23.1
	cv	Covellite	23 (8)
	alb	Alabandite	23
	bn	Bornite	23
	dg	Digenite	22
	ilr	Ilmenorutile	≈ 22? (≤ rt)
	rl	Realgar	21
	mt	Magnetite	21
	ger	Germanite	21
	mng	Manganite	20 (14)
	il	Ilmenite	19.5
	bra	Braunita	19.5
	lpc	Lepidocrocite	19 (9)
	fk	Franklinite	18
	tp	Tapiolite	18
	wf	Wolframite	18.5–16
	gth	Goethite	18–15
	sp	Sphalerite	17
	wz	Wurtzite	17
	van	Vanadinite	≈ 17–16
	urn	Uraninite	16.8
	cbt	Columbotantalite	16.5
	vll	Valleriite	16 (9)
	hub	Hübnerite	16
	um	Umangite	16
	gra	Graphite	16 (6)
	ulv	Ulvöspinel (Ulvite)	15.5
	lim	Limonite	≈ 15–10
	bad	Baddeleyite	14.8–13.4
	pyc	Pyrochlore	13.5
	cst	Cassiterite	13
	chr	Chromite	12.5
Very low R (R < 12)	sch	Scheelite	10
	az	Azurite	9.5 (6)
	mch	Malachite	9.5 (6)
	mz	Monazite	≈ 9–8
	cag	Chlorargyrite	9
	rdl	Ramsdellite	9–41
	bst	Bastnäsite	≈ 8.5–7.2
	spn	Spinel	8
	crc	Chrysocolla	≈ 5.5–5
	q	Quartz	5

(*)Based on data from QDF3-Criddle and Stanley (1993), Picot and Johan (1982), Galopin and Henry (1972) Hagel (1979), Uytenbogaardt and Burke (1971), Schneiderhöhn (1952) and author's observations or measurements

Annex 4.2 Polishing Hardness and Scratch Hardness

Annex 4.2.1 characterizes common ores by their **Polishing Hardness** (***PolH***): in Table 4.2.1.1 they are sorted by direct comparison, and listed according to decreasing *PolH*. If there is no accurate data(*) for a precise position, they are sorted by *PolH* groups (Table 4.2.1.2).

Annex 4.2.2 defines the **Talmage** scratch hardness types (**Talm**) and arranges the documented ores into groups in relation to these types (listed according to Schneiderhöhn 1952).

Annex 4.2.1

Table 4.2.1.1 Compared polishing hardnesses (PolH) of common ores(*)

PolH	Abbrv.	Mineral
PolH Group 5 (very high)	C	Diamond
	SiC	Carborundum
	cor	Corundum
	spn	Spinel
	cst	Cassiterite
	prl	Polianite (xx-prl)
	q	Quartz
	tp	Tapiolite
	cbt	Columbotantalite
	spy	Sperrylite
PolH Group 4 (high)	py	Pyrite
	mc	Marcasite
	ht	Hematite
	rt	Rutile
	urn	Uraninite
	cob	Cobaltite
	asp	Arsenopyrite
	gld	Glaucodot
	sch	Scheelite
	bra	Braunita
	chr	Chromite
	il	Ilmenite
	wf	Wolframite
	ulv	Ulvöspinel (Ulvite)
	mgh	Maghemite
	mt	Magnetite
	lo	Löllingite
	ul	Ullmannite
	gf	Gersdorffite
	po	Pyrrhotite
	Ir	Iridium
PolH Group 3 (medium)	Pt	Platinum
	pn	Pentlandite
	sf	Safflorite
	rm	Rammelsbergite
	sk (Co)	Skutterudite (Co)
	sk (Ni)	Skutterudite (Ni)
	prm	Pararammelsbergite
	mau	Maucherite
	bth	Breithauptite
	nc	Nickeline
	ln	Linneite

(continued)

Table 4.2.1.1 (continued)

PolH	Abbrv.	Mineral
	gud	Gudmundite
	Fe	Iron
	sp	Sphalerite
	wf	Wurtzite
	mch	Malachite
	stn	Stannite
	az	Azurite
	cup	Cuprite
	lim	Limonite(**)
	ml	Millerite
	Cu	Copper
	td	Tetrahedite
	Ag	Silver
	el	Electrum
	Au	Gold
	en	Enargite
	lz	Luzonite
	gra	Graphite
	mol	Molybdenite
	cbn	Cubanite
	ccp	Chalcopyrite
	bn	Bornite
	bnn	Bournonite
	bm	Bismuthinite
	dg	Digenite
	gn	Galena
PolH Group 2 (low)	cc	Chalcocite
	bl	Boulangerite
	clv	Calaverite
	dy	Dyscrasite
	cin	Cinnabar
	clt	Clausthalite
	um	Umangite
	my	Miargyrite
	As	Arsenic
	sba	Stibarsen
	Sb	Antimony
	stm	Stromeyerite
	cv	Covellite
	Bi	Bismuth
	brt	Berthierite
	sbt	Stibnite
PolH Group 1 (very low)	*orp*	Orpiment
	stp	Stephanite
	rl	Rejalgar
	prg	Pyrargyrite
	pru	Proustite
	agp	Argentopyrite
	plb	Polybasite
	pc	Pearceite
	ac	Acanthite
	cag	Chlorargyrite

(*)Compiled after Schneiderhöhn (1952), Ramdohr (1980) and author's observations

(**)**Goethite** should be listed in PolH Group 3 (Medium), but actually it is assumed to be comprised in **lim**: after Schneiderhöhn's German tradition goethite (= Nadeleisenerz) is one of the mixed components of **limonite**

Table 4.2.1.2 PolH of other ores (in Groups)(*)

Abv	Mineral	Group
agl	Aguilarite	1
alb	Alabandite	3
bst	Bastnäsite	3
colo (cld)	Coloradoite	2
cry	Cryptomelane	4
erl	Erlichmanite	5
fm	Famatinite	3
fk	Franklinite	4
flb	Freieslebenite	2
gth	Goethite	3
ilr	Ilmenorutile	4
isc	Isocubanite	3
krn	Krennerite	2
mck	Mackinawite	3
mng	Manganite	4
mz	Monazite	3
nau	Naumannite	2
Os	Osmium	5
pyc	Pyrochlore	3
psi	Psilomelane	4
rdl	Ramsdellite	3
rom	Romanechite	3–4
Te	Tellurium	2
tit	Titanite	4
vll	Valleriite	1

(*)The ores are ordered alphabetically by their abbreviations (**Annex 1**), and their PolH is reported as an estimated value, defined by the numbers of the corresponding groups in Table 4.2.1.1

Annex 4.2.2 Talmage Hardness (*Talm H*)(*)

Group	Type mineral	Load (g)
A	Acanthite	0.105
B	Galena	0.201
C	Chalcopyrite	0.365
D	Tetrahedrite	0.520
E	Nickeline	0.775
F	Magnetite	1.035
G	Ilmenite	1.550

Common ores sorted by *Talm H*: increasing downwards and to the right(**)

Group A Acanthite type	Group B Galena type	Group C Chalcopyrite type	Group D Stibnite type	Group E Nickeline type	Group F Magnetite type	Group G Ilmenite type
ac	cin−	prg−	znc−	tno−	gf−	psi−
agl	Te−	colo−	brt−	lim−	mc−	cob−
hss+	mol−	dy−	in−	ml	py−	urn
clt+	gn	syl−	po−	mau	sk−	il
Bi+	cc	plb−	pn−	rm	lo	rt+
	Au	bnn−	wz−	nc	asp−	cst+
	ker	As	cup−	bth	mt	chr+
	nau	aik	stn−	sf	gld	
	rl	pc	lz	ul	fk+	
	Ag	ccp	td	ln+		
	stp	zk	tn	hm+		
	stm	alb				
	Cu	clv				
	my	cbn				
	Um	jd				
	pru+	krn				
	bl+	fm+				
	sbt+	sp+				
	cv+					
	orp+					
	bn+					
	jm+					
	bm+					
	tdy+					

(*)Scale (A–G) defined by mineral type according to the load (g) of the microsclerometer which produces a standard scratch
(**)The scratch hardness of ores marked +/− may vary, resulting in: DTalm $\geq / \leq$ the type, respectively (after Schneiderhöhn 1952). Mineral abbreviations: cf. **Annex 1**

Annex 4.3 Minerals with Distinct Internal Reflections (I.R.)

IR Colour(*)	Mineral (Abbrv.)	Remarks(**)
Blue	Azurite (az)	m
	Chrysocolla (crc)	m. Turquoise blue
Green	Malachite (mch)	m
	Annabergite (anbg)	m
	Spinel (spn)	m/d
	Alabandite (alb)	d
Pink	Erythrine (ery)	m
Yellow	Oropimente (orp)	m
	Sulfur (S)	m
Yellow to reddish brown	Sphalerite (sp)(***)	m/d
	Wurtzite (wz)	Colorless to *RdBr*
	Cassiterite (cst)	m. Colorless to YlBr
Yellowish red to reddish brown	Rejalgar (rl)	m
	Lepidocrocite (lpc)	m
	Goethite (gth)	m
	Titanite (tit)	m
	Pyrochlore (pyc)	m
	Rutile (rt)	m/d
	Wolframite (wf) (Hübnerite, hub)	d (m, like sp)
	Chromite (chr)	d
	Tetrahedrite (td)-Tenantite (tn)	o/d
	Boulangerite (bl)	o
	Columbotantalite (cbt)	o
	Uraninite (urn)	o
	Ilmenite (il)	o
Red	Cinnabar (cin)	m
	Pyrargyrite (prg)	m
	Proustite (pru)	m
	Kermesite (ker)	m
	Zincite (znc)	m
	Miargirite (my)	m
	Pearceite (pc)–Polybasite (plb)	d/m
	Hematite (ht)	d
	Cuprite (cup)	d
	Manganite (mng)	d
	Franklinite (fk)	d
Orange	Vanadinite (van)	m
Varied	Fluorite (fl)	m
Colorless	Anatase (ana)–Leucoxene (lcx)	m
	Bismite (bim)	m
	Cervantite (cerv)	m
	Senarmontite (sen)	m
	Scheelite (sch)	m
	Zircon (zir)	m
	Gangue (gg, undif.)	m/d

(*)*Yl* yellow; *Br* brown; *RdBr* reddish brown; *YlBr* yellowish brown

(**)Abudance of IR: **m** = massive to abundant; **d** = discrete to frequent; **o** = scarce to occasional

(***)I.R. of sphalerite vary with composition from: colorless or yellowish (m, *blenda acaramelada*: pure ZnS) to *RdBr* and blood red (d, *marmatite*: Fe-rich sp.); even green (Ga-bearing varieties)

Annex 4.4 Distinctly Bireflective or Pleochroic Ores

Very strong to distinct bireflectance versus pleochroism(*)			
Mineral (abbrv.)	B	P	Remarks(**)
Graphite (gra)	+++	++	**B & P** very strong. R minimum ≈ R_{gg}
Covellite (cv)	+++	+++	Dramatic color change: Blu/Vi (d/o)
Molybdenite (mol)	+++	++	Strong birrefl. White-gray, Rmin > R_{gg}
Umangite (um)	++	+++	Strongly pleochroic: *OgViRd-YlGyBr*
Valerıte (vll)	+++	++	**B & P** both strong, *YlBr-Gy*. Rmin > R_{gra}
Stibnite (sbt)	+++	+	Strongly birreflectant. Stays gray
Breithauptite (bth)	+	++	*ViPi to ViGy*
Nickeline (nc)	+	++	*PiWh-YlBrPi*
Kermesite (ker)	+	++	*BluGy-GyBr-OlGr*
Berthierite (brt)	+	++	*WhGy-BrPiGy*
Oropimente (orp)	++	+	Yellowish gray: **B** ≫ **P**
Lepidocrocite (lpc)	++	–	White to gray: **B** ⋙ **P**
Pyrolusite (prl)	++	++	Yellowish white to gray
Stephanite (stp)	+	+	*Gy-BrGy or PiGy*
Bismuthine (bm)	+	+	*YlWh-GyWh*
Stromeyerite (stm)	+	+	*PiGy-ViG/BluGr*
Millerite (ml)	+	+	*Yl-YlWh*
Cubanite (cbn)	+	+	*PiBr-GyBr*
Enargite (en)	+	+	*PiBr-ViGy*, dark
Gudmundite (gud)	+	+	*Wh-PiWh*
Jamesonite (jm)	+	+	Gray to greenish gray
Pyrrhotite (po)	+	+	*PiBr-YlBr*
Luzonite (lz)	+	+	*OgYl-PiYl*
Marcasite (mc)	–	+	*GnWh-PiWh (oil: osp)*
Manganite (mng)	+	–	Shades of gray
Miargirite (my)	+	–	White–gray
Boulangerite (blg)	+	–	Shades of gray
Calaverite (clv)	+	–	Yellowish white
Goethite (gth)	+	–	Shades of gray
Proustite (pru)	+	–	Shades *BluGy*
Pyrargyrite (prg)	+	–	Shades *BluGy*
Pearceite (pc)	+	–	*GnGy* shades
Rutile (rt)	+	–	Light shades of gray
Tellurium (tea)	+	–	White (**B**<, **R**⋙)
Cinnabar (cin)	+	–	*GyWh*
Malachite (mch)	+	–	Carbonates: under reflected light they are gray colored ≈ gg, and show: • **R** low, but > most silicates • **B** ≫ silicate gg
Cerussite (crs)	+	–	
Siderite (sid)	+	–	
Carbonates (Carb)	+	–	

(*)Bireflectance (**B**) or pleochroism (**P**) **intensity**, in air (*dsp*): +++ = very strong; ++ = strong; + = distinct; – = imperceptible
Visible **effects** (*dsp/osp*), according to B/P intensity: **P+**: color changes clearly with stage rotation. **B+/P–**: no color change, but *brightness* varies (lighter–darker) with stage rotation
(**)Color **abbreviations** (see **Annex 1** or § **1-Box 1.1**). *Wh* white, *Gy* gray, *Bk* black, *Blu* blue, *Gn* green, *Yl* yellow, *Og* orange, *Rd red*, *Pu* purple, *Vi* violet, *Pi* pink, *Ol* olive, *Br* brown. Color combinations: *BrRd* brownish red, *GnGy* greenish gray, etc.

Annex 4.5. Isotropic or Very Weakly Anisotropic Minerals(*)

Color	Mineral	Abbrv.	Remarks
White, gray or faintly colored ore minerals(**)			
White	Native Platinum	Pt	Very bright
	Iridium	Ir	Very bright
	Pentlandite	pn	Faint brownish-light yellow tone
	Galena	gn	Triangular pits (*T-pits*), soft
	Clausthalite	clt	T-pits; ≈ gn, but brighter
	Laurite	lrt	PGM, white with slightly bluish tinge, very hard
	Erlichmannite	erl	
	Sperrylite	spy	PGM, very hard
	Gersdorfite	gf	Frequently autom., zoned, T-pits
	Skutterudite	sk	Frequent zoning
	Ullmannite	ul	T-pits
Yellowish white	Pyrite	py	Occas. anomalous anisotropism
	Native Silver	Ag	Very shiny but tarnishes quickly
Pink white	Cobaltite	cob	Frequently automorphic
Grayish white	Coloradoite	colo	Grayish white, slightly brownish
	Maghemite	mgh	Bluish shade
	Tetrahedrite (*s.l.*)	td	Broad group. Olive-gray shades
Gray	Uraninite	urn	Internal reflections
	Spinel	spn	Internal reflections
	Chromite	chr	Internal reflections
	Pyrochlore	pyc	Internal reflections
	Franklinite	fk	Internal reflections
	Alabandite	alb	Internal reflections
	Sphalerite	sp	Internal reflections
Brownish gray	Magnetite	mt	Occas. anomalous anisotropism
	Jacobsite	jac	Occasional IR
Distinctly colored ore minerals(**)			
Yellow	Native Gold	Au	Very bright, imperfect extinction
	Electrum	el	Very bright, imperfect extinction
	Amalgam	am	Very bright (≈ Eugenite, Schachnerite)
Red	Copper	Cu	Quick tarnish (darker patina)
Blue	Digenite	dg	Pale blue color next to cv
Violaceous	Germanite	ger	GrPi hue ≈ bn (often associated)
Pink to yellowish-brown	Linneite (*s.l.*)	ln	Large and varied group
	Bravoite	brv	Zoned variety of py (Ni…)
	Melnicovite	mln	Pyritic gel

(*)Cubic (isotropic, but sometimes with imperfect extinction) or amorphous and other weakly anisotropic ores

(**)These are the natural colors of the mineral, seen **without analyzer** (i.e. ***dsp/osp***). They are the only colors relevant in isotropic ores, which with analyzer (*d+p/o+p*) are always in extinction

They are described by *13 hues* or their combinations: *Wh* white, *Gy* gray, *Bk* black, *Blu* blue, *Gn* green, *Yl* yellow, *Og* orange, *Rd* red, *Pu* purple, *Vi* violet, *Pi* pink, *Ol* olive, *Br* brown

Annex 4.6 Polarization Colors of Distinctly Anisotropic Ores(*)

Dominant colors (+P)	Mineral	Abbrv.	Remarks(**)
Yellow	Millerite	ml	Yl-GnBlueGr, light shades
	Graphite	gra	Anistr. extreme, very bright
	Valleriite	vll	Anistr. extreme, very bright
	Pyrolusite	prl	
	Native Arsenic	As	Very fast and dark tarnishing
	Famatinite	fm	GnYl vivid shades, twins
Orange (to OgBr-RdBr)	Osmium	Os	Very bright anisotropism
	Antimony	Sb	
	Stibarsen	sba	
	Pararammelsbergite	prm	
	Covellite	cv	Very bright golden tones
	Chalcocite	cc	Og-shaded but dull tones
	Gudmundite	gud	
	Ferroselite	fse	Bright anisotropism
	Vonsenite	von	GrViGn vivid, eye-catching tones
	Enargite	en	Strong, multicolored anisotr
Pink	Molybdenite	mol	Anistr. strong, very eye-catching
	Alloclasite	alc	
	Sternbergite	stb	Pi, RdBr, and GnGy shades
	Stibnite	sbt	Pi and bluish shades
Red	Umangite	um	OgRd shades
Brown	Pyrrhotite	po	
	Gratonite	gtt	
	Dyscrasite	dy	
	Maucherite	mau	
	Wolframite	wf	IR dominant in *hubnerite* (Mn)
	Argentopyrite	agp	Tones darker than stb
	Renierite	ren	
	Calaverite	clv	
	Stephanite	stp	Br + Yl and Gn shades
	Freieslebenite	flb	Br + Yl and BluGy shades
	Cosalite	cos	YlBr shades
Blue	Stromeyerite	stm	Luminous shades
	Cubanite	cbn	
	Nickeline	nc	Bright and vivid GnBlu shades
	Goetithe	gth	BluWh to Gy. IR abundant in *lim*
	Polybasite	plb	IR
	Bournonite	bnn	GnBlu anisotropism
	Arsenopyrite	asp	Assorted colors: BluPuVi...
	Löllingite		Blu-YlBr
	Rammelsbergite	rm	ViBlu shades
	Safflorite	sf	Vivid colors, twinning stars
	Miargyrite	my	BluWh, IR
	Jordanite	jd	GnBlu-Vi
	Geochronite	gc	GnBlu
Green	Marcasite	mc	Strong and very bright
	Pearceite	pc	Anistr. stronger than plb, IR
	Berthierite	brt	
	Bismuthinite	bm	
	Native Bismuth	Bi	Glossy, but polishing scratches

(continued)

(continued)

Dominant colors (+P)	Mineral	Abbrv.	Remarks(**)
	Breithauptite	bth	
	Luzonite	lz	
	Cinnabar	cin	Very abundant IR
	Cuprite	cup	Frequent/abundant I.R.
	Jamesonite	jm	GnGr
	Boulangerite	blg	GnGr
Violet	Kermesite	ker	Violet to blue, eye-catching
	Stannite	stn	
Violaceous gray to greenish gray	Acanthite	ac	ViGy to BluGn dominant
	Ilmenite	il	GnGr
	Hematite	ht	GnGr, IR
Gray	Manganite	mng	
	Pyrargyrite	prg	IR masks strong anisotropism
	Sylvanite	syl	BrGy
	Krennerite	krn	
	Tellurium	Te	Gy-BrGy
	Romanechite	rom	
White (very bright)	Mackinawite	mck	Extreme, bright anisotropism
	Lepidocrocite	lpc	Very abundant IR

(*)Only clearly anisotropic ores. Cubic ores are excluded, even if showing anomalous anisotropism (such as pyrite or magnetite) or imperfect extinction (such as native metals: Ag, Au, etc.)

(**)The colors mentioned in this table are not the natural colors of the mineral, i.e. those seen without analyzer (*dsp/osp*), but those characterizing the *effects of anisotropism* (with analyzer crossed at 90°: ***d+p/o+p***): they refer exclusively to the **polarization colors**, especially those dominant in positions of maximum illumination

Color abbreviations (13 *hues* used): *Wh* white, *Gy* gray, *Bk* black, *Blu* blue, *Gn* green, *Yl* yellow, *Og* orange, *Rd* red, *Pu* purple, *Vi* violet, *Pi* pink, *Ol* olive, *Br* brown. Their combinations: *GnGy* greenish gray…

Annex 4.7 Common Ores with Typical Natural Colors

Common colored ores(*)								
Color	Intense		Distinct		Faint		Just a nuance	
	Mineral	Abbrv.	Mineral	Abbrv.	Mineral	Abbrv.	Mineral	Abbrv.
Blue	Covellite	cv	Digenite	dg	Pyrargyrite	prg	Cuprite	cup (BluGy)
					Proustite	pru		
Yellow	Native Gold	Au	Millerite	ml	Pyrite	py	Native Silver	Ag (YlWh)
	Chalcopyrite	ccp			Electrum	el		
Red	Copper	Cu	–		–		–	
Pink	Bornite (fresh)	bn	–		–		Cobaltite	cob
							Maucherita	mau (PiWh)
Brown	–		Pyrrhotite	po	Magnetite	mt	–	
			Enargite	en				
Violet	Germanite	ger	–		–		–	
	Bornite (tarnished)	Bn_T						
OgYl	–		Renierite	ren	–		–	
			Luzonite	lz				
			Nickeline	nc				
rYl	–		–		Melnicovite	mln	–	
PiBr	-		Ilmenite	il	Linneite	ln	–	
					Bravoite	brv		
ViPi	–		Breithauptite	bth	–		–	
			Famatinite	fm				
OlGn-GyBr	–		–		Tetrahedrite	td	–	
					Estannita	stn		

(*)Natural colors of the ore, without polarization. Microscope setting: *dsp/osp* (no analyzer). *White–Gray–Black "colors" excluded*
Color symbols (or abbreviations, § **Annex 1** or § **1-Box 1.1**). *Wh* white, *Gy* gray, *Bk* black, *Blu* blue, *Gn* green, *Yl* yellow, *Og* orange, *Rd red*, *Pu* purple, *Vi* violet, *Pi* pink, *Ol* olive, *Br* brown. Color combinations: *BrRd* brownish red, *GnGy* greenish gray, *YlWh* yellowish white, etc.

References

Criddle AJ, Stanley CJ (eds) (1993) Quantitative data file for ore minerals, 3rd edn. Chapman & Hall, London, 635 p

Galopin R, Henry NFM (1972) Microscopic study of opaque minerals. Heffer, Cambridge, 322 p

Hagel E (1979) Microscopía de Menas. Instº Geología Económica Aplicada, Universidad de Chile, Santiago, 205 p

Picot P, Johan Z (1982) Atlas of ore minerals. B.R.G.M./Elsevier

Ramdohr P (1980) The ore minerals and their intergrowths, 2nd edn, 2 vols. Pergamon Press, Oxford, 1205 p

Schneiderhöhn H (1952) Erzmikroskopisches Praktikum. E. Schweizerbart'sche Verlagsbuchhandlung, Stuttgart, 274 p

Uytenbogaardt W, Burke EAJ (1971) Tables for microscopic identification of ore minerals, 2nd edn. Elsevier, Amsterdam

Annex 5 Determinative Tables

Annex 5.1 Background and Methodology

Since the initial determinative tables published by Schneiderhöhn and Ramdohr (1931), there were several proposals of tables intended to help the microscopist. The aim was to establish effective criteria to achieve an accurate diagnosis of the identity of the ore in practice. They comprised versions in both book (Schouten 1962; Uytenbogaardt and Burke 1971) and magazine format (Tarkian 1974; Spry and Gedlinske 1987; Tarkian and Liessmann 1991).

The first (*Erzmikroskopische Bestimmungstafeln*, Schneiderhöhn and Ramdohr 1931) responds in 47 pages to the objective of systematizing and guiding the observations aimed at identifying ores with the aid of the reflection microscope. They comprised the laboratory tests available then, including physical measurements, structural etching and microchemical tests, among others.

The means available have evolved greatly since then, so this early and valuable contribution is no longer applied in its original form. However, it opened the way to modern approaches. It prefigures and combines the two main determinative lines followed until now: qualitative (observation) and quantitative (determination of physical properties or chemical composition).

The quantitative pathway has diversified enormously, first with the application of precise instrumental techniques for the measurement of properties such as hardness or specular reflectance. They had a great impact through their combination in *key diagrams* or **VHN-R% tables** (Bowie and Taylor 1958;

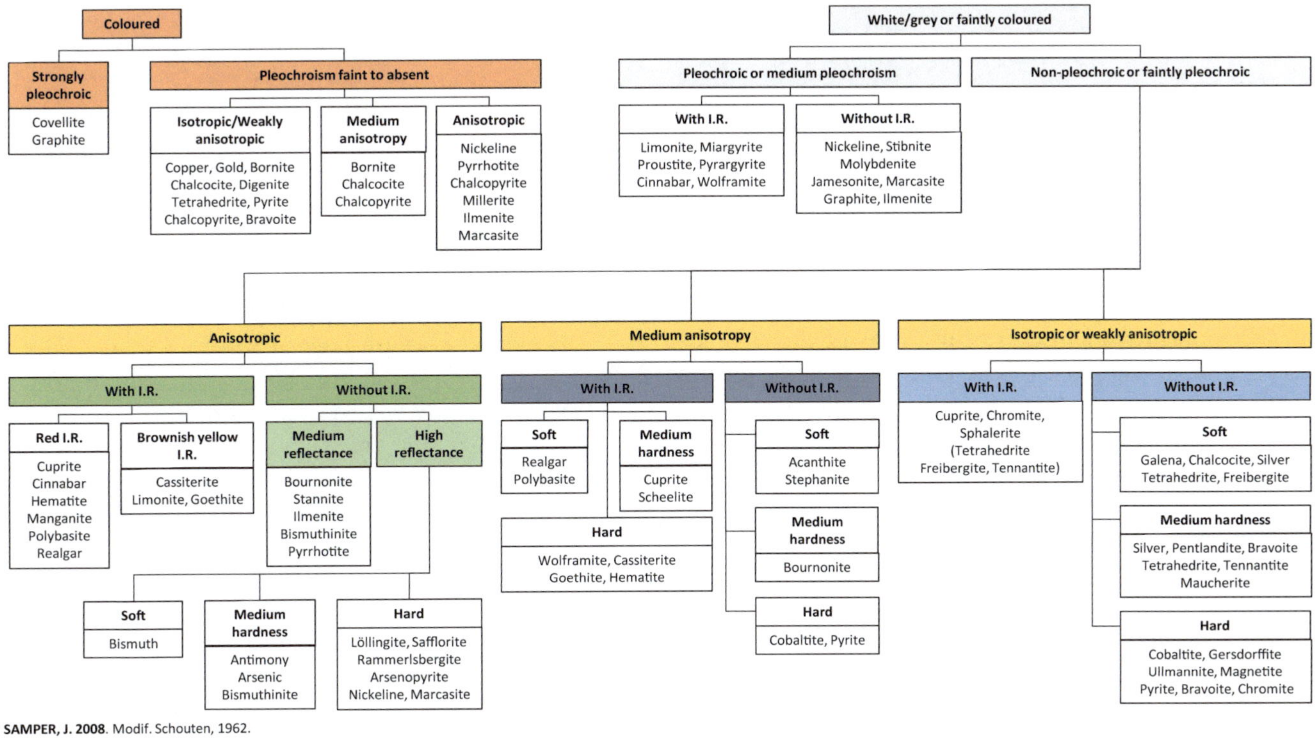

Fig. 5.1 Determinative table (Samper 2008, after Schouten 1962)

R. Castroviejo, *A Practical Guide to Ore Microscopy—Volume 1*,
https://doi.org/10.1007/978-3-031-12654-3

McLeod and Chamberlain 1968; Tarkian 1974; Tarkian and Bernhardt 1984; Tarkian and Liessmann 1991). Then, with the current development of microanalytical techniques (EMPA, SEM, etc.), electronics and computer methods, automation was made possible (cf. § **Chap. 1** and Vol. 2, Chap. 2).

With regard to the tables following the qualitative line, in the most modern versions (e.g. Spry and Gedlinske 1987, after Schouten 1962) an attempt has been made to simplify their use, presenting the information in the form of systematic schemes or *flowcharts,* with decision nodes based on "yes/no" or "±" responses to opposing options. As a result, the number of ores selected for comparison decreases and the amount of information processed is smaller, yet for the beginner they can be a useful support (*vbgr.* Price 2013; Samper 2008: Fig. 5.1), as they help to interpret, compare and critically assess the observations, and to work them out systematically, overcoming the insecurity of the first attempts.

However, as Schouten (1962) already noted, simplifying an identification process based on inevitably varied and complex microscopic observations, to the point of reducing it to a brief outline, limits the rigour of the method and, it has been shown, the accuracy of the results. In comparison, the methodology of Galopin and Henry (1972) is very enlightening. This combines explanatory text with a re-elaborated and abbreviated version of Schneiderhöhn and Ramdohr' tables (1931), reduced to nine items combined as follows:

Ores	Colorless	Slightly colored	Clearly colored
Isotropic	Table I	Table I B	Table I C
Weakly anisotropic	Table II A	Table II B	Table II C
Strongly anisotropic	Table III A	Table III B	Table III C

The use of these tables is related to the follow-up of the text itself (Galopin and Henry 1972) and of the practical exercises that it proposes to boost the beginner's progress.

Methodological approaches of this type are probably the most appropriate strategy. It is also the choice of the present text. It is illusory to think that an inexperienced microscopist can correctly interpret and apply an advanced determinative scheme without having developed the necessary observational skills and mastered the basic concepts. If the practitioner is an experienced microscopist, they will most likely no longer need a simple scheme. Therefore, it does not seem practical to insist on the elaboration of simple schemes of general application and questionable practical results. As a solution, the whole text is conceived with recognition criteria and as a determinative method. Instead of a simple scheme, synthesis or support tables are offered (**Annexes 4** and **5**), which help to progressively reduce the indeterminacy until the focus is on identification.

Annex 5.2 Determinative Strategy and Use of the Tables

The alternative planned in the text is to integrate the determinative keys or diagnostic criteria with the description of each ore, and to provide independently the necessary orientations so that the microscopist can progress in their observations or measurements until able to determine alone and with certainty the identity of the mineral. This cannot be achieved without the observer having acquired the necessary experience; to omit this step, raising false expectations of easy and immediate success, is to lead the beginner to failure. This is why we insist on a two-step strategy:

1. If you are a beginner, you should first of all practice with the appropriate instrumentation and specimens, always with the support of an expert and a conceptual background that you should not neglect, until you acquire basic confidence in your own observation skills with the reflected light microscope (cf. **Box Annex 5.1**).

Box Annex 5.1 Recommendations for the Beginner It is advisable to write down one's observations **orderly** on a *worksheet* for each mineral. This sheet or card is a basic personal document that allows you to remember, follow up, correct and finally memorize the descriptions, as well as to control and evaluate your own progress. It is recommended to select ore samples from available collections in the following sequence:

1. Start with easy samples (comprising one or, at best, several ores selected in the Mineral Indices or in **Annex 1.2** as Learning Priority 1: e.g. pyrite, magnetite, sphalerite, galena, silver or native gold), observe them with a dry objective (in air) and with single polarizer (no analyzer: dsp):
   a. Compare and order the species found by their brightness and color, estimating relative values of their reflectances (to start with, take as a reference quartz gangue as R ≈ 5% and pyrite as R ≈ 55%);
   b. Do the same for a sample containing magnetite (R ≈ 20%);
   c. Do the same for a sample containing galena (R ≈ 43%) and/or sphalerite (R ≈ 17%).

d. Do the same for a sample containing either native silver (reference as R ≈ 90%) or native gold (reference as R ≈ 70%).
e. Do the same on all of them for their polishing hardnesses (*PolH*); compare, particularly, PolH of pyrite, magnetite, sphalerite and galena; describe cleavage, where this is visible (*triangular pits* of galena).

2. Observe the same samples with the analyzer crossed at 90° (***d+p***), and distinguish isotropic from anisotropic species (if any); compare and describe the **IR**, internal reflections of sphalerite and gangue; describe the effects of anisotropism, if anisotropic species are present.
3. Observe the same samples in immersion (**osp**), and note the changes in their appearance (brightness, color); compare the effect of immersion on gangue, magnetite and sphalerite.
4. Do the same with the analyzer crossed at 90° (***o +p***); compare the effect of immersion on gangue and sphalerite IR.
5. Compare your own observations with the descriptions of each mineral given in the text or elsewhere. Resolve any discrepancies with new observations or by consulting an expert.
6. Move now to samples—still *Priority* 1—of anisotropic minerals (such as hematite, covellite, chalcopyrite, arsenopyrite, marcasite, nickelite, pyrrhotite, molybdenite, graphite) and observe them following steps 1–5 for them all (not forgetting any of the four basic settings: ***dsp***, ***d+p***, **osp**, ***o +p***; and using the most suitable objective in each case). Enrich the observations until all outlined properties have been completed (**Boxes 1.1** and **1.2**). For example, describe and compare the bireflectance of hematite and molybdenite; the pleochroism of covellite, pyrrhotite and graphite; the twinning of chalcopyrite; the anisotropism effects (colors) or polarization tints of pyrrhotite, marcasite, molybdenite or graphite.
7. Continue with the same observation methodology until you are familiar with all Priority 1 and then Priority 2 ores (cf. Mineral Indices).

Box Annex 5.2 Most Frequent Minerals in Common Ores

Common ores: basic selection (Priority 1)		
Class	Mineral	Symbol
Elements	Native Gold	Au
	Native Silver	Ag
	Native Platinum	Pt
Sulfides/Arsenides	Acanthite	ac
	Arsenopyrite	asp
	Bornite	bn
	Chalcopyrite	ccp
	Chalcocite	cc
	Cinnabar	cin
	Covellite	cv
	Digenite	dg
	Sphalerite	sp
	Stibnite	sbt
	Galena	gn
	Linneite	ln
	Marcasite	mc
	Molybdenite	mol
	Nickeline	nc
	Pentlandite	pn
	Pyrite	py
	Pyrrhotite	po
Sulfosalts	Tetrahedrite (*s.l.* or fahlore)	td
Oxides, hydroxides, tungstates, etc.	Cassiterite	cst
	Chromite	chr
	Cuprite	cup
	Goethite (and Limonite)	gth
	Hematite	ht
	Ilmenite	il
	Magnetite	mt
	Pyrolusite	prl
	Romanechite (Psilomelane)	rom (psi)
	Rutile	rt
	Uraninite	urn
	Wolframite	wf
Gangue	Quartz	q
	Carbonates	Carb
	Coal and Organic matter	COM

2. If you are an experienced professional or have successfully passed step 1, you may find the following guidelines helpful. In any case, you should first make sure that you recognize the fundamental properties of at least the elementary selection of priority minerals with which you should already be familiar from step 1 (cf. **Box Annex 5.2**).

The **strategy** for microscopic ore recognition has this in common with the strategy for mineral exploration: the need to eliminate uncertainty as soon as possible and at minimum cost. In the case of exploration, it is a matter of discarding barren ground in order to focus attention (and investment) on the land potentially endowed with a deposit (*target*). In microscopic identification, it is a matter of discarding false options and focusing on diagnostic data to determine the identity as soon as possible. For this reason, attention should be paid first and foremost to **general properties such as luster**, **color**, **hardness**, **anisotropism,** and so on.

In addition, if the usual information is available on the origin of the ore, it is wise to pay attention to it from the beginning. Actually, each deposit type has a **mineral association** or **mineral assemblage** (**Annexes 2.1** and **2.2**) related to the ores' parageneses and is therefore meaningful for the sample to be studied. These data provide a powerful tool that can save the beginner false steps and much time, as seen in the examples that follow.

The general properties mentioned above have been synthesized for their application in tables (**Annex 4**), serving to support the identification process. Before discussing them, it is useful to show the method to be followed with an elementary example: the application of the specular reflectance, **R** (in this case in *white light* or, if collected as multispectral data from the ore descriptions, its *proxy* at $\lambda = 546$ nm: R_{546}).

The estimation of the **brightness**—the *proxy* of the **specular reflectance** ***R***—is a good start to examining a polished section, since it can be appreciated at a simple glance (always taking the precaution of working with a constant intensity of illumination—in principle, the nominal one of the microscope—otherwise the appreciation can be much biased).

It is easy to approximate the *Reflectance Group, R* (**Annex 4.1**, values always referred here to white light) if one starts with the darkest components or those which appear gray in reflected light (*Group: very low R*, **Annex 4.1**);

- If the gray level (brightness) is similar to that of the resin ($R \approx 5\%$), the most likely option is gangue minerals, among which the most frequent are silicates and carbonates:
  - Carbonates are distinguished (*d+p*) by their remarkable bireflectance (**Annex 4.4**) and, once this has been checked, can be characterized according to **§ 3.3.2**; otherwise, there is a possibility of silicate or some other type of gg (cf. **§ 3.3**).
  - The IR should also be checked, the absence of which suggests organic or carbonaceous matter (COM: cf. **§ 3.3.6**).
- Attention should be paid if the shade of gray is a bit lighter, because in the upper range of the same group there are certain possible ores like bastnäsite or monazite (both important potential sources of REE), malachite (green IR), azurite (blue IR) or scheelite (colorless IR), which should be contrasted with their descriptions (cf. bst, mz, mch, az, sch).

The *low R Group* starts with a slightly lighter gray color ($R \geq 12.5\%$) (**Annex 4.1**), typically beginning with *oxides* such as:

- Chromite or cassiterite, both gray, dark and very hard, yet easy to distinguish by their parageneses and by their symmetry (the former cubic and therefore isotropic, the latter tetragonal and anisotropic).
- With increasing R, in addition to oxides, the group includes *graphite* (immediately recognizable by its very strong bireflectance, from $R \approx 6\%$ almost like gg to R 27%, and extreme anisotropism) and a few *sulfides*, in general easy to identify, such as:
  - Valleriite (very similar to graphite, but with higher R_{min})
  - Digenite (delicate blue color, isotropic)
  - Bornite (pink color, fast tarnishing)
  - Covellite (extreme and unmistakable pleochroism and anisotropism)
  - Realgar and orpiment (IR abundant and colored).

For the remaining ores with progressively higher R values, a similar analysis is applied, but with a greater variety of species.

Knowledge of the mineral association, in correspondence with the typology, is again a valuable aid as it allows one to anticipate the predictable R-ranges in the sample, starting from the typical paragenesis in each class of deposits, as shown in **Box Annex 5.3**. The beginner soon discovers how useful it is to return again and again to the typological associations.

Box Annex 5.3 Classification of Ores by Reflectance Ranges(*) (R%, White Light) in Typical Mineral Associations(**)

In ultramafic rocks and ophiolithic complexes	
1	**Gold** (72), **PGE** (65)
2	**Arsenopyrite** (52), **Pyrite** (51), Pentlandite (49), Nickeline (50′5–45), **Chalcopyrite** (45)
3	**Pyrrhotite** (39)
4	Martite (29), **Hematite** (29), Rutile (23.5)
5	**Ilmenite** (19.5), **Sphalerite** (19), **Magnetite** (17), **Chromite** (12.5), **Spinel** (8)
6	**Gangue** (6)
Pegmatitic(***)–Pneumatolytic	
1	**Gold** (72), Bismuth (67)
Sulfide phases:	
2	Arsenopyrite (52), **Pyrite** (51), **Chalcopyrite** (45), **Molybdenite** (42-19), **Galena** (43), Stannite (28), **Pyrrhotite** (39)
Oxidized phases:	
3	Rutile (23.5)
4	**Wolframite** (17), **Cassiterite** (13), **Scheelite** (11)
Silicates:	
5	Beryl, Tourmaline (6), Mica (5), Quartz (5)
Hydrothermal Pb–Zn–Ag	
1	**Silver** (85), ***Electrum*** (80)
2	**Arsenopyrite** (52), **Pyrite** (51)
3	**Chalcopyrite** (45), **Galena** (43)
4	Bournonite (36), Acanthite (32), **Tetrahedrite** (31) and other sulfosalts
5	**Sphalerite** (19)
6	Gangue (6)
Volcanogenic-sedimentary massive sulfides (VMS)	
1	**Silver** (85), ***Electrum*** (80), **Gold** (72)
2	**Arsenopyrite** (52), **Pyrite** (51), White Arsenides (51)
3	**Chalcopyrite** (45), **Galena** (43), **Pyrrhotite** (39)
4	**Hematite** (29), Rutile (23.5), **Magnetite** (17)
5	**Goethite**(18–15), **Cassiterite** (13), Jarosite (7)
In metamorphic terrains: orogenic gold deposits	
1	***Electrum*** (80), **Gold** (72)
2	**Arsenopyrite** (52), **Pyrite** (51)
3	Chalcopyrite (45)
4	Galena (43), Stibnite (47–30), Molybdenite (42–19), Pyrrhotite (39)
5	Hematite (29), Rutile (23.5)
6	Sphalerite (19), Graphite (16–6)z
7	Gangue (6)

(*)Average reflectance values (R%), calculated in white light and given in parenthesis after the mineral name, group the ores into categories according to their range; these give a first indication of spectral differences. Important minerals in each type are highlighted in bold: these are not always the most common

(**)According to the typology of mineral deposits presented in **Annexes 2.1** and **2.2**

(***)Complex pegmatites (REM or rare-earth minerals, gems, semi-precious stones, etc.) require individual treatment

It should be stressed that the term of reference here is ***mineral association***, although it might seem at first more appropriate to refer to *paragenesis*; that is, the set of minerals formed simultaneously, in the same place and by the same genetic process. To establish the paragenesis it is indispensable to explain the genesis. For the moment that is not the problem: what the observer has to establish at first is the identity of the minerals present, whatever their genesis.

There is no doubt that, in a single sample, several episodes may have taken place from which as many parageneses may result. The assemblage of them all, that is to say the totality of the minerals present, is the mineral association. Thus, for example, if a massive sulfide ore composed of pyrite, chalcopyrite and galena (*primary minerals*) undergoes a process of supergene enrichment with partial replacement of pyrite and chalcopyrite by chalcocite, covellite and digenite (*secondary minerals*), when studying the samples these six minerals may be found together, and are the ones that the microscopist has to identify. These define the real existing association, although later it will be possible —and it will be very convenient—to establish by applying textural interpretation criteria that this association comprises two superimposed parageneses: the primary (py, ccp, gn); and the supergene (or secondary: cc, cv, dg).

The *medium R* (25–40%) and *high R* (40–60%) groups are the most numerous. Common ores such as hematite (R $\approx$ 29%) or tetrahedrite (R $\approx$ 31%) for the former, and galena (R $\approx$ 43%) or pyrite (R $\approx$ 51%) for the latter can be remembered as references points on the scale of R values.

The *very high* R group ($\geq$ 60%) gathers essentially native elements, mainly metals and precious metal alloys or compounds (Ag, Au, EGP). It is not a very numerous group and the various species can be diagnosed, generally quickly, by readily observable properties (provided that the abundance or grain size allows it, since they are usually very scarce minerals and almost always micro to cryptocrystalline).

For the **common cases** of application of the comparative tables (**Annex 4**), the reflectance (R: **Annex 4.1**) and *polishing hardness* (***PolH***: **Annex 4.2.1**) tables provide a general framework for contrasting the options obtained from other more directly diagnostic properties, such as bireflectance, color, anisotropism, and so on. However, the R and *PolH* tables may also be almost diagnostic in the extreme value ranges (e.g., very high or very low R), as we have just seen when dealing with brightness (native metals, R$\gg$; gangue and some oxides, R$\ll$). Nonetheless, as far as hardness is concerned, although it is undoubtedly an important set of properties, some reservations are unavoidable. The determination of VHN is now rare, due to the lack of instrumentation. As for the polishing, *PolH* and scratch hardnesses, apart from the fact that they may vary with composition or, in anisotropic ores, with orientation, they are not always observable with the precision and rigour required and, consequently, the determinations vary according to the authors.[2] Although the maximum consistency of the data has been sought in the compilation presented (**Annex 4.2.1**) there remains a certain margin of uncertainty; this is normally insignificant, but it may make it advisable to compare to the Talmage hardness (**Annex 4.2.2**), even if it is seldom used today.

The other properties listed in the tables allow for discretionary use suited to the sample under study. As a general guideline, it is recommended to check by a quick initial examination if there are noticeable *internal reflections*, ***IR*** (**Annex 4.3**) or if there are *colored ores* (**Annex 4.7**), because finding any of these properties can eliminate many options and considerably focus the determination. The same is true for the *Bireflectance-B vs. Pleochroism-P* tables (**Annex 4.4**). Properties that are very easily checked can greatly reduce the range of options to be considered: for example, the very strong B/P combination reduces the options to six.[3]

The determination of the *isotropic* (**Annex 4.5**) or anisotropic (**Annex 4.6**) *character* of the mineral provides clues that can be decisive, especially since Picot and Johan (1977, 1982) showed the diagnostic value of *colors of anisotropism* (or polarisation tints) listed in the second of these tables (**Annex 4.6**). This diagnostic value of color requires extreme care in dealing with photomicrographs, either capturing or reproducing the digital images, to ensure the fidelity of the colors obtained (virtual or printed) to those visible under the microscope. For this reason, all images have been processed and defined in the *sRGB color space* and to ensure color fidelity must be viewed on a monitor with color calibration.

References

Bowie SHU, Taylor K (1958) A system of ore mineral identification. Min Mag Lond 99:265–277

McLeod CR, Chamberlain JA (1968) Reflectivity and Vickers microhardness. Paper 68-64, Geological Survey of Canada

Picot P, Johan Z (1977) Atlas des Mineraux Metalliques. Mem. B.R.G. M., 90. Paris, 406 p

Price JG (2013) Reflected light microscopy and minerals associated with ore systems. Short course Overview of ore-forming systems handouts, 7–10 Apr 2015. München University, Germany, 16 p

Ramdohr P (1980) The ore minerals and their intergrowths, 2nd edn, 2 vols. Pergamon Press, Oxford, 1205 p

Samper J (2008) Curso de Microscopía de Menas (con descripción de muestras). ETSI Minas y Energía, Univ. Politécnica de Madrid, 90 p

[2]Actually, notable differences have been found in the various *PolH* tables. Nevertheless, the data of Ramdohr (1980) and those of Schneiderhöhn (1952) are generally consistent with each other, and also with the author's observations, for which reason they have been given preference. The greatest divergences were found in those published by Uytenbogaardt and Burke (1971), and these were discarded when they proved irreconcilable.

[3]Of all ores under study (common ores).

Schneiderhöhn H (1952) Erzmikroskopisches Praktikum. E. Schweizerbart'sche Verlagsbuchhandlung, Stuttgart, 274 p

Schneiderhöhn H, Ramdohr P (1931) Lehrbuch der Erzmikroskopie, I & II Band, mit Anhang: Erzmikroskopische Bestimmungstafeln. Gebr. Borntraeger, Berlin, 714, 313 and 47 p

Schouten C (1962) Determination tables for ore microscopy. Elsevier Publishing Company, Amsterdam, 242 p

Spry PG, Gedlinske BL (1987) Tables for the determination of common opaque minerals. Economic Geology Publ. Co. Littleton, Colo_USA, 52 p

Tarkian M (1974) A key-diagram for the optical determination of common ore minerals. Min Sci Eng 6(2):101–105

Tarkian M, Bernhardt HJ (1984) A key-diagram for the optical determination of platinum-group minerals. Tschermarks Min Petr Mitt 33(2):121–129

Tarkian M, Liessmann W (1991) A guide for optical and analytical identification of ore minerals. Clausthaler Geol. Abh. 47, Verlag Sven von Loga, Köln, 78 p

Uytenbogaardt W, Burke EAJ (1971) Tables for microscopic identification of ore minerals, 2nd edn. Elsevier, Amsterdam

Bibliography

Atencio D, Andrade MB, Christy AG, Giere R, Kartashov PM (2010) The pyrochlore supergroup of minerals: nomenclature. Can Mineral 48:673–698

Atencio D, Andrade MB, Bastos Neto AC, Pereira VP (2017) Ralstonite renamed hydrokenoralstonite, coulsellite renamed fluornatrocoulsellite, and their incorporation into the pyrochlore supergroup. Can Mineral 55:115–120

Barton BP (1991) Ore textures: problems and opportunities. Mineral Mag 55:303–315

Bastin ES (1960) Interpretation of ore textures. Waverly Press Inc., Baltimore, p 101

Berrezueta E, Castroviejo R (2007) Reconocimiento automatizado de menas metálicas mediante análisis digital de imagen: un apoyo al proceso mineralúrgico. I: ensayo metodológico. Rev Metal 43 (4):294–309

Best MG (1982) Igneous and metamorphic petrology. Freeman, New York, p 630

Cabri LJ, Hall R, Szymanski JT, Stewart JM (1973) On the transformation of cubanite. Can Mineral 12:33–38

Cameron EN (1963) Optical symmetry from reflectance measurements. Am Mineral 48:1070–1079

Castroviejo R, Brea C, Pérez-Barnuevo L, Catalina JC, Segundo F, Bernhardt HJ, Pirard E (2009) Using computer vision for microscopic identification of ores with reflected light: preliminary results. In: Williams P et al (eds) Proceedings of 10th biennial SGA meet, vol 2. Townsville, Australia, pp 682–684. ISSN: 9780980558685

Castroviejo R, Quesada C, Soler M (2011) Post-depositional tectonic modifications of VMS deposits in Iberia and its economic significance. Miner Deposita 46:615–637. ISSN: 0026-4598

Catalina JC (2018) Desarrollo de un Sistema Multiespectral para Microscopía de Menas. Tesis doctoral, ETS Ingenieros de Minas y Energía, Universidad Politécnica de Madrid. https://doi.org/10.20868/UPM.thesis.52452; enlace: http://oa.upm.es/52452/

Catalina JC, Castroviejo R (2017) Microscopía de reflectancia multiespectral: aplicación al reconocimiento automatizado de menas metálicas. Rev Metal Madrid 53(4) (Oct-Dic 2017):e107. https://doi.org/10.3989/revmetalm.107. CSIC/CENIM. ISSN-L: 0034-8570

Craig JR (1990a) Textures of the ore minerals. In: Jambor JL, Vaughan DJ (eds) Advanced microscopic studies of ore minerals, chap 9, vol 17. MAC Short Course, pp 213–261

Craig JR (1990b) Ore textures and paragenetic studies—some modern case histories and sources of comparative data. In: Jambor JL, Vaughan DJ (eds) Advanced microscopic studies of ore minerals, chap 10, vol 17. MAC Short Course, pp 263–317

Craig JR (2001) Ore-mineral textures and the tales they tell. Can Mineral 39:937–956

Edwards AB (1965) Textures of the ore minerals and their significance. Brown, Prior, Anderson Pty. Ltd., Australia, 242 p

Fettes D, Desmons J (2007) Metamorphic rocks: a classification and glossary of terms (recommendations of the IUGS/SCMR). Cambridge Univ. Press, Cambridge, 244 p

Frost BRT, Raynor GV (1950) The system silver-magnesium-antimony, with reference to the theory of alloy formation. Proc Roy Soc (Lond) A-203:134–135

Gaudin AM (1939) Principles of mineral dressing. McGraw-Hill, New York, p 554

Grigoriev J (1928) Structures of mineral intergrowths in ores. Mem Russ Min Soc 57:11–56

Head RE (1928) Microscopic study of ore: aid in Cu leaching. Eng Min J 126:13–15

Head RE et al (1932) Statistical microscopic examination of mill products of the Copper Queen concentrates of the Phelps Dodge Corporation, Bisbee, Arizona. US Bur. Mines Tech., 533 p

Huston DL, Sie SH, Suter GF, Cooke DR, Both RA (1995) Trace elements in sulfide minerals from Eastern Australia volcanic-hosted masive sulfide deposits: Part I. Proton microprobe analyses of pyrite, chalcopyrite and sphalerite. Part II. Selenium levels in pyrite. Econ Geol 90:1167–1196

Jones M (1987) Applied mineralogy: a quantitative approach. Graham and Trotman, Londres, 260 p. ISBN: 978-0-86010-510-7

Kojima S, Sugaki A (1985) Phase relations in the Cu-Fe-Zn-S system between 800° and 500 °C. Min J 12:15–28

Korzhinskii DS (1970) Theory of metsomatic zoning. Clarendon Press, Oxford, p 160

Kucha H, Mayer W, Piestrzyński A, Wieczorek A (1981) The replacement of rutile by chalcocite in the Zechstein copper ores of the Fore-Sudetic Monocline. Mineral Polon 12(1):69–76

López-Benito A, Catalina JC, Alarcón D, Grunwald U, Romero P, Castroviejo R (2019) Automated ore microscopy based on multispectral measurements of specular reflectance. I—A comparative study of some supervised classification techniques. Miner Eng 146:106–136. https://doi.org/10.1016/j.mineng.2019.106136

Park CF, MacDiarmid RA (1981) Yacimientos minerales. Omega, Barcelona, p 512

Pérez-Barnuevo L (2014) Caracterización automatizada de texturas de menas mediante análisis digital de imagen para su aplicación geometalúrgica. PhD, ETSI Minas y Energía, Universidad Politécnica de Madrid, 298 p

Pérez-Barnuevo L, Pirard E, Castroviejo R (2013) Automated characterisation of intergrowth textures in mineral particles. A case study. Miner Eng 52:136–142. ISSN: 0892-6875

Petruk W (2000) Applied mineralogy in the mining industry. Elsevier, Amsterdam, p 268

R. Castroviejo, *A Practical Guide to Ore Microscopy—Volume 1*,
https://doi.org/10.1007/978-3-031-12654-3

Petruk W, Harris DC, Cabri LJ, Stewart JM (1969) Native silver and silver-antimony minerals in the Cobalt-Gowganda ores. Can Mineral 10:144ff

Petruk W, Cabri LJ, Harris DC, Stewart JM (1970) Allargentum redefined. Can Mineral 10:163–172

Pirard E (2004) Multispectral imaging of ore minerals in optical microscopy. Mineral Mag 68(2):323–333

Schwartz GM (1923) Solving metallurgical problems with the reflecting microscope. Eng Min J 116:237–238

Schwartz GM (1951) Classification and definitions of textures and mineral structures in ores. Econ Geol 46:578–591

Sociedad Española de Mineralogía (SEM) (2020) Criterios para la nomenclatura de minerales en español. Propuesta presentada a la Asamblea Ordinaria de socios de la SEM, 2020/06/26, Ronda, Málaga

Stanton RL (1972) Ore petrology. McGraw, New York, p 713

Taylor R (2009) Ore textures. Recognition and interpretation. Springer, Heidelberg, 288 p

Websites

AMCO System. https://mmt-systems.com/

IMA & IMA/COM List of Minerals. https://www.ima-mineralogy.org/Minlist.htm (visit. 20220221)

International Mineralogical Association (IMA). https://www.ima-mineralogy.org/about.htm (visit. 20220221)

GPSR Compliance

The European Union's (EU) General Product Safety Regulation (GPSR) is a set of rules that requires consumer products to be safe and our obligations to ensure this.

If you have any concerns about our products, you can contact us on ProductSafety@springernature.com

In case Publisher is established outside the EU, the EU authorized representative is:

Springer Nature Customer Service Center GmbH
Europaplatz 3
69115 Heidelberg, Germany

Batch number: 10756714

Printed by Printforce, the Netherlands